Convexity

Convexity

ROGER WEBSTER

Department of Pure Mathematics,
Sheffield University

Oxford New York Tokyo
OXFORD UNIVERSITY PRESS
1994

Oxford University Press, Walton Street, Oxford OX2 6DP

Oxford New York Toronto
Delhi Bombay Calcutta Madras Karachi
Kuala Lumpur Singapore Hong Kong Tokyo
Nairobi Dar es Salaam Cape Town
Melbourne Auckland Madrid

and associated companies in
Berlin Ibadan

Oxford is a trade mark of Oxford University Press

Published in the United States
by Oxford University Press Inc., New York

A catalogue record for this book is available from the British Library

Library of Congress Cataloging in Publication Data
Webster, Roger.
Convexity / Roger Webster.
Includes bibliographical references and index.
1. Convex sets. 2. Convex functions. I. Title.
QA640.W43 1994 516'.08–dc20 93-34951

ISBN 0 19 853147 8

Typeset by Colset Pte. Ltd., Singapore
Printed in Great Britain on acid-free paper
by Bookcraft Ltd., Midsomer Norton, Avon

This book is dedicated to my wife
PENNY,
and my two little pencil sharpeners
MICHAEL and ANNE

Preface

Although the systematic study of convex sets can be traced back to the end of the nineteenth century, it is only since the mid-twentieth century that convexity has become a well established branch of mathematics. Indeed, *Convex sets and related geometric topics* is one (number 52) of the 95 main subject classifications used by *Mathematical Reviews*. The present text provides a wide-ranging introduction to the theory and applications of convex sets and functions, suitable for final-year undergraduates and graduate students.

Convexity draws upon geometry, analysis, linear algebra, and topology and has a role to play in topics as diverse as number theory, classical extremum problems, combinatorial geometry, linear programming, game theory, polytopes, bodies of constant width, and the theory of inequalities (linear, classical, and matrix). Convex sets are studied here in the setting of the n-dimensional Euclidean space $\mathbb{R}^n$. This has the advantage of allowing an elementary exposition that demands only a modest knowledge of analysis and linear algebra, and enables new concepts to be motivated through simple geometrical examples.

Convexity is an attractive subject to study, for many reasons. Its fundamental concepts are natural and appealing, while the prerequisites it demands will be possessed by any final-year mathematics major. One does not have to journey far along its path before encountering significant, aesthetically pleasing results. Its eclectic nature makes it a showcase for displaying fruitful interactions amongst apparently unrelated parts of mathematics. It helps to develop geometric intuition, a facility so often found wanting in today's students, reared on a diet of algebra and analysis. Over and above applications to pure and applied mathematics, it can boast ties with economics, science, engineering, and technology and, despite being an active field of research, it abounds in easily stated, unsolved problems that have an instant intuitive appeal.

Today convexity is very much a postgraduate preserve, a situation reflected in the advanced treatises written about it, and the dearth of undergraduate courses on the subject. The first text devoted to a general exposition of convex sets was the German *Theorie der konvexen Körper* by Bonnesen and Fenchel (1934); the first in English was *Convexity* by Eggleston (1958), both being at graduate level. In recent years a few introductory textbooks have appeared, and it is hoped that the present work will contribute to the popularization of convexity in the university curriculum.

This book grew out of courses that I have given over many years to undergraduates at Dalhousie University, Nova Scotia, Canada and Sheffield University, England. The lectures varied from one year to the next, but all contained some material from Sections 2.1, 2.2, 2.3, 2.4, and 2.6 on basic properties of convex sets, Sections 5.1, 5.2, and 5.3 on the application of convex functions to real analysis, Section 7.1 on Helly's theorem, and Section 7.4 on the geometry of numbers. It is perhaps worth recording that one year, I invited the students to vote on which had been their favourite, and which their least favourite topic. On the latter, the students were divided. On the former, however, there was complete unanimity: all *thirty-five* students voted for Artin's treatment of the gamma function employing the concept of log-convexity (Section 5.3). Needless to remark, I always find a place for *this* topic in my convexity courses!

Although most of the material presented here was designed with an undergraduate course on convexity in mind, it has served more than one purpose during its evolution. Extracts from it have formed the basis for a reading course. Some sections have provided topics for final year dissertations, while others have been used for undergraduate seminars in which a student prepares and delivers a lecture to fellow students. On one occasion it provided the source for a short series of lectures, entitled *What is a Convex Set?*, that I gave to colleagues at Sheffield University.

The text proceeds at a gentle pace, with new ideas being introduced by means of simple examples in familiar surroundings. The reader is encouraged to draw diagrams to illustrate proofs, wherever possible, and to attempt some of the exercises that conclude each section. These range from easy calculations right through to those demanding a tricky counterexample or an extended proof. Detailed solutions to all but the most trivial of the 241 exercises included in the book are given, a factor which should be particularly helpful for the reader studying in isolation.

The single feature of the present work that distinguishes it from other texts on convexity is the host of diverse applications which it highlights. As far as I am aware, no other general introduction to the subject discusses applications to the Gauss–Lucas theorem, the computation of the volume of the closed unit ball in $\mathbb{R}^n$, the geometry of numbers, the theory of inequalities (linear, classical, and matrix), the structure of polyhedral sets, the gamma function, the convex programming problem, and the minimax approximation of continuous functions by polynomials. A few results appear here for the first time outside research journals. For example, Theorem 5.8.8 on the harmonic means of diagonals of doubly stochastic matrices, the characterization of convexity-preserving mappings given in Section 7.3, and some applications of Brouwer's fixed-point theorem found in Section 7.5. Much of Chapter 3 on convex polytopes has not appeared before in a general text on convexity. Another novel aspect of our account here is the inclusion of a full treatment of the concept of volume for convex sets, thus permitting a rigorous discussion in $\mathbb{R}^n$ of mixed volumes, the isoperimetric and isodiametric inequalities, and the Brunn–Minkowski theorem, material hardly ever covered in elementary treatises.

Any instructor intending to base a lecture course on the book will clearly need to make a careful selection of the topics to be included. In this connection, the following information should be useful. The specialized Chapter 3 and the difficult Chapter 6 can be omitted without seriously impeding the flow of the book. The individual sections of Chapter 7 are independent of one another. The brief notes heading each chapter indicate the interrelations amongst the various sections, and should be of assistance in both designing a course and reading the book.

For an advanced course, appropriate for graduates and able undergraduates, suitable material will be found in Chapters 3, 6, and 7, which could also be used in conjunction with a reading seminar or a student dissertation. The *Notes on further reading* at the end of the book should prove helpful for all the suggested uses of the text.

The courses that I have taught on convexity at Sheffield University in recent years have been encouragingly well supported, with over a hundred students enrolling on one occasion. Large classes, however, present problems in the form of the widely varying backgrounds, abilities, and mathematical interests of those attending. Fortunately, the nature of convexity itself, as a mosaic composed from different facets of mathematics, many themselves being a rich mixture of pure and applied ideas, together with its high degree of accessibility, made the resolution of these difficulties relatively simple.

I based the course upon carefully selected topics from Sections 2.1, 2.2, 2.4, 4.3, 4.4, 4.5, 4.6, 5.1, 5.2, 5.3, 5.8, 7.1, and 7.4, with a little bridging material taken from Sections 2.3, 2.6, 3.1, 3.2, 4.2, and 5.4. Considerations of time and the lack of a solid mathematical background on the part of many students led to a few compromises as regards proof. For example, in Section 7.1 on Helly's theorem, I proved only its *finite* version, merely stating the *infinite* one. This saved much time on difficult topological preliminaries, and allowed me to proceed immediately to some impressive applications of the theorem. Since, by design, most course topics have both a theoretical and an applicable side to them, it was easy to set assignments that included a wide variety of problems ranging from routine manipulations to ones calling for intricate arguments.

To set the course in a wider perspective, I would always include historical and biographical detail, give examples of unsolved problems, and indicate some of the exciting developments in convexity that lie beyond the confines of the lectures. Although such considerations have been largely ignored in the text, the interested reader will find a few helpful references in *Notes on further reading*. For a very brief history of convexity, see Fenchel (1983), and for two personal overviews of the subject, see Klee (1971) and Berger (1990). Valentine (1964), and Klee and Wagon (1991) both contain a wide range of unsolved problems.

Sheffield
May 1994

R. J. W.

Acknowledgements

My sincere thanks are due to a number of friends for their help and encouragement throughout the writing of the book. I am enormously grateful to the late Leon Mirsky, who not only suggested that I undertake this enterprise, but also gave freely of his expert advice. My good friend Victor Bryant made first drafts of a few sections of the book and generously assisted me at every stage of the preparation. It was whilst I was supervising his doctoral dissertation that we investigated axiomatic convexity, and this was the genesis of Section 7.2 on join spaces. To Victor go my heartfelt thanks.

I am delighted to place on record my gratitude to Douglas Northcott and Hazel Perfect for reading the entire manuscript and making many valuable suggestions. Further thanks are due to colleagues Harry Burkill, Roger Cook, Alan Douglas, Michael Edelstein, Mary Hart, Ben Haytock, Chris Knight, Peter McMullen, Michael Piff, David Sharpe, Geoffrey Shephard, and Mike Stannett, all of whom commented on or helped in some way with preliminary versions of the present text. I must also express my appreciation to those many Sheffield University students who have attended my lectures on convexity over the years, for their suggestions, their enthusiasm, and their insistence that I complete this book, started before some of them had even been born. Perhaps no longer shall I be subject to the tease from staff and students alike: will you *ever* finish your book on convexity?

Much work is involved between the completion of the final manuscript and the appearance of the book in print. In this connection, I commend Alex Hart, Rosa Hepworth, and Janet Williams for their quietly proficient, secretarial assistance. I would also like to praise Oxford University Press for its patience, helpfulness, and efficiency.

My greatest thanks must be reserved for my wife Penny, and my children Michael and Anne, who have assisted me in countless ways throughout the twenty-one year gestation period of the book. Indeed, during this time the *book* became a member of the family, with whom we all learned to live. Whereas most kitchens are equipped with food mixers, microwaves, and dishwashers, over a long period ours was unique in having a heavy-duty pencil sharpener clamped to a work surface. How often have visitors to our home enquired: why is there a giant pencil sharpener lurking in your kitchen? The children, having sharpened some thousand pencils for me to write the book, had no difficulty in supplying the answer!

Contents

Introduction

Here we establish the basic notation and conventions used throughout the text and outline the prerequisites needed for our study of convexity. The end of a proof is marked with □. The appearance of □ immediately following the statement of a proposition indicates that the proof is left to the reader.

The empty set is denoted by $\emptyset$. A set consisting of a single element is called a singleton. We use the symbol $\{x_1, \ldots, x_m\}_{\neq}$ to denote the set consisting of *distinct elements* $x_1, \ldots, x_m$. If A and B are sets, then $A \backslash B$ is used to denote the set consisting of those elements belonging to A but not to B. Two sets are said to meet if they have a non-empty intersection, otherwise they are said to be disjoint. If A is a subset of some universal set X, then the set $X \backslash A$ is called the complement of A in X and is denoted by A^c. The intersection (union) of an empty family of subsets of the universal set X is taken to be $X(\emptyset)$. Let $(A_i : i \in I)$ be a family of subsets of X indexed by some index set I. Then the family is said to be pairwise disjoint if A_i and A_j are disjoint whenever $i, j \in I$ with $i \neq j$. De Morgan's complementation laws assert that

$$(\bigcup(A_i : i \in I))^c = \bigcap(A_i^c : i \in I) \quad \text{and} \quad (\bigcap(A_i : i \in I))^c = \bigcup(A_i^c : i \in I).$$

Consider a mapping $f: X \to Y$, where X and Y are non-empty sets. Then f is said to be injective if $f(x) = f(x')$, where $x, x' \in X$, implies that $x = x'$; it is said to be surjective if, for each $y \in Y$, there exists $x \in X$ such that $f(x) = y$. A mapping which is both injective and surjective is said to be bijective. An injective (surjective, bijective) mapping is called an injection (surjection, bijection). Each bijection $f: X \to Y$ gives rise to an inverse mapping $f^{-1}: Y \to X$ defined by the condition that $f^{-1}(y) = x$ if and only if $f(x) = y$, where $x \in X$, $y \in Y$. If $f: X \to Y$, $A \subseteq X$, $B \subseteq Y$, then the image $f(A)$ of A under f is the subset $\{f(a) : a \in A\}$ of Y and the inverse image $f^{-1}(B)$ of B under f is the subset $\{x \in X : f(x) \in B\}$ of X. If $f: X \to Y$ and $g: Y \to Z$ are mappings, then the composite mapping $g \circ f: X \to Z$ is defined by the equation $(g \circ f)(x) = g(f(x))$ for $x \in X$.

The reader is assumed to have a good grasp of the rudiments of real analysis, including a knowledge of sequences, series, and the continuity, differentiability, and integration of functions. One important result needed in the book is the Bolzano–Weierstrass theorem, which asserts that every bounded sequence of real numbers contains a convergent subsequence. The infinite product notation is used in the following naïve sense: if the real sequence $a_1, a_2, \ldots, a_k, \ldots$ is such that the sequence of partial products $a_1, a_1 a_2, \ldots, a_1 a_2 \ldots a_k, \ldots$ converges to the real number a, then this is indicated by writing

$$a = \prod_{k=1}^{\infty} a_k.$$

The first mean value theorem (of the differential calculus) is used in the form: if the function $f: [a, b] \to \mathbb{R}$ is continuous on $[a, b]$ and differentiable on $]a, b[$, then there exists c in $]a, b[$ such that

$$f'(c) = \frac{f(b) - f(a)}{b - a}.$$

If the real function f is such that the one-sided limit

$$\lim_{x \to a_+} \frac{f(x) - f(a)}{x - a}$$

exists, then its value is denoted by $f'_+(a)$ and is called the right derivative of f at a. Similar remarks apply to the left derivative $f'_-(a)$ of f at a. If both $f'_-(a)$ and $f'_+(a)$ exist, then f is continuous at a; if, in addition, $f'_-(a) = f'_+(a)$, then f is differentiable at a. A superficial knowledge of the differentiability of a real function of n real variables is assumed in Section 5.5, where a passing knowledge of the chain rule is required.

The reader is also assumed to have a basic understanding of elementary linear algebra, including a knowledge of vector spaces, bases, dimension, linear transformations, and eigenvectors. The only vector space considered here is the real n-dimensional space $\mathbb{R}^n$ whose points are real n-tuples $\mathbf{x} = (x_1, \ldots, x_n)$, and in which addition and scalar multiplication are defined coordinatewise. One result that will be needed in our discussion is the dimension theorem. This asserts that, if A and B are finite dimensional subspaces of a vector space, then

$$\dim(A + B) + \dim(A \cap B) = \dim A + \dim B,$$

where $\dim A$ etc. denotes the dimension of A. The inner product $\mathbf{x} \cdot \mathbf{y}$ of vectors $\mathbf{x} = (x_1, \ldots, x_n)$ and $\mathbf{y} = (y_1, \ldots, y_n)$ in $\mathbb{R}^n$ is the real number defined by the equation

$$\mathbf{x} \cdot \mathbf{y} = x_1 y_1 + \ldots + x_n y_n.$$

The norm or length $\|\mathbf{x}\|$ of a vector $\mathbf{x} = (x_1, \ldots, x_n)$ in $\mathbb{R}^n$ is the non-negative number defined by the equations

$$\|\mathbf{x}\| = \sqrt{(\mathbf{x} \cdot \mathbf{x})} = \sqrt{(x_1^2 + \ldots + x_n^2)}.$$

Two important inequalities relating to the inner product and the norm are the Cauchy–Schwarz inequality and the triangle inequality, which assert, respectively, that $|\mathbf{x}\cdot\mathbf{y}| \leqslant \|\mathbf{x}\|\,\|\mathbf{y}\|$ and $\|\mathbf{x}+\mathbf{y}\| \leqslant \|\mathbf{x}\| + \|\mathbf{y}\|$. The orthogonal complement $A^{\perp}$ of a subspace A of $\mathbb{R}^n$ is the subspace of $\mathbb{R}^n$ defined by the equation

$$A^{\perp} = \{\mathbf{x}\in\mathbb{R}^n : \mathbf{x}\cdot\mathbf{a} = 0 \text{ for all } \mathbf{a}\in A\}.$$

Moreover, $\dim A + \dim A^{\perp} = n$, and each point of $\mathbb{R}^n$ can be expressed *uniquely* in the form $\mathbf{a}+\mathbf{b}$, where $\mathbf{a}\in A$ and $\mathbf{b}\in A^{\perp}$.

All matrices considered in the book are real. They are denoted by bold, upper case letters and expressed using square brackets. Thus we write $\mathbf{A}$ is the $m \times n$ matrix $[a_{ij}]$ to indicate that $\mathbf{A}$ is the $m \times n$ matrix having the element a_{ij} in its ith row and jth column. A zero matrix, i.e. one all of whose elements are zero, is denoted by $\mathbf{0}$, its size being determined by the context. The identity matrix $\mathbf{I}_n$ is the $n \times n$ matrix having ones along its leading diagonal and zeros elsewhere. The transpose of a matrix $\mathbf{A}$ is denoted by $\mathbf{A}^{\mathrm{T}}$, and the determinant of a square matrix $\mathbf{A}$ by det $\mathbf{A}$. A square matrix $\mathbf{A}$ that has a non-zero determinant is said to be non-singular and has an inverse matrix $\mathbf{A}^{-1}$. Matrices can be used to represent both systems of linear equations, and linear transformations. Thus a general system of m linear equations in n unknowns $x_1, \ldots, x_n$ can be represented by the matrix form $\mathbf{Ax} = \mathbf{b}$, where $\mathbf{A}$ is an $m \times n$ matrix, $\mathbf{x}\in\mathbb{R}^n$ and $\mathbf{b}\in\mathbb{R}^m$, $\mathbf{x}$ and $\mathbf{b}$ being regarded as column matrices. When $\mathbf{b} = \mathbf{0}$, the system is said to be homogeneous. Such a homogeneous system for which $m < n$ always has a non-trivial solution. A transformation $f\colon \mathbb{R}^n \to \mathbb{R}^m$ is linear if and only if it can be expressed in the form $f(\mathbf{x}) = \mathbf{Ax}$, where $\mathbf{A}$ is an $m \times n$ matrix and $\mathbf{x}\in\mathbb{R}^n$. The set of all linear combinations of the columns of an $m \times n$ matrix is a subspace of $\mathbb{R}^m$ called the column space of $\mathbf{A}$, its dimension being called the column rank of $\mathbf{A}$. Similar definitions apply to the row space and the row rank. The column rank and the row rank of a matrix are always equal, and are referred to simply as the rank of the matrix.

Let $\mathbf{A}$ be a symmetric $n \times n$ matrix. Then the eigenvalues of $\mathbf{A}$ are all real and there exists an orthonormal basis for $\mathbb{R}^n$ consisting of eigenvectors of $\mathbf{A}$. The matrix $\mathbf{A}$ gives rise to a quadratic function $q\colon \mathbb{R}^n \to \mathbb{R}$ of n real variables $x_1, \ldots, x_n$ defined by the equation $q(\mathbf{x}) = \mathbf{x}^{\mathrm{T}}\mathbf{Ax}$, where $\mathbf{x} = (x_1, \ldots, x_n)$ and the 1×1 matrix $\mathbf{x}^{\mathrm{T}}\mathbf{Ax}$ is identified with the real number defining it. We say that $\mathbf{A}$ is non-negative semi-definite if $\mathbf{x}^{\mathrm{T}}\mathbf{Ax} \geqslant 0$ for all $\mathbf{x}$, and positive definite if $\mathbf{x}^{\mathrm{T}}\mathbf{Ax} > 0$ when $\mathbf{x} \neq \mathbf{0}$. If $\mathbf{A}$ is non-negative semi-definite (positive definite), then its eigenvalues are non-negative (positive).

1
The Euclidean space $\mathbb{R}^n$

This chapter contains those results about the Euclidean space $\mathbb{R}^n$ that are needed in our discussion of convexity. The reader should already be acquainted with some of these.

Theorem 1.3.12 (used only incidentally in Section 7.3) and that part of Section 1.6 following Theorem 1.6.2 (required in Section 6.2) may be omitted without serious loss.

1.1 THE EUCLIDEAN SPACE $\mathbb{R}^n$

Convexity, as we study it here, is a branch of Euclidean geometry in as much as its main concerns, such as points, lines, planes, polygons, circles, polyhedra, angles, lengths, areas, and volumes, were all discussed in Euclid's *Elements*, written around 300 BC. Our approach, however, differs in two very important aspects from that of Euclid himself. Firstly, we do not restrict our attention to three-dimensional space, and secondly, we make constant use of both algebraic and analytical arguments.

We begin by recalling some results of elementary geometry and seeing how they motivate the definition of the n-dimensional Euclidean space $\mathbb{R}^n$. In three-dimensional coordinate geometry a *point* or *vector* is determined by its *coordinates* x, y, z relative to some rectangular coordinate system, and we can (and do) identify the point or vector with the ordered triple (x, y, z). Vectors are added together according to a *parallelogram law*, which is equivalent to the addition of corresponding coordinates. The word **scalar** is used as a synonym for real number. The product of a scalar and a vector is equivalent to the multiplication of each coordinate of the vector by the scalar. Thus, if (x, y, z) and (u, v, w) are vectors, and λ is a scalar, then

$$(x, y, z) + (u, v, w) = (x + u, y + v, z + w)$$

and

$$\lambda(x, y, z) = (\lambda x, \lambda y, \lambda z).$$

These equations can be extended in the natural way to define vector addition and scalar multiplication of ordered sets of n real numbers, i.e. real *n-tuples*. This observation prompts the definition of $\mathbb{R}^n$.

For each positive integer n, denote by $\mathbb{R}^n$ the set of all n-tuples $(x_1, \ldots, x_n)$ of real numbers. Then $\mathbb{R}^n$ is called the ***n*-dimensional**

Euclidean space. Each element $\mathbf{x} = (x_1, \ldots, x_n)$ of $\mathbb{R}^n$ is called a **point** or **vector** of $\mathbb{R}^n$ and the real numbers $x_1, \ldots, x_n$ are called the **coordinates** of $\mathbf{x}$. For $n = 1$, we identify the 1-tuple $\mathbf{x} = (x_1)$ with the real number x_1 itself, so that $\mathbb{R}^1$ becomes simply $\mathbb{R}$, the set of *real numbers*. For $n = 1, 2, 3$, we often write x, (x, y), (x, y, z) instead of (x_1), (x_1, x_2), (x_1, x_2, x_3). Geometrically, $\mathbb{R}^1$ can be thought of as a line, $\mathbb{R}^2$ as a plane, and $\mathbb{R}^3$ as the set of points in space. Lower case Roman letters such as $\mathbf{a}, \mathbf{b}, \mathbf{c}$, $\mathbf{x}, \mathbf{y}, \mathbf{z}$ will denote points of $\mathbb{R}^n$, lower case Roman and Greek letters such as x, y, z, λ, μ, ν will denote scalars, and capital Roman letters such as A, B, C will denote subsets of $\mathbb{R}^n$. Addition and scalar multiplication in $\mathbb{R}^n$ are defined coordinatewise. Thus, if $\mathbf{x} = (x_1, \ldots, x_n)$, $\mathbf{y} = (y_1, \ldots, y_n)$, and λ is a scalar, then

$$\mathbf{x} + \mathbf{y} = (x_1 + y_1, \ldots, x_n + y_n) \quad \text{and} \quad \lambda\mathbf{x} = (\lambda x_1, \ldots, \lambda x_n).$$

The vector $(0, \ldots, 0)$ of $\mathbb{R}^n$, all of whose coordinates are 0, is denoted by $\mathbf{0}$ and is called the **zero vector** or **origin** of $\mathbb{R}^n$. The vector in $\mathbb{R}^n$ whose only non-zero coordinate is a 1 in the ith position is denoted by $\mathbf{e}_i$ and is called the ith **elementary vector**. A point of $\mathbb{R}^n$ all of whose coordinates are integers is called a **lattice point**. The vector $(-1)\mathbf{x}$ is written simply as $-\mathbf{x}$, and vector subtraction is defined by the rule $\mathbf{x} - \mathbf{y} = \mathbf{x} + (-1)\mathbf{y}$. It is sometimes convenient to write $\mathbf{x}/\lambda$ for $(1/\lambda)\mathbf{x}$. The set $\mathbb{R}^n$, equipped with the above operations of vector addition and scalar multiplication, is a *real vector space*: that is to say, if $\mathbf{x}$, $\mathbf{y}$, $\mathbf{z} \in \mathbb{R}^n$ and $\lambda, \mu \in \mathbb{R}$, then the following relations hold:

(i) $\mathbf{x} + \mathbf{y} = \mathbf{y} + \mathbf{x}$;

(ii) $\mathbf{x} + (\mathbf{y} + \mathbf{z}) = (\mathbf{x} + \mathbf{y}) + \mathbf{z}$;

(iii) $\mathbf{x} + \mathbf{0} = \mathbf{x}$;

(iv) $\mathbf{x} + (-\mathbf{x}) = \mathbf{0}$;

(v) $1\mathbf{x} = \mathbf{x}$;

(vi) $\lambda(\mu\mathbf{x}) = (\lambda\mu)\mathbf{x}$;

(vii) $\lambda(\mathbf{x} + \mathbf{y}) = \lambda\mathbf{x} + \lambda\mathbf{y}$;

(viii) $(\lambda + \mu)\mathbf{x} = \lambda\mathbf{x} + \mu\mathbf{x}$.

We extend the operations of vector addition and scalar multiplication to subsets of $\mathbb{R}^n$ by defining:

$$A + B = \{\mathbf{a} + \mathbf{b} : \mathbf{a} \in A, \mathbf{b} \in B\} \quad \text{and} \quad \lambda A = \{\lambda\mathbf{a} : \mathbf{a} \in A\},$$

where $A, B \subseteq \mathbb{R}^n$ and $\lambda \in \mathbb{R}$. The set $A + B$ is called the **vector sum** of A and B. It follows from the above definitions that both sets $A + B$ and λA are empty when A is empty. We write $-A$ for the set $(-1)A$, and $A - B$ for the set $A + (-B)$. It is sometimes convenient to write A/λ for $(1/\lambda)A$. The set A in $\mathbb{R}^n$ is said to be **0-symmetric**,

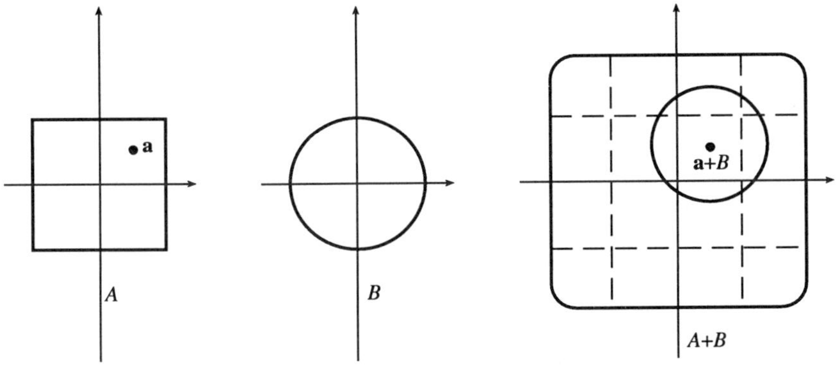

Fig. 1.1

or simply **symmetric**, if $-A = A$. Geometrically, A is symmetric if it is its own reflection in the origin. Examples of symmetric sets in $\mathbb{R}^2$ are ellipses and parallelograms whose centres are at the origin, lines through the origin, and $\mathbb{R}^2$ itself. The set $\{\mathbf{a}\} + B$, where $\mathbf{a} \in \mathbb{R}^n$, is often written as $\mathbf{a} + B$ and is called a **translate** of B or, more precisely, the **translate of B by a**. It is an easy exercise in set theory to show that

$$A + B = \bigcup (\mathbf{a} + B : \mathbf{a} \in A),$$

i.e. $A + B$ is the union of all translates of B by vectors $\mathbf{a}$ in A. This result can help us to visualize $A + B$ in simple cases. For example, suppose that A and B are the square and the circular disc in $\mathbb{R}^2$ defined by the equations

$$A = \{(x, y) : |x|, |y| \leqslant 1\}, \qquad B = \{(x, y) : x^2 + y^2 \leqslant 1\}.$$

Then $\mathbf{a} + B$ is the circular disc with centre $\mathbf{a}$ and radius 1; $A + B$ is the union of all such discs for $\mathbf{a} \in A$. This example is illustrated in Fig. 1.1.

Vector addition and scalar multiplication, when applied to sets in $\mathbb{R}^n$, do not have all the properties one might expect, and the reader is warned to be cautious. For example, it is not always true that $A + A = 2A$. To see this, let A consist of *distinct* points $\mathbf{a}$ and $\mathbf{b}$ in $\mathbb{R}^n$. Then $A + A = \{2\mathbf{a}, 2\mathbf{b}, \mathbf{a} + \mathbf{b}\}$, whereas $2A = \{2\mathbf{a}, 2\mathbf{b}\}$. Properties (i)–(viii) above do, however, partially generalize to give the following easily verified results:

(i)* $\quad A + B = B + A;$

(ii)* $\quad A + (B + C) = (A + B) + C;$

(iii)* $\quad A + \mathbf{0} = A;$

(iv)* $\quad \mathbf{0} \in A + (-A) \quad$ when $A \neq \varnothing$;

(v)* $\quad 1A = A;$

(vi)* $\lambda(\mu A) = (\lambda\mu)A;$

(vii)* $\lambda(A + B) = \lambda A + \lambda B;$

(viii)* $(\lambda + \mu)A \subseteq \lambda A + \mu A.$

Exercises 1.1

1. Show that, for all sets A, B, C in $\mathbb{R}^n$ and all scalars λ, μ:
   (i) $A + (B + C) = (A + B) + C$, and
   (ii) $(\lambda + \mu)A \subseteq \lambda A + \mu A$.
2. For which sets A in $\mathbb{R}^n$ does $(\lambda + \mu)A = \lambda A + \mu A$ for *all* scalars λ, μ?
3. Find the set $A + B$ in each of the following cases:
   (i) $A = \{(x, y): x^2 + y^2 \leqslant 2^2\}$, $B = \{(x, y): x^2 + y^2 \leqslant 3^2\}$;
   (ii) $A = \{(x, y): x + y + 1 = 0\}$, $B = \{(x, y): 2x + y + 1 = 0\}$;
   (iii) $A = \{(x, y): x + y + 1 = 0\}$, $B = \{(x, y): x + y + 2 = 0\}$;
   (iv) $A = \{(x, y): x > 0,\ xy = 1\}$, $B = \{(x, y): x < 0,\ xy = -1\}$;
   (v) $A = B = \{(x, y): x^2 + y^2 = 1\}$.
4. In $\mathbb{R}^n$ a set A is said to be **homothetic** to a set B, written here as $A \sim B$, if $A = \lambda B + \mathbf{u}$ for some $\lambda > 0$ and $\mathbf{u} \in \mathbb{R}^n$. Show that: (i) $A \sim A$; (ii) $A \sim B$ implies $B \sim A$; (iii) $A \sim B$ and $B \sim C$ imply $A \sim C$.

1.2 FLATS

In this section we introduce the concept of a *flat*, which is a generalization to $\mathbb{R}^n$ of the terms *point*, *line*, and *plane* of elementary geometry. Points, lines, and planes share the property that, with each two of their points, they contain the *entire* line through them. This property turns out to be an important one, and prompts the definition of a *flat* in $\mathbb{R}^n$. Since the idea of a *line* in $\mathbb{R}^n$ plays a central role in our discussion, we first examine it in the familiar surroundings of the three-dimensional space $\mathbb{R}^3$.

For each point $\mathbf{x}$ on the line through distinct points $\mathbf{a}$ and $\mathbf{b}$ of $\mathbb{R}^3$, there exists a unique scalar λ such that

$$\mathbf{x} = \mathbf{b} + \lambda(\mathbf{a} - \mathbf{b}) = \lambda\mathbf{a} + (1 - \lambda)\mathbf{b}.$$

See Fig. 1.2. Conversely, each point $\mathbf{x}$ of this form lies on the line through $\mathbf{a}$ and $\mathbf{b}$. Thus the line through $\mathbf{a}$ and $\mathbf{b}$ is the set

$$\{\lambda\mathbf{a} + (1 - \lambda)\mathbf{b} : \lambda \in \mathbb{R}\},$$

which can also be written in the symmetrical form

$$\{\lambda\mathbf{a} + \mu\mathbf{b} : \lambda + \mu = 1\}.$$

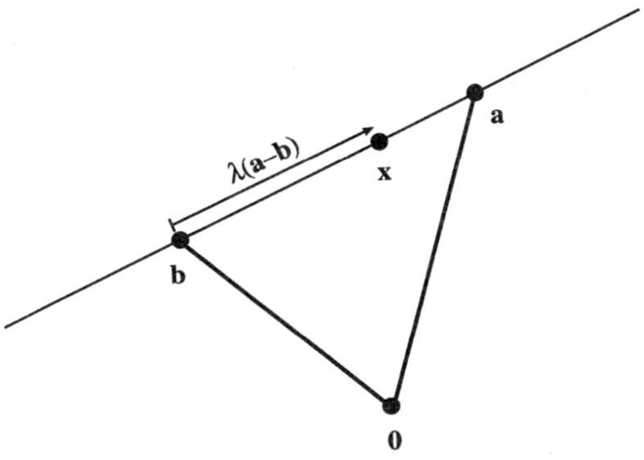

Fig. 1.2

We note that the subset

$$\{\lambda\mathbf{a} + (1-\lambda)\mathbf{b} : 0 \leqslant \lambda \leqslant 1\} = \{\lambda\mathbf{a} + \mu\mathbf{b} : \lambda, \mu \geqslant 0, \lambda + \mu = 1\}$$

of the line through **a** and **b** is the *line segment* joining **a** and **b**, a fact which will be needed when we define a convex set in Section 2.1. The discussion above in $\mathbb{R}^3$ leads us to make the following definition in $\mathbb{R}^n$. The **line** through distinct points **a** and **b** of $\mathbb{R}^n$ is the set

$$\{\lambda\mathbf{a} + \mu\mathbf{b} : \lambda + \mu = 1\}.$$

Clearly this set contains both **a** and **b**, and its points can be placed into a bijective correspondence with the points of the real line $\mathbb{R}$ itself.

In view of our earlier remarks, the following definition of a flat should come as no surprise. The set A in $\mathbb{R}^n$ is called a **flat** if whenever it contains two points, it also contains the entire line through them. Expressed algebraically, A is a flat if $\lambda\mathbf{a} + \mu\mathbf{b} \in A$ whenever $\mathbf{a}, \mathbf{b} \in A$ and $\lambda + \mu = 1$. Equivalently, A is a flat if $\lambda A + \mu A \subseteq A$ whenever $\lambda + \mu = 1$. Synonyms for flat used by other authors are: *affine set*, *affine variety*, *affine manifold*, *linear variety*, and *linear manifold*. The empty set, singletons, lines, and $\mathbb{R}^n$ itself are examples of flats in $\mathbb{R}^n$; planes are flats in $\mathbb{R}^3$.

Flats which contain the origin turn out to be of special importance, and indeed are familiar objects as we shall now see. Let A be a flat in $\mathbb{R}^n$ which contains the origin. Suppose that $\mathbf{a}, \mathbf{b} \in A$ and $\lambda \in \mathbb{R}$. Since A is a flat and $\mathbf{a}, \mathbf{0} \in A$, $\lambda\mathbf{a} + (1-\lambda)\mathbf{0} \in A$, i.e. $\lambda\mathbf{a} \in A$. Thus A is closed under scalar multiplication. Since A is a flat and $\mathbf{a}, \mathbf{b} \in A$, $\frac{1}{2}\mathbf{a} + \frac{1}{2}\mathbf{b} \in A$. But A is closed under scalar multiplication, so $2(\frac{1}{2}\mathbf{a} + \frac{1}{2}\mathbf{b}) \in A$, i.e. $\mathbf{a} + \mathbf{b} \in A$. Thus A is closed under addition. Hence A is a non-empty subset of $\mathbb{R}^n$ which is closed under addition and scalar multiplication, i.e. A is a *subspace* of the real vector space $\mathbb{R}^n$. Trivially, a subspace of $\mathbb{R}^n$ is a flat containing the origin. We have thus shown that flats through the origin in $\mathbb{R}^n$ are precisely the subspaces of $\mathbb{R}^n$. This fact allows us to deduce properties of flats from the corresponding properties of subspaces. Flats are the geometric objects in which we are interested, but subspaces have the simpler algebraic structure.

The following theorem gives the exact relationship between flats and subspaces in $\mathbb{R}^n$.

Theorem 1.2.1 *The non-empty flats in $\mathbb{R}^n$ are precisely the translates of subspaces of $\mathbb{R}^n$.*

Proof Suppose first that A is a non-empty flat in $\mathbb{R}^n$. Let $\mathbf{a} \in A$. We show that $A - \mathbf{a}$ is a flat. Let $\mathbf{x}, \mathbf{y} \in A - \mathbf{a}$ and $\lambda + \mu = 1$. Then $\mathbf{x} + \mathbf{a}$, $\mathbf{y} + \mathbf{a} \in A$, whence

$$\lambda(\mathbf{x} + \mathbf{a}) + \mu(\mathbf{y} + \mathbf{a}) = \lambda\mathbf{x} + \mu\mathbf{y} + \mathbf{a} \in A.$$

Thus $\lambda\mathbf{x} + \mu\mathbf{y} \in A - \mathbf{a}$, and $A - \mathbf{a}$ is a flat. Since $A - \mathbf{a}$ contains the origin, it must be a subspace of $\mathbb{R}^n$. Hence the non-empty flat A is the translate of the subspace $A - \mathbf{a}$ of $\mathbb{R}^n$ by the vector $\mathbf{a}$.

Suppose next that A is a subspace of $\mathbb{R}^n$ and that $\mathbf{u} \in \mathbb{R}^n$. We show that $A + \mathbf{u}$ is a flat. Let $\mathbf{x}, \mathbf{y} \in A + \mathbf{u}$ and $\lambda + \mu = 1$. Then there exist $\mathbf{a}, \mathbf{b} \in A$ such that $\mathbf{x} = \mathbf{a} + \mathbf{u}, \mathbf{y} = \mathbf{b} + \mathbf{u}$. Now

$$\lambda\mathbf{x} + \mu\mathbf{y} = \lambda\mathbf{a} + \mu\mathbf{b} + \mathbf{u} \in A + \mathbf{u},$$

since $\lambda\mathbf{a} + \mu\mathbf{b} \in A$, as A is a subspace of $\mathbb{R}^n$. This shows that $A + \mathbf{u}$ is a flat. □

Corollary 1.2.2 *Each non-empty flat in $\mathbb{R}^n$ is the translate of precisely one subspace of $\mathbb{R}^n$.*

Proof Let A be a non-empty flat in $\mathbb{R}^n$. Suppose that A is a translate of both the subspaces B and C of $\mathbb{R}^n$. Then C must be a translate of B, so there exists $\mathbf{b}$ in $\mathbb{R}^n$ such that $C = B + \mathbf{b}$. Since $\mathbf{0}$ lies in C, it follows that $-\mathbf{b}$, and hence $\mathbf{b}$, lies in B. Thus $C = B + \mathbf{b} \subseteq B$. By symmetry, $B \subseteq C$. Hence $B = C$, and A is the translate of precisely one subspace of $\mathbb{R}^n$. □

The observation that two (distinct) lines in $\mathbb{R}^2$ are parallel if and only if one is a translate of the other prompts the following definition. In $\mathbb{R}^n$ a flat A is said to be **parallel** to a flat B if each is a translate of the other. The relation of parallelism is an equivalence relation on the family of all flats in $\mathbb{R}^n$. This notion of parallelism does not quite accord with that used in elementary geometry on two counts. Firstly, a flat is considered to be parallel to itself. Secondly, it only allows parallelism between flats of the *same dimension*. For example, we cannot speak of a line being parallel to a plane. Corollary 1.2.2 shows that each non-empty flat in $\mathbb{R}^n$ is parallel to precisely one subspace of $\mathbb{R}^n$.

The next result expresses one of the most important properties of the family of flats in $\mathbb{R}^n$, namely that it is closed under the formation of arbitrary intersections.

Theorem 1.2.3 *The intersection of an arbitrary family of flats in $\mathbb{R}^n$ is a flat.*

Proof Let $(A_i : i \in I)$ be a family of flats in $\mathbb{R}^n$. If $\mathbf{a}, \mathbf{b} \in \bigcap (A_i : i \in I)$ and $\lambda + \mu = 1$, then $\mathbf{a}, \mathbf{b} \in A_i$ and, as A_i is a flat, $\lambda\mathbf{a} + \mu\mathbf{b} \in A_i$ for each $i \in I$. Thus $\lambda\mathbf{a} + \mu\mathbf{b} \in \bigcap (A_i : i \in I)$, which shows that the intersection is a flat. □

The **affine hull** aff A of a set A in $\mathbb{R}^n$ is the intersection of all flats in $\mathbb{R}^n$ containing A. Such flats exist, since $\mathbb{R}^n$ is a flat containing A. In view of Theorem 1.2.3, aff A is a flat which contains A. Moreover, if B is any flat in $\mathbb{R}^n$ containing A, then aff $A \subseteq B$. Thus we may refer to aff A as the *smallest* flat in $\mathbb{R}^n$ containing A. Clearly, A is a flat if and only if $A = \text{aff}\, A$, and $\text{aff}(\text{aff}\, A) = \text{aff}\, A$. Another easy result is that $\text{aff}\, A \subseteq \text{aff}\, B$ when $A \subseteq B$. In the space $\mathbb{R}^3$ the affine hull of two distinct points is the line through them, the affine hull of three non-collinear points is the plane which they determine, and the affine hull of four non-coplanar points is the whole space $\mathbb{R}^3$ itself. We return to the topic of affine hulls later in the section, when we show how aff A can be constructed from the points of A itself.

By definition, a set A in $\mathbb{R}^n$ is a flat if $\lambda\mathbf{a} + \mu\mathbf{b} \in A$ whenever $\mathbf{a}, \mathbf{b} \in A$ and $\lambda + \mu = 1$. This defining relation of a flat implies a more general one, as we now establish in the following fundamental theorem.

Theorem 1.2.4 *Let $\mathbf{a}_1, \ldots, \mathbf{a}_m$ be points of a flat A in $\mathbb{R}^n$. Let $\lambda_1 + \ldots + \lambda_m = 1$. Then $\lambda_1\mathbf{a}_1 + \ldots + \lambda_m\mathbf{a}_m \in A$.*

Proof Let $\mathbf{a} \in A$. Then the points $\mathbf{a}_1 - \mathbf{a}, \ldots, \mathbf{a}_m - \mathbf{a}$ lie in the subspace $A - \mathbf{a}$ of $\mathbb{R}^n$, whence so too does the point

$$\lambda_1(\mathbf{a}_1 - \mathbf{a}) + \ldots + \lambda_m(\mathbf{a}_m - \mathbf{a}) = \lambda_1\mathbf{a}_1 + \ldots + \lambda_m\mathbf{a}_m - \mathbf{a}.$$

Hence $\lambda_1\mathbf{a}_1 + \ldots + \lambda_m\mathbf{a}_m \in A$ as desired. □

This last result leads us to make the following definition. A point $\mathbf{x}$ is said to be an **affine combination** of points $\mathbf{a}_1, \ldots, \mathbf{a}_m$ in $\mathbb{R}^n$ if there exist scalars $\lambda_1, \ldots, \lambda_m$ with $\lambda_1 + \ldots + \lambda_m = 1$ such that

$$\mathbf{x} = \lambda_1\mathbf{a}_1 + \ldots + \lambda_m\mathbf{a}_m.$$

Theorem 1.2.4 can now be expressed as follows: *every affine combination of points of a flat in* $\mathbb{R}^n$ *belongs to that flat.*

The affine hull of a set was defined by means of flats containing that set. In contrast, the following theorem expresses the affine hull of a set in terms of points of the set itself.

Theorem 1.2.5 *Let* A *be a set in* $\mathbb{R}^n$. *Then* $\operatorname{aff} A$ *is the set of all affine combinations of points of* A.

Proof Denote by B the set of all affine combinations of points of A. That $B \subseteq \operatorname{aff} A$ follows from Theorem 1.2.4 and the inclusion $A \subseteq \operatorname{aff} A$.

We next show that B is a flat. If $\mathbf{x}, \mathbf{y} \in B$, then

$$\mathbf{x} = \lambda_1 \mathbf{a}_1 + \ldots + \lambda_m \mathbf{a}_m, \quad \mathbf{y} = \mu_1 \mathbf{b}_1 + \ldots + \mu_p \mathbf{b}_p,$$

for some $\mathbf{a}_1, \ldots, \mathbf{a}_m, \mathbf{b}_1, \ldots, \mathbf{b}_p \in A$, and scalars $\lambda_1, \ldots, \lambda_m$, $\mu_1, \ldots, \mu_p$ with $\lambda_1 + \ldots + \lambda_m = 1$, $\mu_1 + \ldots + \mu_p = 1$. Let $\lambda + \mu = 1$. Then

$$\lambda \mathbf{x} + \mu \mathbf{y} = \lambda\lambda_1 \mathbf{a}_1 + \ldots + \lambda\lambda_m \mathbf{a}_m + \mu\mu_1 \mathbf{b}_1 + \ldots + \mu\mu_p \mathbf{b}_p$$

and

$$\begin{aligned} &\lambda\lambda_1 + \ldots + \lambda\lambda_m + \mu\mu_1 + \ldots + \mu\mu_p \\ = \; &\lambda(\lambda_1 + \ldots + \lambda_m) + \mu(\mu_1 + \ldots + \mu_p) = \lambda + \mu = 1. \end{aligned}$$

Thus $\lambda \mathbf{x} + \mu \mathbf{y} \in B$, so B is a flat. Since B is a flat and $B \supseteq A$, it follows that $B \supseteq \operatorname{aff} A$. Hence $B = \operatorname{aff} A$. □

Corollary 1.2.6 *Let* $\mathbf{a}_1, \ldots, \mathbf{a}_m \in \mathbb{R}^n$. *Then*

$$\operatorname{aff}\{\mathbf{a}_1, \ldots, \mathbf{a}_m\} = \{\lambda_1 \mathbf{a}_1 + \ldots + \lambda_m \mathbf{a}_m : \lambda_1 + \ldots + \lambda_m = 1\}. \quad \square$$

Example 1.2.7 Each point $\mathbf{x} = (x_1, \ldots, x_n)$ of $\mathbb{R}^n$ can be expressed as an affine combination of the zero vector $\mathbf{0}$ and the elementary vectors $\mathbf{e}_1, \ldots, \mathbf{e}_n$ as follows:

$$\mathbf{x} = (1 - x_1 - \ldots - x_n)\mathbf{0} + x_1 \mathbf{e}_1 + \ldots + x_n \mathbf{e}_n.$$

Corollary 1.2.6 now shows that $\operatorname{aff}\{\mathbf{0}, \mathbf{e}_1, \ldots, \mathbf{e}_n\} = \mathbb{R}^n$. □

Let A be a non-empty set in $\mathbb{R}^n$. We recall that a point of the form $\lambda_1 \mathbf{a}_1 + \ldots + \lambda_m \mathbf{a}_m$, where $\mathbf{a}_1, \ldots, \mathbf{a}_m \in A$ and $\lambda_1, \ldots, \lambda_m$ are scalars, is said to be a *linear combination* of points of A. The set of all such linear combinations is the *smallest* subspace of $\mathbb{R}^n$ which contains A, and is called here the *linear hull* of A and we denote it by $\operatorname{lin} A$. Since $\operatorname{lin} A$ is a flat containing $A \cup \{\mathbf{0}\}$, it follows that $\operatorname{aff}(A \cup \{\mathbf{0}\}) \subseteq \operatorname{lin} A$. On the other hand, $\operatorname{aff}(A \cup \{\mathbf{0}\})$ is a subspace of $\mathbb{R}^n$ containing A, so $\operatorname{lin} A \subseteq \operatorname{aff}(A \cup \{\mathbf{0}\})$. We conclude that $\operatorname{lin} A = \operatorname{aff}(A \cup \{\mathbf{0}\})$. We define lin $\varnothing = \{\mathbf{0}\}$. This ensures that lin $\varnothing$ is the *smallest* subspace of $\mathbb{R}^n$ which contains $\varnothing$, and that lin $\varnothing = \text{aff}\,(\varnothing \cup \{\mathbf{0}\})$.

We conclude the section by examining how flats behave with respect to the operations of addition and scalar multiplication.

Theorem 1.2.8 *Let A, B be flats in $\mathbb{R}^n$ and let α be a scalar. Then $A + B$ and αA are flats.*

Proof Let $\lambda + \mu = 1$. Since A and B are flats, $\lambda A + \mu A \subseteq A$ and $\lambda B + \mu B \subseteq B$. Thus

$$\lambda(A + B) + \mu(A + B) = (\lambda A + \mu A) + (\lambda B + \mu B) \subseteq A + B$$

and

$$\lambda(\alpha A) + \mu(\alpha A) = \alpha(\lambda A + \mu A) \subseteq \alpha A.$$

This shows that $A + B$ and αA are flats. □

Corollary 1.2.9 *Let $A_1, \ldots, A_m$ be flats in $\mathbb{R}^n$ and let $\lambda_1, \ldots, \lambda_m$ be scalars. Then $\lambda_1 A_1 + \ldots + \lambda_m A_m$ is a flat.* □

We saw in the last section that it is not in general true that $A + A = 2A$. It is true, however, when A is a flat. This, and more, is proved in the next theorem.

Theorem 1.2.10 *Let A be a flat in $\mathbb{R}^n$ and let $\lambda_1, \ldots, \lambda_m$ be scalars with $\lambda_1 + \ldots + \lambda_m \neq 0$. Then*

$$(\lambda_1 + \ldots + \lambda_m)A = \lambda_1 A + \ldots + \lambda_m A.$$

Proof Write $\lambda = \lambda_1 + \ldots + \lambda_m$. Then, using Theorem 1.2.4, we deduce that

$$\begin{aligned}(\lambda_1 + \ldots + \lambda_m)A &\subseteq \lambda_1 A + \ldots + \lambda_m A\\ &= \lambda\left(\frac{\lambda_1}{\lambda}A + \ldots + \frac{\lambda_m}{\lambda}A\right)\\ &\subseteq \lambda A\\ &= (\lambda_1 + \ldots + \lambda_m)A.\end{aligned}$$

Thus $(\lambda_1 + \ldots + \lambda_m)A = \lambda_1 A + \ldots + \lambda_m A$, as desired. □

Exercises 1.2

1. Show that, if $a_0, a_1, \ldots, a_n$ are scalars, then the set

$$\{(x_1, \ldots, x_n) : a_1 x_1 + \ldots + a_n x_n = a_0\}$$

is a flat in $\mathbb{R}^n$. When is this flat a subspace?

2. Let A be a non-empty flat in $\mathbb{R}^n$. Show that $A - A$ is the unique subspace of $\mathbb{R}^n$ which is parallel to A.
3. Show that each point $(x_1, \ldots, x_n)$ of $\mathbb{R}^n$, where $n \geqslant 2$, can be expressed as an affine combination of the points $\mathbf{e}_1, \ldots, \mathbf{e}_n$, $\mathbf{e}_1 + \ldots + \mathbf{e}_n$. What can be deduced from this result?
4. Give an example of a flat A in $\mathbb{R}^n$ and scalars $\lambda_1, \ldots, \lambda_m$ such that

$$(\lambda_1 + \ldots + \lambda_m)A \neq \lambda_1 A + \ldots + \lambda_m A.$$

5. Let $\mathbf{a}, \mathbf{a}_1, \ldots, \mathbf{a}_m \in \mathbb{R}^n$. Prove that

$$\mathbf{a} + \operatorname{aff}\{\mathbf{a}_1, \ldots, \mathbf{a}_m\} = \operatorname{aff}\{\mathbf{a} + \mathbf{a}_1, \ldots, \mathbf{a} + \mathbf{a}_m\}.$$

1.3 DIMENSION

In elementary geometry, each flat has a *dimension* associated with it as follows: a point has dimension zero, a line dimension one, a plane dimension two, and the whole space dimension three. Our aim in this section is to introduce the concept of dimension for flats in $\mathbb{R}^n$, and establish its properties using the fact that $\mathbb{R}^n$ is a real vector space.

Our discussion centres around the notion of *affine dependence*. The term *affinely dependent* is a generalization to $\mathbb{R}^n$ of the familiar terms collinear and coplanar, both of which imply a degree of degeneracy in the sets that they describe. In general, three points of $\mathbb{R}^3$ are *not* collinear; they are so, precisely when one of them lies on the line passing through the other two. Expressed in the language of affine hulls: three points in $\mathbb{R}^3$ are collinear, precisely when one of them belongs to the affine hull of the remaining two. Similarly, four points in $\mathbb{R}^3$ are coplanar, precisely when one of them belongs to the affine hull of the remaining three.

These considerations lead us to make the following definition. The set A in $\mathbb{R}^n$ is said to be **affinely dependent** if there exists $\mathbf{a} \in A$ such that $\mathbf{a} \in \operatorname{aff}(A \backslash \{\mathbf{a}\})$. Thus in $\mathbb{R}^3$, a set of three points is affinely dependent if and only if it is collinear, a set of four points is affinely dependent if and only if it is coplanar, and any set having *more* than four points is affinely dependent. A set in $\mathbb{R}^n$ which is not affinely dependent is said to be **affinely independent**. In $\mathbb{R}^3$, a set of three points is affinely independent precisely when it is the vertex set of a non-degenerate triangle; and a set of four points is affinely independent precisely when it is the vertex set of a non-degenerate tetrahedron. In $\mathbb{R}^n$, the empty set, every singleton, and every set consisting of two points are affinely independent. Since any set in $\mathbb{R}^n$ which contains an affinely dependent set is itself affinely dependent, it follows that every subset of an affinely independent set is affinely independent.

The following theorem provides a working criterion for affine dependence.

Theorem 1.3.1 *Let* A *be a set in* $\mathbb{R}^n$. *Then* A *is affinely dependent if and only if there exist distinct points* $\mathbf{a}_1, \ldots, \mathbf{a}_m$ *of* A *and scalars* $\lambda_1, \ldots, \lambda_m$, *not all zero, such that*

$$\lambda_1 \mathbf{a}_1 + \ldots + \lambda_m \mathbf{a}_m = \mathbf{0} \qquad \text{and} \qquad \lambda_1 + \ldots + \lambda_m = 0. \quad (1.3.1)$$

Proof Suppose that A is affinely dependent. Then there exists $\mathbf{a}_1 \in A$ such that $\mathbf{a}_1 \in \operatorname{aff}(A \backslash \{\mathbf{a}_1\})$. By Theorem 1.2.5, there exist (distinct) points $\mathbf{a}_2, \ldots, \mathbf{a}_m$ of $A \backslash \{\mathbf{a}_1\}$ and scalars $\mu_2, \ldots, \mu_m$ such that

$$\mathbf{a}_1 = \mu_2 \mathbf{a}_2 + \ldots + \mu_m \mathbf{a}_m \qquad \text{and} \qquad \mu_2 + \ldots + \mu_m = 1.$$

Write $\lambda_1 = -1$, $\lambda_2 = \mu_2, \ldots, \lambda_m = \mu_m$. Then $\lambda_1, \ldots, \lambda_m$ are not all zero and satisfy (1.3.1).

Suppose next that there exist *distinct* points $\mathbf{a}_1, \ldots, \mathbf{a}_m$ of A, and scalars $\lambda_1, \ldots, \lambda_m$, not all zero, which satisfy (1.3.1). Suppose that $\lambda_1 \neq 0$. Then

$$\mathbf{a}_1 = -(1/\lambda_1)(\lambda_2 \mathbf{a}_2 + \ldots + \lambda_m \mathbf{a}_m) \quad \text{and} \quad -(1/\lambda_1)(\lambda_2 + \ldots + \lambda_m) = 1,$$

which shows that $\mathbf{a}_1$ is an affine combination of $\mathbf{a}_2, \ldots, \mathbf{a}_m$. Hence

$$\mathbf{a}_1 \in \operatorname{aff}\{\mathbf{a}_2, \ldots, \mathbf{a}_m\} \subseteq \operatorname{aff}(A \backslash \{\mathbf{a}_1\}),$$

and so A is affinely dependent. □

Corollary 1.3.2 *A subset* $\{\mathbf{a}_1, \ldots, \mathbf{a}_m\}_{\neq}$ *of* $\mathbb{R}^n$ *is affinely dependent if and only if there exist scalars* $\lambda_1, \ldots, \lambda_m$, *not all zero, such that*

$$\lambda_1 \mathbf{a}_1 + \ldots + \lambda_m \mathbf{a}_m = \mathbf{0} \qquad \text{and} \qquad \lambda_1 + \ldots + \lambda_m = 0. \ \square$$

Corollary 1.3.3 *Let* $\{\mathbf{a}_1, \ldots, \mathbf{a}_m\}_{\neq}$ *be an affinely independent set in* $\mathbb{R}^n$. *Then each point of* $\operatorname{aff}\{\mathbf{a}_1, \ldots, \mathbf{a}_m\}$ *can be expressed uniquely in the form*

$$\lambda_1 \mathbf{a}_1 + \ldots + \lambda_m \mathbf{a}_m, \qquad \text{where } \lambda_1 + \ldots + \lambda_m = 1.$$

Proof Corollary 1.2.6 shows that each point of $\operatorname{aff}\{\mathbf{a}_1, \ldots, \mathbf{a}_m\}$ can be expressed in the desired form. To establish the uniqueness, suppose that

$$\lambda_1 \mathbf{a}_1 + \ldots + \lambda_m \mathbf{a}_m = \mu_1 \mathbf{a}_1 + \ldots + \mu_m \mathbf{a}_m,$$

where

$$\lambda_1 + \ldots + \lambda_m = \mu_1 + \ldots + \mu_m = 1.$$

Then

$$(\lambda_1 - \mu_1)\mathbf{a}_1 + \ldots + (\lambda_m - \mu_m)\mathbf{a}_m = \mathbf{0} \text{ and } (\lambda_1 - \mu_1) + \ldots + (\lambda_m - \mu_m) = 0.$$

Since $\{\mathbf{a}_1, \ldots, \mathbf{a}_m\}_{\neq}$ is affinely independent, Corollary 1.3.2 shows that the scalars $\lambda_1 - \mu_1, \ldots, \lambda_m - \mu_m$ must be zero. Thus $\lambda_1 = \mu_1, \ldots, \lambda_m = \mu_m$, and the uniqueness is established. □

We mentioned above that any set of *more* than four points in $\mathbb{R}^3$ is affinely dependent. This is a special case of the following corollary.

Corollary 1.3.4 *An affinely independent set in $\mathbb{R}^n$ cannot contain more than $n + 1$ points.*

Proof To prove the result, it will suffice to show that every set of the form $\{\mathbf{a}_1, \ldots, \mathbf{a}_m\}_{\neq}$ in $\mathbb{R}^n$, where $m > n + 1$, is affinely dependent. Let $\{\mathbf{a}_1, \ldots, \mathbf{a}_m\}_{\neq}$ be a set in $\mathbb{R}^n$, where $m > n + 1$. Then the system of the $n + 1$ linear simultaneous equations

$$\lambda_1 \mathbf{a}_1 + \ldots + \lambda_m \mathbf{a}_m = \mathbf{0}, \qquad \lambda_1 + \ldots + \lambda_m = 0$$

in the m unknowns $\lambda_1, \ldots, \lambda_m$ is homogeneous, and so has a *non-trivial* solution, since $m > n + 1$. Hence $\{\mathbf{a}_1, \ldots, \mathbf{a}_m\}_{\neq}$ is affinely dependent by Corollary 1.3.2. □

Corollary 1.3.5 *Let A be an affinely independent subset of $\mathbb{R}^n$. Suppose that* **a** *is a point of $\mathbb{R}^n$ not lying in* aff A. *Then the set $A \cup \{\mathbf{a}\}$ is affinely independent.*

Proof We argue by contradiction. Suppose that $A \cup \{\mathbf{a}\}$ is affinely dependent. Then there exist distinct points $\mathbf{a}_1, \ldots, \mathbf{a}_m$ of A and scalars $\lambda, \lambda_1, \ldots, \lambda_m$, not all zero, such that

$$\lambda \mathbf{a} + \lambda_1 \mathbf{a}_1 + \ldots + \lambda_m \mathbf{a}_m = \mathbf{0} \quad \text{and} \quad \lambda + \lambda_1 + \ldots + \lambda_m = 0. \tag{1.3.2}$$

The scalar λ cannot be zero, for (1.3.2) would then imply that A is affinely dependent. Thus (1.3.2) can be used to express **a** as an affine combination of $\mathbf{a}_1, \ldots, \mathbf{a}_m$, whence $\mathbf{a} \in \operatorname{aff}\{\mathbf{a}_1, \ldots, \mathbf{a}_m\}$. This, however, contradicts the hypothesis that $\mathbf{a} \notin \operatorname{aff} A$. Hence $A \cup \{\mathbf{a}\}$ is affinely independent. □

Example 1.3.6 In $\mathbb{R}^n$ the set $\{\mathbf{0}, \mathbf{e}_1, \ldots, \mathbf{e}_n\}$ is affinely independent. To see this, suppose that the scalars $\lambda, \lambda_1, \ldots, \lambda_n$ satisfy

$$\lambda \mathbf{0} + \lambda_1 \mathbf{e}_1 + \ldots + \lambda_n \mathbf{e}_n = \mathbf{0} \quad \text{and} \quad \lambda + \lambda_1 + \ldots + \lambda_n = 0.$$

The first of these equations shows that $\lambda_1, \ldots, \lambda_n$ are all zero, whence λ must also be zero from the second equation. Corollary 1.3.2 now shows that the set $\{\mathbf{0}, \mathbf{e}_1, \ldots, \mathbf{e}_n\}$ is affinely independent. □

We now continue the informal discussion on dimension with which we began the section. In $\mathbb{R}^3$ as a simple case-by-case consideration shows, each r-dimensional flat ($r = 0, 1, 2, 3$) is the affine hull of some affinely independent set of $r + 1$ points. For example, a plane is the affine hull of any three of its points which are not collinear, while Examples 1.2.7 and 1.3.6 show that $\mathbb{R}^3$ is the affine hull of the affinely independent set $\{\mathbf{0}, \mathbf{e}_1, \mathbf{e}_2, \mathbf{e}_3\}$. This suggests that we might assign a *dimension* r to a flat in $\mathbb{R}^n$ if it is the affine hull of some affinely independent set of $r + 1$ points. Before we can formalize this idea, however, two results need to be established: (i) every flat in $\mathbb{R}^n$ *is* the affine hull of some finite affinely independent set; (ii) if two affinely independent sets in $\mathbb{R}^n$ have the same affine hull, then they have the same number of elements. These results are contained in the following theorem.

Theorem 1.3.7 *Every flat in* $\mathbb{R}^n$ *is the affine hull of some finite affinely independent subset of* $\mathbb{R}^n$. *Moreover, the number of elements in such a subset is determined uniquely by the flat itself.*

Proof Consider the non-trivial case of a flat A in $\mathbb{R}^n$ which is neither empty nor a singleton. Let m be the largest positive integer such that A contains an affinely independent subset of $m + 1$ elements. Such an m exists by Corollary 1.3.4, and $m \geqslant 1$, since A contains at least two points. Let $\{\mathbf{a}_0, \mathbf{a}_1, \ldots, \mathbf{a}_m\}_{\neq}$ be an affinely independent subset of A. Since A is a flat, $\operatorname{aff}\{\mathbf{a}_0, \mathbf{a}_1, \ldots, \mathbf{a}_m\} \subseteq A$. Now $A \subseteq \operatorname{aff}\{\mathbf{a}_0, \mathbf{a}_1, \ldots, \mathbf{a}_m\}$, for otherwise there would exist some point $\mathbf{a}$ of A not lying in $\operatorname{aff}\{\mathbf{a}_0, \mathbf{a}_1, \ldots, \mathbf{a}_m\}$ and, by Corollary 1.3.5, $\{\mathbf{a}, \mathbf{a}_0, \mathbf{a}_1, \ldots, \mathbf{a}_m\}$ would be an affinely independent subset of A having $m + 2$ elements, so contradicting the definition of m. Hence $A \subseteq \operatorname{aff}\{\mathbf{a}_0, \mathbf{a}_1, \ldots, \mathbf{a}_m\}$, and $A = \operatorname{aff}\{\mathbf{a}_0, \mathbf{a}_1, \ldots, \mathbf{a}_m\}$. This establishes the first assertion of the theorem.

We now complete the proof by showing that m is the dimension of the unique subspace B of $\mathbb{R}^n$ that is parallel to A. This we do by proving that the subset $\{\mathbf{a}_1 - \mathbf{a}_0, \ldots, \mathbf{a}_m - \mathbf{a}_0\}$ of B is a basis for B. Let $\mathbf{b} \in B$. Then $\mathbf{b} = \mathbf{x} - \mathbf{a}_0$ for some $\mathbf{x} \in A$, and there exist scalars $\lambda_0, \lambda_1, \ldots, \lambda_m$ such that

$$\mathbf{x} = \lambda_0\mathbf{a}_0 + \lambda_1\mathbf{a}_1 + \ldots + \lambda_m\mathbf{a}_m \qquad \text{and} \qquad \lambda_0 + \lambda_1 + \ldots + \lambda_m = 1.$$

Hence

$$\mathbf{b} = \mathbf{x} - \mathbf{a}_0 = \lambda_1(\mathbf{a}_1 - \mathbf{a}_0) + \ldots + \lambda_m(\mathbf{a}_m - \mathbf{a}_0),$$

which shows that $\{\mathbf{a}_1 - \mathbf{a}_0, \ldots, \mathbf{a}_m - \mathbf{a}_0\}$ spans B. Finally, suppose that scalars $\mu_1, \ldots, \mu_m$ satisfy

$$\mu_1(\mathbf{a}_1 - \mathbf{a}_0) + \ldots + \mu_m(\mathbf{a}_m - \mathbf{a}_0) = \mathbf{0}.$$

Then

$$-(\mu_1 + \ldots + \mu_m)\mathbf{a}_0 + \mu_1\mathbf{a}_1 + \ldots + \mu_m\mathbf{a}_m = \mathbf{0} \quad \text{and}$$
$$-(\mu_1 + \ldots + \mu_m) + \mu_1 + \ldots + \mu_m = 0.$$

But $\{\mathbf{a}_0, \mathbf{a}_1, \ldots, \mathbf{a}_m\}_{\neq}$ is affinely independent, whence all of $\mu_1, \ldots, \mu_m$ are zero. Thus $\{\mathbf{a}_1 - \mathbf{a}_0, \ldots, \mathbf{a}_m - \mathbf{a}_0\}$ is linearly independent. We conclude that $\{\mathbf{a}_1 - \mathbf{a}_0, \ldots, \mathbf{a}_m - \mathbf{a}_0\}$ is a basis for B. Hence m is the dimension of B, and so is uniquely determined by A. □

Theorem 1.3.7 enables us to make the following definition. A flat in $\mathbb{R}^n$ which is the affine hull of some affinely independent set of $r + 1$ points is said to have **dimension** r and is called an r-**flat**. It follows from Theorem 1.3.7 that each flat in $\mathbb{R}^n$ has a unique dimension r attached to it, and from Corollary 1.3.4 that $r \leqslant n$. The empty flat is the affine hull of the (affinely independent) empty set, and so has dimension -1. Clearly every singleton (point) has dimension 0 and every line has dimension 1. We have already seen that $\mathbb{R}^n$ is the affine hull of the affinely independent set $\{\mathbf{0}, \mathbf{e}_1, \ldots, \mathbf{e}_n\}$, whence $\mathbb{R}^n$ has dimension n. The concept of dimension is extended to arbitrary subsets of $\mathbb{R}^n$ by defining the **dimension** dim A of a set A in $\mathbb{R}^n$ to be the dimension of the flat aff A. We note that when a flat in $\mathbb{R}^n$ is also a subspace of $\mathbb{R}^n$ its dimension as defined above coincides with its dimension as a subspace of the real vector space $\mathbb{R}^n$; see the proof of Theorem 1.3.7. Hence we may apply the term *dimension* unambiguously both to flats and subspaces of $\mathbb{R}^n$.

The intersection theory of flats is more complicated than that of subspaces, due to the fact that two flats can have an empty intersection. We do, however, have the following result.

Theorem 1.3.8 *Let A and B be flats in $\mathbb{R}^n$ which have a non-empty intersection. Then*

$$\dim(A + B) + \dim(A \cap B) = \dim A + \dim B.$$

Proof Let $\mathbf{c} \in A \cap B$. Then $A - \mathbf{c}$ and $B - \mathbf{c}$ are subspaces of $\mathbb{R}^n$, and so, by the dimension theorem of elementary linear algebra,

$$\dim((A - \mathbf{c}) + (B - \mathbf{c})) + \dim((A - \mathbf{c}) \cap (B - \mathbf{c}))$$
$$= \dim(A - \mathbf{c}) + \dim(B - \mathbf{c}),$$

that is,

$$\dim(A + B - 2\mathbf{c}) + \dim((A \cap B) - \mathbf{c}) = \dim(A - \mathbf{c}) + \dim(B - \mathbf{c}).$$

The proof of Theorem 1.3.7 shows that the dimension of a non-empty flat in $\mathbb{R}^n$ coincides with the dimension of the unique subspace of $\mathbb{R}^n$ which is parallel to it. It follows from this last result that the dimension of any translate of a flat is the same as the dimension of the flat itself. Thus the last equation above simplifies to

$$\dim(A + B) + \dim(A \cap B) = \dim A + \dim B. \quad \square$$

An **affine basis** for a flat in $\mathbb{R}^n$ is any affinely independent set in $\mathbb{R}^n$ whose affine hull is that flat. Theorem 1.3.7 shows that every flat has an affine basis. By definition, every affine basis for an r-flat has precisely $r + 1$ elements. For example, $\{\mathbf{0}, \mathbf{e}_1, \ldots, \mathbf{e}_n\}$ is an affine basis for $\mathbb{R}^n$. The next result shows that every affinely independent subset of a set in $\mathbb{R}^n$ can be extended to an affine basis for the affine hull of the set.

Theorem 1.3.9 *Let B be an affinely independent subset of a set A in $\mathbb{R}^n$. Then there exists an affine basis for* aff A *that lies in A and contains B.*

Proof Consider the non-empty family $\mathfrak{F}$ of all affinely independent subsets of A which contain B. Since no affinely independent set in $\mathbb{R}^n$ contains more than $n + 1$ points, there must exist some member C of $\mathfrak{F}$ that is not *properly* contained in any other member of $\mathfrak{F}$. Since C is a subset of A, we have $\operatorname{aff} C \subseteq \operatorname{aff} A$. We claim that $\operatorname{aff} C = \operatorname{aff} A$. Suppose that $\operatorname{aff} C \subset \operatorname{aff} A$. Since $\operatorname{aff} A$ is the *smallest* flat containing A, we cannot have $A \subseteq \operatorname{aff} C$, whence there exists some point $\mathbf{a}$ of A not lying in $\operatorname{aff} C$. We can now use Corollary 1.3.5 to deduce that $C \cup \{\mathbf{a}\}$ is a member of $\mathfrak{F}$ which *properly* contains C. This contradicts the choice of C. Thus $\operatorname{aff} C = \operatorname{aff} A$ and C is an affine basis of $\operatorname{aff} A$. $\square$

Corollary 1.3.10 *Let A be a subset of $\mathbb{R}^n$. Then A contains an affine basis for* aff A. $\square$

Let $\{\mathbf{a}_0, \ldots, \mathbf{a}_r\}$ be an affine basis for a non-empty r-flat A in $\mathbb{R}^n$. Then, by Corollary 1.3.3, each point $\mathbf{x}$ of A can be expressed uniquely in the form

$$\mathbf{x} = \lambda_0\mathbf{a}_0 + \ldots + \lambda_r\mathbf{a}_r, \qquad \text{where} \quad \lambda_0 + \ldots + \lambda_r = 1.$$

The scalars $\lambda_0, \ldots, \lambda_r$ are called the **barycentric coordinates** of $\mathbf{x}$ relative to (the *ordered affine basis*) $\mathbf{a}_0, \ldots, \mathbf{a}_r$. Example 1.2.7 shows that the barycentric coordinates of a point $\mathbf{x} = (x_1, \ldots, x_n)$ of $\mathbb{R}^n$ relative to $\mathbf{0}, \mathbf{e}_1, \ldots, \mathbf{e}_n$ are $1 - x_1 - \ldots - x_n, x_1, \ldots, x_n$.

Theorem 1.3.11 *Let $\{\mathbf{a}_0, \ldots, \mathbf{a}_r\}$ be an affine basis for a non-empty r-flat A in $\mathbb{R}^n$. Let $\lambda_0, \ldots, \lambda_r$ be the barycentric coordinates of a point $\mathbf{x} = (x_1, \ldots, x_n)$ of A relative to $\mathbf{a}_0, \ldots, \mathbf{a}_r$. Then there exist scalars a_{ij} $(i = 0, \ldots, r;\ j = 0, \ldots, n)$ such that, for $i = 0, \ldots, r$,*

$$\lambda_i = a_{i0} + a_{i1}x_1 + \ldots + a_{in}x_n.$$

Proof Extend, if necessary, $\{\mathbf{a}_0, \ldots, \mathbf{a}_r\}$ to an affine basis $\{\mathbf{a}_0, \ldots, \mathbf{a}_n\}$ for $\mathbb{R}^n$. Each point $\mathbf{x} = (x_1, \ldots, x_n)$ of $\mathbb{R}^n$ can be written uniquely in the form

$$\mathbf{x} = \lambda_0 \mathbf{a}_0 + \ldots + \lambda_n \mathbf{a}_n, \qquad \text{where} \qquad \lambda_0 + \ldots + \lambda_n = 1. \quad (1.3.3)$$

In particular, each of the points $\mathbf{0}, \mathbf{e}_1, \ldots, \mathbf{e}_n$ can be so expressed. Write $\mathbf{e}_0 = \mathbf{0}$. Then there are scalars $b_{ij} (i = 0, \ldots, n; j = 0, \ldots, n)$ such that, for $i = 0, \ldots, n$,

$$\mathbf{e}_i = b_{0i}\mathbf{a}_0 + \ldots + b_{ni}\mathbf{a}_n \qquad \text{and} \qquad b_{0i} + \ldots + b_{ni} = 1.$$

Write

$$\mathbf{x} = (1 - x_1 - \ldots - x_n)\mathbf{e}_0 + x_1\mathbf{e}_1 + \ldots + x_n\mathbf{e}_n.$$

Then

$$\mathbf{x} = \mu_0 \mathbf{a}_0 + \ldots + \mu_n \mathbf{a}_n,$$

where, for $i = 0, \ldots, n$,

$$\mu_i = b_{i0}(1 - x_1 - \ldots - x_n) + b_{i1}x_1 + \ldots + b_{in}x_n.$$

A routine verification shows that $\mu_0 + \ldots + \mu_n = 1$. Since the representation of $\mathbf{x}$ in the form (1.3.3) is unique, we can deduce that $\lambda_i = \mu_i$ for $i = 0, \ldots, n$. We complete the proof by putting $a_{i0} = b_{i0}$ for $i = 0, \ldots, n$, and $a_{ij} = b_{ij} - b_{i0} (i = 0, \ldots, n; \ j = 1, \ldots, n)$, and noting that $\mathbf{x} \in A$ if and only if $\lambda_{r+1} = 0, \ldots, \lambda_n = 0$. □

In elementary geometry, two lines are said to be *parallel* if they lie in the same plane and do not meet. Because of the slightly different definition of parallelism that we have adopted here, the following result (that will be needed in later work) does require a proof!

Theorem 1.3.12 *Let L and M be two lines that lie in a 2-flat A of $\mathbb{R}^n$ and which do not meet. Then L and M are parallel.*

Proof Let $\mathbf{a}, \mathbf{b}$ be distinct points of L, and let $\mathbf{c}, \mathbf{d}$ be distinct points of M. Since $\{\mathbf{a}, \mathbf{b}, \mathbf{c}\}$ is affinely independent, it will form an affine basis for $\mathbf{A}$. Thus $\mathbf{d} = \alpha\mathbf{a} + \beta\mathbf{b} + \gamma\mathbf{c}$ for some $\alpha, \beta, \gamma \in \mathbb{R}$ with $\alpha + \beta + \gamma = 1$. A typical point on M, the line joining $\mathbf{c}$ and $\mathbf{d}$, has the form

$$(1 - \theta)\mathbf{c} + \theta\mathbf{d} = \theta\alpha\mathbf{a} + \theta\beta\mathbf{b} + (\theta(\gamma - 1) + 1)\mathbf{c}$$

for some $\theta \in \mathbb{R}$. Since the latter point does not lie on L for any θ, we must have $\gamma = 1$ and $\mathbf{d} = \alpha(\mathbf{a} - \mathbf{b}) + \mathbf{c}$, whence

$$\mathbf{d} - (\mathbf{c} - \mathbf{a}) = (\alpha + 1)\mathbf{a} - \alpha\mathbf{b} \in L.$$

Thus $M - (\mathbf{c} - \mathbf{a}) \subseteq L$. Since $M - (\mathbf{c} - \mathbf{a})$ is a line, we must have $M - (\mathbf{c} - \mathbf{a}) = L$. Thus L and M are parallel. □

Exercises 1.3

1. For each point $\mathbf{a} = (a_1, \ldots, a_n)$ of $\mathbb{R}^n$, denote by $\mathbf{a}^*$ the point $(a_1, \ldots, a_n, 1)$ of $\mathbb{R}^{n+1}$. Prove that $\{\mathbf{a}_1, \ldots, \mathbf{a}_m\}_{\neq}$ is affinely independent in $\mathbb{R}^n$ if and only if $\{\mathbf{a}_1^*, \ldots, \mathbf{a}_m^*\}$ is linearly independent in $\mathbb{R}^{n+1}$.
2. Show that $\{(1,0,0), (0,0,1), (1,1,1)\}$ is an affine basis for the flat $\{(x,y,z): x - y + z = 1\}$ in $\mathbb{R}^3$. Find the barycentric coordinates of a general point (x,y,z) of this flat relative to $(1,0,0)$, $(0,0,1)$, $(1,1,1)$.
3. Let A be an r-flat in $\mathbb{R}^n$. Suppose that B and C are sets of $r+1$ points of $\mathbb{R}^n$ such that: (i) B lies in A and is affinely independent; (ii) $\operatorname{aff} C = A$. Show that B and C are affine bases for A.

1.4 HYPERPLANES

Consider the following system of m linear equations in n real variables $x_1, \ldots, x_n$:

$$\left.\begin{array}{l} a_{11}x_1 + \ldots + a_{1n}x_n = a_{10} \\ \cdots\cdots\cdots\cdots\cdots\cdots\cdots\cdots \\ a_{m1}x_1 + \ldots + a_{mn}x_n = a_{m0} \end{array}\right\},$$

where a_{ij} are given scalars. By the **solution set** of this system is meant the set of all n-tuples $(x_1, \ldots, x_n)$ of $\mathbb{R}^n$ that satisfy it. The solution set of the system is clearly the intersection of the solution sets of the m linear equations which comprise it. An easy verification shows that the solution set of any one of the individual linear equations is a flat, whence the solution set of the whole system is a flat. Later in the section, we shall show that every flat is the solution set of some system of linear equations. In general, the solution set of a single linear equation $a_1x_1 + \ldots + a_nx_n = a_0$ is an $(n-1)$-dimensional flat in $\mathbb{R}^n$. In the study of convexity in $\mathbb{R}^n$, flats of dimension $n-1$ play a key role, and are given their own name, **hyperplanes**.

To be precise, we should refer not to a *hyperplane*, but to a *hyperplane in* $\mathbb{R}^n$. When no ambiguity is likely to arise, however, we do speak simply of a hyperplane. A hyperplane in $\mathbb{R}^1$ is a point, in $\mathbb{R}^2$ a line, and in $\mathbb{R}^3$ a plane. Thus a hyperplane in $\mathbb{R}^2$ has an equation of the form $ax + by + c = 0$, where not *both* of a and b are zero; a hyperplane in $\mathbb{R}^3$ has an equation of the form $ax + by + cz + d = 0$, where not *all* of a, b, and c are zero. These results are generalized in the following theorem.

Theorem 1.4.1 *A set H in $\mathbb{R}^n$ is a hyperplane if and only if there exist scalars $c_0, c_1, \ldots, c_n$, where not all $c_1, \ldots, c_n$ are zero, such that*

$$H = \{(x_1, \ldots, x_n) : c_0 + c_1x_1 + \ldots + c_nx_n = 0\}.$$

Proof Let $H = \{(x_1, \ldots, x_n) : c_0 + c_1x_1 + \ldots + c_nx_n = 0\}$, where $c_0, c_1, \ldots, c_n$ are scalars and not all $c_1, \ldots, c_n$ are zero, say $c_1 \neq 0$. Let $\mathbf{u} = (u_1, \ldots, u_n)$, $\mathbf{v} = (v_1, \ldots, v_n)$ lie in H and let $\lambda + \mu = 1$. Then

$$\begin{aligned} &c_0 + c_1(\lambda u_1 + \mu v_1) + \ldots + c_n(\lambda u_n + \mu v_n) \\ &= \lambda(c_0 + c_1u_1 + \ldots + c_nu_n) + \mu(c_0 + c_1v_1 + \ldots + c_nv_n) \\ &= \lambda 0 + \mu 0 \\ &= 0. \end{aligned}$$

Thus $\lambda\mathbf{u} + \mu\mathbf{v} \in H$ and H is a flat.

Define points $\mathbf{a}_1, \ldots, \mathbf{a}_n$ of H by the equations

$$\mathbf{a}_1 = (-c_0/c_1, 0, 0, \ldots, 0)$$

and

$$\mathbf{a}_2 = (-(c_0 + c_2)/c_1, 1, 0, \ldots, 0), \ldots, \mathbf{a}_n = (-(c_0 + c_n)/c_1, 0, 0, \ldots, 1).$$

Since H is a flat, $\operatorname{aff}\{\mathbf{a}_1, \ldots, \mathbf{a}_n\} \subseteq H$. We now establish the opposite inclusion. Let $\mathbf{x} \in H$. Then the equations

$$\mathbf{x} = (x_1, \ldots, x_n) = (1 - x_2 - \ldots - x_n)\mathbf{a}_1 + x_2\mathbf{a}_2 + \ldots + x_n\mathbf{a}_n$$

express $\mathbf{x}$ as an affine combination of $\mathbf{a}_1, \ldots, \mathbf{a}_n$, whence $\mathbf{x} \in \operatorname{aff}\{\mathbf{a}_1, \ldots, \mathbf{a}_n\}$ and $H \subseteq \operatorname{aff}\{\mathbf{a}_1, \ldots, \mathbf{a}_n\}$. Hence $H = \operatorname{aff}\{\mathbf{a}_1, \ldots, \mathbf{a}_n\}$.

To show that the set $\{\mathbf{a}_1, \ldots, \mathbf{a}_n\}$ is affinely independent, suppose that $\lambda_1, \ldots, \lambda_n$ satisfy

$$\lambda_1\mathbf{a}_1 + \ldots + \lambda_n\mathbf{a}_n = \mathbf{0} \qquad \text{and} \qquad \lambda_1 + \ldots + \lambda_n = 0.$$

Comparing the ith coordinates ($i = 2, \ldots, n$) on both sides of the first of these equations, we find that $\lambda_2, \ldots, \lambda_n$ are all zero, whence so too is λ_1, from the second equation. Thus $\{\mathbf{a}_1, \ldots, \mathbf{a}_n\}$ is affinely independent. But $H = \operatorname{aff}\{\mathbf{a}_1, \ldots, \mathbf{a}_n\}$, and so H is an $(n-1)$-dimensional flat, i.e. H is a hyperplane.

Conversely, suppose that H is a hyperplane in $\mathbb{R}^n$. Let $\{\mathbf{b}_1, \ldots, \mathbf{b}_n\}$ be an affine basis for H. Extend this to an affine basis $\{\mathbf{b}_0, \mathbf{b}_1, \ldots, \mathbf{b}_n\}$ for $\mathbb{R}^n$. Then each $\mathbf{x} = (x_1, \ldots, x_n)$ in $\mathbb{R}^n$ can be written uniquely in the form

$$\mathbf{x} = \lambda_0\mathbf{b}_0 + \lambda_1\mathbf{b}_1 + \ldots + \lambda_n\mathbf{b}_n, \qquad \text{where } \lambda_0 + \lambda_1 + \ldots + \lambda_n = 1.$$

Thus $\lambda_0, \lambda_1, \ldots, \lambda_n$ are the barycentric coordinates of $\mathbf{x}$ relative to the (ordered) affine basis $\mathbf{b}_0, \mathbf{b}_1, \ldots, \mathbf{b}_n$. By Theorem 1.3.11, there exist scalars $c_0, c_1, \ldots, c_n$ such that

$$\lambda_0 = c_0 + c_1x_1 + \ldots + c_nx_n.$$

Since $\mathbf{x} \in H$ if and only if $\lambda_0 = 0$,

$$H = \{(x_1, \ldots, x_n) : c_0 + c_1x_1 + \ldots + c_nx_n = 0\}\,.$$

Not all of $c_1, \ldots, c_n$ are zero, for this would imply that either H is empty (if $c_0 \neq 0$) or $\mathbb{R}^n$ (if $c_0 = 0$), both of which contradict the assumption that H is an $(n-1)$-dimensional flat. □

Corollary 1.4.2 *In* $\mathbb{R}^n$ *each r-flat* ($r = -1, \ldots, n$) *can be expressed as the intersection of* $n - r$ *hyperplanes, and so is the solution set of some system of* $n - r$ *linear equations.*

Proof The only (-1)-flat in $\mathbb{R}^n$ is the empty set, which is the intersection of the $n + 1$ hyperplanes $x_1 = 0, \ldots, x_n = 0$, $x_1 + \ldots + x_n = 1$. The only n-flat in $\mathbb{R}^n$ is $\mathbb{R}^n$ itself, which is the intersection of no hyperplanes. Consider now the case of an r-flat A in $\mathbb{R}^n$, where $r = 0, \ldots, n-1$. Let $\{\mathbf{a}_0, \ldots, \mathbf{a}_r\}$ be an affine basis for A. Extend this to an affine basis $\{\mathbf{a}_0, \ldots, \mathbf{a}_n\}$ for $\mathbb{R}^n$. Then each $\mathbf{x}$ in $\mathbb{R}^n$ can be expressed uniquely in the form

$$\mathbf{x} = \lambda_0\mathbf{a}_0 + \ldots + \lambda_n\mathbf{a}_n, \qquad \text{where } \lambda_0 + \ldots + \lambda_n = 1.$$

Now A is the set in $\mathbb{R}^n$ consisting precisely of those $\mathbf{x}$'s whose barycentric coordinates $\lambda_{r+1}, \ldots, \lambda_n$ are all zero. Since each of the sets $\{\mathbf{x} : \lambda_i = 0\}$ is the hyperplane $\operatorname{aff}\{\mathbf{a}_0, \ldots, \mathbf{a}_{i-1}, \mathbf{a}_{i+1}, \ldots, \mathbf{a}_n\}$, it follows that A is the intersection of the $n - r$ hyperplanes with equations $\lambda_{r+1} = 0, \ldots, \lambda_n = 0$. □

Given a hyperplane H in $\mathbb{R}^n$, there exist scalars $c_0, c_1, \ldots, c_n$, with not all $c_1, \ldots, c_n$ zero, such that

$$H = \{(x_1, \ldots, x_n) : c_0 + c_1x_1 + \ldots + c_nx_n = 0\}.$$

We now consider to what extent H determines the scalars $c_0, c_1, \ldots, c_n$. It certainly does not determine them uniquely, for the scalars $\theta c_0, \theta c_1, \ldots, \theta c_n$, where $\theta \neq 0$, serve equally well in the above equation for H. Suppose that $d_0, d_1, \ldots, d_n$ are also scalars such that

$$H = \{(x_1, \ldots, x_n) : d_0 + d_1x_1 + \ldots + d_nx_n = 0\}.$$

Assume that $c_1 \neq 0$, and let $\mathbf{a}_1, \ldots, \mathbf{a}_n$ be the points of H specified as in the first part of the proof of Theorem 1.4.1. Substituting the coordinates of the $\mathbf{a}_i$ into the above equation for H in terms of the d's, we deduce that $d_i = (d_1/c_1)c_i$ for $i = 0, \ldots, n$. Since not all of $d_0, d_1, \ldots, d_n$ can be zero, we deduce that d_1, and hence d_1/c_1, is not zero. Writing $\theta = d_1/c_1$, we find that $d_0 = \theta c_0, d_1 = \theta c_1, \ldots, d_n = \theta c_n$. Thus the hyperplane H determines the scalars $c_0, c_1, \ldots, c_n$ to within a common non-zero scalar multiple.

The importance of hyperplanes in $\mathbb{R}^n$ is that they divide the whole space into two *halfspaces* in a natural way. For example, a line in $\mathbb{R}^2$

with equation $ax + by + c = 0$ divides $\mathbb{R}^2$ into the two *halfplanes* determined by the inequalities $ax + by + c \leqslant 0$ and $ax + by + c \geqslant 0$. The general situation can be described in equally simple terms. A hyperplane in $\mathbb{R}^n$ with equation $c_0 + c_1x_1 + \ldots + c_nx_n = 0$ divides $\mathbb{R}^n$ into the two halfspaces determined by the inequalities

$$c_0 + c_1x_1 + \ldots + c_nx_n \leqslant 0 \qquad \text{and} \qquad c_0 + c_1x_1 + \ldots + c_nx_n \geqslant 0.$$

It turns out to be convenient to define two types of halfspace, the *closed* and the *open*. The reason for this choice of adjectives will become clear in Section 1.7.

Let $c_0, c_1, \ldots, c_n$ be scalars, where not all $c_1, \ldots, c_n$ are zero. Then a set of either of the forms

$$\{(x_1, \ldots, x_n): c_0 + c_1x_1 + \ldots + c_nx_n \leqslant 0\} \qquad \text{or}$$
$$\{(x_1, \ldots, x_n): c_0 + c_1x_1 + \ldots + c_nx_n \geqslant 0\}$$

is called a **closed halfspace** in $\mathbb{R}^n$; a set of either of the forms

$$\{(x_1, \ldots, x_n): c_0 + c_1x_1 + \ldots + c_nx_n < 0\} \qquad \text{or}$$
$$\{(x_1, \ldots, x_n): c_0 + c_1x_1 + \ldots + c_nx_n > 0\}$$

is called an **open halfspace** in $\mathbb{R}^n$. We note that if (in the above discussion) the scalars $c_0, c_1, \ldots, c_n$ are replaced, respectively, by $\theta c_0, \theta c_1, \ldots, \theta c_n$, for some $\theta \neq 0$, then we obtain the same pair of closed halfspaces and the same pair of open halfspaces, although the order of the halfspaces is reversed when $\theta < 0$. Thus, if H is a hyperplane in $\mathbb{R}^n$ with equation $c_0 + c_1x_1 + \ldots + c_nx_n = 0$, then the above pair of closed halfspaces and the above pair of open halfspaces are determined by H, and not by the particular equation chosen to represent H. Hence we may refer unambiguously to the closed halfspaces and the open halfspaces *determined* by H. Using an obvious terminology, we say that the closed (open) halfspaces determined by H are *opposite* to one another.

Example 1.4.3 *Any line through two points lying in opposite halfspaces determined by a hyperplane in $\mathbb{R}^n$ meets the hyperplane.* Suppose that the hyperplane H has equation $c_0 + c_1x_1 + \ldots + c_nx_n = 0$, and that the points $\mathbf{a} = (a_1, \ldots, a_n)$ and $\mathbf{b} = (b_1, \ldots, b_n)$ lie in opposite halfspaces determined by H. We may suppose that

$$c_0 + c_1a_1 + \ldots + c_na_n = \alpha < 0 \quad \text{and} \quad c_0 + c_1b_1 + \ldots + c_nb_n = \beta > 0,$$

thus omitting the trivial case when either of $\mathbf{a}$ or $\mathbf{b}$ lies on H. The points on the line L through $\mathbf{a}$ and $\mathbf{b}$ are precisely those points of the form

$$(\lambda a_1 + (1 - \lambda)b_1, \ldots, \lambda a_n + (1 - \lambda)b_n),$$

where the scalar λ assumes all real values. We find, by substituting these coordinates into the equation of H, that $\lambda = \beta/(\beta - \alpha)$ corresponds to the unique point of intersection of L and H. We note that this value of λ satisfies $0 \leqslant \lambda \leqslant 1$. This means, as will be explained in Section 2.1, that the portion of L lying *between* $\mathbf{a}$ and $\mathbf{b}$, the so-called *line segment* joining $\mathbf{a}$ and $\mathbf{b}$, meets H. □

Theorem 1.4.4 *Let H and H' be hyperplanes in $\mathbb{R}^n$ with respective equations*

$$c_0 + c_1x_1 + \ldots + c_nx_n = 0 \qquad \text{and} \qquad c_0' + c_1'x_1 + \ldots + c_n'x_n = 0.$$

Then H and H' are parallel if and only if there exists a scalar θ such that $c_1' = \theta c_1, \ldots, c_n' = \theta c_n$.

Proof Suppose first that H and H' are parallel, say $H' = H + \mathbf{a}$, where $\mathbf{a} = (a_1, \ldots, a_n)$. Then $(x_1, \ldots, x_n) \in H$ if and only if

$$\begin{aligned} & c_0' + c_1'(x_1 + a_1) + \ldots + c_n'(x_n + a_n) \\ = \quad & c_0' + c_1'a_1 + \ldots + c_n'a_n + c_1'x_1 + \ldots + c_n'x_n = 0. \end{aligned}$$

Thus, by the above remarks on the representation of hyperplanes by linear equations, there exists a scalar θ such that $c_1' = \theta c_1, \ldots, c_n' = \theta c_n$.

Suppose next that $c_1' = \theta c_1, \ldots, c_n' = \theta c_n$, where θ is a (non-zero) scalar. Then H' is represented by the equation $d_0 + c_1x_1 + \ldots + c_nx_n = 0$, where $d_0 = c_0'/\theta$. Let $\mathbf{b} = (b_1, \ldots, b_n)$ satisfy $c_1b_1 + \ldots + c_nb_n = c_0 - d_0$. Then H' also has the equation $c_0 + c_1(x_1 - b_1) + \ldots + c_n(x_n - b_n) = 0$. Thus $\mathbf{x} = (x_1, \ldots, x_n) \in H'$ if and only if $\mathbf{x} - \mathbf{b} \in H$. Hence $H' = H + \mathbf{b}$, which shows that H and H' are parallel. □

Corollary 1.4.5 *Two parallel hyperplanes in $\mathbb{R}^n$ are either identical or disjoint. Two non-parallel hyperplanes in $\mathbb{R}^n$ must meet.*

Proof Let H and H' be parallel hyperplanes in $\mathbb{R}^n$. Then they have respective equations

$$c_0 + c_1x_1 + \ldots + c_nx_n = 0 \qquad \text{and} \qquad c_0' + \theta c_1x_1 + \ldots + \theta c_nx_n = 0$$

say, where θ is a non-zero scalar. If $c_0' = \theta c_0$, then H and H' are identical, otherwise they are disjoint.

Let H and H' be non-parallel hyperplanes in $\mathbb{R}^n$ having respective equations

$$c_0 + c_1x_1 + \ldots + c_nx_n = 0 \qquad \text{and} \qquad c_0' + c_1'x_1 + \ldots + c_n'x_n = 0.$$

Then there is no scalar θ such that $c_1' = \theta c_1, \ldots, c_n' = \theta c_n$. It follows that $n \geqslant 2$. Suppose that $c_1 \neq 0$. Then, for some $j \in \{2, \ldots, n\}$, $c_j' \neq (c_1'/c_1)c_j$, say $c_2' \neq (c_1'/c_1)c_2$. It is easily verified that the point

$$((c_0'c_2 - c_0c_2')/(c_1c_2' - c_1'c_2),\ (c_0c_1' - c_0'c_1)/(c_1c_2' - c_1'c_2), 0, \ldots, 0)$$

lies in $H \cap H'$, and so H and H' meet. □

Exercises 1.4

1. Prove that $H = \text{aff}\{(3, 1, 1), (11, 2, -1), (-2, 0, 2)\}$ is a hyperplane in $\mathbb{R}^3$. Find a linear equation representing H.
2. In $\mathbb{R}^n$ a hyperplane H meets, but does not contain, a flat A. Show that $\dim(A \cap H) = \dim A - 1$.
3. Suppose that in $\mathbb{R}^n$ a flat A lies in the closed halfspace defined by the inequality $c_0 + c_1x_1 + \ldots + c_nx_n \leqslant 0$. Prove that, if $\mathbf{a} = (a_1, \ldots, a_n)$ and $\mathbf{b} = (b_1, \ldots, b_n)$ lie in A, then

$$c_1a_1 + \ldots + c_na_n = c_1b_1 + \ldots + c_nb_n.$$

Deduce that some translate of A lies in the hyperplane with equation $c_0 + c_1x_1 + \ldots + c_nx_n = 0$.

1.5 AFFINE TRANSFORMATIONS

A mapping T from one Euclidean space $\mathbb{R}^n$ to another one $\mathbb{R}^m$ is said to be an *affine transformation* if it preserves affine structure, in the sense that it maps every affine combination of points of $\mathbb{R}^n$ onto the same affine combination of their images under T. Such transformations map flats onto flats. In this short section, we establish those properties of affine transformations which will be needed in our discussion of convexity.

A mapping $T: \mathbb{R}^n \to \mathbb{R}^m$ is called an **affine transformation** if $T(\lambda\mathbf{x} + \mu\mathbf{y}) = \lambda T(\mathbf{x}) + \mu T(\mathbf{y})$ whenever $\mathbf{x}, \mathbf{y} \in \mathbb{R}^n$ and $\lambda + \mu = 1$. A simple example of an affine transformation is the mapping $T: \mathbb{R}^3 \to \mathbb{R}^3$ defined by the equation $T(x, y, z) = (x, y, 1)$. Geometrically, T is the orthogonal projection of $\mathbb{R}^3$ onto the plane with equation $z = 1$. For each vector $\mathbf{q}$ in $\mathbb{R}^n$, the mapping $T: \mathbb{R}^n \to \mathbb{R}^n$ defined by the equation $T(\mathbf{x}) = \mathbf{x} + \mathbf{q}$ is an affine transformation called the **translation** of $\mathbb{R}^n$ through $\mathbf{q}$.

Clearly every linear transformation from $\mathbb{R}^n$ to $\mathbb{R}^m$ is also an affine one. That not every affine transformation from $\mathbb{R}^n$ to $\mathbb{R}^m$ is linear, follows from the observation that it need not map the zero vector of $\mathbb{R}^n$ to the zero vector of $\mathbb{R}^m$. See the two examples of affine transformations given above. The exact relationship between linear and affine transformations is given in the following result.

Theorem 1.5.1 *Let $T: \mathbb{R}^n \to \mathbb{R}^m$ be an affine transformation. Then T is linear if and only if $T(\mathbf{0}) = \mathbf{0}$.*

Proof In view of the remarks above, it will suffice to show that T is linear when $T(\mathbf{0}) = \mathbf{0}$. Suppose, then, that $T(\mathbf{0}) = \mathbf{0}$. Let $\mathbf{x}, \mathbf{y} \in \mathbb{R}^n$ and $\lambda \in \mathbb{R}$. Then

$$T(\lambda\mathbf{x}) = T(\lambda\mathbf{x} + (1 - \lambda)\mathbf{0}) = \lambda T(\mathbf{x}) + (1 - \lambda)T(\mathbf{0}) = \lambda T(\mathbf{x}).$$

Using this last result, we deduce that

$$\begin{aligned} T(\mathbf{x} + \mathbf{y}) &= T(2(\tfrac{1}{2}\mathbf{x} + \tfrac{1}{2}\mathbf{y})) = 2T(\tfrac{1}{2}\mathbf{x} + \tfrac{1}{2}\mathbf{y}) \\ &= 2(\tfrac{1}{2}T(\mathbf{x}) + \tfrac{1}{2}T(\mathbf{y})) = T(\mathbf{x}) + T(\mathbf{y}). \end{aligned}$$

Thus T is linear. □

In the following discussion, all vectors considered will be identified with column vectors in the natural way.

Theorem 1.5.2 *The affine transformations* $T: \mathbb{R}^n \to \mathbb{R}^m$ *are precisely those mappings* $T: \mathbb{R}^n \to \mathbb{R}^m$ *which can be expressed in the form* $T(\mathbf{x}) = \mathbf{Qx} + \mathbf{q}$, *for some real* $m \times n$ *matrix* $\mathbf{Q}$ *and some real* $m \times 1$ *matrix* $\mathbf{q}$.

Proof It is easily verified that a mapping of the type under consideration is an affine transformation. Assume, then, that $T: \mathbb{R}^n \to \mathbb{R}^m$ is an affine transformation. Let $T(\mathbf{0}) = \mathbf{q}$. Then the mapping $T': \mathbb{R}^n \to \mathbb{R}^m$ defined by the equation $T'(\mathbf{x}) = T(\mathbf{x}) - \mathbf{q}$ is readily shown to be an affine transformation with $T'(\mathbf{0}) = \mathbf{0}$. Theorem 1.5.1 shows that T' is linear, whence there is a real $m \times n$ matrix $\mathbf{Q}$ such that $T'(\mathbf{x}) = \mathbf{Qx}$. Thus $T(\mathbf{x}) = \mathbf{Qx} + \mathbf{q}$ as desired. □

We note that the affine transformation $T: \mathbb{R}^n \to \mathbb{R}^m$ determines the matrices $\mathbf{Q}$ and $\mathbf{q}$ *uniquely*, for the jth column of $\mathbf{Q}$ must be $T(\mathbf{e}_j) - T(\mathbf{0})$ and $\mathbf{q}$ must be $\mathrm{T}(\mathbf{0})$. The above representation of an affine transformation in terms of matrices shows easily that, if $T: \mathbb{R}^n \to \mathbb{R}^m$ is an affine transformation, $\mathbf{a}_1, \ldots, \mathbf{a}_r \in \mathbb{R}^n$ and $\lambda_1 + \ldots + \lambda_r = 1$, then

$$T(\lambda_1\mathbf{a}_1 + \ldots + \lambda_r\mathbf{a}_r) = \lambda_1 T(\mathbf{a}_1) + \ldots + \lambda_r T(\mathbf{a}_r),$$

a fact which we use in the proof of the following corollary.

Corollary 1.5.3 *Let* $T: \mathbb{R}^n \to \mathbb{R}^m$ *be an affine transformation and let* A *be a set in* $\mathbb{R}^n$. *Then* $T(\operatorname{aff} A) = \operatorname{aff} T(A)$. *If* A *is a flat, then so too is* $T(A)$.

Proof A point $\mathbf{x}$ lies in $T(\operatorname{aff} A)$ if and only if there exist $\mathbf{a}_1, \ldots, \mathbf{a}_r \in A$ and $\lambda_1, \ldots, \lambda_r$ with $\lambda_1 + \ldots + \lambda_r = 1$ such that

$$\mathbf{x} = T(\lambda_1\mathbf{a}_1 + \ldots + \lambda_r\mathbf{a}_r) = \lambda_1 T(\mathbf{a}_1) + \ldots + \lambda_r T(\mathbf{a}_r),$$

that is, if and only if $\mathbf{x} \in \operatorname{aff} T(A)$. Thus $T(\operatorname{aff} A) = \operatorname{aff} T(A)$. If A is a flat, then aff $T(A) = T(\text{aff } A) = T(A)$, which shows that $T(A)$ is a flat. □

Consider now an affine transformation $T: \mathbb{R}^n \to \mathbb{R}^n$ of $\mathbb{R}^n$ into itself. By Theorem 1.5.2, there exist a real $n \times n$ matrix $\mathbf{Q}$ and a real $n \times 1$ matrix $\mathbf{q}$ such that $T(\mathbf{x}) = \mathbf{Qx} + \mathbf{q}$. The affine transformation T is said to be **non-singular** if the determinant $\det \mathbf{Q}$ of the matrix $\mathbf{Q}$ is non-zero, that is if $\mathbf{Q}$ has an inverse, i.e. is non-singular.

Theorem 1.5.4 *Let $T: \mathbb{R}^n \to \mathbb{R}^n$ be an affine transformation. Then T has an inverse if and only if T is non-singular. When T is non-singular, its inverse $T^{-1}: \mathbb{R}^n \to \mathbb{R}^n$ is an affine transformation.*

Proof Let $\mathbf{Q}$ be a real $n \times n$ matrix and $\mathbf{q}$ a real $n \times 1$ matrix such that $T(\mathbf{x}) = \mathbf{Qx} + \mathbf{q}$ for all $\mathbf{x}$ in $\mathbb{R}^n$. Suppose first that $\mathbf{Q}$ is non-singular. Then $\det \mathbf{Q}$ is non-zero and $\mathbf{Q}$ has an inverse $\mathbf{Q}^{-1}$. For each $\mathbf{y}$ in $\mathbb{R}^n$, the equation $T(\mathbf{x}) = \mathbf{y}$ has the *unique* solution $\mathbf{x} = \mathbf{Q}^{-1}\mathbf{y} - \mathbf{Q}^{-1}\mathbf{q}$. It follows that T has an inverse, which is the affine transformation $T^{-1}: \mathbb{R}^n \to \mathbb{R}^n$ defined by the equation $T^{-1}(\mathbf{y}) = \mathbf{Q}^{-1}\mathbf{y} - \mathbf{Q}^{-1}\mathbf{q}$ for $\mathbf{y}$ in $\mathbb{R}^n$. Suppose next that $\det \mathbf{Q}$ is zero. Then there exists a *non-zero* vector $\mathbf{z}$ in $\mathbb{R}^n$ such that $\mathbf{Q}(\mathbf{z}) = \mathbf{0}$, whence $T(\mathbf{z}) = T(\mathbf{0})$ and T is not injective. Hence T has no inverse. □

Theorem 1.5.5 *Let $\{\mathbf{a}_0, \ldots, \mathbf{a}_r\}_{\neq}$ and $\{\mathbf{b}_0, \ldots, \mathbf{b}_r\}_{\neq}$ be affinely independent sets in $\mathbb{R}^n$. Then there exists a non-singular affine transformation $T: \mathbb{R}^n \to \mathbb{R}^n$ such that $T(\mathbf{a}_i) = \mathbf{b}_i$ for $i = 0, \ldots, r$.*

Proof Extend the sets $\{\mathbf{a}_0, \ldots, \mathbf{a}_r\}_{\neq}$ and $\{\mathbf{b}_0, \ldots, \mathbf{b}_r\}_{\neq}$, respectively, to affine bases $\{\mathbf{a}_0, \ldots, \mathbf{a}_n\}$ and $\{\mathbf{b}_0, \ldots, \mathbf{b}_n\}$ for $\mathbb{R}^n$. Then each $\mathbf{x}$ in $\mathbb{R}^n$ can be written *uniquely* in the form

$$\mathbf{x} = \lambda_0 \mathbf{a}_0 + \ldots + \lambda_n \mathbf{a}_n, \qquad \text{where } \lambda_0 + \ldots + \lambda_n = 1.$$

Define a mapping $T: \mathbb{R}^n \to \mathbb{R}^n$ by the equation

$$T(\mathbf{x}) = \lambda_0 \mathbf{b}_0 + \ldots + \lambda_n \mathbf{b}_n.$$

It is a routine exercise to verify that T is a bijective affine transformation. Hence T is a non-singular affine transformation such that $T(\mathbf{a}_i) = \mathbf{b}_i$ for $i = 0, \ldots, r$. □

Corollary 1.5.6 *Let A and B be flats in $\mathbb{R}^n$ of the same dimension. Then there is a non-singular affine transformation $T: \mathbb{R}^n \to \mathbb{R}^n$ such that $T(A) = B$.*

Proof If A and B are both empty, then T can be taken as the identity mapping of $\mathbb{R}^n$ onto itself. Suppose, then, that A and B are non-empty and have affine bases $\{\mathbf{a}_0, \ldots, \mathbf{a}_r\}_{\neq}$ and $\{\mathbf{b}_0, \ldots, \mathbf{b}_r\}_{\neq}$, respectively. Let T be as in the theorem. Then, by Corollary 1.5.3,

$$T(A) = T(\operatorname{aff}\{\mathbf{a}_0, \ldots, \mathbf{a}_r\}) = \operatorname{aff}\{\mathbf{b}_0, \ldots, \mathbf{b}_r\} = B. \quad \square$$

Suppose that B is an r-dimensional flat ($r \geqslant 1$) in $\mathbb{R}^n$ and that $A = \operatorname{aff}\{\mathbf{0}, \mathbf{e}_1, \ldots, \mathbf{e}_r\}$. Then A and B are flats of the same dimension. By Corollary 1.5.6, there exists a non-singular affine transformation $T: \mathbb{R}^n \to \mathbb{R}^n$ such that $T(A) = B$. The flat A consists precisely of those points $(x_1, \ldots, x_n)$ for which $x_{r+1} = 0, \ldots, x_n = 0$. Hence A can be identified with $\mathbb{R}^r$ by associating the point $(x_1, \ldots, x_n)$ of A with the point $(x_1, \ldots, x_r)$ of $\mathbb{R}^r$. Under this identification $T(\mathbb{R}^r) = B$. Thus every r-dimensional flat ($r \geqslant 1$) can be considered to be an affine *copy* of $\mathbb{R}^r$. This identification is often helpful when working with r-dimensional sets in $\mathbb{R}^n$, for we may consider them as subsets of $\mathbb{R}^r$ and make use of the resulting algebraic simplification.

Exercises 1.5

1. Let $T_1: \mathbb{R}^n \to \mathbb{R}^m$ and $T_2: \mathbb{R}^m \to \mathbb{R}^p$ be affine transformations. Show that their composite $T_2 \circ T_1: \mathbb{R}^n \to \mathbb{R}^p$ defined by the equation $T_2 \circ T_1(\mathbf{x}) = T_2(T_1(\mathbf{x}))$ for $\mathbf{x} \in \mathbb{R}^n$ is an affine transformation.
2. Prove that affine transformations preserve parallelism between flats.
3. Let A be a set in $\mathbb{R}^m$. Show that, for each affine transformation $T: \mathbb{R}^n \to \mathbb{R}^m$, $\operatorname{aff} T^{-1}(A) \subseteq T^{-1}(\operatorname{aff} A)$. Give an example where *strict* inclusion holds. Prove that, if A is a flat, then so too is $T^{-1}(A)$.
4. Show that, if $T: \mathbb{R}^n \to \mathbb{R}^m$ is an injective (surjective, bijective) affine transformation, then $m \geqslant n$ ($m \leqslant n, m = n$).

1.6 LENGTH, DISTANCE, AND ANGLE

Convexity, as it is studied here, is first and foremost a branch of Euclidean geometry. Its fundamental concepts are geometrically inspired and its proofs, although couched in the language of algebra and analysis, are often geometrically motivated. In this section, we define the basic notions of *length*, *distance*, and *angle* in $\mathbb{R}^n$ and examine those of their properties which will be needed in later discussions.

We introduce length, distance, and angle into $\mathbb{R}^n$ by means of an *inner product*. The **inner product** $\mathbf{x} \cdot \mathbf{y}$ of vectors $\mathbf{x} = (x_1, \ldots, x_n)$ and $\mathbf{y} = (y_1, \ldots, y_n)$ in $\mathbb{R}^n$ is the real number defined by the equation

$$\mathbf{x} \cdot \mathbf{y} = x_1 y_1 + \ldots + x_n y_n.$$

The following properties of the inner product are immediate consequences of its definition. For $\mathbf{x}, \mathbf{y}, \mathbf{z} \in \mathbb{R}^n$ and $\lambda, \mu \in \mathbb{R}$:

(i) $\mathbf{x} \cdot \mathbf{x} \geqslant 0$, and $\mathbf{x} \cdot \mathbf{x} = 0$ if and only if $\mathbf{x} = \mathbf{0}$;

(ii) $\mathbf{x} \cdot \mathbf{y} = \mathbf{y} \cdot \mathbf{x}$;

(iii) $(\lambda \mathbf{x} + \mu \mathbf{y}) \cdot \mathbf{z} = \lambda(\mathbf{x} \cdot \mathbf{z}) + \mu(\mathbf{y} \cdot \mathbf{z})$.

The **norm** or **length** $\|\mathbf{x}\|$ of a vector $\mathbf{x} = (x_1, \ldots, x_n)$ in $\mathbb{R}^n$ is the non-negative real number defined by the equation

$$\|\mathbf{x}\| = \sqrt{(\mathbf{x} \cdot \mathbf{x})}, \qquad \text{whence} \qquad \|\mathbf{x}\| = \sqrt{(x_1^2 + \ldots + x_n^2)}.$$

The **distance** between points $\mathbf{x} = (x_1, \ldots, x_n)$ and $\mathbf{y} = (y_1, \ldots, y_n)$ of $\mathbb{R}^n$ is the non-negative real number

$$\|\mathbf{x} - \mathbf{y}\| = \sqrt{((x_1 - y_1)^2 + \ldots + (x_n - y_n)^2)},$$

i.e. the length of the vector $\mathbf{x} - \mathbf{y}$, or $\mathbf{y} - \mathbf{x}$. Since distance is defined in terms of the norm, we concentrate on the properties of the norm. The norm of the zero vector is 0 and the norm of each elementary vector $\mathbf{e}_i$ is 1. In general, any vector in $\mathbb{R}^n$ which has norm 1 is called a **unit vector**. The following properties of the norm are simple consequences of its definition. For $\mathbf{x}, \mathbf{y} \in \mathbb{R}^n$ and $\lambda, \mu \in \mathbb{R}$:

(i) $\|\mathbf{x}\| \geqslant 0$, and $\|\mathbf{x}\| = 0$ if and only if $\mathbf{x} = \mathbf{0}$;

(ii) $\|\lambda \mathbf{x}\| = |\lambda| \, \|\mathbf{x}\|$;

(iii) $\|\lambda \mathbf{x} + \mu \mathbf{y}\|^2 = \lambda^2 \|\mathbf{x}\|^2 + 2\lambda\mu \mathbf{x} \cdot \mathbf{y} + \mu^2 \|\mathbf{y}\|^2$.

Some less immediate, but very important, properties of the norm are given in the theorem following.

Theorem 1.6.1 *Let* $\mathbf{x}, \mathbf{y} \in \mathbb{R}^n$. *Then*

(i) $|\mathbf{x} \cdot \mathbf{y}| \leqslant \|\mathbf{x}\| \, \|\mathbf{y}\|$ (Cauchy–Schwarz inequality);

(ii) $\|\mathbf{x} + \mathbf{y}\| \leqslant \|\mathbf{x}\| + \|\mathbf{y}\|$ (triangle inequality);

(iii) $| \, \|\mathbf{x}\| - \|\mathbf{y}\| \, | \leqslant \|\mathbf{x} - \mathbf{y}\|$;

(iv) if, for some $\alpha > 0$, $\|\mathbf{x} + \lambda \mathbf{y}\| \geqslant \|\mathbf{x}\|$ whenever $0 < \lambda < \alpha$, then $\mathbf{x} \cdot \mathbf{y} \geqslant 0$.

Proof We content ourselves with proving (iv), for (i), (ii), and (iii) are standard results. Let $\alpha > 0$ be such that $\|\mathbf{x} + \lambda \mathbf{y}\| \geqslant \|\mathbf{x}\|$ whenever $0 < \lambda < \alpha$. Then, whenever $0 < \lambda < \alpha$,

$$\|\mathbf{x}\|^2 \leqslant \|\mathbf{x} + \lambda \mathbf{y}\|^2 = \|\mathbf{x}\|^2 + 2\lambda \mathbf{x} \cdot \mathbf{y} + \lambda^2 \|\mathbf{y}\|^2,$$

whence

$$\mathbf{x} \cdot \mathbf{y} + \tfrac{1}{2}\lambda\|\mathbf{y}\|^2 \geqslant 0.$$

Letting $\lambda \to 0_+$ in the last inequality, we deduce that $\mathbf{x}.\mathbf{y} \geqslant 0$. □

The Cauchy–Schwarz inequality allows us to introduce the concept of angle into $\mathbb{R}^n$ as follows. The **angle** between non-zero vectors $\mathbf{x}$ and $\mathbf{y}$ of $\mathbb{R}^n$ is the *unique* real number θ satisfying the conditions

$$\cos\theta = \frac{\mathbf{x} \cdot \mathbf{y}}{\|\mathbf{x}\|\,\|\mathbf{y}\|} \quad \text{and} \quad 0 \leqslant \theta \leqslant \pi.$$

This definition accords with the usual one of elementary geometry. The angle between $\mathbf{x}$ and $\mathbf{y}$ is called **acute** or **obtuse** according as $\mathbf{x} \cdot \mathbf{y}$ is positive or negative. Vectors $\mathbf{x}$ and $\mathbf{y}$, whether zero or not, are said to be **orthogonal** if $\mathbf{x} \cdot \mathbf{y} = 0$.

The ideas of inner product and orthogonality are helpful in giving a simple geometric interpretation to the equation of a hyperplane in $\mathbb{R}^n$. Consider a hyperplane H in $\mathbb{R}^n$ with equation $c_0 + c_1x_1 + \ldots + c_nx_n = 0$. This equation can be written in the form $c_0 + \mathbf{c} \cdot \mathbf{x} = 0$, where $\mathbf{c}$ is the non-zero vector $(c_1, \ldots, c_n)$ and $\mathbf{x}$ is $(x_1, \ldots, x_n)$. Such a vector $\mathbf{c}$ is said to be a **normal vector** to H. In view of the discussion in Section 1.4 on the representation of hyperplanes by means of linear equations, it follows that the normal vectors of H are precisely those vectors of the form $\lambda\mathbf{c}$ for some non-zero scalar λ. Thus H has exactly *two* unit normal vectors, namely $\pm\mathbf{c}/\|\mathbf{c}\|$. Hence, given any hyperplane H in $\mathbb{R}^n$, it may be assumed that it has an equation of the form $c_0 + \mathbf{c} \cdot \mathbf{x} = 0$, where $\mathbf{c}$ is a unit vector. This concept of a normal vector generalizes the one familiar in elementary geometry. To see this, suppose that $\mathbf{v}$ and $\mathbf{w}$ lie in a hyperplane H in $\mathbb{R}^n$ with equation $c_0 + \mathbf{c} \cdot \mathbf{x} = 0$. Then $c_0 + \mathbf{c} \cdot \mathbf{v} = 0$ and $c_0 + \mathbf{c} \cdot \mathbf{w} = 0$, whence $\mathbf{c} \cdot (\mathbf{w} - \mathbf{v}) = 0$. This shows that $\mathbf{c}$ is orthogonal to every vector which is the difference of two vectors in H. See Fig. 1.3.

The idea of orthogonality between vectors can be usefully extended to subspaces of $\mathbb{R}^n$ as we now show. Let A be a subspace of $\mathbb{R}^n$. Then the **orthogonal complement** $A^\perp$ of A is the set of *all* those vectors in $\mathbb{R}^n$ which are orthogonal to all the vectors in A, i.e.

$$A^\perp = \{\mathbf{x} \in \mathbb{R}^n : \mathbf{x} \cdot \mathbf{a} = 0 \text{ for all } \mathbf{a} \in A\}.$$

It follows easily from this definition that $A^\perp$ is a subspace of $\mathbb{R}^n$ which intersects A in the set $\{\mathbf{0}\}$. A standard result of linear algebra asserts that each vector of $\mathbb{R}^n$ can be expressed *uniquely* in the form $\mathbf{a} + \mathbf{b}$, where $\mathbf{a} \in A$ and $\mathbf{b} \in A^\perp$. Thus $A + A^\perp = \mathbb{R}^n$, a fact which will be needed in our discussion of unbounded convex sets in Section 2.5.

A sequence $\mathbf{u}_1, \ldots, \mathbf{u}_m$ of vectors in $\mathbb{R}^n$ is said to be an **orthonormal sequence** if $\mathbf{u}_i \cdot \mathbf{u}_j$ is 1 or 0 according as $i = j$ or $i \neq j$. The simplest

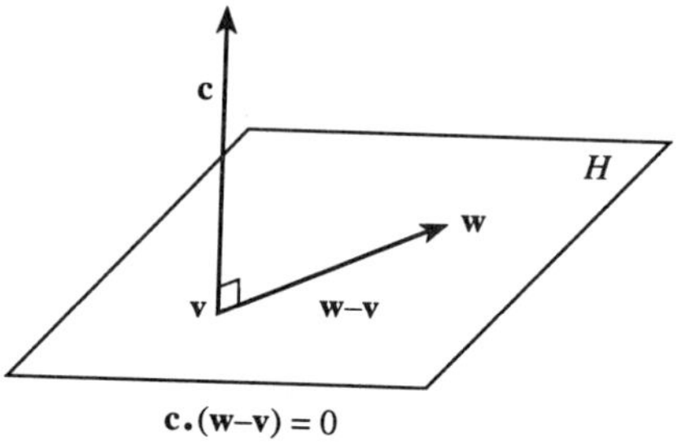

Fig. 1.3

example of such a sequence is the sequence $\mathbf{e}_1, \ldots, \mathbf{e}_n$ of elementary vectors in $\mathbb{R}^n$. In an orthonormal sequence, each term is a unit vector, each two terms are orthogonal, and no two terms are the same. The terms of an orthonormal sequence $\mathbf{u}_1, \ldots, \mathbf{u}_m$ in $\mathbb{R}^n$ form a linearly independent set $\{\mathbf{u}_1, \ldots, \mathbf{u}_m\}$. To see this, suppose that scalars $\lambda_1, \ldots, \lambda_m$ are such that

$$\lambda_1 \mathbf{u}_1 + \ldots + \lambda_m \mathbf{u}_m = \mathbf{0}.$$

Then, for $i = 1, \ldots, m$,

$$\lambda_i = (\lambda_1 \mathbf{u}_1 + \ldots + \lambda_m \mathbf{u}_m) \cdot \mathbf{u}_i = \mathbf{0} \cdot \mathbf{u}_i = 0,$$

which shows that $\{\mathbf{u}_1, \ldots, \mathbf{u}_m\}$ is linearly independent. Hence $\{\mathbf{u}_1, \ldots, \mathbf{u}_m\}$ is an *orthonormal basis* for the subspace $\operatorname{lin}\{\mathbf{u}_1, \ldots, \mathbf{u}_m\}$ of $\mathbb{R}^n$. Thus each point $\mathbf{x}$ of $\operatorname{lin}\{\mathbf{u}_1, \ldots, \mathbf{u}_m\}$ can be written *uniquely* as a linear combination of $\mathbf{u}_1, \ldots, \mathbf{u}_m$, say

$$\mathbf{x} = \mu_1 \mathbf{u}_1 + \ldots + \mu_m \mathbf{u}_m.$$

Then, for $i = 1, \ldots, m$,

$$\mathbf{x} \cdot \mathbf{u}_i = (\mu_1 \mathbf{u}_1 + \ldots + \mu_m \mathbf{u}_m) \cdot \mathbf{u}_i = \mu_i.$$

We conclude that

$$\mathbf{x} = (\mathbf{x} \cdot \mathbf{u}_1)\mathbf{u}_1 + \ldots + (\mathbf{x} \cdot \mathbf{u}_m)\mathbf{u}_m,$$

a fact which will be needed in our discussion of matrix inequalities in Section 5.8.

A congruence transformation in elementary plane geometry is a transformation of the plane which preserves distance. Examples of such transformations are reflections, rotations, translations, and combinations of these. Algebraically, the congruence transformations of $\mathbb{R}^2$ are precisely those affine transformations $T: \mathbb{R}^2 \to \mathbb{R}^2$ that can be expressed in the form $T(\mathbf{x}) = \mathbf{Q}\mathbf{x} + \mathbf{q}$, where $\mathbf{Q}$ is a 2×2 orthogonal matrix and $\mathbf{q}$ is a 2×1 matrix. We now show how these ideas generalize to $\mathbb{R}^n$.

A mapping $T: \mathbb{R}^n \to \mathbb{R}^n$ is said to be a **congruence transformation** of $\mathbb{R}^n$ if $\|T(\mathbf{x}) - T(\mathbf{y})\| = \|\mathbf{x} - \mathbf{y}\|$ for all $\mathbf{x}, \mathbf{y} \in \mathbb{R}^n$, i.e. T preserves distance. Before giving a general description of a congruence transformation of $\mathbb{R}^n$, we need some items of notation. We use a superscript T to denote the transpose of a matrix or a vector. Thus, recalling that we identify a point $\mathbf{x} = (x_1, \ldots, x_n)$ of $\mathbb{R}^n$ with a column vector in the natural way, we see that $\mathbf{x}^{\mathrm{T}}\mathbf{x}$ is the 1×1 matrix whose single element is the scalar $x_1^2 + \ldots + x_n^2$. We identify this scalar with the matrix $\mathbf{x}^{\mathrm{T}}\mathbf{x}$ itself, so that we may write

$$\|\mathbf{x}\|^2 = x_1^2 + \ldots + x_n^2 = \mathbf{x}^{\mathrm{T}}\mathbf{x}.$$

We now show that an affine transformation $T: \mathbb{R}^n \to \mathbb{R}^n$, which is defined by an equation of the form $T(\mathbf{x}) = \mathbf{Q}\mathbf{x} + \mathbf{q}$, where $\mathbf{Q}$ is an $n \times n$ orthogonal matrix and $\mathbf{q}$ is an $n \times 1$ matrix, is a congruence transformation of $\mathbb{R}^n$. Let $\mathbf{x}, \mathbf{y} \in \mathbb{R}^n$. Then

$$\begin{aligned}\|T(\mathbf{x}) - T(\mathbf{y})\|^2 &= \|\mathbf{Q}(\mathbf{x} - \mathbf{y})\|^2 \\ &= (\mathbf{Q}(\mathbf{x} - \mathbf{y}))^{\mathrm{T}}(\mathbf{Q}(\mathbf{x} - \mathbf{y})) \\ &= (\mathbf{x} - \mathbf{y})^{\mathrm{T}}\mathbf{Q}^{\mathrm{T}}\mathbf{Q}(\mathbf{x} - \mathbf{y}) \\ &= (\mathbf{x} - \mathbf{y})^{\mathrm{T}}(\mathbf{x} - \mathbf{y}) \\ &= \|\mathbf{x} - \mathbf{y}\|^2.\end{aligned}$$

Hence $\|T(\mathbf{x}) - T(\mathbf{y})\| = \|\mathbf{x} - \mathbf{y}\|$, which shows that T is a congruence transformation of $\mathbb{R}^n$. In a more substantial result, we now show that every congruence transformation of $\mathbb{R}^n$ is an affine transformation of the type considered above.

Theorem 1.6.2 *Let $T: \mathbb{R}^n \to \mathbb{R}^n$ be a congruence transformation of $\mathbb{R}^n$. Then there exist an $n \times n$ orthogonal matrix $\mathbf{Q}$ and an $n \times 1$ matrix $\mathbf{q}$ such that $T(\mathbf{x}) = \mathbf{Q}\mathbf{x} + \mathbf{q}$ for all $\mathbf{x}$ in $\mathbb{R}^n$.*

Proof Let $\mathbf{x}, \mathbf{y} \in \mathbb{R}^n$. Define a mapping $f: \mathbb{R}^n \to \mathbb{R}^n$ by the equation $f(\mathbf{x}) = T(\mathbf{x}) - T(\mathbf{0})$. Since T preserves distance,

$$\|f(\mathbf{x})\| = \|T(\mathbf{x}) - T(\mathbf{0})\| = \|\mathbf{x} - \mathbf{0}\| = \|\mathbf{x}\|,$$

whence f preserves norms. Also

$$\|f(\mathbf{x}) - f(\mathbf{y})\|^2 = \|T(\mathbf{x}) - T(\mathbf{y})\|^2 = \|\mathbf{x} - \mathbf{y}\|^2,$$

so

$$\|f(\mathbf{x})\|^2 - 2f(\mathbf{x}) \cdot f(\mathbf{y}) + \|f(\mathbf{y})\|^2 = \|\mathbf{x}\|^2 - 2\mathbf{x} \cdot \mathbf{y} + \|\mathbf{y}\|^2.$$

Since $\|f(\mathbf{x})\| = \|\mathbf{x}\|$ and $\|f(\mathbf{y})\| = \|\mathbf{y}\|$, we can deduce from the last equation that $f(\mathbf{x}) \cdot f(\mathbf{y}) = \mathbf{x} \cdot \mathbf{y}$, whence f preserves inner products. It follows that $f(\mathbf{e}_1), \ldots, f(\mathbf{e}_n)$ is an orthonormal sequence in $\mathbb{R}^n$, and hence that

$$f(\mathbf{x}) = (f(\mathbf{x})\cdot f(\mathbf{e}_1))f(\mathbf{e}_1) + \ldots + (f(\mathbf{x})\cdot f(\mathbf{e}_n))f(\mathbf{e}_n).$$

Writing $\mathbf{x}$ for $(x_1, \ldots, x_n)$ and $\mathbf{Q}$ for the $n \times n$ orthogonal matrix whose columns are $f(\mathbf{e}_1), \ldots, f(\mathbf{e}_n)$, we deduce that

$$\begin{aligned} f(\mathbf{x}) &= (\mathbf{x}\cdot\mathbf{e}_1)f(\mathbf{e}_1) + \ldots + (\mathbf{x}\cdot\mathbf{e}_n)f(\mathbf{e}_n) \\ &= x_1 f(\mathbf{e}_1) + \ldots + x_n f(\mathbf{e}_n) \\ &= \mathbf{Qx}. \end{aligned}$$

The proof is completed by putting $\mathbf{q} = T(\mathbf{0})$. □

We have thus identified the congruence transformations $T\colon \mathbb{R}^n \to \mathbb{R}^n$ of $\mathbb{R}^n$ as being precisely those affine transformations $T\colon \mathbb{R}^n \to \mathbb{R}^n$ which can be expressed in the form $T(\mathbf{x}) = \mathbf{Qx} + \mathbf{q}$, where $\mathbf{Q}$ is an $n \times n$ orthogonal matrix and $\mathbf{q}$ is an $n \times 1$ matrix. Sets A and B in $\mathbb{R}^n$ are said to be **congruent** if there is a congruence transformation T of $\mathbb{R}^n$ such that $T(A) = B$. It is easy to verify that congruence is an equivalence relation on the family of all subsets of $\mathbb{R}^n$. In elementary geometry, any two points are congruent, any two lines are congruent, and any two planes are congruent. These results are special cases of the next theorem.

Theorem 1.6.3 *Let A and B be r-flats in $\mathbb{R}^n$. Then A and B are congruent.*

Proof We consider the non-trivial cases when $r \geqslant 1$. First we show that the r-flat A is congruent to the r-flat R_r defined by the equation

$$R_r = \{(x_1, \ldots, x_r, 0, \ldots, 0) : x_1, \ldots, x_r \in \mathbb{R}\}.$$

Let $\mathbf{a} \in A$. Then $A - \mathbf{a}$ is an r-dimensional subspace of $\mathbb{R}^n$. Let $\{\mathbf{u}_1, \ldots, \mathbf{u}_n\}$ be an orthonormal basis for $\mathbb{R}^n$ such that $\{\mathbf{u}_1, \ldots, \mathbf{u}_r\}$ is an orthonormal basis for $A - \mathbf{a}$. Define a congruence transformation T of $\mathbb{R}^n$ by the equation

$$T(\mathbf{x}) = [\mathbf{u}_1, \ldots, \mathbf{u}_n]\mathbf{x} + \mathbf{a}.$$

Then $T(R_r) = A$, so A and R_r are congruent. Similarly, B and R_r are congruent. Thus A and B are congruent. □

We now show how, given any r-dimensional set A in $\mathbb{R}^n$ with $1 \leqslant r \leqslant n$, it is possible to find a *congruent copy* of A in the space $\mathbb{R}^r$. Moreover, we show that any two such congruent copies of A in $\mathbb{R}^r$ are themselves congruent to one another in $\mathbb{R}^r$. As an example of a situation where such an idea might be helpful, consider the case of a triangle located in $\mathbb{R}^3$. If we are interested in some intrinsic property of the triangle, such as its area, then it is natural, and often algebraically simpler, to study not the triangle itself, but some congruent copy of it in $\mathbb{R}^2$.

Let A be an r-dimensional $(1 \leqslant r \leqslant n)$ set in $\mathbb{R}^n$. Then aff$A$ is an r-flat, and so by Theorem 1.6.3 there is a congruence transformation of $\mathbb{R}^n$ which maps aff$A$ onto the r-flat

$$R_r = \{(x_1, \ldots, x_r, 0, \ldots, 0) : x_1, \ldots, x_r \in \mathbb{R}\}.$$

It follows that there is a set B in R_r which is congruent to A. Let $i: R_r \to \mathbb{R}^r$ be the mapping that identifies each point $(x_1, \ldots, x_r, 0, \ldots, 0)$ of R_r with the point $(x_1, \ldots, x_r)$ of $\mathbb{R}^r$. Then $i(B)$ is a set lying in $\mathbb{R}^r$ which is a *congruent copy* of the set A in $\mathbb{R}^n$. In general, there will be an infinite number of such copies. We now see how any two of these copies of A are related.

Let $i(B)$ and $i(C)$ be congruent copies of A in $\mathbb{R}^r$, where B and C are congruent to A in $\mathbb{R}^n$ and lie in R_r. Then there is a congruence transformation T of $\mathbb{R}^n$ such that $T(B) = C$, and which maps R_r onto itself. By considering the images of $\mathbf{0}$ and the elementary vectors $\mathbf{e}_1, \ldots, \mathbf{e}_r$ under T, it follows that T can be expressed in the form

$$T(\mathbf{x}) = \left[\begin{array}{c:c} \mathbf{Q} & \mathbf{0} \\ \hdashline \mathbf{0} & * \end{array}\right] \mathbf{x} + \left[\begin{array}{c} \mathbf{q} \\ \hdashline \mathbf{0} \end{array}\right],$$

where $\mathbf{Q}$ is an $r \times r$ orthogonal matrix, $\mathbf{q}$ is an $r \times 1$ matrix, and $\mathbf{0}$ represents zero matrices of suitable shapes and sizes. Denote by T_r the congruence transformation of $\mathbb{R}^r$ defined by the equation

$$T_r(\mathbf{x}) = \mathbf{Q}\mathbf{x} + \mathbf{q},$$

where $\mathbf{x} = (x_1, \ldots, x_r)$. Then $T_r(i(B)) = i(C)$, which shows that the congruent copies $i(B)$ and $i(C)$ of A in $\mathbb{R}^r$ are themselves congruent to one another in $\mathbb{R}^r$.

Exercises 1.6

1. Let $\mathbf{x}, \mathbf{y} \in \mathbb{R}^n$ and $\lambda + \mu = 1$. Show that

$$\lambda\|\mathbf{x}\|^2 + \mu\|\mathbf{y}\|^2 = \|\lambda\mathbf{x} + \mu\mathbf{y}\|^2 + \lambda\mu\|\mathbf{x} - \mathbf{y}\|^2.$$

Deduce that

$$\|\mathbf{x} + \mathbf{y}\|^2 + \|\mathbf{x} - \mathbf{y}\|^2 = 2\|\mathbf{x}\|^2 + 2\|\mathbf{y}\|^2.$$

Given also that $\lambda, \mu > 0$, $\|\mathbf{x}\| = \|\mathbf{y}\| = 1$ and $\mathbf{x} \neq \mathbf{y}$, prove that $\|\lambda\mathbf{x} + \mu\mathbf{y}\| < 1$.

2. Let a hyperplane H in $\mathbb{R}^n$ have equation $c_0 + \mathbf{c} \cdot \mathbf{x} = 0$. Suppose that $\mathbf{a}$ is a point of $\mathbb{R}^n$ not lying in H. Show that there is a *unique*

point $\mathbf{b}$ of H for which $\mathbf{a} - \mathbf{b}$ is a normal vector to H. Prove that, for all $\mathbf{x} \in H$,

$$\|\mathbf{x} - \mathbf{a}\|^2 = \|\mathbf{x} - \mathbf{b}\|^2 + \|\mathbf{b} - \mathbf{a}\|^2.$$

Deduce that $\mathbf{b}$ is the *unique* nearest point of H to $\mathbf{a}$. Prove that the shortest distance $\|\mathbf{a} - \mathbf{b}\|$ between $\mathbf{a}$ and H is $|c_0 + \mathbf{c} \cdot \mathbf{a}|/\|\mathbf{c}\|$.

3. Find a congruence transformation $T: \mathbb{R}^4 \to \mathbb{R}^4$ which maps the hyperplane H in $\mathbb{R}^4$ with equation $1 + 2x_1 + 2x_2 - x_3 - x_4 = 0$ onto the 3-flat R_3 consisting of all those points of $\mathbb{R}^4$ whose fourth coordinate is zero.

1.7 OPEN SETS AND CLOSED SETS

Whereas so far we have concentrated on the algebraic properties of sets (i.e. those relating to vector addition and scalar multiplication), in the remaining three sections of the chapter we turn our attention to the metric properties of sets (i.e. those relating to distance). The reader with a knowledge of elementary topology will recognize many of the results in these sections as special cases of theorems which hold in arbitrary metric spaces. We begin by introducing the concepts of *open* sets and *closed* sets, ideas which arise naturally in any discussion of the interior and the boundary of a set in $\mathbb{R}^n$.

Just as circles and spheres play a central role in elementary geometry, so their higher-dimensional analogues, known as *balls*, figure prominently in the study of convex sets in $\mathbb{R}^n$. It turns out to be useful to consider two types of ball, namely the *open ball* and the *closed ball*. Roughly speaking, open balls are the analogues of solid spheres with their surfaces excluded, whereas closed balls are the analogues of solid spheres with their surfaces included. The particular relevance of open balls and closed balls here is that they are, respectively, important examples of open sets and closed sets. Moreover, open balls are the building blocks out of which open sets are constructed.

Let $\mathbf{a} \in \mathbb{R}^n$ and $r > 0$. Then the **open ball** $B(\mathbf{a}; r)$ (**closed ball** $B[\mathbf{a}; r]$) with **centre a** and **radius** r is the set of all points of $\mathbb{R}^n$ whose distance from $\mathbf{a}$ is less than (less than or equal to) r, i.e.

$$B(\mathbf{a}; r) = \{\mathbf{x} \in \mathbb{R}^n : \|\mathbf{x} - \mathbf{a}\| < r\} \quad \text{and} \quad B[\mathbf{a}; r] = \{\mathbf{x} \in \mathbb{R}^n : \|\mathbf{x} - \mathbf{a}\| \leqslant r\}.$$

In $\mathbb{R}^1$ the open (closed) ball with centre a and radius r is the open (closed) interval $(a - r, a + r)$ ($[a - r, a + r]$). In $\mathbb{R}^2$ open (closed) balls are referred to as **open (closed) discs**. The open disc $B(\mathbf{a}; r)$ and the closed disc $B[\mathbf{a}; r]$ are illustrated in Fig. 1.4. The balls $B(\mathbf{0}; 1)$ and $B[\mathbf{0}; 1]$ in $\mathbb{R}^n$ are called, respectively, the **open unit ball** and the **closed unit ball**. If we denote them, respectively, by V and U, then

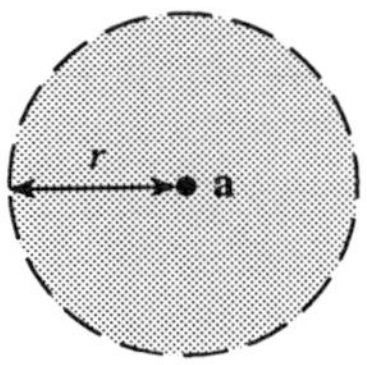

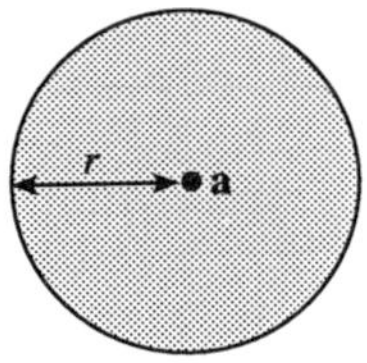

Fig. 1.4

$$V = \{\mathbf{x} \in \mathbb{R}^n : \|\mathbf{x}\| < 1\} \quad \text{and} \quad U = \{\mathbf{x} \in \mathbb{R}^n : \|\mathbf{x}\| \leqslant 1\}.$$

It follows that $B(\mathbf{a}; r) = \mathbf{a} + rV$ and $B[\mathbf{a}; r] = \mathbf{a} + rU$. We adopt U as the standard notation for the closed unit ball.

A point $\mathbf{a}$ of a set A in $\mathbb{R}^n$ is said to be an **interior point** of A if it is the centre of some open ball which lies in A, i.e. if there exists some $r > 0$ such that $B(\mathbf{a}; r) \subseteq A$. The set of interior points of A is called the **interior** of A and is denoted by $\operatorname{int} A$. Clearly $\operatorname{int} B \subseteq \operatorname{int} A$ when $B \subseteq A$. A set in $\mathbb{R}^n$, each of whose points is an interior point of the set, is said to be **open**. Since $\operatorname{int} A \subseteq A$ is always true, A is open if and only if $\operatorname{int} A = A$. Clearly the sets $\varnothing$ and $\mathbb{R}^n$ are open.

Theorem 1.7.1 *In $\mathbb{R}^n$ open balls and open halfspaces are open, and hyperplanes have empty interiors.*

Proof Consider the open ball $B(\mathbf{a}; r)$, where $\mathbf{a} \in \mathbb{R}^n$ and $r > 0$. Let $\mathbf{x} \in B(\mathbf{a}; r)$. We prove that $B(\mathbf{a}; r)$ is open by showing that $B(\mathbf{x}; s) \subseteq B(\mathbf{a}; r)$, where s is the positive number $r - \|\mathbf{x} - \mathbf{a}\|$. The proof is illustrated in Fig. 1.5. Let $\mathbf{y} \in B(\mathbf{x}; s)$. Then $\|\mathbf{y} - \mathbf{x}\| < s$, and so by the triangle inequality

$$\|\mathbf{y} - \mathbf{a}\| \leqslant \|\mathbf{y} - \mathbf{x} + \mathbf{x} - \mathbf{a}\| \leqslant \|\mathbf{y} - \mathbf{x}\| + \|\mathbf{x} - \mathbf{a}\| < s + \|\mathbf{x} - \mathbf{a}\| = r.$$

Thus $\mathbf{y} \in B(\mathbf{a}; r)$, and so $B(\mathbf{x}; s)] \subseteq B(\mathbf{a}; r)$ as required.

Consider the open halfspace A in $\mathbb{R}^n$ which is defined by the inequality $c_0 + \mathbf{c} \cdot \mathbf{x} > 0$, where $\mathbf{c}$ is a unit vector. Let $\mathbf{a} \in A$. We prove that A is open by showing that $B(\mathbf{a}; r) \subseteq A$, where r is the positive number $c_0 + \mathbf{c} \cdot \mathbf{a}$. Let $\mathbf{y} \in B(\mathbf{a}; r)$. Then $\|\mathbf{y} - \mathbf{a}\| < r$ and

$$c_0 + \mathbf{c} \cdot \mathbf{y} = c_0 + \mathbf{c} \cdot \mathbf{a} + \mathbf{c} \cdot (\mathbf{y} - \mathbf{a}) = r + \mathbf{c} \cdot (\mathbf{y} - \mathbf{a}) > 0,$$

since, by the Cauchy–Schwarz inequality, $|\mathbf{c} \cdot (\mathbf{y} - \mathbf{a})| \leqslant \|\mathbf{y} - \mathbf{a}\| < r$. Thus $\mathbf{y} \in A$, and so $B(\mathbf{a}; r) \subseteq A$ as required.

Consider the hyperplane H in $\mathbb{R}^n$ with equation $c_0 + \mathbf{c} \cdot \mathbf{x} = 0$, where $\mathbf{c}$ is a unit vector. We show that no point $\mathbf{a}$ of H is an interior point of H. Let $r > 0$. Then $\mathbf{a} + \frac{1}{2} r\mathbf{c} \notin H$ and $\|\mathbf{a} + \frac{1}{2} r\mathbf{c} - \mathbf{a}\| = \frac{1}{2} r$, whence $\mathbf{a} + \frac{1}{2} r\mathbf{c} \in B(\mathbf{a}; r)$ and $B(\mathbf{a}; r) \nsubseteq A$. Hence $\mathbf{a}$ is not an interior point of H, and H has an empty interior. □

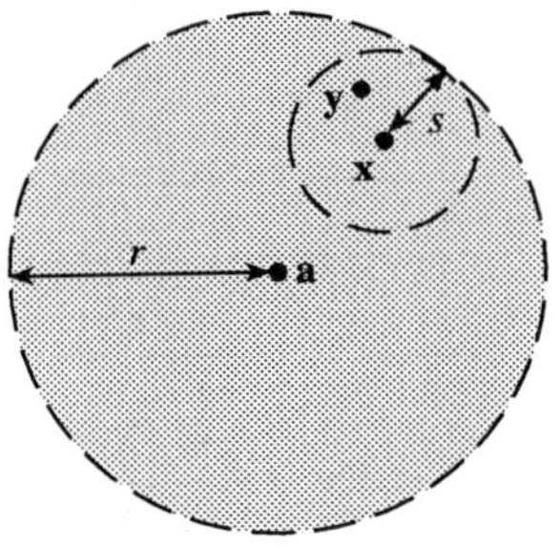

Fig. 1.5

Corollary 1.7.2 *Let A be a set in* $\mathbb{R}^n$. *Then* $\operatorname{int} A$ *is open and* $\operatorname{int}(\operatorname{int} A) = \operatorname{int} A$.

Proof If $\mathbf{a} \in \operatorname{int} A$, then there exists $r > 0$ such that $B(\mathbf{a}; r) \subseteq A$. Since $B(\mathbf{a}; r)$ is open,

$$B(\mathbf{a}; r) = \operatorname{int}(B(\mathbf{a}; r)) \subseteq \operatorname{int} A,$$

whence $\mathbf{a} \in \operatorname{int}(\operatorname{int} A)$ and $\operatorname{int} A \subseteq \operatorname{int}(\operatorname{int} A)$. Thus $\operatorname{int} A$ is open and $\operatorname{int}(\operatorname{int} A) = \operatorname{int} A$. □

Theorem 1.7.3 *In* $\mathbb{R}^n$ *every union and every finite intersection of open sets is open.*

Proof Let A be the union of a family $(A_i : i \in I)$ of open sets in $\mathbb{R}^n$. If $\mathbf{a} \in A$, then $\mathbf{a} \in A_i$ for some $i \in I$. Since A_i is open, there is an $r > 0$ such that $B(\mathbf{a}; r) \subseteq A_i$, whence $B(\mathbf{a}; r) \subseteq A$. Thus A is open.

Let A be the intersection of the open sets $A_1, \ldots, A_m$ in $\mathbb{R}^n$. If $\mathbf{a} \in A$, then $\mathbf{a} \in A_1, \ldots, \mathbf{a} \in A_m$. Since $A_1, \ldots, A_m$ are open, there exist $r_1, \ldots, r_m > 0$ such that $B(\mathbf{a}; r_1) \subseteq A_1, \ldots, B(\mathbf{a}; r_m) \subseteq A_m$. Let $r = \min \{r_1, \ldots, r_m\}$. Then $r > 0$ and

$$B(\mathbf{a}; r) \subseteq B(\mathbf{a}; r_1) \cap \ldots \cap B(\mathbf{a}; r_m) \subseteq A_1 \cap \ldots \cap A_m = A.$$

Thus A is open. □

An arbitrary intersection of open sets in $\mathbb{R}^n$ need not be open. To see this, we note that the intersection of the sequence $V, \frac{1}{2}V, \frac{1}{3}V, \ldots, \frac{1}{k}V, \ldots$ of open balls centred on the origin of $\mathbb{R}^n$ is the singleton set $\{\mathbf{0}\}$, which is *not* open.

A point $\mathbf{a}$ of $\mathbb{R}^n$ is said to be a **closure point** of a set A in $\mathbb{R}^n$ if *every* open ball with centre $\mathbf{a}$ meets A, i.e. if for every $r > 0$ the ball $B(\mathbf{a}; r)$ meets A. The set of closure points of A is called the **closure** of A and is denoted by $\operatorname{cl} A$. Clearly $A \subseteq \operatorname{cl} A$, and $\operatorname{cl} B \subseteq \operatorname{cl} A$ whenever $B \subseteq A$. Roughly speaking, the closure of A is the set of all points in $\mathbb{R}^n$ which either lie in A or are *arbitrarily close to A*. Thus in $\mathbb{R}^1$ the closures of the intervals $(0, 1]$, $(0, 1)$, $[0, 1)$ are all equal to the interval $[0, 1]$. In $\mathbb{R}^2$ the closures of the discs $B(\mathbf{a}; r)$ and $B[\mathbf{a}; r]$ are both equal to the disc $B[\mathbf{a}; r]$. A set in $\mathbb{R}^n$ each of whose closure points lies in the set is said to be **closed**. Thus a set A in $\mathbb{R}^n$ is closed if and only if $\operatorname{cl} A \subseteq A$. Since $A \subseteq \operatorname{cl} A$ is always true, A is closed if and only if $\operatorname{cl} A = A$. Clearly the sets $\varnothing$ and $\mathbb{R}^n$ are closed. Thus the sets $\varnothing$ and $\mathbb{R}^n$ are both open and closed; it can be shown that they are the only sets in $\mathbb{R}^n$ with this property. A set in $\mathbb{R}^n$ may be neither open nor closed. For example, in $\mathbb{R}^1$ the interval $[0, 1)$ is such a set.

Before stating the next theorem, which gives the fundamental relationship between the interior and closure operations, we need an item of notation. For each set A in $\mathbb{R}^n$, we denote by A^c the complement of A in $\mathbb{R}^n$, i.e. the set $\mathbb{R}^n \setminus A$.

Theorem 1.7.4 *Let A be a set in* $\mathbb{R}^n$. *Then* $\operatorname{cl} A = (\operatorname{int} A^c)^c$.

Proof If $\mathbf{x} \in \operatorname{cl} A$, then *each* open ball with centre $\mathbf{x}$ contains a point of A, and so $\mathbf{x}$ cannot belong to $\operatorname{int} A^c$, i.e. $\mathbf{x} \in (\operatorname{int} A^c)^c$. If $\mathbf{x} \in (\operatorname{int} A^c)^c$, then *each* open ball with centre $\mathbf{x}$ must contain a point of A, i.e. $\mathbf{x} \in \operatorname{cl} A$. Thus $\operatorname{cl} A = (\operatorname{int} A^c)^c$. □

Theorem 1.7.5 *A set in* $\mathbb{R}^n$ *is closed if and only if its complement in* $\mathbb{R}^n$ *is open.*

Proof Let A be a set in $\mathbb{R}^n$. Suppose first that A is closed. Then $\operatorname{cl} A = A$, and it follows from Corollary 1.7.2 and Theorem 1.7.4 that A^c is the open set $\operatorname{int} A^c$. Suppose next that A^c is open. Then $\operatorname{int} A^c = A^c$, and it follows from Theorem 1.7.4 that $\operatorname{cl} A = A$, i.e. A is closed. □

Corollary 1.7.6 *Let A be a set in* $\mathbb{R}^n$. *Then* $\operatorname{cl} A$ *is closed and* $\operatorname{cl}(\operatorname{cl} A) = \operatorname{cl} A$.

Proof Now $\operatorname{int} A^c$ is open by Corollary 1.7.2, whence by Theorem 1.7.5 its complement $\operatorname{cl} A$ is closed. □

Theorem 1.7.7 *In* $\mathbb{R}^n$ *every intersection and every finite union of closed sets is closed.*

Proof Let $(A_i : i \in I)$ be a family of closed sets in $\mathbb{R}^n$. Then, for each $i \in I$, A^c is open. By Theorem 1.7.3, $\bigcup (A_i^c : i \in I)$ is open, whence

$$\bigcap (A_i : i \in I) = \left(\bigcup (A_i^c : i \in I)\right)^c$$

is closed. Now let $A_1, \ldots, A_m$ be closed sets in $\mathbb{R}^n$. Then $A_1^c, \ldots, A_m^c$ are open. By Theorem 1.7.3, $A_1^c \cap \ldots \cap A_m^c$ is open, whence

$$A_1 \cup \ldots \cup A_m = (A_1^c \cap \ldots \cap A_m^c)^c$$

is closed. □

Corollary 1.7.8 *Let $A_1, \ldots, A_m$ be sets in $\mathbb{R}^n$. Then*

$$\mathrm{cl}\,(A_1 \cup \ldots \cup A_m) = \mathrm{cl}\,A_1 \cup \ldots \cup \mathrm{cl}\,A_m.$$

Proof Since $A_1 \cup \ldots \cup A_m$ is contained in the closed set $\mathrm{cl}\,A_1 \cup \ldots \cup \mathrm{cl}\,A_m$,

$$\mathrm{cl}\,(A_1 \cup \ldots \cup A_m) \subseteq \mathrm{cl}\,A_1 \cup \ldots \cup \mathrm{cl}\,A_m.$$

Trivially,

$$\mathrm{cl}\,(A_1 \cup \ldots \cup A_m) \supseteq \mathrm{cl}\,A_1 \cup \ldots \cup \mathrm{cl}\,A_m.$$

Thus

$$\mathrm{cl}\,(A_1 \cup \ldots \cup A_m) = \mathrm{cl}\,A_1 \cup \ldots \cup \mathrm{cl}\,A_m. \ \square$$

Theorem 1.7.9 *In $\mathbb{R}^n$ closed balls, closed halfspaces, and flats are closed.*

Proof Let A be the closed ball $B[\mathbf{a}; r]$, where $\mathbf{a} \in \mathbb{R}^n$ and $r > 0$. We prove that A^c is open. Let $\mathbf{x} \in A^c$. Then we show that $B(\mathbf{x}; s) \subseteq A^c$, where s is the positive number $\|\mathbf{x} - \mathbf{a}\| - r$. Suppose that this is not the case. Then there is some point of A, $\mathbf{y}$ say, which lies in $B(\mathbf{x}; s)$. Now

$$\|\mathbf{x} - \mathbf{a}\| = \|\mathbf{x} - \mathbf{y} + \mathbf{y} - \mathbf{a}\| < s + r = \|\mathbf{x} - \mathbf{a}\|,$$

which is impossible. Hence $B(\mathbf{x}; s) \subseteq A^c$, which proves that A^c is open. Thus A, i.e. the closed ball $B[\mathbf{a}; r]$, is closed.

Theorem 1.7.1 shows that open halfspaces in $\mathbb{R}^n$ are open, whence their complements in $\mathbb{R}^n$, i.e. the closed halfspaces, are closed. In $\mathbb{R}^n$ each hyperplane is the intersection of two closed halfspaces, and so is closed. By Corollary 1.4.2, each flat in $\mathbb{R}^n$ is an intersection of hyperplanes, and so is closed. □

A point $\mathbf{a}$ of $\mathbb{R}^n$ is said to be a **boundary point** of a set A in $\mathbb{R}^n$ if *every* open ball with centre $\mathbf{a}$ meets *both* A and its complement A^c. The set of boundary points of A is called the **boundary** of A and is denoted by $\mathrm{bd}\,A$. Thus a boundary point of a set in $\mathbb{R}^n$ is a point of $\mathbb{R}^n$ which is *arbitrarily close* both to the set and its complement. It follows immediately from the above definitions that $\mathrm{bd}\,A = (\mathrm{cl}\,A) \cap (\mathrm{cl}\,A^c)$, whence the boundary of a set in $\mathbb{R}^n$ is always closed, being the intersection of two closed sets. A boundary point of a set in $\mathbb{R}^n$ may or may not belong to the set itself. For example, in $\mathbb{R}^1$ the interval $[0, 1)$ contains its boundary point 0, but not its boundary point 1. We observe that, for any set A in $\mathbb{R}^n$, the sets A and A^c have the same boundary, and the sets $\operatorname{int} A$, $\operatorname{bd} A$, $\operatorname{int} A^c$ form a partition of $\mathbb{R}^n$. Open (closed) sets in $\mathbb{R}^n$ are characterized by the property that they contain *none* (*all*) of their boundary points.

The above definitions of the interior and the boundary of a set do, of course, depend upon the space in which the set is embedded. For example, a closed line segment in $\mathbb{R}^2$ has an empty interior and is its own boundary, whereas the *same* line segment considered as a subset of $\mathbb{R}^1$ has for its interior the set of all of its points with the exception of its two boundary points, these forming its boundary in $\mathbb{R}^1$. The latter interior and boundary, obtained by regarding the one-dimensional line segment as a set in the one-dimensional space $\mathbb{R}^1$, correspond to what may be thought of as the intrinsic interior and boundary of the segment. The idea of studying a set in $\mathbb{R}^n$ in relation to the space of smallest dimension containing it is an important one in the study of convex sets, and leads us to introduce the concepts of the *relative interior* and the *relative boundary* of a set in $\mathbb{R}^n$.

A point $\mathbf{a}$ of a set A in $\mathbb{R}^n$ is said to be a **relative interior point** of A if it is the centre of some open ball whose intersection with $\operatorname{aff} A$ is contained in A, i.e. if there exists $r > 0$ such that

$$B(\mathbf{a}; r) \cap \operatorname{aff} A \subseteq A.$$

The set of all relative interior points of A is called the **relative interior** of A and is denoted by $\operatorname{ri} A$. The relative interior of an n-dimensional set in $\mathbb{R}^n$ coincides with its interior, and the relative interior of any flat in $\mathbb{R}^n$ is itself. A point $\mathbf{a}$ of $\mathbb{R}^n$ is said to be a **relative boundary point** of a set A in $\mathbb{R}^n$ if it lies in the closure of A but not in its relative interior. The set of all relative boundary points of A is called the **relative boundary** of A and is denoted by $\operatorname{rebd} A$. The relative boundary of an n-dimensional set in $\mathbb{R}^n$ coincides with its boundary. It is to be noted that while the inclusion $B \subseteq A$ implies both $\operatorname{int} B \subseteq \operatorname{int} A$ and $\operatorname{cl} A \subseteq \operatorname{cl} B$, it does *not* in general imply $\operatorname{ri} B \subseteq \operatorname{ri} A$. For example, if B is one side of a square A in $\mathbb{R}^2$, then $\operatorname{ri} B$ and $\operatorname{ri} A$ are non-empty but disjoint. If, however, $B \subseteq A$ and $\dim B = \dim A$ or, equivalently, $\operatorname{aff} B = \operatorname{aff} A$, then $\operatorname{ri} B \subseteq \operatorname{ri} A$.

Suppose now that $\mathbf{a}$ is a point of a set A in $\mathbb{R}^n$ and that $\mathbf{x}$ is a point of $\operatorname{aff} A$ *not* lying in A. Define a scalar λ_0 by the equation

$$\lambda_0 = \sup\{\lambda \in [0, 1] : (1 - \lambda)\mathbf{a} + \lambda\mathbf{x} \in A\}.$$

Then $(1 - \lambda_0)\mathbf{a} + \lambda_0\mathbf{x}$ is a relative boundary point of A lying *between* $\mathbf{a}$ and $\mathbf{x}$, a fact that will be used many times in the course of the book. It follows that flats are the only sets in $\mathbb{R}^n$ which have an empty relative boundary.

Exercises 1.7

1. Show that, if $\mathbf{a}, \mathbf{b} \in \mathbb{R}^n$ and $r, s > 0$ are such that $B[\mathbf{a}; r] = B[\mathbf{b}; s]$, then $\mathbf{a} = \mathbf{b}$ and $r = s$.
2. Which of the following assertions are true for *all* subsets A and B of $\mathbb{R}^n$?
   (i) $\operatorname{int}(A \cup B) = \operatorname{int} A \cup \operatorname{int} B$;
   (ii) $\operatorname{cl}(A \cup B) = \operatorname{cl} A \cup \operatorname{cl} B$;
   (iii) $\operatorname{int}(A \cap B) = \operatorname{int} A \cap \operatorname{int} B$;
   (iv) $\operatorname{cl}(A \cap B) = \operatorname{cl} A \cap \operatorname{cl} B$.
3. Give an example to show that an infinite union of closed sets in $\mathbb{R}^n$ need not be closed.
4. Prove that in $\mathbb{R}^n$ the closure of an open ball is a closed ball and the closure of an open halfspace is a closed halfspace.
5. Show that the union of two closed sets in $\mathbb{R}^n$, each of which has an empty interior, itself has an empty interior.

1.8 CONVERGENCE AND COMPACTNESS

We now establish the basic convergence properties of sequences of points in $\mathbb{R}^n$, and introduce the fundamentally important concept of compactness.

In $\mathbb{R}^n$ a sequence $\mathbf{x}_1, \ldots, \mathbf{x}_k, \ldots$ of points is said to **converge** to a point $\mathbf{x}$ if $\|\mathbf{x}_k - \mathbf{x}\| \to 0$ as $k \to \infty$, i.e. if the distance $\|\mathbf{x}_k - \mathbf{x}\|$ between $\mathbf{x}_k$ and $\mathbf{x}$ tends to zero as k tends to infinity. We indicate such convergence by writing $\mathbf{x}_k \to \mathbf{x}$ as $k \to \infty$, or simply $\mathbf{x}_k \to \mathbf{x}$. This convergence for sequences of points in $\mathbb{R}^n$ coincides with that of classical convergence for real sequences. The inequality of Theorem 1.6.1 (iii) shows that

$$\big|\, \|\mathbf{x}_k\| - \|\mathbf{x}\| \,\big| \leqslant \|\mathbf{x}_k - \mathbf{x}\|,$$

whence $\|\mathbf{x}_k\| \to \|\mathbf{x}\|$ as $k \to \infty$ whenever $\mathbf{x}_k \to \mathbf{x}$ as $k \to \infty$. The triangle inequality shows that

$$\|\mathbf{x}_i - \mathbf{x}_j\| \leqslant \|\mathbf{x}_i - \mathbf{x}\| + \|\mathbf{x} - \mathbf{x}_j\|,$$

whence $\|\mathbf{x}_i - \mathbf{x}_j\| \to 0$ as $i, j \to \infty$ whenever $\mathbf{x}_k \to \mathbf{x}$ as $k \to \infty$.

Suppose that $\mathbf{x}_k = (x_{k1}, \ldots, x_{kn})$ for $k = 1, 2, \ldots$ and that $\mathbf{x} = (x_1, \ldots, x_n)$. Then, for $i = 1, \ldots, n$, we have

$$|x_{ki} - x_i|^2 \leqslant (x_{k1} - x_1)^2 + \ldots + (x_{kn} - x_n)^2 = \|\mathbf{x}_k - \mathbf{x}\|^2$$

and

$$\|\mathbf{x}_k - \mathbf{x}\|^2 = (x_{k1} - x_1)^2 + \ldots + (x_{kn} - x_n)^2$$
$$\leqslant (|x_{k1} - x_1| + \ldots + |x_{kn} - x_n|)^2.$$

Hence

$$|x_{ki} - x_i| \leqslant \|\mathbf{x}_k - \mathbf{x}\| \leqslant |x_{k1} - x_1| + \ldots + |x_{kn} - x_n|. \quad (1.8.1)$$

Inequalities (1.8.1) show that $\mathbf{x}_k \to \mathbf{x}$ if and only if $x_{ki} \to x_i$ for $i = 1, \ldots, n$. Thus the convergence of $\mathbf{x}_1, \ldots, \mathbf{x}_k, \ldots$ to $(x_1, \ldots, x_n)$ is equivalent to the convergence of each of the coordinate sequences $x_{1i}, \ldots, x_{ki}, \ldots$ to x_i, for $i = 1, \ldots, n$. One immediate consequence of this result is that a sequence of points in $\mathbb{R}^n$ can converge to at most one point. Others are that, if $\mathbf{x}_k \to \mathbf{x}$, $\mathbf{y}_k \to \mathbf{y}$ in $\mathbb{R}^n$ and $\lambda_k \to \lambda$, $\mu_k \to \mu$ in $\mathbb{R}$, then $\mathbf{x}_k \cdot \mathbf{y}_k \to \mathbf{x} \cdot \mathbf{y}$ in $\mathbb{R}$ and $\lambda_k \mathbf{x}_k + \mu_k \mathbf{y}_k \to \lambda \mathbf{x} + \mu \mathbf{y}$ in $\mathbb{R}^n$.

We recall that a sequence $x_1, \ldots, x_k, \ldots$ of real numbers is said to be *bounded* if there exists a real number r such that $|x_k| \leqslant r$ for $k = 1, 2, \ldots$. Similarly, a sequence $\mathbf{x}_1, \ldots, \mathbf{x}_k, \ldots$ of points in $\mathbb{R}^n$ is defined to be **bounded** if there exists a real number r such that $\|\mathbf{x}_k\| \leqslant r$ for $k = 1, 2, \ldots$. Every convergent sequence of real numbers is bounded, and the same is also true for convergent sequences of points in $\mathbb{R}^n$. To see this, suppose that $\mathbf{x}_k \to \mathbf{x}$ in $\mathbb{R}^n$. By what we proved above, $\|\mathbf{x}_k\| \to \|\mathbf{x}\|$, and so there exists a real number r such that $\|\mathbf{x}_k\| \leqslant r$ for $k = 1, 2, \ldots$. The next theorem generalizes to $\mathbb{R}^n$ the classical result that every bounded sequence of real numbers contains a convergent subsequence.

Theorem 1.8.1 *Every bounded sequence of points of $\mathbb{R}^n$ contains a convergent subsequence.*

Proof Let $\mathbf{x}_1, \ldots, \mathbf{x}_k, \ldots$ be a bounded sequence of points in $\mathbb{R}^n$. Then (1.8.1) with $\mathbf{x} = \mathbf{0}$ shows that each of the n coordinate sequences associated with $\mathbf{x}_1, \ldots, \mathbf{x}_k, \ldots$ is bounded in $\mathbb{R}$. In particular, the sequence of the first coordinates of $\mathbf{x}_1, \ldots, \mathbf{x}_k, \ldots$ is a bounded sequence of real numbers. Thus there exists a subsequence of $\mathbf{x}_1, \ldots, \mathbf{x}_k, \ldots$ such that the sequence of its first coordinates converges. Similarly, there exists a subsequence of this subsequence of $\mathbf{x}_1, \ldots, \mathbf{x}_k, \ldots$ such that the sequence of its second coordinates converges. After performing this subsequence operation n times in all, we arrive at a subsequence of $\mathbf{x}_1, \ldots, \mathbf{x}_k, \ldots$ each of whose n coordinate sequences converges, i.e. we have found a convergent subsequence of $\mathbf{x}_1, \ldots, \mathbf{x}_k, \ldots$. □

The following result provides a useful characterization of the closure of a set in $\mathbb{R}^n$ in terms of convergent sequences.

Theorem 1.8.2 *Let A be a set in $\mathbb{R}^n$. Then $\mathbf{x} \in \operatorname{cl} A$ if and only if there exists a sequence of points of A which converges to $\mathbf{x}$.*

Proof Suppose first that $\mathbf{x}_1, \ldots, \mathbf{x}_k, \ldots$ is a sequence of points of A which converges to a point $\mathbf{x}$ of $\mathbb{R}^n$. Then, for each $r > 0$, there is some point $\mathbf{x}_k$ of the sequence such that $\|\mathbf{x}_k - \mathbf{x}\| < r$, whence the open ball $B(\mathbf{x}; r)$ meets A. This shows that $\mathbf{x} \in \operatorname{cl} A$.

Suppose next that $\mathbf{x} \in \operatorname{cl} A$. Then, for each positive integer k, the ball $B(\mathbf{x}; 1/k)$ meets A, whence there exists $\mathbf{x}_k \in A$ such that $\|\mathbf{x}_k - \mathbf{x}\| < 1/k$. It follows that $\mathbf{x}_1, \ldots, \mathbf{x}_k, \ldots$ converges to $\mathbf{x}$. □

Corollary 1.8.3 *Let A be a set in $\mathbb{R}^n$. Then A is closed if and only if each convergent sequence of points of A converges to a point of A.*

Proof The corollary follows from the theorem and the fact that A is closed if and only if $A = \operatorname{cl} A$. □

Thus far we have only defined the notion of boundedness for sequences of points in $\mathbb{R}^n$; now we extend it to subsets of $\mathbb{R}^n$. The set A in $\mathbb{R}^n$ is said to be **bounded** if there exists a real number r such that $\|\mathbf{a}\| \leqslant r$ for all $\mathbf{a} \in A$. Clearly, a set in $\mathbb{R}^n$ is bounded if and only if each sequence of its points is bounded. In $\mathbb{R}^n$ balls and finite sets are bounded, whereas r-flats ($r \geqslant 1$) are not.

Theorem 1.8.1 and Corollary 1.8.3, taken together, show that each sequence of points of a closed bounded set in $\mathbb{R}^n$ contains some subsequence which converges to a point of the set. This property of the closed bounded sets in $\mathbb{R}^n$, which at first sight might seem to be of little interest, turns out to be of central importance, and we are led to make the following definition. A subset of $\mathbb{R}^n$ is said to be **compact**, if each sequence of its points contains some subsequence that converges to a point of the subset.

Theorem 1.8.4 *Let A be a set in $\mathbb{R}^n$. Then A is compact if and only it it is both closed and bounded.*

Proof We have already noted that closed bounded subsets of $\mathbb{R}^n$ are compact. Suppose, then, that A is compact. We show first that A is closed. If $\mathbf{x} \in \operatorname{cl} A$, then, by Theorem 1.8.2, there is a sequence of points of A which converges to $\mathbf{x}$. Every subsequence of such a sequence also converges to $\mathbf{x}$. The compactness of A and the uniqueness of limits thus show that $\mathbf{x} \in A$, whence A is closed.

Suppose next that A is *not* bounded. Then, for each positive integer k, there must exist a point $\mathbf{x}_k$ of A such that $\|\mathbf{x}_k\| \geqslant k$. Clearly the sequence $\mathbf{x}_1, \ldots, \mathbf{x}_k, \ldots$ of points of A contains no bounded subsequence, and hence no convergent subsequence, contrary to the hypothesis that A is compact. Hence A is both closed and bounded. □

Of the many interesting and important results about compact sets in $\mathbb{R}^n$, we consider only those which we need in our study of convexity.

Theorem 1.8.5 *Let A be a non-empty compact set in $\mathbb{R}^n$ and let $r > 0$. Then there exists a finite number of points $\mathbf{a}_1, \ldots, \mathbf{a}_m$ of A such that*

$$A \subseteq B(\mathbf{a}_1; r) \cup \ldots \cup B(\mathbf{a}_m; r).$$

Proof We argue by contradiction. Suppose that *no* such finite number of points of A exists. Let $\mathbf{x}_1 \in A$. Then $A \nsubseteq B(\mathbf{x}_1; r)$, whence there exists a point $\mathbf{x}_2$ of A such that $\|\mathbf{x}_2 - \mathbf{x}_1\| \geqslant r$. Now $A \nsubseteq B(\mathbf{x}_1; r) \cup B(\mathbf{x}_2; r)$, whence there exists a point $\mathbf{x}_3$ of A such that $\|\mathbf{x}_3 - \mathbf{x}_1\| \geqslant r$ and $\|\mathbf{x}_3 - \mathbf{x}_2\| \geqslant r$. Continuing in this way, we produce a sequence $\mathbf{x}_1, \ldots, \mathbf{x}_k, \ldots$ of points of A with the property that $\|\mathbf{x}_i - \mathbf{x}_j\| \geqslant r$ whenever $i \neq j$. Clearly such a sequence cannot contain a convergent subsequence, thus contradicting the compactness of A. This establishes the theorem. □

Before proving the fundamental property of compact sets expressed below in Theorem 1.8.7, we need the following preparatory lemma.

Lemma 1.8.6 *Let A be a compact set in $\mathbb{R}^n$ and let $(U_i : i \in I)$ be a family of open sets in $\mathbb{R}^n$ whose union contains A. Then there exists $r > 0$ such that, for each $\mathbf{x}$ in A, the open ball $B(\mathbf{x}; r)$ is contained in some U_i.*

Proof We argue by contradiction. Suppose that no such $r > 0$ exists. Then, for each positive integer k, there is some point $\mathbf{a}_k$ of A such that $B(\mathbf{a}_k; 1/k)$ is not contained in any U_i. Since A is compact, the sequence $\mathbf{a}_1, \ldots, \mathbf{a}_k, \ldots$ has a subsequence which converges to a point $\mathbf{a}$ of A. This point $\mathbf{a}$ must belong to one of the U_i's, U^* say. Since U^* is open, there is an $s > 0$ such that $B(\mathbf{a}; 2s) \subseteq U^*$. Because some subsequence of $\mathbf{a}_1, \ldots, \mathbf{a}_k, \ldots$ converges to $\mathbf{a}$, there are infinitely many positive integers k for which $\|\mathbf{a}_k - \mathbf{a}\| < s$. Choose one of these positive integers, m say, so large that $1/m < s$. Let $\mathbf{x} \in B(\mathbf{a}_m; 1/m)$. Then

$$\|\mathbf{x} - \mathbf{a}\| \leqslant \|\mathbf{x} - \mathbf{a}_m\| + \|\mathbf{a}_m - \mathbf{a}\| < s + s = 2s,$$

so $\mathbf{x} \in B(\mathbf{a}; 2s)$. Thus

$$B(\mathbf{a}_m; 1/m) \subseteq B(\mathbf{a}; 2s) \subseteq U^*,$$

which contradicts the assumption that $B(\mathbf{a}_m; 1/m)$ is not contained in any U_i. This establishes the lemma. □

Theorem 1.8.7 *Let A be a compact set in $\mathbb{R}^n$ and let $(U_i : i \in I)$ be a family of open sets in $\mathbb{R}^n$ whose union contains A. Then there exists a finite subset I^* of I such that the union of the family $(U_i : i \in I^*)$ contains A.*

Proof We may suppose that A is non-empty. By Lemma 1.8.6, there is an $r > 0$ such that, for each $\mathbf{x}$ in A, the open ball $B(\mathbf{x}; r)$ is contained in some U_i. By Theorem 1.8.5, there exist points $\mathbf{a}_1, \ldots, \mathbf{a}_m$ in A such that

$$A \subseteq B(\mathbf{a}_1; r) \cup \ldots \cup B(\mathbf{a}_m; r).$$

For each $k = 1, \ldots, m$, there exists $i_k \in I$ such that $B(\mathbf{a}_k; r) \subseteq U_{i_k}$. We complete the proof by taking I^* to be the set $\{i_1, \ldots, i_m\}$. □

Corollary 1.8.8 *Let $(A_i : i \in I)$ be a family of compact sets in $\mathbb{R}^n$ whose intersection is empty. Then there exists a finite subset I^* of I such that the intersection of the family $(A_i : i \in I^*)$ is empty.*

Proof Let $i_0 \in I$ and let $I_0 = I \backslash \{i_0\}$. Then, since $\bigcap (A_i : i \in I)$ is empty, $A_{i_0} \subseteq \bigcup (A_i^c : i \in I_0)$. By the theorem, which is applicable since the sets A_i^c are open, being the complements of closed sets in $\mathbb{R}^n$, there is a finite subset I' of I_0 such that $A_{i_0} \subseteq \bigcup (A_i^c : i \in I')$. It follows that, if I^* denotes the finite subset $I' \cup \{i_0\}$ of I, then $\bigcap (A_i : i \in I^*)$ is empty. □

Corollary 1.8.9 *Let $A_1, \ldots, A_k, \ldots$ be a sequence of non-empty compact sets in $\mathbb{R}^n$ such that $A_1 \supseteq \ldots \supseteq A_k \supseteq \ldots$. Then the intersection $\bigcap (A_k : k = 1, 2, \ldots)$ is non-empty.*

Proof The results follows from Corollary 1.8.8 and the observation that the intersection of any finite number of members of the family is itself a member of the family, and so is non-empty. □

Theorem 1.8.10 *Let A and B be sets in $\mathbb{R}^n$ and let $\lambda, \mu \in \mathbb{R}$. Then $\lambda A + \mu B$ is*

(i) *open when A is open and $\lambda \neq 0$;*

(ii) *closed when A is compact and B is closed;*

(iii) *bounded when A and B are bounded;*

(iv) *compact when A and B are compact.*

Proof (i) Let A be open and let $\lambda \neq 0$. If $\mathbf{x} \in \lambda A + \mu B$, then $\mathbf{x} = \lambda \mathbf{a} + \mu \mathbf{b}$ for some $\mathbf{a} \in A$ and $\mathbf{b} \in B$. Since A is open, there is an $r > 0$ such that $\mathbf{a} + rV \subseteq A$, where V is the open unit ball $\{\mathbf{x} \in \mathbb{R}^n : \|\mathbf{x}\| < 1\}$. Thus

$$\mathbf{x} + \lambda r V = \lambda \mathbf{a} + \mu \mathbf{b} + \lambda r V = \lambda(\mathbf{a} + rV) + \mu \mathbf{b} \subseteq \lambda A + \mu B,$$

which shows that $B(\mathbf{x}; |\lambda| r) \subseteq \lambda A + \mu B$, and hence that $\lambda A + \mu B$ is open.

(ii) Let A be compact and let B be closed. We consider only the non-trivial case $\mu \neq 0$. If $\mathbf{x} \in \mathrm{cl}\,(\lambda A + \mu B)$, then there exist sequences $\mathbf{a}_1, \ldots, \mathbf{a}_k, \ldots$ of points of A, and $\mathbf{b}_1, \ldots, \mathbf{b}_k, \ldots$ of points of B such that $\lambda \mathbf{a}_k + \mu \mathbf{b}_k \to \mathbf{x}$ as $k \to \infty$. Since A is compact, there is a subsequence $\mathbf{a}_{i_1}, \ldots, \mathbf{a}_{i_k}, \ldots$ of $\mathbf{a}_1, \ldots, \mathbf{a}_k, \ldots$ which converges to some point $\mathbf{a}$ of A. Thus $\lambda \mathbf{a}_{i_k} + \mu \mathbf{b}_{i_k} \to \mathbf{x}$ and $\mathbf{b}_{i_k} \to (\mathbf{x} - \lambda \mathbf{a})/\mu$, as $k \to \infty$. But B is closed, and so $(\mathbf{x} - \lambda \mathbf{a})/\mu \in B$, whence $\mathbf{x} \in \lambda A + \mu B$. Thus $\mathbf{x} \in \mathrm{cl}\,(\lambda A + \mu B)$ implies that $\mathbf{x} \in \lambda A + \mu B$, which shows that $\lambda A + \mu B$ is closed.

(iii) Let A and B be bounded. Then there exist real numbers r_1 and r_2 such that $\|\mathbf{a}\| \leqslant r_1$ and $\|\mathbf{b}\| \leqslant r_2$ whenever $\mathbf{a} \in A$ and $\mathbf{b} \in B$. If $\mathbf{x} \in \lambda A + \mu B$, then $\mathbf{x} = \lambda \mathbf{a} + \mu \mathbf{b}$ for some $\mathbf{a} \in A$ and $\mathbf{b} \in B$, whence

$$\|\mathbf{x}\| = \|\lambda \mathbf{a} + \mu \mathbf{b}\| \leqslant |\lambda|\,\|\mathbf{a}\| + |\mu|\,\|\mathbf{b}\| \leqslant |\lambda| r_1 + |\mu| r_2,$$

which shows that $\lambda A + \mu B$ is bounded.

(iv) This follows immediately from (ii) and (iii). □

Exercises 1.8

1. Prove that every r-flat ($r \geqslant 1$) in $\mathbb{R}^n$ is unbounded.
2. Show that the closure of a bounded set in $\mathbb{R}^n$ is always compact.
3. Suppose that a bounded sequence $\mathbf{x}_1, \ldots, \mathbf{x}_k, \ldots$ of points in $\mathbb{R}^n$ is such that *each* of its convergent subsequences converges to the same point $\mathbf{x}$. Prove that the sequence $\mathbf{x}_1, \ldots, \mathbf{x}_k, \ldots$ itself converges to $\mathbf{x}$.
4. Give an example to show that the vector sum of two closed sets in $\mathbb{R}^n$ need not be closed.

1.9 CONTINUITY

In this section we define the concept of continuity for mappings between Euclidean spaces. Of several possible approaches, we have chosen one that uses convergent sequences. We introduce the *distance function* of a non-empty set in $\mathbb{R}^n$, show that it is continuous, and use it to establish the existence of *nearest points* of two non-empty sets in $\mathbb{R}^n$, one of which is compact and the other closed.

Let $f: A \to \mathbb{R}^m$ be a mapping, where A is a non-empty set in $\mathbb{R}^n$. Then f is said to be **continuous** at a point $\mathbf{a}$ of A if, for each sequence $\mathbf{a}_1, \ldots, \mathbf{a}_k, \ldots$ of points of A that converges to $\mathbf{a}$, the sequence $f(\mathbf{a}_1), \ldots, f(\mathbf{a}_k), \ldots$ of points of $\mathbb{R}^m$ converges to $f(\mathbf{a})$. If f is continuous at all points of A, then f is said to be **continuous** on A. An important example of a continuous mapping is the **norm mapping** $\| \ \|:\mathbb{R}^n \to \mathbb{R}$ defined by the equation $\| \ \|(\mathbf{x}) = \|\mathbf{x}\|$ for each point $\mathbf{x}$ of $\mathbb{R}^n$. That $\| \ \|$ is continuous follows immediately from the fact, established at the beginning of Section 1.8, that $\|\mathbf{a}_k\| \to \|\mathbf{a}\|$ as $k \to \infty$ whenever $\mathbf{a}_k \to \mathbf{a}$ as $k \to \infty$.

A mapping $f:A \to \mathbb{R}^m$ defined on a non-empty set A in $\mathbb{R}^n$ is said to satisfy a **Lipschitz condition** on A if there exists a real number s such that, for all $\mathbf{x}, \mathbf{y} \in A$,

$$\|f(\mathbf{x}) - f(\mathbf{y})\| \leqslant s\|\mathbf{x} - \mathbf{y}\|. \tag{1.9.1}$$

If $f:A \to \mathbb{R}^m$ satisfies the Lipschitz condition (1.9.1), then it is continuous on A. To see this, suppose that $\mathbf{a}_1, \ldots, \mathbf{a}_k, \ldots$ is a sequence of points of A that converges to a point $\mathbf{a}$ of A, so that $\|\mathbf{a}_k - \mathbf{a}\| \to 0$ as $k \to \infty$. The Lipschitz condition shows that

$$\|f(\mathbf{a}_k) - f(\mathbf{a})\| \leqslant s\|\mathbf{a}_k - \mathbf{a}\|,$$

whence $\|f(\mathbf{a}_k) - f(\mathbf{a})\| \to 0$ as $k \to \infty$, i.e. the sequence $f(\mathbf{a}_1), \ldots, f(\mathbf{a}_k), \ldots$ converges to $f(\mathbf{a})$. Thus f is continuous at the arbitrary point $\mathbf{a}$ of A, which proves that f is continuous on A.

Several of the mappings that we shall encounter in our study of convexity satisfy Lipschitz conditions. The norm mapping $\| \ \|:\mathbb{R}^n \to \mathbb{R}$ considered above, for example, satisfies the Lipschitz condition

$$\big| \|\mathbf{x}\| - \|\mathbf{y}\| \big| \leqslant \|\mathbf{x} - \mathbf{y}\|.$$

Every affine transformation $T:\mathbb{R}^n \to \mathbb{R}^m$ satisfies a Lipschitz condition on $\mathbb{R}^n$, as we now demonstrate. Suppose that $\mathbf{Q} = [q_{ij}]$ is the real $m \times n$ matrix and $\mathbf{q}$ the real $m \times 1$ matrix such that, for each vector $\mathbf{x}$ in $\mathbb{R}^n$, considered as a column vector,

$$T(\mathbf{x}) = \mathbf{Q}\mathbf{x} + \mathbf{q}.$$

Let $\mathbf{x}, \mathbf{y} \in \mathbb{R}^n$, and write $\mathbf{u} = (u_1, \ldots, u_n) = \mathbf{x} - \mathbf{y}$. Then, using the Cauchy-Schwarz inequality in the form

$$(q_{i1}u_1 + \ldots + q_{in}u_n)^2 \leqslant (q_{i1}^2 + \ldots + q_{in}^2)(u_1^2 + \ldots + u_n^2) \quad \text{for } i = 1, \ldots, m,$$

we deduce that

$$\begin{aligned} \|T(\mathbf{x}) - T(\mathbf{y})\|^2 &= \|\mathbf{Q}\mathbf{u}\|^2 \\ &= \sum_{i=1}^{m} (q_{i1}u_1 + \ldots + q_{in}u_n)^2 \\ &\leqslant \sum_{i=1}^{m} (q_{i1}^2 + \ldots + q_{in}^2)(u_1^2 + \ldots + u_n^2) \\ &= s^2\|\mathbf{u}\|^2, \end{aligned}$$

where

$$s = \sqrt{\left(\sum_{i=1}^{m} \sum_{j=1}^{n} q_{ij}^2\right)}.$$

Thus T satisfies the Lipschitz condition $\|T(\mathbf{x}) - T(\mathbf{y})\| \leqslant s\|\mathbf{x} - \mathbf{y}\|$ on $\mathbb{R}^n$, and so affine transformations are always continuous.

A particularly important example of a mapping satisfying a Lipschitz condition is the **distance function** $d_A : \mathbb{R}^n \to \mathbb{R}$ of a non-empty set A in $\mathbb{R}^n$. In essence, this function d_A associates with each point $\mathbf{x}$ of $\mathbb{R}^n$ its *distance* $d_A(\mathbf{x})$ from A. Formally, d_A is defined by the equation

$$d_A(\mathbf{x}) = \inf\{\|\mathbf{x} - \mathbf{a}\| : \mathbf{a} \in A\} \qquad \text{for } \mathbf{x} \in \mathbb{R}^n.$$

If A is the singleton set $\{\mathbf{a}\}$, then $d_A(\mathbf{x}) = \|\mathbf{x} - \mathbf{a}\|$. In particular, if $\mathbf{a} = \mathbf{0}$, then $d_A(\mathbf{x}) = \|\mathbf{x}\|$. It follows from the definition of d_A and Theorem 1.8.2 that a point $\mathbf{x}$ of $\mathbb{R}^n$ lies in the closure $\operatorname{cl} A$ of A if and only if its *distance* $d_A(\mathbf{x})$ from A is zero. Suppose now that $\mathbf{x}, \mathbf{y}$ lie in $\mathbb{R}^n$. Then, for each $\varepsilon > 0$, there exists $\mathbf{a}$ in A such that

$$\|\mathbf{x} - \mathbf{a}\| < d_A(\mathbf{x}) + \varepsilon.$$

By the triangle inequality,

$$d_A(\mathbf{y}) \leqslant \|\mathbf{y} - \mathbf{a}\| \leqslant \|\mathbf{y} - \mathbf{x}\| + \|\mathbf{x} - \mathbf{a}\| < \|\mathbf{y} - \mathbf{x}\| + d_A(\mathbf{x}) + \varepsilon.$$

Since $\varepsilon > 0$ is arbitrary,

$$d_A(\mathbf{y}) \leqslant \|\mathbf{y} - \mathbf{x}\| + d_A(\mathbf{x}).$$

Interchanging $\mathbf{x}$ and $\mathbf{y}$ in the last inequality, we find that

$$d_A(\mathbf{x}) \leqslant \|\mathbf{x} - \mathbf{y}\| + d_A(\mathbf{y}),$$

whence d_A satisfies the Lipschitz condition

$$|d_A(\mathbf{x}) - d_A(\mathbf{y})| \leqslant \|\mathbf{x} - \mathbf{y}\|,$$

and d_A is continuous on $\mathbb{R}^n$.

In general, the inf in the definition of d_A cannot be replaced by min. To see this, suppose that A is the set $\mathbb{R}^n \setminus \{\mathbf{0}\}$. Then $d_A(\mathbf{0}) = 0$, but there is no $\mathbf{a} \in A$ such that $\|\mathbf{0} - \mathbf{a}\| = 0$. We do, however, have the following result.

Theorem 1.9.1 *Let A be a non-empty closed set in $\mathbb{R}^n$ and let $\mathbf{x} \in \mathbb{R}^n$. Then there exists $\mathbf{a}_0 \in A$ such that $d_A(\mathbf{x}) = \|\mathbf{x} - \mathbf{a}_0\|$.*

Proof It follows easily from the definition of $d_A(\mathbf{x})$ that there exists a sequence $\mathbf{a}_1, \ldots, \mathbf{a}_k, \ldots$ of points of A such that $\|\mathbf{x} - \mathbf{a}_k\| \to d_A(\mathbf{x})$ as $k \to \infty$. Since convergent sequences in $\mathbf{R}$ are bounded, there exists a real number r such that $\|\mathbf{x} - \mathbf{a}_k\| \leqslant r$ for $k = 1, 2, \ldots$. The inequalities

$$\|\mathbf{a}_k\| \leqslant \|\mathbf{a}_k - \mathbf{x}\| + \|\mathbf{x}\| \leqslant r + \|\mathbf{x}\| \qquad \text{for } k = 1, 2, \ldots$$

show that the sequence $\mathbf{a}_1, \ldots, \mathbf{a}_k, \ldots$ is bounded. Hence it contains some subsequence $\mathbf{a}_{i_1}, \ldots, \mathbf{a}_{i_k}, \ldots$, which converges to a point $\mathbf{a}_0$ of $\mathbb{R}^n$. Since A is closed, $\mathbf{a}_0 \in A$. Now $\|\mathbf{x} - \mathbf{a}_{i_k}\| \to \|\mathbf{x} - \mathbf{a}_0\|$ as $k \to \infty$, but we already know that $\|\mathbf{x} - \mathbf{a}_{i_k}\| \to d_A(\mathbf{x})$ as $k \to \infty$. The uniqueness of limits in $\mathbb{R}$ shows that $d_A(\mathbf{x}) = \|\mathbf{x} - \mathbf{a}_0\|$, as required. □

For obvious reasons, the point $\mathbf{a}_0$ in the statement of Theorem 1.9.1 is called a **nearest point** of A to $\mathbf{x}$. The next result shows how fruitfully the concepts of continuity and compactness interact, while its corollary generalizes the well-known result of elementary analysis which asserts that a continuous function $f: [a, b] \to \mathbb{R}$ is bounded and attains its bounds.

Theorem 1.9.2 *Let $f: A \to \mathbb{R}^m$ be a continuous mapping, where A is a non-empty compact set in $\mathbb{R}^n$. Then $f(A)$ is a compact set in $\mathbb{R}^m$.*

Proof Let $f(\mathbf{a}_1), \ldots, f(\mathbf{a}_k), \ldots$ be a sequence of points of $f(A)$, where $\mathbf{a}_1, \ldots, \mathbf{a}_k, \ldots$ is a sequence of points of A. Since A is compact, there is a subsequence $\mathbf{a}_{i_1}, \ldots, \mathbf{a}_{i_k}, \ldots$ of $\mathbf{a}_1, \ldots, \mathbf{a}_k, \ldots$ which converges to some point $\mathbf{a}$ of A. The continuity of f shows that the subsequence $f(\mathbf{a}_{i_1}), \ldots, f(\mathbf{a}_{i_k}), \ldots$ of $f(\mathbf{a}_1), \ldots, f(\mathbf{a}_k), \ldots$ converges to the point $f(\mathbf{a})$ of $f(A)$. Thus $f(A)$ is compact. □

Corollary 1.9.3 *Let $f: A \to \mathbb{R}$ be a continuous mapping, where A is a non-empty compact set in $\mathbb{R}^n$. Then there exist $\mathbf{a}, \mathbf{b} \in A$ such that*

$$f(\mathbf{a}) = \inf\{f(\mathbf{x}) : \mathbf{x} \in A\} \qquad \textit{and} \qquad f(\mathbf{b}) = \sup\{f(\mathbf{x}) : \mathbf{x} \in B\}.$$

Proof The theorem shows that the non-empty set $f(A) = \{f(\mathbf{x}) : \mathbf{x} \in A\}$ of real numbers is compact, and therefore closed and bounded. Thus $f(A)$ possesses both an infimum and supremum. Moreover, the infimum and supremum of $f(A)$ belong to $\operatorname{cl} f(A)$ and hence, since $f(A)$ is closed, to $f(A)$. Hence there exist $\mathbf{a}, \mathbf{b} \in A$ such that $f(\mathbf{a}) = \inf f(A)$ and $f(\mathbf{b}) = \sup f(A)$. □

Theorem 1.9.4 *Let A and B be non-empty sets in $\mathbb{R}^n$ with A closed and B compact. Then there exist $\mathbf{a}_0 \in A$, $\mathbf{b}_0 \in B$ such that*

$$\|\mathbf{a}_0 - \mathbf{b}_0\| = \inf\{\|\mathbf{a} - \mathbf{b}\| : \mathbf{a} \in A, \mathbf{b} \in B\}.$$

Proof The distance function d_A of A is continuous on $\mathbb{R}^n$, and so, by restriction, on B. By Corollary 1.9.3, applicable since B is compact, there exists $\mathbf{b}_0 \in B$ such that

$$d_A(\mathbf{b}_0) = \inf\{d_A(\mathbf{b}) : \mathbf{b} \in B\}.$$

By Theorem 1.9.1, applicable since A is closed, there exists $\mathbf{a}_0 \in A$ such that

$$d_A(\mathbf{b}_0) = \|\mathbf{b}_0 - \mathbf{a}_0\|.$$

For each $\mathbf{a} \in A$, $\mathbf{b} \in B$, we have

$$\|\mathbf{a} - \mathbf{b}\| \geqslant d_A(\mathbf{b}) \geqslant d_A(\mathbf{b}_0) = \|\mathbf{a}_0 - \mathbf{b}_0\|,$$

which, since $\mathbf{a}_0 \in A$, $\mathbf{b}_0 \in B$, shows that

$$\|\mathbf{a}_0 - \mathbf{b}_0\| = \inf\{\|\mathbf{a} - \mathbf{b}\| : \mathbf{a} \in A, \mathbf{b} \in B\}. \quad \square$$

We refer to the points $\mathbf{a}_0$ and $\mathbf{b}_0$ in the statement of Theorem 1.9.4 as **nearest points** of A and B. These points $\mathbf{a}_0$ and $\mathbf{b}_0$ are not, in general, unique.

The next theorem generalizes the result of elementary analysis which asserts that, if a real function is both continuous and positive at some point, then it is positive at all points of its domain sufficiently close to that point.

Theorem 1.9.5 *Let the mapping $f: A \to \mathbb{R}$ be both continuous and positive at some point $\mathbf{a}$ of a set A in $\mathbb{R}^n$. Then there exists an $r > 0$ such that $f(\mathbf{x}) > 0$ whenever $\mathbf{x} \in B(\mathbf{a}; r) \cap A$.*

Proof Suppose that the stated conclusion does not hold. Then, for each $k = 1, 2, \ldots$, there exists $\mathbf{a}_k \in B(\mathbf{a}; 1/k) \cap A$ such that $f(\mathbf{a}_k) \leqslant 0$. Since f is continuous at $\mathbf{a}$ and $\mathbf{a}_k \to \mathbf{a}$ as $k \to \infty$, $f(\mathbf{a}_k) \to f(\mathbf{a})$ as $k \to \infty$. Because $f(\mathbf{a}_k) \leqslant 0$ for $k = 1, 2, \ldots$, it follows that $f(\mathbf{a}) \leqslant 0$. This contradiction establishes the theorem. $\square$

The result of elementary analysis that a *continuous function of a continuous function is itself continuous* generalizes in the following obvious way.

Theorem 1.9.6 *Let $f: A \to \mathbb{R}^m$ and $g: B \to \mathbb{R}^p$ be continuous mappings, where A and B are, respectively, non-empty sets in $\mathbb{R}^n$ and $\mathbb{R}^m$ such that $f(A) \subseteq B$. Then the composite mapping $g \circ f: A \to \mathbb{R}^p$ is continuous.*

Proof Let $\mathbf{a}_1, \ldots, \mathbf{a}_k, \ldots$ be a sequence of points of A that converges to a point $\mathbf{a}$ of A. Since f is continuous, the sequence of points $f(\mathbf{a}_1), \ldots, f(\mathbf{a}_k), \ldots$ of B converges to the point $f(\mathbf{a})$ of B. Since g is continuous, the sequence $g(f(\mathbf{a}_1)), \ldots, g(f(\mathbf{a}_k)), \ldots$ converges to $g(f(\mathbf{a}))$, i.e. the sequence $(g \circ f)(\mathbf{a}_1), \ldots, (g \circ f)(\mathbf{a}_k), \ldots$ converges to $(g \circ f)(a)$. This shows that $g \circ f$ is continuous. $\square$

Theorem 1.9.2 shows that the continuous image of a compact set is compact. It is not true that the continuous image of a closed (open) set is always closed (open). We do, however, have the following important result.

Theorem 1.9.7 *Let $f: \mathbb{R}^n \to \mathbb{R}^m$ be a continuous mapping and let B be a closed (open) subset of $\mathbb{R}^m$. Then $f^{-1}(B)$ is closed (open).*

Proof Suppose first that B is closed, and that $\mathbf{a}_1, \ldots, \mathbf{a}_k, \ldots$ is a sequence of points of $f^{-1}(B)$ that converges to a point $\mathbf{a}$ of $\mathbb{R}^n$. The continuity of f shows that the sequence of points $f(\mathbf{a}_1), \ldots, f(\mathbf{a}_k), \ldots$ of B converges to the point $f(\mathbf{a})$ of $\mathbb{R}^m$. But B is closed, so $f(\mathbf{a}) \in B$ and $\mathbf{a} \in f^{-1}(B)$. This shows that $f^{-1}(B)$ is closed. Suppose next that B is open. Then the complement B^c of B in $\mathbb{R}^m$ is closed, whence, by what has just been proved, $f^{-1}(B^c)$ is closed in $\mathbb{R}^n$. Thus the complement of $f^{-1}(B^c)$ in $\mathbb{R}^n$, i.e. $f^{-1}(B)$, is open. □

Exercises 1.9

1. Give an example of a continuous mapping $f: \mathbb{R}^2 \to \mathbb{R}^2$ and subsets A, B, C of $\mathbb{R}^2$ with A closed, B open, C compact such that $f(A)$ is *not* closed, $f(B)$ is *not* open, $f^{-1}(C)$ is *not* compact.
2. Let $f, g: A \to \mathbb{R}$ be continuous mappings, where A is a non-empty closed set in $\mathbb{R}^n$. Show that the set $\{\mathbf{x} \in A : f(\mathbf{x}) \leqslant g(\mathbf{x})\}$ is closed.
3. Let A be a non-empty bounded set in $\mathbb{R}^n$. Define a mapping $f: \mathbb{R}^n \to \mathbb{R}$ by the equation
$$f(\mathbf{x}) = \sup\{\|\mathbf{x} - \mathbf{a}\| : \mathbf{a} \in A\} \qquad \text{for } \mathbf{x} \in \mathbb{R}^n.$$
Prove that f satisfies a Lipschitz condition on $\mathbb{R}^n$. Find this function f, when A is the closed ball $B[\mathbf{a}; r]$ in $\mathbb{R}^n$.
4. Verify that, for all $\mathbf{x}, \mathbf{y} \in \mathbb{R}^n$,
$$2(\|\mathbf{y}\|\mathbf{x} - \|\mathbf{x}\|\mathbf{y}) = (\|\mathbf{x}\| + \|\mathbf{y}\|)(\mathbf{x} - \mathbf{y}) + (\|\mathbf{y}\| - \|\mathbf{x}\|)(\mathbf{x} + \mathbf{y}).$$
Use this identity to help you prove that the mapping $f: \mathbb{R}^n \to \mathbb{R}^n$, defined by the equation $f(\mathbf{x}) = \mathbf{x}/(1 + \|\mathbf{x}\|)$ for $\mathbf{x} \in \mathbb{R}^n$, satisfies a Lipschitz condition on $\mathbb{R}^n$.
5. Give examples of disjoint non-empty closed sets A and B in $\mathbb{R}^n$ for which inf $\{\|\mathbf{a} - \mathbf{b}\| : \mathbf{a} \in A, \mathbf{b} \in B\}$ is zero.

2
Convex sets

This chapter gives a full account of the fundamental properties of convex sets that will be needed in our subsequent discussion of convexity.

The material in Section 2.6 following Theorem 2.6.18 is not used elsewhere in the body of the book. Section 2.8 on duality is only needed in Sections 3.5 and 5.6 and may be omitted on a first reading.

2.1 BASIC PROPERTIES OF CONVEX SETS

The concept of a convex set is a simple one. A set in space is **convex** if whenever it contains two points, it also contains the line segment joining them. Elementary geometry abounds in convex sets: ellipses, triangles, parallelograms, balls, halfspaces, and cubes are convex. An annulus, a crescent, and the vertex set of a cube are examples of non-convex sets. Convex and non-convex sets are illustrated in Fig. 2.1.

Now for the formal definition of a convex set in $\mathbb{R}^n$. Let $\mathbf{x}$ and $\mathbf{y}$ be *distinct* points of $\mathbb{R}^n$. Then the subset

$$\{\lambda\mathbf{x} + \mu\mathbf{y} : \lambda, \mu \geqslant 0, \lambda + \mu = 1\}$$

of the line through $\mathbf{x}$ and $\mathbf{y}$ is called the **line segment** joining $\mathbf{x}$ and $\mathbf{y}$. The reader will need little convincing that this definition accords with the intuitive idea of a line segment in $\mathbb{R}^3$. The set A in $\mathbb{R}^n$ is said to be **convex** if whenever it contains two points, it also contains the line segment joining them. Expressed algebraically, A is convex if $\lambda\mathbf{x} + \mu\mathbf{y} \in A$ whenever $\mathbf{x}, \mathbf{y} \in A$ and $\lambda, \mu \geqslant 0$ with $\lambda + \mu = 1$. Equivalently, A is convex if $\lambda A + \mu A \subseteq A$ whenever $\lambda, \mu \geqslant 0$ with $\lambda + \mu = 1$.

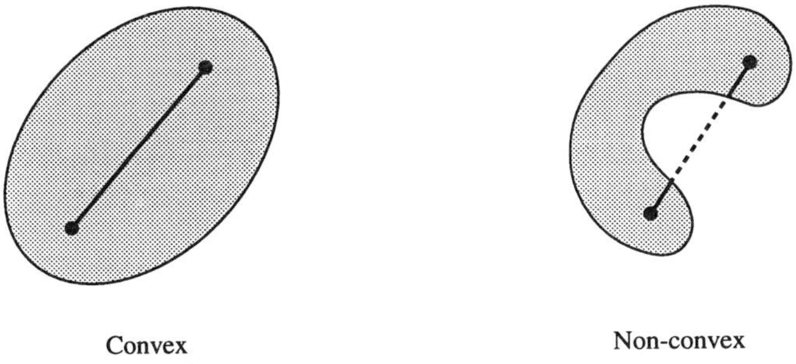

Fig. 2.1

The condition for a set to be convex is less restrictive than for it to be a flat, and so every flat is a convex set. In particular, the empty set, singletons, lines, hyperplanes, and $\mathbb{R}^n$ itself are convex. Other important examples of convex sets are balls, halfspaces, and line segments.

Example 2.1.1 *Balls are convex.* We show that the closed ball $B[\mathbf{a}; r]$ in $\mathbb{R}^n$ is convex. Let $\mathbf{x}, \mathbf{y} \in B[\mathbf{a}; r]$ and let $\lambda, \mu \geqslant 0$ with $\lambda + \mu = 1$. Then $\|\mathbf{x} - \mathbf{a}\| \leqslant r$, $\|\mathbf{y} - \mathbf{a}\| \leqslant r$, and so

$$\begin{aligned}\|\lambda\mathbf{x} + \mu\mathbf{y} - \mathbf{a}\| &= \|\lambda(\mathbf{x} - \mathbf{a}) + \mu(\mathbf{y} - \mathbf{a})\| \\ &\leqslant \lambda\|\mathbf{x} - \mathbf{a}\| + \mu\|\mathbf{y} - \mathbf{a}\| \leqslant \lambda r + \mu r = r.\end{aligned}$$

Thus $\lambda\mathbf{x} + \mu\mathbf{y} \in B[\mathbf{a}; r]$, which proves that $B[\mathbf{a}; r]$ is convex. A similar argument shows that the open ball $B(\mathbf{a}; r)$ is convex. □

Example 2.1.2 *Halfspaces are convex.* We show that the closed halfspace A in $\mathbb{R}^n$ defined by the inequality $\mathbf{u} \cdot \mathbf{x} \leqslant u_0$ is convex. Let $\mathbf{x}, \mathbf{y} \in A$ and let $\lambda, \mu \geqslant 0$ with $\lambda + \mu = 1$. Then $\mathbf{u} \cdot \mathbf{x} \leqslant u_0$, $\mathbf{u} \cdot \mathbf{y} \leqslant u_0$, and so

$$\mathbf{u} \cdot (\lambda\mathbf{x} + \mu\mathbf{y}) = \lambda\mathbf{u} \cdot \mathbf{x} + \mu\mathbf{u} \cdot \mathbf{y} \leqslant \lambda u_0 + \mu u_0 = u_0.$$

Thus $\lambda\mathbf{x} + \mu\mathbf{y} \in A$, which proves that A is convex. A similar argument shows that open halfspaces are convex. □

The next result expresses one of the most important properties of the family of convex sets, namely that it is closed under the formation of arbitrary intersections.

Theorem 2.1.3 *The intersection of an arbitrary family of convex sets in $\mathbb{R}^n$ is convex.*

Proof Let $(A_i : i \in I)$ be a family of convex sets in $\mathbb{R}^n$. If $\mathbf{a}, \mathbf{b} \in \bigcap (A_i : i \in I)$ and $\lambda, \mu \geqslant 0$ with $\lambda + \mu = 1$, then $\mathbf{a}, \mathbf{b} \in A_i$ and, as A_i is convex, $\lambda\mathbf{a} + \mu\mathbf{b} \in A_i$ for each $i \in I$. Thus $\lambda\mathbf{a} + \mu\mathbf{b} \in \bigcap (A_i : i \in I)$, which shows that the intersection is convex. □

By definition, a set A in $\mathbb{R}^n$ is convex if $\lambda\mathbf{x} + \mu\mathbf{y} \in A$ whenever $\mathbf{x}, \mathbf{y} \in A$ and $\lambda, \mu \geqslant 0$ with $\lambda + \mu = 1$. This defining relation of a convex set implies a more general one, as we now establish in the following fundamental theorem.

Theorem 2.1.4 *Let $\mathbf{a}_1, \ldots, \mathbf{a}_m$ be points of a convex set A in $\mathbb{R}^n$. Let $\lambda_1, \ldots, \lambda_m \geqslant 0$ with $\lambda_1 + \ldots + \lambda_m = 1$. Then $\lambda_1 \mathbf{a}_1 + \ldots + \lambda_m \mathbf{a}_m \in A$.*

Proof We argue by induction on m. When $m = 1$ the assertion is trivial. Suppose that the assertion holds when m is some positive integer k. Let

$$\mathbf{x} = \lambda_1 \mathbf{a}_1 + \ldots + \lambda_{k+1} \mathbf{a}_{k+1},$$

where $\mathbf{a}_1, \ldots, \mathbf{a}_{k+1} \in A$ and $\lambda_1, \ldots, \lambda_{k+1} \geqslant 0$ with $\lambda_1 + \ldots + \lambda_{k+1} = 1$. At least one λ_i must be less than 1, say $\lambda_{k+1} < 1$. Write

$$\mathbf{y} = \frac{\lambda_1}{\lambda}\mathbf{a}_1 + \ldots + \frac{\lambda_k}{\lambda}\mathbf{a}_k, \qquad \text{where } \lambda = \lambda_1 + \ldots + \lambda_k = 1 - \lambda_{k+1} > 0.$$

By the induction hypothesis, $\mathbf{y} \in A$. Since A is convex and contains both $\mathbf{y}$ and $\mathbf{a}_{k+1}$, the equation $\mathbf{x} = \lambda\mathbf{y} + \lambda_{k+1}\mathbf{a}_{k+1}$ shows that $\mathbf{x} \in A$. This completes the proof by induction. □

This last result prompts the following definition. A point $\mathbf{x}$ is said to be a **convex combination** of points $\mathbf{a}_1, \ldots, \mathbf{a}_m$ in $\mathbb{R}^n$ if there exist scalars $\lambda_1, \ldots, \lambda_m \geqslant 0$ with $\lambda_1 + \ldots + \lambda_m = 1$ such that

$$\mathbf{x} = \lambda_1 \mathbf{a}_1 + \ldots + \lambda_m \mathbf{a}_m.$$

Theorem 2.1.4 can thus be expressed as: *every convex combination of points of a convex set in $\mathbb{R}^n$ belongs to that set.*

We now show how well the concepts of convexity, vector addition, and scalar multiplication interact with one another.

Theorem 2.1.5 *Let A, B be convex sets in $\mathbb{R}^n$ and let α be a scalar. Then $A + B$ and αA are convex.*

Proof Let $\lambda, \mu \geqslant 0$ with $\lambda + \mu = 1$. Since A, B are convex, $\lambda A + \mu A \subseteq A$ and $\lambda B + \mu B \subseteq B$. Thus

$$\lambda(A + B) + \mu(A + B) = (\lambda A + \mu A) + (\lambda B + \mu B) \subseteq A + B$$

and

$$\lambda(\alpha A) + \mu(\alpha A) = \alpha(\lambda A + \mu A) \subseteq \alpha A.$$

This shows that $A + B$ and αA are convex. □

Corollary 2.1.6 *Let $A_1, \ldots, A_m$ be convex sets in $\mathbb{R}^n$ and let $\lambda_1, \ldots, \lambda_m$ be scalars. Then $\lambda_1 A_1 + \ldots + \lambda_m A_m$ is convex.* □

Theorem 2.1.7 *Let A be a convex set in $\mathbb{R}^n$ and let $\lambda_1, \ldots, \lambda_m \geqslant 0$. Then $(\lambda_1 + \ldots + \lambda_m)A = \lambda_1 A + \ldots + \lambda_m A$.*

Proof The result is trivial when each λ_i is zero, so we may suppose that $\lambda = \lambda_1 + \ldots + \lambda_m > 0$. With the help of Theorem 2.1.4, we can deduce that

$$
\begin{aligned}
(\lambda_1 + \ldots + \lambda_m)A &\subseteq \lambda_1 A + \ldots + \lambda_m A \\
&= \lambda\left(\frac{\lambda_1}{\lambda}A + \ldots + \frac{\lambda_m}{\lambda}A\right) \\
&\subseteq \lambda A \\
&= (\lambda_1 + \ldots + \lambda_m)A.
\end{aligned}
$$

Thus $(\lambda_1 + \ldots + \lambda_m)A = \lambda_1 A + \ldots + \lambda_m A$. □

The next result, while not having the same basic importance as others in this section, does have its own intrinsic interest and will be needed later in the chapter.

Theorem 2.1.8 *Let A, B, C be sets in $\mathbb{R}^n$. Suppose that A is non-empty and bounded, that C is closed and convex, and that $A + B \subseteq A + C$. Then $B \subseteq C$.*

Proof Let $\mathbf{a}_0 \in A$. If $\mathbf{b} \in B$, then $\mathbf{a}_0 + \mathbf{b} \in A + B \subseteq A + C$, and so there exist $\mathbf{a}_1 \in A$, $\mathbf{c}_1 \in C$ such that $\mathbf{a}_0 + \mathbf{b} = \mathbf{a}_1 + \mathbf{c}_1$. Similarly, there exist $\mathbf{a}_2, \ldots, \mathbf{a}_i \in A$ and $\mathbf{c}_2, \ldots, \mathbf{c}_i \in C$ with $\mathbf{a}_1 + \mathbf{b} = \mathbf{a}_2 + \mathbf{c}_2, \ldots, \mathbf{a}_{i-1} + \mathbf{b} = \mathbf{a}_i + \mathbf{c}_i$. We add the i equations above together to deduce that

$$\mathbf{a}_0 + i\mathbf{b} = \mathbf{a}_i + \mathbf{c}_1 + \ldots + \mathbf{c}_i.$$

Since C is convex, the point $\mathbf{x}_i$ defined by the equation

$$\mathbf{x}_i = \frac{1}{i}(\mathbf{c}_1 + \ldots + \mathbf{c}_i)$$

lies in C. Now

$$\|\mathbf{b} - \mathbf{x}_i\| = \frac{1}{i}\|\mathbf{a}_i - \mathbf{a}_0\| \to 0 \qquad \text{as } i \to \infty,$$

since A is bounded. Thus $\mathbf{x}_i \to \mathbf{b}$ as $i \to \infty$. But C is closed, whence $\mathbf{b} \in C$ and $B \subseteq C$. □

Corollary 2.1.9 *Let A, B, C be sets in $\mathbb{R}^n$. Suppose that A is non-empty and bounded, that B and C are closed and convex, and that $A + B = A + C$. Then $B = C$.* □

We conclude this section by showing that affine transformations preserve convex sets. In Section 7.3, we shall show that, if $f: \mathbb{R}^n \to \mathbb{R}^m$ $(n \geqslant 2)$ is an injective mapping which preserves convex sets, then f must be an affine transformation. Thus the affine transformations play a central role in the study of convexity-preserving mappings.

Theorem 2.1.10 *Let $f:\mathbb{R}^n \to \mathbb{R}^m$ be an affine transformation. Then $f(A)$ is convex for each convex set A in $\mathbb{R}^n$, and $f^{-1}(B)$ is convex for each convex set B in $\mathbb{R}^m$.*

Proof Let A be a convex set in $\mathbb{R}^n$. Let $\lambda, \mu \geqslant 0$ with $\lambda + \mu = 1$. If $\mathbf{x}, \mathbf{y} \in f(A)$, *then* $\mathbf{x} = f(\mathbf{a})$, $\mathbf{y} = f(\mathbf{b})$ for some $\mathbf{a}, \mathbf{b} \in A$. Since A is convex, $\lambda\mathbf{a} + \mu\mathbf{b} \in A$. Since f is affine,

$$\lambda\mathbf{x} + \mu\mathbf{y} = \lambda f(\mathbf{a}) + \mu f(\mathbf{b}) = f(\lambda\mathbf{a} + \mu\mathbf{b}).$$

Thus $\lambda\mathbf{x} + \mu\mathbf{y} \in f(A)$, which shows that $f(A)$ is convex.

Let B be a convex set in $\mathbb{R}^m$. Let $\lambda, \mu \geqslant 0$ with $\lambda + \mu = 1$. If $\mathbf{x}, \mathbf{y} \in f^{-1}(B)$, then $f(\mathbf{x}), f(\mathbf{y}) \in B$. Since B is convex, $\lambda f(\mathbf{x}) + \mu f(\mathbf{y}) \in B$. Since f is affine,

$$f(\lambda\mathbf{x} + \mu\mathbf{y}) = \lambda f(\mathbf{x}) + \mu f(\mathbf{y}) \in B.$$

Thus $\lambda\mathbf{x} + \mu\mathbf{y} \in f^{-1}(B)$, which shows that $f^{-1}(B)$ is convex. □

Exercises 2.1

1. Which of the following sets are convex?
   (i) $\{\mathbf{x} \in \mathbb{R}^n : \|\mathbf{x}\| = 1\}$;
   (ii) $\{(x_1, \ldots, x_n) : a_1x_1 + \ldots + a_nx_n > a_0\}$, where $a_0, a_1, \ldots, a_n \in \mathbb{R}$;
   (iii) $\mathbb{R}^n \setminus \{\mathbf{0}\}$;
   (iv) $\{(x_1, \ldots, x_n) : |x_1| + \ldots + |x_n| \leqslant 1\}$;
   (v) $\{(x_1, \ldots, x_n) : (a_{11}x_1 + \ldots + a_{1n}x_n)^2 + \ldots + (a_{n1}x_1 + \ldots + a_{nn}x_n)^2 \leqslant 1\}$, where $a_{ij} \in \mathbb{R}$ for $i, j = 1, \ldots, n$.
2. Let A and B be convex sets in $\mathbb{R}^n$. Show that the set
$$C = \bigcup(\alpha A \cap (1-\alpha)B : 0 \leqslant \alpha \leqslant 1)$$
is convex. Find C when both A and B are the closed unit ball U in $\mathbb{R}^n$.
3. Prove that every union of an increasing sequence of convex sets in $\mathbb{R}^n$ is convex. Deduce that, if $A_1, \ldots, A_i, \ldots$ is *any* sequence of convex sets in $\mathbb{R}^n$, then the set
$$\bigcup_{j=1}^{\infty}\left(\bigcap_{i=j}^{\infty} A_i\right)$$
is convex.
4. Let A be a symmetric convex set in $\mathbb{R}^n$ that contains no non-zero lattice points. Show that, if $\mathbf{a}$ and $\mathbf{b}$ are *distinct* lattice points in $\mathbb{R}^n$, then the sets $\mathbf{a} + \frac{1}{2}A$ and $\mathbf{b} + \frac{1}{2}A$ are disjoint.
5. Give examples of non-empty compact sets A, B, C in $\mathbb{R}^n$ satisfying $A + B \subseteq A + C$ and $B \nsubseteq C$.

6. By expressing a closed ball as a vector sum of a point and a scalar multiple of the closed unit ball U, show that a closed ball in $\mathbb{R}^n$ uniquely determines its centre and radius.

2.2 THE CONVEX HULL

The **convex hull** $\operatorname{conv} A$ of a set A in $\mathbb{R}^n$ is the intersection of all convex sets in $\mathbb{R}^n$ containing A. We show that $\operatorname{conv} A$ is the *smallest* convex set in $\mathbb{R}^n$ containing A, characterizing it as the set of all convex combinations of points of A. Using this characterization, we prove two combinatorial results, one due to Carathéodory, the other to Radon; and a theorem of Gauss on the location of the roots of the derivative of a complex polynomial. We also show that some important properties of a set in $\mathbb{R}^n$ are inherited by its convex hull.

The above definition of $\operatorname{conv} A$, together with Theorem 2.1.3, shows that $\operatorname{conv} A$ is a convex set containing A. Moreover, if C is *any* convex set in $\mathbb{R}^n$ containing A, then $\operatorname{conv} A \subseteq C$. Thus we may refer to $\operatorname{conv} A$ as the *smallest* convex set in $\mathbb{R}^n$ containing A. Clearly, A is convex if and only if $A = \operatorname{conv} A$, and $\operatorname{conv}(\operatorname{conv} A) = \operatorname{conv} A$. Another easy result is that $\operatorname{conv} A \subseteq \operatorname{conv} B$ whenever $A \subseteq B$. In space, the convex hull of two distinct points is the line segment joining them, the convex hull of three non-collinear points is the triangle which they determine, and the convex hull of four non-coplanar points is the tetrahedron which they determine. In $\mathbb{R}^2$ the convex hull of m points symmetrically placed on the circumference of a circle, where $m \geqslant 3$, is a regular m-sided polygon.

Example 2.2.1 *The convex hull of the set* $A = \{\mathbf{x} \in \mathbb{R}^n : \|\mathbf{x}\| = 1\}$ *is the closed unit ball* $U = \{\mathbf{x} \in \mathbb{R}^n : \|\mathbf{x}\| \leqslant 1\}$. The ball U is convex and contains A, so $\operatorname{conv} A \subseteq U$. We now show that $U \subseteq \operatorname{conv} A$. Let $\mathbf{x} \in U$. If $\mathbf{x} = \mathbf{0}$ and $\mathbf{y} \in A$, then $\mathbf{x} = \frac{1}{2}\mathbf{y} + \frac{1}{2}(-\mathbf{y})$, which, since $\operatorname{conv} A$ is convex and contains $\mathbf{y}$ and $-\mathbf{y}$, shows that $\mathbf{x} \in \operatorname{conv} A$. If $\mathbf{x} \neq \mathbf{0}$, then $0 < \|\mathbf{x}\| \leqslant 1$, and the equation

$$\mathbf{x} = \left(\frac{1 + \|\mathbf{x}\|}{2}\right)\frac{\mathbf{x}}{\|\mathbf{x}\|} + \left(\frac{1 - \|\mathbf{x}\|}{2}\right)\frac{-\mathbf{x}}{\|\mathbf{x}\|}$$

shows that $\mathbf{x} \in \operatorname{conv} A$, since $\operatorname{conv} A$ is convex and contains $\mathbf{x}/\|\mathbf{x}\|$ and $-\mathbf{x}/\|\mathbf{x}\|$. Thus $U \subseteq \operatorname{conv} A$ and $U = \operatorname{conv} A$ as desired. □

The convex hull of a set was defined by means of convex sets containing that set. By contrast, the following theorem expresses the convex hull of a set in terms of the points of the set themselves.

Theorem 2.2.2 *Let A be a set in $\mathbb{R}^n$. Then* conv A *is the set of all convex combinations of points of A.*

Proof Denote by B the set of all convex combinations of points of A. That $B \subseteq \operatorname{conv} A$ follows from Theorem 2.1.4 and the inclusion $A \subseteq \operatorname{conv} A$.

We next show that B is convex. If $\mathbf{x}, \mathbf{y} \in B$, then

$$\mathbf{x} = \lambda_1 \mathbf{a}_1 + \ldots + \lambda_m \mathbf{a}_m, \qquad \mathbf{y} = \mu_1 \mathbf{b}_1 + \ldots + \mu_p \mathbf{b}_p,$$

for some $\mathbf{a}_1, \ldots, \mathbf{a}_m, \mathbf{b}_1, \ldots, \mathbf{b}_p \in A$ and $\lambda_1, \ldots, \lambda_m, \mu_1, \ldots, \mu_p \geqslant 0$ with $\lambda_1 + \ldots + \lambda_m = 1$, $\mu_1 + \ldots + \mu_p = 1$. Let $\lambda, \mu \geqslant 0$ with $\lambda + \mu = 1$. Then

$$\lambda\mathbf{x} + \mu\mathbf{y} = \lambda\lambda_1 \mathbf{a}_1 + \ldots + \lambda\lambda_m \mathbf{a}_m + \mu\mu_1 \mathbf{b}_1 + \ldots + \mu\mu_p \mathbf{b}_p$$

and

$$\begin{aligned}\lambda\lambda_1 + \ldots + \lambda\lambda_m + \mu\mu_1 + \ldots + \mu\mu_p &= \lambda(\lambda_1 + \ldots + \lambda_m) + \mu(\mu_1 + \ldots + \mu_p) \\ &= \lambda + \mu = 1.\end{aligned}$$

Thus $\lambda\mathbf{x} + \mu\mathbf{y} \in B$, so B is convex. Since B is convex and $B \supseteq A$, it follows that $B \supseteq \operatorname{conv} A$. Hence $B = \operatorname{conv} A$. □

Corollary 2.2.3 *Let* $\mathbf{a}_1, \ldots, \mathbf{a}_m \in \mathbb{R}^n$. *Then*

$$\begin{aligned}\operatorname{conv}\{\mathbf{a}_1, \ldots, \mathbf{a}_m\} = \{\lambda_1 \mathbf{a}_1 + \ldots + \lambda_m \mathbf{a}_m &: \lambda_1, \ldots, \lambda_m \geqslant 0, \\ &\lambda_1 + \ldots + \lambda_m = 1\}. \quad \square\end{aligned}$$

Theorem 2.2.2 shows that each point of the convex hull of a set in $\mathbb{R}^n$ is a convex combination of points of that set, but makes no reference to the number of points in the combination. Carathéodory's theorem, which is proved below, states that each point of the convex hull of an r-dimensional set can be expressed as a convex combination of $r + 1$ or fewer points of the set. Thus a point in the convex hull of a set in $\mathbb{R}^3$ is either a point of the set or belongs to a line segment, a triangle, or a tetrahedron with vertices in the set.

Theorem 2.2.4 (Carathéodory's theorem) *Let* $\mathbf{a} \in \operatorname{conv} A$, *where A is an r-dimensional set in $\mathbb{R}^n$. Then* $\mathbf{a}$ *can be expressed as a convex combination of $r + 1$ or fewer points of A.*

Proof Theorem 2.2.2 shows the existence of points $\mathbf{a}_1, \ldots, \mathbf{a}_m$ of A and scalars $\lambda_1, \ldots, \lambda_m \geqslant 0$ with $\lambda_1 + \ldots + \lambda_m = 1$ such that

$$\mathbf{a} = \lambda_1 \mathbf{a}_1 + \ldots + \lambda_m \mathbf{a}_m.$$

We assume that this representation of $\mathbf{a}$ is so chosen that $\mathbf{a}$ cannot be expressed as a convex combination of fewer than m points of A. It follows that no two of the points $\mathbf{a}_1, \ldots, \mathbf{a}_m$ are equal and that

$\lambda_1, \ldots, \lambda_m > 0$. We prove the theorem by showing that $m \leqslant r + 1$. This we do using a contradiction argument. Suppose that $m > r + 1$. Then, since A is r-dimensional, the set $\{\mathbf{a}_1, \ldots, \mathbf{a}_m\}$ must be affinely dependent, and so there exist scalars $\mu_1, \ldots, \mu_m$, not all zero, such that

$$\mathbf{0} = \mu_1 \mathbf{a}_1 + \ldots + \mu_m \mathbf{a}_m \qquad \text{and} \qquad \mu_1 + \ldots + \mu_m = 0.$$

Let $t > 0$ be such that the scalars $\lambda_1 + \mu_1 t, \ldots, \lambda_m + \mu_m t$ are non-negative with at least one of them zero; such a t exists since the λ's are all positive and at least one of the μ's is negative. The equation

$$\mathbf{a} = (\lambda_1 + \mu_1 t)\mathbf{a}_1 + \ldots + (\lambda_m + \mu_m t)\mathbf{a}_m,$$

when its terms with zero coefficients are omitted, exhibits $\mathbf{a}$ as a convex combination of fewer than m points of A. This contradiction to the minimality of m shows that $m \leqslant r + 1$. □

Radon's theorem, which we now prove, has been the subject of much research in recent years. We shall need it in our discussions of neighbourly polytopes in Section 3.4 and of Helly's theorem in Section 7.1.

Theorem 2.2.5 (Radon's theorem) *Let* $\mathbf{a}_1, \ldots, \mathbf{a}_m \in \mathbb{R}^n$ $(m \geqslant n + 2)$. *Then the set* $\{1, \ldots, m\}$ *can be partitioned into two subsets I and J such that* $\operatorname{conv}\{\mathbf{a}_i : i \in I\}$ *meets* $\operatorname{conv}\{\mathbf{a}_j : j \in J\}$.

Proof We consider the non-trivial case when the $\mathbf{a}_1, \ldots, \mathbf{a}_m$ are distinct. It follows from Corollary 1.3.4 that there exist scalars $\lambda_1, \ldots, \lambda_m$, not all zero, such that

$$\lambda_1 \mathbf{a}_1 + \ldots + \lambda_m \mathbf{a}_m = \mathbf{0} \qquad \text{and} \qquad \lambda_1 + \ldots + \lambda_m = 0.$$

Some of the λ's will be positive, others negative. Let $I = \{i : \lambda_i \geqslant 0\}$ and $J = \{j : \lambda_j < 0\}$. Then

$$\frac{\sum_{i \in I} \lambda_i \mathbf{a}_i}{\sum_{i \in I} \lambda_i} = \frac{\sum_{j \in J} (-\lambda_j)\mathbf{a}_j}{\sum_{j \in J} (-\lambda_j)} = \mathbf{x}$$

say. Thus $\mathbf{x}$ is a convex combination of points of both $\{\mathbf{a}_i : i \in I\}$ and $\{\mathbf{a}_j : j \in J\}$, whence

$$\mathbf{x} \in \operatorname{conv}\{\mathbf{a}_i : i \in I\} \cap \operatorname{conv}\{\mathbf{a}_j : j \in J\},$$

which shows that I and J satisfy the requirements of the theorem. □

Radon's theorem yields information about the possible configurations of four points in $\mathbb{R}^2$. It shows that either one of the four points

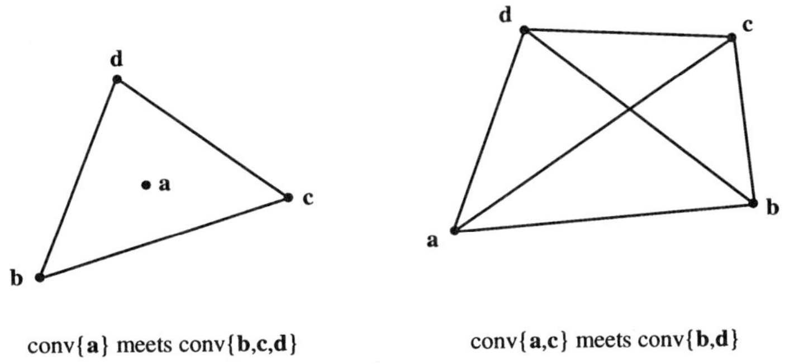

Fig. 2.2

belongs to the (possibly degenerate) triangle determined by the remaining three, or else the four points are the vertices of a convex quadrilateral. These two cases are illustrated in Fig. 2.2.

We now show that some important properties of a set in $\mathbb{R}^n$ are inherited by its convex hull, whereas others are not.

Theorem 2.2.6 *In $\mathbb{R}^n$ the convex hull of an open set is open and the convex hull of a compact set is compact.*

Proof Let A be an open set in $\mathbb{R}^n$. If $\mathbf{a} \in \operatorname{conv} A$, then $\mathbf{a} = \lambda_1 \mathbf{a}_1 + \ldots + \lambda_m \mathbf{a}_m$ for some $\mathbf{a}_1, \ldots, \mathbf{a}_m \in A$ and $\lambda_1, \ldots, \lambda_m \geqslant 0$ with $\lambda_1 + \ldots + \lambda_m = 1$. Since A is open, there exist $r_1, \ldots, r_m > 0$ such that $B(\mathbf{a}_1; r_1) \subseteq A, \ldots, B(\mathbf{a}_m; r_m) \subseteq A$. Let $r = \min\{r_1, \ldots, r_m\}$, so $r > 0$. We show that $B(\mathbf{a}; r) \subseteq \operatorname{conv} A$. Let $\mathbf{x} \in B(\mathbf{a}; r)$. Then $\|\mathbf{x} - \mathbf{a}\| < r$. For $i = 1, \ldots, m$, the point $\mathbf{x}_i = \mathbf{a}_i + \mathbf{x} - \mathbf{a}$ lies in $B(\mathbf{a}_i; r)$, and hence in $B(\mathbf{a}_i; r_i)$ and A. The equation

$$\mathbf{x} = \lambda_1 \mathbf{x}_1 + \ldots + \lambda_m \mathbf{x}_m$$

expresses $\mathbf{x}$ as a convex combination of points of A, and so $\mathbf{x} \in \operatorname{conv} A$. Thus $B(\mathbf{a}; r) \subseteq \operatorname{conv} A$ and each point of $\operatorname{conv} A$ is an interior point of $\operatorname{conv} A$, i.e. $\operatorname{conv} A$ is open.

Now let A be a compact set in $\mathbb{R}^n$. If $\mathbf{x}_1, \ldots, \mathbf{x}_k, \ldots$ is a sequence in $\operatorname{conv} A$, then, by Carathéodory's theorem, $\mathbf{x}_k$ can be expressed in the form

$$\mathbf{x}_k = \lambda_{k0} \mathbf{a}_{k0} + \ldots + \lambda_{kn} \mathbf{a}_{kn},$$

for some $\mathbf{a}_{k0}, \ldots, \mathbf{a}_{kn} \in A$ and $\lambda_{k0}, \ldots, \lambda_{kn} \geqslant 0$ with $\lambda_{k0} + \ldots + \lambda_{kn} = 1$; it may be necessary to include some extra $\mathbf{a}$'s with zero coefficients to bring the number of $\mathbf{a}$'s in the expression for $\mathbf{x}_k$ up to $n + 1$. Each sequence

$$\mathbf{a}_{1j}, \ldots, \mathbf{a}_{kj}, \ldots \qquad (j = 0, \ldots, n) \tag{2.2.1}$$

belongs to the compact set A. Each real sequence

$$\lambda_{1j}, \ldots, \lambda_{kj}, \ldots \qquad (j = 0, \ldots, n) \tag{2.2.2}$$

belongs to the compact interval $[0, 1]$. Since there is only a finite number, namely $2n + 2$, of the sequences (2.2.1) and (2.2.2), we can, by repeatedly forming convergent subsequences of sequences whose members lie in a compact set, find a subsequence $i_1, \ldots, i_k, \ldots$ of $1, \ldots, k, \ldots$, points $\mathbf{a}_0, \ldots, \mathbf{a}_n$ of A and scalars $\lambda_0, \ldots, \lambda_n \geqslant 0$ with $\lambda_0 + \ldots + \lambda_n = 1$, such that

$$\mathbf{a}_{i_k j} \to \mathbf{a}_j \qquad \text{and} \qquad \lambda_{i_k j} \to \lambda_j \qquad (k \to \infty; j = 0, \ldots, n).$$

Thus the subsequence $\mathbf{x}_{i_1}, \ldots, \mathbf{x}_{i_k}, \ldots$ of $\mathbf{x}_1, \ldots, \mathbf{x}_k, \ldots$ converges to the point $\lambda_0 \mathbf{a}_0 + \ldots + \lambda_n \mathbf{a}_n$ of $\operatorname{conv} A$. This shows that $\operatorname{conv} A$ is compact. □

Corollary 2.2.7 *The convex hull of a finite set in* $\mathbb{R}^n$ *is compact.* □

Theorem 2.2.6 makes no reference to the convex hull of a closed set. The reason for this is simple: except in $\mathbb{R}^1$, the convex hull of a closed set need not be closed. For example, in $\mathbb{R}^n$ the union of a line and a point not on it is a closed set whose convex hull is not closed.

Since a set in $\mathbb{R}^n$ is bounded if and only if it lies in some ball, and balls are convex, it follows that the convex hull of a bounded set in $\mathbb{R}^n$ is bounded. Theorem 2.2.8 below can be viewed as a strengthening of this result. Before we can proceed to the theorem, however, we need to introduce the idea of the diameter of a set. The **diameter** of a non-empty bounded set A in $\mathbb{R}^n$ is the non-negative real number

$$\sup \{\|\mathbf{a} - \mathbf{b}\| : \mathbf{a}, \mathbf{b} \in A\}.$$

In $\mathbb{R}^2$ the diameter of a triangle is the length of a longest side, and that of a rectangle is the length of a diagonal. In $\mathbb{R}^n$ the balls $B(\mathbf{a}; r)$ and $B[\mathbf{a}; r]$ both have diameter $2r$.

Theorem 2.2.8 *Let A be a non-empty bounded set in* $\mathbb{R}^n$. *Then A and* $\operatorname{conv} A$ *have the same diameter.*

Proof Suppose that A has diameter s. Let $\mathbf{x}, \mathbf{y} \in \operatorname{conv} A$. Then

$$\mathbf{x} = \lambda_1 \mathbf{a}_1 + \ldots + \lambda_m \mathbf{a}_m, \qquad \mathbf{y} = \mu_1 \mathbf{b}_1 + \ldots + \mu_p \mathbf{b}_p,$$

for some $\mathbf{a}_1, \ldots, \mathbf{a}_m, \mathbf{b}_1, \ldots, \mathbf{b}_p \in A$ and $\lambda_1, \ldots, \lambda_m, \mu_1, \ldots, \mu_p \geqslant 0$ with $\lambda_1 + \ldots + \lambda_m = 1$, $\mu_1 + \ldots + \mu_p = 1$. Thus

$$\mathbf{x} = \sum_{i=1}^{m} \sum_{j=1}^{p} \lambda_i \mu_j \mathbf{a}_i, \qquad \mathbf{y} = \sum_{i=1}^{m} \sum_{j=1}^{p} \lambda_i \mu_j \mathbf{b}_j,$$

whence, by a simple generalization of the triangle inequality,

$$\begin{aligned}\|\mathbf{x}-\mathbf{y}\| &= \left\|\sum_{i=1}^{m}\sum_{j=1}^{p}\lambda_i\mu_j(\mathbf{a}_i-\mathbf{b}_j)\right\| \\ &\leqslant \sum_{i=1}^{m}\sum_{j=1}^{p}\lambda_i\mu_j\|\mathbf{a}_i-\mathbf{b}_j\| \\ &\leqslant \sum_{i=1}^{m}\sum_{j=1}^{p}\lambda_i\mu_j s \\ &= s.\end{aligned}$$

Hence the diameter of conv A does not exceed s. Since $A \subseteq \operatorname{conv} A$, the diameter of conv A is at least s. Thus conv A has diameter s. □

We conclude this section with a delightful result concerning the location of the roots of the derivative of a complex polynomial. In a natural way we can identify the Euclidean space $\mathbb{R}^2$ with the complex plane by identifying each point (x, y) of $\mathbb{R}^2$ with the complex number $x + iy$ and vice versa. This identification allows us to refer to the convex hull of a set of complex numbers or to a convex combination of complex numbers.

To motivate our discussion, consider a complex polynomial $P(z) = az^2 + bz + c$. Then P has roots $(-b \pm \sqrt{(b^2 - 4ac)})/2a$, and its derivative P' has root $-b/2a$. Hence the root of P' lies midway between the roots of P, and so is in the convex hull of these roots. This observation is a special case of the following result.

Theorem 2.2.9 (Gauss–Lucas theorem) *The roots of the derivative of a non-constant complex polynomial belong to the convex hull of the set of roots of the polynomial itself.*

Proof Let P be the complex polynomial defined for complex z by the equation

$$P(z) = a_n z^n + \ldots + a_1 z + a_0,$$

where $n \geqslant 1$ and $a_0, a_1, \ldots, a_n$ are complex numbers with $a_n \neq 0$. Then

$$P(z) = a_n(z - z_1)\ldots(z - z_n),$$

where $z_1, \ldots, z_n$ are the roots of P, each being repeated according to its multiplicity. A routine verification shows that, for $z \neq z_1, \ldots, z_n$,

$$\frac{P'(z)}{P(z)} = \frac{1}{z - z_1} + \ldots + \frac{1}{z - z_n} = \frac{\bar{z} - \bar{z}_1}{|z - z_1|^2} + \ldots + \frac{\bar{z} - \bar{z}_n}{|z - z_n|^2}. \tag{2.2.3}$$

Suppose now that z is a root of P'. We establish the theorem by exhibiting z as a convex combination of $z_1, \ldots, z_n$. This can be done trivially if z is one of $z_1, \ldots, z_n$, so we assume that this is not the case. Putting $P'(z) = 0$ in (2.2.3), we find easily that

$$z = \frac{\dfrac{1}{|z - z_1|^2} z_1 + \ldots + \dfrac{1}{|z - z_n|^2} z_n}{\dfrac{1}{|z - z_1|^2} + \ldots + \dfrac{1}{|z - z_n|^2}},$$

which expresses z as a convex combination of $z_1, \ldots, z_n$. □

Corollary 2.2.10 *Suppose that the roots of a non-constant complex polynomial lie in some given convex set. Then the roots of its derivative lie in the same convex set.* □

As a simple example of this last corollary, we quote the following result, which is often found in courses on complex analysis: *if all the roots of a non-constant complex polynomial have positive imaginary parts, then the same is also true of the roots of its derivative.*

Exercises 2.2

1. Let $A = \{(0,0,0),\ (0,0,1),\ (0,1,0),\ (1,0,0),\ (1,1,0),\ (1,0,1),\ (0,1,1),\ (1,1,1)\}$. Prove that

$$\operatorname{conv} A = \{(x, y, z) : 0 \leqslant x, y, z \leqslant 1\}.$$

2. Let A, B be disjoint non-empty convex sets in $\mathbb{R}^n$ and let $\mathbf{x}$ be a point of $\mathbb{R}^n$. Show that

$$\operatorname{conv}(\{\mathbf{x}\} \cup A) = \{\lambda \mathbf{a} + (1 - \lambda)\mathbf{x} : 0 \leqslant \lambda \leqslant 1, \mathbf{a} \in A\}.$$

   Deduce that either $\operatorname{conv}(\{\mathbf{x}\} \cup A)$ is disjoint from B or $\operatorname{conv}(\{\mathbf{x}\} \cup B)$ is disjoint from A.

3. Prove that $\operatorname{conv}(A + B) = \operatorname{conv} A + \operatorname{conv} B$ for all sets A, B in $\mathbb{R}^n$.
4. Give examples to show that the numbers $n + 2$ in Radon's theorem and $n + 1$ in Carathéodory's theorem cannot be reduced.
5. Show that $\operatorname{conv}(\operatorname{int} A) \subseteq \operatorname{int}(\operatorname{conv} A)$ for every set A *in* $\mathbb{R}^n$. Is it always true that $\operatorname{conv}(\operatorname{int} A) = \operatorname{int}(\operatorname{conv} A)$?
6. Prove that the diameter of the set of the roots of the complex polynomial $z^4 + az^3 + bz^2 + cz + d$ is at least $\sqrt{|(a^2/4) - (2b/3)|}$.

2.3 INTERIORS AND CLOSURES

In this short, but basic, section we establish those properties of the interiors, relative interiors, and closures of convex sets that will be needed in later discussions. For many purposes, the relative interior of a convex set is a more natural concept than that of its interior; one of the reasons for this is given in the following result.

Theorem 2.3.1 *The relative interior of a non-empty convex set in* $\mathbb{R}^n$ *is non-empty.*

Proof Let A be a non-empty r-dimensional convex set in $\mathbb{R}^n$. Then A contains points $\mathbf{a}_0, \ldots, \mathbf{a}_r$ which form an affine basis for the r-flat $\operatorname{aff} A$, and the barycentric coordinates $\lambda_0, \ldots, \lambda_r$ of a point $\mathbf{x}$ of $\operatorname{aff} A$ relative to $\mathbf{a}_0, \ldots, \mathbf{a}_r$ are continuous functions of $\mathbf{x}$, a fact which follows easily from Theorem 1.3.11. Let

$$\mathbf{a} = \frac{1}{r+1}(\mathbf{a}_0 + \ldots + \mathbf{a}_r).$$

Then $\mathbf{a}$ lies in A and its barycentric coordinates $\lambda_0, \ldots, \lambda_r$ are positive, each being $1/(r+1)$. By the continuity of the barycentric coordinates, for each $i = 0, \ldots, r$, there exists $s_i > 0$ such that $\lambda_i > 0$ whenever $\mathbf{x}$ lies in $B(\mathbf{a}; s_i) \cap \operatorname{aff} A$. Let s be the minimum of $s_0, \ldots, s_r$, so $s > 0$. Then if $\mathbf{x}$ lies in $B(\mathbf{a}; s) \cap \operatorname{aff} A$, all its barycentric coordinates $\lambda_0, \ldots, \lambda_r$ are positive, and so, since A is convex, $\mathbf{x}$ lies in A. Thus the relative interior of A contains the point $\mathbf{a}$. □

Corollary 2.3.2 *A convex set in* $\mathbb{R}^n$ *has an empty interior if and only if it lies in some hyperplane of* $\mathbb{R}^n$.

Proof Since a hyperplane in $\mathbb{R}^n$ has an empty interior, so does each of its subsets. A convex set in $\mathbb{R}^n$ which does not lie in any hyperplane of $\mathbb{R}^n$ must be n-dimensional, and therefore its interior coincides with its relative interior; this relative interior, however, is non-empty by the theorem. □

Before proving the fundamental Theorem 2.3.4, we first need to establish the following special case of it.

Lemma 2.3.3 *Let A be a convex set in* $\mathbb{R}^n$. *Let* $\mathbf{a} \in \operatorname{ri} A$ *and* $\mathbf{b} \in A$. *Then* $\lambda\mathbf{a} + (1-\lambda)\mathbf{b} \in \operatorname{ri} A$ *for* $0 < \lambda \leqslant 1$.

Proof Since $\mathbf{a} \in \operatorname{ri} A$, there is an $r > 0$ such that $B(\mathbf{a}; r) \cap \operatorname{aff} A \subseteq A$. Let $\mathbf{c} = \lambda\mathbf{a} + (1-\lambda)\mathbf{b}$, where $0 < \lambda \leqslant 1$. A simple proportionality argument in the two-dimensional case suggests that $B(\mathbf{c}; \lambda r) \cap \operatorname{aff} A \subseteq A$,

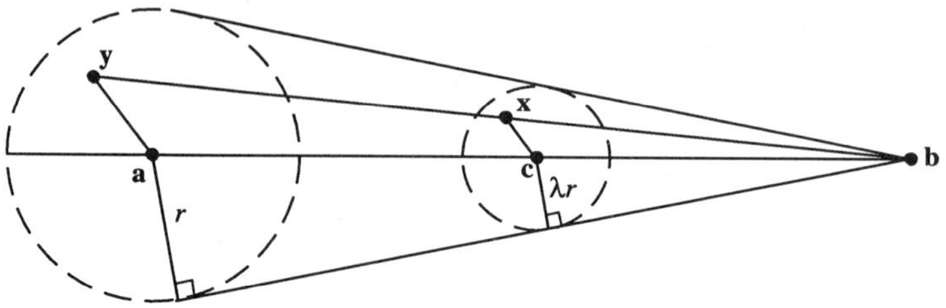

Fig. 2.3

and we now show this to be true in general. See Fig. 2.3. Let $\mathbf{x} \in B(\mathbf{c}; \lambda r) \cap \operatorname{aff} A$, and let

$$\mathbf{y} = \mathbf{a} + (1/\lambda)(\mathbf{x} - \mathbf{c}) = (1/\lambda)\mathbf{x} + (1 - (1/\lambda))\mathbf{b}.$$

Then $\mathbf{y} \in \operatorname{aff} A$ and $\|\mathbf{y} - \mathbf{a}\| = (1/\lambda)\|\mathbf{x} - \mathbf{c}\| < r$. Thus $\mathbf{y}$ lies in $B(\mathbf{a}; r) \cap \operatorname{aff} A$, and hence in A. The equation $\mathbf{x} = \lambda\mathbf{y} + (1 - \lambda)\mathbf{b}$, together with the convexity of A, shows that $\mathbf{x} \in A$. Hence $B(\mathbf{c}; \lambda r) \cap \operatorname{aff} A \subseteq A$, and $\mathbf{c} \in \operatorname{ri} A$ as required. □

Geometrically, the next theorem asserts that, with the possible exception of an endpoint, the entire line segment joining a closure point and a relative interior point of a convex set lies in the relative interior of the set. The remaining results of the section are, for the most part, easy consequences of this theorem.

Theorem 2.3.4 *Let A be a convex set in $\mathbb{R}^n$. Let $\mathbf{a} \in \operatorname{ri} A$ and $\mathbf{b} \in \operatorname{cl} A$. Then $\lambda\mathbf{a} + (1 - \lambda)\mathbf{b} \in \operatorname{ri} A$ for $0 < \lambda \leqslant 1$.*

Proof Since $\mathbf{a} \in \operatorname{ri} A$, there is an $r > 0$ such that $B(\mathbf{a}; r) \cap \operatorname{aff} A \subseteq A$. Let $\mathbf{c} = \lambda\mathbf{a} + (1 - \lambda)\mathbf{b}$, where $0 < \lambda \leqslant 1$. Since $\mathbf{b} \in \operatorname{cl} A$, there exists $\mathbf{d} \in A$ satisfying $(1 - \lambda)\|\mathbf{d} - \mathbf{b}\| < \lambda r$. Let

$$\mathbf{e} = \mathbf{a} + ((1 - \lambda)/\lambda)(\mathbf{b} - \mathbf{d}) = (1/\lambda)\mathbf{c} + (1 - (1/\lambda))\mathbf{d}.$$

Then $\mathbf{e} \in \operatorname{aff} A$ and $\|\mathbf{e} - \mathbf{a}\| = ((1 - \lambda)/\lambda)\|\mathbf{b} - \mathbf{d}\| < r$. Thus $\mathbf{e}$ lies in $B(\mathbf{a}; r) \cap \operatorname{aff} A$, and hence in $\operatorname{ri} A$. The equation $\mathbf{c} = \lambda\mathbf{e} + (1 - \lambda)\mathbf{d}$, together with the lemma, shows that $\mathbf{c} \in \operatorname{ri} A$. □

Theorem 2.3.5 *Let A be a convex set in $\mathbb{R}^n$. Then* $\operatorname{ri} A$, $\operatorname{int} A$, *and* $\operatorname{cl} A$ *are convex.*

Proof If $\mathbf{a}, \mathbf{b} \in \operatorname{ri} A$ and $0 \leqslant \lambda \leqslant 1$, then $\lambda\mathbf{a} + (1 - \lambda)\mathbf{b} \in \operatorname{ri} A$, either trivially, if $\lambda = 0$, or by Theorem 2.3.4 otherwise. Thus $\operatorname{ri} A$ is convex. That $\operatorname{int} A$ is convex follows from the fact that either $\operatorname{int} A$ is empty or coincides with $\operatorname{ri} A$. If $\mathbf{a}, \mathbf{b} \in \operatorname{cl} A$, then there are sequences $\mathbf{a}_1, \ldots,$

$\mathbf{a}_k, \ldots$ and $\mathbf{b}_1, \ldots, \mathbf{b}_k, \ldots$ of points of A such that $\mathbf{a}_k \to \mathbf{a}, \mathbf{b}_k \to \mathbf{b}$ as $k \to \infty$. Let $0 \leqslant \lambda \leqslant 1$. Then $\lambda \mathbf{a}_k + (1 - \lambda)\mathbf{b}_k \in A$ for each k, since A is convex. Now $\lambda \mathbf{a}_k + (1 - \lambda)\mathbf{b}_k \to \lambda \mathbf{a} + (1 - \lambda)\mathbf{b}$ as $k \to \infty$, which shows that $\lambda \mathbf{a} + (1 - \lambda)\mathbf{b} \in \operatorname{cl} A$. Thus $\operatorname{cl} A$ is convex. □

Theorem 2.3.5 shows that, for any set A in $\mathbb{R}^n$, the set $\operatorname{cl}(\operatorname{conv} A)$ is a closed convex set containing A. If B is *any* closed convex set in $\mathbb{R}^n$ containing A, then

$$B = \operatorname{cl}(\operatorname{conv} B) \supseteq \operatorname{cl}(\operatorname{conv} A).$$

This shows that $\operatorname{cl}(\operatorname{conv} A)$ is the *smallest* closed convex set containing A; it is called the **closed convex hull** of A.

The next result provides a useful characterization of the relative interior points of a convex set A in $\mathbb{R}^n$. Geometrically, it asserts that a point $\mathbf{a}_0$ of A is a relative interior point of A if and only if every line segment lying in A, and having $\mathbf{a}_0$ as an endpoint, can be extended some distance beyond $\mathbf{a}_0$ without leaving A.

Theorem 2.3.6 *Let $\mathbf{a}_0$ be a point of a convex set A in $\mathbb{R}^n$. Then $\mathbf{a}_0 \in \operatorname{ri} A$ if and only if, for each $\mathbf{a} \in A$, there exists $\mu > 1$ such that $(1 - \mu)\mathbf{a} + \mu \mathbf{a}_0 \in A$.*

Proof Clearly, if $\mathbf{a}_0 \in \operatorname{ri} A$, then the condition of the theorem is satisfied. Conversely, suppose that $\mathbf{a}_0$ satisfies this condition. Let $\mathbf{a} \in \operatorname{ri} A$. Then there is $\mu > 1$ such that the point $\mathbf{x} = (1 - \mu)\mathbf{a} + \mu \mathbf{a}_0$ lies in A. Hence $\mathbf{a}_0 = \lambda \mathbf{a} + (1 - \lambda)\mathbf{x}$, where $0 < \lambda = 1 - \mu^{-1} < 1$. But $\mathbf{a} \in \operatorname{ri} A$ and $\mathbf{x} \in A$, so $\mathbf{a}_0 \in \operatorname{ri} A$ by Lemma 2.3.3. □

We put Theorem 2.3.6 to immediate use in helping to find a characterization of the relative interior points of the convex hull of a finite set in $\mathbb{R}^n$, a characterization that will be needed in our study of convex polytopes in the next chapter.

Theorem 2.3.7 *Let $A = \operatorname{conv}\{\mathbf{a}_1, \ldots, \mathbf{a}_m\}$, where $\mathbf{a}_1, \ldots, \mathbf{a}_m \in \mathbb{R}^n$. Then*

$$\operatorname{ri} A = \{\lambda_1 \mathbf{a}_1 + \ldots + \lambda_m \mathbf{a}_m : \lambda_1, \ldots, \lambda_m > 0,\ \lambda_1 + \ldots + \lambda_m = 1\}.$$

Proof Suppose first that $\mathbf{a}_0 = \lambda_1 \mathbf{a}_1 + \ldots + \lambda_m \mathbf{a}_m$, where $\lambda_1, \ldots, \lambda_m > 0$ with $\lambda_1 + \ldots + \lambda_m = 1$, and that $\mathbf{a} \in A$. Then $\mathbf{a}_0 \in A$, and $\mathbf{a} = \mu_1 \mathbf{a}_1 + \ldots + \mu_m \mathbf{a}_m$ for some $\mu_1, \ldots, \mu_m \geqslant 0$ with $\mu_1 + \ldots + \mu_m = 1$. Choose $\mu > 1$ such that

$$\mu\lambda_1 + (1 - \mu)\mu_1 \geqslant 0, \ldots, \mu\lambda_m + (1 - \mu)\mu_m \geqslant 0.$$

Then $(1 - \mu)\mathbf{a} + \mu \mathbf{a}_0 \in A$, and so $\mathbf{a}_0 \in \operatorname{ri} A$ by Theorem 2.3.6.

Suppose next that $\mathbf{a}_0 \in \operatorname{ri} A$, and that $\mathbf{a}^* = (1/m)(\mathbf{a}_1 + \ldots + \mathbf{a}_m)$.

Then $\mathbf{a}^* \in A$. By Theorem 2.3.6, there exist $\mu > 1$ and $\mathbf{a} \in A$ such that $\mathbf{a} = (1 - \mu)\mathbf{a}^* + \mu\mathbf{a}_0$; say $\mathbf{a} = \mu_1 \mathbf{a}_1 + \ldots + \mu_m \mathbf{a}_m$, where $\mu_1, \ldots, \mu_m \geqslant 0$ with $\mu_1 + \ldots + \mu_m = 1$. The equation

$$\mathbf{a}_0 = ((\mu_1 + (\mu - 1)/m)/\mu)\mathbf{a}_1 + \ldots + ((\mu_m + (\mu - 1)/m)/\mu)\mathbf{a}_m$$

now expresses $\mathbf{a}_0$ in the form $\lambda_1 \mathbf{a}_1 + \ldots + \lambda_m \mathbf{a}_m$, where $\lambda_1, \ldots, \lambda_m > 0$ with $\lambda_1 + \ldots + \lambda_m = 1$. □

Theorem 2.3.8 *Let A be a convex set in $\mathbb{R}^n$. Then* $\operatorname{ri} A = \operatorname{ri}(\operatorname{cl} A)$ *and* $\operatorname{cl} A = \operatorname{cl}(\operatorname{ri} A)$.

Proof We assume, throughout, that A is non-empty with $\mathbf{a} \in \operatorname{ri} A$.

The inclusion $\operatorname{ri} A \subseteq \operatorname{ri}(\operatorname{cl} A)$ follows from the inclusion $A \subseteq \operatorname{cl} A$ and the fact that the affine hulls of A and $\operatorname{cl} A$ coincide. To establish the inclusion $\operatorname{ri}(\operatorname{cl} A) \subseteq \operatorname{ri} A$, suppose that $\mathbf{b} \in \operatorname{ri}(\operatorname{cl} A)$. By Theorem 2.3.6, there exist $\mu > 1$ and $\mathbf{c} \in \operatorname{cl} A$ such that $\mathbf{c} = (1 - \mu)\mathbf{a} + \mu\mathbf{b}$, whence $\mathbf{b} = (1 - \lambda)\mathbf{a} + \lambda\mathbf{c}$, where $0 < \lambda = 1/\mu < 1$. That $\mathbf{b} \in \operatorname{ri} A$ follows from Theorem 2.3.4. Thus $\operatorname{ri}(\operatorname{cl} A) \subseteq \operatorname{ri} A$, and $\operatorname{ri} A = \operatorname{ri}(\operatorname{cl} A)$ as required.

The inclusion $\operatorname{cl}(\operatorname{ri} A) \subseteq \operatorname{cl} A$ is clear. To establish the inclusion $\operatorname{cl} A \subseteq \operatorname{cl}(\operatorname{ri} A)$, suppose that $\mathbf{b} \in \operatorname{cl} A$. Theorem 2.3.4 shows that $\lambda\mathbf{a} + (1 - \lambda)\mathbf{b} \in \operatorname{ri} A$ for $0 < \lambda \leqslant 1$, whence $\mathbf{b} \in \operatorname{cl}(\operatorname{ri} A)$. Thus $\operatorname{cl} A \subseteq \operatorname{cl}(\operatorname{ri} A)$, and $\operatorname{cl} A = \operatorname{cl}(\operatorname{ri} A)$ as required. □

Corollary 2.3.9 *Let A be a convex set in $\mathbb{R}^n$. Then* $\operatorname{int} A = \operatorname{int}(\operatorname{cl} A)$ *and, when* $\operatorname{int} A$ *is non-empty,* $\operatorname{cl} A = \operatorname{cl}(\operatorname{int} A)$.

Proof If $\operatorname{int} A$ is non-empty, then $\operatorname{ri} A = \operatorname{int} A$ and $\operatorname{ri}(\operatorname{cl} A) = \operatorname{int}(\operatorname{cl} A)$, and the corollary follows from the theorem. If $\operatorname{int} A$ is empty, then A, and hence $\operatorname{cl} A$, lie in a hyperplane of $\mathbb{R}^n$, whence both $\operatorname{int} A$ and $\operatorname{int}(\operatorname{cl} A)$ are empty. □

Corollary 2.3.10 *Let A be a convex set in $\mathbb{R}^n$. Then* $\operatorname{rebd} A = \operatorname{rebd}(\operatorname{cl} A)$ *and* $\operatorname{bd} A = \operatorname{bd}(\operatorname{cl} A)$.

Proof By the theorem and its first corollary,

$$\operatorname{rebd}(\operatorname{cl} A) = \operatorname{cl}(\operatorname{cl} A) \setminus \operatorname{ri}(\operatorname{cl} A) = \operatorname{cl} A \setminus \operatorname{ri} A = \operatorname{rebd} A$$

and

$$\operatorname{bd}(\operatorname{cl} A) = \operatorname{cl}(\operatorname{cl} A) \setminus \operatorname{int}(\operatorname{cl} A) = \operatorname{cl} A \setminus \operatorname{int} A = \operatorname{bd} A. \; \square$$

Exercises 2.3

1. Show that, for any set A in $\mathbb{R}^n$, the sequence

$$A,\ \operatorname{cl} A,\ \operatorname{conv} A,\ \operatorname{cl}(\operatorname{conv} A),\ \operatorname{conv}(\operatorname{cl} A),\ \operatorname{cl}(\operatorname{conv}(\operatorname{cl} A)),\ \operatorname{conv}(\operatorname{cl}(\operatorname{conv} A)), \ldots$$

contains at most five distinct members. Give an example of a set A for which the sequence does contain five distinct members. Why must such an example be unbounded?

2. The convex set A in $\mathbb{R}^n$ intersects each line in $\mathbb{R}^n$ in a closed set. Prove that A is itself closed. Can the convexity condition be omitted in this last result?
3. Prove that $\operatorname{rebd} A = \operatorname{rebd}(\operatorname{ri} A)$ for each convex set A in $\mathbb{R}^n$.
4. Let $\mathbf{a}_0$ be a point of a convex set A in $\mathbb{R}^n$. Suppose that, for each point $\mathbf{x}$ in $\mathbb{R}^n$, there exists a scalar μ with $0 < \mu < 1$ such that $(1 - \mu)\mathbf{x} + \mu\mathbf{a}_0 \in A$. Show that $\mathbf{a}_0 \in \operatorname{int} A$.

2.4 SEPARATION AND SUPPORT

The results that we establish in this section on the separation and support of convex sets are among the most important ones in convexity. From several possible ways of approach, we have chosen that which utilizes the idea of a *nearest point*, because of its simplicity and immediate geometric appeal.

Theorem 1.9.1 shows that, given a non-empty closed set A and a point $\mathbf{x}$ in $\mathbb{R}^n$, there is a point $\mathbf{a}_0$ of A which is *nearest* to $\mathbf{x}$. We show below that, if A is also convex, then this nearest point $\mathbf{a}_0$ of A to $\mathbf{x}$ is *unique* and that, for each point $\mathbf{a}$ of A, the vectors $\mathbf{x} - \mathbf{a}_0$ and $\mathbf{a} - \mathbf{a}_0$ make a non-acute angle with each other.

Theorem 2.4.1 *In $\mathbb{R}^n$ let A be a non-empty closed convex set and let $\mathbf{x}$ be a point. Then there exists a unique point $\mathbf{a}_0$ of A such that*

$$\|\mathbf{x} - \mathbf{a}_0\| = \inf\{\|\mathbf{x} - \mathbf{z}\| : \mathbf{z} \in A\}.$$

Moreover, $(\mathbf{x} - \mathbf{a}_0)\cdot(\mathbf{a} - \mathbf{a}_0) \leqslant 0$ *for each* $\mathbf{a}$ *in* A.

Proof By Theorem 1.9.1, there exists $\mathbf{a}_0 \in A$ such that

$$\|\mathbf{x} - \mathbf{a}_0\| = \inf\{\|\mathbf{x} - \mathbf{z}\| : \mathbf{z} \in A\}.$$

Let $\mathbf{a} \in A$ and $0 < \lambda \leqslant 1$. The convexity of A shows $(1 - \lambda)\mathbf{a}_0 + \lambda\mathbf{a} \in A$. See Fig. 2.4. The choice of $\mathbf{a}_0$ shows that

$$\|\mathbf{x} - ((1 - \lambda)\mathbf{a}_0 + \lambda\mathbf{a})\|^2 = \|(\mathbf{x} - \mathbf{a}_0) + \lambda(\mathbf{a}_0 - \mathbf{a})\|^2 \geqslant \|\mathbf{x} - \mathbf{a}_0\|^2.$$

We deduce, using Theorem 1.6.1(iv), that $(\mathbf{x} - \mathbf{a}_0)\cdot(\mathbf{a} - \mathbf{a}_0) \leqslant 0$.

Suppose that $\mathbf{a}_1 \in A$ also satisfies the equation

$$\|\mathbf{x} - \mathbf{a}_1\| = \inf\{\|\mathbf{x} - \mathbf{z}\| : \mathbf{z} \in A\}.$$

Then, by what we have just proved, $(\mathbf{x} - \mathbf{a}_0)\cdot(\mathbf{a}_1 - \mathbf{a}_0) \leqslant 0$. Because

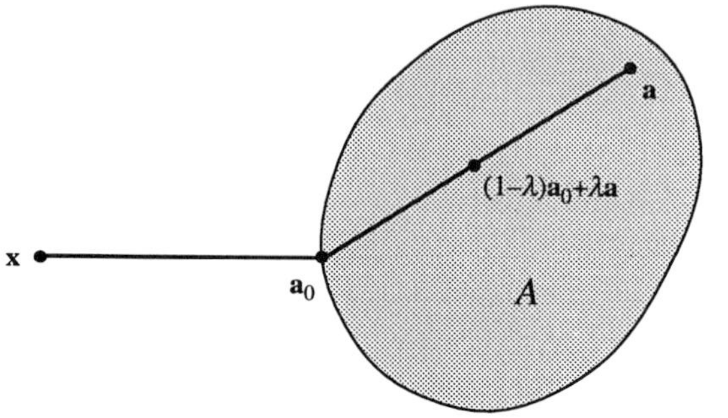

Fig. 2.4

of the symmetry between $\mathbf{a}_0$ and $\mathbf{a}_1$, we have $(\mathbf{x}-\mathbf{a}_1)\cdot(\mathbf{a}_0-\mathbf{a}_1)\leqslant 0$. Adding these last two inequalities together, we deduce that

$$\|\mathbf{a}_1-\mathbf{a}_0\|^2=(\mathbf{a}_1-\mathbf{a}_0)\cdot(\mathbf{a}_1-\mathbf{a}_0)\leqslant 0,$$

which proves that $\mathbf{a}_0=\mathbf{a}_1$. This establishes the uniqueness of the nearest point $\mathbf{a}_0$ of A to $\mathbf{x}$. □

The last theorem shows how each non-empty closed convex set A in $\mathbb{R}^n$ gives rise to a mapping $f\colon\mathbb{R}^n\to A$ defined by $f(\mathbf{x})=\mathbf{a}_0$, where $\mathbf{a}_0$ is the nearest point of A to a point $\mathbf{x}$ of $\mathbb{R}^n$. This mapping f is called the **projection operator** of A.

Corollary 2.4.2 *Let A be a non-empty closed convex set in $\mathbb{R}^n$. Then the projection operator $f\colon\mathbb{R}^n\to A$ of A satisfies the Lipschitz condition $\|f(\mathbf{x})-f(\mathbf{y})\|\leqslant\|\mathbf{x}-\mathbf{y}\|$ for all $\mathbf{x},\mathbf{y}\in\mathbb{R}^n$, and so is continuous.*

Proof Let $\mathbf{x},\mathbf{y}\in\mathbb{R}^n$, and write $\mathbf{u}=\mathbf{x}-f(\mathbf{x})$, $\mathbf{v}=\mathbf{y}-f(\mathbf{y})$. Then, by the theorem,

$$\mathbf{u}\cdot(f(\mathbf{y})-f(\mathbf{x}))\leqslant 0 \qquad\text{and}\qquad \mathbf{v}\cdot(f(\mathbf{x})-f(\mathbf{y}))\leqslant 0,$$

so that

$$(\mathbf{u}-\mathbf{v})\cdot(f(\mathbf{x})-f(\mathbf{y}))\geqslant 0.$$

Thus

$$\begin{aligned}\|\mathbf{x}-\mathbf{y}\|^2&=\|(\mathbf{u}-\mathbf{v})+(f(\mathbf{x})-f(\mathbf{y}))\|^2\\&=\|\mathbf{u}-\mathbf{v}\|^2+2(\mathbf{u}-\mathbf{v})\cdot(f(\mathbf{x})-f(\mathbf{y}))+\|f(\mathbf{x})-f(\mathbf{y})\|^2\\&\geqslant\|f(\mathbf{x})-f(\mathbf{y})\|^2,\end{aligned}$$

and so $\|f(\mathbf{x})-f(\mathbf{y})\|\leqslant\|\mathbf{x}-\mathbf{y}\|$. □

Geometrically, the following corollary states that, if $f(\mathbf{x})$ is the

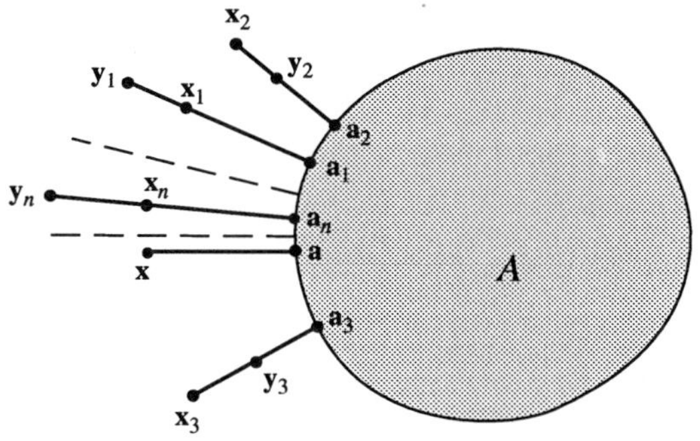

Fig. 2.5

nearest point of a non-empty closed convex set A in $\mathbb{R}^n$ to a point $\mathbf{x}$ of $\mathbb{R}^n$ not belonging to A, then it is also the nearest point of A to any point on the halfline starting at $f(\mathbf{x})$ and passing through $\mathbf{x}$.

Corollary 2.4.3 *Let A be a non-empty closed convex set in* $\mathbb{R}^n$ *with projection operator f. Then, for all* $\mathbf{x} \in \mathbb{R}^n$ *and* $\lambda \geqslant 0$,

$$f(f(\mathbf{x}) + \lambda(\mathbf{x} - f(\mathbf{x}))) = f(\mathbf{x}).$$

Proof Write $\mathbf{y} = f(\mathbf{x}) + \lambda(\mathbf{x} - f(\mathbf{x}))$, where $\mathbf{x} \in \mathbb{R}^n$ and $\lambda \geqslant 0$. By the theorem,

$$(\mathbf{x} - f(\mathbf{x})) \cdot (f(\mathbf{y}) - f(\mathbf{x})) \leqslant 0 \quad \text{and} \quad (\mathbf{y} - f(\mathbf{y})) \cdot (f(\mathbf{x}) - f(\mathbf{y})) \leqslant 0.$$

From these inequalities, we deduce that

$$0 \leqslant (f(\mathbf{y}) - f(\mathbf{x})) \cdot (f(\mathbf{y}) - f(\mathbf{x})) = (\lambda(\mathbf{x} - f(\mathbf{x})) + f(\mathbf{y}) - \mathbf{y}) \cdot (f(\mathbf{y}) - f(\mathbf{x})) \leqslant 0,$$

whence $\|f(\mathbf{y}) - f(\mathbf{x})\|^2 = 0$ and $f(\mathbf{y}) = f(\mathbf{x})$, as required. □

The next theorem shows that every relative boundary point of a non-empty closed convex set A in $\mathbb{R}^n$ is the nearest point of A to some point of aff A not lying in A.

Theorem 2.4.4 *Let* $\mathbf{a}$ *be a relative boundary point of a non-empty closed convex set A in* $\mathbb{R}^n$ *with projection operator f. Then there exists* $\mathbf{x} \in (\operatorname{aff} A) \setminus A$ *such that* $f(\mathbf{x}) = \mathbf{a}$.

Proof Since $\mathbf{a} \in \operatorname{rebd} A$, there exists, for each positive integer m, a point $\mathbf{y}_m$ of $(\operatorname{aff} A) \setminus A$ satisfying $\|\mathbf{y}_m - \mathbf{a}\| \leqslant 1/m$. Write $\mathbf{a}_m = f(\mathbf{y}_m)$ and

$$\mathbf{x}_m = \mathbf{a}_m + (\mathbf{y}_m - \mathbf{a}_m)/\|\mathbf{y}_m - \mathbf{a}_m\|.$$

See Fig. 2.5. Then $\|\mathbf{x}_m - \mathbf{a}_m\| = 1$ and, by Corollary 2.4.3, $f(\mathbf{x}_m) = \mathbf{a}_m$. Corollary 2.4.2 shows that

$$\|\mathbf{a}_m - \mathbf{a}\| = \|f(\mathbf{y}_m) - f(\mathbf{a})\| \leqslant \|\mathbf{y}_m - \mathbf{a}\| \leqslant 1/m,$$

so $\mathbf{a}_m \to \mathbf{a}$ as $m \to \infty$. The inequality

$$\|\mathbf{x}_m\| \leqslant \|\mathbf{x}_m - \mathbf{a}_m\| + \|\mathbf{a}_m - \mathbf{a}\| + \|\mathbf{a}\| \leqslant 1 + (1/m) + \|\mathbf{a}\|$$

shows that the sequence $\mathbf{x}_1, \mathbf{x}_2, \ldots$ is bounded. Thus $\mathbf{x}_1, \mathbf{x}_2, \ldots$ contains a convergent subsequence. Assume, without loss of generality, that $\mathbf{x}_1, \mathbf{x}_2, \ldots$ itself converges to some point $\mathbf{x}$ of $\mathbb{R}^n$. Since $\mathbf{x}_1, \mathbf{x}_2, \ldots$ belong to aff A, we can deduce that $\mathbf{x} \in \operatorname{aff} A$. The continuity of f shows that $f(\mathbf{x}_m) \to f(\mathbf{x})$ as $m \to \infty$, i.e. $\mathbf{a}_m \to f(\mathbf{x})$ as $m \to \infty$. But $\mathbf{a}_m \to \mathbf{a}$ as $m \to \infty$, whence $f(\mathbf{x}) = \mathbf{a}$. Clearly $\|\mathbf{x} - \mathbf{a}\| = 1$, so $\mathbf{x} \notin A$. □

Corollary 2.4.5 *Let A be a non-empty closed convex set in $\mathbb{R}^n$ with projection operator f and let B be a bounded set in $\mathbb{R}^n$ such that $A \subseteq B$. Then $f(\operatorname{bd} B) = \operatorname{bd} A$.*

Proof We show that $\operatorname{bd} A \subseteq f(\operatorname{bd} B)$, the opposite inclusion being obvious. Let $\mathbf{a} \in \operatorname{bd} A$. Then there exists $\mathbf{x} \in \mathbb{R}^n \setminus A$ such that $f(\mathbf{x}) = \mathbf{a}$. This can be proved by substituting bd A for rebd A, and $\mathbb{R}^n$ for aff A, in the proof of Theorem 2.4.4. As B is bounded, there is some $\mu > 0$ such that, for $\lambda \geqslant \mu$, $\mathbf{a} + \lambda(\mathbf{x} - \mathbf{a}) \notin B$. But $\mathbf{a} \in B$, so $\mathbf{a} + \lambda_0(\mathbf{x} - \mathbf{a}) \in \operatorname{bd} B$ for some $\lambda_0 \geqslant 0$. Corollary 2.4.3 shows that $f(\mathbf{a} + \lambda_0(\mathbf{x} - \mathbf{a})) = \mathbf{a}$, whence $\operatorname{bd} A \subseteq f(\operatorname{bd} B)$. □

The concept of separation, which we now discuss, plays a central role in any study of convexity. It is based on the fact that a hyperplane in $\mathbb{R}^n$ divides the space into two halfspaces. Let A and B be sets, and let H be a hyperplane in $\mathbb{R}^n$. Then H is said to **separate** A and B if A lies in one of the closed halfspaces determined by H, and B lies in the other; H is said to **separate** A and B **properly** if it separates them, but not both A and B lie in H. If A and B lie in opposite open halfspaces determined by H, then H is said to **separate** A and B **strictly**. It follows from the convexity of halfspaces that, if a hyperplane separates two sets, then it also separates their convex hulls; for this reason, we consider only the separation of convex sets.

To illustrate the three types of separation that we have just defined, we consider some simple examples. Of course, it is not always possible to separate two convex sets by a hyperplane. For example, there is no line separating the set $\{\mathbf{0}\}$ and the closed unit disc $\{(x, y): x^2 + y^2 \leqslant 1\}$ in $\mathbb{R}^2$. Any two sets that can be strictly separated can be properly separated, unless they are both empty. The convex sets $\{(x, y): x \leqslant 0\}$ and $\{(x, y): x > 0, y \geqslant 1/x\}$ in $\mathbb{R}^2$ cannot be strictly separated, but they are properly separated by the y-axis. A hyperplane in $\mathbb{R}^n$ separates any two of its subsets, but does not separate them properly.

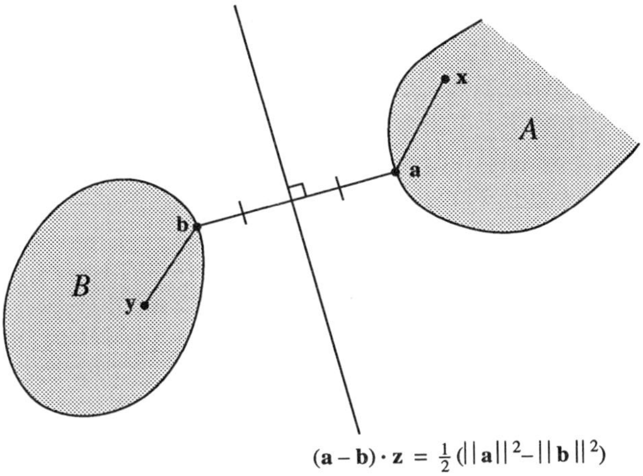

Fig. 2.6

Theorem 2.4.6 *Let A and B be disjoint non-empty convex sets in $\mathbb{R}^n$ with A closed and B compact. Then A and B can be strictly separated by a hyperplane in $\mathbb{R}^n$.*

Proof The geometry of the proof is as follows. Let $\mathbf{a}$ and $\mathbf{b}$ be nearest points of A and B. Then the hyperplane through the midpoint of the line segment joining $\mathbf{a}$ and $\mathbf{b}$ with normal vector $\mathbf{a} - \mathbf{b}$ strictly separates A and B. See Fig. 2.6.

Let $\mathbf{a} \in A$, $\mathbf{b} \in B$ be such that $\mathbf{a}$ is the nearest point of A to $\mathbf{b}$, and $\mathbf{b}$ is the nearest point of B to $\mathbf{a}$. This is possible by Theorem 1.9.4. Since A and B are disjoint, $\mathbf{a} \neq \mathbf{b}$. Let $\mathbf{x} \in A$, $\mathbf{y} \in B$. Then, by Theorem 2.4.1, $(\mathbf{b} - \mathbf{a}) \cdot (\mathbf{x} - \mathbf{a}) \leqslant 0$ and $(\mathbf{a} - \mathbf{b}) \cdot (\mathbf{y} - \mathbf{b}) \leqslant 0$. Thus

$$\begin{aligned}(\mathbf{a} - \mathbf{b}) \cdot \mathbf{x} &\geqslant (\mathbf{a} - \mathbf{b}) \cdot \mathbf{a} \\ &= \tfrac{1}{2}(\|\mathbf{a}\|^2 - \|\mathbf{b}\|^2 + \|\mathbf{a} - \mathbf{b}\|^2) \\ &> \tfrac{1}{2}(\|\mathbf{a}\|^2 - \|\mathbf{b}\|^2) \\ &> \tfrac{1}{2}(\|\mathbf{a}\|^2 - \|\mathbf{b}\|^2 - \|\mathbf{a} - \mathbf{b}\|^2) \\ &= (\mathbf{a} - \mathbf{b}) \cdot \mathbf{b} \\ &\geqslant (\mathbf{a} - \mathbf{b}) \cdot \mathbf{y}.\end{aligned}$$

Write $\mathbf{c} = \mathbf{a} - \mathbf{b}$ and $c_0 = \tfrac{1}{2}(\|\mathbf{a}\|^2 - \|\mathbf{b}\|^2)$. Then we have shown that the hyperplane $\mathbf{c} \cdot \mathbf{z} = c_0$ strictly separates A and B. □

Corollary 2.4.7 *In $\mathbb{R}^n$ let A be a closed convex set and let $\mathbf{b}$ be a point not lying in A. Then A and $\{\mathbf{b}\}$ can be strictly separated by a hyperplane in $\mathbb{R}^n$.* □

Corollary 2.4.8 *Each closed convex set A in $\mathbb{R}^n$ is the intersection of all the closed halfspaces in $\mathbb{R}^n$ containing A.*

Proof Denote by B the intersection of all the closed halfspaces in $\mathbb{R}^n$ containing A. Then B is a closed convex set containing A. If $\mathbf{b} \notin A$, then Corollary 2.4.7 shows that there exists some closed halfspace in $\mathbb{R}^n$ which contains A but not $\mathbf{b}$, whence $\mathbf{b} \notin B$. Thus $B \subseteq A$, and $A = B$ as desired. □

The above results on strict separation have all imposed conditions, such as closedness or compactness, on the convex sets to be separated. We now come to a very general result on the separation of convex sets, Theorem 2.4.10, in which no such extra conditions are assumed. First we need a lemma.

Lemma 2.4.9 *In $\mathbb{R}^n$ let A be a non-empty convex set not containing the origin. Then there exists a hyperplane in $\mathbb{R}^n$ which separates the origin and A, and does not contain A.*

Proof Suppose first that $\mathbf{0} \notin \operatorname{cl} A$. Then the lemma follows from Corollary 2.4.7 applied to the closed convex set $\operatorname{cl} A$. Suppose next that $\mathbf{0} \in \operatorname{cl} A$. Then $\mathbf{0} \in \operatorname{rebd} A$. By Corollary 2.3.10, $\operatorname{rebd} A = \operatorname{rebd}(\operatorname{cl} A)$, so $\mathbf{0} \in \operatorname{rebd}(\operatorname{cl} A)$. Theorem 2.4.4 asserts the existence of a point $\mathbf{x}$ of $(\operatorname{aff}(\operatorname{cl} A)) \setminus \operatorname{cl} A$ whose nearest point in $\operatorname{cl} A$ is $\mathbf{0}$. By Theorem 2.4.1, $\operatorname{cl} A \subseteq \{\mathbf{z} \in \mathbb{R}^n : \mathbf{x} \cdot \mathbf{z} \leqslant 0\}$, which shows that the hyperplane $H = \{\mathbf{z} \in \mathbb{R}^n : \mathbf{x} \cdot \mathbf{z} = 0\}$ separates $\{\mathbf{0}\}$ and A. We cannot have $A \subseteq H$, for this would imply that $\mathbf{x} \in \operatorname{aff}(\operatorname{cl} A) \subseteq H$, which is impossible since $\mathbf{x} \cdot \mathbf{x} > 0$. Thus H separates $\{\mathbf{0}\}$ and A, and does not contain A. □

Theorem 2.4.10 *Each pair of disjoint non-empty convex sets A and B in $\mathbb{R}^n$ can be properly separated by a hyperplane in $\mathbb{R}^n$.*

Proof The non-empty convex set $A - B$ does not contain the origin, and so, by the lemma, there exists a hyperplane in $\mathbb{R}^n$ which separates $\{\mathbf{0}\}$ and $A - B$, and that does not contain $A - B$. Thus there exist $\mathbf{c} \in \mathbb{R}^n$ with $\mathbf{c} \neq \mathbf{0}$ and $c_0 \in \mathbb{R}$ such that

$$0 = \mathbf{c} \cdot \mathbf{0} \leqslant c_0 \qquad \text{and} \qquad \mathbf{c} \cdot (\mathbf{a} - \mathbf{b}) \geqslant c_0 \qquad \text{for } \mathbf{a} \in A, \mathbf{b} \in B;$$

also, for some $\mathbf{a}_0 \in A$, $\mathbf{b}_0 \in B$, we have

$$\mathbf{c} \cdot (\mathbf{a}_0 - \mathbf{b}_0) > c_0 \geqslant 0. \tag{2.4.1}$$

For every $\mathbf{a} \in A$, $\mathbf{b} \in B$,

$$\mathbf{c} \cdot \mathbf{a} \geqslant \mathbf{c} \cdot \mathbf{b} + c_0 \geqslant \mathbf{c} \cdot \mathbf{b}.$$

Thus there is a scalar d satisfying the inequalities

$$\inf\{\mathbf{c}\cdot\mathbf{a}:\mathbf{a}\in A\} \geqslant d \geqslant \sup\{\mathbf{c}\cdot\mathbf{b}:\mathbf{b}\in B\}.$$

For any $\mathbf{a}'\in A$, $\mathbf{b}'\in B$, we have $\mathbf{c}\cdot\mathbf{a}' \geqslant d \geqslant \mathbf{c}\cdot\mathbf{b}'$, so the hyperplane H with equation $\mathbf{c}\cdot\mathbf{z} = d$ separates A and B. Now H cannot contain *both* A and B, for this would imply that $\mathbf{c}\cdot(\mathbf{a}_0 - \mathbf{b}_0) = 0$, which contradicts (2.4.1). Thus H separates A and B properly. □

Corollary 2.4.11 *Each pair of non-empty convex sets A and B in $\mathbb{R}^n$ whose relative interiors are disjoint can be properly separated by a hyperplane in $\mathbb{R}^n$.*

Proof The non-empty convex sets $\operatorname{ri} A$ and $\operatorname{ri} B$ are disjoint, and so, by the theorem, there exists a hyperplane H in $\mathbb{R}^n$ which properly separates them. Since closed halfspaces are closed, H also properly separates $\operatorname{cl}(\operatorname{ri} A) = \operatorname{cl} A$ and $\operatorname{cl}(\operatorname{ri} B) = \operatorname{cl} B$, and hence A and B. □

A notion closely allied with that of the separation of sets is that of a support hyperplane to a set. In $\mathbb{R}^n$ a hyperplane H is called a **support hyperplane** to a set A if H meets $\operatorname{cl} A$ and A lies in one of the closed halfspaces determined by H. Such a hyperplane H is said to **support** A at those points where H meets $\operatorname{cl} A$. A hyperplane H cannot support a set A at an interior point of A, because every ball with centre in H meets both the open halfspaces determined by H. A hyperplane in $\mathbb{R}^n$ is a *trivial* support hyperplane to each of its non-empty subsets. A support hyperplane to a set in $\mathbb{R}^n$ is said to be a **non-trivial support hyperplane** to the set if it does not contain the set itself. The next result expresses what is, arguably, the single most important property of convex sets.

Theorem 2.4.12 *Through each boundary point of a convex set A in $\mathbb{R}^n$ there passes a support hyperplane to A, and through each relative boundary point of A there passes a non-trivial support hyperplane to A.*

Proof Suppose first that $\mathbf{a}$ is a boundary point of A, but not a relative boundary point of A. Then A cannot be n-dimensional, and so lies in some hyperplane H of $\mathbb{R}^n$. Clearly H is a support hyperplane to A passing through $\mathbf{a}$.

Suppose next that $\mathbf{a}$ is a relative boundary point of A. Theorem 2.4.10 shows the existence of a hyperplane H which *properly* separates $\{\mathbf{a}\}$ and $\operatorname{ri} A$. Now H cannot contain $\operatorname{ri} A$, for this would imply that H also contains $\operatorname{cl}(\operatorname{ri} A) = \operatorname{cl} A$, and hence $\mathbf{a}$. By the definition of separation, $\{\mathbf{a}\}$ and $\operatorname{ri} A$, and thus $\{\mathbf{a}\}$ and $\operatorname{cl}(\operatorname{ri} A) = \operatorname{cl} A$, belong to opposite closed halfspaces determined by H. Since $\mathbf{a}\in\operatorname{cl} A$, we must have $\mathbf{a}\in H$. Thus H is a non-trivial support hyperplane to A passing through $\mathbf{a}$. □

We illustrate the above ideas by considering the support properties of a closed ball in space. Geometrically, it is clear that through each boundary point of such a ball there passes a unique support hyperplane, namely the tangent plane at that point; moreover, this hyperplane meets the ball in a single point. This result is generalized in the following example.

Example 2.4.13 Let H be a support hyperplane to a closed ball $B[\mathbf{a}; r]$ in $\mathbb{R}^n$ at some point $\mathbf{c}$. Since H does not meet the interior of the ball, every point of H must be a distance of at least r from $\mathbf{a}$, whence $\mathbf{c}$ must be a nearest point of H to $\mathbf{a}$. By the uniqueness of nearest points of convex sets (Theorem 2.4.1), H can meet the ball only in the point $\mathbf{c}$. The last assertion of Theorem 2.4.1 shows that $(\mathbf{h} - \mathbf{c}) \cdot (\mathbf{a} - \mathbf{c}) \leqslant 0$ for all $\mathbf{h}$ in H. We cannot have $(\mathbf{h} - \mathbf{c}) \cdot (\mathbf{a} - \mathbf{c}) < 0$ for some $\mathbf{h}$ in H, for this would imply that the point $\mathbf{h}' = 2\mathbf{c} - \mathbf{h}$ of H satisfies

$$(\mathbf{h}' - \mathbf{c}) \cdot (\mathbf{a} - \mathbf{c}) = (\mathbf{c} - \mathbf{h}) \cdot (\mathbf{a} - \mathbf{c}) > 0,$$

which is impossible. Thus H must be the hyperplane with equation $(\mathbf{x} - \mathbf{c}) \cdot (\mathbf{a} - \mathbf{c}) = 0$, and so H is the *unique* support hyperplane to $B[\mathbf{a}; r]$ at $\mathbf{c}$. □

We conclude this section by establishing a formula for the distance between a point $\mathbf{a}$ and a hyperplane H with equation $\mathbf{c} \cdot \mathbf{x} = c_0$ in $\mathbb{R}^n$. Denote by $\mathbf{a}_0$ the point defined by the equation

$$\mathbf{a}_0 = \mathbf{a} + \frac{c_0 - \mathbf{c} \cdot \mathbf{a}}{\|\mathbf{c}\|^2} \mathbf{c}.$$

Then $\mathbf{a}_0$ lies in H, and for any $\mathbf{x}$ in H, we have

$$\|\mathbf{a} - \mathbf{x}\|^2 = \|(\mathbf{a} - \mathbf{a}_0) + (\mathbf{a}_0 - \mathbf{x})\|^2 = \|\mathbf{a} - \mathbf{a}_0\|^2 + \|\mathbf{a}_0 - \mathbf{x}\|^2,$$

which shows that $\mathbf{a}_0$ is the unique nearest point of H to $\mathbf{a}$, and that the (shortest) distance between $\mathbf{a}$ and H is

$$\|\mathbf{a} - \mathbf{a}_0\| = \frac{|\mathbf{c} \cdot \mathbf{a} - c_0|}{\|\mathbf{c}\|}.$$

When $\mathbf{c}$ is a unit vector and $\mathbf{a}$ is the origin, this distance becomes $|c_0|$. Clearly, the (shortest) distance between parallel hyperplanes $\mathbf{c} \cdot \mathbf{x} = c_0$ and $\mathbf{c} \cdot \mathbf{x} = d_0$ is

$$\frac{|d_0 - c_0|}{\|\mathbf{c}\|},$$

which becomes $|d_0 - c_0|$ when $\mathbf{c}$ is a unit vector.

Exercises 2.4

1. The point $\mathbf{x}$ of $\mathbb{R}^n$ and the point $\mathbf{a}_0$ of a set A in $\mathbb{R}^n$ are such that $(\mathbf{x}-\mathbf{a}_0)\cdot(\mathbf{a}-\mathbf{a}_0)\leqslant 0$ for *all* $\mathbf{a}$ in A. Prove that $\mathbf{a}_0$ is the nearest point of A to $\mathbf{x}$. Deduce that, if $\|\mathbf{x}\|\geqslant 1$ and $A=\{\mathbf{a}\in\mathbb{R}^n : \|\mathbf{a}\|\leqslant 1\}$, then $\mathbf{x}/\|\mathbf{x}\|$ is the nearest point of A to $\mathbf{x}$.
2. Let f be the projection operator of a non-empty closed convex set A in $\mathbb{R}^n$. Show that if, for some $\mathbf{x},\mathbf{y}$ in $\mathbb{R}^n$, $\|f(\mathbf{x})-f(\mathbf{y})\| = \|\mathbf{x}-\mathbf{y}\|$, then $\mathbf{x}-f(\mathbf{x})=\mathbf{y}-f(\mathbf{y})$. What can be deduced if, for *all* $\mathbf{x},\mathbf{y}$ in $\mathbb{R}^n$, $\|f(\mathbf{x})-f(\mathbf{y})\|=\|\mathbf{x}-\mathbf{y}\|$?
3. Find (i) all closed convex sets in $\mathbb{R}^n$ whose complements are convex, and (ii) all convex sets in $\mathbb{R}^n$ which have no support hyperplanes.
4. Prove that every compact convex set in $\mathbb{R}^n$ can be expressed as an intersection of closed balls in $\mathbb{R}^n$.
5. Let A be a non-empty bounded set in $\mathbb{R}^n$ and let $\mathbf{u}$ be a non-zero vector in $\mathbb{R}^n$. Show that there is a scalar u_0 such that the hyperplane with equation $\mathbf{u}\cdot\mathbf{x}=u_0$ is a support hyperplane to A.
6. Let A and B be disjoint non-empty sets in $\mathbb{R}^n$ with A a flat and B a compact convex set. Prove that there exists a hyperplane in $\mathbb{R}^n$ which contains A and does not meet B.
7. Show that there is only one hyperplane in $\mathbb{R}^3$ which separates the disjoint closed convex sets A and B defined by the equations
$$A=\{(0,y,1):y\in\mathbb{R}\},\qquad B=\{(x,y,z):x,y,z\geqslant 0,\ xy\geqslant z^2\},$$
and that this hyperplane meets both A and B.

2.5 UNBOUNDED CONVEX SETS

We now establish the basic properties of unbounded convex sets, showing that they have a relatively simple structure and that their unboundedness can be usefully described in terms of *recession cones* and *lineality spaces*.

To guide our discussion and to help us discover what properties a closed unbounded convex set might possess, we consider a simple example in the plane $\mathbb{R}^2$. Let A be the closed unbounded convex set in $\mathbb{R}^2$ that is defined by the equation
$$A=\{(x,y):y\geqslant 1/x,\ x>0\},$$
and let $\mathbf{a}\in A$. See Fig. 2.7. Then there are halflines starting at $\mathbf{a}$ which are contained in A. If we denote by $A_{\mathbf{a}}$ the union of all these halflines, then $A_{\mathbf{a}}$ is a closed convex set that is a union of halflines starting at

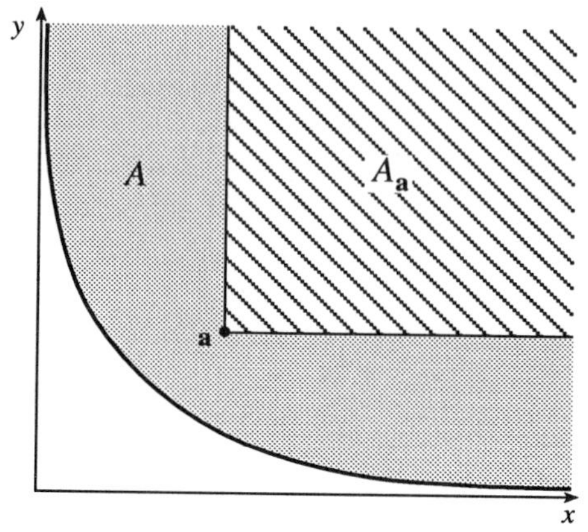

Fig. 2.7

$\mathbf{a}$. Such a set $A_{\mathbf{a}}$ is called a *closed convex cone with apex* $\mathbf{a}$. In fact, $A_{\mathbf{a}} = \mathbf{a} + P$, where P is the **non-negative quadrant** $\{(x, y) : x \geqslant 0, y \geqslant 0\}$ of $\mathbb{R}^2$. The important observation here is that P is determined by the set A alone, being independent of the initial choice of the point $\mathbf{a}$ in A. We refer to P as the *recession cone* of A. Roughly speaking, the recession cone of a convex set indicates in which directions the set *recedes to infinity*.

A **halfline** L^+ in $\mathbb{R}^n$ is a set of the form $\{\mathbf{x}_0 + \lambda\mathbf{y} : \lambda \geqslant 0\}$, where $\mathbf{x}_0, \mathbf{y} \in \mathbb{R}^n$ and $\mathbf{y} \neq \mathbf{0}$. The reason for this name is clear: the line joining the points $\mathbf{x}_0$ and $\mathbf{x}_0 + \mathbf{y}$ is the set

$$\{(1 - \lambda)\mathbf{x}_0 + \lambda(\mathbf{x}_0 + \mathbf{y}) : \lambda \in \mathbb{R}\} = \{\mathbf{x}_0 + \lambda\mathbf{y} : \lambda \in \mathbb{R}\},$$

and so L^+ can, with good reason, be called a halfline. A halfline L_0^+ of the form $\{\lambda\mathbf{y} : \lambda \geqslant 0\}$, where $\mathbf{y} \in \mathbb{R}^n$ and $\mathbf{y} \neq \mathbf{0}$, is called a **ray**. The equation $L^+ = \mathbf{x}_0 + L_0^+$ expresses L^+ as a translate of the ray L_0^+. Since $\mathbf{x}_0$ is the only point of L^+ whose removal from L^+ leaves a convex set, and $L_0^+ = L^+ - \mathbf{x}_0$, it follows that $\mathbf{x}_0$ and L_0^+ are uniquely determined by L^+. We refer to L^+ as the halfline with **direction L_0^+ and initial point** $\mathbf{x}_0$. The word **direction** will be used as a synonym for ray.

Theorem 2.5.1 *Let A be a closed unbounded convex set in $\mathbb{R}^n$. Then A contains a halfline. Moreover, if A contains some halfline with direction L_0^+, then it contains every halfline with direction L_0^+ whose initial point is in A.*

Proof Since A is unbounded, it contains a sequence $\mathbf{a}_1, \ldots, \mathbf{a}_k, \ldots$ of non-zero vectors such that $\|\mathbf{a}_k\| \to \infty$ as $k \to \infty$. Let $\lambda_k = 1/\|\mathbf{a}_k\|$. Then the sequence $\lambda_1\mathbf{a}_1, \ldots, \lambda_k\mathbf{a}_k, \ldots$ lies in the compact set $\{\mathbf{x} : \|\mathbf{x}\| = 1\}$, and so contains some subsequence converging to a point $\mathbf{a}$ with $\|\mathbf{a}\| = 1$.

We may suppose that the sequence itself converges to $\mathbf{a}$. Let L^+ be the direction $\{\lambda\mathbf{a}:\lambda \geqslant 0\}$. Then we show that $\mathbf{a}_0 + L^+ \subseteq A$ for every $\mathbf{a}_0$ in A.

Let $\mathbf{a}_0 \in A$ and let $\lambda \geqslant 0$. Since $\lambda_k \to 0$ as $k \to \infty$, we must have, for all but a finite number of k's,

$$0 \leqslant \lambda\lambda_k \leqslant 1 \qquad \text{and} \qquad (1 - \lambda\lambda_k)\mathbf{a}_0 + \lambda\lambda_k\mathbf{a}_k \in A.$$

Clearly,

$$(1 - \lambda\lambda_k)\mathbf{a}_0 + \lambda\lambda_k\mathbf{a}_k \to \mathbf{a}_0 + \lambda\mathbf{a} \qquad \text{as } k \to \infty,$$

and so $\mathbf{a}_0 + \lambda\mathbf{a} \in A$, since A is closed. Thus the halfline $\mathbf{a}_0 + L^+$ is contained in A.

Suppose next that A contains the halfline $\mathbf{b}_0 + L_0^+$, where L_0^+ is the direction $\{\lambda\mathbf{b}:\lambda \geqslant 0\}$ for some $\mathbf{b} \neq \mathbf{0}$. We show that every halfline $\mathbf{c}_0 + L_0^+$, where $\mathbf{c}_0 \in A$, is contained in A. Let $\mu \geqslant 0$. Then, for all $\lambda > \mu$,

$$\left(1 - \frac{\mu}{\lambda}\right)\mathbf{c}_0 + \frac{\mu}{\lambda}(\mathbf{b}_0 + \lambda\mathbf{b}) \in A.$$

Letting $\lambda \to \infty$ in this last relation and using the fact that A is closed, we deduce that $\mathbf{c}_0 + \mu\mathbf{b} \in A$. Thus $\mathbf{c}_0 + L_0^+ \subseteq A$, as required. □

Corollary 2.5.2 *Let A be an unbounded convex set in $\mathbb{R}^n$. Then* ri A *contains a halfline. Moreover, if* cl A *contains some halfline with direction L_0^+, then* ri A *contains every halfline with direction L_0^+ whose initial point is in* ri A.

Proof We apply the theorem to the closed unbounded convex set cl A. Suppose that cl A contains the halfline $\mathbf{b}_0 + L_0^+$, where L_0^+ is the direction $\{\lambda\mathbf{b}:\lambda \geqslant 0\}$. Let $\mathbf{a}_0 \in \operatorname{ri} A$ and let $\mu \geqslant 0$. Then $\mathbf{a}_0 + 2\mu\mathbf{b} \in \operatorname{cl} A$, and so $\mathbf{a}_0 + \mu\mathbf{b} \in \operatorname{ri} A$, by Theorem 2.3.4. Thus $\mathbf{a}_0 + L_0^+ \subseteq \operatorname{ri} A$, as required. □

Before continuing our discussion of unbounded convex sets, we must pause to introduce the idea of a *cone*. A non-empty set A in $\mathbb{R}^n$ is called a **cone** if $\lambda\mathbf{a} \in A$ whenever $\mathbf{a} \in A$ and $\lambda \geqslant 0$. Examples of cones are subspaces, rays, and the **non-negative orthant**

$$\{(x_1, \ldots, x_n):x_1 \geqslant 0, \ldots, x_n \geqslant 0\}$$

of $\mathbb{R}^n$. All cones contain the origin and are, with the exception of the trivial cone $\{\mathbf{0}\}$, unbounded. Cones need not be convex: the set $\{(x, y): xy \geqslant 0\}$ is a non-convex cone in $\mathbb{R}^2$. Our primary concern, however, will be with convex cones.

Theorem 2.5.3 *Let A be a non-empty set in $\mathbb{R}^n$. Then A is a convex cone if and only if $\lambda\mathbf{a} + \mu\mathbf{b} \in A$ whenever $\mathbf{a}, \mathbf{b} \in A$ and $\lambda, \mu \geqslant 0$.*

Proof Let A be a convex cone. Suppose that $\mathbf{a}, \mathbf{b} \in A$ and $\lambda, \mu \geqslant 0$. If $\lambda + \mu = 0$, then $\lambda = \mu = 0$ and trivially $\lambda\mathbf{a} + \mu\mathbf{b} \in A$. If $\lambda + \mu > 0$, then

$$\frac{\lambda}{\lambda + \mu}\mathbf{a} + \frac{\mu}{\lambda + \mu}\mathbf{b} \in A,$$

since A is convex, and

$$\lambda\mathbf{a} + \mu\mathbf{b} = (\lambda + \mu)\left(\frac{\lambda}{\lambda + \mu}\mathbf{a} + \frac{\mu}{\lambda + \mu}\mathbf{b}\right) \in A,$$

since A is a cone. Thus in all cases $\lambda\mathbf{a} + \mu\mathbf{b} \in A$.

Suppose next that $\lambda\mathbf{a} + \mu\mathbf{b} \in A$ whenever $\mathbf{a}, \mathbf{b} \in A$ and $\lambda, \mu \geqslant 0$. Clearly A is convex. To show that A is a cone, let $\mathbf{a} \in A$ and $\lambda \geqslant 0$. Then, by our hypothesis, $\lambda\mathbf{a} = \lambda\mathbf{a} + 0\mathbf{a} \in A$. □

Corollary 2.5.4 *Let A be a non-empty set in $\mathbb{R}^n$. Then A is a convex cone if and only if $\mathbf{a} + \mathbf{b} \in A$ and $\lambda\mathbf{a} \in A$ whenever $\mathbf{a}, \mathbf{b} \in A$ and $\lambda \geqslant 0$.* □

It is a routine matter to show that the intersection of any family of convex cones in $\mathbb{R}^n$ is a convex cone. Hence cone A, defined as the intersection of all convex cones containing a set A in $\mathbb{R}^n$, is a convex cone; it is called the **convex cone generated by** A. Clearly cone A is the *smallest* convex cone containing A. We note that cone $\varnothing = \{\mathbf{0}\}$. We now characterize cone A, in the case when A is non-empty, as the set of all *non-negative linear combinations* of points of A, i.e. points of the form $\lambda_1\mathbf{a}_1 + \ldots + \lambda_m\mathbf{a}_m$, where $\mathbf{a}_1, \ldots, \mathbf{a}_m \in A$ and $\lambda_1, \ldots, \lambda_m \geqslant 0$.

Theorem 2.5.5 *Let A be a non-empty set in $\mathbb{R}^n$. Then cone A is the set of all non-negative linear combinations of points of A.*

Proof Denote by B the set of all non-negative linear combinations of points of A. Let $\mathbf{x} \in B$. Then

$$\mathbf{x} = \lambda_1\mathbf{a}_1 + \ldots + \lambda_m\mathbf{a}_m \quad \text{for some } \mathbf{a}_1, \ldots, \mathbf{a}_m \in A \text{ and } \lambda_1, \ldots, \lambda_m \geqslant 0;$$

and $\mathbf{x} \in \operatorname{cone} A$ by repeated use of Corollary 2.5.4 to cone A, whence $B \subseteq \operatorname{cone} A$. Corollary 2.5.4 shows that B is a convex cone. Clearly $A \subseteq B$, and so cone $A \subseteq B$. Thus $B = \operatorname{cone} A$. □

A non-empty convex set A in $\mathbb{R}^n$ is said to **recede in a direction** L_0^+, or to have a **direction of recession** L_0^+, if every halfline with initial point in A and direction L_0^+ lies in A, i.e. if $A + L_0^+ \subseteq A$. The union of all directions of recession of A, together with the zero vector, is called the **recession cone** of A. Theorem 2.5.1 shows that a non-empty closed convex set in $\mathbb{R}^n$ is bounded if and only if its recession cone consists of the zero vector alone. The recession cone of a non-empty flat is the unique subspace which is parallel to it. The set

$$\{(x, y): x > 0, y > 0\} \cup \{(0, 0)\}$$

is its own recession cone, and is an example of a set whose recession cone is not closed.

Theorem 2.5.6 *Let A be a non-empty convex set in $\mathbb{R}^n$. Then the recession cone of A consists of all those points $\mathbf{x}$ such that $A + \mathbf{x} \subseteq A$. Moreover, the recession cone of A is a convex cone, which is closed when A is closed.*

Proof If $\mathbf{x}$ belongs to the recession cone of A, then trivially $A + \mathbf{x} \subseteq A$. Conversely, if $A + \mathbf{x} \subseteq A$, then

$$A + 2\mathbf{x} = (A + \mathbf{x}) + \mathbf{x} \subseteq A + \mathbf{x} \subseteq A,$$

and, by repeated application of this argument, $A + m\mathbf{x} \subseteq A$ for each positive integer m. But A is convex, and so $A + \lambda\mathbf{x} \subseteq A$ for all $\lambda \geqslant 0$. Hence $\mathbf{x}$ lies in the recession cone of A.

Denote by C the recession cone of A. Let $\mathbf{x}, \mathbf{y} \in C$ and $\lambda, \mu \geqslant 0$. Then

$$A + \lambda\mathbf{x} + \mu\mathbf{y} = (A + \lambda\mathbf{x}) + \mu\mathbf{y} \subseteq A + \mu\mathbf{y} \subseteq A,$$

and so $\lambda\mathbf{x} + \mu\mathbf{y} \in C$, whence C is a convex cone by Theorem 2.5.3.

Suppose now that A is closed. Let $\mathbf{x}_1, \ldots, \mathbf{x}_k, \ldots$ be a sequence of points of the recession cone C that converges to some point $\mathbf{x}$ of $\mathbb{R}^n$. Then $\mathbf{a} + \mathbf{x}_k \in A$ for each k and for each point $\mathbf{a}$ of A. But A is closed, so $\mathbf{a} + \mathbf{x} \in A$ for each point $\mathbf{a}$ of A, i.e. $A + \mathbf{x} \subseteq A$. Thus $\mathbf{x} \in C$, which shows that C is closed. □

Let A be a non-empty convex set in $\mathbb{R}^n$. We saw above how, by considering the halflines contained in A, we could associate with A a convex cone, namely its recession cone. We shall now show how, by considering the lines contained in A, we can associate with A a subspace, its so-called *lineality space*.

Let L be a line in $\mathbb{R}^n$. Then by the **direction** of L is meant the *unique* line L_0 in $\mathbb{R}^n$, which is parallel to L and passes through the origin. A line is uniquely determined by specifying one of its points and giving its direction. Indeed, if $\mathbf{x}$ lies on a line L in $\mathbb{R}^n$, then the direction L_0 of L is simply the line $L - \mathbf{x}$, and $L = \mathbf{x} + L_0$.

Let L_0 be a line in $\mathbb{R}^n$ that passes through the origin. Then a non-empty convex set A in $\mathbb{R}^n$ is said to be **linear in the direction** L_0, or to have a **direction of linearity** L_0, if every line meeting A which has direction L_0 lies in A, i.e. if $A + L_0 \subseteq A$. The union of all the directions of linearity of A, together with the zero vector, is called the

lineality space of A. Theorem 2.5.1 shows that, if a closed convex set A contains a line with direction L_0, then it contains every line with direction L_0 which meets A, i.e. L_0 is a direction of linearity of A. Thus a non-empty closed convex set contains a line if and only if its lineality space does not consist of the zero vector alone. The lineality space of a non-empty flat is the unique subspace which is parallel to it. The lineality space of the unbounded circular cylinder

$$\{(x, y, z): x^2 + y^2 \leqslant 1\}$$

is the subspace $\{(0, 0, z): z \in \mathbb{R}\}$, i.e. the z-axis.

Theorem 2.5.7 *Let A be a non-empty convex set in $\mathbb{R}^n$. Then the lineality space of A consists of all those points $\mathbf{x}$ of $\mathbb{R}^n$ such that $A + \mathbf{x} = A$, and is a subspace of $\mathbb{R}^n$.*

Proof If $\mathbf{x}$ belongs to the lineality space of A, then trivially $A + \mathbf{x} \subseteq A$ and $A - \mathbf{x} \subseteq A$, whence $A + \mathbf{x} = A$. Conversely, if $A + \mathbf{x} = A$, then, as in the proof of Theorem 2.5.6, $A + m\mathbf{x} = A$ for each positive integer m. If m is a negative integer, then

$$A + m\mathbf{x} = (A + (-m)\mathbf{x}) + m\mathbf{x} = A.$$

Hence $A + m\mathbf{x} = A$ for all integers m. But A is convex, so $A + \lambda\mathbf{x} = A$ for all real λ. Thus $\mathbf{x}$ belongs to the lineality space of A.

Let S be the lineality space of A. Let $\mathbf{x}, \mathbf{y} \in S$ and $\lambda, \mu \in \mathbb{R}$. Then trivially $A + \lambda\mathbf{x} = A$ and $A + \mu\mathbf{y} = A$. Thus

$$A + \lambda\mathbf{x} + \mu\mathbf{y} = (A + \lambda\mathbf{x}) + \mu\mathbf{y} = A + \mu\mathbf{y} = A,$$

and so $\lambda\mathbf{x} + \mu\mathbf{y} \in S$, which shows that S is a subspace of $\mathbb{R}^n$. □

Our next theorem shows that each non-empty closed convex set in $\mathbb{R}^n$ can be expressed as the vector sum of a subspace and a convex set containing no lines, a result which will be needed in the proof of the *finite basis theorem* in Section 4.1.

Theorem 2.5.8 *Let A be a non-empty closed convex set in $\mathbb{R}^n$ with lineality space S. Then*

$$A = S + (A \cap S^{\perp}),$$

and the convex set $A \cap S^{\perp}$ contains no lines.

Proof Let $\mathbf{a} \in A$. Then $\mathbf{a}$ can be expressed *uniquely* in the form $\mathbf{a} = \mathbf{b} + \mathbf{c}$, where $\mathbf{b} \in S$ and $\mathbf{c} \in S^{\perp}$. Since $\mathbf{b} \in S$, $-\mathbf{b} \in S$, whence, by Theorem 2.5.7, $\mathbf{c} = \mathbf{a} - \mathbf{b} \in A$. Thus $\mathbf{c} \in A \cap S^{\perp}$, and so $A \subseteq S + (A \cap S^{\perp})$. The opposite inclusion follows immediately from Theorem 2.5.7. Thus $A = S + (A \cap S^{\perp})$.

Suppose that $A \cap S^{\perp}$ does contain a line. Then there exist $\mathbf{x}, \mathbf{y}$ in $\mathbb{R}^n$ with $\mathbf{y} \neq \mathbf{0}$ such that $\mathbf{x} + \lambda\mathbf{y} \in A \cap S^{\perp}$ for all real λ. Theorem 2.5.1 shows that $\mathbf{y} \in S$. Hence, for all real λ,

$$(\mathbf{x} + \lambda\mathbf{y}) \cdot \mathbf{y} = \mathbf{x} \cdot \mathbf{y} + \lambda\|\mathbf{y}\|^2 = 0,$$

which is clearly impossible. Thus $A \cap S^{\perp}$ contains no lines. □

Exercises 2.5

1. Give an example of an unbounded convex set whose recession cone consists only of the zero vector.
2. Let A be a non-empty convex set in $\mathbb{R}^n$, $\mathbf{a} \in \mathbb{R}^n$, and $\lambda > 0$. How are the recession cones of: (i) $\operatorname{ri} A$ and $\operatorname{cl} A$; (ii) A and $\lambda A + \mathbf{a}$ related?
3. Find the recession cones of the following sets:
   (i) $\{(x, y) : y \geqslant x^4\}$;
   (ii) $\{(x, y) : x \geqslant 1, y \geqslant 1\}$;
   (iii) $\{(x, y) : y > \mathrm{e}^x\}$.
4. Prove that the recession cone of a non-empty flat is the unique subspace which is parallel to it. Give an example of a closed unbounded convex set, which is not a flat, but whose recession cone is a subspace.
5. Give a description of all closed convex sets in $\mathbb{R}^2$ which contain a line.
6. The non-empty convex set A in $\mathbb{R}^n$ has recession cone C and lineality space S. Show that $S = C \cap (-C)$.

2.6 FACIAL STRUCTURE

The reader will no doubt be familiar with the term *face*, as it applies to convex polyhedra such as cubes, triangular prisms, square pyramids, and so on. In the present section we introduce the concept of *face* for arbitrary convex sets, and then investigate the facial structure of such sets. The results that we establish here will play a central role in our discussion of convex polytopes in the next chapter.

Each face F of a three-dimensional convex polyhedron is a convex subset of P with the property that whenever the relative interior of a line segment L lying in P meets F, then the endpoints of L lie in F. This observation motivates the definition of a face of a general convex set. A **face** of a convex set A in $\mathbb{R}^n$ is a convex subset B of A such that whenever $\lambda\mathbf{x} + \mu\mathbf{y} \in B$, where $\mathbf{x}, \mathbf{y} \in A$ and $\lambda, \mu > 0$ with $\lambda + \mu = 1$, then $\mathbf{x}, \mathbf{y} \in B$. Every convex set A in $\mathbb{R}^n$ has the faces $\varnothing$ and A; these are

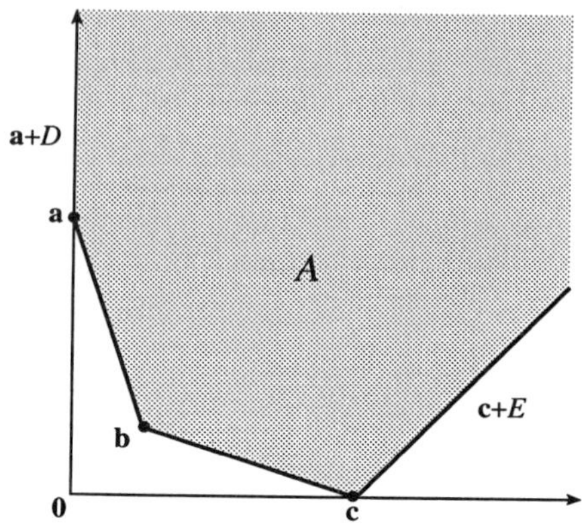

Fig. 2.8

called **improper faces** of A. Faces of A other than $\varnothing$ and A are called **proper faces** of A. The above definition of a face is more comprehensive than the one usually understood in elementary geometry. For example, it ascribes to a cube not only *six* two-dimensional faces, but also *one* three-dimensional face (itself), *twelve* one-dimensional faces (its edges), *eight* zero-dimensional faces (its vertices), and *one* face of dimension -1 (the empty set). In general, we refer to a k-dimensional face of a convex set as a **k-face**.

Certain faces of a convex set are of particular importance and are given special names. The 0-faces of a convex set are called its **extreme points**, and its faces that are halflines are called its **extreme halflines**. The directions of the extreme halflines of a convex set are called its **extreme directions**. Clearly, a point $\mathbf{a}$ of a convex set A in $\mathbb{R}^n$ is an extreme point of A if and only if whenever $\mathbf{a} = \lambda\mathbf{x} + \mu\mathbf{y}$, where $\mathbf{x}, \mathbf{y} \in A$ and $\lambda, \mu > 0$ with $\lambda + \mu = 1$, then $\mathbf{x} = \mathbf{y} = \mathbf{a}$.

Example 2.6.1 Denote by A the set of those points (x, y) in $\mathbb{R}^2$ which satisfy the four inequalities $x \geqslant 0$, $-x + y + 1 \geqslant 0$, $x + 3y - 1 \geqslant 0$, and $3x + y - 1 \geqslant 0$. See Fig. 2.8. The extreme points of A are the points $\mathbf{a} = (0, 1)$, $\mathbf{b} = (\frac{1}{4}, \frac{1}{4})$, and $\mathbf{c} = (1, 0)$, the extreme directions of A are the directions $D = \{(0, \lambda) : \lambda \geqslant 0\}$ and $E = \{(\lambda, \lambda) : \lambda \geqslant 0\}$, and the extreme halflines of A are the halflines $\mathbf{a} + D$ and $\mathbf{c} + E$. □

Some general properties of the faces of a convex set can be discovered by examining simple specific examples. For instance, consider the case of a 2-face B of a cube A in $\mathbb{R}^3$. It is readily seen that the set $A \setminus B$, i.e. the cube A with its face B removed, is convex. Also B is the intersection of its affine hull aff B, i.e. the plane containing B, with the cube A itself. These two results generalize to the situation when B is *any* face of *any* convex set A in $\mathbb{R}^n$, and can be used to obtain a useful characterization of the faces of a convex set, as the next theorem shows.

Theorem 2.6.2 *Let B be a convex subset of a convex set A in $\mathbb{R}^n$. Then B is a face of A if and only if $A \setminus B$ is convex and $B = (\operatorname{aff} B) \cap A$. In particular, a point $\mathbf{a}$ of A is an extreme point of A if and only if $A \setminus \{\mathbf{a}\}$ is convex.*

Proof Suppose that B is a face of A. If $\mathbf{x}, \mathbf{y} \in A \setminus B$ and $\lambda, \mu > 0$ with $\lambda + \mu = 1$, then $\lambda\mathbf{x} + \mu\mathbf{y} \in A$, since $\mathbf{x}, \mathbf{y} \in A$ and A is convex. We cannot have $\lambda\mathbf{x} + \mu\mathbf{y} \in B$, for this would imply that $\mathbf{x}, \mathbf{y} \in B$. Thus $\lambda\mathbf{x} + \mu\mathbf{y} \in A \setminus B$ and $A \setminus B$ is convex. Trivially, $B \subseteq (\operatorname{aff} B) \cap A$. We now establish the opposite inclusion. Suppose that $\mathbf{u} \in (\operatorname{aff} B) \cap A$. Let $\mathbf{b} \in \operatorname{ri} B$. Then there exist $\mathbf{v} \in B$ and $\alpha, \beta > 0$ with $\alpha + \beta = 1$ such that $\mathbf{b} = \alpha\mathbf{u} + \beta\mathbf{v}$. Since $\mathbf{u}, \mathbf{v} \in A$ and B is a face of A, $\mathbf{u} \in B$. Hence $(\operatorname{aff} B) \cap A \subseteq B$, and $B = (\operatorname{aff} B) \cap A$ as desired.

Suppose next that $A \setminus B$ is convex and $B = (\operatorname{aff} B) \cap A$. If $\lambda\mathbf{x} + \mu\mathbf{y} \in B$, where $\mathbf{x}, \mathbf{y} \in A$ and $\lambda, \mu > 0$ with $\lambda + \mu = 1$, then not *both* $\mathbf{x}$ and $\mathbf{y}$ can lie in $A \setminus B$, for the convexity of $A \setminus B$ would imply that $\lambda\mathbf{x} + \mu\mathbf{y} \in A \setminus B$. Suppose that $\mathbf{x} \notin A \setminus B$. Then $\mathbf{x} \in B$, and so $\mathbf{y} \in \operatorname{aff} B$. Thus $\mathbf{y} \in (\operatorname{aff} B) \cap A$, whence $\mathbf{y} \in B$. Hence $\mathbf{x}, \mathbf{y} \in B$ and B is a face of A.

The final assertion of the theorem follows from what we have just proved and the fact that a singleton set is its own affine hull. □

Corollary 2.6.3 *Each face of a closed convex set in $\mathbb{R}^n$ is closed.*

Proof Let B be a face of a closed convex set A in $\mathbb{R}^n$. Then B, being the intersection of the closed sets $\operatorname{aff} B$ and A, is itself closed. □

Corollary 2.6.4 *Let $A = \operatorname{conv} C$, where C is a set in $\mathbb{R}^n$. Then each extreme point of A lies in C.*

Proof Suppose that there is an extreme point $\mathbf{a}$ of A which does not lie in C. Then $A \setminus \{\mathbf{a}\}$ is a *proper* convex subset of A containing C, and hence $\operatorname{conv} C$, i.e. A. This contradiction shows that each extreme point of A lies in C. □

Theorem 2.6.5 *Let A be a convex set in $\mathbb{R}^n$. Then:*

(i) *the intersection of any non-empty family of faces of A is a face of A;*

(ii) *if B is a face of A, and C is a face of B, then C is a face of A;*

(iii) *the intersection of A with each of its support hyperplanes is a face of A.*

Proof (i) Let $(A_i : i \in I)$ be a non-empty family of faces of A. Then $\bigcap (A_i : i \in I)$ is a convex subset of A. If $\lambda\mathbf{x} + \mu\mathbf{y} \in \bigcap (A_i : i \in I)$, where $\mathbf{x}, \mathbf{y} \in A$ and $\lambda, \mu > 0$ with $\lambda + \mu = 1$, then $\lambda\mathbf{x} + \mu\mathbf{y} \in A_i$ for all $i \in I$, and so, since each A_i is a face of A, we have $\mathbf{x}, \mathbf{y} \in A_i$ for all $i \in I$. Hence $\mathbf{x}, \mathbf{y} \in \bigcap (A_i : i \in I)$ and $\bigcap (A_i : i \in I)$ is a face of A.

(ii) Let B be a face of A and let C be a face of B. If $\lambda\mathbf{x} + \mu\mathbf{y} \in C$, where $\mathbf{x}, \mathbf{y} \in A$ and $\lambda, \mu > 0$ with $\lambda + \mu = 1$, then $\lambda\mathbf{x} + \mu\mathbf{y} \in B$, since $C \subseteq B$. Since B is a face of A, $\mathbf{x}, \mathbf{y} \in B$. But C is a face of B, so $\mathbf{x}, \mathbf{y} \in C$. This proves that C is a face of A.

(iii) Let $H = \{z \in \mathbb{R}^n : \mathbf{c} \cdot \mathbf{z} = c_0\}$, where $c_0 \in \mathbb{R}$, $\mathbf{c} \in \mathbb{R}^n$ and $\mathbf{c} \neq \mathbf{0}$, be a support hyperplane to A. Suppose that $\mathbf{c} \cdot \mathbf{a} \leqslant c_0$ whenever $\mathbf{a} \in A$. If $\lambda\mathbf{x} + \mu\mathbf{y} \in A \cap H$, where $\mathbf{x}, \mathbf{y} \in A$ and $\lambda, \mu > 0$ with $\lambda + \mu = 1$, then

$$c_0 = \mathbf{c} \cdot (\lambda\mathbf{x} + \mu\mathbf{y}) = \lambda\mathbf{c} \cdot \mathbf{x} + \mu\mathbf{c} \cdot \mathbf{y} \leqslant \lambda c_0 + \mu c_0 = c_0.$$

Since $\lambda, \mu > 0$, we must have $\mathbf{c} \cdot \mathbf{x} = \mathbf{c} \cdot \mathbf{y} = c_0$, whence $\mathbf{x}, \mathbf{y} \in A \cap H$. Thus $A \cap H$ is a face of A, for clearly it is a convex subset of A. □

We now show that the defining property of a face of a convex set implies a stronger property involving arbitrary subsets of the set, not just line segments.

Theorem 2.6.6 *Let B be a face of a convex set A in $\mathbb{R}^n$. Suppose that C is a subset of A such that* ri C *meets B. Then $C \subseteq B$.*

Proof Let $\mathbf{c} \in C$ and $\mathbf{b} \in B \cap \operatorname{ri} C$. Then there exist $\mathbf{d} \in C$ and $\lambda, \mu > 0$ with $\lambda + \mu = 1$ such that $\mathbf{b} = \lambda\mathbf{c} + \mu\mathbf{d}$. Since $\mathbf{c}, \mathbf{d} \in A$ and B is a face of A, we see that $\mathbf{c} \in B$. Thus $C \subseteq B$. □

Corollary 2.6.7 *Let B and C be faces of a convex set A in $\mathbb{R}^n$ such that* ri B *and* ri C *meet. Then $B = C$.*

Proof Since ri C meets B, we have $C \subseteq B$. Similarly, $B \subseteq C$. Thus $B = C$. □

Corollary 2.6.8 *Let B be a face of a convex set A in $\mathbb{R}^n$, other than A itself. Then* $\dim B < \dim A$.

Proof We suppose that B is non-empty. Clearly, $\operatorname{aff} B \subseteq \operatorname{aff} A$ and $\dim B \leqslant \dim A$. If $\dim B = \dim A$, then $\operatorname{aff} B = \operatorname{aff} A$ and $\varnothing \subset \operatorname{ri} B \subseteq \operatorname{ri} A$, whence $A = B$ by Corollary 2.6.7. This contradiction shows that $\dim B < \dim A$. □

Corollary 2.6.9 *The intersection of any family of faces of a convex set in $\mathbb{R}^n$ can be expressed as an intersection of $n + 1$ or fewer members of the family.*

Proof Suppose that the result is false. Then there exist faces $A_1, \ldots, A_{n+2}$ of some convex set A in $\mathbb{R}^n$ such that

$$A_1 \subset A_2 \subset \ldots \subset A_{n+2} \subset A.$$

Since A_i is a face of A_{i+1} for $i = 1, \ldots, n+1$, Corollary 2.6.8 shows that

$$-1 \leqslant \dim A_1 < \dim A_2 < \ldots < \dim A_{n+2} \leqslant n - 1,$$

which is impossible. This establishes the corollary. □

Each point of a convex set belongs to at least one face of the set, namely the set itself, and in general belongs to several different faces. For example, a vertex of a three-dimensional cube belongs to one 0-face, three 1-faces, three 2-faces, and one 3-face of the cube. Suppose that $\mathbf{a}$ is a point of a convex set A in $\mathbb{R}^n$ and that $F_{\mathbf{a}}$ is the intersection of all faces of A containing $\mathbf{a}$. Then it follows from Theorem 2.6.5 (i) that $F_{\mathbf{a}}$ is the *smallest* face of A containing $\mathbf{a}$. We now characterize $F_{\mathbf{a}}$ as the *unique* face of A which contains $\mathbf{a}$ in its relative interior.

Theorem 2.6.10 *Let* $\mathbf{a}$ *be a point of a convex set* A *in* $\mathbb{R}^n$ *and let* $F_{\mathbf{a}}$ *be the intersection of all faces of* A *containing* $\mathbf{a}$. *Then* $\mathbf{a} \in \operatorname{ri} F_{\mathbf{a}}$ *and the relative interiors of the faces of* A *form a partition of* A.

Proof If $\mathbf{a} \notin \operatorname{ri} F_{\mathbf{a}}$, then $\mathbf{a} \in \operatorname{rebd} F_{\mathbf{a}}$ and so, by Theorem 2.4.2, there exists a support hyperplane H of $F_{\mathbf{a}}$ passing through $\mathbf{a}$ but not containing $F_{\mathbf{a}}$, whence, by Theorem 2.6.5 (iii), $H \cap F_{\mathbf{a}}$ is a face of A containing $\mathbf{a}$ which is *strictly* contained in $F_{\mathbf{a}}$. Since this is impossible, $\mathbf{a} \in \operatorname{ri} F_{\mathbf{a}}$. Thus each point $\mathbf{a}$ of A belongs to the relative interior of the face $F_{\mathbf{a}}$ of A. The relative interiors of two different faces of A are disjoint by Corollary 2.6.7. Hence the relative interiors of the faces of A form a partition of A. □

Corollary 2.6.11 *Let* $\mathbf{a}$ *be a relative boundary point of a convex set* A *in* $\mathbb{R}^n$. *Then* $\dim F_{\mathbf{a}} < \dim A$.

Proof The faces A and $F_{\mathbf{a}}$ of A cannot be equal because $\mathbf{a} \in \operatorname{ri} F_{\mathbf{a}}$. Thus $F_{\mathbf{a}} \subset A$, whence $\dim F_{\mathbf{a}} < \dim A$ by Corollary 2.6.8. □

We now consider the conditions under which a closed convex set is the convex hull of its relative boundary. A closed convex set in $\mathbb{R}^n$ which is *not* the convex hull of its relative boundary is said be **primitive.** The reader should have little difficulty in discovering that the only primitive sets in $\mathbb{R}^2$ are points, lines, halflines, closed halfplanes, and $\mathbb{R}^2$ itself. Before we can extend this last result to $\mathbb{R}^n$, we need to generalize the concepts of halflines and halfplanes from $\mathbb{R}^2$ to $\mathbb{R}^n$.

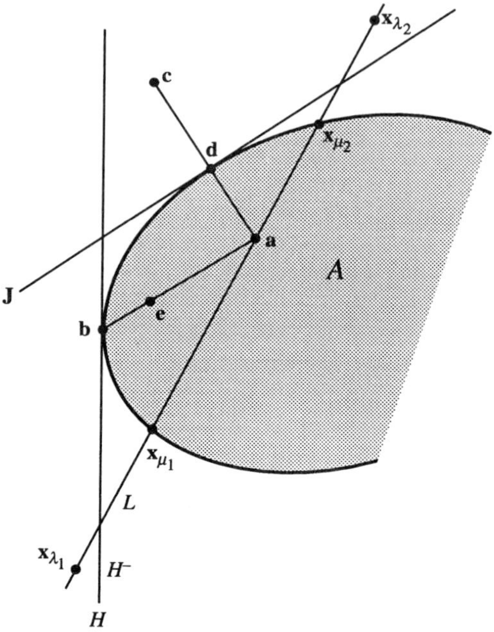

Fig. 2.9

In $\mathbb{R}^n$ a **closed halfflat** is the intersection of a flat with a closed halfspace which meets it, but does not contain it.

Theorem 2.6.12 *A closed convex set in $\mathbb{R}^n$ is primitive if and only if it is either a non-empty flat or a closed halfflat.*

Proof We establish only the non-trivial part of the theorem, that is, for each closed convex set A in $\mathbb{R}^n$ other than a flat or a closed halfflat, $A = \text{conv}(\text{rebd}\,A)$. Since $\text{conv}(\text{rebd}\,A) \subseteq A$, $\text{rebd}\,A \subseteq \text{conv}(\text{rebd}\,A)$ and $A = (\text{rebd}\,A) \cup (\text{ri}\,A)$, we need only show that $\text{ri}\,A \subseteq \text{conv}(\text{rebd}\,A)$. This we now do.

Let $\mathbf{a} \in \text{ri}\,A$. Since A is not a flat, its relative boundary is not empty, say $\mathbf{b} \in \text{rebd}\,A$. Theorem 2.4.12 shows that there is a non-trivial support hyperplane H to A at $\mathbf{b}$. See Fig. 2.9. Since $A \cap H$ is a *proper* face of A, Theorem 2.6.6 shows that $\mathbf{a} \notin H$. Thus there exist $u_0 \in \mathbb{R}$, $\mathbf{u} \in \mathbb{R}^n$ such that H has equation $\mathbf{u} \cdot \mathbf{x} = u_0$ and $\mathbf{u} \cdot \mathbf{a} < u_0$, $\mathbf{u} \cdot \mathbf{b} = u_0$ with A lying in the closed halfspace

$$H^- = \{\mathbf{x} \in \mathbb{R}^n : \mathbf{u} \cdot \mathbf{x} \leqslant u_0\}.$$

By the hypothesis, A is not a closed halfflat, and so cannot be $(\text{aff}\,A) \cap H^-$. Thus there exists a point $\mathbf{c}$ of $(\text{aff}\,A) \cap H^-$ that does not lie in A. Denote by $\mathbf{d}$ the point where the line segment joining $\mathbf{a}$ and $\mathbf{c}$ meets $\operatorname{rebd} A$. Since $\mathbf{u}\cdot\mathbf{a} < u_0$ and $\mathbf{u}\cdot\mathbf{c} \leqslant u_0$, it follows that $\mathbf{u}\cdot\mathbf{d} < u_0$. The existence of a non-trivial support hyperplane to A at $\mathbf{d}$, J say, shows that there exist $v_0 \in \mathbb{R}$, $\mathbf{v} \in \mathbb{R}^n$ such that J has equation $\mathbf{v}\cdot\mathbf{x} = v_0$ and $\mathbf{v}\cdot\mathbf{a} < v_0$, $\mathbf{v}\cdot\mathbf{d} = v_0$ with A lying in the closed halfspace $\{\mathbf{x} \in \mathbb{R}^n : \mathbf{v}\cdot\mathbf{x} \leqslant v_0\}$.

Since $\mathbf{u}\cdot\mathbf{a} < u_0$, $\mathbf{u}\cdot\mathbf{b} = u_0$, and $\mathbf{u}\cdot\mathbf{d} < u_0$, there is a point $\mathbf{e}$ of $\operatorname{ri} A$ lying on the line segment joining $\mathbf{a}$ and $\mathbf{b}$ such that $\mathbf{u}\cdot\mathbf{e} > \mathbf{u}\cdot\mathbf{d}$. Because $\mathbf{e}$ lies in $\operatorname{ri} A$, we must have $\mathbf{v}\cdot\mathbf{e} < v_0$. For each scalar λ, denote by $\mathbf{x}_\lambda$ the point $\mathbf{a} + \lambda(\mathbf{d} - \mathbf{e})$ on the line L joining the points $\mathbf{a}$ and $\mathbf{a} + \mathbf{d} - \mathbf{e}$ of $\operatorname{aff} A$. Choose scalars λ_1, λ_2 such that

$$\lambda_1 < (u_0 - \mathbf{u}\cdot\mathbf{a})/(\mathbf{u}\cdot\mathbf{d} - \mathbf{u}\cdot\mathbf{e}) < 0 < (v_0 - \mathbf{v}\cdot\mathbf{a})/(\mathbf{v}\cdot\mathbf{d} - \mathbf{v}\cdot\mathbf{e}) < \lambda_2.$$

Then $\mathbf{u}\cdot\mathbf{x}_{\lambda_1} > u_0$ and $\mathbf{v}\cdot\mathbf{x}_{\lambda_2} > v_0$, whence neither $\mathbf{x}_{\lambda_1}$ nor $\mathbf{x}_{\lambda_2}$ lies in A. Thus there are scalars μ_1, μ_2 with $\lambda_1 < \mu_1 < 0 < \mu_2 < \lambda_2$ such that $\mathbf{x}_{\mu_1}$, $\mathbf{x}_{\mu_2} \in \operatorname{rebd} A$. Hence $\mathbf{a} \in \operatorname{conv}\{\mathbf{x}_{\mu_1}, \mathbf{x}_{\mu_2}\}$, and so $\operatorname{ri} A \subseteq \operatorname{conv}(\operatorname{rebd} A)$ as desired. □

We are now able to prove the fundamental result on the facial structure of closed convex sets in $\mathbb{R}^n$, deducing from it some equally important corollaries. By the *convex hull* of a family of sets in $\mathbb{R}^n$ is meant the convex hull of its union.

Theorem 2.6.13 *Every closed convex set in $\mathbb{R}^n$ is the convex hull of its primitive faces.*

Proof Let A be a closed convex set in $\mathbb{R}^n$. We argue by induction on the dimension of A. The case $\dim A = -1$ is trivial. Suppose that $\dim A = m$, where $m > -1$, and that the assertion is true for all closed convex sets in $\mathbb{R}^n$ with dimension less than m. The theorem is trivial when A is primitive. Suppose, then, that A is not primitive. Denote by B the convex hull of the primitive faces of A. Then $B \subseteq A$, so we need only show that $A \subseteq B$. Since A is not primitive, we have $A = \operatorname{conv}(\operatorname{rebd} A)$. Let $\mathbf{a} \in \operatorname{rebd} A$. Then $\mathbf{a}$ lies in $F_{\mathbf{a}}$, the smallest face of A containing $\mathbf{a}$. By Corollary 2.6.11, $\dim F_{\mathbf{a}} < \dim A$. The induction hypothesis shows that $F_{\mathbf{a}}$ is the convex hull of its primitive faces. Since each primitive face of $F_{\mathbf{a}}$ is a primitive face of A, $F_{\mathbf{a}} \subseteq B$. Hence $\mathbf{a} \in F_{\mathbf{a}} \subseteq B$ and $\operatorname{rebd} A \subseteq B$, so $A = \operatorname{conv}(\operatorname{rebd} A) \subseteq B$. Thus $A = B$, i.e. A is the convex hull of its primitive faces. This completes the proof by induction. □

Corollary 2.6.14 *Every closed convex set in $\mathbb{R}^n$ is the convex hull of those of its faces which are flats or closed halfflats.*

Proof The result follows from the theorem and Theorem 2.6.12. □

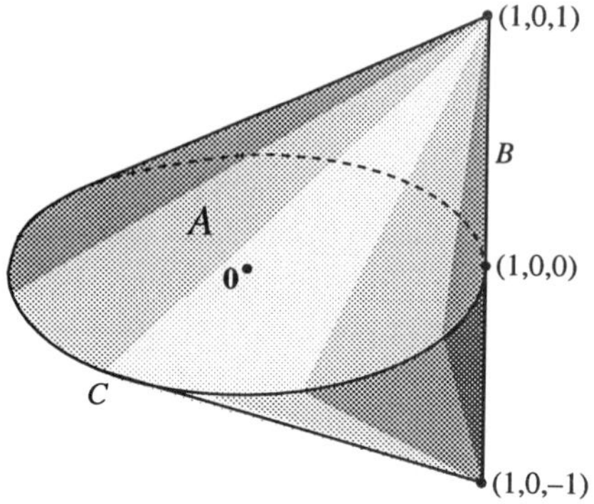

Fig. 2.10

Corollary 2.6.15 *Every closed convex set in* $\mathbb{R}^n$ *that contains no lines is the convex hull of its extreme points and extreme halflines.*

Proof The corollary follows from the theorem and the fact that points and halflines are the only primitive sets which contain no lines. □

Theorem 2.6.16 (Krein–Milman) *Every compact convex set in* $\mathbb{R}^n$ *is the convex hull of its extreme points.*

Proof The theorem follows from Theorem 2.6.13 and the fact that points are the only compact primitive sets. □

The Krein–Milman theorem shows that the convex hull of the extreme points of a compact convex set in $\mathbb{R}^n$ is closed; it is not true, however, that the set of extreme points itself is necessarily closed. To see this, let A and B denote the circular disc and the line segment in $\mathbb{R}^3$ given by the equations

$$A = \{(x, y, 0) : x^2 + y^2 \leqslant 1\} \qquad \text{and} \qquad B = \{(1, 0, z) : -1 \leqslant z \leqslant 1\},$$

and let $C = \operatorname{conv}(A \cup B)$. See Fig. 2.10. Then C is a compact convex set. Its set of extreme points consists of $(1, 0, 1)$ and $(1, 0, -1)$ together with the points on the relative boundary of A with the exception of $(1, 0, 0)$, and this set is *not* closed.

Theorem 2.6.5 (iii) shows that the intersection of a convex set in $\mathbb{R}^n$ with one of its support hyperplanes is a face of the set; a face which arises in this way is called an **exposed face** of the set. It is technically convenient to allow the empty set and the set itself as exposed faces of any convex set in $\mathbb{R}^n$. Thus an exposed face of a convex set in $\mathbb{R}^n$ is either the empty set, the set itself, or the intersection of the set with one of its support hyperplanes. Faces of a convex set are not always

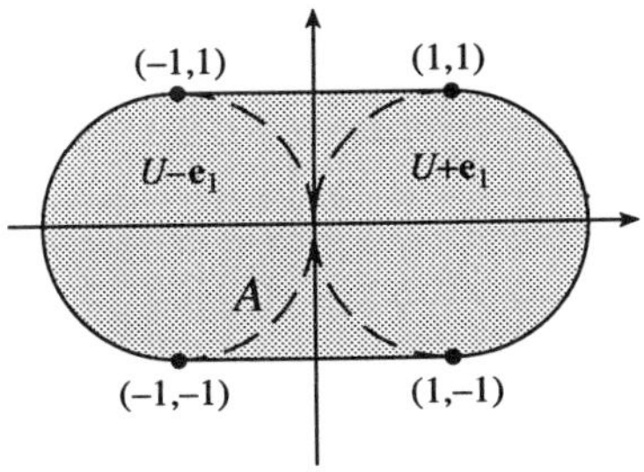

Fig. 2.11

exposed, as the following example shows. Let A be the convex hull of the union of the circular discs $U + \mathbf{e}_1$ and $U - \mathbf{e}_1$ in $\mathbb{R}^2$, where

$$U = \{(x, y) : x^2 + y^2 \leqslant 1\} \qquad \text{and} \qquad \mathbf{e}_1 = (1, 0).$$

See Fig. 2.11. Then the points $(1, 1)$, $(1, -1)$, $(-1, 1)$, $(-1, -1)$ are faces of A that are not exposed. This example also serves to show that an exposed face of an exposed face of a convex set need not be an exposed face of that convex set, for the line segment joining $(-1, 1)$ and $(1, 1)$ is an exposed face of A and $(1, 1)$ is an exposed face of this line segment, but $(1, 1)$ is *not* an exposed face of A.

Theorem 2.6.17 *The intersection of any non-empty family of exposed faces of a convex set in* $\mathbb{R}^n$ *is an exposed face of the set.*

Proof In view of Corollary 2.6.9, we may assume that the family of exposed faces is finite. Let $A_1, \ldots, A_m$ be exposed faces of a convex set A in $\mathbb{R}^n$. We show that $A_1 \cap \ldots \cap A_m$ is an exposed face of A, considering only the non-trivial case when the $A_1, \ldots, A_m$ are proper exposed faces of A, whose intersection is non-empty, containing some point $\mathbf{a}_0$ say.

For each $i = 1, \ldots, m$, there exists $\mathbf{u}_i \in \mathbb{R}^n$ such that

$$A_i = \{\mathbf{a} \in A : \mathbf{u}_i \cdot \mathbf{a} = \mathbf{u}_i \cdot \mathbf{a}_0\} \qquad \text{and} \qquad A \subseteq \{\mathbf{x} \in \mathbb{R}^n : \mathbf{u}_i \cdot \mathbf{x} \leqslant \mathbf{u}_i \cdot \mathbf{a}_0\}.$$

It follows easily that

$$A_1 \cap \ldots \cap A_m = \{\mathbf{a} \in A : (\mathbf{u}_1 + \ldots + \mathbf{u}_m) \cdot \mathbf{a} = (\mathbf{u}_1 + \ldots + \mathbf{u}_m) \cdot \mathbf{a}_0\},$$

and that

$$A \subseteq \{\mathbf{x} \in \mathbb{R}^n : (\mathbf{u}_1 + \ldots + \mathbf{u}_m) \cdot \mathbf{x} \leqslant (\mathbf{u}_1 + \ldots + \mathbf{u}_m) \cdot \mathbf{a}_0\},$$

which shows that $A_1 \cap \ldots \cap A_m$ is an exposed face of A. □

Theorem 2.6.18 *Let A and B be convex sets in $\mathbb{R}^n$. Then each exposed face of $A + B$ has the form $C + D$, where C is an exposed face of A and D is an exposed face of B.*

Proof Suppose that F is a *proper* exposed face of $A + B$. Then there exist $\mathbf{a}_0 \in A$, $\mathbf{b}_0 \in B$, and a non-zero $\mathbf{u}$ in $\mathbb{R}^n$ such that

$$A + B \subseteq \{\mathbf{x} \in \mathbb{R}^n : \mathbf{u} \cdot \mathbf{x} \leqslant \mathbf{u} \cdot (\mathbf{a}_0 + \mathbf{b}_0)\} \qquad \text{and}$$
$$F = \{\mathbf{x} \in A + B : \mathbf{u} \cdot \mathbf{x} = \mathbf{u} \cdot (\mathbf{a}_0 + \mathbf{b}_0)\}.$$

If $\mathbf{a} \in A$, then $\mathbf{a} + \mathbf{b}_0 \in A + B$, whence $\mathbf{u} \cdot (\mathbf{a} + \mathbf{b}_0) \leqslant \mathbf{u} \cdot (\mathbf{a}_0 + \mathbf{b}_0)$ and $\mathbf{u} \cdot \mathbf{a} \leqslant \mathbf{u} \cdot \mathbf{a}_0$. Similarly, if $\mathbf{b} \in B$, then $\mathbf{u} \cdot \mathbf{b} \leqslant \mathbf{u} \cdot \mathbf{b}_0$. Thus

$$C = \{\mathbf{x} \in A : \mathbf{u} \cdot \mathbf{x} = \mathbf{u} \cdot \mathbf{a}_0\} \qquad \text{and} \qquad D = \{\mathbf{x} \in B : \mathbf{u} \cdot \mathbf{x} = \mathbf{u} \cdot \mathbf{b}_0\}$$

are, respectively, exposed faces of A and B. Clearly $C + D \subseteq F$. If $\mathbf{f} \in F$, then $\mathbf{f} = \mathbf{a} + \mathbf{b}$ for some $\mathbf{a} \in A$, $\mathbf{b} \in B$. Now $\mathbf{u} \cdot (\mathbf{a} + \mathbf{b}) = \mathbf{u} \cdot (\mathbf{a}_0 + \mathbf{b}_0)$, $\mathbf{u} \cdot \mathbf{a} \leqslant \mathbf{u} \cdot \mathbf{a}_0$, and $\mathbf{u} \cdot \mathbf{b} \leqslant \mathbf{u} \cdot \mathbf{b}_0$, whence $\mathbf{u} \cdot \mathbf{a} = \mathbf{u} \cdot \mathbf{a}_0$ and $\mathbf{u} \cdot \mathbf{b} = \mathbf{u} \cdot \mathbf{b}_0$. Thus $\mathbf{a} \in C$, $\mathbf{b} \in D$, and $F \subseteq C + D$. Hence $F = C + D$ as desired. □

The zero-dimensional exposed faces of a convex set are called its **exposed points**. Thus a point $\mathbf{a}$ of a convex set A in $\mathbb{R}^n$ is an exposed point of A if and only if there is some support hyperplane to A meeting it in the single point $\mathbf{a}$. Every exposed point of a convex set in $\mathbb{R}^n$ is one of its extreme points, but not necessarily conversely. The point $(1, 1)$ of the set A illustrated in Fig. 2.11 is an extreme, but not an exposed, point of A. In view of Theorem 2.6.16, it is natural to enquire if there is a representation of a compact convex set in terms of its exposed points. The answer, yes, is given in Theorem 2.6.21 below. To help us prove this theorem, we need some interesting results on farthest points.

Let A be a non-empty compact set in $\mathbb{R}^n$ and let $\mathbf{b}$ be a point of $\mathbb{R}^n$. For each point $\mathbf{x}$ of A, denote by $f(\mathbf{x})$ the distance $\|\mathbf{x} - \mathbf{b}\|$ of $\mathbf{x}$ from $\mathbf{b}$. Then f is a continuous real-valued function defined on a non-empty compact set A, and so is bounded and attains its bounds. In particular, f attains its upper bound, so there is a point $\mathbf{a}$ of A such that $\|\mathbf{x} - \mathbf{b}\| \leqslant \|\mathbf{a} - \mathbf{b}\|$ for all $\mathbf{x}$ in A. Each such point $\mathbf{a}$ of A is called a **farthest point** of A from $\mathbf{b}$. The following result states a close relation between farthest points and exposed points of a compact convex set in $\mathbb{R}^n$.

Theorem 2.6.19 *Let $\mathbf{a}$ be a farthest point of a compact convex set A in $\mathbb{R}^n$ from some point $\mathbf{b}$ of $\mathbb{R}^n$. Then $\mathbf{a}$ is an exposed point of A.*

Proof We consider the non-trivial case when $\mathbf{a} \neq \mathbf{b}$. Since $\mathbf{a}$ is a farthest point of A from $\mathbf{b}$, we have, for each point $\mathbf{x}$ of A,

$$\begin{aligned}\|\mathbf{a}-\mathbf{b}\|^2 &\geqslant \|\mathbf{x}-\mathbf{b}\|^2\\ &= \|(\mathbf{x}-\mathbf{a})+(\mathbf{a}-\mathbf{b})\|^2\\ &= \|\mathbf{x}-\mathbf{a}\|^2 + 2(\mathbf{x}-\mathbf{a})\cdot(\mathbf{a}-\mathbf{b}) + \|\mathbf{a}-\mathbf{b}\|^2,\end{aligned}$$

whence

$$0 \geqslant (\mathbf{x}-\mathbf{a})\cdot(\mathbf{a}-\mathbf{b}).$$

Equality occurs in the last inequality if and only if $\mathbf{x} = \mathbf{a}$. It follows that the hyperplane

$$H = \{\mathbf{z} \in \mathbb{R}^n : (\mathbf{a}-\mathbf{b})\cdot(\mathbf{z}-\mathbf{a}) = 0\}$$

supports A at $\mathbf{a}$, and that $H \cap A = \{\mathbf{a}\}$. Thus $\mathbf{a}$ is an exposed point of A. □

Lemma 2.6.20 *A compact convex set in $\mathbb{R}^n$ has an exposed point in every open halfspace which meets it.*

Proof Suppose that the compact convex set A in $\mathbb{R}^n$ meets the open halfspace

$$J = \{\mathbf{z} \in \mathbb{R}^n : \mathbf{u}\cdot\mathbf{z} + u_0 < 0\},$$

where $u_0 \in \mathbb{R}$, $\mathbf{u} \in \mathbb{R}^n$, and $\mathbf{u} \neq \mathbf{0}$; say $\mathbf{a} \in A \cap J$. Let $\lambda > 0$ satisfy

$$s^2 + 2\lambda(\mathbf{u}\cdot\mathbf{a} + u_0) < 0,$$

where s is the diameter of A. Let $\mathbf{c} = \mathbf{a} + \lambda\mathbf{u}$. For each point $\mathbf{x}$ of $A \setminus J$, $\mathbf{u}\cdot\mathbf{x} + u_0 \geqslant 0$ and $\mathbf{u}\cdot(\mathbf{a}-\mathbf{x}) \leqslant \mathbf{u}\cdot\mathbf{a} + u_0$, whence

$$\begin{aligned}\|\mathbf{c}-\mathbf{x}\|^2 &= \|\mathbf{a}-\mathbf{x}+\lambda\mathbf{u}\|^2\\ &= \|\mathbf{a}-\mathbf{x}\|^2 + 2\lambda\mathbf{u}\cdot(\mathbf{a}-\mathbf{x}) + \lambda^2\|\mathbf{u}\|^2\\ &\leqslant s^2 + 2\lambda(\mathbf{u}\cdot\mathbf{a}+u_0) + \lambda^2\|\mathbf{u}\|^2\\ &< \lambda^2\|\mathbf{u}\|^2 = \|\mathbf{c}-\mathbf{a}\|^2.\end{aligned}$$

Thus no point $\mathbf{x}$ of $A \setminus J$ is a farthest point of A from $\mathbf{c}$, and so every farthest point of A from $\mathbf{c}$ is an exposed point of A lying in J. □

Theorem 2.6.21 (Straszewicz) *Every compact convex set in $\mathbb{R}^n$ is the closure of the convex hull of its exposed points.*

Proof Let B be the set of the exposed points of a compact convex set A in $\mathbb{R}^n$. Trivially $\mathrm{cl}(\mathrm{conv}\, B) \subseteq A$, and so we need only show that $A \subseteq \mathrm{cl}(\mathrm{conv}\, B)$. Suppose that this is not so. Then there is a point $\mathbf{a}$ of A which does not belong to the closed convex set $\mathrm{cl}(\mathrm{conv}\, B)$. It follows immediately from Corollary 2.4.7 that there is an open halfspace J in $\mathbb{R}^n$ which contains $\mathbf{a}$ but is disjoint from $\mathrm{cl}(\mathrm{conv}\, B)$. By Lemma 2.6.20, there is a point of B lying in J, which is impossible. Thus $A \subseteq \mathrm{cl}(\mathrm{conv}\, B)$ as desired. □

The two-dimensional set illustrated in Fig. 2.11 shows that the closure requirement in Straszewicz's theorem cannot be omitted.

Exercises 2.6

1. Which closed convex sets in $\mathbb{R}^2$ have no extreme points?
2. Let A be a set in $\mathbb{R}^n$. Show that the extreme points of $\operatorname{conv} A$ are precisely those points $\mathbf{a}$ of A for which $\mathbf{a} \notin \operatorname{conv}(A \setminus \{\mathbf{a}\})$.
3. Prove that the set of extreme points of a closed convex set in $\mathbb{R}^2$ is closed.
4. Show that every $(n-1)$-dimensional face of a convex set in $\mathbb{R}^n$ is an exposed face of the set.
5. Give an example of an exposed point $\mathbf{a}$ of a compact convex set A in $\mathbb{R}^2$ such that there is *no* point $\mathbf{b}$ of $\mathbb{R}^2$ for which $\mathbf{a}$ is a farthest point of A from $\mathbf{b}$.
6. Prove that each extreme point of a closed convex set in $\mathbb{R}^n$ is the limit of some sequence of exposed points of the set.
7. Show that each relative interior point of a compact convex set in $\mathbb{R}^n$ can be expressed as a convex combination of $n+1$ or fewer exposed points of the set.
8. Give an example of a convex set A in $\mathbb{R}^2$ and convex subsets B, C of A which are *not* faces of A, but satisfy: (i) $A \setminus B$ is convex; (ii) $C = (\operatorname{aff} C) \cap A$.

2.7 THE BLASCHKE SELECTION PRINCIPLE

There are many occasions in the study of convexity when it is helpful to have a notion of distance between two sets in $\mathbb{R}^n$. In this section we introduce the *Hausdorff metric*, which assigns a distance between any two non-empty compact sets in $\mathbb{R}^n$. The metric will enable us to make precise statements such as: (i) every non-empty compact convex set in $\mathbb{R}^2$ can be approximated arbitrarily closely by convex polygons; (ii) each circular disc in $\mathbb{R}^2$ is the limit of a sequence of regular polygons inscribed in it; and (iii) the function which attaches to each non-empty compact set in $\mathbb{R}^n$ its diameter is continuous. Our main result, the *Blaschke selection principle*, asserts that every sequence of non-empty compact subsets of some fixed ball in $\mathbb{R}^n$ contains a convergent subsequence. The principle is often used to establish the existence of solutions to extremal problems in geometry. We shall indicate how it can be employed to show that amongst all plane compact convex sets with a given diameter, there exist ones having maximal area.

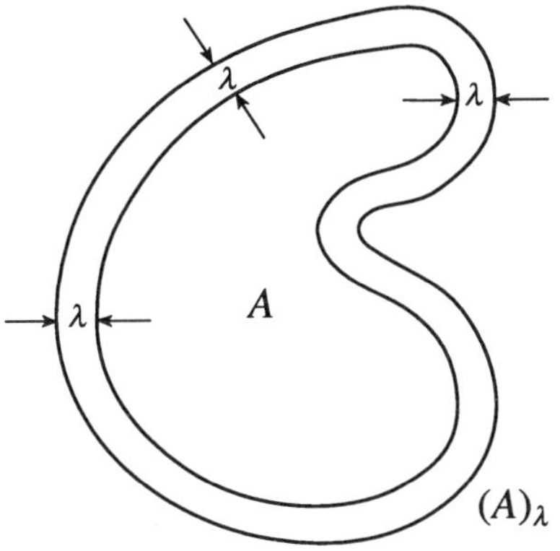

Fig. 2.12

Sets A and B in $\mathbb{R}^n$ are said to be a *finite distance apart* if there exists $\lambda \geqslant 0$ such that, for each point $\mathbf{a}$ of A, there is a point $\mathbf{b}$ of $\mathbf{B}$ whose distance $\|\mathbf{a} - \mathbf{b}\|$ from $\mathbf{a}$ does not exceed λ, and vice versa. In this situation, we say that the *distance between A and B does not exceed λ*. The **distance between** sets A and B in $\mathbb{R}^n$ that are a finite distance apart is defined to be the infimum of the set of all those $\lambda \geqslant 0$ for which the distance between A and B does not exceed λ. This definition does not assign a distance between the empty set and a non-empty set or between a bounded set and an unbounded one. On the other hand, the definition always assigns a distance between two non-empty bounded sets. For our purposes here, it will be sufficient to restrict attention to non-empty compact sets.

To help us express the above definition of distance between sets in a convenient and workable form, we introduce the idea of a λ-neighbourhood of a set in $\mathbb{R}^n$. Let A be a set in $\mathbb{R}^n$ and let $\lambda \geqslant 0$. Then the **λ-neighbourhood** $(A)_\lambda$ of A is the set $A + \lambda U$, where U denotes the closed unit ball $\{\mathbf{x} \in \mathbb{R}^n : \|\mathbf{x}\| \leqslant 1\}$. This definition is illustrated for compact A in Fig. 2.12. The figure makes it clear why the set $(A)_\lambda$ is often referred to as the **outer parallel set of A at distance λ**. Clearly, if $\mathbf{a} \in \mathbb{R}^n$, $r > 0$, and $\lambda \geqslant 0$, then the r-neighbourhood of $\{\mathbf{a}\}$ is the closed ball $B[\mathbf{a}; r]$ and the λ-neighbourhood of $B[\mathbf{a}; r]$ is the closed ball $B[\mathbf{a}; r + \lambda]$.

Theorem 2.7.1 *Let A, B be sets in $\mathbb{R}^n$ and let $\lambda, \mu \geqslant 0$. Then:*

(i) $(A)_0 = A$ *and* $A \subseteq (A)_\lambda$;

(ii) $(A)_\lambda \subseteq (B)_\lambda$ *when* $A \subseteq B$;

(iii) $(A)_\lambda$ *is convex when A is*;

(iv) $((A)_\lambda)_\mu = (A)_{\lambda+\mu}$.

Proof Parts (i) and (ii) are easy consequences of the definition of a λ-neighbourhood. Part (iii) follows from Example 2.1.1 and Theorem 2.1.5. To prove part (iv) we note, using Theorem 2.1.7, that

$$((A)_\lambda)_\mu = (A)_\lambda + \mu U = (A + \lambda U) + \mu U = A + (\lambda + \mu)U = (A)_{\lambda+\mu}. \ \square$$

Clearly the assertions that, for each point **a** of a set A in $\mathbb{R}^n$, there is a point **b** of a set B in $\mathbb{R}^n$ such that $\|\mathbf{a} - \mathbf{b}\| \leqslant \lambda$, and that $A \subseteq (B)_\lambda$, where $\lambda \geqslant 0$, are equivalent. Thus the definition we now give of the distance between non-empty compact sets in $\mathbb{R}^n$ coincides with the one given earlier. The **distance** $\rho(A, B)$ between non-empty compact sets A, B in $\mathbb{R}^n$ is defined by the equation

$$\rho(A, B) = \inf\{\lambda \geqslant 0 : A \subseteq (B)_\lambda \text{ and } B \subseteq (A)_\lambda\}.$$

The assumptions that A and B are non-empty and compact ensure that $\rho(A, B)$ is well-defined. The function ρ is known as the **Hausdorff metric** or **Hausdorff distance**. It is easily seen that the Hausdorff distance $\rho(\{\mathbf{a}\}, \{\mathbf{b}\})$ between the singleton sets $\{\mathbf{a}\}$ and $\{\mathbf{b}\}$ in $\mathbb{R}^n$ is $\|\mathbf{a} - \mathbf{b}\|$, i.e. the distance between the points **a** and **b** themselves. Another readily verified fact is that the Hausdorff distance is invariant under translation in the sense that, if A and B are non-empty compact sets in $\mathbb{R}^n$ and **x** is a point of $\mathbb{R}^n$, then $\rho(A, B) = \rho(A + \mathbf{x}, B + \mathbf{x})$.

Example 2.7.2 Let A and B be, respectively, the closed balls $B[\mathbf{a}; r]$ and $B[\mathbf{b}; s]$ in $\mathbb{R}^n$. Then $\rho(A, B) = \|\mathbf{b} - \mathbf{a}\| + |s - r|$.

Suppose first that $r \leqslant s$. Now

$$A \subseteq B - (\mathbf{b} - \mathbf{a}) \subseteq (B)_{\|\mathbf{b}-\mathbf{a}\|}$$

and

$$B = A + \mathbf{b} - \mathbf{a} + (s - r)U \subseteq (A)_{\|\mathbf{b}-\mathbf{a}\|+s-r},$$

whence $\rho(A, B) \leqslant \|\mathbf{b} - \mathbf{a}\| + s - r$. Now B contains a point whose distance from **a** is $\|\mathbf{b} - \mathbf{a}\| + s$. Thus, if $\lambda \geqslant 0$ and $B \subseteq (A)_\lambda = B[\mathbf{a}; \lambda + r]$, then $\|\mathbf{b} - \mathbf{a}\| + s \leqslant \lambda + r$, whence $\rho(A, B) \geqslant \|\mathbf{b} - \mathbf{a}\| + s - r$. Thus $\rho(A, B) = \|\mathbf{b} - \mathbf{a}\| + s - r$, i.e. $\rho(A, B) = \|\mathbf{b} - \mathbf{a}\| + |s - r|$. The case $s \leqslant r$ is similar. $\square$

Theorem 2.7.3 *Let A and B be non-empty compact sets in $\mathbb{R}^n$ with $\rho(A, B) = \lambda$. Then $A \subseteq (B)_\lambda$ and $B \subseteq (A)_\lambda$.*

Proof Let $\mathbf{a} \in A$. For each $\varepsilon > 0, A \subseteq (B)_{\lambda+\varepsilon}$, whence there is a point $\mathbf{b}_\varepsilon$ of B for which $\|\mathbf{a} - \mathbf{b}_\varepsilon\| \leqslant \lambda + \varepsilon$, and so $\inf\{\|\mathbf{a} - \mathbf{b}\| : \mathbf{b} \in B\} \leqslant \lambda$. Theorem 1.9.1 shows that there exists some point $\mathbf{b}_0$ of B such that $\|\mathbf{a} - \mathbf{b}_0\| \leqslant \lambda$. Thus $\mathbf{a} \in (B)_\lambda$ and $A \subseteq (B)_\lambda$. Similarly $B \subseteq (A)_\lambda$. $\square$

Theorem 2.7.4 *Let A, B, C be non-empty compact sets in $\mathbb{R}^n$ and let $\theta \geqslant 0$. Then*:

(i) $\rho(A, B) \geqslant 0$ *and* $\rho(A, B) = 0$ *if and only if* $A = B$;

(ii) $\rho(A, B) = \rho(B, A)$;

(iii) $\rho(A, C) \leqslant \rho(A, B) + \rho(B, C)$;

(iv) $\rho(\operatorname{conv} A, \operatorname{conv} B) \leqslant \rho(A, B)$;

(v) *if A and B are convex, then* $\rho(A, B) = \rho((A)_\theta, (B)_\theta)$.

Proof We use Theorems 2.7.1 and 2.7.3 throughout this proof.

(i) Trivially $\rho(A, B) \geqslant 0$ and $\rho(A, A) = 0$. If $\rho(A, B) = 0$, then $A \subseteq (B)_0 = B$ and $B \subseteq (A)_0 = A$, whence $A = B$.

(ii) This follows immediately from the definition of ρ.

(iii) Let $\rho(A, B) = \alpha$ and $\rho(B, C) = \beta$. Then

$$A \subseteq (B)_\alpha \subseteq ((C)_\beta)_\alpha = (C)_{\alpha+\beta} \quad \text{and} \quad C \subseteq (B)_\beta \subseteq ((A)_\alpha)_\beta = (A)_{\alpha+\beta},$$

whence

$$\rho(A, C) \leqslant \alpha + \beta = \rho(A, B) + \rho(B, C).$$

(iv) Let $\rho(A, B) = \alpha$. Then $(\operatorname{conv} A)_\alpha$ is convex and

$$B \subseteq (A)_\alpha \subseteq (\operatorname{conv} A)_\alpha,$$

whence $\operatorname{conv} B \subseteq (\operatorname{conv} A)_\alpha$. Similarly, $\operatorname{conv} A \subseteq (\operatorname{conv} B)_\alpha$. Thus

$$\rho(\operatorname{conv} A, \operatorname{conv} B) \leqslant \alpha = \rho(A, B).$$

We note that $\operatorname{conv} A$ and $\operatorname{conv} B$ are compact by Theorem 2.2.6.

(v) Let A and B be convex. The sets $(A)_\theta$ and $(B)_\theta$ are compact by Theorem 1.8.10. Let $\rho(A, B) = \alpha$ and $\rho((A)_\theta, (B)_\theta) = \beta$. Then

$$(A)_\theta \subseteq ((B)_\alpha)_\theta = ((B)_\theta)_\alpha \quad \text{and} \quad (B)_\theta \subseteq ((A)_\alpha)_\theta = ((A)_\theta)_\alpha,$$

which shows that $\beta \leqslant \alpha$. Also

$$A + \theta U \subseteq (B + \theta U) + \beta U \quad \text{and} \quad B + \theta U \subseteq (A + \theta U) + \beta U,$$

i.e.

$$A + \theta U \subseteq (B + \beta U) + \theta U \quad \text{and} \quad B + \theta U \subseteq (A + \beta U) + \theta U,$$

whence $A \subseteq B + \beta U$ and $B \subseteq A + \beta U$ by Theorem 2.1.8. Thus $\alpha \leqslant \beta$, and so $\alpha = \beta$ as desired. □

The sequence $A_1, \ldots, A_j, \ldots$ of non-empty compact sets in $\mathbb{R}^n$ is said to **converge** to the non-empty compact set A in $\mathbb{R}^n$, written $A_j \to A$ as $j \to \infty$, if $\rho(A_j, A) \to 0$ as $j \to \infty$. Such a sequence cannot converge to more than one non-empty compact set A, for if it also converges to a non-empty compact set B in $\mathbb{R}^n$, then

$$0 \leqslant \rho(A, B) \leqslant \rho(A, A_j) + \rho(A_j, B) \to 0 \qquad \text{as } j \to \infty,$$

whence $\rho(A, B) = 0$ and $A = B$. Another easy consequence of the above definition is that, if the sequence $A_1, \ldots, A_j, \ldots$ converges to A and each A_j is convex, then so too is A. To see why this is so, suppose that a sequence $A_1, \ldots, A_j, \ldots$ of non-empty compact convex sets in $\mathbb{R}^n$ converges to a non-empty compact set A in $\mathbb{R}^n$. Theorem 2.7.4 (iv) shows that

$$\rho(A_j, \operatorname{conv} A) = \rho(\operatorname{conv} A_j, \operatorname{conv} A) \leqslant \rho(A_j, A) \to 0 \qquad \text{as } j \to \infty.$$

Thus $A_1, \ldots, A_j, \ldots$ also converges to $\operatorname{conv} A$. But a sequence cannot converge to two *different* limits, whence $A = \operatorname{conv} A$ and A is convex.

Theorem 2.7.5 *Let sequences $A_1, \ldots, A_j, \ldots$ and $B_1, \ldots, B_j, \ldots$ converge, respectively, to A_0 and B_0, where all the A's and B's are non-empty compact sets in $\mathbb{R}^n$. Let real sequences $\alpha_1, \ldots, \alpha_j, \ldots$ and $\beta_1, \ldots, \beta_j, \ldots$ converge, respectively, to α and β. Then the sequence $\alpha_1 A_1 + \beta_1 B_1, \ldots, \alpha_j A_j + \beta_j B_j, \ldots$ converges to $\alpha A_0 + \beta B_0$.*

Proof For $\mathbf{a}, \mathbf{b}, \mathbf{a}_j, \mathbf{b}_j \in \mathbb{R}^n$,

$$\begin{aligned}\|\alpha_j \mathbf{a}_j + \beta_j \mathbf{b}_j - \alpha\mathbf{a} - \beta\mathbf{b}\| &= \|\alpha_j(\mathbf{a}_j - \mathbf{a}) + (\alpha_j - \alpha)\mathbf{a} + \beta_j(\mathbf{b}_j - \mathbf{b}) + (\beta_j - \beta)\mathbf{b}\| \\ &\leqslant |\alpha_j|\|\mathbf{a}_j - \mathbf{a}\| + |\alpha_j - \alpha|\|\mathbf{a}\| + |\beta_j|\|\mathbf{b}_j - \mathbf{b}\| \\ &\quad + |\beta_j - \beta|\|\mathbf{b}\|.\end{aligned}$$

Write $\theta_j = \rho(A_j, A_0)$ and $\varphi_j = \rho(B_j, B_0)$. Let $r > 0$ be such that $\|\mathbf{a}\|$, $\|\mathbf{b}\| \leqslant r$ whenever $\mathbf{a} \in A_0, \mathbf{b} \in B_0$. It follows easily from the above inequality that

$$\rho(\alpha_j A_j + \beta_j B_j, \alpha A_0 + \beta B_0) \leqslant |\alpha_j|\theta_j + |\alpha_j - \alpha| r + |\beta_j|\varphi_j + |\beta_j - \beta| r,$$

whence $\alpha_j A_j + \beta_j B_j \to \alpha A_0 + \beta B_0$ as $j \to \infty$. □

Corollary 2.7.6 *For $i = 1, \ldots, m$, let the sequence $A_1^i, \ldots, A_j^i, \ldots$ converge to A_0^i, where all the A's are non-empty compact sets in $\mathbb{R}^n$, and let the real sequence $\alpha_1^i, \ldots, \alpha_j^i, \ldots$ converge to α_i. Then $\alpha_j^1 A_j^1 + \ldots + \alpha_j^m A_j^m \to \alpha_1 A_0^1 + \ldots + \alpha_m A_0^m$ as $j \to \infty$.* □

We now use ideas about the convergence of a sequence of balls to establish some geometrical properties of convex sets. Let C be a compact convex set in $\mathbb{R}^n$ that has a non-empty interior. Such a set is called a **convex body**. Denote by r the supremum of the set of radii of closed balls lying in C, and by R the infimum of the set of radii of closed balls in $\mathbb{R}^n$ containing C. Then r is called the **inradius** of C and R is called the **circumradius** of C. Clearly both r and R are positive real numbers satisfying $r \leqslant R$.

Theorem 2.7.7 *Let C be a convex body in $\mathbb{R}^n$ with inradius r and circumradius R. Then C contains a closed ball of radius r and is contained in a unique closed ball of radius R.*

Proof The definition of R implies that, for each $j = 1, 2, \ldots$, there exist $\mathbf{a}_j \in \mathbb{R}^n$ and $R_j > 0$ such that $C \subseteq B[\mathbf{a}_j; R_j]$ and $R_j < R + (1/j)$. The sequence $R_1, \ldots, R_j, \ldots$ converges to R and the sequence $\mathbf{a}_1, \ldots, \mathbf{a}_j, \ldots$ is bounded. Thus there is some subsequence $\mathbf{a}_{i_1}, \ldots, \mathbf{a}_{i_j}, \ldots$ of $\mathbf{a}_1, \ldots, \mathbf{a}_j, \ldots$ that converges to some point $\mathbf{a}$ of $\mathbb{R}^n$. It follows from Example 2.7.2 that $B[\mathbf{a}_{i_j}; R_{i_j}] \to B[\mathbf{a}; R]$ as $j \to \infty$. We show that $C \subseteq B[\mathbf{a}; R]$. Let $\mathbf{c} \in C$. Since $C \subseteq B[\mathbf{a}_{i_j}; R_{i_j}]$, $\|\mathbf{c} - \mathbf{a}_{i_j}\| \leqslant R_{i_j}$. Letting $j \to \infty$ in the last inequality, we find that $\|\mathbf{c} - \mathbf{a}\| \leqslant R$. Thus $\mathbf{c} \in B[\mathbf{a}; R]$ and $C \subseteq B[\mathbf{a}; R]$. The proof that C contains a closed ball of radius r is similar to the one which we have just given.

Suppose that C lies in *both* of the closed balls $B[\mathbf{a}; R]$ and $B[\mathbf{b}; R]$ of radius R in $\mathbb{R}^n$. Then, for each $\mathbf{x}$ in C,

$$\|\mathbf{x} - \tfrac{1}{2}(\mathbf{a} + \mathbf{b})\|^2 = \tfrac{1}{2}\|\mathbf{x} - \mathbf{a}\|^2 + \tfrac{1}{2}\|\mathbf{x} - \mathbf{b}\|^2 - \tfrac{1}{4}\|\mathbf{a} - \mathbf{b}\|^2 \leqslant R^2 - \tfrac{1}{4}\|\mathbf{a} - \mathbf{b}\|^2,$$

whence $C \subseteq B[\tfrac{1}{2}(\mathbf{a} + \mathbf{b}); \sqrt{(R^2 - \tfrac{1}{4}\|\mathbf{a} - \mathbf{b}\|^2)}]$. Since C cannot lie in a closed ball of radius *less* than R, we must have $\mathbf{a} = \mathbf{b}$. Thus there is *precisely* one closed ball of radius R in $\mathbb{R}^n$ which contains C. □

Let C be a convex body in $\mathbb{R}^n$ with inradius r and circumradius R. Then any closed ball of radius r lying in C is called an **inball** of C and its centre an **incentre** of C. The unique closed ball of radius R which contains C is called the **circumball** of C and its centre the **circumcentre** of C. A (non-square) rectangle in $\mathbb{R}^2$ is an example of a convex body that does not have a unique incentre. Our definitions of the terms *circumradius*, *circumcircle*, and *circumcentre* as applied to obtuse-angled triangles do not coincide with those used in elementary geometry. For example, consider an isosceles triangle with sides $2, 2, 2\sqrt{3}$. In the parlance of elementary geometry, its circumradius is 2 and its circumcentre lies exterior to the triangle, whereas for us here its circumradius is $\sqrt{3}$ and its circumcentre is the midpoint of its longest side.

Theorem 2.7.7 asserts the existence of solutions to two extremal problems in geometry: the first to find a ball of minimal radius containing a given convex body, the second to find a ball of maximal radius lying in the body. These are particularly simple problems and we virtually solved them with bare hands! Most extremal problems, however, call for more sophisticated machinery. The key step in proving Theorem 2.7.7 was the extraction of a convergent subsequence from a sequence of closed balls, and it is a generalization of this idea that turns out to be useful in finding solutions to many extremal problems. What is needed is a criterion for a sequence of sets to contain a convergent subsequence. One such criterion is that of *uniform boundedness*, which we now introduce. A sequence of sets in $\mathbb{R}^n$ is said to be **uniformly bounded** if there exists some ball in $\mathbb{R}^n$ that contains every member of the sequence. Since the proof of our main result, Theorem 2.7.9, is somewhat involved, we prepare for it with the following preliminary result.

Lemma 2.7.8 *Let $A_1, \ldots, A_j, \ldots$ be a uniformly bounded sequence of non-empty compact sets in $\mathbb{R}^n$. Let $\varepsilon > 0$. Then there exists a subsequence $A_{i_1}, \ldots, A_{i_j}, \ldots$ of $A_1, \ldots, A_j, \ldots$ such that*

$$\rho(A_{i_j}, A_{i_k}) \leqslant \varepsilon \qquad \textit{for all } j, k = 1, 2, \ldots .$$

Proof Since there is a ball in $\mathbb{R}^n$ which contains every member of the given sequence, there is a *finite* set E in $\mathbb{R}^n$ such that $A_j \subseteq (E)_{\frac{1}{2}\varepsilon}$ for $j = 1, 2, \ldots$. For each $j = 1, 2, \ldots$, denote by E_j the non-empty subset $E \cap (A_j)_{\frac{1}{2}\varepsilon}$ of E. It is easily verified that $\rho(E_j, A_j) \leqslant \frac{1}{2}\varepsilon$. Because E is finite, there can only be a finite number of possible *different* sets E_j for $j = 1, 2, \ldots$. Hence the sequence $E_1, \ldots, E_j, \ldots$ must contain some *constant* subsequence, $E_{i_1}, \ldots, E_{i_j}, \ldots$ say. For $j, k = 1, 2, \ldots$, we have

$$\rho(A_{i_j}, A_{i_k}) \leqslant \rho(A_{i_j}, E_{i_j}) + \rho(E_{i_j}, A_{i_k}) = \rho(A_{i_j}, E_{i_j}) + \rho(E_{i_k}, A_{i_k}) \leqslant \tfrac{1}{2}\varepsilon + \tfrac{1}{2}\varepsilon = \varepsilon.$$

Thus the subsequence $A_{i_1}, \ldots, A_{i_j}, \ldots$ of $A_1, \ldots, A_j, \ldots$ has the property demanded by the lemma. □

Theorem 2.7.9 *Every uniformly bounded sequence $A_1, \ldots, A_j, \ldots$ of non-empty compact sets in $\mathbb{R}^n$ contains a subsequence which converges to some non-empty compact set A in $\mathbb{R}^n$.*

Proof It follows, by repeated applications of the lemma with $\varepsilon = 1, 1/2, \ldots, 1/j, \ldots$, that the sequence $A_1, \ldots, A_j, \ldots$ contains a sequence of subsequences

$$\begin{array}{l} A_{11}, A_{12}, \ldots, A_{1j}, \ldots \\ A_{21}, A_{22}, \ldots, A_{2j}, \ldots \\ \cdots\cdots\cdots\cdots\cdots\cdots \\ A_{r1}, A_{r2}, \ldots, A_{rj}, \ldots \\ \cdots\cdots\cdots\cdots\cdots\cdots \end{array}$$

where each subsequence after the first in the list is a subsequence of the preceding one, and $\rho(A_{rj}, A_{rk}) \leqslant 1/r$ for $j, k = 1, 2, \ldots$. The diagonal sequence $A_{11}, A_{22}, \ldots, A_{jj}, \ldots$ is a subsequence of $A_1, A_2, \ldots, A_j, \ldots$ with the property that $\rho(A_{jj}, A_{kk}) \leqslant 1/j$ whenever $j \leqslant k$.

Write $B_j = A_{jj}$ for $j = 1, 2, \ldots$. We complete the proof by showing that the subsequence $B_1, \ldots, B_j, \ldots$ of $A_1, \ldots, A_j, \ldots$ converges to the non-empty compact set B defined by the equation

$$B = \bigcap((B_k)_{1/k}: k = 1, 2, \ldots).$$

Let j be a positive integer and let $\mathbf{b}_j \in B_j$. For $i = 1, 2, \ldots$, choose $\mathbf{b}_{j+i} \in B_{j+i}$ such that $\|\mathbf{b}_{j+i} - \mathbf{b}_j\| \leqslant 1/j$; this is possible because $\rho(B_j, B_{j+i}) \leqslant 1/j$. The sequence $\mathbf{b}_{j+1}, \mathbf{b}_{j+2}, \ldots$ lies in the compact set $(B_1)_1$, and so contains a subsequence converging to some point $\mathbf{b}$ of $\mathbb{R}^n$. Since $B_{j+i} \subseteq (B_k)_{1/k}$ whenever $j + i \geqslant k$, all but a finite number of terms of the sequence $\mathbf{b}_{j+1}, \mathbf{b}_{j+2}, \ldots$ lie in the compact set $(B_k)_{1/k}$ for $k = 1, 2, \ldots$, whence $\mathbf{b} \in (B_k)_{1/k}$ and $\mathbf{b} \in B$. But $\mathbf{b}_j$ is an arbitrary point of B_j and clearly $\|\mathbf{b} - \mathbf{b}_j\| \leqslant 1/j$, so $B_j \subseteq (B)_{1/j}$. Trivially $B \subseteq (B_j)_{1/j}$. Thus $\rho(B_j, B) \leqslant 1/j$ and $B_j \to B$ as $j \to \infty$. □

Theorem 2.7.10 (Blaschke selection principle) *Every uniformly bounded sequence of non-empty compact convex sets in $\mathbb{R}^n$ contains a subsequence which converges to some non-empty compact convex set in $\mathbb{R}^n$.*

Proof The principle is a consequence of Theorem 2.7.9 and the fact that a convergent sequence of convex sets must converge to a convex limit. □

A typical extremal problem of elementary geometry is to maximize or minimize a real-valued function f defined on some family $\mathfrak{F}$ of non-empty compact sets in $\mathbb{R}^n$. It is important to have a concept of continuity for such functions $f: \mathfrak{F} \to \mathbb{R}$, and this we now introduce. The function f is said to be **continuous** on $\mathfrak{F}$ if $f(A_j) \to f(A)$ as $j \to \infty$ whenever $A_j \to A$ as $j \to \infty$, where all the sets under consideration belong to $\mathfrak{F}$.

Example 2.7.11 We show that the diameter function D, which associates with each non-empty compact set A in $\mathbb{R}^n$ its diameter $D(A)$, is continuous on the family $\mathfrak{F}$ of all non-empty compact sets in $\mathbb{R}^n$. It is easily verified that $D((A)_\lambda) = D(A) + 2\lambda$, where $A \in \mathfrak{F}$ and $\lambda \geqslant 0$. Suppose now that $A, B \in \mathfrak{F}$ and that $\rho(A, B) = \lambda$. Then

$$D(A) \leqslant D((B)_\lambda) = D(B) + 2\lambda \quad \text{and} \quad D(B) \leqslant D((A)_\lambda) = D(A) + 2\lambda,$$

whence $|D(A) - D(B)| \leqslant 2\lambda = 2\rho(A, B)$ and the continuity of D is now clear. □

We now indicate how the Blaschke selection principle can be used to show that in the family $\mathfrak{F}$ of all compact convex sets in $\mathbb{R}^2$ with diameter 1, there exist sets having maximal area. For each set A in $\mathfrak{F}$, denote by $f(A)$ the area of A. This area function will be defined formally in Chapter 6, where it will be shown to be continuous on the family of all non-empty compact convex sets in $\mathbb{R}^2$. Let α be the supremum of the set of areas of members of $\mathfrak{F}$. For each positive integer j, there is a member A_j of $\mathfrak{F}$ such that $f(A) > \alpha - 1/j$. We may suppose, by translating the A_j's if necessary, that they are uniformly bounded. The Blaschke selection principle guarantees the existence of a subsequence $A_{i_1}, \ldots, A_{i_j}, \ldots$ of $A_1, \ldots, A_j, \ldots$ which converges to some non-empty compact convex set A_0 in $\mathbb{R}^2$. The continuity of the diameter function shows that A_0 has diameter 1, and so lies in $\mathfrak{F}$. For each j,

$$\alpha \geqslant f(A_{i_j}) > \alpha - 1/i_j \geqslant \alpha - 1/j.$$

Letting $j \to \infty$ in these inequalities, we deduce, using the continuity of f, that $f(A_0) = \alpha$. Thus A_0 is a member of $\mathfrak{F}$ having maximal possible area. We return to this so-called **isodiametric problem** in Section 6.5, where A_0 will be shown to be a closed circular disc with diameter 1.

Exercises 2.7

1. Give an example of a non-convex set A in $\mathbb{R}^n$ for which $(A)_\lambda$ is convex for *all* $\lambda > 0$. Show that, if A is a closed non-convex set in $\mathbb{R}^n$, then there exists $\varepsilon > 0$ such that $(A)_\lambda$ is non-convex whenever $0 \leqslant \lambda < \varepsilon$.
2. Let C_r be the closed circular disc of radius r centred on the origin of $\mathbb{R}^2$ and let P be a regular n-sided polygon inscribed in C_1. Determine the Hausdorff distance $\rho(C_r, P)$ and find its minimal value as r varies.
3. Suppose that A_0 and A_1 are non-empty compact convex sets in $\mathbb{R}^n$ with $\rho(A_0, A_1) = 1$. Prove that, if $A_\theta = (1 - \theta)A_0 + \theta A_1$ for $0 < \theta < 1$, then $\rho(A_\alpha, A_\beta) = \beta - \alpha$ for $0 \leqslant \alpha \leqslant \beta \leqslant 1$.
4. In $\mathbb{R}^n$ the sequence $A_1, A_2, \ldots$ of non-empty compact sets converges to the non-empty compact set A. Show that
$$A = \bigcap_{i=1}^{\infty} \left(\mathrm{cl} \left(\bigcup_{j=i}^{\infty} A_j \right) \right).$$
5. Prove that the inradius function r and the circumradius function R are continuous on the family $\mathfrak{F}$ of all convex bodies in $\mathbb{R}^n$.

2.8 DUALITY

In this section, we show how to associate with each set in $\mathbb{R}^n$ another set called its *polar dual*. Duality, the study of the relationship between a set and its polar dual, proves to be a fruitful concept in several areas of convexity. It can be used to derive new results from ones already known, in addition to sometimes showing how seemingly unconnected results may in fact be closely related.

To motivate our discussion, we consider two ways of viewing a closed convex set A in $\mathbb{R}^n$. On the one hand, A is the union of its points (an internal representation), whilst on the other hand, it is the intersection of all closed halfspaces in $\mathbb{R}^n$ that contain A (an external representation). This observation suggests that we might usefully explore the relationship between points and closed halfspaces in $\mathbb{R}^n$, and this we now do.

For each non-zero vector $\mathbf{u}$ in $\mathbb{R}^n$, the set $H^- = \{\mathbf{x} \in \mathbb{R}^n : \mathbf{u} \cdot \mathbf{x} \leqslant 1\}$ is a closed halfspace in $\mathbb{R}^n$ containing the origin as an interior point. Conversely, for each closed halfspace H^- in $\mathbb{R}^n$ containing the origin as an interior point, there is a *unique* non-zero vector $\mathbf{u}$ in $\mathbb{R}^n$ such that $H^- = \{\mathbf{x} \in \mathbb{R}^n : \mathbf{u} \cdot \mathbf{x} \leqslant 1\}$. Thus there is a bijection between the set of non-zero vectors in $\mathbb{R}^n$ and the set of closed halfspaces in $\mathbb{R}^n$ containing the origin as an interior point. These ideas lead us to define, for each point $\mathbf{u}$ of $\mathbb{R}^n$, a set $\mathbf{u}^*$ in $\mathbb{R}^n$ by the equation

$$\mathbf{u}^* = \{\mathbf{x} \in \mathbb{R}^n : \mathbf{u} \cdot \mathbf{x} \leqslant 1\}.$$

We note that $\mathbf{0}^* = \mathbb{R}^n$.

Recalling our earlier discussion in which we linked unions of points with intersections of closed halfspaces, we now take a bold step forward by defining the **polar dual** A^* of a set A in $\mathbb{R}^n$ to be the intersection of all the sets $\mathbf{a}^*$ for $\mathbf{a} \in A$, i.e.

$$A^* = \bigcap(\mathbf{a}^* : \mathbf{a} \in A) = \{\mathbf{x} \in \mathbb{R}^n : \mathbf{a} \cdot \mathbf{x} \leqslant 1 \text{ for all } \mathbf{a} \in A\}.$$

What can be said about the polar dual of a set? For each set A in $\mathbb{R}^n$, its polar dual A^* is defined as an intersection of closed convex sets containing the origin, and so A^* is itself a closed convex set containing the origin. The polar duals $\varnothing^*$ and $\{\mathbf{0}\}^*$ are both $\mathbb{R}^n$, while the polar dual of $\mathbb{R}^n$ is $\{\mathbf{0}\}$.

It is instructive to examine the sets $\mathbf{u}^*$ for *non-zero* vectors $\mathbf{u}$ in $\mathbb{R}^n$. By definition, $\mathbf{u}^*$ is the closed halfspace which is bounded by the hyperplane $H = \{\mathbf{x} \in \mathbb{R}^n : \mathbf{u} \cdot \mathbf{x} = 1\}$ and contains the origin. This hyperplane H has $\mathbf{u}$ as one of its normal vectors and passes through the point $\mathbf{u}/\|\mathbf{u}\|^2$. The distance of $\mathbf{u}/\|\mathbf{u}\|^2$ from the origin is $1/\|\mathbf{u}\|$, which exceeds, equals, or is less than 1, according as $\|\mathbf{u}\|$ is less

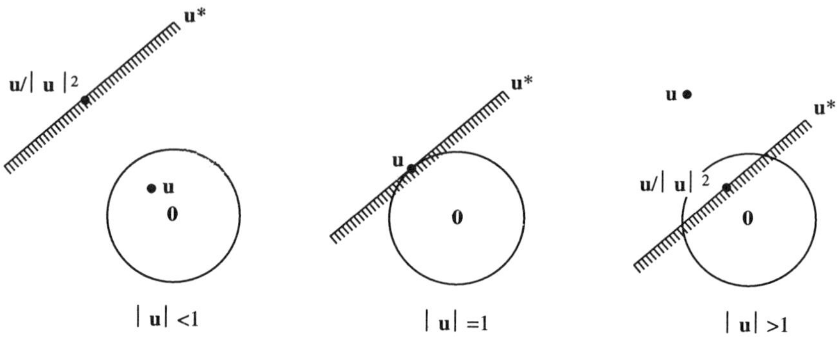

Fig. 2.13

than, equals, or exceeds 1. The relative positions of **u** and **u*** for the cases $\|\mathbf{u}\| < 1$, $\|\mathbf{u}\| = 1$, $\|\mathbf{u}\| > 1$ are illustrated in Fig. 2.13.

Theorem 2.8.1 *Let A, B be sets in $\mathbb{R}^n$, let U be the closed unit ball centred on the origin of $\mathbb{R}^n$ and let λ be a non-zero scalar. Then:*

(i) $A \subseteq B$ *implies that* $B^* \subseteq A^*$;

(ii) $(A \cup B)^* = A^* \cap B^*$;

(iii) $(\lambda A)^* = (1/\lambda)A^*$;

(iv) $U^* = U$.

Proof (i) Suppose that $A \subseteq B$. Then

$$B^* = \{\mathbf{x} \in \mathbb{R}^n : \mathbf{b} \cdot \mathbf{x} \leqslant 1,\ \mathbf{b} \in B\} \subseteq \{\mathbf{x} \in \mathbb{R}^n : \mathbf{a} \cdot \mathbf{x} \leqslant 1,\ \mathbf{a} \in A\} = A^*.$$

(ii) Now $\mathbf{x} \in (A \cup B)^*$ if and only if both $\mathbf{a} \cdot \mathbf{x} \leqslant 1$ for $\mathbf{a} \in A$, and $\mathbf{b} \cdot \mathbf{x} \leqslant 1$ for $\mathbf{b} \in B$, i.e. if and only if $\mathbf{x} \in A^* \cap B^*$. Thus $(A \cup B)^* = A^* \cap B^*$.

(iii) Now $\mathbf{x} \in (\lambda A)^*$ if and only if $\lambda\mathbf{a} \cdot \mathbf{x} = \mathbf{a} \cdot (\lambda\mathbf{x}) \leqslant 1$ for $\mathbf{a} \in A$, i.e. if and only if $\lambda\mathbf{x} \in A^*$ or, equivalently, $\mathbf{x} \in (1/\lambda)A^*$. Thus $(\lambda A)^* = (1/\lambda)A^*$.

(iv) Let $\mathbf{x} \in U$. Then, for all $\mathbf{u} \in U$, $\mathbf{u} \cdot \mathbf{x} \leqslant \|\mathbf{u}\|\,\|\mathbf{x}\| \leqslant 1$, whence $\mathbf{x} \in U^*$ and $U \subseteq U^*$. Conversely, let $\mathbf{x} \in U^*$ with $\mathbf{x} \neq \mathbf{0}$. Then $\mathbf{x}/\|\mathbf{x}\| \in U$, and so $(\mathbf{x}/\|\mathbf{x}\|) \cdot \mathbf{x} = \|\mathbf{x}\| \leqslant 1$, which shows that $\mathbf{x} \in U$. Thus $U^* \subseteq U$ and $U^* = U$. □

Example 2.8.2 We find the polar dual A^* of the *n-cube* A defined by the equation

$$A = \{(a_1, \ldots, a_n) : |a_1| \leqslant 1, \ldots, |a_n| \leqslant 1\}.$$

For each point $(x_1, \ldots, x_n)$ of A^*, define a point $(a_1, \ldots, a_n)$ of A by the conditions that $a_i = 1$ if $x_i \geqslant 0$, and $a_i = -1$ if $x_i < 0$. Then

$$(a_1,\ldots,a_n)\cdot(x_1,\ldots,x_n) = a_1x_1 + \ldots + a_nx_n = |x_1| + \ldots + |x_n| \leqslant 1.$$

Conversely, suppose that $(x_1,\ldots,x_n)$ satisfies $|x_1| + \ldots + |x_n| \leqslant 1$. Then, for any point $(a_1,\ldots,a_n)$ of A,

$$(a_1,\ldots,a_n)\cdot(x_1,\ldots,x_n) = a_1x_1 + \ldots + a_nx_n \leqslant |a_1||x_1| + \ldots + |a_n||x_n|$$
$$\leqslant |x_1| + \ldots + |x_n| \leqslant 1,$$

which shows that $(x_1,\ldots,x_n)\in A^*$. Thus A^* is the set, known as a regular n-crosspolytope, defined by the equation

$$A^* = \{(x_1,\ldots,x_n): |x_1| + \ldots + |x_n| \leqslant 1\}.$$

We now find A^{**} the polar dual of the polar dual A^* of the n-cube A. For each point $(u_1,\ldots,u_n)$ of A^{**}, define a point $(x_1,0,\ldots,0)$ of A^* by the conditions that $x_1 = 1$ if $u_1 \geqslant 0$, and $x_1 = -1$ if $u_1 < 0$. Then

$$(x_1,0,\ldots,0)\cdot(u_1,u_2,\ldots,u_n) = |u_1| \leqslant 1.$$

Similarly, $|u_2| \leqslant 1,\ldots,|u_n| \leqslant 1$, whence $(u_1,\ldots,u_n)\in A$ and $A^{**}\subseteq A$. The inclusion $A\subseteq A^{**}$, which holds for *every* set A in $\mathbb{R}^n$, follows immediately from the definition of the polar dual. Hence $A^{**} = A$. □

This last example suggests that we examine the **double-polar dual** A^{**} of an arbitrary set A in $\mathbb{R}^n$ and see how it is related to A. Since the polar dual of any set in $\mathbb{R}^n$ is always a closed convex set containing the origin, a necessary condition for the equality of the sets A and A^{**} is that A is a closed convex set containing the origin. We now show that this condition is also sufficient, and establish the exact relationship between A^{**} and A.

Theorem 2.8.3 *Let A be a set in $\mathbb{R}^n$. Then $A^{**} = \mathrm{cl}(\mathrm{conv}(A\cup\{\mathbf{0}\}))$. In particular, if A is closed, convex, and contains the origin, then $A^{**} = A$.*

Proof For all $\mathbf{a}\in A$, $\mathbf{x}\in A^*$, we have $\mathbf{a}\cdot\mathbf{x}\leqslant 1$, whence $\mathbf{a}\in A^*$ and $A\subseteq A^{**}$. Since A^{**} is a closed convex set containing $\mathbf{0}$ and A, $\mathrm{cl}(\mathrm{conv}(A\cup\{\mathbf{0}\}))\subseteq A^{**}$. We now establish the opposite inclusion. Suppose that $\mathbf{z}$ is a point of $\mathbb{R}^n$ *not* lying in $\mathrm{cl}(\mathrm{conv}(A\cup\{\mathbf{0}\}))$. By Corollary 2.4.7, there exists a hyperplane *strictly* separating $\{\mathbf{z}\}$ and $\mathrm{cl}(\mathrm{conv}(A\cup\{\mathbf{0}\}))$. Thus, since such a hyperplane cannot pass through the origin, there exists $\mathbf{u}$ in $\mathbb{R}^n$ such that $\mathbf{u}\cdot\mathbf{z} > 1$ and $\mathbf{u}\cdot\mathbf{a} < 1$ for *all* $\mathbf{a}$ in A. This shows that $\mathbf{u}\in A^*$ and $\mathbf{z}\notin A^{**}$. Hence $A^{**}\subseteq \mathrm{cl}(\mathrm{conv}(A\cup\{\mathbf{0}\}))$, and so $A^{**} = \mathrm{cl}(\mathrm{conv}(A\cup\{\mathbf{0}\}))$. □

In view of the last result it is not surprising that duality is most effective when applied to families of closed convex sets containing the origin, and for the remainder of this section we restrict our attention to such families. The reader is asked to note the part played by duality in the next theorem, which comprises two assertions together with their *duals*. Once the two original results have been established, duality both suggests and proves the dual results at one and the same time.

Theorem 2.8.4 *Let A be a closed convex set in $\mathbb{R}^n$ containing the origin. Then A is bounded if and only if the origin is an interior point of A^*, and A^* is bounded if and only if the origin is an interior point of A.*

Proof We use Theorems 2.8.1 and 2.8.3. Suppose first that A is bounded. Then, for some $r > 0$, $A \subseteq rU$, whence $(1/r)U \subseteq A^*$ and the origin is an interior point of A^*. By applying the last result to the set A^*, we deduce that, if A^* is bounded, then the origin is an interior point of $A^{**} = A$. Suppose next that the origin is an interior point of A. Then, for some $s > 0$, $sU \subseteq A$, whence $A^* \subseteq (1/s)U$ and A^* is bounded. By applying the last result to A^*, we deduce that, if the origin is an interior point of A^*, then $A^{**} = A$ is bounded. □

Corollary 2.8.5 *Let $\mathfrak{F}$ be the family of all compact convex sets in $\mathbb{R}^n$ which contain the origin as an interior point. Then the mapping $\theta: \mathfrak{F} \to \mathfrak{F}$ defined for $A \in \mathfrak{F}$ by the equation $\theta(A) = A^*$ is a bijection.*

Proof Let $A, B \in \mathfrak{F}$. The theorem shows that $\theta(A) \in \mathfrak{F}$. Now θ is injective, for $A^* = B^*$ implies $A^{**} = B^{**}$, i.e. $A = B$; it is surjective, since $\theta(A^*) = A$. □

Suppose now that A is a compact convex set in $\mathbb{R}^n$ that contains the origin as an interior point. Let B be an exposed face of A. Then, for each point $\mathbf{b}$ in B, the set $\{\mathbf{x} \in A^* : \mathbf{b} \cdot \mathbf{x} = 1\}$ is an exposed face (possibly empty) of A^*. Thus the set $\varphi(B)$, defined by the equation

$$\varphi(B) = \{\mathbf{x} \in A^* : \mathbf{b} \cdot \mathbf{x} = 1 \text{ for } \mathbf{b} \in B\},$$

being an intersection of exposed faces of A^*, is itself an exposed face of A^*. In this way we have constructed a mapping φ from the set of exposed faces of A to the set of exposed faces of A^*; we call φ the **polar face mapping** of A.

Theorem 2.8.6 *Let A be a compact convex set in $\mathbb{R}^n$ which contains the origin as an interior point. Then the polar face mapping φ of A is an inclusion-reversing bijection.*

Proof That φ is inclusion reversing follows immediately from its definition. Let ψ be the polar face mapping of A^*. We show that, for each

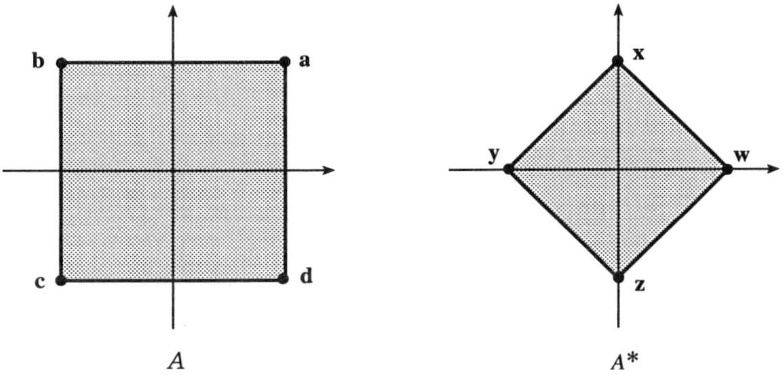

Fig. 2.14

exposed face B of A, $\psi(\varphi(B)) = B$. This is clear when B is either $\varnothing$ or A, and we assume that B is a *proper* exposed face of A. Thus there is $\mathbf{u}$ in $\mathbb{R}^n$ such that

$$B = \{\mathbf{a} \in A : \mathbf{u} \cdot \mathbf{a} = 1\} \qquad \text{and} \qquad \mathbf{u} \cdot \mathbf{a} \leqslant 1 \text{ for } \mathbf{a} \in A,$$

which shows that $\mathbf{u} \in A^*$ and $\mathbf{u} \in \varphi(B)$. Let $\mathbf{v} \in \psi(\varphi(B))$. Then $\mathbf{v} \in A^{**} = A$ and $\mathbf{u} \cdot \mathbf{v} = 1$, whence $\mathbf{v} \in B$ and $\psi(\varphi(B)) \subseteq B$. Conversely, let $\mathbf{b} \in B$. Then $\mathbf{b} \in A = (A^*)^*$ and $\mathbf{b} \cdot \mathbf{w} = 1$ for all $\mathbf{w} \in \varphi(B)$, whence $\mathbf{b} \in \psi(\varphi(B))$ and $B \subseteq \psi(\varphi(B))$. Thus $\psi(\varphi(B)) = B$.

We have just shown that the composite mapping $\psi \circ \varphi$ is the identity mapping on the set of exposed faces of A. By interchanging the roles of A and A^* in the discussion above, we can deduce that $\varphi \circ \psi$ is the identity mapping on the set of exposed faces of A^*. It now follows easily that φ is a bijection. □

Example 2.8.7 Let A be the square $\operatorname{conv}\{\mathbf{a}, \mathbf{b}, \mathbf{c}, \mathbf{d}\}$ in $\mathbb{R}^2$, where $\mathbf{a} = (1, 1)$, $\mathbf{b} = (-1, 1)$, $\mathbf{c} = (-1, -1)$, $\mathbf{d} = (1, -1)$. By Example 2.8.2, the polar dual A^* of A is the square $\operatorname{conv}\{\mathbf{w}, \mathbf{x}, \mathbf{y}, \mathbf{z}\}$, where $\mathbf{w} = (1, 0)$, $\mathbf{x} = (0, 1)$, $\mathbf{y} = (-1, 0)$, $\mathbf{w} = (0, -1)$. See Fig. 2.14. The polar face mapping φ of A is indicated below, where the faces of A and A^* are represented by their extreme points.

	$\varnothing$	**a**	**b**	**c**	**d**	**a, b**	**b, c**	**c, d**	**d, a**	A
$\varphi \downarrow$	↓	↓	↓	↓	↓	↓	↓	↓	↓	↓
	A^*	**w, x**	**x, y**	**y, z**	**z, w**	**x**	**y**	**z**	**w**	$\varnothing$

□

Exercises 2.8

1. Find the polar duals of the following sets in $\mathbb{R}^2$:
   (i) $\{(x, y) : x \geqslant 2\}$;
   (ii) $\{(x, y) : x \leqslant 2\}$;

(iii) $\{(x, y): x = 2\}$;
(iv) $\{(x, y): y \geqslant x^2\}$;
(v) $\{(x, y): (x/a)^2 + (y/b)^2 \leqslant 1\}$, where $a, b > 0$.

2. Show that $A^{***} = A^*$ for *every* set A in $\mathbb{R}^n$.
3. Prove that $(A \cap B)^* = \operatorname{cl}(\operatorname{conv}(A^* \cup B^*))$, whenever A and B are closed convex sets in $\mathbb{R}^n$ that contain the origin.
4. Let φ be the polar face mapping of a compact convex set A in $\mathbb{R}^n$ that has the origin as an interior point. Let B be an exposed face of A and let $\mathbf{b}_0 \in \operatorname{ri} B$. Show that $\varphi(B)$ is the set $\{\mathbf{x} \in A^* : \mathbf{b}_0 \cdot \mathbf{x} = 1\}$.
5. Let B be the set of exposed points of a compact convex set A in $\mathbb{R}^n$. Show that $A^* = B^*$. Find C^* for the set C in $\mathbb{R}^2$ defined by the equation

$$C = \{(x, y): x^2 + y^2 \leqslant 1 \text{ and } x \leqslant 0\} \cup \{(x, y): 0 \leqslant x \leqslant 1 \text{ and } |y| \leqslant 1\}.$$

Specify the polar face mapping of C.

3
Convex polytopes

This chapter gives a much fuller account of convex polytopes than is customary in an elementary text on convexity. Indeed, the first three chapters of the book could serve as a general introduction to convex sets with an emphasis on polytopes.

Sections 3.3–3.6 are not needed in the further development of the book, and so may be omitted on a first reading. Theorems 3.1.3 and 3.1.6, together with Section 3.2, however, should be considered as minimal reading here.

3.1 POLYTOPES

A **convex polytope**, or simply a **polytope**, is the convex hull of a *finite* set of points in $\mathbb{R}^n$. Polytopes were amongst the first convex sets ever to be studied. Points, line segments, polygons, tetrahedra, cubes, octahedra, dodecahedra, and icosahedra are all discussed in Euclid's *Elements*, written around 300 BC. Apart from their own very great intrinsic interest, they play an important role in the measurement of convex sets, in linear programming, the theory of games, and combinatorial geometry. In this section we establish their simplest properties, show that they can approximate compact convex sets to within any degree of accuracy, and introduce the concept of *combinatorial equivalence*.

Since the convex hull of a finite set in $\mathbb{R}^n$ is compact (Corollary 2.2.7), polytopes are compact convex sets, a fact of great importance, and one which we shall assume without comment whenever it is needed. A polytope of dimension r is called an r-**polytope**. The simplest example of an r-polytope is an r-**simplex** ($r = -1, \ldots, n$), which is defined to be the convex hull of an affinely independent set in $\mathbb{R}^n$ consisting of $r + 1$ points. There is precisely one (-1)-simplex, namely the empty set. Not surprisingly, we refer to a 0-simplex as a *point*, a 1-simplex as a line segment, a 2-simplex as a *triangle*, and a 3-simplex as a *tetrahedron*.

An important example of an r-polytope is an r-**crosspolytope** ($r = 1, \ldots, n$), which is defined to be the convex hull of r linearly independent line segments in $\mathbb{R}^n$ whose midpoints coincide, i.e. a translate of a set of the form $\operatorname{conv}\{\pm\mathbf{a}_1, \ldots, \pm\mathbf{a}_r\}$, where $\{\mathbf{a}_1, \ldots, \mathbf{a}_r\}_{\neq}$ is a linearly independent set of vectors in $\mathbb{R}^n$. Such a crosspolytope is called **regular** when the $\mathbf{a}_1, \ldots, \mathbf{a}_r$ have equal lengths and are

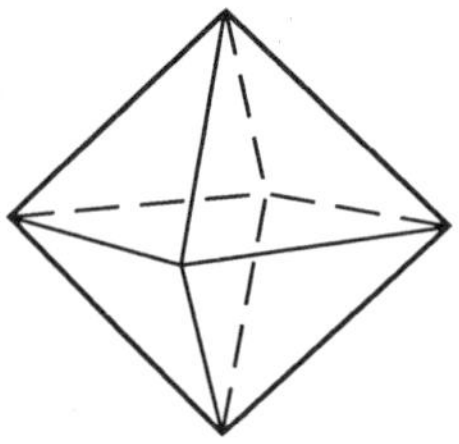

Regular octahedron

Fig. 3.1

mutually orthogonal. Thus conv$\{\pm\mathbf{e}_1, \ldots, \pm\mathbf{e}_r\}$, where $\mathbf{e}_1, \ldots, \mathbf{e}_r$ are elementary vectors in $\mathbb{R}^n$, is a regular r-crosspolytope. In $\mathbb{R}^3$ a regular 2-crosspolytope is a square, and a regular 3-crosspolytope is a *regular octahedron*, which is a regular solid bounded by eight congruent equilateral triangles. See Fig. 3.1.

Theorem 3.1.1 *Let A, B be polytopes in $\mathbb{R}^n$ and let $\alpha \in \mathbb{R}$. Then $A + B$ and αA are polytopes.*

Proof We consider the non-trivial case when neither A nor B is empty. Let $A = \operatorname{conv}\{\mathbf{a}_1, \ldots, \mathbf{a}_k\}$, $B = \operatorname{conv}\{\mathbf{b}_1, \ldots, \mathbf{b}_m\}$, where $\mathbf{a}_1, \ldots, \mathbf{a}_k, \mathbf{b}_1, \ldots, \mathbf{b}_m \in \mathbb{R}^n$. Denote by C the finite set consisting of all those points of the form $\mathbf{a}_i + \mathbf{b}_j$, where $i = 1, \ldots, k$ and $j = 1, \ldots, m$, and by D the finite set whose points are $\alpha\mathbf{a}_1, \ldots, \alpha\mathbf{a}_k$. We prove the theorem by showing that $A + B = \operatorname{conv} C$ and $\alpha A = \operatorname{conv} D$.

Now $A + B$ is a convex set containing C, whence $\operatorname{conv} C \subseteq A + B$. If $\mathbf{x} \in A + B$, then there exist scalars $\lambda_1, \ldots, \lambda_k, \mu_1, \ldots, \mu_m \geqslant 0$ with $\lambda_1 + \ldots + \lambda_k = 1$, $\mu_1 + \ldots + \mu_m = 1$ such that

$$\mathbf{x} = \lambda_1\mathbf{a}_1 + \ldots + \lambda_k\mathbf{a}_k + \mu_1\mathbf{b}_1 + \ldots + \mu_m\mathbf{b}_m = \sum_{i=1}^{k}\sum_{j=1}^{m} \lambda_i\mu_j(\mathbf{a}_i + \mathbf{b}_j),$$

which shows that $\mathbf{x}$ is a convex combination of points of C, whence $\mathbf{x} \in \operatorname{conv} C$ and $A + B \subseteq \operatorname{conv} C$. Thus $A + B = \operatorname{conv} C$.

Now αA is a convex set containing D, whence $\operatorname{conv} D \subseteq \alpha A$. If $\mathbf{x} \in \alpha A$, then there exist $\lambda_1, \ldots, \lambda_k \geqslant 0$ with $\lambda_1 + \ldots + \lambda_k = 1$ such that

$$\mathbf{x} = \alpha(\lambda_1\mathbf{a}_1 + \ldots + \lambda_k\mathbf{a}_k) = \lambda_1(\alpha\mathbf{a}_1) + \ldots + \lambda_k(\alpha\mathbf{a}_k),$$

which shows that $\mathbf{x}$ is a convex combination of points of D, whence $\mathbf{x} \in \operatorname{conv} D$ and $\alpha A \subseteq \operatorname{conv} D$. Thus $\alpha A = \operatorname{conv} D$. □

Corollary 3.1.2 *Let $A_1, \ldots, A_m$ be polytopes in $\mathbb{R}^n$ and let $\alpha_1, \ldots, \alpha_m \in \mathbb{R}$. Then $\alpha_1 A_1 + \ldots + \alpha_m A_m$ is a polytope.* □

Corollary 3.1.2 shows that the vector sum of a finite number of line segments in $\mathbb{R}^n$ is a polytope; such a polytope is called a **zonotope**. An important example of a zonotope is an *r*-**cube** ($r = 1, \ldots, n$) in $\mathbb{R}^n$ which is the vector sum of r mutually orthogonal line segments in $\mathbb{R}^n$, all of equal length, i.e. a set of the form

$$\operatorname{conv}\{\mathbf{a}_1, \mathbf{b}_1\} + \ldots + \operatorname{conv}\{\mathbf{a}_r, \mathbf{b}_r\},$$

where $\mathbf{a}_1, \ldots, \mathbf{a}_r, \mathbf{b}_1, \ldots, \mathbf{b}_r \in \mathbb{R}^n$, $(\mathbf{a}_i - \mathbf{b}_i) \cdot (\mathbf{a}_j - \mathbf{b}_j) = 0$ if and only if $i \neq j$, and $\|\mathbf{a}_i - \mathbf{b}_i\| = \|\mathbf{a}_j - \mathbf{b}_j\|$ for all i, j. A simple example of an n-cube with *edge-length* 1 in $\mathbb{R}^n$ is the polytope

$$\operatorname{conv}\{\mathbf{0}, \mathbf{e}_1\} + \ldots + \operatorname{conv}\{\mathbf{0}, \mathbf{e}_n\} = \{(x_1, \ldots, x_n) : 0 \leqslant x_1, \ldots, x_n \leqslant 1\}.$$

We now turn our attention to the facial structure of a polytope P in $\mathbb{R}^n$. It is customary to call the extreme points of P its **vertices** and its 1-faces its **edges**. The set of all P's vertices is called its **vertex set**. If $P = \operatorname{conv}\{\mathbf{a}_1, \ldots, \mathbf{a}_m\}$, for some $\mathbf{a}_1, \ldots, \mathbf{a}_m \in \mathbb{R}^n$, then Corollary 2.6.4 shows that the vertex set of P is contained in $\{\mathbf{a}_1, \ldots, \mathbf{a}_m\}$, a fact which we use in the proof of the next theorem.

Theorem 3.1.3 *Every polytope in $\mathbb{R}^n$ has only a finite number of faces, and each of these is a polytope.*

Proof Consider a non-empty polytope $A = \operatorname{conv}\{\mathbf{a}_1, \ldots, \mathbf{a}_m\}$, where $\mathbf{a}_1, \ldots, \mathbf{a}_m \in \mathbb{R}^n$. By Theorem 2.6.16, each face F of A is the convex hull of its extreme points. Theorem 2.6.5 (ii) shows that each extreme point of F is also an extreme point of A, whence F is the convex hull of some subset of the vertex set of A. Since $\{\mathbf{a}_1, \ldots, \mathbf{a}_m\}$ contains the vertex set of A, it follows that F is the convex hull of some subset of $\{\mathbf{a}_1, \ldots, \mathbf{a}_m\}$. The desired result is now immediate. □

Suppose that V is the vertex set of a polytope P in $\mathbb{R}^n$. Then the proof of the last theorem shows that each face of P has the form $\operatorname{conv} W$, for some subset W of V. The question naturally arises as to which subsets W of V **determine** a face of P, i.e. are such that $\operatorname{conv} W$ is a face of P. A workable answer is provided by the following result.

Theorem 3.1.4 *Let W be a subset of the vertex set V of a polytope P in $\mathbb{R}^n$. Then $\operatorname{conv} W$ is a face of P if and only if*

$$(\operatorname{aff} W) \cap \operatorname{conv}(V \setminus W) = \varnothing. \tag{3.1.1}$$

Proof Suppose first that $\operatorname{conv} W$ is a face of P. If $\mathbf{v} \in V \setminus W$, then

$P \setminus \{\mathbf{v}\}$ is convex, by Theorem 2.6.2, and contains W. Hence $\operatorname{conv} W \subseteq P \setminus \{\mathbf{v}\}$, $\mathbf{v} \notin \operatorname{conv} W$, and $V \setminus W \subseteq P \setminus \operatorname{conv} W$. Theorem 2.6.2 also shows that $P \setminus \operatorname{conv} W$ is convex, so $\operatorname{conv}(V \setminus W) \subseteq P \setminus \operatorname{conv} W$, and that $(\operatorname{aff} W) \cap P = \operatorname{conv} W$. It follows that (3.1.1) holds.

Suppose next that (3.1.1) is satisfied. Clearly $\operatorname{conv} W$ is a face of P if either W is empty or V, so we assume that this is not the case. Let $V = \{\mathbf{v}_1, \ldots, \mathbf{v}_s\}_{\neq}$ and $W = \{\mathbf{v}_1, \ldots, \mathbf{v}_r\}$, where $1 \leqslant r < s$. Let $\mathbf{w} = \lambda\mathbf{x} + \mu\mathbf{y}$, where $\mathbf{w} \in \operatorname{conv} W$, $\mathbf{x}, \mathbf{y} \in P$, and $\lambda, \mu > 0$ with $\lambda + \mu = 1$. Then $\mathbf{x} = \lambda_1\mathbf{v}_1 + \ldots + \lambda_s\mathbf{v}_s$, $\mathbf{y} = \mu_1\mathbf{v}_1 + \ldots + \mu_s\mathbf{v}_s$, for some $\lambda_1, \ldots, \lambda_s, \mu_1, \ldots, \mu_s \geqslant 0$ with $\lambda_1 + \ldots + \lambda_s = 1$, $\mu_1 + \ldots + \mu_s = 1$. For $i = 1, \ldots, s$, write $\nu_i = \lambda\lambda_i + \mu\mu_i$. Then $\nu_1, \ldots, \nu_s \geqslant 0$, $\nu_1 + \ldots + \nu_s = 1$, and $\mathbf{w} = \nu_1\mathbf{v}_1 + \ldots + \nu_s\mathbf{v}_s$. Write $\alpha = \nu_{r+1} + \ldots + \nu_s$. If $\alpha > 0$, then the point

$$(\mathbf{w} - \nu_1\mathbf{v}_1 - \ldots - \nu_r\mathbf{v}_r)/\alpha = (\nu_{r+1}\mathbf{v}_{r+1} + \ldots + \nu_s\mathbf{v}_s)/\alpha$$

lies *both* in $\operatorname{aff} W$ and $\operatorname{conv}(V \setminus W)$, which contradicts (3.1.1). Thus $\alpha = 0$, which entails $\nu_{r+1}, \ldots, \nu_s = 0$ and $\lambda_{r+1}, \ldots, \lambda_s, \mu_{r+1}, \ldots, \mu_s = 0$. Hence $\mathbf{x}, \mathbf{y} \in \operatorname{conv} W$, and so $\operatorname{conv} W$ is a face of P. □

In proving the *if* part of the last theorem, we used the fact that $\operatorname{conv} V = P$, but not the fact that each element of V was a vertex of P. We thus have the following result.

Corollary 3.1.5 *Let W be a subset of a finite set V in $\mathbb{R}^n$ such that* $(\operatorname{aff} W) \cap \operatorname{conv}(V \setminus W) = \varnothing$. *Then* $\operatorname{conv} W$ *is a face of the polytope* $\operatorname{conv} V$. □

We are now able to give a complete description of the facial structure of a simplex S in $\mathbb{R}^n$. Suppose that $S = \operatorname{conv} V$, where V is an affinely independent set in $\mathbb{R}^n$. We have already seen that each face of S is the convex hull of some subset of V; now we establish the converse of this result. Let $W \subseteq V$. Since V is affinely independent,

$$(\operatorname{aff} W) \cap \operatorname{conv}(\mathrm{V} \setminus W) \subseteq (\operatorname{aff} W) \cap \operatorname{aff}(\mathrm{V} \setminus W) = \varnothing,$$

whence $\operatorname{conv} W$ is a face of S by Corollary 3.1.5. In particular, each point of V is a vertex of S.

We conclude this brief discussion of the facial structure of polytopes by introducing the basic concept of *combinatorial equivalence*. Let P, P' be polytopes, not necessarily lying in the same Euclidean space, with vertex sets V, V', respectively. Then P and P' are said to be **combinatorially equivalent** if there exists a bijection $\varphi: V \to V'$ such that a subset W of V determines a face of P if and only if $\varphi(W)$ determines a face of P'. Since 1-polytopes are simply line segments, they are all combinatorially equivalent to one another. Two 2-polytopes (polygons) are combinatorially equivalent if and only if they have the same number of vertices. Clearly, if two polytopes are combinatorially equivalent, then they must have the same number of vertices. The converse of this result is not true. To see this, in $\mathbb{R}^3$ let P be a square pyramid and let P' be the polytope obtained by taking the union of a regular tetrahedron and its reflection in one of its triangular faces. Both P and P' have five vertices, but they are not combinatorially equivalent, for P has a face with four vertices, but P' does not. We shall show in Section 3.6 that every 3-polytope with five vertices is combinatorially equivalent to either P or P', i.e. there are just two *combinatorial types* for 3-polytopes having five vertices.

There are occasions in the study of convexity when it is expedient to regard a compact convex set as the limit of a sequence of polytopes. For example, one way of defining the perimeter of a compact convex set in $\mathbb{R}^2$ is as the limit of the perimeters of a sequence of polygons which converges to the set. The basis for arguments which involve a compact convex set being approximated by a sequence of polytopes is the next theorem and its corollary. We shall make much use of these results in our account of *mixed volumes* in Section 6.4. In the discussion below, ρ denotes the Hausdorff metric which was introduced in Section 2.7.

Theorem 3.1.6 *Let A be a non-empty compact convex set in $\mathbb{R}^n$ and let $\varepsilon > 0$. Then there exist polytopes P, Q in $\mathbb{R}^n$ such that $P \subseteq A \subseteq Q$ and $\rho(A, P) \leqslant \varepsilon$, $\rho(A, Q) \leqslant \varepsilon$.*

Proof By Theorem 1.8.5, there exists a *finite* set E in $\mathbb{R}^n$ such that $E \subseteq A \subseteq (E)_\varepsilon$. Let $P = \operatorname{conv} E$. Then P is a polytope satisfying $P \subseteq A \subseteq (P)_\varepsilon$, whence $\rho(A, P) \leqslant \varepsilon$. Replacing A by $(A)_\varepsilon$ in the last argument, we deduce the existence of a polytope Q satisfying $Q \subseteq (A)_\varepsilon \subseteq (Q)_\varepsilon$. The inclusion $(A)_\varepsilon \subseteq (Q)_\varepsilon$, i.e. $A + \varepsilon U \subseteq Q + \varepsilon U$, implies $A \subseteq Q$ by Theorem 2.1.8. The inequality $\rho(A, Q) \leqslant \varepsilon$ now follows. □

Corollary 3.1.7. *Let A be a non-empty compact convex set in $\mathbb{R}^n$. Then there exist sequences $P_1, \ldots, P_i, \ldots$ and $Q_1, \ldots, Q_i, \ldots$ of non-empty polytopes in $\mathbb{R}^n$ such that $P_i \subseteq A \subseteq Q_i$ for $i = 1, 2, \ldots,$ and $P_i \to A$ and $Q_i \to A$ as $i \to \infty$.* □

Exercises 3.1

1. Show that the set $\{(x_1, \ldots, x_n) : x_1, \ldots, x_n \geqslant 0$ and $x_1 + \ldots + x_n \leqslant 1\}$ is an n-simplex in $\mathbb{R}^n$.
2. In $\mathbb{R}^n$ let V be the vertex set of a polytope P and let $\mathbf{x}$ be a point

not lying in aff P. Show that conv $(\{\mathbf{x}\} \cup P)$ is a polytope with vertex set $\{\mathbf{x}\} \cup V$.

3. Prove that in $\mathbb{R}^n$ every compact convex set can be expressed as the intersection of a sequence of polytopes.
4. Let $P_1, \ldots, P_i, \ldots,$ be a sequence of non-empty polytopes in $\mathbb{R}^n$ that converges to a non-empty compact convex set A. Denote by $v(P)$ the number of vertices of a polytope P. Show that, if A is *not* a polytope, then $v(P_i) \to \infty$ as $i \to \infty$. Is the converse of this last result true?
5. Prove that a polytope in $\mathbb{R}^n$ is *not* the union of a *strictly* increasing sequence of convex sets.

3.2 POLYHEDRAL SETS

Before continuing with our discussion of polytopes, we pause to introduce the closely related concept of a *polyhedral set*. The definition of a polyhedral set as the solution set of a finite system of linear inequalities immediately makes it clear why such sets play a leading role in linear programming, a topic discussed in the next chapter. Here we establish some basic properties of polyhedral sets, and characterize polytopes as bounded polyhedral sets.

A **polyhedral set** is the intersection of a finite family of closed halfspaces in $\mathbb{R}^n$. Equivalently, a polyhedral set is the set of all points $(x_1, \ldots, x_n)$ in $\mathbb{R}^n$ which satisfy a finite system of linear inequalities of the form

$$\begin{gathered} a_{11}x_1 + \ldots + a_{1n}x_n \leqslant a_{10} \\ \cdots\cdots\cdots\cdots\cdots\cdots\cdots \\ a_{m1}x_1 + \ldots + a_{mn}x_n \leqslant a_{m0}. \end{gathered}$$

Clearly polyhedral sets are closed and convex, and the intersection of any *finite* family of polyhedral sets is a polyhedral set. Each hyperplane in $\mathbb{R}^n$ is an intersection of two closed halfspaces, and so is a polyhedral set. Since each flat in $\mathbb{R}^n$ is a finite intersection of hyperplanes, all flats are polyhedral sets. In particular, the empty set and $\mathbb{R}^n$ itself are polyhedral sets. Later in this section we show that every polytope is a polyhedral set, a fact which can be easily verified for particular polytopes in $\mathbb{R}^3$. The non-negative orthant in $\mathbb{R}^n$ and the set illustrated in Fig. 2.8 are other examples of polyhedral sets.

A concept of special importance in the study of polyhedral sets is that of a *facet*. A **facet** of an r-dimensional polyhedral set in $\mathbb{R}^n$ is a *proper* $(r-1)$-dimensional face of the set. In $\mathbb{R}^3$ the non-negative orthant has three facets, a tetrahedron four, a square pyramid five, and a cube six.

Since flats have no *proper* faces, they have no facets; it is shown in Theorem 3.2.1 that flats are the only polyhedral sets with this property.

Our first result shows how the facets of a polyhedral set can be determined, and how these facets determine the facial structure of the set itself.

Theorem 3.2.1 *Suppose that the polyhedral set A in $\mathbb{R}^n$ is not a flat and that*

$$A = (\operatorname{aff} A) \cap \{\mathbf{x} \in \mathbb{R}^n : \mathbf{a}_1 \cdot \mathbf{x} \leqslant \alpha_1\} \cap \ldots \cap \{\mathbf{x} \in \mathbb{R}^n : \mathbf{a}_m \cdot \mathbf{x} \leqslant \alpha_m\},$$

where $\alpha_1, \ldots, \alpha_m \in \mathbb{R}$, $\mathbf{a}_1, \ldots, \mathbf{a}_m \in \mathbb{R}^n \setminus \{\mathbf{0}\}$ and no one of the closed halfspaces in the intersection can be omitted. For each $i = 1, \ldots, m$, let

$$F_i = A \cap \{\mathbf{x} \in \mathbb{R}^n : \mathbf{a}_i \cdot \mathbf{x} = \alpha_i\}.$$

Then:

(i) $\operatorname{ri} A = \{\mathbf{a} \in A : \mathbf{a}_1 \cdot \mathbf{a} < \alpha_1, \ldots, \mathbf{a}_m \cdot \mathbf{a} < \alpha_m\}$;

(ii) $\operatorname{rebd} A = F_1 \cup \ldots \cup F_m$;

(iii) *the facets of A are precisely the sets $F_1, \ldots, F_m$;*

(iv) *each proper face of A is the intersection of those facets of A that contain it;*

(v) *A has a finite number of faces, each of which is exposed;*

(vi) *each face of A is a polyhedral set;*

(vii) *let B_j, B_k be j- and k-faces, respectively, of A $(0 \leqslant j \leqslant k - 2)$ such that $B_j \subset B_k$. Then there are faces $B_{j+1}, \ldots, B_{k-1}$ of A such that, for each $i = j, \ldots, k - 1$, the face B_i is a facet of B_{i+1}.*

Proof We note that every polyhedral set A in $\mathbb{R}^n$ can be expressed in the form required by the theorem. The assumption that A is not a flat implies that $m \geqslant 1$.

(i) Suppose first that $\mathbf{a} \in A$ and that $\mathbf{a}_1 \cdot \mathbf{a} < \alpha_1, \ldots, \mathbf{a}_m \cdot \mathbf{a} < \alpha_m$. Then $\mathbf{a}$ belongs to the set

$$C = \{\mathbf{x} \in \mathbb{R}^n : \mathbf{a}_1 \cdot \mathbf{x} < \alpha_1, \ldots, \mathbf{a}_m \cdot \mathbf{x} < \alpha_m\},$$

which is open, being a finite intersection of open halfspaces. Thus there exists $r > 0$ such that $B(\mathbf{a}; r) \subseteq C$, whence

$$B(\mathbf{a}; r) \cap \operatorname{aff} A \subseteq C \cap \operatorname{aff} A \subseteq A.$$

Hence $\mathbf{a} \in \operatorname{ri} A$.

Suppose next that $\mathbf{a} \in \operatorname{ri} A$. Since no one of the closed halfspaces in the representation of A given in the statement of the theorem can be omitted, for each $i = 1, \ldots, m$, there exists $\mathbf{z}_i \in \operatorname{aff} A$ such that $\mathbf{a}_j \cdot \mathbf{z}_i \leqslant \alpha_j$ when $j \neq i$ and $\mathbf{a}_i \cdot \mathbf{z}_i > \alpha_i$. Hence, for each $i = 1, \ldots, m$, there exists $\lambda_i \in (0, 1)$ such that $\lambda_i \mathbf{z}_i + (1 - \lambda_i)\mathbf{a} \in A$. Hence

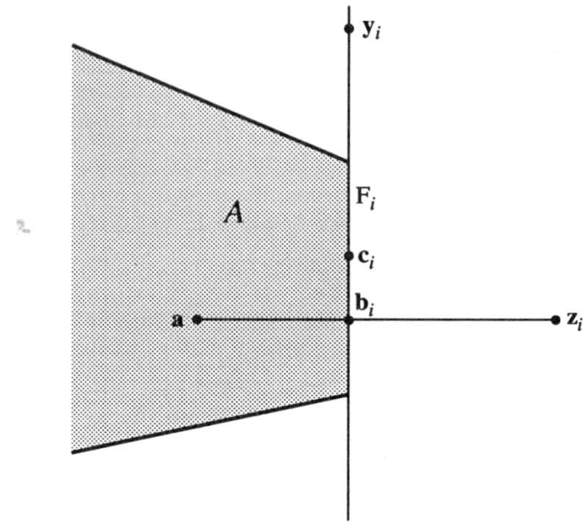

Fig. 3.2

$$\alpha_i \geqslant \mathbf{a}_i \cdot (\lambda_i \mathbf{z}_i + (1 - \lambda_i)\mathbf{a}) = \lambda_i \mathbf{a}_i \cdot \mathbf{z}_i + (1 - \lambda_i)\mathbf{a}_i \cdot \mathbf{a} > \lambda_i \alpha_i + (1 - \lambda_i)\mathbf{a}_i \cdot \mathbf{a},$$

which shows that $\mathbf{a}_i \cdot \mathbf{a} < \alpha_i$. This establishes (i).

(ii) This follows immediately from (i).

(iii) We now show that, for each $i = 1, \ldots, m$, F_i is a facet of A. Let $\mathbf{a} \in \operatorname{ri} A$. Let $\mathbf{z}_i$ be as in (i). Then $\mathbf{a}_i \cdot \mathbf{a} < \alpha_i < \mathbf{a}_i \cdot \mathbf{z}_i$. Write $\mu_i = (\alpha_i - \mathbf{a}_i \cdot \mathbf{a})/(\mathbf{a}_i \cdot \mathbf{z}_i - \mathbf{a}_i \cdot \mathbf{a})$, so that $0 < \mu_i < 1$, and write $\mathbf{b}_i = \mu_i \mathbf{z}_i + (1 - \mu_i)\mathbf{a}$. Then $\mathbf{b}_i \in \operatorname{aff} A$, $\mathbf{a}_i \cdot \mathbf{b}_i = \alpha_i$ and $\mathbf{a}_j \cdot \mathbf{b}_i < \alpha_j$ for $j \neq i$, whence $\mathbf{b}_i \in A$. Thus $\mathbf{b}_i \in F_i$ and $\mathbf{a}_i \cdot \mathbf{x} = \alpha_i$ is a support hyperplane to A at $\mathbf{b}_i$; it follows that F_i is a proper exposed face of A. See Fig. 3.2. We now show that

$$\operatorname{aff} F_i = (\operatorname{aff} A) \cap \{\mathbf{x} \in \mathbb{R}^n : \mathbf{a}_i \cdot \mathbf{x} = \alpha_i\}. \tag{3.2.1}$$

Let $\mathbf{y}_i$ be a point belonging to the set on the right-hand side of (3.2.1). Choose $\theta_i > 0$ such that $\theta_i(\mathbf{a}_j \cdot \mathbf{y}_i - \mathbf{a}_j \cdot \mathbf{b}_i) \leqslant \alpha_j - \mathbf{a}_j \cdot \mathbf{b}_i$ when $j \neq i$. Write $\mathbf{c}_i = \theta_i \mathbf{y}_i + (1 - \theta_i)\mathbf{b}_i$. Then $\mathbf{c}_i \in \operatorname{aff} A$, $\mathbf{a}_i \cdot \mathbf{c}_i = \alpha_i$, and $\mathbf{a}_j \cdot \mathbf{c}_i \leqslant \alpha_j$ when $j \neq i$, whence $\mathbf{c}_i \in F_i$. Since

$$\mathbf{y}_i = (1/\theta_i)\mathbf{c}_i + (1 - (1/\theta_i))\mathbf{b}_i \in \operatorname{aff} F_i,$$

we deduce that

$$(\operatorname{aff} A) \cap \{\mathbf{x} \in \mathbb{R}^n : \mathbf{a}_i \cdot \mathbf{x} = \alpha_i\} \subseteq \operatorname{aff} F_i.$$

The opposite inclusion is trivial, and so (3.2.1) is established. Using (3.2.1) together with Theorem 1.3.8, we conclude that

$$\dim F_i = \dim(\operatorname{aff} F_i) = \dim(\operatorname{aff} A) - 1 = \dim A - 1,$$

and hence that F_i is a facet of A.

We now show that each facet of A is one of the F_i's. Let F be a facet of A. Let $\mathbf{f} \in \operatorname{ri} F$. Since F is a *proper* face of A, $\mathbf{f} \notin \operatorname{ri} A$. Hence, by (ii), $\mathbf{f} \in F_{i_0}$ for some $i_0 \in \{1, \ldots, m\}$. Now the faces F and F_{i_0} of A have the same dimension and $\mathbf{f} \in \operatorname{ri} F$, whence $F = F_{i_0}$.

(iv) Suppose that B is a *proper* face of A. Let $\mathbf{b} \in \operatorname{ri} B$. Denote by I the non-empty set of those i's in $\{1, \ldots, m\}$ for which $\mathbf{a}_i \cdot \mathbf{b} = \alpha_i$, i.e. $\mathbf{b} \in F_i$. Denote by J the set of those j's in $\{1, \ldots, m\}$ for which $\mathbf{a}_j \cdot \mathbf{b} < \alpha_j$. Let E be the intersection of all those facets of A which contain $\mathbf{b}$. Since $\mathbf{b} \in F_i$ if and only if $B \subseteq F_i$, the set E is the intersection of all those facets of A which contain B. Hence E is a face of A which contains B. Choose $r > 0$ such that, for each $j \in J$,

$$B(\mathbf{b}; r) \subseteq \{\mathbf{x} \in \mathbb{R}^n : \mathbf{a}_j \cdot \mathbf{x} < \alpha_j\}.$$

This inclusion, together with the trivial inclusions

$$\operatorname{aff} E \subseteq \operatorname{aff} A \qquad \text{and} \qquad \operatorname{aff} E \subseteq \{\mathbf{x} \in \mathbb{R}^n : \mathbf{a}_i \cdot \mathbf{x} = \alpha_i\} \qquad \text{for } i \in I,$$

shows that

$$B(\mathbf{b}; r) \cap \operatorname{aff} E \subseteq E,$$

whence $\mathbf{b} \in \operatorname{ri} E$. Thus $\mathbf{b} \in \operatorname{ri} B \cap \operatorname{ri} E$, and so $B = E$, as desired.

(v) This follows easily from (iv), since A has only m facets and each of these is an exposed face of A.

(vi) This follows from the facts that each *proper* face of A is the intersection of A with one of its support hyperplanes, and the intersection of two polyhedral sets is itself a polyhedral set.

(vii) Now B_j is a proper face of the polyhedral set B_k, whence by (iv) there is some facet B_{k-1} of B_k which contains B_j. If $j = k - 2$, then the proof is complete. Otherwise, repeat this last argument $k - j - 2$ more times to obtain the desired faces $B_{k-2}, \ldots, B_{j+1}$. □

Theorem 3.2.1 concerns polyhedral sets which are not flats. It is convenient, however, to have a statement of the main properties of general polyhedral sets. This is given in the next result.

Theorem 3.2.2 *Let A be a polyhedral set in $\mathbb{R}^n$. Then A has a finite number of faces, each of which is exposed and a polyhedral set. Every proper face of A is the intersection of those facets of A that contain it, and* $\operatorname{rebd} A$ *is the union of all the facets of A. If A has a non-empty face of dimension s, then A has faces of all dimensions from s to* $\dim A$.

Proof The theorem is trivially true when A is a flat. When A is not a flat, it follows easily from Theorem 3.2.1. □

Theorem 3.2.3 *Let A be a closed convex set in $\mathbb{R}^n$ which has only a finite number of exposed faces. Then A is a polyhedral set.*

Proof If A has no *proper* exposed faces, then it must be a flat, which is polyhedral. Suppose, then, that A has proper *exposed* faces $B_1, \ldots, B_m$. Let $H_1, \ldots, H_m$ be support hyperplanes to A such that $B_1 = A \cap H_1, \ldots, B_m = A \cap H_m$. For each $i = 1, \ldots, m$, let J_i be the closed halfspace of $\mathbb{R}^n$ bounded by H_i which contains A.

Define a polyhedral set P by the equation

$$P = J_1 \cap \ldots \cap J_m \cap \operatorname{aff} A.$$

Clearly, $A \subseteq P$. Suppose that $P \nsubseteq A$. Then there is a point $\mathbf{p}$ lying in $P \setminus A$. Let $\mathbf{a} \in \operatorname{ri} A$. Since A is closed and $\mathbf{p} \in \operatorname{aff} A$, there exists $\lambda \in (0, 1)$ such that the point $\mathbf{b} = \lambda\mathbf{p} + (1 - \lambda)\mathbf{a}$ belongs to $\operatorname{rebd} A$. By Theorem 2.4.12, there is some $i \in \{1, \ldots, m\}$ such that $\mathbf{b} \in B_i$. Now H_i is a face of J_i, $\mathbf{b} \in H_i$, and $\mathbf{p}, \mathbf{a} \in J_i$, whence $\mathbf{a} \in H_i$. Thus $\mathbf{a} \in B_i$, which is impossible, since $\mathbf{a}$ cannot be both a relative interior point of A *and* a member of one of its proper faces! Hence $P \subseteq A$, and A is the polyhedral set P. □

Corollary 3.2.4 *A closed convex set in $\mathbb{R}^n$ which has only a finite number of faces is a polyhedral set.* □

Theorem 3.2.1(v) and Corollary 3.2.4 characterize the polyhedral sets as those closed convex sets which have a finite number of faces. The next theorem characterizes polytopes as bounded polyhedral sets.

Theorem 3.2.5 *A set in $\mathbb{R}^n$ is a polytope if and only if it is a bounded polyhedral set.*

Proof Each polytope in $\mathbb{R}^n$ is compact and has a finite number of faces, and so, by Corollary 3.2.4, must be a bounded polyhedral set. Conversely, every bounded polyhedral set in $\mathbb{R}^n$ is compact and has a finite number of faces. In particular, it has a finite number of extreme points, and so, by Theorem 2.6.16, must be a polytope. □

Corollary 3.2.6 *The intersection of two polytopes in $\mathbb{R}^n$ is a polytope.*

Proof In view of the theorem, the corollary simply states the obvious fact that the intersection of two bounded polyhedral sets is a bounded polyhedral set. □

Exercises 3.2

1. Show that in $\mathbb{R}^n$ the complement of a finite union of polyhedral sets can be expressed as a finite union of open convex sets. Deduce

that, if C is a convex set and $P_1, \ldots, P_m$ are polyhedral sets in $\mathbb{R}^n$ such that

$$C \subseteq P_1 \cup \ldots \cup P_m,$$

then there exists a polyhedral set P such that

$$C \subseteq P \subseteq P_1 \cup \ldots \cup P_m.$$

2. Prove that every n-dimensional polytope in $\mathbb{R}^n$ has at least $n + 1$ facets.
3. Show that the polar dual of a polyhedral set in $\mathbb{R}^n$ which contains the origin in its interior is a polytope.
4. Let A be a convex set in $\mathbb{R}^n$ having a finite, non-zero number of extreme points. Prove that, if the intersection of A with every polytope in $\mathbb{R}^n$ is a polytope, then A is a polyhedral set.

3.3 PYRAMIDS, BIPYRAMIDS, AND PRISMS

We now look at some particular examples of polytopes, introducing the analogues in $\mathbb{R}^n$ of the pyramids, bipyramids, and prisms of elementary geometry, and studying their combinatorial properties. Throughout, we denote by $f_k(P)$ the number of k-faces (faces of dimension k) of an r-polytope P, whence

$$f_{-1}(P) = f_r(P) = 1, \quad \text{and} \quad f_k(P) = 0 \quad \text{when} \quad k < -1 \quad \text{or} \quad k > r.$$

Our results will lead us to anticipate *Euler's relation*, which asserts that, for any non-empty r-polytope P,

$$f_{-1}(P) - f_0(P) + \ldots + (-1)^{r+1} f_r(P) = 0,$$

a fact which will be proved in Section 3.5.

Let S be a non-empty r-simplex in $\mathbb{R}^n$. Then $S = \operatorname{conv} V$ for some affinely independent set V of $r + 1$ points of $\mathbb{R}^n$. We know from Section 3.1 that, for each $k = -1, 0, \ldots, r$, the k-faces of S are precisely those sets of the form $\operatorname{conv} W$, where W is a subset of V having $k + 1$ points. Thus $f_k(S)$ equals the number of ways of choosing $k + 1$ points from a set of $r + 1$ points. Hence, using the standard notation for the binomial coefficients, we see that

$$f_k(S) = \binom{r+1}{k+1} = \frac{(r+1)!}{(k+1)!(r-k)!},$$

whence, for all real x,

$$(1 + x)^{r+1} = f_{-1}(S) + f_0(S)x + \ldots + f_r(S)x^{r+1}.$$

Setting $x = -1$ in this equation, we deduce that

$$f_{-1}(S) - f_0(S) + \ldots + (-1)^{r+1} f_r(S) = 0,$$

which shows that S satisfies Euler's relation.

In elementary geometry a pyramid is usually specified by stating its base and its apex, and this continues to be the case in $\mathbb{R}^n$. Let Q be a non-empty $(r-1)$-polytope in $\mathbb{R}^n$ and let $\mathbf{x}$ be a point of $\mathbb{R}^n$ not lying in aff Q. Then the *r-pyramid* P with *apex* $\mathbf{x}$ and *base* Q is defined to be the r-polytope conv $(\{\mathbf{x}\} \cup Q)$; we say that P is obtained from Q by applying the *cone construction* with apex $\mathbf{x}$.

Theorem 3.3.1 *Let P be an r-pyramid in $\mathbb{R}^n$ with apex $\mathbf{x}$ and base a non-empty $(r-1)$-polytope Q. Then*

$$f_k(P) = f_k(Q) + f_{k-1}(Q) \qquad \textit{for } k = -1, \ldots, r.$$

Proof We show first that, for $A, B \subseteq \text{aff}\, Q$,

$$(\text{aff}\,(\{\mathbf{x}\} \cup A)) \cap B = (\text{aff}\, A) \cap B. \tag{3.3.1}$$

Consider the non-trivial case when A is non-empty. If $\mathbf{b}$ lies in the set on the left-hand side of (3.3.1), then there exist $\mathbf{a}_1, \ldots, \mathbf{a}_m \in A$ and $\lambda, \lambda_1, \ldots, \lambda_m \in \mathbb{R}$ with $\lambda + \lambda_1 + \ldots + \lambda_m = 1$ such that

$$\mathbf{b} = \lambda \mathbf{x} + \lambda_1 \mathbf{a}_1 + \ldots + \lambda_m \mathbf{a}_m.$$

If $\lambda \neq 0$, then the last equation can be rearranged to express $\mathbf{x}$ as an affine combination of points of aff Q, which contradicts the (implied) hypothesis that $\mathbf{x} \notin \text{aff}\, Q$. Thus $\lambda = 0$, $\mathbf{b} \in (\text{aff}\, A) \cap B$ and

$$(\text{aff}\,(\{\mathbf{x}\} \cup A)) \cap B \subseteq (\text{aff}\, A) \cap B.$$

The opposite inclusion is clear, and so (3.3.1) is established.

Denote by V the vertex set of Q. Then $P = \text{conv}\,(\{\mathbf{x}\} \cup V)$. By Corollary 3.1.5, $\{\mathbf{x}\}$ and Q are faces of P. Hence each of the $f_k(Q)$ k-faces of Q is also a k-face of P, whence the set of extreme points of P is $\{\mathbf{x}\} \cup V$. Suppose that $W \subseteq V$ is such that conv W is one of the $f_{k-1}(Q)$ $(k-1)$-faces of Q. Then by (3.3.1),

$$(\text{aff}\,(\{\mathbf{x}\} \cup W)) \cap \text{conv}(V \setminus W) = (\text{aff}\, W) \cap \text{conv}\,(V \setminus W) = \varnothing,$$

which shows that conv $(\{\mathbf{x}\} \cup W)$ is a k-face of P. It follows easily that

$$f_k(P) \geqslant f_k(Q) + f_{k-1}(Q).$$

Suppose next that $W \subseteq V$ is such that either conv W or conv $(\{\mathbf{x}\} \cup W)$ is a face of P—every face of P must be of one of these two forms. Then either

$$(\text{aff}\, W) \cap \text{conv}\,(\{\mathbf{x}\} \cup (V \setminus W)) = \varnothing$$

or

$$(\text{aff}(\{\mathbf{x}\} \cup W)) \cap \text{conv}(V \setminus W) = \varnothing.$$

In both cases

$$(\text{aff } W) \cap \text{conv}(V \setminus W) = \varnothing,$$

which shows that conv W is a face of Q. Thus each face of P is either a face of Q or the convex hull of $\mathbf{x}$ and a face of Q. Hence

$$f_k(P) \leqslant f_k(Q) + f_{k-1}(Q).$$

We conclude that

$$f_k(P) = f_k(Q) + f_{k-1}(Q). \quad \square$$

The formula of Theorem 3.3.1 is easily verified for a 3-pyramid P in $\mathbb{R}^3$ which has for base an m-sided convex polygon. Here $f_0(Q) = m$, $f_1(Q) = m$; $f_0(P) = m + 1$, $f_1(P) = 2m$, $f_2(P) = m + 1$. We note that P satisfies Euler's relation.

Suppose now that P is an r-pyramid with base an $(r-1)$-polytope Q, and that Q is an $(r-1)$-pyramid with base an $(r-2)$-polytope S, so that P is obtained from S by applying the cone construction twice. We say that P is a *2-fold r-pyramid* with *2-base* S, or a *1-fold r-pyramid* with *1-base* Q. Theorem 3.3.1 shows that, for $k = -1, \ldots, r$,

$$\begin{aligned} f_k(P) &= f_k(Q) + f_{k-1}(Q) = f_k(S) + f_{k-1}(S) + f_{k-1}(S) + f_{k-2}(S) \\ &= f_k(S) + 2f_{k-1}(S) + f_{k-2}(S). \end{aligned}$$

The above ideas generalize in a natural way. Let P be an r-polytope in $\mathbb{R}^n$ $(r = 1, \ldots, n)$ and let Q be an $(r-s)$-polytope in $\mathbb{R}^n$ $(s = 1, \ldots, r)$. Then P is said to be an *s-fold r-pyramid* with *s-base* Q if it can be obtained from Q by applying the cone construction s times. A simple induction argument, using Theorem 3.3.1, shows that, for an s-fold r-pyramid P with s-base Q, we have

$$f_k(P) = \sum_{i=0}^{s} \binom{s}{i} f_{k-i}(Q) \qquad \text{for } k = -1, \ldots, r.$$

Clearly, an r-fold r-pyramid is an r-simplex. Since an $(r-1)$-fold r-pyramid has a line segment for an $(r-1)$-base, which is itself a 1-fold 1-pyramid, it follows that each $(r-1)$-fold r-pyramid is an r-fold r-pyramid, i.e. an r-simplex.

We now introduce the idea of a bipyramid. Let I be a line segment in $\mathbb{R}^n$ and let Q be an $(r-1)$-polytope in $\mathbb{R}^n$ such that $I \cap Q$ is a single point which is a relative interior point of *both* I and Q. Then the *r-bipyramid* P with *axis* I and *base* Q is defined to be the r-polytope conv $(I \cup Q)$; we say that P is obtained from Q by applying the *double-cone construction* with axis I. Suppose that $I = \text{conv}\{\mathbf{a}, \mathbf{b}\}$, where $\mathbf{a}$ and

$\mathbf{b}$ are distinct points of $\mathbb{R}^n$. Then an argument similar to that used in the proof of Theorem 3.3.1 shows that the k-faces ($k = -1, \ldots, r-2$) of P are precisely the k-faces of Q and the k-polytopes of the form conv $(\{\mathbf{a}\} \cup F)$ or conv $(\{\mathbf{b}\} \cup F)$, where F is a $(k-1)$-face of Q. The $(r-1)$-faces of P are simply the $(r-1)$-polytopes conv $(\{\mathbf{a}\} \cup F)$ and conv $(\{\mathbf{b}\} \cup F)$, where F is an $(r-2)$-face of Q. We thus arrive at the following result.

Theorem 3.3.2 *Let P be an r-bipyramid in $\mathbb{R}^n$ with axis I and base a non-empty $(r-1)$-polytope Q. Then*

$$f_k(P) = f_k(Q) + 2f_{k-1}(Q) \qquad \textit{for } k = -1, \ldots, r-2$$
$$f_{r-1}(P) = 2f_{r-2}(Q). \ \square$$

The formulas of Theorem 3.3.2 are easily verified for a 3-bipyramid P in $\mathbb{R}^3$ which has for base an m-sided convex polygon Q. Here $f_0(Q) = m$, $f_1(Q) = m$; $f_0(P) = m + 2$, $f_1(P) = 3m$, $f_2(P) = 2m$. We note that P satisfies Euler's relation.

The definition of an s-fold r-bipyramid follows a similar pattern to that of an s-fold r-pyramid given earlier. Let P be an r-polytope in $\mathbb{R}^n$ ($r = 1, \ldots, n$) and let Q be an $(r-s)$-polytope in $\mathbb{R}^n$ ($s = 1, \ldots, r$). Then P is said to be an *s-fold r-bipyramid* with *s-base* Q if it can be obtained from Q by applying the double-cone construction s times. Since an $(r-1)$-fold r-bipyramid has a line segment for an $(r-1)$-base, which is itself a 1-fold 1-bipyramid, it follows that each $(r-1)$-fold r-bipyramid is also an r-fold r-bipyramid.

The simplest example of an r-fold r-bipyramid is the r-crosspolytope, which was introduced in Section 3.1. Consider the r-crosspolytope P in $\mathbb{R}^n$ ($r = 1, \ldots, n$), which is the convex hull of r linearly independent line segments conv $\{\mathbf{a}_1, \mathbf{b}_1\}, \ldots,$ conv $\{\mathbf{a}_r, \mathbf{b}_r\}$ (i.e. the vectors $\mathbf{a}_1 - \mathbf{b}_1, \ldots, \mathbf{a}_r - \mathbf{b}_r$ are linearly independent) whose midpoints coincide. The facial structure of P is easily described. For each $k = 0, \ldots, r-1$, let $I = \{i_1, \ldots, i_{k+1}\}$ be a subset of $\{1, \ldots, r\}$ which has $k+1$ points and let $T = \{\mathbf{x}_{i_1}, \ldots, \mathbf{x}_{i_{k+1}}\}$ be such that each $\mathbf{x}_{i_j}$ is either $\mathbf{a}_{i_j}$ or $\mathbf{b}_{i_j}$ for $j = 1, \ldots, k+1$. Then conv T is a k-face of P and all k-faces of P arise in this way. Since there are $\binom{r}{k+1}$ possibilities for the set I and each I gives rise to 2^{k+1} possibilities for the set T, it follows that

$$f_k(P) = 2^{k+1}\binom{r}{k+1} \qquad \text{for } k = -1, \ldots, r-1.$$

We now introduce the idea of a *prism*. Let Q be a non-empty $(r-1)$-polytope in $\mathbb{R}^n$ and let $\mathbf{x}$ be a point of $\mathbb{R}^n$ which does not lie in the subspace of $\mathbb{R}^n$ which is parallel to aff Q. Let I be the line segment conv $\{\mathbf{0}, \mathbf{x}\}$. Then the *r-prism P* with *axis I* and *base Q* is defined to be the r-polytope $Q + I$ or, equivalently, conv $(Q \cup (Q + \mathbf{x}))$; we say that P is obtained from Q by applying the *prism construction* with *axis I*. An argument similar to that used in the proof of Theorem 3.3.1 shows that the k-faces ($k = 1, \ldots, r$) of P are precisely the k-faces of Q and its translate $Q + \mathbf{x}$, together with k-polytopes of the form $F + I$, where F is a $(k - 1)$-face of Q. We thus arrive at the following result.

Theorem 3.3.3 *Let P be an r-prism in* $\mathbb{R}^n$ *with axis I and base a non-empty* $(r - 1)$*-polytope Q. Then*

$$f_k(P) = 2f_k(Q) + f_{k-1}(Q) \qquad \text{for } k = 1, \ldots, r$$
$$f_0(P) = 2f_0(Q). \quad \square$$

The formulas of Theorem 3.3.3 are easily verified for a 3-prism P in $\mathbb{R}^3$ which has for base an m-sided convex polygon Q. Here $f_0(Q) = m$, $f_1(Q) = m$; $f_0(P) = 2m$, $f_1(P) = 3m$, $f_2(P) = m + 2$. We note that P satisfies Euler's relation.

The reader will not be surprised to learn that we now define an s-fold r-prism! Let P be an r-polytope in $\mathbb{R}^n$ ($r = 1, \ldots, n$) and let Q be an $(r - s)$-polytope in $\mathbb{R}^n$ ($s = 1, \ldots, r$). Then P is said to be an *s-fold r-prism* with *s-base Q* if it can be obtained from Q by applying the prism construction s times. Since an $(r - 1)$-fold r-prism has a line segment for an $(r - 1)$-base, which is itself a 1-fold 1-prism, it follows that each $(r - 1)$-fold r-prism is also an r-fold r-prism.

An r-fold r-prism P in $\mathbb{R}^n$ ($r = 1, \ldots, n$) is called an *r-parallelotope* and has the form

$$P = \mathbf{x} + \{\lambda_1 \mathbf{x}_1 + \ldots + \lambda_r \mathbf{x}_r : 0 \leqslant \lambda_1, \ldots, \lambda_r \leqslant 1\},$$

where $\mathbf{x} \in \mathbb{R}^n$ and $\mathbf{x}_1, \ldots, \mathbf{x}_r$ are linearly independent vectors in $\mathbb{R}^n$. Thus a 2-parallelotope in $\mathbb{R}^2$ is a parallelogram and a 3-parallelotope in $\mathbb{R}^3$ is a parallelepiped. When the vectors $\mathbf{x}_1, \ldots, \mathbf{x}_r$ are pairwise orthogonal, P is known as an *r-orthotope* and when, in addition, all the vectors $\mathbf{x}_1, \ldots, \mathbf{x}_r$ have the same length, P is called on *r-cube*. This definition of a cube agrees with the one given in Section 3.1. A simple induction argument, using Theorem 3.3.3, shows that, for any r-parallelotope P in $\mathbb{R}^n$, we have

$$f_k(P) = 2^{r-k}\binom{r}{k} \qquad \text{for } k = 0, \ldots, r.$$

Exercises 3.3

1. Calculate the number of k-faces ($k = 0, 1, 2, 3, 4$) of (i) a 5-simplex, (ii) a 5-crosspolytope, and (iii) a 5-parallelotope.

2. Find in $\mathbb{R}^4$ (i) a 2-fold 4-pyramid, (ii) a 2-fold 4-bipyramid, and (iii) a 2-fold 4-prism having conv $\{(0, 0, 0, 0), (1, 0, 0, 0), (0, 1, 0, 0)\}$ for a 2-base.
3. Show that all *proper* faces of a crosspolytope are simplexes.
4. Suppose that P is an r-pyramid (r-bipyramid, r-prism) in $\mathbb{R}^n$ with base a non-empty $(r - 1)$-polytope Q. Prove that, if Q satisfies Euler's relation, then so does P. What can you conclude about simplexes, crosspolytopes, and parallelotopes?

3.4 CYCLIC POLYTOPES

The interesting family of *cyclic polytopes*, although discovered in the early 1900s, has been extensively studied only in relatively recent times. These polytopes derive much of their importance from *the upper bound theorem* of 1970, which asserts that no n-dimensional polytope with v vertices has more faces of a given dimension than does an n-dimensional cyclic polytopes with v vertices. Some properties of cyclic polytopes may surprise the reader. We shall see, for example, that all four-dimensional cyclic polytopes (of which there are infinitely many different combinatorial types) are such that any two of their vertices determine an edge; this contrasts sharply with the three-dimensional situation, where only 3-simplexes enjoy this property. In general, any polytope having *more than k* vertices which is such that every k-membered subset of its vertex set determines one of its faces, is said to be k-**neighbourly**. Thus n-simplexes ($n \geqslant 1$) are n-neighbourly and four-dimensional cyclic polytopes are 2-neighbourly. We give a brief discussion of neighbourly polytopes at the end of this section.

Cyclic polytopes are defined using the **moment curve** M_n in $\mathbb{R}^n$ which is determined parametrically by the equation

$$\mathbf{x}(t) = (t, t^2, \ldots, t^n) \qquad \text{for all real } t. \tag{3.4.1}$$

The importance of the moment curve M_n for our present purposes is that every set comprising $n + 1$ of its points is affinely independent; see Theorem 3.4.1 below. Cyclic polytopes can be described by means of curves other than the moment curve, but the moment curve has the advantage of simplifying calculations.

Clearly, equation (3.4.1) sets up a bijection between the set $\mathbb{R}$ of real numbers and the set M_n of points on the moment curve. This bijection induces an ordering on M_n which is isomorphic to the standard ordering on $\mathbb{R}$. Having now made this remark, we shall in future refer to the ordering of points on M_n exactly as if they were real numbers. For example, if points $\mathbf{x}(t_1)$, $\mathbf{x}(t_2)$, $\mathbf{x}(t_3)$ on M_n are such that $t_1 < t_2 < t_3$, then we shall say that $\mathbf{x}(t_2)$ *lies between* $\mathbf{x}(t_1)$ and $\mathbf{x}(t_3)$.

Theorem 3.4.1 *Each set of $n+1$ or fewer points on the moment curve M_n in $\mathbb{R}^n$ is affinely independent.*

Proof For $i = 0, 1, \ldots, n$, let $\mathbf{x}(t_i) = (t_i, t_i^2, \ldots, t_i^n)$, where $t_0 < t_1 < \ldots < t_n$. To prove the theorem it will suffice to show that the set $\{\mathbf{x}(t_0), \mathbf{x}(t_1), \ldots, \mathbf{x}(t_n)\}$ is affinely independent, which is equivalent to the non-vanishing of the $(n+1) \times (n+1)$ determinant

$$\begin{vmatrix} 1 & t_0 & t_0^2 & \ldots & t_0^n \\ 1 & t_1 & t_1^2 & \ldots & t_1^n \\ \ldots & \ldots & \ldots & \ldots & \ldots \\ 1 & t_n & t_n^2 & \ldots & t_n^n \end{vmatrix}$$

It is a well-known result of elementary algebra that this determinant, called *Vandermonde's determinant*, equals

$$\prod_{0 \leqslant i < j \leqslant n} (t_j - t_i),$$

and hence is non-zero. □

A **cyclic polytope** $C(v, n)$ is the convex hull of v $(v \geqslant n+1)$ *distinct* points on the moment curve M_n in $\mathbb{R}^n$. Strictly speaking, $C(v, n)$ is a whole family of polytopes, all of the same combinatorial type. Our first result is that cyclic polytopes are **simplicial**; this means that *all* of their *proper* faces are simplexes. Examples of simplicial polytopes are simplexes, bipyramids with simplicial bases, and crosspolytopes.

Theorem 3.4.2 *Cyclic polytopes are simplicial.*

Proof Let F be a proper face of a cyclic polytope $C(v, n)$ in $\mathbb{R}^n$. Then $F = \operatorname{conv}\{\mathbf{x}_1, \ldots, \mathbf{x}_m\}$ for some distinct points $\mathbf{x}_1, \ldots, \mathbf{x}_m$ $(1 \leqslant m < v)$ on the moment curve M_n. Since the face F is proper, the set $\{\mathbf{x}_1, \ldots, \mathbf{x}_m\}$ cannot contain an affinely independent subset of more than n points. Hence, by Theorem 3.4.1, $m \leqslant n$ and $\{\mathbf{x}_1, \ldots, \mathbf{x}_m\}$ is affinely independent. Thus F is a simplex, which shows that $C(v, n)$ is simplicial. □

The problem of determining the number of faces of each dimension of a cyclic polytope $C(v, n)$ is not trivial. For faces of dimension $k-1$ with $2k \leqslant n$, however, the problem has a simple solution, which follows readily from the next result.

Theorem 3.4.3 *Let $C(v, n)$ be the convex hull of the distinct points $\mathbf{x}_1, \ldots, \mathbf{x}_v (v \geqslant n + 1 \geqslant 3)$ on the moment curve M_n in $\mathbb{R}^n$. Let k be an integer satisfying $1 \leqslant k \leqslant \frac{1}{2}n$. Then each set of k points of $\{\mathbf{x}_1, \ldots, \mathbf{x}_v\}$ determines a $(k-1)$-face of $C(v, n)$ and $\mathbf{x}_1, \ldots, \mathbf{x}_v$ are the vertices of $C(v, n)$.*

Proof It will suffice to show that $\mathbf{x}_1, \ldots, \mathbf{x}_k$ determine a $(k-1)$-face of $C(v, n)$. For each $i = 1, \ldots, k$, let $\mathbf{x}_i = (t_i, t_i^2, \ldots, t_i^n)$. Define a polynomial p for real t by the equation

$$p(t) = (t - t_1)^2(t - t_2)^2 \ldots (t - t_k)^2;$$

say

$$p(t) = t^{2k} + a_{2k-1}t^{2k-1} + \ldots + a_1 t + a_0,$$

where $a_0, a_1, \ldots, a_{2k-1} \in \mathbb{R}$. Clearly, $p(t) \geqslant 0$ for all real t, and $p(t) = 0$ if and only if t has one of the values $t_1, \ldots, t_k$. It follows that the hyperplane with equation

$$a_0 + a_1 x_1 + \ldots + a_{2k-1}x_{2k-1} + x_{2k} = 0$$

is a support hyperplane to $C(v, n)$ which meets $C(v, n)$ in the set $\operatorname{conv}\{\mathbf{x}_1, \ldots, \mathbf{x}_k\}$. Thus $\operatorname{conv}\{\mathbf{x}_1, \ldots, \mathbf{x}_k\}$ is a face of $C(v, n)$. By Theorem 3.4.1, $\{\mathbf{x}_1, \ldots, \mathbf{x}_k\}$ is affinely independent, and so $\operatorname{conv}\{\mathbf{x}_1, \ldots, \mathbf{x}_k\}$ is a $(k-1)$-simplex. That $\mathbf{x}_1, \ldots, \mathbf{x}_v$ are vertices of $C(v, n)$ follows from the result just proved with $k = 1$. □

Corollary 3.4.4 *The cyclic polytope $C(v, n)$ in $\mathbb{R}^n$ $(v \geqslant n + 1 \geqslant 3)$ has $\binom{v}{k}$ $(k-1)$-faces, when k is an integer satisfying $1 \leqslant k \leqslant \frac{1}{2}n$.*

Proof By the theorem, each set of k vertices of $C(v, n)$ determines one of its $(k-1)$-faces. Conversely, by Theorem 3.4.2 each $(k-1)$-face of $C(v, n)$ is the convex hull of some k of its vertices. Thus $C(v, n)$ has as many $(k-1)$-faces as there are ways of choosing a subset of k points from a set of v points, namely $\binom{v}{k}$. □

Theorem 3.2.1 shows that each proper face of a polytope is the intersection of those facets of the polytope which contain that face. Thus the facial structure of a polytope is completely determined by the vertex sets of its facets. We now give a simple criterion for determining which sets of vertices of a cyclic polytope determine one of its facets.

1	2	3	4	5	6	7
*	*	*	*			
*	*	*				*
*	*		*	*		
*	*			*	*	
*	*				*	*
*		*	*			*
*			*	*		*

1	2	3	4	5	6	7
*				*	*	*
	*	*	*	*		
	*	*		*	*	
	*	*			*	*
		*	*	*	*	
		*	*		*	*
			*	*	*	*

Fig. 3.3

Theorem 3.4.5 (Gale's evenness condition) *Let W be a set of n points of the vertex set V of a cyclic polytope $C(v, n)$ in $\mathbb{R}^n$ $(v \geqslant n+1)$. Then* conv *W is a facet of $C(v, n)$* if and only if each two points of $V \setminus W$ *are separated on the moment curve M_n by an even number of points of W.*

Proof Let W consist of the n points $(t_i, t_i^2, \ldots, t_i^n)$ for $i = 1, \ldots, n$. Consider the real polynomial p defined (for real t) by the equation

$$p(t) = (t - t_1) \ldots (t - t_n) = t^n + a_{n-1}t^{n-1} + \ldots + a_1 t + a_0,$$

where $a_0, a_1, \ldots, a_{n-1} \in \mathbb{R}$. Then the hyperplane H in $\mathbb{R}^n$ that contains W has equation

$$a_0 + a_1 x_1 + \ldots + a_{n-1}x_{n-1} + x_n = 0.$$

Now conv W will be a facet of $C(v, n)$ if and only if H is a support hyperplane to $C(v, n)$. This will be the case if and only if *all* the numbers $p(t)$, where t is such that $(t, t^2, \ldots, t^n) \in V \setminus W$, have the same sign. As t increases through all real values, the polynomial p changes sign precisely when t passes through one of the values $t_1, \ldots, t_n$. Thus $p(r)$ and $p(s)$, where r and s are unequal real numbers that are not equal to any of the values $t_1, \ldots, t_n$, will have the same sign if and only if an *even* number of $t_1, \ldots, t_n$ lie between r and s. The desired result is now immediate. □

We now use Gale's evenness condition to calculate the number of facets of the cyclic polytope $C(7, 4)$. This is equivalent to finding how many subsets W of a totally ordered set V of seven elements there are having four elements, and which are such that between any two elements of $V \setminus W$ there is an even number of elements of W. The totality of such subsets W of V is illustrated in Fig. 3.3, where V is represented by the numbers 1,2,3,4,5,6,7 on the real line with their usual ordering, and where the points of W are marked by asterisks. It is readily found that there are 14 such sets W, and so $C(7,4)$ has 14 facets. Since each proper face of $C(7,4)$ is an intersection of facets of $C(7,4)$, we find that $C(7,4)$ has 28 2-faces corresponding to the following subsets of V: $\{1,2,3\}$, $\{1,2,4\}$, $\{1,2,5\}$, $\{1,2,6\}$, $\{1,2,7\}$, $\{1,3,4\}$, $\{1,3,7\}$, $\{1,4,5\}$, $\{1,4,7\}$, $\{1,5,6\}$, $\{1,5,7\}$, $\{1,6,7\}$, $\{2,3,4\}$, $\{2,3,5\}$, $\{2,3,6\}$, $\{2,3,7\}$, $\{2,4,5\}$, $\{2,5,6\}$, $\{2,6,7\}$, $\{3,4,5\}$, $\{3,4,6\}$, $\{3,4,7\}$, $\{3,5,6\}$, $\{3,6,7\}$, $\{4,5,6\}$, $\{4,5,7\}$, $\{4,6,7\}$, $\{5,6,7\}$. The reader who is surprised at such a large number of 2-faces should recall that, by the upper bound theorem, no 4-polytope with 7 vertices has more 2-faces than $C(7,4)$, viz. 28. By Corollary 3.4.4, $C(7,4)$ has $\binom{7}{2}$, i.e. 21, 1-faces. Thus, denoting the polytope $C(7,4)$ by P, we find that

$$f_{-1}(P) - f_0(P) + f_1(P) - f_2(P) + f_3(P) - f_4(P) =$$
$$1 - 7 + 21 - 28 + 14 - 1 = 0,$$

and Euler's relation is verified for $C(7,4)$.

Formulas are known for the number of faces of $C(v,n)$ of each dimension. Here, however, we content ourselves with establishing a formula for the number of facets of $C(v,n)$.

Theorem 3.4.6 *The cyclic polytope $C(v,n)$ in $\mathbb{R}^n$ $(v \geqslant n+1)$ has*

$$\frac{v}{v-d}\binom{v-d}{d} \quad \text{or} \quad 2\binom{v-d-1}{d}$$

facets, according as $n = 2d$ is even or $n = 2d+1$ is odd.

Proof We first establish a simple combinatorial lemma. Let $A = \{1,\ldots,r\}$, $B = \{1,\ldots,r-s\}$, where r, s are integers satisfying $r \geqslant 1$ and $0 \leqslant 2s \leqslant r$. Then a subset of A is said to be *s-paired* if it has the form

$$\{i_1, i_1+1, i_2, i_2+1, \ldots, i_s, i_s+1\},$$

where

$$i_1 < i_1 + 1 < i_2 < i_2 + 1 < \ldots < i_s < i_s + 1;$$

the empty set (corresponding to $s = 0$) is considered to be 0-paired. By associating with each such s-paired set the subset $\{i_1, i_2 - 1, \ldots, i_s - (s-1)\}$ of B, we set up a bijection between the s-paired subsets of A and the subsets of B having s elements. Thus A has $\binom{r-s}{s}$ s-paired subsets.

Gale's evenness condition shows that the number of facets of $C(v,n)$ is the number of subsets W of the set $V = \{1,\ldots,v\}$ having n elements, and which are such that between any two integers of $V \setminus W$ there is an even number of integers of W. For this proof only, we refer to such a subset W of V as a *facet* of V. We need to determine the number of facets W of V. The cases n even and n odd will be treated separately.

Suppose first that $n = 2d$ is even. Then the facets W of V are of two types: either W is a d-paired subset of V, or $W \setminus \{1, v\}$ is a $(d-1)$-paired subset of $\{2, \ldots, v-1\}$. Conversely, each d-paired subset of V is a facet of V, and each $(d-1)$-paired subset of $\{2, \ldots, v-1\}$, when augmented with 1 and v, is a facet of V. By the combinatorial lemma, V has $\binom{v-d}{d}$ facets of the first type and $\binom{v-2-(d-1)}{d-1} = \binom{v-d-1}{d-1}$ facets of the second type. Thus the total number of the facets of V is

$$\binom{v-d}{d} + \binom{v-d-1}{d-1} = \frac{(v-d)!}{(v-2d)!\,d!} + \frac{(v-d-1)!}{(v-2d)!(d-1)!}$$
$$= \frac{v}{v-d}\binom{v-d}{d}.$$

Suppose next that $n = 2d+1$ is odd. Again the facets W of V are of two types: either $W \setminus \{1\}$ is a d-paired subset of $\{2, \ldots, v\}$, or $W \setminus \{v\}$ is a d-paired subset of $\{1, \ldots, v-1\}$. Conversely, each d-paired subset of $\{2, \ldots, v\}$, when augmented with 1, is a facet of V, and each d-paired subset of $\{1, \ldots, v-1\}$, when augmented with v, is a facet of V. Thus the number of facets of V of either type is $\binom{v-1-d}{d}$, whence the total number of facets of V is $2\binom{v-1-d}{d}$. □

Cyclic polytopes are important examples of the neighbourly polytopes that we now discuss. Let k be a positive integer. Then a polytope in $\mathbb{R}^n$ (having more than k vertices) is said to be k-**neighbourly** if every set of k of its vertices determines a face of the polytope. Thus each r-polytope $(r \geqslant 1)$ is 1-neighbourly and each r-simplex $(r \geqslant 1)$ is r-neighbourly. Theorem 3.4.3 shows that the cyclic polytope $C(v, n)$, where $v \geqslant n+1 \geqslant 3$, is $[\frac{1}{2}n]$-neighbourly—here $[\frac{1}{2}n]$ denotes the greatest integer not exceeding $\frac{1}{2}n$.

Theorem 3.4.7 *Let P be a k-neighbourly polytope in $\mathbb{R}^n$. Then every set of k vertices of P is affinely independent and each $(k-1)$-face of P is a $(k-1)$-simplex.*

Proof Suppose that $\mathbf{v}_1, \ldots, \mathbf{v}_k$ are k vertices of P which are affinely dependent, say $\mathbf{v}_k \in \operatorname{aff}\{\mathbf{v}_1, \ldots, \mathbf{v}_{k-1}\}$. Since P has more than k vertices, there is a vertex $\mathbf{v}_0$ of P different from $\mathbf{v}_1, \ldots, \mathbf{v}_k$. But P is

k-neighbourly, and so $\operatorname{conv}\{\mathbf{v}_0, \ldots, \mathbf{v}_{k-1}\}$ is a face of P. By Theorem 3.1.4, $\mathbf{v}_k \notin \operatorname{aff}\{\mathbf{v}_0, \ldots, \mathbf{v}_{k-1}\}$, which contradicts the assumption that $\mathbf{v}_k \in \operatorname{aff}\{\mathbf{v}_1, \ldots, \mathbf{v}_{k-1}\}$. Thus every set of k vertices of P is affinely independent.

Suppose now that F is a $(k-1)$-face of P. Then F must contain an affinely independent subset W consisting of precisely k vertices of P. Since P is k-neighbourly, conv W is a $(k-1)$-face of P, and hence of F. But F has only one $(k-1)$-dimensional face, namely itself. Thus $F = \operatorname{conv} W$, and so F is a $(k-1)$-simplex. □

Corollary 3.4.8 *Let P be a k-neighbourly polytope in $\mathbb{R}^n$ with v vertices. Let $j \in \{1, \ldots, k\}$. Then P is j-neighbourly and has $\binom{v}{j}$ $(j-1)$-faces.*

Proof Let X be a set of j vertices of P. Then $X \subseteq W$ for some set W of k vertices of P. Now conv W is a simplex and a face of P, whence conv X is a face of conv W, and hence of P. This shows that P is j-neighbourly.

The j-neighbourliness of P, together with the theorem, shows that P has as many $(j-1)$-faces as there are ways of choosing a set of j points from a set of v points, namely $\binom{v}{j}$. □

We now show that the only n-polytopes which are *more neighbourly* than the general cyclic polytope $C(v, n)$ are the n-simplexes.

Theorem 3.4.9 *Let P be an n-polytope in $\mathbb{R}^n$ which is k-neighbourly for some k with $k > [\frac{1}{2}n]$. Then P is an n-simplex.*

Proof Suppose that P is *not* an n-simplex. Then the vertex set V of P must contain some subset W of $n+2$ points. By Radon's theorem, W can be partitioned into two subsets X and Y such that

$$(\operatorname{conv} X) \cap (\operatorname{conv} Y) \neq \varnothing. \tag{3.4.2}$$

One of X and Y, X say, has no more than $[\frac{1}{2}n] + 1$ points. Corollary 3.4.8 shows that conv X is a face of P, whence, by Theorem 3.1.4,

$$(\operatorname{conv} X) \cap (\operatorname{conv} Y) \subseteq (\operatorname{aff} X) \cap (\operatorname{conv}(V \setminus X)) = \varnothing,$$

which contradicts (3.4.2). Thus P is an n-simplex. □

Corollary 3.4.10 *Let P be an n-neighbourly $2n$-polytope in $\mathbb{R}^{2n}$. Then P is simplicial.*

Proof Let F be a facet of P. Then F is an n-neighbourly $(2n-1)$-polytope, and so, exactly as in the proof of the theorem, F is a simplex.

Since each proper face of P is a face of some facet of P, each proper face of P must be a simplex, i.e. P is simplicial. □

Exercises 3.4

1. Which of the following regular polyhedra in $\mathbb{R}^3$ are simplicial: tetrahedron, cube, octahedron, icosahedron, dodecahedron?
2. Find the number $f_k (k = -1, 0, \ldots, 5)$ of k-faces of a cyclic polytope $C(7, 5)$. Verify that $f_{-1} - f_0 + f_1 - f_2 + f_3 - f_4 + f_5 = 0$.
3. Show that the vertices of a cyclic polytope $C(2n + 2, 2n)$ in $\mathbb{R}^{2n}$ can be partitioned into two sets of equal size in such a way that $2n$ of the vertices determine a facet if and only if n of them belong to one of the sets and the remaining n belong to the other. Hence calculate the number of facets of $C(2n + 2, 2n)$. Use Theorem 3.4.6 to check the answer.
4. Find the smallest and the largest numbers of facets that a cyclic polytope with 12 vertices can have.
5. Is every n-neighbourly $(2n + 1)$-polytope in $\mathbb{R}^{2n+1}$ simplicial?

3.5 EULER'S RELATION

We now consider the most well-known result in the combinatorial theory of polytopes, *Euler's relation*. This states that, if a non-empty r-polytope P has $f_k(P)$ k-faces, then

$$f_{-1}(P) - f_0(P) + \ldots + (-1)^{r+1} f_r(P) = 0,$$

a result we have already verified for several different polytopes P. Euler discovered this fact for three-dimensional polytopes in 1752—here it assumes the more familiar form $v - e + f = 2$, where v, e, f denote, respectively, the number of vertices, edges, and facets of a 3-polytope. The general case was established by Poincaré in 1899, and for this reason the relation is sometimes referred to as the *Euler–Poincaré relation*. It was attempts to generalize Euler's relation that first drew mathematicians' attention to the importance of the concept of convexity.

Theorem 3.5.1 (Euler's relation) *Let P be a non-empty r-polytope in $\mathbb{R}^n$. Then*

$$f_{-1}(P) - f_0(P) + \ldots + (-1)^{r+1} f_r(P) = 0,$$

where $f_k(P)$ denotes the number of k-faces of P.

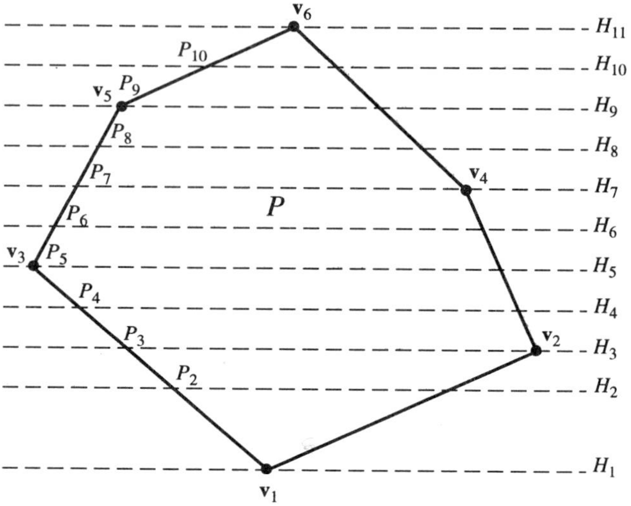

Fig. 3.4

Proof We argue by induction on r. The theorem is trivial when $r = 0$, since $f_{-1}(P) = 1$, $f_0(P) = 1$, and when $r = 1$, since $f_{-1}(P) = 1$, $f_0(P) = 2$, $f_1(P) = 1$. Suppose that the theorem has been established for polytopes of dimension $r - 1$, where $r \geqslant 2$.

Let P be an r-polytope ($r \geqslant 2$) in $\mathbb{R}^n$ with vertices $\mathbf{a}_1, \ldots, \mathbf{a}_v$. Choose a vector $\mathbf{a}$ in $\mathbb{R}^n$ such that the scalars $\mathbf{a} \cdot \mathbf{a}_1, \ldots, \mathbf{a} \cdot \mathbf{a}_v$ are *distinct*. (For the existence of such an $\mathbf{a}$, see the first problem at the end of the section.) Suppose that the vertices of P are labelled so that $\mathbf{a} \cdot \mathbf{a}_1 < \ldots < \mathbf{a} \cdot \mathbf{a}_v$. Define hyperplanes $H_1, H_3, \ldots, H_{2v-1}$ in $\mathbb{R}^n$ by the equations

$$H_{2k-1} = \{\mathbf{x} \in \mathbb{R}^n : \mathbf{a} \cdot \mathbf{x} = \mathbf{a} \cdot \mathbf{a}_k\} \qquad \text{for } k = 1, \ldots, v.$$

Choose scalars $c_1, c_2, \ldots, c_{v-1}$ such that

$$\mathbf{a} \cdot \mathbf{a}_1 < c_1 < \mathbf{a} \cdot \mathbf{a}_2 < c_2 < \ldots < c_{v-1} < \mathbf{a} \cdot \mathbf{a}_v,$$

and define hyperplanes $H_2, H_4, \ldots, H_{2v-2}$ in $\mathbb{R}^n$ by the equations

$$H_{2k} = \{\mathbf{x} \in \mathbb{R}^n : \mathbf{a} \cdot \mathbf{x} = c_k\} \qquad \text{for } k = 1, \ldots, v - 1.$$

This situation for a two-dimensional polytope with six vertices is illustrated in Fig. 3.4.

The following observations about the hyperplanes $H_1, H_2, \ldots, H_{2v-1}$ are immediate: (i) they are distinct and parallel to one another; (ii) each of the hyperplanes $H_1, H_3, \ldots, H_{2v-1}$ contains just one vertex of P; (iii) H_1 and H_{2v-1} are support hyperplanes to P which meet P in a single point; (iv) the set $P \cap H_k$, for $k = 2, 3, \ldots, 2v - 2$, is an

$(r-1)$-polytope, P_k say; and (v) none of the polytopes $P_2, P_4, \ldots, P_{2v-2}$ contains a vertex of P.

For each j-face F_j of P, where $j = 1, \ldots, r$, and for each polytope P_i, where $i = 2, 3, \ldots, 2v-2$, define an integer $\psi(F_j, P_i)$ to be 1 if $\operatorname{ri} F_j$ meets P_i, and 0 otherwise. For each j-face F_j of P, where $j = 1, \ldots, r$, denote by s and t, respectively, the smallest and largest integers i amongst $1, 2, \ldots, 2v-1$ for which H_i meets F_j. Clearly s and t are odd with $s < t$, and $\psi(F_j, P_i) = 1$ precisely when $s < i < t$. Thus

$$\sum_{i=2}^{2v-2} (-1)^i \psi(F_j, P_i) = \sum_{i=s+1}^{t-1} (-1)^i = 1,$$

and so, for each fixed $j = 1, \ldots, r$,

$$\sum_{j\text{-faces}} \left(\sum_{i=2}^{2v-2} (-1)^i \psi(F_j, P_i) \right) = f_j(P),$$

where the summation is over all the j-faces F_j of P. Hence

$$\sum_{j=1}^{r} (-1)^j \left(\sum_{j\text{-faces}} \left(\sum_{i=2}^{2v-2} (-1)^i \psi(F_j, P_i) \right) \right) = \sum_{j=1}^{r} (-1)^j f_j(P). \quad (3.5.1)$$

We now find an alternative expression for the left-hand side of (3.5.1). If i is one of $2, 4, \ldots, 2v-2$ or $1 < j \leqslant r$, then the number of $(j-1)$-faces of P_i is the same as the number of j-faces of P whose relative interiors meet P_i. If i is one of $1, 3, \ldots, 2v-1$, then the number of vertices of P_i is one more than the number of edges of P whose relative interiors meet P_i. These observations are summarized in the following equations, where it is assumed that i is one of $2, 3, \ldots, 2v-2$; j is one of $1, \ldots, r$, and $f_k(P_i)$ denotes the number of k-faces of P_i:

$$\sum_{j\text{-faces}} \psi(F_j, P_i) = \begin{cases} f_{j-1}(P_i), & \text{if } i \text{ is even or } 1 < j \leqslant r \\ -1 + f_{j-1}(P_i), & \text{if } i \text{ is odd and } j = 1. \end{cases}$$

Hence

$$\sum_{j=1}^{r} (-1)^j \left(\sum_{j\text{-faces}} \psi(F_j, P_i) \right) = \begin{cases} \displaystyle\sum_{j=1}^{r} (-1)^j f_{j-1}(P_i), & \text{if } i \text{ is even} \\ \displaystyle 1 + \sum_{j=1}^{r} (-1)^j f_{j-1}(P_i), & \text{if } i \text{ is odd.} \end{cases}$$

By the induction hypothesis,

$$\sum_{j=-1}^{r-1} (-1)^j f_j(P_i) = 0, \qquad \text{so} \qquad 1 + \sum_{j=1}^{r} (-1)^j f_{j-1}(P_i) = 0,$$

whence

$$\sum_{j=1}^{r}(-1)^j\left(\sum_{j\text{-faces}}\psi(F_j,P_i)\right)=\begin{cases}-1, & \text{if } i \text{ is even}\\ 0, & \text{if } i \text{ is odd.}\end{cases}$$

Hence

$$\sum_{i=2}^{2v-2}(-1)^i\left(\sum_{j=1}^{r}(-1)^j\left(\sum_{j\text{-faces}}\psi(F_j,P_i)\right)\right)=1-v. \tag{3.5.2}$$

Comparing equations (3.5.1) and (3.5.2), we deduce that

$$\sum_{j=1}^{r}(-1)^j f_j(P)=1-v=f_{-1}(P)-f_0(P),$$

and so

$$\sum_{j=-1}^{r}(-1)^j f_j(P)=0.$$

This completes the proof by induction. □

We now obtain a striking and little-known generalization of Euler's relation. Suppose that F is a k-face of an r-polytope P $(-1 \leqslant k < r)$ and that $h_i(F)$ denotes the number of i-faces of P containing F. For example, if F is a vertex of a cube P in $\mathbb{R}^3$, then this vertex belongs to three edges and three facets of P, so in this case: $h_0(F)=1$, $h_1(F)=3$, $h_2(F)=3$, $h_3(F)=1$. We note that

$$h_0(F)-h_1(F)+h_2(F)-h_3(F)=1-3+3-1=0,$$

which suggests that we consider the alternating sum

$$h_k(F)-h_{k+1}(F)+\ldots+(-1)^{r-k}h_r(F)$$

in the general case. As an impressive use of a duality argument in combinatorial results, we show that this alternating sum is always zero. This generalizes Euler's relation, which corresponds to the case when F is the empty face of P.

Before proving this generalization of Euler's relation, we need to apply some of the results on polar duality in Section 2.8 to polytopes. Let P be an r-polytope $(r \geqslant 1)$ in $\mathbb{R}^r$ containing the origin as an interior point. Then the polar dual P^* of P is a compact convex set in $\mathbb{R}^r$ containing the origin as an interior point. Suppose that P has extreme points $\mathbf{a}_1,\ldots,\mathbf{a}_m$. Then $P=\operatorname{conv}\{\mathbf{a}_1,\ldots,\mathbf{a}_m\}$ and P^* is the intersection of the m closed halfspaces $\mathbf{a}_i\cdot\mathbf{x}\leqslant 1$ for $i=1,\ldots,m$, whence P^* is a polyhedral set. Thus P^* is a bounded polyhedral set, i.e. a polytope.

Suppose further that F_i is an i-face of P $(i=1,\ldots,r)$. Then Theorem 3.2.1 (vii) shows that there exists a sequence $F_{-1},\ldots,F_i,\ldots,F_r$ of $r+2$ faces of P such that

$$F_{-1} \subset \ldots \subset F_i \subset \ldots \subset F_r.$$

Denote by φ the polar face mapping of P. Recalling that φ is an inclusion-reversing bijection from the family of faces of P to the family of faces of P^*, we see that $\varphi(F_{-1}), \ldots, \varphi(F_i), \ldots, \varphi(F_r)$ is a sequence of $r + 2$ faces of P^* such that

$$\varphi(F_r) \subset \ldots \subset \varphi(F_i) \subset \ldots \subset \varphi(F_{-1}).$$

It now follows easily from Corollary 2.6.8 that $\dim \varphi(F_i) = r - i - 1$. Hence the number of i-faces of P is the same as the number of $(r - i - 1)$-faces of P^*. We can now prove the promised generalization of Euler's relation.

Theorem 3.5.2 *Let F be a k-face of an r-polytope P ($k = 1, \ldots, r - 1$) in $\mathbb{R}^n$. Then*

$$h_k(F) - h_{k+1}(F) + \ldots + (-1)^{r-k} h_r(F) = 0,$$

where $h_i(F)$, for $i = k, \ldots, r$, denotes the number of i-faces of P containing F.

Proof We may assume, without loss of generality, that $r = n$ and that P contains the origin as an interior point. Denote by φ the polar face mapping of P. Then the number $h_i(F)$ of i-faces of P containing F is the same as the number $f_{n-i-1}(\varphi(F))$ of $(n - i - 1)$-faces of $\varphi(F)$. Euler's relation applied to the polytope $\varphi(F)$ shows that

$$\begin{aligned} & f_1(\varphi(F)) - f_0(\varphi(F)) + \ldots + (-1)^{n-k} f_{n-1-k}(\varphi(F)) \\ &= h_n(F) - h_{n-1}(F) + \ldots + (-1)^{n-k} h_k(F) \\ &= 0. \quad \square \end{aligned}$$

Euler's relation shows that, for every r-polytope P ($r \geqslant 1$), the numbers $f_0(P), \ldots, f_{r-1}(P)$ of faces of P of dimensions $0, \ldots, r - 1$, respectively, satisfy the linear equation

$$f_0(P) - f_1(P) + \ldots + (-1)^{r-1} f_{r-1}(P) = 1 - (-1)^r.$$

We now prove that this is essentially the only linear equation which is satisfied by the numbers $f_0(P), \ldots, f_{r-1}(P)$ for *all* r-polytopes $P (r \geqslant 1)$.

Theorem 3.5.3 *Let r be a positive integer. Suppose that $\alpha_0, \ldots, \alpha_r$ are real numbers such that the numbers $f_i(P)$ of the i-faces ($i = 0, \ldots, r - 1$) of any r-polytope P satisfy the equation*

$$\alpha_0 f_0(P) + \alpha_1 f_1(P) + \ldots + \alpha_{r-1} f_{r-1}(P) = \alpha_r.$$

Then $\alpha_1 = -\alpha_0$, $\alpha_2 = \alpha_0, \ldots, \alpha_{r-1} = (-1)^{r-1} \alpha_0$, $\alpha_r = (1 - (-1)^r)\alpha_0$.

Proof We argue by induction on r. The theorem is trivially true when $r = 1$, for in this case $f_0(P) = 2$ for all 1-polytopes. Suppose, then, that the theorem has been proved for the case when r is some positive integer k, and that $\alpha_0, \ldots, \alpha_{k+1}$ are real numbers such that

$$\alpha_0 f_0(P) + \alpha_1 f_1(P) + \ldots + \alpha_k f_k(P) = \alpha_{k+1} \qquad (3.5.3)$$

for all $(k + 1)$-polytopes P.

Let Q be any k-polytope. Let S be a $(k + 1)$-pyramid with base combinatorially equivalent to Q, and let T be a $(k + 1)$-bipyramid with base combinatorially equivalent to Q. Theorems 3.3.1 and 3.3.2 show that (with an obvious notation)

$$f_i(S) = f_{i-1}(Q) + f_i(Q) \qquad (i = 0, \ldots, k)$$

and

$$f_i(T) = 2f_{i-1}(Q) + f_i(Q) \qquad (i = 0, \ldots, k - 1)$$

$$f_k(T) = 2f_{k-1}(Q).$$

Substituting these values for $f_i(S)$ and $f_i(T)$ in equation (3.5.3) and subtracting the results, we find that

$$\alpha_1 f_0(Q) + \alpha_2 f_1(Q) + \ldots + \alpha_k f_{k-1}(Q) = \alpha_k - \alpha_0.$$

Since this last equation holds for *all* k-polytopes Q, the induction hypothesis shows that

$$\alpha_2 = -\alpha_1,\ \alpha_3 = \alpha_1, \ldots,\ \alpha_k = (-1)^{k-1}\alpha_1,\ \alpha_k - \alpha_0 = (1 - (-1)^k)\alpha_1,$$

and so $\alpha_1 = -\alpha_0$. Equation (3.5.3) can thus be written in the form

$$\alpha_0(f_0(P) - f_1(P) + \ldots + (-1)^k f_k(P)) = \alpha_{k+1}.$$

But Euler's relation applied to any $(k + 1)$-polytope P shows that

$$f_0(P) - f_1(P) + \ldots + (-1)^k f_k(P) = 1 - (-1)^{k+1},$$

whence $\alpha_{k+1} = (1 - (-1)^{k+1})\alpha_0$. This completes the proof by induction. □

Although the Euler relation is the only linear equation satisfied by the numbers of faces of various dimensions of *every* polytope with a given dimension, there are other linear equations that are satisfied by the numbers of faces of various dimensions of every *simplicial* polytope with a given dimension—these are the *Dehn–Sommerville equations*.

Theorem 3.5.4 (Dehn–Sommerville equations) *Let P be a simplicial r-polytope* ($r \geqslant 1$) *in* $\mathbb{R}^n$. *Then*

$$\sum_{j=k}^{r-1} (-1)^j \binom{j+1}{k+1} f_j(P) = (-1)^{r-1} f_k(P) \qquad \text{for } k = -1, \ldots, r-2.$$

Proof For each k-face F of P ($k = -1, \ldots, r-2$), consider the equation of Theorem 3.5.2. We add together these equations corresponding to all the k-faces F of P to deduce that

$$h_k - h_{k+1} + \ldots + (-1)^{r-k} h_r = 0,$$

where h_j ($j = k, \ldots, r$) denotes the total number of inclusions of the form $F_k \subseteq F_j$, where F_k and F_j are, respectively, k- and j-faces of P. If $j < r$, then each of the $f_j(P)$ j-faces of P is a j-simplex, and so has $\binom{j+1}{k+1}$ k-faces, whence $h_j = \binom{j+1}{k+1} f_j(P)$. If $j = r$, then the only j-face of P is P itself, which has $f_k(P)$ k-faces, and so $h_r = f_k(P)$. The Dehn–Sommerville equations now follow immediately. □

The Dehn–Sommerville equation corresponding to $k = -1$ is simply the Euler relation. We now derive the Dehn–Sommerville equations corresponding to $k = 0, \ldots, r-2$ for simplicial r-polytopes P with $r = 2, 3, 4$. For $k = 0$ and $r = 2$, we have the equation

$$f_0(P) - 2f_1(P) = -f_0(P),$$

which is the same as the Euler relation. For $k = 0, 1$ and $r = 3$, we have the equations:

$$\begin{aligned} f_0(P) - 2f_1(P) + 3f_2(P) &= f_0(P), \\ -f_1(P) + 3f_2(P) &= f_1(P), \end{aligned}$$

which are the same as one another, but essentially different from the Euler relation. For $k = 0, 1, 2$ and $r = 4$, we have the equations:

$$\begin{aligned} f_0(P) - 2f_1(P) + 3f_2(P) - 4f_3(P) &= -f_0(P), \\ -f_1(P) + 3f_2(P) - 6f_3(P) &= -f_1(P), \\ f_2(P) - 4f_3(P) &= -f_2(P). \end{aligned}$$

The last two of these are the same, while the first one can be deduced from Euler's relation and the second (or third) equation.

We now use some of the ideas developed in this section to investigate the existence of regular 3-polytopes of various types. A 3-polytope P is said to be **regular of type** $(p|q)$ if there exist positive integers p, q with $p, q \geqslant 3$ such that each facet of P is a regular p-gon and each vertex of P belongs to q such facets. Suppose now that P is a regular

3-polytope of type $(p|q)$ which has v vertices, e edges, and f facets. It follows immediately from Theorem 3.5.2 that each edge of a 3-polytope is contained in precisely two of its facets and that each vertex of P belongs to precisely q of its edges. Counting the edges of P by (i) vertices, and (ii) facets, in an obvious way, we find that $qv = 2e$ and $pf = 2e$. Substituting $v = 2e/q$ and $f = 2e/p$ into Euler's relation, and then rearranging the result, we find that

$$(p-2)(q-2) = 4 - \frac{2pq}{e} < 4.$$

Thus the only possible types of regular 3-polytopes are:

$$(3|3), \qquad (3|4), \qquad (4|3), \qquad (3|5), \qquad (5|3).$$

These types do indeed exist: the regular tetrahedron, the regular octahedron, the cube, the regular icosahedron, and the regular dodecahedron serving as examples.

Exercises 3.5

1. Let $\mathbf{a}_1, \ldots, \mathbf{a}_v$ be distinct points of $\mathbb{R}^n$. Show that there exists a vector $\mathbf{a}$ in $\mathbb{R}^n$ such that the scalars $\mathbf{a}\cdot\mathbf{a}_1, \ldots, \mathbf{a}\cdot\mathbf{a}_v$ are distinct.
2. Show that each $(r-2)$-face F of an r-polytope P $(r \geqslant 1)$ is contained in precisely two facets of P.
3. For each r-polytope P $(r \geqslant 1)$, the vector
$$f(P) = (f_0(P), \ldots, f_{r-1}(P)),$$
where $f_i(P)$, for $i = 0, \ldots, r-1$, denotes the number of i-faces of P, is called the **f-vector** of P. Show that the affine hull of the set of f-vectors of *all* r-polytopes P is a hyperplane (the *Euler hyperplane*) in $\mathbb{R}^r$.
4. Write down the Dehn–Sommerville equations for a simplicial r-polytope $(r \geqslant 2)$ corresponding to the values $k = r-3$ and $k = r-2$. What do you observe?
5. For $i = 0, \ldots, 4$, denote by f_i the number of i-faces of a simplicial 5-polytope. Express f_2, f_3, f_4 in terms of f_0, f_1.
6. Give examples to show that no one of the following sets of conditions on a 3-polytope P is sufficient to guarantee its regularity.
   (i) The facets of P are congruent regular polygons.
   (ii) Each facet of P is a regular polygon and each vertex of P belongs to the same number of facets of P.
   (iii) The facets of P are congruent polygons and each vertex of P belongs to the same number of facets of P.

3.6 GALE TRANSFORMS

To conclude our brief study of polytopes, we discuss the elegant and powerful technique of representing polytopes by means of *Gale transforms*. We use this technique to discover which sets of vertices of a polytope determine one of its faces, and to calculate the number of combinatorial types of n-polytopes there are with $n + 2$ vertices.

Before entering into a detailed discussion of Gale transforms, we first give a brief overview of the general theory. A Gale transform of a vertex sequence $\mathbf{a}_1, \ldots, \mathbf{a}_m$ of an n-polytope P in $\mathbb{R}^n$ $(m > n + 1)$ is a sequence $\bar{\mathbf{a}}_1, \ldots, \bar{\mathbf{a}}_m$ in $\mathbb{R}^{m-n-1}$. Such a transform contains full information about the combinatorial structure of P. We shall see that the problem of finding which subsequences of $\mathbf{a}_1, \ldots, \mathbf{a}_m$ are vertex sequences of faces of P is equivalent to that of establishing which subsequences of $\bar{\mathbf{a}}_1, \ldots, \bar{\mathbf{a}}_m$ have the origin as a relative interior point of their convex hulls. The use of Gale transforms is most effective when the dimension $m - n - 1$ is smaller than n.

Two simple ideas which we need in our discussion of Gale transforms are those of an *affine dependence* and a *coface*. An **affine dependence** of a sequence of points $\mathbf{a}_1, \ldots, \mathbf{a}_m$ in $\mathbb{R}^n$ is a point $(\lambda_1, \ldots, \lambda_m)$ of $\mathbb{R}^m$ such that

$$\lambda_1 \mathbf{a}_1 + \ldots + \lambda_m \mathbf{a}_m = \mathbf{0} \qquad \text{and} \qquad \lambda_1 + \ldots + \lambda_m = 0.$$

Clearly the zero vector of $\mathbb{R}^m$ is an affine dependence of any sequence of points $\mathbf{a}_1, \ldots, \mathbf{a}_m$ in $\mathbb{R}^n$. A subset W of the vertex set V of a polytope P in $\mathbb{R}^n$ is called a **coface** of P if $\operatorname{conv}(V \setminus W)$ is a face of P. For example, every set comprising three vertices of a square in $\mathbb{R}^2$ is a coface of that square.

Theorem 3.6.1 *Let $\mathbf{a}_1, \ldots, \mathbf{a}_m$ be the vertices of a polytope P in $\mathbb{R}^n$. Then $\{\mathbf{a}_1, \ldots, \mathbf{a}_r\}$, where $1 \leqslant r \leqslant m$, is a coface of P if and only if there is no affine dependence $(\lambda_1, \ldots, \lambda_m)$ of $\mathbf{a}_1, \ldots, \mathbf{a}_m$ such that $\lambda_1, \ldots, \lambda_r \geqslant 0$ with at least one of $\lambda_1, \ldots, \lambda_r$ positive.*

Proof Suppose that $\{\mathbf{a}_1, \ldots, \mathbf{a}_r\}$ is not a coface of P. Then, by Theorem 3.1.4,

$$\operatorname{conv}\{\mathbf{a}_1, \ldots, \mathbf{a}_r\} \cap \operatorname{aff}\{\mathbf{a}_{r+1}, \ldots, \mathbf{a}_m\} \neq \emptyset,$$

whence there exist scalars $\mu_1, \ldots, \mu_m$ with $\mu_1, \ldots, \mu_r \geqslant 0$, $\mu_1 + \ldots + \mu_r = 1$ and $\mu_{r+1} + \ldots + \mu_m = 1$ such that

$$\mu_1 \mathbf{a}_1 + \ldots + \mu_r \mathbf{a}_r = \mu_{r+1} \mathbf{a}_{r+1} + \ldots + \mu_m \mathbf{a}_m .$$

Let $\lambda_1 = \mu_1, \ldots, \lambda_r = \mu_r$ and $\lambda_{r+1} = -\mu_{r+1}, \ldots, \lambda_m = -\mu_m$. Then $(\lambda_1, \ldots, \lambda_m)$ is an affine dependence of $\mathbf{a}_1, \ldots, \mathbf{a}_m$ for which $\lambda_1, \ldots, \lambda_r \geqslant 0$ and at least one of $\lambda_1, \ldots, \lambda_r$ is positive.

Conversely, suppose that $(\lambda_1, \ldots, \lambda_m)$ is an affine dependence of $\mathbf{a}_1, \ldots, \mathbf{a}_m$ such that $\lambda_1, \ldots, \lambda_r \geqslant 0$ and at least one of $\lambda_1, \ldots, \lambda_r$ is positive. Then

$$\frac{\lambda_1\mathbf{a}_1 + \ldots + \lambda_r\mathbf{a}_r}{\lambda_1 + \ldots + \lambda_r} = \frac{(-\lambda_{r+1})\mathbf{a}_{r+1} + \ldots + (-\lambda_m)\mathbf{a}_m}{(-\lambda_{r+1}) + \ldots + (-\lambda_m)},$$

whence

$$\operatorname{conv}\{\mathbf{a}_1, \ldots, \mathbf{a}_r\} \cap \operatorname{aff}\{\mathbf{a}_{r+1}, \ldots, \mathbf{a}_m\} \neq \varnothing$$

and $\{\mathbf{a}_1, \ldots, \mathbf{a}_r\}$ is *not* a coface of P by Theorem 3.1.4. □

We denote the set of all affine dependences of a sequence $\mathbf{a}_1, \ldots, \mathbf{a}_m$ in $\mathbb{R}^n$ by $\mathcal{A}(\mathbf{a}_1, \ldots, \mathbf{a}_m)$. Theorem 3.6.1 suggests that an exact description of $\mathcal{A}(\mathbf{a}_1, \ldots, \mathbf{a}_m)$ might be helpful in studying the facial structure of the polytope $\operatorname{conv}\{\mathbf{a}_1, \ldots, \mathbf{a}_m\}$, and this is indeed the case. Such a description is given in the following result, in which the statement that the sequence $\mathbf{a}_1, \ldots, \mathbf{a}_m$ in $\mathbb{R}^n$ is *n-dimensional* simply means that the affine hull of its points is $\mathbb{R}^n$.

Theorem 3.6.2 *Let $\mathbf{a}_1, \ldots, \mathbf{a}_m$ be an n-dimensional sequence in $\mathbb{R}^n$. Then $\mathcal{A}(\mathbf{a}_1, \ldots, \mathbf{a}_m)$ is an $(m - n - 1)$-dimensional subspace of $\mathbb{R}^m$.*

Proof Denote the rows of the $(n + 1) \times m$ matrix

$$\begin{bmatrix} \mathbf{a}_1 & \ldots & \mathbf{a}_m \\ 1 & \ldots & 1 \end{bmatrix},$$

in which $\mathbf{a}_1, \ldots, \mathbf{a}_m$ are considered column vectors, by $\mathbf{b}_1, \ldots, \mathbf{b}_{n+1}$, considered as points of $\mathbb{R}^m$. Denote by S the row space of the matrix, i.e. the set of all linear combinations of its rows. Then

$$\mathcal{A}(\mathbf{a}_1, \ldots, \mathbf{a}_m) = \{\boldsymbol{\lambda} \in \mathbb{R}^m : \boldsymbol{\lambda} \cdot \mathbf{b}_i = 0 \quad \text{for} \quad i = 1, \ldots, n + 1\} = S^{\perp}.$$

Since $\mathbf{a}_1, \ldots, \mathbf{a}_m$ is n-dimensional, the column space of the matrix has dimension $n + 1$, whence so too does S. Since

$$\dim S + \dim S^{\perp} = m,$$

it follows that $\mathcal{A}(\mathbf{a}_1, \ldots, \mathbf{a}_m) = S^{\perp}$ is an $(m - n - 1)$-dimensional subspace of $\mathbb{R}^m$. □

We now show how to find *all* the affine dependences of an n-dimensional sequence $\mathbf{a}_1, \ldots, \mathbf{a}_m$ in $\mathbb{R}^n$ $(m > n + 1)$. It follows from Theorem 3.6.2 that $\mathcal{A}(\mathbf{a}_1, \ldots, \mathbf{a}_m)$ has a basis consisting of $m - n - 1$ vectors of $\mathbb{R}^m$, say

$$\mathbf{x}_1 = (x_{11}, \ldots, x_{1m}), \ldots, \mathbf{x}_{m-n-1} = (x_{m-n-1\,1}, \ldots, x_{m-n-1\,m}).$$

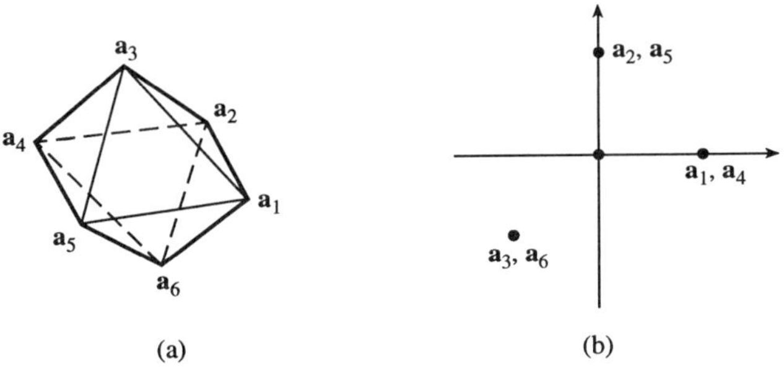

Fig. 3.5

(It is here that we have used the condition $m > n + 1$ to avoid an exceptional, but trivial case.) Now $\boldsymbol{\lambda} \in \mathcal{A}(\mathbf{a}_1, \ldots, \mathbf{a}_m)$ if and only if there exist scalars $c_1, \ldots, c_{m-n-1}$ such that

$$\boldsymbol{\lambda} = c_1(x_{11}, \ldots, x_{1m}) + \ldots + c_{m-n-1}(x_{m-n-1\,1}, \ldots, x_{m-n-1\,m})$$
$$= (c_1x_{11} + \ldots + c_{m-n-1}x_{m-n-1\,1}, \ldots, c_1x_{1m} + \ldots + c_{m-n-1}x_{m-n-1\,m}).$$

Write

$$\bar{\mathbf{a}}_1 = (x_{11}, \ldots, x_{m-n-1\,1}), \ldots, \bar{\mathbf{a}}_m = (x_{1m}, \ldots, x_{m-n-1\,m}).$$

Then we see that $\boldsymbol{\lambda}$ lies in $\mathcal{A}(\mathbf{a}_1, \ldots, \mathbf{a}_m)$ if and only if there exists a vector $\mathbf{c} = (c_1, \ldots, c_{m-n-1})$ in $\mathbb{R}^{m-n-1}$ such that

$$\boldsymbol{\lambda} = (\mathbf{c} \cdot \bar{\mathbf{a}}_1, \ldots, \mathbf{c} \cdot \bar{\mathbf{a}}_m).$$

We have thus found a simple way of expressing *all* of the affine dependences of $\mathbf{a}_1, \ldots, \mathbf{a}_m$ in terms of $\bar{\mathbf{a}}_1, \ldots, \bar{\mathbf{a}}_m$. The sequence of vectors $\bar{\mathbf{a}}_1, \ldots, \bar{\mathbf{a}}_m$ in $\mathbb{R}^{m-n-1}$ is called a **Gale transform** of the sequence $\mathbf{a}_1, \ldots, \mathbf{a}_m$ of vectors in $\mathbb{R}^n$.

Example 3.6.3 We find a Gale transform of the sequence $\mathbf{a}_1 = (1, 0, 0)$, $\mathbf{a}_2 = (0, 1, 0)$, $\mathbf{a}_3 = (0, 0, 1)$, $\mathbf{a}_4 = (-1, 0, 0)$, $\mathbf{a}_5 = (0, -1, 0)$, $\mathbf{a}_6 = (0, 0, -1)$, which lists the vertices of a regular octahedron in $\mathbb{R}^3$. See Fig. 3.5(a).

The subspace $\mathcal{A}(\mathbf{a}_1, \ldots, \mathbf{a}_6)$ of $\mathbb{R}^6$ consists of those points $(\lambda_1, \ldots, \lambda_6)$ which satisfy the simultaneous equations

$$\begin{aligned}
\lambda_1 \qquad\qquad\qquad - \lambda_4 \qquad\qquad\qquad &= 0,\\
\lambda_2 \qquad\qquad\qquad - \lambda_5 \qquad\qquad &= 0,\\
\lambda_3 \qquad\qquad\qquad - \lambda_6 &= 0,\\
\lambda_1 + \lambda_2 + \lambda_3 + \lambda_4 + \lambda_5 + \lambda_6 &= 0.
\end{aligned}$$

The general solution to this system of linear equations can be expressed in the form

$$(\lambda_1, \lambda_2, \lambda_3, \lambda_4, \lambda_5, \lambda_6) = (\alpha, \beta, -(\alpha + \beta), \alpha, \beta, -(\alpha + \beta)),$$

where $\alpha, \beta \in \mathbb{R}$. Thus $\mathbf{x}_1 = (1, 0, -1, 1, 0, -1)$, $\mathbf{x}_2 = (0, 1, -1, 0, 1, -1)$ form a basis for $\mathcal{A}(\mathbf{a}_1, \ldots, \mathbf{a}_6)$, which has dimension $m - n - 1 = 6 - 3 - 1 = 2$. The Gale transform derived from the above basis is the sequence

$$\bar{\mathbf{a}}_1 = (1, 0), \quad \bar{\mathbf{a}}_2 = (0, 1), \quad \bar{\mathbf{a}}_3 = (-1, -1), \quad \bar{\mathbf{a}}_4 = (1, 0),$$
$$\bar{\mathbf{a}}_5 = (0, 1), \quad \bar{\mathbf{a}}_6 = (-1, -1)$$

in $\mathbb{R}^2$. This is illustrated in Fig. 3.5(b). We note that although the six points $\mathbf{a}_1, \ldots, \mathbf{a}_6$ are distinct, the points $\bar{\mathbf{a}}_1, \ldots, \bar{\mathbf{a}}_6$ are not—such situations frequently occur. □

Theorem 3.6.4 *Let $\bar{\mathbf{a}}_1, \ldots, \bar{\mathbf{a}}_m$ be a Gale transform in $\mathbb{R}^{m-n-1}$ of an n-dimensional sequence $\mathbf{a}_1, \ldots, \mathbf{a}_m$ in $\mathbb{R}^n$ $(m > n + 1)$. Then:*

(i) *a vector in $\mathbb{R}^m$ is an affine dependence of $\mathbf{a}_1, \ldots, \mathbf{a}_m$ if and only if it has the form $(\mathbf{c} \cdot \bar{\mathbf{a}}_1, \ldots, \mathbf{c} \cdot \bar{\mathbf{a}}_m)$ for some $\mathbf{c} \in \mathbb{R}^{m-n-1}$;*

(ii) *the sequence $\bar{\mathbf{a}}_1, \ldots, \bar{\mathbf{a}}_m$ is $(m - n - 1)$-dimensional;*

(iii) $\bar{\mathbf{a}}_1 + \ldots + \bar{\mathbf{a}}_m = \mathbf{0}$;

(iv) *the origin of $\mathbb{R}^{m-n-1}$ is an interior point of* $\operatorname{conv}\{\bar{\mathbf{a}}_1, \ldots, \bar{\mathbf{a}}_m\}$;

(v) *every open halfspace of $\mathbb{R}^{m-n-1}$ whose bounding hyperplane passes through the origin contains at least one of the points $\bar{\mathbf{a}}_1, \ldots, \bar{\mathbf{a}}_m$.*

Proof (i) This result was established in the discussion, following Theorem 3.6.2, which motivated the definition of a Gale transform.

(ii), (iii) Let

$$\mathbf{x}_1 = (x_{11}, \ldots, x_{1m}), \ldots, \mathbf{x}_{m-n-1} = (x_{m-n-1\,1}, \ldots, x_{m-n-1\,m})$$

be the basis for $\mathcal{A}(\mathbf{a}_1, \ldots, \mathbf{a}_m)$ for which

$$\bar{\mathbf{a}}_1 = (x_{11}, \ldots, x_{m-n-1\,1}), \ldots, \quad \bar{\mathbf{a}}_m = (x_{1m}, \ldots, x_{m-n-1\,m}).$$

Since $\mathbf{x}_1, \ldots, \mathbf{x}_{m-n-1}$ are affine dependences of $\mathbf{a}_1, \ldots, \mathbf{a}_m$, we have

$$x_{11} + \ldots + x_{1m} = \ldots = x_{m-n-1\,1} + \ldots + x_{m-n-1\,m} = 0,$$

whence $\bar{\mathbf{a}}_1 + \ldots + \bar{\mathbf{a}}_m = \mathbf{0}$. Since $\bar{\mathbf{a}}_1, \ldots, \bar{\mathbf{a}}_m$ can be identified with the rows of the matrix whose columns are $\mathbf{x}_1, \ldots, \mathbf{x}_{m-n-1}$, and these columns are linearly independent, it follows that $\bar{\mathbf{a}}_1, \ldots, \bar{\mathbf{a}}_m$ linearly span $\mathbb{R}^{m-n-1}$. Now $\mathbf{0} = (1/m)(\bar{\mathbf{a}}_1 + \ldots + \bar{\mathbf{a}}_m)$, whence $\mathbf{0} \in \operatorname{aff}\{\bar{\mathbf{a}}_1, \ldots, \bar{\mathbf{a}}_m\}$. Thus $\operatorname{aff}\{\bar{\mathbf{a}}_1, \ldots, \bar{\mathbf{a}}_m\}$ is a subspace of $\mathbb{R}^{m-n-1}$ containing $\bar{\mathbf{a}}_1, \ldots, \bar{\mathbf{a}}_m$, and hence must be $\mathbb{R}^{m-n-1}$. This shows that $\bar{\mathbf{a}}_1, \ldots, \bar{\mathbf{a}}_m$ is $(m - n - 1)$-dimensional.

(iv) Theorem 2.3.7 and the equation $\mathbf{0} = (1/m)(\bar{\mathbf{a}}_1 + \ldots + \bar{\mathbf{a}}_m)$

show that $\mathbf{0} \in \operatorname{ri}(\operatorname{conv}\{\bar{\mathbf{a}}_1, \ldots, \bar{\mathbf{a}}_m\})$, whence from (ii), $\mathbf{0} \in \operatorname{int}(\operatorname{conv}\{\bar{\mathbf{a}}_1, \ldots, \bar{\mathbf{a}}_m\})$.

(v) Let H be a hyperplane in $\mathbb{R}^{m-n-1}$ passing through the origin. Denote by H^- and H^+ the open halfspaces determined by H. Suppose that H^- contains none of the points $\bar{\mathbf{a}}_1, \ldots, \bar{\mathbf{a}}_m$. Then $\bar{\mathbf{a}}_1, \ldots, \bar{\mathbf{a}}_m$ lie in the closed halfspace $H \cup H^+$, whence $\operatorname{conv}\{\bar{\mathbf{a}}_1, \ldots, \bar{\mathbf{a}}_m\} \subseteq H \cup H^+$. This, however, is incompatible with (iv). Thus H^- must contain at least one of the points $\bar{\mathbf{a}}_1, \ldots, \bar{\mathbf{a}}_m$. □

Before we can establish how Gale transforms are used to determine the facial structure of a polytope, we need the following lemma.

Lemma 3.6.5 *Let* $\mathbf{a}_1, \ldots, \mathbf{a}_r \in \mathbb{R}^n$. *Then* $\mathbf{0} \in \operatorname{ri}(\operatorname{conv}\{\mathbf{a}_1, \ldots, \mathbf{a}_r\})$ *if and only if there exists no* $\mathbf{c} \in \mathbb{R}^n$ *such that* $\mathbf{c} \cdot \mathbf{a}_1 \geqslant 0, \ldots, \mathbf{c} \cdot \mathbf{a}_r \geqslant 0$, *with at least one of the inequalities being strict.*

Proof Suppose that $\mathbf{0} \in \operatorname{ri}(\operatorname{conv}\{\mathbf{a}_1, \ldots, \mathbf{a}_r\})$. Then, by Theorem 2.3.7, there exist $\lambda_1, \ldots, \lambda_r > 0$ such that $\mathbf{0} = \lambda_1 \mathbf{a}_1 + \ldots + \lambda_r \mathbf{a}_r$, and clearly there exists no $\mathbf{c} \in \mathbb{R}^n$ for which $\mathbf{c} \cdot \mathbf{a}_1 \geqslant 0, \ldots, \mathbf{c} \cdot \mathbf{a}_r \geqslant 0$, with at least one of these inequalities being strict.

Conversely, suppose that $\mathbf{0} \notin \operatorname{ri}(\operatorname{conv}\{\mathbf{a}_1, \ldots, \mathbf{a}_r\})$. Then $\{\mathbf{0}\}$ and $\operatorname{conv}\{\mathbf{a}_1, \ldots, \mathbf{a}_r\}$ can be properly separated, so there exist $\mathbf{c} \in \mathbb{R}^n$, $c_0 \in \mathbb{R}$ such that $\mathbf{c} \cdot \mathbf{0} = 0 \leqslant c_0$ and $\mathbf{c} \cdot \mathbf{a}_1 \geqslant c_0, \ldots, \mathbf{c} \cdot \mathbf{a}_r \geqslant c_0$, where at least one of these $r + 1$ inequalities is strict. If $c_0 = 0$, then at least one of the inequalities $\mathbf{c} \cdot \mathbf{a}_1 \geqslant 0, \ldots, \mathbf{c} \cdot \mathbf{a}_r \geqslant 0$ must be strict. If $c_0 > 0$, then all of the inequalities $\mathbf{c} \cdot \mathbf{a}_1 \geqslant 0, \ldots, \mathbf{c} \cdot \mathbf{a}_r \geqslant 0$ are strict. Thus, in every case, there is $\mathbf{c} \in \mathbb{R}^n$ such that $\mathbf{c} \cdot \mathbf{a}_1 \geqslant 0, \ldots, \mathbf{c} \cdot \mathbf{a}_r \geqslant 0$, with at least one of these inequalities being strict. □

Let $\bar{\mathbf{a}}_1, \ldots, \bar{\mathbf{a}}_m$ be a Gale transform of a vertex sequence $\mathbf{a}_1, \ldots, \mathbf{a}_m$ of some n-polytope in $\mathbb{R}^n$ $(m > n + 1)$. Then, for each subset W of $\{\mathbf{a}_1, \ldots, \mathbf{a}_m\}$, we define a set $\overline{W}$ by the equation $\overline{W} = \{\bar{\mathbf{a}}_1, \ldots, \bar{\mathbf{a}}_m\}$. This notation is assumed in the following theorem and its proof.

Theorem 3.6.6 *Let* $\bar{\mathbf{a}}_1, \ldots, \bar{\mathbf{a}}_m$ *be a Gale transform in* $\mathbb{R}^{m-n-1}$ *of a vertex sequence* $\mathbf{a}_1, \ldots, \mathbf{a}_m$ *of an n-polytope P in* $\mathbb{R}^n$ $(m > n + 1)$. *Then a subset W of* $\{\mathbf{a}_1, \ldots, \mathbf{a}_m\}$ *is a coface of P if and only if either* $\overline{W}$ *is empty or* $\mathbf{0} \in \operatorname{ri}(\operatorname{conv}\overline{W})$.

Proof We assume throughout the proof that $W = \{\mathbf{a}_1, \ldots, \mathbf{a}_r\}$ for some r with $1 \leqslant r \leqslant m$. Suppose first that W is *not* a coface of P. By Theorem 3.6.1, there exists an affine dependence $(\lambda_1, \ldots, \lambda_m)$ of $\mathbf{a}_1, \ldots, \mathbf{a}_m$ such that $\lambda_1, \ldots, \lambda_r \geqslant 0$, with at least one of $\lambda_1, \ldots, \lambda_r$ positive. By Theorem 3.6.4 (i), $\lambda_1 = \mathbf{c} \cdot \bar{\mathbf{a}}_1, \ldots, \lambda_m = \mathbf{c} \cdot \bar{\mathbf{a}}_m$ for some $\mathbf{c}$ in $\mathbb{R}^{m-n-1}$. The lemma now shows that $\mathbf{0} \notin \operatorname{ri}(\operatorname{conv}\overline{W})$.

Suppose next that $\mathbf{0} \notin \text{ri}(\text{conv}\,\bar{W})$. Then the lemma shows the existence of $\mathbf{c}$ in $\mathbb{R}^{m-n-1}$ such that $\mathbf{c}\cdot\bar{\mathbf{a}}_1 \geqslant 0, \ldots, \mathbf{c}\cdot\bar{\mathbf{a}}_r \geqslant 0$, with at least one of the inequalities being strict. Let $\lambda_1 = \mathbf{c}\cdot\bar{\mathbf{a}}_1, \ldots, \lambda_m = \mathbf{c}\cdot\bar{\mathbf{a}}_m$. By Theorem 3.6.4 (i), $(\lambda_1, \ldots, \lambda_m)$ is an affine dependence of $\mathbf{a}_1, \ldots, \mathbf{a}_m$. It now follows from Theorem 3.6.1 that W is *not* a coface of P. □

The reader is asked to compare the following corollary with Theorem 3.6.4 (v).

Corollary 3.6.7 *Let* $\bar{\mathbf{a}}_1, \ldots, \bar{\mathbf{a}}_m$ *be a Gale transform in* $\mathbb{R}^{m-n-1}$ *of a vertex sequence* $\mathbf{a}_1, \ldots, \mathbf{a}_m$ *of an n-polytope P in* $\mathbb{R}^n$ $(m > n+1)$. *Then every open halfspace in* $\mathbb{R}^{m-n-1}$ *whose bounding hyperplane passes through the origin contains at least two terms of the sequence* $\bar{\mathbf{a}}_1, \ldots, \bar{\mathbf{a}}_m$.

Proof Let H be a hyperplane in $\mathbb{R}^{m-n-1}$ passing through the origin. Denote by H^- and H^+ the open halfspaces determined by H. Suppose that H^- contains *fewer* than two terms of $\bar{\mathbf{a}}_1, \ldots, \bar{\mathbf{a}}_m$. Then Theorem 3.6.4(v) shows that H^- must contain *precisely one* term of $\bar{\mathbf{a}}_1, \ldots, \bar{\mathbf{a}}_m$, say the first one. Since $\mathbf{a}_1$ is a vertex of P, the theorem shows that $\mathbf{0} \in \text{ri}(\text{conv}\{\bar{\mathbf{a}}_2, \ldots, \bar{\mathbf{a}}_m\})$, which is impossible, because $\bar{\mathbf{a}}_2, \ldots, \bar{\mathbf{a}}_m$ lie in the closed halfspace $H \cup H^+$ with at least one of them being in H^+, by Theorem 3.6.4(v). This contradiction shows that H^- must contain at least two terms of $\bar{\mathbf{a}}_1, \ldots, \bar{\mathbf{a}}_m$. □

To see how Theorem 3.6.6 works in practice, we consider again the Gale transform of the octahedron with vertices $\mathbf{a}_1, \ldots, \mathbf{a}_6$ discussed in Example 3.6.3, and illustrated in Fig. 3.5. By Theorem 3.6.6, a subset W of $\{\mathbf{a}_1, \ldots, \mathbf{a}_6\}$ is a coface of the octahedron if and only if $\mathbf{0} \in \text{ri}(\text{conv}\,\bar{W})$. From Fig. 3.5, this is the case if and only if W contains at least one of $\mathbf{a}_1, \mathbf{a}_4$, at least one of $\mathbf{a}_2, \mathbf{a}_5$, and at least one of $\mathbf{a}_3, \mathbf{a}_6$. Thus a non-empty subset of $\{\mathbf{a}_1, \ldots, \mathbf{a}_6\}$ determines a proper face of the octahedron if and only if contains at most one of $\mathbf{a}_1, \mathbf{a}_4$, at most one of $\mathbf{a}_2, \mathbf{a}_5$, and at most one of $\mathbf{a}_3, \mathbf{a}_6$. Here we have used Gale transforms to deduce a known fact. Later in the section, we shall use them to obtain new results.

The next theorem gives precise conditions under which a sequence of m terms of $\mathbb{R}^{m-n-1}$ $(m > n+1)$ is a Gale transform of a vertex sequence of some n-polytope in $\mathbb{R}^n$.

Theorem 3.6.8 *A sequence* $\bar{\mathbf{a}}_1, \ldots, \bar{\mathbf{a}}_m$ *of points in* $\mathbb{R}^{m-n-1}$ $(m > n+1)$ *is a Gale transform of a vertex sequence of some n-polytope in* $\mathbb{R}^n$ *if and only if* (i) $\bar{\mathbf{a}}_1 + \ldots + \bar{\mathbf{a}}_m = \mathbf{0}$, *and* (ii) *every open halfspace in* $\mathbb{R}^{m-n-1}$ *whose bounding hyperplane passes through the origin contains at least two terms of the sequence* $\bar{\mathbf{a}}_1, \ldots, \bar{\mathbf{a}}_m$.

Proof The *only if* part of the theorem follows from Theorem 3.6.4 (iii) and Corollary 3.6.7. Suppose, then, that $\bar{\mathbf{a}}_1, \ldots, \bar{\mathbf{a}}_m$ is a sequence of points in $\mathbb{R}^{m-n-1}$ $(m > n + 1)$ which satisfies conditions (i) and (ii) of the theorem.

First we find an n-dimensional sequence $\bar{\mathbf{a}}_1, \ldots, \bar{\mathbf{a}}_m$ in $\mathbb{R}^n$ of which $\bar{\mathbf{a}}_1, \ldots, \bar{\mathbf{a}}_m$ is a Gale transform; this we do by reversing the procedure whereby the Gale transform of a sequence was constructed. Let

$$\bar{\mathbf{a}}_1 = (x_{11}, \ldots, x_{m-n-1\,1}), \ldots, \bar{\mathbf{a}}_m = (x_{1m}, \ldots, x_{m-n-1\,m}).$$

Define points $\mathbf{x}_1, \ldots, \mathbf{x}_{m-n-1}$ in $\mathbb{R}^m$ by the equations

$$\mathbf{x}_1 = (x_{11}, \ldots, x_{1m}), \ldots, \mathbf{x}_{m-n-1} = (x_{m-n-1\,1}, \ldots, x_{m-n-1\,m}).$$

Condition (ii) ensures that $\bar{\mathbf{a}}_1, \ldots, \bar{\mathbf{a}}_m$ span $\mathbb{R}^{m-n-1}$, whence $\mathbf{x}_1, \ldots, \mathbf{x}_{m-n-1}$ form a basis for some $(m - n - 1)$-dimensional subspace of $\mathbb{R}^m$, S say. Thus $S^\perp$ has dimension $m - (m - n - 1) = n + 1$. Condition (i) shows that $(1, \ldots, 1) \in S^\perp$. Hence $(1, \ldots, 1)$ can be extended by vectors $(a_{11}, \ldots, a_{m1}), \ldots, (a_{1n}, \ldots, a_{mn})$ in $\mathbb{R}^m$ to form a basis for $S^\perp$. Write

$$\mathbf{a}_1 = (a_{11}, \ldots, a_{1n}), \ldots, \mathbf{a}_m = (a_{m1}, \ldots, a_{mn}).$$

Then $\mathbf{a}_1, \ldots, \mathbf{a}_m$ is an n-dimensional sequence in $\mathbb{R}^n$ that has $\bar{\mathbf{a}}_1, \ldots, \bar{\mathbf{a}}_m$ for a Gale transform.

We complete the proof by showing that $\mathbf{a}_1, \ldots, \mathbf{a}_m$ is a vertex sequence of the n-polytope $\operatorname{conv}\{\mathbf{a}_1, \ldots, \mathbf{a}_m\}$. To do this, we show that, for $i = 1, \ldots, m$,

$$\mathbf{a}_i \notin \operatorname{conv}\{\mathbf{a}_1, \ldots, \mathbf{a}_{i-1}, \mathbf{a}_{i+1}, \ldots, \mathbf{a}_m\}.$$

Suppose that this is not so. Then, for some i in $\{1, \ldots, m\}$, there exists an affine dependence $(\lambda_1, \ldots, \lambda_m)$ of $\mathbf{a}_1, \ldots, \mathbf{a}_m$ with $\lambda_i = -1$ and $\lambda_j \geqslant 0$ for $j \in \{1, \ldots, m\} \setminus \{i\}$. By Theorem 3.6.4 (i) there is $\mathbf{c}$ in $\mathbb{R}^{m-n-1}$ such that $\mathbf{c} \cdot \bar{\mathbf{a}}_i = \lambda_i < 0$ and $\mathbf{c} \cdot \bar{\mathbf{a}}_j = \lambda_j$ for $j \in \{1, \ldots, m\} \setminus \{i\}$. Thus the open halfspace $\{\mathbf{z} \in \mathbb{R}^{m-n-1} : \mathbf{c} \cdot \mathbf{z} < 0\}$ in $\mathbb{R}^{m-n-1}$ has the origin on its boundary and contains only one term of the sequence $\bar{\mathbf{a}}_1, \ldots, \bar{\mathbf{a}}_m$, contradicting condition (ii). Thus $\mathbf{a}_1, \ldots, \mathbf{a}_m$ is a vertex sequence of the n-polytope $\operatorname{conv}\{\mathbf{a}_1, \ldots, \mathbf{a}_m\}$. □

We now show how Gale transforms can be used to determine when a polytope is simplicial.

Theorem 3.6.9 *Let $\bar{\mathbf{a}}_1, \ldots, \bar{\mathbf{a}}_m$ be a Gale transform in $\mathbb{R}^{m-n-1}$ of a vertex sequence $\mathbf{a}_1, \ldots, \mathbf{a}_m$ of an n-polytope P in $\mathbb{R}^n$ $(m > n + 1)$. Then P is simplicial if and only if the origin of $\mathbb{R}^{m-n-1}$ cannot be expressed as a positive convex combination of fewer than $m - n$ terms of $\bar{\mathbf{a}}_1, \ldots, \bar{\mathbf{a}}_m$.*

Proof The result follows from Theorem 3.6.6 and the observation that P is simplicial if and only if it has no *proper* face with more than n vertices, and hence no non-empty coface with fewer than $m - n$ vertices. □

Since a Gale transform of a polytope contains full information about its combinatorial structure, the combinatorial type of a polytope can be determined from any one of its Gale transforms. Suppose that $\mathbf{a}_1, \ldots, \mathbf{a}_m$ and $\mathbf{b}_1, \ldots, \mathbf{b}_m$ are, respectively, vertex sequences of n-polytopes P and Q in $\mathbb{R}^n$ $(m > n + 1)$. Then it follows from Theorem 3.6.6 and the definition of combinatorial equivalence that, if $\bar{\mathbf{a}}_1, \ldots, \bar{\mathbf{a}}_m$ and $\bar{\mathbf{b}}_1, \ldots, \bar{\mathbf{b}}_m$ are, respectively, Gale transforms of $\mathbf{a}_1, \ldots, \mathbf{a}_m$ and $\mathbf{b}_1, \ldots, \mathbf{b}_m$, then P and Q are combinatorially equivalent if and only if there is a permutation θ of $\{1, \ldots, m\}$ such that $\mathbf{0} \in \mathrm{ri}\,(\mathrm{conv}\,\{\bar{\mathbf{a}}_j : j \in J\})$ if and only if $\mathbf{0} \in \mathrm{ri}\,(\mathrm{conv}\,\{\bar{\mathbf{b}}_{\theta(j)} : j \in J\})$, for every subset J of $\{1, \ldots, m\}$.

We conclude our discussion of Gale transforms by demonstrating their effectiveness in determining the number of combinatorial types of n-polytopes with $n + 2$ vertices, and by seeing how many of these types are simplicial.

Theorem 3.6.10 *There are $[\frac{1}{4}n^2]$ combinatorial types of n-polytopes with $n + 2$ vertices and $[\frac{1}{2}n]$ of these are simplicial.*

Proof Let $\bar{\mathbf{a}}_1, \ldots, \bar{\mathbf{a}}_{n+2}$ be a Gale transform in $\mathbb{R}^1$ of a vertex sequence $\mathbf{a}_1, \ldots, \mathbf{a}_{n+2}$ of an n-polytope P in $\mathbb{R}^n$. By Theorem 3.6.4 (iii), this transform is a sequence of $n + 2$ real numbers whose sum is zero. Suppose that this sequence has r positive terms and s negative ones, so that $r \geqslant 2$, $s \geqslant 2$ and $r + 2 \leqslant n + 2$; we call such a sequence a G-**sequence of type** (r, s). Suppose next that $\bar{\mathbf{b}}_1, \ldots, \bar{\mathbf{b}}_{n+2}$ is a Gale transform in $\mathbb{R}^1$ of a vertex sequence $\mathbf{b}_1, \ldots, \mathbf{b}_{n+2}$ of an n-polytope Q in $\mathbb{R}^n$. Suppose that $\bar{\mathbf{b}}_1, \ldots, \bar{\mathbf{b}}_{n+2}$ is a G-sequence of type (r', s'). In view of our remarks in the paragraph following Theorem 3.6.9, P and Q are combinatorially equivalent if and only if *either* $r = r'$ and $s = s'$ *or* $r = s'$ and $s = r'$. Theorem 3.6.8 shows that every G-sequence of $n + 2$ terms of $\mathbb{R}^1$ is a Gale transform of some n-polytope in $\mathbb{R}^n$ with $n + 2$ vertices.

Thus the number of combinatorial types of n-polytopes with $n + 2$ vertices equals the number of ordered pairs (r, s) of integers there are satisfying $s \geqslant r \geqslant 2$ and $r + s \leqslant n + 2$; we now calculate this number. When n is even, these ordered pairs are:

$$\begin{array}{llll}(2, n), & (2, n-1), & \ldots, & (2,3), \quad (2,2); \\ (3, n-1), & (3, n-2), & \ldots, & (3,3); \\ \ldots\ldots\ldots\ldots\ldots\ldots ; \\ (\tfrac{1}{2}(n+2), \tfrac{1}{2}(n+2)). \end{array}$$

The total number of these pairs is $(n-1)+(n-3)+\ldots+1$, which equals $\frac{1}{4}n^2$. When n is odd, the ordered pairs which we seek are the same as those listed above except that the last line needs to be replaced by

$$(\tfrac{1}{2}(n+1), \tfrac{1}{2}(n+3)), \quad (\tfrac{1}{2}(n+1), \tfrac{1}{2}(n+1)).$$

The total number of these pairs is $(n-1)+(n-3)+\ldots+2$, which equals $\frac{1}{4}(n^2-1)$. In both cases, the required number is $[\frac{1}{4}n^2]$.

Theorem 3.6.9 shows that a G-sequence of $n+2$ terms which is of type (r, s) corresponds to a simplicial n-polytope with $n+2$ vertices if and only if 0 is not one of its terms, i.e. if and only if $r+s=n+2$. Thus the number of combinatorial types of simplicial n-polytopes with $n+2$ vertices equals the number of ordered pairs (r, s) of integers such that $s \geqslant r \geqslant 2$ and $r+s=n-2$. This number is $\frac{1}{2}n$ when n is even, and $\frac{1}{2}(n-1)$ when n is odd. In both cases it equals $[\frac{1}{2}n]$. □

The last theorem with $n=3$ shows that there are precisely two combinatorial types of 3-polytopes with five vertices, only one type being simplicial. We have already seen, in Section 3.1, examples of these two types: a square pyramid (non-simplicial), and the polytope formed by taking the union of a regular tetrahedron and its reflection in one of its triangular faces (simplicial). Possible Gale transforms for these two examples: $1, -1, 1, -1, 0$ and $2, 2, 2, -3, -3$, themselves make it clear why the two examples are of different combinatorial types, and that the first one (the square pyramid) is non-simplicial, as 0 occurs in its Gale transform. This very simple example serves to show the power and potential of Gale transform techniques in studying the combinatorial properties of polytopes.

Exercises 3.6

1. Find Gale transforms for each of the following sequences in $\mathbb{R}^3$:
   (i) $(1,1,0), (-1,1,0), (-1,-1,0), (1,-1,0), (0,0,1)$—vertices of a square pyramid;
   (ii) $(1,0,0), (0,1,0), (-1,-1,0), (0,0,1), (0,0,-1)$—vertices of a triangular bipyramid;
   (iii) $(0,0,0), (1,0,0), (0,1,0), (0,0,1), (1,0,1), (0,1,1)$—vertices of a triangular prism.

2. Show that any Gale transform $\bar{\mathbf{a}}_1, \ldots, \bar{\mathbf{a}}_m$ of a vertex sequence $\mathbf{a}_1, \ldots, \mathbf{a}_m$ of an s-fold n-pyramid P in $\mathbb{R}^n$ $(m > n + 1)$ contains at least s zero terms.
3. Describe a Gale transform of the cyclic polytope $C(n + 2, n)$ for $n \geqslant 2$. Hence determine the number of its facets. Find the number of faces of each dimension of the cyclic polytope $C(7, 5)$, and verify the Dehn–Sommerville equations for this polytope.
4. Let r, s be integers such that $r, s \geqslant 2$ and $r + s = n + 2$. Construct an n-polytope P in $\mathbb{R}^n$ having $n + 2$ vertices such that any Gale transform in $\mathbb{R}^1$ of one of its vertex sequences has r terms of one sign and s terms of the opposite sign.
5. Let $\bar{\mathbf{a}}_1, \ldots, \bar{\mathbf{a}}_m$ be a Gale transform in $\mathbb{R}^{m-n-1}$ of a vertex sequence $\mathbf{a}_1, \ldots, \mathbf{a}_m$ of an n-polytope P in $\mathbb{R}^n$ $(m > n + 1)$. Then the sequence $\hat{\mathbf{a}}_1, \ldots, \hat{\mathbf{a}}_m$ defined by the equations

$$\hat{\mathbf{a}}_i = \mathbf{0} \quad \text{if} \quad \bar{\mathbf{a}}_i = \mathbf{0} \qquad \text{and} \qquad \hat{\mathbf{a}}_i = \bar{\mathbf{a}}_i / \|\bar{\mathbf{a}}_i\| \quad \text{if} \quad \bar{\mathbf{a}}_i \neq \mathbf{0},$$

is called a **Gale diagram** of P. Show, using an obvious notation, that a set W of vertices of P is a coface of P if and only if either $\hat{W} = \emptyset$ or $\mathbf{0} \in \operatorname{ri}(\operatorname{conv} \hat{W})$.

4
Linear programming

This chapter gives a full introduction to both the theory and practice of linear programming and game theory. The important role played by ideas from convexity is emphasized.

With the exception of Theorem 4.4.2 (needed in Section 5.8), the remainder of the book is independent of this chapter. The *method of the artificial variable*, discussed at the end of Section 4.5, may be omitted on a first reading.

4.1 THE FINITE BASIS THEOREM

We now continue the discussion of polyhedral sets begun in Section 3.2, characterizing them as those sets which can be represented as a vector sum of a polytope and a finitely generated convex cone. This result, known as the *finite basis theorem*, provides a simple internal representation of polyhedral sets that has applications to, amongst other topics, linear programming.

A **finitely generated** convex cone is one that is generated by a finite set, i.e. a convex cone of the form

$$\operatorname{cone}\{\mathbf{a}_1, \ldots, \mathbf{a}_m\} = \{\lambda_1\mathbf{a}_1 + \ldots + \lambda_m\mathbf{a}_m : \lambda_1, \ldots, \lambda_m \geqslant 0\},$$

where $\mathbf{a}_1, \ldots, \mathbf{a}_m \in \mathbb{R}^n$. A convex cone in $\mathbb{R}^n$ which is also a polyhedral set is called a **polyhedral cone**. Clearly, a set in $\mathbb{R}^n$ is a polyhedral cone if and only if it is a finite intersection of closed halfspaces whose bounding hyperplanes pass through the origin.

Our next result, which shows that finitely generated convex cones and polyhedral cones are one and the same thing, can be viewed as the finite basis theorem for polyhedral cones.

Theorem 4.1.1 *A convex cone in $\mathbb{R}^n$ is finitely generated if and only if it is polyhedral.*

Proof Suppose first that C is a polyhedral cone in $\mathbb{R}^n$. Let P be a polytope in $\mathbb{R}^n$ such that $\mathbf{0} \in \operatorname{int} P$. Then $C \cap P$ is a bounded polyhedral set, and hence a polytope. Thus $C \cap P$ is $\operatorname{conv}\{\mathbf{a}_1, \ldots, \mathbf{a}_m\}$ for some points $\mathbf{a}_1, \ldots, \mathbf{a}_m$ of $C \cap P$. We show that C is the finitely generated convex cone $\operatorname{cone}\{\mathbf{a}_1, \ldots, \mathbf{a}_m\}$. Since C is a convex cone containing $\mathbf{a}_1, \ldots, \mathbf{a}_m$, $\operatorname{cone}\{\mathbf{a}_1, \ldots, \mathbf{a}_m\} \subseteq C$. If $\mathbf{c} \in C$, then, since $\mathbf{0} \in \operatorname{int} P$, there is some $\lambda > 0$ such that $\lambda\mathbf{c} \in P$. Thus

$$\lambda\mathbf{c}\in C\cap P=\operatorname{conv}\{\mathbf{a}_1,\ldots,\mathbf{a}_m\}\subseteq\operatorname{cone}\{\mathbf{a}_1,\ldots,\mathbf{a}_m\},$$

and so $\mathbf{c}\in(1/\lambda)\operatorname{cone}\{\mathbf{a}_1,\ldots,\mathbf{a}_m\}=\operatorname{cone}\{\mathbf{a}_1,\ldots,\mathbf{a}_m\}$. Hence we have $C\subseteq\operatorname{cone}\{\mathbf{a}_1,\ldots,\mathbf{a}_m\}$, and $C=\operatorname{cone}\{\mathbf{a}_1,\ldots,\mathbf{a}_m\}$ as desired.

Suppose next that C is the finitely generated convex cone $\operatorname{cone}\{\mathbf{a}_1,\ldots,\mathbf{a}_m\}$, where $\mathbf{a}_1,\ldots,\mathbf{a}_m\in\mathbb{R}^n$. Since polytopes are polyhedral sets, $\operatorname{conv}\{\mathbf{0},\mathbf{a}_1,\ldots,\mathbf{a}_m\}$ can be written as the intersection of some closed halfspaces $J_1,\ldots,J_r$ in $\mathbb{R}^n$. We show that C is the polyhedral cone A formed by the intersection of those J_i's which have the origin on their boundaries. Since $\mathbf{a}_1,\ldots,\mathbf{a}_m\in A$, $C\subseteq A$. If $\mathbf{a}\in A$, then $\lambda\mathbf{a}\in J_i$ for *all* $\lambda>0$ when $\mathbf{0}$ lies on the boundary of J_i, and $\lambda\mathbf{a}\in J_i$ for *all sufficiently small* $\lambda>0$ when $\mathbf{0}$ does not lie on the boundary of J_i. It follows that there is a $\lambda>0$ such that

$$\lambda\mathbf{a}\in J_1\cap\ldots\cap J_r=\operatorname{conv}\{\mathbf{0},\mathbf{a}_1,\ldots,\mathbf{a}_m\}\subseteq C,$$

whence $\mathbf{a}\in(1/\lambda)C=C$. Hence $A\subseteq C$, and $C=A$ as desired. □

Corollary 4.1.2 *Finitely generated convex cones in $\mathbb{R}^n$ are closed.* □

Theorem 4.1.3 (Finite basis theorem) *A set in $\mathbb{R}^n$ is a polyhedral set if and only if it can be expressed as a vector sum of a polytope and a finitely generated convex cone.*

Proof Suppose first that $P=\operatorname{conv}S+\operatorname{cone}T$, where S and T are finite sets in $\mathbb{R}^n$. Then P is the vector sum of the *compact* polytope $\operatorname{conv}S$ and the *closed* finitely generated convex cone $\operatorname{cone}T$, and so is closed by Theorem 1.8.10 (ii). Since $\operatorname{conv}S$ and $\operatorname{cone}T$ are polyhedral sets, they only have a finite number of faces. It follows from Theorem 2.6.18 that P has only a finite number of exposed faces. Thus P is a closed convex set which has only a finite number of exposed faces, and so must be a polyhedral set by Theorem 3.2.3.

Suppose now that P is a polyhedral set in $\mathbb{R}^n$ which contains no lines. If P is bounded, then it is a polytope, and so trivially the vector sum of a polytope and the zero cone. Assume, then, that P is unbounded. By Corollary 2.6.15

$$P=\operatorname{conv}(S\cup L_1\cup\ldots\cup L_r),$$

where S is the set of extreme points of P and $L_1,\ldots,L_r$ are its extreme halflines. Let $T=\{\mathbf{a}_1,\ldots,\mathbf{a}_r\}$, where $\mathbf{a}_1,\ldots,\mathbf{a}_r$ are non-zero vectors belonging to the directions of $L_1,\ldots,L_r$ respectively. Theorem 2.5.1 shows that T lies in the recession cone of P, whence so too does cone T. By Theorem 2.5.6,

$$\operatorname{conv}S+\operatorname{cone}T\subseteq P+\operatorname{cone}T\subseteq P.$$

On the other hand, $\operatorname{conv}S+\operatorname{cone}T$ is a convex set containing

$S \cup L_1 \cup \ldots \cup L_r$, which shows that $P \subseteq \operatorname{conv} S + \operatorname{cone} T$. Thus P is the vector sum of the polytope conv S and the finitely generated convex cone cone T.

Suppose, finally, that P is a polyhedral set in $\mathbb{R}^n$ which contains a line. Theorem 2.5.8 shows that $P = (P \cap L^{\perp}) + L$, where L is the (non-zero) lineality space of P, and $P \cap L^{\perp}$ is a polyhedral set containing no lines. By what we have just proved, $P \cap L^{\perp}$ can be expressed as conv S + cone T for some finite sets S and T in $\mathbb{R}^n$. Let T' be a basis for the subspace L. Then $L = \operatorname{cone}(T' \cup (-T'))$ and

$$\begin{aligned} P &= \operatorname{conv} S + \operatorname{cone} T + \operatorname{cone}(T' \cup (-T')) \\ &= \operatorname{conv} S + \operatorname{cone}(T \cup T' \cup (-T')). \end{aligned}$$

Thus we have expressed P as a vector sum of a polytope and a finitely generated convex cone. □

The finite basis theorem shows that, given any polyhedral set P in $\mathbb{R}^n$, there exist $\mathbf{a}_1, \ldots, \mathbf{a}_m \in \mathbb{R}^n$ and $k \in \{0, 1, \ldots, m\}$ such that

$$P = \{\lambda_1 \mathbf{a}_1 + \ldots + \lambda_m \mathbf{a}_m : \lambda_1, \ldots, \lambda_m \geqslant 0 \quad \text{and} \quad \lambda_1 + \ldots + \lambda_k = 1\};$$

it being understood that P is empty when $m = 0$, and cone $\{\mathbf{a}_1, \ldots, \mathbf{a}_m\}$ when $k = 0$ and $m > 0$. The ordered pair $(\{\mathbf{a}_1, \ldots, \mathbf{a}_k\}, \{\mathbf{a}_{k+1}, \ldots, \mathbf{a}_m\})$ is sometimes referred to as a *finite basis* for P, and P is said to be *finitely generated* by $\mathbf{a}_1, \ldots, \mathbf{a}_k; \mathbf{a}_{k+1}, \ldots, \mathbf{a}_m$.

Theorem 4.1.4 *Let A, B be polyhedral sets in $\mathbb{R}^n$ and let α be a scalar. Then $A + B$ and αA are polyhedral sets.*

Proof By the finite basis theorem, there are finite sets C, D, E, F in $\mathbb{R}^n$ such that $A = \operatorname{conv} C + \operatorname{cone} D$ and $B = \operatorname{conv} E + \operatorname{cone} F$. Now

$$\operatorname{conv} C + \operatorname{conv} E = \operatorname{conv}(C + E)$$

and

$$\operatorname{cone} D + \operatorname{cone} F = \operatorname{cone}(D \cup F).$$

The first of these equations can be established by the argument used in the first part of the proof of Theorem 3.1.1, and the second is trivial. Thus

$$A + B = \operatorname{conv}(C + E) + \operatorname{cone}(D \cup F),$$

which shows that $A + B$ is a polyhedral set. The easily verified equation $\alpha A = \operatorname{conv}(\alpha C) + \operatorname{cone}(\alpha D)$ shows that αA is a polyhedral set. □

It is not always possible to separate two disjoint non-empty closed convex sets in $\mathbb{R}^n$ in such a way that neither lies in the hyperplane of separation itself; when the sets are polyhedral, however, they can always be strictly separated.

Theorem 4.1.5 *Each pair A and B of disjoint non-empty polyhedral sets in $\mathbb{R}^n$ can be strictly separated.*

Proof Theorem 4.1.4 shows that $A - B$ is a polyhedral set. Since $A - B$ is closed and does not contain the origin, it can be strictly separated from the origin. Thus there exist $\mathbf{c} \in \mathbb{R}^n$ and $c_0 \in \mathbb{R}$ such that

$$\mathbf{c} \cdot (\mathbf{a} - \mathbf{b}) > c_0 > 0 \qquad \text{for } \mathbf{a} \in A, \mathbf{b} \in B.$$

It follows easily that there is a scalar d satisfying

$$\inf \{\mathbf{c} \cdot \mathbf{a} : \mathbf{a} \in A\} > d > \sup \{\mathbf{c} \cdot \mathbf{b} : \mathbf{b} \in B\},$$

and that the hyperplane with equation $\mathbf{c} \cdot \mathbf{z} = d$ strictly separates A and B. □

The next theorem anticipates a result (Theorem 4.3.2) in linear programming.

Theorem 4.1.6 *Suppose that $f: P \to \mathbb{R}$ is a linear function which is bounded above on a non-empty line-free polyhedral set P in $\mathbb{R}^n$. Then f attains its upper bound at an extreme point of P.*

Proof We can write

$$P = \operatorname{conv} \{\mathbf{a}_1, \ldots, \mathbf{a}_m\} + \operatorname{cone} \{\mathbf{b}_1, \ldots, \mathbf{b}_p\},$$

where $\mathbf{a}_1, \ldots, \mathbf{a}_m, \mathbf{b}_1, \ldots, \mathbf{b}_p \in \mathbb{R}^n$ and $\mathbf{a}_1, \ldots, \mathbf{a}_m$ are the extreme points of P. A typical point $\mathbf{x}$ of P can be written in the form

$$\mathbf{x} = \lambda_1 \mathbf{a}_1 + \ldots + \lambda_m \mathbf{a}_m + \mu_1 \mathbf{b}_1 + \ldots + \mu_p \mathbf{b}_p,$$

where $\lambda_1, \ldots, \lambda_m, \ \mu_1, \ldots, \mu_p \geqslant 0$ and $\lambda_1 + \ldots + \lambda_m = 1$. Since f is linear,

$$f(\mathbf{x}) = \lambda_1 f(\mathbf{a}_1) + \ldots + \lambda_m f(\mathbf{a}_m) + \mu_1 f(\mathbf{b}_1) + \ldots + \mu_p f(\mathbf{b}_p).$$

The assumption that f is bounded above and the fact that $\mu_1, \ldots, \mu_p$ may assume any positive values combine to show that $f(\mathbf{b}_1), \ldots, f(\mathbf{b}_p) \leqslant 0$. It is now easy to see that f assumes its upper bound (maximal value) at any extreme point $\mathbf{a}_i$ of P for which $f(\mathbf{a}_i) = \max \{f(\mathbf{a}_1), \ldots, f(\mathbf{a}_m)\}$. □

Exercises 4.1

1. Express the finitely generated convex cone

$$\{\lambda_1 \mathbf{a}_1 + \lambda_2 \mathbf{a}_2 + \lambda_3 \mathbf{a}_3 + \lambda_4 \mathbf{a}_4 : \lambda_1, \lambda_2, \lambda_3, \lambda_4 \geqslant 0\},$$

where $\mathbf{a}_1 = (1, 1, 1)$, $\mathbf{a}_2 = (-1, 1, 1)$, $\mathbf{a}_3 = (-1, -1, 1)$, $\mathbf{a}_4 = (1, -1, 1)$ in $\mathbb{R}^3$, as a finite intersection of closed halfspaces.

2. Let P be the polyhedral set consisting of all those points (x, y) in $\mathbb{R}^2$ satisfying the inequalities $x \geqslant 0$, $-x + y + 1 \geqslant 0$, $x + 3y - 1 \geqslant 0$, $3x + y - 1 \geqslant 0$. Express P as a vector sum of a polytope and a finitely generated convex cone.
3. Suppose that $A = B + C$, where B is a non-empty bounded convex set in $\mathbb{R}^n$ and C is a closed convex cone in $\mathbb{R}^n$. Show that C is the recession cone of A. Deduce that the recession cone of a non-empty polyhedral set in $\mathbb{R}^n$ is a polyhedral cone.

4.2 LINEAR INEQUALITIES

Linear inequalities not only play a central role in linear programming and game theory, but are also well worth studying in their own right, as we now hope to show. Beginning with a simple result concerning the separation of a point and a cone, we proceed in easy stages to establish the fundamental results of the subject.

Finite systems of linear inequalities can be conveniently expressed in matrix notation. We denote matrices by square brackets and we identify the points of $\mathbb{R}^n$ with column matrices. Thus we write

$$\mathbf{x} = (x_1, \ldots, x_n) = \begin{bmatrix} x_1 \\ \vdots \\ x_n \end{bmatrix}.$$

We denote the transpose of a matrix $\mathbf{A}$ by $\mathbf{A}^{\mathrm{T}}$ and we identify a square matrix of order one with the number which determines it. Thus we write

$$\mathbf{x}^{\mathrm{T}} = (x_1, \ldots, x_n)^{\mathrm{T}} = [x_1 \ldots x_n]$$

and

$$\mathbf{x} \cdot \mathbf{y} = (x_1, \ldots, x_n) \cdot (y_1, \ldots, y_n) = x_1 y_1 + \ldots + x_n y_n = \mathbf{x}^{\mathrm{T}} \mathbf{y}.$$

When real $m \times n$ matrices $\mathbf{A} = [a_{ij}]$, $\mathbf{B} = [b_{ij}]$ are such that $a_{ij} < b_{ij}$ for $i = 1, \ldots, m$ and $j = 1, \ldots, n$, we write $\mathbf{A} < \mathbf{B}$; similar definitions apply to the inequalities $\mathbf{A} \leqslant \mathbf{B}$, $\mathbf{A} > \mathbf{B}$, $\mathbf{A} \geqslant \mathbf{B}$. In the following discussion, $\mathbf{A}$ denotes a real $m \times n$ matrix $[a_{ij}]$, $\mathbf{b}$ denotes a point $(b_1, \ldots, b_m)$ of $\mathbb{R}^m$, $\mathbf{x}$ denotes a point $(x_1, \ldots, x_n)$ of $\mathbb{R}^n$, $\mathbf{y}$ denotes a point $(y_1, \ldots, y_m)$ of $\mathbb{R}^m$, and $\mathbf{0}$ denotes a zero vector, whose size can be determined from the context.

Our discussion is based on the following result, which asserts that, if $\mathbf{a}$ is a vector not lying in a closed convex cone C, then there is a vector $\mathbf{u}$ making an acute or right angle with each vector in C, but an obtuse angle with $\mathbf{a}$.

Theorem 4.2.1 *In $\mathbb{R}^n$ let $\mathbf{a}$ be a point not lying in a closed convex cone C. Then there exists a point $\mathbf{u}$ in $\mathbb{R}^n$ such that $\mathbf{u}\cdot\mathbf{a} < 0$, and $\mathbf{u}\cdot\mathbf{c} \geqslant 0$ for all points $\mathbf{c}$ in C.*

Proof Since $\mathbf{a}$ does not belong to the closed convex set C, $\mathbf{a}$ can be *strictly* separated from C. Thus there exist $\mathbf{u}\in\mathbb{R}^n$ and $u_0\in\mathbb{R}$ such that

$$\mathbf{u}\cdot\mathbf{a} < u_0 < \mathbf{u}\cdot\mathbf{c} \qquad \text{for } \mathbf{c}\in C.$$

Since $\mathbf{0}\in C$, we have $\mathbf{u}\cdot\mathbf{a} < u_0 < 0$. Let $\mathbf{c}\in C$ and $\lambda > 0$. Then $\lambda\mathbf{c}\in C$, and so $\mathbf{u}\cdot(\lambda\mathbf{c}) > u_0$ and $\mathbf{u}\cdot\mathbf{c} > u_0/\lambda$. Letting $\lambda\to\infty$ in the last inequality, we deduce that $\mathbf{u}\cdot\mathbf{c}\geqslant 0$. □

Consider now the following system of m linear inequalities in n variables:

$$\left.\begin{array}{c} a_{11}x_1 + \ldots + a_{1n}x_n \leqslant b_1 \\ \cdots\cdots\cdots\cdots\cdots\cdots\cdots \\ a_{m1}x_1 + \ldots + a_{mn}x_n \leqslant b_m \end{array}\right\}. \tag{4.2.1}$$

In matrix notation, this system of inequalities assumes the form $\mathbf{Ax}\leqslant\mathbf{b}$. If there exist real numbers $x_1,\ldots,x_n$ which simultaneously satisfy the m linear inequalities of the system (4.2.1), then the system is said to be **consistent**; otherwise it is said to be **inconsistent**. To show that a system is consistent, we only have to find $x_1,\ldots,x_n$ which satisfy it. How can we show that a system is inconsistent? To help us answer the question, consider the following system of three inequalities in three variables:

$$\left.\begin{array}{rl} x - y + 2z \leqslant & -1 \\ -2x + y - 3z \leqslant & 4 \\ 5x - y + 6z \leqslant & -14 \end{array}\right\}.$$

After making several unsuccessful attempts at finding x, y, z which simultaneously satisfy these three inequalities, the reader might correctly conclude that the sytem is inconsistent; but such a lack of success does not, of course, prove the inconsistency. Let us suppose, arguing for a contradiction, that the real numbers x, y, z satisfy the given inequalities. Then, after multiplying the inequalities by $3, 4, 1$, respectively, and adding the resulting inequalities together, we find that

$$0x + 0y + 0z = 0 \leqslant -1.$$

This contradiction shows that the given system of inequalities must be inconsistent. We now see how this method of proving inconsistency applies to the general system (4.2.1). Suppose that $x_1,\ldots,x_n$ satisfy the inequalities (4.2.1) and that $y_1,\ldots,y_m\geqslant 0$. Then

$$\begin{array}{c}(a_{11}y_1 + \ldots + a_{m1}y_m)x_1 + \ldots + (a_{1n}y_1 + \ldots + a_{mn}y_m)x_n \\ \leqslant b_1y_1 + \ldots + b_my_m,\end{array}$$

which in matrix form is $\mathbf{y}^T\mathbf{A}\mathbf{x} \leqslant \mathbf{y}^T\mathbf{b}$. This last inequality cannot be satisfied if

$$a_{11}y_1 + \ldots + a_{m1}y_m = 0, \qquad \ldots, \qquad a_{1n}y_1 + \ldots + a_{mn}y_m = 0,$$
$$b_1y_1 + \ldots + b_my_m < 0,$$

i.e. if $\mathbf{y}^T\mathbf{A} = \mathbf{0}^T$, $\mathbf{y}^T\mathbf{b} < 0$. We have thus shown that, if there exists $\mathbf{y} \geqslant \mathbf{0}$ such that $\mathbf{y}^T\mathbf{A} = \mathbf{0}^T$, $\mathbf{y}^T\mathbf{b} < 0$, then the system $\mathbf{A}\mathbf{x} \leqslant \mathbf{b}$ is inconsistent. The surprise (?) is that the converse of the last assertion is true, i.e. if the system $\mathbf{A}\mathbf{x} \leqslant \mathbf{b}$ is inconsistent, then there exists $\mathbf{y} \geqslant \mathbf{0}$ such that $\mathbf{y}^T\mathbf{A} = \mathbf{0}^T$, $\mathbf{y}^T\mathbf{b} < 0$. Before proving this, we need to establish the following lemma.

Lemma 4.2.2 *Suppose that there exists no* $\mathbf{x} \geqslant \mathbf{0}$ *in* $\mathbb{R}^n$ *such that* $\mathbf{A}\mathbf{x} = \mathbf{b}$. *Then there exists* $\mathbf{y}$ *in* $\mathbb{R}^m$ *such that* $\mathbf{y}^T\mathbf{A} \geqslant \mathbf{0}^T$, $\mathbf{y}^T\mathbf{b} < 0$.

Proof Denote the columns of $\mathbf{A}$ by $\mathbf{c}_1, \ldots, \mathbf{c}_n$. By the hypothesis of the lemma, there is no $\mathbf{x} = (x_1, \ldots, x_n) \geqslant \mathbf{0}$ such that

$$\mathbf{A}\mathbf{x} = x_1\mathbf{c}_1 + \ldots + x_n\mathbf{c}_n = \mathbf{b},$$

i.e. $\mathbf{b} \notin \text{cone}\{\mathbf{c}_1, \ldots, \mathbf{c}_n\}$. Theorem 4.2.1 shows that there exists $\mathbf{y}$ in $\mathbb{R}^m$ such that $\mathbf{y}^T\mathbf{b} < 0$ and

$$\mathbf{y}^T\mathbf{c}_1 \geqslant 0, \ldots, \mathbf{y}^T\mathbf{c}_n \geqslant 0,$$

that is

$$\mathbf{y}^T[\mathbf{c}_1, \ldots, \mathbf{c}_n] = \mathbf{y}^T\mathbf{A} \geqslant \mathbf{0}^T. \; \square$$

Theorem 4.2.3 *The system of inequalities* $\mathbf{A}\mathbf{x} \leqslant \mathbf{b}$ *is inconsistent if and only if there exists* $\mathbf{y} \geqslant \mathbf{0}$ *such that* $\mathbf{y}^T\mathbf{A} = \mathbf{0}^T$, $\mathbf{y}^T\mathbf{b} < 0$.

Proof We have already seen that, if there exists $\mathbf{y} \geqslant \mathbf{0}$ such that $\mathbf{y}^T\mathbf{A} = \mathbf{0}^T$, $\mathbf{y}^T\mathbf{b} < 0$, then the system $\mathbf{A}\mathbf{x} \leqslant \mathbf{b}$ is inconsistent. Suppose, then, that the system $\mathbf{A}\mathbf{x} \leqslant \mathbf{b}$ is inconsistent. Consider the system of m linear equations in $2n + m$ variables represented by the matrix equation

$$[\mathbf{A}, -\mathbf{A}, \mathbf{I}_m]\mathbf{z} = \mathbf{b}, \qquad \text{where } \mathbf{z} = (z_1, \ldots, z_{2n+m}).$$

This system cannot have a solution $\mathbf{z}$ for which $\mathbf{z} \geqslant \mathbf{0}$, since the existence of such a solution would imply that

$$\mathbf{A}(z_1 - z_{n+1}, \ldots, z_n - z_{2n}) \leqslant \mathbf{b},$$

which contradicts the assumption that the system $\mathbf{A}\mathbf{x} \leqslant \mathbf{b}$ is inconsistent. By Lemma 4.2.2, there exists $\mathbf{y}$ such that $\mathbf{y}^T\mathbf{b} < 0$ and

$$\mathbf{y}^T[\mathbf{A}, -\mathbf{A}, \mathbf{I}_m] \geqslant \mathbf{0}^T,$$

that is

$$\mathbf{y}^{\mathrm{T}}\mathbf{A} \geqslant \mathbf{0}^{\mathrm{T}}, \qquad -\mathbf{y}^{\mathrm{T}}\mathbf{A} \geqslant \mathbf{0}^{\mathrm{T}}, \qquad \mathbf{y}^{\mathrm{T}}\mathbf{I}_m \geqslant \mathbf{0}^{\mathrm{T}},$$

whence

$$\mathbf{y} \geqslant \mathbf{0}, \qquad \mathbf{y}^{\mathrm{T}}\mathbf{A} = \mathbf{0}^{\mathrm{T}}, \qquad \mathbf{y}^{\mathrm{T}}\mathbf{b} < 0. \ \square$$

An immediate consequence of the last theorem is that *precisely one* of the following systems of inequalities in $\mathbf{x}$ and $\mathbf{y}$ has a solution:

(i) $\mathbf{Ax} \leqslant \mathbf{b}$;

(ii) $\mathbf{y}^{\mathrm{T}}\mathbf{A} = \mathbf{0}^{\mathrm{T}}, \quad \mathbf{y}^{\mathrm{T}}\mathbf{b} < 0, \quad \mathbf{y} \geqslant \mathbf{0}$.

Two finite systems of linear inequalities such as (i) and (ii), precisely one of which has a solution, are said to form a **dual pair** or to be **dual** to each other. It follows easily from Lemma 4.2.2 that the systems

(i) $\mathbf{Ax} = \mathbf{b}, \quad \mathbf{x} \geqslant \mathbf{0}$;

(ii) $\mathbf{y}^{\mathrm{T}}\mathbf{A} \geqslant \mathbf{0}^{\mathrm{T}}, \quad \mathbf{y}^{\mathrm{T}}\mathbf{b} < 0$

are dual to each other. One interesting application of dual pairs is to the theory of linear equations. It is an easy exercise to show that the systems

(i) $\mathbf{Ax} = \mathbf{b}$;

(ii) $\mathbf{y}^{\mathrm{T}}\mathbf{A} = \mathbf{0}^{\mathrm{T}}, \quad \mathbf{y}^{\mathrm{T}}\mathbf{b} \neq 0$

form a dual pair. Thus, if the system of linear equations $\mathbf{Ax} = \mathbf{b}$ is inconsistent, then some linear combination of its equations yields the contradiction $0 \neq 0$; a result often tacitly assumed, but rarely proved.

We now turn our attention to a mixed system of weak and strict inequalities. Let $\mathbf{A}$ be an $m \times n$ matrix, $\mathbf{C}$ a $p \times n$ matrix, $\mathbf{x}$ an $n \times 1$ matrix, $\mathbf{b}$ an $m \times 1$ matrix, and $\mathbf{d}$ a $p \times 1$ matrix. Consider the system comprising the m weak inequalities $\mathbf{Ax} \leqslant \mathbf{b}$ and the p strict inequalities $\mathbf{Cx} < \mathbf{d}$. When is this mixed system of linear inequalities inconsistent? Two possibilities immediately suggest themselves: (i) if there is $\mathbf{u} = (u_1, \ldots, u_m) \geqslant \mathbf{0}$, $\mathbf{v} = (v_1, \ldots, v_p) \geqslant \mathbf{0}$ with $\mathbf{v} \neq \mathbf{0}$ such that $\mathbf{u}^{\mathrm{T}}\mathbf{A} + \mathbf{v}^{\mathrm{T}}\mathbf{C} = \mathbf{0}^{\mathrm{T}}$ and $\mathbf{u}^{\mathrm{T}}\mathbf{b} + \mathbf{v}^{\mathrm{T}}\mathbf{d} \leqslant 0$, then supposing that $\mathbf{Ax} \leqslant \mathbf{b}$, $\mathbf{Cx} < \mathbf{d}$, we may conclude that

$$(\mathbf{u}^{\mathrm{T}}\mathbf{A} + \mathbf{v}^{\mathrm{T}}\mathbf{C})\mathbf{x} = 0 < \mathbf{u}^{\mathrm{T}}\mathbf{b} + \mathbf{v}^{\mathrm{T}}\mathbf{d} \leqslant 0,$$

which shows that the mixed system is inconsistent; (ii) if there is $\mathbf{u} = (u_1, \ldots, u_m) \geqslant \mathbf{0}$ such that $\mathbf{u}^{\mathrm{T}}\mathbf{A} = \mathbf{0}^{\mathrm{T}}$ and $\mathbf{u}^{\mathrm{T}}\mathbf{b} < 0$, then supposing only that $\mathbf{Ax} \leqslant \mathbf{b}$, we may conclude that

$$\mathbf{u}^{\mathrm{T}}\mathbf{Ax} = 0 \leqslant \mathbf{u}^{\mathrm{T}}\mathbf{b} < 0,$$

which shows that the mixed system is inconsistent. The next theorem, in which the above notation is assumed, shows that these are the only ways in which the mixed system can be inconsistent.

Theorem 4.2.4 *Suppose that the mixed system of inequalities* $\mathbf{Ax} \leqslant \mathbf{b}$, $\mathbf{Cx} < \mathbf{d}$ *is inconsistent. Then either* (i) *there exist* $\mathbf{u} \geqslant \mathbf{0}$, $\mathbf{v} \geqslant \mathbf{0}$ *with* $\mathbf{v} \neq \mathbf{0}$ *such that* $\mathbf{u}^{\mathrm{T}}\mathbf{A} + \mathbf{v}^{\mathrm{T}}\mathbf{C} = \mathbf{0}^{\mathrm{T}}$ *and* $\mathbf{u}^{\mathrm{T}}\mathbf{b} + \mathbf{v}^{\mathrm{T}}\mathbf{d} \leqslant 0$ *or* (ii) *there exists* $\mathbf{u} \geqslant \mathbf{0}$ such that $\mathbf{u}^{\mathrm{T}}\mathbf{A} = \mathbf{0}^{\mathrm{T}}$ *and* $\mathbf{u}^{\mathrm{T}}\mathbf{b} < 0$.

Proof Consider the following system of $m + p + 1$ weak inequalities in the $n + 1$ variables $z_1, \ldots, z_n, z$:

$$\begin{bmatrix} \mathbf{A}, & -\mathbf{b} \\ \mathbf{C}, & -\mathbf{d} \\ \mathbf{0}^{\mathrm{T}}, & -1 \end{bmatrix} \begin{bmatrix} z_1 \\ \vdots \\ z_n \\ z \end{bmatrix} \leqslant \begin{bmatrix} \mathbf{0}_m \\ -\mathbf{1}_p \\ -1 \end{bmatrix},$$

where $\mathbf{0}_m$ is the column vector consisting of m 0's, and $-\mathbf{1}_p$ is the column vector consisting of p -1's. This system is inconsistent, for if it were satisfied by $z_1, \ldots, z_n, z$, then $x_1 = z_1/z, \ldots, x_n = z_n/z$ would satisfy the inconsistent system of the theorem. By Theorem 4.2.3, there exist $\mathbf{u} = (u_1, \ldots, u_m) \geqslant \mathbf{0}$, $\mathbf{v} = (v_1, \ldots, v_p) \geqslant \mathbf{0}$, $w \geqslant 0$ such that

$$\mathbf{u}^{\mathrm{T}}\mathbf{A} + \mathbf{v}^{\mathrm{T}}\mathbf{C} = \mathbf{0}^{\mathrm{T}}, \quad -\mathbf{u}^{\mathrm{T}}\mathbf{b} - \mathbf{v}^{\mathrm{T}}\mathbf{d} - w = 0, \quad -v_1 - \ldots - v_p - w < 0.$$

The alternatives (i) and (ii) of the theorem correspond to the cases $\mathbf{v} \neq \mathbf{0}$ and $\mathbf{v} = \mathbf{0}$ respectively. □

Corollary 4.2.5 *Suppose that the system of strict inequalities* $\mathbf{Cx} < \mathbf{d}$ *is inconsistent. Then there exists* $\mathbf{v} \geqslant \mathbf{0}$ *with* $\mathbf{v} \neq \mathbf{0}$ *such that* $\mathbf{v}^{\mathrm{T}}\mathbf{C} = \mathbf{0}^{\mathrm{T}}$ *and* $\mathbf{v}^{\mathrm{T}}\mathbf{d} \leqslant 0$.

Proof Take $\mathbf{A}$ and $\mathbf{b}$ to be zero matrices in the theorem. □

The reader will, no doubt, be familiar with how to form linear combinations of a given set of linear equations, so as to obtain new equations that are necessary consequences of the given ones. Indeed this is the idea behind the method of Gaussian elimination, which is used to solve a finite system of linear equations. We now consider the corresponding situation for a finite system of linear inequalities. Much more care is needed in dealing with inequalities than with equations. Firstly, there are two types of inequality, *weak* and *strict*; secondly, when an inequality is multiplied by a negative number, the inequality sign must be reversed.

Consider the following system $\mathcal{S}$ of m linear inequalities in n variables $x_1, \ldots, x_n$:

$$\left.\begin{array}{lcc} a_{11}x_1 + \ldots + a_{1n}x_n & \mathsf{r}_1 & b_1 \\ \cdots\cdots\cdots\cdots\cdots & & \\ a_{m1}x_1 + \ldots + a_{mn}x_n & \mathsf{r}_m & b_m \end{array}\right\},$$

where each r_i is either $\leqslant$ or $<$. By a **solution** to $\mathcal{S}$ is meant an n-tuple

$(x_1, \ldots, x_n)$ whose coordinates simultaneously satisfy all the inequalities of $\mathbb{S}$. The system $\mathbb{S}$ is said to be **consistent** it it has a solution; otherwise it is said to be **inconsistent**. An inequality

$$e_1x_1 + \ldots + e_nx_n \quad \mathsf{r} \quad f,$$

where r is either $\leqslant$ or $<$ is called a **consequence** of $\mathbb{S}$ if it is satisfied by all solutions of $\mathbb{S}$. If $\mathbb{S}$ is inconsistent, then every linear inequality in $x_1, \ldots, x_n$ is (vacuously) a consequence of $\mathbb{S}$.

We now describe an obvious way of forming consequences of the system $\mathbb{S}$. Let $y_1, \ldots, y_m \geqslant 0$ and let b be a scalar such that $b_1y_1 + \ldots + b_my_m \leqslant b$. Then an inequality of the form

$$(a_{11}y_1 + \ldots + a_{m1}y_m)x_1 + \ldots + (a_{1n}y_1 + \ldots + a_{mn}y_m)x_m \quad \mathsf{r} \quad b,$$

where *either* r is $\leqslant$; *or* r is $<$ and for some $i \in \{1, \ldots, m\}$, $y_i > 0$ and r_i is $<$; *or* r is $<$ and $b_1y_1 + \ldots + b_my_m < b$; is a consequence of $\mathbb{S}$ called a **legal linear combination** of the inequalities of $\mathbb{S}$. The reason for this choice of name should be clear. We shall prove, later in the section, that every consequence of a consistent system $\mathbb{S}$ must be a legal linear combination of its inequalities.

Three earlier results of this section can now be brought together in the following fundamental theorem on the consistency of a finite system of linear inequalities.

Theorem 4.2.6 *The finite system $\mathbb{S}$ of linear inequalities is consistent if and only if the inequality*

$$0x_1 + \ldots + 0x_n < 0$$

is not a legal linear combination of the inequalities of $\mathbb{S}$.

Proof If $\mathbb{S}$ is consistent, then clearly $0x_1 + \ldots + 0x_n < 0$ is not a consequence of $\mathbb{S}$, and so certainly not a legal linear combination of the inequalities of $\mathbb{S}$.

If $\mathbb{S}$ is inconsistent, then the inequality $0x_1 + \ldots + 0x_n < 0$ can be expressed as a legal linear combination of the inequalities of $\mathbb{S}$ by means of one of Theorem 4.2.3, Theorem 4.2.4, or Corollary 4.2.5, according as the inequalities of $\mathbb{S}$ are weak, mixed, or strict. We give the details for the case when $\mathbb{S}$ is an inconsistent system of weak inequalities, $\mathbf{Ax} \leqslant \mathbf{b}$, say. By Theorem 4.2.3, there exists $\mathbf{y} \geqslant \mathbf{0}$ such that $\mathbf{y}^{\mathrm{T}}\mathbf{A} = \mathbf{0}^{\mathrm{T}}$, $\mathbf{y}^{\mathrm{T}}\mathbf{b} < 0$, so $0x_1 + \ldots + 0x_n < 0$ is a legal linear combination of the inequalities of $\mathbb{S}$. □

Theorem 4.2.7 *Let $\mathbb{S}$ be a finite consistent system of linear inequalities. Then every consequence of $\mathbb{S}$ is a legal linear combination of the inequalities of $\mathbb{S}$.*

Proof Suppose first that $\mathbf{e}^{\mathrm{T}}\mathbf{x} \leqslant f$ is a consequence of $\mathcal{S}$. We consider three cases.

Case (i) Suppose that (in the notation used earlier) $\mathcal{S}$ is the system of inequalities $\mathbf{Ax} \leqslant \mathbf{b}$. Since $\mathbf{e}^{\mathrm{T}}\mathbf{x} \leqslant f$ is a consequence of S, the mixed system of inequalities

$$\left.\begin{aligned} \mathbf{Ax} &\leqslant \ \ \mathbf{b} \\ -\mathbf{e}^{\mathrm{T}}\mathbf{x} &< -f \end{aligned}\right\}$$

must be inconsistent. By Theorem 4.2.4, there exist $\mathbf{u} \geqslant \mathbf{0}$, $v > 0$, such that

$$\mathbf{u}^{\mathrm{T}}\mathbf{A} - v\mathbf{e}^{\mathrm{T}} = \mathbf{0}^{\mathrm{T}} \qquad \text{and} \qquad \mathbf{u}^{\mathrm{T}}\mathbf{b} - vf \leqslant 0;$$

the possibility (ii) of Theorem 4.2.4 cannot occur here for it would imply that $\mathcal{S}$ was inconsistent. Thus

$$\mathbf{e}^{\mathrm{T}} = (\mathbf{u}/v)^{\mathrm{T}}\mathbf{A} \qquad \text{and} \qquad (\mathbf{u}/v)^{\mathrm{T}}\mathbf{b} \leqslant f,$$

which shows that $\mathbf{e}^{\mathrm{T}}\mathbf{x} \leqslant f$ is a legal linear combination of the inequalities of $\mathcal{S}$.

Case (ii) Suppose that $\mathcal{S}$ is the mixed system of inequalities $\mathbf{Ax} \leqslant \mathbf{b}$, $\mathbf{Cx} < \mathbf{d}$. Since $\mathbf{e}^{\mathrm{T}}\mathbf{x} \leqslant f$ is a consequence of $\mathcal{S}$, the mixed system of inequalities

$$\left.\begin{aligned} -\mathbf{Ax} &\leqslant \ \ \mathbf{b} \\ \mathbf{Cx} &< \ \ \mathbf{d} \\ -\mathbf{e}^{\mathrm{T}}\mathbf{x} &< -f \end{aligned}\right\}$$

must be inconsistent. By Theorem 4.2.4, there exist $\mathbf{u} \geqslant \mathbf{0}$, $\mathbf{v} \geqslant 0$, $w > 0$, such that

$$\mathbf{u}^{\mathrm{T}}\mathbf{A} + \mathbf{v}^{\mathrm{T}}\mathbf{C} - w\mathbf{e}^{\mathrm{T}} = \mathbf{0}^{\mathrm{T}} \qquad \text{and} \qquad \mathbf{u}^{\mathrm{T}}\mathbf{b} + \mathbf{v}^{\mathrm{T}}\mathbf{d} - wf \leqslant 0;$$

neither possibility (ii) of Theorem 4.2.4 nor the possibility $w = 0$ can occur for each would imply that $\mathcal{S}$ was inconsistent. Thus

$$\mathbf{e}^{\mathrm{T}} = (\mathbf{u}/w)^{\mathrm{T}}\mathbf{A} + (\mathbf{v}/w)^{\mathrm{T}}\mathbf{C} \qquad \text{and} \qquad (\mathbf{u}/w)^{\mathrm{T}}\mathbf{b} + (\mathbf{v}/w)^{\mathrm{T}}\mathbf{d} \leqslant f,$$

which exhibits $\mathbf{e}^{\mathrm{T}}\mathbf{x} \leqslant f$ as a legal linear combination of the inequalities of $\mathcal{S}$.

Case (iii) Suppose that $\mathcal{S}$ is the system of inequalities $\mathbf{Cx} < \mathbf{d}$. That $\mathbf{e}^{\mathrm{T}}\mathbf{x} \leqslant f$ is a legal linear combination of the inequalities of $\mathcal{S}$ follows from case (ii) by taking $\mathbf{A}$ and $\mathbf{b}$ to be zero matrices.

In the same manner, we can prove that every consequence of $\mathcal{S}$ of the form $\mathbf{e}^{\mathrm{T}}\mathbf{x} < f$ is a legal linear combination of the inequalities of $\mathcal{S}$. □

Corollary 4.2.8 (Farkas' lemma) *Let* $\mathbf{a}, \mathbf{a}_1, \ldots, \mathbf{a}_m \in \mathbb{R}^n$ *be such that* $\mathbf{a} \cdot \mathbf{x} \geqslant 0$ *whenever* $\mathbf{x} \in \mathbb{R}^n$ *and* $\mathbf{a}_1 \cdot \mathbf{x} \geqslant 0, \ldots, \mathbf{a}_m \cdot \mathbf{x} \geqslant 0$. *Then there exist* $\lambda_1, \ldots, \lambda_m \geqslant 0$ *such that* $\mathbf{a} = \lambda_1 \mathbf{a}_1 + \ldots + \lambda_m \mathbf{a}_m$. □

With very few exceptions, the results in this section have been algebraic in character, and may have seemed far removed from convexity. This, however, is not the case; many of these results have simple geometric interpretations in terms of the separation of polyhedral sets. To illustrate this point, consider the dual pair:

(i) $\mathbf{Ax} = \mathbf{b}$;

(ii) $\mathbf{y}^{\mathrm{T}}\mathbf{A} = \mathbf{0}^{\mathrm{T}}, \quad \mathbf{y}^{\mathrm{T}}\mathbf{b} \neq 0$.

Suppose that the equations $\mathbf{Ax} = \mathbf{b}$ are inconsistent. Geometrically, this means that in $\mathbb{R}^m$ the point $\mathbf{b}$ does not belong to the subspace S spanned by the columns of $\mathbf{A}$. The existence of a $\mathbf{y}$ with $\mathbf{y}^{\mathrm{T}}\mathbf{A} = \mathbf{0}^{\mathrm{T}}$ and $\mathbf{y}^{\mathrm{T}}\mathbf{b} \neq 0$ means that the hyperplane $\mathbf{y}^{\mathrm{T}}\mathbf{z} = 0$, which has $\mathbf{y}$ as a normal vector and passes through the origin, contains S but not $\mathbf{b}$. Thus the existence of this dual pair is equivalent to the following result: *a point belongs to a subspace of* $\mathbb{R}^m$ *if and only if there exists no hyperplane containing the subspace but not the point.*

Exercises 4.2

1. Prove that each of the systems (a) and (b) below forms a dual pair:
   (a) (i) $\mathbf{Ax} = \mathbf{b}$; (ii) $\mathbf{y}^{\mathrm{T}}\mathbf{A} = \mathbf{0}^{\mathrm{T}}, \quad \mathbf{y}^{\mathrm{T}}\mathbf{b} \neq 0$.
   (b) (i) $\mathbf{Ax} \leqslant \mathbf{b}, \mathbf{x} \geqslant \mathbf{0}$; (ii) $\mathbf{y}^{\mathrm{T}}\mathbf{A} \geqslant \mathbf{0}^{\mathrm{T}}, \quad \mathbf{y}^{\mathrm{T}}\mathbf{b} < 0, \quad \mathbf{y} \geqslant \mathbf{0}$.
2. Use Theorem 4.2.3 to show that any two disjoint non-empty polyhedral sets in $\mathbb{R}^n$ can be strictly separated by a hyperplane.
3. Show that, if the equation $\mathbf{e}^{\mathrm{T}}\mathbf{x} = f$ is a consequence of the consistent system of equations $\mathbf{Ax} = \mathbf{b}$, then there exists $\mathbf{y}$ such that $\mathbf{e}^{\mathrm{T}} = \mathbf{y}^{\mathrm{T}}\mathbf{A}$, $f = \mathbf{y}^{\mathrm{T}}\mathbf{b}$.
4. Let $\mathbf{A}$ be a skew-symmetric matrix, i.e. $\mathbf{A}^{\mathrm{T}} = -\mathbf{A}$. Show that there exists $\mathbf{x} \geqslant \mathbf{0}$ such that $\mathbf{Ax} \geqslant \mathbf{0}$ and $\mathbf{Ax} + \mathbf{x} > \mathbf{0}$.

4.3 LINEAR PROGRAMMING

One important factor that has stimulated interest in convexity over the last half-century is its application to linear programming, a subject developed after the Second World War in response to the call for the use of mathematical techniques in military strategies. Today it is a dynamic, well-established branch of mathematics that has been successfully applied to such diverse areas as game theory, industry, commerce, economics, agriculture, and traffic analysis. We begin our discussion of linear programming by considering its theory, leaving the methods of solution for particular problems to following sections. The highlight of this section is, without doubt, the *duality theorem*, a pivotal result in the subject and one with great aesthetic appeal.

We now motivate the general linear programming problem. Suppose that a manufacturer produces n products and that he produces, and sells, x_j units of the jth product, it being assumed that $x_j \geqslant 0$. If c_j denotes his income from the sale of one unit of the jth product, then his total income is $c_1x_1 + \ldots + c_nx_n$. Suppose further that each of the n products is made from m raw materials, there being available b_i units of the ith raw material. If the amount of the ith raw material used in producing a unit of the jth product is a_{ij}, then

$$a_{i1}x_1 + \ldots + a_{in}x_n \leqslant b_i \qquad \text{for } i = 1, \ldots, m.$$

We are thus led to define the **standard maximum problem** P as follows:

$$\begin{array}{ll} \text{maximize} & c_1x_1 + \ldots + c_nx_n \\ \text{subject to} & a_{11}x_1 + \ldots + a_{1n}x_n \leqslant b_1 \\ & \cdots\cdots\cdots\cdots\cdots\cdots \\ & a_{m1}x_1 + \ldots + a_{mn}x_n \leqslant b_m \\ & \quad x_1 \geqslant 0, \ldots, x_n \geqslant 0. \end{array}$$

In this definition, no restriction is placed on the signs of the real numbers a_{ij}, b_i, c_j, although in the problem used to motivate it, they are tacitly supposed to be non-negative. This standard maximum problem P can be expressed in matrix notation as follows:

$$\text{maximize} \quad \mathbf{c}^{\mathrm{T}}\mathbf{x} \quad \text{subject to} \quad \mathbf{Ax} \leqslant \mathbf{b}, \quad \mathbf{x} \geqslant \mathbf{0},$$

where $\mathbf{A}$ is the real $m \times n$ matrix $[a_{ij}]$, $\mathbf{b} = (b_1, \ldots, b_m)$, $\mathbf{c} = (c_1, \ldots, c_n)$, and $\mathbf{x} = (x_1, \ldots, x_n)$.

A vector $\mathbf{x}$ satisfying the constraints of the standard maximum problem P, i.e. $\mathbf{Ax} \leqslant \mathbf{b}$ and $\mathbf{x} \geqslant \mathbf{0}$, is said to be a **feasible vector** for P, the set of all such feasible vectors being called the **feasible set** for P. The problem P is called **feasible** or **infeasible** according as its feasible set is non-empty or empty. The feasible set for P is the intersection of the m closed halfspaces represented by the inequalities $\mathbf{Ax} \leqslant \mathbf{b}$ and the n closed halfspaces represented by the inequalities $\mathbf{x} \geqslant \mathbf{0}$, and so is a polyhedral set.

A feasible vector $\mathbf{x}_0$ for P which satisfies $\mathbf{c}^{\mathrm{T}}\mathbf{x} \leqslant \mathbf{c}^{\mathrm{T}}\mathbf{x}_0$ for *all* feasible vectors $\mathbf{x}$ for P is called an **optimal vector** for P, with the scalar $\mathbf{c}^{\mathrm{T}}\mathbf{x}_0$ being called the **value** of P. The problem P is said to be **soluble** or **insoluble** according to whether it has an optimal vector or not. The next theorem shows how to find an optimal vector for P when its feasible set is a non-empty polytope.

Theorem 4.3.1 *Suppose that the feasible set for the standard maximum problem P is a non-empty polytope with extreme points $\mathbf{a}_1, \ldots, \mathbf{a}_k$. Let $i \in \{1, \ldots, k\}$ be such that*

$$\mathbf{c}^{\mathrm{T}}\mathbf{a}_i = \max\{\mathbf{c}^{\mathrm{T}}\mathbf{a}_1, \ldots, \mathbf{c}^{\mathrm{T}}\mathbf{a}_k\}.$$

Then P is soluble having $\mathbf{a}_i$ as an optimal vector and value $\mathbf{c}^{\mathrm{T}}\mathbf{a}_i$.

Proof Let $\mathbf{x}$ lie in the feasible set $\operatorname{conv}\{\mathbf{a}_1, \ldots, \mathbf{a}_k\}$ of P. Then

$$\mathbf{x} = \lambda_1\mathbf{a}_1 + \ldots + \lambda_k\mathbf{a}_k$$

for some $\lambda_1, \ldots, \lambda_k \geqslant 0$ with $\lambda_1 + \ldots + \lambda_k = 1$. Then

$$\mathbf{c}^{\mathrm{T}}\mathbf{x} = \lambda_1\mathbf{c}^{\mathrm{T}}\mathbf{a}_1 + \ldots + \lambda_k\mathbf{c}^{\mathrm{T}}\mathbf{a}_k \leqslant \lambda_1\mathbf{c}^{\mathrm{T}}\mathbf{a}_i + \ldots + \lambda_k\mathbf{c}^{\mathrm{T}}\mathbf{a}_i = \mathbf{c}^{\mathrm{T}}\mathbf{a}_i. \quad \square$$

This last result can be used to solve some linear programming problems graphically. Consider the following problem.

A tailor has 16 units of material A, 11 units of material B, and 15 units of material C from which he cuts suits and dresses. Each suit requires 2 units of A, 1 unit of B, 1 unit of C. Each dress requires 1 unit of A, 2 units of B, 3 units of C. Suits sell at 30 units, dresses at 50 units. How can the tailor maximize his income?

This problem, admittedly an artificial one, does indicate how a similar one could represent a real situation. Suppose that the tailor makes x_1 suits and x_2 dresses. Then the tailor's problem is to

$$\begin{array}{ll} \text{maximize} & 30x_1 + 50x_2 \\ \text{subject to} & 2x_1 + x_2 \leqslant 16 \\ & x_1 + 2x_2 \leqslant 11 \\ & x_1 + 3x_2 \leqslant 15 \\ & x_1 \geqslant 0, x_2 \geqslant 0. \end{array}$$

Perhaps we should add the constraints that x_1 and x_2 are integers! We will, however, suppose that our tailor can produce, and sell, any non-negative number of suits and dresses, subject only to the amount of materials he has at his disposal. We will use this example on several further occasions in this chapter to illustrate points in the discussion, and we will refer to it as the *tailor's problem*. Now to solve the problem! The feasible set F for the problem is the intersection of the closed halfplanes $2x_1 + x_2 \leqslant 16$, $x_1 + 2x_2 \leqslant 11$, $x_1 + 3x_2 \leqslant 15$ with the non-negative quadrant. It is readily verified that F is the pentagon whose extreme points are:

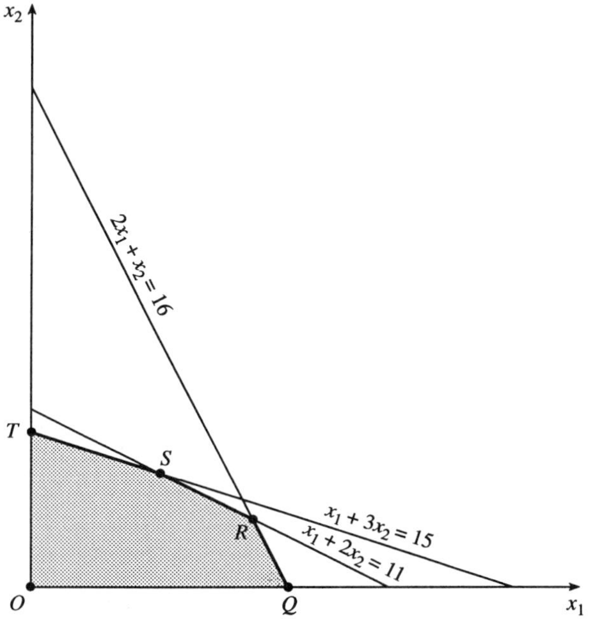

Fig. 4.1

$O = (0, 0)$, $\quad Q = (8, 0)$, $\quad R = (7, 2)$, $\quad S = (3, 4)$, $\quad T = (0, 5)$.

See Fig. 4.1. The values of $30x_1 + 50x_2$ at the points O, Q, R, S, T are, respectively: 0, 240, 310, 290, 250. Thus, by Theorem 4.3.1, the problem has optimal vector (7,2) and value 310: the tailor should make 7 suits and 2 dresses so as to give him a maximal possible income of 310 units.

Theorem 4.3.2 *Suppose that the function* $\mathbf{c}^{\mathrm{T}}\mathbf{x}$ *is bounded above as* $\mathbf{x}$ *ranges over a non-empty feasible set F for the standard maximum problem P. Then P is soluble and at least one of its optimal vectors is an extreme point of F.*

Proof This result follows from Theorem 4.1.6, since F, being a subset of the non-negative orthant of $\mathbb{R}^n$, contains no lines. □

Suppose now that the standard maximum problem P has a non-empty feasible set F. Theorem 4.3.2 shows that P is soluble when the set $\{\mathbf{c}^{\mathrm{T}}\mathbf{x} : \mathbf{x} \in F\}$ of real numbers has an upper bound. How might such an upper bound be obtained? Suppose that, for some $\mathbf{y} = (y_1, \ldots, y_m) \geqslant \mathbf{0}$, $\mathbf{y}^{\mathrm{T}}\mathbf{A} \geqslant \mathbf{c}^{\mathrm{T}}$. Then

$$\mathbf{c}^{\mathrm{T}}\mathbf{x} \leqslant \mathbf{y}^{\mathrm{T}}\mathbf{A}\mathbf{x} \leqslant \mathbf{y}^{\mathrm{T}}\mathbf{b} \qquad \text{for } \mathbf{x} \in F,$$

and P is soluble with value not exceeding $\mathbf{y}^T\mathbf{b}$. The smaller the number $\mathbf{y}^T\mathbf{b}$, the more information we can deduce about the value of P. We are thus led to consider the following problem:

$$\text{minimize} \quad \mathbf{y}^T\mathbf{b} \quad \text{subject to} \quad \mathbf{y}^T\mathbf{A} \geqslant \mathbf{c}^T, \quad \mathbf{y} \geqslant \mathbf{0},$$

which turns out to be closely related to the standard maximum problem P. The relationship between these two problems forms the basis for the concept of *duality* in linear programming, which we now discuss.

The **standard minimum problem** is:

$$\text{minimize} \quad \mathbf{c}^T\mathbf{x} \quad \text{subject to} \quad \mathbf{A}\mathbf{x} \geqslant \mathbf{b}, \quad \mathbf{x} \geqslant \mathbf{0},$$

that is,

$$\begin{array}{ll} \text{minimize} & c_1x_1 + \ldots + c_nx_n \\ \text{subject to} & a_{11}x_1 + \ldots + a_{1n}x_n \geqslant b_1 \\ & \cdots\cdots\cdots\cdots\cdots\cdots \\ & a_{m1}x_1 + \ldots + a_{mn}x_n \geqslant b_m \\ & \quad x_1 \geqslant 0, \ldots, x_n \geqslant 0. \end{array}$$

The definitions of *feasible vector*, *feasible set*, *feasible*, *infeasible*, *optimal vector*, *value*, *soluble*, and *insoluble* are modified in the obvious way so as to apply to the standard minimum problem.

With each standard maximum problem P, we associate a standard minimum problem P^* called the **dual** of P as follows:

P	P^*
maximize $\mathbf{c}^T\mathbf{x}$	minimize $\mathbf{b}^T\mathbf{y}$
subject to $\mathbf{A}\mathbf{x} \leqslant \mathbf{b}$	subject to $\mathbf{A}^T\mathbf{y} \geqslant \mathbf{c}$
$\mathbf{x} \geqslant \mathbf{0}$	$\mathbf{y} \geqslant \mathbf{0}$
that is	that is
maximize $c_1x_1 + \ldots + c_nx_n$	minimize $b_1y_1 + \ldots + b_my_m$
subject to $a_{11}x_1 + \ldots + a_{1n}x_n \leqslant b_1$	subject to $a_{11}y_1 + \ldots + a_{m1}y_m \geqslant c_1$
$\cdots\cdots\cdots\cdots$	$\cdots\cdots\cdots\cdots$
$a_{m1}x_1 + \ldots + a_{mn}x_n \leqslant b_m$	$a_{1n}y_1 + \ldots + a_{mn}y_m \geqslant c_n$
$x_1 \geqslant 0, \ldots, x_n \geqslant 0$	$y_1 \geqslant 0, \ldots, y_m \geqslant 0.$

In the context that we are considering, the problem P is referred to as the **primal problem**. We note that, ignoring the non-negativity constraints on $\mathbf{x}$ and $\mathbf{y}$, the primal problem has m constraints in n variables, whereas the dual problem has n constraints in m variables. As a simple example of these ideas, we note that the dual of the tailor's problem is:

$$\begin{array}{ll} \text{minimize} & 16y_1 + 11y_2 + 15y_3 \\ \text{subject to} & 2y_1 + y_2 + y_3 \geqslant 30 \\ & y_1 + 2y_2 + 3y_3 \geqslant 50 \\ & y_1 \geqslant 0,\ y_2 \geqslant 0,\ y_3 \geqslant 0. \end{array}$$

We now establish two fundamental results concerning duality in linear programming. Together, they give insight into the nature of the subject, as well as providing information which can be helpful in solving particular problems. The first of these results is called the *equilibrium theorem* by some authors, but we prefer to call it the *complementary slackness theorem*.

Theorem 4.3.3 (Complementary slackness theorem) *Let* $\mathbf{x}, \mathbf{y}$ *be feasible vectors for the problems* P, P^*, *respectively. Then* $\mathbf{c}^T\mathbf{x} \leqslant \mathbf{b}^T\mathbf{y}$, *with equality if and only if*

(i) $x_j > 0$ implies $a_{1j}y_1 + \ldots + a_{mj}y_m = c_j$

and

(ii) $y_i > 0$ implies $a_{i1}x_1 + \ldots + a_{in}x_n = b_i$.

Moreover, if $\mathbf{c}^T\mathbf{x} = \mathbf{b}^T\mathbf{y}$, *then* $\mathbf{x}, \mathbf{y}$ *are optimal vectors for their respective problems.*

Proof Since $\mathbf{Ax} \leqslant \mathbf{b}$ and $\mathbf{y} \geqslant \mathbf{0}$, we have $\mathbf{y}^T\mathbf{Ax} \leqslant \mathbf{y}^T\mathbf{b}$, with equality holding if and only if (ii) is true. Similarly, $\mathbf{c}^T\mathbf{x} \leqslant \mathbf{y}^T\mathbf{Ax}$, with equality holding if and only if (i) is true. Thus

$$\mathbf{c}^T\mathbf{x} \leqslant \mathbf{y}^T\mathbf{Ax} \leqslant \mathbf{y}^T\mathbf{b}, \qquad \text{and} \qquad \mathbf{c}^T\mathbf{x} = \mathbf{b}^T\mathbf{y}$$

if and only if both (i) and (ii) hold.

Suppose now that $\mathbf{c}^T\mathbf{x} = \mathbf{b}^T\mathbf{y}$. Let $\mathbf{x}', \mathbf{y}'$ be feasible vectors for the problems P, P^*, respectively. Then, by what we have just proved,

$$\mathbf{c}^T\mathbf{x}' \leqslant \mathbf{b}^T\mathbf{y} = \mathbf{c}^T\mathbf{x} \qquad \text{and} \qquad \mathbf{b}^T\mathbf{y}' \geqslant \mathbf{c}^T\mathbf{x} = \mathbf{b}^T\mathbf{y},$$

which shows that $\mathbf{x}, \mathbf{y}$ are optimal for their respective problems. □

Theorem 4.3.4 (Duality theorem of linear programming) *Denote by* P *the standard maximum problem, and by* P^* *its dual.*

(i) *If either one of* P *and* P^* *is soluble, then so too is the other, and both problems have the same value.*

(ii) *If both* P *and* P^* *are feasible, then they are both soluble.*

Proof (i) Suppose first that P is soluble with optimal vector $\mathbf{x}_0$ and value v. Then the inequality $\mathbf{c}^T\mathbf{x} \leqslant v$ is a consequence of the consistent combined system of inequalities $\mathbf{Ax} \leqslant \mathbf{b}$, $-\mathbf{I}_n\mathbf{x} \leqslant \mathbf{0}$. By Theorem

4.2.7, there exist $\mathbf{y}_0 \geqslant \mathbf{0}$, $\mathbf{u} \geqslant \mathbf{0}$, such that $\mathbf{y}_0^{\mathrm{T}}\mathbf{A} - \mathbf{u}^{\mathrm{T}} = \mathbf{c}^{\mathrm{T}}$ and $\mathbf{y}_0^{\mathrm{T}}\mathbf{b} \leqslant v$. This shows that $\mathbf{y}_0$ is a feasible vector P^*. By Theorem 4.3.3,

$$\mathbf{c}^{\mathrm{T}}\mathbf{x}_0 = v \geqslant \mathbf{y}_0^{\mathrm{T}}\mathbf{b} \geqslant \mathbf{c}^{\mathrm{T}}\mathbf{x}_0,$$

which proves that $\mathbf{c}^{\mathrm{T}}\mathbf{x}_0 = \mathbf{y}_0^{\mathrm{T}}\mathbf{b}$, and $\mathbf{y}_0$ is an optimal vector for P^*. Thus P^* is soluble and has the same value as P, namely v. A similar argument shows that, if P^* is soluble with value v, then so too is P.

(ii) Suppose that both P and P^* are feasible. Let $\mathbf{y}_0$ be a feasible vector for P^*. By Theorem 4.3.3, $\mathbf{c}^{\mathrm{T}}\mathbf{x} \leqslant \mathbf{b}^{\mathrm{T}}\mathbf{y}_0$ for any feasible vector $\mathbf{x}$ of P. Theorem 4.3.2 now shows that P is soluble, and the desired result follows from part (i) of this theorem. □

We now use the complementary slackness theorem to confirm that the vector $(x_1, x_2) = (7, 2)$, obtained earlier by graphical means, is optimal for the tailor's problem, and to obtain an optimal vector for its dual. Certainly (x_1, x_2) is a feasible vector for the problem. Suppose that there is a feasible vector (y_1, y_2, y_3) for the dual which, together with (x_1, x_2), satisfies conditions (i) and (ii) of the complementary slackness theorem. Since $x_1, x_2 > 0$, we have from (i) that:

$$2y_1 + y_2 + y_3 = 30 \qquad \text{and} \qquad y_1 + 2y_2 + 3y_3 = 50.$$

Since the third constraint of the primal, i.e. $x_1 + 3x_2 \leqslant 15$, is *strictly* satisfied, we have from (ii) that $y_3 = 0$. Thus $2y_1 + y_2 = 30$, $y_1 + 2y_2 = 50$, and so $y_1 = 10/3$, $y_2 = 70/3$. A routine verification now shows $(10/3, 70/3, 0)$ is feasible for the dual with

$$30(7) + 50(2) = 16(10/3) + 11(70/3) + 15(0) = 310.$$

The last statement of the complementary slackness theorem now enables us to conclude that $(7, 2)$ is optimal for the tailor's problem, $(10/3, 70/3, 0)$ is optimal for its dual, and both problems have value 310.

Exercises 4.3

1. Solve the following problem:

$$\begin{array}{lrl} \text{minimize} & 2x_1 + 3x_2 + 11x_3 & \\ \text{subject to} & -2x_1 + x_2 + x_3 \geqslant & 2 \\ & x_1 - 2x_2 + 2x_3 \geqslant & -3 \\ & x_1 \geqslant 0,\ x_2 \geqslant 0,\ x_3 \geqslant 0. & \end{array}$$

2. Show that $(7/5, 24/5)$ is an optimal vector for the following problem P, and solve its dual P^*:

$$\begin{aligned} \text{maximize} \quad & -3x_1 + 2x_2 \\ \text{subject to} \quad & -2x_1 + x_2 \leqslant 2 \\ & x_1 - 2x_2 \leqslant 3 \\ & x_1 + 2x_2 \leqslant 11 \\ & x_1 \geqslant 0,\ x_2 \geqslant 0. \end{aligned}$$

3. Let the letters S, F, I stand for *soluble*, *feasible but no soluble*, and *infeasible*, respectively. In each of the nine spaces in the matrix below, put a tick or a cross according as the situation which it represents is possible or not. Justify your answer.

		Dual P^*		
		S	F	I
	S			
Primal P	F			
	I			

4.4 BASIC SOLUTIONS OF LINEAR EQUATIONS

We saw in the last section that the extreme points of a polyhedral set play an important role in the solution of linear programming problems. Here we show how to find these extreme points by solving a system of linear equations. The results which we establish, linking together the geometric concept of an extreme point of a polyhedral set with the algebraic concept of a basic solution to a system of linear equations, will be needed when we discuss the *simplex algorithm* in the next section.

Consider the following system of m linear equations in n variables:

$$\begin{aligned} a_{11}x_1 + \ldots + a_{1n}x_n &= b_1 \\ \ldots\ldots\ldots&\ldots\ldots \\ a_{m1}x_1 + \ldots + a_{mn}x_n &= b_m, \end{aligned}$$

which in matrix notation is $\mathbf{Ax} = \mathbf{b}$, where $\mathbf{A}$ is a real $m \times n$ matrix $[a_{ij}]$, $\mathbf{x} = (x_1, \ldots, x_n)$ and $\mathbf{b} = (b_1, \ldots, b_m)$. To avoid a vacuous discussion, we shall assume throughout, unless stated otherwise, that some m of the columns of $\mathbf{A}$ form a linear basis for $\mathbb{R}^m$, i.e. that $\mathbf{A}$ has rank m; in particular, $m \leqslant n$. Denote the columns of $\mathbf{A}$ by $\mathbf{a}_1, \ldots, \mathbf{a}_n$. Then the system of equations can be written in the form

$$x_1\mathbf{a}_1 + \ldots + x_n\mathbf{a}_n = \mathbf{b}.$$

Suppose that the columns $\mathbf{a}_{i_1}, \ldots, \mathbf{a}_{i_m}$ form a linear basis for $\mathbb{R}^m$. Then there exist unique scalars $x_{i_1}, \ldots, x_{i_m}$ such that

$$x_{i_1}\mathbf{a}_{i_1} + \ldots + x_{i_m}\mathbf{a}_{i_m} = \mathbf{b}.$$

If we put the remaining $n - m$ x_i's equal to zero, we obtain a solution $\mathbf{x} = (x_1, \ldots, x_n)$ of $\mathbf{Ax}=\mathbf{b}$; a solution obtained in this way is called a **basic solution** of $\mathbf{Ax}=\mathbf{b}$.

Example 4.4.1 Find the basic solutions of the system of equations:

$$\begin{aligned} x_1 + x_2 + x_3 &= 3 \\ 3x_1 + 2x_2 + 4x_3 &= 10. \end{aligned}$$

Every two of the columns of the matrix of coefficients on the left-hand side of this system of equations form a basis for $\mathbb{R}^2$, and so the system has three basic solutions. First we put $x_1 = 0$ to obtain the basic solution $(0, 1, 2)$. Next we put $x_2 = 0$ to obtain the basic solution $(2, 0, 1)$. Lastly we put $x_3 = 0$ to obtain the basic solution $(4, -1, 0)$. □

The next theorem makes clear the significance of the non-negative basic solutions of a system of linear equations.

Theorem 4.4.2 *The extreme points of the polyhedral set*

$$C = \{\mathbf{x} \in \mathbb{R}^n : \mathbf{Ax} = \mathbf{b}, \mathbf{x} \geqslant \mathbf{0}\}$$

are precisely the non-negative basic solutions of $\mathbf{Ax} = \mathbf{b}$.

Proof Suppose that $\mathbf{x}_0$ is a non-negative basic solution of $\mathbf{Ax}=\mathbf{b}$, say

$$\mathbf{x}_0 = (x_1, \ldots, x_m, 0, \ldots, 0),$$

where the first m columns $\mathbf{a}_1, \ldots, \mathbf{a}_m$ of $\mathbf{A}$ are linearly independent. Let $\mathbf{x}_0 = \lambda\mathbf{y} + \mu\mathbf{z}$, where $\lambda, \mu > 0$ with $\lambda + \mu = 1$, and $\mathbf{y}, \mathbf{z} \in C$. Since $\mathbf{y}, \mathbf{z} \geqslant \mathbf{0}$ and $\lambda, \mu > 0$, we deduce, on equating the last $n - m$ coordinates on each side of the last expression for $\mathbf{x}_0$, that $\mathbf{y}$ and $\mathbf{z}$ must have the forms

$$\mathbf{y} = (y_1, \ldots, y_m, 0, \ldots, 0), \qquad \mathbf{z} = (z_1, \ldots, z_m, 0, \ldots, 0).$$

Since $\mathbf{y}, \mathbf{z} \in C$, we have

$$y_1\mathbf{a}_1 + \ldots + y_m\mathbf{a}_m = \mathbf{b}, \qquad z_1\mathbf{a}_1 + \ldots + z_m\mathbf{a}_m = \mathbf{b}.$$

But $\mathbf{a}_1, \ldots, \mathbf{a}_m$ are linearly independent, whence $y_1 = z_1, \ldots, y_m = z_m$. Thus $\mathbf{x}_0 = \mathbf{y} = \mathbf{z}$, which shows that $\mathbf{x}_0$ is an extreme point of C.

Suppose next that $\mathbf{x}_0$ is an extreme point of C. If $\mathbf{x}_0 = \mathbf{0}$, then certainly $\mathbf{x}_0$ is a non-negative basic solution of $\mathbf{Ax} = \mathbf{b}$. Assume, then, that $\mathbf{x}_0 \neq \mathbf{0}$; say $\mathbf{x}_0 = (x_1, \ldots, x_r, 0, \ldots, 0)$ for some $r \in \{1, \ldots, n\}$, where $x_1, \ldots, x_r > 0$. Then the first r columns of $\mathbf{A}$, say $\mathbf{a}_1, \ldots, \mathbf{a}_r$, must be linearly independent. To see why this is so, let the scalars $\lambda_1, \ldots, \lambda_r$ be such that

$$\lambda_1 \mathbf{a}_1 + \ldots + \lambda_r \mathbf{a}_r = \mathbf{0}.$$

Choose $\theta > 0$ so small that the points

$$\mathbf{y} = (x_1 + \theta\lambda_1, \ldots, x_r + \theta\lambda_r, 0, \ldots, 0),$$
$$\mathbf{z} = (x_1 - \theta\lambda_1, \ldots, x_r - \theta\lambda_r, 0, \ldots, 0)$$

belong to C. Then $\mathbf{x}_0 = \frac{1}{2}(\mathbf{y} + \mathbf{z})$. But $\mathbf{x}_0$ is an extreme point of C, so $\mathbf{y} = \mathbf{z}$ and $\lambda_1 = 0, \ldots, \lambda_r = 0$. Thus $\mathbf{a}_1, \ldots, \mathbf{a}_r$ are linearly independent. By extending $\{\mathbf{a}_1, \ldots, \mathbf{a}_r\}$ to a linear basis for $\mathbb{R}^m$ using the columns of $\mathbf{A}$, we deduce that $\mathbf{x}_0$ is a non-negative basic solution of $\mathbf{Ax} = \mathbf{b}$. □

We now use this last result in studying the **canonical maximum problem**, which is to

$$\text{maximize} \quad \mathbf{c}^{\mathrm{T}}\mathbf{x} \quad \text{subject to} \quad \mathbf{Ax} = \mathbf{b}, \quad \mathbf{x} \geqslant \mathbf{0}.$$

Here the notation is the same as that used in the last section, except that we now also assume that some m columns of $\mathbf{A}$ are linearly independent. The following terminology for the canonical maximum problem should occasion no surprise. A vector $\mathbf{x} \geqslant \mathbf{0}$ satisfying $\mathbf{Ax} = \mathbf{b}$ is said to be a **feasible vector** for the problem, the set of all such feasible vectors being called the **feasible set** for the problem. A feasible vector $\mathbf{x}_0$ such that $\mathbf{c}^{\mathrm{T}}\mathbf{x} \leqslant \mathbf{c}^{\mathrm{T}}\mathbf{x}_0$ for all feasible vectors $\mathbf{x}$ is called an **optimal vector** for the problem. An optimal vector which is also a basic solution of $\mathbf{Ax} = \mathbf{b}$ is called a **basic optimal vector** for the problem.

Theorem 4.4.3 *Suppose that the canonical maximum problem has an optimal vector. Then it has a basic optimal vector.*

Proof We consider the non-trivial case when $\mathbf{c} \neq \mathbf{0}$. Suppose that the canonical maximum problem has feasible set C and optimal vector $\mathbf{x}_0$. Then the hyperplane H with equation $\mathbf{c} \cdot \mathbf{x} = \mathbf{c} \cdot \mathbf{x}_0$ supports C at $\mathbf{x}_0$. The non-empty polyhedral set $C \cap H$ contains no lines, and so possesses an extreme point, $\mathbf{x}^*$ say. By Theorem 2.6.5, $\mathbf{x}^*$ is an extreme point of C, and by Theorem 4.4.2, $\mathbf{x}^*$ is a basic solution of $\mathbf{Ax} = \mathbf{b}$. Since $\mathbf{x}^* \in C$ and $\mathbf{c} \cdot \mathbf{x}^* = \mathbf{c} \cdot \mathbf{x}_0$, $\mathbf{x}^*$ is a basic optimal vector for the canonical maximum problem. □

We now examine the relationship between the standard and canonical maximum problems. For the present discussion, we shall not insist that some m of the columns of $\mathbf{A}$ form a linear basis for $\mathbb{R}^m$, so that $\mathbf{A}$ will denote *any* real $m \times n$ matrix. We recall that the standard maximum problem P is:

$$\begin{array}{ll} \text{maximize} & c_1x_1 + \ldots + c_nx_n \\ \text{subject to} & a_{11}x_1 + \ldots + a_{1n}x_n \leqslant b_1 \\ & \cdots\cdots\cdots\cdots\cdots\cdots \\ & a_{m1}x_1 + \ldots + a_{mn}x_n \leqslant b_m \\ & x_1 \geqslant 0, \ldots, x_n \geqslant 0. \end{array}$$

We pass from this problem involving m inequalities (excluding the non-negativity constraints on $x_1, \ldots, x_n$) to an equivalent problem involving m equations by introducing m new variables $x_{n+1}, \ldots, x_{n+m}$ as follows:

$$\begin{array}{l} x_{n+1} = b_1 - a_{11}x_1 - \ldots - a_{1n}x_n \\ \cdots\cdots\cdots\cdots\cdots\cdots\cdots \\ x_{n+m} = b_m - a_{m1}x_1 - \ldots - a_{mn}x_n. \end{array}$$

Since each x_{n+i} $(i = 1, \ldots, m)$ measures the amount of *slack* in the inequality

$$a_{i1}x_1 + \ldots + a_{in}x_n \leqslant b_i,$$

the variables $x_{n+1}, \ldots, x_{n+m}$ are referred to as *slack variables*. It is now easy to see that the above standard maximum problem P is equivalent to the following related canonical problem P_{R}:

$$\begin{array}{lll} \text{maximize} & c_1x_1 + \ldots + c_nx_n + 0x_{n+1} + \ldots + 0x_{n+m} & \\ \text{subject to} & a_{11}x_1 + \ldots + a_{1n}x_n + x_{n+1} & = b_1 \\ & \cdots\cdots\cdots\cdots\cdots\cdots\cdots\cdots\cdots & \\ & a_{m1}x_1 + \ldots + a_{mn}x_n \qquad\qquad + x_{n+m} & = b_m \\ & x_1 \geqslant 0, \ldots, x_{n+m} \geqslant 0. & \end{array}$$

Denote by F the feasible set for the above standard maximum problem P, and by F_{R} the feasible set for the related canonical maximum problem P_{R}. Then there is a natural bijection $f: F \to F_{\mathrm{R}}$ defined by the equation

$$f(\mathbf{x}) = f(x_1, \ldots, x_n) = (x_1, \ldots, x_{n+m}) = (\mathbf{x}, \mathbf{b} - \mathbf{A}\mathbf{x}).$$

Clearly f preserves convex combinations of points, and so the extreme points of F and F_{R} correspond under f.

Example 4.4.4 We now solve the tailor's problem using the ideas introduced in this section. The canonical maximum problem related to the tailor's problem is:

$$\begin{array}{ll} \text{maximize} & 30x_1 + 50x_2 + 0x_3 + 0x_4 + 0x_5 \\ \text{subject to} & 2x_1 + x_2 + x_3 = 16 \\ & x_1 + 2x_2 + x_4 = 11 \\ & x_1 + 3x_2 + x_5 = 15 \\ & x_1 \geqslant 0, \ldots, x_5 \geqslant 0. \end{array}$$

Clearly this canonical problem has an optimal vector, and hence a basic optimal vector. Thus to solve the problem, we find at which non-negative basic solutions of the above system of equations the function $30x_1 + 50x_2$ has its maximum. This we now do. Table 4.1 is readily constructed.

Table 4.1

Columns selected	Basic solution	Extreme point of F_R	Extreme point of F	$30x_1 + 50x_2$
1, 2, 3	(3, 4, 6, 0, 0)	(3, 4, 6, 0, 0)	(3, 4)	290
1, 2, 4	$(\frac{33}{5}, \frac{14}{5}, 0, -\frac{6}{5}, 0)$			
1, 2, 5	(7, 2, 0, 0, 2)	(7, 2, 0, 0, 2)	(7, 2)	310
1, 3, 4	(15, 0, −14, −4, 0)			
1, 3, 5	(11, 0, −6, 0, 4)			
1, 4, 5	(8, 0, 0, 3, 7)	(8, 0, 0, 3, 7)	(8, 0)	240
2, 3, 4	(0, 5, 11, 1, 0)	(0, 5, 11, 1, 0)	(0, 5)	250
2, 3, 5	$(0, \frac{11}{2}, \frac{21}{2}, 0, -\frac{3}{2})$			
2, 4, 5	(0, 16, 0, −21, −33)			
3, 4, 5	(0, 0, 16, 11, 15)	(0, 0, 16, 11, 15)	(0, 0)	0

The table shows that the canonical problem has optimal vector (7, 2, 0, 0, 2) and that the tailor's problem has optimal vector (7, 2). □

The method just outlined for solving a linear programming problem is rarely used in practice. Firstly, the method gives no indication as to whether or not the problem has a solution. Secondly, the amount of work in finding a solution is often prohibitive. A system of m equations in $m + n$ unknowns can have as many as $(m + n)!/m!n!$ basic solutions, each one obtained as the solution of a system of m linear equations in m unknowns. Thus a more practical method of solving linear programming problems is required. The most well-known of such methods, the *simplex algorithm*, is the subject of the next section.

Exercises 4.4

1. The polyhedral set P is defined as the intersection of a finite number of closed halfspaces in $\mathbb{R}^n$, and $\mathbf{x}_0$ is an extreme point of P. Show that $\mathbf{x}_0$ lies on at least n of the hyperplanes bounding the closed halfspaces defining P.
2. Suppose that the canonical maximum problem P:

$$\text{maximize} \quad \mathbf{c}^{\mathrm{T}}\mathbf{x} \quad \text{subject to} \quad \mathbf{Ax}=\mathbf{b}, \quad \mathbf{x} \geqslant \mathbf{0},$$

where some m columns of $\mathbf{A}$ are linearly independent, has an optimal vector. Prove that any optimal vector for P that has as many zero coordinates as possible is a basic optimal vector for P.
3. Construct a table, similar to that given in Example 4.4.4, for the following standard maximum problem, and hence solve the problem.

$$\begin{aligned} \text{maximize} \quad & 2x_1 - 3x_2 \\ \text{subject to} \quad & -2x_1 + x_2 \leqslant 2 \\ & x_1 - 2x_2 \leqslant 3 \\ & x_1 + 2x_2 \leqslant 11 \\ & x_1 \geqslant 0, x_2 \geqslant 0. \end{aligned}$$

4. Let P denote the following standard minimum problem:

$$\begin{aligned} \text{minimize} \quad & 3x_1 + 2x_2 \\ \text{subject to} \quad & 5x_1 + x_2 \geqslant 10 \\ & x_1 + x_2 \geqslant 6 \\ & x_1 + 4x_2 \geqslant 12 \\ & x_1 \geqslant 0, x_2 \geqslant 0. \end{aligned}$$

Write down the canonical problem P_{R}, and the standard problem Q whose dual is P. Find the basic feasible solutions for P_{R}, given that there are four of them and that they have the forms

$$(+,+,+,0,0), \quad (+,+,0,0,+), \quad (+,0,+,+,0,), \quad (0,+,0,+,+).$$

Hence solve the problems P and Q.

4.5 THE SIMPLEX ALGORITHM

The methods that we have so far used to solve linear programming problems have only limited applications. The graphical method of

Section 4.3 is restricted to problems in two variables, and the method of finding all basic solutions to a system of linear equations given in Section 4.4 is too uneconomical in all but the simplest cases. What is required is an efficient technique which can solve any linear programming problem and be readily adapted for use on a computer—such a technique is the *simplex algorithm*.

The method of *pivoting*, which is so useful in solving systems of linear equations, forms an important component of the simplex algorithm, and we now recall it. Consider the following system of equations:

$$\begin{aligned} a_{11}x_1 + \ldots + a_{1n}x_n &= b_1 \\ \cdots\cdots&\cdots\cdots \\ a_{m1}x_1 + \ldots + a_{mn}x_n &= b_m. \end{aligned}$$

Suppose that $a_{ij} \neq 0$. Then we obtain a new system equivalent to the given one as follows:

(i) divide the ith equation by a_{ij};

(ii) subtract multiples of the ith equation from the remaining ones in such a way as to remove their x_j term.

The new system that we obtain is:

$$\left(a_{11} - \frac{a_{1j}}{a_{ij}}a_{i1}\right)x_1 + \ldots + 0x_j + \ldots + \left(a_{1n} - \frac{a_{1j}}{a_{ij}}a_{in}\right)x_n = b_1 - \frac{a_{1j}}{a_{ij}}b_i$$

$$\cdots\cdots\cdots\cdots\cdots\cdots\cdots\cdots$$

$$\frac{a_{i1}}{a_{ij}}x_1 + \ldots + x_j + \ldots\ldots\ldots\ldots + \frac{a_{in}}{a_{ij}}x_n = \frac{b_i}{a_{ij}}$$

$$\cdots\cdots\cdots\cdots\cdots\cdots\cdots\cdots$$

$$\left(a_{m1} - \frac{a_{mj}}{a_{ij}}a_{i1}\right)x_1 + \ldots + 0x_j + \ldots + \left(a_{mn} - \frac{a_{mj}}{a_{ij}}a_{in}\right)x_n = b_m - \frac{a_{mj}}{a_{ij}}b_i.$$

We say that this new system has been obtained from the original one by **pivoting** about a_{ij}, this a_{ij} being called the **pivot**.

Before introducing the simplex algorithm formally, we use it informally to solve the tailor's problem. The canonical form of the problem is to maximize $\hat{x}$, subject to the constraints:

$$\begin{array}{l} 30x_1 + 50x_2 + 0x_3 + 0x_4 + 0x_5 = \hat{x} \\ \hline 2x_1 + x_2 + x_3 = 16 \\ x_1 + 2x_2 + x_4 = 11 \\ x_1 + 3x_2 + x_5 = 15, \end{array} \qquad (4.5.1)$$

and the non-negativity contraints $x_1 \geqslant 0, \ldots, x_5 \geqslant 0$. Here we have added the defining equation of the objective function $\hat{x}$ to the constraint equations of the problem.

We seek a basic optimal vector for the problem, beginning at the extreme point (non-negative basic solution) $(0, 0, 16, 11, 15)$, where $\hat{x}$ is 0. Can we find an extreme point where $\hat{x} > 0$? Yes, the system (4.5.1) shows that we can increase $\hat{x}$ by increasing x_1 from 0, while keeping x_2 at 0 and adjusting x_3, x_4, x_5 as required by the equations (4.5.1). As x_1 increases in this way to $8, 11, 15$, x_3, x_4, x_5 decrease, respectively, to 0. Since x_3, x_4, x_5 must be non-negative, we can only increase x_1 to 8, while keeping x_2 at 0, when x_3, x_4, x_5 are $0, 3, 7$ respectively. We have thus arrived at the extreme point $(8, 0, 0, 3, 7)$, where $\hat{x} = 240$. We now express $\hat{x}$ in terms of the new zero variables x_2, x_3. This we do by pivoting the whole system of equations (4.5.1) about the 2 in the second row and the first column to obtain the following system of equations:

$$\begin{array}{lr} 35x_2 - 15x_3 & = \hat{x} - 240 \\ \hline x_1 + \tfrac{1}{2}x_2 + \tfrac{1}{2}x_3 & = 8 \\ \tfrac{3}{2}x_2 - \tfrac{1}{2}x_3 + x_4 & = 3 \\ \tfrac{5}{2}x_2 - \tfrac{1}{2}x_3 \qquad + x_5 & = 7. \end{array} \tag{4.5.2}$$

It is clear from this system of equations that $\hat{x} = 240$ at the extreme point $(8, 0, 0, 3, 7)$.

Can we find an extreme point where $\hat{x} > 240$? Yes, the system (4.5.2) shows that we can increase $\hat{x}$ by increasing x_2 from 0, while keeping x_3 at 0 and adjusting x_1, x_4, x_5 as required by the equations (4.5.2). It is easy to see that x_2 can be increased to 2, when $x_1 = 7$, $x_4 = 0$, $x_5 = 2$. We have thus arrived at the extreme point $(7, 2, 0, 0, 2)$, where $\hat{x} = 310$. We now express $\hat{x}$ in terms of the new zero variables x_3, x_4. This we do by pivoting the whole system of equations (4.5.2) about the $\tfrac{3}{2}$ in the third row and the second column to obtain the following system of equations:

$$\begin{array}{lr} -\tfrac{10}{3}x_3 - \tfrac{70}{3}x_4 & = \hat{x} - 310 \\ \hline x_1 \quad + \tfrac{2}{3}x_3 - \tfrac{1}{3}x_4 & = 7 \\ x_2 - \tfrac{1}{3}x_3 + \tfrac{2}{3}x_4 & = 2 \\ \tfrac{1}{3}x_3 - \tfrac{5}{3}x_4 + x_5 & = 2. \end{array} \tag{4.5.3}$$

It is clear from this system of equations that $\hat{x} = 310$ at the extreme point $(7, 2, 0, 0, 2)$.

Can we find an extreme point where $\hat{x} > 310$? No, we cannot, for the first equation of system (4.5.3) shows that

$$\hat{x} = 310 - \tfrac{10}{3}x_3 - \tfrac{70}{3}x_4 \leqslant 310,$$

since $x_3, x_4 \geqslant 0$ for all feasible vectors $(x_1, x_2, x_3, x_4, x_5)$ for the canonical problem. This ends the search, $(7, 2, 0, 0, 2)$ is a basic optimal vector for the canonical problem, from which it follows that the tailor's problem has optimal vector $(7, 2)$ and value 310.

We now interpret the above solution to the tailor's problem in Fig. 4.1. The search for an optimal vector began at the origin, where $\hat{x}$ was 0. It then moved to the adjacent extreme point $A = (8, 0)$, where $\hat{x}$ was 240. Finally it moved to the adjacent extreme point $(7, 2)$, where $\hat{x}$ assumed its maximum of 310. To summarize: the search started at an extreme point of the feasible set and then moved along the edges of the feasible set, passing from one extreme point to an adjacent one, in such a way that $\hat{x}$ increased at each successive extreme point until it reached its maximum on the feasible set. This is the basic principle that underlies the *simplex algorithm* which we now describe.

The particular argument just used to solve the tailor's problem can be generalized to solve the standard maximum problem, which, in canonical form, is to maximize $\hat{x}$ subject to the constraints:

$$\begin{array}{l}
c_1x_1 + \ldots + c_nx_n + 0x_{n+1} + \ldots + 0x_{n+m} = \hat{x} \\
\hline
a_{11}x_1 + \ldots + a_{1n}x_n + x_{n+1} \qquad\qquad\quad = b_1 \\
\ldots\ldots\ldots\ldots\ldots\ldots\ldots\ldots\ldots\ldots\ldots \\
a_{m1}x_1 + \ldots + a_{mn}x_n \qquad\qquad + x_{n+m} = b_m,
\end{array}$$

and $x_1 \geqslant 0, \ldots, x_{n+m} \geqslant 0$. We denote by F the feasible set for the problem. To simplify our initial discussion, we make two assumptions about the system of equations:

$$[\mathbf{A}, \mathbf{I}_m](x_1, \ldots, x_{n+m}) = \mathbf{b}. \tag{4.5.4}$$

We suppose that: (i) $\mathbf{b} = (b_1, \ldots, b_m) \geqslant \mathbf{0}$; and (ii) every non-negative solution $(x_1, \ldots, x_{n+m})$ of the system (4.5.4) has *at least m* positive coordinates. Condition (ii) is usually referred to as the **non-degeneracy** of the system (4.5.4). Since $(0, \ldots, 0, b_1, \ldots, b_m)$ is a solution of the system (4.5.4), assumptions (i) and (ii) together imply that $b_1, \ldots, b_m > 0$.

The canonical problem can be usefully summarized in tableau form as follows:

x_1	x_2	$\cdots$	x_n	x_{n+1}	x_{n+2}	$\cdots$	x_{n+m}	
c_1	c_2	$\cdots$	c_n	0	0	$\cdots$	0	$\hat{x}$
a_{11}	a_{12}	$\cdots$	a_{1n}	1	0	$\cdots$	$\cdots$	b_1
$\cdots$	$\cdots$	$\cdots$	$\cdots$	$\cdots$	$\cdots$	$\cdots$	$\cdots$	$\cdots$
a_{m1}	a_{m2}	$\cdots$	a_{mn}	0	0	$\cdots$	1	b_m

It is clear from this initial tableau that $(0, \ldots, 0, b_1, \ldots, b_m)$ is an extreme point of the feasible set F at which $\hat{x} = 0$. The variables $x_1, \ldots, x_n$ which are zero at this point are called the **non-basic variables** and the non-zero variables $x_{n+1}, \ldots, x_{n+m}$ are called the **basic variables.**

Can we increase $\hat{x}$ and continue to satisfy the constraints of the problem? Certainly not if $c_1 \leqslant 0, \ldots, c_n \leqslant 0$, for then

$$\hat{x} = c_1 x_1 + \ldots + c_n x_n \leqslant 0,$$

as $x_1, \ldots, x_n \geqslant 0$ for all vectors $(x_1, \ldots, x_{n+m})$ in the feasible set F. Thus in this case $(0, \ldots, 0, b_1, \ldots, b_m)$ is an optimal vector for the problem and the problem has value 0.

Suppose, then, that at least one of $c_1, \ldots, c_n$ is positive, say $c_1 > 0$. If all the numbers in the column below c_1 in the initial tableau are non-positive, then, for any $x_1 \geqslant 0$,

$$(x_1, 0, \ldots, 0, b_1 - a_{11}x_1, \ldots, b_m - a_{m1}x_1) \in F$$

and $\hat{x} = c_1 x_1$ at this point. Thus $\hat{x}$ is not bounded above on F and the problem is insoluble.

Suppose, then, that at least one of $a_{11}, \ldots, a_{m1}$ is positive. For each i such that $a_{i1} > 0$, find b_i/a_{i1} and choose an i which minimizes these quotients; say $a_{11} > 0$ and that b_1/a_{11} is the minimum of the quotients. We now increase x_1 from 0 to b_1/a_{11}, while keeping $x_2, \ldots, x_n$ at 0 and adjusting $x_{n+1}, \ldots, x_{n+m}$ as required by the constraints of the problem. We thus arrive at the extreme point

$$(b_1/a_{11}, 0, \ldots, 0, b_2 - (b_1/a_{11})a_{21}, \ldots, b_m - (b_1/a_{11})a_{m1})$$

of F, where $\hat{x} = (b_1/a_{11})c_1 > 0$. We now express $\hat{x}$ in terms of the new non-basic (zero) variables $x_2, \ldots, x_{n+1}$ by pivoting about the number a_{11} in the first tableau, to obtain a second tableau with the following form:

0	c_2'	$\cdots$	c_n'	c_{n+1}'	0	$\cdots$	0	$\hat{x} - (b_1/a_{11})c_1$
1	a_{12}'	$\cdots$	a_{1n}'	$a_{1\,n+1}'$	0	$\cdots$	0	b_1/a_{11}
0	a_{22}'	$\cdots$	a_{2n}'	$a_{2\,n+1}'$	1	$\cdots$	0	$b_2 - (b_1/a_{11})a_{21}$
$\cdots$	$\cdots$	$\cdots$	$\cdots$	$\cdots$	$\cdots$	$\cdots$	$\cdots$	$\cdots$
0	a_{m2}'	$\cdots$	a_{mn}'	$a_{m\,n+1}'$	0	$\cdots$	1	$b_m - (b_1/a_{11})a_{m1}$

Here we have found it unnecessary to include the x_i's at the heads of the columns of the tableau. This new tableau shows immediately that $\hat{x} = (b_1/a_{11})c_1$ at the new extreme point of F, for here the variables $x_2, \ldots, x_{n+1}$ are zero. Because of the non-degeneracy assumption, the (non-negative) elements in the last column of the tableau under $\hat{x} - (b_1/a_{11})c_1$ cannot be zero—they must be positive. The non-zero coordinates $x_1, x_{n+2}, \ldots, x_{n+m}$ of the new extreme point can be read off immediately from the above tableau. Since $(b_1/a_{11})c_1 > 0$, the value of $\hat{x}$ at the new extreme point is strictly larger than its value at the initial extreme point.

We now study the second tableau. If $c_2' \leqslant 0, \ldots, c_{n+1}' \leqslant 0$, the extreme point just found will be an optimal vector for the problem. Suppose, then, that at least one of $c_2', \ldots, c_{n+1}'$ is positive, say $c_{j_0}' > 0$. If all the numbers in the column below c_{j_0}' in this second tableau are non-positive, then $\hat{x}$ is not bounded above on F and the problem is insoluble. Suppose, then, that at least one of $a_{1j_0}', \ldots, a_{mj_0}'$ is positive. For each i such that $a_{ij_0}' > 0$, consider b_i'/a_{ij_0}', where b_i' is the number in the same row as a_{ij_0}' and in the last column of the tableau; say $a_{i_0j_0}' > 0$ and that $b_{i_0}'/a_{i_0j_0}'$ is the minimum of these quotients. Now pivot about the number $a_{i_0j_0}'$ in the second tableau to obtain a third tableau, which will (as earlier in our discussion) indicate a third extreme point, where the value of $\hat{x}$ exceeds its value at the second extreme point.

We now repeat the procedure. Since F, being a polyhedral set, has only a finite number of extreme points and $\hat{x}$ strictly increases in value at each stage in the algorithm, one of two possibilities must occur: (i) a tableau is reached in which the first $m + n$ numbers on the top row are non-positive; or (ii) a tableau is reached which has one of its first $m + n$ numbers on the top row positive with all the numbers below it non-positive. In case (i) the tableau, which is called a **final tableau**, will yield an optimal vector when the non-basic variables are put equal to zero and the values of the basic variables are read off from the tableau; the value v of the problem is to be found from the last entry on the first row of the tableau which is $\hat{x} - v$. In case (ii) the problem is insoluble.

Example 4.5.1 We use the simplex algorithm to solve the problem:

$$\begin{array}{ll} \text{maximize} & 2x_1 - 3x_2 + x_3 \\ \text{subject to} & 3x_1 + 6x_2 + x_3 \leqslant 6 \\ & 4x_1 + 2x_2 + x_3 \leqslant 4 \\ & x_1 - x_2 + x_3 \leqslant 3 \\ & x_1 \geqslant 0, x_2 \geqslant 0, x_3 \geqslant 0. \end{array}$$

We convert this standard problem to a canonical problem in the usual way to obtain the following initial tableau in the simplex algorithm.

2	−3	1	0	0	0	0
3	6	1	1	0	0	6
④	2	1	0	1	0	4
1	−1	1	0	0	1	3

The reader will observe that we have omitted the $\hat{x}$ in the top right-hand corner of the tableau; this is usual, it being understood that the number in this position is the negative of the value of $\hat{x}$ at the extreme point indicated by the tableau.

We examine the top row of the tableau for positive entries, selecting the 2 in the first column (although the 1 in the third column would serve equally well). According to the simplex algorithm, we next choose the least of the ratios 6/3, 4/4, 3/1, i.e. 4/4, and so pivot about the 4 in the first column, indicating this by circling the 4 in the initial tableau. We thus obtain the second tableau:

0	−4	$\frac{1}{2}$	0	$-\frac{1}{2}$	0	−2
0	$\frac{9}{2}$	$\frac{1}{4}$	1	$-\frac{3}{4}$	0	3
1	$\frac{1}{2}$	$\frac{1}{4}$	0	$\frac{1}{4}$	0	1
0	$-\frac{3}{2}$	($\frac{3}{4}$)	0	$-\frac{1}{4}$	1	2

We examine the top row of the tableau for positive entries, selecting the $\frac{1}{2}$ in the third column. The least of the ratios to be considered, viz. 3/(1/4), 1/(1/4), 2/(3/4), is the last one. Thus we pivot about the $\frac{3}{4}$ in the third column to obtain the third tableau:

0	−3	0	0	$-\frac{1}{3}$	$-\frac{2}{3}$	$-\frac{10}{3}$
0	5	0	1	$-\frac{2}{3}$	$-\frac{1}{3}$	$\frac{7}{3}$
1	1	0	0	$\frac{1}{3}$	$-\frac{1}{3}$	$\frac{1}{3}$
0	−2	1	0	$-\frac{1}{3}$	$\frac{4}{3}$	$\frac{8}{3}$

There are no positive entries on the top row here, so we have a final tableau. The non-basic variables indicated by this tableau are x_2, x_5, x_6, which are zero. The basic variables x_1, x_3, x_4 have values $\frac{1}{3}$, $\frac{8}{3}, \frac{7}{3}$, respectively, which can be easily read off from the above tableau. Hence $(\frac{1}{3}, 0, \frac{8}{3}, \frac{7}{3}, 0, 0)$ is an optimal vector for the canonical problem. Thus the standard problem has optimal vector $(\frac{1}{3}, 0, \frac{8}{3})$ and value $\frac{10}{3}$. □

Applying the simplex algorithm with bare hands can scarcely be counted amongst the delights of mathematics. Nonetheless, for those adventurous enough to follow such a pursuit, there is a reward to be reaped from the final tableau of the algorithm. Some readers may have noticed that (ignoring signs) the coefficients of the slack variables x_3, x_4, x_5 occurring on the top row of the final tableau for the tailor's problem are $\frac{10}{3}, \frac{70}{3}, 0$, which yield the optimal vector to the dual of the tailor's problem, found at the end of Section 4.3. Nor is this just fortuitous, as the following simple argument shows.

Suppose that, in the usual notation, the initial and final tableaux corresponding to the solution of the standard maximum problem by the simplex algorithm are as follows:

$$\begin{bmatrix} \mathbf{c}^{\mathrm{T}}, & \mathbf{0}^{\mathrm{T}}, & 0 \\ \mathbf{A}, & \mathbf{I}_m, & \mathbf{b} \end{bmatrix}, \qquad \begin{bmatrix} -z_1, \ldots, -z_n, & -y_1, \ldots, -y_m, & -v \\ * & * & * \end{bmatrix},$$

Initial tableau **Final tableau**

where $z_1, \ldots, z_n, y_1, \ldots, y_m \geqslant 0$ and v is the value of the problem. The method of operation of the simplex algorithm shows that the first row of the final tableau is obtained from the initial tableau by adding multiples of its last m rows to its first row. In particular, $[-y_1, \ldots, -y_m]$ must be a linear combination of the rows of $\mathbf{I}_m$, so the multiples just referred to above must be $-y_1, \ldots, -y_m$, in that order. Thus, writing $\mathbf{y} = (y_1, \ldots, y_m)$, we deduce that

$$[-z_1, \ldots, -z_n] = \mathbf{c}^{\mathrm{T}} + [-y_1, \ldots, -y_m]\mathbf{A} \quad \text{and}$$
$$-v = [-y_1, \ldots, -y_m]\mathbf{b}.$$

Thus

$$\mathbf{A}^{\mathrm{T}}\mathbf{y} = \mathbf{c} + (z_1, \ldots, z_n) \geqslant \mathbf{c} \qquad \text{and} \qquad v = \mathbf{b}^{\mathrm{T}}\mathbf{y},$$

which shows that $\mathbf{y}$ is a feasible vector for the dual problem and that $\mathbf{b}^{\mathrm{T}}\mathbf{y} = v$, where v is the value of the primal problem. By the complementary slackness theorem, $\mathbf{y}$ is an optimal vector for the dual. Using this result, we can deduce immediately that $(0, \frac{1}{3}, \frac{2}{3})$ is an optimal vector for the dual of the problem considered in Example 4.5.1.

In our discussion of the simplex algorithm we made two initial assumptions: (i) the vector $\mathbf{b}$ was non-negative; and (ii) the system of equations $[\mathbf{A}, \mathbf{I}_m](x_1, \ldots, x_{n+m}) = \mathbf{b}$ was non-degenerate. The first assumption was needed at the outset of the algorithm to show that $(0, \ldots, 0, b_1, \ldots, b_m)$ was an initial extreme point. Without this assumption, it would not have been clear how to find an extreme point with which to begin the simplex algorithm—indeed such an extreme point might not exist. We now describe a method which will tell us if the feasible set of the canonical problem has an extreme point and, if it does, how to find it.

Consider the canonical maximum problem:

$$\text{maximize} \quad c_1x_1 + \ldots + c_nx_n = \hat{x}$$

$$\text{subject to} \quad [\mathbf{A}, \mathbf{I}_m](x_1, \ldots, x_{n+m}) = \mathbf{b}; \quad x_1, \ldots, x_{n+m} \geqslant 0,$$

under the single assumption of non-degeneracy. Since we have discussed the case when $\mathbf{b} \geqslant \mathbf{0}$, we suppose that at least one of $b_1, \ldots, b_m$ is negative. Consider now the following **augmented problem**:

$$\begin{array}{lll} \text{maximize} & -x_0 & = \tilde{x} \\ \text{subject to} & -x_0 + a_{11}x_1 + \ldots + a_{1n}x_n + x_{n+1} & = b_1 \\ & \ldots\ldots\ldots\ldots\ldots\ldots\ldots\ldots\ldots\ldots & \\ & -x_0 + a_{m1}x_1 + \ldots + a_{mn}x_n \qquad + x_{n+m} & = b_m \\ & x_0 \geqslant 0, x_1 \geqslant 0, \ldots, x_{n+m} \geqslant 0. & \end{array} \tag{4.5.5}$$

We can make some immediate observations about this agumented problem:

(i) The problem is feasible, for if x_0 is chosen so that $b_1 + x_0 \geqslant 0, \ldots, b_m + x_0 \geqslant 0$, then

$$(x_0, 0, \ldots, 0, b_1 + x_0, \ldots, b_m + x_0)$$

is a feasible vector.

(ii) The objective function $\tilde{x} = -x_0$ is bounded above by 0, so, in view of (i), the problem is soluble.

(iii) Suppose that the unaugmented problem has a feasible vector $(x_1, \ldots, x_{n+m})$. Then $(0, x_1, \ldots, x_{n+m})$ is an optimal vector for the augmented problem, which has value 0. Conversely, if $(x_0, x_1, \ldots, x_{n+m})$ is a basic optimal vector for the augmented problem giving it value 0, then $x_0 = 0$ and $(x_1, \ldots, x_{n+m})$ is an extreme point of the feasible set for the canonical problem.

Thus what we need first is to solve the augmented problem. If its value is negative, then the canonical problem is insoluble. If its value is zero and $(0, x_1, \ldots, x_{n+m})$ is one of its optimal vectors, then $(x_1, \ldots, x_{n+m})$ is the sought-for extreme point of the feasible set for the canonical problem. Once such an extreme point has been found, the simplex algorithm can be started.

How, then, can we solve the agumented problem? We cannot initially solve this problem by the simplex algorithm, for at least one of $b_1, \ldots, b_m$ is negative. Fortunately, this difficulty is easy to overcome. Suppose, without loss of generality, that b_1 is less than or equal to each of $b_2, \ldots, b_m$, and hence negative. We pivot about the -1 in the first row and the first column of the system of equations (4.5.5) to obtain the following problem, which is equivalent to the augmented problem:

$$\begin{array}{llr}
\text{maximize} & -a_{11}x_1 - \ldots - a_{1n}x_n - x_{n+1} & = \tilde{x} - b_1 \\
\text{subject to} & x_0 - a_{11}x_1 - \ldots - a_{1n}x_n - x_{n+1} & = -b_1 \\
& a'_{21}x_1 + \ldots + a'_{2n}x_n - x_{n+1} + x_{n+2} & = b_2 - b_1 \\
& \cdots\cdots\cdots\cdots\cdots\cdots\cdots\cdots & \\
& a'_{m1}x_1 + \ldots + a'_{mn}x_n - x_{n+1} \qquad + x_{n+m} & = b_m - b_1 \\
& x_0 \geqslant 0, x_1 \geqslant 0, \ldots, x_{n+m} \geqslant 0, &
\end{array}$$

where $a'_{21}, \ldots, a'_{mn}$ are real numbers whose specific values do not interest us here. Since $b_1 < 0$, $b_1 \leqslant b_2, \ldots, b_1 \leqslant b_m$,

$$-b_1 > 0, \qquad b_2 - b_1 \geqslant 0, \qquad \ldots, \qquad b_m - b_1 \geqslant 0.$$

Thus we now have a problem to which we can apply the simplex algorithm, and hence find an extreme point of the feasible set of the unaugmented problem (should such an extreme point exist). This procedure for finding an extreme point by solving the augmented problem is known as the method of the **artificial variable**, because the variable x_0 is artificially introduced, only to disappear before the final solution of the original problem is obtained.

Example 4.5.2 We use the method of the artificial variable to solve the following problem P, which can also be solved graphically:

$$\begin{array}{ll}
\text{maximize} & x_1 + x_2 \\
\text{subject to} & 2x_1 + 3x_2 \leqslant 18 \\
& 4x_1 + x_2 \leqslant 13 \\
& -x_1 - 2x_2 \leqslant -5 \\
& x_1 \geqslant 0, x_2 \geqslant 0.
\end{array}$$

Denote by P_R the canonical maximum problem related to P, and by F_R the feasible set for P_R. The augmented problem associated with P_R is:

$$\begin{array}{llr}
\text{maximize} & -x_0 & \\
\text{subject to} & -x_0 + 2x_1 + 3x_2 + x_3 & = 18 \\
& -x_0 + 4x_1 + x_2 \qquad + x_4 & = 13 \\
& -x_0 - x_1 - 2x_2 \qquad\qquad + x_5 & = -5 \\
& x_0 \geqslant 0, \ldots, x_5 \geqslant 0. &
\end{array}$$

The initial tableau for this problem is:

-1	0	0	0	0	0	0
-1	2	3	1	0	0	18
-1	4	1	0	1	0	13
(-1)	-1	-2	0	0	1	-5

Since -5 is the smallest element in the right-hand column, we pivot about the element in the same row as this and in the column of the artificial variable, to obtain the following tableau:

0	1	2	0	0	-1	5
0	3	5	1	0	-1	23
0	5	3	0	1	-1	18
1	1	(2)	0	0	-1	5

This tableau is not final, because of the 1 and 2 in its top row. We choose a pivot in the column headed by the 2; the pivot is easily seen to be 2. After pivoting about this element, we arrive at the final tableau:

-1	0	0	0	0	0	0
0	$\frac{1}{2}$	0	1	0	$\frac{3}{2}$	$\frac{21}{2}$
$-\frac{3}{2}$	$\frac{7}{2}$	0	0	1	$\frac{1}{2}$	$\frac{21}{2}$
$\frac{1}{2}$	$\frac{1}{2}$	1	0	0	$-\frac{1}{2}$	$\frac{5}{2}$

Thus the augmented problem has value 0 and optimal vector $(0, 0, \frac{5}{2}, \frac{21}{2}, \frac{21}{2}, 0)$, whence $(0, \frac{5}{2}, \frac{21}{2}, \frac{21}{2}, 0)$ is an extreme point of F_R. The non-basic variables at this last extreme point are x_1 and x_5. We now express the objective function $x_1 + x_2$ of P_R in terms of x_1 and x_5. Since $-x_1 - 2x_2 + x_5 = -5$, it follows that $x_1 + x_2 = \frac{1}{2}x_1 + \frac{1}{2}x_5 + \frac{5}{2}$. This enables us to write down an initial tableau for P_R with starting point $(0, \frac{5}{2}, \frac{21}{2}, \frac{21}{2}, 0)$ and then to proceed to a final tableau as follows.

$\frac{1}{2}$	0	0	0	$\frac{1}{2}$	$-\frac{5}{2}$
$\frac{1}{2}$	0	1	0	$\frac{3}{2}$	$\frac{21}{2}$
($\frac{7}{2}$)	0	0	1	$\frac{1}{2}$	$\frac{21}{2}$
$\frac{1}{2}$	1	0	0	$-\frac{1}{2}$	$\frac{5}{2}$

0	0	0	$-\frac{1}{7}$	$\frac{3}{7}$	-4
0	0	1	$-\frac{1}{7}$	$\textcircled{\frac{10}{7}}$	9
1	0	0	$\frac{2}{7}$	$\frac{1}{7}$	3
0	1	0	$-\frac{1}{7}$	$-\frac{4}{7}$	1

0	0	$-\frac{3}{10}$	$-\frac{1}{10}$	0	$-\frac{67}{10}$
0	0	$\frac{7}{10}$	$-\frac{1}{10}$	1	$\frac{63}{10}$
1	0	$-\frac{1}{10}$	$\frac{3}{10}$	0	$\frac{21}{10}$
0	1	$\frac{2}{5}$	$-\frac{1}{5}$	0	$\frac{23}{5}$

We read off from the last tableau that P has optimal vector $(\frac{21}{10}, \frac{23}{5})$ and value $\frac{67}{10}$. □

When we had found the final tableau for the augmented problem in the last example, we needed to express the objective function $x_1 + x_2$ of P_R in terms of the non-basic variables x_1 and x_5. We could have avoided this by adding a row

$$0 \quad 1 \quad 1 \quad 0 \quad 0 \quad 0 \quad 0$$

on top of the initial tableau for the augmented problem (to represent $x_1 + x_2$), and then operating on the enlarged tableau in the natural way, remembering never to pivot about an element in the first row. When the final tableau for the augmented problem is reached, the initial tableau for the second application of the simplex algorithm is obtained from it by deleting the first column and the second row.

Throughout our discussion we have only considered non-degenerate problems; we even made a tacit assumption of non-degeneracy in our account of the method of the artificial variable just described. Where did we need this non-degeneracy assumption? It was in showing that the algorithm terminated after a finite number of steps. Without the assumption, it would be possible to enter into an infinite sequence of pivoting operations without ever reaching a solution (even when one exists!). Such a phenomenon is called **cycling**. This difficulty is more apparent than real, for it can be shown that for any problem, there is a sequence of pivots which will ensure that the simplex algorithm is completed in a finite number of steps. In practice, cycling rarely occurs, although problems have been specially constructed to demonstrate its existence. From a computational point of view, the reader is advised to ignore it.

Exercises 4.5

1. Use the simplex algorithm to investigate, and solve where possible, the following two problems and their duals:

$$\begin{array}{llll} \text{maximize} & 40x + 50y & \text{maximize} & 3x + 2y \\ \text{subject to} & 5x + 2y \leqslant 30 & \text{subject to} & -x + y \leqslant 1 \\ & 5x + 7y \leqslant 35 & & x - 4y \leqslant 2 \\ & 2x + 5y \leqslant 20 & & x \geqslant 0, y \geqslant 0. \\ & x \geqslant 0, y \geqslant 0, & & \end{array}$$

2. Use the simplex algorithm to solve the following problem:

$$\begin{array}{ll} \text{minimize} & 12x + 3y + 2z \\ \text{subject to} & 2x + 2y + 3z \geqslant 9 \\ & 3x - y + z \geqslant 2 \\ & -5x + 3y - 2z \geqslant 5 \\ & x \geqslant 0, y \geqslant 0, z \geqslant 0. \end{array}$$

3. Find a solution to the problem below by the method of the artificial variable:

$$\begin{array}{ll} \text{maximize} & 5x + 3y \\ \text{subject to} & -4x - 5y \leqslant -10 \\ & 5x + 2y \leqslant 10 \\ & 3x + 8y \leqslant 12 \\ & x \geqslant 0, y \geqslant 0. \end{array}$$

4. Solve the problem:

$$\begin{array}{ll} \text{maximize} & 2x_1 + 4x_2 + x_3 + x_4 \\ \text{subject to} & x_1 + 3x_2 + x_4 \leqslant 4 \\ & 2x_1 + x_2 \leqslant 3 \\ & x_2 + 4x_3 + x_4 \leqslant 3 \\ & x_1 \geqslant 0, x_2 \geqslant 0, x_3 \geqslant 0, x_4 \geqslant 0, \end{array}$$

and hence its dual, by the simplex algorithm. For each tableau in your solution, state which extreme point of the feasible set is being examined.

4.6 GAME THEORY

Do not be misled: the matrix games discussed in this section are of a particularly simple type, far removed from such complicated games as bridge and chess. Despite their simplicity, however, they do give rise to interesting mathematical problems. It turns out that there is a close connection between game theory and linear programming; indeed, the central result of the subject, the minimax theorem, will be proved by means of the duality theorem of linear programming.

The optimization problems which we have so far considered have been those of maximizing and minimizing an objective function subject to given constraints. In the games that we now study, the situation is very different from this. Two players confront one another in a competitive situation, each of them attempting to further his own interests. The players being in conflict, it is unlikely that they can realize their objectives simultaneously. Each player, when determining his own strategy, must take into account the possible behaviour of his opponent. The mathematical study of the strategies employed by the two players in such a game, where the decisions of one affect the other, is known as *game theory* or the *theory of games*.

The description of a **matrix game** could hardly be simpler. Two players, a **row player** R and a **column player** C, compete against each other. A game is determined by a real $m \times n$ matrix $\mathbf{A} = [a_{ij}]$, called the **pay-off matrix** of the game. In essence, the row player chooses a row of $\mathbf{A}$ (i.e. one of the numbers $1, \ldots, m$), while at the same time the column player chooses a column of $\mathbf{A}$ (i.e. one of the numbers $1, \ldots, n$). Each players acts in ignorance of his opponent. If R chooses i while C chooses j, then R receives an amount a_{ij} from C. This procedure constitutes one **play** of the game, and the game consists of a large number of plays. The object of each player is to maximize his gains (or to minimize his losses).

Example 4.6.1 Consider the following game.

Player R selects two of the numbers $1, 2, 4$, while C independently selects one of them. For each number chosen by R, but not by C, C pays R that number; and for each number chosen by both R and C, R pays C that number.

This is essentially a matrix game, whose pay-off matrix we now construct. Here R has three choices: (i) $1, 2$; (ii) $1, 4$; (iii) $2, 4$; and C has three choices: (i) 1; (ii) 2; (iii) 4. Suppose that both players play their first choices. Then R pays 1 to C and C pays 2 to R. The net result of this play is a gain of 1 to R, so the element in the first row and the first column of the pay-off matrix is 1. The completed matrix is as follows:

$$\begin{bmatrix} 1 & -1 & 3 \\ 3 & 5 & -3 \\ 6 & 2 & -2 \end{bmatrix}. \qquad \square$$

To motivate the basic definitions of game theory, we discuss, informally, the game determined by a real $m \times n$ matrix $\mathbf{A} = [a_{ij}]$. Consider a *large number* N of plays of the game. Suppose that R chooses $1, \ldots, m$, respectively, $N_1, \ldots, N_m$ times. Then $N_1 + \ldots + N_m = N$, and R has made the choice i ($i = 1, \ldots, m$) with relative frequency $x_i = N_i/N$. Clearly, $x_1, \ldots, x_m \geqslant 0$ and $x_1 + \ldots + x_m = 1$. Suppose, similarly, that C has made the choice j ($j = 1, \ldots, n$) with relative frequency y_j. Then $y_1, \ldots, y_n \geqslant 0$ and $y_1 + \ldots + y_n = 1$. We say that R employs strategy $\mathbf{x} = (x_1, \ldots, x_m)$ and that C employs strategy $\mathbf{y} = (y_1, \ldots, y_n)$. The question which interests us here is: how much can R expect to receive from C during the game? We assume that the players, within their preferred strategies, make their choices in a random way and we proceed in the following naïve fashion. Since R chooses i with relative frequency x_i and C chooses j with relative frequency y_j, the relative frequency with which *both* R chooses i and C chooses j is $x_i y_j$, the number of times on which this occurs being about $x_i y_j N$. Hence the amount which R receives from C as a result of these $x_i y_j N$ plays is $a_{ij} x_i y_j N$. Thus the total amount R receives from C after N plays is

$$\sum_{i=1}^{m} \sum_{j=1}^{n} a_{ij} x_i y_j N,$$

and the average amount R can expect to receive from C for a single play is

$$\sum_{i=1}^{m} \sum_{j=1}^{n} a_{ij} x_i y_j.$$

This last expression, denoted by $E(\mathbf{x}, \mathbf{y})$, is called R's *expected gain* and C's *expected loss*.

We now formalize the above ideas to give *precise* definitions for the game determined by a real $m \times n$ matrix $\mathbf{A} = [a_{ij}]$. A **strategy** for R is a vector $\mathbf{x} = (x_1, \ldots, x_m)$ for which $x_1, \ldots, x_m \geqslant 0$ and $x_1 + \ldots + x_m = 1$, and a **strategy** for C is a vector $\mathbf{y} = (y_1, \ldots, y_n)$ for which $y_1, \ldots, y_n \geqslant 0$ and $y_1 + \ldots + y_n = 1$. The set of all strategies for R is denoted by S_m, while the set of all strategies for C is denoted by S_n. The simplest strategies are the **pure strategies** in which a player consistently chooses a given row or column. The ith **pure strategy** for R is the m-vector $(0, \ldots, 1, \ldots, 0)$, which has a 1 in the ith place and zeros elsewhere; the jth **pure strategy** for C is the n-vector $(0, \ldots, 1, \ldots, 0)$, which has a 1 in the jth place and zeros elsewhere. Clearly, the set S_m of all strategies for R is a polytope in $\mathbb{R}^m$ whose extreme points are R's pure strategies; similar remarks apply to S_n. Suppose now that R and C employ strategies $\mathbf{x}$ and $\mathbf{y}$, respectively. Then R's **expected gain** (which is C's **expected loss**), denoted by $E(\mathbf{x}, \mathbf{y})$, is defined as follows:

$$E(\mathbf{x}, \mathbf{y}) = \sum_{i=1}^{m} \sum_{j=1}^{n} a_{ij} x_i y_j = \mathbf{x}^{\mathrm{T}} \mathbf{A} \mathbf{y}.$$

We observe that $E(\mathbf{x}, \mathbf{y})$ is simply a particular value of the bilinear form associated with the matrix $\mathbf{A}$.

Let $\underline{a}$ and $\bar{a}$ denote, respectively, the minimal and maximal elements of the matrix $\mathbf{A}$. Then it is easily seen that, whatever the strategies adopted by the two players, R's expected gain is at least $\underline{a}$, while C's expected loss is at most $\bar{a}$. What can the players do to improve these expectations? We consider first the use of pure strategies by both players. Which of R's pure strategies will guarantee him the maximal expected gain against *any* pure strategy of his opponent? If R plays his ith pure strategy against a pure strategy of C, his expected gain will be one of the numbers $a_{i1}, \ldots, a_{in}$, and so he can be certain of receiving at least

$$\min_j a_{ij}.$$

Clearly, R should choose i in such a way as to make this minimum as large as possible. Suppose that this minimum is maximized when $i = i_0$, say. Then we have shown that, by suitable choice of a pure strategy, R can guarantee an expected gain of at least

$$\max_i \min_j a_{ij}$$

against any pure strategy of C. Similarly, for some $j = j_0$, the j_0th pure strategy of C will keep his expected loss to at most

$$\min_i \max_j a_{ij}$$

against any pure strategy of R. By considering R's expected gain (C's expected loss) when R chooses his i_0th pure strategy and C chooses his j_0th pure strategy, we can deduce that

$$\underline{a} \leqslant \max_i \min_j a_{ij} \leqslant \min_j \max_i a_{ij} \leqslant \bar{a}.$$

Example 4.6.2. Consider the matrix

$$\begin{bmatrix} 3 & 7 & 2 \\ 8 & 1 & 6 \\ 4 & 9 & 5 \end{bmatrix}.$$

Here

$$1 = \underline{a} \leqslant \max_i \min_j a_{ij} = 4 < 6 = \min_j \max_i a_{ij} \leqslant \bar{a} = 9,$$

and R's best pure strategy is to play his third row, while C's is to play his third column. When both players choose their best strategies, the expected gain (loss) is 5, which lies strictly between the above max-min and min-max. □

We now consider the general matrix game, determined by a real $m \times n$ matrix $\mathbf{A} = [a_{ij}]$, from the point of view of the row player R. Suppose that he decides on some strategy $\mathbf{x} \in S_m$. Then he cannot compute his expectation $E(\mathbf{x}, \mathbf{y})$, since he does not know the strategy $\mathbf{y}$ of C. He can, however, reason as follows to obtain a minimal value of this expectation. Denote by $\mathbf{e}_1, \ldots, \mathbf{e}_n$ the pure strategies of C and let $\mathbf{y} = (y_1, \ldots, y_n) \in S_n$, Then

$$\begin{aligned} E(\mathbf{x}, \mathbf{y}) &= E(\mathbf{x}, y_1\mathbf{e}_1 + \ldots + y_n\mathbf{e}_n) \\ &= y_1 E(\mathbf{x}, \mathbf{e}_1) + \ldots + y_n E(\mathbf{x}, \mathbf{e}_n) \\ &\geqslant \min \{E(\mathbf{x}, \mathbf{e}_1), \ldots, E(\mathbf{x}, \mathbf{e}_n)\}, \end{aligned}$$

and so R can be sure that his expectation is at least equal to $u_R(\mathbf{x})$, where

$$u_R(\mathbf{x}) = \min_{\mathbf{y} \in S_n} E(\mathbf{x}, \mathbf{y}) = \min \{E(\mathbf{x}, \mathbf{e}_1), \ldots, E(\mathbf{x}, \mathbf{e}_n)\}.$$

The number $u_R(\mathbf{x})$ is called R's **security level** for his strategy $\mathbf{x}$. Since the functions $E(\mathbf{x}, \mathbf{e}_1), \ldots, E(\mathbf{x}, \mathbf{e}_n)$ are linear in $\mathbf{x}$, they are continuous. Hence $u_R : S_m \to \mathbb{R}$, being the minimum of a finite number of continuous functions, is itself continuous. Player R will naturally choose a strategy $\mathbf{x}$ in such a way as to make his security level $u_R(\mathbf{x})$ as large as possible. Since u_R is a continuous real-valued function defined on the compact set S_m, its maximal value, v_R say, will be attained at some point $\mathbf{x}_R$ of S_m. Thus

$$v_R = u_R(\mathbf{x}_R) = \max_{\mathbf{x} \in S_m} u_R(\mathbf{x}) = \max_{\mathbf{x} \in S_m} \min_{\mathbf{y} \in S_n} E(\mathbf{x}, \mathbf{y}).$$

The number v_R is called the **value** of R's game, and any strategy such as $\mathbf{x}_R$ which gives R a security level of v_R is called an **optimal strategy** for R.

We now consider briefly the game from the point of view of the column player C. Because of the very nature of a matrix game, C can see his objective as minimizing R's expectation. Suppose C decides on a strategy $\mathbf{y} \in S_n$. Then he can be sure that R's expectation is at most

$$u_C(\mathbf{y}) = \max_{\mathbf{x} \in S_m} E(\mathbf{x}, \mathbf{y}).$$

In perfect analogy to R maximizing $u_R(\mathbf{x})$, C tries to minimize $u_C(\mathbf{y})$. There exist $\mathbf{y}_C \in S_n$ and $v_C \in \mathbb{R}$ such that

$$v_C = u_C(\mathbf{y}_C) = \min_{\mathbf{y} \in S_n} u_C(\mathbf{y}) = \min_{\mathbf{y} \in S_n} \max_{\mathbf{x} \in S_m} E(\mathbf{x}, \mathbf{y}).$$

The number v_C is called the **value** of C's game, and any strategy such as $\mathbf{y}_C$ which give $u_C(\mathbf{y})$ the value v_C is called an **optimal strategy** for C.

Suppose now that R and C have optimal strategies $\mathbf{x}_R, \mathbf{y}_C$, and that the values of their games are v_R, v_C. Then if R uses $\mathbf{x}_R$ he can guarantee himself an expectation of at least v_R, and if C uses $\mathbf{y}_C$ he guarantees that R's expectation will not exceed v_C. Thus

$$v_R \leqslant E(\mathbf{x}_R, \mathbf{y}_C) \leqslant v_C,$$

and so $v_R \leqslant v_C$. The *minimax theorem*, proved below, asserts the equality of the values v_R and v_C. The theorem, therefore, shows that every matrix game is *soluble* in the sense that there exists a number v for which: R can play so that his expectation is at least v, and C can play so that R's expectation is at most v.

Theorem 4.6.3 (von Neumann's minimax theorem) *In the matrix game determined by a real $m \times n$ matrix $\mathbf{A}$, the value of R's game is equal to the value of C's game, i.e.*

$$v_R = \max_{\mathbf{x} \in S_m} \min_{\mathbf{y} \in S_n} \mathbf{x}^{\mathrm{T}}\mathbf{A}\mathbf{y} = \min_{\mathbf{y} \in S_n} \max_{\mathbf{x} \in S_m} \mathbf{x}^{\mathrm{T}}\mathbf{A}\mathbf{y} = v_C.$$

Proof Suppose first that the elements a_{ij} of $\mathbf{A}$ are all positive. Consider the following linear programming problem P and its dual P^*:

P	P^*
maximize $y_1 + \ldots + y_n$	minimize $x_1 + \ldots + x_m$
subject to $a_{11}y_1 + \ldots + a_{1n}y_n \leqslant 1$	subject to $a_{11}x_1 + \ldots + a_{m1}x_m \geqslant 1$
....	....
$a_{m1}y_1 + \ldots + a_{mn}y_n \leqslant 1$	$a_{1n}x_1 + \ldots + a_{mn}x_m \geqslant 1$
$y_1 \geqslant 0, \ldots, y_n \geqslant 0,$	$x_1 \geqslant 0, \ldots, x_m \geqslant 0.$

Since $\mathbf{A}$ has all positive elements, both P and P^* are feasible. Hence, by the duality theorem of linear programming, both P and P^* are soluble and have the same value, v say. Let $\mathbf{X} = (X_1, \ldots, X_m)$, $\mathbf{Y} = (Y_1, \ldots, Y_n)$ be optimal vectors for P^*, P, respectively. Then $\mathbf{X} \geqslant \mathbf{0}$, $\mathbf{Y} \geqslant \mathbf{0}$, $\mathbf{X}^{\mathrm{T}}\mathbf{A} \geqslant [1, \ldots, 1]$, $\mathbf{AY} \leqslant (1, \ldots, 1)$, and

$$X_1 + \ldots + X_m = v = Y_1 + \ldots + Y_n.$$

Write $\mathbf{x}_R = (1/v)\mathbf{X}$ and $\mathbf{y}_C = (1/v)\mathbf{Y}$. Then $\mathbf{x}_R \in S_m$, $\mathbf{y}_C \in S_n$, and, for all $\mathbf{x} \in S_m$, $\mathbf{y} \in S_n$,

$$E(\mathbf{x}_R, \mathbf{y}) = \mathbf{x}_R^{\mathrm{T}}\mathbf{A}\mathbf{y} \geqslant (1/v)[1, \ldots, 1]\mathbf{y} = 1/v$$

and

$$E(\mathbf{x}, \mathbf{y}_C) = \mathbf{x}^{\mathrm{T}}\mathbf{A}\mathbf{y}_C \leqslant (1/v)\mathbf{x}^{\mathrm{T}}(1, \ldots, 1) = 1/v.$$

Thus $v_C \leqslant 1/v \leqslant v_R$. But $v_R \leqslant v_C$, and so $v_R = 1/v = v_C$. Hence $\mathbf{x}_R, \mathbf{y}_C$ are optimal strategies for R, C, respectively.

Consider now the general case when **A** is not assumed to be positive. Let k be any real number such that the matrix **B** obtained by adding k to each element of **A** is positive. By what we have just proved,

$$\max_{\mathbf{x} \in S_m} \min_{\mathbf{y} \in S_n} \mathbf{x}^{\mathrm{T}}\mathbf{B}\mathbf{y} = \min_{\mathbf{y} \in S_n} \max_{\mathbf{x} \in S_m} \mathbf{x}^{\mathrm{T}}\mathbf{B}\mathbf{y},$$

that is

$$k + \max_{\mathbf{x} \in S_m} \min_{\mathbf{y} \in S_n} \mathbf{x}^{\mathrm{T}}\mathbf{A}\mathbf{y} = k + \min_{\mathbf{y} \in S_n} \max_{\mathbf{x} \in S_m} \mathbf{x}^{\mathrm{T}}\mathbf{A}\mathbf{y},$$

and the desired result is immediate. □

Since, in a matrix game, the values v_R and v_C are the same, either of them is referred to simply as the **value** of the game. By a **solution** to a matrix game is meant an optimal strategy for R, an optimal strategy for C, and the value of the game. In the course of proving the minimax theorem, we showed how the solution of a matrix game could be found by solving a certain linear programming problem and its dual. We now illustrate this method.

Example 4.6.4 Consider the matrix game of Example 4.6.1 that has pay-off matrix

$$\begin{bmatrix} 1 & -1 & 3 \\ 3 & 5 & -3 \\ 6 & 2 & -2 \end{bmatrix}.$$

There are some non-positive elements in this matrix, and so we add 4 to each of its elements, and discuss the game with pay-off matrix

$$\begin{bmatrix} 5 & 3 & 7 \\ 7 & 9 & 1 \\ 10 & 6 & 2 \end{bmatrix}.$$

To find a solution to this game, we solve the following linear programming problem and its dual:

$$\begin{array}{ll} \text{maximize} & y_1 + y_2 + y_3 \\ \text{subject to} & 5y_1 + 3y_2 + 7y_3 \leqslant 1 \\ & 7y_1 + 9y_2 + y_3 \leqslant 1 \\ & 10y_1 + 6y_2 + 2y_3 \leqslant 1 \\ & y_1, y_2, y_3 \geqslant 0. \end{array}$$

This we do using the simplex algorithm as follows.

1	1	1	0	0	0	0
5	3	(7)	1	0	0	1
7	9	1	0	1	0	1
10	6	2	0	0	1	1

$\frac{2}{7}$	$\frac{4}{7}$	0	$-\frac{1}{7}$	0	0	$-\frac{1}{7}$
$\frac{5}{7}$	$\frac{3}{7}$	1	$\frac{1}{7}$	0	0	$\frac{1}{7}$
$\frac{44}{7}$	($\frac{60}{7}$)	0	$-\frac{1}{7}$	1	0	$\frac{6}{7}$
$\frac{60}{7}$	$\frac{36}{7}$	0	$-\frac{2}{7}$	0	1	$\frac{5}{7}$

$-\frac{2}{15}$	0	0	$-\frac{2}{15}$	$-\frac{1}{15}$	0	$-\frac{1}{5}$
$\frac{2}{5}$	0	1	$\frac{3}{20}$	$-\frac{1}{20}$	0	$\frac{1}{10}$
$\frac{11}{15}$	1	0	$-\frac{1}{60}$	$\frac{7}{60}$	0	$\frac{1}{10}$
$\frac{24}{5}$	0	0	$-\frac{1}{5}$	$-\frac{3}{5}$	1	$\frac{1}{5}$

Thus $(0, \frac{1}{10}, \frac{1}{10})$ and $(\frac{2}{15}, \frac{1}{15}, 0)$ are optimal vectors for the problem and its dual respectively, the value of both problems is $\frac{1}{5}$. Referring back to the proof of the minimax theorem, we deduce that, for the modified game: an optimal row strategy is $(\frac{2}{3}, \frac{1}{3}, 0)$, an optimal column strategy is $(0, \frac{1}{2}, \frac{1}{2})$, and the value of the game is 5. For the original game: an optimal row strategy is $(\frac{2}{3}, \frac{1}{3}, 0)$, an optimal column strategy is $(0, \frac{1}{2}, \frac{1}{2})$, and its value is 1, i.e. $5 - 4$. □

The complementary slackness theorem of linear programming has a counterpart in game theory—it is Theorem 4.6.5 below. Before coming to this result, we must first introduce the idea of an *essential strategy*. In a matrix game, the ith pure strategy for the row player is said to be **essential** if there is an optimal strategy $(x_1, \ldots, x_m)$ for R in which

$x_i > 0$. A similar definition applies to the pure strategies for the column player. In Example 4.6.4, $(\frac{2}{3}, \frac{1}{3}, 0)$ and $(0, \frac{1}{2}, \frac{1}{2})$ were shown to be optimal strategies for the row and column players, respectively. Thus the first two pure strategies for the row player and the last two pure strategies for the column player are essential.

Theorem 4.6.5 *Suppose that some pure strategy for a player in a matrix game is essential. Then this strategy achieves the value of the game against each optimal strategy of the opponent.*

Proof Suppose that, in a game with $m \times n$ pay-off matrix $\mathbf{A} = [a_{ij}]$ and value v, R's ith pure strategy is essential and $\mathbf{x} = (x_1, \ldots, x_m)$ is an optimal strategy for R in which $x_i > 0$. Let $\mathbf{y} = (y_1, \ldots, y_n)$ be an optimal strategy for C. Then $E(\mathbf{x}, \mathbf{y}) = v$, i.e.

$$x_1(a_{11}y_1 + \ldots + a_{1n}y_n) + \ldots + x_m(a_{m1}y_1 + \ldots + a_{mn}y_n) = v. \tag{4.6.1}$$

Since $(y_1, \ldots, y_n)$ is optimal for C, it will give C an expected loss of at most v against each pure strategy of R, whence

$$a_{11}y_1 + \ldots + a_{1n}y_n \leqslant v, \ldots, a_{m1}y_1 + \ldots + a_{mn}y_n \leqslant v. \tag{4.6.2}$$

It now follows from (4.6.1) and (4.6.2), together with the relations $x_1, \ldots, x_m \geqslant 0$, $x_1 + \ldots + x_m = 1$, $x_i > 0$, that

$$a_{i1}y_1 + \ldots + a_{in}y_n = v.$$

Thus R's ith pure strategy achieves the value v of the game against any optimal strategy $(y_1, \ldots, y_n)$ of C. □

Every matrix game can be solved using the simplex algorithm, as illustrated in Example 4.6.4. Some types of game, however, can be solved by easier methods. We now consider two such types.

First we discuss games which have a *saddle point*. Suppose that the (i_0, j_0)th position in a real $m \times n$ matrix $\mathbf{A} = [a_{ij}]$ is such that $a_{i_0j_0}$ is the least element in its row and the greatest element in its column. Then $\mathbf{A}$ is said to have a **saddle point** at (i_0, j_0) with *value* $a_{i_0j_0}$. Suppose that, in the game with pay-off matrix $\mathbf{A}$, R plays his i_0th pure strategy and C plays an arbitrary strategy $(y_1, \ldots, y_n)$. Then R's expected gain is

$$a_{i_01}y_1 + \ldots + a_{i_0n}y_n \geqslant a_{i_0j_0},$$

and $v \geqslant a_{i_0j_0}$, where v denotes the value of the game. Suppose next that R plays an arbitrary strategy $(x_1, \ldots, x_m)$ and C plays his j_0th pure strategy. Then C's expected loss is

$$a_{1j_0}x_1 + \ldots + a_{mj_0}x_m \leqslant a_{i_0j_0},$$

and $v \leqslant a_{i_0 j_0}$. It follows that $v = a_{i_0 j_0}$, and that R's i_0th pure strategy and C's j_0th pure strategy are both optimal for their respective players. Not all matrices have saddle points; see, for instance, the matrices occurring in Examples 4.6.1 and 4.6.2. The matrix

$$\begin{bmatrix} 7 & 6 & 8 \\ 2 & 4 & 3 \\ 1 & -1 & 8 \end{bmatrix}$$

has a saddle point at $(1, 2)$ with value 6. The game defined by this matrix has value 6; optimal strategies for the row and column players are $(1, 0, 0)$ and $(0, 1, 0)$, respectively.

We now show how games whose pay-off matrices have either just two rows or just two columns can be solved graphically. It will suffice to consider games whose pay-off matrices have two rows. We illustrate the general method by solving the game determined by the 2×3 matrix

$$\begin{bmatrix} 2 & 4 & 3 \\ 4 & 1 & 2 \end{bmatrix}.$$

Suppose that R employs the strategy $\mathbf{x} = (x, 1 - x)$, where $0 \leqslant x \leqslant 1$. Denoting the pure strategies of C by $\mathbf{e}_1, \mathbf{e}_2, \mathbf{e}_3$, we find that:

$$\begin{aligned} E(\mathbf{x}, \mathbf{e}_1) &= 2x + 4(1 - x) = 4 - 2x; \\ E(\mathbf{x}, \mathbf{e}_2) &= 4x + 1(1 - x) = 1 + 3x; \\ E(\mathbf{x}, \mathbf{e}_3) &= 3x + 2(1 - x) = 2 + x. \end{aligned}$$

Thus, recalling the discussion earlier in this section, we see that R's security level for his strategy $\mathbf{x}$ is given by the equation

$$u_R(\mathbf{x}) = \min\{4 - 2x, 1 + 3x, 2 + x\}.$$

The graphs of $E(\mathbf{x}, \mathbf{e}_1)$, $E(\mathbf{x}, \mathbf{e}_2)$, and $E(\mathbf{x}, \mathbf{e}_3)$ are drawn in Fig. 4.2, where the graph of u_R is drawn with a thick line. It is clear from this figure that the value v of the game is given by the equations

$$v = \max\{u_R(\mathbf{x}) : 0 \leqslant x \leqslant 1\} = 2\tfrac{2}{3}$$

and that this maximum occurs when $x = \frac{2}{3}$. Thus $(\frac{2}{3}, \frac{1}{3})$ is an optimal strategy for R. Suppose now that (y_1, y_2, y_3) is an optimal strategy for C. Figure 4.2 shows that C's second pure strategy does not achieve the value of the game against R's optimal strategy $(\frac{2}{3}, \frac{1}{3})$. Thus, by Theorem 4.6.5, this strategy is not essential for C, and so $y_2 = 0$. Since both of R's pure strategies are essential, they must achieve the value of the game against C's optimal strategy $(y_1, 0, y_3)$, whence

$$2y_1 + 3y_3 = 2\tfrac{2}{3} \qquad \text{and} \qquad 4y_1 + 2y_3 = 2\tfrac{2}{3}.$$

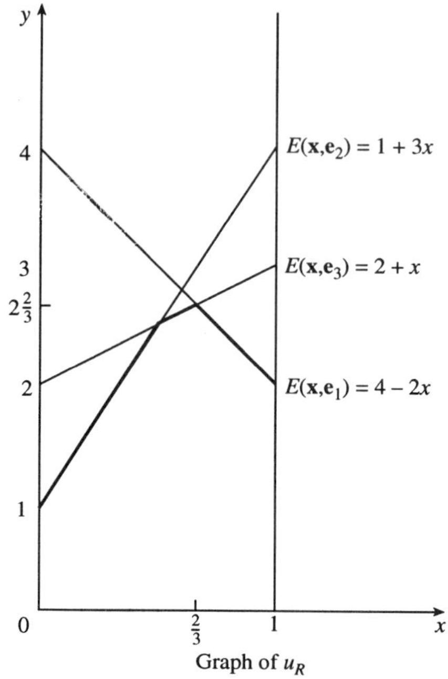

Fig. 4.2 Graph of u_R.

It follows that $(\frac{1}{3}, 0, \frac{2}{3})$ is an optimal strategy for C. This completely solves the game.

We conclude our discussion by mentioning the idea of *dominance*, which allows the simplification of certain games. Consider a game whose pay-off matrix has rows $\mathbf{r}_1, \ldots, \mathbf{r}_m$ and columns $\mathbf{c}_1, \ldots, \mathbf{c}_n$. Suppose that $\mathbf{r}_i \leqslant \mathbf{r}_j$ $(i \neq j)$. Then choosing the ith row offers no advantage to R over choosing the jth row, and so he can exclude it in his search for an optimal strategy. We say that the ith row is **dominated** by the jth row. In this case the ith row can be omitted from the game. Similarly, if $\mathbf{c}_i \leqslant \mathbf{c}_j$ $(i \neq j)$, then choosing the jth column offers no advantage to C over choosing the ith column. We say that the jth column is **dominated** by the ith column. In this case the jth column can be omitted from the game. We illustrate these ideas with an example.

Example 4.6.6 Consider the game with pay-off matrix

$$\begin{bmatrix} 2 & 4 \\ 3 & 1 \\ 2 & 3 \end{bmatrix}.$$

Here the third row is dominated by the first row. Hence we exclude the third row from the game, and consider the reduced game determined by the matrix

$$\begin{bmatrix} 2 & 4 \\ 3 & 1 \end{bmatrix}.$$

This game is easily solved graphically. We leave the reader to verify that its value is $2\frac{1}{2}$, and that optimal strategies for the row and column players are $(\frac{1}{2}, \frac{1}{2})$ and $(\frac{3}{4}, \frac{1}{4})$, respectively. Reverting to the original game, we see that its value is $2\frac{1}{2}$, and that optimal strategies for the row and column players are $(\frac{1}{2}, \frac{1}{2}, 0)$ and $(\frac{3}{4}, \frac{1}{4})$ respectively. □

Exercises 4.6

1. In a matrix game, each player has two red cards and two black cards. Both players lay down two of their cards. R wins 1 if either both players put down two black cards or both put down one card of each colour. When four red cards are played, R wins 2. When two black cards are played against two red cards, C wins 1. Two red cards played against a black card and a red card wins 1 for the player putting down the red cards. Neither player wins when precisely three black cards are played. Find the pay-off matrix and a solution to the game.
2. Consider the game whose pay-off matrix is

$$\begin{bmatrix} a & b \\ c & d \end{bmatrix}.$$

Show that if $a = b$ then the matrix has a saddle point. Prove that if the matrix has no saddle point then *either* $a < b$, $a < c$, $d < b$, $d < c$ *or* $b < a$, $c < a$, $b < d$, $c < d$. Find the value of the game when the matrix has no saddle point.
3. Show that the set of optimal strategies for a player in a matrix games is a polytope. Deduce that, if in an $m \times n$ matrix game every pure strategy for the row player is essential, then he has an optimal strategy $(x_1, \ldots, x_m)$ for which $x_1, \ldots, x_m > 0$.
4. Of the games determined by the three matrices below, one has a saddle point, one has a dominated row, and one has three essential strategies for the row player. Identify these games and find their solutions.

$$\begin{bmatrix} 6 & 0 & 6 \\ 8 & -2 & 0 \\ 4 & 6 & 5 \end{bmatrix}, \quad \begin{bmatrix} 6 & -4 & -2 \\ 1 & -1 & 3 \\ -8 & -3 & 7 \end{bmatrix}, \quad \begin{bmatrix} 3 & 5 & 2 \\ 2 & 1 & 3 \\ 2 & 3 & 1 \end{bmatrix}.$$

5. Solve the games with the following pay-off matrices:

$$\begin{bmatrix} 2 & 6 \\ 5 & 4 \end{bmatrix}, \begin{bmatrix} -1 & 5 & 1 & -2 \\ 1 & -3 & -2 & 5 \end{bmatrix}, \begin{bmatrix} -1 & 1 \\ 1 & -2 \\ 0 & -1 \end{bmatrix}.$$

6. Suppose that, in a game determined by an $m \times n$ matrix $\mathbf{A}$, the first row is dominated by the second one. Denote by $\mathbf{B}$ the $(m-1) \times n$ matrix obtained by deleting the first row from $\mathbf{A}$. Given that the game determined by $\mathbf{B}$ has value v and optimal strategies $(x_2, \ldots, x_m)$ and $(y_1, \ldots, y_n)$ for the row and column players, respectively, prove that the initial game has value v and optimal strategies $(0, x_2, \ldots, x_m)$ and $(y_1, \ldots, y_n)$ for the row and column players, respectively.

7. Suppose that some pure strategy for a player in a matrix game is not essential. Show that there exists some optimal strategy for the opponent, against which this pure strategy does not achieve the value of the game.

5
Convex functions

Here we discuss the theory and applications of convex functions both in $\mathbb{R}^1$ and $\mathbb{R}^n$. Section 5.6 deals with an important convex function associated with a non-empty bounded convex set, its support function.

Sections 5.1, 5.2, and 5.3 (excluding the treatment of Stirling's formula), together with Section 5.4, should be regarded as minimal reading here. Section 5.6 (excluding the discussion of gauge functions) is needed for our account of mixed volumes and extremum problems in Chapter 6.

5.1 CONVEX FUNCTIONS ON THE REAL LINE

Our study of convex functions begins in the familiar setting of real-valued functions whose domains are intervals on the real line. We define convex functions of a single variable and establish their basic properties, thus paving the way for some impressive applications in later sections.

Throughout our discussion here, we shall be concerned with a real-valued function $f: I \to \mathbb{R}$ defined on a non-degenerate (i.e. contains more than one point) interval I of the real line. Such a function f is said to be **convex** if

$$f(\lambda x + \mu y) \leqslant \lambda f(x) + \mu f(y)$$

whenever $x, y \in I$ and $\lambda, \mu \geqslant 0$ with $\lambda + \mu = 1$. Geometrically, f is convex if every chord joining two points on its graph lies on or above the graph. See Fig. 5.1. If $-f: I \to \mathbb{R}$ is convex, then $f: I \to \mathbb{R}$ is said to be **concave**.

As a simple example, we show that the square function $f: \mathbb{R} \to \mathbb{R}$ defined for real x by the equation $f(x) = x^2$ is convex. Let $x, y \in \mathbb{R}$ and let $\lambda, \mu \geqslant 0$ with $\lambda + \mu = 1$. Then

$$\begin{aligned}\lambda f(x) + \mu f(y) - f(\lambda x + \mu y) &= \lambda x^2 + \mu y^2 - (\lambda x + \mu y)^2 \\ &= \lambda\mu(x - y)^2 \geqslant 0,\end{aligned}$$

which establishes the convexity of the square function.

Our first results, concerning the continuity and differentiability of convex functions, appear obvious when interpreted graphically. Figure 5.1 suggests that, if X, Z, Y are points on the graph of a convex function, so listed that their x-coordinates are in increasing order, then

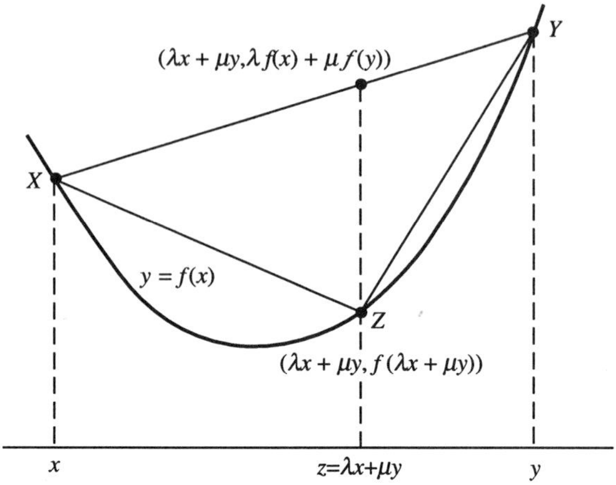

Fig. 5.1

$$\text{slope } XZ \leqslant \text{slope } XY \leqslant \text{slope } ZY.$$

This is indeed the case and the result is proved in the following theorem, which is sometimes referred to as the *three chords lemma.*

Theorem 5.1.1 *Let $f: I \to \mathbb{R}$ be a convex function and let $x, y, z \in I$ satisfy $x < z < y$. Then*

$$\frac{f(z) - f(x)}{z - x} \leqslant \frac{f(y) - f(x)}{y - x} \leqslant \frac{f(y) - f(z)}{y - z}.$$

Proof We express z as a convex combination of x and y as follows:

$$z = \frac{y - z}{y - x} x + \frac{z - x}{y - x} y.$$

By the convexity of f,

$$f(z) \leqslant \frac{y - z}{y - x} f(x) + \frac{z - x}{y - x} f(y).$$

Thus

$$f(z) - f(x) \leqslant \frac{z - x}{y - x} (f(y) - f(x))$$

and

$$\frac{f(z) - f(x)}{z - x} \leqslant \frac{f(y) - f(x)}{y - x}.$$

The other inequality in the theorem follows similarly. □

Corollary 5.1.2 *Let $f: I \to \mathbb{R}$ be a convex function and let $a \in I$. Then the function $g: I \setminus \{a\} \to \mathbb{R}$ defined by the equation*

$$g(x) = \frac{f(x) - f(a)}{x - a} \qquad \text{for } x \in I \setminus \{a\}$$

is increasing.

Proof If $b, c \in I \setminus \{a\}$ with $b < c$, then we must show that $g(b) \leqslant g(c)$. Either $b < c < a$, $b < a < c$, or $a < b < c$. Suppose that $b < c < a$. Then the theorem with $x = b, y = a, z = c$ shows that $g(b) \leqslant g(c)$. The other cases can be proved in a similar fashion. □

The next theorem and its corollaries show how well behaved convex functions are with respect to their continuity and differentiability properties.

Theorem 5.1.3 *Let $f: I \to \mathbb{R}$ be a convex function. Then f posseses left and right derivatives at each interior point of I, and if a, b are interior points of I with $a < b$, then*

$$f'_-(a) \leqslant f'_+(a) \leqslant \frac{f(b) - f(a)}{b - a} \leqslant f'_-(b) \leqslant f'_+(b).$$

Proof Let c be an interior point of I and let x, y be points of I such that $x < c < y$. Corollary 5.1.2 shows that, as x increases to c from below,

$$\frac{f(x) - f(c)}{x - c}$$

increases and is bounded above by

$$\frac{f(y) - f(c)}{y - c}.$$

Thus the left derivative $f'_-(c)$ exists and satisfies the inequality

$$f'_-(c) \leqslant \frac{f(y) - f(c)}{y - c}.$$

Letting y decrease to c in this inequality, we see that the right derivative $f'_+(c)$ exists and satisfies the inequality $f'_-(c) \leqslant f'_+(c)$. Thus, if a, b are interior points of I, then $f'_-(a) \leqslant f'_+(a)$ and $f'_-(b) \leqslant f'_+(b)$. The remaining inequalities of the theorem follow immediately from the definitions of $f'_+(a)$, $f'_-(b)$, and Corollary 5.1.2. □

Corollary 5.1.4 *Let $f: I \to \mathbb{R}$ be a convex function. Then, on the interior of I, f is continuous and f'_-, f'_+ are increasing.*

Proof At each interior point of I, f has both left and right derivatives, and so is continuous from the left and from the right, and hence is continuous. That f'_-, f'_+ are increasing on the interior of I follows immediately from the theorem. □

A convex function need not be continuous at the boundary points of its domain. For example, the convex function $f: [0, 1] \to \mathbb{R}$ defined by the equations

$$f(x) = \begin{cases} 0 & \text{for } 0 < x < 1 \\ 1 & \text{for } x = 0, 1 \end{cases}$$

is not continuous at 0 and 1. Also a convex function need not be differentiable, even at an interior point of its domain; the modulus function is not differentiable at the origin, where its left and right derivatives are -1 and 1, respectively. We do, however, have the following surprising result.

Corollary 5.1.5 *Let $f: I \to \mathbb{R}$ be a convex function. Then the set of those points of I at which f is not differentiable is countable.*

Proof Let C be the set of points of int I at which f is *not* differentiable. With each c in C, we associate a rational r_c such that $f'_-(c) < r_c < f'_+(c)$. It follows from the theorem that, if $c, d \in C$ with $c < d$, then

$$f'_-(c) < r_c < f'_+(c) \leqslant f'_-(d) < r_d < f'_+(d),$$

whence $r_c < r_d$. This shows immediately that the set of points of int I, and hence of I, at which f is not differentiable is countable. □

The definition of a convex function is only of limited use when checking that a particular function is convex. For example, the exponential function can be *seen* to be convex from its graph, but to show this from the definition involves establishing an inequality which is far from obvious, an inequality, in fact, of the very type we intend to prove using the theory of convex functions! Elementary calculus comes to the rescue by providing an effective criterion for the convexity of functions that has a wide range of applications.

Theorem 5.1.6 *Let $f: I \to \mathbb{R}$ be a differentiable function. Then f is convex if and only if f' is increasing.*

Proof Suppose first that f is convex. Let $a, b \in I$ with $a < b$. Then Corollary 5.1.2 shows that

$$f'(a) = \lim_{x \to a_+} \frac{f(x) - f(a)}{x - a} \leqslant \frac{f(b) - f(a)}{b - a} \leqslant \lim_{x \to b_-} \frac{f(x) - f(b)}{x - b} = f'(b),$$

whence $f'(a) \leqslant f'(b)$ and f' is increasing.

Suppose next that f' is increasing. Let $a, b \in I$ with $a < b$ and let $\lambda, \mu > 0$ with $\lambda + \mu = 1$. By the first mean value theorem, there exist real numbers, c, d with $a < c < \lambda a + \mu b < d < b$ such that

$$\frac{f(\lambda a + \mu b) - f(a)}{\lambda a + \mu b - a} = f'(c) \leqslant f'(d) = \frac{f(b) - f(\lambda a + \mu b)}{b - \lambda a - \mu b}.$$

Thus

$$\frac{f(\lambda a + \mu b) - f(a)}{\mu(b - a)} \leqslant \frac{f(b) - f(\lambda a + \mu b)}{\lambda(b - a)}$$

and

$$f(\lambda a + \mu b) \leqslant \lambda f(a) + \mu f(b),$$

which shows that f is convex. □

Corollary 5.1.7 *Let $f: I \to \mathbb{R}$ be a twice differentiable function. Then f is convex if and only if $f''(x) \geqslant 0$ for all x in I.* □

It follows from the last two results that e^x is convex on $\mathbb{R}$, $-\log x$ and $x \log x$ are convex on $(0, \infty)$, and x^p $(p \geqslant 1)$ is convex on $[0, \infty]$.

We now introduce the idea of a *support* of a convex function, which is closely related to its differentiability. Through each point on the graph of a convex function defined on an open interval of the real line, there passes a line which lies on or below the graph at all of its points. Moreover, this line is unique if and only if the function is differentiable at the point concerned. We now make these ideas precise.

Suppose that $f: I \to \mathbb{R}$ is a real-valued function defined on an open interval I of the real line and that $x_0 \in I$. Then an affine transformation $T: \mathbb{R} \to \mathbb{R}$ is said to **support** f at x_0 if $T(x_0) = f(x_0)$ and $T(x) \leqslant f(x)$ for all $x \in I$ and we say that f has **support** T at x_0. This situation is illustrated in Fig. 5.2. Such an affine transformation T can be expressed in the form

$$T(x) = f(x_0) + m(x - x_0)$$

for some real number m; $y = f(x_0) + m(x - x_0)$ is the equation of the line with slope m passing through the point $(x_0, f(x_0))$ on the graph of f. The condition $T(x) \leqslant f(x)$ means that this line lies on or below the graph of f.

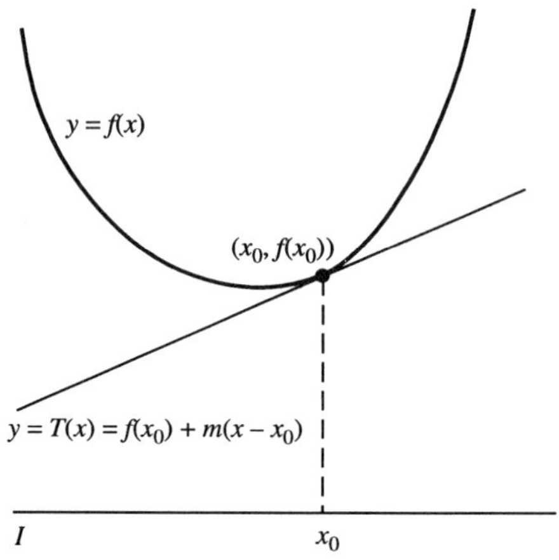

Fig. 5.2

Theorem 5.1.8 *Let $f: I \to \mathbb{R}$ be a real-valued function defined on an open interval I of $\mathbb{R}$. Then f is convex if and only if it has support at each point of I.*

Proof Suppose first that f has support at each point of I. Let $x, y \in I$ and let $\lambda, \mu \geqslant 0$ with $\lambda + \mu = 1$. Let T support f at $\lambda x + \mu y$. Then

$$f(\lambda x + \mu y) = T(\lambda x + \mu y) = \lambda T(x) + \mu T(y) \leqslant \lambda f(x) + \mu f(y),$$

which shows that f is convex.

Suppose next that f is convex. Let $x_0 \in I$ and let m be a real number satisfying the inequalities $f'_-(x_0) \leqslant m \leqslant f'_+(x_0)$. Define an affine transformation $T: \mathbb{R} \to \mathbb{R}$ by the equation

$$T(x) = f(x_0) + m(x - x_0) \qquad \text{for } x \in \mathbb{R}.$$

Let $y, z \in I$ be such that $y < x_0 < z$. Then, by Theorem 5.1.3,

$$\begin{aligned} \frac{f(y) - f(x_0)}{y - x_0} \leqslant f'_-(x_0) \leqslant \frac{T(y) - T(x_0)}{y - x_0} &= m \\ &= \frac{T(z) - T(x_0)}{z - x_0} \\ &\leqslant f'_+(x_0) \leqslant \frac{f(z) - f(x_0)}{z - x_0}, \end{aligned}$$

whence $T(y) \leqslant f(y)$ and $T(z) \leqslant f(z)$. Thus T supports f at x_0. □

Theorem 5.1.9 *Let $f: I \to \mathbb{R}$ be a convex function defined on an open interval I of $\mathbb{R}$. Then f is differentiable at a point x_0 of I if and only if it has unique support at x_0.*

Proof Suppose first that f is differentiable at x_0. Let $T: \mathbb{R} \to \mathbb{R}$ support f at x_0; say

$$T(x) = f(x_0) + m(x - x_0) \qquad \text{for } x \in \mathbb{R},$$

where m is a real number. Let $y, z \in I$ be such that $y < x_0 < z$. Then

$$\begin{aligned}\frac{f(y) - f(x_0)}{y - x_0} &\leqslant \frac{T(y) - T(x_0)}{y - x_0} \\ &= m \\ &= \frac{T(z) - T(x_0)}{z - x_0} \\ &\leqslant \frac{f(z) - f(x_0)}{z - x_0}.\end{aligned}$$

Thus, letting $y \to x_{0_-}$, $z \to x_{0_+}$, we deduce that $m = f'(x_0)$. Hence f has unique support at x_0.

Suppose next that f has unique support at x_0. Let the real number m satisfy $f'_-(x_0) \leqslant m \leqslant f'_+(x_0)$. Then, as in the proof of Theorem 5.1.8, the affine transformation T defined by the equation

$$T(x) = f(x_0) + m(x - x_0)$$

supports f at x_0. But f has *unique* support at x_0, whence m is unique and $f'_-(x_0) = f'_+(x_0)$, i.e. f is differentiable at x_0. □

Exercises 5.1

1. Let $f, g: I \to \mathbb{R}$ be convex functions and let $\alpha \in \mathbb{R}$. Which of the functions $\alpha f, f + g, f - g, |f|, fg, \max\{f, g\}, \min\{f, g\}$ are necessarily convex?
2. Let $f, g: \mathbb{R} \to \mathbb{R}$ be convex functions with g increasing. Show that $g \circ f$ is convex. Deduce that, if $h: \mathbb{R} \to \mathbb{R}$ is a positive function and $\log h$ is convex, then h is convex. Give an example of a positive convex function defined on $\mathbb{R}$ whose logarithm is not convex.
3. Show that the reciprocal of a positive concave function is convex. Is the reciprocal of a positive convex function always concave?
4. Suppose that $f: I \to \mathbb{R}$ is a convex function. Prove that

$$f(x_1 + \delta) - f(x_1) \leqslant f(x_2 + \delta) - f(x_2)$$

whenever $x_1, x_2 \in I$ and $\delta > 0$ are such that $x_1 < x_2$ and $x_2 + \delta \in I$.

5. Let $f: I \to \mathbb{R}$ be a convex function. Show that:
   (i) if f has a global maximum at an interior point of I, then f is constant;
   (ii) if f has a global minimum at c, then f is increasing to the right of c and decreasing to the left of c;
   (iii) if f has a local minimum, then this minimum is a global minimum.
6. Prove that a convex function defined on a bounded interval is bounded below, and that a convex function defined and bounded above on $\mathbb{R}$ must be constant.

5.2 CLASSICAL INEQUALITIES

One impressive application of the theory of convex functions is to the study of classical inequalities. Here we show how the theory provides an elementary, elegant, and unified treatment of some of the best-known inequalities in mathematics.

The defining inequality for a convex function implies a more general one, known as Jensen's inequality, and it is this which we need. To establish an inequality, we first prove that some function is convex (by showing that its second derivative is non-negative) and then apply Jensen's inequality.

Theorem 5.2.1 (Jensen's inequality) *Let $f: I \to \mathbb{R}$ be a convex function. Let $x_1, \ldots, x_m \in I$ and let $\lambda_1, \ldots, \lambda_m \geqslant 0$ with $\lambda_1 + \ldots + \lambda_m = 1$. Then*

$$f(\lambda_1 x_1 + \ldots + \lambda_m x_m) \leqslant \lambda_1 f(x_1) + \ldots + \lambda_m f(x_m).$$

Proof We argue by induction on m. The inequality is trivially true when $m = 1$. Assume, then, that it is true when $m = k$, where $k \geqslant 1$. Let a real number x be defined by the equation

$$x = \lambda_1 x_1 + \ldots + \lambda_{k+1} x_{k+1},$$

where $x_1, \ldots, x_{k+1} \in I$ and $\lambda_1, \ldots, \lambda_{k+1} \geqslant 0$ with $\lambda_1 + \ldots + \lambda_{k+1} = 1$. At least one of $\lambda_1, \ldots, \lambda_{k+1}$ must be less than 1, say $\lambda_{k+1} < 1$. Write

$$\lambda = \lambda_1 + \ldots + \lambda_k = 1 - \lambda_{k+1},$$

so $\lambda > 0$, and write

$$y = \frac{\lambda_1}{\lambda} x_1 + \ldots + \frac{\lambda_k}{\lambda} x_k.$$

The induction hypothesis shows that

$$f(y) \leqslant \frac{\lambda_1}{\lambda} f(x_1) + \ldots + \frac{\lambda_k}{\lambda} f(x_k).$$

Since f is convex,

$$\begin{aligned} f(x) = f(\lambda y + \lambda_{k+1} x_{k+1}) &\leqslant \lambda f(y) + \lambda_{k+1} f(x_{k+1}) \\ &\leqslant \lambda_1 f(x_1) + \ldots + \lambda_{k+1} f(x_{k+1}), \end{aligned}$$

and the inequality is established for $m = k + 1$. This completes the proof by induction. □

In this section the word *number* will be used exclusively to mean *positive* real number. The **arithmetic mean** and the **geometric mean** of numbers x_1 and x_2 are defined to be $\frac{1}{2}(x_1 + x_2)$ and $\sqrt{(x_1 x_2)}$, respectively. The basic inequality between these means is that the geometric mean never exceeds the arithmetic mean, i.e. $\sqrt{(x_1 x_2)} \leqslant \frac{1}{2}(x_1 + x_2)$, a result which follows immediately from the fact that $(\sqrt{x_1} - \sqrt{x_2})^2 \geqslant 0$. The **arithmetic mean** and the **geometric mean** of numbers $x_1, \ldots, x_m$ are defined, respectively, to be

$$\frac{1}{m}(x_1 + \ldots + x_m) \qquad \text{and} \qquad (x_1 \ldots x_m)^{1/m}.$$

Once again the geometric mean never exceeds the arithmetic mean, although the proof of this is appreciably more difficult than the case $m = 2$. The concepts of arithmetic and geometric means can be generalized by attaching *weights* $\alpha_1, \ldots, \alpha_m$ to the numbers as follows. Let $\alpha_1, \ldots, \alpha_m$ be numbers whose sum is 1. Then the numbers

$$\alpha_1 x_1 + \ldots + \alpha_m x_m \qquad \text{and} \qquad x_1^{\alpha_1} \ldots x_m^{\alpha_m}$$

are called, respectively, the **weighted arithmetic mean** and the **weighted geometric mean** of the numbers $x_1, \ldots, x_m$ with respect to the weights $\alpha_1, \ldots, \alpha_m$. These *weighted* means reduce to the usual means when each of the weights $\alpha_1, \ldots, \alpha_m$ is $1/m$. That the weighted geometric mean never exceeds the weighted arithmetic mean is the content of the next theorem.

Theorem 5.2.2 *Let $x_1, \ldots, x_m, \alpha_1, \ldots, \alpha_m > 0$ with $\alpha_1 + \ldots + \alpha_m = 1$. Then*

$$x_1^{\alpha_1} \ldots x_m^{\alpha_m} \leqslant \alpha_1 x_1 + \ldots + \alpha_m x_m.$$

Proof The function $-\log$ is convex on $(0, \infty)$, whence, by Jensen's inequality,

$$\begin{aligned} -\log(\alpha_1 x_1 + \ldots + \alpha_m x_m) &\leqslant -(\alpha_1 \log x_1 + \ldots + \alpha_m \log x_m) \\ &= -\log(x_1^{\alpha_1} \ldots x_m^{\alpha_m}). \end{aligned}$$

Since log is a *strictly* increasing function, we can deduce that

$$x_1^{\alpha_1}\dots x_m^{\alpha_m} \leqslant \alpha_1 x_1 + \dots + \alpha_m x_m. \ \square$$

Corollary 5.2.3 *Let* $x_1, \dots, x_m > 0$. *Then*

$$(x_1 \dots x_m)^{1/m} \leqslant \frac{1}{m}(x_1 + \dots + x_m). \ \square$$

Two of the most important inequalities in analysis are those of Hölder and Minkowski, and both can easily be established using the theory of convex functions. We deduce Hölder's inequality from the following very general result.

Theorem 5.2.4 *Let* $a_{ij} > 0$ $(i = 1, \dots, m; j = 1, \dots, n)$ *and let* $\alpha_1, \dots, \alpha_m > 0$ *with* $\alpha_1 + \dots + \alpha_m = 1$. *Then*

$$a_{11}^{\alpha_1}\dots a_{m1}^{\alpha_m} + \dots + a_{1n}^{\alpha_1}\dots a_{mn}^{\alpha_m} \leqslant (a_{11} + \dots + a_{1n})^{\alpha_1}\dots(a_{m1} + \dots + a_{mn})^{\alpha_m}.$$

Proof We use Theorem 5.2.2 to deduce that, for each $j = 1, \dots n$,

$$\frac{a_{1j}^{\alpha_1}\dots a_{mj}^{\alpha_m}}{(a_{11} + \dots + a_{1n})^{\alpha_1}\dots(a_{m1} + \dots + a_{mn})^{\alpha_m}}$$

$$\leqslant \frac{\alpha_1 a_{1j}}{a_{11} + \dots + a_{1n}} + \dots + \frac{\alpha_m a_{mj}}{a_{m1} + \dots + a_{mn}}.$$

Adding these n inequalities together, we deduce that

$$\sum_{j=1}^{n} \frac{a_{1j}^{\alpha_1}\dots a_{mj}^{\alpha_m}}{(a_{11} + \dots + a_{1n})^{\alpha_1}\dots(a_{m1} + \dots + a_{mn})^{\alpha_m}} \leqslant \alpha_1 + \dots + \alpha_m = 1,$$

and the desired result follows immediately. $\square$

Corollary 5.2.5 *Let* $x_1, \dots, x_m, y_1, \dots, y_m > 0$. *Then*

$$(x_1 \dots x_m)^{1/m} + (y_1 \dots y_m)^{1/m} \leqslant (x_1 + y_1)^{1/m} \dots (x_m + y_m)^{1/m}.$$

Proof Let $n = 2, \alpha_1 = 1/m, \dots, \alpha_m = 1/m, a_{i1} = x_i$ and $a_{i2} = y_i$ in the theorem. $\square$

Corollary 5.2.6 (Hölder's inequality) *Let* $x_1, \dots, x_n, y_1, \dots, y_n > 0$. *Suppose that* $p, q > 0$ *satisfy* $1/p + 1/q = 1$. *Then*

$$\sum_{i=1}^{n} x_i y_i \leqslant \left(\sum_{i=1}^{n} x_i^p\right)^{1/p}\left(\sum_{i=1}^{n} y_i^q\right)^{1/q}.$$

Proof Let $m = 2, \alpha_1 = 1/p, \alpha_2 = 1/q$, and let $a_{1j} = x_j^p, a_{2j} = y_j^q$ in the above theorem. $\square$

Theorem 5.2.7 (Minkowski's inequality) *Let* $x_1, \ldots, x_n, y_1, \ldots, y_n > 0$ *and let* $p \geqslant 1$. *Then*

$$\left(\sum_{i=1}^{n} (x_i + y_i)^p\right)^{1/p} \leqslant \left(\sum_{i=1}^{n} x_i^p\right)^{1/p} + \left(\sum_{i=1}^{n} y_i^p\right)^{1/p}.$$

Proof Write $a = (\Sigma_{i=1}^n x_i^p)^{1/p}$ and $b = (\Sigma_{i=1}^n y_i^p)^{1/p}$. Since x^p $(p \geqslant 1)$ is convex on $(0, \infty)$, we can deduce that, for $i = 1, \ldots, n$,

$$\left(\frac{x_i + y_i}{a+b}\right)^p = \left(\frac{a}{a+b}\left(\frac{x_i}{a}\right) + \frac{b}{a+b}\left(\frac{y_i}{b}\right)\right)^p \leqslant \frac{a}{a+b}\left(\frac{x_i}{a}\right)^p + \frac{b}{a+b}\left(\frac{y_i}{b}\right)^p.$$

Adding these n inequalities together, we deduce that

$$\begin{aligned}\sum_{i=1}^{n}\left(\frac{x_i + y_i}{a+b}\right)^p &\leqslant \frac{a}{a+b}\left(\frac{\sum_{i=1}^{n} x_i^p}{a^p}\right) + \frac{b}{a+b}\left(\frac{\sum_{i=1}^{n} y_i^p}{b^p}\right)\\ &= \frac{a}{a+b} + \frac{b}{a+b}\\ &= 1.\end{aligned}$$

Thus

$$\sum_{i=1}^{n} (x_i + y_i)^p \leqslant (a+b)^p = \left(\left(\sum_{i=1}^{n} x_i^p\right)^{1/p} + \left(\sum_{i=1}^{n} y_i^p\right)^{1/p}\right)^p,$$

and Minkowski's inequality follows immediately. □

In addition to the arithmetic and geometric means, two other means have their own names: the *harmonic mean* and the *root mean square*. Given the numbers $x_1, \ldots, x_m$, their **harmonic mean** is defined to be

$$\frac{1}{\dfrac{1}{m}\left(\dfrac{1}{x_1} + \ldots + \dfrac{1}{x_m}\right)},$$

and their **root mean square** is defined to be

$$\sqrt{\left(\frac{x_1^2 + \ldots + x_m^2}{m}\right)}.$$

The basic inequalities connecting the four means are:

harmonic mean $\leqslant$ geometric mean $\leqslant$ arithmetic mean $\leqslant$ root mean square.

The harmonic mean and the root mean square are generalized in the natural way to the corresponding weighted means. Let $\alpha_1, \ldots, \alpha_m > 0$ with $\alpha_1 + \ldots + \alpha_m = 1$. Then the numbers

$$\frac{1}{\dfrac{\alpha_1}{x_1} + \ldots + \dfrac{\alpha_m}{x_m}} \quad \text{and} \quad \sqrt{(\alpha_1 x_1^2 + \ldots + \alpha_m x_m^2)}$$

are called, respectively, the **weighted harmonic mean** and the **weighted root mean square** of the numbers $x_1, \ldots, x_m$ with respect to the weights $\alpha_1, \ldots, \alpha_m$. Corollary 5.2.9 below shows that the basic inequalities stated above connecting the four unweighted means continue to hold for the weighted means.

The four means that we have so far introduced are special cases of the *mean of order t* which we now define. Let $\boldsymbol{\alpha} = (\alpha_1, \ldots, \alpha_m)$, $\mathbf{x} = (x_1, \ldots, x_m)$, where $\alpha_1, \ldots, \alpha_m, x_1, \ldots, x_m > 0$ and $\alpha_1 + \ldots + \alpha_m = 1$. Then for each *non-zero* real number t the **mean $M_t(\mathbf{x}; \boldsymbol{\alpha})$ of order t** is defined by the equation

$$M_t(\mathbf{x}; \boldsymbol{\alpha}) = (\alpha_1 x_1^t + \ldots + \alpha_m x_m^t)^{1/t}.$$

The values $t = -1, 1, 2$ give rise, respectively, to the weighted harmonic mean, the weighted arithmetic mean, and the weighted root mean square.

The weighted geometric mean is not the mean of order t for any *non-zero* real number t, and so does not immediately fit into the general pattern of means. Is there any meaningful sense in which the weighted geometric mean can be considered to be the mean of order zero? To answer this question, we consider the limit of $M_t(\mathbf{x}; \boldsymbol{\alpha})$ as t tends to zero. This we do by taking logarithms on both sides of the defining equation of $M_t(\mathbf{x}; \boldsymbol{\alpha})$ to obtain the relation

$$\log M_t(\mathbf{x}; \boldsymbol{\alpha}) = \frac{\log(\alpha_1 x_1^t + \ldots + \alpha_m x_m^t)}{t}.$$

By definition,

$$\lim_{t \to 0} \frac{\log(\alpha_1 x_1^t + \ldots + \alpha_m x_m^t)}{t}$$

is the derivative of $\log(\alpha_1 x_1^t + \ldots + \alpha_m x_m^t)$ at $t = 0$, and this is readily found to be $\log x_1^{\alpha_1} \ldots x_m^{\alpha_m}$. Thus

$$M_t(\mathbf{x}; \boldsymbol{\alpha}) = e^{\log M_t(\mathbf{x}; \boldsymbol{\alpha})} \to e^{\log x_1^{\alpha_1} \ldots x_m^{\alpha_m}} = x_1^{\alpha_1} \ldots x_m^{\alpha_m},$$

and so $M_t(\mathbf{x}; \boldsymbol{\alpha})$ tends to the weighted geometric mean $x_1^{\alpha_1} \ldots x_m^{\alpha_m}$ as t tends to zero, which is what we had hoped for. Thus we formally define $M_0(\mathbf{x}; \boldsymbol{\alpha})$, the **mean of order zero**, to be $x_1^{\alpha_1} \ldots x_m^{\alpha_m}$. Thus $M_t(\mathbf{x}; \boldsymbol{\alpha})$ is now defined for every real number t and is continuous on the whole of $\mathbb{R}$, in particular at $t = 0$. We now establish a sweeping generalization of Theorem 5.2.2.

Theorem 5.2.8 *Let* $\mathbf{x} = (x_1, \ldots, x_m)$, $\boldsymbol{\alpha} = (\alpha_1, \ldots, \alpha_m)$, *where* $x_1, \ldots, x_m$, $\alpha_1, \ldots, \alpha_m > 0$ *and* $\alpha_1 + \ldots + \alpha_m = 1$. *Then* $M_t(\mathbf{x}; \boldsymbol{\alpha})$ *is an increasing function of* t.

Proof Since $\mathbf{x}$ and $\boldsymbol{\alpha}$ are fixed, we write $M_t(\mathbf{x}; \boldsymbol{\alpha})$ simply as $M(t)$. We show that $M'(t) \geqslant 0$ for all non-zero real numbers t. Since M is continuous at 0, this will show that M is increasing on $\mathbb{R}$. The equation

$$t \log M(t) = \log (\alpha_1 x_1^t + \ldots + \alpha_m x_m^t),$$

when differentiated, becomes

$$t \frac{M'(t)}{M(t)} + \log M(t) = \frac{\alpha_1 x_1^t \log x_1 + \ldots + \alpha_m x_m^t \log x_m}{\alpha_1 x_1^t + \ldots + \alpha_m x_m^t} \qquad (t \neq 0).$$

Thus, for $t \neq 0$,

$$t^2 \frac{M'(t)}{M(t)} + t \log M(t) = \frac{\alpha_1 x_1^t \log x_1^t + \ldots + \alpha_m x_m^t \log x_m^t}{\alpha_1 x_1^t + \ldots + \alpha_m x_m^t},$$

and so

$$\frac{t^2 M'(t)(\alpha_1 x_1^t + \ldots + \alpha_m x_m^t)}{M(t)}$$

equals

$$\alpha_1 x_1^t \log x_1^t + \ldots + \alpha_m x_m^t \log x_m^t$$
$$- (\alpha_1 x_1^t + \ldots + \alpha_m x_m^t) \log (\alpha_1 x_1^t + \ldots + \alpha_m x_m^t).$$

Jensen's inequality, applied to the convex function $y \log y$ on $(0, \infty)$, shows that, for all $y_1, \ldots, y_m > 0$,

$$(\alpha_1 y_1 + \ldots + \alpha_m y_m) \log (\alpha_1 y_1 + \ldots + \alpha_m y_m)$$
$$\leqslant \alpha_1 y_1 \log y_1 + \ldots + \alpha_m y_m \log y_m.$$

If we put $y_i = x_i^t$ for $i = 1, \ldots, m$ in this inequality, we deduce from the equality previously stated that $M'(t) \geqslant 0$ for $t \neq 0$. □

Corollary 5.2.9 *Let* $x = (x_1, \ldots, x_m)$, $\boldsymbol{\alpha} = (\alpha_1, \ldots, \alpha_m)$, *where* $x_1, \ldots, x_m$, $\alpha_1, \ldots, \alpha_m > 0$ *and* $\alpha_1 + \ldots + \alpha_m = 1$. *Then*

$$M_{-1}(\mathbf{x}; \boldsymbol{\alpha}) \leqslant M_0(\mathbf{x}; \boldsymbol{\alpha}) \leqslant M_1(\mathbf{x}; \boldsymbol{\alpha}) \leqslant M_2(\mathbf{x}; \boldsymbol{\alpha}). \quad \square$$

Exercises 5.2

1. Let $f: I \to \mathbb{R}$ be a convex function defined on an open interval I. Let $x_1, \ldots, x_m \in I$ and let $\lambda_1, \ldots, \lambda_m \geqslant 0$ with $\lambda_1 + \ldots + \lambda_m = 1$. Prove Jensen's inequality by considering a support of f at the point $\lambda_1 x_1 + \ldots + \lambda_m x_m$.
2. Use the inequality between the weighted arithmetic and weighted geometric means to show that
$$x + 1/(xy) + y^2 \geqslant 5(2^{-4/5}) \qquad \text{for } x, y > 0.$$
Find values of x and y which give equality here.
3. Show that the function $-\sin$ is convex on $[0, \pi]$. Deduce that the maximal perimeter of an n-sided convex polygon whose vertices lie on a unit circle is $2n \sin(\pi/n)$.
4. Prove that, if f is convex on $(0, \infty)$ and $x_1, \ldots, x_m, y_1, \ldots, y_m > 0$, then
$$(x_1 + \ldots + x_m) f\left(\frac{y_1 + \ldots + y_m}{x_1 + \ldots + x_m}\right) \leqslant x_1 f\left(\frac{y_1}{x_1}\right) + \ldots + x_m f\left(\frac{y_m}{x_m}\right).$$
Show that $f(x) = (1 + x^p)^{1/p}$ is convex on $(0, \infty)$ when $p \geqslant 1$, and deduce from the first part of the exercise that
$$((x_1 + \ldots + x_m)^p + (y_1 + \ldots + y_m)^p)^{1/p}$$
$$\leqslant (x_1^p + y_1^p)^{1/p} + \ldots + (x_m^p + y_m^p)^{1/p}.$$
5. Let $\mathbf{x} = (x_1, \ldots, x_m)$, $\boldsymbol{\alpha} = (\alpha_1, \ldots, \alpha_m)$, where $x_1, \ldots, x_m, \alpha_1, \ldots, \alpha_m > 0$ and $\alpha_1 + \ldots + \alpha_m = 1$. Let $x_k = \max\{x_1, \ldots, x_m\}$. Show that
$$\alpha_k^{1/t} x_k \leqslant M_t(\mathbf{x}; \boldsymbol{\alpha}) \leqslant x_k \qquad \text{for } t > 0.$$
Deduce the value of the limit
$$\lim_{t \to \infty} M_t(\mathbf{x}; \boldsymbol{\alpha}).$$
Sketch the graph of $M_t(\mathbf{x}; \boldsymbol{\alpha})$ and make clear its behaviour at $-\infty$ and $+\infty$.

5.3 THE GAMMA AND BETA FUNCTIONS

We now introduce the concept of log-convexity and use it to study the gamma and beta functions. Of course, these special functions of analysis can be, and usually are, studied independently of the theory of convex functions. The treatment of the gamma and beta functions given here enables many of their properties to be developed easily from general considerations of log-convexity, without recourse to topics such as Fourier series, double integration, and complex analysis. As a bonus, the context of log-convex functions clarifies the extent to which the gamma function is a *unique* extension of the factorial function.

One result that we shall need in our discussion is Hölder's inequality for integrals, which is Theorem 5.3.1. In the statement of this theorem, $\int_I f$ denotes the (Riemann) integral of a continuous function $f: I \to \mathbb{R}$ over an interval I of the real line.

Theorem 5.3.1 (Hölder's inequality for integrals) *Let $f, g: I \to \mathbb{R}$ be continuous non-negative functions for which the integrals $\int_I f$, $\int_I g$ are positive. Let $\lambda, \mu \geqslant 0$ with $\lambda + \mu = 1$. Then*

$$\int_I f^\lambda g^\mu \leqslant \left(\int_I f\right)^\lambda \left(\int_I g\right)^\mu.$$

Proof By Theorem 5.2.2, for $t \in I$,

$$\left(\frac{f(t)}{\int_I f}\right)^\lambda \left(\frac{g(t)}{\int_I g}\right)^\mu \leqslant \lambda \frac{f(t)}{\int_I f} + \mu \frac{g(t)}{\int_I g}.$$

We integrate both sides of this inequality to deduce that

$$\frac{\int_I f^\lambda(t) g^\mu(t)\,\mathrm{d}t}{(\int_I f)^\lambda (\int_I g)^\mu} \leqslant \lambda \frac{\int_I f}{\int_I f} + \mu \frac{\int_I g}{\int_I g} = \lambda + \mu = 1,$$

whence

$$\int_I f^\lambda g^\mu \leqslant \left(\int_I f\right)^\lambda \left(\int_I g\right)^\mu. \quad \square$$

Let $f: I \to \mathbb{R}$ be a function defined on an interval I of the real line. Then f is said to be **log-convex** if it is positive and its logarithm, $\log f: I \to \mathbb{R}$, is convex. Thus a positive function f is log-convex on an interval I if and only if, whenever $x, y \in I$ and $\lambda, \mu \geqslant 0$ with $\lambda + \mu = 1$, we have

$$\log f(\lambda x + \mu y) \leqslant \lambda \log f(x) + \mu \log f(y) = \log f^\lambda(x) f^\mu(y),$$

that is

$$f(\lambda x + \mu y) \leqslant f^\lambda(x) f^\mu(y).$$

We shall make frequent use of this reformulation of log-convexity. Since $f^\lambda(x) f^\mu(y) \leqslant \lambda f(x) + \mu f(y)$, it follows that every log-convex function is convex. On the other hand, on the interval $(0, \infty)$, the positive function x is convex but not log-convex. Clearly, for any positive number a, the function a^x is log-convex on $\mathbb{R}$.

The class of functions which are log-convex on some interval I is closed under addition and multiplication. To see this, suppose that the functions $f, g: I \to \mathbb{R}$ are log-convex. Let $x, y \in I$ and let $\lambda, \mu \geqslant 0$ with $\lambda + \mu = 1$. By Theorem 5.2.4,

$$\begin{aligned}(f+g)(\lambda x + \mu y) &= f(\lambda x + \mu y) + g(\lambda x + \mu y)\\ &\leqslant f^{\lambda}(x) f^{\mu}(y) + g^{\lambda}(x) g^{\mu}(y)\\ &\leqslant (f(x) + g(x))^{\lambda}(f(y) + g(y))^{\mu}\\ &= (f+g)^{\lambda}(x)(f+g)^{\mu}(y),\end{aligned}$$

which shows that $f + g$ is log-convex. Also

$$\begin{aligned}(fg)(\lambda x + \mu y) &= f(\lambda x + \mu y) g(\lambda x + \mu y)\\ &\leqslant f^{\lambda}(x) f^{\mu}(y) g^{\lambda}(x) g^{\mu}(y)\\ &= (fg)^{\lambda}(x)(fg)^{\mu}(y),\end{aligned}$$

which shows that fg is log-convex.

The **gamma function** $\Gamma: (0, \infty) \to \mathbb{R}$ is defined by the equation

$$\Gamma(x) = \int_0^{\infty} t^{x-1} e^{-t}\, dt \qquad \text{for } x > 0.$$

Elementary analysis show that, for each $x > 0$, $\Gamma(x)$ is a well-defined positive number.

Theorem 5.3.2 *The gamma function has the following properties*:

(i) $\Gamma(x+1) = x\Gamma(x)$ for $x > 0$;

(ii) $\Gamma(1) = 1$;

(iii) Γ is log-convex.

Proof (i) For $x > 0$,

$$\Gamma(x+1) = \int_0^{\infty} t^{x} e^{-t}\, dt = [-t^{x} e^{-t}]_0^{\infty} + x \int_0^{\infty} t^{x-1} e^{-t}\, dt = x\Gamma(x).$$

(ii) $\Gamma(1) = \int_0^{\infty} e^{-t}\, dt = 1.$

(iii) Let $x, y > 0$. Let $\lambda, \mu \geqslant 0$ with $\lambda + \mu = 1$. Then, by Theorem 5.3.1,

$$\begin{aligned}
\Gamma(\lambda x + \mu y) &= \int_0^\infty t^{\lambda x + \mu y - 1} \mathrm{e}^{-t}\, \mathrm{d}t \\
&= \int_0^\infty (t^{x-1}\mathrm{e}^{-t})^\lambda (t^{y-1}\mathrm{e}^{-t})^\mu \, \mathrm{d}t \\
&\leqslant \left(\int_0^\infty t^{x-1}\mathrm{e}^{-t}\,\mathrm{d}t\right)^\lambda \left(\int_0^\infty t^{y-1}\mathrm{e}^{-t}\,\mathrm{d}t\right)^\mu \\
&= \Gamma^\lambda(x)\Gamma^\mu(y),
\end{aligned}$$

which shows that Γ is log-convex. □

Corollary 5.3.3 *For* $n = 0, 1, 2, \ldots, \Gamma(n+1) = n!$.

Proof By the theorem, $\Gamma(1) = 1$. Hence, for $n = 1, 2, \ldots,$

$\Gamma(n+1) = n\Gamma(n) = n(n-1)\Gamma(n-1) = n(n-1)\ldots 1\Gamma(1) = n!$. □

Corollary 5.3.4 *The gamma function is convex, continuous, and* $x\Gamma(x) \to 1$, $\Gamma(x) \to \infty$ *as* $x \to 0_+$.

Proof The gamma function is log-convex, and hence convex. By Corollary 5.1.4, it must also be continuous. The continuity of Γ at 1 shows that

$$x\Gamma(x) = \Gamma(x+1) \to \Gamma(1) = 1 \qquad \text{as } x \to 0_+,$$

whence $\Gamma(x) \to \infty$ as $x \to 0_+$. □

The last theorem and its corollaries are helpful when we come to sketch the graph of the gamma function. See Fig. 5.3. The graph has a minimum between 1 and 2. Gauss found this minimum to be at a number whose decimal expansion begins $1\cdot 461632145\ldots$.

Since $\Gamma(n+1) = n!$ for $n = 0, 1, 2, \ldots$, the gamma function can be considered to be an extension of the factorial function, even if the two functions are one unit out of phase with each other. There are, of course, infinitely many functions $f: (0, \infty) \to \mathbb{R}$ satisfying $f(n+1) = n!$ for $n = 0, 1, 2, \ldots$, and the natural question that arises is: Is there some sense in which the gamma function is a *unique* extension of the factorial function? One answer is given in the following surprising result.

Theorem 5.3.5 (Artin's characterization of the gamma function) *Let the function* $f: (0, \infty) \to \mathbb{R}$ *satisfy: (i)* $f(x+1) = xf(x)$ *for* $x > 0$; *(ii)* $f(1) = 1$; *(iii)* f *is log-convex. Then* $f = \Gamma$.

Proof Conditions (i), (ii) imply that $f(n+1) = n!$ for $n = 0, 1, 2, \ldots$. *Let* $0 < x \leqslant 1$ and let n be any positive integer. Then the log-convexity of f and condition (i) show that

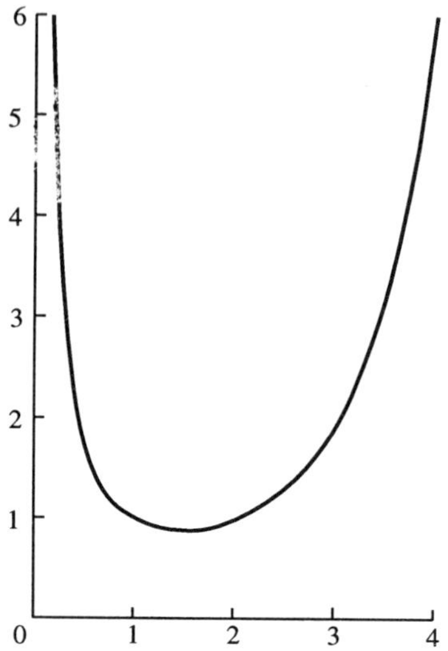

Fig. 5.3

$$\begin{aligned} f(n+1+x) &= f((1-x)(n+1)+x(n+2)) \\ &\leqslant f^{1-x}(n+1)f^{x}(n+2) \\ &= f^{1-x}(n+1)((n+1)f(n+1))^{x} \\ &= (n+1)^{x}f(n+1) = (n+1)^{x}n! \end{aligned}$$

and

$$\begin{aligned} n! = f(n+1) &= f(x(n+x)+(1-x)(n+1+x)) \\ &\leqslant f^{x}(n+x)f^{1-x}(n+1+x) \\ &= (n+x)^{-x}f^{x}(n+1+x)f^{1-x}(n+1+x) \\ &= (n+x)^{-x}f(n+1+x). \end{aligned}$$

Thus, since $f(n+1+x) = (n+x)(n-1+x)\ldots xf(x)$, we can deduce that

$$\left(1+\frac{x}{n}\right)^{x} \leqslant \frac{(n+x)(n-1+x)\ldots xf(x)}{n!n^{x}} \leqslant \left(1+\frac{1}{n}\right)^{x},$$

and hence that

$$f(x) = \lim_{n\to\infty} \frac{n!n^{x}}{(n+x)(n-1+x)\ldots x} \qquad \text{for } 0 < x \leqslant 1.$$

Suppose now that $x > 1$ and that m is the positive integer such that $0 < x - m \leqslant 1$. Then, by condition (i) and what we have just proved,

$$\begin{aligned}
f(x) &= (x-1)\dots(x-m)f(x-m)\\
&= (x-1)\dots(x-m)\lim_{n\to\infty}\frac{n!n^{x-m}}{(n+x-m)(n-1+x-m)\dots(x-m)}\\
&= \lim_{n\to\infty}\left(\frac{n!n^x}{(n+x)(n-1+x)\dots x}\cdot\frac{(n+x)(n+x-1)\dots(n+x-(m-1))}{n^m}\right)\\
&= \lim_{n\to\infty}\frac{n!n^x}{(n+x)(n-1+x)\dots x}\cdot\lim_{n\to\infty}\left(\left(1+\frac{x}{n}\right)\left(1+\frac{x-1}{n}\right)\dots\right.\\
&\qquad\qquad\left.\dots\left(1+\frac{1+x-m}{n}\right)\right)\\
&= \lim_{n\to\infty}\frac{n!n^x}{(n+x)(n-1+x)\dots x}.
\end{aligned}$$

Thus, for all $x > 0$,

$$f(x) = \lim_{n\to\infty}\frac{n!n^x}{(n+x)(n-1+x)\dots x}.$$

This is a remarkable conclusion, since it shows that f is *uniquely determined* by conditions (i), (ii), and (iii). Since Γ satisfies these three conditions, we must have $f = \Gamma$. □

The proof of the theorem yields the following important formula.

Corollary 5.3.6 For $x > 0$, $\Gamma(x) = \lim_{n\to\infty}\dfrac{n!n^x}{(n+x)\dots x}$. □

Not only is Artin's characterization of the gamma function a delightful result in itself, but it also generates some elegantly simple proofs of several properties of the gamma function which are far from obvious. We give two such examples later in this section.

Before establishing a fundamental identity linking the gamma and sine functions, we need to express $\sin x$ as an infinite product. Since the reader may not have encountered this result before, we sketch an elementary proof of it.

Theorem 5.3.7 *For all real* x, $\sin x = x\prod_{k=1}^{\infty}\left(1-\dfrac{x^2}{k^2\pi^2}\right)$.

Proof De Moivre's theorem shows that, for each positive integer n, $\sin(2n+1)\theta$ is a polynomial of degree $2n+1$ in $\sin\theta$. This polynomial has roots $\pm\sin(k\pi/(2n+1))$ for $k = 0, 1, \dots, n$. It follows that

$$\sin(2n+1)\theta = (2n+1)\sin\theta\prod_{k=1}^{n}\left(1-\frac{\sin^2\theta}{\sin^2\dfrac{k\pi}{2n+1}}\right).$$

We suppose that $x > 0$. Let m, n be integers such that $x < m < n$. The last identity shows that

$$\frac{\sin x}{(2n+1)\sin\dfrac{x}{2n+1}} = \prod_{k=1}^{n}\left(1 - \frac{\sin^2\dfrac{x}{2n+1}}{\sin^2\dfrac{k\pi}{2n+1}}\right).$$

Denote by a_k the kth factor in this last product. Since $2\theta/\pi < \sin\theta < \theta$ when $0 < \theta < \pi/2$, we find that

$$0 < 1 - \frac{x^2}{4k^2} < a_k < 1 \qquad \text{for } m < k \leqslant n,$$

whence

$$1 > a_{m+1}\ldots a_n > \prod_{k=m+1}^{n}\left(1 - \frac{x^2}{4k^2}\right) > 1 - \frac{x^2}{4}\sum_{k=m+1}^{n}\frac{1}{k^2}\cdots > 1 - \frac{x^2}{4m}.$$

Hence

$$\frac{\sin x}{(2n+1)\sin\dfrac{x}{2n+1}}$$

lies between

$$\prod_{k=1}^{m}\left(1 - \frac{\sin^2\dfrac{x}{2n+1}}{\sin^2\dfrac{k\pi}{2n+1}}\right) \quad \text{and} \quad \left(1 - \frac{x^2}{4m}\right)\prod_{k=1}^{m}\left(1 - \frac{\sin^2\dfrac{x}{2n+1}}{\sin^2\dfrac{k\pi}{2n+1}}\right),$$

and so letting $n \to \infty$, we deduce that $(\sin x)/x$ lies between

$$\prod_{k=1}^{m}\left(1 - \frac{x^2}{k^2\pi^2}\right) \quad \text{and} \quad \left(1 - \frac{x^2}{4m}\right)\prod_{k=1}^{m}\left(1 - \frac{x^2}{k^2\pi^2}\right).$$

We complete the proof by letting $m \to \infty$ in this stated inequality. □

Theorem 5.3.8 *For every real x with $0 < x < 1$,*

$$\Gamma(x)\Gamma(1-x) = \frac{\pi}{\sin \pi x}.$$

Proof Corollary 5.3.6 shows that, for $0 < x < 1$,

$$\Gamma(x)\Gamma(1-x) = \lim_{n\to\infty} \frac{n!n^x n!n^{1-x}}{(n+x)\dots x(n+1-x)\dots(1-x)}$$
$$= \frac{1}{x\prod_{k=1}^{\infty}\left(1-\frac{x^2}{k^2}\right)}$$
$$= \frac{\pi}{\sin \pi x}. \quad \square$$

Corollary 5.3.9 $\Gamma(\frac{1}{2}) = \sqrt{\pi}$. □

As an example of the effectiveness of Theorem 5.3.5, we use it to provide an elegant proof of a fundamental functional equation satisfied by the gamma function.

Theorem 5.3.10 (Legendre's duplication formula)

$$\Gamma\left(\frac{x}{2}\right)\Gamma\left(\frac{x+1}{2}\right) = \frac{\sqrt{\pi}}{2^{x-1}}\Gamma(x) \qquad \text{for } x > 0.$$

Proof Define a function $f: (0, \infty) \to \mathbb{R}$ by the equation

$$f(x) = \frac{2^{x-1}}{\sqrt{\pi}}\Gamma\left(\frac{x}{2}\right)\Gamma\left(\frac{x+1}{2}\right) \qquad \text{for } x > 0.$$

Then f is a product of log-convex functions, and so is itself log-convex. It is easily verified that $f(x+1) = xf(x)$ for $x > 0$, and $f(1) = 1$. Thus $f = \Gamma$, by Theorem 5.3.5, and the desired result is immediate. □

We now prove Stirling's formula, which is an important tool in analytical probability theory; it asserts that $n! \sim \sqrt{(2\pi)}n^{n+\frac{1}{2}}e^{-n}$, where the sign $\sim$ indicates that the ratio of the two sides tends to 1 as $n \to \infty$. First a lemma.

Lemma 5.3.11 *The sequence whose* nth *term is*

$$\log n! - (n + \tfrac{1}{2})\log n + n$$

converges.

Proof Let $a_n = \log n! - (n + \frac{1}{2})\log n + n$. First we show that the sequence (a_n) is decreasing. Then we show that it is bounded below.

We note that, for $n = 1, 2, \dots,$

$$a_n - a_{n+1} = (n + \tfrac{1}{2})\log\left(1 + \frac{1}{n}\right) - 1. \qquad (5.3.1)$$

Since $1/x$ is convex on $(0, \infty)$, the area bounded by the graph of $y = 1/x$, the x-axis, and the lines $x = n$, $x = n + 1$ exceeds that of the trapezium bounded by the tangent to $y = 1/x$ at the point $(n + \frac{1}{2}, 1/(n + \frac{1}{2}))$, the x-axis, and the lines $x = n$, $x = n + 1$; i.e.

$$\int_n^{n+1} \frac{dx}{x} = \log\left(1 + \frac{1}{n}\right) > \frac{1}{n + \frac{1}{2}}.$$

It now follows from (5.3.1) that $a_n - a_{n+1} > 0$, and hence that the sequence (a_n) is decreasing.

Since $\log x$ is concave on $(0, \infty)$, the area bounded by the graph of $y = \log x$, the x-axis, and the lines $x = r - \frac{1}{2}$, $x = r + \frac{1}{2}$ for $r = 1, 2, \ldots$, is less than that of the trapezium bounded by the tangent to $y = \log x$ at the point $(r, \log r)$, the x-axis, and the lines $x = r - \frac{1}{2}$, $x = r + \frac{1}{2}$, i.e.

$$\int_{r-\frac{1}{2}}^{r+\frac{1}{2}} \log x \, dx < \log r.$$

It follows easily that, for $n \geqslant 3$,

$$\int_1^n \log x \, dx = \int_1^{1\frac{1}{2}} \log x \, dx + \int_{1\frac{1}{2}}^{2\frac{1}{2}} \log x \, dx + \ldots + \int_{n-\frac{3}{2}}^{n-\frac{1}{2}} \log x \, dx + \int_{n-\frac{1}{2}}^{n} \log x \, dx$$

$$< \tfrac{1}{2} \log 1\tfrac{1}{2} + \log 2 + \ldots + \log(n-1) + \tfrac{1}{2} \log n$$

$$< \tfrac{1}{2} + \log n! - \tfrac{1}{2} \log n.$$

Thus

$$\int_1^n \log x \, dx = n \log n - n + 1 < \tfrac{1}{2} + \log n! - \tfrac{1}{2} \log n,$$

whence

$$a_n = \log n! - (n + \tfrac{1}{2}) \log n + n > \tfrac{1}{2}.$$

Thus the decreasing sequence (a_n) is bounded below by $\frac{1}{2}$, and so converges. □

Theorem 5.3.12 (Stirling's formula) $n! \sim \sqrt{(2\pi)} n^{n+\frac{1}{2}} e^{-n}$.

Proof In the notation of the proof of the lemma, let

$$b_n = e^{a_n} = \frac{n!}{n^{n+\frac{1}{2}} e^{-n}} \qquad \text{for } n = 1, 2, \ldots.$$

Then the sequence (b_n) converges to some $b > 0$. Thus

$$\frac{(b_n)^2}{b_{2n}} = \frac{2^{2n+\frac{1}{2}} (n!)^2}{n^{\frac{1}{2}} (2n)!} \to \frac{b^2}{b} = b \qquad \text{as } n \to \infty. \tag{5.3.2}$$

For $n = 1, 2, \ldots$, let

$$c_n = \frac{n!n^{\frac{1}{2}}}{(n+\frac{1}{2})\ldots\frac{3}{2}\frac{1}{2}}.$$

Then, by Corollary 5.3.6, $c_n \to \Gamma(\frac{1}{2}) = \sqrt{\pi}$ as $n \to \infty$. Hence

$$\frac{(b_n)^2}{b_{2n}} = c_n\left(1 + \frac{1}{2n}\right)\sqrt{2} \to \sqrt{(2\pi)} \qquad \text{as } n \to \infty. \tag{5.3.3}$$

Comparing (5.3.2) and (5.3.3), we see that $b = \sqrt{(2\pi)}$. We conclude, as required, that

$$b_n = \frac{n!}{n^{n+\frac{1}{2}}e^{-n}} \to \sqrt{(2\pi)} \qquad \text{as } n \to \infty. \quad \square$$

Closely related to the gamma function is the **beta function** B, which is the real function of two variables defined by the equation

$$B(x, y) = \int_0^1 t^{x-1}(1-t)^{y-1}\,dt \qquad \text{for } x, y > 0.$$

Theorem 5.3.13 *The beta function has the following properties*:

(i) $B(x+1, y) = \dfrac{x}{x+y}B(x, y) \qquad \text{for } x, y > 0.$

(ii) $B(x, y)$ is a log-convex function of x for each fixed $y > 0$.

(iii) $B(x, y) = \dfrac{\Gamma(x)\Gamma(y)}{\Gamma(x+y)} \qquad \text{for } x, y > 0.$

Proof (i)
$$\begin{aligned} B(x+1, y) &= \int_0^1 t^x(1-t)^{y-1}\,dt \\ &= \int_0^1 (1-t)^{x+y-1}(t/(1-t))^x\,dt \\ &= \left[\frac{-(1-t)^{x+y}}{x+y}\left(\frac{t}{1-t}\right)^x\right]_0^1 + \int_0^1 \frac{x}{x+y}t^{x-1}(1-t)^{y-1}\,dt \\ &= \frac{x}{x+y}B(x, y). \end{aligned}$$

(ii) Let $a, b, y > 0$. Let $\lambda, \mu \geqslant 0$ with $\lambda + \mu = 1$. By Theorem 5.3.1,

$$\begin{aligned} B(\lambda a + \mu b, y) &= \int_0^1 (t^{a-1}(1-t)^{y-1})^\lambda (t^{b-1}(1-t)^{y-1})^\mu\,dt \\ &\leqslant \left(\int_0^1 t^{a-1}(1-t)^{y-1}\,dt\right)^\lambda \left(\int_0^1 t^{b-1}(1-t)^{y-1}\,dt\right)^\mu \\ &= B^\lambda(a, y)B^\mu(b, y). \end{aligned}$$

Thus $B(x, y)$ is a log-convex function of x, for fixed $y > 0$.

(iii) Let $y > 0$. Define a function $f_y:(0, \infty) \to \mathbb{R}$ by the equation

$$f_y(x) = \frac{\Gamma(x+y)B(x,y)}{\Gamma(y)} \qquad \text{for } x > 0.$$

Then f_y is a product of log-convex functions, and so is itself log-convex. For $x > 0$,

$$\begin{aligned} f_y(x+1) &= \frac{\Gamma(x+y+1)B(x+1,y)}{\Gamma(y)} \\ &= \frac{\{(x+y)\Gamma(x+y)\}\{x/(x+y)\}B(x,y)}{\Gamma(y)} = xf_y(x), \end{aligned}$$

and

$$f_y(1) = \frac{\Gamma(1+y)B(1,y)}{\Gamma(y)} = y\int_0^1 (1-t)^{y-1}\,\mathrm{d}t = 1.$$

Thus $f_y = \Gamma$, by Theorem 5.3.5, and so

$$B(x,y) = \frac{\Gamma(x)\Gamma(y)}{\Gamma(x+y)} \qquad \text{for } x, y > 0. \ \square$$

Exercises 5.3

1. Prove that the function $t \log M_t(\mathbf{x}; \boldsymbol{\alpha})$, where M_t denotes the mean of order t, is a convex function of t on $\mathbb{R}$.
2. Show that $\Gamma(n + \frac{1}{2}) = ((2n)!\sqrt{\pi})/(n!4^n)$ for $n = 0, 1, 2, \ldots$.
3. Establish *Wallis' formula*:

$$\frac{\pi}{2} = \lim_{n\to\infty} \frac{2.2}{1.3}\cdot\frac{4.4}{3.5}\cdots\frac{(2n)(2n)}{(2n-1)(2n+1)}.$$

4. Prove *Weierstrass' formula* for the gamma function:

$$\Gamma(x) = \frac{e^{-\gamma x}}{x}\prod_{n=1}^{\infty}\left(1+\frac{x}{n}\right)^{-1} e^{x/n} \qquad \text{for } x > 0,$$

where $\gamma = \lim_{n\to\infty}\left(1 + \frac{1}{2} + \ldots + \frac{1}{n} - \log n\right)$ is Euler's constant.

5. Use the log-convexity of Γ to show that

$$\sqrt{\left(\frac{2}{n+1}\right)} \leqslant \frac{\Gamma(\frac{1}{2}(n+1))}{\Gamma(\frac{1}{2}(n+2))} \leqslant \sqrt{\left(\frac{2}{n}\right)} \qquad \text{for } n = 1, 2, 3, \ldots.$$

6. Show that, for all $x > 0$,

$$\Gamma\left(\frac{x}{3}\right)\Gamma\left(\frac{x+1}{3}\right)\Gamma\left(\frac{x+2}{3}\right) = \frac{2\pi\sqrt{3}}{3^x}\Gamma(x).$$

7. Evaluate the following integrals:

$$\int_0^\infty e^{-\frac{1}{2}t^2}\,dt; \qquad \int_0^\infty \sqrt{t}\,e^{-t^3}\,dt;$$

$$\int_0^\infty \frac{\sqrt{t}}{1+t^2}\,dt; \qquad \int_0^1 \frac{dt}{(1-t^{1/3})^{1/3}}.$$

8. Prove that

$$B(x,y) = 2\int_0^{\pi/2} (\sin^{2x-1} t)(\cos^{2y-1} t)\,dt \qquad \text{for } x, y > 0.$$

Deduce that, when m is a non-negative integer,

$$\int_0^{\pi/2} \sin^{2m} t\,dt = \frac{(2m)!\pi}{2^{2m+1}(m!)^2}.$$

9. Establish the following result (due to Euler):

$$\int_0^\infty \frac{t^{x-1}}{1+t}\,dt = \frac{\pi}{\sin \pi x} \qquad \text{for } 0 < x < 1.$$

10. Find the area of the set

$$\left\{(x,y): \left(\frac{|x|}{a}\right)^{1/\alpha} + \left(\frac{|y|}{b}\right)^{1/\beta} \leqslant 1\right\} \qquad \text{for } a, b, \alpha, \beta > 0.$$

5.4 CONVEX FUNCTIONS ON $\mathbb{R}^n$

In this section we study basic properties of convex functions of n real variables. The definition of a convex function on the real line generalizes in an obvious way to $\mathbb{R}^n$. A real-valued function f defined on a non-empty convex set X in $\mathbb{R}^n$ is said to be **convex** if

$$f(\lambda\mathbf{x} + \mu\mathbf{y}) \leqslant \lambda f(\mathbf{x}) + \mu f(\mathbf{y})$$

whenever $\mathbf{x}, \mathbf{y} \in X$ and $\lambda, \mu \geqslant 0$ with $\lambda + \mu = 1$. The convexity of X ensures that $\lambda\mathbf{x} + \mu\mathbf{y} \in X$. A **concave function** is one whose negative is convex. Exactly as in the case of a convex function of a single real variable, each convex function $f: X \to \mathbb{R}^n$ satisfies *Jensen's inequality*, i.e.

$$f(\lambda_1\mathbf{x}_1 + \ldots + \lambda_m\mathbf{x}_m) \leqslant \lambda_1 f(\mathbf{x}_1) + \ldots + \lambda_m f(\mathbf{x}_m),$$

whenever $\mathbf{x}_1, \ldots, \mathbf{x}_m \in X$ and $\lambda_1, \ldots, \lambda_m \geqslant 0$ with $\lambda_1 + \ldots + \lambda_m = 1$. Affine transformations $f: \mathbb{R}^n \to \mathbb{R}$ and their restrictions to non-empty convex subsets of $\mathbb{R}^n$ provide important examples of convex functions.

Many close connections exist between convex functions and convex sets. Here we study two of these; others will be discussed later. In

Section 1.9 the distance function $d_X:\mathbb{R}^n \to \mathbb{R}$ of a non-empty set X in $\mathbb{R}^n$ was defined by the equation

$$d_X(\mathbf{u}) = \inf\{\|\mathbf{u} - \mathbf{x}\| : \mathbf{x} \in X\} \qquad \text{for } \mathbf{u} \in \mathbb{R}^n.$$

We now assume that X is convex and show that in this case the resulting distance function d_X is convex. Let $\mathbf{u}, \mathbf{v} \in \mathbb{R}^n$ and let $\lambda, \mu \geqslant 0$ with $\lambda + \mu = 1$. Then, for each $\varepsilon > 0$, there exist points $\mathbf{x}, \mathbf{y} \in X$ such that

$$\|\mathbf{u} - \mathbf{x}\| \leqslant d_X(\mathbf{u}) + \varepsilon \qquad \text{and} \qquad \|\mathbf{v} - \mathbf{y}\| \leqslant d_X(\mathbf{v}) + \varepsilon.$$

Since X is convex, $\lambda\mathbf{x} + \mu\mathbf{y} \in X$, and so

$$\begin{aligned} d_X(\lambda\mathbf{u} + \mu\mathbf{v}) &\leqslant \|\lambda\mathbf{u} + \mu\mathbf{v} - (\lambda\mathbf{x} + \mu\mathbf{y})\| \\ &\leqslant \lambda\|\mathbf{u} - \mathbf{x}\| + \mu\|\mathbf{v} - \mathbf{y}\| \\ &\leqslant \lambda d_X(u) + \mu d_X(\mathbf{v}) + \varepsilon. \end{aligned}$$

But $\varepsilon > 0$ is arbitrary, whence

$$d_X(\lambda\mathbf{u} + \mu\mathbf{v}) \leqslant \lambda d_X(\mathbf{u}) + \mu d_X(\mathbf{v}),$$

which shows that d_X is a convex function.

To introduce our next connection between convex sets and convex functions, we consider the convex function $f(x_1) = x_1^2$ defined on $\mathbb{R}^1$. The graph of f is the parabola $\{(x_1, x_1^2): x_1 \in \mathbb{R}\}$ in $\mathbb{R}^2$, which is clearly *not* convex; the set of points $\{(x_1, x): x_1 \in \mathbb{R}, x \geqslant x_1^2\}$ in $\mathbb{R}^2$ which lie on or above the graph of f, however, *is* convex. Thus with this particular convex function of a single variable, we have associated a convex set in $\mathbb{R}^2$. We now show how the convexity of a real-valued function of n variables is equivalent to the convexity of a certain subset of $\mathbb{R}^{n+1}$.

Let f be a real-valued function defined on a non-empty convex set X in $\mathbb{R}^n$. Then the **graph** of f is defined to be the subset

$$\{(x_1, \ldots, x_n, f(x_1, \ldots, x_n)) : (x_1, \ldots, x_n) \in X\}$$

of $\mathbb{R}^{n+1}$. The **epigraph** of f, denoted by epi f, is defined to be the subset

$$\{(x_1, \ldots, x_n, x) : (x_1, \ldots, x_n) \in X,\ x \geqslant f(x_1, \ldots, x_n)\}$$

of $\mathbb{R}^{n+1}$.

Theorem 5.4.1 *Let f be a real-valued function defined on a non-empty convex set X in $\mathbb{R}^n$. Then f is convex if and only if its epigraph is convex.*

Proof First an item of notation. For each point $\mathbf{x} = (x_1, \ldots, x_n)$ of $\mathbb{R}^n$ and for each scalar x, we denote by $(\mathbf{x}, x)$ the point $(x_1, \ldots, x_n, x)$ of $\mathbb{R}^{n+1}$.

Suppose that f is convex. Let $(\mathbf{x}, x)$, $(\mathbf{y}, y) \in \operatorname{epi} f$, so $\mathbf{x}, \mathbf{y} \in X$ and

$x \geqslant f(\mathbf{x})$, $y \geqslant f(\mathbf{y})$. Let $\lambda, \mu \geqslant 0$ with $\lambda + \mu = 1$. Then the convexity of f shows that

$$f(\lambda\mathbf{x} + \mu\mathbf{y}) \leqslant \lambda f(\mathbf{x}) + \mu f(\mathbf{y}) \leqslant \lambda x + \mu y.$$

Thus the point

$$\lambda(\mathbf{x}, x) + \mu(\mathbf{y}, y) = (\lambda\mathbf{x} + \mu\mathbf{y}, \lambda x + \mu y)$$

belongs to epi f, and so epi f is convex.

Conversely, suppose that epi f is convex. Let $\mathbf{x}, \mathbf{y} \in X$ and let $\lambda, \mu \geqslant 0$ with $\lambda + \mu = 1$. Since epi f is convex, the point

$$\lambda(\mathbf{x}, f(\mathbf{x})) + \mu(\mathbf{y}, f(\mathbf{y})) = (\lambda\mathbf{x} + \mu\mathbf{y}, \lambda f(\mathbf{x}) + \mu f(\mathbf{y}))$$

belongs to epi f. Hence

$$f(\lambda\mathbf{x} + \mu\mathbf{y}) \leqslant \lambda f(\mathbf{x}) + \mu f(\mathbf{y}),$$

which shows that f is a convex function. □

The last theorem can be used to derive results about convex functions from properties of convex sets. We illustrate this point with a simple example, which can also be easily established directly from the definition of a convex function.

Theorem 5.4.2 *Let $(f_i : i \in I)$ be a non-empty family of convex functions defined on a non-empty convex set X in $\mathbb{R}^n$ such that, for each $\mathbf{x}$ in X, the set $\{f_i(\mathbf{x}) : i \in I\}$ of real numbers is bounded above. Then the function $f : X \to \mathbb{R}$ defined by the equation*

$$f(\mathbf{x}) = \sup\{f_i(\mathbf{x}) : i \in I\} \qquad \text{for } \mathbf{x} \in X$$

is convex.

Proof We observe that

$$\begin{aligned}
\operatorname{epi} f &= \{(x_1, \ldots, x_n, x) : (x_1, \ldots, x_n) \in X,\ x \geqslant f(x_1, \ldots, x_n)\} \\
&= \{(x_1, \ldots, x_n, x) : (x_1, \ldots, x_n) \in X \text{ and } x \geqslant f_i(x_1, \ldots, x_n) \text{ for } i \in I\} \\
&= \bigcap (\{(x_1, \ldots, x_n, x) : (x_1, \ldots, x_n) \in X,\ x \geqslant f_i(x_1, \ldots, x_n)\} : i \in I) \\
&= \bigcap (\operatorname{epi} f_i : i \in I).
\end{aligned}$$

Theorem 5.4.1 shows that all of the sets epi f_i are convex, whence so is their intersection epi f. Thus, by Theorem 5.4.1, f is a convex function. □

Theorem 5.4.2 shows that the class of convex functions is closed under the taking of supremums. We now consider some other operations under which the class of convex functions is closed.

Theorem 5.4.3 *Let f, g be convex functions defined on a non-empty convex subset X of $\mathbb{R}^n$ and let $\alpha, \beta \geqslant 0$. Then the function $\alpha f + \beta g$ is convex.*

Proof Let $\mathbf{x}, \mathbf{y} \in X$ and let $\lambda, \mu \geqslant 0$ with $\lambda + \mu = 1$. Then

$$\begin{aligned}(\alpha f + \beta g)(\lambda \mathbf{x} + \mu \mathbf{y}) &= \alpha f(\lambda \mathbf{x} + \mu \mathbf{y}) + \beta g(\lambda \mathbf{x} + \mu \mathbf{y}) \\ &\leqslant \alpha(\lambda f(\mathbf{x}) + \mu f(\mathbf{y})) + \beta(\lambda g(\mathbf{x}) + \mu g(\mathbf{y})) \\ &= \lambda(\alpha f + \beta g)(\mathbf{x}) + \mu(\alpha f + \beta g)(\mathbf{y}). \quad \square\end{aligned}$$

Theorem 5.4.4 *Let f be a convex function defined on a non-empty convex set X in $\mathbb{R}^n$ and let $g: I \to \mathbb{R}$ be an increasing convex function defined on an interval I of $\mathbb{R}$ which contains the image $f(X)$ of X under f. Then the composite function $g \circ f: X \to \mathbb{R}$ is convex.*

Proof Let $\mathbf{x}, \mathbf{y} \in X$ and let $\lambda, \mu \geqslant 0$ with $\lambda + \mu = 1$. Then

$$\begin{aligned}(g \circ f)(\lambda \mathbf{x} + \mu \mathbf{y}) &= g(f(\lambda \mathbf{x} + \mu \mathbf{y})) \\ &\leqslant g(\lambda f(\mathbf{x}) + \mu f(\mathbf{y})) \\ &\leqslant \lambda g(f(\mathbf{x})) + \mu g(f(\mathbf{y})) \\ &= \lambda(g \circ f)(\mathbf{x}) + \mu(g \circ f)(\mathbf{y}). \quad \square\end{aligned}$$

The ideas concerning the support of real-valued functions of a single variable discussed in Section 5.1 extend in the natural way to real-valued functions of n variables. Let f be a real-valued function defined on a convex set X in $\mathbb{R}^n$ and let $\mathbf{x}_0 \in X$. Then an affine transformation $T: \mathbb{R}^n \to \mathbb{R}$ is said to **support** f at $\mathbf{x}_0$ if $T(\mathbf{x}_0) = f(\mathbf{x}_0)$ and $T(\mathbf{x}) \leqslant f(\mathbf{x})$ for all $\mathbf{x} \in X$. The geometrical interpretation of T supporting f at $\mathbf{x}_0$ is clear: the set

$$\{(x_1, \ldots, x_n, T(x_1, \ldots, x_n)) : (x_1, \ldots, x_n) \in \mathbb{R}^n\}$$

is a hyperplane in $\mathbb{R}^{n+1}$ that passes through the point $(\mathbf{x}_0, f(\mathbf{x}_0))$ and lies *on* or *below* the graph

$$\{(x_1, \ldots, x_n, f(x_1, \ldots, x_n)) : (x_1, \ldots, x_n) \in X\}$$

of f.

Theorem 5.4.5 *Let f be a real-valued function defined on a non-empty open convex set X in $\mathbb{R}^n$. Then f is convex if and only if it has support at each point of X.*

Proof Suppose that f has support at each point of X. Let $\mathbf{x}, \mathbf{y} \in X$ and let $\lambda, \mu \geqslant 0$ with $\lambda + \mu = 1$. Then there is an affine transformation $T: \mathbb{R}^n \to \mathbb{R}$ which supports f at $\lambda \mathbf{x} + \mu \mathbf{y}$. Hence

$$f(\lambda \mathbf{x} + \mu \mathbf{y}) = T(\lambda \mathbf{x} + \mu \mathbf{y}) = \lambda T(\mathbf{x}) + \mu T(\mathbf{y}) \leqslant \lambda f(\mathbf{x}) + \mu f(\mathbf{y}),$$

which shows that f is convex.

Conversely, suppose that f is convex and that $\mathbf{x}_0 \in X$. Since f is convex, its epigraph epi$f$ is a convex set in $\mathbb{R}^{n+1}$. Now $(\mathbf{x}_0, f(\mathbf{x}_0))$ is a boundary point of epi$f$, and so there exists a support hyperplane H to epi$f$ at $(\mathbf{x}_0, f(\mathbf{x}_0))$. Suppose that H has equation

$$a_1x_1 + \ldots + a_nx_n + a_{n+1}x_{n+1} = a_0$$

and that

$$a_1x_1 + \ldots + a_nx_n + a_{n+1}x_{n+1} \geqslant a_0$$

whenever $(x_1, \ldots, x_n) \in X$ and $x_{n+1} \geqslant f(x_1, \ldots, x_n)$. Now $a_{n+1} \neq 0$, for $a_{n+1} = 0$ would imply that the hyperplane in $\mathbb{R}^n$ with equation

$$a_1x_1 + \ldots + a_nx_n = a_0$$

supports X at $\mathbf{x}_0$, but this is impossible because $\mathbf{x}_0$ is an interior point of X. Since, for each $(x_1, \ldots, x_n) \in X$,

$$a_1x_1 + \ldots + a_nx_n + a_{n+1}\lambda \geqslant a_0 \qquad \text{for all } \lambda \geqslant f(x_1, \ldots, x_n),$$

we deduce that $a_{n+1} > 0$. Define an affine transformation $T: \mathbb{R}^n \to \mathbb{R}$ by the equation

$$T(x_1, \ldots, x_n) = \frac{1}{a_{n+1}}(a_0 - a_1x_1 - \ldots - a_nx_n) \qquad \text{for } (x_1, \ldots, x_n) \in \mathbb{R}^n.$$

Since H supports epi$f$ at $(\mathbf{x}_0, f(\mathbf{x}_0))$ and $a_{n+1} > 0$, $T(\mathbf{x}_0) = f(\mathbf{x}_0)$ and $f(\mathbf{x}) \geqslant T(\mathbf{x})$ for all $\mathbf{x} \in X$. Thus T supports f at $\mathbf{x}_0$. □

We continue our discussion of the support of a convex function in the next section, where we show that the *uniqueness* of support to a convex function at a point is equivalent to its differentiability there.

Many of the functions which arise naturally in convexity are real-valued functions f defined on a convex cone x in $\mathbb{R}^n$ (often $\mathbb{R}^n$ itself) that satisfy the equation $f(\lambda\mathbf{x}) = \lambda f(\mathbf{x})$ for *all* $\mathbf{x} \in X$ and *all* $\lambda \geqslant 0$. Such functions are called **positively homogeneous**. The most important example of such a function is the norm mapping $\| \quad \|$, which is defined on the whole of $\mathbb{R}^n$. We shall meet more examples of positively homogeneous functions in the following two sections.

Theorem 5.4.6 *Let f be a positively homogeneous function defined on a convex cone X in $\mathbb{R}^n$. Then f is convex if and only if $f(\mathbf{x} + \mathbf{y}) \leqslant f(\mathbf{x}) + f(\mathbf{y})$ for all $\mathbf{x}, \mathbf{y} \in X$.*

Proof Suppose that f is convex. Let $\mathbf{x}, \mathbf{y} \in X$. Then

$$\tfrac{1}{2}f(\mathbf{x} + \mathbf{y}) = f(\tfrac{1}{2}\mathbf{x} + \tfrac{1}{2}\mathbf{y}) \leqslant \tfrac{1}{2}f(\mathbf{x}) + \tfrac{1}{2}f(\mathbf{y}),$$

and so $f(\mathbf{x} + \mathbf{y}) \leqslant f(\mathbf{x}) + f(\mathbf{y})$.

Conversely, suppose that $f(\mathbf{x} + \mathbf{y}) \leqslant f(\mathbf{x}) + f(\mathbf{y})$ for all $\mathbf{x}, \mathbf{y} \in X$.

Then, for all $\mathbf{x}, \mathbf{y} \in X$ and for all $\lambda, \mu \geqslant 0$ with $\lambda + \mu = 1$,

$$f(\lambda\mathbf{x} + \mu\mathbf{y}) \leqslant f(\lambda\mathbf{x}) + f(\mu\mathbf{y}) = \lambda f(\mathbf{x}) + \mu f(\mathbf{y}),$$

which shows that f is convex. □

Let f be a real-valued function defined on a non-empty convex set X in $\mathbb{R}^n$. Then, for each scalar α, the **level set L_α of f at height α** is the set defined by the equation

$$L_\alpha = \{\mathbf{x} \in X : f(\mathbf{x}) \leqslant \alpha\}.$$

We show that each level set L_α of a convex function $f: X \to \mathbb{R}$ is convex. Let $\mathbf{x}, \mathbf{y} \in L_\alpha$ and let $\lambda, \mu \geqslant 0$ with $\lambda + \mu = 1$. Then, by the convexity of f,

$$f(\lambda\mathbf{x} + \mu\mathbf{y}) \leqslant \lambda f(\mathbf{x}) + \mu f(\mathbf{y}) \leqslant \lambda\alpha + \mu\alpha = \alpha.$$

Thus $\lambda\mathbf{x} + \mu\mathbf{y} \in L_\alpha$ and L_α is convex. There exist non-convex functions *all* of whose level sets are convex: for example, the cube function defined on the real line. For a non-negative positively homogeneous function f, however, the convexity of the single level set L_1 ensures the convexity of f itself.

Theorem 5.4.7 *Let f be a non-negative positively homogeneous function defined on a convex cone X in $\mathbb{R}^n$ such that the level set $\{\mathbf{x} \in X : f(\mathbf{x}) \leqslant 1\}$ is convex. Then f is a convex function.*

Proof We use the criterion established in Theorem 5.4.6 to show that f is convex. Let $\mathbf{x}, \mathbf{y} \in X$. Choose scalars α, β such that $\alpha > f(\mathbf{x})$, $\beta > f(\mathbf{y})$. Since f is non-negative and positively homogeneous, $f(\mathbf{x}/\alpha) \leqslant 1$ and $f(\mathbf{y}/\beta) \leqslant 1$. Thus $\mathbf{x}/\alpha$ and $\mathbf{y}/\beta$ lie in the level set of f at height 1. The assumed convexity of this level set shows that

$$\frac{1}{\alpha + \beta} f(\mathbf{x} + \mathbf{y}) = f\left(\frac{\mathbf{x} + \mathbf{y}}{\alpha + \beta}\right) = f\left(\frac{\alpha}{\alpha + \beta}\frac{\mathbf{x}}{\alpha} + \frac{\beta}{\alpha + \beta}\frac{\mathbf{y}}{\beta}\right) \leqslant 1,$$

whence

$$f(\mathbf{x} + \mathbf{y}) \leqslant \alpha + \beta \qquad \text{whenever } \alpha > f(\mathbf{x}),\ \beta > f(\mathbf{y}).$$

Hence $f(\mathbf{x} + \mathbf{y}) \leqslant f(\mathbf{x}) + f(\mathbf{y})$, which shows that f is convex. □

We conclude this section with an example of the use of positively homogeneous functions. Let $p \geqslant 1$. Define a function f on the non-negative orthant X of $\mathbb{R}^n$ by the equation

$$f(x_1, \ldots, x_n) = (x_1^p + \ldots + x_n^p)^{1/p} \qquad \text{for } x_1, \ldots, x_n \geqslant 0.$$

Then f is non-negative and positively homogeneous. It follows from

Theorem 5.4.3 and the fact that the function x^p is convex on the interval $[0, \infty)$, that the function $f^p : X \to \mathbb{R}^n$ is convex. Hence the level set

$$\{\mathbf{x} \in X : f^p(\mathbf{x}) \leqslant 1\} = \{\mathbf{x} \in X : f(\mathbf{x}) \leqslant 1\}$$

is convex. By Theorem 5.4.7, f is convex. Let $\mathbf{x} = (x_1, \ldots, x_n)$, $\mathbf{y} = (y_1, \ldots, y_n)$ belong to X. Then, by Theorem 5.4.6, $f(\mathbf{x} + \mathbf{y}) \leqslant f(\mathbf{x}) + f(\mathbf{y})$, i.e.

$$((x_1 + y_1)^p + \ldots + (x_n + y_n)^p)^{1/p} \leqslant (x_1^p + \ldots + x_n^p)^{1/p} + (y_1^p + \ldots + y_n^p)^{1/p},$$

and we have re-proved Minkowski's inequality.

Exercises 5.4

1. The convex functions $f_1, \ldots, f_m$ defined on a non-empty convex set X in $\mathbb{R}^n$ are such that there exists no $\mathbf{x}_0 \in X$ such that $f_1(\mathbf{x}_0) < 0, \ldots, f_m(\mathbf{x}_0) < 0$. Show that there exist scalars $\lambda_1, \ldots, \lambda_m \geqslant 0$, not all zero, such that, for all $\mathbf{x} \in X$,

$$\lambda_1 f_1(\mathbf{x}) + \ldots + \lambda_m f_m(\mathbf{x}) \geqslant 0.$$

2. Prove that there exists no convex function $f : (0, \infty) \to \mathbb{R}$ such that $f(x) \leqslant \log x$ for all $x > 0$.
3. Let C be a non-empty convex set in $\mathbb{R}^{n+1}$. Denote by X the convex subset of $\mathbb{R}^n$ defined by the equation

$$X = \{\mathbf{x} \in \mathbb{R}^n : (\mathbf{x}, c) \in C \text{ for some } c \in \mathbb{R}\}.$$

Suppose that, for each $\mathbf{x} \in X$, the set

$$\{c \in \mathbb{R} : (\mathbf{x}, c) \in C\}$$

of real numbers is bounded below. Show that the function $f : X \to \mathbb{R}$, defined by the equation

$$f(\mathbf{x}) = \inf \{c \in \mathbb{R} : (\mathbf{x}, c) \in C\} \qquad \text{for } \mathbf{x} \in X,$$

is convex.
4. Let A be a non-empty bounded set in $\mathbb{R}^n$. Define a function $f : \mathbb{R}^n \to \mathbb{R}$ by the equation

$$f(\mathbf{x}) = \sup \{\mathbf{x} \cdot \mathbf{a} : \mathbf{a} \in A\} \qquad \text{for } \mathbf{x} \in \mathbb{R}^n.$$

Prove that f is convex and positively homogeneous.
5. A function $f : X \to \mathbb{R}$ defined on a non-empty convex set X in $\mathbb{R}^n$ is said to be **quasi-convex** if, for every real number α, the level set $\{\mathbf{x} \in X : f(\mathbf{x}) \leqslant \alpha\}$ is convex. Let $\alpha_1, \ldots, \alpha_n > 0$. Show that the function f defined on the positive orthant of $\mathbb{R}^n$ by the equation

$$f(x_1, \ldots, x_n) = -x_1^{\alpha_1} \ldots x_n^{\alpha_n} \qquad \text{for } x_1, \ldots, x_n > 0$$

is quasi-convex, but not convex, when $\alpha_1 + \ldots + \alpha_n > 1$; and is convex when $\alpha_1 + \ldots + \alpha_n \leqslant 1$.

5.5 CONTINUITY AND DIFFERENTIABILITY OF CONVEX FUNCTIONS

In Section 5.1 we showed that a convex function defined on an open interval of $\mathbb{R}$ is continuous, has one-sided derivatives everywhere on its domain, and is differentiable at a point if and only if it has unique support there. Moreover, we also saw that a twice differentiable function defined on an open interval of $\mathbb{R}$ is convex if and only if its second derivative is non-negative. Here we establish the corresponding results for a real-valued function defined on an open convex set in $\mathbb{R}^n$.

Let f be a convex function defined on an open convex set X in $\mathbb{R}^n$. Let $\mathbf{x} \in X$ and $\mathbf{y} \in \mathbb{R}^n$. Then the set

$$I = \{\lambda \in \mathbb{R} : \mathbf{x} + \lambda\mathbf{y} \in X\}$$

is an open interval of $\mathbb{R}$ which contains the origin, and the function $g: I \to \mathbb{R}$ defined by the equation

$$g(\lambda) = f(\mathbf{x} + \lambda\mathbf{y}) \qquad \text{for } \lambda \in I$$

is convex. To see that g is convex, let $a, b \in I$ and let $\lambda, \mu \geqslant 0$ with $\lambda + \mu = 1$. Then

$$\begin{aligned} g(\lambda a + \mu b) &= f(\mathbf{x} + (\lambda a + \mu b)\mathbf{y}) \\ &= f(\lambda(\mathbf{x} + a\mathbf{y}) + \mu(\mathbf{x} + b\mathbf{y})) \\ &\leqslant \lambda f(\mathbf{x} + a\mathbf{y}) + \mu f(\mathbf{x} + b\mathbf{y}) \\ &= \lambda g(a) + \mu g(b), \end{aligned}$$

which shows that g is convex. Thus

$$g'_+(0) = \lim_{\lambda \to 0_+} \frac{g(\lambda) - g(0)}{\lambda} = \lim_{\lambda \to 0_+} \frac{f(\mathbf{x} + \lambda\mathbf{y}) - f(\mathbf{x})}{\lambda}$$

exists, a result that we shall use frequently in this section.

Theorem 5.5.1 *Let f be a convex function defined on a non-empty open convex set X in $\mathbb{R}^n$. Then f is continuous on X.*

Proof Let $\mathbf{x}_0 \in X$ and let $\mathbf{y}_1, \ldots, \mathbf{y}_m$ be the vertices of some polytope P which is contained in X and has $\mathbf{x}_0$ as an interior point. Choose $r > 0$ such that $B[\mathbf{x}_0; r] \subseteq P$. Each point $\mathbf{x}$ of $B[\mathbf{x}_0; r]$ can be expressed in the form

$$\mathbf{x} = \lambda_1\mathbf{y}_1 + \ldots + \lambda_m\mathbf{y}_m \qquad \text{for some } \lambda_1, \ldots, \lambda_m \geqslant 0 \text{ with } \lambda_1 + \ldots + \lambda_m = 1.$$

By Jensen's inequality applied to f,

$$f(\mathbf{x}) \leqslant \lambda_1 f(\mathbf{y}_1) + \ldots + \lambda_m f(\mathbf{y}_m) \leqslant \lambda_1 M + \ldots + \lambda_m M = M,$$

where $M = \max\{f(\mathbf{y}_1), \ldots, f(\mathbf{y}_m)\}$. Hence f is bounded above by M on the closed ball $B[\mathbf{x}_0; r]$.

Let $\mathbf{x} \in \mathbb{R}^n$ satisfy the inequalities $0 < \|\mathbf{x} - \mathbf{x}_0\| \leqslant r$. Then the function $g: [-r, r] \to \mathbb{R}$ defined by the equation

$$g(t) = f\left(\mathbf{x}_0 + t\frac{\mathbf{x} - \mathbf{x}_0}{\|\mathbf{x} - \mathbf{x}_0\|}\right) \qquad \text{for } -r \leqslant t \leqslant r$$

is convex, and $g(t) \leqslant M$ for $-r \leqslant t \leqslant r$. By Corollary 5.1.2,

$$-\frac{(M-g(0))}{r} \leqslant \frac{g(-r)-g(0)}{-r} \leqslant \frac{g(\|\mathbf{x}-\mathbf{x}_0\|)-g(0)}{\|\mathbf{x}-\mathbf{x}_0\|} \leqslant \frac{g(r)-g(0)}{r} \leqslant \frac{M-g(0)}{r},$$

whence

$$|f(\mathbf{x}) - f(\mathbf{x}_0)| \leqslant \frac{M - f(\mathbf{x}_0)}{r}\|\mathbf{x} - \mathbf{x}_0\|.$$

Thus, if $\mathbf{x}_1, \ldots, \mathbf{x}_k, \ldots$ is a sequence of points of X that converges to $\mathbf{x}_0$, then $f(\mathbf{x}_k) \to f(\mathbf{x}_0)$ as $k \to \infty$ and f is continuous at $\mathbf{x}_0$. □

Before discussing the differentiability of convex functions, we need to introduce the notion of a *directional derivative*. Let f be a real-valued function defined on an open set X in $\mathbb{R}^n$ and let $\mathbf{x} = (x_1, \ldots, x_n)$ be a point of X. We recall that the ith *partial derivative* $\partial f/\partial x_i$ of f at $\mathbf{x}$, when it exists, is the derivative at x_i of the function of a single variable obtained by regarding f as a function of its ith variable only, the remaining $n - 1$ variables being held fixed to their values at $\mathbf{x}$. Thus, for $i = 1, \ldots, n$,

$$\frac{\partial f}{\partial x_i}(\mathbf{x}) = \lim_{\lambda \to 0} \frac{f(x_1, \ldots, x_{i-1}, x_i + \lambda, x_{i+1}, \ldots, x_n) - f(x_1, \ldots, x_n)}{\lambda}$$

or, more succinctly,

$$\frac{\partial f}{\partial x_i}(\mathbf{x}) = \lim_{\lambda \to 0} \frac{f(\mathbf{x} + \lambda \mathbf{e}_i) - f(\mathbf{x})}{\lambda},$$

where $\mathbf{e}_i$ denotes the ith elementary vector in $\mathbb{R}^n$. A directional derivative is a natural generalization of a partial derivative; we simply consider the above limit with an arbitrary vector $\mathbf{y}$ in $\mathbb{R}^n$ replacing the vector $\mathbf{e}_i$. The **directional derivative** of f **at x relative to y** is defined to be the limit

$$\lim_{\lambda \to 0} \frac{f(\mathbf{x} + \lambda \mathbf{y}) - f(\mathbf{x})}{\lambda},$$

whenever this limit exists. Thus the partial derivative $\partial f/\partial x_i$ is simply the directional derivative of f relative to $\mathbf{e}_i$.

A convex function defined on an open interval of $\mathbb{R}$ need not be differentiable, but it always possesses both one-sided derivatives. It is for this reason that we shall be particularly interested in one-sided directional derivatives. With the notation as above, the **one-sided directional derivative of f at $\mathbf{x}$ relative to $\mathbf{y}$** is defined to be the limit

$$f'(\mathbf{x}; \mathbf{y}) = \lim_{\lambda \to 0_+} \frac{f(\mathbf{x} + \lambda \mathbf{y}) - f(\mathbf{x})}{\lambda},$$

provided that this limit exists. Since

$$-f'(\mathbf{x}; -\mathbf{y}) = \lim_{\lambda \to 0_-} \frac{f(\mathbf{x} + \lambda \mathbf{y}) - f(\mathbf{x})}{\lambda},$$

it follows that the directional derivative of f at $\mathbf{x}$ relative to $\mathbf{y}$ exists if and only if both of the one-sided directional derivatives $f'(\mathbf{x}; \mathbf{y})$ and $f'(\mathbf{x}; -\mathbf{y})$ exist and satisfy the relation $f'(\mathbf{x}; \mathbf{y}) = -f'(\mathbf{x}; -\mathbf{y})$. If, for some $\mathbf{x} \in X$, the one-sided directional derivative $f'(\mathbf{x}; \mathbf{y})$ exists for each $\mathbf{y} \in \mathbb{R}^n$, we write $f'(\mathbf{x};\ \ )$ to denote the function $f'(\mathbf{x};\ \ ): \mathbb{R}^n \to \mathbb{R}$ whose value at $\mathbf{y}$ is $f'(\mathbf{x}; \mathbf{y})$. Our remarks immediately preceding Theorem 5.5.1 show that, for each convex function $f: X \to \mathbb{R}^n$, the one-sided directional derivative $f'(\mathbf{x}; \mathbf{y})$ exists for every $\mathbf{x}$ in the interior of X and for all $\mathbf{y}$ in $\mathbb{R}^n$.

We illustrate the above ideas by considering the convex function $f: \mathbb{R}^n \to \mathbb{R}$ defined, for each $\mathbf{x} = (x_1, \ldots, x_n)$, by

$$f(\mathbf{x}) = \|\mathbf{x}\|^2 = x_1^2 + \ldots + x_n^2.$$

Then, for each $\mathbf{y} = (y_1, \ldots, y_n)$ in $\mathbb{R}^n$,

$$\begin{aligned} f'(\mathbf{x}; \mathbf{y}) &= \lim_{\lambda \to 0_+} \frac{f(\mathbf{x} + \lambda \mathbf{y}) - f(\mathbf{x})}{\lambda} \\ &= \lim_{\lambda \to 0_+} \frac{2\lambda(x_1 y_1 + \ldots + x_n y_n) + \lambda^2(y_1^2 + \ldots + y_n^2)}{\lambda} \\ &= 2x_1 y_1 + \ldots + 2x_n y_n \\ &= 2\mathbf{x} \cdot \mathbf{y}. \end{aligned}$$

Thus $f'(\mathbf{x}; \mathbf{y})$ exists and equals $2\mathbf{x} \cdot \mathbf{y}$. We note that, for this particular function, the (two-sided) directional derivative of f at $\mathbf{x}$ relative to $\mathbf{y}$ exists; also the one-sided derivative $f'(\mathbf{x};\ \ ): \mathbb{R}^n \to \mathbb{R}$ is linear for each $\mathbf{x}$ in $\mathbb{R}^n$.

Theorem 5.5.2 *Let f be a convex function defined on an open convex set X in $\mathbb{R}^n$ and let $\mathbf{x} \in X$. Then $f'(\mathbf{x};\quad)$ is a positively homogeneous convex function such that $f'(\mathbf{x};\mathbf{y}) \geqslant -f'(\mathbf{x};-\mathbf{y})$ for all $\mathbf{y}$ in $\mathbb{R}^n$. If f has a directional derivative at $\mathbf{x}$ relative to $\mathbf{y}$, then $f'(\mathbf{x};\lambda\mathbf{y}) = \lambda f'(\mathbf{x};\mathbf{y})$ for all scalars λ.*

Proof Let $\mu > 0$ and let $\mathbf{y} \in \mathbb{R}^n$. Then

$$f'(\mathbf{x};\mu\mathbf{y}) = \lim_{\lambda\to 0_+}\frac{f(\mathbf{x}+\lambda\mu\mathbf{y})-f(\mathbf{x})}{\lambda} = \lim_{\lambda\to 0_+}\mu\cdot\frac{f(\mathbf{x}+\lambda\mu\mathbf{y})-f(\mathbf{x})}{\lambda\mu} = \mu f'(\mathbf{x};\mathbf{y}),$$

which shows that $f'(\mathbf{x};\quad)$ is positively homogeneous.
Let $\mathbf{y},\mathbf{z} \in \mathbb{R}^n$. By the convexity of f,

$$\begin{aligned} f'(\mathbf{x};\mathbf{y}+\mathbf{z}) &= \lim_{\lambda\to 0_+}\frac{f(\mathbf{x}+\lambda(\mathbf{y}+\mathbf{z}))-f(\mathbf{x})}{\lambda} \\ &\leqslant \lim_{\lambda\to 0_+}\left(\frac{1}{2}\frac{f(\mathbf{x}+2\lambda\mathbf{y})-f(\mathbf{x})}{\lambda}+\frac{1}{2}\frac{f(\mathbf{x}+2\lambda\mathbf{z})-f(\mathbf{x})}{\lambda}\right) \\ &= f'(\mathbf{x};\mathbf{y}) + f'(\mathbf{x};\mathbf{z}). \end{aligned}$$

Theorem 5.4.6 shows that $f'(\mathbf{x};\quad)$ is convex.
By what we have just proved, for each $\mathbf{y}$ in $\mathbb{R}^n$,

$$0 = f'(\mathbf{x};\mathbf{0}) = f'(\mathbf{x};\mathbf{y}-\mathbf{y}) \leqslant f'(\mathbf{x};\mathbf{y}) + f'(\mathbf{x};-\mathbf{y}),$$

whence $f'(\mathbf{x};\mathbf{y}) \geqslant -f'(\mathbf{x};-\mathbf{y})$.
Suppose, finally, that f has a directional derivative at $\mathbf{x}$ relative to $\mathbf{y}$. Then $f'(\mathbf{x};\mathbf{y}) = -f'(\mathbf{x};-\mathbf{y})$. If $\lambda < 0$, then, since f is positively homogeneous,

$$f'(\mathbf{x};\lambda\mathbf{y}) = f'(\mathbf{x};(-\lambda)(-\mathbf{y})) = -\lambda f'(\mathbf{x};-\mathbf{y}) = \lambda f'(\mathbf{x};\mathbf{y}),$$

whence $f'(\mathbf{x};\lambda\mathbf{y}) = \lambda f'(\mathbf{x};\mathbf{y})$ for *all* scalars λ. □

Suppose now that f is a real-valued function defined on an open set X in $\mathbb{R}^n$ and that $\mathbf{x}$ is a point of X. We recall that f is *differentiable* at $\mathbf{x}$ if there exists a vector $\mathbf{x}'$ (necessarily unique) such that

$$\lim_{\mathbf{u}\to\mathbf{0}}\frac{f(\mathbf{x}+\mathbf{u})-f(\mathbf{x})-\mathbf{x}'\cdot\mathbf{u}}{\|\mathbf{u}\|} = 0;$$

when such an $\mathbf{x}'$ exists it is called the *gradient* of f at $\mathbf{x}$. Let f be differentiable at $\mathbf{x}$ with gradient $\mathbf{x}'$ there. Then, for any non-zero vector $\mathbf{y}$ in $\mathbb{R}^n$,

$$\lim_{\lambda\to 0}\frac{|f(\mathbf{x}+\lambda\mathbf{y})-f(\mathbf{x})-\mathbf{x}'\cdot(\lambda\mathbf{y})|}{\|\lambda\mathbf{y}\|} = \lim_{\lambda\to 0}\frac{1}{\|\mathbf{y}\|}\left|\frac{f(\mathbf{x}+\lambda\mathbf{y})-f(\mathbf{x})}{\lambda}-\mathbf{x}'\cdot\mathbf{y}\right| = 0,$$

which shows that f possesses a directional derivative at $\mathbf{x}$ relative to $\mathbf{y}$

and that $f'(\mathbf{x};\mathbf{y}) = \mathbf{x}'\cdot\mathbf{y}$, so $f'(\mathbf{x};\quad)$ is linear. The existence of the directional derivatives of f at $\mathbf{x}$ relative to *all* points $\mathbf{y}$ in $\mathbb{R}^n$ neither guarantees that f is differentiable nor that $f'(\mathbf{x};\quad)$ is linear. When f is convex, however, we do have the following result.

Theorem 5.5.3 *Suppose that a convex function f defined on an open convex set X in $\mathbb{R}^n$ possesses all its partial derivatives $\partial f/\partial x_1, \ldots, \partial f/\partial x_n$ at some point $\mathbf{x}$ of X. Then f is differentiable at $\mathbf{x}$.*

Proof Let $r > 0$ be such that $B(\mathbf{x}; r) \subseteq X$. For each $\mathbf{u} = (u_1, \ldots, u_n)$ in $B(\mathbf{0}; r)$, let

$$\psi(\mathbf{u}) = f(\mathbf{x}+\mathbf{u}) - f(\mathbf{x}) - \left(\frac{\partial f}{\partial x_1}u_1 + \ldots + \frac{\partial f}{\partial x_n}u_n\right).$$

Then ψ is convex on $B(\mathbf{0}; r)$. For each $i = 1, \ldots, n$, define a function θ_i on $B(\mathbf{0}; r)$ at a point $\mathbf{u} = (u_1, \ldots, u_n)$ of $B(\mathbf{0}; r)$ as follows:

$$\theta_i(\mathbf{u}) = \begin{cases} (\psi(u_i\mathbf{e}_i))/u_i & \text{for } u_i \neq 0 \\ 0 & \text{for } u_i = 0. \end{cases}$$

Then $\theta_i(\mathbf{u}) \to 0$ as $\mathbf{u} \to \mathbf{0}$. For each $\mathbf{u} = (u_1, \ldots, u_n)$ such that $n\|\mathbf{u}\| < r$, Jensen's inequality applied to the convex function ψ shows that

$$\begin{aligned} \psi(\mathbf{u}) &= \psi\left(\frac{1}{n}(nu_1\mathbf{e}_1) + \ldots + \frac{1}{n}(nu_n\mathbf{e}_n)\right) \\ &\leqslant \frac{1}{n}\psi(nu_1\mathbf{e}_1) + \ldots + \frac{1}{n}\psi(nu_n\mathbf{e}_n) \\ &= u_1\theta_1(n\mathbf{u}) + \ldots + u_n\theta_n(n\mathbf{u}) \\ &\leqslant \|\mathbf{u}\|(|\theta_1(n\mathbf{u})| + \ldots + |\theta_n(n\mathbf{u})|). \end{aligned}$$

Since

$$0 = \psi(\tfrac{1}{2}\mathbf{u} + \tfrac{1}{2}(-\mathbf{u})) \leqslant \tfrac{1}{2}\psi(\mathbf{u}) + \tfrac{1}{2}\psi(-\mathbf{u}), \qquad \psi(\mathbf{u}) \geqslant -\psi(-\mathbf{u}).$$

Thus

$$-\|\mathbf{u}\|(|\theta_1(-n\mathbf{u})| + \ldots + |\theta_n(-n\mathbf{u})|) \leqslant \psi(\mathbf{u}) \leqslant \|\mathbf{u}\|(|\theta_1(n\mathbf{u})| + \ldots + |\theta_n(n\mathbf{u})|),$$

and so

$$(\psi(\mathbf{u}))/\|\mathbf{u}\| \to 0 \qquad \text{as } \mathbf{u} \to \mathbf{0}.$$

This shows that f is differentiable at $\mathbf{x}$ with gradient $(\partial f/\partial x_1, \ldots, \partial f/\partial x_n)$ there. □

The next result is the natural generalization of Theorem 5.1.9 to convex functions of n variables.

Theorem 5.5.4 *Let f be a convex function defined on an open convex set X in $\mathbb{R}^n$. Then f is differentiable at a point $\mathbf{x}_0$ of X if and only if it has unique support at $\mathbf{x}_0$.*

Proof Suppose that f is differentiable at $\mathbf{x}_0$. Let $T:\mathbb{R}^n \to \mathbb{R}$ be a support for f at $\mathbf{x}_0$. Then there exists $\mathbf{x}' \in \mathbb{R}^n$ such that, for all $\mathbf{x} \in \mathbb{R}^n$, $T(\mathbf{x}_0 + \mathbf{x}) = f(\mathbf{x}_0) + \mathbf{x}' \cdot \mathbf{x}$. Let $\mathbf{y} \in \mathbb{R}^n$. Then, for all sufficiently small $\lambda > 0$,

$$f(\mathbf{x}_0 + \lambda\mathbf{y}) - f(\mathbf{x}_0) \geqslant \lambda\mathbf{x}' \cdot \mathbf{y},$$

whence $f'(\mathbf{x}_0; \mathbf{y}) \geqslant \mathbf{x}' \cdot \mathbf{y}$. Replacing $\mathbf{y}$ by $-\mathbf{y}$ in this last inequality and using the fact that f is differentiable at $\mathbf{x}_0$, we deduce that

$$-f'(\mathbf{x}_0; \mathbf{y}) = f'(\mathbf{x}_0; -\mathbf{y}) \geqslant -\mathbf{x}' \cdot \mathbf{y},$$

and hence that $f'(\mathbf{x}_0; \mathbf{y}) = \mathbf{x}' \cdot \mathbf{y}$. It follows that

$$\mathbf{x}' = (f'(\mathbf{x}_0; \mathbf{e}_1), \ldots, f'(\mathbf{x}_0; \mathbf{e}_n)),$$

whence f has *unique* support T at $\mathbf{x}_0$.

Suppose next that f has unique support $T:\mathbb{R}^n \to \mathbb{R}$ at $\mathbf{x}_0$. Let m be any real number satisfying the inequalities

$$-f'(\mathbf{x}_0; -\mathbf{e}_1) \leqslant m \leqslant f'(\mathbf{x}_0; \mathbf{e}_1)$$

and let L be the line in $\mathbb{R}^{n+1}$ defined by the equation

$$L = \{(\mathbf{x}_0 + t\mathbf{e}_1, f(\mathbf{x}_0) + mt): t \in \mathbb{R}\}.$$

Then L meets the epigraph of f at $(\mathbf{x}_0, f(\mathbf{x}_0))$ but does not meet its interior. This last result follows from the inequality

$$f(\mathbf{x}_0) + mt \leqslant f(\mathbf{x}_0 + t\mathbf{e}_1) \qquad \text{for } \mathbf{x}_0 + t\mathbf{e}_1 \in X,$$

which can be established by an argument similar to that used in the second part of the proof of Theorem 5.1.8. Corollary 2.4.11 shows that there is a support hyperplane to the epigraph of f at $(\mathbf{x}_0, f(\mathbf{x}_0))$ which contains L. The uniqueness of the support to f at $\mathbf{x}_0$ shows that this support hyperplane must be the graph of T, whence

$$T(\mathbf{x}_0 + t\mathbf{e}_1) = f(\mathbf{x}_0) + mt = T(\mathbf{x}_0) + mt \qquad \text{for } t \in \mathbb{R}.$$

Thus m is uniquely determined by T, whence, by the choice of m,

$$-f'(\mathbf{x}_0; -\mathbf{e}_1) = f'(\mathbf{x}_0; \mathbf{e}_1),$$

which shows that the partial derivative $\partial f/\partial x_1$ at $\mathbf{x}_0$ exists. Similarly, the partial derivatives $\partial f/\partial x_2, \ldots, \partial f/\partial x_n$ exist. Hence f is differentiable by Theorem 5.5.3. □

We now establish a criterion for the convexity of a function of n variables in terms of its second-order partial derivatives, which generalizes the result that a twice differentiable function of a single variable is convex if and only if its second derivative is non-negative.

Theorem 5.5.5 *Let f be a real-valued function which is defined and has continuous second-order partial derivatives on a non-empty convex set X in $\mathbb{R}^n$. Then f is convex if and only if, for every $\mathbf{x} \in X$,*

$$\sum_{i=1}^{n} \sum_{j=1}^{n} \left[\frac{\partial^2 f}{\partial x_i \partial x_j} \right]_{\mathbf{x}} z_i z_j \geqslant 0$$

for all $(z_1, \ldots, z_n) \in \mathbb{R}^n$.

Proof Let $\mathbf{y} \in X$ and $\mathbf{z} = (z_1, \ldots, z_n) \in \mathbb{R}^n$. Let I be the open interval of $\mathbb{R}$ defined by the equation

$$I = \{\lambda \in \mathbb{R} : \mathbf{y} + \lambda \mathbf{z} \in X\}.$$

We have already seen that the function $g: I \to \mathbb{R}$ defined by the equation

$$g(\lambda) = f(\mathbf{y} + \lambda \mathbf{z}) \qquad \text{for } \lambda \in I$$

is convex when f is. Conversely, suppose that each such function g is convex; we show that this implies that f is convex. Let $\mathbf{x}, \mathbf{y} \in X$ and let $0 \leqslant \lambda \leqslant 1$. Write $\mathbf{z} = \mathbf{x} - \mathbf{y}$. Since g is convex,

$$\begin{aligned} f(\lambda \mathbf{x} + (1 - \lambda)\mathbf{y}) &= g((1 - \lambda)0 + \lambda . 1) \\ &\leqslant (1 - \lambda)g(0) + \lambda g(1) \\ &= \lambda f(\mathbf{x}) + (1 - \lambda) f(\mathbf{y}), \end{aligned}$$

which shows that f is convex. Thus f is convex on X if and only if each function g (as above) is convex on I.

Since f has continuous second-order partial derivatives on X, each function g is differentiable twice on I. The first two derivatives of g can be calculated from the chain rule for functions of n variables; indeed they are:

$$g'(\lambda) = \sum_{j=1}^{n} \frac{\partial f}{\partial x_j} z_j,$$

$$g''(\lambda) = \sum_{i=1}^{n} \sum_{j=1}^{n} \frac{\partial^2 f}{\partial x_i \partial x_j} z_i z_j,$$

where $\lambda \in I$ and the partial derivatives are evaluated at the point $\mathbf{x} = \mathbf{y} + \lambda \mathbf{z}$. The desired result now follows from Corollary 5.1.7. □

Suppose now that f is as in the last theorem. Then the $n \times n$ matrix whose (i, j)th element is $\partial^2 f / \partial x_i \partial x_j$ evaluated at a point $\mathbf{x}$ of X is called the **Hessian matrix** of f at $\mathbf{x}$. The conditions which we have imposed upon f ensure that this matrix is symmetric. We have thus proved that

f is convex on X if and only if its Hessian matrix is *non-negative semidefinite* at each point of X.

Exercises 5.5

1. Let f be a convex function defined on an open convex set X in $\mathbb{R}^n$ and let A be a non-empty compact subset of X. Show that, for some $\lambda > 0$, the λ-neighbourhood $(A)_\lambda$ of A is contained in X. Deduce that $|f(\mathbf{x}) - f(\mathbf{y})| \leqslant (2M/\lambda)\|\mathbf{x} - \mathbf{y}\|$ for $\mathbf{x}, \mathbf{y} \in A$, where $M = \sup\{|f(\mathbf{x})| : \mathbf{x} \in (A)_\lambda\}$.
2. Let f be a differentiable real-valued function defined on a non-empty open convex set X in $\mathbb{R}^n$. Prove that f is convex if and only if, for all $\mathbf{x}, \mathbf{y} \in X$,
$$f(\mathbf{y}) - f(\mathbf{x}) \geqslant f'(\mathbf{x}; \mathbf{y} - \mathbf{x}).$$
3. Write down the Hessian matrix of a real-valued function f, defined on a non-empty open convex set X in $\mathbb{R}^2$, which has continuous second-order partial derivatives. Hence give simple conditions on the second-order partial derivatives of f which will ensure its convexity.
4. The function f is defined on the positive orthant of $\mathbb{R}^n$ by the equation
$$f(x_1, \ldots, x_n) = -x_1^{\alpha_1} \ldots x_n^{\alpha_n} \qquad \text{for } x_1, \ldots, x_n > 0,$$
where $\alpha_1, \ldots, \alpha_n > 0$ and $\alpha_1 + \ldots + \alpha_n \leqslant 1$. Find the Hessian matrix of f and deduce that f is convex.

5.6 SUPPORT FUNCTIONS

We have seen on many occasions what an important role the support properties of convex sets play in convexity. In this section we are led, by considering such properties, to associate with each non-empty compact convex set a positively homogeneous convex function known as its *support function*. One reason for introducing these functions is that it is often easier to deal with them than it is with the sets themselves; we shall see this to be so when we discuss the theory of mixed volumes in the next chapter.

Let A be a non-empty compact convex set in $\mathbb{R}^n$ and let $\mathbf{u}$ be a non-zero vector in $\mathbb{R}^n$. For each real number α, denote by H_α the hyperplane defined by the equation
$$H_\alpha = \{\mathbf{x} \in \mathbb{R}^n : \mathbf{u} \cdot \mathbf{x} = \alpha\},$$
and by H_α^- the closed halfspace defined by the equation

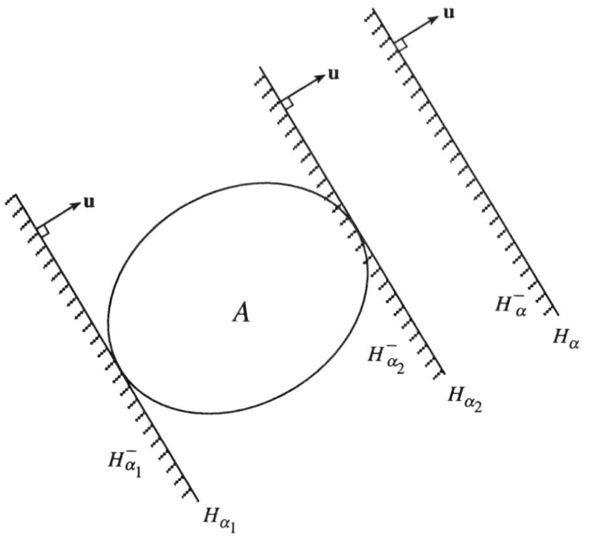

Fig. 5.4

$$H_{\alpha}^{-} = \{\mathbf{x} \in \mathbb{R}^n : \mathbf{u} \cdot \mathbf{x} \leqslant \alpha\}.$$

As α increases, the hyperplane H_{α} describes a family of parallel hyperplanes each having $\mathbf{u}$ as a normal vector. See Fig. 5.4. In general, there will be two values of α for which the hyperplane H_{α} supports A; these values are α_1 and α_2 in Fig. 5.4. Only one of these, α_2 in the figure, will be such that $A \subseteq H_{\alpha}^{-}$. Clearly $A \subseteq H_{\alpha}^{-}$ if and only if $\mathbf{u} \cdot \mathbf{a} \leqslant \alpha$ for all $\mathbf{a}$ in A, i.e. if and only if

$$\sup\{\mathbf{u} \cdot \mathbf{a} : \mathbf{a} \in A\} \leqslant \alpha.$$

If, in addition to the requirement $A \subseteq H_{\alpha}^{-}$, it is also demanded that H_{α} supports A, then, for some point $\mathbf{a}_0$ of A, $\mathbf{u} \cdot \mathbf{a}_0 = \alpha$. Thus H_{α} is a support hyperplane to A such that $A \subseteq H_{\alpha}^{-}$ if and only if

$$\alpha = \sup\{\mathbf{u} \cdot \mathbf{a} : \mathbf{a} \in A\}.$$

The following definition should now seem a natural one. The **support function** h, or more precisely h_A, of a non-empty compact convex set A in $\mathbb{R}^n$ is defined by the equation

$$h(\mathbf{u}) = \sup\{\mathbf{u} \cdot \mathbf{a} : \mathbf{a} \in A\} \qquad \text{for each } \mathbf{u} \text{ in } \mathbb{R}^n.$$

Since A is non-empty and bounded, for each $\mathbf{u}$ in $\mathbb{R}^n$, the subset $\{\mathbf{u} \cdot \mathbf{a} : \mathbf{a} \in A\}$ of $\mathbb{R}$ is non-empty and bounded, whence $h(\mathbf{u})$ is well defined. The above definition of h makes sense even if A is only assumed to be non-empty and bounded. For our purposes here, however, it will suffice to consider the restricted case when A is a non-empty compact convex set.

By way of an example, we find the support function h of the regular n-crosspolytope A defined by the equation

$$A = \{(x_1, \ldots, x_n) \in \mathbb{R}^n : |x_1| + \ldots + |x_n| \leqslant 1\}.$$

Let $\mathbf{u} = (u_1, \ldots, u_n)$. Then

$$\begin{aligned} h(\mathbf{u}) &= \sup \{\mathbf{u} \cdot \mathbf{a} : \mathbf{a} \in A\} \\ &= \sup \{u_1 a_1 + \ldots + u_n a_n : |a_1| + \ldots + |a_n| \leqslant 1\} \\ &\leqslant \sup \{|u_1||a_1| + \ldots + |u_n||a_n| : |a_1| + \ldots + |a_n| \leqslant 1\} \\ &\leqslant \sup \{(\max \{|u_1|, \ldots, |u_n|\})(|a_1| + \ldots + |a_n|) : |a_1| + \ldots + |a_n| \leqslant 1\} \\ &= \max \{|u_1|, \ldots, |u_n|\}. \end{aligned}$$

Let $m \in \{1, \ldots, n\}$ be such that $|u_m| = \max \{|u_1|, \ldots, |u_n|\}$. Define a point $\mathbf{a} = (a_1, \ldots, a_n)$ of A by the conditions $a_i = 0$ when $i \neq m$ and a_m is 1 or -1 according as u_m is non-negative or negative. Then

$$\mathbf{u} \cdot \mathbf{a} = |u_m| = \max \{|u_1|, \ldots, |u_n|\},$$

whence $h(\mathbf{u}) \geqslant \max \{|u_1|, \ldots, |u_n|\}$. We have thus shown that

$$h(\mathbf{u}) = \max \{|u_1|, \ldots, |u_n|\}.$$

We note that this support function is positively homogeneous and convex; this is a special case of the next theorem.

Theorem 5.6.1 *The support function of a non-empty compact convex set in $\mathbb{R}^n$ is positively homogeneous and convex.*

Proof Let h be the support function of a non-empty compact convex set A in $\mathbb{R}^n$. Let $\mathbf{u}, \mathbf{v} \in \mathbb{R}^n$ and let $\lambda \geqslant 0$. Then

$$h(\lambda \mathbf{u}) = \sup \{(\lambda \mathbf{u}) \cdot \mathbf{a} : \mathbf{a} \in A\} = \lambda \sup \{\mathbf{u} \cdot \mathbf{a} : \mathbf{a} \in A\} = \lambda h(\mathbf{u}),$$

which shows that h is positively homogeneous. Also

$$\begin{aligned} h(\mathbf{u} + \mathbf{v}) &= \sup \{(\mathbf{u} + \mathbf{v}) \cdot \mathbf{a} : \mathbf{a} \in A\} \\ &= \sup \{\mathbf{u} \cdot \mathbf{a} + \mathbf{v} \cdot \mathbf{a} : \mathbf{a} \in A\} \\ &\leqslant \sup \{\mathbf{u} \cdot \mathbf{a} : \mathbf{a} \in A\} + \sup \{\mathbf{v} \cdot \mathbf{a} : \mathbf{a} \in A\} \\ &= h(\mathbf{u}) + h(\mathbf{v}). \end{aligned}$$

The convexity of h now follows from Theorem 5.4.6. □

Suppose again that h is the support function of a non-empty compact convex set A in $\mathbb{R}^n$, and that $\mathbf{u}$ is a non-zero vector in $\mathbb{R}^n$. By the definition of h, $\mathbf{u} \cdot \mathbf{a} \leqslant h(\mathbf{u})$ for each $\mathbf{a}$ in A, whence

$$A \subseteq \{\mathbf{x} \in \mathbb{R}^n : \mathbf{u} \cdot \mathbf{x} \leqslant h(\mathbf{u})\}.$$

Consider the function $f: A \to \mathbb{R}$ defined by the rule $f(\mathbf{a}) = \mathbf{u} \cdot \mathbf{a}$ for each point $\mathbf{a}$ in A. Then f is continuous, and so is bounded and attains its bounds on the compact set A. In particular, there exists a point $\mathbf{a}_0$ in A such that

$$\mathbf{u} \cdot \mathbf{a}_0 = \sup\{\mathbf{u} \cdot \mathbf{a} : \mathbf{a} \in A\} = h(\mathbf{u}),$$

which shows that the hyperplane with equation $\mathbf{u} \cdot \mathbf{x} = h(\mathbf{u})$ supports A at $\mathbf{a}_0$. The distance of this support hyperplane from the origin is $|h(\mathbf{u})|/\|\mathbf{u}\|$, which simplifies to $h(\mathbf{u})$ when $\mathbf{u}$ is a unit vector and the origin is a point of A. The earlier discussion shows that the set

$$\{\mathbf{x} \in \mathbb{R}^n : \mathbf{u} \cdot \mathbf{x} = h(\mathbf{u})\} \cap A = \{\mathbf{x} \in A : \mathbf{u} \cdot \mathbf{x} = h(\mathbf{u})\}$$

is a non-empty exposed face of A; it is called the **exposed face** of A with **outward normal u** and is denoted by $A^{\mathbf{u}}$. Since h is positively homogeneous, for $\lambda > 0$,

$$A^{\lambda\mathbf{u}} = \{\mathbf{x} \in A : (\lambda\mathbf{u}) \cdot \mathbf{x} = h(\lambda\mathbf{u})\} = \{x \in A : \mathbf{u} \cdot \mathbf{x} = h(\mathbf{u})\} = A^{\mathbf{u}}.$$

Theorem 5.6.2 *Let A, B be non-empty compact convex sets in $\mathbb{R}^n$ with support functions h_A, h_B, respectively. Then the support functions h_{A+B} of $A + B$ and $h_{\lambda A}$ of λA, where $\lambda \geqslant 0$, are given by the equations $h_{A+B} = h_A + h_B$ and $h_{\lambda A} = \lambda h_A$.*

Proof Let $\mathbf{u} \in \mathbb{R}^n$. Then

$$\begin{aligned} h_{A+B}(\mathbf{u}) &= \sup\{\mathbf{u} \cdot (\mathbf{a} + \mathbf{b}) : \mathbf{a} \in A, \mathbf{b} \in B\} \\ &= \sup\{\mathbf{u} \cdot \mathbf{a} : \mathbf{a} \in A\} + \sup\{\mathbf{u} \cdot \mathbf{b} : \mathbf{b} \in B\} \\ &= h_A(\mathbf{u}) + h_B(\mathbf{u}), \end{aligned}$$

whence $h_{A+B} = h_A + h_B$. Also

$$h_{\lambda A}(\mathbf{u}) = \sup\{\mathbf{u} \cdot (\lambda\mathbf{a}) : \mathbf{a} \in A\} = \lambda \sup\{\mathbf{u} \cdot \mathbf{a} : \mathbf{a} \in A\} = \lambda h_A(\mathbf{u}),$$

whence $h_{\lambda A} = \lambda h_A$. □

Theorem 5.6.3 *Let A, B be non-empty compact convex sets in $\mathbb{R}^n$. Then, for each non-zero vector $\mathbf{u}$ in $\mathbb{R}^n$ and for each $\lambda \geqslant 0$, $(A + B)^{\mathbf{u}} = A^{\mathbf{u}} + B^{\mathbf{u}}$ and $(\lambda A)^{\mathbf{u}} = \lambda A^{\mathbf{u}}$.*

Proof We note that

$$\begin{aligned} (A + B)^{\mathbf{u}} &= \{\mathbf{a} + \mathbf{b} : \mathbf{a} \in A, \mathbf{b} \in B, h_{A+B}(\mathbf{u}) = \mathbf{u} \cdot (\mathbf{a} + \mathbf{b})\} \\ &= \{\mathbf{a} + \mathbf{b} : \mathbf{a} \in A, \mathbf{b} \in B, h_A(\mathbf{u}) + h_B(\mathbf{u}) = \mathbf{u} \cdot \mathbf{a} + \mathbf{u} \cdot \mathbf{b}\} \\ &= \{\mathbf{a} + \mathbf{b} : \mathbf{a} \in A, \mathbf{b} \in B, h_A(\mathbf{u}) = \mathbf{u} \cdot \mathbf{a}, h_B(\mathbf{u}) = \mathbf{u} \cdot \mathbf{b}\} \\ &= \{\mathbf{a} \in A : h_A(\mathbf{u}) = \mathbf{u} \cdot \mathbf{a}\} + \{\mathbf{b} \in B : h_B(\mathbf{u}) = \mathbf{u} \cdot \mathbf{b}\} \\ &= A^{\mathbf{u}} + B^{\mathbf{u}}. \end{aligned}$$

We also have, for $\lambda \geqslant 0$,

$$\begin{aligned}(\lambda A)^{\mathbf{u}} &= \{\lambda \mathbf{a} : \mathbf{a} \in A, h_{\lambda A}(\mathbf{u}) = \mathbf{u} \cdot (\lambda \mathbf{a})\} \\ &= \lambda\{\mathbf{a} \in A : \lambda h_A(\mathbf{u}) = \lambda \mathbf{u} \cdot \mathbf{a}\} \\ &= \lambda\{\mathbf{a} \in A : h_A(\mathbf{u}) = \mathbf{u} \cdot \mathbf{a}\} \\ &= \lambda A^{\mathbf{u}}. \quad \square\end{aligned}$$

The next theorem shows how a non-empty compact convex set can be determined from its support function. In particular, it implies that different non-empty compact convex sets have different support functions.

Theorem 5.6.4 *Let h be the support function of a non-empty compact convex set A in $\mathbb{R}^n$. Then*

$$A = \{\mathbf{x} \in \mathbb{R}^n : \mathbf{u} \cdot \mathbf{x} \leqslant h(\mathbf{u}) \text{ for all } \mathbf{u} \in \mathbb{R}^n\}.$$

Proof We prove the theorem by showing that

(i) if $\mathbf{a} \in A, \mathbf{u} \in \mathbb{R}^n$, then $\mathbf{u} \cdot \mathbf{a} \leqslant h(\mathbf{u})$;

(ii) if $\mathbf{a}_0 \in \mathbb{R}^n \setminus A$, then $\mathbf{u} \cdot \mathbf{a}_0 > h(\mathbf{u})$ for some $\mathbf{u} \in \mathbb{R}^n$.

Statement (i) follows immediately from the definition of h. Suppose that $\mathbf{a}_0 \in \mathbb{R}^n \setminus A$. Then $\{\mathbf{a}_0\}$ and A can be strictly separated by a hyperplane. Thus there exists $\mathbf{u} \in \mathbb{R}^n$ such that

$$h(\mathbf{u}) = \sup\{\mathbf{u} \cdot \mathbf{a} : \mathbf{a} \in A\} < \mathbf{u} \cdot \mathbf{a}_0.$$

This verifies statement (ii). $\square$

We now show that every positively homogeneous convex function is the support function of some non-empty compact convex set.

Theorem 5.6.5 *Let $g : \mathbb{R}^n \to \mathbb{R}$ be a positively homogeneous convex function. Then the set A defined by the equation*

$$A = \{\mathbf{x} \in \mathbb{R}^n : \mathbf{u} \cdot \mathbf{x} \leqslant g(\mathbf{u}) \text{ for all } \mathbf{u} \in \mathbb{R}^n\}$$

is non-empty, compact, convex, and has support function g.

Proof Let $\mathbf{u} \in \mathbb{R}^n$. Since g is convex, it has support at $\mathbf{u}$, so there exist $a_0 \in \mathbb{R}, \mathbf{a} \in \mathbb{R}^n$ such that

$$a_0 + \mathbf{a} \cdot \mathbf{u} = g(\mathbf{u}) \quad \text{and} \quad a_0 + \mathbf{a} \cdot \mathbf{v} \leqslant g(\mathbf{v}) \quad \text{for } \mathbf{v} \in \mathbb{R}^n. \tag{5.6.1}$$

Putting $\mathbf{v} = \lambda \mathbf{u}$ in (5.6.1), we see that, for all $\lambda \geqslant 0$,

$$a_0 + \lambda(\mathbf{a} \cdot \mathbf{u}) \leqslant g(\lambda \mathbf{u}) = \lambda g(\mathbf{u}) = \lambda a_0 + \lambda(\mathbf{a} \cdot \mathbf{u}).$$

Thus $a_0 \leqslant \lambda a_0$ for all $\lambda \geqslant 0$, whence $a_0 = 0$. Putting $a_0 = 0$ in (5.6.1), we find that $\mathbf{a} \cdot \mathbf{u} = g(\mathbf{u})$ and $\mathbf{a} \in A$.

We have just shown that A is non-empty. From its definition, A is an intersection of closed halfspaces, and so is closed and convex. For each $\mathbf{a} = (a_1, \ldots, a_n)$ in A,

$$-g(-\mathbf{e}_i) \leqslant \mathbf{a} \cdot \mathbf{e}_i = a_i \leqslant g(\mathbf{e}_i)$$

for $i = 1, \ldots, n$, which shows that A is bounded. Thus A is a non-empty compact convex set.

Denote by h the support function of A. Let $\mathbf{u} \in \mathbb{R}^n$. By the first part of this proof, there is some $\mathbf{a} \in A$ for which $\mathbf{a} \cdot \mathbf{u} = g(\mathbf{u})$, whence $g(\mathbf{u}) \leqslant h(\mathbf{u})$. For each $\mathbf{a}$ in A, $\mathbf{a} \cdot \mathbf{u} \leqslant g(\mathbf{u})$, and so $h(\mathbf{u}) \leqslant g(\mathbf{u})$. Thus $g = h$ and g is the support function of A. □

Now we introduce another positively homogeneous convex function associated with a convex set, its *gauge function*. We shall show later that the support and gauge functions of a convex set are related through the concept of duality. Whereas we defined the support function of a non-empty compact convex set, we define the gauge function of a closed convex set which has the origin as an interior point.

Let A be a closed convex set in $\mathbb{R}^n$ having the origin as an interior point. Then it follows easily that $\lambda A \subseteq \mu A$ whenever $0 \leqslant \lambda \leqslant \mu$; and that, for each $\mathbf{x}$ in $\mathbb{R}^n$, there is some $\lambda \geqslant 0$ such that $\mathbf{x} \in \lambda A$. Thus $\mathbb{R}^n$ can be expressed as an increasing union of convex sets as follows:

$$\mathbb{R}^n = \bigcup (\lambda A : \lambda \geqslant 0).$$

The **gauge function** g, or more precisely g_A, of A is the function $g : \mathbb{R}^n \to \mathbb{R}$ defined, for each $\mathbf{x}$ in $\mathbb{R}^n$, by the equation

$$g(\mathbf{x}) = \inf\{\lambda \geqslant 0 : \mathbf{x} \in \lambda A\}.$$

In view of our earlier comments, g is well defined. Some immediate consequences of the definition are:

(i) $g(\mathbf{0}) = 0$ and $g(\mathbf{x}) \geqslant 0$ for $\mathbf{x} \in \mathbb{R}^n$;

(ii) $g(\mathbf{x}) \leqslant 1$ when $\mathbf{x} \in A$;

(iii) if $g(\mathbf{x}) = 0$, then $\{\mu \mathbf{x} : \mu \geqslant 0\} \subseteq A$;

(iv) $g(\mathbf{x}) = 0$ for all $\mathbf{x} \in \mathbb{R}^n$ if and only if $A = \mathbb{R}^n$.

Suppose now that $g(\mathbf{x}) > 0$. Then, for each $\varepsilon > 0$, $\mathbf{x} \in (g(\mathbf{x}) + \varepsilon)A$, whence $\mathbf{x}/(g(\mathbf{x}) + \varepsilon) \in A$. Letting $\varepsilon \to 0_+$ and using our assumption that A is closed, we deduce that $\mathbf{x}/g(\mathbf{x}) \in A$, and hence that $\mathbf{x} \in g(\mathbf{x})A$. In particular, if $0 < g(\mathbf{x}) \leqslant 1$, then $\mathbf{x} \in g(\mathbf{x})A \subseteq A$. We have thus established:

(v) $A = \{\mathbf{x} \in \mathbb{R}^n : g(\mathbf{x}) \leqslant 1\}$.

Theorem 5.6.6 *The gauge function of a closed convex set having the origin as an interior point is positively homogeneous and convex.*

Proof Let g be the gauge function of a closed convex set A in $\mathbb{R}^n$ which contains the origin in its interior.

Let $\mathbf{x} \in \mathbb{R}^n$ and let $\lambda > 0$. Then $\lambda\mathbf{x} \in \mu A$ if and only if $\mathbf{x} \in (\mu/\lambda)A$, whence it follows easily from the definition of g that $g(\lambda\mathbf{x}) = \lambda g(\mathbf{x})$. Trivially, $g(0\mathbf{x}) = 0g(\mathbf{x})$. Thus g is positively homogeneous.

Let $\mathbf{x}, \mathbf{y} \in \mathbb{R}^n$ and let $\lambda, \mu \geqslant 0$ with $\lambda + \mu = 1$. Then, for each $\varepsilon > 0$, $\mathbf{x} \in (g(\mathbf{x}) + \varepsilon)A$, $\mathbf{y} \in (g(\mathbf{y}) + \varepsilon)A$, and so

$$\lambda\mathbf{x} + \mu\mathbf{y} \in (\lambda g(\mathbf{x}) + \mu g(\mathbf{y}) + \varepsilon)A.$$

Since $\varepsilon > 0$ is arbitrary, $g(\lambda\mathbf{x} + \mu\mathbf{y}) \leqslant \lambda g(\mathbf{x}) + \mu g(\mathbf{y})$, which shows that g is convex. □

By way of an example, we find the gauge function g of the n-cube A defined by the equation

$$A = \{(x_1, \ldots, x_n) : |x_1|, \ldots, |x_n| \leqslant 1\}.$$

Let $\mathbf{u} = (u_1, \ldots, u_n)$. Then, for $\lambda \geqslant 0$,

$$\lambda A = \{(x_1, \ldots, x_n) : |x_1|, \ldots, |x_n| \leqslant \lambda\},$$

and so $\mathbf{u} \in \lambda A$ if and only if $\max\{|u_1|, \ldots, |u_n|\} \leqslant \lambda$. Thus

$$g(\mathbf{u}) = \max\{|u_1|, \ldots, |u_n|\}.$$

We now show that every non-negative positively homogeneous convex function on $\mathbb{R}^n$ is the gauge function of some closed convex set in $\mathbb{R}^n$ which contains the origin in its interior.

Theorem 5.6.7 *Let $f: \mathbb{R}^n \to \mathbb{R}$ be a non-negative positively homogeneous convex function. Then the set A defined by the equation*

$$A = \{\mathbf{x} \in \mathbb{R}^n : f(\mathbf{x}) \leqslant 1\}$$

is closed, convex, contains the origin in its interior, and has gauge function f.

Proof The function f is continuous by Theorem 5.5.1. Thus A is closed and contains the open set $\{\mathbf{x} \in \mathbb{R}^n : f(\mathbf{x}) < 1\}$, which contains the origin. The set A is convex, being the level set of a convex function. Hence A is a closed convex set containing the origin in its interior.

Denote by g the gauge function of A. Then, as proved earlier,

$$A = \{\mathbf{x} \in \mathbb{R}^n : g(\mathbf{x}) \leqslant 1\}.$$

Suppose that $\mathbf{u} \in \mathbb{R}^n$ satisfies $g(\mathbf{u}) > 0$. Then, since g is positively homogeneous, $g(\mathbf{u}/g(\mathbf{u})) = 1$, whence $\mathbf{u}/g(\mathbf{u}) \in A$. Since f is positively homogeneous and $\mathbf{u}/g(\mathbf{u}) \in A$,

$$f(\mathbf{u}/g(\mathbf{u})) = f(\mathbf{u})/g(\mathbf{u}) \leqslant 1,$$

which shows that $f(\mathbf{u}) \leqslant g(\mathbf{u})$. If $g(\mathbf{u}) = 0$, then, for all $\lambda > 0$, $\lambda\mathbf{u} \in A$, so

$$0 \leqslant f(\lambda\mathbf{u}) = \lambda f(\mathbf{u}) \leqslant 1;$$

it follows that $f(\mathbf{u}) = 0$. Thus $f(\mathbf{u}) \leqslant g(\mathbf{u})$ for all $\mathbf{u} \in \mathbb{R}^n$. By a similar argument, $g(\mathbf{u}) \leqslant f(\mathbf{u})$ for all $\mathbf{u} \in \mathbb{R}^n$. Hence $f = g$ and f is the gauge function of A. □

We conclude this section by showing how the concept of duality provides a link between support and gauge functions. We have already seen that the support function of the regular n-crosspolytope

$$\{(x_1, \ldots, x_n): |x_1| + \ldots + |x_n| \leqslant 1\}$$

and the gauge function of its dual, the n-cube

$$\{(x_1, \ldots, x_n): |x_1|, \ldots, |x_n| \leqslant 1\},$$

are the same, namely the function $f: \mathbb{R}^n \to \mathbb{R}$ defined by the equation

$$f(\mathbf{u}) = \max\{|u_1|, \ldots, |u_n|\} \qquad \text{for } \mathbf{u} = (u_1, \ldots, u_n) \in \mathbb{R}^n.$$

This is a special case of the next theorem.

Theorem 5.6.8 *Suppose that g, h are the gauge and support functions, respectively, of a compact convex set A in $\mathbb{R}^n$ which has the origin as an interior point. Then the gauge and support functions of the dual A^* of A are h, g, respectively.*

Proof If $\mathbf{u} \in A^*$, then $\mathbf{u} \cdot \mathbf{a} \leqslant 1$ for all $\mathbf{a}$ in A, whence $h(\mathbf{u}) \leqslant 1$. Conversely, if $h(\mathbf{u}) \leqslant 1$, then $\mathbf{u} \cdot \mathbf{a} \leqslant 1$ for all $\mathbf{a}$ in A, and so $\mathbf{u} \in A^*$. Thus

$$A^* = \{\mathbf{x} \in \mathbb{R}^n: h(\mathbf{x}) \leqslant 1\}.$$

Since A contains the origin, h is non-negative. Thus h is a non-negative, positively homogeneous convex function, whence, by Theorem 5.6.7, h is the gauge function of A^*. By what we have just proved, the support function of A^* is the gauge function of $A^{**} = A$, viz. g. □

Exercises 5.6

1. Find the support and gauge functions of the following sets:
   (i) $\{(x_1, \ldots, x_n): |x_1|, \ldots, |x_n| \leqslant r\}$, where $r > 0$;
   (ii) $\{(x_1, \ldots, x_n): |x_1| + \ldots + |x_n| \leqslant r\}$, where $r > 0$;
   (iii) $\{(x, y): (x/a)^2 + (y/b)^2 \leqslant 1\}$, where $ab \neq 0$.

2. Let A, B be non-empty compact convex sets in $\mathbb{R}^n$ with support functions h_A, h_B, respectively. Show that the Hausdorff distance $\rho(A, B)$ between A and B, as defined in Section 2.7, is given by the formula
$$\rho(A, B) = \sup\{|h_A(\mathbf{u}) - h_B(\mathbf{u})| : \|\mathbf{u}\| = 1\}.$$
3. Let h be the support function of a non-empty compact convex set A in $\mathbb{R}^n$. Show that, for each $\mathbf{v} \in \mathbb{R}^n$, $h'(\mathbf{v};\quad)$ is the support function of the exposed face
$$B = \{\mathbf{x} \in A : \mathbf{x} \cdot \mathbf{v} = h(\mathbf{v})\}$$
of A. What can be deduced when B is a singleton?
4. Let A and B be compact convex sets in $\mathbb{R}^n$ which contain the origin in their interiors. Denote their support and gauge functions by h_A, h_B and g_A, g_B, respectively. Prove that the support function h of $\operatorname{conv}(A \cup B)$, and the gauge function g of $A \cap B$ are given by the equations
$$h(\mathbf{u}) = \max\{h_A(\mathbf{u}), h_B(\mathbf{u})\}, \qquad g(\mathbf{u}) = \max\{g_A(\mathbf{u}), g_B(\mathbf{u})\}.$$

5.7 THE CONVEX PROGRAMMING PROBLEM

The *convex programming problem* is to minimize a convex function on a convex set defined by means of a finite number of convex functions. First we show the equivalence of this problem to another one, the so-called *saddle-point problem*. Then, assuming the differentiability of the functions concerned, we derive the *Kuhn–Tucker conditions*, which characterize the solutions of the saddle-point problem.

Throughout this section $f, g_1, \ldots, g_m$ will denote convex functions defined on $\mathbb{R}^n$. The **convex programming problem** is to minimize $f(\mathbf{x})$ subject to the constraints $\mathbf{x} \geqslant \mathbf{0}, g_1(\mathbf{x}) \leqslant 0, \ldots, g_m(\mathbf{x}) \leqslant 0$. The **feasible set** for the problem is the convex set X defined by the equation
$$X = \{\mathbf{x} \in \mathbb{R}^n : \mathbf{x} \geqslant \mathbf{0}, g_1(\mathbf{x}) \leqslant 0, \ldots, g_m(\mathbf{x}) \leqslant 0\}.$$
Thus the convex programming problem is to find $\mathbf{x}_0 \in X$ such that $f(\mathbf{x}_0) \leqslant f(\mathbf{x})$ for all $\mathbf{x} \in X$.

Our first result, which is of interest in its own right, will be used to establish the equivalence of the convex programming and saddle-point problems.

Theorem 5.7.1 *Let $f_1, \ldots, f_k$ be convex functions defined on a non-empty convex set Y in $\mathbb{R}^n$. Suppose that there exists no $\mathbf{y}$ in Y such that $f_1(\mathbf{y}) < 0, \ldots, f_k(\mathbf{y}) < 0$. Then there exist $a_1, \ldots, a_k \geqslant 0$, not all zero, such that*

$$a_1 f_1(\mathbf{y}) + \ldots + a_k f_k(\mathbf{y}) \geqslant 0 \qquad \textit{for all } \mathbf{y} \in Y.$$

Proof Define a set C in $\mathbb{R}^k$ by the equation

$$C = \{(z_1, \ldots, z_k)\text{: there is } \mathbf{y} \in Y \text{ such that } f_i(\mathbf{y}) < z_i \text{ for } i = 1, \ldots, k\}.$$

Let $\mathbf{u} = (u_1, \ldots, u_k)$, $\mathbf{v} = (v_1, \ldots, v_k) \in C$, and let $\lambda, \mu \geqslant 0$ with $\lambda + \mu = 1$. Then there exist $\mathbf{a}, \mathbf{b} \in Y$ such that, for $i = 1, \ldots, k$, $f_i(\mathbf{a}) < u_i$ and $f_i(\mathbf{b}) < v_i$. The convexity of $f_1, \ldots, f_k$ shows that, for $i = 1, \ldots, k$,

$$f_i(\lambda\mathbf{a} + \mu\mathbf{b}) \leqslant \lambda f_i(\mathbf{a}) + \mu f_i(\mathbf{b}) < \lambda u_i + \mu v_i,$$

whence, since $\lambda\mathbf{a} + \mu\mathbf{b} \in Y$, $\lambda\mathbf{u} + \mu\mathbf{v} \in C$. Thus C is convex. By hypothesis, C does not contain the origin of $\mathbb{R}^k$, and so the origin and C can be separated by a hyperplane. Thus there exist scalars $a_1, \ldots, a_k$, not all zero, such that, for all $\mathbf{y} \in Y$ and all $\lambda_1, \ldots, \lambda_k > 0$,

$$a_1(f_1(\mathbf{y}) + \lambda_1) + \ldots + a_k(f_k(\mathbf{y}) + \lambda_k) \geqslant 0. \tag{5.7.1}$$

Letting $\lambda_1 \to \infty$, whilst keeping $\lambda_2, \ldots, \lambda_k$ fixed in (5.7.1), we deduce that $a_1 \geqslant 0$. Similarly, $a_2 \geqslant 0, \ldots, a_k \geqslant 0$. Letting $\lambda_1 \to 0_+, \ldots, \lambda_k \to 0_+$ in (5.7.1), we deduce that, for all $\mathbf{y}$ in Y,

$$a_1 f_1(\mathbf{y}) + \ldots + a_k f_k(\mathbf{y}) \geqslant 0. \;\square$$

The method of Lagrange multipliers, which is used to convert a constrained extremal problem into an unconstrained one, at the expense of increasing the number of variables, is also helpful in the discussion of the convex programming problem. The **Lagrangian function** associated with the convex programming problem is the function F of the $m + n$ variables $x_1, \ldots, x_n, y_1, \ldots, y_m$ defined by the equation

$$F(\mathbf{x}, \mathbf{y}) = f(\mathbf{x}) + y_1 g_1(\mathbf{x}) + \ldots + y_m g_m(\mathbf{x}),$$

where $\mathbf{x} = (x_1, \ldots, x_n)$, $\mathbf{y} = (y_1, \ldots, y_m)$. The **saddle-point problem** is to determine a **saddle point** of F, that is, a point $(\mathbf{x}_0, \mathbf{y}_0)$ of $\mathbb{R}^{m+n}$ such that $\mathbf{x}_0 \geqslant \mathbf{0}$, $\mathbf{y}_0 \geqslant \mathbf{0}$ and

$$F(\mathbf{x}_0, \mathbf{y}) \leqslant F(\mathbf{x}_0, \mathbf{y}_0) \leqslant F(\mathbf{x}, \mathbf{y}_0)$$

for all $\mathbf{x} \geqslant \mathbf{0}, \mathbf{y} \geqslant \mathbf{0}$.

Theorem 5.7.2 *Let $(\mathbf{x}_0, \mathbf{y}_0)$ be a saddle point of the Lagrangian function F. Then $\mathbf{x}_0$ is a solution to the convex programming problem and $F(\mathbf{x}_0, \mathbf{y}_0) = f(\mathbf{x}_0)$.*

Proof Let $\mathbf{x}_0 = (x_1^0, \ldots, x_n^0) \geqslant \mathbf{0}$ and $\mathbf{y}_0 = (y_1^0, \ldots, y_m^0) \geqslant \mathbf{0}$. For all $\mathbf{y} = (y_1, \ldots, y_m) \geqslant \mathbf{0}$, $F(\mathbf{x}_0, \mathbf{y}_0) \geqslant F(\mathbf{x}_0, \mathbf{y})$, and so

$$y_1^0 g_1(\mathbf{x}_0) + \ldots + y_m^0 g_m(\mathbf{x}_0) \geqslant y_1 g_1(\mathbf{x}_0) + \ldots + y_m g_m(\mathbf{x}_0).$$

By fixing $y_2, \dots, y_m$ and letting $y_1 \to \infty$ in this last inequality, we deduce that $g_1(\mathbf{x}_0) \leqslant 0$. Similarly, $g_2(\mathbf{x}_0) \leqslant 0, \dots, g_m(\mathbf{x}_0) \leqslant 0$. Thus $\mathbf{x}_0$ is a point of the feasible set X of the convex programming problem.

Putting $\mathbf{y} = \mathbf{0}$ in the saddle-point inequality $F(\mathbf{x}_0, \mathbf{y}_0) \geqslant F(\mathbf{x}_0, \mathbf{y})$ and using the fact that $\mathbf{x}_0 \in X$, we deduce that

$$0 \leqslant y_1^0 g_1(\mathbf{x}_0) + \dots + y_m^0 g_m(\mathbf{x}_0) \leqslant 0,$$

whence

$$y_1^0 g_1(\mathbf{x}_0) + \dots + y_m^0 g_m(\mathbf{x}_0) = 0 \qquad \text{and} \qquad F(\mathbf{x}_0, \mathbf{y}_0) = f(\mathbf{x}_0).$$

Since $F(\mathbf{x}_0, \mathbf{y}_0) \leqslant F(\mathbf{x}, \mathbf{y}_0)$ for all $\mathbf{x} \geqslant \mathbf{0}$, we deduce that, for all $\mathbf{x} \in X$,

$$f(\mathbf{x}_0) \leqslant f(\mathbf{x}) + y_1^0 g_1(\mathbf{x}) + \dots + y_m^0 g_m(\mathbf{x}) \leqslant f(\mathbf{x}),$$

which shows that $\mathbf{x}_0$ is a solution to the convex programming problem. □

It is not true that, given any solution $\mathbf{x}_0$ of the convex programming problem, there is always a $\mathbf{y}_0$ such that $(\mathbf{x}_0, \mathbf{y}_0)$ is a saddle point of the Lagrangian function F. (See Exercise 5.7.1.) We do, however, have the following result.

Theorem 5.7.3 *Suppose that* $\mathbf{x}_0$ *is a solution of the convex programming problem. Suppose also that there exists* $\mathbf{x}^* \geqslant \mathbf{0}$ *such that* $g_1(\mathbf{x}^*) < 0, \dots, g_m(\mathbf{x}^*) < 0$. *Then there exists* $\mathbf{y}_0 \in \mathbb{R}^m$ *for which* $(\mathbf{x}_0, \mathbf{y}_0)$ *is a saddle point of the Lagrangian function F.*

Proof Suppose that $\mathbf{x}$ belongs to the non-negative orthant Y of $\mathbb{R}^n$. Then not all of the following inequalities can hold:

$$g_1(\mathbf{x}) < 0, \dots, g_m(\mathbf{x}) < 0, \qquad f(\mathbf{x}) - f(\mathbf{x}_0) < 0.$$

Thus, by Theorem 5.7.1, there exist $a_1, \dots, a_m, a_0 \geqslant 0$, not all zero, such that

$$a_1 g_1(\mathbf{x}) + \dots + a_m g_m(\mathbf{x}) + a_0(f(\mathbf{x}) - f(\mathbf{x}_0)) \geqslant 0 \qquad (5.7.2)$$

whenever $\mathbf{x} \in Y$, i.e. $\mathbf{x} \geqslant \mathbf{0}$. If $a_0 = 0$, then

$$0 > a_1 g_1(\mathbf{x}^*) + \dots + a_m g_m(\mathbf{x}^*) \geqslant 0,$$

which is impossible. Thus $a_0 > 0$. For $i = 1, \dots, m$, let $y_i^0 = a_i/a_0$ and let $\mathbf{y}_0 = (y_1^0, \dots, y_m^0) \geqslant \mathbf{0}$. Then, for any $\mathbf{x} \geqslant \mathbf{0}$, we deduce from (5.7.2) that

$$f(\mathbf{x}_0) \leqslant f(\mathbf{x}) + y_1^0 g_1(\mathbf{x}) + \dots + y_m^0 g_m(\mathbf{x}) = F(\mathbf{x}, \mathbf{y}_0).$$

Hence

$$f(\mathbf{x}_0) \leqslant f(\mathbf{x}_0) + y_1^0 g_1(\mathbf{x}_0) + \dots + y_m^0 g_m(\mathbf{x}_0) \leqslant f(\mathbf{x}_0),$$

and so

$$y_1^0 g_1(\mathbf{x}_0) + \ldots + y_m^0 g_m(\mathbf{x}_0) = 0.$$

Thus, for all $\mathbf{x} \geqslant \mathbf{0}$, $F(\mathbf{x}_0, \mathbf{y}_0) = f(\mathbf{x}_0) \leqslant F(\mathbf{x}, \mathbf{y}_0)$. For $\mathbf{y} = (y_1, \ldots, y_m) \geqslant \mathbf{0}$,

$$F(\mathbf{x}_0, \mathbf{y}_0) = f(\mathbf{x}_0) \geqslant f(\mathbf{x}_0) + y_1 g_1(\mathbf{x}_0) + \ldots + y_m g_m(\mathbf{x}_0) = F(\mathbf{x}_0, \mathbf{y}).$$

This shows that $(\mathbf{x}_0, \mathbf{y}_0)$ is a saddle point of F. □

Theorem 5.7.4 (Kuhn–Tucker conditions) *Suppose that the convex functions $f, g_1, \ldots, g_m: \mathbb{R}^n \to \mathbb{R}$ are differentiable. Then $(\mathbf{x}_0, \mathbf{y}_0)$, where $\mathbf{x}_0 = (x_1^0, \ldots, x_n^0)$ and $\mathbf{y}_0 = (y_1^0, \ldots, y_m^0)$, is a saddle point of the Lagrangian function F if and only if*

$$\mathbf{x}_0 \geqslant \mathbf{0}, \tag{5.7.3a}$$

$$\frac{\partial F}{\partial x_j}(\mathbf{x}_0, \mathbf{y}_0) = \frac{\partial f}{\partial x_j}(\mathbf{x}_0) + \sum_{i=1}^{m} y_i^0 \frac{\partial g_i}{\partial x_j}(\mathbf{x}_0) \geqslant 0, \tag{5.7.3b}$$

$$\frac{\partial F}{\partial x_j}(\mathbf{x}_0, \mathbf{y}_0) = 0 \qquad \textit{whenever } x_j^0 > 0, \tag{5.7.3c}$$

and

$$\mathbf{y}_0 \geqslant \mathbf{0}, \tag{5.7.4a}$$

$$\frac{\partial F}{\partial y_j}(\mathbf{x}_0, \mathbf{y}_0) = g_j(\mathbf{x}_0) \leqslant 0, \tag{5.7.4b}$$

$$\frac{\partial F}{\partial y_j}(\mathbf{x}_0, \mathbf{y}_0) = 0 \qquad \textit{whenever } y_j^0 > 0. \tag{5.7.4c}$$

Proof Suppose first that $(\mathbf{x}_0, \mathbf{y}_0)$ is a saddle point of F. Then certainly (5.7.3a) and (5.7.4a) are satisfied. For each $j = 1, \ldots, n$,

$$F(\mathbf{x}_0 + \lambda \mathbf{e}_j, \mathbf{y}_0) \geqslant f(\mathbf{x}_0, \mathbf{y}_0) \qquad \text{whenever } \lambda \geqslant -x_j^0.$$

It now follows, by elementary calculus, that

$$\frac{\partial F}{\partial x_j}(\mathbf{x}_0, \mathbf{y}_0) \geqslant 0 \qquad \text{and} \qquad \frac{\partial F}{\partial x_j}(\mathbf{x}_0, \mathbf{y}_0) = 0 \qquad \text{whenever } x_j^0 > 0.$$

Thus (5.7.3b) and (5.7.3c) are satisfied. It follows easily from Theorem 5.7.2 that conditions (5.7.4b) and (5.7.4c) are satisfied.

Suppose next that the six Kuhn–Tucker conditions are satisfied. The function $F(\mathbf{x}, \mathbf{y}_0)$ of $\mathbf{x}$, for fixed $\mathbf{y}_0$, is convex and differentiable, because $f, g_1, \ldots, g_m$ are, and $\mathbf{y}_0 \geqslant \mathbf{0}$. Thus $F(\mathbf{x}, \mathbf{y}_0)$ has unique support at $\mathbf{x}_0$, whence, for all $\mathbf{x} = (x_1, \ldots, x_n) \geqslant \mathbf{0}$,

$$F(\mathbf{x}, \mathbf{y}_0) \geqslant F(\mathbf{x}_0, \mathbf{y}_0) + (x_1 - x_1^0)\frac{\partial F}{\partial x_1}(\mathbf{x}_0, \mathbf{y}_0) + \ldots + (x_n - x_n^0)\frac{\partial F}{\partial x_n}(\mathbf{x}_0, \mathbf{y}_0)$$

$$= F(\mathbf{x}_0, \mathbf{y}_0) + x_1 \frac{\partial F}{\partial x_1}(\mathbf{x}_0, \mathbf{y}_0) + \ldots + x_n \frac{\partial F}{\partial x_n}(\mathbf{x}_0, \mathbf{y}_0)$$

$$\geqslant F(\mathbf{x}_0, \mathbf{y}_0).$$

Here we have used the three conditions (5.7.3).

Finally, for $\mathbf{y} = (y_1, \ldots, y_m) \geqslant \mathbf{0}$, we have

$$F(\mathbf{x}_0, \mathbf{y}) = F(\mathbf{x}_0, \mathbf{y}_0) + (y_1 - y_1^0)g_1(\mathbf{x}_0) + \ldots + (y_m - y_m^0)g_m(\mathbf{x}_0)$$

$$= F(\mathbf{x}_0, \mathbf{y}_0) + y_1 g_1(\mathbf{x}_0) + \ldots + y_m g_m(\mathbf{x}_0)$$

$$\leqslant F(\mathbf{x}_0, \mathbf{y}_0).$$

Here we have used the three conditions (5.7.4). We have thus shown that $(\mathbf{x}_0, \mathbf{y}_0)$ is a saddle point of F. □

Example 5.7.5 Solve the programming problem:

$$\text{minimize} \qquad -6x_1 + 2x_1^2 - 2x_1x_2 + 2x_2^2$$

$$\text{subject to} \qquad x_1 + x_2 \leqslant 2, \quad x_1 \geqslant 0, \quad x_2 \geqslant 0.$$

Solution We write $f(x_1, x_2) = -6x_1 + 2x_1^2 - 2x_1x_2 + 2x_2^2$ and $g_1(x_1, x_2) = x_1 + x_2 - 2$. The Lagrangian function F of the problem is defined by the equation

$$F(\mathbf{x}, \mathbf{y}) = -6x_1 + 2x_1^2 - 2x_1x_2 + 2x_2^2 + y_1(x_1 + x_2 - 2).$$

The Kuhn–Tucker conditions give the three equations:

$$x_1(-6 + 4x_1 - 2x_2 + y_1) = 0,$$
$$x_2(-2x_1 + 4x_2 + y_1) = 0,$$
$$y_1(x_1 + x_2 - 2) = 0;$$

and the inequalities

$$-6 + 4x_1 - 2x_2 + y_1 \geqslant 0,$$
$$-2x_1 + 4x_2 + y_1 \geqslant 0,$$
$$x_1 + x_2 - 2 \leqslant 0,$$
$$x_1 \geqslant 0, \quad x_2 \geqslant 0, \quad y_1 \geqslant 0.$$

The three equations have the following six solutions:

	x_1	x_2	y_1
(i)	0	0	0
(ii)	0	2	-8
(iii)	$1\frac{1}{2}$	0	0
(iv)	2	0	-2
(v)	2	1	0
(vi)	$1\frac{1}{2}$	$\frac{1}{2}$	1.

Of these solutions only (vi) satisfies all the remaining inequalities. Hence f has minimal value $-5\frac{1}{2}$ at $(1\frac{1}{2}, \frac{1}{2})$. □

Exercises 5.7

1. Solve the following convex programming problem:

$$\text{minimize} \quad -x \quad \text{subject to} \quad x^2 \leqslant 0, \quad x \geqslant 0.$$

Write down the Lagrangian function and the Kuhn–Tucker conditions for the problem. Does the Lagrangian function have a saddle point? Comment.

2. Use the Kuhn–Tucker conditions to solve the following convex programming problem:

$$\begin{aligned} &\text{minimize} && (x_1 + 1)^2 + (x_2 - 2)^2 \\ &\text{subject to} && 0 \leqslant x_1 \leqslant 2, \quad 0 \leqslant x_2 \leqslant 1. \end{aligned}$$

3. The **quadratic programming problem** is:

$$\begin{aligned} &\text{minimize} && f(\mathbf{x}) = \mathbf{p}^{\mathrm{T}}\mathbf{x} + \mathbf{x}^{\mathrm{T}}\mathbf{Q}\mathbf{x} \\ &\text{subject to} && \mathbf{A}\mathbf{x} \leqslant \mathbf{b}, \quad \mathbf{x} \geqslant \mathbf{0}, \end{aligned}$$

where $\mathbf{p} \in \mathbb{R}^n$, $\mathbf{Q}$ is a non-negative semidefinite $n \times n$ matrix, $\mathbf{A}$ is an $m \times n$ matrix, and $\mathbf{b} \in \mathbb{R}^m$. Show that f is convex. Write down the Kuhn–Tucker condition for the problem.

4. The standard linear programming problem:

$$\text{minimize} \quad -\mathbf{c}^{\mathrm{T}}\mathbf{x} \quad \text{subject to} \quad \mathbf{A}\mathbf{x} \leqslant \mathbf{b}, \quad \mathbf{x} \geqslant \mathbf{0},$$

is also a convex programming problem. Interpret the Kuhn–Tucker conditions for this problem.

5.8 MATRIX INEQUALITIES

Our principal aim here is to establish a general inequality involving a convex function f and a real symmetric matrix $\mathbf{A}$, and to deduce from it the classical matrix inequalities of Hadamard and Minkowski. To be precise, we shall show that there is some ordering $\lambda_1, \ldots, \lambda_n$ of the eigenvalues of $\mathbf{A}$ such that, for *any* orthonormal sequence $\mathbf{v}_1, \ldots, \mathbf{v}_n$ in $\mathbb{R}^n$,

$$f(\mathbf{v}_1^{\mathrm{T}}\mathbf{A}\mathbf{v}_1, \ldots, \mathbf{v}_n^{\mathrm{T}}\mathbf{A}\mathbf{v}_n) \leqslant f(\lambda_1, \ldots, \lambda_n).$$

Central to our discussion is a celebrated theorem of G. Birkhoff on *doubly stochastic* matrices, i.e. matrices with non-negative elements in which the sum of the elements in each row and column is 1.

To see how doubly stochastic matrices can occur in practice, we consider a simple problem whose solution will be needed later. Associated with each real symmetric square matrix $\mathbf{A}$ of order n, there is a quadratic function $q: \mathbb{R}^n \to \mathbb{R}$ defined for each $\mathbf{x}$ in $\mathbb{R}^n$ by the equation

$$q(\mathbf{x}) = \mathbf{x}^{\mathrm{T}}\mathbf{A}\mathbf{x} = (\mathbf{A}\mathbf{x}) \cdot \mathbf{x}.$$

Let $\mathbf{u}_1, \ldots, \mathbf{u}_n$ be an orthonormal sequence in $\mathbb{R}^n$ consisting of eigenvectors of $\mathbf{A}$ corresponding to the eigenvalues $\lambda_1, \ldots, \lambda_n$ of $\mathbf{A}$. Then, for $i = 1, \ldots, n$,

$$\mathbf{u}_i^{\mathrm{T}}\mathbf{A}\mathbf{u}_i = (\mathbf{A}\mathbf{u}_i) \cdot \mathbf{u}_i = (\lambda_i \mathbf{u}_i) \cdot \mathbf{u}_i = \lambda_i,$$

whence

$$(q(\mathbf{u}_1), \ldots, q(\mathbf{u}_n)) = (\lambda_1, \ldots, \lambda_n).$$

The problem alluded to at the beginning of the paragraph can now be posed as follows:

If $\mathbf{v}_1, \ldots, \mathbf{v}_n$ is any orthonormal sequence in $\mathbb{R}^n$, how are the points

$$\mathbf{u} = (q(\mathbf{u}_1), \ldots, q(\mathbf{u}_n)) \qquad \text{and} \qquad \mathbf{v} = (q(\mathbf{v}_1), \ldots, q(\mathbf{v}_n))$$

related to one another?

To answer the problem, we express each $\mathbf{v}_i$, for $i = 1, \ldots, n$, as a linear combination of $\mathbf{u}_1, \ldots, \mathbf{u}_n$ thus:

$$\mathbf{v}_i = (\mathbf{v}_i \cdot \mathbf{u}_1)\mathbf{u}_1 + \ldots + (\mathbf{v}_i \cdot \mathbf{u}_n)\mathbf{u}_n. \tag{5.8.1}$$

Hence

$$\begin{aligned} q(\mathbf{v}_i) &= ((\mathbf{v}_i \cdot \mathbf{u}_1)\mathbf{A}\mathbf{u}_1 + \ldots + (\mathbf{v}_i \cdot \mathbf{u}_n)\mathbf{A}\mathbf{u}_n) \cdot ((\mathbf{v}_i \cdot \mathbf{u}_1)\mathbf{u}_1 + \ldots + (\mathbf{v}_i \cdot \mathbf{u}_n)\mathbf{u}_n) \\ &= (\lambda_1(\mathbf{v}_i \cdot \mathbf{u}_1)\mathbf{u}_1 + \ldots + \lambda_n(\mathbf{v}_i \cdot \mathbf{u}_n)\mathbf{u}_n) \cdot ((\mathbf{v}_i \cdot \mathbf{u}_1)\mathbf{u}_1 + \ldots + (\mathbf{v}_i \cdot \mathbf{u}_n)\mathbf{u}_n) \\ &= (\mathbf{v}_i \cdot \mathbf{u}_1)^2\lambda_1 + \ldots + (\mathbf{v}_i \cdot \mathbf{u}_n)^2\lambda_n \\ &= (\mathbf{v}_i \cdot \mathbf{u}_1)^2 q(\mathbf{u}_1) + \ldots + (\mathbf{v}_i \cdot \mathbf{u}_n)^2 q(\mathbf{u}_n). \end{aligned}$$

Thus $\mathbf{v} = \mathbf{S}\mathbf{u}$, where $\mathbf{S}$ is the square matrix of order n whose (i,j)th element is $(\mathbf{v}_i \cdot \mathbf{u}_j)^2$. What are the properties of this matrix $\mathbf{S}$? It is certainly a square matrix all of whose elements are non-negative real numbers. Squaring both sides of equation (5.8.1) and using the orthonormality of the sequences $\mathbf{u}_1, \ldots, \mathbf{u}_n$ and $\mathbf{v}_1, \ldots, \mathbf{v}_n$, we deduce that, for $i = 1, \ldots, n$,

$$(\mathbf{v}_i \cdot \mathbf{u}_1)^2 + \ldots + (\mathbf{v}_i \cdot \mathbf{u}_n)^2 = \|\mathbf{v}_i\|^2 = 1.$$

Similarly, for $j = 1, \ldots, n$,

$$(\mathbf{u}_j \cdot \mathbf{v}_1)^2 + \ldots + (\mathbf{u}_j \cdot \mathbf{v}_n)^2 = \|\mathbf{u}_j\|^2 = 1.$$

Thus $\mathbf{S}$ is a square matrix of order n whose elements are non-negative real numbers, and the sum of the elements in each of its rows and columns is equal to 1. Such a matrix is called a **doubly stochastic matrix**. The set of all doubly stochastic $n \times n$ matrices will be denoted by Ω_n.

The simplest example of a doubly stochastic matrix is a **permutation matrix**, which is a square matrix with precisely one 1 in each row and column, all of its other elements being zero. Equivalently, a permutation matrix is one that can be obtained by permuting the rows of an identity matrix. Clearly every convex combination (in the obvious sense) of permutation matrices is a doubly stochastic matrix. The converse of this result, namely that every doubly stochastic matrix is a convex combination of permutation matrices, is also true and it is known as *Birkhoff's theorem*. This theorem, which forms part of Theorem 5.8.2 below, is arguably the most fundamental result in the whole study of doubly stochastic matrices.

Many proofs of Birkhoff's theorem are known; we present the one which is most in keeping with our account of convex sets. In a natural way we may regard each real $n \times n$ matrix $\mathbf{A} = [a_{ij}]$ as a point $\mathbf{a} = (a_{ij})$ of $\mathbb{R}^{n^2}$, the n^2 elements of $\mathbf{A}$ corresponding in some prescribed way to the n^2 coordinates of $\mathbf{a}$. To be definite, we set up the correspondence

$$\mathbf{A} = [a_{ij}] \leftrightarrow (a_{11}, \ldots, a_{1n}, a_{21}, \ldots, a_{2n}, \ldots, a_{n1}, \ldots, a_{nn}) = \mathbf{a}.$$

This correspondence is a bijection between the set of all real $n \times n$ matrices and the set of points in $\mathbb{R}^{n^2}$; it preserves linear combinations, and so we can usefully identify the matrix $\mathbf{A}$ with the point $\mathbf{a}$. Under this identification, we may think of the set Ω_n of doubly stochastic $n \times n$ matrices as a set in $\mathbb{R}^{n^2}$ and refer to some of its members as being permutation matrices. This identification is assumed in Theorem 5.8.2 below. Before proving this theorem, however, we need the following lemma.

Lemma 5.8.1 *Let* $\mathbf{B}$ *be a non-singular square matrix of order n that can be partitioned in the form*

$$\begin{bmatrix} \mathbf{P} \\ \mathbf{Q} \end{bmatrix},$$

where $\mathbf{P}$ *and* $\mathbf{Q}$ *are matrices of* 0*'s and* 1*'s, so arranged that no column of either* $\mathbf{P}$ *or* $\mathbf{Q}$ *contains more than one* 1. *Then* $\det \mathbf{B} = \pm 1$.

Proof We argue by induction on n. The case $n = 2$ is trivial. Suppose that $n \geqslant 3$ and that the assertion is true for square matrices of order $n - 1$. Let $\mathbf{B}$ be as in the statement of the lemma. At least one column of $\mathbf{B}$ contains *precisely* one 1, for otherwise the rows of $\mathbf{P}$ could be added to the negatives of the rows of $\mathbf{Q}$ to produce a zero row, contradicting the assumption that $\mathbf{B}$ is non-singular. Expand $\det \mathbf{B}$ by a column containing precisely one 1 to deduce that $\det \mathbf{B} = \pm \det \mathbf{C}$, where $\mathbf{C}$ is a square matrix of order $n - 1$ of the form considered in the lemma. Hence $\det \mathbf{B} = \pm 1$, since $\det \mathbf{C} = \pm 1$ by the induction hypothesis. This completes the proof by induction. □

Theorem 5.8.2 *The set* Ω_n *is a polytope in* $\mathbb{R}^{n^2}$ *whose extreme points are the permutation matrices in* Ω_n. *Every doubly stochastic matrix is a convex combination of permutation matrices.*

Proof The set Ω_n is polyhedral, since it consists of those points (x_{ij}) in $\mathbb{R}^{n^2}$ satisfying the relations:

$$x_{ij} \geqslant 0 \qquad \text{for } i, j = 1, \ldots, n; \tag{5.8.2}$$

$$\sum_{j=1}^{n} x_{ij} = 1 \qquad \text{for } i = 1, \ldots, n; \tag{5.8.3}$$

$$\sum_{i=1}^{n} x_{ij} = 1 \qquad \text{for } j = 1, \ldots, n - 1. \tag{5.8.4}$$

We note that the equality $x_{1n} + \ldots + x_{nn} = 1$ follows from the $2n - 1$ equations (5.8.3) and (5.8.4). The relations (5.8.2) and (5.8.3) show that, if $(x_{ij}) \in \Omega_n$, then $0 \leqslant x_{ij} \leqslant 1$, whence Ω_n is a bounded polyhedral set, i.e. a polytope.

That each permutation matrix in Ω_n is one of its extreme points follows easily from the definitions of extreme point and permutation matrix. The non-trivial part of the proof is to show that each extreme point of Ω_n is a permutation matrix.

Let (a_{ij}) be an extreme point of Ω_n. Then, by Theorem 4.4.2, (a_{ij}) is a non-negative basic solution for the system of the $2n - 1$ equations

(5.8.3) and (5.8.4), i.e. the system $\mathbf{Ax} = \mathbf{b}$, where $\mathbf{A}$ is the $(2n - 1) \times n^2$ matrix

$$\begin{bmatrix} 111 \ldots 11 & 000 \ldots 00 & \ldots & 000 \ldots 00 \\ 000 \ldots 00 & 111 \ldots 11 & \ldots & 000 \ldots 00 \\ \cdots & \cdots & \cdots & \cdots \\ 000 \ldots 00 & 000 \ldots 00 & \ldots & 111 \ldots 11 \\ 100 \ldots 00 & 100 \ldots 00 & \ldots & 100 \ldots 00 \\ 010 \ldots 00 & 010 \ldots 00 & \ldots & 010 \ldots 00 \\ \cdots & \cdots & \cdots & \cdots \\ 000 \ldots 10 & 000 \ldots 10 & \ldots & 000 \ldots 10 \end{bmatrix}$$

and $\mathbf{b} = (1, \ldots, 1) \in \mathbb{R}^{2n-1}$. At least $n^2 - (2n - 1) = (n - 1)^2$ of the a_{ij} must be zero; the others, $a_1, \ldots, a_{2n-1}$ say, satisfy a system of linear equations of the form

$$\mathbf{B}(a_1, \ldots, a_{2n-1}) = \mathbf{b},$$

where $\mathbf{B}$ is a non-singular $(2n - 1) \times (2n - 1)$ submatrix of $\mathbf{A}$. The matrix $\mathbf{B}$ satisfies the conditions of the lemma, so $\det \mathbf{B} = \pm 1$. Thus the elements of $\mathbf{B}^{-1}$, and hence of $(a_1, \ldots, a_{2n-1})$, are integers. It follows that the doubly stochastic matrix (a_{ij}) has only integer elements, and so must be a permutation matrix.

We complete the proof by noting that a polytope is the convex hull of its extreme points. □

Suppose now that $\boldsymbol{\lambda}$ is an n-tuple of the (necessarily real) eigenvalues, in some order, of a real symmetric $n \times n$ matrix $\mathbf{A}$. The set $\Lambda_{\mathbf{A}}$ of all such n-tuples $\boldsymbol{\lambda}$ is called the **$\boldsymbol{\lambda}$-set** of $\mathbf{A}$. Clearly $\Lambda_{\mathbf{A}}$ is a finite set containing at most $n!$ points.

Theorem 5.8.3 *Let $f: X \to \mathbb{R}$ be a convex function which is defined on a convex set X in $\mathbb{R}^n$ containing the $\boldsymbol{\lambda}$-set $\Lambda_{\mathbf{A}}$ of a real symmetric $n \times n$ matrix $\mathbf{A}$. Let $(\lambda_1, \ldots, \lambda_n)$ be a point of $\Lambda_{\mathbf{A}}$ where f assumes its maximum on $\Lambda_{\mathbf{A}}$. Then, for any orthonormal sequence $\mathbf{v}_1, \ldots, \mathbf{v}_n$ in $\mathbb{R}^n$,*

$$f(\mathbf{v}_1^{\mathrm{T}}\mathbf{A}\mathbf{v}_1, \ldots, \mathbf{v}_n^{\mathrm{T}}\mathbf{A}\mathbf{v}_n) \leqslant f(\lambda_1, \ldots, \lambda_n).$$

Proof We show first that the point $\mathbf{v} = (\mathbf{v}_1^{\mathrm{T}}\mathbf{A}\mathbf{v}_1, \ldots, \mathbf{v}_n^{\mathrm{T}}\mathbf{A}\mathbf{v}_n)$ lies in X, and then that $f(\mathbf{v}) \leqslant f(\boldsymbol{\lambda})$, where $\boldsymbol{\lambda} = (\lambda_1, \ldots, \lambda_n)$. Let $\mathbf{u}_1, \ldots, \mathbf{u}_n$ be an orthonormal sequence of eigenvectors of $\mathbf{A}$ corresponding to the eigenvalues $\lambda_1, \ldots, \lambda_n$. Then, as we proved at the beginning of this section, there is a matrix $\mathbf{S}$ of Ω_n such that $\mathbf{v} = \mathbf{S}\boldsymbol{\lambda}$. By Birkhoff's theorem, there exist $\mu_1, \ldots, \mu_m \geqslant 0$ with $\mu_1 + \ldots + \mu_m = 1$ such that

$$\mathbf{S} = \mu_1\mathbf{P}_1 + \ldots + \mu_m\mathbf{P}_m,$$

where $\mathbf{P}_1, \ldots, \mathbf{P}_m$ are the permutation matrices in Ω_n. Hence

$$\mathbf{v} = \mathbf{S}\boldsymbol{\lambda} = \mu_1(\mathbf{P}_1\boldsymbol{\lambda}) + \ldots + \mu_m(\mathbf{P}_m\boldsymbol{\lambda}) \in \operatorname{conv} \Lambda_{\mathbf{A}} \subseteq X.$$

The convexity of f shows that

$$f(\mathbf{v}) \leqslant \mu_1 f(\mathbf{P}_1\boldsymbol{\lambda}) + \ldots + \mu_m f(\mathbf{P}_m\boldsymbol{\lambda}) \leqslant \mu_1 f(\boldsymbol{\lambda}) + \ldots + \mu_m f(\boldsymbol{\lambda}) = f(\boldsymbol{\lambda}). \quad \square$$

Theorem 5.8.4 *Let* $\mathbf{A}$ *be a non-negative semidefinite* $n \times n$ *matrix with eigenvalues* $\lambda_1, \ldots, \lambda_n$. *Then, for any orthonormal sequence* $\mathbf{v}_1, \ldots, \mathbf{v}_n$ *in* $\mathbb{R}^n$,

$$\det \mathbf{A} = \lambda_1 \ldots \lambda_n \leqslant \prod_{j=1}^{n} \mathbf{v}_j^{\mathrm{T}} \mathbf{A} \mathbf{v}_j.$$

Proof Since $\mathbf{A}$ is non-negative semidefinite, $\lambda_1, \ldots, \lambda_n \geqslant 0$. The function $f: X \to \mathbb{R}$ defined on the non-negative orthant X of $\mathbb{R}^n$ by the equation

$$f(x_1, \ldots, x_n) = -(x_1 \ldots x_n)^{1/n} \qquad \text{for} \quad x_1, \ldots, x_n \geqslant 0$$

is easily seen to be convex from Corollary 5.2.5. The λ-set of $\mathbf{A}$ is clearly contained in X. Theorem 5.8.3 shows that

$$-\left(\prod_{j=1}^{n} \mathbf{v}_j^{\mathrm{T}} \mathbf{A} \mathbf{v}_j\right)^{1/n} \leqslant -(\lambda_1 \ldots \lambda_n)^{1/n} = -(\det \mathbf{A})^{1/n},$$

and the desired result is immediate. $\square$

We now derive the best known of matrix inequalities, which was first proved in 1893, by J. Hadamard.

Theorem 5.8.5 (Hadamard's determinant inequality) *Let* $\mathbf{A} = [a_{ij}]$ *be a real* $n \times n$ *matrix. Then*

$$(\det \mathbf{A})^2 \leqslant (a_{11}^2 + \ldots + a_{n1}^2) \ldots (a_{1n}^2 + \ldots + a_{nn}^2).$$

If $\mathbf{A}$ *is non-negative semidefinite, then*

$$\det \mathbf{A} \leqslant a_{11} \ldots a_{nn}.$$

Proof Let $\mathbf{B} = [b_{ij}]$ denote the non-negative semidefinite matrix $\mathbf{A}^{\mathrm{T}}\mathbf{A}$. Then applying Theorem 5.8.4 to $\mathbf{B}$, and using the orthonormal sequence $\mathbf{e}_1, \ldots, \mathbf{e}_n$ of elementary vectors, we deduce that

$$\det \mathbf{B} = (\det \mathbf{A})^2 \leqslant \prod_{j=1}^{n} \mathbf{e}_j^{\mathrm{T}} \mathbf{B} \mathbf{e}_j = b_{11} \ldots b_{1n}.$$

Hence

$$(\det \mathbf{A})^2 \leqslant (a_{11}^2 + \ldots + a_{n1}^2) \ldots (a_{1n}^2 + \ldots + a_{nn}^2).$$

When $\mathbf{A}$ is itself non-negative semidefinite, we apply Theorem 5.8.4 to $\mathbf{A}$ and the orthonormal sequence $\mathbf{e}_1, \ldots, \mathbf{e}_n$ to deduce the inequality $\det \mathbf{A} \leqslant a_{11} \ldots a_{nn}$. □

Theorem 5.8.6 (Minkowski's determinant inequality) *Let* $\mathbf{A}$, $\mathbf{B}$ *be non-negative semidefinite* $n \times n$ *matrices. Then*

$$(\det(\mathbf{A}+\mathbf{B}))^{1/n} \geqslant (\det \mathbf{A})^{1/n} + (\det \mathbf{B})^{1/n}.$$

Proof Let $\mathbf{v}_1, \ldots, \mathbf{v}_n$ be an orthonormal sequence of eigenvectors of the non-negative semidefinite matrix $\mathbf{A}+\mathbf{B}$ corresponding to eigenvalues $\lambda_1, \ldots, \lambda_n$. Then

$$\begin{aligned}(\det(\mathbf{A}+\mathbf{B}))^{1/n} &= (\lambda_1 \ldots \lambda_n)^{1/n} \\ &= \left(\prod_{j=1}^{n} \mathbf{v}_j^{\mathrm{T}}(\mathbf{A}+\mathbf{B})\mathbf{v}_j\right)^{1/n} \\ &= \left(\prod_{j=1}^{n} (\mathbf{v}_j^{\mathrm{T}}\mathbf{A}\mathbf{v}_j + \mathbf{v}_j^{\mathrm{T}}\mathbf{B}\mathbf{v}_j)\right)^{1/n} \\ &\geqslant \left(\prod_{j=1}^{n} \mathbf{v}_j^{\mathrm{T}}\mathbf{A}\mathbf{v}_j\right)^{1/n} + \left(\prod_{j=1}^{n} \mathbf{v}_j^{\mathrm{T}}\mathbf{B}\mathbf{v}_j\right)^{1/n} \\ &\geqslant (\det \mathbf{A})^{1/n} + (\det \mathbf{B})^{1/n}.\end{aligned}$$

The last two inequalities follow from Corollary 5.2.5 and Theorem 5.8.4, respectively. □

We now establish some inequalities concerning the diagonals of a doubly stochastic matrix. A **diagonal** of a real $n \times n$ matrix $\mathbf{A} = [a_{ij}]$ is a finite sequence $a_{1\sigma(1)}, \ldots, a_{n\sigma(n)}$ of elements of $\mathbf{A}$, where $\sigma(1), \ldots, \sigma(n)$ is a permutation of $1, \ldots, n$. To form such a diagonal, we first choose any element d_1 in the first row of $\mathbf{A}$. Next we choose any element d_2 in the second row of $\mathbf{A}$ not lying in the same column as d_1. Then we choose any element d_3 in the third row of $\mathbf{A}$ not lying in the same column as either d_1 or d_2. Continuing in this way, we produce a diagonal $d_1, \ldots, d_n$ of $\mathbf{A}$. Clearly $\mathbf{A}$ has at most $n!$ different diagonals. The diagonal $a_{11}, \ldots, a_{nn}$ is called the **leading diagonal** of $\mathbf{A}$.

A diagonal $d_1, \ldots, d_n$ of $\mathbf{A}$ is said to be **positive** if $d_1, \ldots, d_n > 0$. It is a non-trivial fact that a doubly stochastic matrix always has a positive diagonal. To see that this is so, we observe that, by Birkhoff's theorem, each doubly stochastic matrix $\mathbf{A}$ in Ω_n can be expressed in the form

$$\mathbf{A} = \lambda_1 \mathbf{P}_1 + \ldots + \lambda_m \mathbf{P}_m,$$

where $\mathbf{P}_1, \ldots, \mathbf{P}_m$ are permutation matrices and $\lambda_1, \ldots, \lambda_m > 0$ with $\lambda_1 + \ldots + \lambda_m = 1$. For each $i = 1, \ldots, n$, let $\mathbf{P}_1$ have a 1 in its

ith row and $\sigma(i)$th column. Then $a_{1\sigma(1)}, \ldots, a_{n\sigma(n)}$ is a positive diagonal of **A**.

Before proving our main result concerning the positive diagonals of a doubly stochastic matrix, we need the following important corollary to Birkhoff's theorem.

Theorem 5.8.7 *Let* $\mathbf{C} = [c_{ij}]$ *be a real* $n \times n$ *matrix. Then there exists a diagonal* $c_{1\sigma(1)}, \ldots, c_{n\sigma(n)}$ *of* **C** *such that*

$$c_{1\sigma(1)} + \ldots + c_{n\sigma(n)} \leqslant \sum_{i,j=1}^{n} c_{ij} s_{ij},$$

for every double stochastic $n \times n$ *matrix* $\mathbf{S} = [s_{ij}]$.

Proof Define a function $f: \Omega_n \to \mathbb{R}$ by the equation

$$f(\mathbf{S}) = \sum_{i,j=1}^{n} c_{ij} s_{ij},$$

for each doubly stochastic matrix $\mathbf{S} = [s_{ij}]$ in Ω_n. Let $\mathbf{P}_1, \ldots, \mathbf{P}_m$ be the permutation matrices in Ω_n. Choose one of these matrices, $\mathbf{P} = [p_{ij}]$ say, for which

$$f(\mathbf{P}) = \min\{f(\mathbf{P}_1), \ldots, f(\mathbf{P}_m)\}.$$

Suppose that the 1 in the ith row of **P** lies in its $\sigma(i)$th column. By Birkhoff's theorem, each doubly stochastic matrix $\mathbf{S} = [s_{ij}]$ in Ω_n can be written in the form

$$\mathbf{S} = \lambda_1 \mathbf{P}_1 + \ldots + \lambda_m \mathbf{P}_m,$$

for some $\lambda_1, \ldots, \lambda_m \geqslant 0$ with $\lambda_1 + \ldots + \lambda_m = 1$. Thus

$$f(\mathbf{S}) = \lambda_1 f(\mathbf{P}_1) + \ldots + \lambda_m f(\mathbf{P}_m) \geqslant f(\mathbf{P})$$

and

$$f(\mathbf{P}) = \sum_{i,j=1}^{n} c_{ij} p_{ij} = c_{1\sigma(1)} + \ldots + c_{n\sigma(n)} \leqslant f(\mathbf{S}) = \sum_{i,j=1}^{n} c_{ij} s_{ij}. \quad \square$$

The next result shows that each doubly stochastic matrix contains a diagonal whose elements are, *on average*, not too small.

Theorem 5.8.8 *Each doubly stochastic* $n \times n$ *matrix has a positive diagonal whose harmonic mean is at least* $1/n$.

Proof Let $\mathbf{A} = [a_{ij}]$ be an $n \times n$ doubly stochastic matrix. Define an $n \times n$ matrix $[c_{ij}]$ by the equations

$$c_{ij} = \begin{cases} 1/a_{ij} & \text{for} \quad a_{ij} > 0 \\ n^2 + 1 & \text{for} \quad a_{ij} = 0. \end{cases}$$

By Theorem 5.8.7, some diagonal $c_{1\sigma(1)}, \ldots, c_{n\sigma(n)}$ of $[c_{ij}]$ satisfies the inequalities

$$c_{1\sigma(1)} + \ldots + c_{n\sigma(n)} \leqslant \sum_{i,j=1}^{n} c_{ij} a_{ij} \leqslant n^2. \tag{5.8.5}$$

Now all the terms on the left-hand side of (5.8.5) are positive, and so no term can be equal to $n^2 + 1$. This implies that, for $i = 1, \ldots, n$, $a_{i\sigma(i)} > 0$ and $c_{i\sigma(i)} = 1/a_{i\sigma(i)}$. Thus, from (5.8.5),

$$1/a_{1\sigma(1)} + \ldots + 1/a_{n\sigma(n)} \leqslant n^2.$$

Consequently, the harmonic mean

$$\left(\frac{1}{n}\left(\frac{1}{a_{1\sigma(1)}} + \ldots + \frac{1}{a_{n\sigma(n)}}\right)\right)^{-1}$$

of the diagonal $a_{1\sigma(1)}, \ldots, a_{n\sigma(n)}$ is at least $1/n$. □

Corollary 5.8.9 *Each doubly stochastic $n \times n$ matrix $[a_{ij}]$ has a positive diagonal $a_{1\sigma(1)}, \ldots, a_{n\sigma(n)}$ satisfying the inequalities*

$$a_{1\sigma(1)} + \ldots + a_{n\sigma(n)} \geqslant 1 \qquad \text{and} \qquad a_{1\sigma(1)} \ldots a_{n\sigma(n)} \geqslant n^{-n}.$$

Proof The corollary follows from the theorem and the fact that the harmonic mean of n positive numbers exceeds neither their arithmetic mean nor their geometric mean. □

Exercises 5.8

1. Show that every doubly stochastic $n \times n$ matrix is a convex combination of $n^2 - 2n + 2$ or fewer permutation matrices.
2. Let $\Lambda_{\mathbf{A}}$ be the $\boldsymbol{\lambda}$-set of a real symmetric $n \times n$ matrix $\mathbf{A}$. Show that, if $\boldsymbol{\lambda}$ is any member of $\Lambda_{\mathbf{A}}$, then

$$\operatorname{conv} \Lambda_{\mathbf{A}} = \{\mathbf{S}\boldsymbol{\lambda} : \mathbf{S} \in \Omega_n\}.$$

3. For each $\mathbf{x} = (x_1, \ldots, x_n)$, denote by $\mathbf{x}^*$ the point $(x_1^*, \ldots, x_n^*)$, where $x_1^*, \ldots, x_n^*$ are the numbers $x_1, \ldots, x_n$ arranged in decreasing order of magnitude. Show that, if $\mathbf{y} = \mathbf{S}\mathbf{x}$, where $\mathbf{y} = (y_1, \ldots, y_n)$ and $\mathbf{S} \in \Omega_n$, then

$$y_1^* + \ldots + y_k^* \leqslant x_1^* + \ldots + x_k^* \qquad \text{for } k = 1, \ldots, n;$$

and

$$y_1^* + \ldots + y_n^* = x_1^* + \ldots + x_n^*.$$

4. Let $\lambda_1, \ldots, \lambda_n$ be the eigenvalues of a real symmetric $n \times n$ matrix $\mathbf{A}$ arranged in increasing order of magnitude. Prove that, for any orthonormal sequence $\mathbf{v}_1, \ldots, \mathbf{v}_n$ in $\mathbb{R}^n$,

$$\sum_{i=1}^{k} \lambda_i \leqslant \sum_{i=1}^{k} \mathbf{v}_i^{\mathrm{T}} \mathbf{A} \mathbf{v}_i \leqslant \sum_{i=1}^{k} \lambda_{n-i+1} \qquad \text{for} \quad k = 1, \ldots, n.$$

What can be deduced about the numbers

$$\sup \{\mathbf{v}^{\mathrm{T}} \mathbf{A} \mathbf{v} : \|\mathbf{v}\| = 1\} \qquad \text{and} \qquad \inf \{\mathbf{v}^{\mathrm{T}} \mathbf{A} \mathbf{v} : \|\mathbf{v}\| = 1\} \text{ ?}$$

5. Let $\lambda_1, \ldots, \lambda_n$ be the eigenvalues of a positive definite $n \times n$ matrix $\mathbf{A}$ arranged in increasing order of magnitude. Show that, for each unit vector $\mathbf{x}$ in $\mathbb{R}^n$, there exist scalars $\alpha_1, \ldots, \alpha_n \geqslant 0$ satisfying $\alpha_1 + \ldots + \alpha_n = 1$ such that, for any $c > 0$,

$$\begin{aligned}(\mathbf{x}^{\mathrm{T}} \mathbf{A} \mathbf{x})(\mathbf{x}^{\mathrm{T}} \mathbf{A}^{-1} \mathbf{x}) &= \left(\sum_{i=1}^{n} \alpha_i \lambda_i\right)\left(\sum_{i=1}^{n} \alpha_i \lambda_i^{-1}\right) \\ &= \left(\sum_{i=1}^{n} \alpha_i \lambda_i c\right)\left(\sum_{i=1}^{n} \alpha_i \lambda_i^{-1} c^{-1}\right).\end{aligned}$$

Deduce, by taking $c = (\lambda_1 \lambda_n)^{-1/2}$, *Kantorovich's inequality:*

$$1 \leqslant (\mathbf{x}^{\mathrm{T}} \mathbf{A} \mathbf{x})(\mathbf{x}^{\mathrm{T}} \mathbf{A}^{-1} \mathbf{x}) \leqslant \frac{1}{4}\left(\left(\frac{\lambda_1}{\lambda_n}\right)^{1/2} + \left(\frac{\lambda_n}{\lambda_1}\right)^{1/2}\right)^2.$$

6. Let $\mathbf{A}, \mathbf{B}$ be real $n \times n$ matrices. Given that no element of $\mathbf{A}$ exceeds M in modulus, and that r of $\mathbf{B}$'s elements are 1's, the others being 0, show that

$$|\det \mathbf{A}| \leqslant M^n n^{n/2} \qquad \text{and} \qquad |\det \mathbf{B}| \leqslant (r/n)^{n/2}.$$

7. Prove or refute the following assertions. Every doubly stochastic $n \times n$ matrix has a diagonal: (i) all of whose elements are at least $1/n$; (ii) all of whose elements are at most $1/n$; (iii) the average of whose elements is at most $1/n$.

8. Show that every doubly stochastic $n \times n$ matrix with *exactly* r positive elements has a diagonal whose harmonic mean is at least n/r.

6
Mixed volumes and extremum problems

The first three sections introduce *volume* for convex sets, establish its properties and show how it can be calculated, while the remaining ones discuss the theory of mixed volumes, surface area, and two of geometry's most famous extremum problems.

The chapter contains some attractive, but quite sophisticated mathematics. It should, perhaps, be omitted on a first reading by those seeking only a basic introduction to convexity. A more acceptable alternative, for some readers, might be to assume familiarity with the properties of volume in $\mathbb{R}^3$, and then proceed through the chapter, omitting proofs and setting the whole discussion against the backdrop of $\mathbb{R}^3$ itself.

6.1 ELEMENTARY SETS

The first two sections of this chapter are devoted to defining the concept of *volume* for bounded convex sets. We begin by introducing the idea of *volume* for a particularly simple class of sets, the so-called *elementary sets*, which although of little interest in themselves, serve as building blocks in the general theory.

The basic elementary set is the *cell.* In $\mathbb{R}^1$ a **cell** is simply a bounded convex subset of the real line, i.e. a set of one of the following forms, in which $a, b \in \mathbb{R}$ with $a < b$:

$$\varnothing, \qquad \{a\}, \qquad [a, b], \qquad [a, b), \qquad (a, b], \qquad (a, b).$$

This definition is generalized to $\mathbb{R}^n$ as follows: a **cell** I in $\mathbb{R}^n$ is a set of the form

$$I = I_1 \times \ldots \times I_n = \{(x_1, \ldots, x_n) : x_1 \in I_1, \ldots, x_n \in I_n\},$$

where $I_1, \ldots, I_n$ are cells in $\mathbb{R}^1$. The empty set and singletons are examples of degenerate cells in $\mathbb{R}^n$. A typical cell in $\mathbb{R}^2$ is a closed rectangle with sides parallel to the coordinate axes, possibly having some or all of its sides removed.

We first mention some properties of cells on the real line. Let I and J be cells in $\mathbb{R}^1$. Then I and J are bounded convex sets, whence so too are $\operatorname{cl} I$, $\operatorname{int} I$, $I \cap J$, and $I + J$. Thus in $\mathbb{R}^1$ the closure and the interior of a cell are cells, as too are the intersection and the vector sum of two cells. In general, the set difference $I \setminus J$ is not a cell; it is, however, easily verified that $I \setminus J$ can be expressed as the union of two disjoint cells (one or both of which may be empty). For example,

$$[3, 7) \setminus (4, 5] = [3, 4] \cup (5, 7).$$

Now let I and J be cells in $\mathbb{R}^n$ specified by the equations

$$I = I_1 \times \ldots \times I_n \quad \text{and} \quad J = J_1 \times \ldots \times J_n,$$

where $I_1, \ldots, I_n, J_1, \ldots, J_n$ are cells in $\mathbb{R}^1$. It is easily verified that

$$\operatorname{cl} I = \operatorname{cl} I_1 \times \ldots \times \operatorname{cl} I_n \quad \text{and} \quad \operatorname{int} I = \operatorname{int} I_1 \times \ldots \times \operatorname{int} I_n,$$

whence the closure and the interior of a cell in $\mathbb{R}^n$ are also cells. The readily established relations

$$I \cap J = (I_1 \cap J_1) \times \ldots \times (I_n \cap J_n)$$

and

$$I + J = (I_1 + J_1) \times \ldots \times (I_n + J_n)$$

show that the intersection and the vector sum of two cells in $\mathbb{R}^n$ are themselves cells. We show that the set difference $I \setminus J$ can be expressed as a finite union of pairwise disjoint cells. For $i = 1, \ldots, n$, $I_i \cap J_i$ is a cell contained in the cell I_i, whence, since cells in $\mathbb{R}^1$ are simply intervals, there exist cells P_i and Q_i in $\mathbb{R}^1$ such that the equation

$$I_i = P_i \cup Q_i \cup (I_i \cap J_i)$$

expresses I_i as a union of three pairwise disjoint cells. It follows that

$$I = (P_1 \cup Q_1 \cup (I_1 \cap J_1)) \times \ldots \times (P_n \cup Q_n \cup (I_n \cap J_n)),$$

whence, by elementary set theory, I can be written as a union of 3^n pairwise disjoint cells in $\mathbb{R}^n$, one of which is

$$(I_1 \cap J_1) \times \ldots \times (I_n \cap J_n) = I \cap J.$$

Thus $I \setminus J$, which equals $I \setminus (I \cap J)$, can be written as the union of $3^n - 1$ pairwise disjoint cells.

A set which can be expressed as a finite union of pairwise disjoint cells in $\mathbb{R}^n$ is called an **elementary set**. Every cell is an elementary set, as also is the set difference of two cells. The next three results show that the class of elementary sets in $\mathbb{R}^n$ is closed under a wide variety of operations: algebraic, topological, and set-theoretic.

Theorem 6.1.1 *Let A and B be elementary sets in $\mathbb{R}^n$. Then $A \cap B$, $A \setminus B$, $A \cup B$, and $A + B$ are elementary sets.*

Proof Suppose that the equations

$$A = \bigcup_{i=1}^{m} I_i \quad \text{and} \quad B = \bigcup_{j=1}^{p} J_j$$

express A and B as finite unions of pairwise disjoint cells in $\mathbb{R}^n$. Then the equation

$$A \cap B = \bigcup_{i=1}^{m} \bigcup_{j=1}^{p} (I_i \cap J_j)$$

expresses $A \cap B$ as a finite union of pairwise disjoint cells in $\mathbb{R}^n$, whence $A \cap B$ is an elementary set. This result easily implies that the intersection of any finite non-zero number of elementary sets is again an elementary set.

Now, for $i = 1, \ldots, m$,

$$I_i \backslash B = I_i \backslash \bigcup_{j=1}^{p} J_j = \bigcap_{j=1}^{p} (I_i \backslash J_j),$$

and so

$$A \backslash B = (\bigcup_{i=1}^{m} I_i) \backslash B = \bigcup_{i=1}^{m} \bigcap_{j=1}^{p} (I_i \backslash J_j).$$

Thus $A \backslash B$ is a finite union of pairwise disjoint elementary sets, and so is itself an elementary set.

The equation

$$A \cup B = (A \backslash B) \cup (B \backslash A) \cup (A \cap B)$$

shows that $A \cup B$ is a finite union of pairwise disjoint elementary sets, and hence is itself an elementary set. This result easily implies that the union of any finite number of elementary sets is an elementary set.

The equation

$$A + B = \bigcup_{i=1}^{m} \bigcup_{j=1}^{p} (I_i + J_j)$$

exhibits $A + B$ as a finite union of elementary sets, and so $A + B$ is an elementary set. □

Corollary 6.1.2 *Every union of a finite number, and every intersection of a finite non-zero number of elementary sets in $\mathbb{R}^n$ is an elementary set.* □

Corollary 6.1.3 *The closure, the interior, and the boundary of an elementary set in $\mathbb{R}^n$ are elementary sets.*

Proof Suppose that in $\mathbb{R}^n$ the elementary set A is the union of the pairwise disjoint cells $I_1, \ldots, I_m$. Then

$$\text{cl}\, A = (\text{cl}\, I_1) \cup \ldots \cup (\text{cl}\, I_m),$$

which shows that $\text{cl}\, A$ is a union of the cells $\text{cl}\, I_1, \ldots, \text{cl}\, I_m$, and so is an elementary set by Corollary 6.1.2.

Let I be an *open* cell in $\mathbb{R}^n$ containing A. Then an argument similar to that used in the proof of Corollary 1.7.4 shows that

$$\text{int}\, A = I \backslash \text{cl}\, (I \backslash A),$$

whence, by the theorem and the first part of this corollary, $\text{int}\, A$ is an elementary set.

Finally, $\text{bd}\, A = \text{cl}\, A \setminus \text{int}\, A$, and so $\text{bd}\, A$ is an elementary set by the theorem, since $\text{cl}\, A$ and $\text{int}\, A$ are elementary sets. □

We now turn our attention to the main objective of this section, the definition of volume for elementary sets. The **length** $l(I)$ of a cell I in $\mathbb{R}^1$ is defined to be zero when I is empty or a singleton, and to be $b - a$ when I is a cell of one of the forms $[a, b]$, $[a, b)$, $(a, b]$, or (a, b), where $a < b$. Suppose next that I is the cell $I_1 \times \ldots \times I_n$ in $\mathbb{R}^n$, where $I_1, \ldots, I_n$ are cells in $\mathbb{R}^1$. Then the **volume** $v(I)$ of I is (uniquely) defined by the equation

$$v(I) = l(I_1) \ldots l(I_n),$$

i.e. $v(I)$ is the product of the lengths of the cells from which I is constructed. This is a natural generalization of the definition of the area of a rectangle and the volume of a rectangular block as encountered in elementary geometry. When I is a cell in $\mathbb{R}^1$, we have $v(I) = l(I)$. Before we can extend the definition of volume to elementary sets, we need the following result.

Theorem 6.1.4 *Let $I_0, I_1, \ldots, I_m$, where $m \geqslant 1$, be cells in $\mathbb{R}^n$ with $I_1, \ldots, I_m$ pairwise disjoint and having union I_0. Then*

$$v(I_0) = \sum_{i=1}^{m} v(I_i).$$

Proof We argue by induction on m. The assertion is trivially true when $m = 1$. Suppose, then, that $m > 1$ and that the assertion is true for all partitions of a cell into fewer than m cells. If one of the cells $I_1, \ldots, I_m$ is empty, the assertion follows from the induction hypothesis and the fact that the empty cell has volume zero. Assume, then, that none of $I_1, \ldots, I_m$ is empty. For $i = 0, 1, \ldots, m$, let

$$I_i = I_{i1} \times \ldots \times I_{in},$$

where $I_{i1}, \ldots, I_{in}$ are cells in $\mathbb{R}^1$. By hypothesis,

$$I_1 \cap I_2 = (I_{11} \cap I_{21}) \times \ldots \times (I_{1n} \cap I_{2n}) = \varnothing,$$

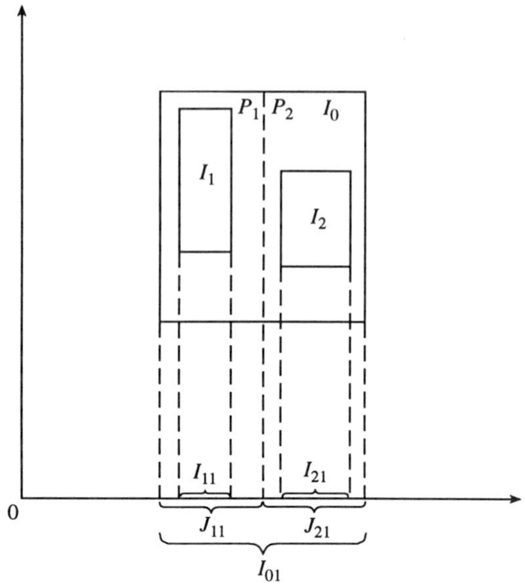

Fig. 6.1

and so one of the cells $I_{11} \cap I_{21}, \ldots, I_{1n} \cap I_{2n}$ must be empty. Suppose that $I_{11} \cap I_{21}$ is empty. Since $I_1 \cup I_2 \subseteq I_0$ and neither of I_1 and I_2 is empty, $I_{11} \cup I_{21} \subseteq I_{01}$. It now follows easily that there exist cells J_{11}, J_{21} in $\mathbb{R}^1$ such that

$$I_{11} \subseteq J_{11}, \qquad I_{21} \subseteq J_{21}, \qquad J_{11} \cup J_{21} = I_{01}, \qquad J_{11} \cap J_{21} = \varnothing.$$

See Fig. 6.1.

Define cells P_1 and P_2 in $\mathbb{R}^n$ by the equations

$$P_1 = J_{11} \times I_{02} \times \ldots \times I_{0n} \qquad \text{and} \qquad P_2 = J_{21} \times I_{02} \times \ldots \times I_{0n}.$$

Then $P_1 \cup P_2 = I_0$, $P_1 \cap P_2 = \varnothing$, and

$$v(P_1) + v(P_2) = (l(J_{11}) + l(J_{21}))l(I_{02})\ldots l(I_{0n}) = l(I_{01})\ldots l(I_{0n}) = v(I_0).$$

Since the cells $P_1 \cap I_2$ and $P_2 \cap I_1$ are empty,

$$P_1 = P_1 \cap I_0 = \bigcup_{i=1}^{m} (P_1 \cap I_i) = (P_1 \cap I_1) \cup (P_1 \cap I_3) \cup \ldots \cup (P_1 \cap I_m),$$

$$P_2 = P_2 \cap I_0 = \bigcup_{i=1}^{m} (P_2 \cap I_i) = (P_2 \cap I_2) \cup (P_2 \cap I_3) \cup \ldots \cup (P_2 \cap I_m).$$

We deduce, using the induction hypothesis, that

$$v(P_1) = v(P_1 \cap I_1) + v(P_2 \cap I_3) + \ldots + v(P_1 \cap I_m) = \sum_{i=1}^{m} v(P_1 \cap I_i),$$

$$v(P_2) = v(P_2 \cap I_2) + v(P_2 \cap I_3) + \ldots + v(P_2 \cap I_m) = \sum_{i=1}^{m} v(P_2 \cap I_i).$$

For $i = 1, \ldots, m$,

$$\begin{aligned} v(P_1 \cap I_i) + v(P_2 \cap I_i) &= (l(J_{11} \cap I_{i1}) + l(J_{21} \cap I_{i1}))l(I_{i2}) \ldots l(I_{in}) \\ &= l(I_{i1})l(I_{i2}) \ldots l(I_{in}) \\ &= v(I_i). \end{aligned}$$

Thus

$$v(I_0) = v(P_1) + v(P_2) = \sum_{i=1}^{m} v(P_1 \cap I_i) + \sum_{i=1}^{m} v(P_2 \cap I_i) = \sum_{i=1}^{m} v(I_i),$$

which shows that the assertion is true for a partition of a cell into m cells. This completes the proof by induction. □

Suppose now that the elementary set A is the pairwise disjoint union of cells $I_1, \ldots, I_m$ in $\mathbb{R}^n$. Then the *obvious* way to define the volume $v(A)$ of A is to put $v(A) = v(I_1) + \ldots + v(I_m)$. Before we can do this, however, we need to show that $v(A)$ is uniquely determined by A, i.e. we must show that, if A is also the pairwise disjoint union of cells $J_1, \ldots, J_p$, then

$$v(I_1) + \ldots + v(I_m) = v(J_1) + \ldots + v(J_p).$$

This we now do.

Corollary 6.1.5 *Suppose that $I_1, \ldots, I_m$ and $J_1, \ldots, J_p$ are partitions of an elementary set A in $\mathbb{R}^n$ into cells. Then*

$$\sum_{i=1}^{m} v(I_i) = \sum_{j=1}^{p} v(J_j).$$

Proof For $i = 1, \ldots, m$, the cell I_i is the union of the pairwise disjoint cells $I_i \cap J_1, \ldots, I_i \cap J_p$. Thus, by the theorem,

$$v(I_i) = \sum_{j=1}^{p} v(I_i \cap J_j),$$

and so

$$\sum_{i=1}^{m} v(I_i) = \sum_{i=1}^{m} \sum_{j=1}^{p} v(I_i \cap J_j) = \sum_{j=1}^{p} \sum_{i=1}^{m} v(I_i \cap J_j) = \sum_{j=1}^{p} v(J_j).$$

Here we have deduced the last equation from the previous ones by interchanging the roles of the I's and the J's. □

We can now extend the volume function v from the class of all cells in $\mathbb{R}^n$ to the class of all elementary sets in $\mathbb{R}^n$. Let A be an elementary set which is the union of pairwise disjoint cells $I_1, \ldots, I_m$ in $\mathbb{R}^n$. Then the **volume** $v(A)$ of A is defined by the equation

$$v(A) = \sum_{i=1}^{m} v(I_i).$$

Corollary 6.1.5 shows that $v(A)$ is uniquely determined by A, i.e. that it is independent of the particular choice of the pairwise disjoint cells $I_1, \ldots, I_m$ whose union is A. A cell I in $\mathbb{R}^n$ is an elementary set, and so is accorded a volume by this last definition, but we defined the volume of a cell earlier in the section. Thus we have defined the volume of a cell in two ways. Corollary 6.1.5, however, shows that the two definitions attach the same volume to I. Hence we may unambiguously refer to the volume $v(I)$ of the cell I.

An immediate consequence of the definition of volume is that, if $A_1, \ldots, A_m$ are pairwise disjoint elementary sets in $\mathbb{R}^n$, then

$$v(A_1 \cup \ldots \cup A_m) = v(A_1) + \ldots + v(A_m).$$

Suppose now that A and B are elementary sets in $\mathbb{R}^n$ such that $A \subseteq B$. Then A and $B \setminus A$ are disjoint elementary sets whose union is B. Thus

$$v(B) = v(A) + v(B \setminus A), \qquad v(B \setminus A) = v(B) - v(A), \qquad v(A) \leqslant v(B).$$

Theorem 6.1.6 *Let A and B be elementary sets in $\mathbb{R}^n$. Then*

$$v(A \cup B) + v(A \cap B) = v(A) + v(B).$$

Proof The set $A \cup B$ is the union of the pairwise disjoint elementary sets $A \setminus (A \cap B)$, $B \setminus (A \cap B)$, and $A \cap B$, so by our comments preceding the theorem,

$$\begin{aligned} v(A \cup B) &= v(A \setminus (A \cap B)) + v(B \setminus (A \cap B)) + v(A \cap B) \\ &= v(A) - v(A \cap B) + v(B) - v(A \cap B) + v(A \cap B) \\ &= v(A) + v(B) - v(A \cap B). \quad \square \end{aligned}$$

Corollary 6.1.7 *Let $A_1, \ldots, A_m$ be elementary sets in $\mathbb{R}^n$. Then*

$$v(A_1 \cup \ldots \cup A_m) \leqslant v(A_1) + \ldots + v(A_m).$$

Proof We argue by induction with respect to m. The case $m = 1$ is trivial. Suppose that $m > 1$ and that the assertion is true for families of fewer than m elementary sets. Then, by Theorem 6.1.6 and the induction hypothesis,

$$\begin{aligned} v(A_1 \cup \ldots \cup A_m) &= v((A_1 \cup \ldots \cup A_{m-1}) \cup A_m) \\ &\leqslant v(A_1 \cup \ldots \cup A_{m-1}) + v(A_m) \\ &\leqslant v(A_1) + \ldots + v(A_{m-1}) + v(A_m). \end{aligned}$$

This completes the proof by induction. □

Corollary 6.1.8 *Let A be an elementary set in $\mathbb{R}^n$. Then*

$$v(\text{int}\, A) = v(A) = v(\text{cl}\, A) \qquad \textit{and} \qquad v(\text{bd}\, A) = 0.$$

Proof We make use of the trivial result that a cell, its interior, and its closure all have the same volume. Let $A = I_1 \cup \ldots \cup I_m$, where $I_1, \ldots, I_m$ are pairwise disjoint cells. Then, by Corollary 6.1.7,

$$\begin{aligned} v(\text{cl}\, A) &= v(\text{cl}\, I_1 \cup \ldots \cup \text{cl}\, I_m) \\ &\leqslant v(\text{cl}\, I_1) + \ldots + v(\text{cl}\, I_m) \\ &= v(I_1) + \ldots + v(I_m) \\ &= v(A) \\ &\leqslant v(\text{cl}\, A), \end{aligned}$$

whence $v(\text{cl}\, A) = v(A)$. Also

$$\begin{aligned} v(\text{int}\, A) &\geqslant v(\text{int}\, I_1 \cup \ldots \cup \text{int}\, I_m) \\ &= v(\text{int}\, I_1) + \ldots + v(\text{int}\, I_m) \\ &= v(I_1) + \ldots + v(I_m) \\ &= v(A) \\ &\geqslant v(\text{int}\, A), \end{aligned}$$

whence $v(\text{int}\, A) = v(A)$. Finally,

$$v(\text{bd}\, A) = v(\text{cl}\, A \backslash \text{int}\, A) = v(\text{cl}\, A) - v(\text{int}\, A) = 0. \ \square$$

Exercises 6.1

1. Prove that the intersection of every non-empty family of cells in $\mathbb{R}^n$ is a cell. Is the intersection of every non-empty family of elementary sets in $\mathbb{R}^n$ always an elementary set?
2. Let C be a bounded subset of a hyperplane in $\mathbb{R}^n$. Show that, for each $\varepsilon > 0$, there is an elementary set A in $\mathbb{R}^n$ such that $C \subseteq A$ and $v(A) < \varepsilon$.
3. Let C be the set $\{(x, y): 0 \leqslant y \leqslant x^2,\ 0 \leqslant x \leqslant 1\}$ in $\mathbb{R}^2$. Show that, for each $\varepsilon > 0$, there are elementary sets $A, B \subseteq \mathbb{R}^2$ with $A \subseteq C \subseteq B$ such that

$$\tfrac{1}{3} - \varepsilon < v(A) < v(B) < \tfrac{1}{3} + \varepsilon.$$

6.2 VOLUME

We now extend the volume function v from the class of elementary sets to a much larger one that includes all bounded convex sets. The method whereby we do this is, in essence, the principle of exhaustion so effectively used by Archimedes in determining the volumes of solids with curved boundaries. Its main feature is the approximation of a set from both within and without by elementary sets.

Denote by $\mathcal{E}$ the class of elementary sets in $\mathbb{R}^n$. Let A be the bounded set in $\mathbb{R}^n$ whose volume we wish to define. We should expect the (as yet undefined) volume of A to be an upper bound for the set of volumes of elementary sets contained in A. This observation leads us to define an **inner-volume** $\underline{v}(A)$ for A by the equation

$$\underline{v}(A) = \sup\{v(E) : E \subseteq A \text{ and } E \in \mathcal{E}\}.$$

The assumption that A is bounded ensures that $\underline{v}(A)$ is a well-defined non-negative real number. Similarly, by considering the volumes of elementary sets containing A, we are led to define an **outer-volume** $\bar{v}(A)$ by the equation

$$\bar{v}(A) = \inf\{v(E) : A \subseteq E \text{ and } E \in \mathcal{E}\}.$$

The basic properties of inner- and outer-volumes are stated in the next theorem.

Theorem 6.2.1 *Let A and B be bounded sets in $\mathbb{R}^n$. Then:*

(i) $\underline{v}(A) \leqslant \bar{v}(A)$;

(ii) $\underline{v}(A) = v(A) = \bar{v}(A)$ *when A is an elementary set;*

(iii) $\underline{v}(A) \leqslant \underline{v}(B)$ *and* $\bar{v}(A) \leqslant \bar{v}(B)$ *whenever* $A \subseteq B$;

(iv) $\underline{v}(A) = \underline{v}(\operatorname{int} A)$ *and* $\bar{v}(A) = \bar{v}(\operatorname{cl} A)$;

(v) $\underline{v}(A \cup B) + \underline{v}(A \cap B) \geqslant \underline{v}(A) + \underline{v}(B)$ *and*
$\bar{v}(A \cup B) + \bar{v}(A \cap B) \leqslant \bar{v}(A) + \bar{v}(B)$.

Proof Both (i) and (ii) follow immediately from the fact that $v(E) \leqslant v(F)$ whenever E and F are elementary sets with $E \subseteq F$, and (iii) is clear from the definitions of $\underline{v}$ and $\bar{v}$.

Suppose now that E is an elementary set with $E \subseteq A$. Then, by Corollary 6.1.3 and Corollary 6.1.8, $\operatorname{int} E$ is an elementary set with $v(\operatorname{int} E) = v(E)$. Also $\operatorname{int} E \subseteq \operatorname{int} A$, and so $\underline{v}(\operatorname{int} A) \geqslant \underline{v}(A)$. But, by (iii), $\underline{v}(\operatorname{int} A) \leqslant \underline{v}(A)$, whence $\underline{v}(\operatorname{int} A) = \underline{v}(A)$. Similarly, $\bar{v}(\operatorname{cl} A) = \bar{v}(A)$.

Finally, let E and F be elementary sets with $E \subseteq A$ and $F \subseteq B$. Then $E \cup F$ and $E \cap F$ are elementary sets with $E \cup F \subseteq A \cup B$ and $E \cap F \subseteq A \cap B$. By Theorem 6.1.6 and (ii), (iii) above,

$$\underline{v}(A \cup B) + \underline{v}(A \cap B) \geqslant v(E \cup F) + v(E \cap F) = v(E) + v(F).$$

Since this inequality holds for all elementary sets E and F with $E \subseteq A$ and $F \subseteq B$, we can deduce that

$$\underline{v}(A \cup B) + \underline{v}(A \cap B) \geqslant \underline{v}(A) + \underline{v}(B).$$

The last part of (v) is proved similarly. □

It can happen that $\underline{v}(A) < \bar{v}(A)$. For example, suppose that A is the set of rational numbers in the interval $[0, 1]$ of the real line. Then $\operatorname{int} A = \emptyset$ and $\operatorname{cl} A = [0, 1]$, whence, by Theorem 6.2.1 (ii) and (iv), $\underline{v}(A) = 0$, whereas $\bar{v}(A) = 1$. Fortunately, however, for all the sets A in which we are interested the numbers $\underline{v}(A)$ and $\bar{v}(A)$ are equal. In particular, this is true when A is a bounded convex set. We say that a set A in $\mathbb{R}^n$ **has volume** if it is bounded and $\underline{v}(A) = \bar{v}(A)$. Theorem 6.2.1 (ii) shows that every elementary set in $\mathbb{R}^n$ has volume. For each set A in $\mathbb{R}^n$ which has volume, we write $v(A)$ for the equal numbers $\underline{v}(A)$ and $\bar{v}(A)$, and we say that A has volume $v(A)$. In this way, we have extended the volume function v from the class of elementary sets to the class of all sets which have volume. The next theorem gives a useful criterion for a set to have volume.

Theorem 6.2.2 *The set A in $\mathbb{R}^n$ has volume if and only if, for each $\varepsilon > 0$, there are elementary sets E and F in $\mathbb{R}^n$ such that $E \subseteq A \subseteq F$ and $v(F \backslash E) < \varepsilon$.*

Proof Suppose that A has volume and that $\varepsilon > 0$. Then there exist elementary sets E and F in $\mathbb{R}^n$ with $E \subseteq A \subseteq F$ such that

$$v(E) > \underline{v}(A) - \tfrac{1}{2}\varepsilon = v(A) - \tfrac{1}{2}\varepsilon \quad \text{and}$$
$$v(F) < \bar{v}(A) + \tfrac{1}{2}\varepsilon = v(A) + \tfrac{1}{2}\varepsilon,$$

whence $v(F \backslash E) = v(F) - v(E) < \varepsilon$.

Conversely, suppose that, for each $\varepsilon > 0$, there are elementary sets E and F in $\mathbb{R}^n$ such that $E \subseteq A \subseteq F$ and $v(F \backslash E) < \varepsilon$. This implies that A is bounded. Let ε, E, F be as described. Then

$$0 \leqslant \bar{v}(A) - \underline{v}(A) \leqslant v(F) - v(E) = v(F \backslash E) < \varepsilon.$$

Since $\varepsilon > 0$ is arbitrary, $\underline{v}(A) = \bar{v}(A)$ and A has volume. □

Theorem 6.2.3 *Let A and B be sets in $\mathbb{R}^n$ which have volume. Then the sets $A \cup B$, $A \cap B$, and $A \backslash B$ have volume.*

Proof The theorem is an easy consequence of Theorem 6.2.2. We content ourselves with showing that $A \backslash B$ has volume; the other two proofs are similar to the one we give. Let $\varepsilon > 0$. Then there exist elementary sets E, F, G, H in $\mathbb{R}^n$ with $E \subseteq A \subseteq F$, $G \subseteq B \subseteq H$ such that $v(F \backslash E) < \frac{1}{2}\varepsilon$ and $v(H \backslash G) < \frac{1}{2}\varepsilon$. Now $E \backslash H$ and $F \backslash G$ are elementary sets with $E \backslash H \subseteq A \backslash B \subseteq F \backslash G$ and

$$(F \backslash G) \backslash (E \backslash H) \subseteq (F \backslash E) \cup (H \backslash G),$$

whence

$$v((F \backslash G) \backslash (E \backslash H)) \leqslant v(F \backslash E) + v(H \backslash G) < \varepsilon.$$

Thus $A \backslash B$ has volume. □

Corollary 6.2.4 *All unions of a finite number, and all intersections of a finite non-zero number, of sets in* $\mathbb{R}^n$ *which have volume also have volume.* □

Theorem 6.2.5 *Let A be a set in $\mathbb{R}^n$ which has volume. Then the sets* $\operatorname{int} A$ *and* $\operatorname{cl} A$ *have volume with* $v(\operatorname{int} A) = v(A) = v(\operatorname{cl} A)$.

Proof By Theorem 6.2.1,

$$\bar{v}(\operatorname{int} A) \leqslant \bar{v}(A) = v(A) = \underline{v}(A) = \underline{v}(\operatorname{int} A) \leqslant \bar{v}(\operatorname{int} A)$$

and

$$\bar{v}(\operatorname{cl} A) = \bar{v}(A) = v(A) = \underline{v}(A) \leqslant \underline{v}(\operatorname{cl} A) \leqslant \bar{v}(\operatorname{cl} A). \quad \square$$

Theorem 6.2.6 *Let A and B be sets in $\mathbb{R}^n$ which have volume. Then*

$$v(A \cup B) + v(A \cap B) = v(A) + v(B).$$

Proof By Theorems 6.2.1 and 6.2.3,

$$\begin{aligned} v(A \cup B) + v(A \cap B) &= \bar{v}(A \cup B) + \bar{v}(A \cap B) \\ &\leqslant \bar{v}(A) + \bar{v}(B) = v(A) + v(B) \end{aligned}$$

and

$$\begin{aligned} v(A \cup B) + v(A \cap B) &= \underline{v}(A \cup B) + \underline{v}(A \cap B) \\ &\geqslant \underline{v}(A) + \underline{v}(B) = v(A) + v(B). \quad \square \end{aligned}$$

Corollary 6.2.7 *Let A, B be sets in $\mathbb{R}^n$ which have volume and are such that $A \subseteq B$. Then $v(B \backslash A) = v(B) - v(A)$ and $v(A) \leqslant v(B)$.*

Proof The first assertion follows by applying the theorem to the sets $B \backslash A$ and A. The second assertion follows immediately from the first. □

Corollary 6.2.8 *Let $A_1, \ldots, A_m$ be sets in $\mathbb{R}^n$ which have volume. Then*

$$v(A_1 \cup \ldots \cup A_m) \leqslant v(A_1) + \ldots + v(A_m),$$

with equality holding when $v(A_i \cap A_j) = 0$ for $1 \leqslant i < j \leqslant m$.

Proof We argue by induction on m. The case $m = 1$ is trivial. Let A_1, A_2 be sets in $\mathbb{R}^n$ which have volume. By the theorem,

$$v(A_1 \cup A_2) + v(A_1 \cap A_2) = v(A_1) + v(A_2),$$

and so $v(A_1 \cup A_2) \leqslant v(A_1) + v(A_2)$ with equality holding when $v(A_1 \cap A_2) = 0$. Suppose that $m > 1$ and that the assertion is true for all sequences of $m - 1$ sets. Then, by the induction hypothesis and the case $m = 2$ just established,

$$\begin{aligned} v(A_1 \cup \ldots \cup A_m) &= v((A_1 \cup \ldots \cup A_{m-1}) \cup A_m) \\ &\leqslant v(A_1 \cup \ldots \cup A_{m-1}) + v(A_m) \\ &\leqslant v(A_1) + \ldots + v(A_{m-1}) + v(A_m). \end{aligned}$$

If $v(A_i \cap A_j) = 0$ when $1 \leqslant i < j \leqslant m$, then

$$\begin{aligned} v((A_1 \cup \ldots \cup A_{m-1}) \cap A_m) &= v((A_1 \cap A_m) \cup \ldots \cup (A_{m-1} \cap A_m)) \\ &\leqslant v(A_1 \cap A_m) + \ldots + v(A_{m-1} \cap A_m) \\ &= 0, \end{aligned}$$

whence

$$\begin{aligned} v(A_1 \cup \ldots \cup A_m) &= v((A_1 \cup \ldots \cup A_{m-1}) \cup A_m) \\ &= v(A_1 \cup \ldots \cup A_{m-1}) + v(A_m) \\ &= v(A_1) + \ldots + v(A_{m-1}) + v(A_m). \end{aligned}$$

Thus the assertion is true for all sequences of m sets. This completes the proof by induction. □

The next result shows that whether or not a bounded set in $\mathbb{R}^n$ has volume depends only upon its boundary.

Theorem 6.2.9 *The bounded set A in $\mathbb{R}^n$ has volume if and only if its boundary* bd A *has volume zero.*

Proof Suppose that A has volume. Then the equation $\operatorname{bd} A = \operatorname{cl} A \setminus \operatorname{int} A$, together with Theorem 6.2.5 and Corollary 6.2.7, shows that $\operatorname{bd} A$ has volume zero.

Conversely, suppose that $\operatorname{bd} A$ has volume zero. Let $\varepsilon > 0$. Then there exists an elementary set E in $\mathbb{R}^n$ with $\operatorname{bd} A \subseteq E$ and $v(E) < \varepsilon$. Let I be a cell in $\mathbb{R}^n$ containing both E and A. Then $I \setminus E$ is an elementary set. Suppose it is the union of the pairwise disjoint cells $I_1, \ldots, I_m$ in $\mathbb{R}^n$. If an I_i meets A, then it must be contained in A, for otherwise, by the convexity of I_i, it would meet $\operatorname{bd} A$, and hence E, which is impossible. Let F be the union of those I_i's which meet A. Then F is an elementary set contained in A, and $E \cup F$ is an elementary set containing A. Also

$$v((E \cup F)\backslash F) = v(E) < \varepsilon,$$

whence, by Theorem 6.2.2, A has volume. □

We are now in a position to prove the main result of this section, namely that every bounded convex set has volume. It is interesting to note that our proof of this result rests upon the fact that every non-empty compact convex set has a projection operator. The following technical lemma, concerning the regular subdivision of the boundary of an n-dimensional cube into $(n-1)$-dimensional cubes, is needed in the proof. The lemma is clear in $\mathbb{R}^3$, where it states essentially that, for any positive integer m, each of the six facets of a cube can be subdivided into m^2 congruent squares, whence the boundary of the cube can be divided into $6m^2$ congruent squares.

Lemma 6.2.10 *Let $a > 0$ and let I be the n-cube in $\mathbb{R}^n$ defined by the equation*

$$I = \{(x_1, \ldots, x_n): -a \leqslant x_i \leqslant a \text{ for } i = 1, \ldots, n\}.$$

Then, for each positive integer m, there exists a subset S of $2nm^{n-1}$ points of $\operatorname{bd} I$ *such that, for each* $\mathbf{x} \in \operatorname{bd} I$, *there is* $\mathbf{s} \in S$ *with*

$$\|\mathbf{x} - \mathbf{s}\| \leqslant a\sqrt{(n-1)}/m.$$

Proof Let J denote the set of midpoints of the intervals obtained by subdividing the interval $[-a, a]$ on the real line into m equal sub-intervals in the obvious way. Then J is a subset of $[-a, a]$ which has m points and the property that, for each x in $[-a, a]$, there is a point t of J such that $|x - t| \leqslant a/m$. Let S be that set in $\mathbb{R}^n$ consisting of all those points exactly one of whose coordinates is either a or $-a$ and whose remaining coordinates belong to the set J. Then S is a subset of $\operatorname{bd} I$ having $2nm^{n-1}$ points.

Now let $\mathbf{x} = (x_1, \ldots, x_n)$ be a point of $\operatorname{bd} I$. Then one of the coordinates of $\mathbf{x}$ must be a or $-a$ and all of its coordinates must belong to $[-a, a]$. Suppose, without loss of generality, that $x_1 = a$. By the construction of J, there exist points $s_2, \ldots, s_n$ of J (supposing $n \geqslant 2$) such that

$$|s_i - x_i| \leqslant a/m \qquad \text{for } i = 2, \ldots, n.$$

Put $\mathbf{s} = (a, s_2, \ldots, s_n)$. Then $\mathbf{s} \in S$ and

$$\begin{aligned}\|\mathbf{x} - \mathbf{s}\|^2 &= (a - a)^2 + (x_2 - s_2)^2 + \ldots + (x_n - s_n)^2 \\ &\leqslant (n-1)a^2/m^2.\end{aligned}$$

The desired result now follows. □

Theorem 6.2.11 *Every bounded convex set in* $\mathbb{R}^n$ *has volume.*

Proof We show that the boundary of a bounded convex set has volume zero, whence the set has volume by Theorem 6.2.9. Since, by Corollary 2.3.10, a convex set and its closure have the same boundary, it will suffice to prove the theorem for a compact convex set.

Let A be a non-empty compact convex set in $\mathbb{R}^n$ with projection operator f. Let $\varepsilon > 0$. Since A is bounded, there exists $a > 0$ such that A is contained in the cube I as defined in the statement of the lemma. Choose an integer m such that $m > 2^{n+1}a^n n^{n+1}/\varepsilon$, and let the set S be as in the lemma. For each $\mathbf{s} \in S$, let $I(\mathbf{s})$ be the cube in $\mathbb{R}^n$ with centre $f(\mathbf{s})$ defined by the equation

$$I(\mathbf{s}) = \{(x_1, \ldots, x_n) : |x_i - y_i| \leqslant na/m \text{ for } i = 1, \ldots, n\},$$

where $f(\mathbf{s}) = (y_1, \ldots, y_n)$. We show that the union of the cubes $I(\mathbf{s})$ for $\mathbf{s} \in S$ contains $\operatorname{bd} A$. To see why this is so, suppose that $\mathbf{a} \in \operatorname{bd} A$. By Corollary 2.4.5, there exists $\mathbf{x} \in \operatorname{bd} I$ with $f(\mathbf{x}) = \mathbf{a}$. The construction of S shows that there is $\mathbf{s} \in S$ such that

$$\|\mathbf{x} - \mathbf{s}\| \leqslant a\sqrt{(n-1)}/m \leqslant na/m.$$

Since f is a projection operator, we have

$$\|\mathbf{a} - f(\mathbf{s})\| = \|f(\mathbf{x}) - f(\mathbf{s})\| \leqslant \|\mathbf{x} - \mathbf{s}\| \leqslant na/m,$$

and it follows that $\mathbf{a} \in I(\mathbf{s})$. Thus $\operatorname{bd} A$ is contained in the union of the (at most) $2nm^{n-1}$ cubes $I(\mathbf{s})$, each of which has volume $(2na/m)^n$. By Corollary 6.2.8, the volume of this union does not exceed

$$2nm^{n-1}(2na/m)^n = 2^{n+1}a^n n^{n+1}/m < \varepsilon.$$

This shows that $\bar{v}(\operatorname{bd} A) < \varepsilon$, whence, since $\varepsilon > 0$ is arbitrary, $\bar{v}(\operatorname{bd} A) = 0$. Thus $\operatorname{bd} A$ has volume zero. □

Corollary 6.2.12 *Every bounded subset of a hyperplane in* $\mathbb{R}^n$ *has volume zero.*

Proof Let A be a bounded subset of a hyperplane in $\mathbb{R}^n$. The theorem shows that $\operatorname{conv} A$ has volume. By Theorem 6.2.5,

$$v(\operatorname{conv} A) = v(\operatorname{int}(\operatorname{conv} A)) = v(\varnothing) = 0.$$

It now follows easily that A has volume zero. □

We now consider the effect that an affine transformation has on volume. To be precise, we show that, if A is a set in $\mathbb{R}^n$ which has volume and $T: \mathbb{R}^n \to \mathbb{R}^n$ is an affine transformation with associated matrix $\mathbf{Q}$, then the image $T(A)$ of A under T has volume given by the formula

$$v(T(A)) = |\det \mathbf{Q}|\, v(A).$$

First we dispose of the simplest type of affine transformation, the translation. Let A be a set in $\mathbb{R}^n$ which has volume and let $\mathbf{a}$ be a point of $\mathbb{R}^n$. If E is an elementary set contained in A, then it is easily verified that $E + \mathbf{a}$ is an elementary set contained in $A + \mathbf{a}$ with $v(E + \mathbf{a}) = v(E)$. It follows that $\underline{v}(A + \mathbf{a}) \geqslant \underline{v}(A)$. Similarly, we have $\bar{v}(A + \mathbf{a}) \leqslant \bar{v}(A)$. But A has volume, so

$$\bar{v}(A + \mathbf{a}) \geqslant \underline{v}(A + \mathbf{a}) \geqslant \underline{v}(A) = \bar{v}(A) \geqslant \bar{v}(A + \mathbf{a}),$$

which shows that $A + \mathbf{a}$ has volume $v(A)$. We have thus proved that every translate of a set having volume has a volume equal to that of the set itself.

Before we can deal with the general affine transformation, we need to recall some basic matrix theory. A real $n \times n$ matrix is said to be an *elementary matrix* if it can be obtained from the identity matrix $\mathbf{I}_n$ by one of the following operations:

(i) the multiplication of a row by a non-zero scalar;

(ii) the interchange of two rows;

(iii) the addition of one row to another one.

The matrices

$$\begin{bmatrix} 1 & 0 & 0 \\ 0 & 1 & 0 \\ 0 & 0 & -8 \end{bmatrix}, \quad \begin{bmatrix} 0 & 0 & 1 \\ 0 & 1 & 0 \\ 1 & 0 & 0 \end{bmatrix}, \quad \begin{bmatrix} 1 & 0 & 0 \\ 0 & 1 & 1 \\ 0 & 0 & 1 \end{bmatrix}$$

illustrate the types (i), (ii), and (iii) of elementary matrix for 3×3 matrices. We assume the result that every non-singular matrix can be expressed as a product of elementary matrices.

Lemma 6.2.13 *Let $T: \mathbb{R}^n \to \mathbb{R}^n$ be the linear transformation given by $T(\mathbf{x}) = \mathbf{Qx}$ for $\mathbf{x} \in \mathbb{R}^n$, where $\mathbf{Q}$ is an elementary matrix. Then, for each cell I in $\mathbb{R}^n$, the set $T(I)$ has volume $|\det \mathbf{Q}|\, v(I)$.*

Proof Let $I = I_1 \times \ldots \times I_n$, where $I_1, \ldots, I_n$ are cells in $\mathbb{R}^1$. That $T(I)$ has volume is clear, since it is bounded and convex. The proof falls naturally into three parts, corresponding to the three types of elementary matrix.

Suppose first that $\mathbf{Q}$ is an elementary matrix of type (i); say $\mathbf{Q}$ is obtained from $\mathbf{I}_n$ by multiplying its rth row by a non-zero scalar λ. Then

$$T(I) = I_1 \times \ldots \times \lambda I_r \times \ldots \times I_n$$

and

$$v(T(I)) = l(I_1)\ldots l(\lambda I_r)\ldots l(I_n) = |\lambda|\, l(I_1)\ldots l(I_n) = |\det \mathbf{Q}|\, v(I).$$

Suppose next that $\mathbf{Q}$ is an elementary matrix of type (ii), say $\mathbf{Q}$ is obtained from $\mathbf{I}_n$ by interchanging its rth and sth rows, where $r < s$. Then $\det \mathbf{Q} = -1$,

$$T(I) = I_1 \times \ldots \times I_s \times \ldots \times I_r \times \ldots \times I_n,$$

i.e. the cells I_r and I_s are transposed from their natural order, and

$$v(T(I)) = l(I_1)\ldots l(I_s)\ldots l(I_r)\ldots l(I_n) = l(I_1)\ldots l(I_n) = |\det \mathbf{Q}|\, v(I).$$

Suppose finally that $\mathbf{Q}$ is an elementary matrix of type (iii), say $\mathbf{Q}$ is obtained from $\mathbf{I}_n$ by adding its second row to its first. Then $\det \mathbf{Q} = 1$. For notational simplicity, we assume that

$$I = \{(x_1, \ldots, x_n) : a_i \leqslant x_i \leqslant b_i \text{ for } i = 1, \ldots, n\},$$

where $a_i < b_i$ for $i = 1, \ldots, n$. Then

$$T(I) = \{(x_1, \ldots, x_n) : a_1 + x_2 \leqslant x_1 \leqslant b_1 + x_2 \quad \text{and} \\ a_i \leqslant x_i \leqslant b_i \,(i = 2, \ldots, n)\}.$$

Let bounded convex sets A and B be defined by the equations

$$A = \{(x_1, \ldots, x_n) : a_1 + a_2 \leqslant x_1 < a_1 + x_2 \text{ and } a_i \leqslant x_i \leqslant b_i \,(i = 2, \ldots, n)\}$$

and

$$B = \{(x_1, \ldots, x_n) : b_1 + x_2 < x_1 \leqslant b_1 + b_2 \text{ and } a_i \leqslant x_i \leqslant b_i \,(i=2, \ldots, n)\}.$$

Then the cell

$$[a_1 + a_2, b_1 + b_2] \times [a_2, b_2] \times \ldots \times [a_n, b_n]$$

is the pairwise disjoint union of the sets A, $T(I)$, and B, whence

$$(b_1 - a_1 + b_2 - a_2)(b_2 - a_2)\ldots(b_n - a_n) = v(A) + v(T(I)) + v(B). \tag{6.2.1}$$

The two-dimensional case is illustrated in Fig. 6.2. The cell

$$[a_1 + a_2, a_1 + b_2] \times [a_2, b_2] \times \ldots \times [a_n, b_n]$$

is easily seen to be the disjoint union of the bounded convex sets $\operatorname{cl} A$ and $B - (b_1 - a_1)\mathbf{e}_1$, whence

$$(b_2 - a_2)(b_2 - a_2)\ldots(b_n - a_n) = v(\operatorname{cl} A) + v(B - (b_1 - a_1)\mathbf{e}_1) \\ = v(A) + v(B). \tag{6.2.2}$$

Equations 6.2.1 and 6.2.2 taken together show that

$$v(T(I)) = (b_1 - a_1)\ldots(b_n - a_n) = v(I) = |\det \mathbf{Q}|\, v(I). \quad \square$$

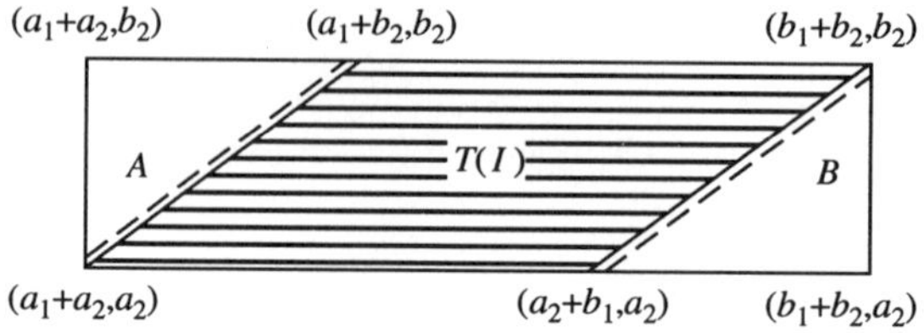

Fig. 6.2

Theorem 6.2.14 *Let $T: \mathbb{R}^n \to \mathbb{R}^n$ be the affine transformation given by $T(\mathbf{x}) = \mathbf{Qx} + \mathbf{q}$ for $\mathbf{x} \in \mathbb{R}^n$, where $\mathbf{Q}$ is an $n \times n$ real matrix and $\mathbf{q} \in \mathbb{R}^n$. Then, for each set A in $\mathbb{R}^n$ that has volume, the set $T(A)$ has volume $|\det \mathbf{Q}|\, v(A)$.*

Proof We consider first the case when $\mathbf{Q}$ is an elementary matrix and $\mathbf{q} = \mathbf{0}$. Let $\varepsilon > 0$. Then there exist pairwise disjoint cells $I_1, \ldots, I_m$ in $\mathbb{R}^n$ such that

$$I_1 \cup \ldots \cup I_m \subseteq A \text{ and } v(I_1) + \ldots + v(I_m) = v(I_1 \cup \ldots \cup I_m) > v(A) - \varepsilon.$$

Now

$$T(I_1) \cup \ldots \cup T(I_m) = T(I_1 \cup \ldots \cup I_m) \subseteq T(A).$$

Using the lemma and the fact that T is a bijection (as $\mathbf{Q}$ is non-singular), we deduce that

$$\begin{aligned} \underline{v}(T(A)) &\geqslant v(T(I_1) \cup \ldots \cup T(I_m)) \\ &= v(T(I_1)) + \ldots + v(T(I_m)) \\ &= |\det \mathbf{Q}|(v(I_1) + \ldots + v(I_m)) \\ &\geqslant |\det \mathbf{Q}|(v(A) - \varepsilon). \end{aligned}$$

Since $\varepsilon > 0$ is arbitrary, $\underline{v}(T(A)) \geqslant |\det \mathbf{Q}|\, v(A)$. Similarly, we have $\bar{v}(T(A)) \leqslant |\det \mathbf{Q}|\, v(A)$. Hence

$$|\det \mathbf{Q}|\, v(A) \leqslant \underline{v}(T(A)) \leqslant \bar{v}(T(A)) \leqslant |\det \mathbf{Q}|\, v(A),$$

which shows that $v(T(A))$ has volume $|\det \mathbf{Q}|\, v(A)$.

We now consider the case when $\mathbf{Q}$ is an arbitrary $n \times n$ real matrix and $\mathbf{q} = \mathbf{0}$. The theorem is obvious when $\mathbf{Q}$ is singular, for then $T(A)$ is a bounded subset of some hyperplane of $\mathbb{R}^n$, so both $v(T(A))$ and $|\det \mathbf{Q}|\, v(A)$ are zero. Suppose, then, that $\mathbf{Q}$ is non-singular, and that $\mathbf{Q} = \mathbf{Q}_1 \ldots \mathbf{Q}_m$, where $\mathbf{Q}_1, \ldots, \mathbf{Q}_m$ are elementary matrices. By repeated applications of the special case of the theorem just proved, we deduce that

$$\begin{aligned} v(T(A)) &= |\det \mathbf{Q}_1| \ldots |\det \mathbf{Q}_m| v(A) \\ &= |\det \mathbf{Q}_1 \ldots \mathbf{Q}_m| v(A) \\ &= |\det \mathbf{Q}| v(A). \end{aligned}$$

Finally, the removal of the restriction $\mathbf{q} = \mathbf{0}$ occasions no difficulty, since, as we have already seen, translations leave volumes unchanged. □

Corollary 6.2.15 *Let A be a set in $\mathbb{R}^n$ which has volume. Then $v(\lambda A + \mathbf{a}) = \lambda^n v(A)$ for all $\lambda \geqslant 0$ and $\mathbf{a} \in \mathbb{R}^n$.* □

Corollary 6.2.16 *Let A and B be congruent sets in $\mathbb{R}^n$ with A having volume. Then $v(B) = v(A)$.*

Proof Since A and B are congruent, there exists an affine transformation $T: \mathbb{R}^n \to \mathbb{R}^n$ given by $T(\mathbf{x}) = \mathbf{Qx} + \mathbf{q}$ for $\mathbf{x} \in \mathbb{R}^n$, where $\mathbf{Q}$ is an $n \times n$ orthogonal matrix and $\mathbf{q} \in \mathbb{R}^n$, such that $T(A) = B$. The result now follows from the theorem, since the determinant of an orthogonal matrix is ± 1. □

We now investigate how the volume function v behaves with respect to the Hausdorff distance ρ. Since each non-empty compact convex set in $\mathbb{R}^n$ has volume, it makes sense to ask: Is v continuous with respect to ρ on the class of all non-empty compact convex sets in $\mathbb{R}^n$? The answer, yes, is given in the next theorem.

Theorem 6.2.17 *Let $A, A_1, \ldots, A_k, \ldots$ be non-empty compact convex sets in $\mathbb{R}^n$ such that $A_k \to A$ as $k \to \infty$. Then $v(A_k) \to v(A)$ as $k \to \infty$.*

Proof Throughout the proof we denote the Hausdorff distance $\rho(A_k, A)$ between A_k and A by θ_k. Consider first the case when A has non-empty interior. Since both volume and the Hausdorff distance are unchanged by translations, we can assume that the origin is an interior point of A, say $rU \subseteq A$ for some $r > 0$. Choose k so large that $\theta_k < r$. Then, by the definition of ρ,

$$A_k \subseteq A + \theta_k U \subseteq A + \frac{\theta_k}{r} A = \left(1 + \frac{\theta_k}{r}\right) A$$

and

$$\left(1 - \frac{\theta_k}{r}\right) A + \frac{\theta_k}{r} A = A \subseteq A_k + \theta_k U \subseteq A_k + \frac{\theta_k}{k} A,$$

whence, by Theorem 2.1.8,

$$\left(1 - \frac{\theta_k}{r}\right) A \subseteq A_k.$$

Thus

$$\left(1-\frac{\theta_k}{r}\right)A \subseteq A_k \subseteq \left(1+\frac{\theta_k}{r}\right)A \qquad \text{and}$$

$$\left(1-\frac{\theta_k}{r}\right)^n v(A) \leqslant v(A_k) \leqslant \left(1+\frac{\theta_k}{r}\right)^n v(A),$$

so $v(A_k) \to v(A)$ as $k \to \infty$.

Suppose now that A has empty interior. Then A lies in some hyperplane of $\mathbb{R}^n$. If $n = 1$, then A is a singleton and $v(A_k) \leqslant 2\theta_k$, so $v(A_k) \to 0 = v(A)$ as $k \to \infty$. Suppose, then, that $n \geqslant 2$. Since both volume and the Hausdorff distance are unchanged by congruence transformations, we can assume that, for some $R > 0$,

$$A \subseteq \{(x_1, \ldots, x_{n-1}, 0) : |x_1|, \ldots, |x_{n-1}| \leqslant R\}.$$

Now

$$A_k \subseteq A + \theta_k U \subseteq \{(x_1, \ldots, x_n) : |x_1|, \ldots, |x_{n-1}| \leqslant R + \theta_k, |x_n| \leqslant \theta_k\},$$

and so

$$v(A_k) \leqslant 2(2R + 2\theta_k)^{n-1}\theta_k \to 0 = v(A) \qquad \text{as } k \to \infty. \ \square$$

We conclude this section with a technical result that will be needed in our discussion of mixed volumes in Section 6.4.

Theorem 6.2.18 *Let A be a union of a finite number of bounded convex sets in $\mathbb{R}^n$ each of which has dimension at most $n - 2$. Then*

$$v((A)_\lambda)/\lambda \to 0 \qquad \textit{as } \lambda \to 0_+.$$

Proof Consider first the case of a bounded convex set A in $\mathbb{R}^n$ having dimension at most $n - 2$. We can assume that $n \geqslant 3$ and that, for some $R > 0$,

$$A \subseteq \{(0, 0, x_3, \ldots, x_n) : |x_3|, \ldots, |x_n| \leqslant R\}.$$

Thus, for $\lambda > 0$,

$$(A)_\lambda \subseteq \{(x_1, \ldots, x_n) : |x_1|, |x_2| \leqslant \lambda;\ |x_3|, \ldots, |x_n| \leqslant R + \lambda\},$$

and so

$$v((A)_\lambda) \leqslant 4\lambda^2(2R + 2\lambda)^{n-2} = 2^n\lambda^2(R + \lambda)^{n-2}.$$

Hence

$$v((A)_\lambda)/\lambda \leqslant 2^n\lambda(R + \lambda)^{n-2} \to 0 \qquad \text{as } \lambda \to 0_+.$$

Consider now the general case when A is the union of bounded convex sets $A_1, \ldots, A_m$ in $\mathbb{R}^n$, each of which has dimension at most $n-2$. Then

$$v((A)_\lambda) = v((A_1)_\lambda \cup \ldots \cup (A_m)_\lambda) \leqslant v((A_1)_\lambda) + \ldots + v((A_m)_\lambda),$$

and so, by what we have just proved,

$$v((A)_\lambda)/\lambda \leqslant (v((A_1)_\lambda))/\lambda + \ldots + (v((A_m)_\lambda))/\lambda \to 0 \qquad \text{as } \lambda \to 0_+. \ \square$$

Exercises 6.2

1. Let $A \subseteq X$, where X is an elementary set in $\mathbb{R}^n$. Prove that
$$\underline{v}(A) + \bar{v}(X \backslash A) = v(X).$$
Deduce that this equation continues to hold when X is any set in $\mathbb{R}^n$ which has volume and contains A.
2. For each ordered pair (A, B) of convex bodies in $\mathbb{R}^n$, define
$$\delta(A, B) = v(A \cup B) - v(A \cap B).$$
Show that, for any convex bodies A, B, C in $\mathbb{R}^n$:
(i) $\delta(A, B) \geqslant 0$ with $\delta(A, B) = 0$ if and only if $A = B$;
(ii) $\delta(A, B) = \delta(B, A)$;
(iii) $\delta(A, C) \leqslant \delta(A, B) + \delta(B, C)$.
3. Let $I_1, \ldots, I_m, \ldots$ be a sequence of pairwise disjoint cells in $\mathbb{R}^n$ whose union is the cell I_0. Show that
$$v(I_0) = \sum_{m=1}^{\infty} v(I_m).$$
4. Find the volume of the *parallelotope*
$$P = \mathbf{a} + \{\lambda_1 \mathbf{a}_1 + \ldots + \lambda_n \mathbf{a}_n : 0 \leqslant \lambda_1, \ldots, \lambda_n \leqslant 1\},$$
where $\mathbf{a}, \mathbf{a}_1, \ldots, \mathbf{a}_n \in \mathbb{R}^n$.
5. Show that the convexity condition on the sets in Theorem 6.2.17 cannot be omitted.

6.3 THE DETERMINATION OF VOLUME

In the previous two sections, we defined volume and established its basic properties. Now we turn our attention to the determination of the volumes of particular convex bodies such as *parallelotopes*, *simplexes*, *balls*, and *ellipsoids*. To help us perform some of the calculations involved, we express the volume of a convex set in terms of an integral.

A set in $\mathbb{R}^n$ is called a **parallelotope** if it is the image of the unit n-cube

$$\{(x_1, \ldots, x_n): 0 \leqslant x_1, \ldots, x_n \leqslant 1\} = [0, 1] \times \ldots \times [0, 1]$$

under a non-singular affine transformation. We find the volume of the n-parallelotope

$$P = \{\lambda_1 \mathbf{a}_1 + \ldots + \lambda_n \mathbf{a}_n : 0 \leqslant \lambda_1, \ldots, \lambda_n \leqslant 1\} + \mathbf{a},$$

where $\mathbf{a}, \mathbf{a}_1, \ldots, \mathbf{a}_n \in \mathbb{R}^n$ with $\mathbf{a}_1, \ldots, \mathbf{a}_n$ linearly independent. Let $[\mathbf{a}_1, \ldots, \mathbf{a}_n]$ be the matrix with columns $\mathbf{a}_1, \ldots, \mathbf{a}_n$. Then P is the image of the n-cube

$$\{(x_1, \ldots, x_n): 0 \leqslant x_1, \ldots, x_n \leqslant 1\} = [0, 1] \times \ldots \times [0, 1]$$

under the affine transformation which maps $\mathbf{x}$ to $[\mathbf{a}_1, \ldots, \mathbf{a}_n]\mathbf{x} + \mathbf{a}$, whence, by Theorem 6.2.14, P has volume $|\det [\mathbf{a}_1, \ldots, \mathbf{a}_n]|$.

So far we have used v to denote the volume function in $\mathbb{R}^n$, realizing, of course, that different values of n give rise to different volume functions v. When we need to indicate the dependence of v upon n, we write v_n for the volume function in $\mathbb{R}^n$ and refer to it as the **n-volume**. Thus $v_1 (= l)$, v_2, v_3 denote, respectively, length in $\mathbb{R}^1$, area in $\mathbb{R}^2$, and volume in $\mathbb{R}^3$. It turns out to be necessary to enlarge the domain of definition of v_{n-1}, for $n \geqslant 2$, to include those sets in $\mathbb{R}^n$ which are congruent to sets in $\mathbb{R}^{n-1}$ having $(n-1)$-volume. This we do in a natural way. Let A be a set in $\mathbb{R}^n$ $(n \geqslant 2)$ that is congruent to some set B in $\mathbb{R}^{n-1}$ having $(n-1)$-volume $v_{n-1}(B)$. Then we define $v_{n-1}(A)$ to be $v_{n-1}(B)$. This defines $v_{n-1}(A)$ uniquely, for if A is also congruent to a set C in $\mathbb{R}^{n-1}$, then B and C are congruent, which shows that $v_{n-1}(B) = v_{n-1}(C)$. It is also helpful to define a volume function v_0 in $\mathbb{R}^1$ by putting $v_0(\varnothing) = 0$, and $v_0(\{a\}) = 1$ for each real number a. Clearly, if A is a set in $\mathbb{R}^n$ for which $v_{n-1}(A)$ is defined, then A must be a bounded subset of some hyperplane of $\mathbb{R}^n$. Also, if A is a bounded subset of some $(n-2)$-flat in $\mathbb{R}^n$ $(n \geqslant 2)$, then it is congruent to some bounded subset of a hyperplane in $\mathbb{R}^{n-1}$, whence $v_{n-1}(A) = 0$. If $T: \mathbb{R}^n \to \mathbb{R}^n$ is a congruence transformation and $\lambda \geqslant 0$, then

$$v_{n-1}(T(A)) = v_{n-1}(A) \qquad \text{and} \qquad v_{n-1}(\lambda A) = \lambda^{n-1} v_{n-1}(A),$$

where A is a subset of $\mathbb{R}^n$ for which $v_{n-1}(A)$ is defined.

Before expressing the volume of a convex body as an integral, we recall the definition of the *Riemann integral* for real functions of a single variable. Let $f: [a, b] \to \mathbb{R}$, where $a < b$, be a bounded real-valued function defined on the compact interval $[a, b]$ of the real line. For each subdivision $\mathfrak{D}$ of $[a, b]$, where $\mathfrak{D}$ is given by

$$a = \xi_0 < \xi_1 < \ldots < \xi_m = b,$$

lower and *upper sums* $s(\mathfrak{D})$ and $S(\mathfrak{D})$ of f with respect to $\mathfrak{D}$ are defined by the equations

$$s(\mathfrak{D}) = \sum_{i=0}^{m-1} (\xi_{i+1} - \xi_i) \inf \{f(x) : \xi_i \leqslant x \leqslant \xi_{i+1}\},$$

$$S(\mathfrak{D}) = \sum_{i=0}^{m-1} (\xi_{i+1} - \xi_i) \sup \{f(x) : \xi_i \leqslant x \leqslant \xi_{i+1}\}.$$

Lower and *upper integrals* ${}_*\int_a^b f(x)\,\mathrm{d}x$ and ${}^*\int_a^b f(x)\,\mathrm{d}x$ of f on $[a, b]$ are defined by the equations:

$${}_*\int_a^b f(x)\,\mathrm{d}x = \sup \{s(\mathfrak{D}) : \mathfrak{D} \text{ is a subdivision of } [a, b]\},$$

$${}^*\int_a^b f(x)\,\mathrm{d}x = \inf \{S(\mathfrak{D}) : \mathfrak{D} \text{ is a subdivision of } [a, b]\}.$$

The inequality ${}_*\int_a^b f(x)\,\mathrm{d}x \leqslant {}^*\int_a^b f(x)\,\mathrm{d}x$ holds between the lower and upper integrals. If ${}_*\int_a^b f(x)\,\mathrm{d}x = {}^*\int_a^b f(x)\,\mathrm{d}x$, then f is said to be *Riemann integrable* on $[a, b]$ and the common value of the two integrals is denoted by $\int_a^b f(x)\,\mathrm{d}x$.

Theorem 6.3.1 *Let A be a bounded convex set in $\mathbb{R}^n$. For each real number x, denote by A_x the intersection of A with the hyperplane $x_1 = x$ in $\mathbb{R}^n$. Let a and b be real numbers such that $a < b$ and A_x is empty whenever $x < a$ or $x > b$. Then*

$$v_n(A) = \int_a^b v_{n-1}(A_x)\,\mathrm{d}x.$$

Proof Let E be *any* elementary set contained in A. For each real number x, denote by E_x the intersection of E with the hyperplane $x_1 = x$ in $\mathbb{R}^n$. The function $v_{n-1}(E_x)$ is a step function, and so is Riemann integrable. Clearly $\int_a^b v_{n-1}(E_x)\,\mathrm{d}x = v_n(E)$ and $v_{n-1}(E_x) \leqslant v_{n-1}(A_x)$. Thus

$$v_n(E) = {}_*\int_a^b v_{n-1}(E_x)\,\mathrm{d}x \leqslant {}_*\int_a^b v_{n-1}(A_x)\,\mathrm{d}x.$$

Since E is *any* elementary set contained in A,

$$v_n(A) \leqslant {}_*\int_a^b v_{n-1}(A_x)\,\mathrm{d}x.$$

A similar argument shows that

$${}^*\int_a^b v_{n-1}(A_x)\,\mathrm{d}x \leqslant v_n(A).$$

Thus $v_{n-1}(A_x)$ is Riemann integrable on $[a, b]$ and

$$v_n(A) = \int_a^b v_{n-1}(A_x)\,dx. \quad \square$$

The formula of Theorem 6.3.1 can be written

$$v_n(A) = \int_{-\infty}^{\infty} v_{n-1}(A_x)\,dx,$$

since $v_{n-1}(A_x) = 0$ when either $x < a$ or $x > b$. This result is used in the following generalization of Theorem 6.3.1.

Corollary 6.3.2 *Let A be a bounded convex set in $\mathbb{R}^n$ and let $\mathbf{u}$ be a unit vector in $\mathbb{R}^n$. For each real number x, denote by A_x the intersection of A with the hyperplane $\mathbf{u} \cdot \mathbf{x} = x$. Then*

$$v_n(A) = \int_{-\infty}^{\infty} v_{n-1}(A_x)\,dx.$$

Proof Let $T: \mathbb{R}^n \to \mathbb{R}^n$ be a congruence transformation such that $T(\mathbf{0}) = \mathbf{0}$ and $T(\mathbf{u}) = \mathbf{e}_1$. By the theorem,

$$\begin{aligned} v_n(A) = v_n(T(A)) &= \int_{-\infty}^{\infty} v_{n-1}(T(A) \cap \{\mathbf{x}:\mathbf{x} \cdot \mathbf{e}_1 = x\})\,dx \\ &= \int_{-\infty}^{\infty} v_{n-1}(T(A_x))\,dx \\ &= \int_{-\infty}^{\infty} v_{n-1}(A_x)\,dx. \quad \square \end{aligned}$$

Let A be a bounded convex subset of a hyperplane $\mathbf{u} \cdot \mathbf{x} = u_0$ in $\mathbb{R}^n$, where $u_0 \in \mathbb{R}, \mathbf{u} \in \mathbb{R}^n$, and $\|\mathbf{u}\| = 1$. Let $\mathbf{c} \in \mathbb{R}^n$. We compute the volumes of the *cylindrical set* B and the *conical set* C defined by the equations

$$B = A + \{\lambda\mathbf{c}:0 \leqslant \lambda \leqslant 1\} \qquad \text{and} \qquad C = \operatorname{conv}(A \cup \{\mathbf{c}\}).$$

See Fig. 6.3.

In calculating $v_n(B)$, we assume initially that $\mathbf{c} \cdot \mathbf{u} > 0$. The hyperplane $\mathbf{u} \cdot \mathbf{x} = x$ meets B in a translate of A if $u_0 \leqslant x \leqslant u_0 + \mathbf{c} \cdot \mathbf{u}$, and in the empty set for other real values of x. Since each translate of A has the same $(n-1)$-volume as A itself, Corollary 6.3.2 shows that

$$v_n(B) = \int_{u_0}^{u_0 + \mathbf{c}\cdot\mathbf{u}} v_{n-1}(A)\,dx = (\mathbf{c} \cdot \mathbf{u})\,v_{n-1}(A).$$

In the general case, i.e. when $\mathbf{c} \cdot \mathbf{u}$ is unrestricted, we have

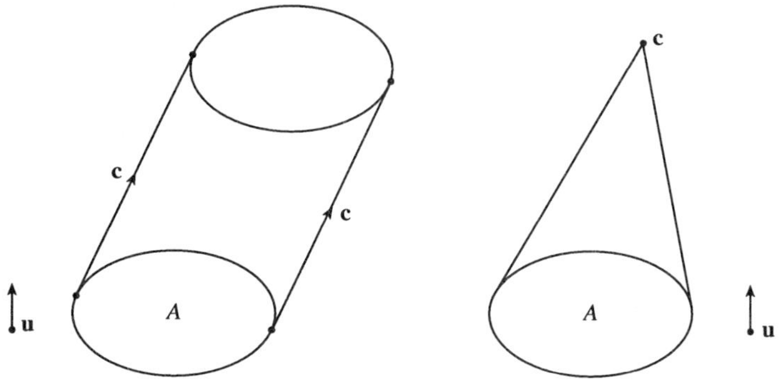

Fig. 6.3

$$v_n(B) = |\mathbf{c} \cdot \mathbf{u}|\, v_{n-1}(A).$$

This formula generalizes the result that the volume of a three-dimensional cylinder is the product of the area of its base with its height. If $\mathbf{c}$ is normal to the hyperplane $\mathbf{u} \cdot \mathbf{x} = u_0$, then $\mathbf{c} \cdot \mathbf{u} = \pm \|\mathbf{c}\|$ and the above formula reduces to

$$v_n(B) = \|\mathbf{c}\|\, v_{n-1}(A).$$

We now calculate $v_n(C)$, assuming initially that A is non-empty and that $u_0 < \mathbf{c} \cdot \mathbf{u}$. Clearly

$$C = \operatorname{conv}(A \cup \{\mathbf{c}\}) = \cup\, (\lambda \mathbf{c} + (1 - \lambda)A : 0 \leqslant \lambda \leqslant 1).$$

The hyperplane $\mathbf{u} \cdot \mathbf{x} = x$ meets C in the set $\lambda \mathbf{c} + (1 - \lambda)A$, for $u_0 \leqslant x \leqslant \mathbf{c} \cdot \mathbf{u}$, where

$$\lambda = \frac{x - u_0}{\mathbf{c} \cdot \mathbf{u} - u_0},$$

and in the empty set for other real values of x. Since

$$v_{n-1}(\lambda \mathbf{c} + (1 - \lambda)A) = v_{n-1}((1 - \lambda)A) = (1 - \lambda)^{n-1} v_{n-1}(A),$$

we deduce from Corollary 6.3.2 that

$$v_n(C) = \int_{u_0}^{\mathbf{c} \cdot \mathbf{u}} \left(\frac{\mathbf{c} \cdot \mathbf{u} - x}{\mathbf{c} \cdot \mathbf{u} - u_0} \right)^{n-1} v_{n-1}(A)\, \mathrm{d}x = \frac{1}{n} (\mathbf{c} \cdot \mathbf{u} - u_0)\, v_{n-1}(A).$$

In the general case, when $\mathbf{c} \cdot \mathbf{u}$ is unrestricted and A may be empty, we have

$$v_n(C) = \frac{1}{n} |\mathbf{c} \cdot \mathbf{u} - u_0|\, v_{n-1}(A).$$

This formula generalizes the result that the volume of a three-dimensional cone is one third the product of the area of its base with its height.

Now we proceed to calculate the volume of a general simplex in $\mathbb{R}^n$. Consider first the simplex S_n in $\mathbb{R}^n$ which is the polytope $\operatorname{conv}\{\mathbf{0}, \mathbf{e}_1, \ldots, \mathbf{e}_n\}$. Let α_n denote the n-volume of S_n. Since, for $n \geqslant 2$,

$$S_n = \operatorname{conv}\{\mathbf{0}, \mathbf{e}_1, \ldots, \mathbf{e}_n\} = \operatorname{conv}(\operatorname{conv}\{\mathbf{0}, \mathbf{e}_1, \ldots, \mathbf{e}_{n-1}\} \cup \{\mathbf{e}_n\}),$$

we deduce, using the formula established above for the volume of a conical set, that $\alpha_n = \alpha_{n-1}/n$. Since $\alpha_1 = 1$, we conclude that $\alpha_n = 1/n!$ for $n \geqslant 1$. Consider next the general n-simplex which is the convex hull of some affinely independent set $\{\mathbf{a}_0, \ldots, \mathbf{a}_n\}$ in $\mathbb{R}^n$, where $\mathbf{a}_i = (a_{i1}, \ldots, a_{in})$ for $i = 0, \ldots, n$. This simplex is the image of S_n under the affine transformation $T: \mathbb{R}^n \to \mathbb{R}^n$ defined by the equation $T(\mathbf{x}) = \mathbf{Q}\mathbf{x} + \mathbf{q}$ for $\mathbf{x} \in \mathbb{R}^n$, where $\mathbf{Q}$ is the $n \times n$ matrix with columns $\mathbf{a}_1 - \mathbf{a}_0, \ldots, \mathbf{a}_n - \mathbf{a}_0$ and $\mathbf{q} = \mathbf{a}_0$. Thus $\operatorname{conv}\{\mathbf{a}_0, \ldots, \mathbf{a}_n\}$ has n-volume $|\det \mathbf{Q}|/n!$, i.e. the absolute value of

$$\frac{1}{n!} \det \begin{bmatrix} a_{11} - a_{01} & \ldots & a_{n1} - a_{01} \\ \ldots & \ldots & \ldots \\ a_{1n} - a_{0n} & \ldots & a_{nn} - a_{0n} \end{bmatrix},$$

which is the same as the absolute value of

$$\frac{1}{n!} \det \begin{bmatrix} a_{01} & a_{11} & \ldots & a_{n1} \\ \ldots & \ldots & \ldots & \ldots \\ a_{0n} & a_{1n} & \ldots & a_{nn} \\ 1 & 1 & \ldots & 1 \end{bmatrix}.$$

Now to find a formula for ω_n, the n-volume of the closed unit ball U in $\mathbb{R}^n$. It is well known that $\omega_1 = 2, \omega_2 = \pi, \omega_3 = 4\pi/3$. By Theorem 6.3.1,

$$\omega_n = \int_{-1}^{1} v_{n-1}(U_x)\,\mathrm{d}x \qquad \text{for } n = 2, 3, \ldots,$$

where

$$U_x = \{(x, x_2, \ldots, x_n) : x_2^2 + \ldots + x_n^2 = 1 - x^2\} \qquad \text{for } -1 \leqslant x \leqslant 1.$$

For $-1 < x < 1$, U_x is congruent to a closed ball in $\mathbb{R}^{n-1}$ of radius $\sqrt{(1 - x^2)}$, and so

$$v_{n-1}(U_x) = \omega_{n-1}(1 - x^2)^{(n-1)/2} \qquad \text{for } -1 \leqslant x \leqslant 1.$$

Thus

$$\omega_n = \int_{-1}^{1} \omega_{n-1}(1 - x^2)^{(n-1)/2}\,dx = 2\int_0^1 \omega_{n-1}(1 - x^2)^{(n-1)/2}\,dx.$$

It is readily verified that

$$2\int_0^1 (1 - x^2)^{(n-1)/2}\,dx = B\left(\frac{1}{2}, \frac{n+1}{2}\right) = \frac{\sqrt{\pi}\,\Gamma\left(\frac{n+1}{2}\right)}{\Gamma\left(\frac{n+2}{2}\right)},$$

whence

$$\omega_n = \frac{\sqrt{\pi}\,\Gamma\left(\frac{n+1}{2}\right)}{\Gamma\left(\frac{n+2}{2}\right)} \cdot \frac{\sqrt{\pi}\,\Gamma\left(\frac{n}{2}\right)}{\Gamma\left(\frac{n+1}{2}\right)} \cdots \frac{\sqrt{\pi}\,\Gamma\left(\frac{3}{2}\right)\omega_1}{\Gamma\left(\frac{4}{2}\right)} = \frac{2\pi^{n/2}}{n\Gamma\left(\frac{n}{2}\right)},$$

a formula that holds for all positive integers n. Thus $\omega_4 = \pi^2/2$.

An ellipse can be defined as the image of a closed circular disc under a non-singular affine transformation. This observation leads to the following generalization of an ellipse to the space $\mathbb{R}^n$. A set in $\mathbb{R}^n$ is called an **ellipsoid** if it is the image of a closed ball under a non-singular affine transformation. Clearly every ellipsoid is a convex body. We find the volume of the symmetric ellipsoid

$$E = \{(x_1, \ldots, x_n) : (a_{11}x_1 + \ldots + a_{1n}x_n)^2 + \ldots + (a_{n1}x_1 + \ldots + a_{nn}x_n)^2 \leqslant r^2\},$$

where $\mathbf{A} = [a_{ij}]$ is a real $n \times n$ matrix with non-zero determinant and $r > 0$. The image of E under the linear transformation that maps $\mathbf{x}$ in $\mathbb{R}^n$ to $\mathbf{Ax}$ is the closed ball rU. Thus, by Theorem 6.2.14,

$$|\det \mathbf{A}|\, v_n(E) = \omega_n r^n, \qquad \text{whence} \qquad v_n(E) = \omega_n r^n / |\det \mathbf{A}|.$$

Finally, we derive a formula for the volume of an n-polytope in terms of the $(n-1)$-volumes of its facets and its support function, which will be used in our discussion of mixed volumes in the next section.

Theorem 6.3.3 *Let $\mathbf{u}_1, \ldots, \mathbf{u}_m$ be the outward unit normals to the facets of an n-polytope P in $\mathbb{R}^n$ corresponding to the facets $F_1, \ldots, F_m$. Let h be the support function of P. Then*

$$v_n(P) = \frac{1}{n}\sum_{i=1}^{m} h(\mathbf{u}_i)v_{n-1}(F_i) \quad \textit{and} \quad \sum_{i=1}^{m} v_{n-1}(F_i)\mathbf{u}_i = \mathbf{0}.$$

Proof Suppose first that the origin is an interior point of P. For each $i = 1, \ldots, m$, denote by C_i the convex set $\mathrm{conv}(\{\mathbf{0}\} \cup F_i)$. Then $P = C_1 \cup \ldots \cup C_m$ and

$$C_i \cap C_j = \operatorname{conv}(\{\mathbf{0}\} \cup (F_i \cap F_j)) \qquad \text{for } i, j = 1, \ldots, m,$$

which shows that $C_i \cap C_j$ $(i \neq j)$ is at most $(n-1)$-dimensional, so $v_n(C_i \cap C_j) = 0$ $(i \neq j)$. By Corollary 6.2.8,

$$v_n(P) = v_n(C_1 \cup \ldots \cup C_m) = v_n(C_1) + \ldots + v_n(C_m).$$

But, by the formula obtained earlier for the volume of a conical set, for $i = 1, \ldots, m$,

$$v_n(C_i) = \frac{1}{n} h(\mathbf{u}_i) v_{n-1}(F_i).$$

Hence

$$v_n(P) = \frac{1}{n} \sum_{i=1}^{m} h(\mathbf{u}_i) v_{n-1}(F_i).$$

Denote by $\mathbf{a}$ the vector $\Sigma_{i=1}^{m} v_{n-1}(F_i)\mathbf{u}_i$. Choose $\lambda > 0$ small enough to ensure that the origin is an interior point of the polytope $P + \lambda\mathbf{a}$. Applying the formula established above for the volume of a polytope having the origin as an interior point, we deduce that

$$\begin{aligned} v_n(P) = v_n(P + \lambda\mathbf{a}) &= \frac{1}{n} \sum_{i=1}^{m} (h_{P+\lambda\mathbf{a}}(\mathbf{u}_i)) v_{n-1}(F_i + \lambda\mathbf{a}) \\ &= \frac{1}{n} \sum_{i=1}^{m} (h(\mathbf{u}_i) + \lambda\mathbf{a} \cdot \mathbf{u}_i) v_{n-1}(F_i) \\ &= v_n(P) + \frac{\lambda}{n} \|\mathbf{a}\|^2, \end{aligned}$$

which shows that $\mathbf{a} = \Sigma_{i=1}^{m} v_{n-1}(F_i)\mathbf{u}_i = \mathbf{0}$, as required.

Consider now the general case when it is not assumed that the origin is an interior point of P. With each n-polytope P associate the vector $\Sigma_{i=1}^{m} v_{n-1}(F_i)\mathbf{u}_i$. Clearly, this vector is the same for all translates of P. But for any translate of P which has the origin as an interior point, this associated vector is the zero vector, by what we have just proved. Thus

$$\sum_{i=1}^{m} v_{n-1}(F_i)\mathbf{u}_i = \mathbf{0}.$$

Finally, let $\mathbf{c} \in \mathbb{R}^n$ be such that the polytope $P + \mathbf{c}$ has the origin in its interior. Then, by the first part of the proof,

$$\begin{aligned} v_n(P) = v_n(P + \mathbf{c}) &= \frac{1}{n} \sum_{i=1}^{m} (h_{P+\mathbf{c}}(\mathbf{u}_i)) v_{n-1}(F_i + \mathbf{c}) \\ &= \frac{1}{n} \sum_{i=1}^{m} (h(\mathbf{u}_i) + \mathbf{c} \cdot \mathbf{u}_i) v_{n-1}(F_i) \end{aligned}$$

$$= \frac{1}{n} \sum_{i=1}^{m} h(\mathbf{u}_i) v_{n-1}(F_i) + \frac{\mathbf{c}}{n} \cdot \left(\sum_{i=1}^{m} v_{n-1}(F_i) \mathbf{u}_i \right)$$

$$= \frac{1}{n} \sum_{i=1}^{m} h(\mathbf{u}_i) v_{n-1}(F_i). \quad \square$$

Exercises 6.3

1. Find the volumes of the following convex sets:
   (i) $\{(x_1, \ldots, x_n): (x_1/a_1)^2 + \ldots + (x_n/a_n)^2 \leqslant r^2\}$, where $a_1, \ldots, a_n$, $r > 0$;
   (ii) $\operatorname{conv}\{(0,0,0,0), (1,2,3,4), (2,3,4,1), (3,4,1,2), (4,1,2,3)\}$;
   (iii) $\operatorname{conv}(A \cup B)$, where

$$A = \{(0, x_2, \ldots, x_n) : x_2^2 + \ldots + x_n^2 \leqslant a^2\},$$
$$B = \{(c, x_2, \ldots, x_n) : x_2^2 + \ldots + x_n^2 \leqslant b^2\},$$

   and $a, b, c > 0, n \geqslant 2$.
2. Investigate the behaviour of the sequence $\omega_1, \ldots, \omega_n, \ldots$, where $\omega_n = v_n(U)$.
3. Give an *ad hoc* proof of the second assertion of Theorem 6.3.3 in $\mathbb{R}^2$, i.e. show that, if $\mathbf{u}_1, \ldots, \mathbf{u}_m$ are outward unit normals to the sides of lengths $a_1, \ldots, a_m$, respectively, of a convex polygon in $\mathbb{R}^2$, then $a_1\mathbf{u}_1 + \ldots + a_m\mathbf{u}_m = \mathbf{0}$.
4. Given that $F_0, F_1, \ldots, F_m$ are the facets of some n-polytope in $\mathbb{R}^n$, show that

$$v_{n-1}(F_1) + \ldots + v_{n-1}(F_m) \geqslant v_{n-1}(F_0).$$

5. Let $O = (0,0,0)$, $A = (\alpha, 0, 0)$, $B = (0, \beta, 0)$, $C = (0, 0, \gamma)$, where $\alpha, \beta, \gamma > 0$, be points in $\mathbb{R}^3$. Denote by a, b, c, respectively, the lengths of the sides BC, CA, AB of the triangle ABC. Apply the second assertion of Theorem 6.3.3 to the 3-polytope $OABC$ to express the area Δ of triangle ABC in terms of α, β, γ. Deduce that, if $2s = a + b + c$, then

$$\Delta = \sqrt{s(s-a)(s-b)(s-c)}.$$

6.4 MIXED VOLUMES AND SURFACE AREA

In this section we prove that the volume $v_n(\lambda_1 A_1 + \ldots + \lambda_r A_r)$, where $A_1, \ldots, A_r$ are given compact convex sets in $\mathbb{R}^n$ and $\lambda_1, \ldots, \lambda_r$ are non-negative variables, is a homogeneous polynomial of degree n in $\lambda_1, \ldots, \lambda_r$. The theory of mixed volumes is concerned with the

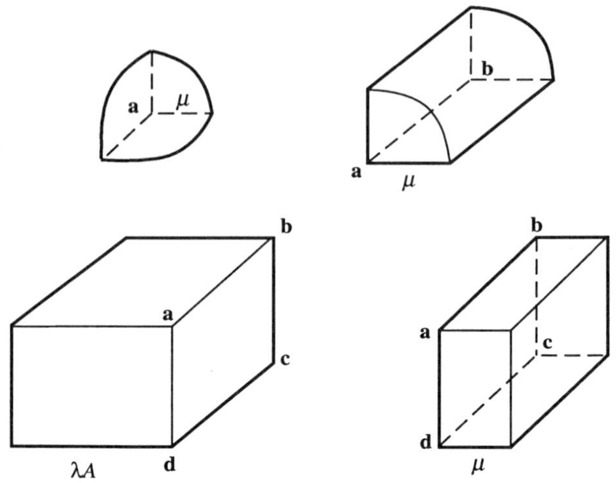

Fig. 6.4

relationship between the coefficients of the polynomial and the sets $A_1, \ldots, A_r$ themselves. This is a hard theoretical section, but we trust that the reader will find the effort amply repaid when some impressive applications of the theory are presented in the next section.

We begin our discussion by considering the following simple problem:

> What is the volume of the convex body $\lambda A + \mu B$, where A is the rectangular block (i.e. 3-orthotope) with edge lengths a, b, c defined by the equation
> $$A = \{(x, y, z) : 0 \leqslant x \leqslant a, 0 \leqslant y \leqslant b, 0 \leqslant z \leqslant c\},$$
> B is the closed unit ball in $\mathbb{R}^3$, and λ,μ are positive scalars?

Now λA is a rectangular block with edge lengths $\lambda a, \lambda b, \lambda c$ and μB is the closed ball of radius μ centred at the origin in $\mathbb{R}^3$. Thus $\lambda A + \mu B$ is the outer parallel body of λA at distance μ. The reader will easily verify that $\lambda A + \mu B$ is the union of λA together with six rectangular blocks (each with height μ and having a facet of λA as base), twelve quadrants of circular cylinders (each with base radius μ and having an edge of λA as axis), and eight octants of balls (each with radius μ and having a vertex of λA as centre). Moreover, any two *different* sets in this union meet in a set of volume zero. Figure 6.4 shows one example each of a rectangular block, a quadrant of a cylinder, and an octant of a ball, as considered above; their positions relative to λA are also indicated. It is readily found that $v_3(\lambda A + \mu B)$ equals

$$(abc)\lambda^3 + 2(ab + bc + ca)\lambda^2\mu + \pi(a + b + c)\lambda\mu^2 + \tfrac{4\pi}{3}\mu^3,$$

the four terms representing in order, the volumes of λA, the union of the six rectangular blocks, the union of the twelve quadrants of circular cylinders, and the union of the eight octants of balls. Thus $v_3(\lambda A + \mu B)$ is a homogeneous polynomial of degree three in λ and μ with non-negative coefficients, which is a special case of the fundamental Theorem 6.4.3. We prepare for this theorem with the following two lemmas.

Lemma 6.4.1 *Let $C_1, \ldots, C_r$ be polytopes in $\mathbb{R}^n$ and let $\alpha_1, \ldots, \alpha_r > 0$. Then $\alpha_1 C_1 + \ldots + \alpha_r C_r$ and $C_1 + \ldots + C_r$ have the same dimension, and the sets of outward unit normals to the $(n-1)$-faces of the two polytopes are equal.*

Proof The result is trivial when one of $C_1, \ldots, C_r$ is empty. Suppose, then, that $\mathbf{c}_1 \in C_1, \ldots, \mathbf{c}_r \in C_r$. Let A be the flat aff $(C_1 + \ldots + C_r)$. Then $A - (\mathbf{c}_1 + \ldots + \mathbf{c}_r)$ is a subspace of $\mathbb{R}^n$ containing each of the sets $C_1 - \mathbf{c}_1, \ldots, C_r - \mathbf{c}_r$, and hence the set $\alpha_1(C_1 - \mathbf{c}_1) + \ldots + \alpha_r(C_r - \mathbf{c}_r)$. It follows that $\alpha_1 C_1 + \ldots + \alpha_r C_r$ lies in the translate

$$A + (\alpha_1 - 1)\mathbf{c}_1 + \ldots + (\alpha_r - 1)\mathbf{c}_r$$

of A, whence the dimension of $\alpha_1 C_1 + \ldots + \alpha_r C_r$ does not exceed that of $C_1 + \ldots + C_r$. It follows, from what we have just proved, that the dimension of the set

$$\alpha_1^{-1}(\alpha_1 C_1) + \ldots + \alpha_r^{-1}(\alpha_r C_r), \qquad \text{i.e. the set } C_1 + \ldots + C_r,$$

does not exceed that of $\alpha_1 C_1 + \ldots + \alpha_r C_r$. Thus the polytopes $\alpha_1 C_1 + \ldots + \alpha_r C_r$ and $C_1 + \ldots + C_r$ have the same dimension.

A unit vector $\mathbf{u}$ is an outward normal to some $(n-1)$-face of $\alpha_1 C_1 + \ldots + \alpha_r C_r$ if and only if the set

$$(\alpha_1 C_1 + \ldots + \alpha_r C_r)^{\mathbf{u}} = \alpha_1 C_1^{\mathbf{u}} + \ldots + \alpha_r C_r^{\mathbf{u}}$$

has dimension $n - 1$. By the first part of the proof, this occurs precisely when

$$C_1^{\mathbf{u}} + \ldots + C_r^{\mathbf{u}} = (C_1 + \ldots + C_r)^{\mathbf{u}}$$

has dimension $n - 1$, i.e. when $\mathbf{u}$ is an outward unit normal to some $(n-1)$-face of $C_1 + \ldots + C_r$. □

Lemma 6.4.2 *Let $A_1, \ldots, A_r$ be polytopes in $\mathbb{R}^n$. Then $v_n(\lambda_1 A_1 + \ldots + \lambda_r A_r)$ is, for all $\lambda_1, \ldots, \lambda_r > 0$, a homogeneous polynomial of degree n in $\lambda_1, \ldots, \lambda_r$ with non-negative coefficients.*

Proof We argue by induction on n. If $n = 1$, then

$$v_1(\lambda_1 A_1 + \ldots + \lambda_r A_r) = \lambda_1 v_1(A_1) + \ldots + \lambda_r v_1(A_r) \quad \text{for } \lambda_1, \ldots, \lambda_r > 0,$$

when none of $A_1, \ldots, A_r$ is empty, and is zero otherwise. This proves the lemma for the case $n = 1$. Suppose, then, that the assertion is true in $\mathbb{R}^{n-1}$, where $n \geqslant 2$. If $A_1 + \ldots + A_r$ has dimension less than n, then, by Lemma 6.4.1, so too does $\lambda_1 A_1 + \ldots + \lambda_r A_r$, whence $v_n(\lambda_1 A_1 + \ldots + \lambda_r A_r)$ is zero for all $\lambda_1, \ldots, \lambda_r > 0$, and the assertion is true in this case.

Suppose now that $A_1 + \ldots + A_r$ has dimension n. Since v_n-volumes are preserved by translations, we can assume that each of the polytopes $A_1, \ldots, A_r$ contains the origin. Let $\mathbf{u}_1, \ldots, \mathbf{u}_m$ be the outward unit normals to the facets of $A_1 + \ldots + A_r$. For each $i = 1, \ldots, m$, the polytopes $A_1^{\mathbf{u}_i}, \ldots, A_r^{\mathbf{u}_i}$ lie in parallel hyperplanes of $\mathbb{R}^n$. Since v_{n-1}-volumes are preserved by translations, we can, for the purposes of calculating

$$v_{n-1}(\lambda_1 A_1^{\mathbf{u}_i} + \ldots + \lambda_r A_r^{\mathbf{u}_i}),$$

assume that $A_1^{\mathbf{u}_i}, \ldots, A_r^{\mathbf{u}_i}$ lie in the same hyperplane of $\mathbb{R}^n$. By identifying this hyperplane with $\mathbb{R}^{n-1}$ and using the induction hypothesis, we deduce the existence of a homogeneous polynomial p_i of degree $n-1$ in $\lambda_1, \ldots, \lambda_r$ with non-negative coefficients such that, for all $\lambda_1, \ldots, \lambda_r > 0$,

$$v_{n-1}(\lambda_1 A_1^{\mathbf{u}_i} + \ldots + \lambda_r A_r^{\mathbf{u}_i}) = v_{n-1}((\lambda_1 A_1 + \ldots + \lambda_r A_r)^{\mathbf{u}_i}) = p_i(\lambda_1, \ldots, \lambda_r).$$

Let $\lambda_1, \ldots, \lambda_r > 0$. By Lemma 6.4.1, the facets of $\lambda_1 A_1 + \ldots + \lambda_r A_r$ are $(\lambda_1 A_1 + \ldots + \lambda_r A_r)^{\mathbf{u}_i}$ with corresponding outward unit normals $\mathbf{u}_i$. Theorem 6.3.3 shows that

$$\begin{aligned} v_n(\lambda_1 A_1 + \ldots + \lambda_r A_r) &= \frac{1}{n} \sum_{i=1}^{m} (h_{\lambda_1 A_1 + \ldots + \lambda_r A_r}(\mathbf{u}_i))\, v_{n-1}((\lambda_1 A_1 + \ldots + \lambda_r A_r)^{\mathbf{u}_i}) \\ &= \frac{1}{n} \sum_{i=1}^{m} (\lambda_1 h_{A_1}(\mathbf{u}_i) + \ldots + \lambda_r h_{A_r}(\mathbf{u}_i))\, p_i(\lambda_1, \ldots, \lambda_r). \end{aligned}$$

Thus $v_n(\lambda_1 A_1 + \ldots + \lambda_r A_r)$ is a homogeneous polynomial of degree n in $\lambda_1, \ldots, \lambda_r$ with non-negative coefficients. This completes the proof by induction. □

Theorem 6.4.3 *Let $A_1, \ldots, A_r$ be compact convex sets in $\mathbb{R}^n$. Then $v_n(\lambda_1 A_1 + \ldots + \lambda_r A_r)$ is, for all $\lambda_1, \ldots, \lambda_r \geqslant 0$, a homogeneous polynomial of degree n in $\lambda_1, \ldots, \lambda_r$ with non-negative coefficients.*

Proof We assume that the sets $A_1, \ldots, A_r$ are non-empty. For each $i = 1, \ldots, r$, let

$$A_i^1, A_i^2, \ldots, A_i^j, \ldots$$

be a sequence of polytopes converging to A_i. By Lemma 6.4.2, for each $j = 1, 2, \ldots,$ there exist non-negative scalars

$$a^j_{i_1\ldots i_n} \qquad \text{for } i_1, \ldots, i_n = 1, \ldots, r$$

such that, for all $\lambda_1, \ldots, \lambda_r > 0$,

$$v_n(\lambda_1 A^j_1 + \ldots + \lambda_r A^j_r) = \sum_{i_1=1}^{r} \ldots \sum_{i_n=1}^{r} a^j_{i_1\ldots i_n} \lambda_{i_1} \ldots \lambda_{i_n}. \qquad (6.4.1)$$

Since the r sequences of polytopes considered above are convergent, there is a closed ball B in $\mathbb{R}^n$ such that $A^j_i \subseteq B$ for $i = 1, \ldots, r$ and $j = 1, 2, \ldots$. We can deduce easily from this, by putting $\lambda_1 = 1, \ldots, \lambda_r = 1$ in (6.4.1), that no $a^j_{i_1\ldots i_n}$ exceeds $r^n v_n(B)$. Since every bounded sequence of real numbers contains a convergent subsequence, it follows that there is a subsequence $k_1, k_2, \ldots, k_j, \ldots$ of $1, 2, \ldots, j, \ldots$ and non-negative scalars

$$a_{i_1\ldots i_n} \qquad \text{for } i_1, \ldots, i_n = 1, \ldots, r$$

such that

$$a^{k_j}_{i_1\ldots i_n} \to a_{i_1\ldots i_n} \qquad \text{as } j \to \infty \text{ for } i_1, \ldots, i_n = 1, \ldots, r.$$

Corollary 2.7.6 shows that, for $\lambda_1, \ldots, \lambda_r > 0$,

$$\lambda_1 A^{k_j}_1 + \ldots + \lambda_r A^{k_j}_r \to \lambda_1 A_1 + \ldots + \lambda_r A_r \qquad \text{as } j \to \infty.$$

The continuity of v_n now shows that

$$v_n(\lambda_1 A^{k_j}_1 + \ldots + \lambda_r A^{k_j}_r) \to v_n(\lambda_1 A_1 + \ldots + \lambda_r A_r) \qquad \text{as } j \to \infty.$$

But from (6.4.1)

$$v_n(\lambda_1 A^{k_j}_1 + \ldots + \lambda_r A^{k_j}_r) \to \sum_{i_1=1}^{r} \ldots \sum_{i_n=1}^{r} a_{i_1\ldots i_n} \lambda_{i_1} \ldots \lambda_{i_n} \qquad \text{as } j \to \infty.$$

Thus, for all $\lambda_1, \ldots, \lambda_r > 0$,

$$v_n(\lambda_1 A_1 + \ldots + \lambda_r A_r) = \sum_{i_1=1}^{r} \ldots \sum_{i_n=1}^{r} a_{i_1\ldots i_n} \lambda_{i_1} \ldots \lambda_{i_n}.$$

Suppose finally that $\lambda_1, \ldots, \lambda_r \geqslant 0$. By what we have just proved, for each $\varepsilon > 0$,

$$v_n((\lambda_1 + \varepsilon) A_1 + \ldots + (\lambda_r + \varepsilon) A_r) = \sum_{i_1=1}^{r} \ldots \sum_{i_n=1}^{r} a_{i_1\ldots i_n} (\lambda_{i_1} + \varepsilon) \ldots (\lambda_{i_n} + \varepsilon).$$

Letting $\varepsilon \to 0_+$ in the last equation, we find that

$$v_n(\lambda_1 A_1 + \ldots + \lambda_r A_r) = \sum_{i_1=1}^{r} \ldots \sum_{i_n=1}^{r} a_{i_1\ldots i_n} \lambda_{i_1} \ldots \lambda_{i_n}. \ \square$$

Each homogeneous polynomial $p(\lambda_1, \ldots, \lambda_r)$ of degree n can be uniquely represented in the form

$$p(\lambda_1, \ldots, \lambda_r) = \sum_{\alpha_1 + \ldots + \alpha_r = n} \frac{n!}{\alpha_1! \ldots \alpha_r!} a_{\alpha_1 \ldots \alpha_r} \lambda_1^{\alpha_1} \ldots \lambda_r^{\alpha_r}.$$

For integers $i_1, i_2, \ldots, i_n$ lying in $\{1, \ldots, r\}$, put

$$v_{i_1 i_2 \ldots i_n} = a_{\alpha_1 \alpha_2 \ldots \alpha_r}, \qquad \text{where } \lambda_{i_1}\lambda_{i_2} \ldots \lambda_{i_n} = \lambda_1^{\alpha_1}\lambda_2^{\alpha_2} \ldots \lambda_r^{\alpha_r}.$$

Then:

(i) $v_{i_1 i_2 \ldots i_n}$ remains unchanged when $i_1, i_2, \ldots, i_n$ are permuted, and

(ii) $p(\lambda_1, \ldots, \lambda_r) = \sum_{i_1=1}^{r} \ldots \sum_{i_n=1}^{r} v_{i_1 \ldots i_n} \lambda_{i_1} \ldots \lambda_{i_n}$;

moreover, the $v_{i_1 \ldots i_n}$ are uniquely determined by (i) and (ii).

When

$$p(\lambda_1, \ldots, \lambda_r) = v_n(\lambda_1 A_1 + \ldots + \lambda_r A_r),$$

where $A_1, \ldots, A_r$ are compact convex sets in $\mathbb{R}^n$ and $\lambda_1, \ldots, \lambda_r \geqslant 0$, the numbers $v_{i_1 \ldots i_n}$ are called the **mixed volumes** of $A_1, \ldots, A_r$. Consider the example studied at the beginning of this section in which A was a rectangular block with edge lengths a, b, c and B was the closed unit ball. We found that, for $\lambda, \mu \geqslant 0$,

$$v_3(\lambda A + \mu B) = (abc)\lambda^3 + 2(ab + bc + ca)\lambda^2\mu + (a + b + c)\pi\lambda\mu^2 + \tfrac{4\pi}{3}\mu^3.$$

It follows easily from this equation that the mixed volumes of A, B are:

$$v_{111} = abc,\ v_{222} = \tfrac{4\pi}{3},\ v_{112} = v_{121} = v_{211} = \tfrac{2}{3}(ab + bc + ca), \text{ and}$$
$$v_{122} = v_{212} = v_{221} = \tfrac{1}{3}(a + b + c)\pi.$$

The mixed volumes of compact convex sets $A_1, \ldots, A_r$ are determined by the function $v_n(\lambda_1 A_1 + \ldots + \lambda_r A_r)$ of $\lambda_1, \ldots, \lambda_r$. It is tempting to assume that any particular mixed volume $v_{i_1 \ldots i_n}$ depends only upon the sets $A_{i_1}, \ldots, A_{i_n}$. For example, when none of the sets $A_1, \ldots, A_r$ is empty, it is easy to see that $v_{1 \ldots 1} = v_n(A_1)$, which only depends upon A_1. If, however, even one of the sets $A_1, \ldots, A_r$ is empty, then all the mixed volumes $v_{i_1 \ldots i_n}$ are zero. We shall show below that, when none of the sets $A_1, \ldots, A_r$ is empty, the mixed volume $v_{i_1 \ldots i_n}$ does indeed depend only upon the sets $A_{i_1}, \ldots, A_{i_n}$.

With each n-tuple $(A_1, \ldots, A_n)$ of non-empty compact convex sets in $\mathbb{R}^n$ we associate a non-negative scalar $v(A_1, \ldots, A_n)$ as follows. Suppose that there are *exactly s distinct* sets occurring in $(A_1, \ldots, A_n)$. We can assume, relabelling the A_i's if necessary, that the sets $A_1, \ldots, A_s$ are distinct. For $i = 1, \ldots, s$, let α_i be the number of times which the set A_i occurs in $(A_1, \ldots, A_n)$. Then $\alpha_1 + \ldots + \alpha_s = n$. For all $\lambda_1, \ldots, \lambda_s \geqslant 0$,

$$v_n(\lambda_1 A_1 + \ldots + \lambda_s A_s) = \sum_{i_1=1}^{s} \ldots \sum_{i_n=1}^{s} v_{i_1 \ldots i_n} \lambda_{i_1} \ldots \lambda_{i_n},$$

where the $v_{i_1 \ldots i_n}$ are the mixed volumes of $A_1, \ldots, A_s$. We now define

$$v(A_1, \ldots, A_n) = v_{\underbrace{1\ldots 1}_{\alpha_1} \ldots \underbrace{s \ldots s}_{\alpha_s}}.$$

This determines $v(A_1, \ldots, A_n)$ uniquely and in such a way that it remains unchanged when $A_1, \ldots, A_n$ are permuted. The next result identifies $v(A_1, \ldots, A_n)$ as a mixed volume in the general case.

Theorem 6.4.4. *Let $A_1, \ldots, A_r$ be non-empty compact convex sets in $\mathbb{R}^n$. Then, for all $\lambda_1, \ldots, \lambda_r \geqslant 0$,*

$$v_n(\lambda_1 A_1 + \ldots + \lambda_r A_r) = \sum_{i_1=1}^{r} \ldots \sum_{i_n=1}^{r} v(A_{i_1}, \ldots, A_{i_n}) \lambda_{i_1} \ldots \lambda_{i_n}.$$

Proof We argue by induction on the *redundancy number* of the finite sequence $A_1, \ldots, A_r$; this is defined to be the non-negative integer $r - s$, where s is the number of *distinct* sets in the sequence. The sequence has redundancy number zero when all of its terms are different, and has redundancy number $r - 1$ when all of its terms are the same.

Suppose first that the sequence $A_1, \ldots, A_r$ has redundancy number zero, i.e. all of its terms are different. For all $\lambda_1, \ldots, \lambda_r \geqslant 0$,

$$v_n(\lambda_1 A_1 + \ldots + \lambda_r A_r) = \sum_{i_1=1}^{r} \ldots \sum_{i_n=1}^{r} v_{i_1 \ldots i_n} \lambda_{i_1} \ldots \lambda_{i_n},$$

where the $v_{i_1 \ldots i_n}$ are the mixed volumes of $A_1, \ldots, A_r$. Consider a particular mixed volume $v_{i_1 \ldots i_n}$ and the corresponding n-tuple $(A_{i_1}, \ldots, A_{i_n})$. Suppose that there are exactly s distinct sets occurring in this last n-tuple, say $A_1, \ldots, A_s$. For $i = 1, \ldots, s$, let α_i be the number of times which the set A_i occurs in $(A_{i_1}, \ldots, A_{i_n})$. Then $\alpha_1 + \ldots + \alpha_s = n$. For all $\lambda_1, \ldots, \lambda_s \geqslant 0$,

$$v_n(\lambda_1 A_1 + \ldots + \lambda_s A_s) = \sum_{i_1=1}^{s} \ldots \sum_{i_n=1}^{s} v_{i_1 \ldots i_n} \lambda_{i_1} \ldots \lambda_{i_n}.$$

By the definition of $v(A_{i_1}, \ldots, A_{i_n})$,

$$v(A_{i_1}, \ldots, A_{i_n}) = v_{\underbrace{1\ldots 1}_{\alpha_1} \ldots \underbrace{s \ldots s}_{\alpha_s}} = v_{i_1 \ldots i_n}.$$

Here we have used the fact that all the r sets $A_1, \ldots, A_r$ are different. Thus the assertion is true for the case of redundancy number zero.

Suppose next that the assertion is true for sequences with redundancy number m, where $m \geqslant 0$. Let the sequence $A_1, \ldots, A_r$ have redundancy number $m+1$. Then at least two terms of this sequence must be equal, say $A_{r-1} = A_r$. Since $A_1, \ldots, A_{r-1}$ has redundancy number m, the induction hypothesis shows that, for all $\lambda_1, \ldots, \lambda_{r-1} \geqslant 0$,

$$v_n(\lambda_1 A_1 + \ldots + \lambda_{r-1} A_{r-1}) = \sum_{i_1=1}^{r-1} \ldots \sum_{i_n=1}^{r-1} v(A_{i_1}, \ldots, A_{i_n}) \lambda_{i_1} \ldots \lambda_{i_n}.$$

Let $v_{i_1 \ldots i_n}$ be a typical mixed volume for the sequence $A_1, \ldots, A_r$. Let $\alpha_1, \ldots, \alpha_r$ be non-negative integers such that $i_1, \ldots, i_n$ is a rearrangement of the sequence

$$\underbrace{1, \ldots, 1}_{\alpha_1}, \ldots, \underbrace{r, \ldots, r}_{\alpha_r}.$$

Then the coefficient of the term $\lambda_1^{\alpha_1} \ldots \lambda_r^{\alpha_r}$ in the polynomial $v_n(\lambda_1 A_1 + \ldots + \lambda_r A_r)$ is

$$\frac{n!}{\alpha_1! \ldots \alpha_r!} v_{\underbrace{1 \ldots 1}_{\alpha_1} \ldots \underbrace{r \ldots r}_{\alpha_r}}.$$

Now $v_n(\lambda_1 A_1 + \ldots + \lambda_r A_r)$ can be obtained from $v_n(\lambda_1 A_1 + \ldots + \lambda_{r-1} A_{r-1})$ by replacing λ_{r-1} with $\lambda_{r-1} + \lambda_r$. Thus the coefficient of $\lambda_1^{\alpha_1} \ldots \lambda_r^{\alpha_r}$ in $v_n(\lambda_1 A_1 + \ldots + \lambda_r A_r)$ is also the product of the coefficient of the term

$$\lambda_1^{\alpha_1} \ldots \lambda_{r-2}^{\alpha_{r-2}} \lambda_{r-1}^{\alpha_{r-1}+\alpha_r}$$

in $v_n(\lambda_1 A_1 + \ldots + \lambda_{r-1} A_{r-1})$ with the coefficient of the term $\lambda_{r-1}^{\alpha_{r-1}} \lambda_r^{\alpha_r}$ in $(\lambda_{r-1} + \lambda_r)^{\alpha_{r-1} + \alpha_r}$, i.e. the product

$$\frac{n!}{\alpha_1! \ldots \alpha_{r-2}!(\alpha_{r-1} + \alpha_r)!} v(\underbrace{A_1}_{\alpha_1}, \ldots, \underbrace{A_{r-2}}_{\alpha_{r-2}}, \underbrace{A_{r-1}}_{\alpha_{r-1}+\alpha_r}) \cdot \frac{(\alpha_{r-1} + \alpha_r)!}{(\alpha_{r-1})!(\alpha_r)!}$$

$$= \frac{n!}{\alpha_1! \ldots \alpha_r!} v(\underbrace{A_1}_{\alpha_1}, \ldots, \underbrace{A_r}_{\alpha_r}).$$

The two expressions which we have found for the coefficient of $\lambda_1^{\alpha_1} \ldots \lambda_r^{\alpha_r}$ in $v_n(\lambda_1 A_1 + \ldots + \lambda_r A_r)$ must be equal, so

$$v_{i_1 \ldots i_n} = v_{\underbrace{1 \ldots 1}_{\alpha_1} \ldots \underbrace{r \ldots r}_{\alpha_r}} = v(\underbrace{A_1}_{\alpha_1}, \ldots, \underbrace{A_r}_{\alpha_r}) = v(A_{i_1}, \ldots, A_{i_n}).$$

Thus the assertion is true for the case of redundancy number $m + 1$. This completes the proof by induction. □

The next result shows that, in some restricted sense, the mixed volume function is linear in each of its arguments.

Theorem 6.4.5 *Let $A_1', A_1, A_2, \ldots, A_n$ be non-empty compact convex sets in $\mathbb{R}^n$. Let $\alpha_1', \alpha_1 \geqslant 0$. Then*

$$v(\alpha_1' A_1' + \alpha_1 A_1, A_2, \ldots, A_n) = \alpha_1' v(A_1', A_2, \ldots, A_n) + \alpha_1 v(A_1, A_2, \ldots, A_n).$$

Proof Equating the coefficients of the term $\lambda_1 \ldots \lambda_n$ in the identical polynomials

$$v_n(\lambda_1(\alpha_1' A_1' + \alpha_1 A_1) + \lambda_2 A_2 + \ldots + \lambda_n A_n), \quad \text{and}$$
$$v_n((\lambda_1 \alpha_1')A_1' + (\lambda_1 \alpha_1)A_1 + \lambda_2 A_2 + \ldots + \lambda_n A_n),$$

we deduce that

$$n! v(\alpha_1' A_1' + \alpha_1 A_1, A_2, \ldots, A_n) = n! \alpha_1' v(A_1', A_2, \ldots, A_n) + n! \alpha_1 v(A_1, A_2, \ldots, A_n).$$

The desired result is now immediate. □

The following lemma will be needed in the proof that mixed volumes depend continuously upon their arguments.

Lemma 6.4.6 *Let m be a positive integer. For each $i = 0, 1, 2, \ldots$, let*

$$P_i(x) = a_{im}x^m + \ldots + a_{i1}x + a_{io}$$

be a real polynomial. Suppose that, for each $x \geqslant 0$, $P_i(x) \to P_0(x)$ as $i \to \infty$. Then $a_{ij} \to a_{0j}$ as $i \to \infty$, for $j = 0, 1, \ldots, m$.

Proof The $m + 1$ vectors $\mathbf{a}_\lambda = (\lambda^m, \ldots, \lambda, 1)$ for $\lambda = 0, 1, \ldots, m$ are linearly independent, and so form a basis for $\mathbb{R}^{m+1}$. Thus there are scalars $\mu_0, \mu_1, \ldots, \mu_m$ such that

$$(1, 0, \ldots, 0) = \mu_0 \mathbf{a}_0 + \mu_1 \mathbf{a}_1 + \ldots + \mu_m \mathbf{a}_m.$$

By the hypothesis,

$$\mu_0 P_i(0) + \mu_1 P_i(1) + \ldots + \mu_m P_i(m) \to \mu_0 P_0(0) + \mu_1 P_0(1) + \ldots + \mu_m P_0(m) \quad \text{as } i \to \infty,$$

i.e. $a_{im} \to a_{0m}$ as $i \to \infty$. Similarly, $a_{ij} \to a_{0j}$ as $i \to \infty$ for $j = 0, 1, \ldots, m - 1$. □

Theorem 6.4.7 (Continuity of mixed volumes) *For each $j = 1, \ldots, n$, let $A_j^1, A_j^2, \ldots, A_j^i, \ldots$ be a sequence of non-empty compact convex sets converging to a non-empty compact convex set A_j^0 in $\mathbb{R}^n$. Then*

$$v(A_1^i, \ldots, A_n^i) \to v(A_1^0, \ldots, A_n^0) \qquad \textit{as } i \to \infty.$$

Proof For $i = 0, 1, 2, \ldots$, and for $\lambda_1, \ldots, \lambda_n \geqslant 0$,

$$v_n(\lambda_1 A_1^i + \ldots + \lambda_n A_n^i) = Q_i(\lambda_1, \ldots, \lambda_n) \text{ say,}$$

is a real homogeneous polynomial of degree n in $\lambda_1, \ldots, \lambda_n$. Since v_n is continuous, for all $\lambda_1, \ldots, \lambda_n \geqslant 0$,

$$Q_i(\lambda_1, \ldots, \lambda_n) \to Q_0(\lambda_1, \ldots, \lambda_n) \qquad \text{as } i \to \infty.$$

Choose a positive integer r so large that the coefficient of $x^r x^{r^2} \ldots x^{r^n}$ in the real polynomial

$$P_i(x) = Q_i(x^r, x^{r^2}, \ldots, x^{r^n})$$

of the single variable x is $n!v(A_1^i, \ldots, A_n^i)$. For each $x \geqslant 0$, $P_i(x) \to P_0(x)$ as $i \to \infty$. We deduce from the lemma that

$$n!v(A_1^i, \ldots, A_n^i) \to n!v(A_1^0, \ldots, A_n^0) \qquad \text{as } i \to \infty,$$

whence the desired result is immediate. □

Mixed volumes are increasing in the sense that $v(A_1, \ldots, A_n) \leqslant v(B_1, \ldots, B_n)$ whenever $A_1 \subseteq B_1, \ldots, A_n \subseteq B_n$. This fact is easier to guess than it is to establish. Before proving it, we need some preliminary results.

Theorem 6.4.8 *Let* $\mathbf{u}_1, \ldots, \mathbf{u}_m$ *be the outward unit normals to the* $(n-1)$*-faces of a polytope* P *in* $\mathbb{R}^n$ *corresponding to faces* $F_1, \ldots, F_m$*, respectively. Then, for any non-empty compact convex set* A *in* $\mathbb{R}^n$ *with support function* h,

$$\lim_{\lambda \to 0_+} \frac{v_n(P + \lambda A) - v_n(P)}{\lambda} = \sum_{i=1}^{m} h(\mathbf{u}_i) v_{n-1}(F_i).$$

Proof Both sides of the above equation are unchanged in value if A is replaced by one of its translates; the non-trivial part of *this* assertion follows from the second part of Theorem 6.3.3. We can, therefore, assume that A contains the origin, and this we do.

Consider first the case when P has dimension n. We begin by showing that, for all $\lambda > 0$,

$$v_n(P + \lambda A) - v_n(P) \geqslant \sum_{i=1}^{m} \lambda h(\mathbf{u}_i) v_{n-1}(F_i).$$

For each $i = 1, \ldots, m$, let $\mathbf{a}_i \in A$ satisfy $\mathbf{u}_i \cdot \mathbf{a}_i = h(\mathbf{u}_i)$, and define a convex subset C_i of $P + \lambda A$ by the equation

$$C_i = \operatorname{ri} F_i + \lambda\{\mu \mathbf{a}_i : 0 \leqslant \mu \leqslant 1\}.$$

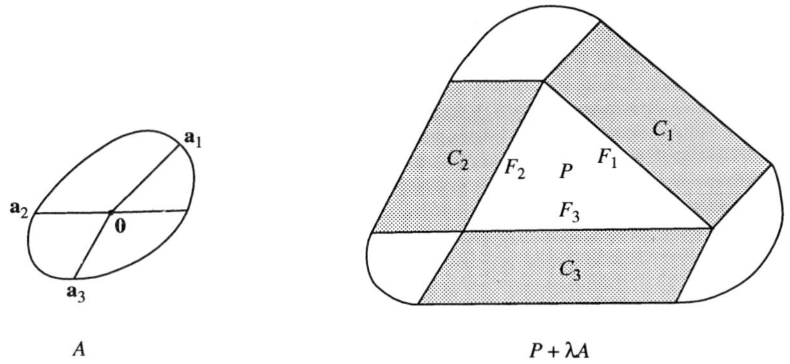

Fig. 6.5

A simple case of this situation is illustrated in Fig. 6.5. The sets $C_1, \ldots, C_m$ are pairwise disjoint. If they were not, there would exist points $\mathbf{f}_i$ in $\operatorname{ri} F_i$, $\mathbf{f}_j$ in $\operatorname{ri} F_j$, and scalars $\theta_i, \theta_j \geqslant 0$ with $i \neq j$ satisfying

$$\mathbf{f}_i + \theta_i \mathbf{a}_i = \mathbf{f}_j + \theta_j \mathbf{a}_j.$$

Since P is n-dimensional, $\mathbf{f}_j \notin F_i$, whence $\mathbf{u}_i \cdot \mathbf{f}_j < \mathbf{u}_i \cdot \mathbf{f}_i$. It follows easily from the definition of $h(\mathbf{u}_i)$ that $\mathbf{u}_i \cdot \mathbf{a}_j \leqslant \mathbf{u}_i \cdot \mathbf{a}_i$. Hence

$$\mathbf{u}_i \cdot \mathbf{f}_i + \theta_i \mathbf{u}_i \cdot \mathbf{a}_i = \mathbf{u}_i \cdot \mathbf{f}_j + \theta_j \mathbf{u}_i \cdot \mathbf{a}_j < \mathbf{u}_i \cdot \mathbf{f}_i + \theta_j \mathbf{u}_i \cdot \mathbf{a}_i.$$

Since A contains the origin, $h(\mathbf{u}_i) = \mathbf{u}_i \cdot \mathbf{a}_i \geqslant 0$, and we can conclude that $\theta_i < \theta_j$. By symmetry, $\theta_j < \theta_i$. This contradiction shows that the sets $C_1, \ldots, C_m$ are pairwise disjoint. For each $i = 1, \ldots, m$, $v_n(C_i \cap P) = 0$ and

$$v_n(C_i) = v_n(\operatorname{cl} C_i) = |\lambda(\mathbf{u}_i \cdot \mathbf{a}_i) v_{n-1}(F_i)| = \lambda h(\mathbf{u}_i) v_{n-1}(F_i).$$

We can thus deduce that

$$v_n(P + \lambda A) - v_n(P) \geqslant v_n(C_1) + \ldots + v_n(C_m) = \sum_{i=1}^{m} \lambda h(\mathbf{u}_i) v_{n-1}(F_i).$$

We find an upper bound for $v_n(P + \lambda A) - v_n(P)$ by showing that, for $\lambda > 0$,

$$P + \lambda A \subseteq P \cup \operatorname{cl} C_1 \cup \ldots \cup \operatorname{cl} C_m \cup (S)_{\lambda s}, \qquad (6.4.2)$$

where S is the union of the $(n-2)$-dimensional faces of P and s is the diameter of A. To establish (6.4.2), we suppose that $\mathbf{x}$ is a point of $P + \lambda A$, but not of any of the sets P, $\operatorname{cl} C_1, \ldots, \operatorname{cl} C_m$, and show that it belongs to $(S)_{\lambda s}$. Let $\mathbf{f}$ be the nearest point of P to $\mathbf{x}$. Then $\mathbf{f} \in F_i$ for some $i = 1, \ldots, m$. If $\mathbf{f} \in \operatorname{rebd} F_i$, then $\mathbf{f} \in S$ and $\mathbf{x} \in (S)_{\lambda s}$. If $\mathbf{f} \in \operatorname{ri} F_i$, $\mathbf{x} = \mathbf{f} + \alpha \mathbf{u}_i$ for some $\alpha > 0$. Since $\mathbf{x} \in P + \lambda A$, $\alpha \leqslant \lambda h(\mathbf{u}_i)$.

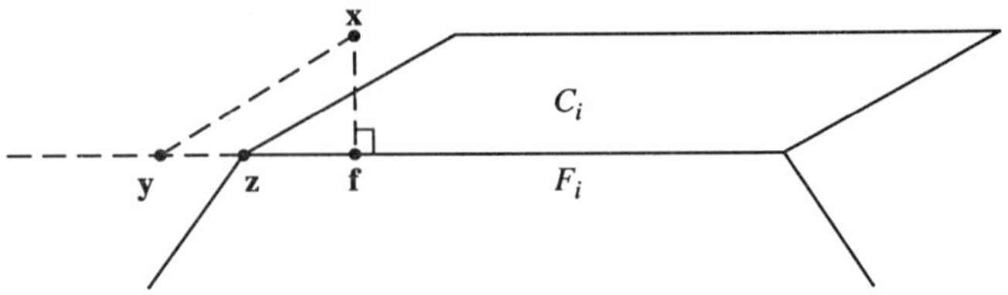

Fig. 6.6

Let $\mathbf{y} = \mathbf{x} - (\alpha\mathbf{a}_i)/h(\mathbf{u}_i)$. Our assumption that $\mathbf{x} \notin \operatorname{cl} C_i$ shows that $\mathbf{y} \notin F_i$. Now $\mathbf{u}_i \cdot \mathbf{y} = \mathbf{u}_i \cdot \mathbf{f}$, and so $\mathbf{y} \in \operatorname{aff} F_i$. Thus, for some β with $0 < \beta < 1$, the point $\mathbf{z} = (1 - \beta)\mathbf{y} + \beta\mathbf{f}$ lies in $\operatorname{rebd} F_i$, and hence in S. See Fig. 6.6. Now

$$\|\mathbf{x} - \mathbf{z}\| \leqslant (1 - \beta)\|\mathbf{x} - \mathbf{y}\| + \beta\|\mathbf{x} - \mathbf{f}\| \leqslant (1 - \beta)\lambda s + \beta\lambda s = \lambda s,$$

whence $\mathbf{x} \in (S)_{\lambda s}$. This establishes (6.4.2). Thus

$$v_n(P + \lambda A) \leqslant v_n(P) + v_n(\operatorname{cl} C_1) + \ldots + v_n(\operatorname{cl} C_m) + v_n((S)_{\lambda s}).$$

Combining this inequality with the one obtained at the end of the last paragraph, we deduce that, for $\lambda > 0$,

$$\sum_{i=1}^{m} h(\mathbf{u}_i)v_{n-1}(F_i) \leqslant \frac{v_n(P + \lambda A) - v_n(P)}{\lambda} \leqslant \sum_{i=1}^{m} h(\mathbf{u}_i)v_{n-1}(F_i) + \frac{1}{\lambda}v_n((S)_{\lambda s}).$$

By Theorem 6.2.18, $(v_n((S)_{\lambda s}))/\lambda \to 0$ as $\lambda \to 0_+$. Thus

$$\lim_{\lambda \to 0_+} \frac{v_n(P + \lambda A) - v_n(P)}{\lambda} = \sum_{i=1}^{m} h(\mathbf{u}_i)v_{n-1}(F_i).$$

This completes the proof for the case when P is n-dimensional.

Suppose next that P has dimension $n - 1$. Then $m = 2$, $\mathbf{u}_1 = \mathbf{u}$, $\mathbf{u}_2 = -\mathbf{u}$, $F_1 = P$, and $F_2 = P$, where $\mathbf{u}$ is a unit normal to the hyperplane containing P. The proof in this case is the same as that just given, except that the sets C_1 and C_2 are not disjoint but meet in the set $\operatorname{ri} P$, which has v_n-volume zero.

When P has dimension less than $n - 1$, the assertion of the theorem is assumed to mean that

$$\lim_{\lambda \to 0_+} \frac{v_n(P + \lambda A) - v_n(P)}{\lambda} = \lim_{\lambda \to 0_+} \frac{v_n(P + \lambda A)}{\lambda} = 0.$$

This is clear, since $P + \lambda A \subseteq (P)_{\lambda s}$, where s is the diameter of A, and by Theorem 6.2.18

$$\lim_{\lambda \to 0_+} \frac{v_n((P)_{\lambda s})}{\lambda} = 0. \ \square$$

Corollary 6.4.9 *Let $\mathbf{u}_1, \ldots, \mathbf{u}_m$ be the outward unit normals to the $(n-1)$-faces of a non-empty polytope P in $\mathbb{R}^n$ corresponding to faces $F_1, \ldots, F_m$, respectively. Then, for any non-empty compact convex set A in $\mathbb{R}^n$ with support function h,*

$$v(A, P, \ldots, P) = \frac{1}{n} \sum_{i=1}^{m} h(\mathbf{u}_i) v_{n-1}(F_i).$$

Proof For all $\lambda > 0$,

$$v_n(P + \lambda A) = \sum_{i=0}^{n} \binom{n}{i} v(\underbrace{A, \ldots, A}_{i}, \underbrace{P, \ldots, P}_{n-i}) \lambda^i,$$

and so

$$v_n(P+\lambda A) - v_n(P) = nv(A, P, \ldots, P)\lambda + \sum_{i=2}^{n} \binom{n}{i} v(\underbrace{A, \ldots, A}_{i}, \underbrace{P, \ldots, P}_{n-i}) \lambda^i.$$

Thus

$$\lim_{\lambda \to 0_+} \frac{v_n(P + \lambda A) - v_n(P)}{\lambda} = nv(A, P, \ldots, P).$$

The corollary now follows from the theorem. □

Theorem 6.4.10 *Let $P_2, \ldots, P_n$ be non-empty polytopes in $\mathbb{R}^n$ $(n \geqslant 2)$. Let $\mathbf{u}_1, \ldots, \mathbf{u}_m$ be the outward unit normals to the $(n-1)$-faces of $P_2 + \ldots + P_n$. Then there are scalars $\alpha_1, \ldots, \alpha_m \geqslant 0$ such that, for every non-empty compact convex set A in $\mathbb{R}^n$ with support function h,*

$$v(A, P_2, \ldots, P_n) = \frac{1}{n} \sum_{i=1}^{m} \alpha_i h(\mathbf{u}_i).$$

Proof Let $Q = \lambda_2 P_2 + \ldots + \lambda_n P_n$ for $\lambda_2, \ldots, \lambda_n > 0$. By repeated applications of Theorem 6.4.5,

$$v(A, Q, \ldots, Q) = \sum_{i_2=2}^{n} \cdots \sum_{i_n=2}^{n} v(A, P_{i_2}, \ldots, P_{i_n}) \lambda_{i_2} \ldots \lambda_{i_n},$$

which is a homogeneous polynomial of degree $n-1$ in $\lambda_2, \ldots, \lambda_n$, the coefficient of $\lambda_2 \ldots \lambda_n$ being $(n-1)! v(A, P_2, \ldots, P_n)$.

By Lemma 6.4.1, the set of outward unit normals to the $(n-1)$-faces of Q is $\{\mathbf{u}_1, \ldots, \mathbf{u}_m\}$. Corollary 6.4.9 shows that

$$v(A, Q, \ldots, Q) = \frac{1}{n} \sum_{i=1}^{m} h(\mathbf{u}_i) v_{n-1}(Q^{\mathbf{u}_i})$$

$$= \frac{1}{n} \sum_{i=1}^{m} h(\mathbf{u}_i) v_{n-1}(\lambda_2 P_2^{\mathbf{u}_i} + \ldots + \lambda_n P_n^{\mathbf{u}_i})$$

$$= \frac{1}{n} \sum_{i=1}^{m} h(\mathbf{u}_i) \left(\sum_{j_2=2}^{n} \cdots \sum_{j_n=2}^{n} u(P_{j_2}^{\mathbf{u}_i}, \ldots, P_{j_n}^{\mathbf{u}_i}) \lambda_{j_2} \ldots \lambda_{j_n} \right),$$

where $u(P_{j_2}^{\mathbf{u}_i}, \ldots, P_{j_n}^{\mathbf{u}_i})$ denotes an $(n-1)$-dimensional mixed volume. This shows again that $v(A, Q, \ldots, Q)$ is a homogeneous polynomial of degree $n-1$ in $\lambda_2, \ldots, \lambda_n$, the coefficient of $\lambda_2 \ldots \lambda_n$ being

$$\frac{1}{n} \sum_{i=1}^{m} h(\mathbf{u}_i)(n-1)!\, u(P_2^{\mathbf{u}_i}, \ldots, P_n^{\mathbf{u}_i}).$$

Equating the two coefficients that we have found for the term $\lambda_2 \ldots \lambda_n$ in $v(A, Q, \ldots, Q)$, we deduce that

$$v(A, P_2, \ldots, P_n) = \frac{1}{n} \sum_{i=1}^{m} h(\mathbf{u}_i) u(P_2^{\mathbf{u}_i}, \ldots, P_n^{\mathbf{u}_i}).$$

The proof is completed by putting $\alpha_i = u(P_2^{\mathbf{u}_i}, \ldots, P_n^{\mathbf{u}_i})$. □

Theorem 6.4.11 *Let $A_1, \ldots, A_n$, $B_1, \ldots, B_n$ be non-empty compact convex sets in $\mathbb{R}^n$ with $A_1 \subseteq B_1, \ldots, A_n \subseteq B_n$. Then $v(A_1, \ldots, A_n) \leqslant v(B_1, \ldots, B_n)$.*

Proof We assume that $n \geqslant 2$. Consider first the case when $A_1, \ldots, A_n$, $B_1, \ldots, B_n$ are polytopes. Let h_{A_1}, h_{B_1} be the support functions of A_1, B_1, respectively. Since $A_1 \subseteq B_1$, $h_{A_1}(\mathbf{u}) \leqslant h_{B_1}(\mathbf{u})$ for all $\mathbf{u}$ in $\mathbb{R}^n$. Using Theorem 6.4.10 and an obvious notation, we deduce that

$$v(A_1, A_2, \ldots, A_n) = \frac{1}{n} \sum_{i=1}^{m} h_{A_1}(\mathbf{u}_i)\alpha_i \leqslant \frac{1}{n} \sum_{i=1}^{m} h_{B_1}(\mathbf{u}_i)\alpha_i = v(B_1, A_2, \ldots, A_n).$$

Repeating this argument $n-1$ times, we find that

$$v(A_1, A_2, \ldots, A_n) \leqslant v(B_1, B_2, \ldots, B_n).$$

Consider now the general case. For each $i = 1, \ldots, n$, let

$$P_i^1, \ldots, P_i^j, \ldots \qquad \text{and} \qquad Q_i^1, \ldots, Q_i^j, \ldots$$

be sequences of non-empty polytopes in $\mathbb{R}^n$ such that $P_i^j \to A_i$, $Q_i^j \to B_i$ as $j \to \infty$, and $P_i^j \subseteq Q_i^j$ for $j = 1, 2, \ldots$. Using the first part of the proof and the continuity of the mixed volumes, we deduce that

$$v(A_1, \ldots, A_n) = \lim_{j\to\infty} v(P_1^j, \ldots, P_n^j) \leqslant \lim_{j\to\infty} v(Q_1^j, \ldots, Q_n^j) = v(B_1, \ldots, B_n). \quad \square$$

Now we introduce the concept of surface area for compact convex sets. Perhaps the natural way of doing this is first to define the surface area of an n-polytope to be the sum of the v_{n-1}-volumes of its facets, and then to extend this definition to compact convex sets by an approximation argument. We, however, choose a different approach. The theory of surface area for convex sets forms part of the theory of mixed volumes. We prefer, therefore, to define surface area in terms of mixed volumes, and to deduce its properties from the results obtained earlier in this section.

To motivate our definition, consider Theorem 6.4.8 for the special case when P is an n-polytope and A is the closed unit ball U. Here the theorem asserts that

$$\sum_{i=1}^{m} v_{n-1}(F_i) = \lim_{\lambda \to 0_+} \frac{v_n((P)_\lambda) - v_n(P)}{\lambda}.$$

The left-hand side of this equation is what we intuitively regard as the surface area of the polytope P, i.e. the sum of the v_{n-1}-volumes of its facets. This observation prompts us to define the **surface area** $s_n(A)$ of a compact convex set A in $\mathbb{R}^n$ by the equation

$$s_n(A) = \lim_{\lambda \to 0_+} \frac{v_n((A)_\lambda) - v_n(A)}{\lambda}.$$

For non-empty A and $\lambda > 0$, we have

$$v_n((A)_\lambda) = v_n(A + \lambda U) = v_n(A) + nv(A, \ldots, A, U)\lambda + \ldots + v_n(U)\lambda^n,$$

whence $s_n(A)$ is well defined and equals $nv(A, \ldots, A, U)$; in $\mathbb{R}^1$ this last assertion is taken to mean that $s_1(A) = v_1(U) = 2$. Thus we can define the surface area $s_n(A)$ of a compact convex set A in $\mathbb{R}^n$ to be $nv(A, \ldots, A, U)$ when A is non-empty, and to be zero when A is empty.

To illustrate the above definition and to obtain an interesting and useful result, we evaluate the surface area $s_n(U)$ of the closed unit ball U in $\mathbb{R}^n$. We know that, for any $\lambda > 0$, $v_n((U)_\lambda) = \omega_n(1 + \lambda)^n$, whence

$$s_n(U) = \lim_{\lambda \to 0_+} \frac{\omega_n(1 + \lambda)^n - \omega_n}{\lambda} = n\omega_n.$$

Thus $s_2(U) = 2\omega_2 = 2\pi$, i.e. the perimeter of a circle of unit radius is 2π, and $s_3(U) = 3\omega_3 = 4\pi$, i.e. the surface area of a closed ball of unit radius in $\mathbb{R}^3$ is 4π.

Since surface area is a multiple of a mixed volume, some of its properties can be deduced immediately; we mention a few of these. The value of $s_n(A)$, when A is a compact convex set in $\mathbb{R}^n$ of dimension at most $n - 1$, is $2v_{n-1}(A)$. Surface area is increasing in the sense that $s_n(A) \leqslant s_n(B)$ whenever A, B are compact convex sets in $\mathbb{R}^n$ with $A \subseteq B$, and it is continuous in the sense that $s_n(A_i) \to s_n(A)$ as $i \to \infty$, whenever $A_1, \ldots, A_i, \ldots$ is a sequence of non-empty compact convex sets which converges to the non-empty compact convex set A in $\mathbb{R}^n$.

We conclude our discussion of mixed volumes by remarking on an interesting property of the surface area function s_2 in $\mathbb{R}^2$; for obvious reasons, s_2 is referred to as the *perimeter function*. Let A, B be non-empty compact convex sets in $\mathbb{R}^2$. Then, by Theorem 6.4.5,

$$s_2(A + B) = 2v(A + B, U) = 2v(A, U) + 2v(B, U) = s_2(A) + s_2(B),$$

i.e. the perimeter of $A + B$ is the sum of the perimeters of A and B.

Exercises 6.4

1. In $\mathbb{R}^2$ let A, B, C be the square, triangle, and closed circular disc defined by the equations:

$$A = \text{conv}\{(0,0), (1,0), (1,1), (0,1)\}, B = \text{conv}\{(0,0), (3,0), (0,4)\},$$
$$C = \{(x, y) : x^2 + y^2 \leqslant 1\}.$$

Find the polynomial $v_2(\lambda A + \mu B + \gamma C)$, where $\lambda, \mu, \nu \geqslant 0$, and hence find the mixed volumes $v(A, B)$, $v(B, C)$, $v(A, C)$.

2. Verify that, if $A = \{(x, y) : |x|, |y| \leqslant 1\}$, $B = \{(x, y) : x^2 + y^2 \leqslant 1\}$, then the function $f: [0, 1] \to \mathbb{R}$ defined below is concave:

$$f(\lambda) = \sqrt{(v_2((1 - \lambda)A + \lambda B))} \qquad \text{for } 0 \leqslant \lambda \leqslant 1.$$

3. Suppose that $A_1, \ldots, A_r$ are sets in $\mathbb{R}^n$ such that, for all $\lambda_1, \ldots, \lambda_r \geqslant 0$, the set $\lambda_1 A_1 + \ldots + \lambda_r A_r$ has volume. Show that $v_n(\lambda_1 A_1 + \ldots + \lambda_r A_r)$ need not be a homogeneous polynomial of degree n in $\lambda_1, \ldots, \lambda_r$ for $\lambda_1, \ldots, \lambda_r \geqslant 0$.

4. Let A, B, C be non-empty compact convex sets in $\mathbb{R}^n$ such that $v_n(A + \lambda C) \geqslant v_n(B + \lambda C)$ for $\lambda > 0$. Show that $v_n(A) \geqslant v_n(B)$, but that it is not necessarily true that $s_n(A) \geqslant s_n(B)$. Prove, however, that if $v_n(A) = v_n(B)$ and $C = U$, then $s_n(A) \geqslant s_n(B)$.

5. Suppose that A, B are compact convex sets in $\mathbb{R}^n$ which have a non-empty intersection and are such that $A \cup B$ is convex. Prove that, for each $i = 0, \ldots, n$,

$$v(\underbrace{A \cup B, \ldots, A \cup B}_{n-i}, \underbrace{U, \ldots, U}_{i}) + v(\underbrace{A \cap B, \ldots, A \cap B}_{n-i}, \underbrace{U, \ldots, U}_{i})$$

equals

$$v(\underbrace{A, \ldots, A}_{n-i}, \underbrace{U, \ldots, U}_{i}) + v(\underbrace{B, \ldots, B}_{n-i}, \underbrace{U, \ldots, U}_{i}).$$

6.5 THE BRUNN-MINKOWSKI THEOREM

The **Brunn–Minkowski theorem** asserts that: *if A, B are convex bodies in $\mathbb{R}^n$, then*

$$v_n^{1/n}(A + B) \geqslant v_n^{1/n}(A) + v_n^{1/n}(B),$$

with equality holding if and only if A and B are homothetic, i.e. if and only if $B = \lambda A + \mathbf{a}$, for some $\lambda > 0$ and $\mathbf{a} \in \mathbb{R}^n$. In the present section, we establish this important result, thereby also solving the most famous of extremum problems, the **isoperimetric problem**: *of all convex bodies in $\mathbb{R}^n$ with a given volume, which have the smallest surface area*?

Theorem 6.1.1 shows that the vector sum of two elementary sets is an elementary set. The vector sum of two sets each of which has volume, however, need not itself have volume. To see this, consider the sets A, B in $\mathbb{R}^2$ defined by the equations:

$$A = \{(x, 0) : 0 \leqslant x \leqslant 1, x \text{ rational}\}, \quad B = \{(0, y) : 0 \leqslant y \leqslant 1, y \text{ rational}\}.$$

Then the sets A and B have zero volume, but the set

$$A + B = \{(x, y) : 0 \leqslant x, y \leqslant 1; x, y \text{ rational}\}$$

does not have volume; its inner-volume is *zero*, its outer-volume is *one*. When, however, $A, B, A + B$ are non-empty sets in $\mathbb{R}^n$, all of which do have volume, then the following inequality holds:

$$v_n^{1/n}(A + B) \geqslant v_n^{1/n}(A) + v_n^{1/n}(B).$$

This was proved for the special case when A and B are convex bodies, by H. Brunn in 1887. The above formulation was established (in even greater generality) by L. A. Lyusternik in 1935. The proof of Brunn's inequality which we give below is due to H. Hadwiger and D. Ohmann in 1956. In the following two lemmas, we isolate some details which are needed in our proof of the inequality. We revert to our earlier practice of writing v_n simply as v, when no confusion is possible; we also write s_n simply as s.

Lemma 6.5.1 *Let A be a set in $\mathbb{R}^n$ which has volume. Let $\theta \geqslant 0$. Then, for $i = 1, \ldots, n$, there is a scalar λ_i such that*

$$v(\{(x_1, \ldots, x_n) \in A : x_i < \lambda_i\}) = \theta v(\{(x_1, \ldots, x_n) \in A : x_i > \lambda_i\}).$$

Proof Since A is bounded, there is $a > 0$ such that

$$A \subseteq \{(x_1, \ldots, x_n) : -a \leqslant x_i \leqslant a \quad \text{for } i = 1, \ldots, n\}.$$

Define a function $f_i : \mathbb{R} \to \mathbb{R}$ by the equation

$$f_i(x) = v(\{(x_1, \ldots, x_n) \in A : x_i < x\}) \qquad \text{for } x \in \mathbb{R}.$$

Then, for $x < y$,

$$0 \leqslant f_i(y) - f_i(x) = v(\{(x_1, \ldots, x_n) \in A : x \leqslant x_i < y\}) \leqslant 2^{n-1}a^{n-1}(y - x).$$

This shows that f_i is continuous. Now $f_i(-a) = 0$, $f_i(a) = v(A)$, and so, for some $\lambda_i \in [-a, a]$, $f_i(\lambda_i) = \theta v(A)/(1 + \theta)$. The desired result now follows easily. □

In the following lemma, the term *non-degenerate cell* is used to mean a cell which has non-empty interior.

Lemma 6.5.2 *Let* $A = I_1 \cup \ldots \cup I_m$, $B = J_1 \cup \ldots \cup J_p$, *where* $I_1, \ldots, I_m$ *and* $J_1, \ldots, J_p$ *are sequences of pairwise disjoint non-degenerate cells in* $\mathbb{R}^n$. *Then*

$$v^{1/n}(A + B) \geqslant v^{1/n}(A) + v^{1/n}(B).$$

Proof We argue by induction on $m + p$. Suppose first that $m + p = 2$, so that $m = 1, p = 1$. Let $A = S_1 \times \ldots \times S_n$, $B = T_1 \times \ldots \times T_n$, where $S_1, \ldots, S_n, T_1, \ldots, T_n$ are cells in $\mathbb{R}^1$ with positive lengths $a_1, \ldots, a_n$, $b_1, \ldots, b_n$, respectively. Corollary 5.2.5 justifies the inequality

$$\begin{aligned} v^{1/n}(A + B) &= ((a_1 + b_1)\ldots(a_n + b_n))^{1/n} \\ &\geqslant (a_1 \ldots a_n)^{1/n} + (b_1 \ldots b_n)^{1/n} = v^{1/n}(A) + v^{1/n}(B). \end{aligned}$$

This proves the lemma in the case $m + p = 2$.

Suppose next that $m + p > 2$ and that the assertion is true for all cases in which the induction variable is less than $m + p$. We can assume that $m \geqslant 2$. Since the cells I_1 and I_2 are disjoint, there is some $i \in \{1, \ldots, n\}$ and some scalar μ such that I_1 lies in the closed halfspace $x_i \leqslant \mu$ and I_2 lies in the closed halfspace $x_i \geqslant \mu$, or vice versa; a similar situation occurs in the proof of Theorem 6.1.4. Denote by A^- and A^+ the intersections of A with the open halfspaces $x_i < \mu$ and $x_i > \mu$, respectively. Then each of A^- and A^+ is non-empty and is a union of *fewer* than m pairwise disjoint non-degenerate cells. Since A is the pairwise disjoint union of A^-, A^+ and a set of volume zero, $v(A)$ equals $v(A^-) + v(A^+)$. Lemma 6.5.1 shows that there is a scalar λ such that the hyperplane $x_i = \lambda$ divides B (in a fashion similar to that considered above for A) into disjoint sets B^-, B^+ and a set of volume zero such that

$$\frac{v(B^-)}{v(A^-)} = \frac{v(B^+)}{v(A^+)} = \alpha, \text{ say.}$$

Each of the sets B^- and B^+ is a union of p or fewer pairwise disjoint non-degenerate cells, and $v(B)$ equals $v(B^-) + v(B^+)$. The sets $A^- + B^-$ and $A^+ + B^+$ lie in opposite open halfspaces bounded by the hyperplane $x_i = \lambda + \mu$, and so are disjoint; their union is a subset of $A + B$. We deduce, applying the induction hypothesis to the pairs (A^-, B^-) and (A^+, B^+), that

$$\begin{aligned}
v(A + B) &\geqslant v(A^- + B^-) + v(A^+ + B^+) \\
&\geqslant (v^{1/n}(A^-) + v^{1/n}(B^-))^n + (v^{1/n}(A^+) + v^{1/n}(B^+))^n \\
&= (v(A^-) + v(A^+))(1 + \alpha^{1/n})^n \\
&= v(A)(1 + \alpha^{1/n})^n \\
&= (v^{1/n}(A) + \alpha^{1/n}v^{1/n}(A))^n \\
&= (v^{1/n}(A) + v^{1/n}(B))^n.
\end{aligned}$$

Thus $v^{1/n}(A + B) \geqslant v^{1/n}(A) + v^{1/n}(B)$. This completes the proof by induction. □

Theorem 6.5.3 (Brunn's inequality) *Let $A, B, A + B$ be non-empty sets in $\mathbb{R}^n$ all of which have volume. Then*

$$v^{1/n}(A + B) \geqslant v^{1/n}(A) + v^{1/n}(B).$$

Proof The inequality is trivial if either A or B has zero volume. We assume, therefore, that $v(A) > 0$ and $v(B) > 0$. There are sequences $A_1, \ldots, A_i, \ldots$ and $B_1, \ldots, B_i, \ldots$ of non-empty elementary sets in $\mathbb{R}^n$ such that $A_i \subseteq A$, $B_i \subseteq B$ for $i = 1, 2, \ldots$, and $v(A_i) \to v(A)$, $v(B_i) \to v(B)$ as $i \to \infty$; moreover, we can suppose that all of these sets A_i and B_i are finite unions of pairwise disjoint non-degenerate cells. By Lemma 6.5.2,

$$v^{1/n}(A + B) \geqslant v^{1/n}(A_i + B_i) \geqslant v^{1/n}(A_i) + v^{1/n}(B_i).$$

Letting $i \to \infty$ in these inequalities, we deduce that

$$v^{1/n}(A + B) \geqslant v^{1/n}(A) + v^{1/n}(B). \quad \square$$

Corollary 6.5.4 *Let A, B be non-empty bounded convex sets in $\mathbb{R}^n$. Then the function $f: [0, 1] \to \mathbb{R}$ defined by the equation*

$$f(t) = v^{1/n}((1 - t)A + tB) \qquad \text{for } 0 \leqslant t \leqslant 1$$

is concave.

Proof Let $x, y \in [0, 1]$. Let $\lambda, \mu \geqslant 0$ with $\lambda + \mu = 1$. We apply the theorem to the sets $\lambda((1 - x)A + xB)$ and $\mu((1 - y)A + yB)$ to deduce that

$$\begin{aligned}
f(\lambda x + \mu y) &= v^{1/n}((1 - (\lambda x + \mu y))A + (\lambda x + \mu y)B) \\
&= v^{1/n}(\lambda((1 - x)A + xB) + \mu((1 - y)A + yB)) \\
&\geqslant \lambda v^{1/n}((1 - x)A + xB) + \mu v^{1/n}((1 - y)A + yB) \\
&= \lambda f(x) + \mu f(y).
\end{aligned}$$

Thus the function f is concave. □

Theorem 6.5.5 (Minkowski's inequality for mixed volumes) *Let A and B be convex bodies in* $\mathbb{R}^n$. *Then*

$$v(A, \ldots, A, B) \geqslant v^{(n-1)/n}(A)v^{1/n}(B),$$

with equality holding if and only if

$$v^{1/n}(A + B) = v^{1/n}(A) + v^{1/n}(B).$$

Proof Define a function $f: [0, 1] \to \mathbb{R}$ by the equation

$$f(t) = v^{1/n}((1 - t)A + tB) \qquad \text{for } 0 \leqslant t \leqslant 1.$$

Then

$$f^n(t) = v(A)(1 - t)^n + nv(A, \ldots, A, B)(1 - t)^{n-1}t + \ldots + v(B)t^n,$$

whence

$$f'(0) = \frac{v(A, \ldots, A, B) - v(A)}{v^{(n-1)/n}(A)}.$$

Now f is concave by Corollary 6.5.4, and Corollary 5.1.2 can be used to show that $f'(0) \geqslant f(1) - f(0)$, from which the desired inequality is immediate. We note that Minkowski's inequality becomes an equality if and only if $f'(0) = f(1) - f(0)$.

We now show that $f'(0) = f(1) - f(0)$ if and only if $f(\frac{1}{2}) = \frac{1}{2}(f(0) + f(1))$. Suppose first that $f'(0) = f(1) - f(0)$. Then Corollary 5.1.2 with $a = 0$ shows that $(f(x) - f(0))/x$ must be constantly equal to $f(1) - f(0)$ for $0 < x \leqslant 1$, and so putting $x = \frac{1}{2}$, we deduce that $f(\frac{1}{2}) = \frac{1}{2}(f(0) + f(1))$. Suppose next that $f(\frac{1}{2}) = \frac{1}{2}(f(0) + f(1))$. Then

$$\frac{f(1) - f(\frac{1}{2})}{1 - \frac{1}{2}} = \frac{f(0) - f(\frac{1}{2})}{0 - \frac{1}{2}} = f(1) - f(0),$$

and Corollary 5.1.2 with $a = \frac{1}{2}$ shows that $(f(x) - f(\frac{1}{2}))/(x - \frac{1}{2})$ must be constantly equal to $f(1) - f(0)$ for $x \in [0, 1] \setminus \{\frac{1}{2}\}$, whence $f'(0) = f(1) - f(0)$. We complete the proof by noting that $f(\frac{1}{2}) = \frac{1}{2}(f(0) + f(1))$ if and only if

$$v^{1/n}(A + B) = v^{1/n}(A) + v^{1/n}(B). \quad \square$$

We have thus established the inequalities of Brunn and Minkowski for convex bodies, and have shown that the conditions under which they reduce to equalities are identical. Now we proceed to give a complete description of what these conditions are. Since our argument is somewhat involved, we choose to isolate its different strands and present them in the following lemma.

Lemma 6.5.6 (i) *Let S be an n-simplex and let T be a convex body in* $\mathbb{R}^n$. *Suppose that* $v(S) = v(T)$ *and that*

$$v(S, \ldots, S, T) = v^{(n-1)/n}(S)v^{1/n}(T).$$

Then T is a translate of S.

(ii) *Let A, B be convex bodies in $\mathbb{R}^n$ such that, for each n-simplex contained in either one of them, there is some translate of it which is contained in the other. Then B is a translate of A.*

Proof (i) Let $F_0, \ldots, F_n$ be the facets of S and let $\mathbf{u}_0, \ldots, \mathbf{u}_n$ be the corresponding outward unit normals. Let h_T be the support function of T. Let C be the simplex which is homothetic to S, and circumscribes T, i.e. $T \subseteq C$ and T meets each facet of C. Suppose that $C = \lambda S + \mathbf{a}$ for some $\lambda > 0$ and $\mathbf{a} \in \mathbb{R}^n$.

Corollary 6.4.9 shows that

$$v^{(n-1)/n}(S)v^{1/n}(T) = v(S, \ldots, S, T) = \frac{1}{n}\sum_{i=0}^{n} h_T(\mathbf{u}_i)v_{n-1}(F_i).$$

Also

$$\lambda^n v(S) = v(C) = \frac{1}{n}\sum_{i=0}^{n} h_C(\mathbf{u}_i)v_{n-1}(\lambda F_i + \mathbf{a}) = \frac{1}{n}\sum_{i=0}^{n} \lambda^{n-1}h_T(\mathbf{u}_i)v_{n-1}(F_i),$$

where h_C is the support function of C. Thus

$$v^{(n-1)/n}(S)v^{1/n}(T) = \lambda v(S), \qquad \text{whence} \qquad v(T) = \lambda^n v(S) = v(C).$$

But $T \subseteq C$, and so $T = C$. Since $v(S) = v(T)$, λ must be 1, and T is the translate $S + \mathbf{a}$ of S.

(ii) Clearly A and B must have the same diameter, s, say. Let $\mathbf{a}, \mathbf{a}' \in A$ be such that $\|\mathbf{a} - \mathbf{a}'\| = s$. Then there is some $\mathbf{x} \in \mathbb{R}^n$ such that $\mathbf{a} + \mathbf{x}, \mathbf{a}' + \mathbf{x} \in B$. Let $\mathbf{c} \in A$. Then $\mathbf{a}, \mathbf{a}', \mathbf{c}$ belong to some n-simplex of A, whence there is some $\mathbf{y} \in \mathbb{R}^n$ such that $\mathbf{a} + \mathbf{y}$, $\mathbf{a}' + \mathbf{y}$, $\mathbf{c} + \mathbf{y} \in B$. Now

$$\begin{aligned}
2\|\mathbf{y} - \mathbf{x}\|^2 + 2s^2 &= 2\|\mathbf{y} - \mathbf{x}\|^2 + 2\|\mathbf{a} - \mathbf{a}'\|^2 \\
&= \|\mathbf{y} - \mathbf{x} + \mathbf{a} - \mathbf{a}'\|^2 + \|\mathbf{y} - \mathbf{x} - \mathbf{a} + \mathbf{a}'\|^2 \\
&= \|(\mathbf{a} + \mathbf{y}) - (\mathbf{a}' + \mathbf{x})\|^2 + \|(\mathbf{a}' + \mathbf{y}) - (\mathbf{a} + \mathbf{x})\|^2 \\
&\leqslant 2s^2,
\end{aligned}$$

since B has diameter s. This shows that $\mathbf{x} = \mathbf{y}$. Hence $\mathbf{c} + \mathbf{x} \in B$ for all $\mathbf{c}$ in A, and so $A + \mathbf{x} \subseteq B$. A similar argument shows that $B - \mathbf{x} \subseteq A$. Thus B is the translate $A + \mathbf{x}$ of A. □

Theorem 6.5.7 (Brunn–Minkowski theorem) *Let A, B be convex bodies in $\mathbb{R}^n$. Then*

$$v^{1/n}(A + B) \geqslant v^{1/n}(A) + v^{1/n}(B) \tag{6.5.1}$$

with equality holding if and only if A and B are homothetic.

Proof We have already established Brunn's contribution to the theorem, the inequality of Theorem 6.5.3. Now we establish Minkowski's contribution, the conditions under which equality occurs. If A and B are homothetic, say $B = \mu A + \mathbf{a}$, where $\mu > 0$ and $\mathbf{a} \in \mathbb{R}^n$, then equality holds in (6.5.1), both sides being equal to $(1 + \mu)v^{1/n}(A)$.

Conversely, suppose that A, B give equality in (6.5.1). Choose $\lambda > 0$ so that λB and A have the same volume. The second assertion of Theorem 6.5.5 shows that the sets $A, \lambda B$ also give equality in Brunn's inequality, i.e.

$$v^{1/n}(A + \lambda B) = v^{1/n}(A) + v^{1/n}(\lambda B).$$

Let S be any n-simplex contained in A. Then $S = J_0 \cap \ldots \cap J_n$ for some closed halfspaces $J_0, \ldots, J_n$ in $\mathbb{R}^n$. Denote by K_0 the translate of J_0 which makes the volumes $v(A \cap J_0)$ and $v((\lambda B) \cap K_0)$ equal. We show that the sets $A \cap J_0$, $(\lambda B) \cap K_0$ give equality in Brunn's inequality. To see why this is so, consider the sets $A^- = A \cap J_0$, $A^+ = A \backslash A^-$, $B^- = (\lambda B) \cap K_0$, $B^+ = (\lambda B) \backslash B^-$. Suppose that A^-, B^- do not give equality in Brunn's inequality. Then

$$v^{1/n}(A^- + B^-) > v^{1/n}(A^-) + v^{1/n}(B^-).$$

The sets $A^- + B^-$, $A^+ + B^+$ are disjoint and are contained in $A + \lambda B$, whence, by Brunn's inequality and the equalities $v(A^-) = v(B^-)$, $v(A^+) = v(B^+)$,

$$\begin{aligned} v^{1/n}(A + \lambda B) &\geqslant (v(A^- + B^-) + v(A^+ + B^+))^{1/n} \\ &> ((v^{1/n}(A^-) + v^{1/n}(B^-))^n + (v^{1/n}(A^+) + v^{1/n}(B^+))^n)^{1/n} \\ &= v^{1/n}(A) + v^{1/n}(\lambda B). \end{aligned}$$

This contradicts our assumption that $A, \lambda B$ give equality in Brunn's inequality. Thus the convex bodies $A \cap J_0$ and $(\lambda B) \cap K_0$ yield equality in Brunn's inequality. We repeat the argument just given n more times to deduce the existence of closed halfspaces $K_1, \ldots, K_n$ in $\mathbb{R}^n$ such that the convex bodies

$$S = A \cap J_0 \cap \ldots \cap J_n, \qquad T = (\lambda B) \cap K_0 \cap \ldots \cap K_n$$

have the same volume and produce equality in Brunn's inequality. We deduce, from the second assertion of Theorem 6.5.5 and the first part of the lemma, that T must be a translate of S. It follows, by symmetry, that, for each n-simplex contained in either one of A and λB, there is some translate of it which is contained in the other. Hence, by the second part of the lemma, A is a translate of λB, whence A and B are homothetic. □

Theorem 6.5.8 *Let A, B be convex bodies in $\mathbb{R}^n$. Then*

$$v(A, \ldots, A, B) \geqslant v^{(n-1)/n}(A)v^{1/n}(B)$$

with equality holding if and only if A and B are homothetic.

Proof The result follows immediately from Theorems 6.5.5 and 6.5.7. □

We recall that the isoperimetric problem is to determine which convex bodies have the smallest surface area for a given volume. The problem has a long, interesting history and its solution is far from trivial. For us, however, the solution is a direct consequence of the last theorem. Since the solution gives a lower bound on the possible surface areas of convex bodies with a given volume, it can be expressed by an inequality; this inequality, which is stated in the next theorem, is called the **isoperimetric inequality**.

Theorem 6.5.9 (Isoperimetric inequality) *Every convex body A in $\mathbb{R}^n$ has a surface area greater than that of a closed ball with the same volume, unless it is itself a closed ball. More specifically,*

$$s^n(A) \geqslant \omega_n n^n v^{n-1}(A)$$

with equality holding if and only if A is a closed ball.

Proof Let B be the closed unit ball U in Theorem 6.5.8. Then we deduce that

$$s(A) = nv(A, \ldots, A, U) \geqslant nv^{(n-1)/n}(A)v^{1/n}(U) = nv^{(n-1)/n}(A)\omega_n^{1/n},$$

and so

$$s^n(A) \geqslant \omega_n n^n v^{n-1}(A),$$

with equality holding if and only if A is homothetic to U, i.e. if and only if A is a closed ball. □

We now apply the Brunn–Minkowski theorem to the **isodiametric problem**: *of all convex bodies in $\mathbb{R}^n$ with a given diameter, which ones have the greatest volume*? The reader's attention is drawn to the use of the Blaschke selection theorem to establish the existence of an extremal body—this is a feature common to the solutions of many extremum problems.

Theorem 6.5.10 (Isodiametric inequality) *Every convex body A in $\mathbb{R}^n$ with diameter d has a volume less than that of a closed ball with diameter d, unless it is itself a closed ball. More specifically,*

$$v(A) \leqslant \omega_n(\tfrac{1}{2}d)^n$$

with equality holding if and only if A is a closed ball.

Proof Denote by $\mathfrak{F}$ the family of all convex bodies in $\mathbb{R}^n$ with diameter d. Let $\alpha = \sup\{v(A): A \in \mathfrak{F}\}$. Then there is a sequence $A_1, \ldots, A_k, \ldots$ of members of $\mathfrak{F}$ which lie in the closed ball dU such that $v(A_k) \to \alpha$ as $k \to \infty$. By the Blaschke selection theorem, there exists a subsequence $i_1, \ldots, i_k, \ldots$ of $1, \ldots, k, \ldots$ and a convex body A_0 such that $A_{i_k} \to A_0$ as $k \to \infty$. Since both volume and diameter are continuous with respect to Hausdorff distance, it follows that $A_0 \in \mathfrak{F}$, $v(A) = \alpha$. Thus A_0 is a member of $\mathfrak{F}$ having maximal possible volume, so A_0 is a solution of the isodiametric problem.

Let C be any member of $\mathfrak{F}$ having maximal volume. It is easily verified that the convex body $C' = \frac{1}{2}(C - C)$ belongs to $\mathfrak{F}$. The Brunn–Minkowski theorem shows that $v(C') \geqslant v(C)$ with equality holding if and only if C is homothetic to $-C$. By the choice of C, $v(C) \geqslant v(C')$. Thus $v(C') = v(C)$, whence $C = -\lambda C + \mathbf{c}$ for some $\lambda > 0$ and $\mathbf{c} \in \mathbb{R}^n$. Since C has the same volume as $-C$, $\lambda = 1$ and $C = -C + \mathbf{c}$. Hence $C - \frac{1}{2}\mathbf{c} = -(C - \frac{1}{2}\mathbf{c})$, and $C - \frac{1}{2}\mathbf{c}$ is a symmetric member of $\mathfrak{F}$ having maximal volume. The symmetry of $C - \frac{1}{2}\mathbf{c}$ together with the fact that its diameter is d shows that $C - \frac{1}{2}\mathbf{c} \subseteq \frac{1}{2}dU$. But $\frac{1}{2}dU \in \mathfrak{F}$ and $v(C - \frac{1}{2}\mathbf{c}) \leqslant v(\frac{1}{2}dU)$, whence $C - \frac{1}{2}\mathbf{c} = \frac{1}{2}dU$. Thus C is the closed ball $\frac{1}{2}\mathbf{c} + \frac{1}{2}dU$, and the desired result is immediate. □

We now describe a construction invented by H. A. Schwarz to solve the isoperimetric problem. The construction associates with each given convex body and each line in $\mathbb{R}^n$ another convex body having the same volume as the given one, and for which the given line is an axis of rotation. Before describing the construction formally, we first consider a simple example of its use.

Suppose that the given convex body is the square pyramid

$$A = \operatorname{conv}\{(0,0,0), (1,1,1), (1,1,-1), (1,-1,1), (1,-1,-1)\}.$$

Then A has for its base a square of side 2 lying in the plane $x_1 = 1$, for its vertex the origin, and its height is 1. For each x with $0 < x \leqslant 1$, denote by A_x the intersection of A with the hyperplane $x_1 = x$, and by C_x the closed circular disc which lies in the hyperplane $x_1 = x$, has its centre on the x_1-axis, and has the same area as that of A_x. Clearly A_x is a square of side $2x$ and C_x has radius $r_x = 2\pi^{-\frac{1}{2}}x$. We write $C_0 = \{(0,0,0)\}$ and $r_0 = 0$. The union

$$C = \bigcup\,(C_x : 0 \leqslant x \leqslant 1)$$

of the circular discs C_x is called the *Schwarz rotation-symmetral* of A in the x_1-axis. Here C is a right circular cone with base a closed disc of radius $2\pi^{-\frac{1}{2}}$, with axis the x_1-axis, and vertex the origin.

We now describe the Schwarz rotation-symmetral of a convex body A about a line in $\mathbb{R}^n$, where $n \geqslant 2$. For simplicity of notation, we suppose that the line is the x_1-axis and that A lies between parallel support hyperplanes $x_1 = a$ and $x_1 = b$ to A, where $a < b$. For each x with $a \leqslant x \leqslant b$, denote by A_x the intersection of A with the hyperplane $x_1 = x$, and define r_x by the equation

$$\omega_{n-1} r_x^{n-1} = v_{n-1}(A_x).$$

Thus, for $a < x < b$, r_x is the radius of an $(n-1)$-ball whose v_{n-1}-volume is the same as that of A_x. For each x with $a \leqslant x \leqslant b$, define a convex set C_x (indeed an $(n-1)$-ball when $a < x < b$) by the equation

$$C_x = \{(x, x_2, \ldots, x_n): x_2^2 + \ldots + x_n^2 \leqslant r_x^2\}.$$

Then the set

$$C = \bigcup (C_x : a \leqslant x \leqslant b)$$

is called the **Schwarz rotation-symmetral** of A in the x_1-axis.

Theorem 6.5.11 *Let A be a convex body in $\mathbb{R}^n$ $(n \geqslant 2)$ whose Schwarz rotation-symmetral in the x_1-axis is C. Then C is a convex body having the same volume as A.*

Proof We assume the notation introduced for the definition of the Schwarz rotation-symmetral. First we show that $r:[a, b] \to \mathbb{R}$ is a concave function. Let $x, y \in [a, b]$ and let $\lambda, \mu \geqslant 0$ with $\lambda + \mu = 1$. By the convexity of A,

$$A_{\lambda x + \mu y} \supseteq \lambda A_x + \mu A_y.$$

Applying Brunn's inequality in $\mathbb{R}^{n-1}$, we find that

$$v_{n-1}^{1/(n-1)}(A_{\lambda x+\mu y}) \geqslant v_{n-1}^{1/(n-1)}(\lambda A_x + \mu A_y) \geqslant \lambda v_{n-1}^{1/(n-1)}(A_x) + \mu v_{n-1}^{1/(n-1)}(A_y),$$

whence

$$r_{\lambda x + \mu y} \geqslant \lambda r_x + \mu r_y.$$

We now establish the convexity of C, leaving the reader to verify that C satisfies the other requirements of a convex body. Let $\mathbf{u}, \mathbf{v} \in C$ and let $\lambda, \mu \geqslant 0$ with $\lambda + \mu = 1$. Then $\mathbf{u} \in C_x$, $\mathbf{v} \in C_y$, for some $x, y \in [a, b]$. Thus

$$\|\mathbf{u} - (x, 0, \ldots, 0)\| \leqslant r_x, \qquad \|\mathbf{v} - (y, 0, \ldots, 0)\| \leqslant r_y.$$

Now $a \leqslant \lambda x + \mu y \leqslant b$ and

$$\begin{aligned}\|\lambda\mathbf{u} + \mu\mathbf{v} - (\lambda x + \mu y, 0, \ldots, 0)\| &\leqslant \lambda\|\mathbf{u} - (x, 0, \ldots, 0)\| + \mu\|\mathbf{v} - (y, 0, \ldots, 0)\| \\ &\leqslant \lambda r_x + \mu r_y \\ &\leqslant r_{\lambda x + \mu y}.\end{aligned}$$

Hence $\lambda\mathbf{u} + \mu\mathbf{v} \in C_{\lambda x+\mu y}$, which shows that $\lambda\mathbf{u} + \mu\mathbf{v} \in C$. Thus C is convex. It follows from Theorem 6.3.1 that

$$v_n(A) = \int_a^b v_{n-1}(A_x)\,dx = \int_a^b v_{n-1}(C_x)\,dx = v_n(C). \quad \square$$

Consider our earlier example in which A was a square pyramid in $\mathbb{R}^3$ and its Schwarz rotation-symmetral C was a circular cone. Here A and C have the same volume 4/3, A has surface area $4 + 4\sqrt{2}$, and C has the smaller surface area $4 + 2\sqrt{(\pi + 4)}$. It is a property of the Schwarz rotation-symmetral of a convex body that its surface area never exceeds that of the body itself, a fact used by Schwarz in his investigations into the isoperimetric problem.

Exercises 6.5

1. Given that A, B are compact convex sets in $\mathbb{R}^n$, state under what conditions $v^{1/n}(A + B) = v^{1/n}(A) + v^{1/n}(B)$.
2. A builder measures the desirability of the space $\mathbb{R}^n$ $(n \geqslant 2)$ in which he builds, by the largest convex volume σ_n he can bound by a unit of surface area. Find a formula for σ_n. In which $\mathbb{R}^n$ might he choose to build?
3. Let A, B be convex bodies in $\mathbb{R}^n$ $(n \geqslant 2)$. Let $f\colon[0, 1] \to \mathbb{R}$ be defined by the equation
$$f(t) = v^{1/n}((1 - t)A + tB) \qquad \text{for } 0 \leqslant t \leqslant 1.$$
Find $f''(0)$ and deduce that
$$v(A)v(A, \ldots, A, B, B) \leqslant v^2(A, \ldots, A, B).$$
4. The convex bodies A, B in $\mathbb{R}^2$ have projections on the x-axis of lengths a, b, respectively. Prove that, for $0 \leqslant t \leqslant 1$,
$$v(tbA + (1 - t)aB) \geqslant tb^2v(A) + (1 - t)a^2v(B).$$
Deduce, by putting $t = \lambda a/(\lambda a + (1 - \lambda)b)$, that, for $0 \leqslant \lambda \leqslant 1$,
$$\begin{aligned} v(\lambda A + (1 - \lambda)B) &\geqslant (\lambda + (1 - \lambda)c)(\lambda v(A) + (1 - \lambda)c^{-1}v(B)) \\ &\geqslant (\lambda v^{1/2}(A) + (1 - \lambda)v^{1/2}(B))^2, \end{aligned}$$
where $c = b/a$.

6.6 STEINER SYMMETRIZATION

In this section we introduce a construction invented by the celebrated Swiss geometer Jakob Steiner (1796–1863) that has impressive applica-

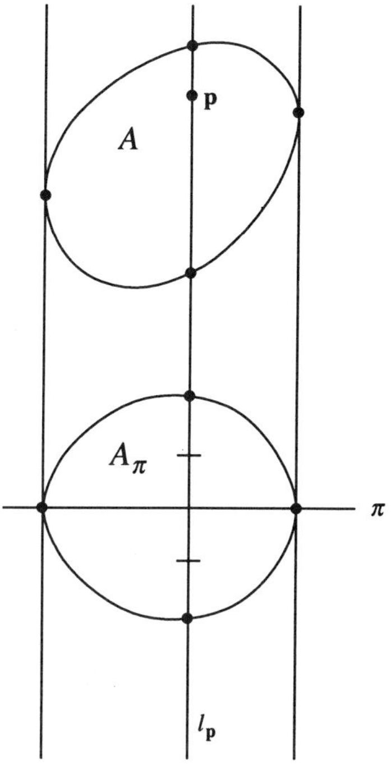

Fig. 6.7

tions to geometrical extremum problems. Here we use it to give simple proofs of the Brunn, isoperimetric, and isodiametric inequalities.

Let A be a non-empty compact convex set and π a hyperplane in $\mathbb{R}^n$. Steiner's construction produces from A and π a convex set A_π in $\mathbb{R}^n$ called the *Steiner symmetral* of A about π. The following informal description should make it clear how A_π is obtained from A and π. For each point $\mathbf{p}$ of A, denote by $\mathbf{l_p}$ the line through $\mathbf{p}$ perpendicular to the hyperplane π. Translate the chord $A \cap \mathbf{l_p}$ of A along $\mathbf{l_p}$ until its midpoint lies on π. The union A_π of all such translated chords is called the *Steiner symmetral* of A about π. This construction is illustrated in Fig. 6.7.

Before giving the formal definition of the Steiner symmetral A_π, we first show how the set A and the hyperplane π give rise to a concave function γ defined on the *projection* $\pi(A)$ of A on π. Essentially γ measures the lengths of the chords of A which are perpendicular to π. Let $\mathbf{u}$ be a unit normal vector to π. Then the **projection** $\pi(A)$ of A on π is the subset of π defined by the equation

$$\pi(A) = \{\mathbf{p} \in \pi : \mathbf{p} + \theta\mathbf{u} \in A \text{ for some } \theta \in \mathbb{R}\}.$$

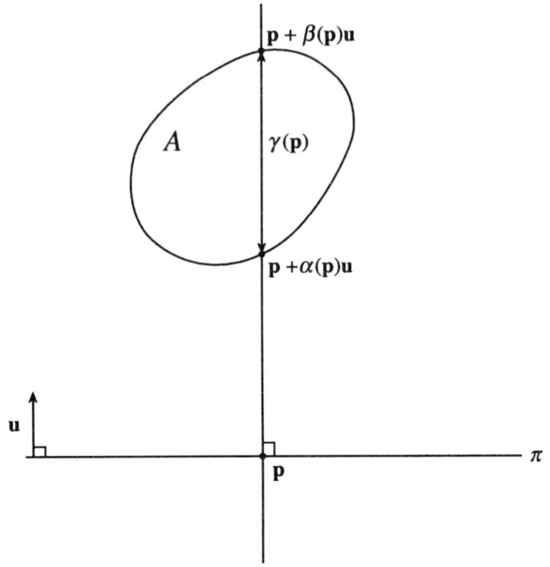

Fig. 6.8

We show that $\pi(A)$ is convex. Let $\mathbf{p}, \mathbf{q} \in \pi(A)$ and let $\lambda, \mu \geqslant 0$ with $\lambda + \mu = 1$. Then there exist $\theta, \varphi \in \mathbb{R}$ such that $\mathbf{p} + \theta\mathbf{u}$, $\mathbf{q} + \varphi\mathbf{u} \in A$. Since A is convex,

$$\lambda(\mathbf{p} + \theta\mathbf{u}) + \mu(\mathbf{q} + \varphi\mathbf{u}) = \lambda\mathbf{p} + \mu\mathbf{q} + (\lambda\theta + \mu\varphi)\mathbf{u} \in A.$$

Hence $\lambda\mathbf{p} + \mu\mathbf{q} \in \pi(A)$, and so $\pi(A)$ is convex. For each $\mathbf{p}$ in $\pi(A)$, denote by $I_\mathbf{p}$ the non-empty compact interval of $\mathbb{R}$ defined by the equation

$$I_\mathbf{p} = \{\theta \in \mathbb{R} : \mathbf{p} + \theta\mathbf{u} \in A\}.$$

Define functions $\alpha, \beta, \gamma : \pi(A) \to \mathbb{R}$ as follows:

$$\alpha(\mathbf{p}) = \min I_\mathbf{p}, \quad \beta(\mathbf{p}) = \max I_\mathbf{p}, \quad \gamma(\mathbf{p}) = \beta(\mathbf{p}) - \alpha(\mathbf{p}) \qquad \text{for } \mathbf{p} \in \pi(A).$$

Thus $\gamma(\mathbf{p})$ is the length of the chord of A which is the intersection of A with the line through $\mathbf{p}$ normal to π. If we choose $-\mathbf{u}$ instead of $\mathbf{u}$ for a unit normal to π, then γ (unlike α and β) remains unchanged. Thus γ is uniquely determined by A and π. For the remainder of this section, unless stated otherwise, $A, \pi, \mathbf{u}, \alpha, \beta, \gamma$ will be as above. See Fig. 6.8.

Theorem 6.6.1 *The function $\gamma : \pi(A) \to \mathbb{R}$ is concave.*

Proof Let $\mathbf{p}, \mathbf{q} \in \pi(A)$ and let $\lambda, \mu \geqslant 0$ with $\lambda + \mu = 1$. Then $\mathbf{p} + \alpha(\mathbf{p})\mathbf{u}$, $\mathbf{q} + \alpha(\mathbf{q})\mathbf{u} \in A$. The convexity of A shows that

$$\lambda(\mathbf{p} + \alpha(\mathbf{p})\mathbf{u}) + \mu(\mathbf{q} + \alpha(\mathbf{q})\mathbf{u}) = \lambda\mathbf{p} + \mu\mathbf{q} + (\lambda\alpha(\mathbf{p}) + \mu\alpha(\mathbf{q}))\mathbf{u} \in A,$$

whence

$$\alpha(\lambda\mathbf{p} + \mu\mathbf{q}) \leqslant \lambda\alpha(\mathbf{p}) + \mu\alpha(\mathbf{q}).$$

Similarly,

$$\beta(\lambda\mathbf{p} + \mu\mathbf{q}) \geqslant \lambda\beta(\mathbf{p}) + \mu\beta(\mathbf{q}).$$

Thus

$$\begin{aligned}\gamma(\lambda\mathbf{p} + \mu\mathbf{q}) &= \beta(\lambda\mathbf{p} + \mu\mathbf{q}) - \alpha(\lambda\mathbf{p} + \mu\mathbf{q})\\ &\geqslant \lambda(\beta(\mathbf{p}) - \alpha(\mathbf{p})) + \mu(\beta(\mathbf{q}) - \alpha(\mathbf{q}))\\ &= \lambda\gamma(\mathbf{p}) + \mu\gamma(\mathbf{q}),\end{aligned}$$

which shows that γ is concave. □

We now formally define the **Steiner symmetral** A_π of A about π by the equation

$$A_\pi = \{\mathbf{p} + \theta\mathbf{u} : \mathbf{p} \in \pi(A) \quad \text{and} \quad |\theta| \leqslant \tfrac{1}{2}\gamma(\mathbf{p})\}.$$

Some easy consequences of the definition are:

(i) A_π is (in an obvious sense) symmetric about π;

(ii) if B is a closed ball with centre on π, then $B_\pi = B$;

(iii) if C is a compact convex set with $A \subseteq C$, then $A_\pi \subseteq C_\pi$.

Theorem 6.6.2 *Let A be a non-empty compact convex set and let π be a hyperplane in $\mathbb{R}^n$. Then A_π is a non-empty compact convex set, which is a convex body when A is.*

Proof Let $\mathbf{a}, \mathbf{b} \in A_\pi$. Then there are $\mathbf{p}, \mathbf{q} \in \pi(A)$ and scalars θ, φ such that $\mathbf{a} = \mathbf{p} + \theta\mathbf{u}$, $\mathbf{b} = \mathbf{q} + \varphi\mathbf{u}$, where $|\theta| \leqslant \frac{1}{2}\gamma(\mathbf{p})$, $|\varphi| \leqslant \frac{1}{2}\gamma(\mathbf{q})$. Let $\lambda, \mu \geqslant 0$ with $\lambda + \mu = 1$. Then

$$\lambda\mathbf{a} + \mu\mathbf{b} = \lambda\mathbf{p} + \mu\mathbf{q} + (\lambda\theta + \mu\varphi)\mathbf{u}.$$

By the concavity of γ,

$$|\lambda\theta + \mu\varphi| \leqslant \lambda|\theta| + \mu|\varphi| \leqslant \tfrac{1}{2}\lambda\gamma(\mathbf{p}) + \tfrac{1}{2}\mu\gamma(\mathbf{q}) \leqslant \tfrac{1}{2}\gamma(\lambda\mathbf{p} + \mu\mathbf{q}).$$

Thus $\lambda\mathbf{a} + \mu\mathbf{b} \in A_\pi$, which shows that A_π is convex.

We now show that A_π is closed. Let $\mathbf{x}_1, \ldots, \mathbf{x}_k, \ldots$ be a sequence of points of A_π that converges to a point $\mathbf{x}$ of $\mathbb{R}^n$. For each $k = 1, 2, \ldots$, there exist $\mathbf{p}_k \in \pi(A)$, $\theta_k \in \mathbb{R}$ such that

$$\mathbf{x}_k = \mathbf{p}_k + \theta_k\mathbf{u} \qquad \text{and} \qquad |\theta_k| \leqslant \tfrac{1}{2}\gamma(\mathbf{p}_k).$$

Now the point $\mathbf{x}$ can be written in the form $\mathbf{p} + \theta\mathbf{u}$, where $\mathbf{p} \in \pi$ and $\theta \in \mathbb{R}$. Since $\|\mathbf{p}_k - \mathbf{p}\| \leqslant \|\mathbf{x}_k - \mathbf{x}\|$ and $\mathbf{x}_k \to \mathbf{x}$ as $k \to \infty$, $\mathbf{p}_k \to \mathbf{p}$ as $k \to \infty$. The points

$$\mathbf{y}_k = \mathbf{p}_k + \alpha(\mathbf{p}_k)\mathbf{u}, \qquad \mathbf{z}_k = \mathbf{p}_k + \beta(\mathbf{p}_k)\mathbf{u}$$

lie in the compact set A, whence there is a subsequence $i_1, \ldots, i_k, \ldots$ of $1, \ldots, k, \ldots$ and points $\mathbf{y}, \mathbf{z} \in A$ such that

$$\mathbf{y}_{i_k} \to \mathbf{y}, \qquad \mathbf{z}_{i_k} \to \mathbf{z} \qquad \text{as } k \to \infty.$$

A simple argument shows that $\mathbf{y} = \mathbf{p} + a\mathbf{u}$, $\mathbf{z} = \mathbf{p} + b\mathbf{u}$, where $a, b \in \mathbb{R}$ are such that $a \leqslant b$ and

$$\alpha(\mathbf{p}_{i_k}) \to a, \qquad \beta(\mathbf{p}_{i_k}) \to b \qquad \text{as } k \to \infty.$$

Thus $\mathbf{p} \in \pi(A)$ and

$$|\theta| = \lim_{k \to \infty} |\theta_{i_k}| \leqslant \lim_{k \to \infty} \tfrac{1}{2}(\beta(\mathbf{p}_{i_k}) - \alpha(\mathbf{p}_{i_k})) = \tfrac{1}{2}(b - a) \leqslant \tfrac{1}{2}\gamma(\mathbf{p}),$$

which shows that $\mathbf{x} \in A_\pi$. Hence A_π is closed.

Since A is bounded, it lies in some closed ball C, whence $A_\pi \subseteq C_\pi$. But C_π is clearly a closed ball, and so A_π is bounded. We have thus shown that A_π is both closed and bounded, i.e. A_π is compact.

If A is a convex body, then it contains some closed ball B, whence $B_\pi \subseteq A_\pi$. But B_π is a closed ball, and so the compact convex set A_π has a non-empty interior, i.e. A_π is a convex body. □

Theorem 6.6.3 *In* $\mathbb{R}^n$ *let* A, B *be non-empty compact convex sets and let* π *be a hyperplane passing through the origin. Then*

$$A_\pi + B_\pi \subseteq (A + B)_\pi.$$

Proof Let $\mathbf{x} \in A_\pi + B_\pi$. Then $\mathbf{x} = \mathbf{a} + \mathbf{b}$ for some $\mathbf{a} \in A_\pi$, $\mathbf{b} \in B_\pi$. We can write, using an obvious notation, $\mathbf{a} = \mathbf{p} + \theta\mathbf{u}$, $\mathbf{b} = \mathbf{q} + \varphi\mathbf{u}$, where $\mathbf{p} \in \pi(A)$, $\mathbf{q} \in \pi(B)$ and $|\theta| \leqslant \frac{1}{2}\gamma_A(\mathbf{p})$, $|\varphi| \leqslant \frac{1}{2}\gamma_B(\mathbf{q})$. Since π is a subspace of $\mathbb{R}^n$, $\mathbf{p} + \mathbf{q} \in \pi$, from which it follows easily that $\mathbf{p} + \mathbf{q} \in \pi(A + B)$. Clearly

$$\gamma_{A+B}(\mathbf{p} + \mathbf{q}) \geqslant \gamma_A(\mathbf{p}) + \gamma_B(\mathbf{q}),$$

whence

$$|\theta + \varphi| \leqslant |\theta| + |\varphi| \leqslant \tfrac{1}{2}\gamma_A(\mathbf{p}) + \tfrac{1}{2}\gamma_B(\mathbf{q}) \leqslant \tfrac{1}{2}\gamma_{A+B}(\mathbf{p} + \mathbf{q}).$$

Thus

$$\mathbf{x} = \mathbf{p} + \mathbf{q} + (\theta + \varphi)\mathbf{u} \in (A + B)_\pi, \quad \text{and so} \quad A_\pi + B_\pi \subseteq (A + B)_\pi. \ □$$

In the following basic result about Steiner symmetrization, D and R denote, respectively, diameter and circumradius.

Theorem 6.6.4 *In* $\mathbb{R}^n$ *let* A *be a non-empty compact convex set and let* π *be a hyperplane. Then* $v_n(A_\pi) = v_n(A)$, $s_n(A_\pi) \leqslant s_n(A)$, $D(A_\pi) \leqslant D(A)$, $R(A_\pi) \leqslant R(A)$.

Proof We suppose throughout that π contains the origin.

To show that $v_n(A_\pi) = v_n(A)$, we argue by induction on n. The case $n = 1$ is trivial. Suppose that $n \geqslant 2$ and that the assertion is known to be true in $\mathbb{R}^{n-1}$. Let $\mathbf{u}$ be a unit vector lying in π. Corollary 6.3.2 shows that, in an obvious notation,

$$v_n(A) = \int_{-\infty}^{\infty} v_{n-1}(A_x)\,dx \qquad \text{and} \qquad v_n(A_\pi) = \int_{-\infty}^{\infty} v_{n-1}((A_\pi)_x)\,dx.$$

It is readily verified that $(A_\pi)_x = (A_x)_\pi$, and it follows easily from the induction hypothesis that $v_{n-1}((A_x)_\pi) = v_{n-1}(A_x)$. Hence $v_n(A) = v_n(A_\pi)$. This completes the proof by induction.

Theorem 6.6.3 shows that, for each $\lambda > 0$,

$$A_\pi + \lambda U = A_\pi + (\lambda U)_\pi \subseteq (A + \lambda U)_\pi.$$

Thus, by the first part of this proof,

$$v_n(A_\pi + \lambda U) \leqslant v_n((A + \lambda U)_\pi) = v_n(A + \lambda U).$$

Hence

$$\lim_{\lambda \to 0_+} \frac{v_n(A_\pi + \lambda U) - v_n(A_\pi)}{\lambda} \leqslant \lim_{\lambda \to 0_+} \frac{v_n(A + \lambda U) - v_n(A)}{\lambda},$$

i.e. $s_n(A_\pi) \leqslant s_n(A)$.

Suppose now that $\mathbf{u}$ is a unit normal to π. Let $\mathbf{x}, \mathbf{y} \in A_\pi$. Then $\mathbf{x} = \mathbf{p} + \theta\mathbf{u}$, $\mathbf{y} = \mathbf{q} + \varphi\mathbf{u}$ for some $\mathbf{p}, \mathbf{q} \in \pi(A)$ and $\theta, \varphi \in \mathbb{R}$ with $|\theta| \leqslant \frac{1}{2}\gamma(\mathbf{p})$, $|\varphi| \leqslant \frac{1}{2}\gamma(\mathbf{q})$. The points

$$\mathbf{x}_\alpha = \mathbf{p} + \alpha(\mathbf{p})\mathbf{u}, \quad \mathbf{x}_\beta = \mathbf{p} + \beta(\mathbf{p})\mathbf{u}, \quad \mathbf{y}_\alpha = \mathbf{q} + \alpha(\mathbf{q})\mathbf{u}, \quad \mathbf{y}_\beta = \mathbf{q} + \beta(\mathbf{q})\mathbf{u}$$

belong to A, and

$$\begin{aligned} \|\mathbf{x}_\alpha - \mathbf{y}_\beta\|^2 &= \|\mathbf{p} - \mathbf{q}\|^2 + |\alpha(\mathbf{p}) - \beta(\mathbf{q})|^2, \\ \|\mathbf{x}_\beta - \mathbf{y}_\alpha\|^2 &= \|\mathbf{p} - \mathbf{q}\|^2 + |\beta(\mathbf{p}) - \alpha(\mathbf{q})|^2. \end{aligned}$$

Now

$$\begin{aligned} \|\mathbf{x} - \mathbf{y}\|^2 &= \|\mathbf{p} - \mathbf{q}\|^2 + |\theta - \varphi|^2 \\ &\leqslant \|\mathbf{p} - \mathbf{q}\|^2 + \tfrac{1}{4}(\gamma(\mathbf{p}) + \gamma(\mathbf{q}))^2 \\ &= \|\mathbf{p} - \mathbf{q}\|^2 + \tfrac{1}{4}(\beta(\mathbf{p}) - \alpha(\mathbf{q}) + \beta(\mathbf{q}) - \alpha(\mathbf{p}))^2 \\ &\leqslant \|\mathbf{p} - \mathbf{q}\|^2 + \tfrac{1}{2}|\beta(\mathbf{p}) - \alpha(\mathbf{q})|^2 + \tfrac{1}{2}|\beta(\mathbf{q}) - \alpha(\mathbf{p})|^2 \\ &= \tfrac{1}{2}\|\mathbf{x}_\alpha - \mathbf{y}_\beta\|^2 + \tfrac{1}{2}\|\mathbf{x}_\beta - \mathbf{y}_\alpha\|^2 \\ &\leqslant D^2(A). \end{aligned}$$

This shows that $D(A_\pi) \leqslant D(A)$.

If C is a closed ball containing A, then C_π is a translate of C containing A_π. Thus $R(A_\pi) \leqslant R(A)$. □

We now use Steiner symmetrization to establish the isodiametric inequality. The proof of Theorem 6.5.10 shows that, for each $d > 0$, there exists amongst all convex bodies in $\mathbb{R}^n$ of diameter d some convex body C which has maximal volume. Let C_0 be the convex body obtained from C by successive Steiner symmetrizations in the hyperplanes $x_1 = 0, \ldots, x_n = 0$. It is a simple exercise to show that C_0 is a symmetric convex body, which has the same volume, and no larger diameter than C_0. Since C_0 is symmetric with diameter less than or equal to d, it must lie in the ball $\frac{1}{2}dU$. Thus, for any convex body A in $\mathbb{R}^n$ with diameter d, we have

$$v_n(A) \leqslant v_n(C) = v_n(C_0) \leqslant \omega_n(\tfrac{1}{2}d)^n,$$

which is the isodiametric inequality.

An important property of Steiner symmetrization, so far as its use in extremum problems is concerned, is that it is continuous on the class of all convex bodies. This we now prove.

Theorem 6.6.5 *Let $A_1, \ldots, A_k, \ldots$ be a sequence of convex bodies that converges to a convex body A in $\mathbb{R}^n$. Then the sequence $(A_1)_\pi, \ldots, (A_k)_\pi, \ldots$ of its Steiner symmetrals about any hyperplane π of $\mathbb{R}^n$ converges to the Steiner symmetral A_π of A about π.*

Proof We assume that the origin is an interior point of A lying on π. Thus there exist $r, s > 0$ and a positive integer N_1 such that

$$rU \subseteq A \subseteq sU \qquad \text{and} \qquad rU \subseteq A_k \subseteq sU \qquad \text{for } k > N_1.$$

Hence

$$rU \subseteq A_\pi \subseteq sU \qquad \text{and} \qquad rU \subseteq (A_k)_\pi \subseteq sU \qquad \text{for } k > N_1.$$

Let $\varepsilon > 0$. Since $A_k \to A$ as $k \to \infty$, there is a positive integer N_2 such that, for $k > N_2$,

$$A_k \subseteq A + \frac{r\varepsilon}{s}U \qquad \text{and} \qquad A \subseteq A_k + \frac{r\varepsilon}{s}U.$$

Let $k > \max\{N_1, N_2\}$. Then

$$A_k \subseteq A + \frac{r\varepsilon}{s}U \subseteq A + \frac{\varepsilon}{s}A = \left(1 + \frac{\varepsilon}{s}\right)A,$$

and so

$$(A_k)_\pi \subseteq \left(1 + \frac{\varepsilon}{s}\right)A_\pi = A_\pi + \frac{\varepsilon}{s}A_\pi \subseteq A_\pi + \varepsilon U.$$

Similarly,

$$A_\pi \subseteq (A_k)_\pi + \varepsilon U.$$

Thus

$$\rho((A_k)_\pi, A_\pi) \leqslant \varepsilon \qquad \text{and} \qquad (A_k)_\pi \to A_\pi \quad \text{as } k \to \infty. \ \square$$

Perhaps the most important feature of Steiner symmetrization is that, given any convex body, it is possible, by applying a suitable sequence of symmetrizations to it, to obtain a convex body which approximates to a closed ball to within any prescribed degree of accuracy. Before stating this result formally, we need an item of notation. Let A be a convex body in $\mathbb{R}^n$. Then we denote by $\mathcal{S}(A)$ the family of all convex bodies which can be obtained from A by a finite number of symmetrizations about hyperplanes through the origin.

Theorem 6.6.6 *Let A be a convex body in $\mathbb{R}^n$. Then there is a sequence of members of $\mathcal{S}(A)$ which converges to the closed ball of volume $v_n(A)$ whose centre is the origin.*

Proof Let

$$r_0 = \inf\{r > 0 : \text{there is } C \text{ in } \mathcal{S}(A) \text{ such that } C \subseteq rU\}.$$

Then, for each $k = 1, 2, \ldots,$ there exists A_k in $\mathcal{S}(A)$ such that $A_k \subseteq (r_0 + k^{-1})U$. By the Blaschke selection theorem, there is a subsequence of $A_1, A_2, \ldots$ which converges to some convex body, B, say. We assume that the sequence itself converges to B. Clearly $B \subseteq r_0 U$ and $v_n(B) = v_n(A)$.

We complete the proof by showing that $B = r_0 U$. Suppose that $B \neq r_0 U$. Then there exist $\mathbf{x}_0 \in \operatorname{bd} r_0 U$ and $s > 0$ such that $B(\mathbf{x}_0; s) \cap B = \emptyset$. Since $\operatorname{bd} r_0 U$ is compact, there exist distinct points $\mathbf{x}_0, \ldots, \mathbf{x}_m$ $(m \geqslant 1)$ of $\operatorname{bd} r_0 U$ such that

$$\operatorname{bd} r_0 U \subseteq B(\mathbf{x}_0; s) \cup \ldots \cup B(\mathbf{x}_m; s).$$

For $i = 0, \ldots, m$, denote by C_i the *cap* $B(\mathbf{x}_i; s) \cap \operatorname{bd} r_0 U$. Then

$$\operatorname{bd} r_0 U = C_0 \cup \ldots \cup C_m.$$

For $i = 1, \ldots, m$, let π_i be the hyperplane through the origin which has $\mathbf{x}_i - \mathbf{x}_0$ for a normal vector. Then C_0 and C_i are *mirror images* of one another in π_i. It follows immediately from the fact that $B(\mathbf{x}_0; s) \cap B = \emptyset$ and the definition of Steiner symmetrization that B_{π_1} is disjoint from $C_0 \cup C_1$. Similarly, $(B_{\pi_1})_{\pi_2}$ is disjoint from $C_0 \cup C_1 \cup C_2$. Continuing in this fashion, we find that the convex body $B^\square$ obtained from B by successive symmetrizations about $\pi_1, \ldots, \pi_m$ is disjoint from $C_0 \cup \ldots \cup C_m$, and hence from $\operatorname{bd} r_0 U$. Since $B^\square$ is a convex body lying in $r_0 U$, there exists ε with $0 < \varepsilon < r_0$

such that $B^{\square} \subseteq (r_0 - \varepsilon)U$. For $k = 1, 2, \ldots,$ denote by $A_k^{\square}$ the convex body obtained from A_k by successive symmetrizations about $\pi_1, \ldots,$ π_m. Then $A_k^{\square} \in \mathcal{S}(A)$ and, by Theorem 6.6.5, $A_k^{\square} \to B^{\square}$ as $k \to \infty$. But $B^{\square} \subseteq (r_0 - \varepsilon)U$, so there is a k such that $A_k^{\square} \subseteq (r_0 - \frac{1}{2}\varepsilon)U$. This, however, contradicts the definition of r_0. Thus $B = r_0 U$. □

We now use the last theorem to give us a new proof of the isoperimetric inequality. Let A be a convex body in $\mathbb{R}^n$ and let r be the radius of a ball having the same volume as that of A, i.e. $\omega_n r^n = v_n(A)$. The theorem shows the existence of a sequence of convex bodies converging to rU, each of whose members has surface area not exceeding $s_n(A)$, whence

$$s_n(rU) = n\omega_n r^{n-1} \leqslant s_n(A).$$

We can thus deduce the isoperimetric inequality:

$$s_n^n(A) \geqslant \omega_n n^n v_n^{n-1}(A).$$

Our final application of Steiner symmetrization is to re-prove Brunn's inequality. Let A, B be convex bodies in $\mathbb{R}^n$ and let $r, s > 0$ be such that $v_n(A) = \omega_n r^n$, $v_n(B) = \omega_n s^n$. Let $0 < k < 1$. It follows, by applying Theorem 6.6.6 twice, that there exists a finite sequence of symmetrizations about hyperplanes through the origin which sends A, B to convex bodies $A^{\square}, B^{\square}$, respectively, such that $A^{\square} \supseteq krU$, $B^{\square} \supseteq ksU$. Theorem 6.6.3 shows, that for any hyperplane π through the origin,

$$v_n(A_\pi + B_\pi) \leqslant v_n((A + B)_\pi) = v_n(A + B).$$

We can deduce, by repeated applications of this result, that

$$\omega_n k^n (r + s)^n = v_n(k(r + s)U) \leqslant v_n(A^{\square} + B^{\square}) \leqslant v_n(A + B).$$

Letting $k \to 1_-$ in these inequalities, we deduce that $\omega_n(r + s)^n \leqslant v_n(A + B)$, whence $v_n^{1/n}(A) + v_n^{1/n}(B) \leqslant v_n^{1/n}(A + B)$. □

We conclude our discussion of Steiner symmetrization by remarking that it can be used to establish the conditions under which equality holds in the Brunn, isoperimetric and isodiametric inequalities, but we shall not do this here.

Exercises 6.6

1. Prove that the Steiner symmetral A_π of a non-empty polytope A about a hyperplane π in $\mathbb{R}^n$ is itself a polytope.
2. Give examples in $\mathbb{R}^2$ of convex bodies A, B and a line π through the origin such that $A_\pi + B_\pi$ is a *proper* subset of $(A + B)_\pi$.

3. Show that the maximal area of a triangle contained in a Steiner symmetral of a plane convex body does not exceed the maximal area of a triangle contained in the body itself.
4. Which Steiner symmetrals of a triangle in $\mathbb{R}^2$ are themselves triangles? Use your answer to prove that, amongst all triangles with a given area, the equilateral triangles, and only these, have minimal perimeter (diameter). Show also that, amongst all triangles with a given area, the equilateral triangles, and only these, have maximal inradius.

7
Selected topics

The topics in this final chapter reflect the author's own personal preferences.

The six sections are independent of one another and draw largely upon the first two chapters. Section 7.4, however, uses ideas relating to volume, such as are discussed in Chapter 6. By assuming that the familiar *volume* in $\mathbb{R}^3$ extends naturally to $\mathbb{R}^n$, the reader should have little difficulty in following this introduction to the geometry of numbers. Not *all* the applications of Helly's theorem in Section 7.1 should be needed to show that it is one of the most versatile in the whole of convexity.

7.1 HELLY'S THEOREM

Helly's theorem on the intersection of convex sets in $\mathbb{R}^n$ is one of convexity's great treasures. Since its discovery in 1913 by the Austrian mathematician Eduard Helly, it has been the focal point of literally hundreds of research articles. Perhaps the most delightful aspect of the theorem is the ease with which it lends itself to a wide variety of impressive applications, enabling new discoveries to be made and providing the basis for the first short proofs of some classical results. Interestingly enough, many theorems that can be proved using it make no mention of convex sets in their statements. In the first half of this section, we indicate the leading role that Helly's theorem plays in combinatorial geometry; in the second, we use it to establish a celebrated result of Chebyshev on the approximation of continuous functions by polynomials.

Theorem 7.1.1 (Helly's theorem) *Let $\mathfrak{F}$ be a finite family of convex sets in $\mathbb{R}^n$ containing at least $n+1$ members. Suppose that every $n+1$ members of $\mathfrak{F}$ have a non-empty intersection. Then $\mathfrak{F}$ has a non-empty intersection.*

Proof We argue by induction on the number of members in $\mathfrak{F}$. The theorem is trivial when $\mathfrak{F}$ has $n+1$ members. Suppose that the assertion is true for families of m convex sets, where $m \geqslant n+1$. Let $A_0, \ldots, A_m$ be a family of $m+1$ convex sets in $\mathbb{R}^n$, every $n+1$ of whose members have a non-empty intersection. By the induction hypothesis, for each $i = 0, \ldots, m$, there exists a point $\mathbf{a}_i$ such that

$$\mathbf{a}_i \in A_0 \cap \ldots \cap A_{i-1} \cap A_{i+1} \cap \ldots \cap A_m.$$

By Radon's theorem of Section 2.2, the set $\{0, \ldots, m\}$ can be partitioned into two sets J, K such that the sets $\operatorname{conv}\{\mathbf{a}_j : j \in J\}$ and $\operatorname{conv}\{\mathbf{a}_k : k \in K\}$ meet. Suppose that

$$\mathbf{a} \in \operatorname{conv}\{\mathbf{a}_j : j \in J\} \cap \operatorname{conv}\{\mathbf{a}_k : k \in K\}.$$

Clearly, for each $j \in J$,

$$\mathbf{a}_j \in \cap (A_k : k \in K).$$

Thus, by the convexity of $\cap (A_k : k \in K)$,

$$\operatorname{conv}\{\mathbf{a}_j : j \in J\} \subseteq \cap (A_k : k \in K).$$

Similarly,

$$\operatorname{conv}\{\mathbf{a}_k : k \in K\} \subseteq \cap (A_j : j \in J).$$

Thus

$$\mathbf{a} \in \operatorname{conv}\{\mathbf{a}_j : j \in J\} \cap \operatorname{conv}\{\mathbf{a}_k : k \in K\} \subseteq A_0 \cap \ldots \cap A_m,$$

which shows that $A_0 \cap \ldots \cap A_m$ is non-empty. This completes the proof by induction. □

Helly's theorem is often used to prove results of the form: *If a family of sets is such that every one of its subfamilies consisting of k members possesses a certain property, then so too does the entire family.* Our first two applications of the theorem are ingenious examples of this type.

By a *transversal* for a family of sets in $\mathbb{R}^n$ is meant a line which meets every member of the family.

Theorem 7.1.2 *Let $\mathfrak{F}$ be a finite family of parallel line segments in $\mathbb{R}^2$ every three of which have a transversal. Then the family $\mathfrak{F}$ has a transversal.*

Proof We let it suffice to prove the theorem when $\mathfrak{F}$ comprises *vertical* line segments S_i of the form

$$S_i = \{(x, y) : x = u_i,\ v_i \leqslant y \leqslant w_i\}$$

for $i = 1, \ldots, r$, where $r \geqslant 3$ and the u_i's are distinct. For each $i = 1, \ldots, r$, denote by C_i the set of those points (m, c) of $\mathbb{R}^2$ for which the line $y = mx + c$ meets the segment S_i, i.e. C_i is the convex set

$$\{(m, c) : v_i \leqslant mu_i + c \leqslant w_i\}.$$

By hypothesis, every three of the sets C_i have a common point, so by Helly's theorem there exists a point (m_0, c_0) belonging to all the sets C_i. Thus the line $y = m_0 x + c_0$ is a transversal for $\mathfrak{F}$. □

In his book *Convex sets*, F. A. Valentine (1964) introduces our next result as follows:

> If a field contains a number of stationary black and white sheep, what simple condition will guarantee that a straight line exists which separates the black sheep from the white sheep? The answer is simple. If every four sheep can be thus separated, then all can be separated in such a fashion.

This result, stated formally in $\mathbb{R}^n$, was established by P. Kirchberger in 1902. His proof, which made no use of Helly's theorem, was nearly 24 pages long. Our proof, discovered by H. Rademacher and I. J. Schoenberg in 1950, is the epitome of brevity and elegance, and demonstrates the power of Helly's theorem in combinatorial geometry.

Theorem 7.1.3 (Kirchberger's theorem) *Let A, B be finite sets in $\mathbb{R}^n$ whose union contains at least $n+2$ points. Suppose that, for each set C of $n+2$ points of $A \cup B$, the sets $A \cap C$ and $B \cap C$ can be strictly separated. Then A and B can be strictly separated.*

Proof For each $\mathbf{a} = (a_1, \ldots, a_n)$ in A and for each $\mathbf{b} = (b_1, \ldots, b_n)$ in B, define open halfspaces $P(\mathbf{a})$ and $Q(\mathbf{b})$ in $\mathbb{R}^{n+1}$ by the equations

$$P(\mathbf{a}) = \{(x_1, \ldots, x_{n+1}) : a_1x_1 + \ldots + a_nx_n + x_{n+1} < 0\},$$
$$Q(\mathbf{b}) = \{(x_1, \ldots, x_{n+1}) : b_1x_1 + \ldots + b_nx_n + x_{n+1} > 0\}.$$

Let $\mathfrak{F}$ be the family of these open halfspaces $P(\mathbf{a})$ and $Q(\mathbf{b})$. Let C be a set of $n+2$ points of $A \cup B$. Since $A \cap C$ and $B \cap C$ can be strictly separated, there exist scalars $c_1, \ldots, c_{n+1}$ such that

$$c_1a_1 + \ldots + c_na_n + c_{n+1} < 0 \qquad \text{for } (a_1, \ldots, a_n) \in A \cap C,$$
$$c_1b_1 + \ldots + c_nb_n + c_{n+1} > 0 \qquad \text{for } (b_1, \ldots, b_n) \in B \cap C.$$

Thus $(c_1, \ldots, c_{n+1}) \in P(\mathbf{a})$ when $\mathbf{a} \in A \cap C$ and $(c_1, \ldots, c_{n+1}) \in Q(\mathbf{b})$ when $\mathbf{b} \in B \cap C$. Hence $\mathfrak{F}$ is a finite family of convex sets in $\mathbb{R}^{n+1}$, which contains at least $n+2$ members and every $n+2$ of whose members have a non-empty intersection. By Helly's theorem in $\mathbb{R}^{n+1}$, there is a point $(d_1, \ldots, d_{n+1})$ belonging to all members of $\mathfrak{F}$. Since the intersection of all members of $\mathfrak{F}$ is open, we can suppose that not all of $d_1, \ldots, d_n$ are zero. Thus

$$d_1a_1 + \ldots + d_na_n + d_{n+1} < 0 \qquad \text{for } (a_1, \ldots, a_n) \in A,$$
$$d_1b_1 + \ldots + d_nb_n + d_{n+1} > 0 \qquad \text{for } (b_1, \ldots, b_n) \in B.$$

Hence the hyperplane $d_1x_1 + \ldots + d_nx_n + d_{n+1} = 0$ strictly separates A and B. □

The finiteness condition in Helly's theorem *cannot* be omitted. To see this, consider the family of closed halfspaces

$$\{(x_1, \ldots, x_n) \in \mathbb{R}^n : x_1 \geqslant i\} \qquad \text{for } i = 1, 2, \ldots.$$

Every finite collection of these sets has a common point, but the intersection of the whole family is empty. When dealing with compact convex sets, however, we can drop the finiteness condition, as we now prove in a second version of Helly's theorem. We shall use this form of the theorem in the remaining examples of the section.

Theorem 7.1.4 (Helly's theorem for infinite families) *Let $\mathfrak{F}$ be an infinite family of compact convex sets in $\mathbb{R}^n$. Suppose that every $n + 1$ members of $\mathfrak{F}$ have a non-empty intersection. Then $\mathfrak{F}$ has a non-empty intersection.*

Proof This result follows from the finite version of Helly's theorem and the fact (Corollary 1.8.8) that, if a family of compact sets in $\mathbb{R}^n$ has an empty intersection, then so too does one of its finite subfamilies. □

We now discuss to what extent a convex body can be considered to have a *centre*. Let A be a symmetric convex body. Then the origin bisects every chord of A on which it lies. More generally, every chord of a translate $A + \mathbf{a}$ of A passing through $\mathbf{a}$ is bisected by it; we call $\mathbf{a}$ the *centre* of $A + \mathbf{a}$. Most convex bodies do not possess a centre in this sense; for example, an n-simplex (for $n \geqslant 2$) has no such centre. However, we shall now show that *every* convex body A in $\mathbb{R}^n$ contains a point $\mathbf{c}$ such that every chord of A through $\mathbf{c}$ has at most $n/(n + 1)$ of its length lying on either side of $\mathbf{c}$.

Before proving the next theorem, we give a formal definition of a *chord* of a convex body. The line segment $\operatorname{conv}\{\mathbf{u}, \mathbf{v}\}$ joining distinct points $\mathbf{u}, \mathbf{v}$ of a convex body A in $\mathbb{R}^n$ is called a **chord** of A if it is the intersection of A with the line $\operatorname{aff}\{\mathbf{u}, \mathbf{v}\}$ through $\mathbf{u}$ and $\mathbf{v}$, i.e. if

$$\operatorname{conv}\{\mathbf{u}, \mathbf{v}\} = A \cap \operatorname{aff}\{\mathbf{u}, \mathbf{v}\}.$$

Theorem 7.1.5 *Every convex body A in $\mathbb{R}^n$ contains a point $\mathbf{c}$ such that, for every chord* $\operatorname{conv}\{\mathbf{u}, \mathbf{v}\}$ *of A through $\mathbf{c}$,*

$$\frac{\|\mathbf{u} - \mathbf{c}\|}{\|\mathbf{u} - \mathbf{v}\|} \leqslant \frac{n}{n + 1}.$$

Proof For each point $\mathbf{x}$ in A, denote by $A_{\mathbf{x}}$ the compact convex set

$$\mathbf{x} + \frac{n}{n + 1}(A - \mathbf{x}).$$

Thus $A_{\mathbf{x}}$ is obtained from A by a dilation at $\mathbf{x}$ with ratio $n/(n+1)$. Let $\mathbf{x}_0, \ldots, \mathbf{x}_n \in A$. Then, for $i = 0, \ldots, n$, the point

$$\mathbf{x}_i + \frac{n}{n+1}\left(\frac{\mathbf{x}_0 + \ldots + \mathbf{x}_n - \mathbf{x}_i}{n} - \mathbf{x}_i\right) = \frac{\mathbf{x}_0 + \ldots + \mathbf{x}_n}{n+1}$$

belongs to $A_{\mathbf{x}_i}$. Thus any $n+1$ of the sets $A_{\mathbf{x}}$ have a common point. By Theorem 7.1.4, there exists a point $\mathbf{c}$ which belongs to $A_{\mathbf{x}}$ for each $\mathbf{x}$ in A. Let $\operatorname{conv}\{\mathbf{u}, \mathbf{v}\}$ be a chord of A passing through $\mathbf{c}$. Since $\mathbf{c} \in A_{\mathbf{u}}$, there is some point $\mathbf{a}$ in A for which

$$\mathbf{c} = \mathbf{u} + \frac{n}{n+1}(\mathbf{a} - \mathbf{u}).$$

Thus

$$\mathbf{a} = \frac{n+1}{n}\mathbf{c} - \frac{1}{n}\mathbf{u} \in A \cap \operatorname{aff}\{\mathbf{c}, \mathbf{u}\} \subseteq A \cap \operatorname{aff}\{\mathbf{u}, \mathbf{v}\}.$$

Since $\operatorname{conv}\{\mathbf{u}, \mathbf{v}\}$ is a chord of A, we can deduce that $\mathbf{a} \in \operatorname{conv}\{\mathbf{u}, \mathbf{v}\}$. Hence $\mathbf{a} = (1-\lambda)\mathbf{u} + \lambda\mathbf{v}$ for some scalar λ with $0 \leqslant \lambda \leqslant 1$. Thus

$$\mathbf{c} = \mathbf{u} + \frac{\lambda n}{n+1}(\mathbf{v} - \mathbf{u})$$

and

$$\frac{\|\mathbf{u} - \mathbf{c}\|}{\|\mathbf{u} - \mathbf{v}\|} = \frac{\lambda n}{n+1} \leqslant \frac{n}{n+1}. \quad \square$$

The next result, first proved by H. W. E. Jung in 1901, is one of the oldest and most well known in combinatorial geometry. It asserts that every set with diameter 1 in $\mathbb{R}^n$ is contained in some ball of radius $\sqrt{(n/2(n+1))}$. Our proof, depending on Helly's theorem, was devised by L. M. Blumenthal and G. E. Wahlin in 1941.

Theorem 7.1.6 (Jung's theorem) *Every set A with diameter* 1 *in* $\mathbb{R}^n$ *lies in some closed ball of radius* $\sqrt{(n/2(n+1))}$.

Proof Suppose first that A has at most $n+1$ points. We assume, without loss of generality, that the origin is the circumcentre of A. Let $\mathbf{a}_0, \ldots, \mathbf{a}_m$ be those points of A whose distance from the origin equals the circumradius R of A. Then $0 < m \leqslant n$. Also $\mathbf{0} \in \operatorname{conv}\{\mathbf{a}_0, \ldots, \mathbf{a}_m\}$, for otherwise the point $\lambda\mathbf{z}$, where $\mathbf{z}$ is the nearest point of $\operatorname{conv}\{\mathbf{a}_0, \ldots, \mathbf{a}_m\}$ to $\mathbf{0}$ and $\lambda > 0$ is sufficiently small, would be the centre of a closed ball containing A with radius less than R. Thus there exist $\lambda_0, \ldots, \lambda_m \geqslant 0$ with $\lambda_0 + \ldots + \lambda_m = 1$ such that

$$\mathbf{0} = \lambda_0\mathbf{a}_0 + \ldots + \lambda_m\mathbf{a}_m.$$

For each $j = 0, \ldots, m$,

$$\begin{aligned} 1 - \lambda_j &= \sum_{i=0}^{m} \lambda_i - \lambda_j \\ &\geqslant \sum_{i=0}^{m} \lambda_i \|\mathbf{a}_i - \mathbf{a}_j\|^2 \\ &= \sum_{i=0}^{m} \lambda_i (\|\mathbf{a}_i\|^2 - 2\mathbf{a}_i \cdot \mathbf{a}_j + \|\mathbf{a}_j\|^2) \\ &= \sum_{i=0}^{m} \lambda_i (2R^2 - 2\mathbf{a}_i \cdot \mathbf{a}_j) \\ &= 2R^2 - 2\left(\sum_{i=0}^{m} \lambda_i \mathbf{a}_i\right) \cdot \mathbf{a}_j \\ &= 2R^2. \end{aligned}$$

Adding the inequalities $1 - \lambda_j \geqslant 2R^2$, for $j = 0, \ldots, m$, together, we deduce that $m \geqslant 2(m + 1)R^2$. Thus

$$R \leqslant \sqrt{(m/2(m + 1))} \leqslant \sqrt{(n/2(n + 1))},$$

and the proof is complete in this case.

Suppose next that A has more than $n + 1$ points. Denote by $\mathfrak{F}$ the family of all closed balls of radius $\sqrt{(n/2(n + 1))}$ having their centres in A. By the first part of this proof, every $n + 1$ members of $\mathfrak{F}$ have a common point, namely the circumcentre of the set of their centres. Helly's theorem, in one of its two versions, shows that there is some point $\mathbf{a}$ belonging to every member of $\mathfrak{F}$.

Thus A lies in the closed ball with centre $\mathbf{a}$ and radius $\sqrt{(n/2(n + 1))}$. □

Our next application of Helly's theorem is to a remarkable result of Krasnosselsky. Before we can state this, we need to introduce the concept of a *starshaped set*. A point $\mathbf{a}_0$ of a set A in $\mathbb{R}^n$ is called a **starcentre** of A if $(1 - \lambda)\mathbf{a}_0 + \lambda\mathbf{a} \in A$ whenever $\mathbf{a} \in A$ and $0 \leqslant \lambda \leqslant 1$. A set which has a starcentre is said to be **starshaped**. Geometrically, a point of a set in $\mathbb{R}^n$ is a starcentre of that set if every line segment joining it to another point of the set lies in the set. Every convex set is starshaped, each of its points serving as one of its starcentres. The set

$$\{(x_1, \ldots, x_n) : \sqrt{|x_1|} + \ldots + \sqrt{|x_n|} \leqslant 1\}$$

has the origin for a starcentre, but it is not convex when $n \geqslant 2$. If $\mathbf{a}$ and $\mathbf{b}$ are points of a set A in $\mathbb{R}^n$, then they are said to be *visible* from each other in A if $\lambda\mathbf{a} + \mu\mathbf{b} \in A$ whenever $\lambda, \mu \geqslant 0$ with $\lambda + \mu = 1$. Thus

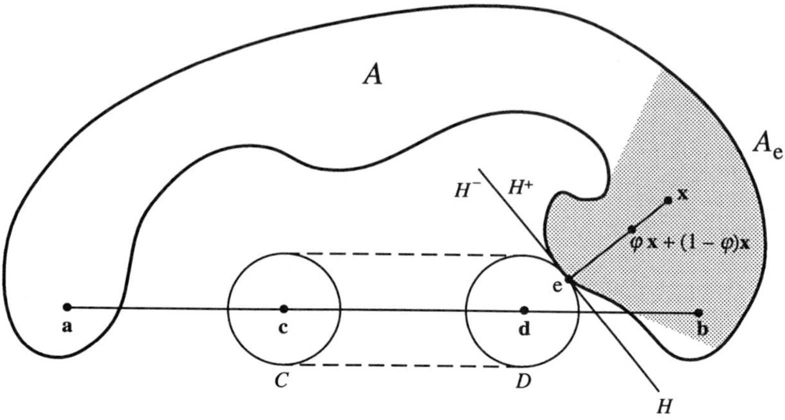

Fig. 7.1

a point $\mathbf{a}$ of a set A is a starcentre of A precisely when the whole of A is visible from $\mathbf{a}$ in A. In the following lemma and theorem, for each point $\mathbf{x}$ of a set A in $\mathbb{R}^n$, the set of those points of A which are visible from $\mathbf{x}$ in A will be denoted by $A_{\mathbf{x}}$, i.e.

$$A_{\mathbf{x}} = \{\mathbf{a} \in A : (1 - \lambda)\mathbf{x} + \lambda\mathbf{a} \in A \text{ for } 0 \leqslant \lambda \leqslant 1\}.$$

Lemma 7.1.7 *The point* $\mathbf{a}$ *is a starcentre of a non-empty closed set A in* $\mathbb{R}^n$ *if and only if* $\mathbf{a} \in \operatorname{conv} A_{\mathbf{x}}$ *for all* $\mathbf{x} \in A$.

Proof If $\mathbf{a}$ is a starcentre of A, then trivially $\mathbf{a} \in A_{\mathbf{x}} \subseteq \operatorname{conv} A_{\mathbf{x}}$ for all $\mathbf{x} \in A$. We complete the proof by showing that, if $\mathbf{a}$ is *not* a starcentre of A, then there is some $\mathbf{e} \in A$ such that $\mathbf{a} \notin \operatorname{conv} A_{\mathbf{e}}$.

Suppose now that $\mathbf{a}$ is *not* a starcentre of A. Then there exists $\mathbf{b} \in A$ such that the line segment joining $\mathbf{a}$ and $\mathbf{b}$ is not contained in A. Thus there is a point $\mathbf{c}$ on this segment that does not belong to A. Since A is closed, there is a closed ball C with centre $\mathbf{c}$ which does not meet A. Define a scalar α by the equation

$$\alpha = \inf\{\lambda \geqslant 0 : A \cap (C + \lambda(\mathbf{b} - \mathbf{c})) \neq \varnothing\}.$$

Denote the closed ball $C + \alpha(\mathbf{b} - \mathbf{c})$ by D, and its centre by $\mathbf{d}$. The definition of α ensures that A meets the boundary, but not the interior, of D. Suppose that $\mathbf{e} \in A \cap D$. See Fig. 7.1.

The hyperplane H with equation $(\mathbf{z} - \mathbf{e}) \cdot (\mathbf{d} - \mathbf{e}) = 0$ supports D at $\mathbf{e}$. Define complementary halfspaces H^- and H^+ as follows:

$$H^- = \{\mathbf{z} : (\mathbf{z} - \mathbf{e}) \cdot (\mathbf{e} - \mathbf{d}) < 0\}, \qquad H^+ = \{\mathbf{z} : (\mathbf{z} - \mathbf{e}) \cdot (\mathbf{e} - \mathbf{d}) \geqslant 0\}.$$

The definition of α ensures that, for all sufficiently small $\theta > 0$, the translate of C centred at $\mathbf{d} - \theta(\mathbf{b} - \mathbf{c})$ does *not* meet A, whence

$$\|\mathbf{e}-\mathbf{d}+\theta(\mathbf{b}-\mathbf{c})\|^2=\|\mathbf{e}-(\mathbf{d}-\theta(\mathbf{b}-\mathbf{c}))\|^2>\|\mathbf{e}-\mathbf{d}\|^2.$$

It follows from Theorem 1.6.1(iv) that

$$(\mathbf{e}-\mathbf{d})\cdot(\mathbf{b}-\mathbf{c})\geqslant 0,\qquad \text{whence}\qquad (\mathbf{e}-\mathbf{d})\cdot(\mathbf{a}-\mathbf{d})\leqslant 0.$$

Hence

$$(\mathbf{a}-\mathbf{e})\cdot(\mathbf{e}-\mathbf{d})=(\mathbf{a}-\mathbf{d})\cdot(\mathbf{e}-\mathbf{d})-\|\mathbf{e}-\mathbf{d}\|^2<0,$$

which shows that $\mathbf{a}\in H^-$.

We now show that $A_\mathbf{e}\subseteq H^+$. Let $\mathbf{x}\in A_\mathbf{e}$. Since A is disjoint from the interior of D and $\varphi\mathbf{x}+(1-\varphi)\mathbf{e}\in A$ for $0\leqslant\varphi\leqslant 1$,

$$\|\mathbf{e}-\mathbf{d}+\varphi(\mathbf{x}-\mathbf{e})\|^2=\|\varphi\mathbf{x}+(1-\varphi)\mathbf{e}-\mathbf{d}\|^2\geqslant\|\mathbf{e}-\mathbf{d}\|^2.$$

Theorem 1.6.1(iv) shows that $(\mathbf{x}-\mathbf{e})\cdot(\mathbf{e}-\mathbf{d})\geqslant 0$, whence $\mathbf{x}\in H^+$. Thus $A_\mathbf{e}$, and hence $\operatorname{conv}A_\mathbf{e}$, is contained in H^+. Since $\mathbf{a}\in H^-$ and $H^-\cap H^+=\varnothing$, we deduce that $\mathbf{a}\notin\operatorname{conv}A_\mathbf{e}$. □

Theorem 7.1.8 (Krasnosselsky's theorem) *Let A be an infinite compact set in $\mathbb{R}^n$. Suppose that, for each $n+1$ points of A, there is some point of A from which all of these points are visible in A. Then A is starshaped.*

Proof A routine argument, which makes use of the fact that A is closed, shows that, for each $\mathbf{x}$ in A, the set $A_\mathbf{x}$ is closed. Thus $A_\mathbf{x}$, being a closed subset of the compact set A, is compact, whence so too is the convex set $\operatorname{conv}A_\mathbf{x}$.

Let $\mathbf{x}_0,\ldots,\mathbf{x}_n$ be any $n+1$ points of A. By the hypothesis, there exists some point $\mathbf{y}$ of $\mathbf{A}$ from which the points $\mathbf{x}_0,\ldots,\mathbf{x}_n$ are visible in A, i.e. $\mathbf{x}_0,\ldots,\mathbf{x}_n\in A_\mathbf{y}$. Thus

$$\mathbf{y}\in A_{\mathbf{x}_0}\cap\ldots\cap A_{\mathbf{x}_n}\subseteq(\operatorname{conv}A_{\mathbf{x}_0})\cap\ldots\cap(\operatorname{conv}A_{\mathbf{x}_n}),$$

and so each $n+1$ members of the infinite family $(\operatorname{conv}A_\mathbf{x}:\mathbf{x}\in A)$ of compact convex sets in $\mathbb{R}^n$ have a common point. Therefore, by Helly's theorem for infinite families, there exists a point $\mathbf{a}$ which belongs to every member of the above family. In view of the lemma, $\mathbf{a}$ is a starcentre of A. Thus A is starshaped. □

We now turn our attention away from combinatorial geometry to give an impressive application of Helly's theorem to the classical problem of approximating a continuous function $f:[a,b]\to\mathbb{R}$, where $a<b$, by real polynomials P whose degrees do not exceed some given non-negative integer n. To be precise, we shall show that amongst all such polynomials P there is a unique one P^* which minimizes the expression

$$\max\{|f(x)-P(x)|:a\leqslant x\leqslant b\}.$$

Moreover, as we shall we see, if ε^* is the minimal value of this expression (obtained when $P = P^*$), then there exist $x_1, \ldots, x_{n+2}$ with $a \leqslant x_1 < \ldots < x_{n+2} \leqslant b$ such that the numbers

$$\varepsilon_1 = f(x_1) - P^*(x_1), \;\ldots,\; \varepsilon_{n+2} = f(x_{n+2}) - P^*(x_{n+2})$$

have the same absolute value ε^* and alternate in sign, i.e. either $\varepsilon_1 \geqslant 0$, $\varepsilon_2 \leqslant 0$, $\varepsilon_3 \geqslant 0, \ldots$ etc., or $\varepsilon_1 \leqslant 0$, $\varepsilon_2 \geqslant 0$, $\varepsilon_3 \leqslant 0, \ldots$ etc.

The special case of the above result when $n = 0$ is readily established. Here the problem is: given a continuous function $f\colon [a, b] \to \mathbb{R}$, where $a < b$, to find a real constant c such that the expression

$$\max\{|f(x) - c| : a \leqslant x \leqslant b\}$$

is a minimum. Clearly, the constant c sought is $\frac{1}{2}(m + M)$, where m, M are, respectively, the minimum and maximum values of f on $[a, b]$, and the ε^* is $\frac{1}{2}(M - m)$. Let $x_1, x_2 \in [a, b]$ with $x_1 \neq x_2$ satisfy $f(x_1) = m$, $f(x_2) = M$. Then, as desired,

$$f(x_1) - \tfrac{1}{2}(m + M) = -(f(x_2) - \tfrac{1}{2}(m + M)).$$

Consider the following finite problem. Let $(x_1, y_1), \ldots, (x_{n+2}, y_{n+2})$, where $x_1 < \ldots < x_{n+2}$, be points in $\mathbb{R}^2$. We seek to approximate these $n + 2$ points by a real polynomial P of degree not exceeding n. (The reader is reminded that, by Lagrange's interpolation formula, given any $n + 1$ points in $\mathbb{R}^2$ having distinct x-coordinates, there is a unique polynomial of degree not exceeding n whose graph passes through these points.) The **error** of such a polynomial P for the given $n + 2$ points is defined to be the non-negative number

$$\max\{|y_i - P(x_i)| : i = 1, \ldots, n + 2\}.$$

A real polynomial P^* of degree n or less is called a **best approximation of degree n to** the given $n + 2$ points if its error does not exceed that of any other polynomial of degree n or less for the given points. More accurately, such a best approximation is called a **minimax approximation**, since it minimizes the expression

$$\max\{|y_i - P(x_i)| : i = 1, \ldots, n + 2\}.$$

Lemma 7.1.9 *Let $(x_1, y_1), \ldots, (x_{n+2}, y_{n+2})$, where $x_1 < \ldots < x_{n+2}$, be points in $\mathbb{R}^2$. Then there exists a unique polynomial P^* of degree at most n such that the real numbers*

$$y_1 - P^*(x_1), \;\ldots,\; y_{n+2} - P^*(x_{n+2})$$

have the same absolute value and alternate in sign. Moreover, P^* is the unique best approximation of degree n to the given points and its associated error ε^* is the absolute value of

$$\begin{vmatrix} x_1^n & x_1^{n-1} & \dots & 1 & y_1 \\ x_2^n & x_2^{n-1} & \dots & 1 & y_2 \\ \dots & \dots & \dots & \dots & \dots \\ x_{n+2}^n & x_{n+2}^{n-1} & \dots & 1 & y_{n+2} \end{vmatrix} \div \begin{vmatrix} x_1^n & x_1^{n-1} & \dots & 1 & 1 \\ x_2^n & x_2^{n-1} & \dots & 1 & -1 \\ \dots & \dots & \dots & \dots & \dots \\ x_{n+2}^n & x_{n+2}^{n-1} & \dots & 1 & (-1)^{n+1} \end{vmatrix}.$$

Proof Consider the following system of $n + 2$ linear equations in the $n + 2$ unknowns $a, a_0, \dots, a_n$:

$$\left.\begin{array}{l} a_n x_1^n \quad + a_{n-1}x_1^{n-1} + \dots + a_0 + \qquad\quad a = y_1 \\ a_n x_2^n \quad + a_{n-1}x_2^{n-1} + \dots + a_0 - \qquad\quad a = y_2 \\ \dots\dots\dots\dots\dots\dots\dots\dots\dots\dots\dots \\ a_n x_{n+2}^n + a_{n-1}x_{n+2}^{n-1} + \dots + a_0 + (-1)^n a = y_{n+2} \end{array}\right\}.$$

The determinant of the $(n + 2) \times (n + 2)$ coefficient matrix on the left-hand side of the above system of equations is non-zero. To see that this is so, expand the determinant by its last column to express its value as a sum of non-zero terms having the same sign. Thus $a, a_0, \dots, a_n$ are uniquely determined by the equations above. Denote by P^* the polynomial defined by the equation

$$P^*(x) = a_n x^n + a_{n-1}x^{n-1} + \dots + a_0.$$

Then clearly P^* is the unique polynomial of degree at most n having the properties demanded by the first assertion of the lemma; the error ε^* of P^* for the given points is $|a|$. Thus the formula for ε^* in the statement of the lemma follows immediately from Cramer's rule.

Before proving that P^* is the unique best approximation of degree n to the given points, we need the following result: *Let Q be a polynomial of degree at most n. Suppose that there exist $z_1, \dots, z_{n+2}$ with $z_1 < \dots < z_{n+2}$ such that the numbers $Q(z_1), \dots, Q(z_{n+2})$ alternate in sign. Then Q is the zero polynomial.* We argue by induction on n. The case $n = 0$ is trivial. Suppose that the assertion is true for polynomials of degree at most $n - 1$, where $n \geqslant 1$. Let $Q, z_1, \dots, z_{n+2}$ be as above. The first mean value theorem applied to Q on the intervals $[z_1, z_2], \dots, [z_{n+1}, z_{n+2}]$ shows the existence of $c_1, \dots, c_{n+1}$ with

$$z_1 < c_1 < z_2 < c_2 < z_3 < \dots < z_{n+1} < c_{n+1} < z_{n+2}$$

such that the numbers $Q'(c_1), \dots, Q'(c_{n+1})$ alternate in sign. But Q' has degree at most $n - 1$, so by the induction hypothesis Q' must be the zero polynomial. But $Q(z_1)$, $Q(z_2)$ alternate in sign, whence Q is the zero polynomial. This completes the proof by induction.

We now show that the polynomial P^* found earlier in the proof is the unique best approximation of degree n to the given $n + 2$ points. Without loss of generality, we may suppose that (in the earlier discussion) $a \geqslant 0$. Let P be *any* polynomial of degree at most n whose error for the given points does not exceed a, the error of P^* for the points. By the definition of the error of P, we must have

$$|y_1 - P(x_1)| \leqslant a = y_1 - P^*(x_1),$$

whence $(P^* - P)(x_1) \leqslant 0$. Similarly, $(P^* - P)(x_2) \geqslant 0$, $(P^* - P)(x_3) \leqslant 0$, etc. But $P^* - P$ is a polynomial of degree at most n, and so, by the result just proved, $P^* - P$ is the zero polynomial. Thus $P = P^*$ and P^* must be the unique best approximation of degree n to the given points. □

Suppose now that n is a fixed non-negative integer and that $f:S \to \mathbb{R}$ is a bounded function defined on a bounded subset S of $\mathbb{R}$ having at least $n+2$ points. For each polynomial P of degree at most n, we define the **error** of P for the function f to be the non-negative number

$$\sup\{|f(x) - P(x)| : x \in S\}.$$

We seek to minimize this error by choosing P suitably. It is easy to obtain a lower bound for the error. For each $(n+2)$-tuple $(x_1, \ldots, x_{n+2})$ of numbers in S with $x_1 < \ldots < x_{n+2}$, denote by $\varepsilon_f(x_1, \ldots, x_{n+2})$ the error of the best approximation of degree n to the points $(x_1, f(x_1)), \ldots, (x_{n+2}, f(x_{n+2}))$; an explicit formula for $\varepsilon_f(x_1, \ldots, x_{n+2})$ can be readily found using the lemma. Let

$$\varepsilon^* = \sup \varepsilon_f(x_1, \ldots, x_{n+2}),$$

the supremum being taken over all $(n+2)$-tuples of the type just considered. Certainly, no polynomial of degree at most n can have error for f *less than* ε^*. The remarkable fact is that there always exists some polynomial of degree at most n whose error for f is *precisely* ε^*.

Theorem 7.1.10 *Let $f:S \to \mathbb{R}$ be a bounded function defined on a bounded set S in $\mathbb{R}$ containing at least $n+2$ points. Let*

$$\varepsilon^* = \sup\{\varepsilon_f(x_1, \ldots, x_{n+2}) : x_1 < \ldots < x_{n+2} \text{ and } x_1, \ldots, x_{n+2} \in S\}.$$

Then there exists a polynomial P^* of degree at most n such that

$$\sup\{|f(x) - P^*(x)| : x \in S\} = \varepsilon^*.$$

Proof For each x in S, denote by C_x the closed convex set

$$\{(y_0, \ldots, y_{n-1}, y_n) : |f(x) - y_n x^n - y_{n-1}x^{n-1} - \ldots - y_0| \leqslant \varepsilon^*\}$$

in $\mathbb{R}^{n+1}$. Let $s_1, \ldots, s_{n+1}$ be distinct points of S. Then it is not a difficult exercise to show that the closed convex set

$$C = C_{s_1} \cap \ldots \cap C_{s_{n+1}}$$

is bounded, and hence compact.

Suppose that $x_1, \ldots, x_{n+2} \in S$ and that $x_1 < \ldots < x_{n+2}$. Since $\varepsilon_f(x_1, \ldots, x_{n+2}) \leqslant \varepsilon^*$, there exists a polynomial

$$P(x) = a_n x^n + a_{n-1}x^{n-1} + \ldots + a_0$$

such that

$$|f(x_i) - a_n x_i^n - a_{n-1}x_i^{n-1} - \ldots - a_0| \leqslant \varepsilon^* \qquad \text{for } i = 1, \ldots, n+2.$$

This shows that

$$(a_0, \ldots, a_{n-1}, a_n) \in C_{x_1} \cap \ldots \cap C_{x_{n+2}}.$$

Thus every $n + 2$ members of the family $(C_x : x \in S)$ have a non-empty intersection. Hence, by Helly's theorem, the intersection of any finite number of members of the family is non-empty.

We now show that the family $(C_x : x \in S)$ itself has a non-empty intersection—we cannot appeal directly to Helly's theorem, for the C_x's are not compact and S might be infinite. Consider the family $(C \cap C_x : x \in S)$ of compact convex sets in $\mathbb{R}^{n+1}$, where C is the set defined in the first paragraph of this proof. Every intersection of a finite number of members of this family is also an intersection of a finite number of members of the family $(C_x : x \in S)$, and so is non-empty. By Helly's theorem, there is a point $(a_0^*, \ldots, a_{n-1}^*, a_n^*)$ belonging to $C \cap C_x$ for each x in S. It follows that the polynomial

$$P^*(x) = a_n^* x^n + a_{n-1}^* x^{n-1} + \ldots + a_0^*$$

satisfies

$$\sup\{|f(x) - P^*(x)| : x \in S\} = \varepsilon^*. \quad \square$$

Theorem 7.1.10 shows that amongst all polynomials of degree at most n, there is one having least possible error for f; such a polynomial is called a **best approximation of degree** n **to** f. Such a best approximation need not be unique. To see that this is so, consider the function $f : [0, 1] \to \mathbb{R}$ defined by the equations

$$f(x) = \begin{cases} 0 & \text{for } 0 \leqslant x < 1 \\ 1 & \text{for } x = 1 \end{cases}.$$

It is easily verified that, for any real m with $0 \leqslant m \leqslant 1$, the polynomial $m(x - 1) + \frac{1}{2}$ is a best approximation of degree 1 to f.

We conclude this section by establishing the *classical equioscillation properties* of the best approximation of degree n to a continuous function defined on a compact interval.

Theorem 7.1.11 *Let $a < b$ and let $f:[a, b] \to \mathbb{R}$ be a continuous function. Then, for each $n = 0, 1, 2, \ldots$, there is a unique best approximation P^* of degree n to f. If ε^* is the error of P^* for f, then there exist $x_1^*, \ldots, x_{n+2}^*$ in $[a, b]$ with $x_1^* < \ldots < x_{n+2}^*$ such that the numbers*

$$f(x_1^*) - P^*(x_1^*), \ldots, f(x_{n+2}^*) - P^*(x_{n+2}^*)$$

have absolute value ε^* and alternate in sign.

Proof So far we have defined $\varepsilon_f(x_1, \ldots, x_{n+2})$ for any *strictly* increasing $(n + 2)$-tuple of numbers in $[a, b]$. We now extend this domain to include any increasing $(n + 2)$-tuple $(x_1, \ldots, x_{n+2})$ of numbers in $[a, b]$ by defining $\varepsilon_f(x_1, \ldots, x_{n+2})$ to be zero when the x_i's are not all different. We now show that ε_f is continuous on its extended domain. It is clear, from the explicit formula for ε_f mentioned earlier, that ε_f is continuous at any $(n + 2)$-tuple satisfying $x_1 < \ldots < x_{n+2}$. Consider, then, any $(n + 2)$-tuple $(x_1, \ldots, x_{n+2})$ in which not all of the x_i's are different. Then there is a polynomial P of degree at most n such that $P(x_i) = f(x_i)$ for $i = 1, \ldots, n + 2$. Suppose that $(x_1, \ldots, x_{n+2})$ is the limit of a sequence of points $(x_1^{(k)}, \ldots, x_{n+2}^{(k)})$ of the extended domain of ε_f. The continuity of f and P show that

$$\varepsilon_f(x_1^{(k)}, \ldots, x_{n+2}^{(k)}) \leqslant \max\{|f(x_i^{(k)}) - P(x_i^{(k)})| : i = 1, \ldots, n + 2\} \to 0$$
$$\text{as } k \to \infty,$$

which establishes the continuity of ε_f at $(x_1, \ldots, x_{n+2})$.

Suppose now that ε^* is the error of a best approximation P^* of degree n to f. If $\varepsilon^* = 0$, then $P^* = f$ and the theorem is trivial. Suppose, then, that $\varepsilon^* > 0$. Since ε^* is the supremum of the continuous function ε_f on the compact set

$$C = \{(x_1, \ldots, x_{n+2}) : a \leqslant x_1 \leqslant \ldots \leqslant x_{n+2} \leqslant b\}$$

in $\mathbb{R}^{n+2}$, it is attained by ε_f at some point, $(x_1^*, \ldots, x_{n+2}^*)$ say, of C. Thus P^* is the unique best approximation of degree n to the points $(x_1^*, f(x_1^*)), \ldots, (x_{n+2}^*, f(x_{n+2}^*))$. The final assertion follows from Lemma 7.1.9 □

Example 7.1.12 We use the last theorem to help us find the best approximation $mx + c$ of degree 1 to $\sqrt{x}$ on the compact interval $[1, 4]$. Let x_1^*, x_2^*, x_3^* be the points of $[1, 4]$ whose existence is guaranteed by the theorem. Since x_1^*, x_2^*, x_3^* are either minimum points or maximum points for $\sqrt{x} - mx - c$ on $[1, 4]$, and this last function can have at most one stationary point (at $1/4m^2$) on $[1, 4]$, we may deduce that $x_1^* = 1$, $x_2^* = 1/4m^2$, $x_3^* = 4$. The equioscillation properties of $\sqrt{x} - mx - c$ at x_1^*, x_2^*, x_3^* show that

$$\sqrt{1 - m - c} = \sqrt{4 - 4m - c} = -(1/2m - 1/4m - c).$$

We find from these equations that the desired best approximation of degree 1 to $\sqrt{x}$ on $[1, 4]$ is $x/3 + 17/24$ with associated error $1/24$. □

Exercises 7.1

1. Show that, if a convex set is contained in the union of a finite family of halfspaces in $\mathbb{R}^n$, then it is contained in the union of some $n + 1$ or fewer halfspaces of the family. Give examples to show that neither the finiteness nor the convexity conditions can be omitted in this result.
2. Let $\mathfrak{F}$ be a finite family of convex sets and let C be a convex set in $\mathbb{R}^n$. Suppose that, for any $n + 1$ or fewer members of $\mathfrak{F}$, there is a translate of C which meets these members. Deduce the existence of some translate of C which meets *all* the members of $\mathfrak{F}$. Why does this result generalize Helly's theorem?
3. In $\mathbb{R}^2$ let A be a set of m $(m \geqslant 1)$ points, let B be a closed disc containing A and let $\mathfrak{F}$ be the family of closed halfplanes which contain *more than* $2m/3$ points of A. Prove that every three members of the family $\mathcal{G} = \{F \cap B : F \in \mathfrak{F}\}$ have a point of A in common. Deduce that there exists a point $\mathbf{p}$ common to all members of $\mathcal{G}$ and that each closed halfplane whose boundary line passes through $\mathbf{p}$ contains at least $m/3$ points of A.
4. A finite family of convex sets in $\mathbb{R}^2$, having at least two members, is such that, given any two of its sets, there is some line parallel to the x-axis which meets them both. Prove the existence of a line parallel to the x-axis which meets all members of the family. Suggest generalizations of this result to $\mathbb{R}^3$.
5. Show that, if R_n denotes the circumradius of a regular n-simplex of diameter 1 in $\mathbb{R}^n$, then $R_n = (4(1 - R_{n-1}^2))^{-\frac{1}{2}}$ for $n \geqslant 2$. Deduce the value of R_n. Comment on your answer.
6. Show that the word *compact* in the statement of Krasnosselsky's theorem cannot be replaced by either of the words *closed* or *bounded*.
7. Let $(x_1, y_1), \ldots, (x_n, y_n)$ be points in $\mathbb{R}^2$, where $x_1 < \ldots < x_n$ and $n \geqslant 3$. Show that the best approximation of degree 1 to these points is unique. Find this approximation for the points $(0, 1)$, $(1, 3)$, $(2, 0)$, $(3, 1)$.
8. Find the best approximations of degree 1 to: (i) x^2 on $[0, 1]$; (ii) $1/x$ on $[1, 4]$; (iii) $f: [0, 2] \to \mathbb{R}$, where

$$f(x) = \begin{cases} 0 & \text{for } 0 \leqslant x < 1 \\ 1 & \text{for } 1 \leqslant x < 2 \\ 2 & \text{for } x = 2. \end{cases}$$

9. For $n = 0, 1, 2, \ldots$, the nth *Chebyshev polynomial* $T_n(x)$ is defined by the requirement $\cos nx = T_n(\cos x)$. Thus $T_0(x) = 1$, $T_1(x) = x$, $T_2(x) = 2x^2 - 1$, $T_3(x) = 4x^3 - 3x$, etc. Each polynomial $P(x)$ of degree n can be expressed uniquely in the form

$$P(x) = a_n T_n(x) + a_{n-1} T_{n-1}(x) + \ldots + a_0 T_0(x).$$

Show that, for $n \geqslant 1$, the best approximation of degree $n - 1$ to $P(x)$ on $[-1, 1]$ is

$$a_{n-1} T_{n-1}(x) + \ldots + a_0 T_0(x)$$

with associated error $|a_n|$.

Find the best approximation of degree 2 to $x^3 - x^2 + 1$ on $[-1, 1]$.

7.2 JOIN SPACES

This section is unlike any other in the book in that it discusses convexity not in the familiar context of $\mathbb{R}^n$, but in a structure known as a *join space*. We indicate how an attractive theory of convexity can be developed within such a framework. The spirit of the section is similar to that of Euclid's *Elements*, with its emphasis on a geometry founded on a small number of carefully chosen axioms.

In several respects $\mathbb{R}^n$ is not always the most suitable setting for studying elementary convexity. For example, its structure is far richer than is necessary to support a rudimentary theory of convex sets, thereby often obscuring the innate geometry of the subject in a maze of algebraic, analytical, and topological detail. To avoid this state of affairs, we now pursue an axiomatic approach to convexity, based on a few geometrically motivated, but algebraically formulated assumptions. Our account, which is coordinate free, applies to geometries of arbitrary dimension, both finite and infinite.

The definition of a convex set, as one containing each line segment joining a pair of its points, at once highlights the key role played by the notion of a line segment in the study of convexity in $\mathbb{R}^n$, and it is this concept which we abstract when coming to define a join space. To help prepare for this, we introduce an idea closely related to that of a line segment, that of a *join*. The **join** of points $\mathbf{a}$ and $\mathbf{b}$ of $\mathbb{R}^n$ is the subset $\mathbf{ab}$ of $\mathbb{R}^n$ defined by the equation

$$\mathbf{ab} = \{\lambda\mathbf{a} + \mu\mathbf{b} : \lambda, \mu > 0,\ \lambda + \mu = 1\}.$$

Clearly, if $\mathbf{a} \neq \mathbf{b}$, then $\mathbf{ab}$ is the line segment joining $\mathbf{a}$ and $\mathbf{b}$ *excluding* the points $\mathbf{a}$ and $\mathbf{b}$ themselves; if $\mathbf{a} = \mathbf{b}$, then $\mathbf{ab}$ consists of the point $\mathbf{a}$ alone. In all cases, $\mathbf{ab}$ equals $\mathbf{ba}$ and is non-empty. The join operation is extended to subsets A, B of $\mathbb{R}^n$ by defining the **join** of A and B to be the subset AB of $\mathbb{R}^n$ given by the equation

$$AB = \bigcup (\mathbf{ab} : \mathbf{a} \in A, \mathbf{b} \in B).$$

Clearly, $A \subseteq AA$, $AB = BA$ and AB is non-empty when both A and B are. We note that AB is the union of all joins $\mathbf{ab}$ for $\mathbf{a} \in A$, $\mathbf{b} \in B$. Thus, as an illustration of the join of two sets, if A and B are (distinct) parallel lines in $\mathbb{R}^2$, then AB is the open parallel strip bounded by A and B.

Before investigating the properties of the join operation, we introduce some items of notation. Throughout this section we shall not distinguish notationally between a singleton set and the element defining it. Thus, if in a join AB, A is the singleton set $\{\mathbf{a}\}$, we write $\mathbf{a}B$ instead of $\{\mathbf{a}\}B$; this is unlikely to lead to confusion as $\{\mathbf{a}\}\{\mathbf{b}\} = \mathbf{ab}$. We write $A \approx B$ (read *A meets B*) to indicate that the sets A and B have a non-empty intersection. Thus we may write $a \approx A$ to mean that $a \in A$.

Suppose that A, B, C are subsets of $\mathbb{R}^n$. Then we can form the sets $A(BC)$ and $(AB)C$. The easiest of calculations shows that both of these sets equal

$$\{\lambda\mathbf{a} + \mu\mathbf{b} + \nu\mathbf{c} : \mathbf{a} \in A, \mathbf{b} \in B, \mathbf{c} \in C, \lambda > 0, \mu > 0, \nu > 0, \lambda + \mu + \nu = 1\}.$$

Hence the join operation is associative in the sense that $A(BC) = (AB)C$. To appreciate the geometrical significance of this algebraic law, we consider the case when A, B, C are, respectively, singleton sets $\mathbf{a}, \mathbf{b}, \mathbf{c}$ in $\mathbb{R}^2$. Here $\mathbf{a}(\mathbf{bc})$ is the union of all the joins of $\mathbf{a}$ to a point of $\mathbf{bc}$, and so is, in the non-degenerate case, the interior of $\Delta\mathbf{abc}$. Similarly, $(\mathbf{ab})\mathbf{c}$ is the union of all the joins of a point of $\mathbf{ab}$ to $\mathbf{c}$, and as before is the interior of $\Delta\mathbf{abc}$. These ideas are illustrated in Fig. 7.2.

Although the basic operation in convexity is that of joining two points, there is another one which is of importance, that of *extending* one point from another. The **extension** of one point $\mathbf{a}$ of $\mathbb{R}^n$ from another $\mathbf{b}$ is the subset $\mathbf{a}/\mathbf{b}$ of $\mathbb{R}^n$ defined by the equation

$$\mathbf{a}/\mathbf{b} = \{\lambda\mathbf{a} + \mu\mathbf{b} : \lambda > 1,\ \lambda + \mu = 1\}.$$

Clearly, if $\mathbf{a} \neq \mathbf{b}$, then $\mathbf{a}/\mathbf{b}$ is the halfline emanating from $\mathbf{a}$ away from $\mathbf{b}$ *excluding* the point $\mathbf{a}$ itself; if $\mathbf{a} = \mathbf{b}$, then $\mathbf{a}/\mathbf{b}$ consists of the point $\mathbf{a}$ alone. See Fig. 7.3. In all cases $\mathbf{a}/\mathbf{b}$ is non-empty and

$$\mathbf{a}/\mathbf{b} = \{\mathbf{x} \in \mathbb{R}^n : \mathbf{a} \approx \mathbf{bx}\}.$$

To see why this equation holds, we note that $\mathbf{x} \in \mathbf{a}/\mathbf{b}$ (or $\mathbf{x} \approx \mathbf{a}/\mathbf{b}$) if and

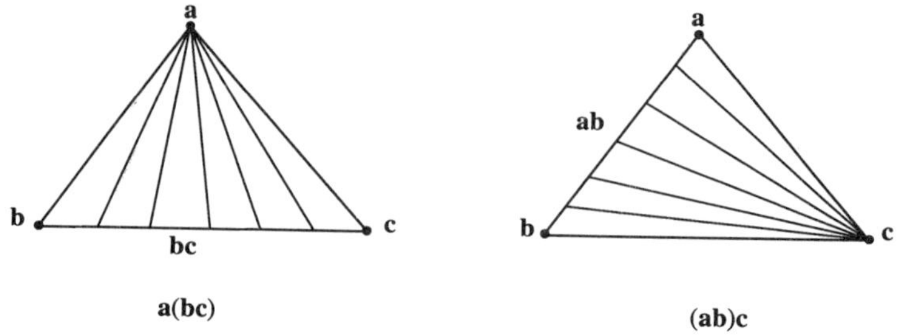

Fig. 7.2

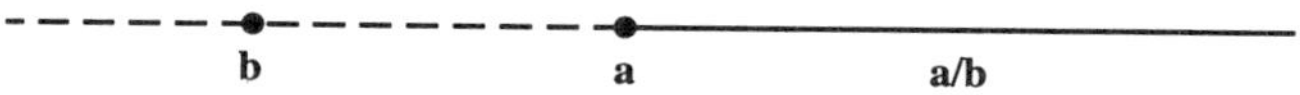

The extension **a/b** of **a** from **b**

Fig. 7.3

only if there are scalars λ, μ with $\lambda > 1$ and $\lambda + \mu = 1$ such that $\mathbf{x} = \lambda\mathbf{a} + \mu\mathbf{b}$, i.e. $\mathbf{a} = (1/\lambda)\mathbf{x} + (-\mu/\lambda)\mathbf{b}$, and this holds precisely when $\mathbf{a} \approx \mathbf{bx}$. The fact that $\mathbf{x} \approx \mathbf{a}/\mathbf{b}$ implies $\mathbf{a} \approx \mathbf{bx}$ shows that a kind of *cross multiplication* operates with respect to $\approx$.

We now pursue this idea of cross multiplication. Let $\mathbf{a}, \mathbf{b}, \mathbf{c}, \mathbf{d} \in \mathbb{R}^n$ be such that $\mathbf{a}/\mathbf{b} \approx \mathbf{c}/\mathbf{d}$. Then, for some scalars $\alpha, \beta, \gamma, \delta$ with $\alpha > 1$, $\gamma > 1$, $\alpha + \beta = 1$, $\gamma + \delta = 1$,

$$\alpha\mathbf{a} + \beta\mathbf{b} = \gamma\mathbf{c} + \delta\mathbf{d}, \qquad \text{whence} \qquad \frac{\alpha\mathbf{a} - \delta\mathbf{d}}{\alpha - \delta} = \frac{\gamma\mathbf{c} - \beta\mathbf{b}}{\gamma - \beta} = \mathbf{e},$$

say. Thus the point **e** lies on both **ad** and **bc**, i.e. $\mathbf{ad} \approx \mathbf{bc}$. Hence $\mathbf{a}/\mathbf{b} \approx \mathbf{c}/\mathbf{d}$ implies $\mathbf{ad} \approx \mathbf{bc}$, which again shows that a cross multiplication operates with respect to $\approx$. The geometrical significance of the last result should be clear from Fig. 7.4. The result itself is, in essence, a postulate of Peano which may be stated as follows: *segments which join two vertices of a triangle to respective points of their opposite sides intersect.*

This completes our discussion of the join and extension operations in $\mathbb{R}^n$ and we now summarize the basic laws which they obey.

Existence Laws. The join **ab** and the extension **a/b** are non-empty.

Commutative Law. The join **ab** equals the join **ba**.

Associative Law. The join **a(bc)** equals the join **(ab)c**.

Transposition Law. If the extensions **a/b** and **c/d** meet, then so do the joins **ad** and **bc**.

Idempotent Laws. The join **aa** and the extension **a/a** both equal **a**.

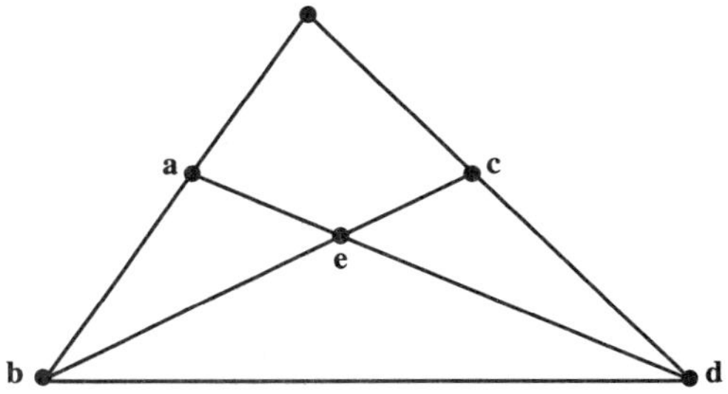

Fig. 7.4

We now carry over the above ideas on the join operation in $\mathbb{R}^n$ to a completely abstract setting. Suppose that X is a non-empty set and that $\cdot$ is a two-term operation which associates with each ordered pair of elements a, b of X a subset of X called the *join* of a and b, denoted by $a \cdot b$ or simply ab. The join operation $\cdot$ gives rise to an inverse operation / as follows: the *extension* a/b of a from b, where $a, b \in X$, is the set of all those elements x in X for which $a \approx bx$. The join and extension operations are extended to subsets A, B of X by the equations

$$AB = \cup\,(ab : a \in A, b \in B) \qquad \text{and} \qquad A/B = \cup\,(a/b : a \in A, b \in B).$$

Immediate consequences of these definitions are: if $A' \subseteq A$, $B' \subseteq B$, then $A'B' \subseteq AB$ and $A'/B' \subseteq A/B$; $A \approx BC$ if and only if $A/B \approx C$; if either A or B is empty, then so too are AB and A/B. The pair $(X, \cdot)$ is called a **join space** if, for all $a, b, c, d \in X$:

(i) $ab \neq \varnothing$, $a/b \neq \varnothing$;

(ii) $ab = ba$;

(iii) $a(bc) = (ab)c$;

(iv) $a/b \approx c/d$ implies $ad \approx bc$;

(v) $aa = a = a/a$.

The following assertions, in which A, B, C, D denote subsets of a join space $(X, \cdot)$, follow directly from the definition of a join space:

(i)* AB and A/B are non-empty when A and B are;

(ii)* $AB = BA$;

(iii)* $A(BC) = (AB)C$;

(iv)* $A/B \approx C/D$ implies $AD \approx BC$;

(v)* $A \subseteq AA$ and $A \subseteq A/A$.

Since the join operation is associative, we can, in the natural way, refer unambiguously to the join $A_1 \ldots A_n$ of sets $A_1, \ldots, A_n$ in a join space. In particular, when all of the sets in the join equal a set A, we may refer to the join $A_1 \dots A_n$ simply as A^n. Since the join operation is commutative, the join $A_1 \dots A_n$ is invariant under any permutation of its members.

The most important example of a join space for us here is, of course, $\mathbb{R}^n$ with the join operation discussed in detail above; we refer to this join space as the *join space* $\mathbb{R}^n$ *with standard join*. There are *other* examples, two of which will be mentioned at the end of this section, where we make a few general observations about join spaces. For the moment, we concentrate on developing some basic ideas of convexity in the framework of an arbitrary join space.

We begin by establishing the fundamental laws obeyed by the join and extension operations in a join space. Since the axioms of a join space are couched in simple algebraic terms, it is perhaps not too surprising that our first results are closely linked in form to ideas from elementary algebra. In the context of a join space, however, these algebraic ideas assume a geometrical significance of their own. Suppose that a, b, c are real numbers. Then with juxtaposition and / denoting the multiplication and division of real numbers, respectively, we know that (when no question of division by zero arises):

(i) $(a/b)/c = a/bc$;

(ii) $a/(b/c) = ac/b$;

(iii) $a(b/c) = ab/c$.

Suppose now that a, b, c are points of a join space $(X, \cdot)$. Then both $(a/b)/c$ and a/bc are well-defined subsets of X, and it is meaningful to ask if (i) holds in this abstract setting. Similar remarks apply to equations (ii) and (iii). The answers to these questions are given in the following result.

Theorem 7.2.1 *Let a, b, c be points in a join space $(X, \cdot)$. Then*

(i) $(a/b)/c = a/bc$;

(ii) $a/(b/c) \subseteq ac/b$;

(iii) $a(b/c) \subseteq ab/c$.

Proof (i) Suppose that $x \approx (a/b)/c$. Then $xc \approx a/b$ and $(xc)b = x(bc) \approx a$, so $x \approx a/bc$. Thus $(a/b)/c \subseteq a/bc$. Suppose next that $x \approx a/bc$. Then $x(bc) = (xc)b \approx a$, so $xc \approx a/b$ and $x \approx (a/b)/c$. Thus $a/bc \subseteq (a/b)/c$. We conclude that $(a/b)/c = a/bc$.

(ii) Suppose that $x \approx a/(b/c)$. Then $x(b/c) = (b/c)x \approx a$, so $b/c \approx a/x$ and $bx \approx ca = ac$. Thus $x \approx ac/b$ and $a/(b/c) \subseteq ac/b$.

(iii) Suppose that $x \approx a(b/c)$. Then $x/a \approx b/c$, whence $cx = xc \approx ab$ and $x \approx ab/c$. Thus $a(b/c) \subseteq (ab)/c$. □

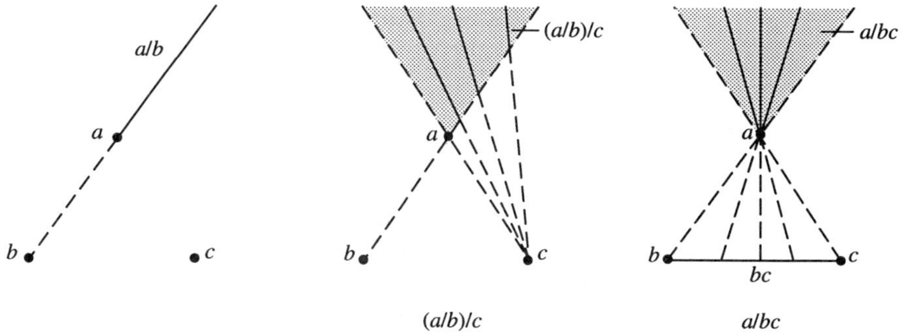

Fig. 7.5

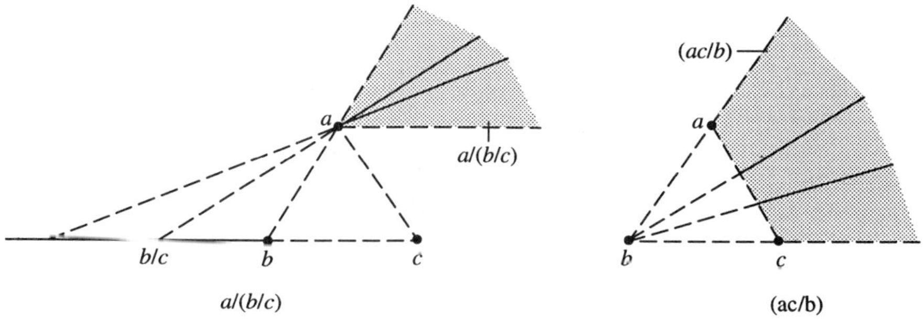

Fig. 7.6

Corollary 7.2.2 *Let* A, B, C *be sets in a join space* $(X, \cdot)$. *Then*:

(i) $(A/B)/C = A/BC$;

(ii) $A/(B/C) \subseteq AC/B$;

(iii) $A(B/C) \subseteq AB/C$. □

We now examine the geometrical significance of Theorem 7.2.1 in the familiar join space $\mathbb{R}^2$. Suppose that a, b, c are the vertices of a non-degenerate triangle in $\mathbb{R}^2$. Then the sets a/b, bc, $(a/b)/c$, a/bc and their *constructions* are illustrated in Fig. 7.5, which should make clear the equation $(a/b)/c = a/bc$ in this case. Figures 7.6 and 7.7 show why the inclusions in Theorem 7.2.1 hold in $\mathbb{R}^2$, and why they cannot be replaced by equations.

The definition of a convex set in a join space $(X, \cdot)$ could hardly be simpler: a set A in $(X, \cdot)$ is **convex** if $ab \subseteq A$ whenever $a, b \in A$ or, equivalently, if $AA \subseteq A$. Here we are simply mimicking the definition of a convex set in $\mathbb{R}^n$. Since, for *any* set A in $(X, \cdot)$, $A \subseteq AA$, it

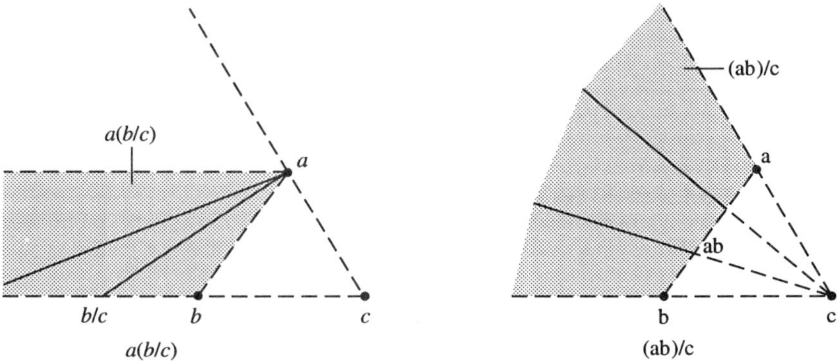

Fig. 7.7

follows that A is convex if and only if $AA = A$, i.e. $A^2 = A$. Clearly, the empty set, singletons, and the set X itself are convex. The next theorem shows that in a join space the join of two convex sets is convex, as is the extension of one convex set from another.

Theorem 7.2.3 *Let A and B be convex sets in a join space. Then AB and A/B are convex.*

Proof Since A and B are convex, $AA \subseteq A$ and $BB \subseteq B$. Thus

$$(AB)(AB) = (AA)(BB) \subseteq AB,$$

so AB is convex. Corollary 7.2.2 (i), (iii) enable us to deduce that

$$(A/B)(A/B) \subseteq ((A/B)A)/B \subseteq (AA/B)/B = AA/BB \subseteq A/B,$$

so A/B is convex. □

Corollary 7.2.4 *Let $A_1, \ldots, A_r$, $B_1, \ldots, B_s$ be convex sets in a join space. Then $A_1 \ldots A_r$ and $A_1 \ldots A_r/B_1 \ldots B_s$ are convex.* □

We now interrupt our general line of development to present a single result which is the example *par excellence* for displaying the advantages of studying convexity in a join space. A similar result was first established by H. Brunn in 1913. In a join space $(X, \cdot)$ a point a of a set A is called a **starcentre** of A if each join of a to a point of A lies in A, i.e. if $aA \subseteq A$. The set of all such starcentres is called the **kernel** of A and is denoted by A^*. Thus

$$A^* = \{a \in A : aA \subseteq A\}.$$

Theorem 7.2.5 (Brunn) *The kernel A^* of any set A in a join space is convex.*

Proof Suppose that $a, b \in A^*$. Then $aA \subseteq A$ and $bA \subseteq A$. If $x \in ab$, then $x \in aA \subseteq A$ and

$$xA \subseteq (ab)A = a(bA) \subseteq aA \subseteq A.$$

Thus $x \in A^*$ and A^* is convex. □

As an illustration of Theorem 7.2.5, we note that the kernel A^* of a set A in $\mathbb{R}^2$ which is the union of two closed circular discs which meet in a single point, together with their common tangent passing through their point of contact, consists of that single point.

We observe that only the associative property of the join was used in proving Brunn's theorem, and it is easy to see why this is so geometrically, a fact which could be so readily obscured in an algebraic proof in $\mathbb{R}^n$.

Many topics discussed in this book within the framework of $\mathbb{R}^n$ can be developed in a join space. To do so, however, in a general work such as this would be inappropriate, and we content ourselves by indicating how just one of these topics can be transferred to the more general setting. The topic that we have chosen is the study of the interior and closure operations on convex sets or, as we refer to them in a join space, the *core* and *linear access* operations.

Suppose that $\mathbf{a}$ is an interior point of a set A in $\mathbb{R}^n$. Then, given any point $\mathbf{x}$ of $\mathbb{R}^n$, there is a point $\mathbf{b}$ of $\mathbf{ax}$ (using the standard join in $\mathbb{R}^n$) such that $\mathbf{ab} \subseteq A$. On the other hand, if $\mathbf{a}$ is a point and A a *convex* set in $\mathbb{R}^n$ such that, given *any* point $\mathbf{x}$ in $\mathbb{R}^n$, there is a point $\mathbf{b}$ of $\mathbf{ax}$ such that $\mathbf{ab} \subseteq A$, then $\mathbf{a}$ is an interior point of A (see Exercise 2.3.4). These observations prompt the following definition. Let a be a point and A a set in a join space $(X, \cdot)$. Then a is said to be a **core point** of A if, given any point x of X, there is a point b of ax such that $ab \subseteq A$. The set of all core points of A is called the **core** of A and is denoted by $\tilde{A}$. It follows easily from the definition of a core point that each core point of A is a point of A, i.e. $\tilde{A} \subseteq A$. In view of our observations at the beginning of the paragraph, the core of a convex set in $\mathbb{R}^n$ (with its standard join) coincides with its interior. To see that this state of affairs does not hold for *all* sets, consider the case of a set A in $\mathbb{R}^2$ which is the union of two closed circular discs which meet in a point, together with their common tangent line passing through their point of contact. Then the interior of A is the union of the two interiors of the discs, whereas the core of A in addition contains the point of contact of the discs. See Fig. 7.8.

Our first result on the core of a convex set is a direct analogue

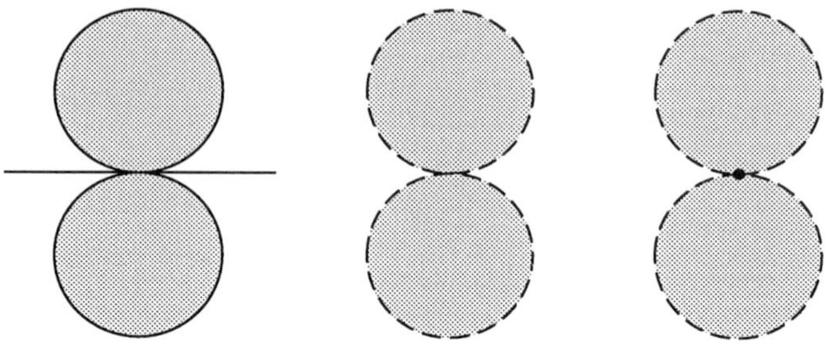

Fig. 7.8

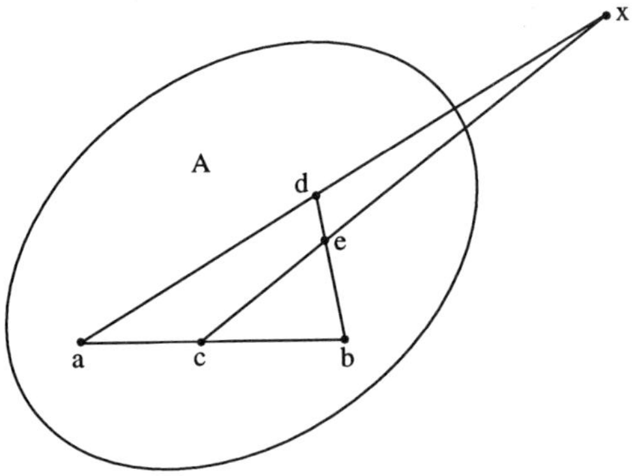

Fig. 7.9

of Lemma 2.3.3; the proofs of the two results are, however, very different.

Theorem 7.2.6 *Let A be a convex set in a join space $(X, \cdot)$. Then $\tilde{A}A \subseteq \tilde{A}$.*

Proof Suppose that $a \in \tilde{A}$, $b \in A$, and $c \approx ab$. Then we must show that $c \in \tilde{A}$. Let $x \in X$. Since $a \in \tilde{A}$, there exists $d \approx ax$ such that $ad \subseteq A$. Now $a \approx (c/b) \cap (d/x)$, whence $c/b \approx d/x$ and $cx \approx bd$. Thus there exists $e \approx (cx) \cap (bd)$. See Fig. 7.9. Since A is convex,

$$ce \subseteq (ab)(bd) = (ad)b \subseteq AA = A.$$

Thus $c \in \tilde{A}$ and $\tilde{A}A \subseteq \tilde{A}$. □

Corollary 7.2.7 *Let A be a convex set in a join space $(X, \cdot)$. Then $\tilde{\tilde{A}} = \tilde{A}$.*

Proof Trivially $\tilde{\tilde{A}} \subseteq \tilde{A}$, so we need only show that $\tilde{A} \subseteq \tilde{\tilde{A}}$. Suppose that $a \in \tilde{A}$ and that $x \in X$. Then there exists $b \approx ax$ such that $ab \subseteq A$. Let $c \approx ab$. Then $c \subseteq ab \subseteq a(ax) = ax$ and, by the theorem, $ac \subseteq \tilde{A}A \subseteq \tilde{A}$. Thus $a \in \tilde{\tilde{A}}$ and $\tilde{A} \subseteq \tilde{\tilde{A}}$ as required. □

Corollary 7.2.8 *Let A be a convex set in a join space $(X, \cdot)$. Then $\tilde{A}$ is convex.*

Proof Theorem 7.2.6 shows that $\tilde{A}\tilde{A} \subseteq \tilde{A}A \subseteq \tilde{A}$, whence $\tilde{A}$ is convex. □

We now turn our attention to the *linear access* operation in a join space, which resembles the closure operation in $\mathbb{R}^n$. To motivate our discussion, suppose that $\mathbf{a}, \mathbf{b}$ are points and A is a set in $\mathbb{R}^n$ such that $\mathbf{ab} \subseteq A$. Then clearly $\mathbf{a}$ is a closure point of A. On the other hand, if $\mathbf{a}$ is a closure point of a *convex* set A in $\mathbb{R}^n$, then (by Theorem 2.3.4) $\mathbf{ab} \subseteq \operatorname{ri} A \subseteq A$, for any point $\mathbf{b}$ of $\operatorname{ri} A$. These observations prompt the following definition. In a join space $(X, \cdot)$, a point a is said to be a **linear access point** of a set A if there exists a point b of X such that $ab \subseteq A$. The set of all linear access points of A is called the **linear access** of A and it is denoted by $\hat{A}$. It follows immediately from the definition of a linear access point that each point of A is a linear access point of A, i.e. $A \subseteq \hat{A}$. In view of our earlier remarks, the linear access of a convex set in $\mathbb{R}^n$ (with its standard join) coincides with its closure. To see that this state of affairs does not hold for *all* sets in $\mathbb{R}^n$, consider the case of a set A which is a semicircular arc in $\mathbb{R}^2$ with its two endpoints removed. Then these two endpoints are closure points but not linear access points of A.

Theorem 7.2.9 *Let A be a convex set in a join space $(X, \cdot)$. Then $\hat{A}$ is convex.*

Proof Suppose that $a, b \in \hat{A}$. Then there exist $c, d \in X$ such that $ac, bd \subseteq A$. Since A is convex, we can deduce that

$$(ab)(cd) = (ac)(bd) \subseteq AA = A.$$

Thus $ab \subseteq \hat{A}$ and $\hat{A}$ is convex. □

The next result is the (partial) analogue in a join space of Theorem 2.3.4.

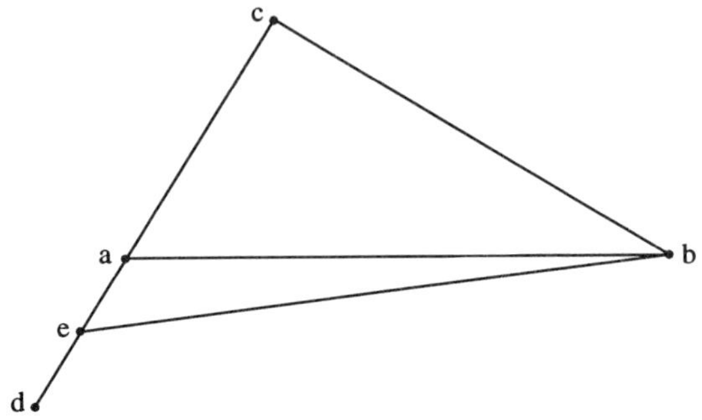

Fig. 7.10

Theorem 7.2.10 *Let A be a convex set in a join space $(X, \cdot)$. Then $\tilde{A}\hat{A} \subseteq \tilde{A}$.*

Proof Suppose that $a \in \tilde{A}$ and $b \in \hat{A}$. Then there exists $c \in X$ such that $bc \subseteq A$. Let $d \approx a/c$. By Corollary 7.2.7, $a \in \tilde{A} = \tilde{\tilde{A}}$, and so there exists $e \approx ad$ such that $ae \subseteq \tilde{A}$. Now $d \approx (a/c) \cap (e/a)$, whence $a/c \approx e/a$ and $aa \approx ce$, i.e. $a \approx ce$. See Fig. 7.10. By Theorem 7.2.6,

$$ab = (aa)b = a(ab) \subseteq (ce)(ab) = (ae)(bc) \subseteq \tilde{A}A \subseteq \tilde{A}. \quad \square$$

Our brief discussion of the core and linear access operations should have served to illustrate the advantages of considering $\mathbb{R}^n$ as a join space. In particular, we mention the close interplay between geometry and algebra, the ease with which some basic ideas in convexity can be translated into the language of joins, and the operation-effectiveness of the algebra of joins. Although we have only given *one* example of a join space, namely $\mathbb{R}^n$ with its standard join, there are others, and much work has been done in recent years on developing convexity in general join spaces. We conclude our discussion by introducing two new join spaces, both of which differ in essential ways from the join space $\mathbb{R}^n$ with its standard join.

Our first example is a natural generalization of $\mathbb{R}^n$. Denote by $\mathbb{R}^*$ the collection of all real sequences of the form $\mathbf{x} = (x_1, \ldots, x_n, \ldots)$ in which *at most* a finite number of the terms are non-zero. The addition and scalar multiplication of points of $\mathbb{R}^*$ are defined coordinatewise as follows:

$$\mathbf{x} + \mathbf{y} = (x_1, \ldots, x_n, \ldots) + (y_1, \ldots, y_n, \ldots) = (x_1 + y_1, \ldots x_n + y_n, \ldots);$$
$$\lambda\mathbf{x} = \lambda(x_1, \ldots, x_n, \ldots) = (\lambda x_1, \ldots, \lambda x_n, \ldots) \qquad \text{for } \lambda \in \mathbb{R}.$$

The *join* $\mathbf{xy}$ of points $\mathbf{x}, \mathbf{y}$ of $\mathbb{R}^*$ is defined by the equation:

$$\mathbf{xy} = \{\lambda\mathbf{x} + \mu\mathbf{y} : \lambda, \mu > 0, \lambda + \mu = 1\}.$$

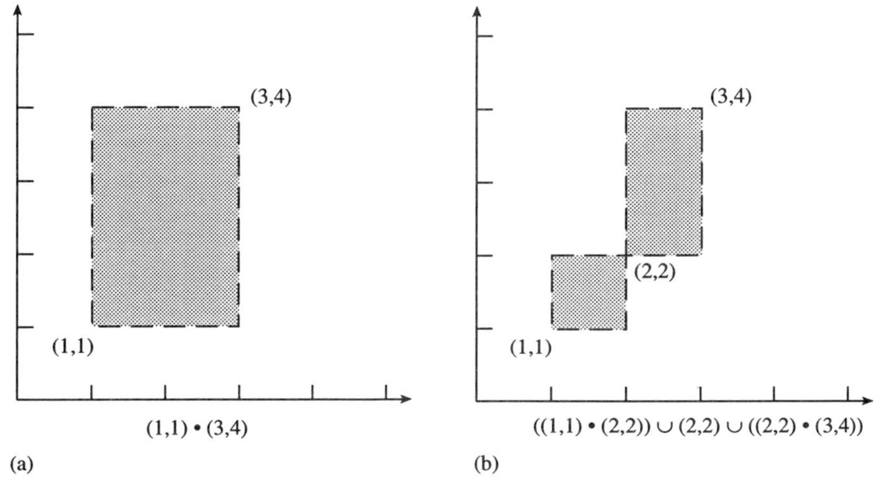

Fig. 7.11

It is verified, exactly as in $\mathbb{R}^n$, that $\mathbb{R}^*$ together with the join operation just defined constitute a join space. Whilst many results in the join space $\mathbb{R}^n$ carry over to the join space $\mathbb{R}^*$, many others do not, due to the *infinite-dimensional* nature of $\mathbb{R}^*$. We give one example of such a result. It follows immediately from the fact that in $\mathbb{R}^n$ a convex set can always be separated from a point not lying in it, that the only convex set in $\mathbb{R}^n$ whose closure is $\mathbb{R}^n$ is $\mathbb{R}^n$ itself. Since the closure and the linear access of a convex set in $\mathbb{R}^n$ coincide, it follows that *the only convex set in* $\mathbb{R}^n$ *whose linear access is* $\mathbb{R}^n$ *is* $\mathbb{R}^n$ *itself.* To see that this result does not transfer to $\mathbb{R}^*$, consider the convex set A consisting of all the non-zero elements of $\mathbb{R}^*$ whose last non-zero term is *positive.* Let $\mathbf{x} = (x_1, \ldots, x_n, 0, 0, \ldots)$ be a typical member of $\mathbb{R}^*$ and let $\mathbf{a} = (x_1, \ldots, x_n, 1, 0, 0, \ldots)$. Then $\mathbf{a} \in A$ and $\mathbf{xa} \subseteq A$, which shows that $\mathbf{x} \in \hat{A}$. Thus $\hat{A} = \mathbb{R}^*$ and A is a convex set in $\mathbb{R}^*$, other than $\mathbb{R}^*$ itself, whose linear access is $\mathbb{R}^*$.

For our second example, we take X to be the set $\mathbb{R}^2$ with the *non-standard join* $(x_1, x_2) \cdot (y_1, y_2)$ of the points (x_1, x_2), (y_1, y_2) of $\mathbb{R}^2$ defined by the equation

$$(x_1, x_2) \cdot (y_1, y_2) = \{(z_1, z_2) : z_1 \in x_1 y_1,\ z_2 \in x_2 y_2\},$$

where $x_1 y_1$, $x_2 y_2$ denote, respectively, the standard joins of x_1 and y_1, and of x_2 and y_2 in $\mathbb{R}^1$. See Fig. 7.11(a), where the particular join $(1, 1) \cdot (3, 4)$ is illustrated. It is a short and straightforward exercise to verify that $(X, \cdot)$ so defined is a join space. Perhaps it is not too surprising that this join space does not completely mimic $\mathbb{R}^2$ with its standard join. Consider, for example, three distinct points $\mathbf{a}, \mathbf{b}, \mathbf{c}$ in $\mathbb{R}^2$ (with its standard join) such that $\mathbf{c} \approx \mathbf{ab}$. Then $\mathbf{ab} = \mathbf{ac} \cup \mathbf{c} \cup \mathbf{cb}$. In the join space $(X, \cdot)$ just defined, this last equation *never* holds; see the special case when $\mathbf{a} = (1, 1)$, $\mathbf{b} = (3, 4)$ and $\mathbf{c} = (2, 2)$, which is illustrated in Fig. 7.11(b).

In this section we have indicated how, on the one hand, by studying convexity in join spaces it is possible to obtain new, simple, geometrically motivated proofs of some familiar results in $\mathbb{R}^n$, and how, on the other hand, join spaces are much more general structures than the Euclidean spaces $\mathbb{R}^n$, sometimes exhibiting characteristics that are completely alien to the classical spaces themselves.

For our final word on the topic, we return to the definition of a join space itself, showing that the five axioms defining it fail to be independent in the following surprising (?) way.

Theorem 7.2.11 *The commutative axiom for a join space is a consequence of the other four axioms.*

Proof Suppose that the pair $(X, \cdot)$ satisfies conditions (i), (iii), (iv), (v) of a join space, i.e. the commutative condition $ab = ba$ is omitted. Let $a, b \in X$. We show first that $ab \subseteq ba$. Let $x \approx ab$. Choose $y \approx xb$. Then $b \approx (y/x) \cap (x/a)$, whence $y/x \approx x/a$ and $x = xx \approx ya \subseteq (xb)a = x(ba)$. Thus $x = x/x \approx ba$. Hence $ab \subseteq ba$. By symmetry, $ba \subseteq ab$. Hence $ab = ba$, and the commutative axiom follows from the other four axioms. $\square$

Exercises 7.2

1. Let a, b, c be points in a join space. Let d lie on the extension b/a and let e lie on the extension c/a. Show that, for each point x on the join bc, the extension x/a meets the join de.
2. The *convex hull* $\operatorname{conv} A$ of a set A in a join space $(X, \cdot)$ is defined to be the intersection of all convex sets in X which contain A. Prove that $\operatorname{conv} A = \bigcup (A^n : n = 1, 2, \ldots)$.
3. The set A in a join space $(X, \cdot)$ is said to be a *flat* if $A/A \subseteq A$. (This definition coincides with the usual one in $\mathbb{R}^n$ having its standard join.) Show that, if A, B are sets in X, then:
   (i) if A is a flat, then A is convex;
   (ii) if A is a flat and $x \in A$, then $A = A/x = x/A$;
   (iii) if A is convex, then A/A is a flat;
   (iv) if A and B are intersecting flats, then A/B is a flat.
4. Let A be a set in a join space $(X, \cdot)$. Show that, if A is convex and $\tilde{A} \neq \varnothing$, then $\hat{\hat{A}} = \hat{A}$.
5. *Let A* be the set of those points $\mathbf{x} = (x_1, \ldots, x_n, \ldots)$ in the join

space $\mathbb{R}^*$ for which there exists a positive integer m such that $x_1 > 1/m$, $x_2 > 0, \ldots, x_m > 0$ and $x_n = 0$ for $n > m$. Prove that: (i) A is convex; (ii) $(t, 0, 0, \ldots) \in \hat{A}$ for $t > 0$; (iii) $(0, 0, \ldots) \notin \hat{A}$. Deduce that $\hat{\hat{A}} \neq \hat{A}$.

7.3 CONVEXITY-PRESERVING MAPPINGS

A mapping $f: \mathbb{R}^n \to \mathbb{R}^m$ is said to be **convexity-preserving** if the image under f of each convex set in $\mathbb{R}^n$ is a convex set in $\mathbb{R}^m$. The most familiar type of such a mapping is the affine transformation. Not every convexity-preserving mapping is affine. To see that this is so, we note that every continuous mapping from $\mathbb{R}^n$ to $\mathbb{R}^1$ is convexity-preserving. The aim of this section is to show that, *if* $f: \mathbb{R}^n \to \mathbb{R}^m$ $(n \geqslant 2)$ *is an injective convexity-preserving mapping, then* f *is an affine transformation*, a result that was proved in 1971 by W. Meyer and D. C. Kay.

Our proof of the result is neither short nor trivial. Fortunately, however, it has been possible to divide the proof into three parts, each consisting of a result worthy of consideration in its own right. The reader is asked to bear this in mind when reading through the section.

A mapping $f: \mathbb{R}^n \to \mathbb{R}^m$ is said to be **collinearity preserving** if whenever $\mathbf{x}, \mathbf{y}, \mathbf{z}$ lie on a line in $\mathbb{R}^n$, then $f(\mathbf{x}), f(\mathbf{y}), f(\mathbf{z})$ lie on a line in $\mathbb{R}^m$. Affine transformations are collinearity-preserving. The first of our three preparatory results asserts that *every injective convexity-preserving mapping is collinearity-preserving*. Even before we can prove this result, we need to establish two lemmas. The first of these is simple and geometrically appealing, the second is of a somewhat technical nature.

Lemma 7.3.1 *Let* $\mathbf{c}$ *be an interior point of a set* C *in* $\mathbb{R}^2$. *Suppose that* $C \setminus \{\mathbf{c}\}$ *is the union of disjoint convex sets* A, B. *Then there is a unique line* L *in* $\mathbb{R}^2$ *which separates* A *and* B. *Moreover,* L *passes through* $\mathbf{c}$ *and is the unique support line to* A *through* $\mathbf{c}$ *and to* B *through* $\mathbf{c}$.

Proof The situation considered in the lemma is illustrated for compact convex C in Fig. 7.12. Since $\mathbf{c}$ is interior to C, and A, B are convex, it follows that $\mathbf{c} \in (\operatorname{cl} A) \cap (\operatorname{cl} B)$. Thus A, B are non-empty disjoint convex sets, and so can be separated by a line L. Since $\mathbf{c} \in (\operatorname{cl} A) \cap (\operatorname{cl} B)$, the line L must pass through $\mathbf{c}$. Hence L supports both A and B at $\mathbf{c}$. Let L' be any line which supports A at $\mathbf{c}$. If L' is *different* from L, then it must meet $\operatorname{int} A$, but this is impossible. Thus L is the *unique* support line to A through $\mathbf{c}$. Similarly, L is the unique support line to B through $\mathbf{c}$. It follows that L is the unique line which separates A and B. $\square$

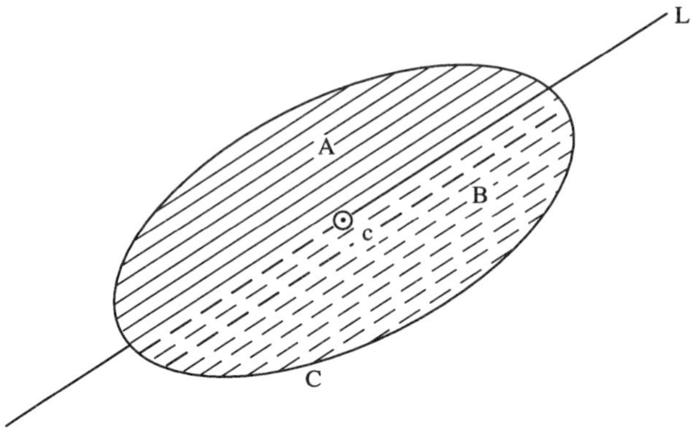

Fig. 7.12

Lemma 7.3.2 *Let $f: S \to C$ be a bijection from a non-empty set S in $\mathbb{R}^1$ onto an open set C in $\mathbb{R}^2$. Then there is some interval I in $\mathbb{R}^1$ such that $f(I \cap S)$ is not convex.*

Proof We argue by contradiction, assuming that $f(I \cap S)$ is convex for *every* interval I in $\mathbb{R}^1$. For each $s \in S$, denote by s^- and s^+ the sets $(-\infty, s) \cap S$ and $(s, \infty) \cap S$, respectively. Now $f(s^-)$ and $f(s^+)$ are disjoint convex sets with union $C \backslash \{f(s)\}$. By Lemma 7.3.1, there is a unique line separating $f(s^-)$ and $f(s^+)$. We denote this line by L^s. Then L^s passes through $f(s)$; also L^s is the unique support line through $f(s)$ to both $f(s^-)$ and $f(s^+)$. We define an equivalence relation $\sim$ on S by writing $s \sim t$ whenever $s, t \in S$ and $L^s = L^t$, and we denote the equivalence class of s in S by $[s]$. In (i), (ii), (iii) below, we establish (amongst other things) that each equivalence class is infinite, conv $[s]$ and conv $[t]$ are disjoint when $[s]$ and $[t]$ are distinct equivalence classes, and there is an uncountable number of equivalence classes.

(i) Let $s, t \in S$ be such that $f(t) \in L^s$. We show that $L^t = L^s$. Suppose that $s \leqslant t$; the case $t < s$ is similar. Since L^s passes through $f(t)$, supports $f(s^+)$, and $f(t^+) \subseteq f(s^+)$, it follows that L^s supports $f(t^+)$ at $f(t)$, whence $L^s = L^t$. But $C \cap L^s$ is infinite and f is bijective, so the equivalence class $[s]$ must be infinite.

(ii) Suppose that $r, s, t \in S$ are such that $r < s < t$ and $L^r = L^t = L$, say. We show that $L^s = L$. Now $f(s^-) \subseteq f(t^-)$, so $f(s^-)$ lies in one of the closed halfplanes determined by $L(=L^t)$. Similarly, $f(s^+)$ also lies in one of the closed halfplanes determined by $L(=L^r)$. It follows easily that L separates $f(s^-)$ and $f(s^+)$, whence $L^s = L$.

Now let $[s]$, $[t]$, where $s, t \in S$, be *distinct* equivalence classes. By what we proved in the last paragraph, no member of $[s]$ can lie between two members of $[t]$ and vice versa. A simple exercise shows that either every

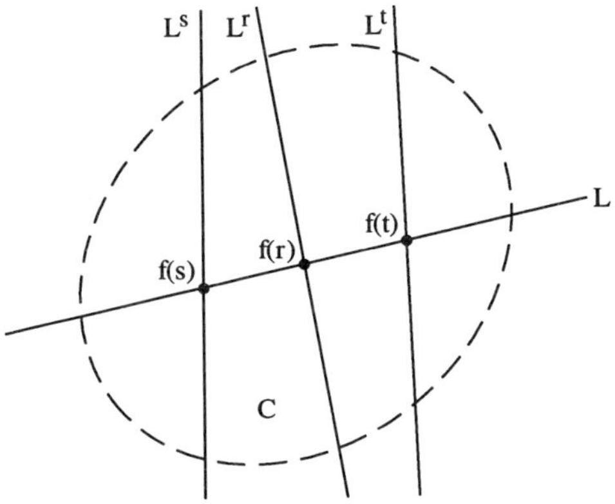

Fig. 7.13

number in $[s]$ is less than every number in $[t]$ or vice versa, whence $\operatorname{conv}[s]$ and $\operatorname{conv}[t]$ are disjoint. Thus the convex hulls of distinct equivalence classes are disjoint.

(iii) Let $s \in S$. Let L be any line through $f(s)$ other than L^s. Choose r, t in S such that $r \neq t$ and $f(r), f(t) \in L$. We argue by contradiction to show that $L^r \neq L^t$ (see Fig. 7.13). Suppose that $L^r = L^t$. Since L^r, L^t pass through $f(r), f(t)$, respectively, both L^r and L^t must be L. But $f(s) \in L$, whence by the result established in (i) above, $L^s = L$, contradicting the choice of L. Thus $L^r \neq L^t$, and so there is an uncountable number of equivalence classes, since $C \cap L$ is uncountable.

We can now easily obtain the contradiction for which we have been striving. For each equivalence class, it is possible to choose a rational number lying in its convex hull. The choices of such rationals arising from different equivalence classes are distinct. This shows that the number of equivalence classes must be countable, so contradicting the result at the end of the last paragraph. □

Theorem 7.3.3 *Let $g: \mathbb{R}^n \to \mathbb{R}^m$ be an injective convexity-preserving mapping. Then g is collinearity-preserving.*

Proof We argue by contradiction. Suppose that g is *not* collinearity-preserving. Then there is a line L in $\mathbb{R}^n$ such that $g(L)$ is not contained in any line of $\mathbb{R}^m$. Thus $g(L)$ contains three points whose affine hull is some 2-flat, P, say. Denote by A the relative interior of the convex set $g(L) \cap P$. Let $h: \mathbb{R}^1 \to L$ be an isometry from $\mathbb{R}^1$ *onto* L and let $j: P \to \mathbb{R}^2$ be an isometry from P *onto* $\mathbb{R}^2$. Then h and j are convexity-preserving bijections. Let $S = \{x \in \mathbb{R}^1 : g(h(x)) \in A\}$ and let $C = j(A)$. Then $S \subseteq \mathbb{R}^1$ and C is an open convex set in $\mathbb{R}^2$. Define a bijection $f: S \to C$, for each s in S, by the equation $f(s) = j(g(h(s)))$. For *any* interval I of $\mathbb{R}^1$, $f(I \cap S)$ is the convex set $j(g(h(I)) \cap A)$. But by Lemma 7.3.2, there is no bijection $f: S \to C$ such that $f(I \cap S)$ is convex for every interval I of $\mathbb{R}^1$. This gives the desired contradiction. □

One important property of injective affine mappings is that they map parallel lines *onto* parallel lines. Using the last theorem, we now show that most injective convexity-preserving mappings share this property.

Theorem 7.3.4 *Let $f: \mathbb{R}^n \to \mathbb{R}^m$ $(n \geqslant 2)$ be an injective convexity-preserving mapping. Then f maps parallel lines in $\mathbb{R}^n$ onto parallel lines in $\mathbb{R}^m$.*

Proof Let L, M be parallel lines lying in some 2-flat A of $\mathbb{R}^n$. We establish the theorem by showing that $f(A)$ is two-dimensional and that $f(L), f(M)$ are lines. See Theorem 1.3.12.

Let $\mathbf{a}, \mathbf{b}, \mathbf{c}$ be non-collinear points in A. Then $f(\mathbf{a}), f(\mathbf{b}), f(\mathbf{c})$ are non-collinear, for if they were, then one of them would lie in the convex hull of the other two, say $f(\mathbf{a}) \in \operatorname{conv}\{f(\mathbf{b}), f(\mathbf{c})\}$. Since $f(\operatorname{conv}\{\mathbf{b}, \mathbf{c}\})$ is a convex set containing $f(\mathbf{b}), f(\mathbf{c})$, we can deduce that

$$f(\mathbf{a}) \in \operatorname{conv}\{f(\mathbf{b}), f(\mathbf{c})\} \subseteq f(\operatorname{conv}\{\mathbf{b}, \mathbf{c}\}),$$

which, because f is injective, enables us to conclude that $\mathbf{a} \in \operatorname{conv}\{\mathbf{b}, \mathbf{c}\}$. This contradicts the non-collinearity of $\mathbf{a}, \mathbf{b}, \mathbf{c}$. Thus $f(\mathbf{a}), f(\mathbf{b}), f(\mathbf{c})$ are *not* collinear.

We show that $f(A)$ is two-dimensional by proving that it lies in the 2-flat $B = \operatorname{aff}\{f(\mathbf{a}), f(\mathbf{b}), f(\mathbf{c})\}$. We use the fact that f preserves collinearity. Denote by S the union of the lines $\operatorname{aff}\{\mathbf{a}, \mathbf{b}\}$, $\operatorname{aff}\{\mathbf{b}, \mathbf{c}\}$, $\operatorname{aff}\{\mathbf{c}, \mathbf{a}\}$. Then $f(S) \subseteq B$. Since each point of A lies on some line joining two points of S, $f(A) \subseteq B$.

Finally, suppose that $f(L)$ is *not* a line in $\mathbb{R}^m$, so that it must be a *proper* subset of a line. Let $\mathbf{u}$ be a point of A such that $f(\mathbf{u})$ is a relative interior point of $f(A)$ not lying on $f(L)$. Choose points $\mathbf{v}, \mathbf{w}$ of A such that $\operatorname{aff}\{f(\mathbf{u}), f(\mathbf{v})\}$, $\operatorname{aff}\{f(\mathbf{u}), f(\mathbf{w})\}$ are *distinct* lines in B which do not meet $f(L)$. It follows easily that $\operatorname{aff}\{\mathbf{u}, \mathbf{v}\}$, $\operatorname{aff}\{\mathbf{u}, \mathbf{w}\}$ are *distinct* lines in A through $\mathbf{u}$ which are parallel to L. This is clearly impossible (a contradiction to Euclid's parallel postulate), and so $f(L)$ must be a line. See Fig. 7.14. Similarly $f(M)$ is a line. Since $f(L), f(M)$ lie in a 2-flat, and f is injective, they must be parallel. □

Before continuing with our geometrical arguments, we need to prove a delightful result about the real numbers. In the language of algebra, it states that the real number field has no non-trivial automorphism.

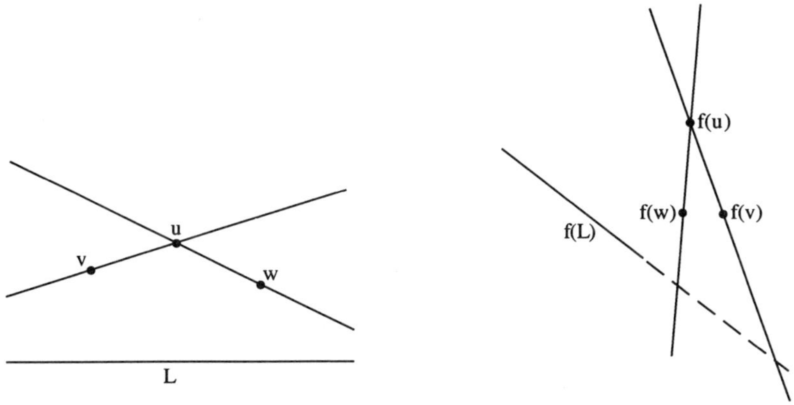

Two parallels to L through u in A

Two lines in B through f(u) not meeting f(L)

Fig. 7.14

Lemma 7.3.5 *Let* $F:\mathbb{R}\to\mathbb{R}$ *be a bijection satisfying*

$$F(\lambda+\mu)=F(\lambda)+F(\mu) \qquad \text{and} \qquad F(\lambda\mu)=F(\lambda)F(\mu)$$

for all $\lambda,\mu\in\mathbb{R}$. *Then* F *is the identity mapping.*

Proof The equation $F(0)+F(0)=F(0)$ shows that $F(0)=0$. Since F is injective, $F(1)\neq 0$. The equation $F(1)F(1)=F(1)$ now implies that $F(1)=1$. A simple argument shows that, for all positive integers n, $F(n)=n$ and $F(n)+F(-n)=0$; it follows that $F(n)=n$ for *all* integers n. Thus, for integers m,n with $n\neq 0$,

$$nF(m/n)=F(n)F(m/n)=F(m)=m \qquad \text{and} \qquad F(m/n)=m/n.$$

Suppose that $\sigma=\sqrt{(\lambda-\mu)}$, where $\mu<\lambda$. Then

$$F(\lambda)-F(\mu)=F(\lambda)+F(-1)F(\mu)=F(\lambda-\mu)=F(\sigma^2)=(F(\sigma))^2>0.$$

Thus F preserves the ordering of the real numbers and leaves each rational number fixed. Suppose finally that c is any real number. Let $a_1,\ldots,a_k,\ldots$ and $b_1,\ldots,b_k,\ldots$ be sequences of rational numbers converging to c such that $a_1<c<b_1,\ldots,a_k<c<b_k,\ldots$. Then

$$a_k=F(a_k)<F(c)<F(b_k)=b_k.$$

Letting $k\to\infty$ in these inequalities, we deduce that $F(c)=c$. Thus F is the identity mapping. □

Theorem 7.3.6 *Let* $g:\mathbb{R}^n\to\mathbb{R}^m$ $(n\geqslant 2)$ *be an injective mapping which maps parallel lines onto parallel lines. Then* g *is an affine transformation.*

Proof Define a mapping $f: \mathbb{R}^n \to \mathbb{R}^m$ by the equation $f(\mathbf{x}) = g(\mathbf{x}) - g(\mathbf{0})$ for $\mathbf{x}$ in $\mathbb{R}^n$. Then f is injective and maps parallel lines onto parallel lines, and leaves the origin fixed. Also $f(\mathbf{x}), f(\mathbf{y})$ are linearly independent whenever $\mathbf{x}, \mathbf{y}$ are. We prove the theorem by showing that f is linear.

Let $\mathbf{x}, \mathbf{y} \in \mathbb{R}^n$. We show that $f(\mathbf{x}+\mathbf{y}) = f(\mathbf{x}) + f(\mathbf{y})$. Suppose first that $\mathbf{x}, \mathbf{y}$ are linearly independent. Since the line through $\mathbf{0}$ and $\mathbf{x}$ is parallel to the line through $\mathbf{y}$ and $\mathbf{x}+\mathbf{y}$, the line through $f(\mathbf{0}) = \mathbf{0}$ and $f(\mathbf{x})$ must be parallel to the line through $f(\mathbf{y})$ and $f(\mathbf{x}+\mathbf{y})$. Thus there exists a scalar α, and similarly a scalar β, such that

$$f(\mathbf{x}+\mathbf{y}) - f(\mathbf{y}) = \alpha f(\mathbf{x}) \qquad \text{and} \qquad f(\mathbf{x}+\mathbf{y}) - f(\mathbf{x}) = \beta f(\mathbf{y}).$$

Hence $f(\mathbf{x}) + \beta f(\mathbf{y}) = \alpha f(\mathbf{x}) + f(\mathbf{y})$. But $f(\mathbf{x}), f(\mathbf{y})$ are linearly independent, whence $\alpha = \beta = 1$ and $f(\mathbf{x}+\mathbf{y}) = f(\mathbf{x}) + f(\mathbf{y})$.

Since the result that we are proving is clearly true when either $\mathbf{x}$ or $\mathbf{y}$ is the zero vector, we need consider only the case of $\mathbf{x}, \mathbf{y}$ linearly dependent and non-zero. Since $n \geqslant 2$, there exists $\mathbf{z}$ in $\mathbb{R}^2$ which is *not* a scalar multiple of $\mathbf{x}$. If $\mathbf{x}+\mathbf{y} \neq \mathbf{0}$, then the pairs of vectors $\mathbf{y}, \mathbf{z}$ and $\mathbf{x}, \mathbf{y}+\mathbf{z}$, and $\mathbf{x}+\mathbf{y}, \mathbf{z}$ are linearly independent. By what we proved in the last paragraph,

$$f(\mathbf{x}) + f(\mathbf{y}) + f(\mathbf{z}) = f(\mathbf{x}) + f(\mathbf{y}+\mathbf{z}) = f(\mathbf{x}+\mathbf{y}+\mathbf{z}) = f(\mathbf{x}+\mathbf{y}) + f(\mathbf{z}),$$

whence $f(\mathbf{x}+\mathbf{y}) = f(\mathbf{x}) + f(\mathbf{y})$. Finally, suppose that $\mathbf{x}+\mathbf{y} = \mathbf{0}$. Then, as $2\mathbf{x} + (-\mathbf{x}) \neq \mathbf{0}$, we can deduce from the case just established that

$$f(\mathbf{x}) = f(2\mathbf{x} + (-\mathbf{x})) = f(2\mathbf{x}) + f(-\mathbf{x}) = f(\mathbf{x}) + f(\mathbf{x}) + f(-\mathbf{x}),$$

whence

$$f(\mathbf{x}) + f(\mathbf{y}) = f(\mathbf{x}) + f(-\mathbf{x}) = \mathbf{0} = f(\mathbf{x}+\mathbf{y}).$$

Hence $f(\mathbf{x}+\mathbf{y}) = f(\mathbf{x}) + f(\mathbf{y})$ in all cases.

Suppose next that $\mathbf{x}$ is a non-zero vector in $\mathbb{R}^n$. Since f maps lines onto lines and $f(\mathbf{0}) = \mathbf{0}$, for each real λ, there exists a unique real number λ' such that $f(\lambda\mathbf{x}) = \lambda' f(\mathbf{x})$. Thus each non-zero $\mathbf{x}$ in $\mathbb{R}^n$ gives rise to a bijection $F_{\mathbf{x}}: \mathbb{R} \to \mathbb{R}$ given by $f(\lambda\mathbf{x}) = F_{\mathbf{x}}(\lambda) f(\mathbf{x})$ for each real λ. We show that $F_{\mathbf{x}}$ is independent of the choice of $\mathbf{x}$. To prove this, suppose that $\mathbf{x}, \mathbf{y}$ are linearly independent vectors in $\mathbb{R}^n$. Then, for any real λ,

$$\begin{aligned}
F_{\mathbf{x}+\mathbf{y}}(\lambda) f(\mathbf{x}) + F_{\mathbf{x}+\mathbf{y}}(\lambda) f(\mathbf{y}) &= F_{\mathbf{x}+\mathbf{y}}(\lambda)(f(\mathbf{x}) + f(\mathbf{y})) \\
&= F_{\mathbf{x}+\mathbf{y}}(\lambda)(f(\mathbf{x}+\mathbf{y})) \\
&= f(\lambda(\mathbf{x}+\mathbf{y})) \\
&= f(\lambda\mathbf{x}) + f(\lambda\mathbf{y}) \\
&= F_{\mathbf{x}}(\lambda) f(\mathbf{x}) + F_{\mathbf{y}}(\lambda) f(\mathbf{y}).
\end{aligned}$$

Since $f(\mathbf{x}), f(\mathbf{y})$ are linearly independent, $F_{\mathbf{x}}(\lambda) = F_{\mathbf{y}}(\lambda)$, and so $F_{\mathbf{x}} = F_{\mathbf{y}}$. When $\mathbf{x}, \mathbf{y}$ are linearly dependent non-zero vectors in $\mathbb{R}^n$, choose $\mathbf{z}$ in $\mathbb{R}^n$ such that both $\mathbf{x}, \mathbf{z}$ and $\mathbf{y}, \mathbf{z}$ are linearly independent pairs of vectors. By what has just been proved, $F_{\mathbf{x}} = F_{\mathbf{z}}$ and $F_{\mathbf{y}} = F_{\mathbf{z}}$, whence $F_{\mathbf{x}} = F_{\mathbf{y}}$. Thus each non-zero $\mathbf{x}$ in $\mathbb{R}^n$ determines the same mapping $F_{\mathbf{x}}$, which we now write simply as F.

For each non-zero $\mathbf{x}$ in $\mathbb{R}^n$ and for all real λ, μ,

$$F(\lambda + \mu)f(\mathbf{x}) = f((\lambda + \mu)\mathbf{x}) = f(\lambda\mathbf{x}) + f(\mu\mathbf{x}) = F(\lambda)f(\mathbf{x}) + F(\mu)f(\mathbf{x}),$$

and

$$F(\lambda\mu)f(\mathbf{x}) = f(\lambda\mu\mathbf{x}) = f(\lambda(\mu\mathbf{x})) = F(\lambda)f(\mu\mathbf{x}) = F(\lambda)F(\mu)f(\mathbf{x}).$$

Since $f(\mathbf{x}) \neq \mathbf{0}$, we may conclude that

$$F(\lambda + \mu) = F(\lambda) + F(\mu) \qquad \text{and} \qquad F(\lambda\mu) = F(\lambda)F(\mu).$$

The lemma now shows that $F: \mathbb{R} \to \mathbb{R}$ is the identity mapping. Thus, for each non-zero vector $\mathbf{x}$ in $\mathbb{R}^n$ and for each real λ, $f(\lambda\mathbf{x}) = \lambda f(\mathbf{x})$. Trivially, $f(\lambda\mathbf{0}) = \lambda f(\mathbf{0})$ for all real λ. Hence f is linear, as desired. □

Theorems 7.3.4 and 7.3.6 now yield the result for which we have been striving throughout this section.

Theorem 7.3.7 *Let $f: \mathbb{R}^n \to \mathbb{R}^m$ $(n \geqslant 2)$ be an injective convexity-preserving mapping. Then f is an affine transformation.* □

Exercises 7.3

1. Prove that an injective mapping $f: \mathbb{R}^n \to \mathbb{R}^m$ is convexity-preserving if and only if $f(\mathbb{R}^n)$ is convex, and f^{-1} preserves betweenness on $f(\mathbb{R}^n)$.
2. Show that, if A is an r-dimensional convex set $(r \geqslant 1)$ in $\mathbb{R}^n$, then there exist convex subsets $A_0, \ldots, A_r$ of A such that any r of them have a non-empty intersection, whereas the intersection of all $r + 1$ of them is empty. Deduce that, if $f: \mathbb{R}^n \to \mathbb{R}^m$ is an injective convexity-preserving mapping, then f does not decrease the dimension of convex sets.
3. Give examples, one in each case, to demonstrate the existence of mappings with the following properties:
   (i) a non-affine bijection $f: \mathbb{R}^1 \to \mathbb{R}^1$ which preserves convex sets;
   (ii) a non-affine mapping $f: \mathbb{R}^2 \to \mathbb{R}^1$ which maps parallel lines *onto* parallel lines;
   (iii) a non-affine, collinearity-preserving mapping $f: \mathbb{R}^2 \to \mathbb{R}^2$ such that dim $(f(\mathbb{R}^2))$ is 2.

4. Use the following existence result to construct a convexity-preserving surjection $f:\mathbb{R} \to \mathbb{R}^n$ which is not affine. *There exists a set S of non-zero real numbers such that: (i) there is a bijection $\theta:S \to \mathbb{R}^n$; (ii) for each non-zero real number r, there is a unique s in S such that $r = qs$ for some rational number q.*

7.4 THE GEOMETRY OF NUMBERS

We now come to a delightful theorem of Minkowski (1910) that gave birth to a whole new branch of number theory, the *geometry of numbers*. The theorem, which combines ideas from geometry, analysis, and number theory asserts that every symmetric convex set in $\mathbb{R}^n$ which has volume exceeding 2^n must contain a non-zero lattice point. We illustrate the power of this deceptively simple result by applying it to such diverse topics as Farey sequences, the approximation of irrational numbers by rational ones, the minimum of a quadratic form, and the representation of a non-negative integer as a sum of four squares.

Lemma 7.4.1 *Let A be a bounded symmetric convex set in $\mathbb{R}^n$ that contains no non-zero lattice points. Then $v(A) \leqslant 2^n$.*

Proof Since A is bounded, there exists $r > 0$ such that $A \subseteq rU$. Let m be a positive integer. Denote by $\mathfrak{F}$ the family of the m^n translates $\mathbf{z} + \frac{1}{2}A$ of $\frac{1}{2}A$, where $\mathbf{z}$ ranges over all the lattice points in the cell

$$\{(x_1, \ldots, x_n) : 0 \leqslant x_i < m \text{ for } i = 1, \ldots, n\}.$$

Then $\mathfrak{F}$ is a family of pairwise disjoint sets. To see that this is so, suppose that the translates $\mathbf{z} + \frac{1}{2}A$ and $\mathbf{z}' + \frac{1}{2}A$ meet for *different* lattice points $\mathbf{z}$ and $\mathbf{z}'$. Then the symmetry and convexity of $\frac{1}{2}A$ show that

$$\mathbf{z} - \mathbf{z}' \in \tfrac{1}{2}A - \tfrac{1}{2}A = \tfrac{1}{2}A + \tfrac{1}{2}A = A,$$

so contradicting the assumption that A contains no non-zero lattice points. The union of the members of $\mathfrak{F}$ lies in the cell

$$\{(x_1, \ldots, x_n) : -\tfrac{1}{2}r \leqslant x_i < m + \tfrac{1}{2}r \text{ for } i = 1, \ldots, n\},$$

whence

$$m^n v(\tfrac{1}{2}A) \leqslant (m + r)^n,$$

and so

$$v(\tfrac{1}{2}A) \leqslant (1 + (r/m))^n.$$

Letting $m \to \infty$ in the last inequality, we deduce that $v(\frac{1}{2}A) \leqslant 1$, whence $v(A) = 2^n v(\frac{1}{2}A) \leqslant 2^n$. □

Theorem 7.4.2 (Minkowski) *Let A be a bounded symmetric convex set in $\mathbb{R}^n$. Then A contains a non-zero lattice point if either* (*i*) $v(A) > 2^n$ *or* (*ii*) $v(A) = 2^n$ *and A is compact.*

Proof Statement (i) follows immediately from Lemma 7.4.1. To establish statement (ii), we argue by contradiction. Suppose that A is compact with $v(A) = 2^n$ and does *not* contain a non-zero lattice point. Let $d > 0$ be the least distance between a non-zero lattice point of $\mathbb{R}^n$ and a point of A. Such a d exists, since the set of non-zero lattice points in $\mathbb{R}^n$ is closed, and disjoint from the compact set A. By (i), the bounded symmetric convex set $A + \frac{1}{2}dU$ contains a non-zero lattice point. The distance of this point from some point of A does not exceed $\frac{1}{2}d$, thus contradicting the definition of d. Hence A must contain a non-zero lattice point. □

Our first application of Minkowski's theorem shows how a non-obvious property of a Farey sequence has a simple geometrical interpretation in terms of lattice points in $\mathbb{R}^2$, hence the name *geometry of numbers*. The **Farey sequence** $\mathfrak{F}_n$ **of order** n is the finite ascending sequence of rational numbers p/q, where p and q are positive integers with $p < q \leqslant n$ and p/q in its lowest terms. Thus $\mathfrak{F}_5$ is the finite sequence

$$\tfrac{1}{5}, \tfrac{1}{4}, \tfrac{1}{3}, \tfrac{2}{5}, \tfrac{1}{2}, \tfrac{3}{5}, \tfrac{2}{3}, \tfrac{3}{4}, \tfrac{4}{5}.$$

For the purpose of our next result only, we say that a triangle or a parallelogram in $\mathbb{R}^2$ is *basic* if its vertices are lattice points, and these are the only lattice points that it contains. Thus the triangle with vertices $(0, 0)$, $(1, 0)$, $(0, 1)$ is basic, whereas the triangle with vertices $(0, 0)$, $(2, 0)$, $(2, 1)$ is not. Suppose now that a/b, c/d are consecutive terms of the Farey sequence $\mathfrak{F}_n$, where a, b, c, d are positive integers and a/b, c/d are in their lowest terms. Consider the triangle $\Delta\mathbf{0uv}$ in $\mathbb{R}^2$, where $\mathbf{u} = (b, a)$ and $\mathbf{v} = (d, c)$. See Fig. 7.15. Since a/b, c/d are in their lowest terms, the only lattice points on the sides $\mathbf{0u}, \mathbf{0v}$ of $\Delta\mathbf{0uv}$ are $\mathbf{0}, \mathbf{u}, \mathbf{v}$. Hence, if $\Delta\mathbf{0uv}$ does contain a lattice point other than $\mathbf{0}, \mathbf{u}, \mathbf{v}$, say $\mathbf{w} = (f, e)$, then the line segment $\mathbf{0w}$ must be interior to the angle $\angle\mathbf{v0u}$, whence the slope of the line segment $\mathbf{0w}$ must lie between the slopes of the line segments $\mathbf{0u}$ and $\mathbf{0v}$, i.e. $a/b < e/f < c/d$. Since $e < f \leqslant n$, it follows that e/f, when reduced to its lowest terms, is a term of $\mathfrak{F}_n$, which lies *strictly* between two of its consecutive terms. This impossibility shows that the triangle $\Delta\mathbf{0uv}$ must be basic. It follows (see Fig. 7.15) that the triangle

$$\Delta\mathbf{uv}(\mathbf{u} + \mathbf{v}) = (-\Delta\mathbf{0uv}) + (\mathbf{u} + \mathbf{v}),$$

the parallelogram P with vertices $\mathbf{0}, \mathbf{u}, \mathbf{v}, \mathbf{u} + \mathbf{v}$, and the parallelograms

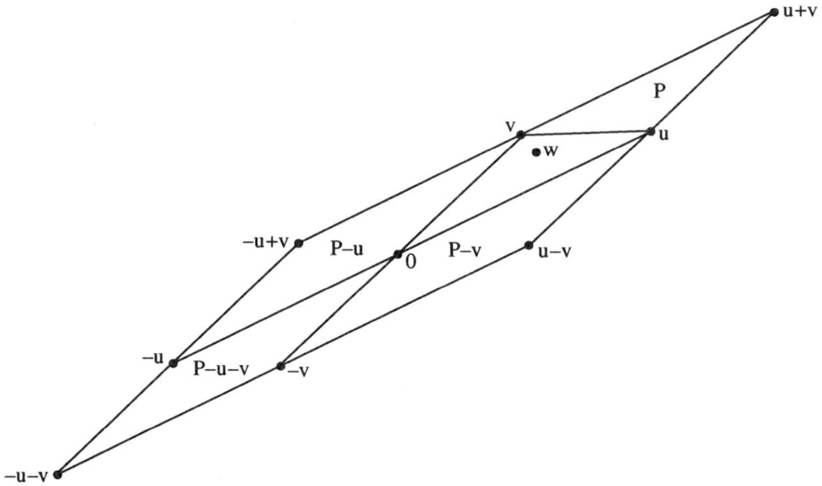

Fig. 7.15

$P - \mathbf{u}$, $P - \mathbf{v}$, $P - \mathbf{u} - \mathbf{v}$ are also basic. Hence the only lattice point in the interior of the parallelogram with vertices $\mathbf{u} + \mathbf{v}$, $\mathbf{u} - \mathbf{v}$, $-\mathbf{u} + \mathbf{v}$, $-\mathbf{u} - \mathbf{v}$ is the origin. Minkowski's theorem shows that the area $4(bc - ad)$ of this parallelogram cannot exceed 4, whence $bc - ad \leqslant 1$. But $bc - ad$ is a positive integer, and so we deduce that $bc - ad = 1$. We have thus established the following non-obvious, non-trivial result about Farey sequences.

Theorem 7.4.3 *Let a/b, c/d be consecutive terms of some Farey sequence. Then $bc - ad = 1$.* □

We now discuss the approximation of irrational numbers by rational ones. Let α be an irrational number and let q be a positive integer. We seek rational approximations to α of the form p/q, where p is an integer. Certainly an integer p can always be found such that $|\alpha - p/q| < 1/2q$. It is not always possible, however, to find an integer p such that $|\alpha - p/q| < 1/q^2$. For example, no such p exists when $\alpha = \pi$, $q = 3$. On the other hand, when $\alpha = \pi$, $q = 7$, we can take $p = 22$ to find that $|\pi - 22/7| < 1/7^2$. We say that a rational number p/q, assumed to be in its lowest terms, is a *good approximation* to α if $|\alpha - p/q| < 1/q^2$. Good approximations to π are not readily found; the *next* one after 22/7 is 333/106. The following theorem guarantees an endless supply of good approximations to any irrational number.

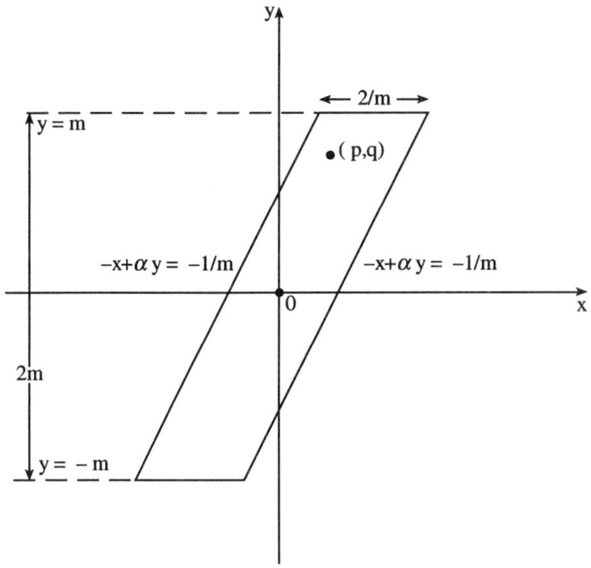

Fig. 7.16

Theorem 7.4.4 *Let α be an irrational number. Then, for any $\varepsilon > 0$, there is a good rational approximation β to α such that $|\alpha - \beta| < \varepsilon$.*

Proof Let m be a non-integral real number such that $m > \max\{1, 1/\varepsilon\}$. The parallelogram

$$\{(x, y): |-x + \alpha y| \leqslant 1/m,\ |y| \leqslant m\}$$

in $\mathbb{R}^2$ is a symmetric convex body with area 4. See Fig. 7.16. By Minkowski's theorem, there exist integers p, q, not both zero, such that $|-p + \alpha q| \leqslant 1/m$ and $|q| \leqslant m$. Since m is not an integer, $|q| < m$. Now q cannot be zero, for this would imply that $|-p| \leqslant 1/m < 1$, and hence that $p = 0$. Thus

$$\left|\alpha - \frac{p}{q}\right| \leqslant \frac{1}{m|q|} < \begin{cases} \varepsilon \\ 1/q^2 \end{cases}.$$

Hence $\beta = p/q$, which may not be in lowest terms, is a good approximation to α satisfying $|\alpha - \beta| < \varepsilon$. □

The problem that we now discuss is the one which led Minkowski to discover the geometry of numbers: *the minimization of positive definite quadratic forms*. Let $f: \mathbb{R}^n \to \mathbb{R}$ be the positive definite quadratic form defined by the equation

$$f(x_1, \ldots, x_n) = (a_{11}x_1 + \ldots + a_{1n}x_n)^2 + \ldots + (a_{n1}x_1 + \ldots + a_{nn}x_n)^2,$$

where the real $n \times n$ matrix $[a_{ij}]$ has non-zero determinant. Then f is positive unless all of $x_1, \ldots, x_n$ are zero. The problem considered by Minkowski is: *how small can* $f(x_1, \ldots, x_n)$ *be made as* $(x_1, \ldots, x_n)$ *ranges over the non-zero lattice points in* $\mathbb{R}^n$? Minkowski's theorem quickly yields the following result.

Theorem 7.4.5 *Let* $f: \mathbb{R}^n \to \mathbb{R}$ *be defined by the equation*

$$f(x_1, \ldots, x_n) = (a_{11}x_1 + \ldots + a_{1n}x_n)^2 + \ldots + (a_{n1}x_1 + \ldots + a_{nn}x_n)^2,$$

where the real $n \times n$ *matrix* $[a_{ij}]$ *has non-zero determinant* Δ. *Then there exist integers* $x_1, \ldots, x_n$, *not all zero, such that*

$$f(x_1, \ldots, x_n) \leqslant 4(|\Delta|/\omega_n)^{2/n}.$$

Proof For $r > 0$, the ellipsoid $\{(x_1, \ldots, x_n): f(x_1, \ldots, x_n) \leqslant r^2\}$ is a symmetric convex body with volume $\omega_n r^n/|\Delta|$. Hence this ellipsoid contains a non-zero lattice point if $\omega_n r^n/|\Delta| = 2^n$, i.e. if $r = 2(|\Delta|/\omega_n)^{1/n}$. It follows that there is a non-zero lattice point $(x_1, \ldots, x_n)$ such that $f(x_1, \ldots, x_n) \leqslant 4(|\Delta|/\omega_n)^{2/n}$. □

We conclude the section in style by proving a celebrated theorem of Lagrange which asserts that every non-negative integer is a sum of four perfect squares. First, however, we need the following result of elementary number theory: *for each odd prime p, there exist integers a, b such that* $a^2 + b^2 + 1 \equiv 0 \pmod{p}$. We outline a proof of the result. A simple argument shows that the $\frac{1}{2}(p+1)$ integers x^2 for $x = 0, 1, \ldots, \frac{1}{2}(p-1)$ are incongruent modulo p (i.e. they have *different* remainders when divided by p), as are the $\frac{1}{2}(p+1)$ integers $-1 - y^2$ for $y = 0, 1, \ldots, \frac{1}{2}(p-1)$. Since there are only p *different* remainders possible on division by p, it follows that there are integers a, b such that a^2 and $-1 - b^2$ give the same remainder on division by p, i.e. $a^2 \equiv -1 - b^2 \pmod{p}$, so $a^2 + b^2 + 1 \equiv 0 \pmod{p}$.

Theorem 7.4.6 (Lagrange's four square theorem) *Every non-negative integer is a sum of four perfect squares.*

Proof Since $(a^2 + b^2 + c^2 + d^2)(u^2 + v^2 + w^2 + t^2)$ equals

$$(au - bv - cw - dt)^2 + (av + bu + ct - dw)^2 + (aw - bt + cu + dv)^2 + (at + bw - cv + du)^2,$$

any product of integers, each of which is a sum of four perfect squares, is itself a sum of four perfect squares. Since every integer exceeding 1 is a product of primes and

$$0 = 0^2 + 0^2 + 0^2 + 0^2, \quad 1 = 1^2 + 0^2 + 0^2 + 0^2, \quad 2 = 1^2 + 1^2 + 0^2 + 0^2,$$

it suffices to prove the theorem for an odd prime p.

Let p be an odd prime and let a, b be integers such that $a^2 + b^2 + 1 \equiv 0 \pmod{p}$. Now the ellipsoid

$$\{(x_1, x_2, x_3, x_4): (px_1 + ax_3 + bx_4)^2 + (px_2 + bx_3 - ax_4)^2 + x_3^2 + x_4^2 \leqslant 2p\}$$

in $\mathbb{R}^4$ has volume

$$\frac{\omega_4(\sqrt{(2p)})^4}{p^2} = \frac{(\pi^2/2)(4p^2)}{p^2} = 2\pi^2 > 2^4,$$

whence, by Minkowski's theorem, there exist integers x_1, x_2, x_3, x_4, not all zero, such that

$$0 < (px_1 + ax_3 + bx_4)^2 + (px_2 + bx_3 - ax_4)^2 + x_3^2 + x_4^2 < 2p.$$

But the integer in the centre of these inequalities is

$$p^2(x_1^2 + x_2^2) + (a^2 + b^2 + 1)(x_3^2 + x_4^2) + 2p(ax_1x_3 + bx_1x_4 + bx_2x_3 - ax_2x_4),$$

which is divisible by p. Thus

$$(px_1 + ax_3 + bx_4)^2 + (px_2 + bx_3 - ax_4)^2 + x_3^2 + x_4^2 = p,$$

and p is a sum of four perfect squares. □

Exercises 7.4

1. The line segment L of length 2λ is centred on the origin of $\mathbb{R}^2$. Prove that there is a non-zero lattice point whose distance from L does not exceed $2(\sqrt{(\lambda^2 + \pi)} - \lambda)/\pi$.
2. Denote by A_r the union of all closed circular discs of radius $r > 0$ centred on non-zero lattice points of $\mathbb{R}^2$ whose distance from the origin does not exceed 50. Show that, if $r < 1/\sqrt{(2501)}$, then there exists a line through the origin which does not meet A_r, but that this is not so when $r = 1/50$.
3. Let a/b, c/d, e/f be consecutive terms of some Farey sequence. Prove that $c/d = (a + e)/(b + f)$. Check this result for the Farey sequence $\mathfrak{F}_7$.
4. A Chinese engineer, Tsu Ch'ung-chi (ca. 480), gave the approximation 355/113 to π. Is this a good approximation to π?
5. For $i = 1, \ldots, n$, let $\xi_i = a_{i1}x_1 + \ldots + a_{in}x_n$ be real linear forms with $\Delta = \det[a_{ij}]$ non-zero. Show that there exist integers $x_1, \ldots, x_n$, not all zero, such that

$$|\xi_1| + \ldots + |\xi_n| \leqslant (n!|\Delta|)^{1/n} \quad \text{and} \quad |\xi_1 \ldots \xi_n| \leqslant n^{-n}n!|\Delta|.$$

6. Let p be a prime of the form $4m + 1$. Let q be an integer such that $q^2 + 1 \equiv 0 \pmod{p}$; such a q always exists. Find for which value of $r > 0$ the ellipse

$$\{(x, y) : x^2 + (qx + py)^2 \leqslant r^2\}$$

in $\mathbb{R}^2$ has area 4. Deduce that p is a sum of two perfect squares.

7. Let α and β be irrational numbers. For each $m > 0$, denote by Q_m the parallelepiped defined by the equation

$$Q_m = \{(x, y, z) : |-x + \alpha z| \leqslant 1/\sqrt{m}, \quad |-y + \beta z| \leqslant 1/\sqrt{m}, \quad |z| \leqslant m\}.$$

Show, by considering Q_m, that, for each $\varepsilon > 0$, there exist integers p, q, r with $r > 0$ such that *both* $|\alpha - p/r|$ and $|\beta - q/r|$ are *less than* $\min\{\varepsilon, r^{-3/2}\}$.

7.5 BROUWER'S FIXED-POINT THEOREM

Brouwer's fixed-point theorem is one of mathematics' most celebrated results. It asserts that every continuous mapping $f: C \to C$ of a non-empty compact convex set C in $\mathbb{R}^n$ into itself has a *fixed point*, i.e. there is a point $\mathbf{c}$ of C such that $f(\mathbf{c}) = \mathbf{c}$. The simplicity of the statement of the theorem belies the difficulty of its proof. Since the discovery of the result in 1912, several proofs of it have been given, each one bearing testimony to the ingenuity of mathematicians when chasing a worthy goal. One very attractive feature of the theorem is that it has a host of impressive applications both in pure and applied mathematics.

This section is devoted to a detailed proof of Brouwer's fixed-point theorem and three of its applications. The first of these shows that every non-negative square matrix has a non-negative eigenvalue. The second and third characterize, respectively, those sets in $\mathbb{R}^n$ having the *unique nearest-point property* and those having the *unique farthest-point property*.

As a naïve example of the theorem, consider the following situation. A convex mass of liquid (in a flask) is shaken and then returned to occupy its original position. The overall effect of the shaking can be regarded as a continuous mapping of a compact convex set (the liquid) into itself, whence, by the fixed-point theorem, there exists at least one particle of liquid which will occupy precisely the same position after the shaking that it did originally, a truly remarkable conclusion.

The one-dimensional form of Brouwer's theorem is easily established. Let $f: [a, b] \to [a, b]$ be a continuous mapping of the compact interval $[a, b]$ of the real line, where $a \leqslant b$, into itself. Then the function $g: [a, b] \to \mathbb{R}$ defined by the equation $g(x) = f(x) - x$ (for $a \leqslant x \leqslant b$) is continuous with $g(a) \geqslant 0$, $g(b) \leqslant 0$, whence by the intermediate value theorem there is some real number c in $[a, b]$ satisfying $g(c) = 0$, i.e. $f(c) = c$. No such simple proof exists in dimensions higher than one. Our chosen proof of Brouwer's theorem makes use of barycentric coordinates and is based on a combinatorial lemma of E. Sperner concerning the subdivision of a simplex into smaller simplexes.

We first establish Brouwer's theorem for the *n-simplex* S in $\mathbb{R}^n$, which is the convex hull of the $n+1$ affinely independent points

$$\mathbf{u}_0 = (0, \ldots, 0), \qquad \mathbf{u}_1 = (1, 0, \ldots, 0), \qquad \mathbf{u}_2 = (1, 1, 0, \ldots, 0), \qquad \ldots, \qquad \mathbf{u}_n = (1, \ldots, 1).$$

For each $i = 0, 1, \ldots, n$, denote by F_i the facet of S that does *not* contain the vertex $\mathbf{u}_i$. It is easily verified that

$$S = \{(x_1, \ldots, x_n) : 1 \geqslant x_1 \geqslant \ldots \geqslant x_n \geqslant 0\}, \tag{7.5.1}$$

and that

$$\begin{aligned} F_0 &= \{(x_1, \ldots, x_n) \in S : x_1 = 1\}; \\ F_i &= \{(x_1, \ldots, x_n) \in S : x_i = x_{i+1}\} \qquad \text{for } i = 1, \ldots, n-1; \\ F_n &= \{(x_1, \ldots, x_n) \in S : x_n = 0\}. \end{aligned}$$

For each positive integer m, we subdivide the simplex S into congruent n-simplexes of diameter $\varepsilon\sqrt{n}$, where $\varepsilon = 1/m$. The set V_m of the vertices of these n-simplexes is defined to consist of those points of S of the form $(m_1\varepsilon, \ldots, m_n\varepsilon)$, where $m_1, \ldots, m_n$ are integers. To help us describe which n-simplexes occur in our subdivision of S, we introduce a relation $\prec$ on the vertex set V_m. If $\mathbf{u}, \mathbf{v} \in V_m$, then we write $\mathbf{u} \prec \mathbf{v}$ to mean that $\mathbf{v} - \mathbf{u} = \varepsilon\mathbf{e}_i$ for some elementary vector $\mathbf{e}_i$, i.e. the vectors $\mathbf{u}$ and $\mathbf{v}$ differ only in their ith coordinates, and the ith coordinate of $\mathbf{v}$ exceeds that of $\mathbf{u}$ by ε. Suppose that the points $\mathbf{v}_0, \ldots, \mathbf{v}_n$ of V_m satisfy the relations $\mathbf{v}_0 \prec \mathbf{v}_1, \mathbf{v}_1 \prec \mathbf{v}_2, \ldots, \mathbf{v}_{n-1} \prec \mathbf{v}_n$ and that $\mathbf{v}_n - \mathbf{v}_0 = (\varepsilon, \ldots, \varepsilon)$. Then the set $\{\mathbf{v}_0, \ldots, \mathbf{v}_n\}$ is affinely independent and has diameter $\varepsilon\sqrt{n}$, whence $\operatorname{conv}\{\mathbf{v}_0, \ldots, \mathbf{v}_n\}$ is an n-simplex with diameter $\varepsilon\sqrt{n}$. The collection of all n-simplexes obtained in this way is denoted by D_m, and it is this subdivision of S which interests us here. In $\mathbb{R}^2$ the subdivision D_m of S consists of m^2 isosceles right-angled triangles whose *equal* sides are parallel to the coordinate axes and have length $\varepsilon = 1/m$, and whose hypotenuses have slope $+1$. The subdivision D_4 of S in $\mathbb{R}^2$ is illustrated in Fig. 7.17(a).

We now investigate how the simplexes forming the subdivision D_m of S fit together. Let τ be a facet of some simplex σ of D_m. We ask the question: how many simplexes of D_m have τ as a facet? Figure 7.17 suggests that, in the two-dimensional case, the answer is two unless τ lies in a facet of S, in which case the anwer is one. This is indeed true as we show in Theorem 7.5.1 below. We warn the reader that the proof of the theorem, although by no means difficult, is somewhat tedious and has little intrinsic interest.

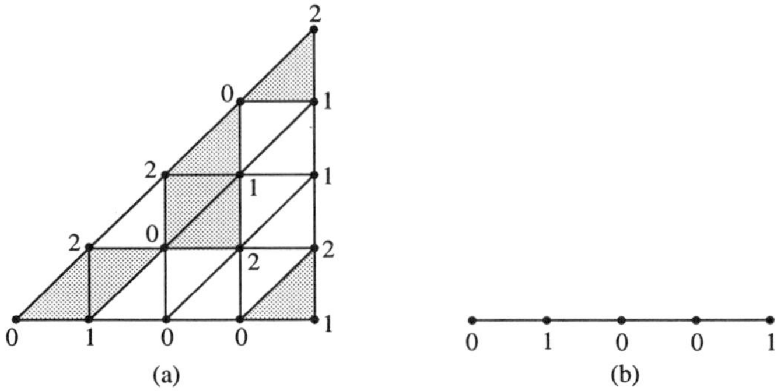

Fig. 7.17

Theorem 7.5.1 *Let τ be a facet of some simplex σ of D_m. Then τ is a facet of precisely two simplexes of D_m, unless τ lies in a facet of S, in which case it is only a facet of σ.*

Proof Let σ have vertices $\mathbf{v}_0, \ldots, \mathbf{v}_n$ in V_m, where $\mathbf{v}_0 \prec \mathbf{v}_1, \ldots, \mathbf{v}_{n-1} \prec \mathbf{v}_n$ and $\mathbf{v}_n - \mathbf{v}_0 = (\varepsilon, \ldots, \varepsilon)$. Now τ must be one of the following forms:

(i) $\operatorname{conv}\{\mathbf{v}_1, \ldots, \mathbf{v}_n\}$;

(ii) $\operatorname{conv}\{\mathbf{v}_0, \ldots, \mathbf{v}_{i-1}, \mathbf{v}_{i+1}, \ldots, \mathbf{v}_n\}$ for $i = 1, \ldots, n-1$;

(iii) $\operatorname{conv}\{\mathbf{v}_0, \ldots, \mathbf{v}_{n-1}\}$.

We consider each case in turn, making frequent use of the characterization of S given by (7.5.1). We denote by $(\mathbf{v})_k$ the kth coordinate of $\mathbf{v}$.

Case (i) Now τ will be a facet of precisely two members of D_m if $\mathbf{v}_1 + (\varepsilon, \ldots, \varepsilon) \in V_m$ and will be a facet of just one member of D_m if $\mathbf{v}_1 + (\varepsilon, \ldots, \varepsilon) \notin V_m$. If τ lies in a facet of S, then this facet must clearly be F_0, whence $\mathbf{v}_1 + (\varepsilon, \ldots, \varepsilon) \notin V_m$. If $\mathbf{v}_1 + (\varepsilon, \ldots, \varepsilon) \notin V_m$, then its first coordinate must exceed 1, which implies that $(\mathbf{v}_k)_1 = 1$ for $k = 1, \ldots, n$, whence τ lies in F_0.

Case (ii) Now τ will be a facet of precisely two members of D_m if $\mathbf{v}_{i-1} + \mathbf{v}_{i+1} - \mathbf{v}_i \in V_m$ and it will be a facet of just one member of D_m if $\mathbf{v}_{i-1} + \mathbf{v}_{i+1} - \mathbf{v}_i \notin V_m$.

Suppose that $\tau \subseteq F_j$ for some j. Since $\mathbf{v}_0, \mathbf{v}_n \in F_j$ and $\mathbf{v}_n - \mathbf{v}_0 = (\varepsilon, \ldots, \varepsilon)$, we must have $j \in \{1, \ldots, n-1\}$. Thus each of $\mathbf{v}_0, \ldots, \mathbf{v}_{i-1}$ has its jth and $(j+1)$th coordinates equal. Since, as we progress through this sequence, only one coordinate changes at a time, it follows that

$$(\mathbf{v}_0)_j = (\mathbf{v}_0)_{j+1} = \ldots = (\mathbf{v}_{i-1})_j = (\mathbf{v}_{i-1})_{j+1}.$$

Similarly,

$$(\mathbf{v}_{i+1})_j = (\mathbf{v}_{i+1})_{j+1} = \ldots = (\mathbf{v}_n)_j = (\mathbf{v}_n)_{j+1}.$$

This shows that either $\mathbf{v}_i - \mathbf{v}_{i-1} = \varepsilon\mathbf{e}_j$ and $\mathbf{v}_{i+1} - \mathbf{v}_i = \varepsilon\mathbf{e}_{j+1}$ or else $\mathbf{v}_i - \mathbf{v}_{i-1} = \varepsilon\mathbf{e}_{j+1}$ and $\mathbf{v}_{i+1} - \mathbf{v}_i = \varepsilon\mathbf{e}_j$. The second of these possibilities can be discounted, because it leads to $(\mathbf{v}_i)_{j+1} > (\mathbf{v}_i)_j$. We thus have $\mathbf{v}_{i-1} + \mathbf{v}_{i+1} - \mathbf{v}_i = \mathbf{v}_{i-1} + \varepsilon\mathbf{e}_{j+1}$ and the $(j+1)$th coordinate of this vector exceeds its jth. Thus $\mathbf{v}_{i-1} + \mathbf{v}_{i+1} - \mathbf{v}_i \notin V_m$.

Suppose now that $\mathbf{v}_{i-1} + \mathbf{v}_{i+1} - \mathbf{v}_i \notin V_m$. Let $\mathbf{v}_i - \mathbf{v}_{i-1} = \varepsilon\mathbf{e}_j$. We show that $\tau \subseteq F_j$. To see this, we note that $\mathbf{v}_{i+1} \in V_m$, but $\mathbf{v}_{i+1} - \varepsilon\mathbf{e}_j \notin V_m$. If $j < n$, this tells us that $(\mathbf{v}_{i+1})_j = (\mathbf{v}_{i+1})_{j+1}$. On the other hand, if $j = n$, we can infer that $(\mathbf{v}_{i+1})_n = 0$, whence

$$(v_{i-1})_n < (v_i)_n = (v_{i+1})_n = 0,$$

which shows that the situation with $j = n$ cannot occur. Thus $j < n$. Now $(\mathbf{v}_{i-1})_j \geqslant (\mathbf{v}_{i-1})_{j+1}$, and therefore, since $\mathbf{v}_i = \mathbf{v}_{i-1} + \varepsilon\mathbf{e}_j$, we have $(\mathbf{v}_i)_j > (\mathbf{v}_i)_{j+1}$. But $(\mathbf{v}_{i+1})_j = (\mathbf{v}_{i+1})_{j+1}$, so we conclude that $\mathbf{v}_{i+1} = \mathbf{v}_i + \varepsilon\mathbf{e}_{j+1}$. Thus $\mathbf{v}_i = \mathbf{v}_{i-1} + \varepsilon\mathbf{e}_j$, $\mathbf{v}_{i+1} = \mathbf{v}_i + \varepsilon\mathbf{e}_{j+1}$, $(\mathbf{v}_{i+1})_j = (\mathbf{v}_{i+1})_{j+1}$. It follows that $(\mathbf{v}_k)_j = (\mathbf{v}_k)_{j+1}$ for all $k \neq i$, whence $\tau \subseteq F_j$. This disposes of case (ii).

Case (iii) This is similar to case (i). It will be found that τ is a facet of two members of D_m unless τ lies in F_n, in which case it is a facet of just one member of D_m. □

Let $\lambda_0, \ldots, \lambda_n : \mathbb{R}^n \to \mathbb{R}$ be the barycentric coordinate functions determined by the vertices $\mathbf{u}_0, \ldots, \mathbf{u}_n$ of the n-simplex S. Then each point $\mathbf{x}$ of $\mathbb{R}^n$ can be written *uniquely* in the form

$$\mathbf{x} = \lambda_0(\mathbf{x})\mathbf{u}_0 + \ldots + \lambda_n(\mathbf{x})\mathbf{u}_n, \qquad \text{where} \qquad \lambda_0(\mathbf{x}) + \ldots + \lambda_n(\mathbf{x}) = 1,$$

and $\mathbf{x} \in S$ if and only if $\lambda_0(\mathbf{x}), \ldots, \lambda_n(\mathbf{x}) \geqslant 0$. Let $\Lambda : V_m \to \{0, 1, \ldots, n\}$ be a mapping such that, if $\Lambda(\mathbf{v}) = i$ for some $\mathbf{v} \in V_m$, then $\lambda_i(\mathbf{v}) > 0$. Then Λ is called a *labelling* of V_m. Thus a labelling of V_m is a rule which attaches to each vertex $\mathbf{v}$ of V_m an integer i such that the ith barycentric coordinate $\lambda_i(\mathbf{v})$ of $\mathbf{v}$ is positive. It follows immediately from the definition of Λ that $\Lambda(\mathbf{u}_0) = 0$, $\Lambda(\mathbf{u}_1) = 1, \ldots, \Lambda(\mathbf{u}_n) = n$. An example of a labelling of V_4 in $\mathbb{R}^2$ is illustrated in Fig. 7.17 (a). By the *label set* of a simplex with vertices in V_m is meant the set of labels of its vertices, i.e. the set of images of the vertices of the simplex for the labelling under consideration. We are interested in the number of simplexes of D_m which have $\{0, 1, \ldots, n\}$ for their label set. The simplexes in Fig. 7.17 (a) which have label set $\{0, 1, 2\}$ are shaded—there are seven of them. In proving Brouwer's theorem, we use the fact that, for any labelling of V_m, there is always a simplex of D_m whose label set is $\{0, 1, \ldots, n\}$. This result follows from the following combinatorial lemma.

Theorem 7.5.2 (Sperner's lemma) *Let* $\Lambda: V_m \to \{0, 1, \ldots, n\}$ *be a labelling of* V_m. *Then there is an odd number of simplexes of* D_m *having label set* $\{0, 1, \ldots, n\}$.

Proof We argue by induction on the dimension n. When $n = 1$, $S = [0, 1] \subseteq \mathbb{R}$ and

$$V_m = \{0, 1/m, \ldots, (m-1)/m, 1\}, \; D_m = \{[0, 1/m], \ldots, [(m-1)/m, 1]\}.$$

The number of 1-simplexes in D_m having label set $\{0, 1\}$ is the number of times there is a change from 0 to 1 or 1 to 0 in the finite sequence of labels $\Lambda(0), \Lambda(1/m), \ldots, \Lambda(1)$. But $\Lambda(0) = 0$, $\Lambda(1) = 1$, so this number must be odd. This proves the lemma when $n = 1$.

Suppose now that $n \geqslant 2$ and that the assertion is true for dimension $n - 1$. For each simplex σ of D_m, denote by $\Phi(\sigma)$ the number of facets of σ having label set $\{0, 1, \ldots, n-1\}$. Then

$$\Phi(\sigma) = \begin{cases} 1 & \text{if } \sigma \text{ has label set } \{0, 1, \ldots, n\} \\ 0 \text{ or } 2 & \text{otherwise.} \end{cases}$$

Hence

$$\sum_{\sigma \in D_m} \Phi(\sigma) = a + 2b,$$

where a is the number of simplexes of D_m having label set $\{0, 1, \ldots, n\}$ and b is an integer. We now evaluate this sum in a different way. Suppose that τ is a facet of some simplex of D_m and that τ has label set $\{0, 1, \ldots, n-1\}$. Then either: (i) τ *lies in the facet* F_n *of* S in which case τ is a facet of only one simplex of D_m, and so is counted just once in the summation above; or (ii) τ *lies in no facet of* S in which case τ is a facet of two simplexes of D_m, and so τ is counted twice in the summation above. Here we have used Theorem 7.5.1 and the observation that the only facet of S which could contain τ is F_n.

Suppose that there are c facets of type (i) and d facets of type (ii). It follows from the remarks in the last paragraph that

$$\sum_{\sigma \in D_m} \Phi(\sigma) = a + 2b = c + 2d.$$

We complete the proof by showing that the number $a = c + 2d - 2b$ of simplexes of D_m with label set $\{0, 1, \ldots, n\}$ is odd. To do this, we use the induction hypothesis to show that the number c is odd.

We denote by S', V'_m, D'_m the counterparts in $\mathbb{R}^{n-1}$ of S, V_m, D_m. The labelling Λ of V_m induces, in the natural way, a labelling Λ' of V'_m. For example, the labelling illustrated in Fig. 7.17 (a) induces the labelling of V'_m in $\mathbb{R}^1$ shown in Figure 7.17 (b). We now make these ideas precise. The bijection $\theta: S' \to F_n$ defined by the equation

$$\theta(x_1, \ldots, x_{n-1}) = (x_1, \ldots, x_{n-1}, 0) \qquad \text{for } (x_1, \ldots, x_{n-1}) \in S'$$

preserves vertices in the sense that $\mathbf{y} \in V'_m$ if and only if $\theta(\mathbf{y}) \in V_m$. Also $\operatorname{conv}\{\mathbf{y}_0, \ldots, \mathbf{y}_{n-1}\} \in D'_m$ if and only if $\operatorname{conv}\{\theta(\mathbf{y}_0), \ldots, \theta(\mathbf{y}_{n-1})\}$ is a facet of some simplex of D_m. The labelling Λ of V_m induces a labelling Λ' of V'_m by defining $\Lambda'(\mathbf{y}) = \Lambda(\theta(\mathbf{y}))$ for each $\mathbf{y} \in V'_m$. Thus the number c of facets of simplexes of D_m which lie in F_n and have label set $\{0, 1, \ldots, n-1\}$ is the same as the number of simplexes of D'_m which have label set $\{0, 1, \ldots, n-1\}$ for the labelling Λ' of V'_m. The latter number c is odd by the induction hypothesis, as desired. This completes the proof by induction. □

Theorem 7.5.3 (Brouwer's fixed-point theorem) *Let $f: C \to C$ be a continuous mapping of a non-empty compact convex set C in $\mathbb{R}^n$ into itself. Then there exists a point $\mathbf{c}$ of C such that $f(\mathbf{c}) = \mathbf{c}$.*

Proof We first discuss the special case when $f: S \to S$ is a continuous mapping of the n-simplex S (considered above) into itself. For each $i = 0, 1, \ldots, n$, denote by A_i the set of those points of S whose ith barycentric coordinates are not increased by f, i.e. (using our earlier notation)

$$A_i = \{\mathbf{x} \in S : (\lambda_i \circ f)(\mathbf{x}) \leqslant \lambda_i(\mathbf{x})\}.$$

Since the barycentric coordinate functions λ_i and the function f are continuous, the sets A_i are closed.

We now use the sets A_i to define a labelling Λ of V_m. Let $\mathbf{v} \in V_m$. Then the barycentric coordinates of $\mathbf{v}$ which are positive have sum 1, and the barycentric coordinates of $f(\mathbf{v})$ are non-negative and have sum 1. Thus there exists an integer i such that *both* $\lambda_i(\mathbf{v}) > 0$ and $\lambda_i(f(\mathbf{v})) \leqslant \lambda_i(\mathbf{v})$. We define $\Lambda(\mathbf{v})$ to be any integer i for which the latter inequalities hold, so $\mathbf{v} \in A_i$. By Sperner's lemma, there is some simplex σ_m of D_m having label set $\{0, 1, \ldots, n\}$.

For each positive integer m, let $\mathbf{x}_m \in \sigma_m$. Then the sequence $\mathbf{x}_1, \ldots, \mathbf{x}_m, \ldots$ lies in the compact simplex S, and so it contains some subsequence, $\mathbf{x}_{i_1}, \ldots, \mathbf{x}_{i_m}, \ldots$, say, which converges to a point $\mathbf{s}$ of S. Since the simplex σ_{i_m} has diameter $\sqrt{n}/i_m$ and has label set $\{0, 1, \ldots, n\}$, it follows that, for each $j = 0, 1, \ldots, n$, there is a vertex $\mathbf{v}_{i_m j}$ of σ_{i_m} lying in A_j such that $\|\mathbf{x}_{i_m} - \mathbf{v}_{i_m j}\| \leqslant \sqrt{n}/i_m$. Thus, for each $j = 0, 1, \ldots, n$, $\mathbf{v}_{i_m j} \to \mathbf{s}$ as $m \to \infty$. Since each set A_j is closed, each A_j contains $\mathbf{s}$. Hence

$$\lambda_0(f(\mathbf{s})) + \ldots + \lambda_n(f(\mathbf{s})) = \lambda_0(\mathbf{s}) + \ldots + \lambda_n(\mathbf{s}) = 1$$

and

$$\lambda_0(f(\mathbf{s})) \leqslant \lambda_0(\mathbf{s}), \ldots, \lambda_n(f(\mathbf{s})) \leqslant \lambda_n(\mathbf{s}),$$

whence

$$\lambda_0(\mathbf{s}) = \lambda_0(f(\mathbf{s})), \ldots, \lambda_n(\mathbf{s}) = \lambda_n(f(\mathbf{s})).$$

Thus $f(\mathbf{s}) = \mathbf{s}$, which proves the theorem for the special case when C is the n-simplex S.

Consider now the general case of a continuous mapping $f: C \to C$ of a non-empty compact convex set C into itself. Let $g: \mathbb{R}^n \to \mathbb{R}^n$ be a *homothety*, i.e. a mapping of the form $g(\mathbf{x}) = \lambda\mathbf{x} + \mathbf{a}$, where $\lambda > 0$ and $\mathbf{a} \in \mathbb{R}^n$, such that $g(C) \subseteq S$. Let $h: S \to g(C)$ be the restriction to S of the projection operator of $g(C)$. Then $g \circ f \circ g^{-1} \circ h: S \to S$ is a continuous mapping of S into itself. By the first part of the proof, there is an $\mathbf{x} \in S$ such that $g(f(g^{-1}(h(\mathbf{x})))) = \mathbf{x}$. It follows that $\mathbf{x} \in g(C)$, whence $h(\mathbf{x}) = \mathbf{x}$ and $f(g^{-1}(\mathbf{x})) = g^{-1}(\mathbf{x})$. Thus f has a fixed point $\mathbf{c} = g^{-1}(\mathbf{x})$. □

We should perhaps mention that there are many non-convex sets A in $\mathbb{R}^n$ which have the property that, for every continuous mapping $f: A \to A$ of A into itself, there is a point $\mathbf{a}$ of A such that $f(\mathbf{a}) = \mathbf{a}$. This idea is taken up in the first two exercises at the end of this section.

We now use Brouwer's fixed-point theorem to prove a classical result of O. Perron and G. Frobenius on non-negative square matrices.

Theorem 7.5.4 (Perron–Frobenius) *Let* $\mathbf{A}$ *be a non-negative square matrix of order n. Then* $\mathbf{A}$ *has a non-negative eigenvalue.*

Proof The theorem is trivial when $\det \mathbf{A} = 0$, for then 0 is an eigenvalue of $\mathbf{A}$. Suppose, then, that $\det \mathbf{A} \neq 0$. Let C be the $(n-1)$-simplex in $\mathbb{R}^n$ defined by the equation

$$C = \{(x_1 \ldots, x_n) : x_1, \ldots, x_n \geqslant 0, \quad x_1 + \ldots + x_n = 1\}.$$

Let continuous linear mappings $g: \mathbb{R}^n \to \mathbb{R}^n$ and $h: \mathbb{R}^n \to \mathbb{R}$ be specified by $g(\mathbf{x}) = \mathbf{A}\mathbf{x}$, and $h(\mathbf{x})$ equals the sum of the coordinates of $g(\mathbf{x})$ for $\mathbf{x} \in \mathbb{R}^n$. If $\mathbf{x} \geqslant \mathbf{0}$ and $h(\mathbf{x}) = 0$, then, using the fact that $\mathbf{A}$ is non-negative, we can deduce that $\mathbf{A}\mathbf{x} = \mathbf{0}$; but $\det \mathbf{A} \neq 0$, so $\mathbf{x} = \mathbf{0}$. This shows that h is *positive* on C. It now follows that the equation $f(\mathbf{x}) = g(\mathbf{x})/h(\mathbf{x})$ for $\mathbf{x} \in C$ defines a continuous mapping $f: C \to C$ of C into itself. By Brouwer's theorem, there exists a $\mathbf{c} \in C$ with $f(\mathbf{c}) = \mathbf{c}$, i.e. $\mathbf{A}\mathbf{c} = h(\mathbf{c})\mathbf{c}$ and $\mathbf{c}$ is an eigenvector of $\mathbf{A}$ associated with the positive eigenvalue $h(\mathbf{c})$. □

As a second application of Brouwer's theorem, we prove a classical result about convex sets discovered by T. S. Motzkin in 1935. A set C in $\mathbb{R}^n$ is called a **Chebyshev set** if, for each point $\mathbf{x}$ of $\mathbb{R}^n$, there is a *unique* nearest point of C to $\mathbf{x}$. In Section 2.4 we showed that every non-empty closed convex set in $\mathbb{R}^n$ is a Chebyshev set. Motzkin's theorem asserts that the converse of this result is true, namely that *every Chebyshev set in* $\mathbb{R}^n$ *is non-empty, closed, and convex.*

Before proving Motzkin's theorem, we establish some properties of

Chebyshev sets. Let C be a Chebyshev set in $\mathbb{R}^n$. Certainly C is non-empty. For each point $\mathbf{x}$ in $\mathbb{R}^n$, denote by $f(\mathbf{x})$ the unique nearest point of C to $\mathbf{x}$. In this way we define a mapping $f\colon \mathbb{R}^n \to C$ called the *projection operator* of C, referring to the point $f(\mathbf{x})$ as the *projection* of $\mathbf{x}$ onto C. We show first that C is closed. Let $\mathbf{x}_1 \ldots, \mathbf{x}_k, \ldots$ be a sequence of points of C which converges to some point $\mathbf{x}$ of $\mathbb{R}^n$. Then, since $f(\mathbf{x})$ is the nearest point of C to $\mathbf{x}$,

$$\|\mathbf{x} - f(\mathbf{x})\| \leqslant \|\mathbf{x} - \mathbf{x}_k\| \qquad \text{for } k = 1, 2, \ldots.$$

Letting $k \to \infty$ in this inequality, we deduce that $\|\mathbf{x} - f(\mathbf{x})\| = 0$, whence $f(\mathbf{x}) = \mathbf{x}$. But $f(\mathbf{x}) \in C$ by the definition of f, so $\mathbf{x} \in C$ and C is closed. We now show that the projection operator f of the Chebyshev set C is continuous. First we note that the distance function $d_C\colon \mathbb{R}^n \to \mathbb{R}$ of C is continuous, as was proved in Section 1.9; here $d_C(\mathbf{x}) = \|\mathbf{x} - f(\mathbf{x})\|$ for each $\mathbf{x}$ in $\mathbb{R}^n$. Let $\mathbf{x}_1, \ldots, \mathbf{x}_k, \ldots$ be a sequence of points of $\mathbb{R}^n$ that converges to some point $\mathbf{x}$ of $\mathbb{R}^n$. The inequality

$$\|f(\mathbf{x}_k)\| \leqslant \|f(\mathbf{x}_k) - \mathbf{x}_k\| + \|\mathbf{x}_k\| = d_C(\mathbf{x}_k) + \|\mathbf{x}_k\|$$

shows that the sequence $f(\mathbf{x}_1), \ldots, f(\mathbf{x}_k), \ldots$ is bounded, since $d_C(\mathbf{x}_k) \to d_C(\mathbf{x})$, $\|\mathbf{x}_k\| \to \|\mathbf{x}\|$ as $k \to \infty$ and convergent sequences of real numbers are bounded. We prove the continuity of f by showing that $f(\mathbf{x}_k) \to f(\mathbf{x})$ as $k \to \infty$. Suppose that this is not the case. Then the boundedness of the sequence $f(\mathbf{x}_1), \ldots, f(\mathbf{x}_k), \ldots$ shows that there exists a subsequence $\mathbf{x}_{i_1}, \ldots, \mathbf{x}_{i_k}, \ldots$ of $\mathbf{x}_1, \ldots, \mathbf{x}_k, \ldots$ and a point $\mathbf{y}$ of $C \backslash \{f(\mathbf{x})\}$ such that the sequence $f(\mathbf{x}_{i_1}), \ldots, f(\mathbf{x}_{i_k}), \ldots$ converges to $\mathbf{y}$. Since d_C is continuous,

$$\|\mathbf{x} - f(\mathbf{x})\| = \lim_{k \to \infty} \|\mathbf{x}_{i_k} - f(\mathbf{x}_{i_k})\| = \|\mathbf{x} - \mathbf{y}\|.$$

But C is a Chebyshev set and $\mathbf{y} \in C$, so $\mathbf{y} = f(\mathbf{x})$. This contradiction shows that $f(\mathbf{x}_k) \to f(\mathbf{x})$ as $k \to \infty$, thus establishing the continuity of f.

Again let C be a Chebyshev set in $\mathbb{R}^n$ with projection operator f. Suppose that $\mathbf{x}$ is a point of $\mathbb{R}^n$ *not* lying in C and that $\mathbf{y} = f(\mathbf{x})$ is the projection of $\mathbf{x}$ onto C. Let $\mathbf{z}$ be any point on the line segment joining $\mathbf{x}$ and $\mathbf{y}$. Then $\mathbf{z} = \lambda\mathbf{x} + (1 - \lambda)\mathbf{y}$ for some λ satisfying $0 \leqslant \lambda \leqslant 1$; see Fig. 7.18. Now

$$\begin{aligned}
\|\mathbf{x} - f(\mathbf{x})\| &\leqslant \|\mathbf{x} - f(\mathbf{z})\| \\
&\leqslant \|\mathbf{x} - \mathbf{z}\| + \|\mathbf{z} - f(\mathbf{z})\| \\
&\leqslant \|\mathbf{x} - \mathbf{z}\| + \|\mathbf{z} - f(\mathbf{x})\| \\
&= (1 - \lambda)\|\mathbf{x} - f(\mathbf{x})\| + \lambda\|\mathbf{x} - f(\mathbf{x})\| \\
&= \|\mathbf{x} - f(\mathbf{x})\|,
\end{aligned}$$

from which we can deduce that $\|\mathbf{x} - f(\mathbf{x})\| = \|\mathbf{x} - f(\mathbf{z})\|$, whence $f(\mathbf{x}) = f(\mathbf{z})$, since C is a Chebyshev set. We have thus shown that all

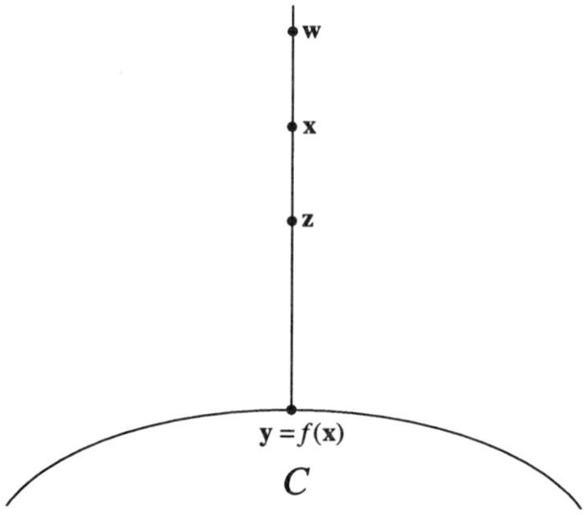

Fig. 7.18

points on the line segment joining $\mathbf{x}$ and $\mathbf{y}$ have the same projection onto C as $\mathbf{x}$ does. We now consider the projections onto C of points on the halfline starting at $\mathbf{x}$ away from $\mathbf{y}$. A Chebyshev set C is called a **sun** if, for each point $\mathbf{x}$ of $\mathbb{R}^n$ not in C, all points $\mathbf{w}$ on the halfline starting at $\mathbf{x}$ away from $\mathbf{y} = f(\mathbf{x})$ have the same projection onto C as $\mathbf{x}$ does. See Fig. 7.18. Corollary 2.4.3 shows that every non-empty closed convex set in $\mathbb{R}^n$ is a sun. In the proof of Theorem 7.5.5 below, we show that every Chebyshev set is a sun and that every sun is convex, thus establishing Motzkin's theorem.

Theorem 7.5.5 (Motzkin's theorem) *Every Chebyshev set in* $\mathbb{R}^n$ *is non-empty, closed, and convex,*

Proof Let C be a Chebyshev set in $\mathbb{R}^n$ with projection operator f. We already know that C is non-empty and closed , so we need only show that C is convex.

First we show that C is a sun. To do this we argue by contradiction. Suppose that C is not a sun. Then it follows easily from the properties of f established earlier, that there is a point $\mathbf{x}$ of $\mathbb{R}^n \setminus C$ such that *it* is the only point on the halfline L starting at $\mathbf{x}$ away from $\mathbf{y} = f(\mathbf{x})$ which has $\mathbf{y}$ for its projection onto C. Denote by B the closed ball with centre $\mathbf{x}$ and radius $\|\mathbf{x} - \mathbf{y}\|$. Define a mapping $g: B \to B$ by the equation

$$g(\mathbf{z}) = \mathbf{x} + \frac{\|\mathbf{x} - \mathbf{y}\|}{\|\mathbf{x} - f(\mathbf{z})\|} (\mathbf{x} - f(\mathbf{z})) \qquad \text{for } \mathbf{z} \in B.$$

The continuity of f shows that g is continuous. By Brouwer's theorem, there is a point $\mathbf{z}_0$ of B such that $g(\mathbf{z}_0) = \mathbf{z}_0$. The definition of g shows that

$$\mathbf{x} = \frac{\|\mathbf{x} - f(\mathbf{z}_0)\|\mathbf{z}_0 + \|\mathbf{x} - \mathbf{y}\| f(\mathbf{z}_0)}{\|\mathbf{x} - f(\mathbf{z}_0)\| + \|\mathbf{x} - \mathbf{y}\|}. \tag{7.5.2}$$

This equation shows that $\mathbf{x}$ lies on the line segment joining $\mathbf{z}_0$ and its projection $f(\mathbf{z}_0)$ onto C, whence $\mathbf{x}$ and $\mathbf{z}_0$ have the same projection onto C. Thus $\mathbf{y} = f(\mathbf{z}_0)$. Putting $\mathbf{y} = f(\mathbf{z}_0)$ in (7.5.2), we find that $\mathbf{z}_0 = 2\mathbf{x} - \mathbf{y}$. Thus $\mathbf{z}_0$ is a point on the halfline L other than $\mathbf{x}$ which has $\mathbf{y}$ for its projection onto C. This contradicts our choice of $\mathbf{x}$. Thus C must be a sun.

We complete the proof by showing that the sun C is convex. Let $\mathbf{x}$, $\mathbf{y} \in C$ and let $\mathbf{z} = \lambda\mathbf{x} + (1 - \lambda)\mathbf{y}$, where $0 \leqslant \lambda \leqslant 1$. Since C is a sun, each point $\mathbf{z} + \theta(\mathbf{z} - f(\mathbf{z}))$ for $\theta > 0$ has $f(\mathbf{z})$ for its projection on C. Hence

$$\|\mathbf{z} + \theta(\mathbf{z} - f(\mathbf{z})) - \mathbf{x}\|^2 \geqslant \|\mathbf{z} + \theta(\mathbf{z} - f(\mathbf{z})) - f(\mathbf{z})\|^2$$

and

$$\|\mathbf{z} - \mathbf{x}\|^2/\theta + 2(\mathbf{z} - f(\mathbf{z}))\cdot(\mathbf{z} - \mathbf{x}) \geqslant \|\mathbf{z} - f(\mathbf{z})\|^2/\theta + 2(\mathbf{z} - f(\mathbf{z}))\cdot(\mathbf{z} - f(\mathbf{z})).$$

Letting $\theta \to \infty$ in this inequality, we deduce that

$$(\mathbf{z} - f(\mathbf{z})) \cdot (\mathbf{z} - \mathbf{x}) \geqslant \|\mathbf{z} - f(\mathbf{z})\|^2.$$

Similarly,

$$(\mathbf{z} - f(\mathbf{z})) \cdot (\mathbf{z} - \mathbf{y}) \geqslant \|\mathbf{z} - f(\mathbf{z})\|^2.$$

Multiplying the last two inequalities, respectively, by λ and $1 - \lambda$, and adding the results together, we find that

$$0 \geqslant \|\mathbf{z} - f(\mathbf{z})\|^2.$$

Thus $\mathbf{z} = f(\mathbf{z}) \in C$, and so C is convex. □

A point $\mathbf{a}_0$ of a set A in $\mathbb{R}^n$ is called a **farthest point** of A from a point $\mathbf{x}$ of $\mathbb{R}^n$ if

$$\|\mathbf{x} - \mathbf{a}_0\| = \sup\{\|\mathbf{x} - \mathbf{a}\| : \mathbf{a} \in A\}.$$

The set A is said to have the **unique farthest-point property** if, for each point $\mathbf{x}$ of $\mathbb{R}^n$, there exists a *unique* farthest point of A from $\mathbf{x}$. Suppose that a set A in $\mathbb{R}^n$ has the unique farthest-point property and that, for each point $\mathbf{x}$ of $\mathbb{R}^n$, the unique farthest point of A from $\mathbf{x}$ is denoted by $f(\mathbf{x})$. In this way we define a mapping $f: \mathbb{R}^n \to A$, which is called the **farthest-point mapping** associated with A. The reader will easily find one example of a set having the unique farthest-point property, namely any singleton set. That other examples are difficult to find is explained by the next result.

Theorem 7.5.6 *The only sets in $\mathbb{R}^n$ that have the unique farthest-point property are the singleton sets.*

Proof Let A be a set in $\mathbb{R}^n$ which has the unique farthest-point property. Let $f:\mathbb{R}^n \to A$ be the farthest-point mapping associated with A. We prove the theorem by showing that f is continuous and then applying Brouwer's theorem to the restriction of f to the set conv (cl A).

Suppose that f is *not* continuous. Then there is a sequence $\mathbf{x}_1,\ldots,\mathbf{x}_k,\ldots$ of points of $\mathbb{R}^n$ converging to a point $\mathbf{x}$ such that the sequence $f(\mathbf{x}_1),\ldots,f(\mathbf{x}_k),\ldots$ does *not* converge to $f(\mathbf{x})$. Since the set A is clearly bounded, there exists a subsequence $\mathbf{x}_{i_1},\ldots,\mathbf{x}_{i_k},\ldots$ of $\mathbf{x}_1,\ldots,\mathbf{x}_k,\ldots$ such that the sequence $f(\mathbf{x}_{i_1}),\ldots f(\mathbf{x}_{i_k}),\ldots$ converges to some point $\mathbf{y}$ *other than* $f(\mathbf{x})$. Clearly $\mathbf{y}\in \mathrm{cl}\,A$. For each $k = 1,2,\ldots,$

$$\|\mathbf{x}-f(\mathbf{x}_{i_k})\| \leqslant \|\mathbf{x}-f(\mathbf{x})\| \qquad \text{and} \qquad \|\mathbf{x}_{i_k}-f(\mathbf{x})\| \leqslant \|\mathbf{x}_{i_k}-f(\mathbf{x}_{i_k})\|.$$

Letting $k\to\infty$ in these inequalities, we find that

$$\|\mathbf{x}-\mathbf{y}\| \leqslant \|\mathbf{x}-f(\mathbf{x})\| \leqslant \|\mathbf{x}-\mathbf{y}\|,$$

whence $\|\mathbf{x}-f(\mathbf{x})\| = \|\mathbf{x}-\mathbf{y}\|$. Since $f(\mathbf{x})$ and $\mathbf{y}$ are *distinct* equidistant points from $\mathbf{x}$, and $f(\mathbf{x})$ is the unique farthest point of A from $\mathbf{x}$, it follows that $\mathbf{y}\notin A$. Let $\mathbf{z} = 2\mathbf{x}-\mathbf{y}$, the *diametrically opposite* point to $\mathbf{y}$ in the closed ball $B[\mathbf{x};r]$, where $r = \|\mathbf{x}-f(\mathbf{x})\|$. Now $A \subseteq B[\mathbf{x};r]$, and the unique farthest point of $B[\mathbf{x};r]$ from $\mathbf{z}$ is $\mathbf{y}$. Since $y\notin A$, $\|\mathbf{z}-f(\mathbf{z})\| < 2r$. Thus A, and hence cl A, lies in $B[\mathbf{z};s]$, where $s = \|\mathbf{z}-f(\mathbf{z})\|$. But $\mathbf{y}\in\mathrm{cl}\,A$, and so $\|\mathbf{y}-\mathbf{z}\| \leqslant s < 2r$. This, however, contradicts the fact that $\|\mathbf{y}-\mathbf{z}\| = 2r$. Hence f must be continuous.

The restriction of f to the non-empty compact convex set conv (cl A) is continuous and maps the set to itself. By Brouwer's theorem, there is a point $\mathbf{a}$ of conv (cl A) such that $f(\mathbf{a}) = \mathbf{a}$. Thus $\mathbf{a}$ is the farthest point of A from $\mathbf{a}$, which shows that A must be the singleton set $\{\mathbf{a}\}$. □

Exercises 7.5

1. A set C in $\mathbb{R}^n$ is said to have the *fixed-point property* if every continuous mapping $f:C\to C$ of C into itself has a fixed point, i.e. a point $\mathbf{c}$ of C such that $f(\mathbf{c}) = \mathbf{c}$. Which of the following sets have the fixed-point property?
   (i) $\{\mathbf{x}\in\mathbb{R}^n:\|\mathbf{x}\| \leqslant 1\}$;
   (ii) $\{\mathbf{x}\in\mathbb{R}^n:\|\mathbf{x}\| < 1\}$;
   (iii) $\{\mathbf{x}\in\mathbb{R}^n:\|\mathbf{x}\| = 1\}$.
2. (a) Two sets A and B in $\mathbb{R}^n$ are said to be *topologically equivalent* if there exists a continuous bijection $f:A\to B$ which has a continuous inverse $f^{-1}:B\to A$. Show that, if A and B are topologically equivalent sets in $\mathbb{R}^n$ and A has the fixed-point property, then so too does B.

   (b) A set B is said to be a *retract* of a set A in $\mathbb{R}^n$ if $B\subseteq A$ and there exists a continuous mapping $f:A\to B$ such that $f(\mathbf{b}) = \mathbf{b}$ for all $\mathbf{b}$ in B. Prove that every retract of a set in $\mathbb{R}^n$ which has the fixed-point property, itself has the fixed-point property.

(c) Use (a), (b) above to: (i) give an example of a non-convex set which has the fixed-point property; (ii) show that the set $\{\mathbf{x} \in \mathbb{R}^n : \|\mathbf{x}\| = 1\}$ is not a retract of the unit ball $\{\mathbf{x} \in \mathbb{R}^n : \|\mathbf{x}\| \leqslant 1\}$.

3. Show that every convex set in $\mathbb{R}^n$ which has the fixed-point property must be compact. Give an example of a set in $\mathbb{R}^n$ which has the fixed-point property, but which is neither closed, bounded, nor convex.

7.6 CONVEX BODIES OF CONSTANT WIDTH

A convex body has *constant width* λ if the distance between any two of its parallel support hyperplanes is λ. The simplest example of such a body is a closed ball with diameter λ. Plane convex bodies of constant width (other than circular discs) were first studied by Euler in the 1770s, and by many mathematicians since. In addition to their own intrinsic interest, these bodies have physical applications to the design of coins, the gear mechanisms of film projectors, and bits that drill nearly square holes! We begin our discussion by introducing two naturally occurring functions associated with any convex body, its *width* and *maximal chord-length functions*.

Let A be a convex body in $\mathbb{R}^n$ and let $\mathbf{u}$ be a unit vector in $\mathbb{R}^n$. Then there are *precisely* two support hyperplanes to A having $\mathbf{u}$ for a normal vector. The distance between these parallel support hyperplanes is

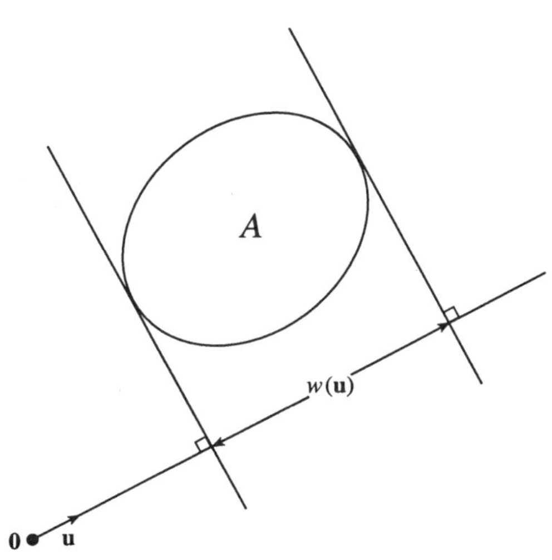

Fig. 7.19

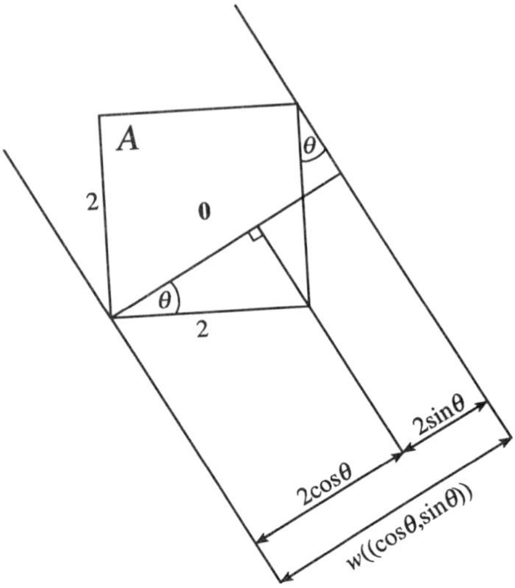

Fig. 7.20

called the **width of** A **in the direction of u and is denoted by** $w(\mathbf{u})$. See Fig. 7.19. Clearly, for all unit vectors $\mathbf{u}$ in $\mathbb{R}^n$, $w(-\mathbf{u}) = w(\mathbf{u})$. The width function of a closed ball of radius r has constant value $2r$. As a less trivial example, consider the square

$$A = \{(x, y): |x|, |y| \leqslant 1\}$$

in $\mathbb{R}^2$. A simple calculation shows that

$$w((\cos\theta, \sin\theta)) = 2(|\sin\theta| + |\cos\theta|) \qquad \text{for } \theta \in \mathbb{R}.$$

The case $0 < \theta < \frac{1}{4}\pi$ is illustrated in Fig. 7.20.

We now show that the width function w of a convex body A in $\mathbb{R}^n$ is nothing more than a restriction of the support function of the *difference body* $A - A$ of A. Denote by h the support function of A. Then the support hyperplanes to A having normal vector $\mathbf{u}$ have equations $\mathbf{u}\cdot\mathbf{x} = h(\mathbf{u})$ and $\mathbf{u}\cdot\mathbf{x} = -h(-\mathbf{u})$, and the distance between them is $h(\mathbf{u}) + h(-\mathbf{u})$. Thus

$$\begin{aligned} w(\mathbf{u}) = h(\mathbf{u}) + h(-\mathbf{u}) &= \sup\{\mathbf{u}\cdot\mathbf{a}:\mathbf{a}\in A\} + \sup\{-\mathbf{u}\cdot\mathbf{b}:\mathbf{b}\in A\} \\ &= \sup\{\mathbf{u}\cdot(\mathbf{a}-\mathbf{b}):\mathbf{a},\mathbf{b}\in A\} \\ &= \sup\{\mathbf{u}\cdot\mathbf{c}:\mathbf{c}\in A - A\}, \end{aligned}$$

which shows that w is the restriction of the support function of $A - A$ to the compact set $\{\mathbf{u}: \|\mathbf{u}\| = 1\}$. Since support functions are continuous, w is continuous and attains its bounds on $\{\mathbf{u}: \|\mathbf{u}\| = 1\}$. The lower bound w_* of w is called the **minimal width** of A, and the upper bound w^* of w is called the **maximal width** of A. The minimal and maximal widths of a square of side a are, respectively, a and $\sqrt{2}a$.

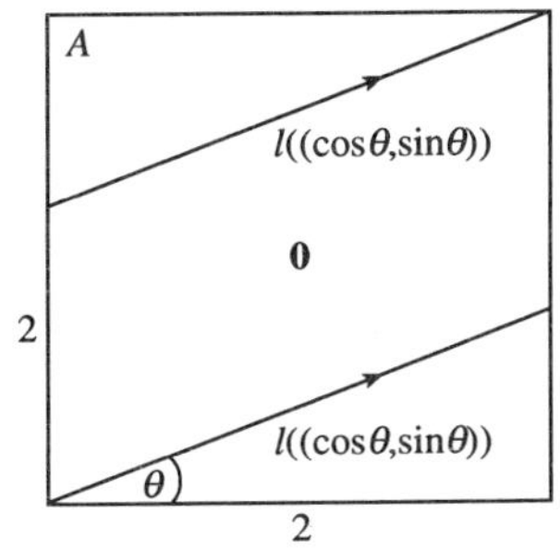

Fig. 7.21

Closely related to the width function of a convex body is the *maximal chord-length function* of the body. Let A be a convex body in $\mathbb{R}^n$ and let $\mathbf{u}$ be a unit vector in $\mathbb{R}^n$. Then the positive number $l(\mathbf{u})$, defined by the equation

$$l(\mathbf{u}) = \sup\{\|\mathbf{b} - \mathbf{a}\| : \mathbf{a}, \mathbf{b} \in A \text{ and } \mathbf{b} - \mathbf{a} = \alpha\mathbf{u} \text{ for some } \alpha \in \mathbb{R}\},$$

is (for obvious reasons) called the **maximal chord length of A in the direction of $\mathbf{u}$**. Clearly, for all unit vectors $\mathbf{u}$ in $\mathbb{R}^n$, $l(-\mathbf{u}) = l(\mathbf{u})$. The maximal chord-length function of a closed ball of radius r has constant value $2r$. As a less trivial example, consider the square

$$A = \{(x, y) : |x|, |y| \leqslant 1\}$$

in $\mathbb{R}^2$. Then a simple calculation shows that

$$l((\cos\theta, \sin\theta)) = 2/\max\{|\cos\theta|, |\sin\theta|\} \qquad \text{for } \theta \in \mathbb{R}.$$

The case $0 < \theta < \frac{1}{4}\pi$ is illustrated in Fig. 7.21. The definition of l can be usefully reformulated as follows:

$$l(\mathbf{u}) = \sup\{\alpha \in \mathbb{R} : \alpha\mathbf{u} \in A - A\}.$$

Some important properties of l follow easily from this observation. Firstly, for each unit vector $\mathbf{u}$ in $\mathbb{R}^n$, the point $l(\mathbf{u})\mathbf{u}$ lies on the boundary of $A - A$ and there exist $\mathbf{a}, \mathbf{b} \in A$ such that $\mathbf{b} - \mathbf{a} = l(\mathbf{u})\mathbf{u}$. Secondly, as $\mathbf{u}$ ranges over all unit vectors in $\mathbb{R}^n$, l attains both its lower bound l_* and its upper bound l^*; moreover, $l_* U \subseteq A - A \subseteq l^* U$, where U is the closed unit ball $\{\mathbf{x} \in \mathbb{R}^n : \|\mathbf{x}\| \leqslant 1\}$. It follows that l_* and l^* are, respectively, the inradius and the circumradius of $A - A$. If l denotes the maximal chord-length function of a square of side a, then $l_* = a$ and $l^* = \sqrt{2}a$. An observant reader might have noticed that, if w denotes the width function of a square of side a, then $l_* = w_*$ and $l^* = w^*$. The next result shows that this is no mere coincidence.

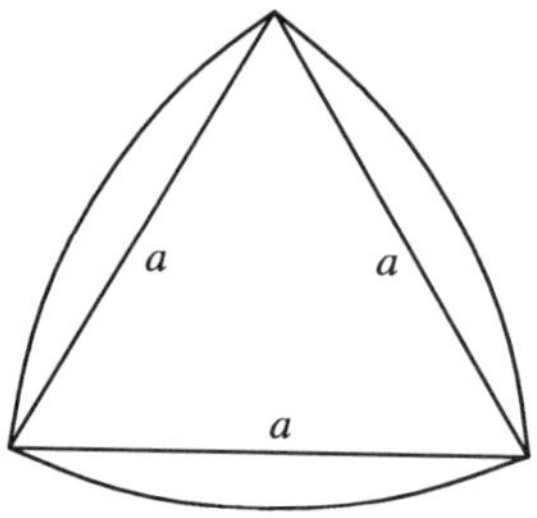

Fig. 7.22

Theorem 7.6.1 *Let l be the maximal chord-length function and w the width function of a convex body A in $\mathbb{R}^n$ with diameter D. Then, for any unit vector* $\mathbf{u}$ *in* $\mathbb{R}^n$, $l(\mathbf{u}) \leqslant w(\mathbf{u})$. *Also* $l_* = w_*$ *and* $l^* = w^* = D$.

Proof For any unit vector $\mathbf{u}$ in $\mathbb{R}^n$, $l(\mathbf{u})\mathbf{u} \in A - A$, whence

$$l(\mathbf{u}) = l(\mathbf{u})\mathbf{u} \cdot \mathbf{u} \leqslant \sup\{\mathbf{u} \cdot \mathbf{x}: \mathbf{x} \in A - A\} = w(\mathbf{u}).$$

Suppose now that $\mathbf{u}$ is a unit vector in $\mathbb{R}^n$ for which $l(\mathbf{u}) = l_*$. Since $A - A$ contains the closed ball $l_* U$, the hyperplane $\mathbf{u} \cdot \mathbf{x} = l_*$ supports $A - A$ at its boundary point $l_*\mathbf{u}$. Thus

$$w_* \leqslant w(\mathbf{u}) = \sup\{\mathbf{u} \cdot \mathbf{x}: \mathbf{x} \in A - A\} = l_* \leqslant w_*,$$

whence $l_* = w_*$.

Since $A - A$ lies in the closed ball $l^* U$, it follows that, for any unit vector $\mathbf{v}$ in $\mathbb{R}^n$,

$$w(\mathbf{v}) = \sup\{\mathbf{v} \cdot \mathbf{x}: \mathbf{x} \in A - A\} \leqslant l^*,$$

and so $w^* \leqslant l^*$. Trivially $l^* \leqslant w^*$. That $l^* = D$ is an easy consequence of the definition of l^*. Hence $l^* = w^* = D$. □

We now come to our main interest in this section: convex bodies of constant width. A convex body is said to have **constant width** if its width function is constant. Thus a closed ball of radius r has constant width $2r$. There are, perhaps surprisingly, convex bodies of constant width other than closed balls. The simplest of these is the *Reuleaux triangle*. This is a plane figure obtained by intersecting three closed circular discs of radius a centred at the vertices of an equilateral triangle with sides of length a. See Fig. 7.22. Reuleaux pentagons, heptagons, nonagons etc. can be constructed in a similar way. Since the vector sum of two convex bodies of constant width is itself a convex body of constant width, other examples of plane convex bodies of constant width can easily be constructed.

Theorem 7.6.1 shows that the following three conditions on a

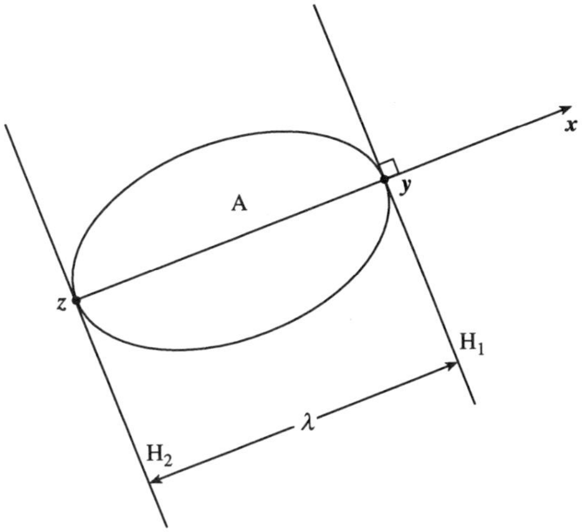

Fig. 7.23

convex body A in $\mathbb{R}^n$ are equivalent: (i) A has constant width λ; (ii) A has constant maximal chord length λ; (iii) $A - A$ is the closed ball λU.

Suppose now that A is a convex body in $\mathbb{R}^n$ of constant width λ that is symmetric, i.e. $-A = A$. Then condition (iii) of the last paragraph shows that $A = \frac{1}{2}\lambda U$. Thus a symmetric convex body of constant width λ must be the closed ball of radius $\frac{1}{2}\lambda$ centred on the origin.

Consider now a convex body A in $\mathbb{R}^n$ of constant width λ, and a point $\mathbf{x}$ of $\mathbb{R}^n$ *not* belonging to A. Let $\mathbf{y}$ be the nearest point of A to $\mathbf{x}$. Then the hyperplane H_1 through $\mathbf{y}$ with $\mathbf{x} - \mathbf{y}$ as a normal vector supports A at $\mathbf{y}$. Denote by H_2 the *other* support hyperplane to A having $\mathbf{x} - \mathbf{y}$ as a normal vector. Let $\mathbf{z}$ be a point of A lying on H_2. See Fig. 7.23. Since H_1 and H_2 are a distance λ apart, the distance between $\mathbf{x}$ and $\mathbf{z}$ must exceed λ. Thus the diameter of $A \cup \{\mathbf{x}\}$ is greater than that of A.

We have just shown that the adjunction of any extra point to a convex body of constant width results in a set of increased diameter. Any set with the property that it is not *properly* contained in another set of the same diameter is said to be **complete**. Hence every convex body of constant width is complete. The converse of this result, that every convex body which is complete has constant width, is also true. Before establishing this non-trivial converse, we need some preliminary results.

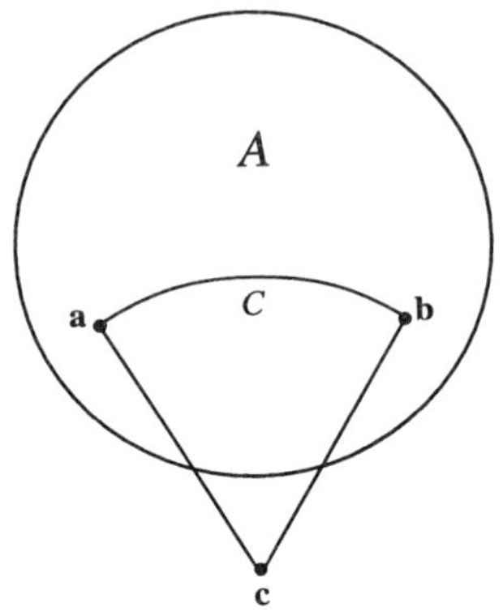

Fig. 7.24

Theorem 7.6.2 *Let* $\mathbf{b}$ *be a boundary point of a complete convex body* A *in* $\mathbb{R}^n$ *with diameter* λ. *Then there exists a point* $\mathbf{c}$ *of* A *such that* $\|\mathbf{c}-\mathbf{b}\| = \lambda$.

Proof Since A is compact, there is a point $\mathbf{c}$ of A whose distance from $\mathbf{b}$ is maximal. If $\|\mathbf{c}-\mathbf{b}\| < \lambda$, then there is a point $\mathbf{y}$ of $\mathbb{R}^n \setminus A$ such that $\|\mathbf{y}-\mathbf{b}\| < \lambda - \|\mathbf{c}-\mathbf{b}\|$. Let $\mathbf{a} \in A$. By the choices of $\mathbf{c}$ and $\mathbf{y}$,

$$\|\mathbf{y}-\mathbf{a}\| \leqslant \|\mathbf{y}-\mathbf{b}\| + \|\mathbf{b}-\mathbf{a}\| \leqslant \|\mathbf{y}-\mathbf{b}\| + \|\mathbf{b}-\mathbf{c}\| < \lambda,$$

which shows that $A \cup \{\mathbf{y}\}$ has diameter λ. This contradicts the completeness of A, and so shows that $\|\mathbf{c}-\mathbf{b}\| = \lambda$. □

Theorem 7.6.3 *The set* A *in* $\mathbb{R}^n$ *with positive diameter* λ *is complete if and only if it is the intersection of all closed balls of radius* λ *with centres in* A.

Proof Denote by C the intersection of all closed balls of radius λ with centres in A. Suppose first that $A = C$. Let $\mathbf{x} \in \mathbb{R}^n \setminus A$. Then there is some ball $\mathbf{B}[\mathbf{a}; \lambda]$ with centre $\mathbf{a}$ in A that does not contain $\mathbf{x}$, i.e. $\|\mathbf{x}-\mathbf{a}\| > \lambda$. Thus the diameter of $A \cup \{\mathbf{x}\}$ exceeds λ, which shows that A is complete.

Suppose next that A is complete. Since A has diameter λ, $A \subseteq C$. Let $\mathbf{x} \in C$. Then $A \cup \{\mathbf{x}\}$ has diameter λ. Since A is complete, $\mathbf{x}$ must lie in A. Hence $C \subseteq A$, and $A = C$ as desired. □

We now consider a simple property of a closed circular disc A of radius r in $\mathbb{R}^2$. Let $\mathbf{a}$, $\mathbf{b}$, $\mathbf{c}$ be non-collinear points of $\mathbb{R}^2$ such that $\mathbf{a}$, $\mathbf{b} \in A$ and $\|\mathbf{a}-\mathbf{c}\| = r$, $\|\mathbf{b}-\mathbf{c}\| = r$. Then it is pictorially clear that the (minor) arc C of the circle with centre $\mathbf{c}$ and radius r that joins $\mathbf{a}$ and $\mathbf{b}$ lies in A. See Fig. 7.24. We now generalize this result to $\mathbb{R}^n$.

First we must see how to express points on the arc C in terms of **a,b,c** and r. The points of C are precisely those points of $\mathbb{R}^2$ that lie in the angle $\angle\mathbf{acb}$ and whose distance from $\mathbf{c}$ is r, i.e. points of the form

$$\mathbf{c} + \alpha(\mathbf{a} - \mathbf{c}) + \beta(\mathbf{b} - \mathbf{c}),$$

where α, $\beta \geqslant 0$, which satisfy

$$\|\alpha(\mathbf{a} - \mathbf{c}) + \beta(\mathbf{b} - \mathbf{c})\| = r.$$

This last remark motivates the following definition. Let $\mathbf{a}, \mathbf{b}, \mathbf{c}$ be non-collinear points in $\mathbb{R}^n$ and let $r > 0$ be such that $\|\mathbf{a} - \mathbf{c}\| = r$, $\|\mathbf{b} - \mathbf{c}\| = r$. Then the *circular arc of radius r with centre* **c** *joining* **a** *and* **b** is defined to be the set

$$\{\mathbf{c} + \alpha(\mathbf{a} - \mathbf{c}) + \beta(\mathbf{b} - \mathbf{c}) : \alpha, \beta \geqslant 0 \text{ and } \|\alpha(\mathbf{a} - \mathbf{c}) + \beta(\mathbf{b} - \mathbf{c})\| = r\}.$$

Theorem 7.6.4 *A complete convex body in* $\mathbb{R}^n$ *of diameter* λ *contains every circular arc of radius* λ joining two of its points.

Proof Since, by Theorem 7.6.3, every complete convex body of diameter λ is an intersection of closed balls of radius λ, it will suffice to prove the theorem for a closed ball of radius λ.

Let B be a closed ball in $\mathbb{R}^n$ with centre $\mathbf{x}$ and radius λ. Let $\mathbf{a}, \mathbf{b}, \mathbf{c}$ be non-collinear points in $\mathbb{R}^n$ such that $\mathbf{a}, \mathbf{b} \in B$, $\|\mathbf{a} - \mathbf{c}\| = \lambda$, and $\|\mathbf{b} - \mathbf{c}\| = \lambda$. Then

$$\|\mathbf{a} - \mathbf{x}\|^2 = \|\mathbf{a} - \mathbf{c} + \mathbf{c} - \mathbf{x}\|^2 = \|\mathbf{a} - \mathbf{c}\|^2 + 2(\mathbf{a} - \mathbf{c}) \cdot (\mathbf{c} - \mathbf{x}) + \|\mathbf{c} - \mathbf{x}\|^2 \leqslant \lambda^2,$$

whence

$$2(\mathbf{a} - \mathbf{c}) \cdot (\mathbf{c} - \mathbf{x}) + \|\mathbf{c} - \mathbf{x}\|^2 \leqslant 0. \tag{7.6.1}$$

Similarly,

$$2(\mathbf{b} - \mathbf{c}) \cdot (\mathbf{c} - \mathbf{x}) + \|\mathbf{c} - \mathbf{x}\|^2 \leqslant 0. \tag{7.6.2}$$

Now let $\mathbf{z}$ lie on the circular arc of radius λ with centre $\mathbf{c}$ joining $\mathbf{a}$ and $\mathbf{b}$. Then $\mathbf{z} = \mathbf{c} + \alpha(\mathbf{a} - \mathbf{c}) + \beta(\mathbf{b} - \mathbf{c})$ for some $\alpha, \beta \geqslant 0$, and $\|\mathbf{z} - \mathbf{c}\| = \lambda$. Since $\|\mathbf{a} - \mathbf{c}\| = \lambda$, $\|\mathbf{b} - \mathbf{c}\| = \lambda$, the equation $\mathbf{z} - \mathbf{c} = \alpha(\mathbf{a} - \mathbf{c}) + \beta(\mathbf{b} - \mathbf{c})$ shows that $\alpha + \beta \geqslant 1$. Multiplying the inequalities (7.6.1) and (7.6.2), respectively, by α and β, and then adding the resulting inequalities together, we deduce that

$$2(\mathbf{z} - \mathbf{c}) \cdot (\mathbf{c} - \mathbf{x}) + \|\mathbf{c} - \mathbf{x}\|^2 \leqslant 2(\mathbf{z} - \mathbf{c}) \cdot (\mathbf{c} - \mathbf{x}) + (\alpha + \beta)\|\mathbf{c} - \mathbf{x}\|^2 \leqslant 0.$$

Hence

$$\begin{aligned}\|\mathbf{z} - \mathbf{x}\|^2 &= \|\mathbf{z} - \mathbf{c} + \mathbf{c} - \mathbf{x}\|^2 \\ &= \|\mathbf{z} - \mathbf{c}\|^2 + 2(\mathbf{z} - \mathbf{c}) \cdot (\mathbf{c} - \mathbf{x}) + \|\mathbf{c} - \mathbf{x}\|^2 \leqslant \|\mathbf{z} - \mathbf{c}\|^2 = \lambda^2,\end{aligned}$$

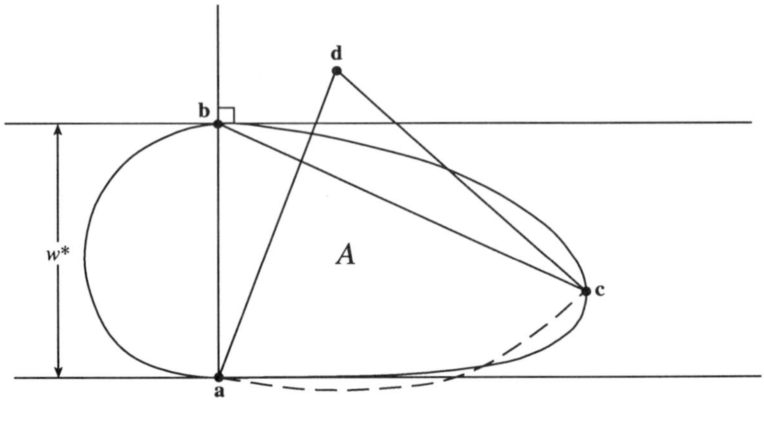

Fig. 7.25

and so $\|\mathbf{z} - \mathbf{x}\| \leqslant \lambda$. Thus $\mathbf{z} \in B$ and the circular arc of radius λ with centre $\mathbf{c}$ joining $\mathbf{a}$ and $\mathbf{b}$ lies in B. □

Theorem 7.6.5 *A convex body in* $\mathbb{R}^n$ *is complete if and only if it has constant width.*

Proof We already know that a convex body of constant width is complete, so we need only show that a complete convex body has constant width. To do this, we argue by contradiction.

Suppose that A is a complete convex body in $\mathbb{R}^n$ with diameter λ that does *not* have constant width. Let $\mathbf{u}$ be a unit vector in whose direction A has minimal width w_*. Then $w_* < w^* = \lambda$. Since $l_* = w_*$, there exist $\mathbf{a}, \mathbf{b} \in A$ such that $\mathbf{b} - \mathbf{a} = w_* \mathbf{u}$ and

$$A \subseteq \{\mathbf{x} \in \mathbb{R}^n : \mathbf{u} \cdot \mathbf{a} \leqslant \mathbf{u} \cdot \mathbf{x} \leqslant \mathbf{u} \cdot \mathbf{b}\}. \tag{7.6.3}$$

See Fig. 7.25.

By Theorem 7.6.2, there is a point $\mathbf{c}$ of A such that $\|\mathbf{c} - \mathbf{b}\| = \lambda$. We establish the existence of a point $\mathbf{d}$ of the form $(1 - \theta)\mathbf{a} + \theta\mathbf{c} + \varepsilon\mathbf{u}$, where $0 \leqslant \theta \leqslant 1$ and $\varepsilon > 0$, which is at a distance λ from both $\mathbf{a}$ and $\mathbf{c}$, distinguishing the cases $\mathbf{u} \cdot \mathbf{a} = \mathbf{u} \cdot \mathbf{c}$ and $\mathbf{u} \cdot \mathbf{a} < \mathbf{u} \cdot \mathbf{c}$.

If $\mathbf{u} \cdot \mathbf{a} = \mathbf{u} \cdot \mathbf{c}$, then we can take $\mathbf{d}$ to be $\frac{1}{2}\mathbf{a} + \frac{1}{2}\mathbf{c} + \varepsilon\mathbf{u}$, where ε is $\sqrt{(\lambda^2 - \frac{1}{4}\|\mathbf{a} - \mathbf{c}\|^2)}$. Suppose, then, that $\mathbf{u} \cdot \mathbf{a} < \mathbf{u} \cdot \mathbf{c}$. The point $\mathbf{f} = \mathbf{b} + \varphi\mathbf{u}$, where $\varphi = (\lambda^2 - w_*^2)/2\mathbf{u} \cdot (\mathbf{c} - \mathbf{a})$, is equidistant from $\mathbf{a}$ and $\mathbf{c}$, the common distance exceeding λ. The point $\frac{1}{2}(\mathbf{a} + \mathbf{c})$ is equidistant from $\mathbf{a}$ and $\mathbf{c}$, the common distance being less than λ. Thus there is a point $\mathbf{d}$ on the line segment joining $\mathbf{f}$ and $\frac{1}{2}(\mathbf{a} + \mathbf{c})$ which is at a distance of λ from both $\mathbf{a}$ and $\mathbf{c}$. Moreover, $\mathbf{d} = (1 - \theta)\mathbf{a} + \theta\mathbf{c} + \varepsilon\mathbf{u}$, for some θ and ε satisfying $0 \leqslant \theta \leqslant 1$ and $\varepsilon > 0$.

Consider now the point $\mathbf{p} = \mathbf{d} - \lambda\mathbf{u}$. The equation

$$\mathbf{p} = \mathbf{d} + \lambda(1 - \theta)\varepsilon^{-1}(\mathbf{a} - \mathbf{d}) + \lambda\theta\varepsilon^{-1}(\mathbf{c} - \mathbf{d})$$

shows that $\mathbf{p}$ lies on the circular arc of radius λ with centre $\mathbf{d}$ joining $\mathbf{a}$ and $\mathbf{c}$. Since A is complete, $\mathbf{p} \in A$. We obtain the desired contradiction by now showing that $\mathbf{p} \notin A$.

The only point on the halfline $\{\mathbf{a} + \mu\mathbf{u} : \mu \geqslant 0\}$ which is at a distance λ from $\mathbf{a}$ is $\mathbf{a} + \lambda\mathbf{u}$, its distance from $\mathbf{c}$ being $\|(\lambda - w_*)\mathbf{u} + (\mathbf{b} - \mathbf{c})\|$, which exceeds $\|\mathbf{b} - \mathbf{c}\| = \lambda$. Thus $\mathbf{d}$ is not of the form $\mathbf{a} + \mu\mathbf{u}$ for any $\mu \geqslant 0$. Hence the nearest point $\mathbf{d} - ((\mathbf{d} - \mathbf{a}) \cdot \mathbf{u})\mathbf{u}$ of the hyperplane H with equation $\mathbf{u} \cdot \mathbf{x} = \mathbf{u} \cdot \mathbf{a}$ to the point $\mathbf{d}$ is *not* $\mathbf{a}$. Since $\mathbf{a}$ lies on H, its distance from $\mathbf{d}$ must exceed the (shortest) distance of $\mathbf{d}$ from H, i.e. $\lambda > (\mathbf{d} - \mathbf{a}) \cdot \mathbf{u}$. Thus

$$\mathbf{u} \cdot \mathbf{p} = \mathbf{u} \cdot \mathbf{d} - \lambda < \mathbf{u} \cdot \mathbf{a}.$$

The inclusion (7.6.3) now shows that $\mathbf{p} \notin A$, as desired. □

Before stating the next result, we recall that some convex bodies, for example a rectangle, do not have a *unique* inball.

Theorem 7.6.6 *Let r be the inradius and R the circumradius of a convex body A in $\mathbb{R}^n$ of constant width λ. Then A has a unique inball, and this is concentric with its circumball. Moreover, $\lambda = R + r$.*

Proof Let $B[\mathbf{a}; r]$ be an inball of A. Then $A \subseteq B[\mathbf{a}; \lambda - r]$. For otherwise, there would exist $\mathbf{a}' \in A$ such that $\|\mathbf{a}' - \mathbf{a}\| > \lambda - r$, and the distance $\|\mathbf{a}' - \mathbf{a}\| + r$ between the points $\mathbf{a} - r(\mathbf{a}' - \mathbf{a})/\|\mathbf{a}' - \mathbf{a}\|$ and $\mathbf{a}'$ of A would exceed the diameter λ of A. Thus $R \leqslant \lambda - r$ and $r + R \leqslant \lambda$.

Let $B[\mathbf{c}; R]$ be the circumball of A. Then $B[\mathbf{c}; \lambda - R] \subseteq A$. For otherwise there would exist $\mathbf{b} \in B[\mathbf{c}; \lambda - R]$ such that $\mathbf{b} \notin A$. But then $\mathbf{b}$ and A could be strictly separated, whence there would exist a unit vector $\mathbf{u}$ such that, for all $\mathbf{x} \in A$,

$$\mathbf{u} \cdot \mathbf{c} - R \leqslant \mathbf{u} \cdot \mathbf{x} < \mathbf{u} \cdot \mathbf{b} \leqslant \mathbf{u} \cdot \mathbf{c} + \lambda - R,$$

and the width of A in the direction of $\mathbf{u}$ would be less than λ. Thus $r \geqslant \lambda - R$ and $r + R \geqslant \lambda$.

The inequalities $\lambda \leqslant r + R \leqslant \lambda$ show that $r + R = \lambda$. The first part of the proof shows that, if $B[\mathbf{a}; r]$ is an inball of A, then $B[\mathbf{a}; R]$ is the circumball of A. Thus A has a unique inball, and this is concentric with its circumball. □

We conclude this section, and also the book, with the following remarkable theorem, which generalizes the result that the circumference of a circle is π times its diameter,

Theorem 7.6.7 (Barbier's theorem) *All convex bodies in* $\mathbb{R}^2$ *of constant width* λ *have perimeter* $\pi\lambda$.

Proof Let A be a convex body in $\mathbb{R}^2$ of constant width λ. Then $A - A = \lambda U$. Since A and $-A$ have the same perimeter, and in $\mathbb{R}^2$ the perimeters of convex bodies are additive with respect to vector addition (see the end of Section 6.4), the perimeter of A must be half that of λU, i.e. $\pi\lambda$. □

Exercises 7.6

1. Find the maximal chord-length function l and the width function w of the ellipse $\{(x, y): x^2/a^2 + y^2/b^2 \leqslant 1\}$ in $\mathbb{R}^2$, where $a \geqslant b > 0$. Verify that $l_* = w_*$ and $l^* = w^*$.
2. Prove that each boundary point of a convex body of constant width is an exposed point of the body.
3. The convex body A in $\mathbb{R}^n$ of diameter λ is such that, for each boundary point $\mathbf{b}$ of A, there exists a point $\mathbf{c}$ of A such that $\|\mathbf{c} - \mathbf{b}\| = \lambda$. Prove that A has constant width λ.
4. Show that every set in $\mathbb{R}^n$ of positive diameter λ is contained in some convex body of constant width λ.
5. Prove that, if a plane convex body of constant width λ is rotated about an axis of symmetry, then the resulting solid of revolution has constant width λ.

8
Solutions to exercises

Exercises 1.1

1. (i) $$\begin{aligned} A + (B + C) &= \{\mathbf{a} + (\mathbf{b} + \mathbf{c}) : \mathbf{a} \in A, \mathbf{b} \in B, \mathbf{c} \in C\} \\ &= \{(\mathbf{a} + \mathbf{b}) + \mathbf{c} : \mathbf{a} \in A, \mathbf{b} \in B, \mathbf{c} \in C\} \\ &= (A + B) + C. \end{aligned}$$
   (ii) $(\lambda + \mu)A = \{(\lambda + \mu)\mathbf{a} : \mathbf{a} \in A\} = \{\lambda\mathbf{a} + \mu\mathbf{a} : \mathbf{a} \in A\} \subseteq \lambda A + \mu A$.
2. It follows, by considering the special case $\lambda = 1$, $\mu = -1$, that A must either be empty or a singleton.
3. (i) $\{(x, y) : x^2 + y^2 \leqslant 5^2\}$; (ii) $\mathbb{R}^2$; (iii) $\{(x, y) : x + y + 3 = 0\}$; (iv) $\{(x, y) : y > 0\}$; (v) $\{(x, y) : x^2 + y^2 \leqslant 2^2\}$.
4. (i) $A = 1A + \mathbf{0}$; (ii) $A = \lambda B + \mathbf{u}$, where $\lambda > 0$, implies $B = A/\lambda - \mathbf{u}/\lambda$ and $1/\lambda > 0$; (iii) $A = \lambda B + \mathbf{u}$ and $B = \mu C + \mathbf{v}$, where $\lambda, \mu > 0$, imply $A = \lambda\mu C + \lambda\mathbf{v} + \mathbf{u}$ and $\lambda\mu > 0$.

Exercises 1.2

1. If $(x_1, \ldots, x_n)$, $(y_1, \ldots, y_n)$ lie in the set and $\lambda + \mu = 1$, then
$$\begin{aligned} & a_1(\lambda x_1 + \mu y_1) + \ldots + a_n(\lambda x_n + \mu y_n) \\ &= \lambda(a_1 x_1 + \ldots + a_n x_n) + \mu(a_1 y_1 + \ldots + a_n y_n) \\ &= \lambda a_0 + \mu a_0 = a_0, \end{aligned}$$
which shows that the set is a flat. The flat is a subspace if and only if it contains the origin, i.e. if and only if $a_0 = 0$.
2. Let $\mathbf{a} \in A$. Then $A - \mathbf{a}$ is the unique subspace of $\mathbb{R}^n$ parallel to A, whence $A - \mathbf{x} = A - \mathbf{a}$ for all $\mathbf{x} \in A$. Thus
$$A - A = \bigcup (A - \mathbf{x} : \mathbf{x} \in A) = A - \mathbf{a}.$$
3. We note that $(x_1, \ldots, x_n) = \lambda_1\mathbf{e}_1 + \ldots + \lambda_n\mathbf{e}_n + \lambda(\mathbf{e}_1 + \ldots + \mathbf{e}_n)$, where $\lambda = (x_1 + \ldots + x_n - 1)/(n - 1)$, $\lambda_1 = x_1 - \lambda, \ldots, \lambda_n = x_n - \lambda$. Thus $\operatorname{aff}\{\mathbf{e}_1, \ldots, \mathbf{e}_n, \mathbf{e}_1 + \ldots + \mathbf{e}_n\} = \mathbb{R}^n$.
4. Let $A = \mathbb{R}^n$, $\lambda_1 = 1$, $\lambda_2 = -1$. Then $\{\mathbf{0}\} = (\lambda_1 + \lambda_2)A \subset \lambda_1 A + \lambda_2 A = \mathbb{R}^n$.
5. The stated result follows from the equation
$$\mathbf{a} + \lambda_1\mathbf{a}_1 + \ldots + \lambda_m\mathbf{a}_m = \lambda_1(\mathbf{a} + \mathbf{a}_1) + \ldots + \lambda_m(\mathbf{a} + \mathbf{a}_m),$$
where $\lambda_1 + \ldots + \lambda_m = 1$.

Exercises 1.3

1. The result follows immediately from the fact that, if $\mathbf{a}_1, \ldots, \mathbf{a}_m$ lie in $\mathbb{R}^n$ and $\lambda_1, \ldots, \lambda_m$ are scalars, then the two equations

$$\lambda_1 \mathbf{a}_1 + \ldots + \lambda_m \mathbf{a}_m = \mathbf{0} \quad \text{and} \quad \lambda_1 + \ldots + \lambda_m = 0$$

are equivalent to the single equation $\lambda_1 \mathbf{a}_1^* + \ldots + \lambda_m \mathbf{a}_m^* = \mathbf{0}$.

2. The given set lies in the flat and is affinely independent. Also, if (x, y, z) lies in the flat, then

$$(x, y, z) = (x - y)(1, 0, 0) + (z - y)(0, 0, 1) + y(1, 1, 1)$$

and $(x - y) + (z - y) + y = x - y + z = 1$. This shows that the given set is an affine basis for the flat, and that the barycentric coordinates of a general point (x, y, z) of the flat relative to $(1, 0, 0)$, $(0, 0, 1)$, $(1, 1, 1)$ are $x - y, z - y, y$.

3. By Theorem 1.3.9, B lies in some affine basis for A. But B contains $r + 1$ elements, so it must itself be an affine basis for A. By Theorem 1.3.9 (with the C of the question serving as the A of the theorem, and the B of the theorem as the empty set), C must contain an affine basis for A. But C contains $r + 1$ elements, so it must itself be an affine basis for A.

Exercises 1.4

1. The three given points are affinely independent, so H is two-dimensional, i.e. a hyperplane. H has equation $x - 2y + 3z = 4$.

2. By Theorem 1.3.8,

$$\dim (A + H) + \dim (A \cap H) = \dim A + \dim H.$$

Let $\{\mathbf{a}_1, \ldots, \mathbf{a}_n\}$ be an affine basis for H and let $\mathbf{a}_0$ be a point of A not lying in H. Let $\mathbf{a} \in A \cap H$. Then $\{\mathbf{a} + \mathbf{a}_0, \ldots, \mathbf{a} + \mathbf{a}_n\}$ is an affinely independent subset of $A + H$, whence $\dim (A + H) = n$. Since $\dim H = n - 1$, it follows from the above equation that $\dim (A \cap H) = \dim A - 1$.

3. Since A is a flat, $\lambda \mathbf{a} + (1 - \lambda)\mathbf{b} \in A$ for *all* scalars λ. Thus, for all scalars λ,

$$c_0 + c_1 b_1 + \ldots + c_n b_n + \lambda(c_1 a_1 + \ldots + c_n a_n - c_1 b_1 - \ldots - c_n b_n) \leqslant 0.$$

Letting $\lambda \to \infty$, we see that $c_1 a_1 + \ldots + c_n a_n - c_1 b_1 - \ldots - c_n b_n \leqslant 0$. By symmetry, $c_1 a_1 + \ldots + c_n a_n - c_1 b_1 - \ldots - c_n b_n \geqslant 0$, and the desired equality follows. Choose $\mathbf{z} = (z_1, \ldots, z_n)$ such that

$$c_0 + c_1 a_1 + \ldots + c_n a_n + c_1 z_1 + \ldots + c_n z_n = 0.$$

It then follows easily that $A + \mathbf{z}$ lies in the hyperplane with equation $c_0 + c_1x_1 + \ldots + c_nx_n = 0$.

Exercises 1.5

1. Let $T_1(\mathbf{x}) = \mathbf{Q}_1\mathbf{x} + \mathbf{q}_1$ and $T_2(\mathbf{y}) = \mathbf{Q}_2\mathbf{y} + \mathbf{q}_2$. Then $T_2 \circ T_1$ is the affine transformation given by $T_2 \circ T_1(\mathbf{x}) = \mathbf{Q}_2\mathbf{Q}_1\mathbf{x} + \mathbf{Q}_2\mathbf{q}_1 + \mathbf{q}_2$.
2. Let $T: \mathbb{R}^n \to \mathbb{R}^m$ be an affine transformation. Let A and B be parallel flats in $\mathbb{R}^n$ related by the equation $A = B + \mathbf{c}$. Then $T(A) = T(B) + (T(\mathbf{c}) - T(\mathbf{0}))$, which shows that $T(A)$ and $T(B)$ are parallel.
3. Let $\mathbf{x} \in \text{aff}\, T^{-1}(A)$. Then $\mathbf{x} = \lambda_1\mathbf{x}_1 + \ldots + \lambda_m\mathbf{x}_m$, where $T(\mathbf{x}_1), \ldots, T(\mathbf{x}_m) \in A$ and $\lambda_1 + \ldots + \lambda_m = 1$. Thus

$$T(\mathbf{x}) = \lambda_1 T(\mathbf{x}_1) + \ldots + \lambda_m T(\mathbf{x}_m) \in \text{aff}\, A,$$

whence $\mathbf{x} \in T^{-1}(\text{aff}\, A)$ and $\text{aff}\, T^{-1}(A) \subseteq T^{-1}(\text{aff}\, A)$. Define an affine transformation $T: \mathbb{R}^1 \to \mathbb{R}^2$ by the equation $T(x) = (x, 0)$. Take $A = \{(0, 1), (0, -1)\}$. Then $\text{aff}\, T^{-1}(A) = \varnothing$, whereas $T^{-1}(\text{aff}\, A) = \{0\}$. Suppose that A is a flat. Then

$$T^{-1}(A) \subseteq \text{aff}\, T^{-1}(A) \subseteq T^{-1}(\text{aff}\, A) = T^{-1}(A).$$

Thus $T^{-1}(A) = \text{aff}\, T^{-1}(A)$, and so $T^{-1}(A)$ is a flat.
4. Let $T: \mathbb{R}^n \to \mathbb{R}^m$ be the affine transformation given by $T(\mathbf{x}) = \mathbf{Q}\mathbf{x} + \mathbf{q}$. Suppose *first* that T is injective. Let $\{\mathbf{a}_0, \ldots, \mathbf{a}_n\}$ be an affine basis for $\mathbb{R}^n$. Then $\{T(\mathbf{a}_0), \ldots, T(\mathbf{a}_n)\}_{\neq}$ is affinely independent. To see this, suppose that

$$\lambda_0 T(\mathbf{a}_0) + \ldots + \lambda_n T(\mathbf{a}_n) = \mathbf{0} \qquad \text{and} \qquad \lambda_0 + \ldots + \lambda_n = 0.$$

Then $\mathbf{Q}(\lambda_0\mathbf{a}_0 + \ldots + \lambda_n\mathbf{a}_n) = \mathbf{0}$, so $T(\lambda_0\mathbf{a}_0 + \ldots + \lambda_n\mathbf{a}_n) = T(\mathbf{0})$. But T is injective, so $\lambda_0\mathbf{a}_0 + \ldots + \lambda_n\mathbf{a}_n = \mathbf{0}$. Since $\{\mathbf{a}_0, \ldots, \mathbf{a}_n\}$ is affinely independent, all of $\lambda_0, \ldots, \lambda_n$ are zero. Thus $\{T(\mathbf{a}_0), \ldots, T(\mathbf{a}_n)\}$ is affinely independent in $\mathbb{R}^m$, so $n + 1 \leqslant m + 1$ and $n \leqslant m$. Suppose *next* that T is surjective. Let $\{\mathbf{b}_0, \ldots, \mathbf{b}_m\}$ be an affine basis for $\mathbb{R}^m$. Since T is surjective, there exist $\mathbf{a}_0, \ldots, \mathbf{a}_m$ in $\mathbb{R}^n$ such that $T(\mathbf{a}_0) = \mathbf{b}_0, \ldots, T(\mathbf{a}_m) = \mathbf{b}_m$. Then $\{\mathbf{a}_0, \ldots, \mathbf{a}_m\}_{\neq}$ is affinely independent. To see this, suppose that

$$\lambda_0\mathbf{a}_0 + \ldots + \lambda_m\mathbf{a}_m = \mathbf{0} \qquad \text{and} \qquad \lambda_0 + \ldots + \lambda_m = 0.$$

Then $\lambda_0\mathbf{b}_0 + \ldots + \lambda_m\mathbf{b}_m = \mathbf{0}$. But $\{\mathbf{b}_0, \ldots, \mathbf{b}_m\}$ is affinely independent, whence all of $\lambda_0, \ldots, \lambda_m$ are zero. Thus $\{\mathbf{a}_0, \ldots, \mathbf{a}_m\}$ is affinely independent, so $m + 1 \leqslant n + 1$ and $m \leqslant n$. The last part follows immediately from the first two parts.

Exercises 1.6

1. We have, since $\lambda + \mu = 1$,

$$\begin{aligned}\|\lambda\mathbf{x} + \mu\mathbf{y}\|^2 + \lambda\mu\|\mathbf{x} - \mathbf{y}\|^2 &= \lambda^2\|\mathbf{x}\|^2 + 2\lambda\mu\mathbf{x}\cdot\mathbf{y} + \mu^2\|\mathbf{y}\|^2 + \\ &\qquad \lambda\mu(\|\mathbf{x}\|^2 - 2\mathbf{x}\cdot\mathbf{y} + \|\mathbf{y}\|^2) \\ &= \lambda\|\mathbf{x}\|^2 + \mu\|\mathbf{y}\|^2.\end{aligned}$$

Putting $\lambda = \mu = \frac{1}{2}$ in the given equation, we find that $\|\mathbf{x} + \mathbf{y}\|^2 + \|\mathbf{x} - \mathbf{y}\|^2 = 2\|\mathbf{x}\|^2 + 2\|\mathbf{y}\|^2$. If $\lambda, \mu > 0$, $\|\mathbf{x}\| = \|\mathbf{y}\| = 1$ and $\mathbf{x} \neq \mathbf{y}$, then $\lambda\mu\|\mathbf{x} - \mathbf{y}\|^2 > 0$, whence

$$\|\lambda\mathbf{x} + \mu\mathbf{y}\|^2 = \lambda\|\mathbf{x}\|^2 + \mu\|\mathbf{y}\|^2 - \lambda\mu\|\mathbf{x} - \mathbf{y}\|^2 = 1 - \lambda\mu\|\mathbf{x} - \mathbf{y}\|^2 < 1$$

and $\|\lambda\mathbf{x} + \mu\mathbf{y}\| < 1$.

2. If $\mathbf{b}$ lies in H and is such that $\mathbf{a} - \mathbf{b}$ is a normal vector to H, then $c_0 + \mathbf{c}\cdot\mathbf{b} = 0$ and $\mathbf{a} - \mathbf{b} = \lambda\mathbf{c}$ for some scalar λ. These two simultaneous equations have unique solution $\lambda = (c_0 + \mathbf{c}\cdot\mathbf{a})/\|\mathbf{c}\|^2$ and $\mathbf{b} = \mathbf{a} - \lambda\mathbf{c}$. Now let $\mathbf{x}$ be any point of H. Then $(\mathbf{a} - \mathbf{b})\cdot(\mathbf{x} - \mathbf{b}) = 0$, and so

$$\begin{aligned}\|\mathbf{x}-\mathbf{a}\|^2 = \|(\mathbf{x}-\mathbf{b}) + (\mathbf{b}-\mathbf{a})\|^2 &= \|\mathbf{x}-\mathbf{b}\|^2 + 2(\mathbf{x}-\mathbf{b})\cdot(\mathbf{b}-\mathbf{a}) + \|\mathbf{b}-\mathbf{a}\|^2 \\ &= \|\mathbf{x}-\mathbf{b}\|^2 + \|\mathbf{b}-\mathbf{a}\|^2.\end{aligned}$$

Hence $\|\mathbf{x} - \mathbf{a}\| \geqslant \|\mathbf{b} - \mathbf{a}\|$ with equality if and only if $\mathbf{x} = \mathbf{b}$. This shows $\mathbf{b}$ to be the unique nearest point of H to $\mathbf{a}$. Also

$$\|\mathbf{a} - \mathbf{b}\| = |\lambda|\,\|\mathbf{c}\| = |c_0 + \mathbf{c}\cdot\mathbf{a}|/\|\mathbf{c}\|.$$

3. We note that $(0, 0, 0, 1) \in H$ and that the subspace $H - (0, 0, 0, 1)$ has equation $2x_1 + 2x_2 - x_3 - x_4 = 0$. An orthonormal basis for this subspace consists of the vectors $(1, 0, 0, 2)/\sqrt{5}$, $(0, 1, 2, 0)/\sqrt{5}$, $(2, -2, 1, -1)/\sqrt{10}$. This can be extended to an orthonormal basis for $\mathbb{R}^4$ by the addition of the vector $(2, 2, -1, -1)/\sqrt{10}$. Thus a congruence transformation T of $\mathbb{R}^4$ which maps R_3 onto H is given by the equation

$$T(\mathbf{x}) = \begin{bmatrix} 1/\sqrt{5} & 0 & 2/\sqrt{10} & 2/\sqrt{10} \\ 0 & 1/\sqrt{5} & -2/\sqrt{10} & 2/\sqrt{10} \\ 0 & 2/\sqrt{5} & 1/\sqrt{10} & -1/\sqrt{10} \\ 2/\sqrt{5} & 0 & -1/\sqrt{10} & -1/\sqrt{10} \end{bmatrix}\mathbf{x} + \begin{bmatrix} 0 \\ 0 \\ 0 \\ 1 \end{bmatrix}.$$

The inverse transformation T^{-1}, given as below, is one possible answer to the question:

$$T^{-1}(\mathbf{x}) = \begin{bmatrix} 1/\sqrt{5} & 0 & 0 & 2/\sqrt{5} \\ 0 & 1/\sqrt{5} & 2/\sqrt{5} & 0 \\ 2/\sqrt{10} & -2/\sqrt{10} & 1/\sqrt{10} & -1/\sqrt{10} \\ 2/\sqrt{10} & 2/\sqrt{10} & -1/\sqrt{10} & -1/\sqrt{10} \end{bmatrix}\mathbf{x} - \begin{bmatrix} 2/\sqrt{5} \\ 0 \\ -1/\sqrt{10} \\ -1/\sqrt{10} \end{bmatrix}.$$

Exercises 1.7

1. By the triangle inequality, the distance apart of two points of $B[\mathbf{a}; r]$ never exceeds $2r$. The points $\mathbf{b} \pm s\mathbf{e}_1$ lie in $B[\mathbf{b}; s]$, and hence $B[\mathbf{a}; r]$, and are at a distance $2s$ apart, whence $2s \leqslant 2r$, i.e. $s \leqslant r$. By symmetry $r \leqslant s$, so $r = s$. Suppose that $\mathbf{a} \neq \mathbf{b}$. Then the points $\mathbf{b} + r(\mathbf{b} - \mathbf{a})/\|\mathbf{b} - \mathbf{a}\|$ and $\mathbf{a} - r(\mathbf{b} - \mathbf{a})/\|\mathbf{b} - \mathbf{a}\|$ lie in $B[\mathbf{a}; r]$ and their distance apart is $2r + \|\mathbf{b} - \mathbf{a}\|$, which exceeds $2r$. This contradiction shows that $\mathbf{a} = \mathbf{b}$.

2. (i) is not always true. For example, take A to be the set of points in $\mathbb{R}^n$, all of whose coordinates are rational, and B to be the complement of A in $\mathbb{R}^n$. Then $\operatorname{int}(A \cup B) = \mathbb{R}^n$, whereas $\operatorname{int} A \cup \operatorname{int} B = \varnothing$.

   (ii) is a special case of Corollary 1.7.8.

   (iii) is true. Trivially $\operatorname{int}(A \cap B) \subseteq \operatorname{int} A \cap \operatorname{int} B$. Since $A \cap B$ contains the open set $\operatorname{int} A \cap \operatorname{int} B$, we see that $\operatorname{int}(A \cap B) \supseteq (\operatorname{int} A) \cap (\operatorname{int} B)$. Hence (iii) is true.

   (iv) is not always true. See the counterexample in (i).

3. The union of the sequence $\frac{1}{2}U, \frac{2}{3}U, \frac{3}{4}U, \ldots$ of closed balls centred on the origin of $\mathbb{R}^n$ is the open unit ball $\{\mathbf{x} \in \mathbb{R}^n : \|\mathbf{x}\| < 1\}$, which is *not* closed.

4. The closure of the open ball $B(\mathbf{a}; r)$ is the closed ball $B[\mathbf{a}; r]$, and the closure of the open halfspace $c_0 + \mathbf{c} \cdot \mathbf{x} > 0$ is the closed halfspace $c_0 + \mathbf{c} \cdot \mathbf{x} \geqslant 0$. We establish the second of these results. Let A and B denote, respectively, the open halfspace $c_0 + \mathbf{c} \cdot \mathbf{x} > 0$ and the closed halfspace $c_0 + \mathbf{c} \cdot \mathbf{x} \geqslant 0$, where $\mathbf{c}$ is a unit vector. Let $\mathbf{b} \in B$ and $r > 0$. Then $\mathbf{b} + \frac{1}{2}r\mathbf{c} \in A$ and $\|\mathbf{b} + \frac{1}{2}r\mathbf{c} - \mathbf{b}\| < r$, which shows that $\mathbf{b} \in \operatorname{cl} A$. Hence $B \subseteq \operatorname{cl} A$. Since B is closed and contains A, $B \supseteq \operatorname{cl} A$. Thus $B = \operatorname{cl} A$.

5. Let A and B be closed sets in $\mathbb{R}^n$ each of which has an empty interior. Let C be *any* open ball in $\mathbb{R}^n$. Since A has empty interior, there exists a point $\mathbf{c}$ of C which does not lie in A, i.e. $\mathbf{c} \in A^c$. Since A is closed, A^c is open and there exists $r > 0$ such that $B(\mathbf{c}; r) \subseteq A^c \cap C$. Thus $B(\mathbf{c}; r)$ is an open ball which lies in C and does not meet A. Repeating this argument, we can show that there exist a point $\mathbf{d}$ of $B(\mathbf{c}; r)$ and $s > 0$ such that $B(\mathbf{d}; s)$ lies in $B(\mathbf{c}; r)$ and does not meet B. Thus $\mathbf{d} \notin A \cup B$, whence $A \cup B$ does not contain C. Since C was an arbitrary open ball in $\mathbb{R}^n$, $A \cup B$ cannot have any interior points, i.e. it has empty interior.

Exercises 1.8

1. Suppose that some r-flat $A\,(r \geqslant 1)$ in $\mathbb{R}^n$ *is* bounded. Then there exist *distinct* points $\mathbf{a}$ and $\mathbf{b}$ of A and a real number s such

that $\|\mathbf{b} + \lambda(\mathbf{a} - \mathbf{b})\| \leqslant s$ for *all* real λ. Dividing through the last inequality by $\lambda > 0$ and letting $\lambda \to \infty$, we deduce that $\|\mathbf{a} - \mathbf{b}\| \leqslant 0$, whence $\mathbf{a} = \mathbf{b}$. This contradiction gives the desired result.

2. Let A be a bounded set in $\mathbb{R}^n$. Then there is $r > 0$ such that $A \subseteq B[\mathbf{0}; r]$. Since closed balls are closed, $\operatorname{cl} A \subseteq B[\mathbf{0}; r]$. Thus $\operatorname{cl} A$ is bounded, in addition to being closed, and so is compact.

3. Suppose that $\mathbf{x}_1, \ldots, \mathbf{x}_k, \ldots$ does *not* converge to $\mathbf{x}$. Then there exists $\varepsilon > 0$ and some subsequence $\mathbf{x}_{i_1}, \ldots, \mathbf{x}_{i_k}, \ldots$ of $\mathbf{x}_1, \ldots, \mathbf{x}_k, \ldots$ such that $\|\mathbf{x}_{i_k} - \mathbf{x}\| \geqslant \varepsilon$ for $k = 1, 2, \ldots$. This subsequence is bounded, and so contains a subsequence, $\mathbf{x}_{j_1}, \ldots, \mathbf{x}_{j_k}, \ldots$ say, which converges to some point $\mathbf{y}$ of $\mathbb{R}^n$. The given hypotheses show that $\mathbf{y} = \mathbf{x}$, which is impossible, since $\|\mathbf{x}_{j_k} - \mathbf{x}\| \geqslant \varepsilon$ for $k = 1, 2, \ldots$. This contradiction shows that $\mathbf{x}_1, \ldots, \mathbf{x}_k, \ldots$ converges to $\mathbf{x}$.

4. Let $\mathbf{e} = (1, 0, \ldots, 0)$ be the first elementary vector in $\mathbb{R}^n$. Define closed sets A and B in $\mathbb{R}^n$ by the equations:

$$A = \{k\mathbf{e} : k = 2, 3, \ldots\} \quad \text{and} \quad B = \{(-k + k^{-1})\mathbf{e} : k = 2, 3, \ldots\}.$$

Clearly, $k^{-1}\mathbf{e} \in A + B$ and $k^{-1}\mathbf{e} \to \mathbf{0}$ as $k \to \infty$, whence $\mathbf{0} \in \operatorname{cl}(A + B)$. But $\mathbf{0} \notin A + B$, so $A + B$ is not closed.

Exercises 1.9

1. Define $f: \mathbb{R}^2 \to \mathbb{R}^2$ by the equation $f(x, y) = (\mathrm{e}^x, 0)$. Let $A = B = \mathbb{R}^2$ and $C = \{(1, 0)\}$. Then f is continuous, A is closed, B is open, and C is compact, but $f(A) = \{(x, 0) : x > 0\}$ is not closed, $f(B) = \{(x, 0) : x > 0\}$ is not open, and $f^{-1}(C) = \{(0, y) : y \in \mathbb{R}\}$ is not compact.

2. Denote the set $\{\mathbf{x} \in A : f(\mathbf{x}) \leqslant g(\mathbf{x})\}$ by B. Suppose that $\mathbf{b}_1, \ldots, \mathbf{b}_k, \ldots$ is a sequence of points of B that converges to a point $\mathbf{b}$ of $\mathbb{R}^n$. Since $B \subseteq A$ and A is closed, $\mathbf{b} \in A$. Since f and g are continuous at $\mathbf{b}$, $f(\mathbf{b}_k) \to f(\mathbf{b})$ and $g(\mathbf{b}_k) \to g(\mathbf{b})$. But $f(\mathbf{b}_k) \leqslant g(\mathbf{b}_k)$, whence letting $k \to \infty$ in this inequality, we deduce that $f(\mathbf{b}) \leqslant g(\mathbf{b})$. Thus $\mathbf{b} \in B$ and B is closed.

3. Let $\mathbf{x}, \mathbf{y} \in \mathbb{R}^n$. Then, for each $\varepsilon > 0$, there exists $\mathbf{a} \in A$ such that $\|\mathbf{x} - \mathbf{a}\| > f(\mathbf{x}) - \varepsilon$. By the triangle inequality,

$$f(\mathbf{x}) - \varepsilon < \|\mathbf{x} - \mathbf{a}\| \leqslant \|\mathbf{x} - \mathbf{y}\| + \|\mathbf{y} - \mathbf{a}\| \leqslant \|\mathbf{x} - \mathbf{y}\| + f(\mathbf{y}).$$

Since $\varepsilon > 0$ is arbitrary, $f(\mathbf{x}) - f(\mathbf{y}) \leqslant \|\mathbf{x} - \mathbf{y}\|$. Similarly, we have $f(\mathbf{y}) - f(\mathbf{x}) \leqslant \|\mathbf{y} - \mathbf{x}\|$. Thus f satisfies the Lipschitz condition $|f(\mathbf{x}) - f(\mathbf{y})| \leqslant \|\mathbf{x} - \mathbf{y}\|$ on $\mathbb{R}^n$. When A is $B[\mathbf{a}; r]$, $f(\mathbf{x}) = \|\mathbf{x} - \mathbf{a}\| + r$ for $\mathbf{x} \in \mathbb{R}^n$.

4. It follows from the given identity that, for $\mathbf{x}, \mathbf{y} \in \mathbb{R}^n$,

$$(1 + \|\mathbf{x}\|)(1 + \|\mathbf{y}\|)\|f(\mathbf{x}) - f(\mathbf{y})\| \leqslant (1 + \|\mathbf{x}\| + \|\mathbf{y}\|)\|\mathbf{x} - \mathbf{y}\|,$$

whence f satisfies the Lipschitz condition $\|f(\mathbf{x}) - f(\mathbf{y})\| \leqslant \|\mathbf{x} - \mathbf{y}\|$ on $\mathbb{R}^n$.

5. Take $A = \{k\mathbf{a} : k = 2, 3, \ldots\}$ and $B = \{(k + k^{-1})\mathbf{a} : k = 2, 3, \ldots\}$, where $\mathbf{a}$ is any non-zero vector in $\mathbb{R}^n$.

Exercises 2.1

1. Only the sets (ii), (iv), (v) are convex.

2. If $\mathbf{x}, \mathbf{y} \in C$, then, for some $\alpha, \beta \in [0, 1]$, $\mathbf{x} \in \alpha A \cap (1 - \alpha)B$ and $\mathbf{y} \in \beta A \cap (1 - \beta)B$. Thus $\mathbf{x} = \alpha\mathbf{a} = (1 - \alpha)\mathbf{b}$, $\mathbf{y} = \beta\mathbf{a}' = (1 - \beta)\mathbf{b}'$ for some $\mathbf{a}, \mathbf{a}' \in A$ and $\mathbf{b}, \mathbf{b}' \in B$. Let $\lambda, \mu \geqslant 0$ with $\lambda + \mu = 1$. Then $0 \leqslant \lambda\alpha + \mu\beta \leqslant 1$ and

$$\lambda\mathbf{x} + \mu\mathbf{y} \in (\lambda\alpha + \mu\beta)A \cap (1 - \lambda\alpha - \mu\beta)B \subseteq C,$$

which shows that C is convex. When $A = B = U$, $C = \frac{1}{2}U$.

3. Let C be the union of an increasing sequence $C_1, C_2, \ldots$ of convex sets in $\mathbb{R}^n$. If $\mathbf{x}, \mathbf{y} \in C$, then $\mathbf{x} \in C_i$ and $\mathbf{y} \in C_j$ for some i and j. Suppose that $i \leqslant j$. Then $C_i, C_j \subseteq C_j$, and so $\mathbf{x}, \mathbf{y} \in C_j$. Let $\lambda, \mu \geqslant 0$ with $\lambda + \mu = 1$. Since C_j is convex, $\lambda\mathbf{x} + \mu\mathbf{y} \in C_j \subseteq C$, which shows that C is convex. The stated set is convex, since it is a union of an increasing sequence of convex sets.

4. Suppose that, for some *distinct* lattice points $\mathbf{a}$ and $\mathbf{b}$, the sets $\mathbf{a} + \frac{1}{2}A$ and $\mathbf{b} + \frac{1}{2}A$ meet. Then

$$\mathbf{a} - \mathbf{b} \in \tfrac{1}{2}A - \tfrac{1}{2}A = \tfrac{1}{2}A + \tfrac{1}{2}A = A,$$

which contradicts the assumption that A contains no non-zero lattice point. This establishes the stated result.

5. Take $A = B = U$ and $C = \{\mathbf{x} \in \mathbb{R}^n : \|\mathbf{x}\| = 1\}$.

6. Note that $B[\mathbf{a}; r] = \mathbf{a} + rU$. Suppose now that $B[\mathbf{a}; r] = B[\mathbf{b}; s]$, where $r \leqslant s$. Then $\mathbf{a} + rU = \mathbf{b} + sU$, so

$$rU + \mathbf{a} = rU + \mathbf{b} + (s - r)U.$$

Corollary 2.1.9 shows that $\mathbf{a} = \mathbf{b} + (s - r)U$, whence $\mathbf{a} - \mathbf{b} = (s - r)U$ and $\mathbf{a} = \mathbf{b}$, $r = s$.

Exercises 2.2

1. The set $B = \{(x, y, z) : 0 \leqslant x, y, z \leqslant 1\}$ is convex and contains A, so $B \supseteq \operatorname{conv} A$. Each point (x, y, z) of B can be expressed as

$$xyz(1,1,1) + xy(1-z)(1,1,0) + x(1-y)z(1,0,1) +$$
$$x(1-y)(1-z)(1,0,0) + (1-x)yz(0,1,1) +$$
$$(1-x)y(1-z)(0,1,0) + (1-x)(1-y)z(0,0,1) +$$
$$(1-x)(1-y)(1-z)(0,0,0),$$

so (x, y, z) is a convex combination of points of A. Thus $B \subseteq \operatorname{conv} A$, and $B = \operatorname{conv} A$ as required.

2. Each point $\mathbf{p}$ of $\operatorname{conv}(\{\mathbf{x}\} \cup A)$, with the possible exception of $\mathbf{x}$, can be expressed in the form $\lambda_0\mathbf{x} + \lambda_1\mathbf{a}_1 + \ldots + \lambda_m\mathbf{a}_m$, where $\mathbf{a}_1, \ldots, \mathbf{a}_m \in A$ and $\lambda_0, \lambda_1, \ldots, \lambda_m \geqslant 0$ with $\lambda = \lambda_1 + \ldots + \lambda_m = 1 - \lambda_0 > 0$. Thus $\mathbf{p} = \lambda\mathbf{a}_0 + (1-\lambda)\mathbf{x}$, where $\mathbf{a}_0 = (\lambda_1\mathbf{a}_1 + \ldots + \lambda_m\mathbf{a}_m)/\lambda \in A$, whence

$$\operatorname{conv}(\{\mathbf{x}\} \cup A) \subseteq \{\lambda\mathbf{a} + (1-\lambda)\mathbf{x} : 0 \leqslant \lambda \leqslant 1, \mathbf{a} \in A\}.$$

The opposite inclusion is clear, so the stated equality is proved.
Suppose next that *both* $\operatorname{conv}(\{\mathbf{x}\} \cup A)$ meets B *and* that $\operatorname{conv}(\{\mathbf{x}\} \cup B)$ meets A. Then there exist $\mathbf{a}, \mathbf{a}' \in A$, $\mathbf{b}, \mathbf{b}' \in B$, $0 \leqslant \lambda, \mu \leqslant 1$, such that $\lambda\mathbf{a} + (1-\lambda)\mathbf{x} = \mathbf{b}$, $\mu\mathbf{b}' + (1-\mu)\mathbf{x} = \mathbf{a}'$. Since A and B are disjoint, $\lambda, \mu < 1$. On eliminating $\mathbf{x}$ from the last two equations, we find that

$$\frac{(1-\mu)\mathbf{b} + (1-\lambda)\mu\mathbf{b}'}{1-\lambda\mu} = \frac{(1-\lambda)\mathbf{a}' + (1-\mu)\lambda\mathbf{a}}{1-\lambda\mu} = \mathbf{c},$$

say. Thus A and B have a common point $\mathbf{c}$, which contradicts the hypothesis. This establishes the stated assertion.

3. The set $\operatorname{conv} A + \operatorname{conv} B$ is convex and contains $A + B$, whence $\operatorname{conv} A + \operatorname{conv} B \supseteq \operatorname{conv}(A + B)$. If $\mathbf{x} \in \operatorname{conv} A + \operatorname{conv} B$, then

$$\mathbf{x} = \lambda_1\mathbf{a}_1 + \ldots + \lambda_m\mathbf{a}_m + \mu_1\mathbf{b}_1 + \ldots \mu_p\mathbf{b}_p$$

for some $\mathbf{a}_1, \ldots, \mathbf{a}_m \in A$, $\mathbf{b}_1, \ldots, \mathbf{b}_p \in B$, $\lambda_1, \ldots, \lambda_m$, $\mu_1, \ldots, \mu_p \geqslant 0$ with $\lambda_1 + \ldots + \lambda_m = 1$, $\mu_1 + \ldots + \mu_p = 1$. The equation

$$\mathbf{x} = \sum_{i=1}^{m} \sum_{j=1}^{p} \lambda_i\mu_j(\mathbf{a}_i + \mathbf{b}_j)$$

expresses $\mathbf{x}$ as a convex combination of points of $A + B$, which shows that $\operatorname{conv} A + \operatorname{conv} B \subseteq \operatorname{conv}(A + B)$. Hence we have $\operatorname{conv} A + \operatorname{conv} B = \operatorname{conv}(A + B)$.

4. Let $A = \{\mathbf{0}, \mathbf{e}_1, \ldots, \mathbf{e}_n\} \subseteq \mathbb{R}^n$ and let $\mathbf{a} = (\mathbf{e}_1 + \ldots + \mathbf{e}_n)/(n+1)$. Then $\mathbf{a} \in \operatorname{conv} A$, but $\mathbf{a}$ cannot be expressed as a convex combination of fewer than $n + 1$ points of A. This shows that the number $n + 1$ in Carathéodory's theorem cannot be reduced. The set A of $n + 1$ points cannot be partitioned into two sets whose convex hulls

meet. This shows that the number $n + 2$ in Radon's theorem cannot be reduced.

5. Now conv (int A) is open by Theorem 2.2.6, and is contained in conv A, whence conv (int A) $\subseteq$ int (conv A). The given equation does not hold when $A = \{\mathbf{x} \in \mathbb{R}^n : \|\mathbf{x}\| = 1\}$, for example.
6. The diameter of the set of roots of the given polynomial, P, say, equals the diameter of the convex hull of its roots and this, by two applications of the Gauss–Lucas theorem, cannot be less than the diameter of the set of roots of P'', which is easily found to be $\sqrt{|(a^2/4) - (2b/3)|}$.

Exercises 2.3

1. Since the sets cl (conv A) and cl (conv (cl A)) are both closed and convex, there cannot be more than six different members in the sequence, namely A, cl A, conv A, conv (cl A), cl (conv A), and cl(conv (cl A)). We show that there cannot be more than five different members in the sequence by establishing the identity cl (conv (cl A)) = cl (conv A). Now $A \subseteq$ conv A, so cl $A \subseteq$ cl (conv A), conv (cl A) $\subseteq$ cl (conv A) and cl (conv (cl A)) $\subseteq$ cl (conv A). The inclusion cl (conv A) $\subseteq$ cl (conv (cl A)) is trivial. Thus the sets cl(conv (cl A)) and cl (conv A) are equal. The set $A = \{(x, x^2) : x > 0\}$ will generate five different sets. Here cl $A = \{(x, x^2) : x \geqslant 0\}$, conv $A = \{(x, y) : y \geqslant x^2, x > 0\}$, conv (cl A) $= \{(x, y) : y \geqslant x^2, x > 0\} \cup \{(0, 0)\}$, and cl (conv A) $= \{(x, y) : y \geqslant x^2, x \geqslant 0\}$. For a bounded set A, the set cl A is compact, whence so is conv (cl A), and the sets conv (cl A) and cl (conv (cl A)) are equal.
2. Suppose that A is not closed. Then A is non-empty, so there exists $\mathbf{a} \in$ ri A; also there exists $\mathbf{b} \in$ cl $A \setminus A$. Let L be the line through $\mathbf{a}$ and $\mathbf{b}$. The set $\{\lambda\mathbf{a} + (1 - \lambda)\mathbf{b} : 0 < \lambda \leqslant 1\}$ lies in $A \cap L$ and $\mathbf{b} \in$ cl $(A \cap L)$. By hypothesis, $A \cap L$ is closed, so $\mathbf{b} \in A \cap L$. Thus $\mathbf{b} \in A$. This contradiction shows that A is closed. No, consider a semicircular arc in $\mathbb{R}^2$ with its 'endpoints' removed.
3. We have rebd A = cl $A \setminus$ri A = cl (ri A)$\setminus$ri (ri A) = rebd (ri A).
4. The given condition implies that A is n-dimensional, and so has a non-empty interior. Let $\mathbf{a} \in$ int A and let $\mathbf{x} = 2\mathbf{a}_0 - \mathbf{a}$. Then there exists μ with $0 < \mu < 1$ such that

$$\mathbf{b} = (1 - \mu)(2\mathbf{a}_0 - \mathbf{a}) + \mu\mathbf{a}_0 \in A.$$

Now

$$\mathbf{a}_0 = \frac{1}{2 - \mu}\mathbf{b} + \frac{1 - \mu}{2 - \mu}\mathbf{a}_0,$$

whence $\mathbf{a}_0 \in$ int A by Lemma 2.3.3.

Exercises 2.4

1. For all $\mathbf{a} \in A$,

$$\|\mathbf{x}-\mathbf{a}\|^2 = \|\mathbf{x}-\mathbf{a}_0\|^2 + 2(\mathbf{x}-\mathbf{a}_0)\cdot(\mathbf{a}_0-\mathbf{a}) + \|\mathbf{a}_0-\mathbf{a}\|^2 \geqslant \|\mathbf{x}-\mathbf{a}_0\|^2.$$

This shows that $\mathbf{a}_0$ is the nearest point of A to $\mathbf{x}$. Suppose that $\|\mathbf{x}\| \geqslant 1$. Then $\mathbf{x}/\|\mathbf{x}\| \in A$. For all $\mathbf{a} \in A$, we have

$$(\mathbf{x}-(\mathbf{x}/\|\mathbf{x}\|))\cdot(\mathbf{a}-(\mathbf{x}/\|\mathbf{x}\|)) = (\|\mathbf{x}\|-1)(\mathbf{a}\cdot\mathbf{x}-\|\mathbf{x}\|)/\|\mathbf{x}\| \leqslant 0.$$

Here we have used the Cauchy–Schwarz inequality to deduce that $\mathbf{a}\cdot\mathbf{x} \leqslant \|\mathbf{a}\|\,\|\mathbf{x}\| \leqslant \|\mathbf{x}\|$. Thus $\mathbf{x}/\|\mathbf{x}\|$ is the nearest point of A to $\mathbf{x}$.

2. The proof of Corollary 2.4.2 shows that $\|\mathbf{x}-\mathbf{y}\| = \|f(\mathbf{x})-f(\mathbf{y})\|$ implies $\mathbf{x}-f(\mathbf{x}) = \mathbf{y}-f(\mathbf{y})$. If $\|\mathbf{x}-\mathbf{y}\| = \|f(\mathbf{x})-f(\mathbf{y})\|$ for *all* $\mathbf{x}, \mathbf{y}$ in $\mathbb{R}^n$, then $\mathbf{x}-f(\mathbf{x}) = \mathbf{0}$ for all $\mathbf{x} \in \mathbb{R}^n$, whence $A = \mathbb{R}^n$.

3. (i) $\varnothing$, $\mathbb{R}^n$ and closed halfspaces. To see why this is so, suppose that A is a closed convex set in $\mathbb{R}^n$ that is neither the empty set nor $\mathbb{R}^n$, and is such that its complement A^c is convex. Then it follows easily from Theorem 2.4.10, applied to the sets A and A^c, that there is a closed halfspace H^+ such that $A \subseteq H^+$ and $\operatorname{int} H^+ \subseteq A$, whence

$$H^+ = \operatorname{cl}(\operatorname{int} H^+) \subseteq \operatorname{cl} A = A \subseteq H^+$$

and A is the closed halfspace H^+.

(ii) $\varnothing$ and $\mathbb{R}^n$. To see why this is so, use Theorem 2.4.12 and note that the only sets in $\mathbb{R}^n$ having empty boundaries are $\varnothing$ and $\mathbb{R}^n$.

4. In $\mathbb{R}^n$ let A be a non-empty compact convex set with diameter d. Denote by B the intersection of all closed balls in $\mathbb{R}^n$ containing A. Then $B \supseteq A$. Suppose that $\mathbf{z} \notin A$ and that $\mathbf{a}$ is the nearest point of A to $\mathbf{z}$. Let $\mathbf{c} = \frac{1}{2}(\mathbf{a}+\mathbf{z})$, $\mathbf{u} = (\mathbf{a}-\mathbf{z})/\|\mathbf{a}-\mathbf{z}\|$ and $\theta = \|\mathbf{a}-\mathbf{c}\|$. If $\mathbf{x} \in A$, then $(\mathbf{x}-\mathbf{a})\cdot\mathbf{u} \geqslant 0$, by Theorem 2.4.1, and so, for all $r > 0$,

$$\|\mathbf{x}-(\mathbf{a}+r\mathbf{u})\|^2 = \|\mathbf{x}-\mathbf{a}\|^2 - 2r(\mathbf{x}-\mathbf{a})\cdot\mathbf{u} + r^2 \leqslant d^2 + r^2.$$

It follows that, if $r \geqslant d^2/2\theta$, then

$$A \subseteq B[\mathbf{a}+r\mathbf{u}; \sqrt{(d^2+r^2)}] \subseteq B[\mathbf{a}+r\mathbf{u}; r+\theta].$$

Since the ball $B[\mathbf{a}+r\mathbf{u}; r+\theta]$ contains A but not $\mathbf{z}$, it follows that $\mathbf{z} \notin B$. Hence $B \subseteq A$, and so $A = B$.

5. One possibility is to take $u_0 = \sup\{\mathbf{u}\cdot\mathbf{a} : \mathbf{a} \in A\}$.

6. By Theorem 2.4.6, there is a hyperplane $\mathbf{u}\cdot\mathbf{x} = u_0$, which strictly separates A and B. Using the fact that A is a flat, it follows that $\mathbf{u}\cdot\mathbf{a}$ is the same real number, v_0, say, for each point $\mathbf{a}$ of A. The hyperplane $\mathbf{u}\cdot\mathbf{x} = v_0$ contains A and does not meet B.

7. Let $a, b, c, d \in \mathbb{R}$ satisfy

$$ax + by + cz \leqslant d \ \ ((x, y, z) \in A), \quad ax + by + cz \geqslant d \ \ ((x, y, z) \in B).$$

The first inequality shows $b = 0$ and $c \leqslant d$. Since $(1/m, m, 1) \in B$, for $m > 0$, the second inequality shows that $c \geqslant d$, whence $c = d$. Since $(0, 0, 0) \in B$ and $(1, m^2, m) \in B$ for $m \geqslant 0$, the second inequality shows that $0 \geqslant d$ and $c \geqslant 0$. Hence $b = c = d = 0$. It follows easily that $x = 0$ is the only hyperplane which separates A and B; moreover, this hyperplane contains A and meets B.

Exercises 2.5

1. The set $\{(x, y): -1 < y < 1\} \cup \{(0, 1)\}$ has recession cone $\{(0, 0)\}$.
2. (i) Theorem 2.5.1 and Corollary 2.5.2 show that $\operatorname{ri} A$ and $\operatorname{cl} A$ have the same recession cones. (ii) Theorem 2.5.6 shows that A and $\lambda A + \mathbf{a}$ have the same recession cones.
3. (i) $\{(0, y): y \geqslant 0\}$; (ii) $\{(x, y): x \geqslant 0, y \geqslant 0\}$; (iii) $\{(x, y): x \leqslant 0, y \geqslant 0\}$.
4. Let $\mathbf{a}$ be a point of a flat A. Then $A - \mathbf{a}$ is the unique subspace parallel to A. Now A has the same recession cone as the subspace $A - \mathbf{a}$, i.e. $A - \mathbf{a}$ itself. The set $\{(x, y): -1 \leqslant y \leqslant 1\}$ has recession cone the subspace $\{(x, 0): x \in \mathbb{R}\}$, i.e. the x-axis.
5. $\mathbb{R}^2$, closed halfplanes, unbounded parallel strips (convex hulls of two distinct parallel lines) and lines.
6. We note that

$$S = \{\mathbf{x}: A + \mathbf{x} = A\} = \{\mathbf{x}: A + \mathbf{x} \subseteq A\} \cap \{\mathbf{x}: A - \mathbf{x} \subseteq A\} = C \cap (-C).$$

Exercises 2.6

1. The empty set, lines, unbounded parallel strips, closed halfplanes, and $\mathbb{R}^2$.
2. If $\mathbf{a}$ is an extreme point of $\operatorname{conv} A$, then $(\operatorname{conv} A) \setminus \{\mathbf{a}\}$ is convex and contains $A \setminus \{\mathbf{a}\}$, so $(\operatorname{conv} A) \setminus \{\mathbf{a}\} \supseteq \operatorname{conv}(A \setminus \{\mathbf{a}\})$. Thus $\mathbf{a} \in A$ and $\mathbf{a} \notin \operatorname{conv}(A \setminus \{\mathbf{a}\})$. Conversely, if $\mathbf{a} \in A$ and $\mathbf{a} \notin \operatorname{conv}(A \setminus \{\mathbf{a}\})$, then it is easily verified that

$$(\operatorname{conv} A) \setminus \{\mathbf{a}\} = \{\lambda \mathbf{a} + (1 - \lambda)\mathbf{b}: 0 \leqslant \lambda < 1 \text{ and } \mathbf{b} \in \operatorname{conv}(A \setminus \{\mathbf{a}\})\},$$

and that this set is convex. Hence $\mathbf{a}$ is an extreme point of $\operatorname{conv} A$.
3. We consider the non-trivial case of a two-dimensional closed convex set A in $\mathbb{R}^2$. Let $\mathbf{a}_1, \ldots, \mathbf{a}_k, \ldots$ be a sequence of extreme points of A that converges to some point $\mathbf{a}$ of A. We prove that $\mathbf{a}$ is an extreme point of A. Suppose that this is not so. Clearly $\mathbf{a}$

must be a boundary point of A, and so there exists a support line L to A at $\mathbf{a}$. Since $\mathbf{a}$ is not an extreme point of the face $A \cap L$ of A, there exist *distinct* points $\mathbf{b}$ and $\mathbf{c}$ of A such that $\mathbf{a}$ lies in $\operatorname{ri}(\operatorname{conv}\{\mathbf{b}, \mathbf{c}\})$. Let $\mathbf{d}$ be an interior point of A. For some sufficiently large k, the point $\mathbf{a}_k$ will lie either in the interior of the triangle $\operatorname{conv}\{\mathbf{b}, \mathbf{c}, \mathbf{d}\}$ or in the relative interior of the line segment $\operatorname{conv}\{\mathbf{b}, \mathbf{c}\}$. In neither of these cases can $\mathbf{a}_k$ be an extreme point of A. This contradiction shows that $\mathbf{a}$ must be an extreme point of A.

4. Let B be an $(n-1)$-dimensional face of a convex set A in $\mathbb{R}^n$. Let $\mathbf{b} \in \operatorname{ri} B$ and let H be a support hyperplane to A at $\mathbf{b}$. Then $A \cap H$ is an exposed face of A containing $\mathbf{b}$, whence $B \subseteq A \cap H$. But B is $(n-1)$-dimensional, whence $\operatorname{aff} B = H$ and $\mathbf{b} \in \operatorname{ri}(A \cap H)$. Corollary 2.6.7 now shows that B must be the exposed face $A \cap H$ of A.
5. Take $A = \{(x, y) : x^4 \leqslant y \leqslant 1\}$ and $\mathbf{a} = (0, 0)$ in $\mathbb{R}^2$.
6. Suppose *first* that $\mathbf{x}$ is an extreme point of a *compact* convex set A in $\mathbb{R}^n$. Denote by B the closure of the set of exposed points of A. Then $\operatorname{conv} B$ is a compact convex set containing all the exposed points of A, so by Straszewicz's theorem $\operatorname{conv} B = A$. Hence $\mathbf{x} = \lambda_1\mathbf{b}_1 + \ldots + \lambda_m\mathbf{b}_m$ for some $\mathbf{b}_1, \ldots, \mathbf{b}_m \in B$ and some $\lambda_1, \ldots, \lambda_m > 0$ with $\lambda_1 + \ldots + \lambda_m = 1$. But $\mathbf{x}$ is an extreme point of A, whence all the $\mathbf{b}_1, \ldots, \mathbf{b}_m$ equal $\mathbf{x}$. Thus $\mathbf{x} \in B$, as required.

   Suppose *next* that $\mathbf{x}$ is an extreme point of a *closed* convex set A in $\mathbb{R}^n$. Let C be a closed ball in $\mathbb{R}^n$ having $\mathbf{x}$ as a centre. By the last paragraph, there is a sequence $\mathbf{x}_1, \ldots, \mathbf{x}_k, \ldots$ of exposed points of $A \cap C$ that converges to $\mathbf{x}$ and lies in the *interior* of C. Since each $\mathbf{x}_k$ must be an exposed point of A, the desired result is clear.
7. Let B be the set of exposed points of a compact convex set A in $\mathbb{R}^n$. Then

$$\operatorname{ri} A = \operatorname{ri}(\operatorname{cl}(\operatorname{conv} B)) = \operatorname{ri}(\operatorname{conv} B) \subseteq \operatorname{conv} B.$$

   We now use Carathéodory's theorem to deduce that each point of $\operatorname{ri} A$ can be expressed as a convex combination of $n + 1$ or fewer points of B.
8. Take A to be a square in $\mathbb{R}^2$, B a line segment joining a midpoint of one of the square's sides to an adjacent corner, and C a diagonal of the square.

Exercises 2.7

1. Take $A = \mathbb{R}^n \backslash \{\mathbf{0}\}$. For a closed non-convex set A in $\mathbb{R}^n$, choose $\mathbf{a} \in (\operatorname{conv} A) \backslash A$ and $\varepsilon > 0$ such that $B(\mathbf{a}; \varepsilon) \cap A = \varnothing$.

2. $\rho(C_r, P)$ is $1 - r$ when $0 < r \leqslant \cos(\pi/n)$, $\max\{1 - r, r - \cos(\pi/n)\}$ when $\cos(\pi/n) \leqslant r \leqslant 1$, and $r - \cos(\pi/n)$ when $r > 1$. The minimal value of $\rho(C_r, P)$ is $\sin^2(\pi/2n)$.
3. We note that

$$A_\alpha + (\beta - \alpha)U = (1 - \beta)A_0 + \alpha A_1 + (\beta - \alpha)(A_0 + U) \supseteq$$
$$(1 - \beta)A_0 + \alpha A_1 + (\beta - \alpha)A_1 = A_\beta,$$

and similarly $A_\beta + (\beta - \alpha)U \supseteq A_\alpha$, whence $\rho(A_\alpha, A_\beta) \leqslant \beta - \alpha$. Now

$$1 = \rho(A_0, A_1) \leqslant \rho(A_0, A_\alpha) + \rho(A_\alpha, A_\beta) + \rho(A_\beta, A_1)$$
$$\leqslant \alpha + (\beta - \alpha) + 1 - \beta = 1,$$

and equality must hold throughout. In particular, $\rho(A_\alpha, \mathrm{A}_\beta) = \beta - \alpha$.
4. Let $\mathbf{x} \in A$ and $\varepsilon > 0$. Then the ball $B(\mathbf{x}; \varepsilon)$ meets all but a finite number of members of $A_1, A_2, \ldots$, whence it also meets $\bigcup_{j=i}^{\infty} A_j$ for each $i = 1, 2, \ldots$. Hence $\mathbf{x} \in \mathrm{cl}(\bigcup_{j=i}^{\infty} A_j)$ for $i = 1, 2, \ldots$, and so $\mathbf{x} \in \bigcap_{i=1}^{\infty}(\mathrm{cl}(\bigcup_{j=i}^{\infty} A_j))$. Conversely, suppose that $\mathbf{x} \notin A$. Then there exists $\varepsilon > 0$ such that the ball $B(\mathbf{x}; 2\varepsilon) \cap A = \varnothing$, whence $B(\mathbf{x}; \varepsilon)$ meets only a finite number of members of $A_1, A_2, \ldots$. Thus $\mathbf{x} \notin \mathrm{cl}(\bigcup_{j=i_0}^{\infty} A_j)$ for some i_0, and so $\mathbf{x} \notin \bigcap_{i=1}^{\infty}(\mathrm{cl}(\bigcup_{j=i}^{\infty} A_j))$.
5. Let $A, B \in \mathfrak{F}$ with $\rho(A, B) = \lambda$. Then $A \subseteq (B)_\lambda$ and $B \subseteq (A)_\lambda$, so

$$R(A) \leqslant R((B)_\lambda) \leqslant R(B) + \lambda \quad \text{and} \quad R(B) \leqslant R((A)_\lambda) \leqslant R(A) + \lambda.$$

Thus $|R(A) - R(B)| \leqslant \lambda = \rho(A, B)$, which proves that R is continuous on $\mathfrak{F}$. A similar argument shows that r is continuous on $\mathfrak{F}$.

Exercises 2.8

1. (i) $\{(x, 0): x \leqslant 0\}$; (ii) $\{(x, 0): 0 \leqslant x \leqslant \frac{1}{2}\}$; (iii) $\{(x, 0): x \leqslant \frac{1}{2}\}$; (iv) $\{(x, y): y \leqslant -\frac{1}{4}x^2\}$; (v) $\{(x, y): (ax)^2 + (by)^2 \leqslant 1\}$.
2. The set A^* is closed, convex, and contains the origin, whence, by Theorem 2.8.3, $A^{***} = (A^*)^{**} = A^*$.
3. Since $(A \cap B)^*$ is closed, convex, and contains $A^* \cup B^*$, $(A \cap B)^* \supseteq \mathrm{cl}(\mathrm{conv}(A^* \cup B^*))$. If $\mathbf{x} \notin \mathrm{cl}(\mathrm{conv}(A^* \cup B^*))$, then there is $\mathbf{u}$ in $\mathbb{R}^n$ such that $\mathbf{u} \cdot \mathbf{x} > 1$ and $\mathbf{u} \cdot \mathbf{y} < 1$ when $\mathbf{y} \in A^* \cup B^*$. Thus $\mathbf{u} \in A^{**} \cap B^{**} = A \cap B$ and $\mathbf{x} \notin (A \cap B)^*$, whence $(A \cap B)^* \subseteq \mathrm{cl}(\mathrm{conv}(A^* \cup B^*))$ and $(A \cap B)^* = \mathrm{cl}(\mathrm{conv}\, A^* \cup B^*))$.
4. Trivially, $\varphi(B) \subseteq \{\mathbf{x} \in A^*: \mathbf{b}_0 \cdot \mathbf{x} = 1\}$. Suppose that $\mathbf{x} \in A^*$ satisfies $\mathbf{b}_0 \cdot \mathbf{x} = 1$. Let $\mathbf{b} \in B$. Since $\mathbf{b}_0 \in \mathrm{ri}\, B$, there exist $\mathbf{b}_1 \in B$ and $\lambda \in (0, 1)$ such that $\mathbf{b}_0 = \lambda \mathbf{b} + (1 - \lambda)\mathbf{b}_1$. Hence $\mathbf{b}_0 \cdot \mathbf{x} = \lambda \mathbf{b} \cdot \mathbf{x} + (1 - \lambda)\mathbf{b}_1 \cdot \mathbf{x}$. But $\mathbf{b} \cdot \mathbf{x}$, $\mathbf{b}_1 \cdot \mathbf{x} \leqslant 1$, whence $\mathbf{b} \cdot \mathbf{x} = 1$ and $\{\mathbf{x} \in A^*: \mathbf{b}_0 \cdot \mathbf{x} = 1\} \subseteq \varphi(B)$.

5. It is easily verified that, for any set S in $\mathbb{R}^n$, $S^* = (\operatorname{conv} S)^*$ and $S^* = (\operatorname{cl} S)^*$, whence $S^* = (\operatorname{cl}(\operatorname{conv} S))^*$. Thus $A^* = (\operatorname{cl}(\operatorname{conv} B))^* = B^*$, by Straszewicz's theorem. The polar dual C^* is given by the equation

$$C^* = \{(x, y): x^2 + y^2 \leqslant 1 \text{ and } x \leqslant 0\} \cup \{(x, y): |y| \leqslant 1 - x \text{ and } x \geqslant 0\}.$$

The exposed faces of C are: $\varnothing$, $\{(1, 1)\}$, $\{(1, -1)\}$, $\{(\cos\theta, \sin\theta)\}$ for $\frac{\pi}{2} < \theta < \frac{3\pi}{2}$, $\{(x, 1): 0 \leqslant x \leqslant 1\}$, $\{(x, -1): 0 \leqslant x \leqslant 1\}$, $\{(1, y): |y| \leqslant 1\}$, and C. The polar face mapping of C associates with these faces, respectively, the following exposed faces of C^*: C^*, $\{(x, 1-x): 0 \leqslant x \leqslant 1\}$, $\{(x, x-1): 0 \leqslant x \leqslant 1\}$, $\{(\cos\theta, \sin\theta)\}$ for $\frac{\pi}{2} < \theta < \frac{3\pi}{2}$, $\{(0, 1)\}$, $\{(0, -1)\}$, $\{(1, 0)\}$, and $\varnothing$.

Exercises 3.1

1. A routine calculation shows that the given set is the convex hull of the zero vector and the elementary vectors $\mathbf{e}_1, \ldots, \mathbf{e}_n$.
2. Let $Q = \operatorname{conv}(\{\mathbf{x}\} \cup P)$. Then $Q = \operatorname{conv}(\{\mathbf{x}\} \cup V)$, and so Q is a polytope whose vertex set is contained in $\{\mathbf{x}\} \cup V$. Corollary 3.1.5 shows that $\{\mathbf{x}\}$ and P are faces of Q, whence, by Theorem 2.6.5 (ii), each vertex of P is a vertex of Q. Hence Q has vertex set $\{\mathbf{x}\} \cup V$.
3. Let A be a non-empty compact convex set in $\mathbb{R}^n$. For each $i = 1, 2, \ldots$, let P_i be a polytope such that $\rho(P_i, A) \leqslant 1/i$ and $P_i \supseteq A$. Then $A = \bigcap(P_i: i = 1, 2, \ldots)$.
4. Suppose that $v(P_i) \nrightarrow \infty$ as $i \to \infty$. Then there exists (after much taking of subsequences) a positive integer m and a subsequence $Q_1, \ldots, Q_i, \ldots$ of $P_1, \ldots, P_i, \ldots$ such that $Q_i = \operatorname{conv}\{\mathbf{a}_{i1}, \ldots, \mathbf{a}_{im}\}$ say, and the sequence $\mathbf{a}_{1j}, \ldots, \mathbf{a}_{ij}, \ldots$ converges to some $\mathbf{a}_j$ for $j = 1, \ldots, m$. Thus $Q_1, \ldots, Q_i, \ldots$, and hence $P_1, \ldots, P_i, \ldots$, converges to the polytope $\operatorname{conv}\{\mathbf{a}_1, \ldots, \mathbf{a}_m\}$. This contradiction shows that $v(P_i) \to \infty$. The converse is false. To see why this is so, for $i = 1, 2, \ldots$, let P_i be the convex hull of i points whose distance from the origin $\mathbf{0}$ in $\mathbb{R}^n$ is $1/i$. Then $P_i \to \{\mathbf{0}\}$ and $v(P_i) = i \to \infty$ as $i \to \infty$.
5. Suppose that in $\mathbb{R}^n$ the polytope P is the union of an increasing sequence $A_1, \ldots, A_i, \ldots$ of convex sets. Since P has only a finite number of vertices, there is a positive integer m such that A_m contains all these vertices, and hence P itself. Thus $A_i = P$ for $i = m, m+1, \ldots$, and the sequence $A_1, \ldots, A_i, \ldots$ is *not* strictly increasing.

Exercises 3.2

1. The first assertion follows from the facts that in $\mathbb{R}^n$: (i) the complement of a polyhedral set is a finite union of open halfspaces; (ii) a finite intersection of sets, each of which is a finite union of open halfspaces, is itself a finite union of sets, each of which is a finite intersection of open halfspaces.

   Suppose that C is non-empty and that $P_1 \cup \ldots \cup P_m$ is not $\mathbb{R}^n$. Then the complement of $P_1 \cup \ldots \cup P_m$ in $\mathbb{R}^n$ has the form $C_1 \cup \ldots \cup C_r$, where $C_1, \ldots, C_r$ are non-empty open convex sets. Thus, for each $i = 1, \ldots, r$, there exists a closed halfspace J_i in $\mathbb{R}^n$ which contains C and is disjoint from C_i. Now let P be the polyhedral set $J_1 \cap \ldots \cap J_r$.

2. Suppose that A is an n-dimensional polytope in $\mathbb{R}^n$ (n $\geqslant 2$). Then A is a polyhedral set, not a flat, and so can be expressed as in Theorem 3.2.1, i.e.

$$A = \{\mathbf{x} \in \mathbb{R}^n : \mathbf{a}_1 \cdot \mathbf{x} \leqslant \alpha_1\} \cap \ldots \cap \{\mathbf{x} \in \mathbb{R}^n : \mathbf{a}_m \cdot \mathbf{x} \leqslant \alpha_m\},$$

   where m is minimal. Suppose that $m \leqslant n$. Then $\mathbf{a}_1, \ldots, \mathbf{a}_{m-1}$ lie in some hyperplane of $\mathbb{R}^n$ containing the origin, whence there exists $\mathbf{u} \neq \mathbf{0}$ such that $\mathbf{u} \cdot \mathbf{a}_1 = 0, \ldots, \mathbf{u} \cdot \mathbf{a}_{m-1} = 0$. Suppose that $\mathbf{u} \cdot \mathbf{a}_m \leqslant 0$. Then $A + \lambda\mathbf{u} \subseteq A$ for all $\lambda \geqslant 0$, which implies that A is unbounded. This contradiction shows that $m \geqslant n + 1$. By Theorem 3.2.1(iii), A has m distinct facets.

3. Let P be a polyhedral set in $\mathbb{R}^n$ which contains the origin in its interior, say P is the intersection of closed halfspaces $\{\mathbf{x} \in \mathbb{R}^n : \mathbf{a}_1 \cdot \mathbf{x} \leqslant 1\}, \ldots, \{\mathbf{x} \in \mathbb{R}^n : \mathbf{a}_m \cdot \mathbf{x} \leqslant 1\}$. Then $\operatorname{conv}\{\mathbf{0}, \mathbf{a}_1, \ldots, \mathbf{a}_m\} \subseteq P^*$. If $\mathbf{x} \notin \operatorname{conv}\{\mathbf{0}, \mathbf{a}_1, \ldots, \mathbf{a}_m\}$, then $\{\mathbf{x}\}$ and $\operatorname{conv}\{\mathbf{0}, \mathbf{a}_1, \ldots, \mathbf{a}_m\}$ can be strictly separated, so there exists some $\mathbf{y}$ in $\mathbb{R}^n$ such that $\mathbf{y} \cdot \mathbf{a}_1 < 1, \ldots, \mathbf{y} \cdot \mathbf{a}_m < 1$ and $\mathbf{y} \cdot \mathbf{x} > 1$. Thus $\mathbf{y} \in P$ and $\mathbf{x} \notin P^*$. Hence $P^* \subseteq \operatorname{conv}\{\mathbf{0}, \mathbf{a}_1, \ldots, \mathbf{a}_m\}$, and P^* is the polytope $\operatorname{conv}\{\mathbf{0}, \mathbf{a}_1, \ldots, \mathbf{a}_m\}$.

4. Suppose that A is *not* a polyhedral set. Since A must be closed (why?), it must have an infinite number of faces, each of which contains an extreme point of A. Thus there is some extreme point $\mathbf{a}$ of A and some infinite family $(F_i : i \in I)$ of distinct faces of A such that $\mathbf{a} \in F_i$ for $i \in I$. Let P be a polytope in $\mathbb{R}^n$ which has $\mathbf{a}$ as an interior point. Then $(P \cap F_i : i \in I)$ is an infinite family of distinct faces of $P \cap A$, which contradicts the hypothesis that $P \cap A$ is a polytope. Thus A is a polyhedral set.

Exercises 3.3

1. (i) 6, 15, 20, 15, 6; (ii) 10, 40, 80, 80, 32; (iii) 32, 80, 80, 40, 10.
2. Convex hulls of (i) (0, 0, 0, 0), (1, 0, 0, 0), (0, 1, 0, 0), (0, 0, 1, 0), (0, 0, 0, 1); (ii) (0, 0, 0, 0), (1, 0, 0, 0), (0, 1, 0, 0), $(\frac{1}{3}, \frac{1}{3}, 1, 0)$, $(\frac{1}{3}, \frac{1}{3}, -1, 0)$, $(\frac{1}{3}, \frac{1}{3}, 0, 1)$, $(\frac{1}{3}, \frac{1}{3}, 0, -1)$; (iii) (0, 0, 0, 0), (1, 0, 0, 0), (0, 1, 0, 0), (0, 0, 1, 0), (1, 0, 1, 0), (0, 1, 1, 0), (0, 0, 0, 1), (1, 0, 0, 1), (0, 1, 0, 1), (0, 0, 1, 1), (1, 0, 1, 1), (0, 1, 1, 1).
3. Use induction on the dimension of the crosspolytope.
4. Consider the case of an r-prism P; the other cases can be treated similarly. By Theorem 3.3.3,

$$\sum_{i=-1}^{r} (-1)^{i+1} f_i(P) = 1 - 2f_0(Q) + 2\sum_{i=1}^{r} (-1)^{i+1} f_i(Q) +$$
$$\sum_{i=1}^{r} (-1)^{i+1} f_{i-1}(Q) = \sum_{i=-1}^{r-1} (-1)^{i+1} f_i(Q).$$

Thus if Q satisfies Euler's relation, so does P. This result and the other two similar results, together with the fact that a point satisfies Euler's relation, show that simplexes (except the empty one), cross-polytopes, and parallelotopes all satisfy Euler's relation.

Exercises 3.4

1. The tetrahedron, the octahedron, and the icosahedron.
2. $f_{-1} = 1$, $f_0 = 7$, $f_1 = 21$, $f_2 = 34$, $f_3 = 30$, $f_4 = 12$, $f_5 = 1$.
3. Let $C(2n + 2, 2n)$ have vertices $\mathbf{v}_i = (t_i, \ldots, t_i^n)$, where $t_1 < \ldots < t_{2n+2}$. Gale's evenness condition shows that $\{\mathbf{v}_1, \ldots, \mathbf{v}_{2n+2}\} \backslash \{\mathbf{v}_i, \mathbf{v}_j\}$ determines a facet if and only if i and j have opposite parity. Thus a set of vertices determines a facet if and only if it has the form $A \cup B$, where A is a set of n vertices with even suffixes and B is a set of n vertices with odd suffixes. There are $n + 1$ possible choices for each of the sets A and B, so $C(2n + 2, 2n)$ has $(n + 1)^2$ facets.
4. $C(12, 2)$ has 12 facets (smallest). $C(12, 6)$ and $C(12, 7)$ have 112 facets (largest).
5. No. A pyramid in $\mathbb{R}^{2n+1}$ which has for a base an n-neighbourly $2n$-polytope (which is not a simplex) is n-neighbourly but not simplicial.

Exercises 3.5

1. Let $\mathbf{b}_{ij} = \mathbf{a}_i - \mathbf{a}_j$ for $i \neq j$ and let H_{ij} be the hyperplane with equation $\mathbf{b}_{ij} \cdot \mathbf{x} = 0$. Each of these $\frac{1}{2}v(v - 1)$ hyperplanes is closed and

has empty interior. It follows easily from Problem 5 of Exercises 1.7 that the union S of these hyperplanes is not $\mathbb{R}^n$. Choose $\mathbf{a} \in \mathbb{R}^n \setminus S$.

2. Theorem 3.5.2 with $k = r - 2$ shows that

$$h_{r-2}(F) - h_{r-1}(F) + h_r(F) = 1 - h_{r-1}(F) + 1 = 0,$$

whence $h_{r-1}(F) = 2$.

3. Denote by A the set of f-vectors of all r-polytopes. By Euler's relation, A lies in the hyperplane with equation

$$x_1 - x_2 + \ldots + (-1)^{r-1}x_r = 1 - (-1)^r.$$

Theorem 3.5.3 shows that this is the only hyperplane in $\mathbb{R}^r$ containing A. Thus this hyperplane is the affine hull of A.

4. Both equations are $2f_{r-2} = rf_{r-1}$.

5. The Dehn–Sommerville equations, $f_0 - f_1 + f_2 - f_3 + f_4 = 2$, $-2f_1 + 3f_2 - 4f_3 + 5f_4 = 0$, $2f_3 = 5f_4$, show that

$$f_2 = 20 - 10f_0 + 4f_1, \quad f_3 = 30 - 15f_0 + 5f_1, \quad f_4 = 12 - 6f_0 + 2f_1.$$

6. (i) The bipyramid formed from two regular tetrahedra with equal edge lengths has congruent equilateral triangles for its facets, but is not regular.

   (ii) A right prism of unit height with an equilateral triangle of unit length for its base has regular facets (three squares and two triangles) and each of its vertices belongs to three facets, but is not regular.

   (iii) It is easy to construct an irregular tetrahedron whose facets are congruent; and each vertex of a tetrahedron always belongs to three of its facets.

Exercises 3.6

1. (i) The subspace of affine dependences is the subspace $\{\alpha(1, -1, 1, -1, 0) : \alpha \in \mathbb{R}\}$ of $\mathbb{R}^5$, which has $(1, -1, 1, -1, 0)$ as a basis. The corresponding Gale transform is $1, -1, 1, -1, 0$ in $\mathbb{R}^1$.

   (ii) The subspace of affine dependences is the subspace $\{\alpha(2, 2, 2, -3, -3) : \alpha \in \mathbb{R}\}$ of $\mathbb{R}^5$, which has $(2, 2, 2, -3, -3)$ as a basis. The corresponding Gale transform is $2, 2, 2, -3, -3$ in $\mathbb{R}^1$.

   (iii) The subspace of affine dependences is the subspace $\{\alpha(-1, 1, 0, 1, -1, 0) + \beta(-1, 0, 1, 1, 0, -1) : \alpha, \beta \in \mathbb{R}\}$ of $\mathbb{R}^6$, which has $(-1, 1, 0, 1, -1, 0)$, $(-1, 0, 1, 1, 0, -1)$ as a basis. The corresponding Gale transform is $(-1, -1)$, $(1, 0)$, $(0, 1)$, $(1, 1)$, $(-1, 0)$, $(0, -1)$ in $\mathbb{R}^2$.

2. Suppose that P is an s-fold n-pyramid $(1 \leqslant s \leqslant n - 2)$ with s-base $\operatorname{conv}\{\mathbf{a}_{s+1}, \ldots, \mathbf{a}_m\}$ and that P is obtained from this base by

s applications of the cone construction using apexes $\mathbf{a}_s, \ldots, \mathbf{a}_1$ in turn. Then $\operatorname{conv}\{\mathbf{a}_2, \ldots, \mathbf{a}_m\}$, $\operatorname{conv}\{\mathbf{a}_3, \ldots, \mathbf{a}_m\}, \ldots, \operatorname{conv}\{\mathbf{a}_{s+1}, \ldots, \mathbf{a}_m\}$ are faces of P, whence $\mathbf{0}$ is a relative interior point of each of the sets $\operatorname{conv}\{\bar{\mathbf{a}}_1\}$, $\operatorname{conv}\{\bar{\mathbf{a}}_1, \bar{\mathbf{a}}_2\}, \ldots, \operatorname{conv}\{\bar{\mathbf{a}}_1, \ldots, \bar{\mathbf{a}}_s\}$. This shows that $\bar{\mathbf{a}}_1 = \bar{\mathbf{a}}_2 = \ldots = \bar{\mathbf{a}}_s = \mathbf{0}$.

3. Gale's evenness condition shows that the Gale transforms of vertex sequences of the cyclic polytope $C(n+2, n)$ consist of $n+2$ non-zero real numbers with zero sum, having an equal number of positive and negative terms when n is even, and $\frac{1}{2}(n+2)$ terms of one sign and $\frac{1}{2}(n+3)$ terms of the opposite sign when n is odd. The number of facets of $C(n+2, n)$ is the product of the number of negative terms with the number of positive terms in one of its Gale transforms, i.e. $\frac{1}{4}(n+2)^2$ when n is even and $\frac{1}{4}(n+1)(n+3)$ when n is odd. Denoting by f_i the number of i-faces of $C(7, 5)$, we calculate (from any Gale transform of $C(7, 5)$) that $f_0 = 7$, $f_1 = 21$, $f_2 = 34$, $f_3 = 30$, $f_4 = 12$. These f_i satisfy the Dehn–Sommerville equations:

$$f_0 - f_1 + f_2 - f_3 + f_4 = 2, \quad 2f_1 - 3f_2 + 4f_3 - 5f_4 = 0, \quad 2f_3 = 5f_4.$$

4. Let P be the convex hull of the points

$$\mathbf{e}_1, \ldots, \mathbf{e}_{r-1}, \quad -(\mathbf{e}_1 + \ldots + \mathbf{e}_{r-1}), \ \mathbf{e}_r, \ldots, \mathbf{e}_n, \quad -(\mathbf{e}_r + \ldots + \mathbf{e}_n),$$

where $\mathbf{e}_1, \ldots, \mathbf{e}_n$ are the vectors $(1, 0, \ldots, 0, 0), \ldots, (0, 0, \ldots, 0, 1)$ in $\mathbb{R}^n$. Then any Gale transform of a vertex sequence of P has r terms of one sign and s terms of the opposite sign.

5. Suppose that $W = \{\mathbf{a}_1, \ldots, \mathbf{a}_r\}$ is a non-empty coface of P. Then, by Theorem 3.6.6, there exist $\lambda_1, \ldots, \lambda_r > 0$ such that

$$\mathbf{0} = \lambda_1 \bar{\mathbf{a}}_1 + \ldots + \lambda_r \bar{\mathbf{a}}_r.$$

Write $\mu_i = \lambda_i \|\bar{\mathbf{a}}_i\|$ if $\bar{\mathbf{a}}_i \neq \mathbf{0}$, and $\mu_i = \lambda_i$ if $\bar{\mathbf{a}}_i = \mathbf{0}$. Then

$$\mathbf{0} = \mu_1 \hat{\mathbf{a}}_1 + \ldots + \mu_r \hat{\mathbf{a}}_r.$$

Since $\mu_1, \ldots, \mu_r > 0$, $\mathbf{0} \in \operatorname{ri}(\operatorname{conv} \hat{W})$. This argument can be reversed to show that, if $\mathbf{0} \in \operatorname{ri}(\operatorname{conv} \hat{W})$, then W is a coface of P.

Exercises 4.1

1. Take the closed halfspaces $z - x \geqslant 0$, $z - y \geqslant 0$, $z + x \geqslant 0$, $z + y \geqslant 0$.
2. $P = \operatorname{conv}\{(0, 1), (1, 0), (\frac{1}{4}, \frac{1}{4})\} + \operatorname{cone}\{(0, 1), (1, 1)\}$.
3. Since $A + C = B + C + C = B + C = A$, C is contained in the recession cone of A. Let $\mathbf{b} \in B$ and let $\mathbf{x}$ lie in the recession cone of A. Then, for each positive integer m, $\mathbf{b} + m\mathbf{x} \in A$, whence $\mathbf{b} + m\mathbf{x} = \mathbf{b}_m + m\mathbf{c}_m$ for some $\mathbf{b}_m \in B$, $\mathbf{c}_m \in C$. Since B is bounded,

$\mathbf{c}_m \to \mathbf{x}$ as $m \to \infty$. But C is closed, so $\mathbf{x} \in C$. Thus the recession cone of A is contained in C, and so must be C.

Let A be a non-empty polyhedral set in $\mathbb{R}^n$. Write $A = B + C$, where B is a polytope and C a finitely generated convex cone. By the first part, the polyhedral cone C is the recession cone of A.

Exercises 4.2

1. (a) Suppose that both (i) and (ii) hold. Then $0 = \mathbf{y}^{\mathrm{T}}\mathbf{A}\mathbf{x} = \mathbf{y}^{\mathrm{T}}\mathbf{b} \neq 0$, which is impossible. If (i) has no solution, then the combined system $\mathbf{A}\mathbf{x} \leqslant \mathbf{b}$, $-\mathbf{A}\mathbf{x} \leqslant -\mathbf{b}$ has no solution, whence by Theorem 4.2.3 there exist $\mathbf{u}, \mathbf{v}$ such that $\mathbf{u}^{\mathrm{T}}\mathbf{A} - \mathbf{v}^{\mathrm{T}}\mathbf{A} = \mathbf{0}^{\mathrm{T}}$, $\mathbf{u}^{\mathrm{T}}\mathbf{b} - \mathbf{v}^{\mathrm{T}}\mathbf{b} < 0$. Let $\mathbf{y} = \mathbf{u} - \mathbf{v}$. Then $\mathbf{y}^{\mathrm{T}}\mathbf{A} = \mathbf{0}^{\mathrm{T}}$, $\mathbf{y}^{\mathrm{T}}\mathbf{b} \neq 0$.

   (b) Suppose that both (i) and (ii) hold. Then $0 \leqslant \mathbf{y}^{\mathrm{T}}\mathbf{A}\mathbf{x} \leqslant \mathbf{y}^{\mathrm{T}}\mathbf{b} < 0$, which is impossible. If (i) has no solution, then the system $[\mathbf{A}, \mathbf{I}_m]\mathbf{z} = \mathbf{b}$, where $\mathbf{z} = (z_1, \ldots, z_{n+m})$, has no solution for which $\mathbf{z} \geqslant \mathbf{0}$. By Lemma 4.2.2, there exists $\mathbf{y}$ such that $\mathbf{y}^{\mathrm{T}}\mathbf{A} \geqslant \mathbf{0}^{\mathrm{T}}$, $\mathbf{y}^{\mathrm{T}}\mathbf{I}_{\mathrm{m}} = \mathbf{y}^{\mathrm{T}} \geqslant \mathbf{0}^{\mathrm{T}}$, $\mathbf{y}^{\mathrm{T}}\mathbf{b} < 0$.
2. Let P, Q be disjoint non-empty polyhedral sets given by the systems of inequalities $\mathbf{A}\mathbf{x} \leqslant \mathbf{b}$, $\mathbf{C}\mathbf{x} \leqslant \mathbf{d}$, respectively. Then the combined system $\mathbf{A}\mathbf{x} \leqslant \mathbf{b}$, $\mathbf{C}\mathbf{x} \leqslant \mathbf{d}$ is inconsistent. By Theorem 4.2.3, there exist $\mathbf{u} \geqslant \mathbf{0}$, $\mathbf{v} \geqslant \mathbf{0}$ such that $\mathbf{u}^{\mathrm{T}}\mathbf{A} + \mathbf{v}^{\mathrm{T}}\mathbf{C} = \mathbf{0}^{\mathrm{T}}$, $\mathbf{u}^{\mathrm{T}}\mathbf{b} + \mathbf{v}^{\mathrm{T}}\mathbf{d} < 0$. The sets P and Q are strictly separated by the hyperplane $(\mathbf{u}^{\mathrm{T}}\mathbf{A})\mathbf{z} = \frac{1}{2}(\mathbf{u}^{\mathrm{T}}\mathbf{b} - \mathbf{v}^{\mathrm{T}}\mathbf{d})$.
3. Let $\mathbf{e}^{\mathrm{T}}\mathbf{x} = f$ be a consequence of the consistent system $\mathbf{A}\mathbf{x} = \mathbf{b}$. Then $\mathbf{e}^{\mathrm{T}}\mathbf{x} \leqslant f$ is a consequence of the consistent system formed by combining the systems $\mathbf{A}\mathbf{x} \leqslant \mathbf{b}$, $-\mathbf{A}\mathbf{x} \leqslant -\mathbf{b}$. Thus there exist $\mathbf{u} \geqslant \mathbf{0}$, $\mathbf{v} \geqslant \mathbf{0}$ such that $\mathbf{u}^{\mathrm{T}}\mathbf{A} - \mathbf{v}^{\mathrm{T}}\mathbf{A} = \mathbf{e}^{\mathrm{T}}$. Let $\mathbf{y} = \mathbf{u} - \mathbf{v}$. Then $\mathbf{y}^{\mathrm{T}}\mathbf{A} = \mathbf{e}^{\mathrm{T}}$ and $\mathbf{e}^{\mathrm{T}}\mathbf{x} = \mathbf{y}^{\mathrm{T}}\mathbf{b}$ is a consequence of $\mathbf{A}\mathbf{x} = \mathbf{b}$, whence $f = \mathbf{y}^{\mathrm{T}}\mathbf{b}$.
4. Suppose that the mixed system of inequalities $\mathbf{I}_n\mathbf{x} \geqslant \mathbf{0}$, $\mathbf{A}\mathbf{x} \geqslant \mathbf{0}$, $(\mathbf{A} + \mathbf{I}_n)\mathbf{x} > \mathbf{0}$ is inconsistent. Then, by Theorem 4.2.4, there exist $\mathbf{u} \geqslant \mathbf{0}$, $\mathbf{v} \geqslant \mathbf{0}$, $\mathbf{w} \geqslant \mathbf{0}$, $\mathbf{w} \neq \mathbf{0}$ such that $\mathbf{u}^{\mathrm{T}}\mathbf{I}_n + \mathbf{v}^{\mathrm{T}}\mathbf{A} + \mathbf{w}^{\mathrm{T}}(\mathbf{A} + \mathbf{I}_n) = \mathbf{0}^{\mathrm{T}}$. Let $\mathbf{y} = \mathbf{u} + \mathbf{w}$, $\mathbf{z} = \mathbf{v} + \mathbf{w}$. Then $\mathbf{y}^{\mathrm{T}}\mathbf{z} > 0$ and $\mathbf{y}^{\mathrm{T}} + \mathbf{z}^{\mathrm{T}}\mathbf{A} = \mathbf{0}^{\mathrm{T}}$. Since $\mathbf{A}^{\mathrm{T}} = -\mathbf{A}$, the last equation implies that $\mathbf{y}^{\mathrm{T}}\mathbf{z} = -\mathbf{z}^{\mathrm{T}}\mathbf{y}$, which is impossible since $\mathbf{y}^{\mathrm{T}}\mathbf{z} = \mathbf{z}^{\mathrm{T}}\mathbf{y} > 0$. Thus the given system of inequalities is consistent.

Exercises 4.3

1. Solve the primal problem, of which the stated problem is the dual, graphically to find the optimal vector $(7, 2)$. The complementary slackness theorem can be used to obtain the optimal vector $(0, 7/4, 1/4)$ for the given problem; its value is 8.

2. The vector $(x_1, x_2) = (7/5, 24/5)$ is feasible for P. Suppose that (y_1, y_2, y_3) is a feasible vector for P^* which, together with (x_1, x_2), satisfies conditions (i) and (ii) of the complementary slackness theorem. Since $x_1, x_2 > 0$, we have from (i) that:

$$-2y_1 + y_2 + y_3 = -3 \quad \text{and} \quad y_1 - 2y_2 + 2y_3 = 2.$$

Since the second constraint of P is *strictly* satisfied, i.e. $x_1 - 2x_2 < 3$, we have from (ii) that $y_2 = 0$. Thus $-2y_1 + y_3 = -3$, $y_1 + 2y_3 = 2$, so $y_1 = 8/5$, $y_3 = 1/5$. It now follows easily that $(8/5, 0, 1/5)$ is feasible for P^* with

$$-3(7/5) + 2(24/5) = 2(8/5) + 3(0) + 11(1/5) = 27/5.$$

It follows that $(7/5, 24/5)$ is optimal for P, $(8/5, 0, 1/5)$ is optimal for P^*, and their common value is $27/5$.

3.

$$\begin{bmatrix} \surd & \times & \times \\ \times & \times & \surd \\ \times & \surd & \surd \end{bmatrix}$$

The crosses are justified by the duality theorem. The tailor's problem justifies the tick on the first row. The ticks in the positions (F, I), (I, F), (I, I) are justified by the following problems in $\mathbb{R}^2$, respectively:

$$\begin{array}{ll} \text{maximize} & 1x_1 + 0x_2 \\ \text{subject to} & 0x_1 + 1x_2 \leqslant 2 \\ & 0x_1 - 1x_2 \leqslant -1 \\ & x_1 \geqslant 0, x_2 \geqslant 0, \end{array}$$

$$\begin{array}{ll} \text{maximize} & -2x_1 + 1x_2 \\ \text{subject to} & 0x_1 + 0x_2 \leqslant -1 \\ & -1x_1 + 1x_2 \leqslant 0 \\ & x_1 \geqslant 0, x_2 \geqslant 0, \end{array}$$

$$\begin{array}{ll} \text{maximize} & 1x_1 + 1x_2 \\ \text{subject to} & 1x_1 + 0x_2 \leqslant -1 \\ & 0x_1 - 1x_2 \leqslant -1 \\ & x_1 \geqslant 0, x_2 \geqslant 0. \end{array}$$

Exercises 4.4

1. Let P be the intersection of the closed halfspaces $\mathbf{a}_i \cdot \mathbf{x} \leqslant b_i$ for $i = 1, \ldots, m$. Suppose that, for some $r \in \{1, \ldots, m\}$, $\mathbf{a}_i \cdot \mathbf{x}_0 = b_i$ for $i = 1, \ldots, r$ and $\mathbf{a}_i \cdot \mathbf{x}_0 < b_i$ for $i = r + 1, \ldots, m$. If $r < n$, then there exists $\mathbf{c} \neq \mathbf{0}$ such that $\mathbf{a}_i \cdot \mathbf{c} = 0$ for $i = 1, \ldots, r$. Thus, for *all sufficiently small* $\lambda > 0$, $\mathbf{x}_0 \pm \lambda\mathbf{c} \in P$. For such λ,

$$\mathbf{x}_0 = \tfrac{1}{2}(\mathbf{x}_0 + \lambda\mathbf{c}) + \tfrac{1}{2}(\mathbf{x}_0 - \lambda\mathbf{c}),$$

which contradicts the fact that $\mathbf{x}_0$ is an extreme point of P. Thus $r \geqslant n$ as desired.

2. Let $\mathbf{x}$ be an optimal vector of P which has as many zero coordinates as possible. If $\mathbf{x} = \mathbf{0}$, then it is a basic optimal vector. Suppose, then, that $\mathbf{x} = (x_1, \ldots, x_r, 0, \ldots, 0)$, where $1 \leqslant r \leqslant n$ and $x_1, \ldots, x_r > 0$. We show that $\mathbf{x}$ is a basic optimal vector by proving that the first columns $\mathbf{a}_1, \ldots, \mathbf{a}_r$ of $\mathbf{A}$ are linearly independent.

Suppose that $\mathbf{a}_1, \ldots, \mathbf{a}_r$ are linearly dependent. Then there exist $\lambda_1, \ldots, \lambda_r$, at least one negative, such that

$$\lambda_1 \mathbf{a}_1 + \ldots + \lambda_r \mathbf{a}_r = \mathbf{0}.$$

Choose $\varepsilon > 0$ such that the two vectors

$$(x_1 \pm \varepsilon\lambda_1, \ldots, x_r \pm \varepsilon\lambda_r, 0, \ldots, 0)$$

are feasible for P. By the optimality of $\mathbf{x}$,

$$c_1(x_1 \pm \varepsilon\lambda_1) + \ldots + c_r(x_r \pm \varepsilon\lambda_r) \leqslant c_1x_1 + \ldots + c_rx_r,$$

whence $c_1\lambda_1 + \ldots + c_r\lambda_r = 0$. Choose $\theta > 0$ such that the scalars

$$x_1 + \theta\lambda_1, \ldots, x_r + \theta\lambda_r$$

are non-negative with at least one of them zero. Then

$$\mathbf{x}^* = (x_1 + \theta\lambda_1, \ldots, x_r + \theta\lambda_r, 0, \ldots, 0)$$

is feasible for P and has more zeros than $\mathbf{x}$; also

$$c_1(x_1 + \theta\lambda_1) + \ldots + c_r(x_r + \theta\lambda_r) = c_1x_1 + \ldots + c_rx_r,$$

so $\mathbf{x}^*$ is optimal for P. This contradiction shows that $\mathbf{a}_1, \ldots, \mathbf{a}_r$ are linearly independent, and hence that $\mathbf{x}$ is a basic optimal vector.

3.

Columns selected	Basic solution	Extreme point of F_R	Extreme point of F	$2x_1 - 3x_2$
1, 2, 3	(7, 2, 14, 0, 0)	(7, 2, 14, 0, 0)	(7, 2)	8
1, 2, 4	$(\frac{7}{5}, \frac{24}{5}, 0, \frac{56}{5}, 0)$	$(\frac{7}{5}, \frac{24}{5}, 0, \frac{56}{5}, 0)$	$(\frac{7}{5}, \frac{24}{5})$	$-\frac{58}{5}$
1, 2, 5	$(-\frac{7}{3}, -\frac{8}{3}, 0, 0, \frac{56}{3})$			
1, 3, 4	(11, 0, 24, −8, 0)			
1, 3, 5	(3, 0, 8, 0, 8)	(3, 0, 8, 0, 8)	(3, 0)	6
1, 4, 5	(−1, 0, 0, 4, 12)			
2, 3, 4	$(0, \frac{11}{2}, -\frac{7}{2}, 14, 0)$			
2, 3, 5	$(0, -\frac{3}{2}, \frac{7}{2}, 0, 14)$			
2, 4, 5	(0, 2, 0, 7, 7)	(0, 2, 0, 7, 7)	(0, 2)	−6
3, 4, 5	(0, 0, 2, 3, 11)	(0, 0, 2, 3, 11)	(0, 0)	0

The value of the problem is 8 and an optimal vector is (7, 2).

4. P_R: minimize $3x_1 + 2x_2$
subject to
$$\begin{aligned} 5x_1 + x_2 - x_3 &= 10 \\ x_1 + x_2 - x_4 &= 6 \\ x_1 + 4x_2 - x_5 &= 12 \\ x_1, \ldots, x_5 \geqslant 0. \end{aligned}$$

Q: maximize $10y_1 + 6y_2 + 12y_3$
subject to
$$\begin{aligned} 5y_1 + y_2 + y_3 &\leqslant 3 \\ y_1 + y_2 + 4y_3 &\leqslant 2 \\ y_1, y_2, y_3 &\geqslant 0. \end{aligned}$$

The basic feasible solutions for P_R are: $(4, 2, 12, 0, 0)$, $(1, 5, 0, 0, 9)$, $(12, 0, 50, 6, 0)$, $(0, 10, 0, 4, 28)$. The problem P has optimal vector $(1, 5)$ and value 13, the problem Q has optimal vector $(\frac{1}{4}, \frac{7}{4}, 0)$.

Exercises 4.5

1. One sequence of tableaux for the first problem is:

Initial tableau

40	50	0	0	0	0
⑤	2	1	0	0	30
5	7	0	1	0	35
2	5	0	0	1	20

0	34	-8	0	0	-240
1	$\frac{2}{5}$	$\frac{1}{5}$	0	0	6
0	⑤	-1	1	0	5
0	$\frac{21}{5}$	$-\frac{2}{5}$	0	1	8

Final tableau

0	0	$-\frac{6}{5}$	$-\frac{34}{5}$	0	-274
1	0	$\frac{7}{25}$	$-\frac{2}{25}$	0	$\frac{28}{5}$
0	1	$-\frac{1}{5}$	$\frac{1}{5}$	0	1
0	0	$\frac{11}{25}$	$-\frac{21}{25}$	1	$\frac{19}{5}$

This problem has optimal vector $(\frac{28}{5}, 1)$ and value 274; an optimal vector for the dual is $(\frac{6}{5}, \frac{34}{5}, 0)$.

For the second problem we have:

Initial tableau

3	2	0	0	0
−1	(1)	1	0	1
1	−4	0	1	2

5	0	−2	0	−2
−1	1	1	0	1
−3	0	4	1	6

The negative column below the 5 in the last tableau shows that the objective function is unbounded on the feasible set. The problem is insoluble.

2. Solve the dual problem first. The given problem has optimal vector (0, 9, 11) and value 49.
3. The auxiliary problem is:

$$\begin{array}{ll} \text{maximize} & -x_0 \\ \text{subject to} & -x_0 - 4x_1 - 5x_2 + x_3 = -10 \\ & -x_0 + 5x_1 + 2x_2 + x_4 = 10 \\ & -x_0 + 3x_1 + 8x_2 + x_5 = 12 \\ & x_0 \geqslant 0, \ldots, x_5 \geqslant 0 \end{array}$$

One possible sequence of tableaux (each increased by an extra initial row to keep track of the original objective function) for this problem is:

0	5	3	0	0	0	0
−1	0	0	0	0	0	0
(−1)	−4	−5	1	0	0	−10
−1	5	2	0	1	0	10
−1	3	8	0	0	1	12

0	5	3	0	0	0	0
0	4	5	−1	0	0	10
1	4	5	−1	0	0	10
0	(9)	7	−1	1	0	20
0	7	13	−1	0	1	22

0	0	$-\frac{8}{9}$	$\frac{5}{9}$	$-\frac{5}{9}$	0	$-\frac{100}{9}$
0	0	$\frac{17}{9}$	$-\frac{5}{9}$	$-\frac{4}{9}$	0	$\frac{10}{9}$
1	0	$\textcircled{\frac{17}{9}}$	$-\frac{5}{9}$	$-\frac{4}{9}$	0	$\frac{10}{9}$
0	1	$\frac{7}{9}$	$-\frac{1}{9}$	$\frac{1}{9}$	0	$\frac{20}{9}$
0	0	$\frac{68}{9}$	$-\frac{2}{9}$	$-\frac{7}{9}$	1	$\frac{58}{9}$

$\frac{8}{17}$	0	0	$\frac{5}{17}$	$-\frac{13}{17}$	0	$-\frac{180}{17}$
-1	0	0	0	0	0	0
$\frac{9}{17}$	0	1	$-\frac{5}{17}$	$-\frac{4}{17}$	0	$\frac{10}{17}$
$-\frac{7}{17}$	1	0	$\frac{2}{17}$	$\frac{5}{17}$	0	$\frac{30}{17}$
-4	0	0	2	1	1	2

This is a final tableau for the auxiliary problem, which therefore has optimal vector $(0, \frac{30}{17}, \frac{10}{17}, 0, 0, 2)$ and value 0. Thus $(\frac{30}{17}, \frac{10}{17}, 0, 0, 2)$ is an extreme point of the feasible set for the canonical problem. Omitting the first column and the second row of the above final tableau, we obtain an initial tableau for the given problem and proceed as follows:

0	0	$\frac{5}{17}$	$-\frac{13}{17}$	0	$-\frac{180}{17}$
0	1	$-\frac{5}{17}$	$-\frac{4}{17}$	0	$\frac{10}{17}$
1	0	$\frac{2}{17}$	$\frac{5}{17}$	0	$\frac{30}{17}$
0	0	$\textcircled{2}$	1	1	2

0	0	0	$-\frac{31}{34}$	$-\frac{5}{34}$	$-\frac{185}{17}$
0	1	0	$-\frac{3}{34}$	$\frac{5}{34}$	$\frac{15}{17}$
1	0	0	$\frac{4}{17}$	$-\frac{1}{17}$	$\frac{28}{17}$
0	0	1	$\frac{1}{2}$	$\frac{1}{2}$	1

We deduce from this final tableau that the given problem has optimal vector $(\frac{28}{17}, \frac{15}{17})$ and value $\frac{185}{17}$.

4. One possible sequence of tableaux is:

2	4	1	1	0	0	0	0
1	3	0	1	1	0	0	4
2	1	0	0	0	1	0	3
0	1	4	①	0	0	1	3

2	3	-3	0	0	0	-1	-3
①	2	-4	0	1	0	-1	1
2	1	0	0	0	1	0	3
0	1	4	1	0	0	1	3

0	-1	5	0	-2	0	1	-5
1	2	-4	0	1	0	-1	1
0	-3	8	0	-2	1	②	1
0	1	4	1	0	0	1	3

0	$\frac{1}{2}$	1	0	-1	$-\frac{1}{2}$	0	$-\frac{11}{2}$
1	$\frac{1}{2}$	0	0	0	$\frac{1}{2}$	0	$\frac{3}{2}$
0	$-\frac{3}{2}$	④	0	-1	$\frac{1}{2}$	1	$\frac{1}{2}$
0	$\frac{5}{2}$	0	1	1	$-\frac{1}{2}$	0	$\frac{5}{2}$

0	$\frac{7}{8}$	0	0	$-\frac{3}{4}$	$-\frac{5}{8}$	$-\frac{1}{4}$	$-\frac{45}{8}$
1	$\frac{1}{2}$	0	0	0	$\frac{1}{2}$	0	$\frac{3}{2}$
0	$-\frac{3}{8}$	1	0	$-\frac{1}{4}$	$\frac{1}{8}$	$\frac{1}{4}$	$\frac{1}{8}$
0	($\frac{5}{2}$)	0	1	1	$-\frac{1}{2}$	0	$\frac{5}{2}$

0	0	0	$-\frac{7}{20}$	$-\frac{11}{10}$	$-\frac{9}{20}$	$-\frac{1}{4}$	$-\frac{13}{2}$
1	0	0	$-\frac{1}{5}$	$-\frac{1}{5}$	$\frac{3}{5}$	0	1
0	0	1	$\frac{3}{20}$	$-\frac{1}{10}$	$\frac{1}{20}$	$\frac{1}{4}$	$\frac{1}{2}$
0	1	0	$\frac{2}{5}$	$\frac{2}{5}$	$-\frac{1}{5}$	0	1

Problem has optimal vector $(1,1,\frac{1}{2},0)$ and value $\frac{13}{2}$. Dual has optimal vector $(\frac{11}{10},\frac{9}{20},\frac{1}{4})$. The extreme points corresponding to the above tableaux are, in order: $(0,0,0,0)$, $(0,0,0,3)$, $(1,0,0,3)$, $(\frac{3}{2},0,0,\frac{5}{2})$, $(\frac{3}{2},0,\frac{1}{8},\frac{5}{2})$, $(1,1,\frac{1}{2},0)$.

Exercises 4.6

1. With an obvious notation, the pay-off matrix is

$$\begin{array}{c} \\ \text{RR} \\ \text{BB} \\ \text{RB} \end{array}\begin{array}{c} \begin{array}{ccc}\text{RR} & \text{BB} & \text{RB}\end{array} \\ \left[\begin{array}{ccc} 2 & -1 & 1 \\ -1 & 1 & 0 \\ -1 & 0 & 1 \end{array}\right]\end{array}$$

We solve the game by the simplex algorithm. First we add 2 to each element of the above matrix to obtain a positive matrix. The following tableaux are then obtained.

1	1	1	0	0	0	0
④	1	3	1	0	0	1
1	3	2	0	1	0	1
1	2	3	0	0	1	1

0	$\frac{3}{4}$	$\frac{1}{4}$	$-\frac{1}{4}$	0	0	$-\frac{1}{4}$
1	$\frac{1}{4}$	$\frac{3}{4}$	$\frac{1}{4}$	0	0	$\frac{1}{4}$
0	$\textcircled{\frac{11}{4}}$	$\frac{5}{4}$	$-\frac{1}{4}$	1	0	$\frac{3}{4}$
0	$\frac{7}{4}$	$\frac{9}{4}$	$-\frac{1}{4}$	0	1	$\frac{3}{4}$

0	0	$-\frac{1}{11}$	$-\frac{2}{11}$	$-\frac{3}{11}$	0	$-\frac{5}{11}$
1	0	$\frac{7}{11}$	$\frac{3}{11}$	$-\frac{1}{11}$	0	$\frac{2}{11}$
0	1	$\frac{5}{11}$	$-\frac{1}{11}$	$\frac{4}{11}$	0	$\frac{3}{11}$
0	0	$\frac{16}{11}$	$-\frac{1}{11}$	$-\frac{7}{11}$	1	$\frac{3}{11}$

Thus the value of the game is $\frac{11}{5}-2=\frac{1}{5}$. Optimal strategies for R and C are, respectively, $(\frac{2}{5},\frac{3}{5},0)$ and $(\frac{2}{5},\frac{3}{5}.0)$.

2. *Either* $a \geqslant c$, with $(1,1)$ a saddle point, *or* $a \geqslant d$ with $(1,2)$ a saddle point, *or* $a<c$, $a<d$, $c \leqslant d$ with $(2,1)$ a saddle point *or* $a<c$, $a<d$, $d<c$ with $(2,2)$ a saddle point. By what we have just

proved, if the matrix has no saddle point, then its elements must be distinct. Suppose that the matrix has no saddle point. Let $a < b$. Since $(1, 1)$ is not a saddle point, $a < c$; since $(2, 1)$ is not a saddle point, $c > d$; and since $(2, 2)$ is not a saddle point, $b > d$; thus $a < b$, $a < c$, $d < b$, $d < c$. Similarly, if $b < a$, then $c < a$, $b < d$, $c < d$. The value required is $(ad - bc)/(a - b - c + d)$.

3. Consider the game with $m \times n$ pay-off matrix $\mathbf{A}$ and value v. Then $\mathbf{x} = (x_1, \ldots, x_m)$ is an optimal strategy for R if and only if $\mathbf{x} \geqslant \mathbf{0}$, $x_1 + \ldots + x_m = 1$ and $\mathbf{x}$ achieves an expected gain of at least v against each pure strategy of C, i.e. if and only if

$$\mathbf{x} \geqslant \mathbf{0}, \qquad x_1 + \ldots + x_m = 1, \qquad \mathbf{x}^{\mathrm{T}}\mathbf{c}_1 \geqslant v, \ldots, \mathbf{x}^{\mathrm{T}}\mathbf{c}_n \geqslant v,$$

where $\mathbf{c}_1, \ldots, \mathbf{c}_n$ denote the columns of $\mathbf{A}$. This shows that the set of R's optimal strategies is a bounded polyhedral set, i.e. a polytope.

Suppose that each pure strategy for the row player is essential. For $i = 1, \ldots, m$, let $\mathbf{z}_i$ be an optimal strategy for R whose ith coordinate is positive. Then, by what we have just proved, $(\mathbf{z}_1 + \ldots + \mathbf{z}_m)/m$ is an optimal strategy for R all of whose coordinates are positive.

4. First game has three essential strategies for the row player; a solution is: $v = \frac{23}{5}$, $\mathbf{x} = (\frac{1}{10}, \frac{1}{10}, \frac{4}{5})$, $\mathbf{y} = (\frac{19}{30}, \frac{7}{30}, \frac{4}{30})$. Second game has a saddle point; a solution is: $v = -1$, $\mathbf{x} = (0, 1, 0)$, $\mathbf{y} = (0, 1, 0)$. In the third game, the third row is dominated by the first one; a solution is $v = \frac{5}{2}$, $\mathbf{x} = (\frac{1}{2}, \frac{1}{2}, 0)$, $\mathbf{y} = (\frac{1}{2}, 0, \frac{1}{2})$.

5. The three games can easily be solved graphically. Their solutions are: $\frac{22}{5}$, $(\frac{1}{5}, \frac{4}{5})$, $(\frac{2}{5}, \frac{3}{5})$; $-\frac{1}{5}$, $(\frac{3}{5}, \frac{2}{5})$, $(\frac{3}{5}, 0, \frac{2}{5}, 0)$; $-\frac{1}{5}$, $(\frac{3}{5}, \frac{2}{5}, 0)$, $(\frac{3}{5}, \frac{2}{5})$.

6. Clearly, $(x_2, \ldots, x_m)^{\mathrm{T}}\mathbf{B} \geqslant (v, \ldots, v)^{\mathrm{T}}$ and $\mathbf{B}(y_1, \ldots, y_n) \leqslant (v, \ldots, v)$. We must show that $(0, x_2, \ldots, x_m)^{\mathrm{T}}\mathbf{A} \geqslant (v, \ldots, v)^{\mathrm{T}}$ and $\mathbf{A}(y_1, \ldots, y_n) \leqslant (v, \ldots, v)$. The first of these two inequalities is immediate from the first stated inequality involving $\mathbf{B}$. The second of these two inequalities follows from the second stated inequality involving $\mathbf{B}$ and the fact that the first row is dominated by the second one.

7. Suppose C's first pure strategy is not essential. Then the following system of inequalities is inconsistent:

$$\begin{array}{rcr}
a_{11}y_1 + \ldots + a_{1n}y_n & \leqslant & v \\
\cdots\cdots\cdots\cdots\cdots\cdots & & \\
a_{m1}y_1 + \ldots + a_{mn}y_n & \leqslant & v \\
-y_1 \qquad\qquad\qquad & \leqslant & 0 \\
\cdots\cdots\cdots\cdots\cdots\cdots & & \\
-y_n & \leqslant & 0 \\
y_1 + \ldots + \qquad y_n & \leqslant & 1 \\
-y_1 - \ldots - \qquad y_n & \leqslant & -1 \\
-y_1 \qquad\qquad\qquad & & 0,
\end{array}$$

where v denotes the value of the game. By Theorem 4.2.4, there exist $u_1, \ldots, u_m, v_1, \ldots, v_n, r, s \geqslant 0$ and $t > 0$ such that

$$\begin{array}{l} a_{11}u_1 + \ldots + a_{m1}u_m - v_1 + r - s - t = 0 \\ a_{12}u_1 + \ldots + a_{m2}u_m - v_2 + r - s \qquad = 0 \\ \cdots\cdots\cdots\cdots\cdots\cdots\cdots\cdots\cdots \\ a_{1n}u_1 + \ldots + a_{mn}u_m - v_n + r - s \qquad = 0 \\ vu_1 + \ldots + \quad vu_m \qquad\quad + r - s \quad \leqslant 0. \end{array}$$

It follows that

$$\begin{array}{l} a_{11}u_1 + \ldots + a_{m1}u_m > v(u_1 + \ldots + u_m) \\ a_{12}u_1 + \ldots + a_{m2}u_m \geqslant v(u_1 + \ldots + u_m) \\ \cdots\cdots\cdots\cdots\cdots\cdots\cdots\cdots\cdots \\ a_{1n}u_1 + \ldots + a_{mn}u_m \geqslant v(u_1 + \ldots + u_m). \end{array}$$

Let $\mathbf{x} = (x_1, \ldots, x_m)$, where $x_i = u_i/(u_1 + \ldots + u_m)$. Then $\mathbf{x}$ is an optimal strategy for R against which C's first pure strategy does not achieve the value of the game.

Exercises 5.1

1. Only $f + g$ and $\max\{f, g\}$ are necessarily convex.
2. Let $x, y \in \mathbb{R}$ and let $\lambda, \mu \geqslant 0$ with $\lambda + \mu = 1$. Then

$$\begin{aligned}(g \circ f)(\lambda x + \mu y) = g(f(\lambda x + \mu y)) &\leqslant g(\lambda f(x) + \mu f(y)) \\ &\leqslant \lambda(g \circ f)(x) + \mu(g \circ f)(y),\end{aligned}$$

so $g \circ f$ is convex. The exponential function is convex and increasing, whence $h = \exp(\log h)$ is convex. The function $x^2 + 1$ is convex, but $\log(x^2 + 1)$ is not.

3. Let $g: I \to \mathbb{R}$ be a positive concave function and let $f = 1/g$. Let $x, y \in I$ and let $\lambda, \mu \geqslant 0$ with $\lambda + \mu = 1$. Denote by c the positive number $g(x)g(y)g(\lambda x + \mu y)$. Then

$$\begin{aligned}c(\lambda f(x) + \mu f(y) - f(\lambda x + \mu y)) &= (\lambda g(y) + \mu g(x))g(\lambda x + \mu y) - g(x)g(y) \\ &\geqslant (\lambda g(y) + \mu g(x))(\lambda g(x) + \mu g(y)) - g(x)g(y) \\ &= \lambda\mu(g(x) - g(y))^2 \\ &\geqslant 0.\end{aligned}$$

Thus $\lambda f(x) + \mu f(y) \geqslant f(\lambda x + \mu y)$, and so f is convex. No, the reciprocal of the exponential function is not concave.

4. Now $x_1 < x_1 + \delta < x_2 + \delta$ and $x_1 < x_2 < x_2 + \delta$, so by the three chords lemma,

$$\frac{f(x_1+\delta)-f(x_1)}{x_1+\delta-x_1} \leqslant \frac{f(x_2+\delta)-f(x_1)}{x_2+\delta-x_1},$$
$$\frac{f(x_2+\delta)-f(x_1)}{x_2+\delta-x_1} \leqslant \frac{f(x_2+\delta)-f(x_2)}{x_2+\delta-x_2}.$$

Hence $f(x_1+\delta)-f(x_1) \leqslant f(x_2+\delta)-f(x_2)$.

5. (i) Suppose f has a global maximum at an interior point c of I. Let $x, y \in I$ with $x < c < y$. Then $f(x) \leqslant f(c)$, $f(y) \leqslant f(c)$. Since f is convex and has a maximum at c, $0 \leqslant f'_-(c) \leqslant f'_+(c) \leqslant 0$, and so $f'(c) = 0$. Hence

$$(f(x)-f(c))/(x-c) \leqslant 0 \leqslant (f(y)-f(c))/(y-c),$$

whence $f(x) \geqslant f(c)$, $f(y) \geqslant f(c)$. Thus $f(x) = f(y) = f(c)$, i.e. f is constant.

(ii) Suppose f has a global minimum at a point c of I. If $x, y \in I$ with $c < x < y$, then by the three chords lemma

$$0 \leqslant (f(x)-f(c))/(x-c) \leqslant (f(y)-f(x))/(y-x),$$

whence $f(x) \leqslant f(y)$ and f is increasing to the right of c.

(iii) Suppose f has a local minimum at a point c of I. Suppose $x \in I\backslash\{c\}$; say $c < x$. Since f has a local minimum at c, there exists y with $c < y < x$ such that $f(y) \geqslant f(c)$. Thus

$$0 \leqslant (f(y)-f(c))/(y-c) \leqslant (f(x)-f(c))/(x-c),$$

which shows that $f(x) \geqslant f(c)$. Hence f has a global minimum at c.

6. That a convex function defined on a bounded interval is bounded below follows easily from the fact that it has support at each interior point of its domain. By what we have just proved, a convex function $f: [a, b] \to \mathbb{R}$ is bounded below and f is bounded above by $\max\{f(a), f(b)\}$, whence f is bounded on $[a, b]$. Let $f: \mathbb{R} \to \mathbb{R}$ be a convex function which is bounded above by M. Let $x, y \in \mathbb{R}$ with $x < y$ and let $\lambda > 0$. By the three chords lemma,

$$\frac{f(x)-M}{\lambda} \leqslant \frac{f(x)-f(x-\lambda)}{\lambda} \leqslant \frac{f(y)-f(x)}{y-x} \leqslant \frac{f(y+\lambda)-f(y)}{\lambda} \leqslant \frac{M-f(y)}{\lambda}.$$

Letting $\lambda \to \infty$ in these inequalities, we deduce that $f(y) = f(x)$, which shows that f is constant.

Exercises 5.2

1. Let $x_0 = \lambda_1 x_1 + \ldots + \lambda_m x_m$. Since f has a support at x_0, there exists some real M such that, for all $x \in I$,

$$f(x) \geqslant f(x_0) + M(x - x_0).$$

Let $x = x_1, \ldots, x_m$ in turn in this inequality. Multiply the resulting inequalities, respectively, by $\lambda_1, \ldots, \lambda_m$ and add to obtain $f(x_0) \leqslant \lambda_1 f(x_1) + \ldots + \lambda_m f(x_m)$, i.e. Jensen's inequality.

2. We use the inequality between the weighted arithmetic and geometric means to deduce that

$$5(2^{-4/5}) = (5x/2)^{2/5}\,(5/2xy)^{2/5}\,(5y^2)^{1/5} \leqslant x + 1/xy + y^2$$

for $x, y > 0$. Equality will occur if $5x/2 = 5/2xy = 5y^2$, i.e. when $x = 2^{1/5}$, $y = 2^{-2/5}$.

3. Since $(-\sin)''(x) = \sin x \geqslant 0$ for $0 \leqslant x \leqslant \pi$, $-\sin$ is convex on $[0, \pi]$. Let the *central* angles of the polygon (assumed to contain the centre of the circle) be $\theta_1, \ldots, \theta_n$. Then $\theta_1 + \ldots + \theta_n = 2\pi$ and the perimeter of the polygon is

$$2n\left(\frac{1}{n}\sin\tfrac{1}{2}\theta_1 + \ldots + \frac{1}{n}\sin\tfrac{1}{2}\theta_n\right),$$

which, by Jensen's inequality, never exceeds

$$2n\left(\sin\frac{1}{n}(\tfrac{1}{2}\theta_1 + \ldots + \tfrac{1}{2}\theta_n)\right) = 2n\sin\frac{\pi}{n}.$$

4. Apply Jensen's inequality to the equation

$$\frac{y_1 + \ldots + y_m}{x_1 + \ldots + x_m} = \frac{x_1}{x_1 + \ldots + x_m}\left(\frac{y_1}{x_1}\right) + \ldots + \frac{x_m}{x_1 + \ldots + x_m}\left(\frac{y_m}{x_m}\right).$$

Note $f''(x) = (p-1)x^{p-2}(1 + x^p)^{(1-2p)/p} \geqslant 0$ for $x > 0$ and $p > 1$.

5. Clearly, for $t > 0$,

$$\alpha_k^{1/t} x_k \leqslant (\alpha_1 x_1^t + \ldots + \alpha_m x_m^t)^{1/t} \leqslant (\alpha_1 x_k^t + \ldots + \alpha_m x_k^t)^{1/t} = x_k.$$

Since $\alpha_\kappa^{1/t} \to 1$ as $t \to \infty$, $M_t(\mathbf{x}; \boldsymbol{\alpha}) \to x_k$ as $t \to \infty$. The graph should be that of an increasing function that is asymptotic to $\min\{x_1, \ldots, x_m\}$ at $-\infty$ and $\max\{x_1, \ldots, x_m\}$ at ∞.

Exercises 5.3

1. The function $\alpha_1 x_1^t + \ldots + \alpha_m x_m^t$ is a sum of log-convex functions, and so is log-convex. Thus $\log(\alpha_1 x_1^t + \ldots + \alpha_m x_m^t) = t \log M_t(\mathbf{x}; \boldsymbol{\alpha})$ is convex.

2. $\Gamma(n+\frac{1}{2}) = (n-\frac{1}{2})(n-\frac{3}{2})\ldots\frac{1}{2}\Gamma(\frac{1}{2}) = ((2n)!\sqrt{\pi})/(n!\,4^n)$.

3. $\dfrac{\pi}{2} = \frac{1}{2}\Gamma^2(\frac{1}{2}) = \frac{1}{2}\lim\limits_{n\to\infty}\left(\dfrac{n!\sqrt{n}}{(n+\frac{1}{2})\ldots\frac{1}{2}}\right)^2 = \lim\limits_{n\to\infty}\dfrac{2.2}{1.3}\cdot\dfrac{4.4}{3.5}\cdots\dfrac{(2n)(2n)}{(2n-1)(2n+1)}$.

4. $\Gamma(x) = \lim\limits_{n\to\infty}\dfrac{n!\,n^x}{(n+x)\ldots x} = \dfrac{1}{x}\lim\limits_{n\to\infty}(e^{x(\log n - (\frac{1}{1}+\ldots+\frac{1}{n}))})\dfrac{e^{x/1}}{1+\frac{x}{1}}\cdots\dfrac{e^{x/n}}{1+\frac{x}{n}}$.

5. Now $\Gamma(\frac{1}{2}(n+1)) \leqslant \sqrt{\Gamma(\frac{1}{2}n)}\sqrt{\Gamma(\frac{1}{2}(n+2))}$, so $\Gamma(\frac{1}{2}(n+1))/\Gamma(\frac{1}{2}(n+2)) \leqslant \sqrt{(2/n)}$. Replacing n by $n+1$, we find that $\Gamma(\frac{1}{2}(n+2))/\Gamma(\frac{1}{2}(n+3)) \leqslant \sqrt{(2/(n+1))}$, and so $\Gamma(\frac{1}{2}(n+1))/\Gamma(\frac{1}{2}(n+2)) \geqslant \sqrt{(2/(n+1))}$.

6. Define $f\colon(0,\infty)\to\mathbb{R}$ by the equation

$$f(x) = \frac{3^x}{2\pi\sqrt{3}}\Gamma\left(\frac{x}{3}\right)\Gamma\left(\frac{x+1}{3}\right)\Gamma\left(\frac{x+2}{3}\right) \qquad \text{for } x>0.$$

Then f is log-convex, satisfies $f(x+1) = xf(x)$, and $f(1) = (3/2\pi\sqrt{3})\Gamma(\frac{1}{3})\Gamma(\frac{2}{3})\Gamma(1) = (3/2\pi\sqrt{3})(\pi/\sin(\pi/3)) = 1$, so $f = \Gamma$.

7. $\sqrt{(\pi/2)}$; $\sqrt{\pi}/3$; $\pi/\sqrt{2}$; 81/40.

8. $B(x,y) = \displaystyle\int_0^1 u^{x-1}(1-u)^{y-1}\,du = 2\int_0^{\pi/2}(\sin^{2x-1}t)(\cos^{2y-1}t)\,dt$, where $u = \sin^2 t$, and

$$\int_0^{\pi/2}\sin^{2m}t\,dt = \tfrac{1}{2}B(m+\tfrac{1}{2},\tfrac{1}{2}) = \tfrac{1}{2}\frac{\Gamma(m+\frac{1}{2})\Gamma(\frac{1}{2})}{\Gamma(m+1)} = \frac{(2m)!\,\pi}{2^{2m+1}(m!)^2}.$$

9. Let $u = t/(1+t)$. Then

$$\frac{\pi}{\sin \pi x} = \Gamma(x)\Gamma(1-x)$$
$$= B(x,1-x) = \int_0^1 u^{x-1}(1-u)^{-x}\,du = \int_0^\infty \frac{t^{x-1}}{1+t}\,dt.$$

10. Area is

$$4\int_0^a b(1-(x/a)^{1/\alpha})^\beta\,dx = 4ab\alpha\beta\,B(\alpha,\beta)/(\alpha+\beta).$$

Exercises 5.4

1. Let

$C = \{(x_1,\ldots,x_m)\text{: there is } \mathbf{x}\in X \text{ such that } x_1 > f_1(\mathbf{x}),\ldots,x_m > f_m(\mathbf{x})\}$.

Then C is a convex set in $\mathbb{R}^m$ not containing the origin. Thus there exist scalars $\lambda_1,\ldots,\lambda_m$, not all zero, such that

$$\lambda_1 x_1 + \ldots + \lambda_m x_m \geqslant 0 \qquad \text{for } (x_1, \ldots, x_m) \in C.$$

Since each x_i can be chosen as large as we please, $\lambda_i \geqslant 0$. For each $\mathbf{x} \in X$ and for each $\varepsilon > 0$, $(f_1(\mathbf{x}) + \varepsilon, \ldots, f_m(\mathbf{x}) + \varepsilon) \in C$, whence

$$\lambda_1(f_1(\mathbf{x}) + \varepsilon) + \ldots + \lambda_m(f_m(\mathbf{x}) + \varepsilon) \geqslant 0.$$

Letting $\varepsilon \to 0_+$ in this inequality, we deduce that

$$\lambda_1 f_1(\mathbf{x}) + \ldots + \lambda_m f_m(\mathbf{x}) \geqslant 0 \qquad \text{for } \mathbf{x} \in X.$$

2. Suppose that such a convex function f does exist. Then epi$f$ is a convex set in $\mathbb{R}^2$ containing the epigraph of the log function, whence epi$f$ must be the open halfplane $x > 0$, which is clearly impossible.

3. Let $\mathbf{x}, \mathbf{y} \in X$ and let $\lambda, \mu \geqslant 0$ with $\lambda + \mu = 1$. Let $\varepsilon > 0$. Then there exist $c_x, c_y \in \mathbb{R}$ such that $0 \leqslant c_x, c_y < \varepsilon$ and

$$(\mathbf{x}, f(\mathbf{x}) + c_x),\ (\mathbf{y}, f(\mathbf{y}) + c_y) \in C.$$

Since C is convex,

$$(\lambda\mathbf{x} + \mu\mathbf{y}, \lambda f(\mathbf{x}) + \mu f(\mathbf{y}) + \lambda c_x + \mu c_y) \in C.$$

This shows that

$$f(\lambda\mathbf{x} + \mu\mathbf{y}) \leqslant \lambda f(\mathbf{x}) + \mu f(\mathbf{y}) + \lambda c_x + \mu c_y < \lambda f(\mathbf{x}) + \mu f(\mathbf{y}) + \varepsilon.$$

Since $\varepsilon > 0$ is arbitrary, $f(\lambda\mathbf{x} + \mu\mathbf{y}) \leqslant \lambda f(\mathbf{x}) + \mu f(\mathbf{y})$ and f is convex.

4. Let $\mathbf{x}, \mathbf{y} \in \mathbb{R}^n$ and let $\lambda \geqslant 0$. Then

$$f(\lambda\mathbf{x}) = \sup\{\lambda\mathbf{x}\cdot\mathbf{a} : \mathbf{a} \in A\} = \lambda \sup\{\mathbf{x}\cdot\mathbf{a} : \mathbf{a} \in A\} = \lambda f(\mathbf{x})$$

and

$$\begin{aligned} f(\mathbf{x} + \mathbf{y}) &= \sup\{(\mathbf{x} + \mathbf{y})\cdot\mathbf{a} : \mathbf{a} \in A\} \\ &\leqslant \sup\{\mathbf{x}\cdot\mathbf{a} : \mathbf{a} \in A\} + \sup\{\mathbf{y}\cdot\mathbf{a} : \mathbf{a} \in A\} = f(\mathbf{x}) + f(\mathbf{y}). \end{aligned}$$

This shows that f is both convex and positively homogeneous.

5. We first show that each level set $\{\mathbf{x} : f(\mathbf{x}) \leqslant \alpha\}$ is convex. This is trivial if $\alpha \geqslant 0$, so we assume that $\alpha < 0$. Let $f(\mathbf{x}) \leqslant \alpha$, $f(\mathbf{y}) \leqslant \alpha$, where $\mathbf{x} = (x_1, \ldots, x_n)$, $\mathbf{y} = (y_1, \ldots, y_n) > \mathbf{0}$. Let $\lambda, \mu \geqslant 0$ with $\lambda + \mu = 1$. Then

$$-x_1^{\alpha_1} \ldots x_n^{\alpha_n} \leqslant \alpha \qquad \text{and} \qquad -y_1^{\alpha_1} \ldots y_n^{\alpha_n} \leqslant \alpha,$$

whence

$$\begin{aligned} &\lambda \log(-\alpha) \leqslant \lambda(\alpha_1 \log x_1 + \ldots + \alpha_n \log x_n) \qquad \text{and} \\ &\mu \log(-\alpha) \leqslant \mu(\alpha_1 \log y_1 + \ldots + \alpha_n \log y_n). \end{aligned}$$

Thus

$$\log(-\alpha) \leqslant \alpha_1 \log x_1^{\lambda} y_1^{\mu} + \ldots + \alpha_n \log x_n^{\lambda} y_n^{\mu}$$
$$\leqslant \alpha_1 \log(\lambda x_1 + \mu y_1) + \ldots + \alpha_n \log(\lambda x_n + \mu y_n)$$
$$= \log(\lambda x_1 + \mu y_1)^{\alpha_1} \ldots (\lambda x_n + \mu y_n)^{\alpha_n},$$

which shows that $f(\lambda\mathbf{x} + \mu\mathbf{y}) \leqslant \alpha$, and hence that the level set $\{\mathbf{x}: f(\mathbf{x}) \leqslant \alpha\}$ is convex. This proves that f is quasi-convex.

Now f is not convex when $\alpha_1 + \ldots + \alpha_n > 1$, for if it were convex, then the function $g: (0, \infty) \to \mathbb{R}$ defined by the equation

$$g(x) = f(x, \ldots, x) = -x^{\alpha_1 + \ldots + \alpha_n} \qquad \text{for } x > 0$$

would be convex, which it is not.

Suppose, finally, that $\alpha_1 + \ldots + \alpha_n = \theta \leqslant 1$. Let $\mathbf{x}, \mathbf{y} > \mathbf{0}$ and let $\lambda, \mu \geqslant 0$ with $\lambda + \mu = 1$. By Theorem 5.2.4,

$$(-f(\lambda\mathbf{x} + \mu\mathbf{y}))^{1/\theta} \geqslant \lambda(-f(\mathbf{x}))^{1/\theta} + \mu(-f(\mathbf{y}))^{1/\theta}.$$

Since x^θ is concave on $(0, \infty)$,

$$-f(\lambda\mathbf{x} + \mu\mathbf{y}) \geqslant (\lambda(-f(\mathbf{x}))^{1/\theta} + \mu(-f(\mathbf{y}))^{1/\theta})^\theta \geqslant -\lambda f(\mathbf{x}) - \mu f(\mathbf{y}),$$

which shows that f is convex.

Exercises 5.5

1. Suppose that no such $\lambda > 0$ exists. Then there is a sequence of points of $\mathbb{R}^n \backslash X$ which converges to a point of A, which is impossible. Thus such a $\lambda > 0$ does exist. Let $\mathbf{x}, \mathbf{y} \in A$ with $\mathbf{x} \neq \mathbf{y}$. If $\mathbf{z} = \mathbf{y} + \lambda(\mathbf{y} - \mathbf{x})/\|\mathbf{y} - \mathbf{x}\|$, then $\mathbf{z} \in (A)_\lambda$, and so, by the convexity of f,
$$f(\mathbf{y}) - f(\mathbf{x}) \leqslant \frac{\|\mathbf{x} - \mathbf{y}\|}{\|\mathbf{x} - \mathbf{y}\| + \lambda}(f(\mathbf{z}) - f(\mathbf{x})) \leqslant \frac{2M\|\mathbf{x} - \mathbf{y}\|}{\lambda}.$$
We complete the proof by interchanging $\mathbf{x}$ and $\mathbf{y}$.
2. Suppose that $f(\mathbf{y}) - f(\mathbf{x}) \geqslant f'(\mathbf{x}; \mathbf{y} - \mathbf{x})$ for $\mathbf{x}, \mathbf{y} \in X$. Then the affine transformation $T: \mathbb{R}^n \to \mathbb{R}$ given by $T(\mathbf{y}) = f(\mathbf{x}) + f'(\mathbf{x}; \mathbf{y} - \mathbf{x})$ supports f at $\mathbf{x}$, and so f is convex by Theorem 5.4.5. Suppose next that f is convex. By Theorem 5.5.4, f has unique support T at each point $\mathbf{x}$ of X, where $T(\mathbf{y}) = f(\mathbf{x}) + f'(\mathbf{x}; \mathbf{y} - \mathbf{x})$ for $\mathbf{y} \in \mathbb{R}^n$, and so, for $\mathbf{y} \in X$, $f(\mathbf{x}) + f'(\mathbf{x}; \mathbf{y} - \mathbf{x}) \leqslant f(\mathbf{y})$.
3. Let $a = \partial^2 f/\partial x_1^2$, $b = \partial^2 f/\partial x_1 \partial x_2$, $c = \partial^2 f/\partial x_2^2$, the partial derivatives being evaluated at some point $\mathbf{x}$ of X. The Hessian matrix of f at $\mathbf{x}$ is
$$\begin{bmatrix} a & b \\ b & c \end{bmatrix}.$$
The convexity of f is equivalent to: $a \geqslant 0$, $c \geqslant 0$, $b^2 \leqslant ac$ for all $\mathbf{x}$ in X.

4. The Hessian matrix of f at $\mathbf{x}$ is the $n \times n$ matrix $[a_{ij}]$, where

$$a_{ii} = \frac{\alpha_i(\alpha_i - 1)f(\mathbf{x})}{x_i^2}, \qquad a_{ij} = \frac{\alpha_i \alpha_j f(\mathbf{x})}{x_i x_j} \qquad (i \neq j).$$

We prove that f is convex by showing that, for all real $z_1, \ldots, z_n$ and for each $\mathbf{x} = (x_1, \ldots, x_n) > \mathbf{0}$,

$$\sum_{i=1}^{n} \sum_{j=1}^{n} a_{ij} z_i z_j = f(\mathbf{x}) \left(\left(\sum_{i=1}^{n} \frac{\alpha_i z_i}{x_i} \right)^2 - \sum_{i=1}^{n} \frac{\alpha_i z_i^2}{x_i^2} \right) \geqslant 0.$$

Let $\theta = \alpha_1 + \ldots + \alpha_n$. By Jensen's inequality, applied to the square function,

$$\left(\sum_{i=1}^{n} \frac{\alpha_i}{\theta} \left(\frac{z_i}{x_i} \right) \right)^2 \leqslant \sum_{i=1}^{n} \frac{\alpha_i}{\theta} \left(\frac{z_i}{x_i} \right)^2, \quad \text{so} \quad \left(\sum_{i=1}^{n} \frac{\alpha_i z_i}{x_i} \right)^2 \leqslant \sum_{i=1}^{n} \frac{\alpha_i z_i^2}{x_i^2},$$

and the desired result now follows.

Exercises 5.6

1. The support and gauge functions are, respectively:
   (i) $h(u_1, \ldots, u_n) = r(|u_1| + \ldots + |u_n|)$ and
   $g(u_1, \ldots, u_n) = r^{-1} \max\{|u_1|, \ldots, |u_n|\}$;
   (ii) $h(u_1, \ldots, u_n) = r \max\{|u_1|, \ldots, |u_n|\}$ and
   $g(u_1, \ldots, u_n) = r^{-1}(|u_1| + \ldots + |u_n|)$;
   (iii) $h(u_1, u_2) = \sqrt{(a^2 u_1^2 + b^2 u_2^2)}$ and $g(u_1, u_2) = \sqrt{((u_1/a)^2 + (u_2/b)^2)}$.
2. The result follows from the following observations:
   (i) $(A)_\eta \supseteq B$ implies that $h_B(\mathbf{u}) - h_A(\mathbf{u}) \leqslant \eta$ whenever $\|\mathbf{u}\| = 1$;
   (ii) if $h_B(\mathbf{u}) - h_A(\mathbf{u}) \leqslant \eta$ whenever $\|\mathbf{u}\| = 1$, then $(A)_\eta \supseteq B$;
   (iii) $(A)_\eta \supseteq B$ and $(B)_\eta \supseteq A$ if and only if $|h_B(\mathbf{u}) - h_A(\mathbf{u})| \leqslant \eta$ whenever $\|\mathbf{u}\| = 1$.
3. We show first that

$$B = \{\mathbf{x} \in \mathbb{R}^n : \mathbf{u} . \mathbf{x} \leqslant h'(\mathbf{v}; \mathbf{u}) \text{ for all } \mathbf{u} \in \mathbb{R}^n\}.$$

It will then follow from Theorem 5.6.5 that $h'(\mathbf{v};\ \ )$ is the support function of B. Let $\mathbf{b} \in B$. Then, for every $\mathbf{u} \in \mathbb{R}^n$ and $\lambda > 0$,

$$\mathbf{u} \cdot \mathbf{b} = \frac{\mathbf{b} \cdot (\mathbf{v} + \lambda \mathbf{u}) - \mathbf{b} \cdot \mathbf{v}}{\lambda} \leqslant \frac{h(\mathbf{v} + \lambda \mathbf{u}) - h(\mathbf{v})}{\lambda},$$

whence $\mathbf{u} \cdot \mathbf{b} \leqslant h'(\mathbf{v}; \mathbf{u})$. Conversely, suppose that

$$\mathbf{u} \cdot \mathbf{b} \leqslant h'(\mathbf{v}; \mathbf{u}) \qquad \text{for all } \mathbf{u} \in \mathbb{R}^n.$$

Then $\mathbf{u} \cdot \mathbf{b} \leqslant h'(\mathbf{v}; \mathbf{u}) \leqslant h(\mathbf{u})$ for all $\mathbf{u} \in \mathbb{R}^n$, so $\mathbf{b} \in A$. Also

$$\mathbf{v}\cdot\mathbf{b} \leqslant h'(\mathbf{v};\mathbf{v}) = h(\mathbf{v}) \qquad \text{and} \qquad -\mathbf{v}\cdot\mathbf{b} \leqslant h'(\mathbf{v};-\mathbf{v}) = -h(\mathbf{v}).$$

Hence $\mathbf{v}\cdot\mathbf{b} = h(\mathbf{v})$. Thus $\mathbf{b}\in B$, and the desired expression for B is established.

If $B = \{(b_1, \ldots, b_n)\}$, then $h'(\mathbf{v};(u_1,\ldots,u_n)) = b_1u_1 + \ldots + b_nu_n$, and so $h'(\mathbf{v};\ \ )$ is linear. Thus h is differentiable at $\mathbf{v}$ with $\partial h/\partial x_i = b_i$ for $i = 1, \ldots, n$.

4. Clearly $h(\mathbf{u}) \geqslant \max\{h_A(\mathbf{u}), h_B(\mathbf{u})\}$. If $\mathbf{x}\in\operatorname{conv}(A\cup B)$, then $\mathbf{x} = \lambda\mathbf{a} + \mu\mathbf{b}$ for some $\mathbf{a}\in A$, $\mathbf{b}\in B$ and some $\lambda, \mu \geqslant 0$ with $\lambda + \mu = 1$. Thus

$$\mathbf{u}\cdot\mathbf{x} = \lambda\mathbf{u}\cdot\mathbf{a} + \mu\mathbf{u}\cdot\mathbf{b} \leqslant \lambda h_A(\mathbf{u}) + \mu h_B(\mathbf{u}) \leqslant \max\{h_A(\mathbf{u}), h_B(\mathbf{u})\},$$

and so $h(\mathbf{u}) \leqslant \max\{h_A(\mathbf{u}), h_B(\mathbf{u})\}$. Thus $h(\mathbf{u}) = \max\{h_A(\mathbf{u}), h_B(\mathbf{u})\}$.

Clearly $g(\mathbf{u}) \geqslant \max\{g_A(\mathbf{u}), g_B(\mathbf{u})\}$. Also

$$\mathbf{u}\in(g_A(\mathbf{u})A)\cap(g_B(\mathbf{u})B) \subseteq \max\{g_A(\mathbf{u}), g_B(\mathbf{u})\}(A\cap B),$$

and so $g(\mathbf{u}) \leqslant \max\{g_A(\mathbf{u}), g_B(\mathbf{u})\}$. Thus $g(\mathbf{u}) = \max\{g_A(\mathbf{u}), g_B(\mathbf{u})\}$.

Exercises 5.7

1. The problem has minimal value 0 when $x = 0$. The Lagrangian function is $-x + yx^2$. The Kuhn–Tucker conditions are: $-1 + 2yx \geqslant 0$; $-1 + 2yx = 0$ when $x > 0$; $x^2 \leqslant 0$; $x^2 = 0$ when $y > 0$; $x \geqslant 0$; $y \geqslant 0$. These conditions cannot be simultaneously satisfied, and so the Lagrangian function has no saddle point. This example shows that the existence of a solution to the convex programming problem does not guarantee a saddle point for the Lagrangian function.
2. The Kuhn–Tucker conditions are: $2(x_1 + 1) + y_1 \geqslant 0$; $2(x_1 + 1) + y_1 = 0$ when $x_1 > 0$; $2(x_2 - 2) + y_2 \geqslant 0$; $2(x_2 - 2) + y_2 = 0$ when $x_2 > 0$; $x_1 - 2 \leqslant 0$; $x_1 - 2 = 0$ when $y_1 > 0$; $x_2 - 1 \leqslant 0$; $x_2 - 1 = 0$ when $y_2 > 0$; $x_1, x_2, y_1, y_2 \geqslant 0$. The conditions are only satisfied by $x_1 = 0$, $x_2 = 1$, $y_1 = 0$, $y_2 = 2$. The problem has minimal value 2 when $x_1 = 0$, $x_2 = 1$.
3. We show that the function $g:\mathbb{R}^n \to \mathbb{R}$ defined by the equation $g(\mathbf{x}) = \mathbf{x}^{\mathrm{T}}\mathbf{Q}\mathbf{x}$ is convex. Let $\mathbf{x}, \mathbf{y}\in\mathbb{R}^n$ and let $\lambda, \mu \geqslant 0$ with $\lambda + \mu = 1$. Then

$$\lambda g(\mathbf{x}) + \mu g(\mathbf{y}) - g(\lambda\mathbf{x} + \mu\mathbf{y}) = \lambda\mu(\mathbf{x} - \mathbf{y})^{\mathrm{T}}\mathbf{Q}(\mathbf{x} - \mathbf{y}) \geqslant 0,$$

since $\mathbf{Q}$ is non-negative semidefinite. This shows that g is convex. The Kuhn–Tucker conditions are (with an obvious notation): $\mathbf{x} \geqslant \mathbf{0}$; $\mathbf{y} \geqslant \mathbf{0}$;

$$p_j + 2\sum_{i=1}^{n} q_{ij}x_i + \sum_{i=1}^{m} a_{ij}y_i \geqslant 0 \qquad \text{with equality if } x_j > 0;$$

$$\sum_{j=1}^{n} a_{ij}x_j - b_i \leqslant 0 \qquad \text{with equality if } y_i > 0.$$

4. The Kuhn–Tucker conditions for the problem are the complementary slackness conditions for the linear programming problem.

Exercises 5.8

1. The set Ω_n lies in the intersection of $2n - 1$ *independent* hyperplanes in $\mathbb{R}^{n^2}$, and so has dimension at most $n^2 - (2n - 1)$. The result now follows from Birkhoff's and Carathéodory's theorems.
2. The set $\{\mathbf{S\lambda}:\mathbf{S} \in \Omega_n\}$ is convex and contains $\Lambda_\mathbf{A}$, whence $\operatorname{conv} \Lambda_\mathbf{A} \subseteq \{\mathbf{S\lambda}:\mathbf{S} \in \Omega_n\}$. The inclusion $\{\mathbf{S\lambda}:\mathbf{S} \in \Omega_n\} \subseteq \operatorname{conv} \Lambda_\mathbf{A}$ follows from Birkhoff's theorem.
3. For some doubly stochastic matrix $\mathbf{S}^* = [s_{ij}]$, $\mathbf{y}^* = \mathbf{S}^*\mathbf{x}^*$. For $j = 1, \ldots, n$, let $s_j = s_{1j} + \ldots + s_{kj}$. Then $s_j \leqslant 1$ and

$$\begin{aligned} y_1^* + \ldots + y_k^* &= s_1x_1^* + \ldots + s_nx_n^* \\ &\leqslant s_1x_1^* + \ldots + s_kx_k^* + (k - s_1 - \ldots - s_k)x_k^* \\ &= s_1(x_1^* - x_k^*) + \ldots + s_k(x_k^* - x_k^*) + kx_k^* \\ &\leqslant x_1^* - x_k^* + \ldots + x_k^* - x_k^* + kx_k^* \\ &= x_1^* + \ldots + x_k^*. \end{aligned}$$

Also

$$y_1^* + \ldots + y_n^* = \sum_{i=1}^{n} (s_{i1}x_1^* + \ldots + s_{in}x_n^*) = x_1^* + \ldots + x_n^*.$$

4. The inequalities follow from Theorem 5.8.3 with $f(x_1, \ldots, x_n) = \pm(x_1 + \ldots + x_k)$. Letting $k = 1$ in the inequalities, we can easily deduce that

$$\lambda_1 = \inf\{\mathbf{v}^\mathrm{T}\mathbf{A}\mathbf{v}:\|\mathbf{v}\| = 1\} \leqslant \sup\{\mathbf{v}^\mathrm{T}\mathbf{A}\mathbf{v}:\|\mathbf{v}\| = 1\} \leqslant \lambda_n.$$

5. Let $\mathbf{u}_1, \ldots, \mathbf{u}_n$ be an orthonormal sequence of eigenvectors of $\mathbf{A}$ corresponding to the eigenvalues $\lambda_1, \ldots, \lambda_n$. Let $\alpha_i = (\mathbf{x} \cdot \mathbf{u}_i)^2$. Since the geometric mean of two numbers never exceeds their arithmetic mean,

$$\begin{aligned} \left(\sum_{i=1}^{n} \alpha_i\lambda_i c\right)\left(\sum_{i=1}^{n} \alpha_i\lambda_i^{-1}c^{-1}\right) &\leqslant \tfrac{1}{4}\left(\sum_{i=1}^{n} \alpha_i(\lambda_i c + \lambda_i^{-1}c^{-1})\right)^2 \\ &\leqslant \tfrac{1}{4}(\max\{\lambda_i c + \lambda_i^{-1}c^{-1}:i = 1, \ldots, n\})^2 \\ &= \tfrac{1}{4}(\max\{\lambda_1 c + \lambda_1^{-1}c^{-1}, \lambda_n c + \lambda_n^{-1}c^{-1}\})^2 \end{aligned}$$

$$= \tfrac{1}{4}\left(\left(\frac{\lambda_1}{\lambda_n}\right)^{1/2} + \left(\frac{\lambda_n}{\lambda_1}\right)^{1/2}\right)^2.$$

In the last but one of the above inequalities, we have used the fact that the convex function $cx + (cx)^{-1}$ on $[\lambda_1, \lambda_n]$ attains its maximal value either at λ_1 or λ_n. By the Cauchy–Schwarz inequality,

$$\left(\sum_{i=1}^{n} \alpha_i\lambda_i c\right)\left(\sum_{i=1}^{n} \alpha_i\lambda_i^{-1}c^{-1}\right) \geqslant \left(\sum_{i=1}^{n} \sqrt{(\alpha_i\lambda_i c\alpha_i\lambda_i^{-1}c^{-1})}\right)^2 = 1.$$

6. Use Hadamard's inequality and Theorem 5.2.2.
7. Assertions (i), (ii) are false as the following matrices show:

$$\begin{bmatrix} \frac{1}{4} & \frac{1}{4} & \frac{1}{2} \\ \frac{1}{4} & \frac{1}{4} & \frac{1}{2} \\ \frac{1}{2} & \frac{1}{2} & 0 \end{bmatrix}, \qquad \begin{bmatrix} \frac{1}{2} & \frac{1}{2} & 0 \\ \frac{1}{2} & \frac{1}{2} & 0 \\ 0 & 0 & 1 \end{bmatrix}.$$

Assertion (iii) is true, the sum of the sums of the elements on the $n!$ diagonals equals $n!$, so there is some diagonal the sum of whose elements is at most 1, and the average of the elements on such a diagonal is at most $1/n$.
8. Replace $\Sigma c_{ij} a_{ij}$ in the proof of Theorem 5.8.8 by r.

Exercises 6.1

1. For each i in some non-empty index set, let I_i be the cell $I_{i1} \times \ldots \times I_{in}$ in $\mathbb{R}^n$, where $I_{i1}, \ldots, I_{in}$ are cells in $\mathbb{R}^1$. Then $\cap I_i$ is the cell

$$(\cap I_{i1}) \times \ldots \times (\cap I_{in}),$$

where all the intersections are over the index set.

No. In $\mathbb{R}^1$ let A_i be the elementary set $[0,1]\backslash\{1/i\}$ for $i = 1, 2, \ldots$. Then $\cap(A_i : i = 1, 2, \ldots)$ is *not* an elementary set.
2. Let C be contained in the ball $\{\mathbf{x} : \|\mathbf{x}\| \leqslant r\}$, where $r > 0$, and in the hyperplane $x_n = a_0 + a_1x_1 + \ldots + a_{n-1}x_{n-1}$ $(n \geqslant 2)$. Let $I_1, \ldots, I_m$ be subintervals of $[-r, r]$ of length $2r/m$ whose union is $[-r, r]$. Let $J_1, \ldots, J_{m^{n-1}}$ be the cells in $\mathbb{R}^n$ of the form

$$I_{i_1} \times \ldots \times I_{i_{n-1}} \times \{0\},$$

where $1 \leqslant i_1, \ldots, i_{n-1} \leqslant m$. For $i = 1, \ldots, m^{n-1}$, let

$$m_i = \inf\{a_0 + a_1x_1 + \ldots + a_{n-1}x_{n-1} : (x_1, \ldots, x_{n-1}, 0) \in J_i\},$$

with M_i being defined by the same equation but with inf replaced by sup. Then

$$M_i - m_i \leqslant 2r(|a_1| + \ldots + |a_{n-1}|)/m.$$

For $i = 1, \ldots, m^{n-1}$, let K_i be the cell

$$K_i = \{(x_1, \ldots, x_n) : (x_1, \ldots, x_{n-1}, 0) \in J_i \text{ and } m_i \leqslant x_n \leqslant M_i\}.$$

Then $A = \cup (K_i : i = 1, \ldots, m^{n-1})$ is an elementary set containing C. Moreover, the volume of A does not exceed

$$(2r)^n(|a_1| + \ldots + |a_{n-1}|)/m.$$

Now choose $m > (2r)^n(|a_1| + \ldots + |a_{n-1}|)/\varepsilon$.

3. Let m be a positive integer such that $m > 1/\varepsilon$. For $i = 1, \ldots, m$, let

$$I_i = [(i-1)/m, i/m], \quad J_i = I_i \times [0, (i-1)^2/m^2], \quad K_i = I_i \times [0, i^2/m^2].$$

Let $A = J_1 \cup \ldots \cup J_m$, $B = K_1 \cup \ldots \cup K_m$. Then $A \subseteq C \subseteq B$ and $v(A) < \frac{1}{3} < v(B)$. Finally, $v(B) - v(A) = \frac{1}{m} < \varepsilon$, and the stated inequality follows.

Exercises 6.2

1. Let $\varepsilon > 0$. Then there is an elementary set $E \subseteq A$ with $v(E) > \underline{v}(A) - \varepsilon$. Now the elementary set $X \backslash E \supseteq X \backslash A$ and

$$v(X) = v(X \backslash E) + v(E) > \bar{v}(X \backslash A) + \underline{v}(A) - \varepsilon.$$

Since $\varepsilon > 0$ is arbitrary, $v(X) \geqslant \bar{v}(X \backslash A) + \underline{v}(A)$. Similarly, $v(X) \leqslant \bar{v}(X \backslash A) + \underline{v}(A)$, so $v(X) = \bar{v}(X \backslash A) + \underline{v}(A)$.

Suppose now that X has volume, $A \subseteq X$ and that F is an elementary set with $X \subseteq F$. By the above result,

$$\underline{v}(A) + \bar{v}(X \backslash A) \leqslant \underline{v}(A) + \bar{v}(F \backslash A) = v(F),$$

whence $\bar{v}(X) \geqslant \underline{v}(A) + \bar{v}(X \backslash A)$. Similarly, $\underline{v}(X) \leqslant \underline{v}(A) + \bar{v}(X \backslash A)$, whence, since $\bar{v}(X) = \underline{v}(X)$, $v(X) = \underline{v}(A) + \bar{v}(X \backslash A)$.

2. (i) Trivially $\delta(A, B) \geqslant 0$. Suppose $A \neq B$, say $\mathbf{a} \in A \backslash B$. Since B is closed, there is $r > 0$ such that $B(\mathbf{a}; r) \cap B = \varnothing$. Since A is convex and has non-empty interior,

$$\delta(A, B) \geqslant v(B(\mathbf{a}; r) \cap A) > 0.$$

(ii) Trivially $\delta(A, B) = \delta(B, A)$. (iii) This follows from the inclusion

$$(A \cup C) \backslash (A \cap C) \subseteq ((A \cup B) \backslash (A \cap B)) \cup ((B \cup C) \backslash (B \cap C)).$$

3. Suppose first that I_0 is closed. Let $\varepsilon > 0$. For $m = 1, 2, \ldots$, let I_m^* be an open cell with $I_m \subseteq I_m^*$ and $v(I_m^*) < v(I_m) + \frac{1}{2^m}\varepsilon$. Then the cells $I_1^*, \ldots, I_m^*, \ldots$ form an open cover of the compact cell I_0, and so there is m_0 such that $I_0 \subseteq I_1^* \cup \ldots \cup I_{m_0}^*$. For $m > m_0$,

$$v(I_1)+\ldots+v(I_m)\leqslant v(I_0)\leqslant v(I_1^*)+\ldots+v(I_m^*)<v(I_1)+\ldots+v(I_m)+\varepsilon,$$

and so

$$|v(I_0)-(v(I_1)+\ldots+v(I_m))|<\varepsilon,$$

which shows that $v(I_0)=\Sigma_{m=1}^{\infty}v(I_m)$. Finally, suppose that the cell I_0 is not closed. Then the elementary set $(\operatorname{cl} I_0)\setminus I_0$ can be expressed in the form $J_1\cup\ldots\cup J_k$, where $J_1,\ldots,J_k$ are pairwise disjoint cells in $\mathbb{R}^n$, each cell having zero volume. Now

$$\operatorname{cl} I_0=J_1\cup\ldots\cup J_k\cup I_1\cup\ldots\cup I_m\cup\ldots,$$

so, by the first part of the solution,

$$v(I_0)=v(\operatorname{cl} I_0)=v(J_1)+\ldots+v(J_k)+\sum_{m=1}^{\infty}v(I_m)=\sum_{m=1}^{\infty}v(I_m).$$

4. Let $[\mathbf{a}_1,\ldots,\mathbf{a}_n]$ denote the matrix with columns $\mathbf{a}_1,\ldots,\mathbf{a}_n$. Then P is the image of the cube

$$\{(x_1,\ldots,x_n):0\leqslant x_1,\ldots,x_n\leqslant 1\}=[0,1]\times\ldots\times[0,1]$$

under the affine transformation which maps $\mathbf{x}$ to $[\mathbf{a}_1,\ldots,\mathbf{a}_n]\mathbf{x}+\mathbf{a}$, whence P has volume $|\det[\mathbf{a}_1,\ldots,\mathbf{a}_n]|$.

5. In $\mathbb{R}^1$, let $A=[0,1]$ and $A_k=\{0/2^k,1/2^k,\ldots,2^k/2^k\}$. Then $A_k\to A$ as $k\to\infty$, yet $v(A_k)=0\nrightarrow 1=v(A)$ as $k\to\infty$.

Exercises 6.3

1. (i) $a_1\ldots a_n\omega_n r^n$; (ii) $6\frac{2}{3}$; (iii) $\omega_{n-1}c(b^{n-1}+ab^{n-2}+\ldots+a^{n-1})/n$.

2. Since $\omega_n/\omega_{n-1}=2\int_0^1(1-x^2)v^{\frac{n-1}{2}}\,\mathrm{d}x$ decreases with n and since $\omega_4<\omega_5$, $\omega_5>\omega_6$, the sequence increases to its fifth term $8\pi^2/15$ and then decreases to zero.

3. Let the convex polygon be $A_0A_1\ldots A_m$ and let $B_0B_1\ldots B_m$ be the polygon obtained from it by a rotation of a quarter turn about the origin. Then the assertion $a_1\mathbf{u}_1+\ldots+a_m\mathbf{u}_m=\mathbf{0}$ is equivalent to the trivial vector identity $\overrightarrow{B_0B_1}+\ldots+\overrightarrow{B_mB_0}=\mathbf{0}$.

4. Let $\mathbf{u}_0,\ldots,\mathbf{u}_m$ be the outward unit normals to the facets $F_0,\ldots,F_m$, respectively. Then

$$v_{n-1}(F_0)\mathbf{u}_0+\ldots+v_{n-1}(F_m)\mathbf{u}_m=\mathbf{0},$$

and so

$$\begin{aligned}v_{n-1}(F_0)&=v_{n-1}(F_0)\mathbf{u}_0\cdot\mathbf{u}_0\\&=-(v_{n-1}(F_1)\mathbf{u}_0\cdot\mathbf{u}_1+\ldots+v_{n-1}(F_m)\mathbf{u}_0\cdot\mathbf{u}_m)\\&\leqslant v_{n-1}(F_1)+\ldots+v_{n-1}(F_m).\end{aligned}$$

5. The area Δ is $\frac{1}{2}\sqrt{(\alpha^2\beta^2 + \beta^2\gamma^2 + \gamma^2\alpha^2)}$. Also $2\alpha^2 = b^2 + c^2 - a^2$, $2\beta^2 = c^2 + a^2 - b^2$, $2\gamma^2 = a^2 + b^2 - c^2$, and the given formula for Δ follows after some algebraic manipulation.

Exercises 6.4

1. $\lambda^2 + 6\mu^2 + \pi\nu^2 + 7\lambda\mu + 12\mu\nu + 4\lambda\nu$; $v(A, B) = 3\frac{1}{2}$, $v(B, C) = 6$, $v(A, C) = 2$.
2. Here $f(\lambda) = \sqrt{((\pi - 4)\lambda^2 + 4)}$ and $f''(\lambda) = 4(\pi - 4)/f^3(\lambda) < 0$.
3. Let $A_1 = \{0, 1\}$, $A_2 = [0, 1]$ in $\mathbb{R}^1$. Were $v_1(\lambda_1 A_1 + \lambda_2 A_2)$ a homogeneous polynomial of degree 1 in λ_1, λ_2, it would have to be $\lambda_1 v_1(A_1) + \lambda_2 v_1(A_2)$, i.e. λ_2. When $\lambda_1 = 1$, $\lambda_2 = 1$, however, $v_1(A_1 + A_2) = 2 \neq \lambda_2$.
4. Letting $\lambda \to 0_+$ in the given inequality, we find that $v_n(A) \geqslant v_n(B)$. Let

$$A = \{(x, 0): 0 \leqslant x \leqslant 1\}, \qquad B = \{(0, y): 0 \leqslant y \leqslant 2\},$$
$$C = \{(0, y): 0 \leqslant y \leqslant 1\}$$

in $\mathbb{R}^2$. Then $v_2(A + \lambda C) = \lambda \geqslant 0 = v_2(B + \lambda C)$ for all $\lambda > 0$, yet $s_2(A) = 2$, $s_2(B) = 4$. If $v_n(A) = v_n(B)$ and $C = U$, then

$$s_n(A) = \lim_{\lambda \to 0_+} \frac{v_n(A + \lambda U) - v_n(A)}{\lambda} \geqslant \lim_{\lambda \to 0_+} \frac{v_n(B + \lambda U) - v_n(B)}{\lambda} = s_n(B).$$

5. For $\lambda > 0$,

$$(A \cup B) + \lambda U = (A + \lambda U) \cup (B + \lambda U),$$
$$(A \cap B) + \lambda U = (A + \lambda U) \cap (B + \lambda U).$$

Thus

$$v_n(A + \lambda U) + v_n(B + \lambda U) = v_n((A \cup B) + \lambda U) + v_n((A \cap B) + \lambda U).$$

We deduce the desired result by expressing both sides of the last equation as polynomials λ, and then comparing their coefficients.

Exercises 6.5

1. (i) $v(A)$, $v(B) > 0$ with A and B homothetic; (ii) $v(A) > 0$ and B a singleton or vice versa; (iii) A and B lie in parallel hyperplanes.
2. It is readily found that $\sigma_n = (\omega_n n^n)^{-1/(n-1)}$ which has maximal value when $n = 3$.
3. The positive function f is concave, and so

$$f^{2n-1}(0)f''(0) = (n-1)(v(A)v(A, \ldots, A, B, B) - v^2(A, \ldots, A, B)) \leqslant 0.$$

4. Suppose that the projections of both bA and aB on the x-axis are the real interval $[0, ab]$. Then (with an obvious notation), for $0 \leqslant t \leqslant 1$,

$$\begin{aligned} v(tbA + (1-t)aB) &= \int_0^{ab} l((tbA + (1-t)aB)_x)\,\mathrm{d}x \\ &\geqslant \int_0^{ab} (t(l(bA)_x) + (1-t)l(aB)_x)\,\mathrm{d}x \\ &= tb^2 v(A) + (1-t)a^2 v(B). \end{aligned}$$

The stated inequalities follow by straightforward algebraic manipulation.

Exercises 6.6

1. Let $\mathbf{c}$ be an extreme point of A_π and let $\mathbf{u}$ be a unit normal to π. Then $\mathbf{c} = \mathbf{p} \pm \frac{1}{2}\gamma(\mathbf{p})\mathbf{u}$ for some $\mathbf{p} \in \pi(A)$. Denote by $F_\mathbf{a}, F_\mathbf{b}$ the smallest faces of A containing the points $\mathbf{a} = \mathbf{p} + \alpha(\mathbf{p})\mathbf{u}$, $\mathbf{b} = \mathbf{p} + \beta(\mathbf{p})\mathbf{u}$ of A, respectively. Since $\mathbf{c}$ is an extreme point of A_π, the projections $\pi(F_\mathbf{a})$, $\pi(F_\mathbf{b})$ of $F_\mathbf{a}$, $F_\mathbf{b}$ on π must have intersection $\{\mathbf{p}\}$. But A has only a finite number of faces. Thus there is only a finite number of choices for $\mathbf{p}$, and hence for $\mathbf{c}$. This shows that A_π has only a finite number of extreme points, and so is a polytope.
2. Let π be the x-axis, and let A, B be the parallelograms defined by the equations

$$\begin{aligned} A &= \operatorname{conv}\{\mathbf{0}, \tfrac{1}{2}\mathbf{e}_2, \mathbf{e}_1 + \tfrac{1}{2}\mathbf{e}_2, \mathbf{e}_1 + \mathbf{e}_2\}, \\ B &= \operatorname{conv}\{\mathbf{0}, \tfrac{1}{2}\mathbf{e}_2, -\mathbf{e}_1 + \tfrac{1}{2}\mathbf{e}_2, -\mathbf{e}_1 + \mathbf{e}_2\}. \end{aligned}$$

3. Suppose that A_π is the Steiner symmetral of a plane convex body A in a line π, which we take to be the x_1-axis. Denote by α, α_π the maximal areas of triangles lying in A, A_π, respectively. Let (x_i, y_i), for $i = 1, 2, 3$, be the vertices of a triangle of area α_π lying in A_π. By the definition of A_π, there exists z_i, for $i = 1, 2, 3$, such that the points $(x_i, z_i \pm y_i)$ lie in A. For each real t, denote by $f(t)$ the area of the triangle with vertices $(x_i, y_i + tz_i)$. Then $f(t) = |mt + \alpha_\pi|$, for some constant m. By considering the triangles with vertices $(x_i, y_i + z_i)$ and $(x_i, -y_i + z_i)$, we find that $f(1) \leqslant \alpha$, $f(-1) \leqslant \alpha$, whence $\alpha_\pi \leqslant \alpha$.
4. Those symmetrals about lines perpendicular to the sides of the triangle. Let T be a non-equilateral triangle whose sides have lengths a, b, c. The assertion about the perimeter follows from the fact that, if $a \neq b$, then the Steiner symmetral T about a line

perpendicular to the side of length c has a perimeter smaller than that of T. The assertion about the diameter follows from the facts: (i) if $a > b$ and $a > c$, then the Steiner symmetral of T about a line perpendicular to the side of length b has diameter less than that of T; (ii) if $a = b > c$, then the Steiner symmetral of T about a line perpendicular to a side of length a has precisely one longest side. The final assertion follows from the first of the two results just proved, together with the fact that the product of the perimeter and the inradius of a triangle is twice its area.

Exercises 7.1

1. Suppose that the convex set C is contained in the union $H_1 \cup \ldots \cup H_m$ of halfspaces in $\mathbb{R}^n$, where $m \geqslant n + 1$. Then
$$C \cap (H_1 \cup \ldots \cup H_m)^c = (C \cap H_1^c) \cap \ldots \cap (C \cap H_m^c) = \varnothing.$$
By the finite version of Helly's theorem, some $n + 1$ of the convex sets $C \cap H_1^c, \ldots, C \cap H_m^c$ have an empty intersection, say
$$(C \cap H_1^c) \cap \ldots \cap (C \cap H_{n+1}^c) = \varnothing, \text{ whence } C \subseteq H_1 \cup \ldots \cup H_{n+1}.$$
In $\mathbb{R}^2$ let C be the convex set $\{(x, y): y > 0\}$ and let H_i be the open halfplane $y > 1/i$ for $i = 1, 2, \ldots$. Then $C \subseteq \bigcup(H_i: i = 1, 2, \ldots)$, but C is not contained in any finite union of the H_i's. The set $\{(1, 0), (0, 1), (-1, 0), (0, -1)\}$ is contained in the union of the four halfplanes $x \geqslant 1, y \geqslant 1, x \leqslant -1, y \leqslant -1$, but is *not* contained in the union of any three of them.
2. Helly's theorem applied to the family $\mathcal{G} = \{A - C: A \in \mathcal{F}\}$ shows the existence of a point $\mathbf{x}$ belonging to all members of $\mathcal{G}$. The translate $C + \mathbf{x}$ of C meets all members of $\mathcal{F}$. The result reduces to Helly's theorem when C is a singleton.
3. Let $C_1, C_2, C_3 \in \mathcal{G}$. The number of points of A in $C_1^c \cup C_2^c \cup C_3^c$ is fewer than $\frac{1}{3}m + \frac{1}{3}m + \frac{1}{3}m = m$, whence C_1, C_2, C_3 have a point of A in common. By Helly's theorem, there exists a point $\mathbf{p}$ common to all members of $\mathcal{G}$. Suppose that a closed halfplane L^+ bounded by a line L through $\mathbf{p}$ contains fewer than $\frac{1}{3}m$ points of A. Then there is a line M parallel to L, one of whose associated closed halfplanes M^+ contains more than $\frac{2}{3}m$ points of A, but does *not* contain $\mathbf{p}$. Thus $M^+ \cap B \in \mathcal{G}$, and so $\mathbf{p} \in M^+ \cap B$. This contradiction proves the desired result.
4. For each set A in the family, denote by A_y the orthogonal projection of A on the y-axis. Any two of the convex sets A_y meet. By Helly's theorem in $\mathbb{R}^1$, there is a point $\mathbf{y}_0$ common to all of the A_y's. The line through $\mathbf{y}_0$ parallel to the x-axis meets all members of the given family.

One possible generalization is to replace *line parallel to the x-axis* by *plane parallel to the xy-plane.*

5. Let $S = \operatorname{conv}\{\mathbf{a}_0, \ldots, \mathbf{a}_n\}$ be a regular n-simplex in $\mathbb{R}^n$ with diameter 1. Suppose that the $(n-1)$-simplex $\operatorname{conv}\{\mathbf{a}_1, \ldots, \mathbf{a}_n\}$ lies in the hyperplane $x_n = 0$ and has its circumcentre at the origin. Suppose that the circumcentre of S is $(0, \ldots, 0, h)$. Then $h = \sqrt{(R_n^2 - R_{n-1}^2)}$. Consideration of the right-angled triangle with vertices $\mathbf{a}_0, \mathbf{0}, \mathbf{a}_1$ shows that $1^2 = R_{n-1}^2 + (h + R_n)^2$. It follows that $R_n = (4(1 - R_{n-1}^2))^{-\frac{1}{2}}$. Since $R_1 = \frac{1}{2}$, we can deduce that $R_n = \sqrt{(n/2(n+1))}$. This shows that the constant in Jung's theorem cannot be reduced.
6. Let A be a ball in $\mathbb{R}^n$ $(n \geqslant 2)$ with its centre removed. Then A is bounded, and given any finite number of points of A, there is a point of A from which all of the given points are visible in A, but A is not starshaped.

   Let B be the set in $\mathbb{R}^n$ $(n \geqslant 2)$ defined by the equation
$$B = \bigcup(X + (m, m, 0, \ldots, 0) : m = 1, 2, \ldots),$$
   where
$$X = \{(x_1, \ldots, x_n) : x_1 \geqslant 0 \text{ and } 0 \leqslant x_2 \leqslant 1\}.$$
   Then B is closed, and given any finite number of points of B, there is a point of B from which all of the given points are visible in B, but B is not starshaped.
7. The uniqueness follows from the fact that the error ε^* of the best approximation asserted by Theorem 7.1.10 must be attained. The best approximation of degree 1 for the given points is $-\frac{1}{2}x + \frac{9}{4}$.
8. (i) $x - \frac{1}{8}$; (ii) $-\frac{1}{4}x + \frac{9}{8}$; (iii) $x - \frac{1}{2}$.
9. The first assertion follows from ideas used in the proof of Lemma 7.1.9 and the following (easily established) results: (i) $|T_n(x)| \leqslant 1$ for $-1 \leqslant x \leqslant 1$; (ii) there exist $x_1, x_2, \ldots, x_{n+1}$ in $[-1, 1]$ such that $x_{n+1} < \ldots < x_2 < x_1$ and $T(x_1) = 1$, $T(x_2) = -1, \ldots, T(x_{n+1}) = (-1)^n$.

   The best approximation of degree 2 to $x^3 - x^2 + 1$ on $[-1, 1]$ (found by the method just indicated) is $-x^2 + \frac{3}{4}x + 1$.

Exercises 7.2

1. We note that $x \approx bc \subseteq (ad)(ae) = ade$, whence $x/a \approx de$.
2. Clearly $\operatorname{conv} A$ is the *smallest* convex set in X containing A. Let $B = \bigcup(A^n : n = 1, 2, \ldots)$. Since $A^m A^n = A^{m+n}$ for all positive

integers m and n, the set B is convex; it also contains A, so $\operatorname{conv} A \subseteq B$. On the other hand, $\operatorname{conv} A$ contains B, whence $\operatorname{conv} A = B$.

3. (i) Suppose that A is a flat. Let $x \approx ab$, where $a, b \in A$. Choose $c \approx a/b$. Then $c \approx A/A \subseteq A$ and $b \approx (a/c) \cap (x/a)$. Thus $a/c \approx x/a$ and $x \approx a/c \subseteq A/A \subseteq A$. This shows that A is convex.

   (ii) Suppose that A is a flat and that $x \in A$. Then $A/x \subseteq A/A \subseteq A$. For $a \in A$, choose $y \approx ax \subseteq A$. Then $a \approx A/x$, so $A \subseteq A/x$. Thus $A/x = A$. A similar argument shows that $x/A = A$.

   (iii) Suppose that A is convex. Using Corollary 7.2.2, we deduce that

$$(A/A)/(A/A) = A/A(A/A) \subseteq A/((AA)/A) \subseteq AA/AA = A/A,$$

which shows that A/A is a flat.

   (iv) Suppose that A, B are flats and that $x \approx A \cap B$. Using Corollary 7.2.2 and (i), (ii) above, we deduce that

$$\begin{aligned}(A/B)/(A/B) &= (A/B)/((x/A)/(x/B)) \\ &\subseteq (A/B)/(xB/xA) \\ &= (A/B)/(B/A) \\ &\subseteq AA/BB \\ &= A/B.\end{aligned}$$

This shows that A/B is a flat.

4. Suppose that A is convex and $\tilde{A} \neq \varnothing$. Trivially $\hat{A} \subseteq \hat{\hat{A}}$. Let $a \in \tilde{A}$ and $x \in \hat{\hat{A}}$. Then there exists $y \in X$ such that $xy \subseteq \hat{A}$. By Theorem 7.2.10,

$$x(ay) = a(xy) \subseteq \tilde{A}\hat{A} \subseteq \tilde{A} \subseteq A.$$

Thus $x \in \hat{A}$ and $\hat{\hat{A}} \subseteq \hat{A}$. Hence $\hat{\hat{A}} = \hat{A}$.

5. (i) Suppose that $\mathbf{x}, \mathbf{y} \in A$. Then there exist positive integers r, s such that $\mathbf{x} = (x_1, \ldots, x_r, 0, 0, \ldots)$, $\mathbf{y} = (y_1, \ldots, y_s, 0, 0, \ldots)$, where $x_1 > 1/r$, $x_2, \ldots, x_r > 0$ and $y_1 > 1/s$, $y_2, \ldots, y_s > 0$. Suppose that $r \geqslant s$. Let $\lambda, \mu > 0$ with $\lambda + \mu = 1$. Then

$$\lambda\mathbf{x} + \mu\mathbf{y} = (\lambda x_1 + \mu y_1, \ldots, \lambda x_r + \mu y_r, 0, 0, \ldots).$$

Clearly $\lambda x_1 + \mu y_1, \ldots, \lambda x_r + \mu y_r > 0$ and $\lambda x_1 + \mu y_1 \geqslant \lambda/r + \mu/s \geqslant 1/r$. Thus $\lambda\mathbf{x} + \mu\mathbf{y} \in A$ and A is convex.

   (ii) Let $\mathbf{a} = (t, 0, 0, \ldots)$, where $t > 0$. Choose a positive integer m such that $1/m < t$. Let $\mathbf{b} = (t, 1, \ldots, 1, 0, 0, \ldots)$ have m positive terms. Then $\mathbf{ab} \subseteq A$, and so $\mathbf{a} \in \hat{A}$.

   (iii) Let $\mathbf{x}$ be a fixed point of $\mathbb{R}^*$. Then, for all sufficiently small $\lambda > 0$, the point $(1 - \lambda)\mathbf{0} + \lambda\mathbf{x}$ does not lie in A. Thus $\mathbf{0} \notin \hat{A}$.

Part (ii) shows that $\mathbf{0} \in \hat{\hat{A}}$ and part (iii) states that $\mathbf{0} \notin \hat{A}$. Hence $\hat{\hat{A}} \neq \hat{A}$.

Exercises 7.3

1. Suppose that f is convexity-preserving and that $f(\mathbf{b})$ lies between $f(\mathbf{a})$ and $f(\mathbf{c})$, where $\mathbf{a}, \mathbf{b}, \mathbf{c} \in \mathbb{R}^n$. Since $f(\operatorname{conv}\{\mathbf{a}, \mathbf{c}\})$ is a convex set containing $f(\mathbf{a}), f(\mathbf{c})$, we have $f(\mathbf{b}) \in f(\operatorname{conv}\{\mathbf{a}, \mathbf{c}\})$. But f is injective, so $\mathbf{b} \in \operatorname{conv}\{\mathbf{a}, \mathbf{c}\}$ and f^{-1} preserves betweenness. Suppose, conversely, that f^{-1} preserves betweenness on the convex set $f(\mathbb{R}^n)$. Let $\mathbf{a}, \mathbf{c}$ be distinct points of a convex set A in $\mathbb{R}^n$. Let $f(\mathbf{b})$ lie between $f(\mathbf{a})$ and $f(\mathbf{c})$, where $\mathbf{b} \in \mathbb{R}^n$. Since f^{-1} preserves betweenness, $\mathbf{b} \in \operatorname{conv}\{\mathbf{a}, \mathbf{c}\}$ and $\mathbf{b} \in A$. Thus $f(\mathbf{b}) \in f(A)$ and $f(A)$ is convex.
2. Let $\mathbf{a}_0, \ldots, \mathbf{a}_r$ be affinely independent points of A. For $i = 0, \ldots, r$, write $A_i = \operatorname{conv}(\{\mathbf{a}_0, \ldots, \mathbf{a}_r\} \setminus \{\mathbf{a}_i\})$. If $f = \mathbb{R}^n \to \mathbb{R}^m$ is injective and convexity-preserving, then $f(A_0), \ldots, f(A_r)$ are convex subsets of $f(A)$; any r of them have a non-empty intersection, but the intersection of all of them is empty. Helly's theorem adapted to the flat aff $(f(A))$ shows that such a situation cannot occur unless $f(A)$ has dimension at least r.
3. (i) $f(x) = x^3$; (ii) $f(x, y) = x + y^3$; (iii) $f(x, y) = (\operatorname{sgn}(x^2 + y^2), \operatorname{sgn} x^2y^2)$, where $\operatorname{sgn} r$ is the sign, either $-1, 0, 1$, of the real number r.
4. For each non-zero real number r, define $f(r) = \theta(s)$, where s is the unique element of S such that $r = qs$ for some rational number q. Define $f(0) = \mathbf{0}$. Then $f: \mathbb{R} \to \mathbb{R}^n$ maps each convex set in $\mathbb{R}$ containing more than one point to $\mathbb{R}^n$ itself.

Exercises 7.4

1. The set of points whose distance from L does not exceed $r > 0$ is a symmetric convex body with area $4\lambda r + \pi r^2$. This body will contain a non-zero lattice point if $4\lambda r + \pi r^2 = 4$, i.e. if $r = 2(\sqrt{(\lambda^2 + \pi)} - \lambda)/\pi$.
2. The line through $(0, 0)$ and $(50, 1)$ does not meet A_r when $r < 1/\sqrt{(2501)}$. Suppose $r = 1/50$. Let L be any line in $\mathbb{R}^2$ through the origin. Let $\mathbf{a}, \mathbf{b}$ be points on L at a distance 50 from the origin. Let R be the rectangle with sides 100 and 1/25 such that $\mathbf{a}, \mathbf{b}$ are the midpoints of opposite sides of R. By Minkowski's theorem, R contains a non-zero lattice point $\mathbf{c}$. The closed disc with centre $\mathbf{c}$ and radius 1/50 lies in A_r and meets L.
3. By Theorem 7.4.3, $bc - ad = ed - cf$, so $c/d = (a + e)/(b + f)$. Now $\mathfrak{F}_7$ is

1/7, 1/6, 1/5, 1/4, 2/7, 1/3, 2/5, 3/7, 1/2, 4/7, 3/5, 2/3, 5/7, 3/4, 4/5, 5/6, 6/7

and the stated equality is readily verified.

4. Yes, a very good approximation!
5. The volume of the symmetric convex body

$$\{(x_1, \ldots, x_n) : |\xi_1| + \ldots + |\xi_n| \leqslant (n!|\Delta|)^{1/n}\}$$

is 2^n and the first of the results now follows from Minkowski's theorem. The second result follows from the first one and the inequality $n|\xi_1 \ldots \xi_n|^{1/n} \leqslant |\xi_1| + \ldots + |\xi_n|$.
6. The desired value of r is $\sqrt{(4p/\pi)}$. Thus there is a non-zero lattice point (x, y) such that

$$0 < x^2 + (qx + py)^2 = (q^2 + 1)x^2 + 2pqxy + p^2y^2 \leqslant 4p/\pi < 2p,$$

and so $x^2 + (qx + py)^2 = p$.
7. Choose a non-integral $m > \max\{1, 1/\varepsilon^2\}$. The given parallelepiped has volume 8, and so by Minkowski's theorem contains a non-zero lattice point (p, q, r), where $r \geqslant 0$. We cannot have $r = 0$, for this would imply that p and q are also zero. Thus $r > 0$ and

$$|\alpha - p/r|, |\beta - q/r| \leqslant 1/r\sqrt{m}.$$

The desired result is now immediate, since (as m is not an integer) we have $r < m$.

Exercises 7.5

1. Set (i) has the fixed-point property by Brouwer's theorem. Sets (ii), (iii) do not have the fixed-point property. The mapping $f(\mathbf{x}) = \frac{1}{2}(\mathbf{x} + \mathbf{e}_1)$ is a continuous mapping of set (ii) into itself that has no fixed point. The mapping $g(\mathbf{x}) = -\mathbf{x}$ is a continuous mapping of set (iii) into itself that has no fixed point.
2. (a) Suppose that $B = f(A)$, where A, B, f are as in the exercise, and that A has the fixed-point property. Let $g: B \to B$ be a continuous mapping of B into itself. Then $f^{-1} \circ g \circ f: A \to A$ is a continuous mapping of A into itself, so there exists $\mathbf{a}$ in A such that $(f^{-1} \circ g \circ f)(\mathbf{a}) = \mathbf{a}$, whence $g(f(\mathbf{a})) = f(\mathbf{a})$ and $f(\mathbf{a})$ is a fixed point of g. Thus B has the fixed-point property.

   (b) Suppose that B is a retract of a set A in $\mathbb{R}^n$ and that A has the fixed-point property. Let $f: A \to B$ be a continuous mapping such that $f(\mathbf{b}) = \mathbf{b}$ for all $\mathbf{b}$ in B. Let $g: B \to B$ be a continuous mapping of B into itself. Then $g \circ f: A \to A$ is continuous, and so there exists $\mathbf{a}$ in A such that $(g \circ f)(\mathbf{a}) = \mathbf{a}$. Thus $\mathbf{a} \in B$, $f(\mathbf{a}) = \mathbf{a}$, $g(\mathbf{a}) = \mathbf{a}$, which shows that B has the fixed-point property.

(c) (i) The union of two sides of a triangle in $\mathbb{R}^2$ is topologically equivalent to a closed line segment, and so has the fixed-point property. (ii) The set $\{\mathbf{x} \in \mathbb{R}^n : \|\mathbf{x}\| = 1\}$ does not have the fixed-point property, and so is not a retract of the unit ball $\{\mathbf{x} \in \mathbb{R}^n : \|\mathbf{x}\| \leqslant 1\}$, which does.

3. Let C be a convex set in $\mathbb{R}^n$ having the fixed-point property. Suppose that C is not closed. Then there exist $\mathbf{a} \in \operatorname{cl} C \backslash C$ and $\mathbf{b} \in \operatorname{ri} C$ such that $\|\mathbf{a} - \mathbf{b}\| < 1$. The mapping $f: C \to C$ defined for $\mathbf{c} \in C$ by the equation

$$f(\mathbf{c}) = \mathbf{a} + \min\{1, \|\mathbf{c} - \mathbf{a}\|\}(\mathbf{b} - \mathbf{a})$$

is continuous, but has no fixed point. This contradiction shows that C is closed. Suppose next that C is unbounded. Then C contains some halfline L, say $L = \{\mathbf{a} + \lambda\mathbf{u} : \lambda \geqslant 0\}$, where $\mathbf{a}, \mathbf{u} \in \mathbb{R}^n$ and $\mathbf{u} \neq \mathbf{0}$. Denote by g the restriction to C of the projection operator of L. Then the mapping $f: C \to C$ defined for $\mathbf{c} \in C$ by the equation $f(\mathbf{c}) = g(\mathbf{c}) + \mathbf{u}$ is continuous, but has no fixed point. This contradiction shows that C is bounded. We have thus shown that C is compact.

Let $g:(0, 1] \to \mathbb{R}$ be defined by $g(x) = x^{-1} \sin(x^{-1})$ for $0 < x \leqslant 1$. Let $A = \{(0, 0)\} \cup \{(x, g(x)) : 0 < x \leqslant 1\}$. Then A is neither closed, bounded, nor convex. We show that A has the fixed-point property. Let $f: A \to A$ be a continuous mapping of A into itself. If either $f(0, 0) = (0, 0)$ or $f(1, \sin 1) = (1, \sin 1)$, then f has a fixed point, so we assume that this is not the case. Define continuous functions $h: A \to [0, 1]$ and $j:(0, 1] \to \mathbb{R}$ by the equations: $h(x, g(x)) = x$ for $0 < x \leqslant 1$, $h(0, 0) = 0$ and $j(x) = h(f(x, g(x))) - x$ for $0 < x \leqslant 1$. The continuity of f at $(0, 0)$ and the fact that $f(0, 0) \neq (0, 0)$ shows that there exists α with $0 < \alpha < 1$ such that $j(\alpha) > 0$. Clearly $j(1) < 0$. Thus there exists β with $\alpha < \beta < 1$ such that $j(\beta) = 0$, whence $h(f(\beta, g(\beta))) = \beta$ and $f(\beta, g(\beta)) = (\beta, g(\beta))$. Hence f has fixed point $(\beta, g(\beta))$.

Exercises 7.6

1. Let $\mathbf{u} = (\cos\theta, \sin\theta)$. Thus $l(\mathbf{u}) = 2ab/\sqrt{(a^2 \sin^2\theta + b^2 \cos^2\theta)}$, $w(\mathbf{u}) = 2\sqrt{(a^2 \cos^2\theta + b^2 \sin^2\theta)}$, $l_* = w_* = 2b$ and $l^* = w^* = 2a$.

2. In $\mathbb{R}^n$ let $\mathbf{b}$ be a boundary point of a convex body A of constant width λ. Let $\mathbf{c}$ be a point of A such that $\|\mathbf{c} - \mathbf{b}\| = \lambda$. Then the hyperplane through $\mathbf{b}$ with $\mathbf{c} - \mathbf{b}$ as a normal vector is a support hyperplane to A which meets it in $\{\mathbf{b}\}$, whence $\mathbf{b}$ is an exposed point of A.

3. Suppose that $\mathbf{a} \in \mathbb{R}^n \backslash A$ and that $\mathbf{b}$ is the nearest point of A to $\mathbf{a}$. Let $\mathbf{c} \in A$ be such that $\|\mathbf{c} - \mathbf{b}\| = \lambda$. Then $\|\mathbf{c} - \mathbf{a}\| > \lambda$, whence $A \cup \{\mathbf{a}\}$ has diameter greater than λ. Thus A is complete, and so of constant width.

4. Let A be a set in $\mathbb{R}^n$ of positive diameter λ. Let $\mathfrak{F}$ be the family of all convex bodies in $\mathbb{R}^n$ which have diameter λ and contain A. Then $\mathfrak{F}$ is non-empty and uniformly bounded. Thus $\mathfrak{F}$ contains a member, C say, of maximum volume. We show that C is complete. If it is not, then for some point $\mathbf{x}$ in $\mathbb{R}^n \backslash C$ the set $C \cup \{\mathbf{x}\}$ has diameter λ, whence conv $(C \cup \{\mathbf{x}\})$ belongs to $\mathfrak{F}$ and has volume exceeding that of C. Thus C is complete, and hence of constant width.

5. In $\mathbb{R}^3$ let A be the convex body of diameter λ obtained by rotating a plane convex body of constant width λ about an axis of symmetry $\mathfrak{l}$. Let $\mathbf{x} \in \mathbb{R}^3 \backslash A$ and let π be a plane containing both $\mathfrak{l}$ and $\mathbf{x}$. Then $\pi \cap A$ is a plane convex body of constant width λ, and hence complete. Thus the diameter of $(\pi \cap A) \cup \{\mathbf{x}\}$, and hence of $A \cup \{\mathbf{x}\}$, exceeds λ. Hence A is complete, and so has constant width λ.

9
Notes on further reading

General accounts of convexity are given in Eggleston (1958), Valentine (1964), Benson (1966), Roberts and Varberg (1973), Kelly and Weiss (1979), Lay (1982), and van Tiel (1984). Of these, Eggleston and Valentine can be considered graduate texts, the remaining ones (primarily) undergraduate. A comprehensive survey of the subject, written for the expert, is the two volume *Handbook of convex geometry* by Gruber and Wills (1993).

Chapter 1: The Euclidean space $\mathbb{R}^n$

§1.1. This introductory material on $\mathbb{R}^n$ is in nearly all textbooks on convexity. See, for example, the first few pages of Lay (1982).

§1.2. Rockafellar (1970) has a detailed account of the basic properties of flats.

§1.3. Our discussion of affine dimension is more extensive than that found in most texts. Kelly and Weiss (1979) gives a careful treatment of the topic.

§1.4. Hyperplanes are studied in both Rockafellar (1970) and Lay (1982).

§1.5. This approach to affine transformations is similar to that taken by Rockafellar (1970), and by Kelly and Weiss (1979), which calls them affine *functions*.

§1.6. Kelly and Weiss (1979), and Lay (1982) have their own accounts of this material.

§1.7. Kelly and Weiss (1979) contains some, but by no means all, of the results presented here. For more background, see Mirsky (1955) for algebraic aspects of the material, and Burkill and Burkill (1970) for further discussion about open sets and closed sets.

§1.8. Burkill and Burkill (1970) has a full account of the standard material relating to convergence and compactness. Theorem 1.8.10 is hardly ever mentioned in books on convexity. At best, it is stated without proof.

§1.9. While the results on continuous functions are routine, our sequential definition of continuity is rarely adopted by fellow authors. For example, Kelly and Weiss (1979) uses an ϵ, δ definition, whereas Lay (1982) *deems* $f: \mathbb{R}^n \to \mathbb{R}^m$ to be continuous if $f^{-1}(B)$ is open in $\mathbb{R}^n$ whenever B is open in $\mathbb{R}^m$.

Chapter 2: Convex sets

§2.1. With the exception of Theorem 2.1.8 and its corollary, the fundamental results presented here are to be found in any general account of convexity, for example, Rockafellar (1970). The *cancellation properties* of convex sets, expressed in the two exceptions just detailed, are due to Rådström (1952).

§2.2. Lay (1982) covers most of the material in this section. The Gauss–Lucas theorem, rarely mentioned in treatises on convexity, frequently appears in books on complex analysis; see Hille (1959). For a generalization of carathéodory's theorem, see Cook and Webster (1972).

§2.3. The basic results concerning the interiors and closures of convex sets are considered in Eggleston (1958) and Rockafellar (1970), the latter also gives a detailed discussion of relative interiors.

§2.4. The concepts of separation and support are amongst the most important ideas in convexity, having a wide range of applicability. It is not, therefore, surprising that there are a variety of ways of establishing the basic results concerning them. An excellent account, using nearest-point arguments, is given by Botts (1942). For a selection of other treatments, see Eggleston (1958), Valentine (1964), and Lay (1982).

§2.5. Rockafellar (1970), and Stoer and Witzgall (1970) treat the results of this section, although the latter's terminology differs from ours in places.

§2.6. Our account blends ideas from Rockafellar (1970), and Stoer and Witzgall (1970). Of historical interest are Straszewicz (1935), and Krein and Milman (1940). That the set of extreme points of a closed convex set in $\mathbb{R}^2$ is closed (Exercise 3) seems to be almost unrecognized in the mathematical literature.

§2.7. Kelly and Weiss (1979), and Lay (1982) give careful treatments of the Blaschke selection theorem and the Hausdorff metric. For a discussion of other metrics on spaces of convex sets, see Webster (1965), Shephard and Webster (1965), and Webster (1967).

§2.8. A gentle, well motivated introduction to duality is given in Lay (1982). See also Grünbaum (1967).

Chapter 3: Convex polytopes

§3.1. Most of the topics touched upon in this section are considered in Grünbaum (1967), and McMullen and Shephard (1971). Eggleston (1958) discusses the approximation of compact convex sets by polytopes. One interesting and historically important class of polytopes, not studied in this book, is that of the regular polyhedra (tetrahedron, cube, octahedron, dodecahedron, and icosahedron) and their analogues in $\mathbb{R}^n$. For a detailed account of these polytopes, see Coxeter (1963) and Fejes-Tóth (1964).

§3.2. Alternative treatments to that given in the text can be found in McMullen and Shephard (1971), and Lay (1982).

§3.3. Chapter 4 of Grünbaum (1967) is devoted to a study of particular examples of polytopes.

§3.4. Both Gale (1963) and Grünbaum (1967) contain a discussion of cyclic and neighbourly polytopes, as does the book *Convex polytopes and the upper bound conjecture* by McMullen and Shephard (1971). During the preparation of the latter, one of the authors (Peter McMullen) succeeded in proving the conjecture, thus changing its status to that of a theorem. His proof is given in the final chapter of the book.

§3.5. The proof of Euler's relation given in the text is taken from McMullen and Shephard (1971), that of the Dehn–Sommerville equations from Grünbaum (1967). The latter, in its notes to Chapters 8 and 9, details the interesting history of those two celebrated results.

§3.6. The idea of representing a polytope by a Gale transform is due to Gale (1956). For a discussion of such transforms, see Grünbaum (1967), and McMullen and Shephard (1971).

Chapter 4: Linear programming

§4.1. Rockafellar (1970), and Stoer and Witzgall (1970) both give detailed accounts of the structure of polyhedral sets, including the result that we have called the *finite basis theorem*.

§4.2. Our approach is motivated by Kuhn (1956), and Stoer and Witzgall (1970). The lemma of Farkas seems to have been first discovered by Minkowski, and then proved by Farkas (1902).

§4.3. There is a vast amount of literature on linear programming. Gass' (1969) book, *Linear programming*, can be recommended as a comprehensive introduction. For shorter accounts, read Lipschutz (1966) and Mital (1976). The duality theorem of linear programming is due to Gale, Klee, and Rockafellar (1951).

§4.4. Gass (1969), and Roberts and Varberg (1973) both cover this fundamental topic of linear programming.

§4.5. The simplex algorithm was invented by Dantzig (1951) while investigating military logistic problems. Gass (1969), Roberts and Varberg (1973), and Mital (1976) all have clear accounts of the simplex algorithm. Gass gives an example of cycling, and Roberts and Varberg detail the method of the artificial variable, which is due to Rockafellar (1964).

§4.6. The minimax theorem, the fundamental result in game theory, was discovered by von Neumann in 1928, and in 1944 von Neumann and Morgenstern published their classic book, *Theory of games and economics behaviour*. The latter contains a proof of the minimax theorem based on the idea of separating a point from a convex set by a hyperplane. Since then, game theory has been applied to a variety of competitive situations. Both Gass (1969) and Mital (1976) have clear accounts of the subject and its relationship to linear programming. Lipschutz (1966) can be recommended for its elementary approach and wealth of numerical examples.

Chapter 5: Convex functions

§5.1. Roberts and Varberg (1973) is an excellent, wide-ranging introduction to convex functions. Equally good, but more advanced, is Rockafellar (1970).

§5.2. Our account follows Roberts and Varberg (1973). The classic work on inequalities is Hardy, Littlewood, and Pólya (1952). More recent accounts are given in Kazarinoff (1961), Beckenbach and Bellman (1965), and Mitrinović (1970).

§5.3. Artin (1964) develops the theory of the gamma function through the concept of log-convexity, in a book that has been described as a little classic of mathematics. His work is based on ideas of Bohr and Mollerup (1922).

§5.4. Convex functions on $\mathbb{R}^n$ are discussed extensively in Berge (1963), Rockafellar (1970), Roberts and Varberg (1973), and van Tiel (1984).

§5.5. The continuity and differentiability properties of convex functions are discussed in Eggleston (1958), Rockafellar (1970), and Roberts and Varberg (1973).

§5.6. The introduction of support functions is one of many significant contributions made by Minkowski (1911) to the study of convex sets. Accounts of these functions can be found in most books on convexity, for example, Eggleston (1958), Valentine (1964), and Lay (1982).

§5.7. The books by Gass (1969) and Mital (1976), in addition to having good accounts of the convex programming problem, both give emphasis to the important role played by convexity in optimization problems.

§5.8. Our treatment follows Roberts and Varberg (1973). For an alternative approach, see Marcus and Minc (1964). The survey article by Mirsky (1963) is a useful source of our information on doubly stochastic matrices. Mirsky (1971) and Webster (1983*a*) discuss the diagonals of doubly stochastic matrices.

Chapter 6: Mixed volumes and extremum problems

§6.1. Elementary sets receive brisk treatments in Rogosinski (1952), where they are called *interval sums*, and in Kolmogorov and Fomin (1961).

§6.2. Most books on convexity adopt one of two approaches to the concept of volume for convex bodies. *Either* (like Eggleston (1958)) they assume a knowledge of Lebesgue measure, in which case little more remains to be done, *or* (like Lay (1982)) they assume the concept for polytopes, and then extend it to general convex bodies by a limiting argument. Our treatment is more basic than either of these. Rogosinski (1952) is close to the spirit of our account, if not the detail. Eggleston discusses the interaction between volume and the Hausdorff distance. The whole of Hadwiger's (1957) book is concerned with the volume and surface area of convex bodies. Mirsky (1955) gives a proof that a non-singular matrix is a product of elementary matrices.

§6.3. Few elementary texts on convexity give the details (or even state the results) of calculating the volumes of specific convex bodies. Eggleston (1958) derives the volume of a polytope in terms of its support function and the surface area of its facets.

§6.4. The theory of mixed volumes was studied intensively around the turn of the last century. Modern treatments of the subject are given in Eggleston (1958) and Leichtweiss (1980). Elementary accounts, in two and three dimensions, can be found in Hadwiger (1955), Lyusternik (1963), and Benson (1966).

§6.5. The works alluded to in the text are Schwarz (1884), Brunn (1887), Minkowski (1911), Lyusternik (1935), and Hadwiger and Ohmann (1956). There is a wealth of literature on the Brunn–Minkowski theorem, one of convexity's most celebrated results. See, for example, Eggleston (1958) and Benson (1966). An elementary account in the two-dimensional case can be found in Lyusternik (1963), while Schneider (1992) is a highly detailed, 544 page treatise on the topic.

§6.6. Blaschke (1949), Eggleston (1958), and Benson (1966) each have their own detailed accounts of Steiner symmetrization. Steiner (1838) is of historical interest.

Chapter 7: Selected topics

§7.1. The survey article by Danzer, Grünbaum, and Klee (1963) contains a wealth of material on Helly's theorem and related topics, including biographical detail on Helly himself. The books Eggleston (1958), Yaglom and Boltyanskii (1961), and Cadwell (1970) each have a chapter on the theorem. Webster (1983*b*) gives a simple proof of Kirchberger's theorem that does not use Helly's theorem. For a discussion of combinatorial geometry, including many unsolved problems, see Hadwiger, Debrunner, and Klee (1964), and Klee and Wagon (1991). Articles referred to in the text are Blumenthal and Wahlin (1941), and Rademacher and Schoenberg

(1950). Of historical interest are Jung (1901), Kirchberger (1903), Helly (1923), and Krasnosselsky (1946). A significant generalization of Helly's theorem can be found in Alon and Kleitman (1992). For Chebyshev polynomials and the equioscillation properties of best approximations, see Davis (1963).

§7.2. An excellent treatment of join geometries is given in Prenowitz and Jantosciak (1979). Related ideas are discussed in Bryant and Webster (1969, 1972, 1973, 1977), and van de Vel (1993). Brunn (1913) is of historical interest.

§7.3. There is little literature on this topic, and the reader is referred to the original papers of Meyer and Kay (1973), and Schöpf (1981).

§7.4. Books on convexity rarely discuss its applications to number theory. Davenport (1947), in his presidential address to the Mathematical Association, provides an excellent introduction to the geometry of numbers. For full-blown treatments of the subject, see Lekkerkerker (1969) and Cassels (1971). See also Hardy and Wright (1960). Baker (1984) devotes a section of his book to Minkowski's theorem, providing an alternative proof to that given in the text.

§7.5. Brouwer's (1912) fixed-point theorem has been the subject of much scrutiny. For a proof using Sperner's lemma, see Knaster, Kuratowski, and Mazurkiewicz (1929). Chapter 12 of Klee and Wagon (1991) contains an interesting discussion of ideas related to the fixed-point theorem. Motzkin (1935), and Motzkin, Straus, and Valentine (1953) are the original papers characterizing those sets in $\mathbb{R}^n$ which have, respectively, the unique nearest-point and the unique farthest-point properties. Klee (1961) and Valentine (1964) are good sources of reference for the latter two topics. Information concerning the Perron–Frobenius theorem can be found in Perron (1907) and Frobenius (1912). For a fixed-point proof of the theorem, see Debreu and Herstein (1953).

§7.6. For a modern survey article on convex bodies of constant width, see Chakerian and Groemer (1983). Some books dealing with the general area of convexity that discuss the topic are: Eggleston (1958), Yaglom and Boltyanskii (1961), Benson (1966) and Lay (1982). Popular accounts can be found in Gardner (1963), Cadwell (1970), and Rademacher and Toeplitz (1990). Of historical interest are Barbier (1860) and Reuleaux (1875).

Bibliography

Alon, N. and Kleitman, D. J. (1992). Piercing convex sets. *Bulletin of the American Mathematical Society*, **27**, 252–6.

Artin, E. (1964). *The gamma function*, (trans. M. Butler). Holt, Rinehart and Winston, New York.

Baker, A. (1984). *A concise introduction to the theory of numbers.* Cambridge University Press.

Barbier, E. (1860). Note sur le problème de l'aiguille et le jeu joint couvert. *Journal de Mathématiques Pures et Appliquées*, series 2, **5**, 273–86.

Beckenbach, E. F., and Bellman, R. (1965). *Inequalities*, 2nd revised printing. Springer-Verlag, Berlin.

Benson, R. V. (1966). *Euclidean geometry and convexity.* McGraw-Hill, New York.

Berge, C. (1963). *Topological spaces*, (trans. E. M. Patterson). Oliver & Boyd, Edinburgh.

Berger, M. (1990). Convexity. *American Mathematical Monthly*, **97**, 650–78.

Blaschke, W. (1949). *Kreis und Kugel.* Chelsea Publishing Company, New York.

Blumenthal, L. M. and Wahlin, G. E. (1941). On the spherical surface of smallest radius enclosing a bounded subset of n-dimensional Euclidean space. *Bulletin of the American Mathematical Society*, **47**, 771–7.

Bohr, H. and Mollerup, J. (1922). *Laerebogi i matematisk analyse* (Copenhagen) III, 149–64.

Bonnesen, T. and Fenchel, W. (1948). *Theorie der konvexen Körper.* Chelsea Publishing Company, New York.

Botts, T. (1942). *Convex sets.* American Mathematical Monthly, **49**, 527–35.

Brouwer, L. E. J. (1912). Über Abbildung von Mannigfaltigkeiten. *Mathematische Annalen*, **71**, 97–115.

Brunn, H. (1887). *Über Ovale und Eiflächen.* Inaugural dissertation, Munich.

Brunn, H. (1913). Über Kerneigebiete. *Mathematische Annalen*, **73**, 430–40.

Bryant, V. W. and Webster, R. J. (1969). Generalizations of the theorems of Radon, Helly, and Carathéodory. *Monatshefte für Mathematik*, **73**, 309–15.

Bryant, V. W. and Webster, R. J. (1972). Convexity spaces I: The basic properties. *Journal of Mathematical Analysis and its Applications*, **37**, 206–13.

Bryant, V. W. and Webster, R. J. (1973). Convexity spaces II: Separation. *Journal of Mathematical Analysis and its Applications*, **43**, 321–7.

Bryant, V. M. and Webster, R. J. (1977). Convexity spaces III: Dimension. *Journal of Mathematical Analysis and its Applications*, **57**, 382–92.

Burkill, J. C. and Burkill, H. (1970). *A second course in mathematical analysis.* Cambridge University Press.

Cadwell, J. H. (1970). *Topics in recreational mathematics.* Cambridge University Press.

Cassels, J. W. S. (1971). *An introduction to the geometry of numbers.* Springer-Verlag, Berlin.

Chakerian, G. D. and Groemer, H. (1983). Convex bodies of constant width.

Convexity and its applications, (ed. P. M. Gruber and J. M. Wills). Birkhäuser Verlag, Basel.

Cook, W. D. and Webster, R. J. (1972). Carathéodory's theorem. *Canadian Mathematical Bulletin*, **15**, 293.

Coxeter, H. S. M. (1963). *Regular polytopes*, (2nd edn). Macmillan, New York.

Dantzig, G. B. (1951). Maximization of a linear function of variables subject to linear inequalities. In *Activity analysis of production and allocation*. Cowles Commission Monograph 13. Wiley, New York.

Danzer, L. W., Grünbaum, B., and Klee, V. (1963). Helly's theorem and its relatives. American Mathematical Society Symposium on Convexity, *Proceedings of Symposia in Pure Mathematics*, **7**, 101–80.

Davenport, H. (1947). The geometry of numbers. *The Mathematical Gazette*, **31**, 206–10.

Davis, P. J. (1963). *Interpolation and approximation*. Blaisdell, New York.

Debreu, G. and Herstein, I. N. (1953). Nonnegative square matrices. *Econometrica*, **21**, 597–607. Chicago.

Eggleston, H. G. (1958). *Convexity*, Cambridge Tracts in Mathematics and Mathematical Physics, No. 47. Cambridge University Press.

Farkas, J. (1902). Theorie der einfachen Ungleichungen. *Journal für die Reine und Angewandte Mathematik*, **124**, 1–27.

Fejes-Tóth, L. (1964). *Regular figures*. Pergamon, London.

Fenchel, W. (1983). Convexity through the ages. *Convexity and its applications*, (ed. P. M. Gruber and J. M. Willis). Birkhäuser Verlag, Basel.

Frobenius, G. (1912). Über Matrizen aus nicht negativen Elementen. *Sitzungsberichte der königlich preussischen Akademie der Wissenschaften*, 456–77.

Gale, D. (1956). Neighboring vertices on a convex polyhedron. In *Linear inequalities and related systems*, (ed. H. W. Kuhn and A. W. Tucker). Princeton University Press.

Gale, D. (1963). Neighborly and cyclic polytopes. American Mathematical Society Symposium on Convexity, *Proceedings of Symposia in Pure Mathematics*, **7**, 225–32.

Gale, D., Klee, V. L. and Rockafellar, R. T. (1951). Linear programming and the theory of games. In *Activity analysis of production and allocation*, Cowles Commission Monograph 13. Wiley, New York.

Gardner, M. (1963). Curves of constant width, one of which makes it possible to drill square holes. *Scientific American*, February, pp. 148–56.

Gass, S. I. (1969). *Linear programming*. McGraw-Hill, New York.

Gruber, P. M. and Willis, J. M. (ed.) (1983). *Convexity and its applications*. Birkhäuser Verlag, Basel.

Gruber, P. M. and Willis, J. M. (ed.) (1993). *Handbook of convex geometry*. North-Holland, Amsterdam.

Grünbaum, B. (1967). *Convex polytopes*. Wiley, London.

Hadwiger, H. (1955). *Altes und Neues über konvexe Körper*. Birkhauser Verlag, Basel.

Hadwiger, H. (1957). *Vorlesungen über Inhalt, Oberfläche und Isoperimetrie*. Springer-Verlag, Berlin.

Hadwiger, H. and Ohmann, D. (1956). Brunn-Minkowskischer Satz und Isoperimetrie. *Mathematische Zeitschrift*, **66**, 1–8.

Hadwiger, H., Debrunner, H., and Klee, V. (1964). *Combinatorial geometry in the plane*, (trans. V. Klee). Holt, Rinehart and Winston, New York.

Hardy, G. H., Littlewood, J. E., and Pólya, G. (1952). *Inequalities*, (2nd edn). Cambridge University Press.

Hardy, G. H. and Wright, E. M. (1960). *An introduction to the theory of numbers*, (4th edn). Oxford University Press.

Helly, E. (1923). Über Mengen konvexer Körper mit gemeinschaftlichen Punkten. *Jahresbericht der Deutscher Mathematiker—Vereingigung*, **32**, 175–6.

Hille, E. (1959). *Analytic function theory I*. Chelsea Publishing Company, New York.

Jung, H. W. E. (1901). Über die kleinste Kugel, die eine räumliche Figur einschliesst. *Journal für die Reine und Angewandte Mathematik*, **123**, 241–57.

Kazarinoff, N. D. (1961). *Analytic inequalities*. Holt, Rinehart and Winston, New York.

Kelly, P. J. and Weiss, M. L. (1979). *Geometry and convexity*. Wiley, New York.

Kirchberger, P. (1903). Über Tschebyschefsche Annäherungsmethoden. *Mathematische Annalen*, **57**, 509–40.

Klee, V. (1961). Convexity of Chebyshev sets. *Mathematische Annalen*, **142**, 292–304.

Klee, V. (1971). What is a convex set? *American Mathematical Monthly*, **78**, 616–31.

Klee, V. and Wagon, S. (1991). *Old and new unsolved problems in plane geometry and number theory*, Dolciani Mathematical Expositions, No. 11. The Mathematical Association of America.

Knaster, B., Kuratowski, C., and Mazurkiewicz, S. (1929). Ein Beweis des Fixpunktsatzes für n-dimensionale Simplexe. *Fundamenta Mathematicae*, **14**, 132–7.

Kolmogorov, A. N. and Fomin, S. V. (1961). *Elements of the theory of functions and functional analysis*: Volume 2, (trans. H. Kamel and H. Komm). Graylock Press, Albany, New York.

Krasnosselsky, M. A. (1946). Sur un critère pour qu'un domain soit étoilé (Russian with French summary). *Matematičeskiĭ Sbornik*, Novaja Serija, **19** (61), 309–10.

Krein, M. and Milman, D. (1940). On the extreme points of regularly convex sets. *Studia Mathematica*, **9**, 133–8.

Kuhn, H. W. (1956). Solvability and consistency for linear equations and inequalities. *American Mathematical Monthly*, **63**, 217–32.

Lay, S. R. (1982). *Convex sets and their applications*. Wiley, New York.

Lekkerkerker, C. G. (1969). *Geometry of numbers*. North-Holland, Amsterdam.

Leichtweiss, K. (1980). *Konvexe Mengen*. Springer-Verlag, Berlin.

Lipschutz, S. (1966). *Finite mathematics*. Schaum Publishing, New York.

Lyusternik, L. A. (1935). Die Brunn—Minkowskische Ungleichung für beliebige messbare Mengen. *Comptes Rendus de l'Académie des Sciences* URSS, **8**, 55–8.

Lyusternik, L. A. (1963). *Convex figures and polyhedra*, (trans. T. J. Smith). Dover Publications, New York.

McMullen, P. and Shephard, G. C. (1971). *Convex polytopes and the Upper Bound Conjecture*, London Mathematical Society Lecture Note Series 3. Cambridge University Press.

Marcus, M. and Minc, H. (1964). *A survey of matrix theory and matrix inequalities.* Allyn and Bacon, Rockleigh, New Jersey.

Meyer, W. and Kay, D. C. (1973). A convexity structure admits but one real linearization of dimension greater than one. *Journal of the London Mathematical Society*, (2) **7**, 124–30.

Minkowski, H. (1910). Geometrie der Zahlen. Leipzig and Berlin.

Minkowski, H. (1911). *Theorie der konvexen Körper, inbesondere Begründung ihres Oberflächenbegriffs*, Gesammelte Abhandlungen II, Leipzig.

Mirsky, L. (1955). *An introduction to linear algebra.* Oxford University Press.

Mirsky, L. (1963). Results and problems in the theory of doubly-stochastic matrices. *Zeitschrift für Wahrscheinlichkeitstheorie und Verwandte Gebiete*, **1**, 319–34.

Mirsky, L. (1971). *Transversal theory.* Academic Press, New York.

Mital, K. V. (1976). *Optimization methods.* Wiley Eastern Limited, New Delhi.

Mitrinović, D. S. (1970). *Analytic inequalities.* Springer-Verlag, Berlin.

Motzkin, T. S. (1935). Sur quelques propriétés caractéristiques des ensembles convexes. *Atti della Accademia Nazionale dei Lincei. Rendiconti. Classi di Scienze Fisiche, Matematiche e Naturali*, (8) **21**, 773–9.

Motzkin, T. S., Straus, E. G., and Valentine F. A. (1953). The number of farthest points. *Pacific Journal of Mathematics*, **3**, 221–32.

von Neumann, J. and Morgenstern, O. (1944). *Theory of games and economic behaviour.* Princeton University Press.

Perron, O. (1907). Grundlagen für eine Theorie des Jacobischen Kettenbruchalgorithmus, *Mathematische Annalen*, **64**, 1–76.

Prenowitz, W. and Jantosciak, J. (1979). *Join geometries.* Springer-Verlag, New York.

Rademacher, H. and Schoenberg, I. J. (1950). Helly's theorem on convex domains and Tchebycheff's approximation problem. *Canadian Journal of Mathematics*, **2**, 245–56.

Rademacher, H. and Toeplitz, O. (1990). *The enjoyment of mathematics*, (trans. H. Zuckerman). Dover Publications, New York.

Rådström, H. (1952). An embedding theorem for spaces of convex sets. *Proceedings of the American Mathematical Society*, **3**, 165–9.

Reuleaux, F. (1875). *Lehrbuch der Kinematik.* Brunswick (Braunschweig, Germany).

Roberts, A. W. and Varberg, D. E. (1973). *Convex functions.* Academic Press, New York.

Rockafellar, R. T. (1964). A combinatorial algorithm for linear programs in the general mixed form. *SIAM Journal on Applied Mathematics*, **12**, 215–25.

Rockafellar, R. T. (1970). *Convex analysis.* Princeton University Press.

Rogosinski, W. W. (1952). *Volume and integral.* Oliver and Boyd, Edinburgh.

Schneider, R. G. (1992). *Convex bodies: the Brunn-Minkowski theory.* Cambridge University Press.

Schöpf, P. (1981). Konvexitätstreue unde Linearität von Abbildungen. *Mathematische Zeitschrift*, **177**, 533–40.

Schwarz, H. A. (1884). Beweis des Satzes, dass die Kugel kleinere Oberfläche besitzt als jeder andere Körper gleichen Volumens. *Nachrichten der Akademie der Wissenchaften in Göttingen*, 1–13.

Shephard, G. C. and Webster, R. J. (1965). Metrics for sets of convex bodies. *Mathematika*, **12**, 73–88.

Steiner, J. (1838). Einfache Beweise der isoperimetrischen Hauptsätze. *Journal für die Reine unde Angewandte Mathematik*, **18**, 281–96.

Stoer, J. and Witzgall, C. (1970). *Convexity and optimization in finite dimensions I.* Springer-Verlag, New York.

Straszewicz, S. (1935). Über exponierte Punkte abgescchlossener Punktmengen. *Fundamenta Mathematicae*, **24**, 139–43.

van Tiel, J. (1984). *Convex analysis.* Wiley, Chichester, England.

Valentine, F. A. (1964). *Convex sets.* McGraw-Hill, New York.

van de Vel, M. L. J. (1993). *Theory of convexity structures.* North-Holland. Mathematical Library, No. 50.

Webster, R. J. (1965). The space of affine-equivalence classes of compact subsets of Euclidean space. *Journal of London Mathematical Society*, **40**, 425–32.

Webster, R. J. (1967). An embedding of a subset space in the Hilbert cube. *Journal of London Mathematical Society*, **42**, 344–8.

Webster, R. J. (1983*a*). The harmonic means of diagonals of doubly-stochastic matrices. *Linear and Multilinear Algebra*, **13**, 367–9.

Webster, R. J. (1983*b*). Another simple proof of Kirchberger's theorem. *Journal of Mathematical Analysis and its Applications*, **92**, 299–300.

Yaglom, I. M. and Boltyanskii, V. G. (1961). *Convex figures.* Holt, Rinehart and Winston, New York.

Index of symbols and notation

General index

COUPLE, MARRIAGE, AND FAMILY THERAPY SUPERVISION

Karin Jordan, PhD, LMFT
EDITOR

SPRINGER PUBLISHING COMPANY
NEW YORK

Springer Publishing Company, LLC
11 West 42nd Street
New York, NY 10036
www.springerpub.com

Acquisitions Editor: Nancy S. Hale
Composition: S4Carlisle Publishing Services

ISBN: 978-0-8261-2678-8
e-book ISBN: 978-0-8261-2679-5

15 16 17 18 / 5 4 3 2 1

Library of Congress Cataloging-in-Publication Data

Couple, marriage, and family therapy supervision / Karin Jordan, editor.
pages cm
Includes bibliographical references and index.
ISBN 978-0-8261-2678-8
1. Psychotherapists—Supervision of. 2. Psychotherapy—Study and teaching—Supervision.
3. Couples therapy. 4. Marital psychotherapy. 5. Family psychotherapy. I. Jordan, Karin, editor.
RC459.C67 2016
616.89'1562—dc23

2015032982

Printed in the United States of America by Gasch Printing.

To my family,

my teachers and supervisors,

and all those I supervised.

Contents

Contributors

Ashley J. Blount, PhD
Assistant Professor
Department of Counseling
University of Nebraska, Omaha
Omaha, NE

Rebecca Boyle, PhD, IMFT, PCC-S
Director, Clinic for Individual and Family Counseling
School of Counseling
The University of Akron
Akron, OH

Sharon Apel Bursky, MA, IMFT, PhD Candidate
Department of Counseling
The University of Akron
Akron, OH

Benjamin E. Caldwell, PsyD
Associate Professor
Alliant International University
California School of Professional Psychology
Los Angeles, CA

Marj Castronova, PhD
MFT Program Director
Assistant Professor
School of Professional Studies
Lincoln Christian University, Las Vegas
Las Vegas, NV

Jessica ChenFeng, PhD
Assistant Professor
Department of Educational Psychology and Counseling
California State University, Northridge
Northridge, CA

Ryan M. Cook, LPCC-S, MFT, AAMFT Approved Supervisor, PhD Candidate
Department of Counseling
The University of Akron
Akron, OH

Angela DeCandia-Vitoria, MSW
Founding Member
New York Center for Emotionally Focused Therapy
New York, NY

George Faller, MS
President
New York Center for Emotionally Focused Therapy
New York, NY

Ulia Fisher, MA
The University of Akron
Akron, OH

Jerry Gale, PhD
Department of Human Development and Family Science
Doctoral Family Therapy Program
University of Georgia
Athens, GA

Joanne Holbert, PhD
Marriage and Family Therapist
Family Connection of Wadsworth
Wadsworth, OH

Karin Jordan, PhD, LMFT
Professor and Director, School of Counseling
Interim Associate Dean, College of Health Professions
The University of Akron
Akron, OH

Eli A. Karam, PhD, LMFT
Assistant Professor
Family Therapy Program
Kent School of Social Work
University of Louisville
Louisville, KY

Heather Katafiasz, PhD
Assistant Professor
The University of Akron
Akron, OH

Glenn W. Lambie, PhD
Professor and Department Chair
Department of Child, Family, and Community Sciences
University of Central Florida
Orlando, FL

Megan A. McCoy, MA
Department of Human Development and Family Science
Doctoral Family Therapy Program
University of Georgia
Athens, GA

Susan McDaniel, PhD
Professor
University of Rochester School of Medicine and Dentistry
Department of Psychiatry
Rochester, NY

Molly McDowell-Burns, MS
The University of Akron
School of Counseling
Akron, OH

Megan Miranda, PhD
The Relationship Center
North Canton, OH

Lisa Palmer-Olsen, PsyD
Director
Alliant Couple and Family Clinic
Alliant International University
San Diego, CA

Rikki Patton, PhD
Assistant Professor
School of Counseling
College of Health Professions
The University of Akron
Akron, OH

Anne M. Prouty, PhD, LMFT
Associate Professor
Department of Community, Family and Addiction Services—MFT Programs
Texas Tech University
Lubbock, TX

Michael Redd, MA
School of Counseling
The University of Akron
Akron, OH

Sandra A. Rigazio-DiGilio, PhD
Professor
University of Connecticut
Department of Human Development and Family Studies
Marriage and Family Therapy Program
Storrs, CT

Thomas W. Roberts, PhD
Department of Child and Family Development
San Diego State University
San Diego, CA

Teresa Rogers, PhD, IMFT-S, PCC
The Relationship Center
North Canton, Ohio

D. Bruce Ross III, MA
Department of Human Development and Family Science
Doctoral Family Therapy Program
University of Georgia
Athens, GA

Jenny Speice, PhD
Associate Professor
University of Rochester School of Medicine and Dentistry
Department of Psychiatry
Rochester, NY

Laura Tejada, PhD
Assistant Professor
Department of Counselor Education
Northeastern Illinois University
Chicago, IL

Markie L. C. Twist, PhD
Associate Professor
Certificate in Sex Therapy Program Coordinator
Department of Human Development and Family Studies
University of Wisconsin-Stout
Marriage and Family Therapy Program
Menomonie, WI

Scott R. Woolley, PhD
Distinguished Professor
Couple and Family Therapy Program
Alliant International University
San Diego, CA

Katie Wootton, MA, PhD Candidate
The University of Akron
Akron, OH

Toni Zimmerman, PhD
Professor
Department of Human Development and Family Studies
Director, Marriage and Family Therapy Graduate Program
Colorado State University
Fort Collins, CO

Preface

The intent of this book is to provide a comprehensive resource guide for Marriage and Family Therapists (MFTs), Approved Supervisors, and Supervisors in training. The book is comprehensive, looking not only at theories used in American Association for Marriage and Family Therapy (AAMFT) supervision, but also at other important aspects of supervision. There are three sections that organize the content logically and make it easy to follow. Section I focuses on supervision across time. Why is it important to look at the history, basic concepts, and theory training of supervision? The history and today's trends have laid the foundation for the development of the process of supervision and show that the supervision process has to be flexible as the field changes and must be reflective of the field as it currently exists.

Section II focuses on the nuts and bolts of supervision. Basic concepts such as how to get started in supervision, the first meeting, the supervision contract, supervision logs, and supervision documentation are all important and are addressed in Chapter 2, which lays the foundation for how to work together. These are also important from a liability perspective. AAMFT Approved Supervisors use various forms of supervision, which are addressed in Chapter 3.

It is important to focus on the developmental readiness of the supervisor-in-training; this is addressed in Chapter 4, which looks at tailoring supervision to the supervisee's developmental readiness. Common factors have been well addressed in the field of Marriage and Family Therapy by such researchers as Doug Sprenkle. These must also be addressed when looking at AAMFT Approved Supervision, as investigated in Chapter 5. It is important that AAMFT Approved Supervisors, those in training, and supervisees follow the AAMFT Code of Ethics. Clinicians and researchers in the field are looking more critically, through empirical and other research, at how culture, race, and gender should be considered and addressed in the process of supervision. This important topic, which must be continually revisited in an ever-changing, diverse, and global society, is addressed in Chapters 6 and 7.

Section III focuses on theory-specific supervision. The Marriage and Family Therapy field, including AAMFT Approved Supervision, has recognized the importance of looking at different theoretical orientations, as empirical research has been growing and more attention has been given to empirical and other research on diversity (which is addressed in the theory specific chapters). To get more clarity on how the various theories would be applied by the AAMFT Approved Supervisor and Supervisor-in-Training, each theory chapter has been written to focus on teaching theoretical context (founders, philosophical foundation, the process of change, role of the therapist, techniques to be used, and culture/race and gender issues/research to consider) and the role of the supervisor and supervisee (supervisor-supervisee relationship and learning goals/objectives). In addition, the same case example is used for each theory chapter, with the following information:

A. AAMFT Approved Supervisor—Dr. Bob
   The supervisor is a 53-year-old Caucasian male. He is a seasoned supervisor who has been credentialed as an AAMFT Approved Supervisor for more than 20 years. He provides supervision for both students (in practicum and internship as an adjunct professor) and postgraduates at his agency.
B. Supervisee—Brittney
   The supervisee is a 33-year-old Caucasian female in her first semester of internship at a local nonprofit agency. She is particularly interested in doing family therapy and is excited to have the opportunity to work with a broad range of families.
C. Family
   Travis (29 years old) and Tekia (27 years old) are an African American couple with two children, Tyrone (5 years old) and Tommy (4 years old). Tekia has been a stay-at-home mom. Travis was in the military (as has been every male in his family) and served two tours of duty in Afghanistan. He was wounded when an improvised explosive device (IED) exploded and left him severely injured. After spending many months in hospitals and rehabilitation, Travis was discharged from the military (which he had wanted to make his career) due to his injuries and now has difficulty finding a job. He now works two part-time jobs and hopes to find a full-time job.

Presenting problem:

- Couple relationship and parenting issues, as Travis reportedly has a "short fuse."
- There are no reported safety concerns.
- Travis's continued recovery from his injuries.
- Job issues—inability to find full-time employment/financial strain.
- Travis's older brother is still in the military and was recently re-deployed to Afghanistan.

This case example, used in all the theoretically specific chapters, can help clarify how the various theories overlap and differ.

Section IV deals with population-specific supervision. Because the field of Marriage and Family Therapy has been changing, there should be an emphasis on health and mental health care for veterans, which Marriage and Family Therapists have been authorized to provide. One chapter has been devoted to Medical Family Therapy Supervision and another to Trauma Supervision.

This book provides guidance for AAMFT Approved Supervisors and those in training as they provide supervision for Marriage and Family Therapists at various developmental levels. It should serve as a valuable resource for the seasoned AAMFT Approved Supervisor but also can be used as a text for the training of AAMFT Approved Supervisors. It is a complex book that provides a wide range of information, from the basics on how to get started to descriptions of the various theories. The case example helps clarify how the various theories differ and overlap, something that has not been seen in other supervision texts.

SECTION I

Supervision Across Time

ONE

History and Future Trends

Karin Jordan
Ulia Fisher

The field of Marriage and Family Therapy (MFT) is on the fast track to becoming a prominent and competitive mental health profession, based on an expected growth in MFT jobs (United States Department of Labor, 2012), as well as a growing body of clinical research (Sprenkle, 2010). The number of programs across the country, ranging from master's to doctoral programs, is continuously growing.

The Commission on Accreditation of Marriage and Family Therapy Education (COAMFTE) develops accreditation standards for graduate and postgraduate Marriage and Family Therapy training programs to ensure that students are getting a strong educational foundation in MFT (see the COAMFTE *Accreditation Standards for Graduate & Post-Graduate Marriage and Family Therapy Training Programs,* Version 12.0 Draft). The COAMFTE has identified its function as fivefold:

> First, the COAMFTE establishes and maintains professional standards in the education and training of CMFTPs.
>
> Second, the COAMFTE is committed to establishing and maintaining professional standards that support diversity and inclusion within MFT programs, and ensuring through accreditation excellence in multi-culturally/internationally informed education.
>
> Third, the COAMFTE accredits and ensures excellence in educational programs that educate CMFTPs who are relationally/systemically, multi-culturally/internationally and ethically informed.
>
> Fourth, the COAMFTE provides leadership and advocates for change in the larger practice and regulatory communities in defining competent CMFTPs.

> Finally, the COAMFTE promotes consistency of qualifications and competency of CMFTPs to the public (COAMFTE Accreditation Standards, 2014).

The COAMFTE focuses on the various aspects of MFT training, including the educational and practice (clinical training) regulatory requirements. Clinical training is an important aspect of training MFTs, which not only requires significant face-to-face contact with individuals, couples, families, and other systems, but also must include relationally oriented supervision. The focus should be on the developmental needs of master's and doctoral students, as the supervisor provides supervision in an ever-changing environment. Supervision for MFT students and postgraduates working toward licensure is generally provided by American Association for Marriage and Family Therapy (AAMFT) Approved Supervisors.

AAMFT Approved Supervisors are an integral part of training MFT students and postgraduates, focusing on clinical growth and development. Because AAMFT Approved Supervisors are so important in the growth and development of today's and future MFTs, it is important to understand the history of AAMFT supervision; this is a building block for today's standards and will be the foundation for future trends and directions. AAMFT Approved Supervision is now and has continuously been one of the most active and fastest-growing subsystems of the MFT field (Liddle, Breunlin, & Schwartz, 1988). Before looking at the history, present trends, and future direction, it is important to understand what an AAMFT Approved Supervisor is, as defined in the *Approved Supervisor Designation: Standards Handbook* (AAMFT, 2014a):

> AAMFT Approved Supervisor (AS) is a marriage and family therapist who has completed the supervisor candidate training requirements established by AAMFT as described below and who has applied for and been awarded the Approved Supervisor designation. The Approved Supervisor designation is not an advanced clinical credential but rather identifies, for the mental health community, those professionals who have met the AAMFT requirements to provide MFT supervision. Therefore, Approved Supervisors must also be Clinical Fellows of AAMFT. Approved Supervisors complete an AAMFT approved refresher course every five years to remain current on supervision standards, literature, and practices, and to renew their designation. (p. 5)

The AAMFT Approved Supervisor definition has evolved for more than 50 years and is continuously changing as the field of MFT grows and changes.

HISTORY

The history of AAMFT Approved Supervision goes back to David Mace in 1961. Mace was the Executive Director of the American Association for Marriage Counseling (AAMC), which changed its name to the American Association for Marriage and Family Therapy (AAMFT) when designated marriage counseling internship sites in one county could not be accessed, and a group of members were asked to serve as supervisors. However, the topic of supervision was something that the organization spent more than a decade discussing, even though Mace strongly advocated for supervision as a means of quality control and increasing membership. Others, however, feared that supervision would create an opposite effect, discouraging practitioners from seeking the credential in marriage counseling, and therefore negatively affecting membership (Benningfield, 1985). It was during this time that a consensus gradually evolved as to the definition of *supervision* and as to what supervisor qualifications should be, and in 1971 the Approved Supervisor designation was established after several years of naming individuals to supervise candidates for membership. Also during this time, a definition of *supervision* was adopted and 67 persons (20 females and 47 males) were grandfathered in and named Approved Supervisors. The number of Approved Supervisors grew quickly, rising to 200 in 1974. During this time, a Committee on Supervision (COS) was developed by the board. Fred Humphry was the chair of the newly formed COS and was charged with reviewing and approving applications for Approved Supervisor status. In 1975, the COS was also charged with setting standards for Approved Supervisors. The COS developed both supervision practices and procedures for Approved Supervisor designation. In 1976, there were 233 Approved Supervisors in the United States, constituting 14% of the AAMFT clinical members (Lee, Nichols, Nichols, & Odom, 2004). In support of Marriage and Family Therapy (MFT) being a unique profession, in 1977 the COS identified that those who sought the Approved Supervisor designation needed to demonstrate systemic conceptualization.

One of the salient components of MFT supervision is live supervision, which sets it apart from other disciplines (Montalvo, 1973). In this supervision medium, the MFT student and/or MFT postgraduate (or therapist) is observed through one-way glass and occasionally receives suggestions via a phone call (Montalvo, 1973) as well as videotapes and use of the "bug in the ear" (Birchler, 1975). This allows provision of immediate feedback to the MFT student or postgraduate. Live supervision goes back to the seminal research conducted by Bateson (1972) in collaboration with Haley, Weakland, and later Jackson, focusing on observation and teamwork. The Milan team, practicing from Milan, Italy (Selvini Palazzoli, Boscolo, Cecchin, & Prata, 1978), utilized both a team behind the one-way glass and a co-therapy team

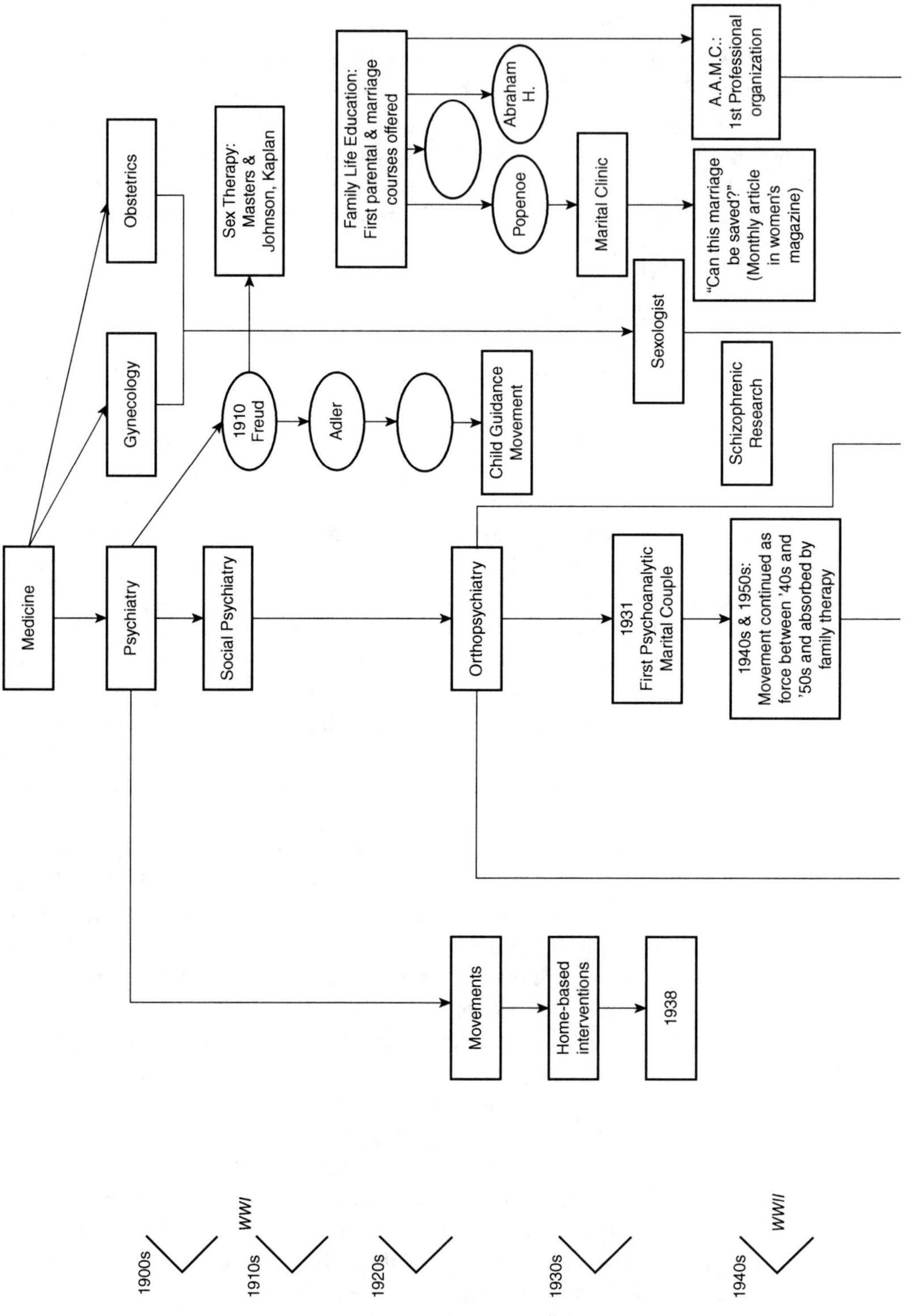
Medicine
Psychiatry
Social Psychiatry
Orthopsychiatry
Gynecology
Obstetrics
1910 Freud
Adler
Child Guidance Movement
Sex Therapy: Masters & Johnson, Kaplan
Family Life Education: First parental & marriage courses offered
Abraham H.
Popenoe
Marital Clinic
A.A.M.C.: 1st Professional organization
"Can this marriage be saved?" (Monthly article in women's magazine)
Sexologist
Schizophrenic Research
1931 First Psychoanalytic Marital Couple
1940s & 1950s: Movement continued as force between '40s and '50s and absorbed by family therapy
Movements
Home-based interventions
1938
1900s
1910s
WWI
1920s
1930s
1940s
WWII

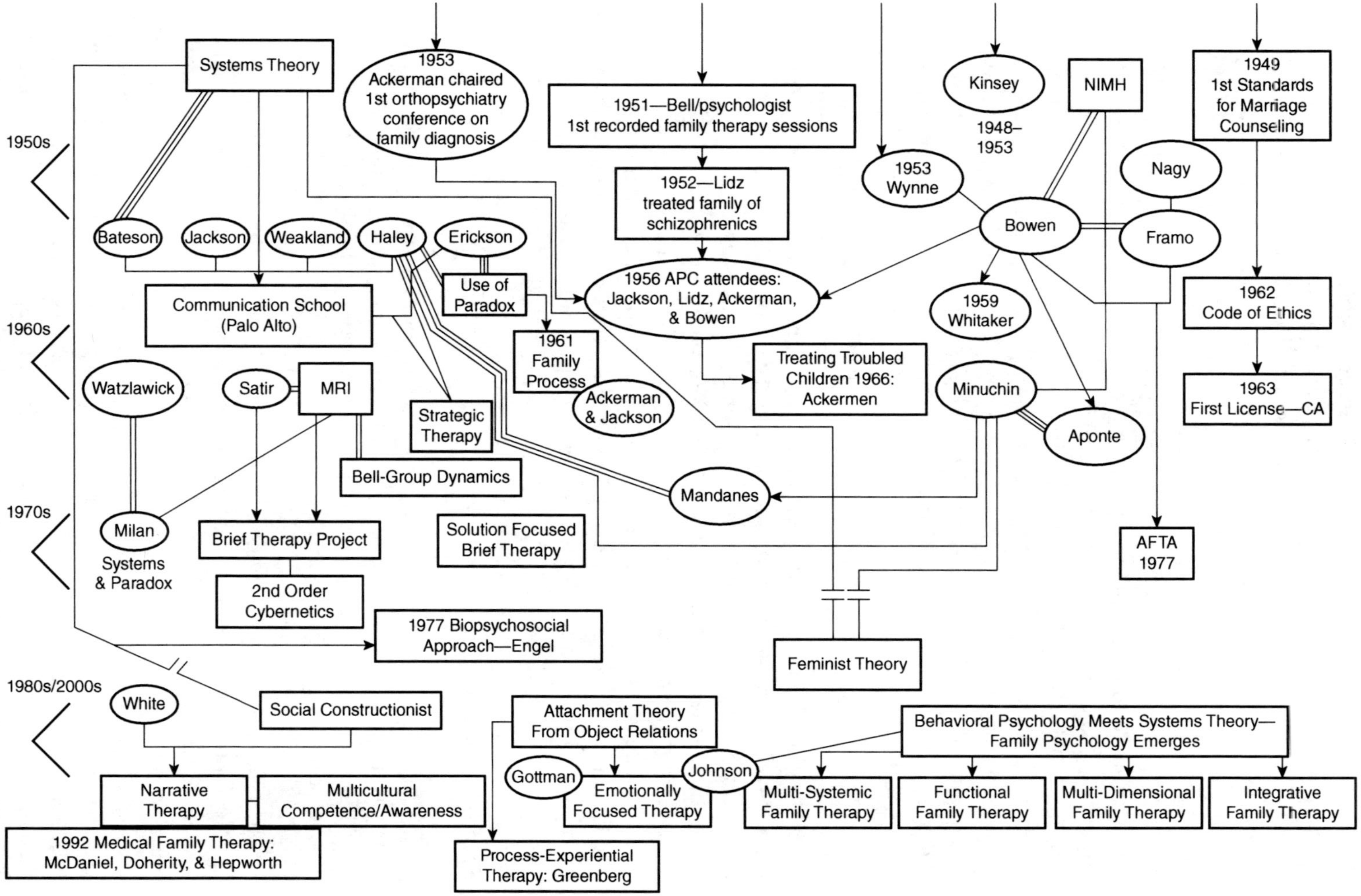

This historical supervision genogram has been included to better demonstrate the MFT supervision evolution.

Source: Parr, Boyle, Tejada, & Perosa (2008), revised by Boyle, Tejada, & McDowell-Burns (2015).

when working with seriously disturbed families. Their model included five stages: (1) pre-session, (2) session, (3) intersession, (4) intervention, and (5) post-session. During the intersession, the whole team (a team behind the one-way glass and a co-therapy team) discussed what they saw and devised a strategy that was delivered as a team message. Computer-assisted supervision was used to provide information to the supervisee (Smith, Mead, & Kinsella, 1981).

In 1983, the AAMFT Board changed the COS to a Commission, to function semi-autonomously. During the next decade, the number of Approved Supervisors kept increasing, and a course on supervision was added to the training of Approved Supervisors (with parts of the course being offered at the annual AAMFT conference, thus initiating the supervision track). Also during this time, the *Supervision Bulletin*, a newsletter for sharing information and supervision trends/issues, was started. The first editor of the *Supervision Bulletin* was Tony Heath. By 1986 the number of Approved Supervisors had grown to 1,286, constituting 14% of the AAMFT clinical members (Lee et al., 2004). According to Everett (1980), the population of AAMFT Approved Supervisors had changed since the 1970s: There were more women, a shift toward systemic theoretical orientation, a decline in the popularity of personal psychotherapy, an increased popularity of the use of video recording, and an increase in persons identifying themselves as MFTs.

A further development arrived in 1986, when Fred Piercy edited the book *Family Therapy Education and Supervision*. One of the chapters, written by Robert Beavers, was titled "Family Therapy Supervision: An Introduction and Consumer Guide." According to Beavers, "Supervision in marriage and family therapy is both a legitimate offspring of individual psychotherapy supervision and a mutant, representing qualitative differences from the parent" (p. 15). The conclusion of the chapter contributed by Piercy and Sprenkle (1986) states:

> The key figures of family therapy were revolutionaries. They took strong, often unpopular, theoretical stands that [ran] counter to the Zeitgeist of their time and paved the way for the theoretical models taught today. (p. 12)

In 1988, another supervision book was published, authored by Howard Liddle, Doug Breunlin, and Richard Schwartz and titled *Handbook of Family Therapy Training and Supervision*. This book focuses on thinking structurally and thinking strategically, using the concepts of cybernetics. It covers the various systemic models to choose from, interconnected relationships in isomorphic proportions, and various forms of supervision (live and video). Neither the Piercy book nor the Liddle et al. book focuses on what constitutes supervision; instead, they focus on systemic thinking as well as the training for and supervision of marriage and family systems work. The requirement from AAMFT was the notion that training in systemic thinking was a prerequisite for supervising other MFTs (MFTs in training). In addition, AAMFT required that supervisees practice systemically.

One of the first recognitions of the importance of cultural influences in supervision was found in the Liddle et al. (1988) book, specifically the chapter written by Falicov, titled "Learning to Think Culturally" (Falicov, 1988). The COS recognized the importance of diversity and, more specifically, the diversity of Approved Supervisors and diversity sensitivity when doing supervision; its focus was on contextual variables such as gender, ethnicity, race, and so on. A minority stipend for supervisors-in-training was also set up, and the COS diversified by filling vacancies with diverse members to create a balanced member representation of minority and majority cultures. Later in 1991, the COS again became a committee. The field of MFT, as well as AAMFT, were out in front, recognizing the importance of cultural influences on couple and family systems. However, the movement from cultural competence to cultural equity and humility has fallen short in AAMFT supervision and will have to be given more attention.

More research in the area of training MFTs and supervision emerged in the 1990s, by researchers such as Avis and Sprenkle (1990), Frankel and Piercy (1990), and Liddle (1991), to list only a few. However, research in the area of supervision was sparse. Liddle (1991) wrote:

> Supervisors must be formally trained. . . . Being a skilled therapist is not enough. Clinical skill and knowledge are indispensable for, but no guarantee of, supervisory success. Without exaggeration, the success of the family therapy field depends on the next generation(s) of supervisors. Our field can progress no further than do those who define it and teach it to others. . . . It is they who carry the torch. (p. 688)

Although supervision research was occurring during the 1990s, there was a lack of research focusing on the effectiveness of AAMFT Approved Supervision for the MFT student and MFT postgraduate.

In 1995, the COS and AAMFT Membership Committee were combined into the Standards Committee. The charge for this new committee was twofold: (1) AAMFT membership procedures, and (2) procedures for Approved Supervisor designation. The Approved Supervisor designation has been developing continuously since that time.

In 2001, a national survey of AAMFT Approved Supervisors showed that major MFT theories were no longer dominated by allegiance to a single theory (Blow & Sprenkle, 2001). In the early 2000s, the focus in the MFT supervision literature was on the importance of common practice, sometimes called *best practices* in AAMFT Approved Supervision (Storm, Sprenkle, & Morgan, 2001). The latest COAMFTE (2002) standards require that accredited programs have a minimum of three faculty members; however, only two of them must be AAMFT Approved Supervisors—one can have an "equivalence" (Lee et al., 2004). During this time, an equivilance could include being an independent licensed Marriage and Family Therapist (IMFT) who has 2 to 3 years experience as an IMFT and has some supervision training (Lee et al., 2004). The focus was on refining the standard of practice to define best practice for Approved Supervisors. Best practice was to provide AAMFT-approved supervision for the MFT after graduation and to include topics such as ethical responsibilities, the gatekeeping role, quality client care, and so on. In 2001, the number of Approved Supervisors had grown to 2,046, constituting 13% of the AAMFT clinical members (Lee et al., 2004).

In 2004, Lee and Everett wrote a book titled *The Integrative Family Therapy Supervisor: A Primer,* in which they identified 12 principles of supervision:

1. Supervision must be respectful.
2. Supervision, like therapy, must be a safe place.
3. A working alliance must be developed.
4. A supervisor does not offer therapy to the clinical family.
5. A supervisor does not offer therapy to the therapist-in-training.
6. The dynamic of supervision includes hierarchy and power.
7. Supervision develops through predictable stages.
8. Supervision interventions are driven by theory.
9. Supervision should be competency based.
10. The supervisor has simultaneous responsibilities to the therapist, the clinical setting/institution, and the self.
11. The supervisor, like the therapist, follows clear ethical principles of conduct and practice.
12. Supervision is unique within each training system. (p. 4)

More specifically, their book referenced the integration of the various systemic concepts and theories (such as Structural Theory and Bowen Theory). It was written using postmodern thinking, attempting to "identify, and appreciate the unique qualities, resources, and constructions of reality of the many therapists and their clients" (p. 4).

In 2007, the AAMFT developed its own book on training AAMFT Approved Supervisors, titled *Approved Supervisor Designation and Standards Handbook.* The book provides information about training requirements, guidelines, application forms, and other tools. It was developed to help AAMFT Approved Supervisors-in-Training gain knowledge about the training and requirements for becoming an AAMFT Approved Supervisor. The book focuses on three steps: (1) prepare to train, (2) complete training, and (3) submit application.

In 2012, Carlson and Lambie presented a systemic developmental approach called *Systemic Developmental Supervision.* This model was developed specifically to govern supervision for MFT students and postgraduates, focusing on the developmental path these students go through as they learn more about functioning as an MFT, integrate skills, and become familiar with the various systemic models that will cumulate in the emergence of the MFT identity. More specifically, this developmental model incorporates aspects of the integrated developmental model (Stoltenberg, 1981; Stoltenberg & McNeill, 2010) and the life-span developmental model (Ronnestad & Skovholt, 1993, 2003).

TODAY'S TRENDS

Today, in a complex and ever-changing world in which communication and education have become easier than ever, technology is rapidly expanding the possibilities for supervision and supervisors. For example, AAMFT Approved Supervisors looking for a refresher workshop can, as they have in the past, sign up for the refresher course offered at the AAMFT annual conference, and/or (at the state level) attend a 5-hour in-person workshop. As of 2014, however, AAMFT Approved Supervisors can also sign up for an online refresher course. This has created convenience and flexibility for AAMFT Approved Supervisors. In addition, the AAMFT has also put its 30-hour fundamentals of supervision course for clinical fellows and preclinical fellows online. This fundamental training is comprised of a 15-hour didactic course, a 15-hour interactive portion, and a personal paper on one's philosophy of supervision. After all three parts have been completed for the fundamentals of supervision, a certificate of completion is awarded. There is also an option to take only one part, either the didactic or interactive portion, of the training online, in addition to the ability to take both portions online.

Today's fast-growing technology is affecting both Marriage and Family Therapy and supervision. Today, there is no AAMFT Approved Supervision Code of Ethics; rather, ethical guidelines for AAMFT Approved Supervisors are addressed in the AAMFT Code of Ethics (2012):

> Principle IV: Responsibility to students and supervisees. Various aspects of supervision are addressed in the literature, however complex client and student issues dealing with technology are non-existent, despite tech's impact and seeming omnipresence—e.g., social media (Facebook, Twitter, etc.), e-mail, texting, web conferencing, etc. It is important to recognize that technology presents opportunities and challenges for supervision that must be addressed in the AAMFT Code of Ethics.

In an era in which technology use is rapidly growing, with clients as well as students and supervisees employing many and various forms of technology in their daily lives, there is an expectation that technology will also be used in the process of marriage and family therapy. This can range from such formal measures as online therapy and supervision to the use of smartphones, Twitter, Skype, and the like. Unfortunately, the Code of Ethics, especially in the area of AAMFT Approved Supervision, has not kept up with the rapid growth in technology.

Today, the field of AAMFT supervision has expanded from theory-specific AAMFT approved supervision to include (and focus on) population-specific supervision. Special attention has turned to such areas as medical family therapy and trauma therapy supervision. The field of medical family therapy (MedFT), which goes back to the 1990s, is the application of systems theory in the form of a bio-psycho-social systems approach to conducting systemic therapy with patients and their families who experience physical health problems, including illness and disability, among other things. MedFT supervision is rooted in systems theory, bio-psycho-social thinking, and collaborative (community agencies, hospitals, etc.) thinking (Doherty, McDaniel, & Baird, 1996; Pratt & Lamson, 2012). Supervision in the field of Trauma Informed Marriage and Family Therapy (T-IMFT) has been addressed in the past by such researchers as Jordan (2003, 2005) and often focused on those working with survivors of abuse. At first, it was realized that marriage and family therapists working with survivors of abuse may experience vicarious traumatization (Jordan, 2010): The theory was that vicarious traumatization is the cumulative effect of the caring marriage and family therapist repeatedly hearing stories of abuse while working with abuse survivors. It is also believed that vicarious traumatization is a job hazard for those working with clients affected by other types of trauma (e.g., veterans, disaster survivors, accident survivors). Even though much research today has focused on working with abuse survivors, the cumulative effect of hearing trauma stories can affect marriage and family therapists on a professional and personal level of functioning as they work with other trauma-affected clients.

Multiculturalism in a globalized society is important for marriage and family therapy students and supervisees as well as AAMFT Approved Supervisors. According to Version 12.0 of the COAMFTE Accreditation Standards (published July 15, 2014; effective January 1, 2016), Standard II: Commitment to Diversity and Inclusion:

> Programs demonstrate their commitment throughout the program to diversity and inclusion. This includes providing a multiculturally-informed education that addresses a range of diversity; a safe, respectful, inclusive learning climate; student experiences with diverse, marginalized, and/or underserved communities; and a commitment to the ethical and social responsibility to diverse, marginalized, and/or underserved communities. (p. 16)

Supervisors must address the issues of multiculturalism, diversity, marginalization, and globalization with students and supervisees. These issues have to be brought into both the therapy process and the supervision process, providing an opportunity for open dialogue and exploration of values, beliefs, and perception. More specifically, marriage and family therapy students and supervisees can benefit from doing a cultural self-reflection, not to evaluate other cultures, but to recognize and evaluate their own prejudices, racism, stereotypes, and personal reactions. The cultural self-assessment will promote the student's/supervisee's awareness of his or her internalized culture. Marriage and family therapy students and supervisees should be knowledgeable about the integrative systemic multicultural approach, as this will help them (a) to understand the individual/couple/family's internal reality and context, as well as (b) to develop an inclusive and effective treatment plan. Multiculturalism and diversity are important in a globalized world; however, of equal importance is the acculturation process of international marriage and family therapy students and supervisees, as they go through adjustments in their personal and professional beliefs and values. The AAMFT Approved Supervisor will need to understand each student's/supervisee's cultural background by being curious and interested. In addition, language can be challenging; both supervisors and supervisees must learn to understand colloquial terms, slang, phrases, stories, and metaphors. Cultural concepts and phenomena can also impact the marriage and family therapy student and supervisee. AAMFT Approved Supervisors are challenged with helping international students and supervisees conceptualize client issues and discuss possible treatment plans and also with helping them prepare to return to their native countries. This kind of AAMFT Approved Supervision can be both challenging (e.g., cultural differences, language differences, etc.) and rewarding for the supervisors, as they expand their own worldviews and gain more cultural knowledge.

The field of AAMFT Approved Supervision appears to be growing parallel to the field of marriage and family therapy. Today there is more focus on the use of technology, population-specific treatment and supervision, and ethics, as well as multiculturalism and globalization. The process of supervision is becoming more effective in an era of changing technology, meeting the demands of today's marriage and family therapy students and supervisees, as well as their clients.

FUTURE DIRECTION: THE USE OF TECHNOLOGY

COAMFTE has charged programs to: "Ensure [that] your program has exposure to the newest innovations and strategies for educating marriage and family therapy students" (COAMFTE Accreditation Standards, Version 11, p. 5, 2005). Technology, which has become an integral component in today's education system, is also used to deliver supervision to mental health clinicians (Lux & Sivakumaran, 2010). Various mental health professionals use technology to provide supervision in clinical settings and training programs when supervising students and postgraduates; however, these uses have raised a question as to whether supervisors and supervisees must meet face-to-face for the entire required number of hours. Other similar questions were raised when technological advances were first implemented in the sphere of mental health services (such as, "Are clinicians able to receive an adequate amount of supervision?"). One of the obstacles for delivering mental health services is the physical distance between clients and clinicians (Layne & Hohenshil, 2005); research is still determining whether technology-enabled services are appropriate and equivalent to in-person contact.

AAMFT Approved Supervisors, similar to supervisors of other mental health professionals, are dealing with the use of technology to provide supervision for MFT students and postgraduates. Today there are approximately 2,000 AAMFT Approved Supervisors in the United States, 200 in Canada, and 40 in other countries (AAMFT, 2014b), while the number of MFT clinicians exceeds 29,000 in the United States alone (Bureau of Labor Statistics, 2013). As described earlier, MFT programs are charged with "innovation" in the delivery of training programs, fulfilling another COAMFTE tenet of accredited MFT programs (COAMFTE Accreditation Standards, Version 11, p. 5, 2005). As the number of MFT student trainees working toward the MFT license grows, the need for access to supervisors grows as well. Furthermore, internship opportunities may also become more varied (Watson, 2003). The use of technology may facilitate the delivery of AAMFT Approved Supervision to MFT students and postgraduates who are not able to meet face-to-face with supervisors (Kanz, 2001) in their licensure jurisdiction, or in cases where licensure jurisdiction of the supervisor varies from that of the student. Also, when technology is used, the cost of AAMFT Approved Supervision may be lower, the convenience and scheduling of supervision meetings may increase, and the supervision process may become more accommodating to all parties (Vaccaro & Lambie, 2007). For example, for an MFT student or postgraduate needing urgent consultation, the AAMFT Approved Supervisor might be more accessible via technological means than by a face-to-face meeting (McAdams & Wyatt, 2010), thus giving both the AAMFT Approved Supervisor and the MFT student/postgraduate peace of mind, as supervision via technological means can be more immediately available in emergency situations. It is also a way to lower potential liability issues, as urgent issues can be quickly addressed. Furthermore, easier access to supervision may lead to higher job satisfaction and lower rates of clinician burnout (Kanz, 2001).

Types of Technologies Used

Technology has been transforming the field of mental health since the 1930s, when computers were first used to assist in the scoring of a vocational assessment, the Strong Vocational Interest Blank. The influx of personal computers into the business and public domains continues to improve the delivery of mental health assessments. By 1990, a variety of technology-assisted methods had been developed and used to connect supervisors, clinicians, and clients. Researchers reported these methods to be electronic mail, teleconferencing (Harvey & Carlson, 2003), chat-rooms (Vaccaro & Lambie, 2007), and videoconferencing (Watson, 2003).

In recent years, telehealth—the use of communication technology to provide mental health and consultation services—has become widespread (Himle et al., 2012). Although e-mail correspondence has improved the efficiency of communication, it has also been shown to decrease communication clarity, as body language and other subtle communication cues are missing

(Watson, 2003). Due to the need for live observation, videoconferencing has become central in connecting individuals in the mental health industry. Videoconferencing has been the subject of recent studies and has shown to be effective in delivering successful therapeutic services to clients suffering from Tic Disorder (Himle et al., 2012), Obsessive Compulsive Disorder (Vogel et al., 2012), substance abuse (King, Brooner, Peirce, Kolodner, & Kidorf, 2014), and anxiety and depression (Dunstan & Tooth, 2012).

Clinical supervisors have also been utilizing videoconferencing technology successfully. A literature review of studies conducted on the efficacy of using videoconferencing in supervision revealed that supervision provided via technology does not differ greatly from face-to-face supervision (Abbass et al., 2011). In fact, one study revealed that some clinicians were not able to meet the hourly requirement for meeting with a supervisor, and videoconferencing was the only means for them to receive supervision (Xavier, Shepherd, & Goldstein, 2007). Videoconferencing can be delivered via a variety of methods, including proprietary platforms such as eGetgoing, offered by the CRC Health Group, Inc. (King et al., 2014), and Skype (Armfield, Gray, & Smith, 2012; Krampe & Musterman, 2013); more recently, Blackboard (Elluminate) Web conferencing has been used to connect Internet users at different locations (Blackboard.com, n.d.). This becomes increasingly important in a global society, as international students come to the United States to be trained in the field of MFT. Supervision for these students can be done via Skype or the use of other technology. For example:

> A Sri Lankan student in a COAMFTE-approved program returned to his home country after the tsunami in 2004, wanting to assist with the aftermath of that disaster. For 1 year he was busy helping out, but after the year ended he contacted his MFT program in the United States and indicated that he wanted to finish his internship experience. He found an AAMFT Approved Supervisor who was able to provide the supervision in Sri Lanka, but needed to get university group supervision. Skype became the technology of choice and the student Skyped in each class meeting and was part of the group supervision provided by the university. He contributed to the class discussions and also presented his clinical cases and received both peer and AAMFT-approved faculty supervision. His supervision by the AAMFT Approved Supervisor in Sri Lanka was helpful, as some of his clients spoke Tamil, although the majority of them spoke English. There were also some cultural aspects that the in-country AAMFT Approved Supervisor seemed to be better equipped to address, as he was very familiar with the local cultural values and customs, having been born and raised in Sri Lanka. These issues were also addressed within the context of group supervision and became a valuable learning opportunity for all of the MFT students in the group supervision class.

This case example shows that technology can be a valuable tool for dealing with AAMFT Approved Supervision needs, at national and international levels, which will become increasingly important in an increasingly globalized society. In addition, most university training programs, including MFT programs, are charged with providing "Global Preparedness" for students. In the preceding case example, not only did the student from Sri Lanka benefit from the use of technology, but so did the students in his internship class who were part of the AAMFT Approved Group Supervision, as issues of cultural diversity were addressed.

Ethical Concerns

Technology in the field of mental health, specifically in MFT, has become more widespread, and therefore defining the ethical standards for services rendered thereby takes on a more prominent role (Mallen, Vogel, & Rochlen, 2005). When one looks at the various mental health professions and how the use of technology in supervision is addressed, it becomes obvious that very little guidance is provided. The use of technology in mental health, and specifically supervision, is growing very rapidly, making it difficult for the various professional codes of ethics to remain up to date. Because the AAMFT Code of Ethics serves as a guideline for what AAMFT Approved Supervisors, MFT students, and postgraduates can do, it leaves them with little guidance as to how to behave prudently and do no harm. The use of technology, because it is constantly changing, is basically uncharted; any uses should be looked at with a critical eye and chosen carefully to ensure safety and protection for the AAMFT Approved Supervisor, the MFT student/or postgraduate, and—most importantly—the clients.

The MFT profession entails an added dimension, as MFT students and postgraduates deal not only with individual clients, but also with couples and families. Providing MFT Approved Supervision tends to be more complex than other kinds of supervision, as the client modality can involve more than one client in the therapy session. Due to the complex process of connecting several family members via technological tools, issues of confidentiality, information disclosure, identity protection, and clinicians' abilities all become major concerns that must be addressed (Baltimore, 2000). For example, an Internet tool might be monitored, or content might be recorded by the provider (Wilson, 1995). This could be a concern not only for the MFT student or postgraduate who is conducting the session, but also for the AAMFT Approved Supervisor who is using technology to conduct supervision.

The use of technology by AAMFT Approved Supervisors raises a variety of ethical questions:

- How can client confidentiality be maintained when using technology such as Skype or Elluminate?

- How can confidentiality be ensured when dealing with transmission and recordkeeping on Internet-based platforms?
- How safe is it to e-mail client and/or supervision documentation? What measures are taken so that these documents do not end up in the wrong hands?
- Who is the owner of electronic transmissions received?
- To what degree does state law influence some of these questions and answers? For example, in the state of Ohio, the records belong to the entity that owns the equipment. So, if it is the agency's computer, then all the e-mail and document correspondence belongs to the agency. What implications might this have for AAMFT Approved Supervisors and MFT students and/or postgraduates? Is there a conflict between state law and the AAMFT Code of Ethics?

These and other questions must be raised by COAMFTE and the AAMFT, so that guidance and standards are available for AAMFT Approved Supervisors, MFT students, and postgraduates, to assure the provision of ethical services and ensure that clients are well served and—at the least—not harmed.

Diversity and Technology Use

Although the use of technology in the MFT field is growing, it is important to identify population sectors that either do not have access to it, or for whom technology is currently cost-prohibitive. A gap exists among those who are able to use information technology (e.g., Internet and social networking, cell phones, e-mail, etc.) and those who are not. Underrepresented populations include Native Americans, African Americans, Hispanics, and others, as well as individuals with disabilities (Taylor, 2010). Likewise, the elderly population may not be well equipped to effectively use new and innovative technology tools (Soares, Jacobs, Callari, Ciairano, & Re, 2012). Although using such technologies is an important part of integrating into the modern society, the elderly may not be able to do so without extensive training (González, Ramírez, & Viadel, 2012). AAMFT Approved Supervisors who depend on the use of technology for supervision meetings with MFT students and postgraduates may encounter barriers with and from these populations. Thus, it is important to assess the use of technology by MFT students and postgraduates, assessing their level of comfort with technology, as well as their skill/proficiency in technology and commitment and desire to grow and learn in this area.

CONCLUSION

MFT is still a young profession, but despite that, recognition of the importance of providing clinical training through relationally oriented supervision arose very early in its history. As seen in this chapter, the process of supervision in the MFT field evolved almost parallel to the growth of MFT theory development. Today there are unique challenges for AAMFT Approved Supervisors, who need to keep abreast of ever-changing technology and seek new ways to become part of a technology-based behavioral health care delivery system in which telehealth creates accessibility and new opportunities for service delivery. It also creates new challenges in training future MFTs to become technologically equipped, knowing when it is and when it is not appropriate. Given the limited guidance in the AAMFT Code of Ethics, supervisory guidance has to be sought out and provided through consultation.

Other areas with which AAMFT Approved Supervisors need to become acquainted and knowledgeable are the various specializations, such as MedFt and T-IMFT. These create new opportunities for MFTs to collaborate and work as part of a team in often complex systems. However, as noted in this chapter, as we become an increasingly global society, it becomes imperative for AAMFT Approved Supervisors to be well versed in working with international students who seek out MFT training. As the field of MFT is continuously growing and maturing, the AAMFT Approved Supervisor needs to grow in a parallel process, which means that AAMFT Approved Supervisor training (initial and refresher courses) must be updated regularly to reflect the changes that are occurring in the profession, mental health field, and global society.

REFERENCES

Abbass, A., Arthey, S., Elliott, J., Fedak, T., Nowoweiski, D., Markovski, J., & Nowoweiski, S. (2011). Web-conference supervision for advanced psychotherapy training: A practical guide. *Psychotherapy, 48*(2), 109–118. doi:10.1037/a0022427

American Association for Marriage and Family Therapy (AAMFT). (2007). *Approved supervisor designation and standards handbook.*

American Association for Marriage and Family Therapy (AAMFT). (2012). *Code of ethics.* Retrieved from https://www.aamft.org/iMIS15/documents/AAMFT%20code_11_2012_secured.pdf

American Association for Marriage and Family Therapy (AAMFT). (2014a). *Approved supervisor designation: Standards handbook.* Retrieved from https://http://aamft.org/imis15/documents/approved_supervisor_handbook_2014.pdf

American Association for Marriage and Family Therapy (AAMFT). (2014b). *Approved supervisor directory.* Retrieved from http://www.aamft.org/iMIS15/AAMFT/Directories/Approved_Supervisors/Content/Directories/supervisor_search_us.aspx

Armfield, N. R., Gray, L. C., & Smith, A. C. (2012). Clinical use of Skype: A review of the evidence base. *Journal of Telemedicine & Telecare, 18*(3), 125–127. doi:10.1258/jtt.2012.SFT101

Avis, J. M., & Sprenkle, D. H. Outcome research on family therapy training: A substantive and methodological review. *Journal of Marital & Family Therapy, 16*(3), 241–264.

Baltimore, M. L. (2000). Ethical considerations in the use of technology for marriage and family counselors. *The Family Journal, 8*(4), 390–393. doi:10.1177/1066480700084010

Bateson, G. (1972). *Steps to an ecology of mind*. New York, NY: Balantine Books.

Beavers, R. (1986). Family therapy supervision: An introduction and consumer guide. In F. Piercy (Ed.), *Family therapy education and supervision* (pp. 15–24).

Benningfield, M. (1988). The commission on supervision: A historical review. (unpublished).

Birchler, G. R. (1975). Live supervision and instant feedback in marriage and family therapy. *Journal of Marriage and Family Counselors, 1*(4), 331–342.

Blackboard.com. (n.d.). *Blackboard reimagine education education technology & services.* Retrieved from http://www.blackboard.com/; http://www.blackboard.com/platforms/collaborate/products/blackboard-collaborate/web-conferencing.aspx

Blow, A. J. & Sprenkle, D. H. (2001). *Journal of Marital & Family Therapy, 27*(3), 385–401.

Boyle, R., Tejada, L., & McDowell-Burns, M. (2015). The historical supervision genogram (unpublished).

Bureau of Labor Statistics, United States Department of Labor. (2013). *Occupational employment and wages, May 2013; 21-1013 Marriage and Family Therapists.* Retrieved from http://www.bls.gov/oes/current/oes211013.htm

Carlson, R. G., & Lambie, G. W. (2012). Systemic-developmental supervision: A clinical supervisory approach for family counseling student interns. *The Family Journal: Counseling Therapy for Couples and Families, 20*, 29–36. doi:10.1177/1066480711419809

Commission on Accreditation of Marriage and Family Therapy Education (COAMFTE). (2002). Retrieved from http://www.aamft.org/iMIS15/Content/COAMFTE/Accreditation_Resources.aspx

Commission on Accreditation of Marriage and Family Therapy Education (COAMFTE). (2005). *Accreditation standards, version 11.0.* Retrieved from http://www.aamft.org/iMIS15/AAMFT/Education_and_Training/Accreditation/Content/COAMFTE/Accreditation_Resources.aspx?hkey=5cb74a42-815a-4b75-9c29-64e5e70f02d5

Commission on Accreditation for Marriage and Family Therapy Education (COAMFTE). (2013). Accreditation Standards. Retrieved from http://www.aamft.org/iMIS15/Content/COAMFTE/Accreditation_Resources.aspx

Commission on Accreditation of Marriage and Family Therapy Education (COAMFTE). (2015). Accreditation standards for graduate & post-graduate marriage and family therapy training programs, Version 12.0. Retrieved from https://www.AAMFT.org/iMI515/Documents/COAMFTE/COAMFTE_Accreditation_Standards_Version_1

Doherty, W. J. & Baird, M. A. (1996). Five levels of primary care/behavioral healthcare collaboration. *Behavioral Healthcare Tomorrow, 5*(5), 25–27.

Dunstan, D. A., & Tooth, S. M. (2012). Treatment via videoconferencing: A pilot study of delivery by clinical psychology trainees. *Australian Journal of Rural Health, 20*(2), 88–94. doi:10.1111/j.1440-1584.2012.01260.x

Everett, C. A. (1980). *Journal of Marital and Family Therapy, 6*(2), 215–226.

Falicov, C. J. (1988). Learning to think culturally. In H. A. Liddle, D. C. Breunlin, & R. C. Schwartz (Eds.), *Handbook of family therapy training and supervision.* pp. 335–357. New York, NY: Guilford Press.

Frankel, B. R. & Piercy, F. P. (1990). The relationship among selected supervisor, therapist, and client behaviors, *Journal of Marital & Family Therapy, 16*(4), 407–421.

González, A., Ramírez, M., & Viadel, V. (2012). Attitudes of the elderly toward information and communications technologies. *Educational Gerontology, 38*(9), 585–594. doi:10.1080/03601277.2011.595314

Harvey, V. S., & Carlson, J. F. (2003). Ethical and professional issues with computer-related technology. *School Psychology Review, 32*, 92–104. Retrieved from Questia database.

Himle, M. B., Freitag, M., Walther, M., Franklin, S. A., Ely, L., & Woods, D. W. (2012). A randomized pilot trial comparing videoconference versus face-to-face delivery of behavior therapy for childhood tic disorders. *Behavior Research & Therapy, 50*(9), 565–570. doi:10.1016/j.brat.2012.05.009

Jordan, K. (2003). Assessment and treatment of children with post-traumatic stress disorder. *Marriage and Family, 6*(2), 249–259.

Jordan, K. (2005). What we learned from 9/11: A terrorism grief and recovery process model. *Brief Treatment and Crisis Intervention, 5*(4), 1–16.

Jordan, K. (2010). An ethical decision making model for crisis counselors. *Vistas 2010*, 1–6.

Kanz, J. E. (2001). Clinical-supervision.com: Issues in the provision of online supervision. *Professional Psychology: Research and Practice, 32*, 415–420. doi:10.1037/0735-7028.32.4.415

King, V. L., Brooner, R. K., Peirce, J. M., Kolodner, K., & Kidorf, M. S. (2014). A randomized trial of Web-based videoconferencing for substance abuse counseling. *Journal of Substance Abuse Treatment, 46*(1), 36–42. doi:10.1016/j.jsat.2013.08.009

Krampe, J., & Musterman, K. (2013). Shall we Skype dance: Connecting nursing students with older adults via Skype for danced-based therapy, cine computers, informatics, *Nursing, 31*(4), 151–154.

Layne, C. M., & Hohenshil, T. H. (2005). High tech counseling: Revisited. *Journal of Counseling & Development, 83*, 222–226.

Lee, R. E., & Everett, C. A. (2004). *The integrative family therapy supervisor: A primer.* New York, NY: Brunner Routledge.

Lee, R. E., Nichols, D. P., Nichols, W. C., & Odom, T. (2004). Trends in family therapy supervision: The past 25 years and into the future, *Journal of Marital Therapy, 30*(1), 61–69.

Liddle, H. A., Breunlin, D. C., & Schwartz, R. C. (1988). *Handbook of family therapy training and supervision.* New York, NY: Guilford Press.

Liddle, H. A. (1991). In A. S. Gurman and D. P. Kniskern (eds.), *Handbook of Family Therapy*, Vol 2, 638–697.

Lux, A., & Sivakumaran, T. (2010). Video integration in graduate counseling programs: Considerations for training and supervision. *Journal of Technology Integration in the Classroom, 2*(2), 181–189.

Mallen, M. J., Vogel, D. L., & Rochlen, A. B. (2005). The practical aspects of online counseling: Ethics, training, technology, and competency. *Counseling Psychologist, 33*(6), 776–818. doi:10.1177/0011000005278625

McAdams III, C. R., & Wyatt, K. (2010). The regulation of technology-assisted distance counseling and supervision in the United States: An analysis of current extent, trends, and implications. *Counselor Education & Supervision, 49*(3), 179–192.

Montalvo, B. (1973). Aspects of live supervision. *Family Process, 12*, 343–359.

Parr, P., Boyle, R., Tejada, L., & Perosa, L. Teaching family history of marriage & family therapists through the use of the genogram. (unpublished).

Piercy, F. (1986). *Family therapy education and supervision.* New York, NY: Haworth Press.

Piercy, F. P., & Sprenkel, D. H. (1986). Family therapy theory building: An integrative training approach. In F. Piercy (Ed.), *Family therapy education and supervision*, 5–14.

Pratt, K. J., & Lamson, A. L. (2012). Supervision in behavioral health: Implications for students, interns, and new professionals. *Journal of Behavioral Health Services & Research*, 285–294.

Ronnestad, M. H., & Skovholt, T. M. (1993). Supervision of beginning and advanced graduate students of counseling and psychotherapy. *Journal of Counseling and Development, 71*, 396–405.

Ronnestad, M. H., & Skovholt, T. M. (2013). The developing practitioner growth and stagnation of therapists & counselors.

Selvini Palazzoli, M., Boscolo, L., Cecchin, G., & Prata, G. (1978). Paradox and counterparadox, New York, NY: Jason Aronson.

Smith, R. C., Mead, D. E., & Kinsella, J. A. (1988). *Journal of Marital and Family Therapy, 24*(1), 113–125.

Soares, M. M., Jacobs, K., Callari, T. C., Ciairano, S., & Re, A. (2012). Elderly-technology interaction: Accessibility and acceptability of technological devices promoting motor and cognitive training. *Work, 41*, 362–369. doi:10.3233/WOR-2012-0183-362

Sprenkle, D. H. (2010). The present and future of MFT doctoral education in research-focused universities. *Journal of Marital and Family Therapy, 36*(3), 270–281. doi:10.1111/j.1752-0606.2009.00159.x

Stoltenberg, C. (1981). Approaching supervision from a developmental perspective: The counselor complexity model. *Journal of Counseling Psychology, 27*, 84–88.

Stoltenberg, C. D., & McNeill, B. W. (2010). IDM supervision: An integrative developmental model for supervising counselors and therapists (3rd ed). New York, NY: Routledge.

Storm, C. L., Todd, T. C. Sprenkle, D. H., & Morgan, M. M. (2001). *Journal of Marital & Family Therapy, 27*(2), 227–239.

Taylor, V. E. (2010). Broadening participation: Cultivating cultural diversity in information technology. *Communications of the ACM, 53*(7), 35–36. doi:10.1145/1785414.1785430

United States Department of Labor. (2012). *Mental health counselors and marriage and family therapists*. Retrieved from http://www.bls.gov/ooh/community-and-social-service/mental-health-counselors-and-marriage-and-family-therapists.htm

Vaccaro, N., & Lambie, G. W. (2007). Computer-based counselor-in-training supervision: Ethical and practical implications for counselor educators and supervisors. *Counselor Education & Supervision, 47*(1), 46–57.

Vogel, P. A., Launes, G., Moen, E. M., Solem, S., Hansen, B., Håland, Å., & Himle, J. A. (2012). Videoconference- and cell phone-based cognitive-behavioral therapy of obsessive-compulsive disorder: A case series. *Journal of Anxiety Disorders, 26*(1), 158–164. doi:10.1016/j.janxdis.2011.10.009

Watson, J. C. (2003). Computer-based supervision: Implementing computer technology into the delivery of counseling supervision. *Journal of Technology in Counseling, 3*(1). Retrieved from http://jtc.colstate.edu/vol3_1/Watson/Watson.htm

Wilson, D. L. (1995). The network has eyes. *Chronicle of Higher Education, 41*(45), 17–18.

Xavier, K., Shepherd, L., & Goldstein, D. (2007). Clinical supervision and education via videoconference: A feasibility project. *Journal of Telemedicine and Telecare, 13*(4), 206–209. doi:10.1258/135763307780907996

SECTION II

Nuts and Bolts of Supervision

TWO

MFT Supervision: An Overview

Sandra A. Rigazio-DiGilio

As noted in Chapter 1, the Marriage and Family Therapy (MFT) profession has long recognized the pivotal role of supervision in preparing and socializing future generations of MFT professionals and in fostering the ongoing personal and professional development of both novice and seasoned MFT practitioners (Todd & Storm, 2014). Two essential proficiencies that MFT supervisors cultivate, regardless of whom they supervise, are the dedication and discipline for building competencies necessary to provide effective and relevant treatment, and the commitment and capability for the ongoing development of these and additional competencies to remain viable MFT practitioners (Watkins & Milne, 2014).

Cultivating these proficiencies is among a host of responsibilities that MFT supervisors assume. Among these responsibilities are: (a) facilitating supervisees' development of self-awareness, inclusive cultural and social empathy, and theoretical, technical, cultural, and contextual competence; (b) monitoring quality of service provision; (c) identifying supervisees' areas of proficiency and necessary development; and (d) serving as gatekeepers for the profession. In their survey of various sources across disciplines and countries, Watkins and Wang (2014) found international and interdisciplinary consensus about the practice competencies supervisors must acquire to carry out these responsibilities. These include the knowledge and skills for: (a) integrating supervision models, methods, and interventions; (b) attending to ethical, legal, and professional concerns; (c) managing supervisory relationships; (d) conducting supervisory assessments and evaluations; (e) fostering attention to diversity; and (f) maintaining a self-reflective stance in supervision. These competencies can be classified into four essential functions of MFT supervision (detailed in Figure 2.1) that, together, balance supervisors' responsibilities to represent the profession, advance supervisees' proficiencies, and ensure the provision of quality services that are responsive to knowledge advancements and to the various sociohistorical forces and current demographic and societal conditions cultivating and challenging clients'[1] lives.

[1]The term *clients* refers to individuals, families, and wider institutional, community, and social systems directly participating in the therapeutic process.

Normative Functions	Socialization Functions
Supervisors ensure that supervisees are cognizant of and uphold procedural practices congruent with the: (a) MFT profession's educational and training requirements, ethical codes, and practice standards, (b) federal and state statutes governing licensure requirements, professional mandates, and best practices, and (c) relevant institutional, agency, and third-party regulations. Supervisors establish supervision parameters such as roles and responsibilities. acceptable standards of practice, and lines of accountability regarding documentation, contracting, evaluations, and gatekeeping. Given supervisees enter supervision with some understanding of procedures specific to their developmental trajectories and training / work settings, this information can serve as a common point of reference for initial supervisory conversations. Preparing for and engaging in such dialogues early on can be an initial step toward establishing safe supervisory environments by defining transparent structures for learning and quality monitoring when relationships are still in their ledging states. Conversely, the absence of such dialogues can increase anxiety and impede the development of solid foundations from which to build safe environments.	Supervision provides a primal socialization matrix for supervisees to integrate aspects of their professional and personal development into their identities as members of the MFT profession. Early on, supervisors ensure that viable aspects of supervisees' personal and professional selves are inculcated within this matrix. As the matrix extends to include relevant indigenous and institutional systems, supervisors ensure that supervisees gain an understanding of these systems and directly experience what dimensions of their professional identities are held as prominent, as well as the varying positions of power, credibility, and influence they are afforded within each system. Thus, supervisees are assisted to build a professional identify, to extend this identify to include the multiple ways it is perceived and responded to in wider systems, and to gain comfort and expertise with the extended roles they can assume as participants in multifaceted service approaches that go well beyond the boundaries of the therapy room(e.g., collaborating with service providers, consulting with natural helping networks, and advocating for accessible and quality resource distribution within institutional and political domains).

Educative Functions	Evaluative Functions
Supervisors employ various learning principles to invite supervisees to reciprocally integrate self-awareness, inclusive cultural empathy, theoretical and technical skills, and historical and current conditions that contribute to real and perceived distress and disorder. In line with best practice recommendations that clinicians draw from techniques consistent with clients' cultural multidimensionality, past and current lived experiences, needs, and preferences, supervisees are provided opportunities to build core competencies and, later, to expand the repertoire of skills they can draw from to complement and extend clients' sense of themselves and their ability to participate in and influence transactions within and across contexts primarily defining and managing issues considered necessary to address. Opportunities to practice, consolidate, extend, and integrate this broad skill set are considered most effective when supervisory environments and styles (e.g., directive, modeling, coaching, collaborating, consulting) are tailored to developmental trajectories, settings, and supervision phases, and when supervisors respectfully and responsively attend to cultural equity, power differentials, and power sharing.	Supervisors use formal and informal methods to facilitate and evaluate supervisee development, functions inextricably linked to gatekeeping and quality assurance responsibilities. The proficiencies addressed and evaluative methods used are informed by multiple factors including: (a) professional, training, and work expectancies, (b) lines of accountability, (c) procedures for establishing remediation plans and for addressing the effects and ramifications of unethical practice, and (d) collaboratively generated goals and objectives. Similarly, proficiencies addressed and evaluative methods used inform every aspect of supervision. Research indicates that collaboratively negotiated contracts and frequent reciprocal evaluations strengthen the supervisory alliance and invite supervisors and supervisees to evaluate and modify proficiencies being addressed and supervisory methods being utilized, examine ambiguities or problems that arise, negotiate remediation or restorative plans to propel supervisee development in agreed-upon areas, and moderate power imbalances. These exchanges enable supervisees to assume more responsibility for self-directed learning and to consider myriad factors in their clinical work.

FIGURE 2.1 Core MFT supervisor functions.

Source: Reproduced with permission of Sandra A. Rigazio-DiGilio, copyright 2014.

The normative, socialization, educative, and evaluative functions detailed in Figure 2.1 are interrelated and intersect over the course of supervision. The salience of each changes over time and in response to the developmental trajectories, learning styles, needs, and goals of supervisees; the relevant expectations within supervision and the various contexts and communities within which it takes place; and the nature and scope of clinical issues being addressed, populations being served, and professional services being provided. Thus, the functions represent a translation of the MFT profession's ecosystemic perspective into a contextualized framework for supervision.

CONTEXTUALIZING MFT SUPERVISION

Supervisors and supervisees bring their personal, family, community, cultural, and professional backgrounds to the supervisory encounter. This exchange takes place in particular institutional, professional, community, cultural, and political contexts. Hence, MFT supervision is subject to the recursive forces that influence the interactive discourse occurring within and across supervisory systems and the broader arenas in which supervision takes place (Rigazio-DiGilio, 2014). Figure 2.2 explains this contextualized view of supervision, which encompasses four interdependent domains: personal and professional worldviews, MFT supervision practice, local communities, and global societies.

MFT Supervisors' and Supervisees' Personal and Professional Worldviews

Our worldviews emerge and acquire multiple meanings through our differential lived personal and professional experiences in various contexts over time. Myriad factors influence the differential nature of these experiences. These include: (a) the various positions of power and relatedness we have held and now occupy within relevant social and intercultural milieus, (b) the degree of congruence across definitions of what is acceptable, and (c) the ways dominant groups make sense of and address incongruences in their own group and within other less dominant groups.

We rely on our worldviews to understand and participate in the multiple learning and practice environments we engage in throughout our professional development and careers. Specific to supervision, participants bring their unique worldviews to the encounter. As such, even when participants come together around basic normative expectancies (detailed in Figure 2.1), each may carry idiosyncratic understandings of the ways these will be upheld, the legitimacy of the actual practices, the time that should be devoted to these practices, and the weight that should be given to evaluating these practices. The complexity increases when supervisory dialogues consider ways to incorporate supervisees' developmental trajectories, learning styles, needs, and

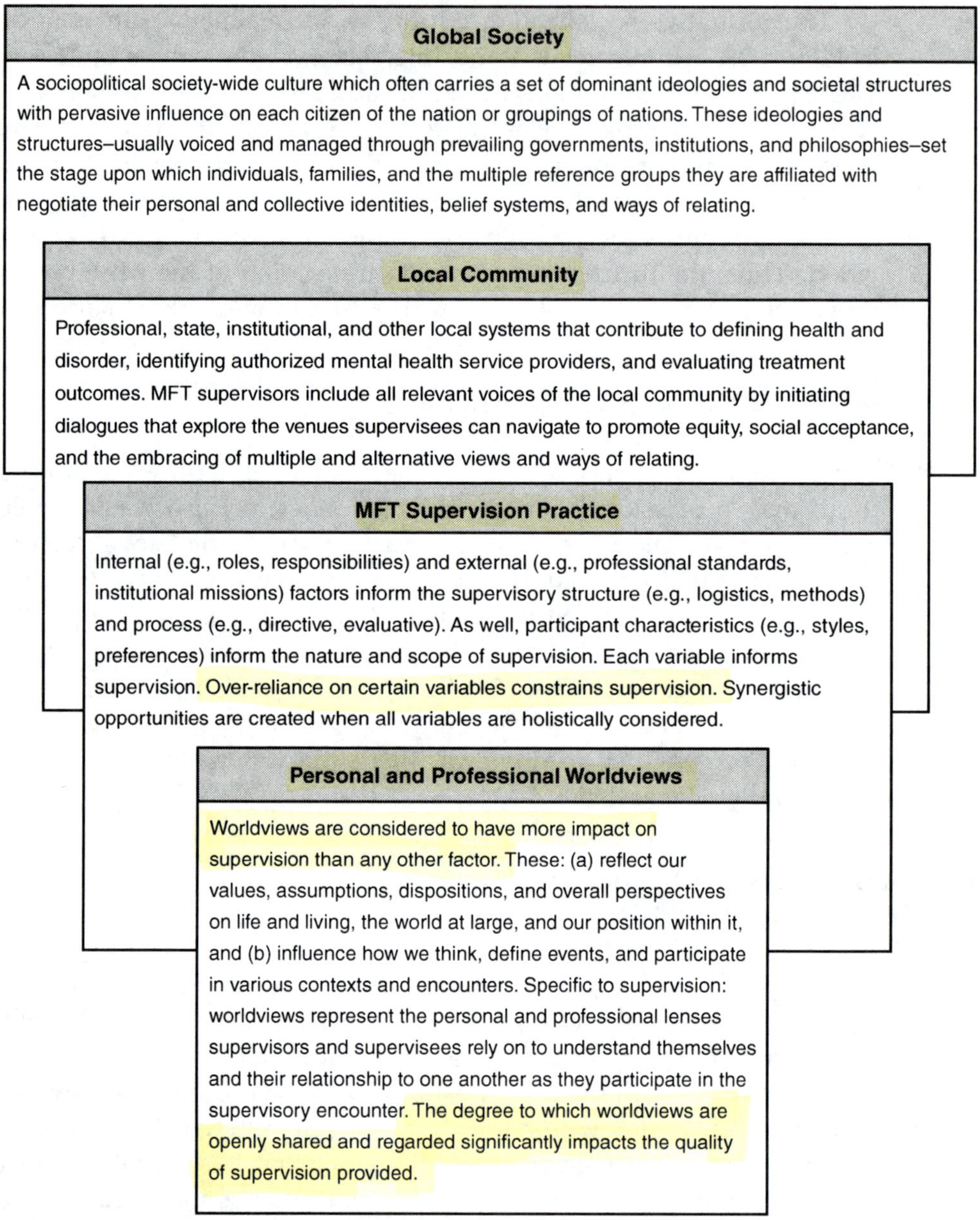

FIGURE 2.2 Contextualizing MFT supervision: An ecosystemic perspective.

An ecosystemic perspective is foundational to the MFT profession. As such, MFT supervisors are cognizant of and responsive to the recursive transactions that take place within the supervisory alliance and the ways in which these are informed by and inform the broader arenas in which supervision takes place.

Source: Reproduced with permission of Sandra A. Rigazio-DiGilio, copyright 2014.

preferences into supervision. Therefore, supervisors should be cognizant of and responsive to the recursive and interdependent factors influencing how both participants are construing themselves independently and in relation to the work of supervision and the supervisory encounter (see Chapter 7).

Initially, this can take the form of co-constructing foundational contracts, shaped by dialogues that evolve in the opening phases of supervision. For example, exploring and clarifying the purposes for and opportunities provided by the normative procedural practices participants reviewed in preparing for their work together might add common points of reference to this component of the foundational contract. Similarly, mutually sharing the nature and scope of personal and professional knowledge, experience, and expertise participants bring to supervision, and the ways each participant can contribute to and benefit from this particular supervision relationship, might inform other components of the foundational contract specific to supervisory styles and methods considered best tailored to the supervisee's learning style. These two examples emphasize core factors that significantly shape supervision at the level of worldviews. That is, initial supervisory exchanges that promote educational clarity, specificity, and shared understandings and that emphasize mutual sharing are considered prerequisites for developing a solid and safe foundation for learning. Conversely, initial supervisory exchanges that fail to promote clarity, focus, and common understandings; that discourage mutual sharing; and that devalue the reciprocal benefits of different experiences and perspectives are considered to impede the development of this essential foundation. They also raise anxiety for supervisees and contribute to role ambiguities and ruptures in already tenuous supervisory relationships. Recent research on worldviews and supervision indicate that these latter types of exchanges remain common experiences for supervisees and often replicate lived experiences and reinforce marginalized worldviews, particularly for those who are minimally represented in the MFT profession. Generational training gaps and existing demographic differences between MFT supervisors and those entering the profession continue to be implicated as contributory factors, particularly for seasoned White supervisors who are less aware of/responsive to the additional social power they are afforded as members of the dominant racial group.

MFT Supervision Practice

The structure and process of supervision are significantly shaped by both internal factors (e.g., assigned roles, supervisees' learning needs, supervisors' inherent power) and external factors (e.g., professional standards, institutional missions, agency resources). Useful examples of the influence of internal factors on supervision are the distinctions between supervision environments and methods best tailored to: (a) training student supervisees toward building core executive competencies (AAMFT, 2004), (b) assisting seasoned practitioners to extend their repertoire of therapeutic and technical competencies, or (c) addressing remediation plans for supervisees who have demonstrated substandard practices with particular clinic populations. In turn, distinctions between expectations of practicum supervisors assisting supervisees to translate constructs and techniques to actual practice, of agency supervisors assisting supervisees to develop the proficiencies of a specific manualized approach, and of supervisors working with supervisees who carry high caseloads in under-resourced agencies are useful examples of the influence of external factors on supervision.

The nature and scope of supervision are significantly determined by supervisor characteristics (e.g., preferred supervision methods) and supervisee characteristics (e.g., paradigms and trends that informed the supervisee's socialization into the profession). A particularly significant characteristic is supervisors' preferred theories and approaches. At the foundation of MFT supervision and therapy models, such as those presented in Section III of this book, is a focus on the systemic dimensions and characteristics of individuals, couples, families, and wider systems participating in (or in other ways influencing) the therapeutic and supervisory process. Although our profession emphasizes balanced and holistic consideration of these dimensions and characteristics, it has been difficult to reach a consensus about just what components warrant our primary focus. This can be illustrated by a review of the territory and perimeters of our profession's models and approaches, factors that tend to predetermine the focus of our assessments and plans, the methods and avenues for promoting client health and supervisee development, and the criteria for evaluating the effectiveness of our work. While each model offers useful and unique perspectives on client health and supervisee development and on ways to achieve each, the worldviews espoused by any one model lead to different determinations of client and supervisee needs, ways to address these needs, and conceptualizations of successful or unsuccessful progress and outcomes. Thus, it is our ethical and professional responsibility, as MFT supervisors, to be aware of (and to clarify) the ways in which our preferred theories and approaches influence what we accentuate and disregard in supervision, with regard to those we supervise and the clients they serve.

A look at inherent power and extended power illustrates how internal and external factors and participants' characteristics come together to shape supervision and the supervisory alliance. For example, although MFT supervisors have the inherent power (legitimized by the profession) required to uphold their ethical responsibilities, how they use this power and the degree to which they extend their definitions of power significantly influences the supervisory relationship. Specific to the evaluative functions detailed in Figure 2.1, supervisors can use power in myriad ways, ranging from unilateral and hierarchical evaluations to recursive and reciprocal evaluations. Supervisors can narrow or extend their definition of power by the degree to which they address such factors as professionally sanctioned power, socially inherent power, the power of each participant's knowledge and experience, and the benefits of power sharing.

Local Community

Supervision takes place in geopolitical communities, significantly shaped by dominant norms and cultural mores regarding functional/acceptable beliefs, practices, lifestyles, and ways of relating, all of which are established as credible through local and regional political, legal, institutional, and professional systems. Methods presumed credible for managing those who deviate from such norms are legitimized through these same systems. Specific to mental health, established norms and legitimized practices are often dictated by long-held beliefs about mental health professionals and those we serve, and by prevailing sociopolitical beliefs held by those in power at any given time. Illustrative examples include: (a) ongoing disparities affecting the mental health care of populations least affiliated with the systems defining and managing mental disorders, (b) differences in treatments provided in resourced and under-resourced communities, and (c) variations in resource allocations.

The MFT profession also is a geopolitical community, shaped by an ecosystemic worldview that is based on a common set of beliefs, norms, and values that influences how members approach assessment and treatment. Additionally, the MFT profession has established itself as a credible participant in the wider health and mental health enterprise by developing professional and ethical standards of practice, licensure requirements, and the like. As part of the socialization functions detailed in Figure 2.1, MFT supervisors assist supervisees to develop their identities both as members of a distinct profession and as participants in the wider mental health care enterprise.

Regarding the former, supervisors help supervisees incorporate MFT's unique ecosystemic perspective, which recognizes that psychological distress and disorder are experienced, expressed, and labeled through interactions among clients' biological and social histories and the cultural contexts that have influenced them over time. Supervisors assist supervisees to become increasingly aware of the ways in which practicing from this perspective influences how they view and work with clients. For example, supervisees may need support to consider multiple perspectives when their worldview significantly diverges from those of the clients they serve and to modify treatment plans and approaches found to be incongruent with the multiple realities of clients' lived experiences. Given that the effectiveness of therapy is most likely enhanced when therapists use modalities and define goals consistent with the life experiences and cultural values of clients (e.g., Sue, Ivey, & Pedersen, 1996/2007), effective supervision provides opportunities for supervisees to expand the repertoire of helping responses available to them, regardless of theoretical orientation(s).

Regarding the latter, supervisors provide opportunities for supervisees to gain comfort and expertise with the extended roles they can assume as MFT professionals when participating in multifaceted service approaches that go well beyond the boundaries of the therapy room. For example, supervisors assist supervisees to use MFT's ecosystemic lens to understand the interactive processes occurring among individuals, families, and the wider society that contribute to how "disorders" are defined and managed, as well as the impact of divergent expectations for treatment outcomes that can exist across supervisory, therapeutic, and wider systems of participants and institutions involved with clients. For example, supervisors assist supervisees to consider a broader scope of assessment when determining the ways clients' presenting difficulties are inextricably linked to, and quickly become a part of, a transactional narrative upheld by relevant systems within the geopolitical community. Additionally, supervisors assist supervisees to consider a broader territory of intervention when considering options for collaborating with wider networks to elicit a sense of shared responsibility for understanding client difficulties and for identifying resources that can be accessed in this wider territory for the resolution of difficulties (Rigazio-DiGilio & Kang, 2015).

By helping supervisees to build a professional identity as members of the MFT profession, and to gain comfort and expertise with the extended roles they can assume in local communities, MFT supervisors are ensuring that current and future generations can exponentially increase the perspectives, options, and resources available to them when assisting clients. MFT practitioners can then assist clients to become and to be perceived as equal participants in this wider network. Over time, it becomes possible for clients to more readily: (a) serve the psychosocial needs of family members while balancing the demands of wider systems, (b) shift the constraints and incongruities experienced when they are perceived as dysfunctional or deviant, and (c) coordinate key players to co-construct congruent images and plans for identifying and managing stress.

Global Society

Preparing today's supervisees for global service involves advancing their understanding of mental health issues at the widest boundary of human existence. Two illustrations of the work required to prepare supervisees as global professionals are exchanges that assist supervisees to consider the hegemonic impact of dominant multinational ideologies and structures on the MFT profession; and the ways in which exported U.S.-based MFT training, supervision, and practice standards impact indigenous healing practices and mental health treatment protocols of other nations.

SUPERVISION FOR A GLOBAL SOCIETY

Since the emergence of supervision as a specialty, based on U.S. cultural values, MFT supervisors have been working with trainees and practitioners who serve diverse client populations with varying degrees of power and relatedness with respect to the systems defining and managing mental distress and disorders. More recently, MFT supervisors have been working with increasingly diverse trainee populations that similarly have varying degrees of power and relatedness with respect to the systems defining standards for MFT training, supervision, and practice. Growing diversity within these two populations increases demands for supervisors to build multicultural, cross-national, and international proficiencies.

Inviting supervisees to use a cultural and contextual lens to understand clients unaffiliated with the dominant systems is illustrative of the types of exchanges that supervisors—with multicultural, cross-national, and international proficiencies—bring to the foreground of supervision. Over time, such exchanges increase the capacity of supervisees to assess the degree to which seemingly idiosyncratic client behaviors and beliefs are in fact expressions of larger heritages and practices, and to determine ways these can be accounted for in treatment (Rigazio-DiGilio & Kang, 2015).

Multicultural, cross-national, and international proficiencies are particularly relevant when working with supervisees who attend institutions external to their nations of origin. Such proficiencies help supervisors expand beyond their awareness of and responsiveness to participant worldviews to consider the ways in which the organizational mission of the host institution intersects with and impacts the supervisor–supervisee exchange. These organizational missions (which are reflected in academic standards, curricular requirements, and other related policies and practices) commonly: (a) reinforce the credibility of supervisors' worldviews; (b) give lesser credence to supervisees' worldviews; and (c) legitimize the universal identification of benchmarks for all students, regardless of nation of origin. Given the ease with which this set of circumstances can privilege the worldviews and practices of supervisors and relevant others also affiliated with the host institution, the notion of culturally encapsulated supervision, and its impact on supervisees from other nations, must be recognized and accounted for. Unfortunately, this is often difficult to acknowledge and address, specifically because such cultural encapsulation occurs as part of a collective meaning-making process. Therefore, the need for supervisors working within host institutions to build multicultural, cross-national, and international proficiencies rises in significance.

Useful illustrations of supervisory practices informed by multicultural, cross-national, and international proficiencies include developing an awareness of: (a) predominant ideologies and systems significantly and differentially impacting the lives of individuals, families, communities, and population subgroups within supervisees' nations of origin; and (b) the nature and scope of mental health services and professions with these nations. Collaborating with students and host institutions to determine ways to prepare supervisees to meet their career goals is another. A final example involves understanding supervisees' intentions once they return home (e.g., to work within already defined systems or to influence change within these systems) and accounting for these when tailoring plans of study within the host institution.

As international professionals, supervisors help supervisees develop an understanding of the clinical implications of seeing nations as interconnected subsystems. By drawing from MFT's ecosystemic perspective, supervisors assist supervisees to build an informed understanding of the ways in which current policies and phenomena (such as migration patterns, economic practices, racial and ethnic genocide, institutionalized racism and oppression), and major natural and human-made disasters are related to their work as mental health professionals. Over time, these exchanges increase the capacity of supervisees to recognize how events and practices at the global level can be linked to the general types of family problems they will deal with in their own communities, such as poverty, family violence, alienation, and substance abuse, to name only a few.

Summary

Each of these four interdependent domains is one part of the larger backdrop you attend to in supervision, ranging from factors of genetic endowment to the forces of social reality. Together, these provide a broad conceptual map of the territory you can draw on to inform your work at any given time or in response to particular circumstances. For example, sometimes you may draw from the worldview of the supervisee to generate understandings of how cultural heritage informs theory and practice. At other times, you may address the implications of your work on the MFT field or the community at large. As you develop your approach, you will have experiences that prompt you to consider how you might attend to, learn from, participate with, and influence these four interdependent domains.

GETTING STARTED: THE PROFESSIONAL JOURNEY OF MFT SUPERVISORS

When viewed from a contextualized framework, MFT supervision is a multifaceted and demanding professional responsibility. Choosing this path conveys your commitment to examine what you believe, what you know, and how you practice. It reflects your willingness to maintain a viable supervisory approach informed by: (a) your worldviews, (b) your knowledge of advances in the field, (c) your understanding of our profession's work within diversifying demographic landscapes and changing geopolitical communities, and (d) your recognition of the dominant ideologies that have perpetuated definitions of mental illness and its management for centuries.

The very way you approach these responsibilities influences those you supervise, eliciting their commitment to provide informed, respectful, and reflective practice and to engage in career-long learning. Ideally, your work models for supervisees how to be ready rather than reluctant to expand the perimeters of their worldviews, knowledge, professional identities, practice competencies, and cultural and contextual proficiencies so that they are prepared to address the persistent and evolving mental health needs of all families—including those underrepresented in our theories and research, and underserved in our practices.

Maintaining a responsive supervisory approach is a formidable challenge, but this is required to ensure that our practices are in synchrony with our foundational and evolving values and beliefs and responsive to the needs engendered by sociohistorical forces and current/changing conditions within geopolitical communities worldwide. For new supervisors, a central question is how to construct a supervisory approach that propels you to act wisely, to believe in the integrity of what you do, to learn from your experiences, and to be ready to embrace future changes in our evolving profession and societies. For veteran supervisors, the central question is how to balance continuity and change in ways that maintain or elicit your sense of synergy, vitality, and validity as an MFT supervisor. A core challenge for all supervisors is recognizing the unique issues particular to time, context, and societal trends when determining the competencies required to meet the changing needs of clients and communities, identifying supervisory exchanges that advance ongoing development of these competencies, and determining the roles supervisors can take to ensure professional preparedness and to promote social justice and equity.

Engaging in multilayered self-appraisals helps to ensure the continued relevance of your supervisory approach. This involves regularly examining how: (a) your cultural and contextual histories and experiences, (b) the predominant paradigms and trends that informed your foundational training, and (c) your participation in professional, sociocultural, and political domains have shaped the worldviews you bring to every supervisory encounter. Additionally, this involves routinely evaluating the effectiveness of the positions and perspectives you tend toward in supervision and the ways you might be inclined to readjust both when tailoring supervision to specific supervisees and supervisory contexts.

Today's supervisors must grapple with the degree to which they can ethically rely on supervision practices that emerged within the parameters of a Western, Euro-American context (Ng, 2005) and, as such, possess certain generic characteristics that are incongruent with and have exclusionary effects with regard to meeting the diverse needs of supervisees, especially those situated outside dominant mainstreams. As well, supervisors must consider ways to respond to long-standing calls for training a contemporary workforce that can competently serve increasingly diverse populations living within both resourced and under-resourced communities and societies, given that supervision frameworks have yet to be significantly shaped by coherent and integrated theories and research centrally informed by cultural and contextual factors. To address such issues, this book provides knowledge supervisors can draw from as they form, extend, or transform their approaches to tailor their work to those they supervise, the contexts where supervision and therapy take place, and some of today's most pressing issues (see Chapters 4 and 7, and sections addressing cultural multidimensionality in Sections III and 4). Additionally, supervisors can use the chapters in Sections III and IV to examine proficiencies attained in specific models and to consider ways to draw from other models when tailoring supervision to the unique needs of supervisees, clients, and wider systems.

Multilayered self-appraisals also involve critically examining our responsiveness to: (a) the differential impact of sociohistorical forces and rapidly changing demographic, social, economic, political, and technological conditions experienced by individuals, families, and the temporary and lifelong groups and subgroups they belong to or affiliate with; and (b) the long-standing calls for professions to examine the many levels of stress that affect clients and to be aware of and responsive to the wider geopolitical community forces that perpetuate oppression, violence, and trauma. Effective responsiveness requires visionary, thoughtful planning and involves using existing data about demographic, generational, social, economic, political, and technological issues that affect people's lives to understand current conditions and emerging social issues that our profession and its members must be prepared to address in their practices (Arredondo & Lewis, 2001).

CONCEPTUALIZING THE MFT SUPERVISORY PROCESS

The theory-specific and integrative supervision approaches in Section III and the population-specific approaches in Section IV represent some of the various models and approaches MFT supervisors can draw from when conducting supervision. Although no single, agreed-upon MFT supervision framework exists, there are generic frameworks that identify core phases generally considered important to cover over the course of supervision, regardless of the models and approaches supervisors draw from when working with supervisees. Figure 2.3 illustrates one generic, seven-phase supervisory framework that is based on the four supervisor functions detailed in Figure 2.1.

Figure 2.3 is not intended to represent supervision as a seamless sequence of seven discrete phases, but rather to identify tasks and activities addressed within each phase, which could quite naturally occur in recursive and cyclical processes, sometimes simultaneously and often in various sequences. The time required to address phase-specific tasks and activities and the circumstances that inform the constellation of phases vary according to the functions being addressed (see Figure 2.1) and the contexts within which supervision takes place (see Figure 2.2).

Phase 1: Co-Constructing Safe Supervision Environments

The co-construction of safe supervisory environments is essential to effective supervision. Research suggests that supervisees appreciate early conversations that clarify the structures for learning and quality monitoring. Specifically, discussions that establish a shared sense of the nature and purpose of supervision; clarify roles and responsibilities with respect to supervision and client care; identify relevant professional, institutional, and wider contextual

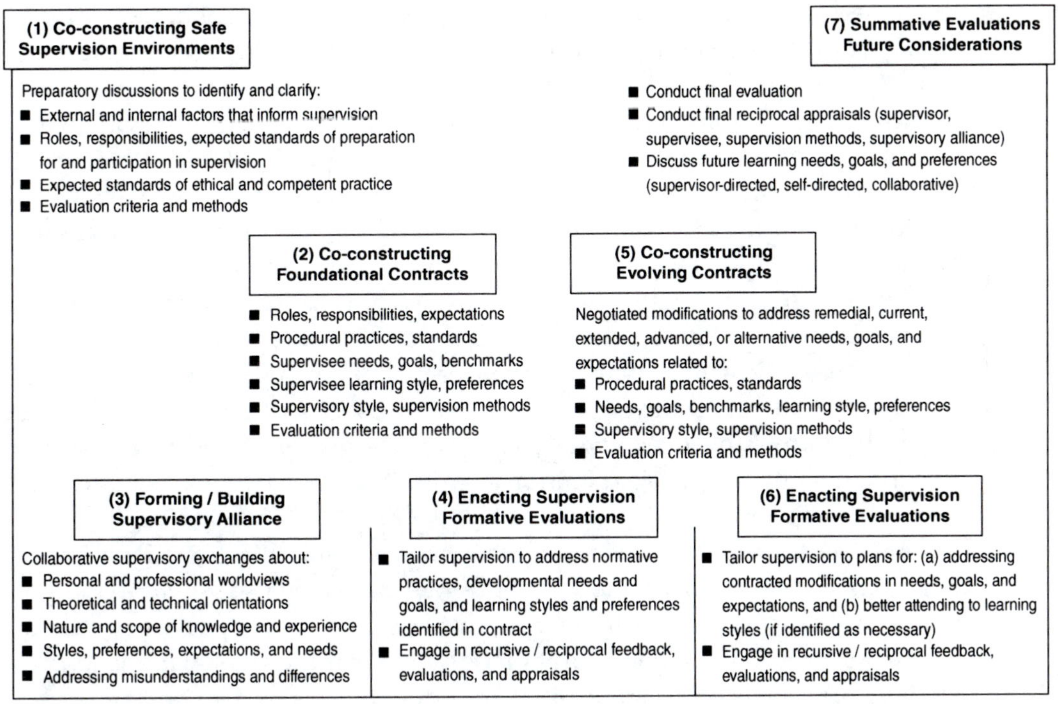

FIGURE 2.3 Seven phases of MFT supervision.

Source: Reproduced with permission of Sandra A. Rigazio-DiGilio, copyright 2014.

regulations and practices governing supervision; and define procedures for addressing emergencies and contacting supervisors are all perceived as providing foundations from which safe environments can evolve. These initial discussions should naturally expand to include dialogues about lines of accountability regarding documentation, contracting, evaluations, and gatekeeping; the boundaries of confidentiality; and the management of conflicts, misunderstandings, and grievances.

As foundations are established, supervisory dialogues tend to shift toward learning about supervisees' developmental trajectories, learning styles, needs, and preferences and exploring ways to incorporate these into the supervision to be provided. These initial discussions significantly impact the nature and scope of supervision that will take form. For example, supervisees who are invited to share, in general terms, their hopes, goals, and preferred learning styles, and who are provided opportunities to invite supervisors to do the same, report a sense that their expectations are heard and will be accounted for in the foundational contract that will inform how supervision will take place.

Phase 2: Building the Supervisory Alliance

This phase overlaps with the first, and, in accordance with the explicit and implicit assumptions across many supervision models and approaches, shifts the supervisory exchange, to varying degrees, toward collaborative discussions about the personal and professional lenses supervisors and supervisees rely on to understand themselves, their relationship with one another, and their perceptions of and relationships with clients. The importance of sharing personal and professional worldviews, exploring the ways in which these intersect with one another and within the collective worldview of the institutions within which supervision and therapy take place, and discussing the ways these intersections inform perceptions of and work with clients is increasingly recognized. Specific to phase two, collaborative and respectful dialogues about: (a) perceptions of and approaches to learning, supervision, and therapy; (b) differential experiences and knowledge; (c) supervisees' developmental needs and supervisors' supervisory styles and methods best suited to addressing these; and (d) ways to be aware of and responsive to misunderstandings serve to position examinations of worldviews in the foreground of supervision.

Because traditional supervision is cast as a hierarchical relationship, supervisors must be knowledgeable about the ways they use inherent and personal power, especially with regard to worldviews. The supervisory alliance is co-constructed through the interaction between the personal and professional worldviews of the supervisor, the supervisee, and relevant contexts. It is viewed as a cultural exchange process wherein participants share aspects of their personal and professional identities in the service of supervisee development and quality care. By seeking supervisees' perspectives about the clinical issues and about the quality of the supervisory alliance, supervisors demonstrate a commitment to collaboration, respect, contextual sharing, reciprocity, coherence, and justice—all of which engender trust. Although client issues and concerns command primary attention, the quality of the supervisory alliance is a primary factor that influences the overall sense of achievement and satisfaction.

Phase 3: Co-Constructing Foundational Contracts

Supervisors and supervisees synthesize information from the first two phases, along with any information from supervised clinical work, to hold specific conversations about a supervisory contract that enumerates important learning goals, working conditions, and evaluative criteria for success. Contracts specify the professional, clinical, cultural, and contextual competencies that will be the focus of supervision, thereby upholding ethical principles and legal expectations. Supervisors use the foundational contract to ensure ethical practice by aligning supervision with therapy, scheduling bidirectional feedback opportunities, positioning evaluation as a key function, and situating all of this within a specific institutional setting.

Building on the positive alliance created in supervision, the (a) expected roles, responsibilities, and practice standards; (b) agreed-upon supervisee learning goals and benchmarks; and (c) supervisory methods can be clearly identified in the foundational contract. Based on the specific learning goals and objectives, the supervisory style and methods, as well as the specific evaluative criteria to be used, are then incorporated into the contract. Identification of the particular methods and a tentative sequence of activities are also contained in the foundational contract. Supervisors understand and explain to supervisees that this initial contract will be revised over the course of supervision as the needs of the supervisee shift. It will be at designated formative evaluation conferences, which should be explicitly stated in the contract, that revisions to the contract will be considered. In this manner, supervisors use contracts as benchmarks to monitor their own progress and to model a reflexive process for supervisees.

Phase 4: Enacting Supervision and Conducting Formative Evaluations

This phase focuses on using professional knowledge the supervisor has gleaned about the supervisee's level of competence and confidence, in general and in particular, to achieve the identified goals of the contract. It is a time to monitor how effective and technically proficient the supervisee is. There is a difference between talking about what the goals and objectives in supervision should be and actually seeing how a supervisee executes learnings from supervision in treatment. Evidence of supervisee work should at first be collected in low-inference formats, such as live supervision, co-therapy, and electronic recording formats. This is especially important for first-time supervisees. When working with veteran practitioners, appropriate supervisory formats might tend more toward consultative rather than direct modalities, methods, and styles.

The performance gap between a supervisee's self-report, actual implementation, and effect on the client must be assessed and discussed by the supervisor and the supervisee. Supervisees need assistance combining technical skills and higher-level skills such as rebalancing the family's sense of power in their community. The skillful supervisor will closely watch the development of both foundational and more advanced competencies. Once a supervisee can confidently and accurately assess and report events from treatment, supervision sessions can rely on descriptions and reflections about the impact of treatment. Live supervision and enactment of co-therapy are methods that promote direct teaching and observation of new therapeutic skills and techniques. Direct methods are useful when introducing new therapeutic skills. Critical incident and general reflective methods where supervisees report what happened in treatment can be used once initial mastery of a new skill is achieved and confirmed. This phase is like an accordion; it can expand or compact to fit the needs of the supervisee and the time limits of the supervision arrangement. There are no particular time limits on this phase, other than institutional or regulatory schedules. It may last a couple of sessions or extend for many. The length of phase 4 is determined by the time needed for the supervisee to practice and master new skills and competencies.

Formative evaluation is the spontaneous and focused assessment of specific skills in specific contexts. In other words, the vast majority of our interactions with supervisees are formative. Sometimes, specific formative assessment opportunities are delineated, such as a mid-contract review of progress explicitly using the standards and criteria identified in the contract; at other times, an evaluation or assessment may consist of the immediate personal and professional insights brought into a nuanced application of one of the skills you are working on with the supervisee.

During announced formative evaluation conferences, both the supervisee and the supervisor should provide evidence on progress, or the lack of progress, and ideas as to how best to achieve the contracted goals at this time. Once information and perceptions from both are furnished, the degree of fit (e.g., similar perceptions) can be explored and strategies to move toward the goals or to revise the goals can be considered. Whether you are engaged in formal or spontaneous assessment conversations, the learning function of professional evaluation in supervision is to provide feedback that stimulates self-appraisal, reflection, analysis, growth, and action.

Phase 5: Co-Constructing Evolving Contracts

In this phase, supervisors and supervisees collaborate in identifying and refining learning goals and objectives. Periodically, formal formative evaluation conferences should happen at designated times for consideration of any modifications to the contract. Based on work to date, supervisors and supervisees use the specified goals and professional standards to determine the progress made thus far. At times, when unsatisfactory progress is being made, that fact must be discussed and documented in a way that supervisees understand what is going to be written about their performance, and how problems will be addressed. Remedial plans may identify other professionals from the training network to be involved, or timelines may have to be extended to achieve competency. In these cases, the terms of the contract should be revised and the future actions that will be taken should be stated.

Similarly, if a supervisee is demonstrating strong progress, the goals and objectives, timelines, and methods may also have to be modified. Even in cases where supervisees are making expected progress, the goals, methods, timelines, expectations, and responsibilities contained in the contract should be reviewed and affirmed. Sometimes shifts in the wider agency happen that require a change in supervisory goals and/or processes. Supervisors will also use these formative assessment opportunities to solicit feedback from the supervisee about the quality of supervision to date and revise the contract accordingly. The supervisor can use this new information to co-construct supervisory experiences and conversations tailored to the current needs of the supervisee.

Phase 6: Enacting Supervision and Conducting Formative Evaluations

This is a second period of practice and provides learning opportunities for supervisees to refine their clinical skills. This phase is marked by the use of supervisory strategies that help consolidate practiced techniques and build on strategies that expand to new concepts and techniques. This is done in service of expanding the supervisee's understanding of case conceptualization, identifying and using community resources, and researching and applying more nuanced client-centered practices. It is also a time to review the quality of supervision to date and make adjustments for the future that are based upon these new goals. Supervisors often focus on the metacognitive skills of evidence-based analysis and dialectic reflection during this phase. In evidence-based analysis, supervisees are expected to support their perceptions and reactions in therapy with evidence of how the family-client is behaviorally, emotionally, and relationally reacting to the specific interventions enacted in treatment. Data about the family's reaction as a whole and individual family members' reactions are all considered in evidence-based analyses. Dialectic reflection asks the supervisee to consider wider influences outside the family, the family's reaction over time, and the differences and similarities of reactions and perceptions between the supervisee and the family.

This phase is punctuated with formative evaluations that may result in making changes to the contract. Practice-oriented formative assessment cycles may happen numerous times in supervision, depending on the length of time of the supervisory contract and the personal and professional skill development of the supervisee. Again, it is also a time for reciprocal feedback from the supervisee to the supervisor that may lead to other modifications of the contract.

Phase 7: Conducting Summative Evaluations

Summative evaluations are assessments that happen when decisions must be made. This is a more formal conversation in which the supervisee's growth and development are assessed over a longer period. During this phase, a decision is being made to move the supervisee on or to identify significant standards-based concerns and forestall any pending professional advancement. During summative evaluations, both members assess the supervisee's growth and current level of proficiency regarding technical, conceptual, executive, metacognitive, and multicultural competencies. The goals and objectives identified in the latest revised contract should be at the center of this evaluation conversation. Appropriate professional standards should be addressed and recommendations as to future learning and professional growth should be identified.

At this time, ethically speaking, there should be no surprises in the summative appraisal. Any and all concerns supervisors may have about the competency and progress of the supervisee should be shared long before the summative evaluation. By proceeding in this fashion, the supervisee is made aware of the concerns and has time to work on the areas identified as needing improvement. Summative evaluations can happen at any point along the supervisory process, especially if a major concern or breach of ethics is encountered and decisions about the supervisee's status have to be addressed.

In a reflexive fashion, both members should have the opportunity to assess the quality of the supervisory experience. What supervisory strategies were most productive? What MFT content and practice skills were most productive? In addition, what qualities of the supervisory alliance helped co-construct an optimal supervisory environment? Recommendations for future learning for the supervisor also should be provided at the time of the summative evaluation.

At the professional level, summative evaluations happen when considering a supervisee for licensure or promotion. At an institution of higher education, the determination of a grade at the close of a semester is a form of summative evaluation. When appropriate, the results of summative evaluations should be communicated to other supervisors, trainers, and staff who may be working next with the supervisee to ensure continuity of learning.

THE OPENING SESSION

Ideally, before the first face-to-face supervision session, an orientation to the work, the institution, and the community has occurred so that the supervisee is mentally prepared to engage in the learning process. The first supervision session will then help establish the priorities and foci of the supervision encounter. There are four tasks to achieve: (a) starting the contracting process, (b) initiating a professional relationship, (c) identifying boundaries around what MFT theories will be addressed, and (d) suggesting possible goals for supervisee development.

The first task is to initiate the professional contract. Clarification of the number of hours of supervision, the meeting times, the types of clinical data required, and other logistics that must be agreed upon early in the relationship should be discussed. The lines of accountability, the boundaries of confidentiality, and the importance of informed consent and due process rights for clients and supervisees should be presented. Communication processes to follow for non-crisis and crisis situations should be made clear at this time. Supervisors understand that contracts are living documents that evolve over the course of supervision, but some hard-and-fast agreements, such as logistics, confidentiality, and emergency procedures, have to be made in the first session.

The second task focuses on sharing salient issues of the worldviews of both supervisee and supervisor in regard to the policies and procedures governing supervision (see Chapter 6), the preferred methods of supervising and learning (see Chapter 3), the cultural and contextual foundations of our work (see Chapter 6), and self of the therapist issues (see Chapters 4 and 7). These initial conversations are as much educative—that is, the supervisor is providing knowledge to the supervisee—as they are informative.

The third task is to chart out, in broad terms, what MFT theories and practices will be the focus of supervision. Some institutions will have preferred delivery styles and practices, including multicultural skills and evidence-based protocols that the supervisee is expected to master. Other institutions will leave the decision-making process concerning practice open-ended and permit the supervisee and the supervisor to chart out the particular technical skills and concepts to be mastered. This conversation can focus on particular MFT school-specific theories and practices, or it may move the supervisee to considered integrative models. It may focus on a particular need of the clientele, or it may address more universal aspects of human healing.

The final task is to begin the process of identifying the unique goals and objectives for supervision. Using pertinent federal, state, and local statutes governing licensure and specific core professional competencies, and best-practice standards and criteria, the supervisor begins exploring the degree of knowledge the supervisee possesses about this work. Asking supervisees to conduct a self-appraisal at this time jump-starts the goal identification process. Areas of strength, areas that need further development, and areas that are missing should be noted as possible topics for future sessions when assessments that are more concrete and goal-specific may be employed.

Sometimes, the supervisory goals and objectives may include particular procedures the organization uses to process clinical work. The supervisor needs to understand everything from the initial intake form, to all the monitoring and billing procedures, to the final documentation and referral processes in order to effectively orient supervisees. In other instances, the supervisee may be a veteran practitioner who is seeking a new skill set, has been required to earn more continuing education credits, or has been referred by an ethics board in order to maintain a license. During the opening session, the operating goals and objectives will be outlined, thus beginning the work toward clearly specifying the actual goals and objectives that will be included in the upcoming contract-setting process.

Throughout all tasks of this first session, the supervisor needs to pay attention to the learning process and be sure that the supervisee is actively engaged in the conversation and contributes to the emerging contract. It is easy to develop a specific agenda to be sure all key aspects of initiating supervision are covered and then simply hand it to the supervisee; unfortunately, this approach communicates the subordinate position of the supervisee. The first session does set the tone for the relationship; if it is going to be a collaborative process, the supervisee's voice needs to stand in the center of the interaction right from the start. Balancing supervisor and supervisee voices is difficult. If it is a top-down encounter, the voice of the supervisee may be silenced. If it is a bottom-up encounter, important structural and professional elements may not be surfaced or attended to. The skill of the effective supervisor is to find the balance so all key concepts and activities happen in a way that supervisees feel heard and see themselves as contributors to the process, not just as passive learners.

CO-CONSTRUCTING FOUNDATIONAL AND EVOLVING CONTRACTS

Contracts serve multiple functions in MFT supervision. Client protection, supervisee growth, and supervisor responsibilities are all covered by a comprehensive and evolving contract process. The supervision contract and goals of supervision must be tailored to the cultural, gender-specific, and developmental needs of supervisees. The ongoing nature of contracting conversations allows the participants to (a) clarify the methods, goals, and expectations of supervision; (b) encourage professional collaboration between the supervisor and supervisee; (c) ensure that ethical principles are upheld; (d) document services to be provided; and (e) align supervision with therapy. Clarity on each element of the contract, especially who and upon what criteria evaluation will be conducted, helps minimize conflicts and frustration for both supervisee and supervisor. The more vague the contract, the more differences of memory (What did we say about that? Mean by this?) may emerge, thus compromising the quality of supervision. A clear contract will have a stabilizing effect on the learning process by identifying how to work through problems as they surface. All supervisory contracts should be signed and dated by both supervisors and supervisees.

The contract details the primary goals, methods, and boundaries of supervision. The contract is a powerful tool supervisors used to co-generate a shared understanding of the purpose, processes, professional goals, and evaluation criteria. It unites both participants in clarifying the tasks to be accomplished, the expectations for each other, and the means of handling conflicts. A mutually developed, clearly articulated contract significantly increases the likelihood that both participants will conform to agreed-upon expectations, will take action to meet those expectations, and will seek and use feedback concerning their efforts and progress. Effective supervisors use the contract to ensure that supervisees clearly understand the evaluative structure, goals, and criteria for success that will be applied in the supervision arrangement. At the initial session, supervisees should be made aware of the boundaries of confidentiality and to whom evaluative data will be sent.

According to Scaife (2009), contracts help avoid misunderstandings, initiate a collaborative supervisory alliance, identify the interests of third parties, and inspire supervisees to clarify their learning needs. Through the contracting work, discussions can focus on how feedback to and from the supervisee is managed and situate the supervisory alliance as a topic of negotiation, reflection, and action. Reflexively, contract conversations are viewed as opportunities to air multiple perspectives about differences in assumptions and observations that may be constraining the supervisory alliance.

As noted previously, the contract is introduced in the first session and then brought up and modified all throughout the supervisory experience to meet the unique needs of supervisees as they evolve with more experience and knowledge. As supervisees increase their technical skills and confidence, the contract will have to be modified to reflect their emerging learning needs.

From the very first meeting, it is important for the supervisor to provide the supervisee with a road map of what to expect from supervision. Research studies confirm that beginning supervisees greatly benefit from making role expectations clear and detailing competency-based expectations at various intermediary stages of evaluation. They also benefit when they know that conversations about how salient issues within the wider community and cultural and geopolitical domains that affect the lived experience of their clients and themselves will be incorporated into the supervision. Supervisees value conversations that expand their understanding about the influence of historic, systemic, and current forms of prejudice, oppression, and marginalization.

Research indicates that although supervisees appreciate knowing what will be discussed and what is expected in supervision, they may actually be worried about what might go wrong. Todd and Rastogi (2014) remind us that breaches of the contract are what supervisees worry most about: gross neglect of the contract (e.g., failure to meet at agreed-upon times; insufficient, vague, and confusing feedback), being marginalized, and experiencing discrimination (e.g., workplace hostility, sexual harassment, and avoidance of conversations about issues salient to the supervisee). Further, alliances that are perceived as autocratic, bureaucratic, impersonal, and focused primarily on technical skills and reviews of case-related materials are viewed as inauthentic and resulting in shallow learning and are perceived as unhelpful and demeaning.

How difficult conversations will be structured and when they will occur should be outlined in the contracting process. Ways for both supervisors and supervisees to bring up touchy and not-so-pleasant issues, and when to do so, must be included in the contracting process. One way to ensure that this happens is to allocate time at the beginning and end of each supervisory session for either person to bring up insights and/or questions and request feedback. Formal formative and summative evaluation conferences scheduled throughout the encounter provide opportunities to raise specific, and sometimes difficult, observations and questions about the quality of the supervisory relationship. Supervisors need to provide safeguards that enable persons with less power in a relationship to know that they have methods to redress unproductive, intimidating, or harmful supervisory behaviors.

DOCUMENTING SUPERVISORY AND THERAPEUTIC ACTIVITIES

Beyond the institutional policies for documenting treatment, there are three other aspects of supervision that must be monitored carefully. The first is to document the number of hours engaged in clinical work; second is to monitor the work of supervision from the supervisee's perspective; and third is to record the supervisor's observations and interventions. A log of hours will suffice to record the number of hours a supervisee accumulates under supervision. Monitoring the quality of supervision is not so simple. In terms of capturing supervisees' insights into their own development, there is a wide variety of standardized and nonstandardized reflective and self-assessment tools that can be used.

Supervisors also need to maintain their own process notes of supervision. A process note should be completed by the supervisor after every single session with a supervisee. These notes should describe supervisee activities in treatment with the client and in supervision, and they should also include supervisor activities (e.g., supervision interventions and training recommendations). By maintaining a record of supervisory activities, you minimize your professional liability. As Falvey (2002) noted, the professional goal of timely, specific, and transparent documentation is intended to focus more on supervisor legal and ethical misconduct than on supervisee growth and development.

Supervisor notes focus on what was observed and the feedback provided. Information for notes can be drawn from any face-to-face or electronic conversation, analysis of clinical artifacts and documents, observations of performance, communiqués from other professionals involved in the case, or any significant incident. In terms of quantity, the more notes and documented evidence of progress, the better. Supervisory logs uphold the ethical responsibilities of supervisees and supervisors and can be used to identify instances when concern is warranted.

All other required documentation should also be shared in a transparent fashion. Communication with third parties should be shared with the supervisee. All written reports should include contextual information about the length, setting, and purpose of the supervision, and the limits of the report's content, in addition to the provision of particular information required in the report.

CONDUCTING COLLABORATIVE AND REFLEXIVE EVALUATIONS

MFT supervision evaluation is the ongoing process of addressing personal and professional strengths and limitations of the supervisee that may enhance or jeopardize the ethical treatment of the client. Through effective evaluation, supervisors help supervisees identify the boundaries of their technical and multicultural competence. The guiding questions are: Can supervisees describe their work (technical skill), and can they apply that knowledge in a culturally responsive fashion (multicultural skill) with a wide array of clientele? What personal, professional, institutional, and global assumptions do supervisees unwittingly bring into treatment that may reflect stereotypical and discriminatory biases toward the clients and the community they are serving? In which domains (see Figure 2.2) can supervisees function as skillful advocates for the client and the profession? What are the gaps and next levels of technical knowledge advancement for the supervisee?

In traditional supervisory situations, evaluative conversations are structured by the supervisor and usually flow from supervisor to supervisee. In contemporary supervisory practice, the responsibility for conducting evaluation and the conversations required to assess, reflect, and plan for supervisee growth and development is shared by both participants. To monitor the quality of the supervisory alliance, both participants require the skills of self-criticism and reflexivity. *Reflexivity* involves "turning one's critical gaze back on oneself as well as the professional, historical, and cultural discourses that empower and constrain one's capabilities to think and act in the context of a relationship" (Hawes, 1998, pp. 97–98). For the supervisees, the self-critical and reflexive evaluation of their work is based on supervisor, supervisee, and client feedback. These conversations are focused on using that feedback to co-generate multiple perspectives, and identify the supervisory support necessary to realize the constraining nature of supervisees' prejudices, to build upon their strengths, and to try new styles and techniques to work with diverse clientele.

Supervisors derive the same benefits of professional and personal growth and development from this type of feedback from the supervisee about the effectiveness of the supervision. At these times, the imbalance of power in the relationship has to be remembered; supervisors need to intentionally help create conditions in which the supervisees feels safe to speak about their real concerns and not just repeat what the supervisor wants to hear. Because supervisee feedback counteracts the usual flow of evaluative conversations, the supervisor needs to clearly identify, in the contract, formal and informal opportunities to engage in supervisee feedback conversations. Relying on the end of the semester/supervision relationship is insufficient in terms of gathering feedback from supervisees to improve the supervisory relationship. Short, specific, and frequent formative conversations about what's working and what isn't working about therapy and supervision ensure that the flow of evaluative information is reciprocal. In this self-critical and reflexive process, participants not only share with each other the questioning of their own preferred ideas/beliefs, but also track the ways in which power relations related to various contexts affect their supervisory relationship, the therapeutic relationship, and relationships within the agency where they work.

Supervision is an extended form of professional development and assessment. All professional behavior and ethical personal behavior is under the microscope, and the supervisor and supervisee may generate an opportunity to engage in feedback and reflection at any time. In any supervision situation that relates to licensure and certification, the role of evaluation is heightened, as the supervisor performs the gatekeeping function. In these instances, supervisors must ensure that there is clarity of expectations; early, regular, and frequent evaluations; and documentation at the first hint of a concern. To avoid the shock of a surprising evaluation conversation, particularly when a concern is going to be discussed and documented, it is recommended that formal evaluation conversations be arranged to specifically review supervisee progress throughout the supervision arrangement. By announcing and holding to these appraisal dates, the supervisor helps create a trusting and respectful environment and can ask the supervisee to come prepared with his or her insights and observations about the work. Because of the long-term, developmental nature of this evaluation process, supervisees are encouraged to maintain an assessment portfolio so that they can chart their growth over time and in relation to the professional standards they are working to achieve. Finally, to model reflexive evaluation, after each appraisal conversation a second conversation should be held in which the supervisee provides authentic feedback to the supervisor about the quality of supervision.

CONCLUSION

In this chapter, an overview of MFT supervision was presented in a noncontextual format. The specific personal and professional worldviews that each participant in the supervisory relationship brings into the encounter will color the specific path supervision will follow. Further, the particular technical knowledge that is the content of supervision was outlined but not specifically identified. The required tasks and structures to be developed were identified, with a focus on multiculturally appropriate service at both the supervisory and therapeutic levels. The significance of an evolving contract and clear evaluation plan based upon self-assessment and reflectivity was emphasized. This chapter presents one portrait of MFT supervision and is intended to be used as a broad overview when planning for specific supervisory encounters.

REFERENCES

AAMFT. (2004). *Marriage and family therapy core competencies.* Retrieved from http://www.aamft.org/imis15/Documents/MFT_Core_Competencie.pdf

Arredondo, P., & Lewis, J. (2001). Counselor roles in the 21st century. In D. C. Locke, J. E. Myers, & E. L. Herr (Eds.), *The handbook of counseling* (pp. 257–268). Thousand Oaks, CA: Sage.

Falvey , J. E. (2002). *Managing clinical supervision.* St. Paul, MN: Brooks/Cole.

Hawes, S. E. (1998). Positioning a dialogic reflexivity in the practice of feminist supervision. In B. M. Bayer & J. Shotter (Eds.), *Reconstructing the psychological subject: Bodies, practices and technologies* (pp. 94–110). London, UK: Sage.

Ng, K. (2005). The development of family therapy around the world. *The Family Journal, 13*(1), 35–42. doi:10.1177/1066480704270264

Rigazio-DiGilio, S. A. (2014). Supervising couple and family therapy practitioners. In C. Watkins & E. Milne (Eds.), *International handbook of clinical supervision* (pp. 662–647). London, UK: John Wiley & Sons.

Rigazio-DiGilio, S. A., & Kang, H. (2015). Using community genograms to position culture and context in family therapy. In M. Stauffer and D. Capuzzi (Eds.), *Fundamentals of couples, marriage, and family counseling* (pp. 25–52). Hoboken, NJ: John Wiley & Sons, Inc.

Scaife, J. (2009). *Supervision in clinical practice: A practitioner's guide.* Hove, UK: Routledge.

Sue, D., Ivey, A., & Pedersen, P. (1996/2007). *Theory of multicultural counseling and therapy.* Belmont, CA: Thomson Brooks/Cole.

Todd, T., & Rastogi, M. (2014). Listening to supervisees about problems in systemic supervision. In T. Todd & C. Storm (Eds.), *The complete systemic supervisor: Context, philosophy, and pragmatics* (2nd ed., pp. 371–394). Chichester, West Sussex [England]: John Wiley & Son, Inc.

Todd, T., & Storm, C. (2014). *The complete systemic supervisor: Context, philosophy, and pragmatics* (2nd ed.). Chichester, West Sussex [England]: John Wiley & Son, Inc.

Watkins, C., Jr., & Milne, E. (2014). *International handbook of clinical supervision.* Oxford, UK: John Wiley & Sons, Inc.

Watkins, C., Jr., & Wang, C. (2014). On the education of clinical supervisors. In C. Watkins, Jr., & D. Milne (Eds.), *The Wiley international handbook of clinical supervision* (pp. 177–203). Oxford, UK: John Wiley & Sons, Inc.

THREE

Modalities of Marriage and Family Therapy Supervision

Rebecca Boyle
Molly McDowell-Burns

MODALITIES OF SUPERVISION

Clinical supervision is a crucial and necessary element in the helping professions (Vaccaro & Lambie, 2007). Supervision is the primary resource that trains therapists/counselors and helps them gain practical skills as well as knowledge that will assist them in becoming ethical and effective helping professionals (Vespia, Heckman-Stone, & Delworth, 2002). One of the hallmarks of Marriage and Family Therapy (MFT) training is supervision, specifically "live" or "raw" supervision. A master's program accredited by the Commission on Accreditation of Marriage and Family Therapy Education (COAMFTE) requires 50 of its required 100 supervision hours to be "raw" supervision. This may include live, videotape, or audiotape supervision. No other mental health discipline requires the use of live supervision in its training, and none requires as many hours of supervised experience as MFTs (American Psychological Association [APA], 2013; Council for Accreditation of Counseling & Related Educational Programs [CACREP], 2009; Council on Social Work Education [CSWE], 2012). This chapter outlines four modalities of supervision, including case consultation, online supervision, videotaped supervision, and live supervision.

CASE CONSULTATION

Case consultation is a broad methodology used to better understand the process of therapy. A 2004 study completed by Lee, Nichols, Nichols, and Odom finds that case consultation continues to be the primary method used for supervision. The survey, which was administered in 2001 (N = 330), reports that 42% used this method of supervision compared to only 21% in 1986 (Lee et al., 2004).

Case consultation requires a therapist to discuss cases with a supervisor in a group or individual format. The therapist describes sessions, interventions used, and his or her perspective on the problem. The supervisor then provides feedback to the therapist and helps the therapist construct informal or formal plans for future sessions (Stewart, 1997). McCollum and Wetchler (1995) define *case consultation* as "an individual supervisor and supervisee discussing a particular case without the use of raw data." Schon (1983, 1987) describes this process of supervision as the construction of the "virtual world" in which therapists conceptualizes their cases.

Development of Case Consultation

McCollum and Wetchler (1995) identify four categories necessary to reap the full benefits of case consultation: (1) understanding the architecture of the therapy process; (2) helping supervisees build theoretical models of change; (3) understanding the clients' broader context; and (4) understanding the therapist's broader context.

The *architecture of therapy* is defined as the context in which the therapist works across multiple sessions with a client and how the therapist puts the parts together to make a comprehensible whole (McCollum & Wetchler, 1995). The process of treatment includes the assessment process, treatment goals, and hypothesis about the nature of the problem (McCollum & Wetchler, 1995). As therapists become more comfortable with and aware of the process, they are better able to conceptualize a case by connecting the past, present, and future. Case consultation allows therapists to explore the construction process of therapy by sharing their experiences with their supervisors and to organize their therapeutic plans in an appropriate manner. It also gives the supervisees time needed to formulate goals for treatment and to ensure that their interventions and direction in therapy are consistent with the treatment goals (McCollum & Wetchler, 1995).

The second category proposed by McCollum and Wetchler (1995) is building theoretical models. This category focuses on helping the supervisees view cases with broader perspectives. In order to build a theoretical model, supervisees work to generalize their theoretical orientation across cases (Liddle, 1982; McCollum & Wetchler, 1995). This developmental process helps supervisees provide meaning to the interventions being used and encourages them to relate this meaning to their theoretical orientation.

The third category identified is understanding the client's broader context. This idea emphasizes the importance of larger system involvement with the client ((Bronfenbrenner, 1977; Imber-Black, 1988; McCollum & Wetchler, 1995; Schwartzman, 1985). Examples of other system involvement are probation/courts, children's services, medical professionals, education services, substance abuse counselors, and/or other mental health professionals. "Supervisees must learn to include these systems in their thinking about the case and not operate in isolation from the larger system" (McCollum & Wetchler,

1995, p. 161). Case consultation allows the supervisor to guide the supervisee during the process of collaborating with other professionals to ensure overall quality of care to the client(s).

The last category recognized by McCollum and Wetchler (1995) is the supervisee's broader context. This category emphasizes therapists' ability to be cognizant of their own influences, motivations, and biases that accompany them or are influenced in the therapy sessions. Supervisors can help therapists to identify issues as they arise and to work through alternative perspectives as needed (Guanipa, 2003; Storm et al., 2001; Sue, 1998).

Advantages and Limitations of Case Consultation

There are many advantages to utilizing case consultation in supervision. Case consultation gives both beginning and experienced therapists time to process and reflect on cases with supervisors, as opposed to live supervision, which does not allot time for processing (Stewart, 1997; Todd & Storm, 1997). It also allows the therapist and supervisor to discuss how the therapy is constructed and at what pace the therapy moves (McCollum & Wetchler, 1995; Stewart, 1997; Todd & Storm, 1997). In addition, it gives supervisees a chance to discuss cases in a broader, socially constructed context: social, political, economic, and/or cultural (Stewart, 1997; Todd & Storm, 1997).

Case consultation also gives supervisors a chance to ask questions that influence the thinking of the therapist. It permits them to share their expertise and experiences with the supervisee in a way that live supervision does not always allow. The narrative dialogue also helps the supervisor to question the social construction of ideas or paradigms the therapist may be experiencing. This may influence sensitivity to cultural, religious, ethnic, or sexual implications (Guanipa, 2003; Stewart, 1997; Sue, 1998).

Another advantage of case consultation is the ability to discuss ethical or legal implications in a controlled setting. Sometimes live supervision encourages a supervisor to act quickly and reduces the amount of time spent on specific situations. Case consultation provides time for the supervisee to process the situation and respond appropriately. An example cited in McCollum and Wetchler (1995) is that of reporting child abuse. Case consultation provides the supervisor with ample time to give guidance to the supervisee and discuss how the supervisee can rejoin with the family after reporting child abuse.

Although there are positives to using case consultation as a supervisory tool, there are also limitations. The supervisor is not able to observe body language, tone of voice, or any other idiosyncrasy that live supervision allows (Stewart, 1997; Todd & Storm, 1997). When observed live, the supervisor also gains a better understanding of how the therapist actually conducts a therapy session, as opposed to having the therapist attempt to reconstruct the session for the supervisor.

Case consultation is still the most widely used methodology for supervising clinicians. Like anything else, there are advantages and disadvantages to utilizing this particular method. As the counseling profession continues to grow, there will always be a need for supervision. More research should be conducted in this area to understand exactly why case consultation is important to the growth of clinicians.

ONLINE SUPERVISION

Technology is an ever-growing entity that continues to prove its necessity in society. As technology becomes more universal and easier to access, it will infiltrate multiple professions and change the ways in which professionals work. Technology has gained attention in the counseling field as it has proved to be an innovative and potentially useful strategy in supervision. Layne and Hohenshil (2005) emphasize that technology in counseling is a trend that will remain constant and believe that supervisees and supervisors need to learn to use it effectively. Research suggests that the full potential of computer-based supervision has not been fully recognized in counselor education programs (Trolley & Silliker, 2005; Vaccaro & Lambie, 2007, p. 47; Watson 2003). Watson (2003) defined supervision interactions that occur over the World Wide Web as "cybersupervision"; we will use this term throughout this chapter.

Several different types of technologies can be utilized for computer-based supervision. Examples are: (1) e-mail; (2) computer-based teleconferencing; (3) electronic mailing lists; (4) chat rooms; and (5) computer-assisted live supervision (Harvey & Carlson, 2003; Vaccaro & Lambie, 2007; Watson, 2003). The technologies mentioned are used in a plethora of ways; however, they come with an intricate set of advantages and disadvantages to the counseling profession.

Advantages and Disadvantages of Online Supervision

Advantage

One of the major attractions of cybersupervision is the ability to meet with supervisors regardless of their geographical locations (Bacigalupe, 2010; Bloom & Walz, 2000; Coursol & Lewis, 2000; Gainor & Constantine, 2002; Oravec, 2000; Watson, 2003). This also gives opportunities for more diverse supervision experiences, because it permits unique individuals to connect with other unique individuals (Bacigalupe, 2010; Bloom & Walz, 2000; Coursol & Lewis, 2000; Gainer et al., 2002; Oravec, 2000). Furthermore, online supervision can also reach those individuals living and working in more rural settings (Bacigalupe, 2010; Parker & Parker, 1998). An intern no longer risks having to travel an extensive amount to get the proper supervision needed to obtain licensure.

In addition to the convenience of avoiding geographic constraints, online supervision also allows more flexibility in scheduling (Bloom & Walz, 2000;

Coursol & Lewis, 2000; Gainor & Constantine, 2002; Oravec, 2000; Parker & Parker, 1998). One no longer has to schedule meetings in specific locations—one need only be accessible to a computer or smartphone.

Although there are advantages to cybersupervision, there are also limitations to these technology innovations. Technology is always changing as people gain a better understanding of how to make things better and more efficient. This presents a challenge because technological advances can become costly. Always striving for the best and most innovative technology can impede the supervision experience as the cost to utilize this technology becomes too much for supervisees and their supervisors (Hildebrand, 1995; Vaccaro & Lambie, 2007).

In addition to cost, technology is not always the most reliable method. Supervisors and supervisees should formulate a backup plan for actions to take if technology fails during supervision (McCarty & Clancy, 2002; Oravec, 2000; Watson, 2003). This can be a time-consuming and also overwhelming (and expensive) task.

Research has suggested that the supervisor-supervisee relationship is the most critical component of supervision (Bernard, Goodyear, & Bernard, 2004; Bloom & Walz, 2000; Harvey & Carlson, 2003; Watson, 2003). The usage of cybersupervision jeopardizes the nuances that face-to-face supervision permits. Nonverbal communication is one of the elements affected by online supervision because it is not as easily observed through technology (Vaccaro & Lambie, 2007). Unfortunately, without these nonverbal cues, the interpersonal relationship between the supervisor and supervisee may lack "depth" and "richness" (Vaccaro & Lambie, 2007, p. 50). Supervisors may not be able to interpret the subtleties that body language conveys to ensure that the supervisee is feeling confident in his or her abilities. This becomes problematic because it could potentially thwart the professional development of the clinician-in-training.

Cybersupervision also brings up ethical considerations, in addition to the other limitations previously mentioned. Corey, Corey, and Callanan (2003, p. 351) suggest that the "greatest clinical and ethical challenge of supervision is that supervisors must attend to the best interests of the client and supervisee simultaneously. Online supervision challenges that assumption as the implications of autonomy, nonmaleficence, and fidelity may be debatable."

Autonomy

Autonomy is the freedom individuals have to make their own decisions and to possess the ability to act upon those decisions without intrusion from anyone else (Corey, Corey, & Callanan, 2007; Gladding, Remley, & Huber, 2001; Welfel, 1998; Wilcoxon, Remley, Gladding, & Huber, 2007). Clients have a right to know how their confidential information is going to be handled (Greenwalt, 2001). A proper informed consent should explain the purpose of cybersupervision, and the clients have the right to decide whether or not this is something they would like to continue. "Therapists should not press the issue, but allow the client the right of autonomy in deciding how the therapist handles their confidential information" (Greenwalt, 2001, p. 12). This can become difficult for supervisees because one client's disapproval could potentially hinder the supervision process. Upholding this ethical standard becomes much more intricate when technology is involved.

Nonmaleficence

Nonmaleficence means to do no harm (Corey et al., 2007; Gladding et al., 2001; Welfel, 1998; Wilcoxon et al., 2007). Specifically, it means to do no harm to the supervisee or the client. If confidentiality is compromised in any way during treatment, it could be potentially harmful to the client. It is not impossible to share things over the Internet without breaking the confidentiality of clients. In fact, there is strict software that promises privacy; however, the risk of having confidential information compromised continues to loom.

Fidelity

Fidelity is a trusting relationship between the therapist and the client (Welfel, 1998). Gladding et al. (2001) define fidelity in a supervisor-supervisee relationship as a supervisor sustaining all commitments and communications through an honest and trusting manner while preserving integrity to the supervisee and clients. Because there are few definitive answers with regard to cybersupervision, supervisors must inform supervisees that their supervision hours may be rejected by state licensing boards. Currently, the American Association for Marriage and Family Therapy (AAMFT) does not have specific ethical guidelines on online supervision, and the American Counseling Association (ACA) guidelines are vague.

> F.2.c Online Supervision:
>
> When using technology in supervision, counselor supervisors are competent in the use of those technologies. Supervisors take the necessary precautions to protect the confidentiality of all information transmitted through any electronic means. (ACA, 2014)

Also, the ability to utilize online supervision may vary from state to state. Laws may prohibit supervisees from seeking supervision online. Based on the ambiguity of the ethical guidelines and the nebulous nature of state laws, one should err on the side of caution when utilizing cybersupervision as the only modality for supervision.

Cybersupervision is a new frontier that promises many innovative and exciting contributions to supervision. As technology continues to advance, there will be a greater need for individuals to grow with technology. Even as the benefits of online supervision are appealing, the ambiguity and limitations are concerning. While cybersupervision may be realistic for supervisees, it should not be the only type of supervision utilized. It is best to use face-to-face supervision and, if needed, accompany it with online supervision.

RAW DATA SUPERVISION

As one can see from the technological advances discussed previously, case report is no longer the only means by which a supervisor can gain information about a supervisee's case. By utilizing just case report, the supervisor is deprived of the ability to hear the interactions and see the nonverbals. Case report only gives one perspective of the session, the supervisee's. With the use of audiotaped, videotaped, or live supervision, multiple perspectives of the therapy session are made available.

Even though it is an integral piece of MFT training, very little research has been conducted as to the efficacy of video and live supervision. Anderson, Schlossberg, and Rigazio-DiGilio (2000) find that supervisees associated better supervisory experiences with live observations and videotape review. Silverthorne, Bartle-Haring, Meyer, and Toviessi (2009) examine the correlation of live observations and the supervisees' and clients' perceptions of therapy progress: Just one instance of live supervision improved perceptions of progress for the supervisee. Another study finds that there was no difference in gaining and retaining MFT skills between supervisees receiving live supervision and supervisees receiving delayed feedback (Fennel, Hovestadt, & Harvey, 1986). More research should be conducted to support the wide use of live supervision in MFT training, especially focusing on the different types of live supervision (such as phone-ins, mid-session breaks, and pre/post session planning).

Audiotape

Audiotaping was one of the first forms of "raw" supervision. It is currently not widely used as a form of raw data supervision. However, according to Protinsky (1997), audiotape supervision may even be the chosen method of supervision, depending on the supervisee's approach. Audiotaping a therapy session and playing it back during supervision may offer the supervisor the opportunity to listen to verbal interactions without being focused on the nonverbals. This can give a different perspective on the session. The Mental Research Institute group (Fisch, 1988) used this method of supervision to focus on the verbal dialogue and between-session homework.

Supervisees may need to tape sessions at their outside sites and thus may have to rely on audiotape, since agencies do not always allow videotaping of clients. While conceptualizing a case presentation, the use of the audiotape can be very enlightening. For example, the supervisor will be able to key into the pitch, rate, and speed of the session. By stopping and replaying portions of the tape, certain sequences can be highlighted and studied in detail. A summary of months of therapy can be condensed into a few sections of tape. Although this is not as rich as video data, it does offer a glimpse into the session and moves farther along the continuum from case report toward live supervision.

Videotape Supervision

In MFT training clinics, videotaped supervision is common. Unlike case consultation, videotape gives the supervisor a glimpse into the session. Carl Rogers utilized videotape and is considered a pioneer in this field (Protinsky, 1997). It became more commonplace in the 1970s as the technology in this area became readily available. This form of supervision is required in COAMFTE accredited programs (COAMFTE, 2014) and in other mental health professional training as well.

Advantages

Videotaped supervision allows the supervisor the time to review and conceptualize the case without the time constraints of live supervision. The supervisor can observe the session including the interactions between the family members and the family and the trainee. It can provide the supervisor and supervisee an in-depth look at any particular moment within the session. The modality includes the ability to rewind and dissect the moments in question. This allows some distance from the actual session to reflect. It also allows for other perspectives or realities of the session. The supervisor can question, reflect on, challenge, or agree with the trainee's perspective while viewing the tape. The supervisor can also find out the trainee's state of mind when watching certain portions of tape. This may help to clarify why certain interventions were done during the session.

Supervisors can get a much richer picture of the client-therapist relationship with the use of videotape. Techniques and theory can be implemented and supervised much differently with the use of tape. This benefits not only the trainee but the client as well. Using the tapes to highlight earlier sessions with the client may help highlight the changes that occur over time. These changes may have gone unnoticed without the ability to play back sessions over time. Other advantages include the ability to see the nonverbals in the room. Structural Family Therapy is particularly interested in where everyone sits and how they interact. Video recording gives the supervisor and the trainee the ability to see nonverbal behavior of the family and the therapist.

Using videotape can also help with honing one's theory and the evaluative process. For example, supervisees may present a case highlighting their theory of choice. This would be shown through a series of small segments of sessions over time. This helps the trainee to incorporate theory and see it implemented in practice over time. Having supervisees focus on theory is a main goal of supervision, and being able to view tapes gives feedback about the progress. It also gives the supervisor the ability to see changes in the trainee's progress over time. Supervisors can use it as an evaluative tool when needed. Having the tape as a reference as one evaluates the supervisees helps to ensure a more in-depth evaluation.

CASE EXAMPLE

Supervisor and supervisee watch video together. Supervisor pays close attention to body language of couple and body language of therapist.

Dr. Bob: [Pauses tape] *Tell me about your body language in this moment.*

Brittney: *Well, I can see that I am fidgeting quite a bit and I continuously tap my foot.*

Dr. Bob: *What seems to be making you uncomfortable? Do you move like this any other time during the session?*

Brittney: *I don't think I do. I think it is just here. We were discussing Travis's time in the military and his injuries.*

Dr. Bob: *Hmm. I wonder how that topic is impacting you as a therapist?*

Brittney: *I guess I hadn't really thought about it, because this happened so many years ago, but my husband was injured while serving in Iraq. Our children were about the same age as Tommy and Tyrone are now.*

Dr. Bob: *How may the couple be impacted by your body language?*

Brittney: *Now that you point that out, I can see that Travis really shut down. He stopped making eye contact and Tekia discussed how his injuries impacted the family. He really didn't speak much the rest of the session.*

Dr. Bob: *You may be on to something; how can you address his body language in future sessions?*

Disadvantages and Ethical Considerations

With all of these advantages, it is easy to overlook some possible disadvantages of taped sessions. First, the supervisee may feel uncomfortable at first when looking at the sessions. Making sure to address this with the supervisee can help facilitate the supervision session. If the supervisee is too worried about how she or he looks or sounds, she or he may not be fully engaged in the supervision session and may miss valuable feedback from the supervisor.

DeRoma, Hickey, and Stanek (2007) report that only 57.4% of trainee and intern MFTs in California use videotaping as a direct form of supervision. On the one hand, this is surprising given the benefits of videotaped supervision; on the other hand, it could also highlight the lack of available, reliable equipment for taping trainees. In addition, the ethical concerns over client confidentiality may be a concern.

Where does one keep the taped sessions? How long does one keep the taped sessions? As with all new technology, ethical codes and laws pertaining to the storage and destruction of these materials have lagged behind the technology (AAMFT Code of Ethics, Standard VI, 2015). AAMFT and ACA both require clients to sign an informed consent for video recording. Their accrediting bodies, COAMFTE and CACREP, require videotaping as part of their training, but clients need to sign an informed consent to allow the videotaping. According to Protinsky (1997; Storm 2014), the client needs to be given information regarding the rationale for taping, who will view the tapes, when the tapes will be erased, and where/how the tapes will be stored. These must be stated clearly in the informed consent. Clients should be given the option to opt out of videotaping; if it is a requirement of a training facility, unwilling clients should be given referrals elsewhere. As with any technology, there will be challenges, but the richness added to the supervision experience and the benefits to the client should outweigh the difficulties that come with videotaping client sessions.

Live Supervision

Live supervision is synonymous with Marriage and Family Therapy training and supervision. It typically involves the supervisor behind a one-way mirror or watching a live feed on a monitor while in a different room. It can include a pre-session meeting, phone calls, a planned mid-session break, a knock on the door, delayed feedback with a post-session debriefing, and use of a "bug-in-ear." Its roots began with therapists demonstrating their techniques while being observed by their colleagues (Storm, 1997). They felt that the intrusion on client privacy was justified because it was such a new way to conduct therapy. Many of these early therapists experimented with live supervision. Jay Haley concluded that a directive given to a therapist while in session could be both an intervention and supervision. The pioneers of family therapy were also the pioneers of live supervision. Haley, Minuchin, and Montalvo began live supervision in the 1950s, and it is still an integral part of training today (Storm, 1997).

Throughout the years, several authors have offered suggestions for best practices for live supervision (Birchler, 1975; Liddle & Schwartz, 1983; Lowe, Hunt, & Simmons, 2008; Montalvo, 1973; Schwartz, Liddle, & Breunlin, 1988). One of the biggest assumptions is that supervisors possess the skills to effectively observe a live session. Schwartz et al. (1988) describe two main issues that occur during live observation. The first is that the trainee follows the instructions verbatim with little understanding or creativity. This creates a dependence on the supervisor because follow-up questions are not asked. The trainee may feel anxious and fear making a mistake, so the instructions are followed word for word. The second is *isomorphism*, the replication of patterns between the two levels: The patterns between the team members may be similar to those occurring between the family members. Role-playing and discussion of what should happen after the supervisor intervenes will help to mitigate some of these issues.

Supervisees should understand that this type of supervision may feel mechanical. Supervisors should try to minimize this by matching their style to that of the supervisees. Prior to the start of the session, the supervisor and supervisee agree to the terms of the supervision. Supervisors agree to call or knock, and the supervisee agrees to take the call or step out. All input is framed as a suggestion, unless client welfare is an issue. Of course, the client needs to fully understand what live supervision is. Typically, this is delineated in the informed consent. The supervisee should also be well versed in how to explain this to the client. The supervisee does have the right to question the demands of the supervisor. The case planning and debriefing should occur prior to and after the session. The supervisor needs to consider the urgency of the intervention and consider that the supervisee may be successful without the interruption. The supervisors need to strike a balance between supervisee dependency and growth. It can be very unnerving for both the client and supervisee when the knock, voice in the ear, or phone call happens. Therefore, if both are well prepared, the intrusion into the session will be more welcome.

Phone-Ins

One way to intervene while doing live observation is to make a phone call into the therapy room during the session. Wright (1986) analyzed approximately 150 phone-ins during live supervision and came up with a number of recommendations for training family therapy supervisors.

1. Supervisors need to be trained in how to utilize the phone-in technique.
2. The supervisor and supervisee need to practice phone-ins prior to seeing clients. This will help the supervisee practice how to transition from the phone call back to therapy.
3. Phone-ins should be used sparingly. They should not be used within the first 10 minutes of a session and no more than 5 should be made during the session. If more assistance is needed, then it may be best for the supervisee to step out and consult with the supervisor.
4. Make sure to use positive reinforcement with the supervisee during the phone call.
5. Phone calls should vary based upon supervisee experience. Beginning supervisees will need more explicit, script-like statements, whereas more advanced supervisees can be given less direction.

6. Supervisors need to be clear and concise with beginners, because those supervisees will use the same tone and phrases as the supervisor when they speak to the family.
7. Supervisors should only give one process statement or none at all. Process statements can be confusing to the beginning therapist. Phone-ins should not exceed 25 seconds.
8. Supervisors should not give more than two instructions per phone call.
9. Even if there is a team behind the mirror, there should only be one designated person conducting the phone calls.

This list is not exhaustive, but it does provide the supervisor with several suggestions to make phone-ins as successful as possible.

CASE EXAMPLE

As Dr. Bob watches a session, he notices that the father has become disengaged. He phones into Brittney's session.

Dr. Bob: *You are doing a really nice job engaging the mother, but I'm wondering about the father. He looks disengaged and I want you to direct a few questions to him.*
Brittney: *OK.*
Dr. Bob: *Please ask him about a time when he felt in charge of the house since he returned. Make sure to explain this phone call to the clients. They look confused.*
Brittney: *OK.*

Brittney hangs up the phone and explains to the family that the supervisor wanted to make sure she was asking everyone about Travis's reintegration into the family. The phone ringing actually seemed to bring everyone back into the session.

The phone-in can be a very effective means of improving the supervisee and client experience in the moment. If the supervisor uses it as outlined here, the supervisee should feel empowered throughout the session. As stated previously, though, the phone-in should not be used more than 5 times throughout the session; in fact, this author feels that it should not be used more than twice during a session. It does disrupt the flow, and depending on the type of phone used, can be very startling if loud. It may be more detrimental to session success when the phone has a loud, jarring ring. For this reason, a mid-session break may be more advantageous.

Mid-session Break and Knock

Before beginning a session, the supervisor and supervisee may decide to use a mid-session break. This can take two forms. The first can be a preplanned break midway through the session. The therapist informs the family that she or he will exit the room halfway through the session to consult with the supervisor or supervision team. This break may have a specific preplanned goal or may be used to give feedback based on the content of the session. The break may be longer than a phone-in, but several of the previously outlined suggestions still apply. The supervisees should have a good script for when they return to the sessions. They need to transition back into the sessions without too much disruption. It is generally acceptable to have the supervisee summarize what has happened thus far in the session when he or she returns. During the break, the supervisor needs to give positive reinforcement for the supervisee on the session. Again, the supervisor should only give two to three directives for when the supervisee returns to the room. The supervisor should keep the process statements to a minimum. If the supervisee is more advanced, then some process statements can be given for the supervisee to implement when the break is over. If there is a supervision team watching from behind the mirror, there are two ways to proceed: either each member can offer one piece of feedback for the supervisee, or these can be summarized and presented by one person so as not to confuse or overwhelm the supervisee.

Another way to intervene is with a knock on the door, whereupon the supervisee steps out into the hall. This is similar to the phone-in, as the intrusion is kept short with only a few directives being given to the supervisee. Even though the supervisee is stepping out of the session, it may be less intrusive than a phone-in, because a phone ringing can be very loud and the supervisee still needs to engage the client while listening to the supervisor. When out in the hall, the focus is on the supervisor. The supervisor can also assess for signs of distress in the supervisee and have him or her take a deep breath if needed. The knock can be preplanned, or the supervisor can knock when the direction of the session should be changed.

CASE EXAMPLE

Dr. Bob and Brittney plan a mid-session break at the 30-minute mark. She informs the family at the beginning of the session that she will be taking a break to consult with her supervisor. She tells the family that this might also be a good time to take the children to the restroom, since they are so young. Brittney returns to the supervision room at the 30-minute mark.

(*continued*)

Dr. Bob: *Tell me what you think you are doing well.*

Brittney: *I feel that I have been able to include Travis much more in this session.*

Dr. Bob: *I agree; you have engaged the parents and elicited a lot of information from them. I think he is beginning to feel more comfortable. I notice that the children are starting to act up and interrupt a lot. No one seems comfortable redirecting them or bringing them into the session.*

Brittney: *I was waiting for the parents to speak with them. I didn't think it was my place.*

Dr. Bob: *I think this is a good time to bring in the parenting piece—ask about how this would look at home. I worry that Travis may get overloaded with the noise. I think you should notice out loud that the children seem to be seeking attention. You can ask if this is typical and how they would handle this at home.*

Brittney: *OK.*

Dr. Bob: *You can then speak openly about the parenting strategies we have covered in supervision and ask if it is OK if you demonstrate them for the family.*

Brittney: *OK.*

Brittney heads back into the room and notices immediately that the children seem calmer after the break and are sitting with the parents. She notices this out loud and begins the parenting discussion.

Delayed Feedback

In some instances, live supervision is conducted but feedback is delayed. Birchler (1975) described a model of live supervision with delayed feedback. Essentially, the supervisor observes the session live but does not intervene during the session. The supervisee and supervisor meet immediately after the session, and feedback is provided at that time. An advantage is that the session is not interrupted and the supervisor can give much more detailed feedback immediately after the session. Another possibility is for the therapist to tape his or her feedback as the session is being conducted so that the supervisee can hear the feedback as she or he listens to the session (Birchler, 1975). As there is no interruption to the session, the supervisee does not become dependent upon the supervisor, but the supervisor still has the option of intervening if needed.

Bug-in-Ear

Still another way to conduct live supervision is with a bug-in-ear. This form of direct supervision occurs in the moment without the intrusion of a knock or a phone call. There is no delay and the supervisor can change the course of the session by giving the supervisee feedback. As with the other forms of intervention, there must be training before an actual session is attempted. The supervisee needs to be able to listen to the supervisor's voice in his or her ear and incorporate the feedback without becoming distracted by the earpiece.

In summary, the advantages of live supervision include immediacy of feedback, client care, and support to the supervisee. Live supervision offers support "in the moment" to the supervisee. This gives the supervisor the ability to see the interactions between the clients and the therapists in real time. It also gives the supervisor the ability to intervene immediately and change the direction of the session if necessary. It improves the quality of care because it allows supervisees to rework a mistake and can keep them from getting stuck (McCollum & Wetchler, 1995). Theoretically, the client benefits from the added layer of supervision during the therapy session. The supervisor is able to intervene if certain ethical issues arise. For example, a family member mentions suicide, but the supervisee does not explore this further. The supervisor is able to speak with the therapist in "real-time" to make sure this is addressed in the session. Supervisees can learn therapy skills while in session with the support of the supervisor (Storm, 1997). If a supervisee is stuck, the supervisor may provide an option not considered by the supervisee.

Despite the many advantages of live supervision, there are possible disadvantages. Several authors have outlined the disadvantages (Liddle, 1991; McCollum & Wetchler, 1995). First, the supervisee may become dependent upon the supervisor's intervention. There may be passivity as the supervisee waits for the supervisor to intervene instead of taking initiative during a session. This may in turn undermine the therapist's authority with the client family. The family may have negative reactions to call-ins or to the therapist leaving a session to consult with the supervisor.

Second, the boundaries between the therapist, the family, and the supervisor can become blurred. The therapist may not take responsibility for the session, and the family may not understand who is ultimately responsible for their well-being. Finally, the rhythm of the session may be disrupted, with the therapist being unable to make a seamless transition back into the therapy session after speaking with the supervisor.

Without the technology, many of these options are not viable. Some can cost a lot of money, which may limit the options to behind-the-mirror supervision. As discussed earlier, many of these disadvantages can be addressed to mitigate the downside of any interruptions.

CONCLUSION

Supervision is a vital and indispensable component of the training experience of MFTs. Supervision prepares therapists by giving them rich learning experiences that foster personal and professional growth. It encourages supervisors to serve as gatekeepers to the profession and to guide therapists ethically, clinically, and professionally. The advances in technology have offered means that greatly enhance the supervision experience but have also added some ethical issues. With the ability to stop or "freeze" time when using a recording, the session can be dissected in a way that makes room for different lenses and realities to come into the session. Technology has also allowed supervisors and supervisees to engage in supervision without being in the same room. The task of making sure that the technology is secure and provides the client with protection falls to the supervisor. With the use of any supervision modality, keeping the client's needs first is of upmost importance. Using a variety of supervision modalities will help to ensure that both client and supervisee are benefiting from the counseling experience.

REFERENCES

American Association for Marriage and Family Therapy (AAMFT). (2015). *AAMFT code of ethics.* Alexandria, VA: Author.

American Counseling Association (ACA). (2014). *ACA code of ethics.* Alexandria, VA: Author.

American Psychological Association (APA). (2013). *Guidelines and principles for accreditation of programs in professional psychology.* Retrieved from https://www.apa.org/ed/accreditation/about/policies/guiding-principles.pdf

Anderson, S. A., Schlossberg, M., & Rigazio-DiGilio, S. (2000). Family therapy trainees' evaluations of their best and worst supervision experiences. *Journal of Marital and Family Therapy, 26*(1), 79–91.

Bacigalupe, G. (2010). Supervision 2.0: E-supervision a decade later. *Family Therapy Supervisor Bulletin, 9*(1), 38–41.

Bernard, J. M., Goodyear, R. K., & Bernard, J. M. (2004). *Fundamentals of clinical supervision* (3rd ed.). Boston, MA: Allyn and Bacon.

Birchler, G. R. (1975). Live supervision and instant feedback in marriage and family therapy. *Journal of Marriage and Family Counseling, 1*, 333–342.

Bloom, J. W., & Walz, G. R. (2000). *Cybercounseling and cyberlearning: Strategies and resources for the millennium.* Alexandria, VA: American Counseling Association.

Bronfenbrenner, U. (1977). Toward an experimental ecology of human development. *American Psychologist, 32*, 513–531.

Commission on Accreditation for Marriage and Family Therapy Education (COAMFTE). (2014). *Accreditation standards: Graduate and post-graduate marriage and family therapy training programs.* Retrieved from https://www.aamft.org/imis15/Documents/COAMFTE/COAMFTE_Accreditation_Standards_Version_12.pdf

Corey, G., Corey, M. S., & Callanan, P. (2003). *Theory and practice of counseling and psychotherapy* (6th ed.). Pacific Grove, CA: Brooks/Cole.

Corey, G., Corey, M. S., & Callanan, P. (2007). *Issues and ethics in the helping professions* (8th ed.). Belmont, CA: Brooks/Cole.

Council for Accreditation of Counseling and Related Educational Programs (CACREP). (2009). *Accreditation standards.* Retrieved from http://www.cacrep.org/wp-content/uploads/2013/12/2009-Standards.pdf

Council on Social Work Education (CSWE). (2012). *Educational policy and accreditation standards.* Retrieved from http://www.cswe.org/File.aspx?id=13780

Coursol, D. H., & Lewis, J. (2000). Cybersupervision: Close encounters in the new millennium. *American Counseling Association/ERIC/CASS.* Retrieved from http://cybercounsel.uncg.edu

DeRoma, V. M., Hickey, D. A., & Stanek, K. M. (2007). Methods of supervision in marriage and family therapist training: A brief report. *North American Journal of Psychology, 9*(3), 415–422.

Fennel, D. L., Hovestadt, A. J., & Harvey, S. J. (1986). A comparison of delayed feedback and live supervision models of marriage and family therapist clinical training. *Journal of Marital and Family Therapy, 12*(2), 181–186.

Fisch, R. (1988). Training in the brief therapy model. In H. Liddle, D. Breunlin, & R. Schwartz (Eds.), *Handbook of family therapy training and supervision* (pp. 78–92). New York, NY: Guilford Press.

Gainor, K. A., & Constantine, M. G. (2002). Multicultural group supervision: A comparison of in-person versus web-based formats. *Professional School Counseling, 6*(2), 104–111.

Gladding, S. T., Remley, T. P., & Huber, C. H. (2001). *Ethical, legal, and professional issues in the practice of marriage and family therapy* (3rd ed.). Columbus, OH: Merrill Prentice Hall.

Greenwalt, B. C. (2001). Cybersupervision: Some ethical issues. *AAMFT Supervision Bulletin,* 12–14.

Guanipa, C. (2003). A preliminary instrument to evaluate multicultural issues in marriage and family therapy supervision. *The Clinical Supervisor, 21*(1), 59–75.

Harvey, V. S., & Carlson, J. F. (2003). Ethical and professional issues with computer-related technology. *School Psychology Review, 32*(1), 92–107.

Hildebrand, J. E. (1995). Videoconferencing in the business curriculum. *Journal of Business and Technical Communication, 9*(2), 228–240.

Imber-Black, E. (1988). *Families and larger systems: A family therapist's guide through the labyrinth.* New York, NY: Guilford Press.

Layne, C. M., & Hohenshil, T. H. (2005). High tech counseling: Revisited. *Journal of Counseling & Development, 83*(2), 222–226.

Lee, R. E., Nichols, D. P., Nichols, W. C., & Odom, T. (2004). Trends in family therapy supervision: The past 25 years and into the future. *Journal of Marital and Family Therapy, 30*(1), 61–69.

Liddle, H. (1982). On the problem of eclecticism: A call for epistemologic clarification and human-scale theories. *Family Process, 21,* 243–247.

Liddle, H. (1991). Training and supervision in family therapy: A comprehensive and critical analysis. In A. S. Gurman & D. P. Kniskern (Eds.), *Handbook of family therapy* (Vol. 2, pp. 638–697). New York, NY: Brunner/Mazel.

Liddle, H., & Schwartz, R. (1983). Live supervision/consultation: Conceptual and pragmatic guidelines for family therapy training. *Family Process, 22,* 477–490.

Lowe, R., Hunt, C., & Simmons, P. (2008). Towards multi-positioned live supervision in family therapy: Combining treatment and observation teams with first-and second-order perspectives. *Contemporary Family Therapy, 30*(1), 3–14.

McCarty, D., & Clancy, C. (2002). Telehealth: Implications for social work practice. *Social Work, 47*(2), 153–161.

McCollum, E. E., & Wetchler, J. L. (1995). In defense of case consultation: Maybe "dead" supervision isn't dead after all. *Journal of Marital and Family Therapy, 21*(2), 155–166.

Montalvo, B. (1973). Aspects of live supervision. *Family Process, 12*, 343–359.

Oravec, J. A. (2000). Online counselling and the Internet: Perspectives for mental health care supervision and education. *Journal of Mental Health, 9*(2), 121–135.

Parker, A., & Parker, L. E. (1998). Distance education vs. in-class instruction. In S. F. Viegas & K. Dunn (Eds.), *Telemedicine: Practicing in the information age* (pp. 135–152). Philadelphia, PA: Lippincott-Raven.

Protinsky, H. (1997). Dismounting the tiger: Using tape in supervision. In T. Todd & C. Storm (Eds.), *The complete systemic supervisor: Context, philosophy, and pragmatics.* Boston, MA: Allyn and Bacon.

Schon, D. A. (1983). *The reflective practitioner: How professionals think in action.* New York, NY: Basic Books.

Schon, D. A. (1987). *Educating the reflective practitioner.* San Francisco, CA: Jossey-Bass, Inc.

Schwartz, R., Liddle, H., & Breunlin, D. (1988). Muddles in live supervision. In H. Liddle, D. Breunlin, & R. Schwartz (Eds.), *Handbook of family therapy training and supervision* (pp. 183–193). New York, NY: Guilford Press.

Schwartzman, J. E. (1985). *Families and other systems: The macrosystemic context of family therapy.* New York, NY: Guilford Press.

Silverthorne, B. C., Bartle-Haring, S., Meyer, K., & Toviessi, P. (2009). Does live supervision make a difference? A multilevel analysis. *Journal of Marital and Family Therapy, 35*(4), 406–414.

Stewart, K. (1997). Case consultation: Stories told about stories. In T. Todd & C. Storm (Eds.). *The complete systemic supervisor: Context philosophy and pragmatics* (pp. 308–318). Boston, MA: Allyn & Bacon.

Storm, C. (1997). Back to the future: A review through time. In T. Todd & C. Storm (Eds.), *The complete systemic supervisor: Context, philosophy, and pragmatics.* Boston, MA: Allyn and Bacon.

Storm, C. (2014). Guidelines for selecting and using systemic supervision methods. In T. Todd & C. Storm (Eds.), *The complete systemic supervisor: Context, philosophy and pragmatics* (2nd ed.; pp. 357–380). John Wiley & Sons.

Storm, C. L., Todd, T. C., Sprenkle, D. H., & Morgan, M. M. (2001). Gaps between MFT supervision assumptions and common practice: Suggested best practices. *Journal of Marital and Family Therapy, 27*(2), 227–239.

Sue, S. (1998). In search of cultural competence in psychotherapy and counseling. *American Psychologist, 53*(4), 440.

Todd, T. C., & Storm, C. L. (1997). *The complete systemic supervisor: Context, philosophy, and pragmatics.* Boston, MA: Allyn and Bacon.

Trolley, B., & Silliker, A. (2005). The use of WebCT in the supervision of counseling interns. *Journal of Technology in Counseling, 4*(1). Retrieved from http://jtc.colstate.edu/Vol4_1/Trolley/Trolley.htm

Vaccaro, N., & Lambie, G. W. (2007). Computer-based counselor-in-training supervision: Ethical and practical implications for counselor educators and supervisors. *Counselor Education and Supervision, 47*(1), 46–57.

Vespia, K. M., Heckman-Stone, C., & Delworth, U. (2002). Describing and facilitating effective supervision behavior in counseling supervisees. *Psychotherapy: Theory, Research, Practice, Training, 39*(1), 56.

Watson, J. C. (2003). Computer-based supervision: Implementing computer technology into the delivery of counseling supervision. *Journal of Technology in Counseling, 3*(1), 1–13.

Welfel, R. E. (1998). *Ethics in counseling and psychotherapy: Standards, research, and emerging issues.* Pacific Grove, CA: Brooks/Cole.

Wilcoxon, S. A., Remley, T. P., Jr., Gladding, S. T., & Huber, C. H. (2007). *Ethical, legal, and professional issues in the practice of marriage and family therapy* (4th ed.). Columbus, OH: Merrill Prentice Hall.

Wright, L. M. (1986). An analysis of live supervision "phone-ins" in family therapy. *Journal of Marital and Family Therapy, 12*(2), 187–190.

FOUR

Tailoring Supervision to Supervisees' Developmental Level

Glenn W. Lambie
Ashley J. Blount

Working in the helping professions (e.g., psychology; marriage, couples, and family therapy; counseling; social work) involves a unique human factor: Individuals in helping professions work with real clientele. Individuals training to become couples, marriage, and family therapists are afforded opportunities to acquire knowledge relating to helping their future clients and applying their acquired knowledge in their practices. As a result, these therapists-in-training need to gain the abilities to apply their knowledge and skills in therapeutic settings with diverse clients. Thus, supervisors of therapists-in-training are faced with the task of imparting information that helps their supervisees learn and develop effective therapeutic skills as well as promoting supervisees' confidence in their ability to be effective couple, marriage, and family therapists.

Couples, marriage, and family therapy (CMFT) "supervision is the process of evaluating, training, and providing oversight to trainees using relational or systemic approaches for the purpose of helping them attain systemic clinical skills" (American Association for Marriage and Family Therapy [AAMFT], 2014, p. 5). In addition, supervision (a) promotes therapists' skill acquisition and development of counseling competencies, (b) socializes novice therapists into their professional identity and ethical practice, (c) protects client welfare, and (d) monitors supervisees' readiness to enter into the counseling profession (Bernard & Goodyear, 2014; Lambie & Sias, 2009). Therefore, a primary focus of effective supervision is to support therapists' continuous development of therapeutic skills and professional dispositions and behaviors, necessitating that supervisors match their supervisees' developmental needs to increase their effectiveness with diverse clients.

Supervisors employ different methods and theories of supervision (e.g., Bernard, 1997; Patterson, 1983; Rønnestad & Skovholt, 2003; Stoltenberg & McNeill, 2010) to enhance supervisees' counseling competence and to evaluate supervisees' therapeutic skills (American Counseling Association [ACA], 2014; Bernard & Goodyear, 2014). In addition, the ACA (2014) *Code of Ethics* states that supervision involves a process of monitoring "client welfare and supervisee clinical performance and professional development" (Standard F.1.a). Furthermore, the Association for Counselor Education and Supervision (ACES, 2011) *Standards for Best Practices Guidelines* notes 12 supervision best practices:

(a) initiating supervision (e.g., supervisory contract);
(b) goal-setting (e.g., collaborative development of specific goals for supervision);
(c) giving feedback (e.g., balance of challenging and supportive feedback);
(d) conducting supervision (e.g., provides a safe, supportive, and structured supervisory climate);
(e) the supervisory relationship (e.g., supervisory relationship is key to effective supervision);
(f) diversity and advocacy considerations (e.g., supervisors infuse multicultural considerations into their supervision);
(g) ethical considerations (e.g., supervisor provides supervisees with a professional disclosure statement);
(h) documentation (e.g., supervisors document supervision sessions to protect client welfare);
(i) evaluation (e.g., supervisor communicate all supervisory evaluation processes);
(j) supervision format (e.g., supervisors employ various supervision formats, addressing supervisees' needs);
(k) the supervisor (e.g., the supervisor is competent in providing clinical supervision); and
(l) supervision preparation, supervision training, and supervision of supervision (e.g., supervisors' training is based on developmental perspective).

Therefore, therapist associations and accrediting bodies (e.g., AAMFT, 2014; ACA, 2014; ACES, 2011; CACREP, 2009) identify the significance of supervision as a tool to promote supervisees' therapeutic skills, knowledge, and professional behavior development.

DEVELOPMENTAL SUPERVISION

Developmental supervision is a supervisory approach in which supervisors attempt to facilitate the personal and professional growth of their supervisees (Blocher, 1983). Likewise, supervision is a primary component of therapists' training and involves a relationship in which an experienced therapist (e.g., supervisor) aids in the development of therapeutic competencies in a less experienced therapist-in-training (Loganbill, Hardy, & Delworth, 1982). In order to foster supervisees' professional development, supervisors attempt to enhance their supervisees' professional functioning in a supportive and educational environment (Bernard & Goodyear, 2014). Specifically, developmental supervision is "psychological education in the fullest and most complete sense of the term" (Blocher, 1983, p. 28). Hence, supervisees need different supervisory environments at diverse developmental levels, requiring supervisors to read, recognize, and respond to their supervisees' changing developmental needs, adapting the supervisory environment to best match—and mismatch—supervisees' needs (Bernard & Goodyear, 2014; Lambie & Sias, 2009).

For the purpose of this chapter, we integrated theoretical tenets from multiple developmental models of supervision (Blocher, 1983; Carlson & Lambie, 2012; Lambie & Sias, 2009; Loganbill et al., 1982; Rønnestad & Skovholt, 2003; Stoltenberg, 1981; Stoltenberg & McNeill, 2010). The five primary tenets of developmental supervision are that (a) supervisees move through a series of developmental levels that are qualitatively different from one another (e.g., Stoltenberg & McNeill, 2010); (b) supervisees' developmental levels each require a qualitatively different supervisory environment to promote growth (e.g., Blocher, 1983; Stoltenberg, 1981); (c) supervisors first match their supervisees' developmental level and then mismatch (support and challenge) their developmental level to foster growth (e.g., Lambie & Sias, 2009; Loganbill et al., 1982); (d) higher levels of therapists' development equate to increased therapist functionality with diverse client configurations (e.g., Lambie & Sias, 2009; Rønnestad & Skovholt, 2003); and (e) effective developmental supervision involves personal salience and emotional engagement for the supervisees (e.g., Lambie & Sias, 2009).

Qualitative Different Levels of Supervisee Development

Supervisees at different levels of development tend to exhibit diverse qualities (Stoltenberg & McNeill, 2010). Specifically, new therapists (*Introductory Therapists*: initial experience as a therapist with clients) are anxious and motivated, are highly dependent on their supervisors, and desire concrete instruction. In addition, Introductory Therapists have limited confidence in their therapeutic abilities and seek approval from their supervisors. As an example, Introductory Therapists might seek out information from supervisors and seek approval at higher rates than more advanced therapists-in-training. In the same sense, Introductory Therapists might *not* ask for help or additional information due to lack of awareness (i.e., not realizing they need help), lack of confidence (i.e., not knowing how to approach the supervisor), or lack of self-efficacy (i.e., not wanting to appear unknowledgeable).

Intermediate Therapists, in contrast, have some experience working with clients; however, their clinical experience is limited (therapist interns). At the second development level, therapists are transitioning from being dependent on their supervisor to becoming more independent practitioners. They struggle to feel confident in working with some clients while feeling secure in working with others. Intermediate Therapists have developed basic counseling competencies (e.g., reflection of content); however, they struggle with more advanced therapist competencies such as case conceptualization and reflection of meaning.

At the third developmental level (*Proficient Therapist*; new practitioner), the therapists are more autonomous clinicians, yet know when they should consult and seek supervision. Proficient Therapists have an increased ability to integrate multiple perspectives regarding their work with clients, resulting in increased case conceptualization. In addition, Proficient Therapists have self-awareness of their limitations as therapists and have the confidence to ask for help or consultation when needed. Essentially, the Proficient Therapist worries less about how he or she is perceived by supervisors and feels more comfortable discussing areas of concern in a collegial manner.

At the fourth developmental level (*Advanced Therapist*; therapist with significant supervised therapeutic experience), therapists are accepting of their strengths and limitations as clinicians. Therapists at this developmental level are open to diverse therapeutic perspectives, possessing the ability and confidence to work with diverse client configurations and presenting problems.

Balance of Challenge and Support

Developmental supervision involves a dance between support and challenge. If supervisees are to learn and grow, they must be both challenged and supported (Lambie & Sias, 2009). Developmental supervisors challenge and support their supervisees so that the supervisees progress to higher developmental levels. Specifically, developmental supervisors facilitate a supervisory environment that first matches the supervisees' needs (supportive supervisory relationship) and then challenges the supervisees' current understandings and skills, promoting disequilibrium and fostering an accommodative response, which leads to developmental growth of the therapist (Lambie & Sias, 2009). In addition, challenging Introductory and Intermediate Therapists may promote learning by providing a new experience (e.g., therapy with diverse client configurations; Kolb, 1984). Nevertheless, supervisors need to be cognizant to provide enough support for their supervisees' developmental growth rather than facilitating a damaging and "miseducative" (Dewey, 1938/1973, p. 25) supervisory experience.

According to Lambie and Sias (2009), supervisors need to continuously assess supervisees' needs for structure, challenge, and support. Following assessment, the supervisory environment can be changed accordingly, so as to meet the therapists-in-training where they are and match their developmental need. An example of providing consistent support and structure is offering a weekly supervisory e-mail to provide immediate support for supervisees (Lambie & Sias, 2009). In order to provide challenge, supervisors attempt to help supervisees gain experience through exploration of personal values and discussions related to the counseling and supervision process (Lambie & Sias, 2009).

Therapists at Higher Developmental Levels Are More Effective With Diverse Clients

A central belief in developmental supervision is that therapists functioning at higher levels of therapeutic maturity are more effective clinicians with diverse client populations (Blocher, 1983; Carlson & Lambie, 2012; Lambie & Sias, 2009; Loganbill et al., 1982; Rønnestad & Skovholt, 2003; Stoltenberg & McNeill, 2010). Therefore, a primary goal of developmental supervision is to promote supervisee growth across the different developmental levels. This means that Introductory Therapists are *not* as effective in working with different therapeutic presenting problems as Intermediate Therapists; thus, the task of the developmental supervisors is to move their supervisees along the supervisee developmental continuum. Two ways in which supervisors can promote supervisee growth and development are by supporting personal salience and keeping supervision emotionally engaging for supervisees.

Personal Salience and Emotional Engagement

Engagement in supervision involves an agreement between supervisors and their supervisees in the supervision process (Bernard & Goodyear, 2014). Specifically, developmental supervision requires both cognitive and affective engagement to support supervisees' growth. "Supervisors need to be both engaging educators and facilitators; supervision is not just a cognitive exercise" (Lambie & Sias, 2009, p. 353). For example, when a supervisor and supervisee are discussing issues regarding domestic violence within a couple, the supervisor fosters both the cognitive and affective engagement of the supervisee by facilitating a dialogue about the supervisee's thoughts and feelings related to domestic violence and the implication of these feelings and thoughts in the supervisee's work with the couple.

Anxiety may influence supervisees' engagement in supervision. Moreover, supervisees' anxiety relates to their work with clients, their interactions with their supervisors, and their openness to supervisory feedback (Ladany, Hill, Corbett, & Nutt, 1996; Mehr, Ladany, & Caskie, 2010). In addition, anxious supervisees tend to discuss more positive interactions with clients, areas they feel confident in, and things they are doing well (Rønnestad & Skovholt, 2003). As a result, a task of developmental supervisors is to facilitate a safe supervisory alliance with their supervisees, mitigating their anxiety and increasing the effectiveness of supervision (Falender & Shafranske, 2004).

To promote supervisees' development, supervision needs to be personally relevant to the supervisees (Lambie & Sias, 2009). Therefore, developmental supervisors maintain a *supervisee-centered style* (Lambie & Sias, 2009, p. 353) in their work with therapists-in-training, while monitoring the therapists' ethical practice and client welfare (ACA, 2014). Supervisors should tailor their supervision structure to meet their supervisees' needs. For example, supervisors may begin supervision by "checking in" with their supervisees about their current feelings, and then adapt the supervision activities to the supervisees' pertinent concerns.

Supervisory Approaches for Supervisees at Different Developmental Levels

As supervisees at different levels of development (Introductory, Intermediate, Proficient, and Advanced Therapists) exhibit distinct qualities (anxious and rigid to confident and flexible), supervisors need to tailor the supervisory environment and strategies to the supervisees' changing needs. Therefore, supervisors assess a supervisee's developmental phase through active inquiry. The assessment of supervisees' developmental level may be attained through active listening, reflection, questioning, and challenging incongruencies (Young et al., 2011). For example, supervisors may inquire how their supervisees feel in working with specific clients: supervisees at lower levels of development will present with high anxiety and a need for concrete direction from their supervisor, whereas more advanced supervisees will display more confidence in their ability to work with the clients and increased acceptance of the ambiguity within therapeutic processes.

Additionally, supervisors may assess their supervisees' developmental function through measures that rate therapists' levels of competencies in diverse areas. Specifically, the *Counseling Competencies Scale* (CCS; Swank, Lambie, & Witta, 2012) and the *Counseling Competencies Scale—Revised* (CCS-R; Lambie, Mullen, Swank, & Blount, in review) assesses therapists' competencies in the domains of counseling skills and therapeutic conditions and professional dispositions and behaviors. Assessing supervisees' development and the importance of matching supervisees' developmental needs is well established (Falender & Shafranske, 2004; Stoltenberg & McNeill, 2010). See Figure 4.1 for a sample of the CCS-R.

Introductory Therapists

New therapists (Introductory Therapists) are anxious, dependent on their supervisors, and have limited confidence in their therapeutic abilities. Therefore, supervisors working with supervisees at this level need to facilitate a safe supervisory environment that alleviates the supervisees' anxiety. To mitigate Introductory Therapists' anxiety, supervisors should employ their therapeutic skills (e.g., active listening, empathy), fostering a supervisory alliance of trust and mutual respect. This process of facilitating a strong supervisor-supervisee relationship is similar to how therapists facilitate a strong therapeutic alliance with clients. In addition, supervisors should normalize their supervisees' anxiety as a normal reaction to a new experience within an evaluative context.

Introductory Therapists need structure from their supervisor (Rønnestad & Skovholt, 2003; Stoltenberg & McNeill, 2010). Therefore, the supervisor should encourage supervisees to be more autonomous while providing support and

Structure, support, normalize

Part I: Counseling Skills and Therapeutic Conditions

#	SCORE	PRIMARY COUNSELING SKILL(S)	SPECIFIC COUNSELING SKILLS AND THERAPEUTIC CONDITIONS DESCRIPTORS	EXCEEDS EXPECTATIONS / DEMONSTRATES COMPETENCIES (5)	MEETS EXPECTATIONS / DEMONSTRATES COMPETENCIES (4)	NEAR EXPECTATIONS / DEVELOPING TOWARD COMPETENCIES (3)	BELOW EXPECTATIONS / UNACCEPTABLE (2)	HARMFUL (1)
1.A		**Nonverbal Skills**	**Includes Body Position, Eye Contact, Posture, Distance From Client, Voice Tone, Rate of Speech, Use of Silence, etc.** (*attuned to the emotional state and cultural norms of the clients*)	Demonstrates effective nonverbal communication skills, conveying connectedness and empathy (85%).	Demonstrates effective nonverbal communication skills for the majority of counseling sessions (70%).	Demonstrates inconsistency in his or her nonverbal communication skills.	Demonstrates limited nonverbal communication skills.	Demonstrates poor nonverbal communication skills, such as ignores client and/or gives judgmental looks.
1.B		**Encouragers**	**Includes Minimal Encouragers and Door Openers such as "Tell me more about . . .," "Hmm"**	Demonstrates appropriate use of encouragers, which supports development of a therapeutic relationship (85%).	Demonstrates appropriate use of encouragers for the majority of counseling sessions, which supports development of a therapeutic relationship (70%).	Demonstrates inconsistency in his or her use of appropriate encouragers.	Demonstrates limited ability to use appropriate encouragers.	Demonstrates poor ability to use appropriate encouragers, such as using skills in a judgmental manner.
1.C		**Questions**	**Use of Appropriate Open and Closed Questioning** (*e.g., avoidance of double questions*)	Demonstrates appropriate use of open and close-ended questions, with an emphasis on open-ended question (85%).	Demonstrates appropriate use of open and close-ended questions for the majority of counseling sessions (70%).	Demonstrates inconsistency in using open-ended questions and may use closed questions for prolonged periods.	Demonstrates limited ability to use open-ended questions with restricted effectiveness.	Demonstrates poor ability to use open-ended questions, such as questions tend to confuse clients or restrict the counseling process.
1.D		**Reflecting$_a$** ***Paraphrasing***	**Basic Reflection of Content–Paraphrasing** (*With couples and families, paraphrasing the different clients' multiple perspectives*)	Demonstrates appropriate use of paraphrasing as a primary therapeutic approach (85%).	Demonstrates appropriate use of paraphrasing (majority of counseling sessions; 70%).	Demonstrates paraphrasing inconsistently and inaccurately or mechanical or parroted responses.	Demonstrates limited proficiency in paraphrasing or is often inaccurate.	Demonstrates poor ability to paraphrase, such as being judgmental and/or dismissive.

FIGURE 4.1 Counseling Competencies Scale-Revised (CCS-R). (*continued*)

#	SCORE	PRIMARY COUNSELING SKILL(S)	SPECIFIC COUNSELING SKILLS AND THERAPEUTIC CONDITIONS DESCRIPTORS	EXCEEDS EXPECTATIONS/ DEMONSTRATES COMPETENCIES (5)	MEETS EXPECTATIONS/ DEMONSTRATES COMPETENCIES (4)	NEAR EXPECTATIONS/ DEVELOPING TOWARD COMPETENCIES (3)	BELOW EXPECTATIONS/ UNACCEPTABLE (2)	HARMFUL (1)
1.E		**Reflecting$_b$** ***Reflection of Feelings***	**Reflection of Feelings** (*With couples and families, reflection of each clients' feelings*)	Demonstrates appropriate use of reflection of feelings as a primary approach (85%).	Demonstrates appropriate use of reflection of feelings (majority of counseling sessions; 70%).	Demonstrates reflection of feelings inconsistently and is *not* matching the client.	Demonstrates limited proficiency in reflecting feelings and/or is often inaccurate.	Demonstrates poor ability to reflective feelings, such as being judgmental and/ or dismissive.
1.F		**Reflecting$_c$** ***Summarizing***	**Summarizing Content, Feelings, Behaviors, and Future Plans** (*With couples and families, summarizing relational patterns of interaction*)	Demonstrates consistent ability to use summarization to include content, feelings, behaviors, and future plans (85%).	Demonstrates ability to appropriately use summarization to include content, feelings, behaviors, and future plans (majority of counseling sessions; 70%).	Demonstrates inconsistent and inaccurate ability to use summarization.	Demonstrates limited ability to use summarization (e.g., summary suggests counselor did *not* understand clients or is overly focused on content rather than process).	Demonstrates poor ability to summarize, such as being judgmental and/ or dismissive.
1.G		**Advanced Reflection** **(*Meaning*)**	**Advanced Reflection of Meaning, Including Values and Core Beliefs** (*Taking counseling to a deeper level*)	Demonstrates consistent use of advanced reflection and promotes discussions of greater depth during counseling sessions (85%).	Demonstrates ability to appropriately use advanced reflection, supporting increased exploration in session (majority of counseling sessions; 70%).	Demonstrates inconsistent and inaccurate ability to use advanced reflection. Counseling sessions appear superficial.	Demonstrates limited ability to use advanced reflection and/or switches topics in counseling often.	Demonstrates poor ability to use advance reflection, such as being judgmental and/or dismissive.
1.H		**Confrontation**	**Counselor Challenges Clients to Recognize and Evaluate Inconsistencies**	Demonstrates the ability to challenge clients through verbalizing inconsistencies and discrepancies in the clients' words and/or actions in a supportive fashion. Balance of challenge and support (85%).	Demonstrates the ability to challenge clients through verbalizing inconsistencies and discrepancies in the clients' words and/or actions in a supportive fashion (can confront, but hesitant) or was *not* needed; therefore, appropriately *not* used (majority of counseling sessions; 70%).	Demonstrates inconsistent ability to challenge clients through verbalizing inconsistencies and discrepancies in clients' words and/or actions in a supportive fashion. Used minimally/ missed opportunity.	Demonstrates limited ability to challenge clients through verbalizing discrepancies in the client's words and/or actions in a supportive and caring fashion, and/or skill is lacking.	Demonstrates poor ability to use confrontation, such as degrading client, harsh, judgmental, and/ or aggressive.

FIGURE 4.1 Counseling Competencies Scale-Revised (CCS-R). (*continued*)

1.I	**Goal Setting**	**Counselor Collaborates With Clients to Establish Realistic, Appropriate, and Attainable Therapeutic Goals** (*With couples and families, goal setting supports clients in establishing common therapeutic goals*)	Demonstrates consistent ability to establish collaborative and appropriate therapeutic goals with clients (85%).	Demonstrates ability to establish collaborative and appropriate therapeutic goals with client (majority of counseling sessions; 70%).	Demonstrates inconsistent ability to establish collaborative and appropriate therapeutic goals with clients.	Demonstrates limited ability to establish collaborative, appropriate therapeutic goals with clients.	Demonstrates poor ability to develop collaborative therapeutic goals, such as identifying unattainable goals, and agreeing with goals that may be harmful to the clients.
1.J	**Focus of Counseling**	**Counselor Focuses (or Refocuses) Clients on Their Therapeutic Goals** (i.e., *purposeful counseling*)	Demonstrates consistent ability to focus and/or refocus counseling on clients' goal attainment (85%).	Demonstrates ability to focus and/or refocus counseling on clients' goal attainment (majority of counseling sessions; 70%).	Demonstrates inconsistent ability to focus and/or refocus counseling on clients' therapeutic goal attainment.	Demonstrates limited ability to focus and/ or refocus counseling on clients' therapeutic goal attainment.	Demonstrates poor ability to maintain focus in counseling, such as counseling moves focus away from clients' goals.
1.K	**Facilitate Therapeutic Environment$_a$:** ***Empathy and Caring***	**Expresses Accurate Empathy and Care. Counselor Is "Present" and Open to Clients** (*Includes immediacy and concreteness*)	Demonstrates consistent ability to be empathic and uses appropriate responses (85%).	Demonstrates ability to be empathic and uses appropriate responses (majority of counseling sessions; 70%).	Demonstrates inconsistent ability to be empathic and/ or use appropriate responses.	Demonstrates limited ability to be empathic and/or uses appropriate responses.	Demonstrates poor ability to be empathic and caring, such as creating an unsafe space for clients.
1.L	**Facilitate Therapeutic Environment$_b$:** ***Respect and Compassion***	**Counselor Expresses Appropriate Respect and Compassion for Clients**	Demonstrates consistent ability to be respectful, accepting, and compassionate with clients (85%).	Demonstrates ability to be respectful, accepting, and compassionate with clients (majority of counseling sessions; 70%).	Demonstrates inconsistent ability to be respectful, accepting, and compassionate with clients.	Demonstrates limited ability to be respectful, accepting, and/ or compassionate with clients.	Demonstrates poor ability to be respectful and compassionate with clients, such as having conditional respect.

_______: **Total Score** (***out of a possible 60 points***)

FIGURE 4.1 Counseling Competencies Scale-Revised (CCS-R). (*continued*)

structure in supervision (e.g., Lambie & Sias, 2009). Supervisory structure may come in the form of supervisory handbook and contracts, which delineate *all* supervisory processes and expectations, supporting Introductory Therapists' need for concrete, consistent, and tangible supervisory guidance.

Intermediate Therapists (Interns)

Intermediate Therapists "are in a state of confusion because they yearn for more freedom but still feel dependent on the supervisor's direction" (Carlson & Lambie, 2012, p. 33). In addition, Intermediate Therapists fluctuate between having some confidence in their therapeutic abilities and displaying self-doubt and anxiety. In other words, the Intermediate Therapist level is akin to adolescence, where the supervisee desires autonomy and at the same time the security of leaning on his or her supervisor. Consequentially, supervisors should balance challenging their supervisees to identify their own solutions for therapeutic issues and providing support in reassuring the supervisees that they are able to be effective clinicians. Supervisors supervising Intermediate Therapists should "provide less structure in supervision . . . and focus more on interpersonal interactions of the supervisee and the client, as well as supervisor and supervisee, through the use of process-oriented techniques, such as the parallel process" (Carlson & Lambie, 2012, p. 33). Thus, Intermediate Therapist supervisees begin to work through more complex issues of personal development and parallel processes or replicate the therapeutic relationship in a variety of settings (i.e., counseling, teaching, supervision; Ekstein & Wallerstein, 1972).

Proficient Therapists (New Professionals)

Proficient Therapists have increased therapeutic insight, self-confidence, and self-awareness (Stoltenberg & McNeill, 2010). Proficient Therapists demonstrate sound therapeutic decision making and understand their limitations. Therefore, supervisors supervising Proficient Therapists take a collegial approach, offering consultation regarding presenting clients' challenges. In addition, "active listening and appropriate self-disclosure can be effective supervisory interventions" (Carlson & Lambie, 2012, p. 34) with a Proficient Therapist. Nevertheless, when Proficient Therapists are confronted with new therapeutic experiences and crisis situations (e.g., a suicidal client), they necessitate more "structured, directive, and supportive supervision" (Lambie & Sias, 2009, p. 352). Supervisees may recycle through previous developmental levels as counseling experiences become more complicated or multifaceted.

Advanced Therapists

Advanced Therapists possess awareness and acceptance of their strengths and limitations as clinicians. They are flexible and open to multiple perspectives in working with clients, making them potentially strong supervisors if given appropriate training in supervision. Advanced Therapists necessitate supervision that is egalitarian and collegial, aligning to the consultant supervisory role (Bernard, 1997). Further, therapists in this developmental stage exhibit advanced multicultural perspectives and have the ability to work with diverse client case conceptualizations. Table 4.1 presents descriptors of the supervisees at the four developmental levels, along with supervisory strategies tailored to match the supervisees' needs.

Supervisor Roles and Matching

Supervisors often change their supervisory style throughout the developmental supervision process. Supervisory role changes are made to match specific supervisee need. Some of the roles a supervisor might take on include those of a counselor, a teacher (psychoeducational role), a consultant, or a monitor (Bernard, 1979, 1997; Bernard & Goodyear, 2014; Blocher, 1983). For instance, supervisors might take on the counselor role if their supervisees are having issues working with clients because their clients remind them of a sibling or a friend. In addition, a counselor role is appropriate if supervisees are having a difficult time maintaining their personal wellness because of outside influences (e.g., illness, relationship issues). Thus, the supervisor takes on the counselor role to help facilitate supervisee growth and to match the supervisee's need. An example of supervisors in the teaching role is when supervisees require knowledge and training in specific therapeutic interventions in order to serve their clients (teaching supervisees about enactments and demonstrating the intervention). Similarly, if a supervisee has the knowledge in a certain situation but lacks the case conceptualization skills to look at clients' concerns from a different perspective, the supervisor may take on a consultant role and facilitate a discussion with the supervisee to promote increased awareness. Finally, supervisees may simply want support or to be observed during a session; so, in order to match supervisees developmentally, the supervisor could play a monitor role (Blocher, 1983). Thus, supervisors tailoring their supervisory styles/roles to meet supervisee developmental needs is a primary component in developmental supervision.

Brittney and Dr. Bob have been working together for the past 3 months and have established a strong supervisory alliance. Brittney trusts Dr. Bob's judgment and feels supported through his supervisory style. When Brittney brings information about the family she is working with into her individual supervision session, she appears to have some anxiety about the direction in which to take the family in the therapeutic sessions. In addition, Brittney appears frustrated by the fact that she has worked with numerous families before but is struggling to connect with this new family. In addition, Brittney discloses that she is having a difficult time with the parents (Travis and Tekia) because she tends to side with Travis, agreeing with his side of the story.

TABLE 4.1 Application of Developmental Supervision per Supervisee Level

	SUPERVISEE DESCRIPTORS	SUPERVISORY ENVIRONMENT	SUPERVISORY INTERVENTIONS	SUPERVISOR CONSIDERATIONS
Introductory Therapists	High anxiety Low self-confidence Low autonomy	Supportive Encouraging Structured Prescriptive	Supervisor as directive educator (e.g., teaching supervisees about enactments) Live supervision with immediate feedback Modeling desired therapist behaviors Role-plays	Provide concrete expectation of supervisees (e.g., supervisor contract) Introduce supervision style Assess supervisees' level of anxiety
Intermediate Therapists (Interns)	Some anxiety Some self-confidence Desire autonomy and security from supervisor	Supportive Encouraging Reflective Insight-oriented	Supervisor as both supportive and challenge counselor (e.g., discussion of family-of-origin on current work with clients) Interpersonal Process Recall Discussion of issues of transference and countertransference	Review concrete expectations for supervisees (e.g., internship handbook) Maintain process-oriented approach Focus of supervision is on supervisee
Proficient Therapists (New Professionals)	Minimal anxiety Self-confident Autonomous	Supportive Collegial Reciprocal Consultative	Supervisor as collaborative colleague (e.g., discussion of diverse approach to working with clients with no definitive best method) Active listening Parallel process Consultation	May use self-disclosure Supportive and validating statements Challenge supervisees to consider diverse therapeutic perspectives, increase their case conceptualization
Advanced Therapists	Insightful Acknowledge limitations Self-aware	Supportive Collegial Consultative Discussion-based	Supervisor as collaborative colleague Peer-to-peer conversation Advanced techniques	Challenging statements Offer different viewpoint

CASE EXAMPLE OF DEVELOPMENTAL SUPERVISION

Dr. Bob is a 53-years-old, Caucasian male, who has been supervising for over 20 years. He is a clinical supervisor (credentialed AAMFT Approved Supervisor) and has worked with a wide range of supervisees throughout his career as an adjunct instructor and in his practice. Brittney is a 33-year-old, Caucasian female who is working through her master's degree in MCFT. She is currently completing her internship experience at a local nonprofit agency. Brittney is interested in conducting family therapy and excited about the opportunity to work with families with diverse needs.

The family that Brittany is working with is of African American descent and includes (a) Travis, 29-year-old father; (b) Tekia, 27-year-old mother; (c) Tyrone, 5-year-old son; and (d) Tommy, 4-year-old son. Travis was in the military (as has been every male in his family) and served two tours of duty in Afghanistan. He was wounded when an improvised explosive device (IED) exploded and left him severely injured. He has spent many months in the hospital and rehabilitation. He was discharged from the military (which he had wanted to make his career) as a result of his injuries and now has difficulty finding a job. He currently works two part-time jobs, hoping to find full-time employment. The primary presenting problems for the family include: (a) couple relationship and parenting issues, as Travis reportedly has a "short fuse"; (b) there are *no* reported safety concerns; (c) Travis continuing recovery from his injury; and (d) job issues related to Travis's inability to find full-time employment and the resulting financial strain. In addition, Travis's older brother is still in the military and was recently redeployed to Afghanistan.

Utilizing a developmental supervisory approach, Dr. Bob recognized that Brittney is at the Intermediate Therapist (Intern) developmental stage because she demonstrated some anxiety; she has increased self-confidence but is still self-doubting (as indicated by her frustration that what had worked with families in the past did *not* seem to be working with her new family clients); and possible transference issues (between Brittney and the parents). Dr. Bob begins by allowing Brittney to discuss the case and give her perspectives on the client concerns, an example of maintaining trust and a safe supervisory environment in order to promote confidence in Brittney. Following Brittney's explanation of the clients and her discussion about her frustrations working with the family (being stuck), Dr. Bob challenges Brittney to think about (a) why she might be struggling with this particular family, (b) why she tends to side with Travis rather than Tekia, and (c) what personal areas may be influencing her therapy with the family. Here, Dr. Bob is using intentionality to focus the supervision session on Brittney. By focusing the session on the supervisee, Dr. Bob is encouraging Brittney to explore the "bigger picture" and promoting an insight-oriented discussion.

Brittney evaluates her struggles with the family and decides that her biggest issue is that she feels Tekia should stop being a stay-at-home mother and find a professional position to reduce the family's financial challenges. Brittney's mother worked a full-time job and Brittney feels that a lot of this family's problems could be solved if Tekia pulled her weight and brought in some money. In addition to her opinions about Tekia, Brittney realizes that as the therapist she was making mental excuses for Travis (e.g., he probably needs more time to heal; he shouldn't have to work so hard; of course he has a short fuse, he was in a war) based on personal opinions rather than family-stated concerns. As Dr. Bob asks Brittney about external areas that might be influencing her feelings, she feels embarrassed that she let her own feelings into her work as a therapist. Though she trusts Dr. Bob, she does *not* divulge her embarrassment but, rather, tells him that she has some "stuff" that might have influenced her work with the family. Brittney and Dr. Bob continue to discuss areas that might be influencing her work with the family and are able to come up with multiple ways in which Brittney could help the family achieve their goals.

In addition to facilitating a discussion with Brittney, Dr. Bob uses interpersonal process recall (IPR; Kagan & Kagan, 1997) in order to provide concrete feedback and to bring the supervision session into the here-and-now (i.e., allow for a live discussion while watching Brittney's session). A video recording of Brittney counseling the family is displayed, and Dr. Bob assures Brittney that the purpose of watching the session is to reflect on feelings and thoughts of the therapist (Brittney) and the client (the family) during the session. Dr. Bob and Brittney watch the video, periodically stopping at points so that questions such as "What else could you have said to the family?" "Were you aware of any feelings during this part?" and "Did Tekia remind you of anyone in your life?" could be discussed. Dr. Bob aimed the questions at opening up a conversation about Brittney's internal reactions to her clients (in this case, the family) and external reactions (e.g., techniques used) during the counseling session. Following the use of IPR, Dr. Bob and Brittney process the experience.

During the depicted supervision session, Dr. Bob offered a supportive yet challenging atmosphere for Brittney to explore her relationship with the family: In addition, Dr. Bob appropriately matched Brittney's developmental level by operating from the counselor supervisory role to help Brittney evaluate personal areas that could have been influencing her work as a therapist. Brittney and Dr. Bob were able to have a discussion about difficult topics such as transference, because of the safe, trusting relationship they shared. Subsequently, Dr. Bob used techniques to facilitate development (e.g., insight-oriented discussion, challenging questions) and supported Brittney's autonomy by allowing her to provide information on the family, discuss the case conceptualization, and offer her perspectives on their concerns.

REFERENCES

American Association for Marriage and Family Therapy (AAMFT). (2014). *Approved supervision designation: Standards handbook*. Retrieved from http://www.aamft.org/imis15/Documents/Approved_Supervisor_Handbook_2014.pdf

American Counseling Association (ACA). (2014). *Code of ethics*. Alexandria, VA: Author.

Association for Counselor Education and Supervision (ACES). (2011). *Standards for best practices guidelines*. Retrieved from http://www.acesonline.net/wp-content/uploads/2011/10/ACES-Best-Practices-in-clinical-supervision-document-FINAL.pdf

Bernard, J. M. (1979). Supervisor training: A discrimination model. *Counselor Education & Supervision, 19*, 60–68.

Bernard, J. M. (1997). The discrimination model. In C. E. Watkins, *Handbook of psychotherapy supervision* (pp. 310–327). New York, NY: Wiley.

Bernard, J. M., & Goodyear, R. K. (2014). *Fundamentals of clinical supervision* (5th ed.). Upper Saddle River, NJ: Pearson Education.

Blocher, D. H. (1983). Toward a cognitive developmental approach to counseling supervision. *The Counseling Psychologist, 11*, 27–34.

Carlson, R. G., & Lambie, G. W. (2012). Systemic-developmental supervision: A clinical supervisory approach for family counseling student-interns. *The Family Journal: Counseling and Therapy for Couples and Families, 20*(1), 25–32.

Council for Accreditation of Counseling and Related Educational Programs (CACREP). (2009). *CACREP accreditation standards and procedures manual*. Alexandria, VA: Author.

Dewey, J. (1938). *Experience and education*. New York, NY: Macmillan.

Dewey, J. (1973). *Experience and education*. New York, NY: Macmillan.

Ekstein, R., & Wallerstein, R. S. (1972). *The teaching and learning of psychotherapy* (2nd ed.). New York, NY: International Universities Press.

Falender, C. A., & Shafranske, E. P. (2004). *Clinical supervision: A competency-based approach*. Washington, DC: American Psychological Association.

Kagan, H. K., & Kagan, N. I. (1997). Interpersonal process recall: Influencing human interaction. In C. E. Watkins, Jr. (Ed.), *Handbook of psychotherapy supervision* (pp. 296–309). New York, NY: Wiley.

Kolb, D. (1984). *Experiential learning: Experience as the source of learning and development*. Upper Saddle River, NJ: Prentice-Hall.

Ladany, N., Hill, C. E., Corbett, M. M., & Nutt, E. A. (1996). Nature, extent, and importance of what psychotherapy trainees do not disclose to their supervisors. *Journal of Counseling Psychology, 43*, 10–24.

Lambie, G. W., Mullen, P. R., Swank, J. M., & Blount, A. J. (2015). *The Counseling Competencies Scale: Validation and refinement*. Unpublished manuscript; Department of Child, Family, and Community Sciences; University of Central Florida, Orlando, Florida.

Lambie, G. W., & Sias, S. M. (2009). An integrative psychological developmental model of supervision for professional school counselor-in-training. *Journal of Counseling & Development, 87*, 384–355.

Loganbill, C., Hardy, E., & Delworth, U. (1982). Supervision, a conceptual model. *The Counseling Psychologist, 10*(1), 3–42.

Mehr, K. E., Ladany, N., & Caskie, G. I. L. (2010). Trainee nondisclosure in supervisión: What are they not telling you? *Counselling and Psythotherapy Research, 10,* 103–113.

Patterson, C. H. (1983). A client-centered approach to supervision. *The Counseling Psychologist, 11*(1), 21–25.

Rønnestad, M. H., & Skovholt, T. M. (2003). The journey of the counselor and therapist: Research finding and perspective on professional development. *Journal of Career Development, 30,* 5–44.

Stoltenberg, C. (1981). Approaching supervision from a developmental perspective: The counselor-complexity model. *Journal of Counseling Psychology, 28,* 59–65.

Stoltenberg, C. D., & McNeill, B. W. (2010). *IDM supervision: An integrative developmental model for supervising counselors and therapists* (3rd ed.). New York, NY: Routledge Taylor & Francis Group.

Swank, J. M., Lambie, G. W., & Witta, E. L. (2012). An exploratory investigation of the Counseling Competencies Scale: A measure of counseling skills, dispositions, and behaviors. *Counselor Education and Supervision, 51,* 189–206. doi:10.1002/j.1556-6978.2012.00014.x

Young, T. L., Lambie, G. W., Hutchinson, T., & Thurston-Dyer, J. (2011). The integration of reflectivity in developmental supervision: Implications for clinical supervisors. *The Clinical Supervisor, 30*(1), 1–18.

FIVE

Integrating Common Factors Into MFT Supervision

Eli A. Karam

In this chapter, I encourage those who train marital, couple, and family therapists (MFTs) to integrate established common factors and empirically supported principles of change into supervision strategies. *Common factors* refers to all techniques and therapeutic change mechanisms that transcend various models and are related to successful outcomes. First I will examine both the history and benefits of teaching specific models in Marriage and Family Therapy (MFT) supervision, before arguing for an integrative, moderate common-factors approach. While I do not propose throwing out the classic MFT theories that have historically guided supervision, I do believe that a core focus on single approaches does not reflect the reality of MFT practice. I believe that learning about common factors is an excellent way to bring about a theoretical integration, not otherwise possible when the focus is on competing models.

THE HISTORY OF SUPERVISING USING SPECIFIC MODELS

MFT as a profession originated in the late 1950s and 1960s with a series of rebellious pioneers who rejected the dominant individual-based behavior and psychoanalytic models of the time in favor of something new. These prominent early figures, many from disparate disciplines, banded together as systemic soldiers against the linear mental health establishment to find a common ground, focusing on similar themes like centrality of the family system, the interconnection between people and things, and changing problematic communication. In tracing the evolution of our profession, Jay Lebow points out how this harmony was short lived, stating that "family therapy had an exciting beginning in which a shared vision was emphasized, followed by a period of conflict between models" (Lebow, 2013, p. 15).

In something akin to a civil war-like period in MFT history, the 1970s centered on promoting distinction and brand differentiation among models. This emphasis on "difference" was influenced by the personas of charismatic model developers, each trying to brand what they did as a unique and pure form of family therapy. These psychotherapeutic "rock stars" toured the country, looking for new fans from the worlds of social work, psychiatry, and other related mental health disciplines who could be recruited to become the first generation of MFT students. At this time, empirical evidence was not necessary in the sales pitch, as model popularity relied primarily on word of mouth, emotional appeal, and the powerful live demonstrations of family therapy techniques. Leaders supervised their fervent followers, not on college campuses, but at free-standing institutes throughout the United States in single-school, pure-model approaches. If you were in the market for a family therapy supervisor, you might have felt both excited and confused, as there were many establishments to choose from, each purporting to teach "the absolute truth," including the Philadelphia Guild Guidance Clinic; the Bowen Center for the Study of the Family in Washington, D.C.; the MRI Brief Therapy Center in Palo Alto, California; the Ackerman Institute of New York; and Haley & Madanes's Family Therapy Institute of Washington, D.C., to name just a few of the early meccas of family therapy training. As a student at any of these centers, however, you were only taught one model, the proprietary approach of the model developer. For example, a family therapy trainee at the Philadelphia Guild Guidance Clinic would be trained entirely in Structural Family Therapy, without having any exposure to popular alternatives of the time, like the experiential models of Satir and Whitaker or the transgenerational theories of Bowen or Nagy.

As the profession and practice of family therapy became more mature, training programs were soon situated primarily in university settings. Then, in 1978, the United States Department of Education (USDE) recognized the Commission on Accreditation for Marriage and Family Therapy Education (COAMFTE) as the national accrediting body for MFT. The standards of the COAMFTE focus on grounding clinicians in a systems/relational perspective as a basis for their treatment of client problems. From this time forward in MFT educational history, students were introduced to different classic models and approaches throughout the curriculum. Even though the exposure to a myriad of models was now available, MFT training still tended to focus on highlighting differences, rather than similarities, between the models (Sprenkle, Davis, & Lebow, 2009). Rather than presenting an integrative framework, many training programs encouraged a "choose your favorite model" approach, resulting in a depth-over-breadth mentality in clinical training.

MAKING A CASE FOR SUPERVISING USING COMMON FACTORS ALONGSIDE SPECIFIC MODELS

Believing that it is better to be passionate about *theory* rather than *a* theory, I argue that the traditional "choose your favorite model" approach is not appropriate for the evolution of MFT supervision. This learning style suggests that if you master one model, you are by definition choosing *not* to master another. Instead, I offer the following reasons to include emphasis on the importance of common factors alongside the teaching of specific models.

The Empirical Proof

Specific models and techniques, used alone, have little influence on the outcome of treatment (Shadish, Ragsdale, Glaser, & Montgomery, 1995). In the largest meta-analysis in the MFT field to date, Shadish and colleagues make the following conclusions: (a) marital and family therapy works; (b) the majority of those who participate in marital and/or family therapy are better off than those who do not; (c) the odds are 2 out of 3 that a randomly chosen client who received MFT is better off than a randomly chosen control client at posttest; and (d) no orientation of MFT was shown to be demonstrably superior to any other MFT orientation (Shadish & Baldwin, 2003; Shadish et al., 1995).

Although clinical experience and this meta-analytic research evidence suggest that there is no single model or theory capable of fitting all clients or family systems, many model developers and their followers continue to perpetuate the myth that there is (Sprenkle et al., 2009). This myth could confuse novice therapists, as they may feel pressured early in their training to pick the "right" model, or they may pick an approach without being fully informed about all of its complexities. Although there is clear evidence for the effectiveness of certain MFT approaches, there is not yet strong evidence for the relative effectiveness of the various models as compared to one another. Karam and Sprenkle (2010) explain how learning this information helps students realize that they do not have to prematurely pledge allegiance to the superiority of any one model, even though there may be value in choosing a model that is a good fit for their own worldview (Simon, 2006). Believing in a model (i.e., allegiance) is itself a common factor associated with positive outcome (Wampold, 2001).

The Practical Proof: The Integrative Reality of Many Practicing MFTs and Supervisors

While it is natural that some models will resonate with students more than others, it is unrealistic to believe that the therapists (or supervisors) will stay with one pure model throughout the duration of their careers, especially as they work with a wide range of clientele and presenting problems. Jay Lebow (1997) has labeled this movement—away from relying on a single model in favor of adapting a more integrative style—as the "quiet revolution" in couple and family therapy.

The majority of seasoned therapists do not utilize only one or two approaches in their work, a notion supported by comprehensive studies which conclude that MFTs generally do not professionally practice this way

(Northey, 2002; Orlinsky & Rønnestad, 2005). Therapists-in-training, especially, must experiment with ideas and techniques on their own so that they can discover for themselves what works and makes sense, given their own particular personality and caseload. Understanding this integrative reality, Sprenkle (1999) contends that MFT supervisors in university training settings must be generalists, capable of exposing students to a variety of techniques and theories, instead of supervision specialists who focus on one model exclusively with a supervisee.

Alignment With AAMFT Core Competencies

Whereas MFT training was once known for input-oriented education (e.g., the accumulation of 500 clinical hours as a graduation requirement), in recent years the focus of skill development for therapists-in-training has shifted to output-oriented education by articulating profession-specific core competencies aimed at a wide variety of common assessment and intervention techniques as well as other conceptual, executive, and professional skills. *Core competencies* refers to the basic or minimum skill set that each practitioner should possess in order to provide safe and effective care before graduating from a COAMFTE program (Nelson et al., 2007).

Most MFT educators and supervisors can relate to the experience of watching students struggle in session when trying to apply a pure model to a specific problem or population. Students may also often encounter frustration or diminished self-confidence as they try to implement complex interventions that do not fit with the client system. These AAMFT Core Competencies are not model specific, but instead are designed to help students think critically about both the strengths and limitations of what they are implementing in session and to match the technique(s) used to the client's needs, goal, and values (AAMFT, 2004). However, if you are only proficient in one model—even if you know it extremely well—it is unlikely that you will be able to excel in all of these core areas. The common-factors approach reinforces the fact that some of the skills students typically already feel good about (e.g., ability to "engage each family member" in order to establish the therapeutic alliance) are both strong evidence-based contributors to change and important core competencies for beginners to master. Nelson and colleagues (2007) credit MFT common-factors research for laying the foundation for the core competencies project by giving a better understanding about what really works in therapy.

TASKS OF THE COMMON-FACTORS SUPERVISOR

Supervisors who intend to integrate common factors into the supervisory experience and education have several tasks, possibly the most important of which is to provide education to supervisees about the factors that truly are common to all psychotherapy modalities. In addition, they should educate supervisees about factors unique to MFT; ask questions to stimulate critical thinking by supervisees about the common factors; coach supervisees on repairing alliance ruptures; and help supervisees develop and cultivate their own theories of change based on common factors. Supervisors should also engage in self-reflection on the role of common factors in their own clinical and supervisory work and integrate feedback instruments that capture common elements into therapy and supervision.

Provide Education on Common Factors Inherent in All Psychotherapy Modalities

What common factors should MFT trainers strive to highlight and infuse into supervision? Almost all that has been documented in individual therapy common-factors research could also hold true for systemic therapies. Michael Lambert (1992) originally proposed categories, which were later modified by Hubble, Duncan, and Miller (1999) and then by Wampold (2001), as pertaining to the field of individual psychotherapy. (Although the space limitations of a book chapter like this do not permit an in-depth review, please see Sprenkle and colleagues' publication (2009) for a detailed historical research and clinical perspective on common factors.)

Client Factors

Client factors are ingredients in the life and environment of the client that contribute to change largely separate from the therapeutic process. Examples of these are family support, chance events such as winning the lottery, meeting a new person, or a religious experience. They also include personality or inherent trait characteristics of clients that allow them to gain far more from the process of therapy than other clients. We have all had the privilege and experience of working with clients who are psychologically minded, who take the process of therapy seriously, and who actively work at their goals outside of the therapy room. What these clients bring to the therapy experience is highly instrumental in their change. Another important change factor to consider related to the client is the natural process of change and the stage the client is in related to change (Prochaska & DiClemente, 1992).

Therapist Factors

In general, specific models deemphasize therapist factors even in the face of empirical and anecdotal evidence that some therapists are more effective than others, and that some therapists are certainly more effective with certain client characteristics and with certain presenting problems. In an excellent review, Wampold (2001) suggests that therapist effects are larger than treatment effects and that in many cases the therapist does precious little—but what she or he does is very precious. Change is greatest when the therapist is skillful and provides trust, acceptance, acknowledgment, collaboration, and respect for the client in an environment that supports risk and maximizes safety.

The Therapeutic Alliance

The client-therapist working alliance has been conceptualized in a variety of ways. Bordin (1979) suggests that the alliance is composed of three elements: bonds (the affective quality of the client-therapist relationship, which includes dimensions like trust, caring, and involvement); tasks (the extent to which the client and therapists are both comfortable with the major activities in therapy and to which the client finds them credible); and goals (the extent to which the client and therapist are working toward compatible goals).

Hope and Expectancy

Clients come to therapy because they have lost hope or have a depleted morale. Hope, or helping clients to think differently about their selves and family systems, can help open the space necessary for change to occur. Howard, Moras, Brill, Martinovich, and Lutz (1996) stressed the importance of moving a client from demoralization to remoralization by tapping into hope and positive expectations early in treatment.

Interventions (Behavioral, Cognitive, Affective) That Cut Across All Models

Behavioral interventions essentially consist of changing the "doing." They occur when therapists facilitate clients' changing of interactional patterns or dysfunctional sequences, modifying boundaries and changing family structures, learning new skills, becoming more supportive of each other, and learning to empower self and others. Cognitive interventions change the ways in which clients view problems and their life situations. Cognitive interventions occur when therapists facilitate clients' gaining new perspectives (new meanings) about interactional processes within themselves and the family, between the family and other systems, and across generations. Affective interventions impact emotional experiencing and regulation. In MFT, they occur when therapists facilitate clients' regulation or experiencing of emotions and making emotional connections with themselves, the therapist, and (most importantly) each other (Davis & Piercy, 2007a, 2007b).

Allegiance of the Therapist

As long as it does not keep the therapist from being flexible and responsive to the needs of the client system, allegiance to an approach may bring hope and confidence to the clinical work. You must fully believe in or "buy into" what you are doing before you can credibly "sell" it to a client. In his meta-analytic research, Wampold (2001) presents compelling empirical proof that allegiance effects account for significantly more of the outcome variance in psychotherapy than does the choice of the actual treatment model.

Feedback

In addition to the preceding factors, it has been recently argued that providing organized feedback to clients is yet another common element associated with successful psychotherapy (Halford et al., 2012). Multiple studies document significant improvements in both retention of and outcome from treatment when therapists have access to formal, real-time feedback from clients regarding the process and outcome of therapy (Anker, Duncan, & Sparks, 2009; Harmon et al., 2007). Michael Lambert's research (2010a, 2010b) also demonstrates that utilizing systematic feedback that potentiates common factors (like the therapeutic alliance and other client variables) consistently improves therapy outcomes and can greatly improve treatment effectiveness for clients at risk of treatment failure. Given the importance of these findings for both initial engagement and overall therapeutic outcome, it makes sense to teach MFTs early in their careers to become proficient in receiving and processing feedback with clients around these important common factors, in order to strengthen the therapeutic alliance and keep therapy focused.

Provide Education on Common Factors Unique to MFT

While an understanding of the seven common factors just described is inclusive enough to be applied to all modalities and types of psychotherapy approaches, there are certain common factors associated only with MFT. Specifically, I believe it is important throughout the curriculum to emphasize the following four common factors unique to MFT: (1) conceptualizing difficulties in relational terms, (2) disrupting dysfunctional relational patterns, (3) expanding the direct treatment system, and (4) expanding the therapeutic alliance (Sprenkle & Blow, 2004; Sprenkle et al., 2009).

Conceptualizing Difficulties in Relational Terms

While not denying the role of biology or intrapsychic causation, most MFTs would conceptualize a client problem within a complex web of reciprocal influences. Essentially, the translation of human difficulties into relational problems involves seeing problems through a systemic lens—a core foundation of the field.

Disrupting Dysfunctional Relational Patterns

The flip side of seeing a conflict in relational terms is subsequently intervening while in session to disrupt the dysfunctional cycle. Davis and Piercy (2007a, 2007b) conclude that therapists help clients to both disrupt and later process their respective parts in negative interactional sequences by using a combination of behavioral, cognitive, and affective interventions.

Expanding the Direct Treatment System

Many family therapists tend to push to involve more people than the immediate client(s) directly in treatment (Pinsof, 1995). These include persons not physically present in treatment but who affect the problem in important ways. For example, after doing family-of-origin work with an adult son, the client is ready to work on his relationship with his aging mother. The therapist expands the system to conduct this dyadic session, as the mother moves from the indirect into the direct client system.

Expanding the Therapeutic Alliance

The expanded therapeutic alliance accounts for the complexity of the different interpersonal subsystems relevant to the therapy. For example, in the case of family therapy with two parents and two teen siblings, the therapist forms an alliance not only with each member of the family individually, but also with certain subsystems (e.g., co-parents, children), and with the family as a whole (Pinsof, 1995).

Ask Questions to Stimulate Supervisee Critical Thinking Around the Common Factors

Common-factors supervision is based on the belief that no one theory is always effective when working with a client system. Rather than focusing on mastering a specific model in first-year supervision, the common-factors framework is used primarily to get students comfortable with hypothesizing and conceptualizing about the various client, therapist, and relationship factors at play in a particular case (Karam, 2011). The following series of questions, when incorporated into the regular routine and structure between supervisor and supervisee, will enhance both awareness and integration of common factors in a therapist-in-training's clinical repertoire.

Therapist factors

1. What personal characteristic do you bring into this therapy that you think will benefit your client system?
2. How do you know if your conceptualization of the problem matches that of the understanding of your client system?

3. How did you decide how active or how passive to be in the session?
4. How did you adapt to the client system during the session? What caused your shift?

Client factors

1. What are the client system strengths (self-identified versus other identified) in this therapy?
2. What has been helpful in the past for your client system in dealing with the issue(s) that brought them to therapy?
3. What stage of change is each member of the system in at the onset of therapy (Pre-Contemplation, Contemplation, Preparation, Action, or Maintenance)?
4. What resources exist to benefit the client in both the direct and indirect systems?

Therapeutic alliance factors

1. What are the goals for this therapy?
2. Please explain how the tasks or structure of the therapy is a good fit for the client system.
3. Did you find yourself liking the client system, or were they difficult to tolerate?
4. Do you feel that the client system felt comforted in your presence and safe enough to disclose important information?

Hope/expectancy factors

1. Describe your client sense of hope about this therapy.
2. Describe your own sense of hope about this therapy.
3. How did you convey a sense of hope during the session?
4. What are the client system's expectations about the situation and the therapy?

Feedback factors

1. What feedback did the client system give you?
2. What feedback did you share with the client system?
3. How did you use assessment instruments/tracking instruments to provide feedback?
4. What other sources of information will you need to gather feedback from the indirect system?

Allegiance/coherence factors

1. Is your approach coherent and organized enough to give you confidence that you know what you are doing with this client system?
2. Why do you believe your approach will work with this client system?
3. What modifications, if any, will you need to make to your approach in order to work with this client system?
4. How did you present your approach to instill confidence in the client system?

Conceptualizing difficulties in relational terms (MFT unique factor)

1. What is this client system's problem cycle?
2. How do you see each member in the system contributing to the problem cycle?

Disrupting dysfunctional relational patterns (MFT unique factor)

1. What techniques and interventions have you used to disrupt the client system's dysfunctional relational patterns?

Expanding the direct treatment system (MFT unique factor)

1. Who are the important players/participants in the indirect client system?
2. How do you believe these people will impact the therapy?
3. Who else should be involved directly in the therapy? Why now?

Expanding the therapeutic alliance (MFT unique factor)

1. Describe the alliance you feel you have with each individual member of the client system.
2. Describe the alliance you feel you have with all family members directly involved in the therapy (i.e., the direct system).
3. Describe the alliance you feel you have with the client system's support network not directly involved in therapy (i.e., the indirect system).
4. Describe the alliance you feel you have with our therapeutic support network assisting you with this case (e.g., supervisors, supervision members, co-therapist).

Self-Reflect on the Role of Common Factors in Your Own Clinical Work and Supervisory Style

In order to be a competent, common-factors-based, systemic supervisor, you must first be able to understand and then articulate how this perspective has informed your own clinical work with individuals, couples, and families. Answering all the same questions that you would ask your supervisee from the preceding section about a client system in your current caseload is one way to accomplish this goal. All AAMFT Approved Supervisors are required to develop a mission statement, also known as an *epistemological declaration*, about their personal beliefs regarding the process and content of supervision (Liddle, 1982). Thus, rewriting your own epistemological declaration, highlighting how common factors and generic mechanisms of change impact your work with supervisees, is another way to move toward this style of supervision.

Integrate Feedback Instruments Into Therapy and Supervision to Capture Common Factors

Soliciting constructive feedback, whether from a client system or a supervisor, is an essential skill for all beginning therapists. Whipple and colleagues

(2003) examine the impact of providing therapists with feedback on various client-rated dimensions of the therapy process and recommend clinical support tools to address problems in the therapeutic relationship. As compared with the no-feedback control group, clients in the feedback-plus-clinical-support-tools group attended more sessions, displayed less deterioration, and demonstrated higher recovery rates. Similarly, Lambert, Hansen, and Finch (2001) determined that outcomes improved when therapists were provided with client-focused feedback that alerted the therapists to potential treatment failure with at-risk clients. Due to the influence of the managed care movement (Wampold, 2001), it appears that an increasing number of agency settings that employ MFTs will begin to rely on some kind of feedback mechanism. Educators and supervisors should prepare therapists-in-training for this eventuality—especially since there is evidence that this type of common-factors-based feedback improves their therapy.

For example, MFT researchers/educators at the Family Institute at Northwestern University developed two linked, common-factors-based systems for measuring client change; providing feedback regarding that change to the therapist; and identifying therapist, client, and relationship-level common factors. The Systemic Therapy Inventory of Change (STIC) tracks clinical change through the use of online self-report questionnaires in multiple systemic domains: individual adult, couple, family/household, and child functioning. Lastly, STIC addresses another universal common factor by monitoring the client's experience of the therapeutic alliance. Designed as a companion instrument to the STIC, the Integrative Therapy Session Report (ITSR) is a therapist self-report instrument that provides a clinically meaningful and statistically reliable picture of what occurred in a recently completed session and could provide students and clinical supervisors with data about both therapist common factors and interventions that cut across models (Pinsof et al., 2009).

The Internet-based system through which therapists access their clients' STIC and ITSR data and receive real-time feedback is currently being used in clinical practice and supervision by more than 75 MFT students being trained at The Family Institute at Northwestern as well as a number of faculty and other therapists in surrounding community mental health agencies. The feedback technology provides bar graphs with clients' absolute initial scores and change profiles on clinically relevant scales and subscales over the course of therapy. Supervisors and supervisees may access STIC and ITSR data about their clients at any point during therapy to use either in session or supervision (Pinsof, Goldsmith, & Latta, 2012).

Although access to empirically validated measures and accompanying feedback methodologies may not be feasible for every training program or practicing supervisor, there are other viable and accessible alternatives that can be used to track the therapeutic relationship and other important common factors. As an example, Duncan and Miller (2008) have developed readily available measures of alliance and outcome that can be used to track therapeutic progress (www.scottdmiller.com). These measures are brief, and scoring is quick and easy, making them well-suited for clinical practice.

Using a Feedback Instrument to Track Change Over Time

For this assignment, supervisees are instructed to implement a feedback instrument on five or more occasions in their work with at least two of their cases, one of which should be a couple or family. Supervisees are required to videotape both how they explain the purpose of the instrument to a client system and how they give and receive feedback using the measure. Finally, supervisees write a critical reflection about using the instrument, addressing the following questions: (a) How did the alliance change throughout treatment? (b) How did client responses either confirm or disconfirm your clinical impressions of the therapy? (c) How were you different as a therapist as you used the feedback forms? (d) What surprised you in using these forms? and (e) What are your recommendations for using this type of in-session feedback instrument based upon what you learned through this process?

Common-Factors Feedback Interview

Initiating dialogue with a client system on the process of therapy is another crucial (but often overlooked) skill in the training and development of an MFT. The common-factors feedback interview assignment was designed specifically to develop this skill, focusing on ways for the supervisee to discuss the therapeutic process with a couple or family system. Some of the questions that students are required to ask include: (a) What has been the pivotal moment or most important part of this therapy for you thus far? (b) What about my therapeutic style or approach works well for you? (c) What have we done in this therapy that hasn't worked as well for you? (d) If you feel misunderstood in this therapy or believe I have sided with another family member, how would you address it with me? After conducting the interview, supervisees are instructed to reflect critically on their capability to solicit feedback and how aspects of the experience will be integrated into their emerging clinical repertoire.

Research on therapist behaviors indicates that personal characteristics like warmth, flexibility, and accurate interpretation are positively associated with strong therapeutic bonds (Ackerman & Hilsenroth, 2003). Humor and levity have also been demonstrated to increase the bond with clients who can laugh about their problems with therapists who are comfortable displaying their light-hearted side. Humor may also reduce tension and facilitate emotional release in the therapy room (Dimmer, Carroll, & Wyatt, 1990). Conversely, certain therapist characteristics and behaviors may constrain the ability to form healthy bonds (e.g., rigidity, criticalness, inappropriate self-disclosure; see Ackerman & Hilsenroth, 2001).

Fife, Whiting, Bradford, and Davis (2014) recently introduced the concept of *way of being* to the common-factors literature. Therapists can either be in the moment and genuinely engaged with the humanity of the client system or removed and impersonal when in the therapy room. If young therapists are trying too hard to be professional or appear more experienced than they really are, to make up for their perceived inadequacies, they may lose the ability to remain present, relaxed, and responsive to the needs of the client system. They start to view their clients like case studies in their textbooks, rather than as real people who may have many of the same fears and insecurities as they do.

Way-of-Being Exercise

To address this potential problem and help supervisees gauge their way of being, the supervisor may structure this self-examination by asking students to select segments of videotaped therapy footage where they felt fully present and connected with their clients. The supervisor may also ask the trainees to highlight portions of the session when they were too much in their own heads, overfocused on their anxiety or performance, instead of connected to the client's experience.

In a variation on this exercise, during live supervision, the supervisor instructs the team members behind the mirror to code for vital therapist characteristics like humor, warmth, and empathy. By providing encouraging feedback, group members may help to build supervisee confidence and comfort. If there is corrective feedback to share during or after the live session concerning the student's way of being, the supervisor may be best able to frame the comments in a way that does not threaten or further damage therapist confidence. Both supervisors and group members should always pay close attention to see if way of being changes as a function of the therapy modality. For example, a student therapist who is in a comfort zone while working with individuals may become more reticent or removed from the experience when conducting couple or family therapy.

Coach Supervisees on How to Repair an Alliance Rupture

Although some types of alliance ruptures (i.e., feeling ignored, judged, deceived) are inherent to every psychotherapy modality, MFT faculty and supervisors should prepare their students for specific relational ruptures that occur only in a conjoint therapy setting. These types of split-alliance ruptures may stem from spousal, co-parental, parent-child, sibling, or extended family therapeutic relationships (Pinsof, 1995). For example, a teenager is no longer willing to participate in family therapy after she discovers that her therapist shared with her mother that she secretly snuck out of the house to attend a party in the interim between sessions.

Research supports the clinical belief that a successful tear and repair will strengthen the overall therapeutic alliance by allowing clients to safely explore their relational problems with the therapist in the here-and-now of the therapeutic relationship (Eubanks-Carter, Muran, & Safran, 2010; Pinsof, 1995). Therefore, it is paramount for supervisors to focus on the therapeutic alliance as a potent common factor and coach their supervisees on effective repair strategies. Safran, Muran, Samstag, and Stevens (2001) suggest the following four steps to repair the therapeutic alliance after a rupture: (1) Encourage and validate client emotional reaction to the breach; (2) show appreciation for the client's openness and ability to acknowledge the rupture; (3) explore the impact and consequences of the rupture on the rest of the therapeutic and client system; and (4) acknowledge the therapist's own role in creating the tear. Safran and Muran (1996) contend that if therapists are willing to accept criticism and take responsibility for their actions, then the sharing of negative client feelings can lead to an increased alliance and positive outcomes. A logical place to begin training students on this repair processes would be after a supervisor either hears about an alliance rupture via supervisee case report, or sees a breach in the therapeutic relationship on videotape or during live supervision.

Help Supervisees Cultivate Their Own, Common-Factors–Based Theory of Change

Many MFT programs have required theory-of-change papers or projects, holding steadfastly to the belief that those students who can integrate and explain "why they do what they do" in session will become better therapists (Taibbi, 2007). This project typically is an opportunity for supervisees or students to integrate their understanding of various theories and skills into a personal model that reflects and drives the manner in which they conduct therapy. This model is theory driven and allows young therapists to articulate and demonstrate their ideas about the relationship between theory and intervention and intervention and outcome or change.

Common-Factors Theory of Change Project

Although many of these MFT training programs want their students to link together the students' theories of therapy and their actual practice of therapy, Nelson and Prior's research (2003) demonstrates that supervisors are sometimes deficient in evaluating this congruence. To address this gap and assist supervisees to actually practice what they preach, a common-factors supervisor may modify this traditional theory-of-change assignment. Take, for example, the current assignment built into a supervision course in an MFT training program: As the culminating project prior to ending advanced MFT supervision, students are required to develop an integrative theory-of-change paper and videotaped representation of their work, using the common factors as a template in which to think about their preferred theoretical orientation. This capstone assignment is designed to be a critical evaluation of how the common factors have influenced the student's clinical development (Karam, 2011).

CONCLUSION

What do our supervisees need to know in order to grow into effective therapists, remain clinically vital, and think critically about what really works in therapy? In an attempt to respond to these questions, I have presented a rationale and given specific examples of how common factors could be infused throughout MFT supervision. What I have provided here are some of our suggestions; I readily acknowledge that there may be many other useful supervisory skills and teaching techniques available to promote this type of integration. If instilled early in training and supervision process, however, this common-factors approach may lay a strong foundation for personal integration and critical thinking that will deepen as the therapist matures, keeping both the supervisor and supervisee responsive to change and open to clinical growth and development.

REFERENCES

Ackerman, S. J., & Hilsenroth, M. J. (2003). A review of therapist characteristics and techniques positively impacting the therapeutic alliance. *Clinical Psychology Review, 23*(1), 1–33.

American Association for Marriage and Family Therapy (AAMFT). (2004). *Marriage and therapy core competencies*. Alexandria, VA: Author.

Anker, M., Duncan, B., & Sparks, J. (2009). Using client feedback to improve couple therapy outcomes: A randomized clinical trial in a naturalistic setting. *Journal of Consulting and Clinical Psychology, 77*(4), 693–704. http://dx.doi.org/10.1037/a0016062

Bordin, E. S. (1979). The generalizability of the psychoanalytic concept of the working alliance. *Psychotherapy, 16*, 252–260.

Davis, S. D., & Piercy, F. P. (2007a). What clients of couple therapy model developers and their former students say about change, part I: Model-dependent common factors across three models. *Journal of Marital and Family Therapy, 33*(3), 318–343.

Davis, S. D., & Piercy, F. P. (2007b). What clients of couple therapy model developers and their former students say about change, part II: Model-independent common factors and an integrative framework. *Journal of Marital and Family Therapy, 33*(3), 344–363.

Dimmer, S. A., Carroll, J. L., & Wyatt, G. K. (1990). Uses of humor in psychotherapy. *Psychological Reports, 66*(3), 795–801.

Duncan, B. L., & Miller, S. D. (2008). The outcome and session rating scales: The revised administration and scoring manual, including the child outcome rating scale. Chicago, IL: Institute for the Study of Therapeutic Change.

Fife, S. T., Whiting, J. B., Bradford, K., & Davis, S. (2014). The therapeutic pyramid: A common factors synthesis of techniques, alliance, and way of being. *Journal of Marital and Family Therapy, 40*, 20–33.

Halford, K., Hayes, S., Christensen, A., Lambert, M., Baucom, D., & Atkins, D. (2012). Toward making progress feedback an effective common factor in couple therapy. *Behavior Therapy, 43*, 49–60.

Harmon, S., Lambert, M., Smart, D., Hawkins, E., Nielsen, S., Slade, K., & Lutz, W. (2007). Enhancing outcome for potential treatment failures: Therapist-client feedback and clinical support tools. *Psychotherapy Research, 17*, 379–392.

Howard, K. I., Moras, K., Brill, P. L., Martinovich, Z., & Lutz, W. (1996). Evaluation of psychotherapy: Efficacy, effectiveness, and patient progress. *American Psychologist, 51*(10), 1059–1064.

Hubble, M. A., Duncan, B. L., & Miller, S. (1999). Directing attention to what works. In M. A. Hubble, B. L. Duncan, & S. Miller (Eds.), *The heart and soul of change: What works in therapy*. Washington, DC: American Psychological Association.

Karam, E. A. (2011). Integrating common factors into a MFT curriculum. *Family Therapy Magazine, 10*(5), 32–34.

Karam, E. A., & Sprenkle, D. H. (2010). The research-informed clinician: A guide to training the next-generation MFT. *Journal of Marital and Family Therapy, 36*(3), 307–319.

Lambert, M. J. (1992). Psychotherapy outcome research: Implications for integrative and eclectic therapists. In J. C. Norcross & M. R. Goldfried (Eds.), *Handbook of psychotherapy integration* (pp. 94–129). New York, NY: Basic Books.

Lambert, M. J. (2010a). *Prevention of treatment failure: The use of measuring, monitoring, and feedback in clinical practice.* Washington, DC: American Psychological Association.

Lambert, M. (2010b). Yes, it is time for clinicians to routinely monitor treatment outcome. In B. L. Duncan, S. D. Miller, B. E. Wampold, & M. A. Hubble (Eds.), *The heart and soul of change: Delivering what works in therapy* (2nd ed., pp. 239–268). Washington, DC: American Psychological Association.

Lambert, M. J., Hansen, N. B., & Finch, A. E. (2001). Patient-focused research: Using patient outcome data to enhance treatment effects. *Journal of Consulting and Clinical Psychology, 69*(2), 159–172.

Lebow, J. (1997). The integrative revolution in couple and family therapy. *Family Process, 36*(1), 1–17.

Lebow, J. L. (2013). *Couple and family therapy: An integrative map of the territory.* Washington, DC: American Psychological Association.

Liddle, H. A. (1982). On the problems of eclecticism: A call for epistemologic clarification and human-scale theories. *Family Process, 21*(2), 243–250.

Nelson, T. S., Chenail, R. J., Alexander, J. F., Crane, D. R., Johnson, S. M., & Schwallie, L. (2007). The development of core competencies for the practice of marriage and family therapy. *Journal of Marital and Family Therapy, 33*(4), 417–438.

Nelson, T. S., & Prior, D. (2003). Theory of change projects in MFT programs. *Contemporary Family Therapy, 25*(2), 133–151.

Northey, W. F. (2002). Characteristics and clinical practices of marriage and family therapists: A national survey. *Journal of Marital and Family Therapy, 28*(4), 487–494.

Orlinsky, D. E., & Rønnestad, M. H. (2005). *How psychotherapists develop: A study of therapeutic work and professional growth.* Washington, DC: American Psychological Association.

Pinsof, W. M. (1995). *Integrative problem-centered therapy: A synthesis of family, individual, and biological therapies.* New York, NY: Basic Books.

Pinsof, W. M., Goldsmith, J. Z., & Latta, T. A. (2012). Information technology and feedback research can bridge the scientist–practitioner gap: A couple therapy example. *Couple and Family Psychology: Research and Practice, 1*(4), 253–273.

Pinsof, W. M., Zinbarg, R. E., Lebow, J. L., Knobloch-Fedders, L. M., Durbin, E., Chambers, A., Latta, . . . Friedman, G. (2009). Laying the foundation for progress research in family, couple, and individual therapy: The development and psychometric features of the Initial Systemic Therapy Inventory of Change. *Psychotherapy Research, 19*(2), 143–156.

Prochaska, J. O., & DiClemente, C. C. (1992). The transtheoretical approach. In J. C. Norcross & M. R. Goldfried (Eds.), Handbook of psychotherapy integration (pp. 300–334). New York: Basic Books.

Safran, J. D., & Muran, J. C. (1996). The resolution of ruptures in the therapeutic alliance. *Journal of Consulting and Clinical Psychology, 64*(3), 447–458.

Safran, J. D., Muran, J. C., & Eubanks-Carter, C. (2011). Repairing alliance ruptures. *Psychotherapy, 48*(1), 80–87.

Safran, J. D., Muran, J. C., Samstag, L. W., & Stevens, C. (2001). Repairing alliance ruptures. *Psychotherapy: Theory, Research, Practice, Training, 38*(4), 406.

Shadish, W. R., & Baldwin, S. A. (2003). Meta-analysis of MFT interventions. *Journal of Marital and Family Therapy, 29,* 547–570.

Shadish, W. R., Ragsdale, K., Glaser, R. R., & Montgomery, L. M. (1995). The efficacy and effectiveness of marital and family therapy: A perspective from meta-analysis. *Journal of Marital and Family Therapy, 21,* 345–360.

Simon, G. M. (2006). The heart of the matter: A proposal for placing the self of the therapist at the center of family therapy research and training. *Family Process, 45,* 331–344.

Sprenkle, D. H. (1999). Toward a general model of family therapy supervision: Comment on Roberts, Winek, and Mulgrew. *Contemporary Family Therapy, 21*(3), 309–315.

Sprenkle, D. H., & Blow, A. J. (2004). Common factors and our sacred models. *Journal of Marital and Family Therapy, 30,* 113–129.

Sprenkle, D. H., Davis, S., & Lebow, J. (2009). *Common factors in couple and family therapy: The overlooked foundation for effective practice.* New York, NY: Guilford Press.

Taibbi, R. (2007). *Doing family therapy: Craft and creativity in clinical practice.* New York, NY: Guilford Press.

Wampold, B. E. (2001). The *great psychotherapy debate: Models, methods, and findings.* Mahwah, NJ: Erlbaum.

Whipple, J. L., Lambert, M. J., Vermeersch, D. A., Smart, D. W., Nielsen, S. L., & Hawkins, E. J. (2003). Improving the effects of psychotherapy: The use of early identification of treatment and problem-solving strategies in routine practice. *Journal of Counseling Psychology, 50*(1), 59–68.

SIX

Ethics and Supervision

Benjamin E. Caldwell

Supervision in family therapy is an inherently complex enterprise. In any supervision of psychotherapy work, the supervisor must navigate the hierarchy of supervisor, therapist, and client; in family therapy, the client unit may include its own complex hierarchy as well. In addition, we recognize that both therapy and supervision take place within a social and historical context. Therapists and their supervisors must balance the needs and goals of multiple individuals in a couple or family, and balance the family's interests with those of their community. Ethical decision making in such a complex environment is understandably challenging for therapist and supervisor alike. It is perhaps because of this complexity that family therapy has been at the forefront when it comes to identifying and codifying the ethical responsibilities of supervisors.

The American Association for Marriage and Family Therapy (AAMFT) Code of Ethics (2014a) was among the first to detail specific supervisor responsibilities and remains a leader in this regard (Thomas, 2014). In addition, the AAMFT's specific training requirements for supervisors make it clear that the supervisory role comes with significant responsibility (AAMFT, 2014b).

THE ROLE OF THE SUPERVISOR

There are very different approaches to the supervision relationship. These different conceptualizations are not necessarily better or worse than one another but do represent meaningful differences in how supervisors will interact with their supervisees. The following elements are commonly considered part of the supervisor's role, though different supervisors give them very different levels of emphasis.

Teaching

The very term *teaching* suggests a power imbalance, where the supervisor has more knowledge about some key elements of doing therapy than the supervisee does, and the supervision process becomes a mechanism for instilling that knowledge. Notably, research data are mixed on how MFT supervisees value teaching in supervision. Some have argued that as the developing therapists move out of their degree programs and into their postdegree, prelicense hours, they value teaching less and collegial relationships more. Supervisors are thus encouraged to use a developmental lens in considering how best to supervise, recognizing that the supervisor's own developmental stage in that role can also impact the process (Lee & Nelson, 2013). However, one study of interns finds that the teaching of specific skills is almost always a part of what the interns label as their best supervision experience (Hildebrandt, 2009).

Building Confidence

Supervisees in psychotherapy regularly report that good supervision makes them more confident in their work (Wheeler & Richards, 2007), and authors in MFT have developed specific supervision practices to improve confidence among supervisees (e.g., Senediak, 2013). Although confidence building is a key concern in supervision, it is important that supervisors not inadvertently risk promoting *false confidence* in supervisees. This can lead to therapists acting beyond their scope of competence or willfully ignoring key assessment information. Ideally, supervisees will carry an *appropriate level* of confidence—balancing the assuredness to trust their clinical instincts and act calmly in tense situations with the openness to be corrected by clients and supervisors as needed—understanding that the appropriate level changes as they develop into professionals. From my own experience as a supervisor, I hold concern about the clinical effectiveness of underconfident supervisees and about legal and ethical issues arising with overconfident supervisees.

Preventing and Fixing Mistakes

Supervisors walk a difficult line between building the confidence of their supervisees and providing the feedback necessary for supervisees to improve. Supervisees are commonly asked to bring in their most difficult cases and most challenging dilemmas, including any perceived mistakes they have made with their cases, to best ensure quality of care. When mistakes are welcomed and treated as learning experiences, the supervision experience for MFT interns can be quite positive (Hildebrandt, 2009). However, when discussion of mistakes results in feedback that is perceived as negative or critical, supervisees experience anxiety and reduced confidence (Daniels & Larson, 2001). Supervisees are also often keenly aware of when the supervisor is responsible for evaluating the quality of their work; as those evaluations may affect their academic standing or continued employment (Perosa & Perosa, 2010), supervisees may understandably want to avoid bringing in such concerns, preferring to focus only on their most successful cases in order to earn the most favorable evaluations. This hinders the supervisor's ability to aid the supervisee in necessary course corrections when therapy is not progressing or when there has been some kind of a rupture in the client-therapist relationship. Supervisors in Marriage and Family Therapy (MFT) are encouraged to track not just some but all of a supervisee's caseload, reducing the possibility that the supervisee will bring only the best-performing clients to the supervisor's attention.

Preparing the Supervisee for Independent Practice

Modern licensing laws use required periods of clinical supervision alongside licensure exams to ensure public safety. In this context, supervision is not presumed to make a therapist more effective necessarily, but instead to ensure that by the end of their supervised experience, the supervised therapists can be trusted to practice independently without endangering the public.

Understanding and Adapting to Difference

As the MFT profession and the clients we serve have become more diverse, much attention has rightly been paid to areas of cultural and contextual difference among clients, therapists, and supervisors. Supervisors in MFT who recognize such differences and respond appropriately have stronger alliances with their supervisees and supervisees who are more satisfied with supervision. Notably, such cultural competence also appears to make case conceptualization more challenging for supervisees (Inman, 2006). Of course, understanding and integrating the role of culture and context in treatment increases the complexity of cases, so this added challenge makes some sense.

Acknowledging Power

Many ethical issues in supervision are ultimately power issues. "It is important to recognize the power a supervisor holds, often to the point of influencing whether a supervisee continues to work in their chosen profession. This power should be openly acknowledged and not hidden. Power can be used in positive ways to help influence and develop supervisees, but it is also important to acknowledge that with power comes responsibility and the potential to do harm" (Hughes, 2012, p. 203). One way of interpreting the seemingly contradictory findings that supervisees want to be *challenged* but vehemently dislike being *criticized* (Hildebrandt, 2009; Reichelt & Skjerve, 2000) is through the lens of power in supervision, as viewed through the eyes of the supervisee: A supervisor's disagreement seems much more likely to be viewed as challenging, rather than critical, if the supervisee feels empowered within the relationship.

SPECIFIC ETHICAL REQUIREMENTS OF SUPERVISORS

Standard IV of the 2015 AAMFT Code of Ethics (AAMFT, 2014a) offers the most direct guidance to supervisors. Violations of these principles would be considered serious enough to threaten one's professional standing.[1] Obviously, supervisors need to be familiar with the full Code to ensure that they and their supervisees are adhering to it. The Code addresses the following supervisor behaviors (citations here refer to specific standards within the Code).

Avoiding Exploitation

As would be expected, supervisors are prohibited from exploiting supervisees or taking advantage of their dependency (4.1). The Code also includes a more specific prohibition against deceiving or exploiting supervisees when making financial arrangements with them (4.8), which is particularly relevant when setting supervision fees. I know of some clinics where MFT interns and trainees volunteer their services to gain hours of experience toward licensure, and the clinic charges the therapist a training or supervision fee—essentially requiring the volunteer to pay to be there. The financial realities of running a low-fee clinic, and the need to pay supervisors to ensure that supervision is available, certainly create a challenging context for many such clinics. However, such arrangements strike me as unsettling at best. Some states prohibit supervisees from directly paying their supervisors (Haug, 1997), so make sure, in any payment-for-supervision arrangement, that you are familiar with the laws of your state.

Avoiding Harassment

It should go without saying that supervisors should not harass their supervisees, sexually or otherwise (3.7). Though it is thankfully rare, such treatment does sometimes occur. In a qualitative examination of 101 MFT interns' best and worst supervision experiences, three reported that part of their worst supervision experience involved supervisors making direct sexual advances toward them, and seven reported that their supervisors routinely used sexist

---

[1]Notably, there is language in the AAMFT Code of Ethics that addresses situations in which supervisors may not be able to fully adhere to the Code. This can arise when workplace policies or practices obligate the supervisor to behave one way, and the Code obligates the supervisor to behave in a different way that is mutually exclusive. While such circumstances should be rare, the Code does allow supervisors caught in such a bind to "make known to the organization their commitment to the AAMFT Code of Ethics and take reasonable steps to resolve the conflict in a way that allows the fullest adherence to the Code of Ethics" (AAMFT, 2014a, Preamble).

language in supervision (Hildebrandt, 2009). Such behaviors can create an unpleasant or even hostile work environment for supervisees.

Some evidence from other mental health professions suggests that supervisors may be hesitant to discuss issues of supervisees' sexual attraction to clients in supervision, for fear of being accused of harassment or otherwise breaching ethical boundaries (McMurtery, Webb, & Arnold, 2011). However, such attractions arise with some frequency, so raising the issue in an appropriate manner has the potential to be a very helpful part of a supervisee's ethical training (Ladany et al., 1997).

how would this be done?

Avoiding or Addressing Multiple Relationships

Supervisors should avoid multiple relationships with supervisees that could impair professional judgment or increase the risk of exploitation (4.6). This area is among the most difficult for supervisors to navigate, especially in rural areas, in clinics that require supervisors to hold both clinical and administrative roles, and in academic settings where the supervisor may also be a teacher. Even the clear prohibitions in the Code can seem muddied upon close examination. For example, a supervisor may not provide therapy to a current supervisee (4.2). While the Code seems to make clear that you should not have separate meetings with a supervisee for whom you are also in the role of his or her therapist, it is less clear how far you should go in working with a supervisee on personal struggles within a supervision context (Aponte, 1994).

No!

The most significant dual relationship concern arises when a supervisor also maintains administrative responsibility over the supervisee; that is, when the supervisor is also the supervisee's manager. Although some argue for strong walls between these two roles, it is unrealistic, in my opinion, to presume that a supervisor would (or should) accept that a supervisee's behavior in supervision would be inconsequential when it comes to evaluating—and, ultimately, keeping or letting go of—the supervisee. It is important for all involved to be clear on the evaluative process and their roles within it, but given the appropriate need for information sharing, it is naïve to presume that the concerns related to supervisors managing their supervisees can be alleviated simply by putting a different person in the managerial role.

The Code includes a clear and specific prohibition against sexual relationships with supervisees (4.3). Just as the power imbalance between therapist and client makes a sexual relationship between the two inherently problematic, the power differential between supervisor and supervisee makes the risk of exploitation and harm significant in a sexual relationship.

Monitoring

Because supervision involves taking responsibility for the welfare of clients of their supervisees, it is vital for supervisors to monitor the supervisees' work, both in therapy and in how the supervisees conduct themselves more generally. The larger community trusts that MFT supervisors will ensure that their supervisees are acting responsibly and professionally in accordance with the standards of the field (Lee & Everett, 2003). A supervisor can be held responsible if he or she allows supervisees to perform services beyond their scope of competence or advertises that they can perform such services (4.4). More generally, a supervisor can be held responsible if the supervisee behaves in any unprofessional manner and the supervisor allows it (4.5).

Confidentiality

Simply put, supervisors cannot violate the confidentiality of supervisees (4.7). Some supervisors are surprised to learn that confidentiality in supervision is very similar to confidentiality in therapy. In settings with multiple supervisors, it is appropriate to share information between supervisors and others who share responsibility for the supervisee's training. However, any other disclosures require written consent from the supervisee. This can become challenging when working with the same supervisee in individual and group supervision contexts; if the supervisor is going to disclose information from a supervisee's individual supervision to the supervision group, the supervisor needs consent to do so. Similar to the arguments concerning a "no secrets" policy in therapy, there are reasonable arguments for having supervisees in such an instance sign a blanket consent allowing the supervisor to share whatever arises during individual supervision within the group, as deemed appropriate by the supervisor. For example, such a release keeps the supervisor from having to keep careful track of what was disclosed to him or her individually so as to avoid unintentional breaches of confidentiality. There also are reasonable arguments against such a policy, as it may erode a supervisee's willingness to share perceived failures or personal problems in individual supervision for fear that these will be shared with the group.

Informed Consent/Contracting

Informed consent for supervision is not directly required by the AAMFT Code of Ethics. However, it is increasingly viewed as a vital component of supervision in psychotherapy (e.g., Thomas, 2007) and is commonly addressed in discussion of MFT supervision (Haug, 1997). Atkinson (1997) and the AAMFT (2014b) offer sample contracts specific to MFT supervision that provide informed consent to and from the supervisee. While informed consent for supervision is not a perfect parallel with informed consent for therapy, many of the same principles apply (Thomas, 2007). Supervisees benefit from understanding their rights and responsibilities as well as the possible risks and benefits of supervision.

Documentation

Related to informed consent is the broader issue of documenting the supervision process. The notion that supervision needs to be documented at all is fairly new, and specific standards for the content of such documentation do not exist. However, writings in the MFT field are increasingly recognizing an expectation that supervision be well documented (e.g., Lee & Everett, 2003) for a number of good reasons: It protects the supervisor against claims of inadequate supervision and can help supervisors recall the current standing of supervisees' cases when there are many supervisees or many cases to track (Riemersma, 2009). Documentation of supervision also may protect supervisees from claims that they failed to appropriately consult on their cases, help to ensure continuity of supervision when supervisors change, and provide justification for any corrective actions that may have to be taken with a supervisee.[2]

Riemersma (2009) offers a list of more than 20 pieces of documentation that a supervisor might consider keeping in supervisees' files, including the supervisee's resume, proof of malpractice insurance coverage, the supervision contract, forms related to the supervisee's state license or registration (including copies of any forms tracking hours of experience toward licensure), a log of all supervision meetings, notes of any recommendations or instructions given to the supervisee, and copies of all evaluations of the supervisee. If any documentation of supervision is stored electronically, the supervisor should follow specific guidelines surrounding the use of technology in the storage or transfer of such records (see the "Technology" subsection later in this chapter).

TROUBLESHOOTING

Ethical issues can and should be a frequent topic of conversation in supervision. If a supervisee tells me he has a case with no legal or ethical concerns whatsoever, my immediate response is that he probably is not looking hard enough. However, what happens when supervisor and supervisee disagree about the most ethical course of action to take? Or worse, when the supervisee's behavior seems to be falling short of ethical standards? There are a number of ways that ethical problems can arise in supervision, each requiring careful attention to be resolved in the best interests of all involved: client, therapist, supervisor, and setting.

Ineffective Supervisees

MFTs are not to continue providing treatment if it is clear that the client is not benefiting from therapy (1.9). In practice, of course, this can be difficult to

[2]Even here, you can see the tension between clinical and evaluative roles.

assess; in systemic work, it is common for families to plateau or even regress at times along the path toward becoming a stable and well-functioning family. An ethical concern arises for the supervisor, however, if it appears that a supervisee is holding clients in therapy while the clients are clearly and consistently failing to make improvements.[3]

Interestingly, there is no similar requirement in the ethics code for supervisors. However, if a supervisee does not appear to be benefiting from the supervision relationship (beyond getting paid, if applicable, and gaining experience toward licensure), both supervisor and supervisee share responsibility for examining the supervision process to see where it might be improved.

unethical vs. less ethical

Unethical Behavior Among Supervisees

There is a meaningful difference between behavior that is *un*ethical—that is, behavior that falls below the defined standards of the profession—and behavior that might be labeled as *differently* ethical or even *less* ethical but that is still above the bar of being minimally acceptable. This is key in addressing concerns about supervisees' ethical behavior; it is one thing if the supervisee is risking her own professional standing and that of the supervisor, and something else if the supervisee is simply engaging in a behavior that the supervisor or other colleagues would *prefer* she do differently.

Russell, DuPree, Beggs, Peterson, and Anderson (2007) assessed supervisors' gatekeeping responses to a series of vignettes. When supervisees were displaying problems with honesty or personal integrity, supervisors were most likely to impose severe consequences; problems with confidentiality were also seen by some as an appropriate reason for probation or dismissal.

Resistant Supervisees

Not all supervisees are open to corrective feedback. Burgess (1994) used the memorable term "impervious to feedback" in describing such supervisees. Obviously, resistance comes in many forms, and resistance is not the same as absolute imperviousness. If a supervisee's resistance appears to be related to personal issues, then the supervisor must exercise careful judgment in determining the degree to which it is appropriate to address those issues in supervision as opposed to directing such supervisees to their own therapy (Aponte, 1994).

---

[3]Naturally, the treatment context here is important in defining *improvement*; if a supervisee is working with clients whose symptoms are so severe that the therapist's job is primarily to keep people stable enough that they stay out of hospitalization, then "improvement" might be considered in relative terms (i.e., comparing the impact of treatment against what would likely be happening if the client were not in therapy) rather than absolute terms (i.e., symptom relief).

Poor Matching

There are times when a supervisor and supervisee are simply not a good fit for one another, due to personality types, teaching styles, preferred models of therapy, or other reasons. Ideally, such mismatches would be easily identifiable before a supervision relationship has even begun. However, sometimes supervisors and supervisees are simply assigned to one another and told to do their best; at other times, there may not be another supervisor readily available. More often, such issues of mismatch arise in a context in which they are more challenges than "dealbreakers": They do not rise to the level where it would be appropriate to end the supervision relationship, but they can create tension in the supervision room. Unfortunately, little guidance regarding such situations is available in the literature. Supervisors in such situations should make sure that their personal distaste for the supervisee does not interfere with the effective provision of supervision and (to whatever degree the pair is open to acknowledging and discussing their differences directly) may even be able to use the situation to demonstrate to supervisees how to remain effective in their interactions with clients that the supervisee may not especially like.

Remediation and Dismissal

It speaks well of the field that supervisors in MFT take their gatekeeping role seriously. Supervisors clearly feel a responsibility to protect both the public and the profession from incompetent clinicians and often report that they have encouraged supervisees to leave the profession. However, such decisions obviously do not come easily. Although we often are reluctant to do so, there are times when our responsibilities as trainers are outweighed by our responsibilities as gatekeepers.

When concerns are raised about supervisees' conduct or performance, going beyond behavior that is simply not likable and into areas of actual incompetence, it is important that they be afforded due process to ensure fair treatment and the opportunity to develop any skills that are lacking, as long as they can do so without endangering their clients (Hughes, 2012). Sampson, Kelly-Trombley, Zubatsky, and Harris (2013) offer guidance for attempting to remediate the problem; the guidance fits nicely for any supervision relationship, inside or outside of an academic context. They suggest that supervisors (1) document the supervisee's demonstrations of incompetence; (2) be direct and concrete when discussing their concerns in supervision; (3) note the supervisee's responses to these discussions'; and (4) develop a plan of action to address the issues, which may include videotaping or audiotaping sessions (in addition to any other remedial steps the supervisor deems necessary). Once such a plan is in place, the supervisor should continue to offer feedback on the supervisee's improvement or lack thereof.

When a supervisor does make the difficult decision to terminate the relationship with a supervisee, having gone through these steps of remediation can help the supervisees to understand the reasons for their dismissal and should help the supervisor have the dismissal conversation in a manner that is at once clear and empathetic to the students' struggles.

EVALUATION OF SUPERVISEES

Many of the trouble areas just described can be avoided through the use of regular feedback processes in supervision. Regular assessment and feedback regarding supervisees' development is now considered an essential element of supervision across the psychotherapy professions (Bernard & Goodyear, 2014). Perosa and Perosa (2010) provide a useful review of various instruments that can be used for supervisors to assess clinical skill development with a particular focus on systemic treatment. Critics of such measures are quick to note, however, that specific skill development does not necessarily equate with more effective practice, and a therapist who is overly focused on proving his or her ability could even harm the relationship with the client (Lambert & Hawkins, 2001; Lambert & Ogles, 2004). As researchers on common factors (those elements of psychotherapy that appear to be present and impactful regardless of the particular theoretical model) argue, a therapist's *way of being* is a much more foundational and powerful presence in the therapy room than his or her development of specific clinical skills (Fife, Whiting, Bradford, & Davis, 2014).

Such concerns may be overcome through the use of direct client feedback measures, which provide objective information about client responses to therapy (Sapyta, Reimer, & Bickman, 2005). Integration of client feedback measures into the therapy process has clinical benefits, appearing to increase retention and improve overall outcomes, especially when a continuous feedback process is chosen over an intermittent one (Lambert, 2010). Positive effects of client feedback have been shown with couple cases (Anker, Duncan, & Sparks, 2009) and with trainees in MFT and counseling (Reese et al., 2009).

However an evaluation process is done, a skilled supervisor will present the results to supervisees in an artful way, focusing on their professional growth. Supervisees who have experienced supervision that includes such data have reported appreciating the clarity it offers in terms of areas on which to focus their own skill development, and minimizing their guesswork about client satisfaction and change. The use of evaluative instruments for therapy and supervision can contribute to an overall culture of accountability in MFT training—one that is likely to ultimately benefit clients, therapists, and the MFT profession as a whole (Sparks, Kisler, Adams, & Blumen, 2011).

CONTEMPORARY ETHICAL ISSUES IN SUPERVISION

Supervision occurs in a constantly changing cultural and professional context. Expectations of supervisors and guidelines for supervision evolve and adapt over time in response to this changing context; the changes in AAMFT's Code of Ethics for 2015, with many new mentions of technology specifically applied to supervision as well as therapy, are a strong example. There is also growing concern about the appropriate balance between therapists' right to choose their clients, and clients' right to quality and nondiscriminatory care.

Technology

Just as therapy itself is in the midst of significant change due to technology, supervision is also being affected by technological advances. As one example, the use of videoconferencing technology in MFT supervision holds great promise: For therapists who wish to specialize in a particular model of family therapy, they now could be supervised by a leader in use of the model, or perhaps by one of the original developers of the model, even if that person lives far away. For therapists working with a specific cultural or language group, videoconferencing in supervision can ensure that they receive culturally competent supervision. For MFTs working in rural areas, videoconference supervision can make it possible for one supervisor to work with multiple supervisees across several sites, ensuring an adequate supply of local services across sparsely populated land. However, transitioning to videoconference supervision, like any other transition into greater reliance on technology, should be done thoughtfully.

The AAMFT Code of Ethics recognized this, and in the 2015 version of the Code added Standard VI, on technology-assisted professional services. This standard is designed to govern both therapy *and* supervision processes that use technology in the delivery of service. Here, the ethical requirements are phrased in terms of what one must minimally do, rather than what behaviors to avoid:

- **Ensure compliance with all relevant laws before providing supervision through electronic means (6.1).** State laws such as the California Telemedicine Act and federal laws such as the Health Insurance Portability and Accountability Act (HIPAA) impose specific requirements on therapists and other health care providers who are using technology in the delivery of services. It is vital that practitioners be aware of these laws. Supervisors who allow their supervisees to practice using technology without ensuring that the supervisees are familiar with and abiding by such laws may themselves be considered to be falling short of ethical standards (4.5).
- **Use professional judgment before starting supervision through electronic means (6.1).** This includes a number of specific responsibilities defined in the Code, each of which should be documented: (a) assess whether the service is appropriate to the supervisee's needs and development; (b) inform supervisees of the potential benefits and risks; (c) ensure, to the degree feasible, that the means of communication are secure; and (d) ensure

that you as the supervisor and they as the supervisees have had appropriate preparation (training, education, or experience) to use the technology.
- **Inform supervisees *in writing* about the risks associated with technology-assisted services and the responsibilities placed on both you and the supervisee to minimize those risks (6.2).**
- **Choose a platform that follows best practices for confidentiality and quality of services, and inform supervisees in writing of the limitations and safeguards involved in the platform you have chosen (6.3).** This is something of a balancing act, as platforms that are the most secure do not necessarily offer the highest quality of services. Consider Skype as an example: although its wide distribution and availability mean that many clients would have access to this videoconferencing software, perhaps improving the quality of care in relation to services provided via telephone, therapists who use Skype may be falling short of the privacy and security standards required by HIPAA (APA, 2014). Supervisors thus may decide that even though the quality of service may be a bit lower, supervision by telephone is preferable to supervision by Skype.[4]
- **Make sure that documentation storage and transfer practices are compliant with applicable law (6.4).** Again, supervisees must be made aware of the limitations of the supervisor's chosen technologies in writing.
- **Do not supervise outside of your legal jurisdiction (6.5).** This could be a chapter unto itself. Generally speaking, your license only enables you to supervise someone who is physically located in the state where you are licensed and is appropriately registered as an intern or associate (if necessary) in that same state.
- **Maintain current training in technology-assisted services (6.6).** The technology we can use to provide supervision is constantly evolving, as are the standards for such services. The Code specifically requires adherence to best practices for the delivery of technology-assisted services, which of course means maintaining awareness of what those best practices are.

Referral and Refusal of Clients

Some religious therapists have asserted a right to refuse to treat gay and lesbian clients, citing a conflict with the therapist's religious belief (Caldwell, 2011). This is especially challenging for supervisors, who have responsibility for the therapist's training as well as ensuring that clients receive appropriate care. The American Psychological Association has offered guidance on this issue,

---

[4]Supervisors should remain aware of all relevant laws when making such decisions. In California, state law clearly allows supervisors in nonprofit settings to use videoconferencing for supervising interns, though such an allowance is not made for private practice settings or for the supervision of trainees. The law is not clear when it comes to supervision by phone, which would seem to be discouraged by language requiring "face-to-face" supervision. Your state laws may be different.

arguing that it may be appropriate for supervisors to transfer clients to another therapist in the same clinic if the original therapist raises such concerns. However, the American Psychological Association guidance is clear that this does not mean a supervisee should expect to be able to avoid a particular kind of client for the duration of his or her training; they maintain responsibility for building their clinical skills in treating clients regardless of sexual orientation, and supervisees do not have an ethical right to opt out of any training deemed appropriate by the supervisor or training program (American Psychological Association, 2013). I have echoed this stance for MFTs (Caldwell, 2014).

WHAT DOES IT MEAN TO BE A *GOOD* SUPERVISOR?

In 2001, there was "a surprising lack of research support, either quantitative or qualitative, for many of the basic assumptions that underlie the supervisory enterprise," wrote Storm, Todd, Sprenkle, and Morgan (p. 227). Studies had not demonstrated that supervision makes for more effective therapists or better protection for their clients, in spite of widespread assumptions that it did. "Indeed, it would not be overstating the case to assert that the field's belief in the importance of supervision rests mostly on faith."

Now, almost 15 years later, not much has changed. The research base on supervision in marriage and family therapy has certainly expanded and has gone much farther in demonstrating that supervised experience improves the *confidence* of developing therapists. But, much to our chagrin, it has yet to convincingly show that supervision impacts much else. It is still true that our perspectives on the need for, and preferred ways of doing, supervision rest mostly on faith.

Thus, writing about supervision in our field can sometimes feel like writing about religion. Questions of practice quickly blend into the questions of philosophy that, ideally, would provide guidance for practice: Why are we here? What are we aiming to achieve? In short, what does it mean to be *good* as a supervisor?

These are not small or inconsequential questions. Indeed, they have major impacts on how we think about the ethics that govern supervision work. We cannot answer the question of what makes a *good* supervisor based on what research tells us makes an *effective* supervisor because studies have so far failed to answer that question. (We now know a great deal about what supervisees *like* [e.g., Hildebrandt, 2009; Lizzio, Wilson, & Que, 2009] but still very little about what makes a supervisor *effective*.)

Experienced supervisors are rightly quick to point out that there can be harsh realities of supervising: We often have a year or less with our supervisees; we are working directly with them for perhaps as little as an hour each week; we may be sharing supervisory responsibility with other supervisors and administrators; we have significant other demands on our professional time. Ultimately, we only have so much time, attention, knowledge, and competence to offer. Defining your role as a supervisor can become a balancing act between what you would *like* to do and what you are reasonably *able* to do. It is important to recognize the value in being a *good enough* supervisor, even when circumstances prevent you from being able to do all you would like.

While we have agreed-upon minimal standards of behavior for MFT supervisors, there is little guidance available for supervisors about how to be *good*—how to pursue a higher calling in supervision than simply gatekeeping, for those supervisors who want and are able to do so. You may experience that lack of guidance as either freeing or terrifying. It certainly grants a *lot* of leeway for supervisors to define themselves and their supervisory practices as they see fit. Although there are minimal standards for supervisory conduct, as described here, there is no more consensus on the *best* way to do supervision than there is on the *best* religion. As you will see in the forthcoming chapters on specific supervision models, there is tremendous variety in the philosophies, goals, and tasks of various models of supervision.

This makes it all the more important that supervisors have a clear sense of what they honor and value in the supervisory relationship and that this be a key component of early supervisory discussions. Some crucial questions to consider:

- Do you believe your role to be one primarily of gatekeeping, or of mentorship? How do you see the balance between these roles?
- In what ways does your supervision process differ from that of other supervisors? How do these differences reflect differences in what you value?
- What are your ideals as a supervisor? What are you striving for, both for yourself and for your supervisees, in your supervisory work?

There is an established line of distinction in the field between *principle ethics*—the stated minimum standards of conduct, specifically detailed in ethical codes and often phrased in terms of what one *must* and *must not* do—and *virtue ethics*, the process of determining what is the *best* course of action in a given set of circumstances, where many different courses of action may meet the standard of being minimally acceptable. While it is fairly easy to get over the bar of being a "good enough" supervisor, there is much room for debate about what it means to be a *good* supervisor. If you have come to a definition of that term for yourself, if you have practices (or, if you're just starting to supervise, a plan) in place for living up to that definition, and if your supervisees have been made aware of your definition (and, hopefully, agree with it), you're off to a good start.

REFERENCES

American Association for Marriage and Family Therapy (AAMFT). (2014a). 2015 Code of Ethics. Alexandria, VA: Author.

American Association for Marriage and Family Therapy (AAMFT). (2014b). *Approved supervision designation: Standards handbook.* Alexandria, VA: Author.

American Psychological Association. (2013). *Preparing professional psychologists to serve a diverse public.* Retrieved from http://www.apa.org/pi/lgbt/resources/policy/diversity-preparation.aspx

American Psychological Association. (2014). *Practitioner pointer: Does the use of Skype raise HIPAA compliance issues?* Retrieved from http://www.apapracticecentral.org/update/2014/04-24/skype-hipaa.aspx

Anker, M. G., Duncan, B. L., & Sparks, J. A. (2009). Using client feedback to improve couple therapy outcomes: A randomized clinical trial in a naturalistic setting. *Journal of Consulting and Clinical Psychology, 77*(4), 693–704.

Aponte, H. J. (1994). How personal can training get? *Journal of Marital and Family Therapy, 20*(1), 3–15.

Atkinson, B. J. (1997). Informed consent form. In C. L. Storm & T. C. Todd (Eds.), *The reasonably complete systemic supervisor resource guide* (pp. 11–15). Boston, MA: Allyn & Bacon.

Bernard, J. M., & Goodyear, R. K. (2014). *Fundamentals of clinical supervision* (5th ed.). Boston, MA: Allyn & Bacon.

Burgess, S. L. (1994). The impaired clinical and counseling psychology doctoral student. Unpublished doctoral dissertation, California School of Professional Psychology, Alameda, CA.

Caldwell, B. E. (2011, September/October). The dilemma: Can a religious therapist refuse to treat gay and lesbian clients? *Family Therapy Magazine, 10*(5), 50–52.

Caldwell, B. E. (2014, July/August). A proposed framework for considering conscience-based referrals. *Family Therapy Magazine, 13*(4), 54–55.

Daniels, J. A., & Larson, L. M. (2001). The impact of performance feedback on counseling self-efficacy and counselor anxiety. *Counselor Education and Supervision, 41*, 120–130.

Fife, S. T., Whiting, J. B., Bradford, K., & Davis, S. (2014). The therapeutic pyramid: A common factors synthesis of techniques, alliance, and way of being. *Journal of Marital and Family Therapy, 40*(1), 20–33.

Haug, I. (1997). Supervisory ethics and standards of practice. In C. L. Storm & T. C. Todd (Eds.), *The reasonably complete systemic supervisor resource guide* (pp. 2–9). Needham Heights, MA: Allyn & Bacon.

Hildebrandt, C. (2009). *Marriage and family therapy interns' best and worst supervision experience* (Doctoral dissertation). Alliant International University, San Diego, CA.

Hughes, J. (2012). Practical aspects of supervision. In I. Fleming & L. Steen (Eds.), *Supervision and clinical psychology: Theory, practice, and perspectives* (2nd ed., pp. 184–206). New York, NY: Routledge.

Inman, A. G. (2006). Supervisor multicultural competence and its relation to supervisory process and outcome. *Journal of Marital and Family Therapy, 32*(1), 73–85.

Ladany, N., Melincoff, D. S., O'Brien, K. M., Hill, C. E., Knox, S., & Petersen, D. A. (1997). Sexual attraction towards clients, use of supervision, and prior training: A qualitative study of predoctoral psychology interns. *Journal of Counseling Psychology, 44*, 413–424.

Lambert, M. J. (2010). Yes, it is time for clinicians to routinely monitor treatment outcome. In B. L. Duncan, S. D. Miller, B. E. Wampold, & M. A. Hubble (Eds.), *The heart and soul of change. Delivering what works* (2nd ed., pp. 239–266). Washington, DC: American Psychological Association Press.

Lambert, M. J., & Hawkins, E. J. (2001). Using information about patient progress in supervision: Are outcomes enhanced? *Australian Psychologist, 36*, 131–138.

Lambert, M. J., & Ogles, B. (2004). The efficacy and effectiveness of psychotherapy. In M. J. Lambert (Ed.), *Bergin and Garfield's handbook of psychotherapy and behavioral change* (5th ed., pp. 139–193). New York, NY: Wiley.

Lee, R. E., & Everett, C. A. (2003). *The integrative family therapy supervisor: A primer.* New York, NY: Routledge.

Lee, R. E., & Nelson, T. S. (2013). *The contemporary relational supervisor.* New York, NY: Routledge.

Lizzio, A., Wilson, K., & Que, J. (2009). Relationship dimensions in the professional supervision of psychology graduates: Supervisee perceptions of process and outcome. *Studies in Continuing Education, 31*(2), 127–140.

McMurtery, R. F., Webb, T. T., & Arnold, R. D. (2011). Assessing perceptions and attitudes of intimate behaviors in clinical supervision among licensed professional counselors, licensed social workers, and licensed psychologists. *The Researcher: An Independent Journal, 24*(2), 57–78.

Perosa, L. M., & Perosa, S. L. (2010). Assessing competencies in couples and family therapy/counseling: A call to the profession. *Journal of Marital and Family Therapy, 36*(2), 126–143.

Reese, R. J., Usher, E. L., Bowman, D., Norsworthy, L., Halstead, J., Rowlands, S., & Chisholm, R. (2009). Using client feedback in psychotherapy training: An analysis of its influence on supervision and counselor self-efficacy. *Training and Education in Professional Psychology, 3*(3), 157–168.

Reichelt, S., & Skjerve, J. (2000). Supervision of inexperienced therapists: A qualitative analysis. *The Clinical Supervisor, 19*(2), 25–43.

Riemersma, M. (2009 September/October). Tips on supervision: Supervisor record-keeping. *The Therapist.* Accessed online August 1, 2014 at https://www.camft.org/COS/Resources/Attorney_Articles/Mary/Tips_on_Supervision_Supervisor_Record_Keeping.aspx

Russell, C. S., DuPree, W. J., Beggs, M. A., Peterson, C. M., & Anderson, M. P. (2007). Responding to remediation and gatekeeping challenges in supervision. *Journal of Marital and Family Therapy, 33*(2), 227–244.

Sampson, J. M., Kelly-Trombley, K. M., Zubatsky, J. M., & Harris, S. M. (2013). Breaking up is hard to do: Dismissing students from MFT training programs. *American Journal of Family Therapy, 41*(1), 26–33.

Sapyta, J., Reimer, M., & Bickman, L. (2005). Feedback to clinicians: Theory, research, and practice. *Journal of Clinical Psychology, 61*, 145–153.

Senediak, C. (2013). Integrating reflective practice in family therapy supervision. *Australian and New Zealand Journal of Family Therapy, 34*(4), 338–351.

Sparks, J. A., Kisler, T. S., Adams, J. F., & Blumen, D. G. (2011). Teaching accountability: Using client feedback to train effective family therapists. *Journal of Marital and Family Therapy, 37*(4), 452–467.

Storm, C. L., Todd, T. C., Sprenkle, D. H., & Morgan, M. M. (2001). Gaps between MFT supervision assumptions and common practice: Suggested best practices. *Journal of Marital and Family Therapy, 27*(1), 227–239.

Thomas, J. T. (2007). Informed consent through contracting for supervision: Minimizing risks, enhancing benefits. *Professional Psychology: Research and Practice, 38*(3), 221–231.

Thomas, J. T. (2014). International ethics for psychotherapy supervisors: Principles, practices, and future directions. In C. E. Watkins Jr. & D. L. Milne (Eds.), *The Wiley international handbook of clinical supervision* (pp. 131–154). Chichester, West Sussex, UK: John Wiley & Sons.

Wheeler, S., & Richards, K. (2007). The impact of clinical supervision on counsellors and therapists, their practice and their clients: A systematic review of the literature. *Counselling and Psychotherapy Research, 7*(1), 54–65.

SEVEN

Diversity and Social Justice in Supervision

Toni Zimmerman
Marj Castronova
Jessica ChenFeng

It is well established that to be a competent supervisor, you must be attentive to diversity and social justice in the process of supervision and in attending to clinical cases (Ali & Bachicha, 2012; Arnold, 1993; Bernard, 2014; Bernard & Goodyear, 1998; Inman, 2006; Ladany, Inman, Constantine, & Hofheinz, 1997; Lawless, Gale, & Bacigalupe, 2001; Pendry, 2012). Therapy training programs are required to prepare trainees to work with diverse populations and to understand the personal and institutional impact of marginalization, power, and privilege as part of their accreditation through regulatory agencies such as the Commission on Accreditation for Marriage and Family Therapy Education (COAMFTE, 2005) and the Council for Accreditation of Counseling and Related Educational Programs (CACREP, 2009). Understanding the human experience based on social location is imperative to building alliances with our supervisees and our clients. *Social location* refers to a person's place in society with respect to certain categories of identity (Berger, 1963). It is understood to be socially constructed as a perception of how things work, what is real, where things belong, and how they fit together. One's social location in a social system can be privileged or marginalized and oppressed (Macintosh, 1989). Nieto and Boyer (2006) point out that as small children we learn a particular set of social rules that we follow in a mostly unconscious way. We learn these based on our social location and status and that of those around us. We learn how members of groups are supposed to act as well as how to sort and categorize people based on their social location. Recognizing this and resisting the unconscious categorizing of people is central to less-biased supervision and therapy.

The American Association for Marriage and Family Therapy's (AAMFT) Code of Ethics (2015) requires that therapists practice in an affirmative and inclusive way in order to be ethical. Supervisors in all settings (e.g., universities, agencies, private practice, schools, military, and medical settings) must prepare supervisees to work with a wide variety of clients from a variety of backgrounds and values in less-biased ways in order to practice ethically. Supervisors must keep the conversation of social location and issues of power and privilege front and center in the process and content of supervision. This is imperative in order to work with supervisees and their clients in affirming and inclusive ways with great sensitivity to oppression and privilege differences in the human experience. In this chapter, we present some ways in which supervisors can keep these conversations central in supervision so that discussions of diversity and oppression remain vibrant, intentional, and ever-present. We outline four primary areas that can be attended to in the supervisor-supervisee relationship in order to do this:

1. Establishing a safe space for supervisory connection (who are we?).
2. Working toward cultural knowledge/awareness and cultural humility (who are the clients?).
3. Recognizing and managing privilege, power, and bias associated with social location for all involved (supervisee, supervisor, and clients).
4. Embracing our role as agents of change in the areas of social justice and advocacy with our clients and in our profession.

For each of these four areas, we discuss relevant goals as well as offer techniques and activities for addressing them in supervision. Our hope is that supervisors will find this chapter to be a guide in bringing and keeping diversity in the forefront of their supervision and in helping trainees bring and keep diversity considerations in the forefront of therapy with their clients.

Establishing a Safe Space for Supervisory Connection (Who Are We?)

Goals for Establishing a Safe Space

1. Understand that the supervisor has significant influence on creating safety so that strong supervisory alliances can be established.
2. Recognize the ways in which the supervisor's and supervisee's backgrounds and values serve as a lens for how they view the world and how they view each client they are treating.
3. Challenge the ways in which the backgrounds of the supervisee and the supervisor limit or enhance the supervisory relationship as well as therapy with the client.

ESTABLISHING A SAFE SPACE

As systems thinkers, we are well aware that we are not blank slates taking in information about people and the world in an objective way (Flemons, Green, & Rambo, 1996; Montgomery, Hendricks, & Bradley, 2001; Polnik,

2013). Instead, we have lenses through which we view the world that affect how and what we see. Understanding the lenses of the supervisor and the supervisee and being able to critically explore how those lenses may be different or similar to those of our clients and each other is important (Zimmerman, 2001). Our lenses affect all components of therapy with clients, including identification of the problem, creation of a treatment plan, assessment and diagnosis, and the process of therapy. If we leave diversity considerations unexamined, we can inadvertently do harm to our clients and supervisees in all of these areas.

Ellis (2010) discusses the myths of supervision. One of the myths is that supervision is about right techniques and right theory. Rather, solid supervision is about the relationships (Deihl & Ellis, 2009a, 2009b; Fama & Ellis, 2005; Inman, 2006). The supervisory relationship is negatively impacted by miscommunication of verbal and nonverbal cues, lack of interpersonal awareness, and the inability to ask questions and elicit responses that are valuable (Cook, 1994; Killian, 2001). When the relationship is built on a solid alliance between the supervisor and the supervisee, it is a major predictor of supervision satisfaction (Inman, 2006; Ladany, Ellis, & Friedlander, 1999). Given the sensitive nature of discussing race, gender, class, sexual orientation, and other categories of social location, the supervisor-supervisee relationship must have a strong alliance for these conversations to be safe and productive. The supervisory alliance is a critical component in diversity conversations (Inman, 2006; Ladany et al., 1997). For instance, when the supervisor and supervisee hold common belief systems about diversity (high racial identity levels), those beliefs and diversity are more likely to be discussed in supervision (Killian, 2001; Ladany et al., 1997). Yet, supervision sessions should be safe so that diversity topics can be discussed regardless of the backgrounds of those involved. The degree to which the supervisor creates a sense of safety for supervisees to self-disclose influences the quality of the supervisory relationship (Killian, 2001). This is particularly true for discussions of social location and marginalization.

Factors that negatively affect supervisory relationships are gender bias, undiscussed racial/ethnic issues, and overemphasis on psychological problems within the context of culture (Fong & Lease, 1997; Killian, 2001). Supervisory alliances within different cultures is a major predictor of satisfaction in supervision (Inman, 2006; Son, Ellis, & Yoo, 2007, 2009). When supervisors are aware, open, and sincere, they are able to facilitate culturally responsive supervision via attending to cultural and racial factors, providing guidance and discussion of culturally specific issues, and creating multicultural activities as well as being vulnerable about their own struggles (Fukuyama, 1994; Helms & Cook, 1999, Killian, 2001; Lawless et al., 2001). The supervisor's multicultural focus is directly related to the working alliance (Silvestri, 2003) and the supervisor-supervisee alliance is a significant positive mediator between multicultural competence and supervision satisfaction (Inman, 2006). It is apparent that "[g]iven the sensitivities and vulnerabilities involved in cultural exploration at both intrapersonal and interpersonal levels, having a supervisory relationship that is based in trust and respect is vital to an environment that allows individuals to take risks in relationships" (Inman, 2006, p. 81). Supervisory relationships can have great impact on the work with supervisees of different ethnicities if the supervisor is skilled in the area of alliance-building and is open and aware (Inman, 2006). Research consistently shows that supervisors who hold authentic and active interest in other cultures and who value differences are seen as creating safe places for conversations on cultural differences (Killian, 2001). Following are a few examples of how supervisors can create space for diversity dialogue and create safe spaces to share and explore.

Techniques and Activities for Establishing Safe Spaces

As a supervisor, you may or may not have done a cultural genogram (Hardy & Laszloffy, 1995, 2002) in your academic program. If you have not, we recommend completing one and using it to share your values and background with your supervisees. If you have completed one, we suggest you share it with your supervisees. You may want to occasionally update your original cultural genogram because over the years, you may have reflected more on the meaning of your background and its impact on your work. Updating can be valuable because, over time, our values can shift and change; we may want to reflect on what changed and what held steady and share this with our supervisees. In doing so, you are role-modeling being reflective and self-aware (Ali & Bachicha, 2012).

You can give examples of how you have learned to use your awareness in therapy. For example, an Asian American supervisor shared that she is from a culture where arguing is frowned upon because of the cultural emphasis on peace and harmony and that she used to be triggered by clients who showed anger in the therapy room. Once she recognized this as a cultural difference between her and some of her clients, she was able to manage her triggers. An African American supervisor shared with his supervisees who were female, White, and young that he worried about supervisees' feeling that they couldn't speak up to or disagree with him given his race, gender, and size (a tall and large man). Sharing these types of concerns and investigating the way they are embedded in societal messages about race, gender, and size creates an open environment that allows attentiveness to these barriers if they present themselves. Asking supervisees to complete and share cultural genograms with supervisors can open conversations about the similarities and differences between supervisees and supervisors. Supervisees frequently report that their supervision experiences would have been more valuable if more time had been spent on issues surrounding cultural differences (Constantine, 1997).

Learning about how both the supervisee and the supervisor have experienced marginalization or privilege, given their backgrounds and social locations, is helpful for self- reflection. Having each share the ways in which their backgrounds may differ in terms of oppression, bias, stereotypes, and privilege can lead to a rich and personal self-reflection of difference and increased understanding of each other (Nolte, 2007). For instance, one author shared a story from her childhood with her Taiwanese American supervisee. She had moved to a small farming community when she was in third grade. She was a different religion than most in the community and her father was in law enforcement. On her first day of school, she and her little brother got on the bus and two senior boys grabbed each of them and wrote "pigs' kids" on the back of their jackets. This opened the door for the supervisor and supervisee to have a rich conversation about the impact of marginalizing experiences and how they inform different perspectives.

From the beginning of the supervisory relationship (Lee & Nelson, 2014), supervisors can set an expectation that when staffing cases, supervisees will revisit the background and values presented in the cultural genograms in order to stay vigilant of the ways their lenses may limit their understanding of the client's worldview. Simon (2010) refers to this as "system supervision as a transgressive partnership," where we are breaking new ground with our supervisees and creating permissions that do not exist within the dominant culture (p. 309). When we give permission, we create room to question the "unspoken censorship" that exists when persons deviate from the accepted culture, gender, age, lifestyle, and so forth (Simon, 2010, p. 310). In these curious questions, we are able to wonder how experiences of oppression or marginalization have impacted all of us and what the consequences might be for acting outside the mainstream (Simon, 2010). Revisiting who we are, what our lenses might be, and how they might influence our case planning is an essential and ongoing dialogue.

These techniques and activities for exploring supervisor and supervisee backgrounds have been of great benefit in our experience as supervisors. We hope you experience as much growth and transformation as we have in utilizing them.

Working Toward Cultural Competency With Cultural Humility (Who Are the Clients?)

Goals for Working Toward Cultural Knowledge With Cultural Humility

1. Have knowledge of a variety of cultures and backgrounds (cultural knowledge and competency).
2. Don't assume that what you know applies to your client (cultural humility).

WORKING TOWARD CULTURAL KNOWLEDGE AND AWARENESS

Although therapists certainly can and do learn about a variety of backgrounds from their clients, it is important to have knowledge of a wide range of cultures and cultural practices as a starting place and not rely solely on the clients to teach the therapist. Adams (1995) draws a distinction between cultural knowledge and cultural awareness. *Knowledge* is when we are familiar with a variety of characteristics, history, values, beliefs, and behaviors of a group; *awareness* is when we have internally changed our attitudes and values because we have developed a sensitivity and understanding. Many scholars have written extensively about the cultures of a variety of clients and supervisees with whom we work. For instance, the textbook *Ethnicity in Family Therapy* (McGoldrick, Pearce, & Giordano, 2005) is an excellent resource that provides more than 50 different ethnic summaries for learning about working with families of diverse backgrounds. Authors such as Dupree, Bhakta, Patel, and Dupree (2013) provide the supervisee with information about themes identified in work with Asian Indian American couples, such as (a) impact of intergenerational/extended family relationships on marriages, (b) acculturation issues, and (c) the influence of gender roles on marital and parental patterns. When therapists begin to see clients from backgrounds unfamiliar to their own, supervision should include learning about that background in a variety of ways.

Techniques and Activities for Working Toward Cultural Knowledge and Awareness

Supervisors can introduce supervisees to relevant resources that match the various clients with whom the supervisee is working. For instance, if a supervisee is working with an Hispanic, Catholic, single mother, the supervisor can suggest readings about the common struggles and resiliency of single mothers and the culture of Catholic, Hispanic families. If a supervisee is working with a Hmong professional woman who is struggling to align her cultural values and her professional self, the supervisor can suggest reading a book such as *The Spirit Catches You and You Fall Down* (Fadiman, 1997). This book provides a rich narrative of one Hmong family's experience with the American medical community. Lor's (2013) biographical/ethnographic approach to a professional Hmong woman's experience sheds light on how culture and gender can limit opportunities. As one woman said:

> Traditionally and culturally, women are viewed as inferiors to males and there are still certain social and public contexts where Hmong women are still expected to behave in this manner.
> For example, though I am very much an independent, strong-minded, bold, and capable Hmong woman, there are times in public and/or social functions that I have to remain submissive and not supersede my husband's role and character. And, if there are situations where I am placed in preference over him, he still is acknowledged and consulted first prior to recognizing me and/or approaching me. These examples add to the concept of a culture value whereby all people have a place in society. (Lor, 2013, p. 45)

Having some working knowledge of a variety of cultural practices will provide the supervisee with an introduction into the culture with which her client identifies. The supervisee can begin to ask herself questions about how this client's background might be similar and/or different from her own and begin to wonder how what she has learned matches or does not match the experience of her client. We often use the phrase "click for clients," meaning to search for reliable resources (e.g., books, chapters, journal articles, videos) to learn about the backgrounds of clients. Looking for resources that address how clients' backgrounds intersect with each other's or with the supervisor's is of particular benefit.

Crenshaw (1993) defines *intersectionality* as a process where various social locations, such as religion, culture, and gender, can co-construct our sense of self, thus influencing our identities, choices, and opportunities, often within the voice of the dominant culture (DeReus, Few, & Blume, 2005). For instance, being gay as a Black man from a conservative religious background will be significantly different from being a White man coming out in a progressive religious culture. Sometimes, resources offer advice about how to tailor a therapy approach and/or technique to a particular background. For example, Richeport-Haley (1998) offers ideas to integrate cultural issues when utilizing strategic family therapy. Lim and Nakamoto (2008) discuss ways to integrate genograms with clients from Asian cultures, paying special attention to sociocultural and political processes. When we consider the various intersects that may be impacting our clients and supervisees, a quick search can assist us in asking questions we may not have considered given our own lenses.

Addressing Cultural Humility

Supervisees should be reminded to exercise cultural humility (sometimes referred to as *cultural curiosity*) with their clients. *Cultural humility* has been defined as "the ability to maintain an interpersonal stance that is other-oriented (or open to the other) in relation to aspects of cultural identity that are most important to the client" (Hook, Davis, Owen, Worthington, & Utsey, 2013, p. 354). Supervisees should be reminded not to assume that what they have learned and read is relevant to the specific client with whom they are working. To be respectful of clients, supervisees should know something about the backgrounds of their clients but stay humble and curious about their clients' actual lived experiences. It means that even when cultural differences threaten to weaken the therapeutic alliance, the therapist still expresses respect and does not assume competency based on prior knowledge (Hook et al., 2013).

Techniques and Activities for Working With Cultural Humility

The Cultural Humility Scale (CHS; Hook et al., 2013) is a measure that has clients share their perception of their therapist's cultural humility. Hook and colleagues found that the higher the client reported perception of the therapist's cultural humility, the stronger the alliance in the therapy process (Hook et al., 2013). In addition, there was a large effect size for the client's improvement in therapy explained by the mediated effect of cultural humility through the working alliance. In working with supervisees, supervisors can use the CHS as a way to assess supervisees with regard to cultural humility and the effect it is having on the working alliance and clients' improvement.

In addition to gaining knowledge through resources, we also suggest that all supervisees gain lived experiences with cultures different from their own. This can be accomplished by traveling, working, or volunteering in various places around the globe to see firsthand how people live and work. Supervisees can also look locally for opportunities to attend community events such as powwows, Hanukah celebrations, Asian New Year's celebrations, a southern crawfish boil, a northern lutefisk dinner, or celebration of Kwanzaa. The point is for supervisors to encourage their supervisees to get out of their comfort zones by urging them to go to places they generally would not go and meet people with whom they do not generally interact. In order for supervisees to have personal encounters with many people who are different from themselves, we suggest that supervisors assist in "brokering opportunities." For instance, if a supervisee is from a small, homogeneous town and has limited interactions with other persons, his first meaningful encounter with this type of diversity perhaps should not be working with clients from unfamiliar groups. In this case, the supervisor might call a friend or colleague who identifies with the particular identity to ask if the supervisee can join them for dinner to interact with the family, meet for lunch, or conduct an interview. Over the years, we have done this for many supervisees, and it has led to honest and meaningful exchanges through which the friend/colleague was happy to introduce the supervisee to his or her culture and background, knowing that the therapist will benefit from it as he encounters clients from that culture. On numerous occasions we have heard that these meetings have been filled with open discussion about bias and misconceptions and deep levels of awareness and sensitivity that are hard to encounter or access in any another way.

Recognizing and Managing Privilege, Power, and Bias Associated With Social Location

Goals for Recognizing and Managing Privilege, Power, and Bias

1. Understand the differences in power and privilege based on social location, and develop sensitivity to the lived experience of having less power and privilege.
2. Understand institutionalized racism and be sensitive to the lived experience of institutionalized isms.
3. Understand one's own biases and stereotypes and be sensitive to the experience of recipients of stereotypes and biases.

ADDRESSING DIFFERENCES IN POWER AND PRIVILEGE BASED ON SOCIAL LOCATION AND INSTITUTIONALIZED "ISMs"

When we refer to *social location*, we are referring to 11 dimensions or categories: gender, race, religion, age, abilities, class, culture, ethnicity, education, sexual orientation, and spirituality. Burnham created an acronym for these, called social GRRAACCEESS (Burnham, 1992, 1993; Burnham, Palma, & Whitehouse, 2008; Roper-Hall, 1998, 2008). Social GRRAACCESS are a "position that says these aspects of experience are important, interrelated, and constitutive of a person's experiences in all aspects of life, and therefore important for all participants in the endeavor of therapy, training, and supervision" (Burnham et al., 2008, p. 530). In the United States, the most personal and institutionalized "isms" occur within these 11 categories: sexism, racism, religious intolerance, ageism, ableism, classism, ethnocentrisms, educational bias, heterosexism, and spiritual bias.

We all have hidden, unconscious, and/or implicit bias in these 11 areas. We have expectations and make judgments based on people's social locations and categories along these dimensions. We perceive and treat people based on these, often without even realizing it. This judgment happens on a personal level as well as at an institutional level. Institutionalized "isms" are found within all major systems and institutions, including educational, corporate, justice, and health systems. An example of sexism in hiring and promotion practices in education is when a male and a female university professor are presented and rated identically on their hiring application packages, and the candidate with the male name (e.g., Brian) is preferred two to one over the candidate with the female name (e.g., Karen; Steinpreis, Anders, & Ritzke, 1999). This same study found that when reviewing identical tenure and promotion packages, reservations were expressed four times more often when the name was female versus when the name was male. In relation to heterosexism, when a male university guest lecturer referred to his partner as Jessica in some class sections and Jason in others, he received 22% more positive comments and 81% fewer negative comments when students thought he was "straight" compared to when they thought he was gay (Steinpreis et al., 1999). Bertrand and Mullainathan (2004) found that when applying for a job using identical resumes, the resume with an African American-sounding name (Jamal) had to be sent out 15 times to get a call back, compared to the resume with the White-sounding name (Greg), which had to be sent out 10 times to get a call back. In addition, for African American-sounding names to yield the same return call rate as White-sounding names, the candidate had to have an additional 8 years of experience. These are institutional examples in which well-meaning people, and in fact people who would most likely care about equality, have implicit bias that works to privilege some categories of people and oppress or marginalize others. If therapists and supervisors do not make issues of disparity central in therapy and supervision, solid alliances are unlikely.

Supervisors must discuss the nanosecond reaction (implicit bias) that may occur in the minds of supervisees who go to the therapy waiting room to meet clients for the first time, such as when seeing an Hispanic man who is dressed in a way that may indicate lower socioeconomic status than the trainee and has an accent indicates that English is his second language. Compare this to the category our minds often create when the client is a well-dressed, articulate White man who appears comfortable in this setting. We often give the White male the "benefit of the doubt" upon hearing his name or meeting him for the first time, rather than acting upon our subconscious implicit bias. The Hispanic man, given his lifelong experience with "isms," is likely aware of the therapist's implicit bias even if the therapist is not. Conversations with trainees about these implicit biases are only possible in safe, open, and caring supervision environments.

The social location form (see Table 7.1) shows which categories of people along the 11 dimensions typically hold more or less privilege and power in an American context.

TABLE 7.1 Social Location Form

SOCIAL LOCATION		
	MOST PRIVILEGED	**LESS PRIVILEGED**
Gender	Men	Women, transgendered, other (e.g., queer-identified)
Race	White	Black, brown, other
Religion	Mainstream Christianity	Non-Christian (e.g., Muslim, Jewish, Hindu)
Age	Middle-aged and young adults	Children, teens, elderly
Ability	Fully able cognitively and physically	Cognitive and/or physically disabled
Class	Upper-middle and middle income	Lower income, poor, homeless
Culture	Western cultural values (e.g., individualism, capitalism)	Non-Western cultural values (e.g., collectivism)
Ethnicity	European American	African American, Asian American, Hispanic American, Native American, biracial, and nonresident and resident immigrants
Education	College education	Trade school, community college, military, high school, GED, or less
Sexuality	Heterosexual	Bisexual, gay, lesbian, other nonheterosexual identity
Spirituality	Moderate expression	High, low, or no expression

This social location chart assumes that the categories are obvious. For instance, a person may be Muslim but assumed to be Christian until learned otherwise. A woman may be a lesbian but assumed to be straight. A person with a disability may be assumed to be able-bodied. Many times, people are categorized or stereotyped based on the dominant culture or what they look like even if that is not how they actually identify. Finding out the identities of supervisee, supervisor, and clients in these 11 salient areas often takes dialogue, rather than a simple visual scan or assumption based on dominancy of an identity in the local culture.

Another important concept is *intersectionality*, defined as the ways in which various dimensions of each person's GRRAACCEESS interact in nuanced ways to impact experiences of marginalization and privilege. An example referenced earlier is the Black, gay man embedded in a religious community that is not accepting of gay people. When he is coming out to his family and church religious community, he will have a far different experience than a White, gay man coming out whose family has had other members come out and is attend a gay-affirming church. A White, high school educated single mother with a biracial son will experience something far different from a Black, college-educated single-parent mother with a biracial daughter. Brown (2008) tells of his journey as a Black therapist in training and the various dimensions and frustrations he encountered in understanding intersectionality: "As a straight, young, educated, handsome, financially independent Black male without a criminal record, I intuitively knew that I had 'options'" (p. 5). He was confronted with his male privilege, realizing that when he partnered in relationships he did so with patriarchal power. This type of privileged power allows a male to structure the interactions of men and women in every sphere of life; it affects prospects and treatment in public life, in civil society, in the workplace, the street, the home, and the bedroom; it conditions our deepest understandings, affects, and habits; and it benefits men and harms women on the whole (Brown, 2008, p. 5).

Crenshaw (1993) notes that while identity-based politics has been a source of empowerment, we must be careful not to ignore intragroup differences. Each individual's experience is unique.

Techniques and Activities for Addressing Differences in Power and Privilege Based on Social Location and Institutionalized "ISMs"

It is critical that supervisees and supervisors discuss clients (and themselves) in terms of their social-location dimensions in order to recognize and be attentive to power and privilege and how those factors may affect the various relationships. One way to do this is to use the chart in Table 7.1 as a "privilege map." Map (put an X) where the supervisor and the supervisee and the clients fall along the 11 dimensions. This assists in keeping possible biases at the forefront of supervision and allows all participants to stay intentional in managing the power and privilege differences involved. Each time a case is discussed, the map can be brought out (from the case chart) to remind the supervisor and the supervisee of social location issues to keep in mind and attend to. It also supports sensitivity to the biases that clients have experienced in their lives from privileged people and institutions as well as biases the clients may hold.

Addressing Our Own Biases and Stereotypes

Although the values and beliefs of most supervisees and supervisors could be described as "fairness for all along all the dimensions," most of us hold biases and prejudices that must be examined and managed. The biggest challenge is for White therapists "who believe that they are just, unbiased and nonracist" (Sue et al., 2007, p. 280) to admit and be honest about their biases. Furthermore, "[t]he harm they perpetrate against their clients of color is either unknown or minimized" (Sue et al., 2007, p. 280). Unknown biases "not only oppress and harm, but they place clients of color in the unenviable position of a catch-22" (Sue et al., 2007, p. 280). The inherent power and transference position of the therapist puts a client in a place of having to sort through the intent. In a relationship that is intended to be built on trust, the client is sorting through if the therapist is aware or unaware, prejudiced or not prejudiced—and all these choices potentially lead the client to a position of not trusting. This type of harm is often referred to as microaggression. *Microaggressions* are acts of discrimination and are identified in three forms: microassaults, microinsults, and microinvalidation (Nadal, Griffin, Wong, Hamit, & Rasmus, 2014). *Microassaults* are more overt than the other forms of microaggression and come in the form of verbal or nonverbal attacks or avoiding behaviors, such as a White woman not getting on an elevator with a Black man. *Microinsults* are behaviors or statements that are rude or insensitive by degrading a person's racial heritage or identity, such as telling an Asian American woman that she is so articulate or following a teenage African American boy around a store. *Microinvalidations* are negations or denials of another person's thoughts, feelings, or experiences, such as when a Caucasian person tells a person of color that racism no longer exists or to stop complaining about their race and move forward.

It is critical to recognize that, despite our best intentions and stated values, most of us hold biases and stereotypes, and we commit forms of microaggressions to marginalized persons. When people are different from us, we are primed to have some sort of biased, prejudiced, or stereotyped response. This unintentional aggression comes from our invisible *implicit bias.* This happens when the neurons that process information in our brains embed certain images and schemas about people and things, based on the society in which we grew up (Stanley, Phelps, & Banaji, 2008). Greenwald and Banaji (1995) define *implicit attitudes* as "introspectively unidentified (or inaccurately identified) traces of past experience that mediate favorable or unfavorable feeling, thought, or action toward social objects" (p. 8).

When we are from privileged domains (e.g., White, educated, male), our privilege is often invisible to us. However, when we are from marginalized domains (e.g., Hispanic, poor, female), we are well aware that we have less privilege, and we know who has more. For example, a White male who is middle class and Christian may never have thought of himself as privileged, and therefore certainly has not considered how to manage that power in a therapy session. His client, who may be a Black, lower-middle-class woman, is well aware of the stereotypes imposed upon her and the lower privilege and status she holds as well as the privilege the therapist holds and may not be aware of. In order for him to effectively build an alliance, the therapist must understand the client's experiences, be aware, and manage his power and privilege.

Techniques and Activities for Addressing Our Own Biases and Stereotypes

There are a number of techniques and activities for addressing our own biases and stereotypes. One way of supporting supervisees in raising their awareness of implicit bias is to have them (and the supervisor) take the Implicit Association Test (IAT; Greenwald, McGhee, & Schwartz, 1998) developed by researchers at Harvard University. This test measures implicit, unconscious attitudes by assessing the underlying automatic process of one's associations. When considering socially sensitive issues, those who take the IAT report a greater degree of bias compared to when they take explicit tests about biases (Livingston & Drwecki, 2007; Teachman, Gapinski, Brownell, Rawlins, & Jeyaram, 2003). The test can be found at implicit.harvard.edu/implicit. In or out of the supervision session, you can view some of the YouTube videos of scholars, teachers, and even the test creators. In all these videos, it is clear that despite our values and beliefs, we must attend to our automatic implicit bias.

In order to normalize, recognize, and manage our automatic thoughts and biases, it is essential to learn to be comfortable sharing them in supervision. Again, the neurons in our brains that process information have been embedded with certain images and schemas about people based on the society in which we grew up (Stanley et al., 2008). This type of processing happens so quickly that often it is hard, if not nearly impossible, to prevent. We call this type of automatic biased thought a "first thought." What we encourage supervisees to do is recognize the first thought and bring forward a "second thought." Our second thought is reminding ourselves that we must recognize our power and privilege to manage it effectively. When we recognize our biases and challenge them, we minimize the chance of doing harm.

Second thoughts can be explored in supervision. For example, after being assigned a particular client, a supervisee might say she thought the client would be uneducated, poor, Hispanic, and have issues with alcohol. She also thought the client might have an illegal status. This was all before she had even met the client and was based on minimal information such as her name. The supervisor can normalize this as an experience of implicit bias and compliment the honesty of sharing and recognizing it. Together, they can strategize questions the therapist can ask the client, such as what does she do for fun, what is she most proud of, what stereotypes has she had to deal with, and what is it she most wants the therapist to know about her. This can help to expand a limited understanding of a client. Ellis (2010) wrote, "The question is, where are you on the continuum of racism or sexism, and what are you doing about it?" as opposed to "Do you have biases?" (p. 96). Our brains are primed to keep our own privileges invisible to us and for our biased, stereotyped thinking to permeate us. Rather than having diversity and the "isms" be the background of our supervision, it is critical and imperative that we make them the foreground of our work with supervisees and their clients.

A popular metaphor known as the "knapsack" can be a useful tool in talking with supervisees about invisible/hidden privilege (Macintosh, 1989). This metaphor supposes that each of us is born with a "knapsack" that contains our privileges (e.g., White, male, upper class). Some people begin life with many more privileges in their "knapsack" than others. We suggest that we (metaphorically) bring our knapsacks around to the front where we can see them because, when our knapsacks are on our backs, we can walk around "bumping" people with them (our privilege) and not be aware of or managing our bias and power. We cannot give up our privilege (empty our knapsacks or take them off), but we can manage them and intentionally work to prevent the potential harm they can inflict on others. We can become more aware of the privileges we have and learn to be more sensitive to others who may not have those same privileges.

Laszloffy and Habekost (2010) provide a list of experiential activities to encourage trainees to understand and experience the impact of marginalization. For each activity, participants write a reflective journal entry of what was experienced, including their thoughts, feelings, the impact of the experience, and its implications. One activity is to have the supervisee go to the local welfare office and fill out an application. Another is having supervisees attend an event where they are in the minority position. In teaching supervisees about confronting sexual orientation bias, supervisees are asked to wear a rainbow pin/button for a week or to walk hand-in-hand in public with someone of their same gender or sex.

Other activities that can be used to promote sensitivity and awareness of power and privilege are included in McGeorge and Carlson's (2010) suggestions to start each supervision session by asking supervisees to reflect on how their social location and privilege is influencing their assumptions and understanding of their clients. Mio and Barker-Hackett (2003) suggest that supervisees journal in between supervision sessions about how they recognize their social location and privilege in therapy sessions as well as the times they were unable to recognize them. In the next supervision session, these experiences are shared and discussed. Lee and Vennum (2010) propose the "cultural bump" exercise, where for 6 weeks supervisees keep a daily journal of critical events that cause them to stop and learn or recognize something about their own culture. Divac and Heaphy (2005) suggest the "Line of Privilege" exercise (p. 281) where supervisees stand on an invisible line of privilege with one end representing the highest privileged and the other end being the least privileged, based on one of the dimensions of GRRAACCEESS. Each supervisee (in a group supervision setting) speaks from his or her social location and shares a story of what it is like to be in that position. Supervisees then change social location positions and speak from this new position and compare and reflect on these experiences. Divac and Heaphy share "Monica's Fifty Questions" from a workshop with McGoldrick (2004), where she posed a series of questions that begin with "Would you stand up if you . . ." and end with statements like "remember the day Martin Luther King died? If you earn more than your partner? If you changed class/social status through marriage?" (Divac & Heaphy, 2005, p. 282). In all these activities, the purpose is to raise awareness about diversity and its many complexities that are ever-shifting and ever-active in society.

We also suggest a few activities to address biases when presenting cases. The first one is called "Dimensions Withheld." In this activity, the discussion of a case intentionally does not begin with the demographic or social location information. After the case is discussed, the supervisor explores what sort of client the group envisioned (e.g., was the client Native American? wealthy? Catholic?). Then the social location information is integrated and the case is discussed again to see if anyone feels differently about the case or treatment plan. Some differences could be positive, based on having cultural competencies, whereas others could be negative if they are embedded in bias. It is interesting to see if additional ideas surface in efforts to be culturally sensitive as well as if supervisees think differently about the case given the new information and dimensions. For instance, a couple comes into therapy for relational difficulties. Their presenting complaint is communication difficulties as well as one person in the relationship wanting to relocate back to their hometown so they can be around their sibling who is having a new baby. The group can discuss some of their hypotheses about the case as well as some ideas about how to work with the couple. Next, the supervisor would have the supervisee talk about the withheld dimensions of social location. In this version of the case, a Caucasian, gay couple comes in for relational difficulties. Their presenting complaints are communication difficulties, an inability to negotiate on whether to move closer to family, and limited employment options for one of the partners. Now reconsider the case and notice what would be different based on replacing the withheld dimensions of social location: for instance, such as a heterosexual, Hispanic couple, where the husband has limited employment opportunities and the wife wants to be closer to family.

Another activity that can be done with a case is "dimension replacement." In this activity, the social location dimensions are intentionally changed when the case is presented. For instance, in our previous example of the Caucasian, gay couple, the social location positions of race and sexual preference could be changed. The couple could be biracial and coming in for relational difficulties due to communication and differing priorities about family. Their presenting complaint is that the husband comes from a close Hispanic family and his brother's family is having its first child. The wife, a Caucasian woman, has investment in her career and it would be deeply and negatively impacted by moving to where her husband grew up. Changing the dimensions in our cases creates the opportunity to look through different lenses as well as to become aware of our implicit bias that can blind us. This activity ends with saying what the social locations actually are and discussing what differences consequently arise in the case planning.

Movies are another avenue through which we can explore our implicit bias. It can be beneficial to assign supervisees to watch movies that include various GRRAACCEESS and then discuss them. For instance, the movie *Crash* (Moresco et al., 2004) portrays the intersections and complexities of gender, race, ability, age, culture, class, education, and ethnicity. Implicit bias is present throughout the movie, revealing its destructive and devastating consequences. The movie *Gran Torino* (Eastwood, 2008) is about the Hmong culture assimilating into America and the changing demographics of a once blue-collar, White neighborhood amidst overt bias and racisms. *Frozen River* (Rae & Hourihan, 2008) is the story of two single-parent mothers' encounters with implicit bias around gender, race, culture, ethnicity, class, and discrimination. While both women confront issues of social location discrimination in regard to gender, class, and education, there is still a clear distinction of privilege for the Caucasian woman versus the Native American woman.

Our thoughts in this section can best be summed up by referring readers to the YouTube video by Chimamanda Adichie, a Nigerian writer, titled *The Danger of a Single Story*. We suggest that supervisors and supervisees watch together. Through the metaphor of "a single story," Adichie eloquently teaches us all about how quickly our minds sum up marginalized persons (e.g., the immigrant, the Jew, the disabled)—yet our minds do not do this with people in categories of power (e.g., White, male). Simply put, we reduce those with less power to a single story without getting to know them or explore who they are. Our mantra in supervision is "don't single-story the clients or each other." Everyone's story is filled with variations of the GRRAACCEESS and, therefore, everyone's story is different.

We need to be willing to share our own unique story with our supervisees. For example, a supervisor might share how she grew up in a conservative religious background where the values included monogamous, covenant marriages between a male and female. The supervisor can share that over the years, as she has grown and learned about diversity, cultural humility, implicit bias, and intersectionality, she does her best to continually keep these values and biases in check by realizing the way they minimize and devalue others. We also need to hear our supervisees' stories in order to help them draw out and hear their clients' unique stories.

Embracing Our Roles in Social Justice and Advocacy

Goals to Embrace Our Roles in Social Justice and Advocacy

1. Identify areas where advocacy can be a part of treatment.
2. Know what advocacy means and how to take action with clients and in the profession.

EMBRACING OUR ROLES IN SOCIAL JUSTICE AND ADVOCACY

It is not enough to address diversity issues in our work with clients in the therapy room. We must also intervene wherever and whenever possible outside of the therapy room. Direct interventions on behalf of our clients must be done within legal and ethical guidelines. For example, if a poor female client reports that she cannot get a callback from a doctor's office, we can empower her to continue to advocate for herself; however, we could also advocate for her by calling the doctor's office for her. The client can sign an authorization to release information and we can call her doctor with her in the room. When we make that call, we might consider the most therapeutic way in which to let the doctor know that we are concerned, given our client's marginalized status, that she may not be getting the care and attention needed. Our hope with advocacy is not only to assist our clients, but also to educate and challenge the many institutions that function in inequitable ways. We can also write letters to courts on our clients' behalf and intervene with schools and legal systems. Again, it is always important to attend to ethical and legal as well as clinical issues.

Techniques and Activities to Embrace Our Roles in Social Justice and Advocacy

In addition to our direct advocacy work on behalf of clients, we can also work to end injustice wherever we encounter it. We can make a commitment to both personally and professionally take action in our areas of influence. The entire July/August 2014 issue of *Family Therapy Magazine* was devoted to social responsibility. This is a great resource for supervisees.

The website, *It's Pronounced Metrosexual* (itspronouncedmetrosexual.com), has great resources for how to be intentional in advocacy and social justice on many levels. One of its pages suggests "Things you can do today for social justice" (itspronouncedmetrosexual.com/2012/11/things-you-can-do-today-for-social-justice).

Other ways of being intentional include paying attention in work settings. For instance, when serving on a search, tenure, or promotion committee for a university, we can keep an eye out for hidden biases and injustice. We can also do this at our agencies and community mental health systems as we notice how clients are treated in different ways. We can also make intentional decisions to have our research benefit marginalized communities.

CONCLUSION

It is our hope that this discussion of supervision and diversity will help supervisors recognize and develop practical ways in which they can move diversity to the forefront of supervision. As we equip the next generation of couple and family therapists, we must prepare them not only to be effective in our changing and diverse world, but also to be effective at changing the world. In doing so, Markovic (2008) noted:

> Systemic ideas can lead us to be seduced by our position on the margin (or meta-position), making us vulnerable to what I sometimes call "systemic arrogance." By privileging our critical ability to question, deconstruct, take risks and encompass multiple context levels, we can put ourselves above it all and ironically, become blind to the limitations of our practice. Being systemic does not protect us from being compliant to the very practices we purport to combat. (as cited in Simon, 2010, p. 323)

Given this, we must be diligent in our pursuit of cultural humility and be aware of and manage our implicit biases. Our privilege as supervisors puts us in a position of being responsible for creating safety in supervision and for clients. In our privileged place, we need to be aware of and manage how that privilege affects the supervision process and ultimately the therapy room. We need to remember that the complexity created by the various intersections of social location leaves us in a place of remaining culturally humble in our stance so we can be curious, sensitive, and aware of the lived experiences of those we supervise.

CASE EXAMPLE

A. AAMFT Approved Supervisor—Dr. Bob
The supervisor is a 53-year-old Caucasian male. He is a seasoned supervisor who has been credentialed as an AAMFT Approved Supervisor for more than 20 years. He provides supervision for both students (in practicum and internship as an adjunct professor), and postgraduates (at his agency).

B. Supervisee—Brittney
The supervisee is a 33-year-old Caucasian female, in her first semester of internship at a local nonprofit agency. She is particularly interested in doing family therapy and is excited to have the opportunity to work with a broad range of families.

(continued)

C. Family

Travis (29) and Tekia (27) are an African American couple with two children, Tyrone (5) and Tommy (4). Tekia has been a stay-at-home mom. Travis was in the military (as has been every male in his family) and "served two tours of duty in Afghanistan." He was injured when an improvised explosive device (IED) exploded and left him severely injured. He has spent many months in the hospital and rehabilitation. He was discharged from the military (which he had wanted to make his career) as a result of his injuries, and now has difficulty finding a job. He is working two part-time jobs, while hoping to find a full-time job.

Presenting Problem:

- Couple relationship and parenting issues, as Travis reportedly has a "short fuse."
- There are no reported safety concerns.
- Continuing recovery from his injury.
- Job issues—inability to find full-time employment/financial strain.

Travis's older brother is still in the military and was recently redeployed to Afghanistan.

Considering the Case Example From a Diversity Supervision Position

In Dr. Bob's first supervision session with Brittney, he discussed his own social location privileges, including his race, gender, educational level, income status, and so on. He wanted to create a supervision environment that was safe, and together he and Brittney made the decision that she would refer to him as Dr. Bob. She wasn't comfortable calling him by his first name, and calling him Dr. Jones seemed too formal.

When Brittney began talking about this case with Dr. Bob, she provided him with her completed Social Location Form, as she does for each case (see Table 7.2). It included her clients' positions as well as Dr. Bob's and her own social locations (SLs). She and Dr. Bob discussed and responded to the questions on the bottom of the form: (a) Given the SL of the client, what factors might you need to pay attention to in terms of culture, oppression, and bias? (b) Intersectionality that you notice; (c) Given the SL of the therapists and the supervisor, what blind spots might occur? (d) What areas might benefit from your advocacy?

(*continued*)

TABLE 7.2 Completed Social Location Form

	MOST PRIVILEGED	LESS PRIVILEGED
Gender	Men AM, S	Women AF, T Transgendered
Race	White S, T	Black, brown, other AM, AF
Religion	Mainstream Christianity S, T, AM, AF	Non-Christian (e.g., Muslim, Jewish)
Age	Middle-aged & young adults	Children, teens, elderly
Ability	Fully able Cognitive & physical S, T, AF	Cognitive and/or physical disabilities AM
Class	Upper-middle & middle income	Lower income, poor, homeless
Culture	Western cultural values (e.g., Individualism, capitalism, etc.) AM	Non-Western cultural values (e.g., collectivism, etc.)
Ethnicity	European-American S, T	African American, Asian American, Hispanic American, Native American, biracial & nonresident and resident immigrants AM, AF
Education	College education S, T	Graduate degree, trade school, community college, military, high school, GED, or less
Sexuality	Heterosexual S, T, AM, AF	Bisexual, gay/lesbian, transgender, etc.
Spirituality	Moderate expression	High or low expression

Note. AF = adult female, AM = adult male, S = supervisor, T = therapist.

Given the SL and Intersectionality of the Clients, What Should Be Considered?

The supervisor and supervisee consider the social location and intersectionality of Travis and his family. They note that they want to be

(*continued*)

aware of possible cultural oppressions or bias embedded within the presenting problem. Travis is Black and has disabilities, which are both societal barriers to getting employment. He reports having two part-time jobs and being unable to find full-time work. This has led to financial strain and demoralization. He is likely up against institutionalized ableism and racism in his search for a job. Tekia is female and Black. Given that she has been a stay-at-home mom, she may be struggling with entering the workforce and the need for the family to obtain two incomes. This may also be an adjustment for their marriage, depending on why they chose for Tekia to stay at home and not work outside the home. Black women and military wives are often the heads of their households and, given Travis's and Tekia's new situation, they will need to make some adjustments in how authority and responsibility for the home and children are managed. These adjustments may include gender equity implications, cultural issues, or religious beliefs or spiritual practices. Travis went from being a fully able-bodied male with a career in the military to having disabilities and difficulties with employment. He may be struggling with societal, cultural, religious, and his own and others' gender expectations. He may also be struggling with the spiritual meaning behind what happened to him and how it affects his religious expression and faith. He is likely very frustrated with the multiple changes in his life given his injuries and therefore has difficulty managing his emotions, particularly anger. Couple and parenting issues will be addressed, keeping in mind these social context factors.

Given the SL of the Therapists and the Supervisor, What Blind Spots Might Exist?

Both the therapist and the supervisor are fully able-bodied and therefore may be blind to the many implications, both physical and emotional, that disability presents for Travis. Neither therapist nor supervisor has lived with a spouse who is not fully able-bodied and therefore may be blind to the struggles and challenges that Tekia faces. Although both the therapist and supervisor identify as Christian, neither sees this as a major force in decision making in their lives. Dr. Bob considers whether religion may in fact be a major force for the couple given the importance they have placed on Tekia's staying at home with the children. Exploring how religion plays a role in what the couple values and in their possible community support is important. Furthermore, Dr. Bob discusses Nancy Boyd-Franklin's work concerning the African American culture. Specifically, Boyd-Franklin and Lockwood (2009) discussed the skepticism that can

(*continued*)

arise in the African American community when a therapist ignores the deep spiritual roots of the African American culture. This is especially significant given the life-changing events that are in process for Tekia and Travis. Pargament (2007), who has done extensive research and training in helping mental health workers know how to explore spirituality with their clients, wrote, "Traumatic events have been associated with changes in religious beliefs, perhaps in part because these events represent spiritual violations" (p. 97). In a study of persons who had been subjected to trauma, Falsetti, Resick, and Davis (2003) found that when a person was diagnosed with posttraumatic stress disorder, 50% of the participants either became more religious or less religious after the traumatic event. These events can either deepen the client's spiritual resources or challenge them. How therapists choose to address these issues (or ignore them) can have a significant impact on clients (Pargament, 2007). Even when therapists represent themselves as secular, empirical studies reveal that there still are changes in the client's images of God (Tisdale et al., 1997). Aponte and Carlsen (2009) suggest that:

> [T]he deprived and disadvantaged need a renewal of spirit that is not dependent on socioeconomic conditions. In this crucible of society's failure, spirituality offers hope—the ability of the most disadvantaged to transcend the turbulence of life and find meaning in every personal struggle. (p. 129)

Both Brittney and Dr. Bob are White and privileged with regard to race. The clients are Black and therefore have certainly experienced both micro- and macroaggressions due to their race. Given institutionalized racism, it will be important that the mental health system strive to not be another area in which they experience bias. In addition, Brittney and Dr. Bob need to be sensitive to the fact that in the African American community, there is a reluctance to discuss "family business in public"—and therapy is considered very public (Boyd-Franklin & Lockwood, 2009). Both Brittney and Dr. Bob are also economically middle- to upper-middle class, whereas the clients are struggling financially. It would be very easy not to appreciate the effort that everyday life tasks take when money is an issue. Given the strong military background family-of-origin that Travis comes from, he may feel disconnected from his family and older brother who was just redeployed. He may feel disempowered and "less than," given his discharge and injuries. All of these issues should be explored within the context of the GRRAACCEESS.

(*continued*)

What Areas Might Benefit From Your Advocacy?

Social advocacy may be needed in areas such as employment, health care, and childcare services. If clients report frustrations with other systems in which they are trying to access services, therapists (within ethical and legal guidelines) can make calls, write letters, and so forth to advocate for the clients. For example, if the VA is not returning calls or helping with filing paperwork needed to access treatment, the therapists can be an additional voice to move this process along. Given race and ability, clients may experience institutionalized racism and ableism in these areas.

Another consideration is the implicit bias Dr. Bob and Brittney may hold in their neurological processing systems and how those biases might come into play in this case. These biases do not necessarily reflect values and beliefs they both hold but are the hidden, invisible biases that we are socialized into holding. Brittney shares that when she thinks about a Black man in the military with a "short fuse," she immediately sees someone very angry and violent. She also has stereotypes about injured military men all having PTSD and being dangerous to others. Dr. Bob asks her to "click for clients" to learn about these topics from the literature in order to bring in knowledge to address this bias. For example, a click on "soldiers and loss of limbs" brought up the following link (journalistsresource.org/studies/government/security-military/combat-incurred-bilateral-transfemoral-limb-loss-comparison-vietnam-wars-afghanistan-iraq). This link leads to information that provides quality-of-life and mental health information from soldiers who lost limbs and served in Vietnam, Afghanistan, or Iraq (Dougherty, McFarland, Smith, & Reiber, 2014). For soldiers who served in Afghanistan or Iraq and as a result of their service wear a prosthesis, 70% experienced some form of mental health issue. Specifically, 20% suffered from depression and 10% suffered from PTSD. In addition, only 60% of soldiers wearing prostheses from the Afghanistan or Iraq wars reported being employed (Dougherty et al., 2014).

Next, Dr. Bob shares that he used to have a bias that if women did not stay home with young children, the children would get inferior care in daycare; however, he notes that his bias is less strong when the woman is Black or Hispanic rather than White. He has read extensively on this subject and shared with Brittney that it is inadequate daycare (poverty) rather than daycare itself that leads to poor outcomes for children. In fact, children in quality daycare generally have excellent outcomes and having a mother in the workforce is a family protective factor that increases financial stability. Dr. Bob and

(*continued*)

Brittney discussed that both of their stereotypes are present in the media and in politics. For instance, Black women are often told to "get off welfare and get to work and not be lazy" and White women are told to "get in the minivan and stay home" or be called selfish. Obviously, this message has embedded in it judgments as to which children are more important in society and speaks to the intersectionality of race, gender, and economics.

Case Planning

Brittney and Dr. Bob collectively decide to use a solution-focused approach with this case in order to focus on resiliency and strengths, especially given the social location challenges of race and ability. Brittney will look for exceptions to the conflicts in Travis's and Tekia's couple and parenting issues and highlight them. Now that she is more aware of the deep spiritual role of religion and the church in African American culture, Brittney will also explore the possibility of the church community resource as an ongoing support system for the couple. In guiding Brittney through this unfamiliar territory, Dr. Bob refers Brittney to read Nancy Boyd-Franklin and Tonya Walker Lockwood's chapter in Froma Walsh's *Spiritual Resources in Family Therapy* (2nd ed., 2009) as a guide to therapeutic work with this couple's spiritual resources.

REFERENCES

Adams, D. (Ed.). (1995). *Health issues for women of color: A cultural diversity perspective.* Thousand Oaks, CA: Sage Publishing.

Adichie, C. (2009). *The Danger of a Single Story* [Video file]. https://www.youtube.com/watch?v=D9Ihs241zeg

Ali, R. K., & Bachicha, D. L. (2012). Systemic supervision practices compared: A closer look at "reflection" and "self" in multisystemic therapy and family therapy supervision. *Clinical Child Psychology and Psychiatry, 17*(2), 192–207.

American Association for Marriage and Family Therapy (AAMFT). (2015). *AAMFT code of ethics.* Alexandria, VA: Author.

Aponte, H. J., & Carlsen, J. C. (2009). An instrument for person-of-the-therapist supervision. *Journal of Marital and Family Therapy, 35*, 395–405.

Arnold, M. S. (1993). Ethnicity and training marital and family therapists. *Counselor Education and Supervision, 33*, 139–147.

Berger, P. (1963). *Invitation to sociology.* New York, NY: Doubleday.

Bernard, J. M. (2014). The use of supervision notes as a targeted training strategy. *American Journal of Psychotherapy, 68*(2), 195–212.

Bernard, J. M., & Goodyear, R. K. (1998). *Fundamentals of clinical supervision* (2nd ed.). Boston, MA: Allyn & Bacon.

Bertrand, M., & Mullainathan, S. (2004). Are Emily and Greg more employable than Lakisha and Jamal? A field experiment on labor market discrimination. *American Economic Review, 94*, 991–1013.

Boyd-Franklin, N., & Lockwood, T. W. (2009). Spirituality and religion: Implications for psychotherapy with African American families. In F. Walsh (Ed.), *Spiritual resources in family therapy* (2nd ed., pp. 141–155). New York, NY: Guilford Press.

Brown, A. L. (2008). I too am feminist: The journey of a Black male in transformative feminist family therapy. *Journal of Feminist Family Therapy, 20*(1), 1–71. doi:10.1080/0895280801907036

Burnham, J. (1992). Approach-method-techniques: Making distinctions and creating connections. *Human Systems, 3*, 3–27.

Burnham, J. (1993). Systemic supervision: The evolution of reflexivity in the context of the supervisory relationship. *Human Systems, 4*, 349–381.

Burnham, J., Palma, D. A., & Whitehouse, L. (2008). Learning as a context for differences and difference as a context for learning. *Journal of Family Therapy, 30*, 529–542.

Commission on Accreditation for Marriage and Family Therapy Education (COAMFTE). (2005). *Accreditation standards: Graduate & post-graduate marriage and family therapy programs* (version 11.0). Alexandria, VA: Author.

Constantine, M. G. (1997). Facilitating multicultural competency in counseling supervision: Operationalizing a practical framework. In D. B. Pope-Davis & H. L. Coleman (Eds.), *Multicultural counseling competencies: Assessment, education and training, and supervision* (vol. 7, pp. 310–324). Thousand Oaks, CA: Sage.

Cook, D. A. (1994). Racial identity in supervision. *Counselor Education and Supervision, 34*, 132–141.

Council for Accreditation of Counseling and Related Educational Programs (CACREP). (2009). *2009 CACREP accreditation manual.* Alexandria, VA: Author.

Crenshaw, K. (1993). Demarginalizing the interaction of race and sex: A Black feminist critique of antidiscrimination doctrine, feminist theory, and antiracist politics. In D. Wiesberg (Ed.), *Feminist legal theory: Foundations* (pp. 255–287). Stanford, CA: Stanford University Press.

Deihl, L., & Ellis, M. V. (2009a). *The contribution of the supervisory working alliance to burnout and vigor among residential counseling paraprofessionals* (Unpublished doctoral dissertation). University at Albany, NY.

Deihl, L., & Ellis, M. V. (2009b, June). *The relevance of the supervisory working alliance in the supervision of residential frontline staff.* Paper presented at the Fifth International Conference on Clinical Supervision, Buffalo, NY.

De Reus, L. A., Few, A. L., & Blume, L. B. (2005). Multicultural and critical race feminisms: Theorizing families in the third wave. In V. L. Bengtson, A. C. Acock, K. R. Allen, P. Dilworth-Anderson, & D. M. Klein (Eds.), *Sourcebook of family theory and research* (2nd ed., pp. 447–467). Thousand Oaks, CA: Sage.

Divac, A., & Heaphy, G. (2005). Spaces for GRRAACCEESS: Training for cultural competence in supervision. *Journal of Family Therapy, 27*, 280–284.

Dougherty, P. J., McFarland, L. V, Smith, D. G., & Reiber, G. E. (2014). Bilateral transfemoral/transtibial amputations due to battle injuries: A comparison of Vietnam veterans and Iraq and Afghanistan service members. *Clinical Orthopedics and Related Research, 472*(10), 3010–3016. doi:10.1007/s11999-014-3534-9

Dupree, W. J., Bhakta, K. A., Patel, P. S., & Dupree, D. G. (2013). Developing culturally competent marriage and family therapists: Guidelines for working with Asian Indian American couples. *American Journal of Family Therapy, 41*, 311–329.

Eastwood, C. (Producer & Director). (2008). *Gran Torino* [Motion picture]. United States: Warner Brothers.

Ellis, M. V. (2010). Bridging the science and practice of clinical supervision: Some discoveries, some misconceptions. *The Clinical Supervisor, 29*, 95–116.

Fadiman, A. (1997). *The spirit catches you and you fall down.* New York, NY: Farrar, Straus, & Giroux.

Falsetti, S. A., Resick, P. A., & Davis, J. L. (2003). Changes in religious beliefs following trauma. *Journal of Traumatic Stress, 16*(4), 391–398.

Fama, L., & Ellis, M. V. (2005, August). Vicarious traumatization: A concern for doctoral level psychology trainees? Paper presented at the 113th Annual Convention of the American Psychological Association, Washington, DC.

Flemons, D. G., Green, S. K., & Rambo, A. H. (1996). Evaluating therapists' practices in the postmodern world: A discussion and a scheme. *Family Process, 35*, 43–56.

Fong, M. L., & Lease, S. H. (1997). Cross-cultural supervision: Issues for the White supervisor. In D. B. Pope-Davis & H. L. Coleman (Eds.), *Multicultural counseling competencies: Assessment, education and training, and supervision* (pp. 387–405). Thousand Oaks, CA: Sage.

Fukuyama, M. A. (1994). Critical incidents in multicultural counseling supervision: A phenomenological approach to supervision research. *Counselor Education and Supervision, 34*, 142–151.

Greenwald, A. G., & Banaji, M. R. (1995). Implicit social cognition: Attitudes, self-esteem, and stereotypes. *Journal of Personality and Social Psychology, 102*, 4–27.

Greenwald, A. G., McGhee, D. E., & Schwartz, J. L. K. (1998). *Measuring individual differences in implicit cognition: The Implicit Association Test, 74*(6), 1464–1480.

Hardy, K. V., & Laszloffy, T. A. (1995). The cultural genogram: Key to training culturally competent family therapists. *Journal of Marital and Family Therapy, 21*, 227–237.

Hardy, K. V., & Laszloffy, T. A. (2002). The cultural genogram: An application. In C. Storm & T. Todd (Eds.), *The reasonably complete systemic supervisor resources guide* (pp. 34–39). New York, NY: iUniverse.

Helms, J. E., & Cook, D. A. (1999). *Using race and culture in counseling and psychotherapy: Theory and process.* Boston, MA: Allyn & Bacon.

Hook, J. N., Davis, D. E., Owen, J., Worthington, E. L., & Utsey, S. O. (2013). Cultural humility: Measuring openness to culturally diverse clients. *Journal of Counseling Psychology, 60*(3), 353–366.

Inman, A. G. (2006). Supervisor multicultural competence and its relation to supervisory process and outcome. *Journal of Marital and Family Therapy, 32*(1), 73–85.

Killian, K. D. (2001). Differences making a difference: Cross-cultural interactions in supervisory relationships. *Journal of Feminist Family Therapy, 30*, 113–129.

Ladany, N., Ellis, M. V., & Friedlander, M. L. (1999). The working alliance in clinical supervision: Relations with trainees' self-efficacy and satisfaction. *Journal of Counseling and Development, 77*, 447–455.

Ladany, N., Inman, A. G., Constantine, M. G., & Hofheinz, E. W. (1997). Supervisee multicultural case conceptualization ability and self-reported multicultural competence as functions of supervisee racial identity and supervisor focus. *Journal of Counseling Psychology, 44*, 284–293.

Laszloffy, T., & Habekost, J. (2010). Using experiential tasks to enhance cultural sensitivity among MFT trainees. *Journal of Marital and Family Therapy, 36*(3), 333–346. doi:10.1111/j.1752-0606.2010.00213.x

Lawless, J. J., Gale, J. E., & Bacigalupe, G. (2001). The discourse of race and culture in family therapy supervision: A conversation analysis. *Contemporary Family Therapy, 23*, 181–197.

Lee, M. M., & Vennum, A. V. (2010). Using critical incident journaling to encourage cultural awareness in doctoral marriage and family therapy students. *Journal of Family Psychotherapy, 21*(4), 238–252.

Lee, R. E., & Nelson, T. S. (2014). *The contemporary relational supervisor*. New York, NY: Routledge.

Lim, S., & Nakamoto, T. (2008). Genograms: Use in therapy with Asian families with diverse cultural heritages. *Contemporary Family Therapy, 30*(4), 199–219.

Livingston, R. W., & Drwecki, B. B. (2007). Why are some individuals not racially biased? Susceptibility to affective conditioning predicts nonprejudice toward blacks. *Psychology Science, 18*(9), 816–823.

Lor, P. (2013). A Hmong professional woman's reflections and perspectives on the influences affecting the changing roles of Hmong women in America. *Journal of Cultural Diversity, 20*(1), 4–48.

Macintosh, P. (1989, July/August). White privilege: Unpacking the invisible knapsack. *Peace and Freedom*, 10–12.

Markovic, D. (2008), Personal communication as cited in Simon, G. (2010). Self-supervision, surveillance and transgression. *Journal of Family Therapy, 32*, 308–325. DOI: 10.1111/j.1467-6427.2010.00505.x

McGeorge, C. R., & Carlson, T. S. (2010). Social justice mentoring: Preparing family therapists for social justice advocacy work. *Michigan Family Review, 14*(1), 42–59.

McGoldrick, M. (2004). Exercise presented at a workshop hosted by the Institute of Family Therapy, London.

McGoldrick, M., Pearce, J. W., & Giordano, J. (2005). *Ethnicity and family therapy* (3rd ed.). New York, NY: Guilford.

Mio, J. S., & Barker-Hackett, L. (2003). Reaction papers and journal writing as techniques for assessing resistance in multicultural courses. *Journal of Multicultural Counseling and Development, 13*, 12–19. Retrieved from http://www.multiculturalcenter.org/jmcd/

Montgomery, C., Hendricks, B., & Bradley, L. J. (2001). Using systems perspective in supervision. *The Family Journal, 9*, 305–311. doi:10.1177/1066480701093011

Moresco, R., Schulman, C., Yari, B., Cheadle, D., & Harris, M. R. (Producers), & Haggis, P. (Director). (2004). *Crash* [Motion picture]. United States: Lion's Gate.

Nadal, K. L., Griffin, K. E., Wong, Y., Hamit, S., & Rasmus, M. (2014). The impact of racial microaggressions on mental health: Counseling implications for clients of color. *Journal of Counseling & Development, 92*, 57–66. doi:10:1002/j.1556-6676.2014.00130.x

Nieto, L., & Boyer, M. F. (2006). Understanding oppression: Strategies in addressing power and privilege. *Colors NW Magazine, 5*, 30–51.

Nolte, L. (2007). White is a colour too: Engaging actively with the risks, challenges and rewards of cross-cultural family therapy training and practice. *Journal of Family Therapy, 29*, 378–388.

Pargament, K. I. (2007). *Spiritually integrated psychotherapy.* New York, NY: Guilford.

Pendry, N. (2012). Race, racism and systemic supervision. *Journal of Family Therapy, 34*, 504–418. doi:10.1111/j.1467.2011.00576.x

Polnik, M. (2013). Conscience? *Family Therapy Magazine, 12*(5), 17–19.

Rae, H., & Hourihan, C. (Producers), & Hunt, C. (Director). (2008). *Frozen river* [Motion picture]. United States: Sony.

Richeport-Haley, M. (1998). Ethnicity in family therapy: A comparison of brief strategic therapy and cultural focused therapy. *American Journal of Family Therapy, 26*(1), 77–90.

Roper-Hall, A. (1998). Working systemically with older people and their families who have "come to grief." In P. Sutcliffe, G. Tufnell, & U. Cornish (Eds.), *Working with the dying and bereaved: Systemic approaches to therapeutic work*. London, UK: Macmillan.

Roper-Hall, A. (2008). Systemic interventions and older people. In R. Woods & L. Clare (Eds.), *Handbook of the clinical psychology of aging* (2nd ed.). London, UK: Wiley.

Silvestri, T. J. (2003). The temporal effect of supervisor focus, the supervisory working alliance, and the graduate training environment upon supervisee multicultural competence. *Dissertation Abstracts International: Section B: The Sciences and Engineering, 63*(12), 6108.

Simon, G. (2010). Self-supervision, surveillance and transgression. *Journal of Family Therapy, 32*, 308–325.

Social responsibility. (2014). *Family Therapy Magazine, 13*(4), 1–72.

Son, E. J., Ellis, M. V., & Yoo, S. K. (2007). The relations among supervisory working alliance, role difficulties, and supervision satisfaction: A cross-cultural comparison. *Korean Journal of Psychology, 26*, 161–182.

Son, E. J., Ellis, M. V., & Yoo, S. K. (2009, August). Clinical supervision in South Korea and the U.S.: A comparison of supervisees' perceptions. Paper presented at the 117th Annual Convention of the American Psychological Association, Toronto, Canada.

Stanley, D., Phelps, E., & Banaji, M. (2008). The neural basis of implicit attitudes. *Current Directions in Psychological Science, 17*(2), 164–170.

Steinpreis, R. E., Anders, K. A., & Ritzke, D. (1999). The impact of gender on the review of the curricula vitae of job applicants and tenure candidates: A national empirical study. *Sex Roles, 41*(7/8), 509–528.

Sue, D. W., Capodilupo, C. M., Torino, G. C., Bucceri, J. M., Holder, A. M. B., Nadal, K. L., & Esquilin, M. (2007). Racial microaggressions in everyday life: Implications for clinical practice. *American Psychologist, 62*(4), 271–286. doi:10.1037/0003-066X.62.4.271

Teachman, B. A., Gapinski, K. D., Brownell, K. D., Rawlins, M., & Jeyaram, S. (2003). Demonstrations of implicit anti-fat: The impact of providing causal information and evoking empathy, *Health Psychology, 22*, 68–78.

Tisdale, T. C., Key, T. L., Edwards, K., J., Brokaw, B. F., Kemperman, S. R., & Cloud, H. (1997). Impact of God image and personal adjustment, and correlations of the God image to personal adjustment and object relations development. *Journal of Psychology and Theology, 5*, 227–239.

Walsh, F. (Ed.). (2009). *Spiritual resources in family therapy* (2nd ed.). New York, NY: Guilford Press.

Zimmerman, T. S. (Ed.) (2001). *Integrating gender and culture in family therapy training*. Binghamton, NY: Haworth Press.

SECTION III

Theory-Specific Supervision

EIGHT

Training the Structural Therapist

Sharon Apel Bursky
Ryan M. Cook

Structural Supervision, adapted from the Structural Family Therapy model developed by Salvador Minuchin, is an effective supervision model used in Marriage and Family Therapy (MFT) training. Similar to the model used in therapy, this supervision model focuses on the present hierarchy that exists between supervisor and supervisee as well as providing the supervisee with the necessary tools and skills needed to develop into a competent and effective Marriage and Family Therapist. A productive supervision process involves engagement from both the supervisor and supervisee. Structural supervision highlights that the structure is to be established and encouraged by the supervisor by maintaining a hierarchical, yet supportive, position. This chapter discusses the components of using Structural Theory in supervision; the philosophical foundation of the theory; the use of techniques and modalities in supervision; the role of the therapist and supervisor; the process of change; and culture, gender, and diversity issues related to the use of this theory in supervision. It also includes a case example.

TEACHING THEORETICAL CONTEXT

Providing a theoretical foundation for supervisees is essential for the growth and development of Marriage and Family Therapists as well as for setting a standard for the profession. Marriage and Family Therapists are known for using strong theoretical approaches during therapy; however, the supervision experience sets the stage for a foundation in therapy. As crucial as the use of theory is in therapy, using a theoretical approach in supervision is equally important. This chapter focuses on the use of Structural Theory in supervision and how this theory is used to provide a theoretical foundation for the supervisee as well as to facilitate an effective supervisory relationship.

FOUNDERS OF STRUCTURAL FAMILY THERAPY AND SUPERVISION

Structural Family Therapy was developed by Salvador Minuchin and his colleagues/collaborators (e.g., Montalvo, Guerney, Rosman, Schumer, Haley, and Fishman) in response to their perception of a need to work with a specific population that was not responding to other available treatment approaches (Jones & Lindblad-Goldberg, 2008). Minuchin and his colleagues noticed that families with behavioral-disordered children who lived in the inner city, were members of minority groups, and who were impoverished needed a more action-oriented, concrete treatment (Jones & Lindblad-Goldberg, 2008). This theory was initially developed as a structure for therapy with families; however, some lead contributors to the theory adapted this model for use as a supervision model. Lead contributors to Structural supervision include Haley, Madanes (McCollum, 1990), and Montalvo (1973).

Live supervision, which seeks to create a new sense of reality, evolved from the Structural approach to treating families (Liddle & Halpin, 1978). Liddle and Halpin (1978) eloquently compare Structural Family Therapy and Structural supervision, saying, "whereas Minuchin attempts to create a sense of competence in family members by reshaping and reframing reality through the creation of therapeutic events, Montalvo accomplishes similar goals with his supervisees through supervisory interventions during the session" (p. 79). This comparison illustrates the isomorphic nature of the therapy and supervision processes through the use of Structural Theory.

Madanes and Haley (1977) believe that supervision should be done from a problem-solving approach and suggest that training in family therapy should focus on teaching students what to do in therapy. To successfully accomplish this, Minuchin pioneered new therapy interventions and training techniques through utilizing the one-way mirror and videotaping of sessions (Rockinson-Szapkiw, Payne, & West, 2011). At that time, the one-way mirror and video playback in supervision were being used exclusively in Structural supervision and not utilized in any other theory (or practice) of supervision. Therefore, this model of supervision set the stage and standards for live and video supervision. Montalvo (1973) worked from the Structural approach and addressed supervisory skills and goals through the use of a live supervisory model. According to Montalvo (1973), the task of the supervisor is to ensure that the therapist does not get caught in the family's unproductive patterns and to encourage the therapist to take control and direction of the therapy session (Liddle & Halpin, 1978). Use of the one-way mirror and video playback allowed supervisors to intervene and guide the supervisee when necessary.

PHILOSOPHICAL FOUNDATION

The primary goal in the Structural model of supervision is correcting the hierarchy (Todd & Storm, 2002). When working from a Structural approach, goals are more cognitively based: They focus on defining therapeutic skills and ways supervisees can intervene in dysfunctional systems (Liddle & Halpin, 1978). The Structural approach implies that an expert's knowledge and skills can be demonstrated and transmitted to a supervisee, and as long as the supervisees can trust the receipt of this knowledge, they can test it, integrate it, and eventually make it their own (Liddle & Halpin, 1978). This model is especially useful with beginning clinicians because it focuses on skills, directives, and working with clients. These tangible skills learned in Structural supervision provide the supervisees with guidance and allows them to develop the necessary skills needed to become clinicians. The Structural supervisor models the use of an appropriate therapeutic stance and use of therapeutic skills within the presence of a hierarchy to help the therapists-in-training understand what role they play when working with families.

Minuchin empowered supervisees working with and under him to become better therapists by what many called his "*servant leadership*" stance (Katzenbach & Smith, 1994; Rockinson-Szapkiw et al., 2011), where he encouraged intellectual stimulation and collaboration (Cooklin, Asen, Ayers, & Glienke, 2006). Many confuse the hierarchical stance with a "what I say goes" position. In actuality, Minuchin took a teaching and coaching position with supervisees in order to help them reach their potential. Rockinson-Szapkiw and colleagues (2011) describe the stance of servant leadership as one who embodies four factors: idealized influence, inspirational motivation, intellectual stimulation, and individual concern. *Idealized influence* is the aspect of the supervisor demonstrating high moral and ethical standards while challenging the supervisee to take responsibility and be active in the supervision and therapy process (Warren, 1977). *Inspirational motivation* in the Structural supervision model is the supervisor communicating expectations for the supervisees while helping the supervisees know their role in achieving therapeutic and supervision goals (Rockinson-Szapkiw et al., 2011). Minuchin knew the importance of a "network" of providers. This led him to support and encourage supervisees to join with the family and become a part of that network before assisting the family in utilizing this network to achieve change. Minuchin describes *intellectual stimulation* as encouraging creativity and challenging the norms in therapy and supervision (Rockinson-Szapkiw et al., 2011). Minuchin would inspire supervisees by sharing his thoughts on ways to challenge the established system, discussing ways to empower others, and implementing strategies that are necessary to bring about change. Intellectual stimulation has always been a goal since the initial development of the Structural supervision therapy model (Cooklin et al., 2006). Lastly, *individual concern* is described as the component of supervision where the supervisor will collaborate and encourage, as well as challenge and confront, the supervisee with flexibility and fluidity that allow for individual growth (Minuchin, 1974). When the servant leadership stance is utilized in supervision, it allows the supervisee to learn from the modeling of the supervisor. The Structural supervision model focuses on establishing a structure that is both encouraging and empowering for supervisees while simultaneously challenging the status quo in order to produce change (Minuchin, Montalvo, Guerney, Rosman, & Schumer, 1967).

An important aspect of the Structural model is called *joining*, although the literature portrays it as less important in the supervisor-supervisee relationship (Todd & Storm, 2002). Joining in the supervision relationship is done through matching the supervisee's body language, verbal language, and energy level in an effort to most effectively connect with the supervisee. Joining can also be accomplished through the appropriate use of self; for example, Minuchin states that "joining has nothing to do with pretending to be what you are not. It means tuning into people and responding to the way they move you" (Minuchin & Nichols, 1993, p. 42). The authors of this chapter argue that joining and the use of self are important to the supervisor-supervisee relationship. They permit the creation of a safe environment in which the supervisee can develop skills and share this intimate journey.

A key principle of Structural Family Therapy is the assumption that families are open systems that are organized so that members of the family relate to one another in structured and patterned ways that can be observed and predicted (Jones & Lindblad-Goldberg, 2008). *Family structure*, which is defined as "the recurring, enduring, and regulating patterns of interaction that develop over time as family members manage the common day-to-day tasks, needs, and connections of family members," is a crucial element in Structural Family Therapy (Jones & Lindblad-Goldberg, 2008, p. 333). Structural Family Therapy operates under the assumption that although variations exist, families tend to organize into subsystems with hierarchies that are usually organized generationally (Jones & Lindblad-Goldberg, 2008). This principle is also an important assumption for supervision, and it should mirror the type of relationship that occurs between the supervisor and supervisee. First, supervision should operate as an open, organized system in which the supervisor and supervisee relate in a structured and patterned manner. For example, directives, skill teaching, and feedback should be consistent. Second, Structural supervision should have a hierarchical organization. This allows for fewer power struggles and better delivery of directives as well as facilitating more of a teaching environment.

THE PROCESS OF CHANGE

Structural Family Therapy posits that change and growth can occur when a trusting relationship is developed within the therapist-client relationship as well as the supervisor-supervisee relationship. According to Structural Family Therapy, change is also thought to occur when transactional patterns are lived out in enactments and reenactments. These enactments and reenactments allow the development of alternative and productive communication patterns. Creating healthy boundaries and an effective hierarchy are also an important process of change in Structural Theory. When relationships are established and patterns become visible, change is created by fostering the ability to manage psychological distance and space while defining and strengthening the individual and subsystem boundaries (Vetere, 2001).

Structural supervisors focus on competencies and strengths of the supervisee, which the supervisor hopes will also encourage a parallel process between the supervisee and the families with which the supervisee works. Focusing on supervisee competencies and strengths creates a safe environment to assist them to work through situations where they may feel therapeutically hindered or stuck (Minuchin & Nichols, 1993; Simon, 1995). Structural supervision attempts to remove constraints that hinder the strengths of the supervisee. This can be done by the supervisor normalizing the "mistakes" that are made in the early stages of a therapist's development as well as through encouragement and continued assessment of the strengths and competencies the supervisee possesses.

A component of change in supervision occurs when the supervisor empowers supervisees through modeling and encouraging a balance between being challenging and supporting (Daft, 2008; Hodgkinson, 1991; Kouzes & Posner, 2007). The Structural supervisor will challenge supervisees to be aware of their own impact on the therapeutic process with clients. Just like the isomorphic process that occurs within supervision and therapy, the supervisor affects the supervisee and supervision process. Conversation about and awareness of the isomorphic process is encouraged to stimulate insights that can be utilized in the therapy sessions, specifically in the realm of joining and appropriate disclosure of self (Cooklin et al., 2006).

ROLE OF THE THERAPIST IN WORKING WITH FAMILIES

The Structural therapist works to bring about a structural change: namely, providing clients with alternative ways of interacting (Colapinto, 1988). Colapinto states that the therapist should assist in creating a context for the family to experience alternative patterns, which will assist the family in structuring and improving familial functioning. This is done through reworking of the system's boundaries and hierarchy that may have been contributing to the conflicts. The therapist's role is to teach the family how to modify their communication and interactions and assist in strengthening the family structure. The goal is to help families create a new way of communicating and interacting but also, and more importantly, to establish a more functional familial structure (Colapinto, 1988). The Structural therapist often takes the stance of a teacher and coach, encouraging and challenging the system. While providing this coaching, the therapist is simultaneously joining with the family in order to create a strong working relationship with the family system. As this process is occurring, the therapist encourages the system to discover strengths that will allow the change that has occurred to be maintained.

ROLE OF THE THERAPIST IN SUPERVISION

The role of the supervisee therapist in the supervision process is to work within the hierarchy established by the supervisor and be open to discuss topics and issues such as techniques, interventions, transference, countertransference, and parallel processes. The hierarchy in supervision is especially important because it sets the stage for modeling the use of hierarchies when the supervisee works with families and for families to create their own hierarchies. The role of the supervisee therapist is to develop the techniques and skills needed to become an effective therapist by being open to guidance and coaching from the supervisor and by being willing to discuss difficult topics.

TECHNIQUES IN STRUCTURAL SUPERVISION

As noted earlier, *joining* is an important concept in Structural supervision as well as an intervention. Joining is considered an intervention, as it is an action the supervisor takes in order to build the relationship between the supervisor and the supervisee, which is typically done in the initial stages of supervision. Jones and Lindblad-Goldberg (2008) discuss how creating growth through promoting interpersonal experiences encouraged by the therapist or supervisor involves two actions: the actions the therapist/supervisor takes with families/supervisees (e.g., doing with) and through the way the therapist/supervisor relates with families/supervisees (e.g., being with). The first, "doing with," refers to the use of concrete tools or methods/techniques that are taught; the second, "being with," refers to the emotional stance the therapist/supervisor takes with the family/supervisee (Jones & Lindblad-Goldberg, 2008). Both the "doing with" and the "being with" are equally important steps in joining and illustrate how this intervention is delivered and maintained throughout the supervision process.

Boundary making is another important intervention in Structural supervision, as it models appropriate behavior for the supervisees as they work with clients. Boundaries can be described on a continuum from diffuse (too close, overinvolved) to rigid (too closed-off) (Jones & Lindblad-Goldberg, 2008). In Structural supervision, it is essential for the supervisor and supervisee to have clear boundaries that are neither too diffuse nor too rigid. Boundaries enhance the supervisor's ability to teach necessary skills to supervisees, and assist in directing the supervisee appropriately, without having dependency or autonomy issues. It is important to be alert to potential issues or problems within the hierarchy and to problems with boundaries; preparation should be made to intervene if and when problems arise (Todd & Storm, 2002).

Enactments are yet another important intervention used in Structural supervision, which are typically utilized during live supervision. This format allows observation of the patterns and enables the supervisor to direct supervisees and deliver interventions that facilitate change (Todd & Storm, 2002). Role-playing can be a useful intervention during a therapeutic situation in which the supervisee feels "stuck," as it allows the supervisor to intervene and redirect (Todd & Storm, 2002, p. 178). Enactments and role-playing are the most common methods therapists use to create change in Structural Family Therapy (Jones & Lindblad-Goldberg, 2008); therefore, they are also two of the most effective interventions used in Structural supervision. Using enactments and role-plays in supervision allows the supervisor to create rich opportunities to help the supervisee become more aware of self-defeating patterns and to practice relating in new ways (Jones & Lindblad-Goldberg, 2008). Enactments and role-plays are thus critical parts of Structural supervision (Todd & Storm, 2014); these techniques allow the supervisor to present difficult or challenging situations that can mirror similar questions, stances, reactions, or emotions a client may have when addressing a specific situation.

A technique that is well known to Structural supervision is the use of *live supervision*. In live supervision, there are times when the supervisor may step into the therapy room and address a hierarchy or boundary to adjust the unbalanced aspect of the relationship. The supervisor may also interrupt a therapy session and discuss with the therapist his or her observations of the situation. The supervisor and supervisee therapist will then develop interventions to implement when the therapist returns to the session. The supervisor may also use the technique of *intensifying*, whereby the supervisor increases the intensity of a situation to help the supervisee become more aware of when she or he is being too tolerant of extreme behaviors in a session (Todd & Storm, 2014).

The *supervisor's use of self* can be a powerful and useful technique but should be approached with caution equal to that exercised when using the technique of the therapist's use of self. A supervisor can employ this technique—for example, by sharing and using past or current experiences (life or therapeutic)—to motivate the supervisee to act (Todd & Storm, 2014). Minuchin also looked at this concept from the angle of the therapist or supervisor recognizing one's self and what one brings to therapy and supervision. Minuchin encouraged the supervisor/supervisee/therapist to take responsibility for using the self to initiate the changing of the status quo in the client family while becoming a change agent in the places in which one works and lives (Lee & Everett, 2004).

CULTURE/RACE AND GENDER ISSUES—RESEARCH TO CONSIDER

Gender

Although gender is a fundamental dimension of supervision in Marriage and Family Therapy, little research has examined gender and Structural supervision (Nelson, 1991). Although there is no literature that specifically addresses gender and Structural supervision, many authors discuss gender generally within the supervision of MFT. Nelson (1991) suggests that all supervisors are gendered and have been socialized based on gender norms.

This author suggests that the field of family therapy does not provide an adequate definition of *gender* and that the social definitions of gender tend to place women as being subordinate to men in power hierarchies (Nelson, 1991). Nelson (1991) further suggests that it is important for supervisors and therapists to remember that the power dimension in society is real and has a strong impact on families, therapy, and supervision. Issues of gender in family therapy training should focus on the interactions between clients and therapists in terms of gender (Nelson, 1991). Brodsky (1980) reports that because sex-role stereotypical attitudes influence the process of therapy, one's perspective will affect the interpretation and perception of interventions based on whether the therapist/supervisor is male or female. Nelson (1991) asserts that it is extremely important for supervisors to attend to their own gendered attitudes and biases and to examine these fully for the effects they may have in and on supervision.

McInnes Miller and Ivey (2006) discuss gender in supervision and how female and male supervisors in the field typically do not fit gender stereotypes, which results in men often being perceived as more open than their female counterparts. It is suggested that training in MFT may limit the degree to which supervisors and clinicians function within the traditional gender roles, such as male supervisors tending to work from a more collaborative style (McInnes Miller & Ivey, 2006). The American Association for Marriage and Family Therapy (AAMFT) guidelines for Approved Supervisors require that supervisors be sensitive to contextual variables such as culture, gender, ethnicity, and economics (AAMFT, 1999).

Diversity

As noted earlier, Structural supervision does not address issues of gender or culture directly. The model was developed to be used within a variety of family dyads; due to its versatility, it can be implemented within a variety of diverse cultures (Connell, 2010). Minuchin, in the origination of Structural supervision, utilized his training techniques to train individuals from the community (urban and Hispanic community) to become therapists and to utilize Structural Family Therapy (Bowen & Lawler, 1995). This theory, from its inception, intended supervision and training techniques to work with supervisees from different minority backgrounds; however, there is limited literature regarding Structural supervision's effectiveness across cross-cultural populations (Weaver et al., 2013). Waldegrave (2009) states that when working from a Structural framework, it is important to be sensitive to multicultural perspectives and to the beliefs, practices, race, ethnicity, religion, gender, level of acculturation, special needs, socioeconomic status, and behavioral expectations.

The literature on the topic of supervision and diversity tends to report that MFT training programs lack diversity in regard to students, faculty, and supervisor populations (Lawless, Gale, & Bacigalupe, 2001). These authors state that there is an absence of African American students and faculty, and if this is paralleled for other ethnic minority groups, such as Hispanics and Asians, this lack of diversity may create an environment where discourse on race, ethnicity, and culture does not take place (Lawless et al., 2001). Brown and Brown (1995) emphasize the need to understand the cross-cultural supervision dynamics. They also state that ways are needed to incorporate cross-cultural dynamics into supervisory models, as supervisors directly influence the knowledge, attitudes, and skills of a supervisee through their own knowledge, attitudes, and skills.

Weiling and Marshall (1999) conducted a study that asked marriage and family supervisors and students to compare and contrast their experiences working with ethnic minority and White persons. Findings from this study indicate that both the supervisors and the students placed great value on having cross-cultural supervisory experiences; however, they report that these opportunities are limited (Weiling & Marshall, 1999). The empirical studies on supervision and diversity that have been conducted deal mostly with race and exclude other cultural and ethnic factors that may influence the supervisory relationship (Weiling & Marshall, 1999). Tolhurst-Christiansen et al. (2011) discuss four ways a supervisor can address these multicultural issues in practice. First, discuss the importance of a supervisor addressing the emotional reactions of the supervisor, supervisee, and client to the situation(s); second, understand that many multicultural instances come up unplanned; third, understand that supervisees need their supervisors to provide support and validation to the experiences, including their negative reactions; and fourth, recognize that a safe space must be available where the experience can be processed at the supervisor level as opposed to the therapeutic level. Stone (1997) reports that it is incredibly easy to avoid cultural factors and discussions in supervision unless they are actively sought out and addressed. He continues by noting that the failure to discuss cultural issues is due either to such issues not being readily apparent to the supervisor, or to the supervisor being uncomfortable with the topic, which leaves a wide gap in the supervisory process (Stone, 1997). The multicultural literature specific to the field of family therapy expresses that a supervisor and supervisee should continue to become competent in diversity through immersion in the culture, purposefully gaining knowledge and experience, and through continued research (McDowell, Goessling, & Melendez, 2012).

Limitations of Diversity Literature in Structural Supervision

In the early stages of Structural Family Therapy and supervision (1970s and 1980s), and particularly during the feminist movement in the 1980s, this theory was criticized and challenged to address the issues of power differential, diversity, and multiculturalism (Goldner, 1998; Vetere & Cooper, 2000). The feminist movement at the time did not feel that these issues were properly addressed within Structural theories (Hare-Musten, 1994). Since that time, more research and literature has been pursued to address these issues within Structural theory. Specific work by Goldner (1998) and Vetere and Cooper (2000) addresses the issues of power, safety, race, and culture by discussing the responsibility the therapist and supervisor have in addressing these issues within family therapy and supervision relationships. Due to the lack of literature on Structural supervision and gender and diversity, it is important for future studies to continue to investigate these issues and the impact gender and diversity issues have on supervision. Future studies should discuss the effectiveness of Structural supervision with diverse supervisees, including supervisees of opposite genders and diverse ethnicities as well as supervisees who hold different sexual orientations. This would provide valuable information regarding the effectiveness of this supervision model across cultures and belief systems. In addition, it would be interesting to see how issues of structure and hierarchy are seen across different cultures and ethnicities and whether this affects the supervision relationship.

ROLE OF THE SUPERVISOR AND SUPERVISEE IN STRUCTURAL SUPERVISION

Haley (1974) identifies four basic tasks for supervisors: teaching supervisees that family therapy is a new orientation to human problems; teaching supervisees the specific skills of family interviewing and how to conduct the initial session; teaching supervisees to develop goals in their work and how to give directives and use strategies in their therapy; and teaching supervisees that the responsibility for therapeutic success or failures is theirs and not to label their client as untreatable or resistant. These tasks are especially helpful in understanding the goals of supervision for beginning clinicians. As a supervisor, it is important to recognize that one must often instruct supervisees when needed and that it is the supervisor's responsibility to teach techniques, skills, interventions, and systemic thinking to supervisees. This chapter argues that the use of a hierarchy and working as a teacher within supervision fosters growth and skill development for supervisees.

Structural supervision has been described as a successful framework for working with students who are beginning their training (McCollum, 1990). McCollum discusses the dual obligation of a Structural supervisor: first, an obligation to the family to ensure that they receive competent treatment; and second, an obligation to attend to the supervisee's needs (such as reducing anxiety) and teach skills that provide a foundation so the supervisee is competent as a marriage and family therapist. In addition to this dual obligation, the Structural model recommends a clear hierarchy in supervision, as it does in the family structure (McCollum, 1990). McCollum also suggests that the supervisor take responsibility for the therapy and actively direct the work of the supervisee, who, in turn, directs the family. This hierarchy and structure increase the supervisee's sense of competence while reducing anxiety (McCollum, 1990).

A Structural supervisor takes an active role through development of interventions, interacting during live observations (Storm, 2012), and helping the therapist to develop interventions to be delivered in an optimal manner. The supervisor works to help the supervisee maintain an appropriate hierarchical position with the family. This is done by discussing the family's structure and hierarchy, and then encouraging the therapist to assess the "dance" in and out of the family's structure and system. The supervisor encourages the therapist to be aware of not getting overly enmeshed with or disengaged from the family and to be able to "dance" between being a part of the system yet an outside supporter of the system. The supervisor encourages discussion and assessment to increase awareness of dangerous or destructive cross-generational coalitions, such as being overly aligned with a child or favoring a spouse (Todd & Storm, 2014). Growth in the supervision relationship occurs when the supervisor and supervisee are aware of their individual selves and biases. The next step of growth occurs when each utilizes therapeutic and supervision techniques to alter and encourage adjustments to improve the system and the supervision relationship.

Supervisor-Supervisee Relationship

The supervisor is responsible for monitoring his or her own tendency to undermine the hierarchy and enter into covert alliances that constrain therapy (Storm, 2012). This means that the supervisor, similar to the supervisee with a family, must work to "dance" in and out of the system while working with the supervisee. A supervisor takes on the task of encouraging change through direct confrontation and techniques (Rockinson-Szapkiw et al., 2011)—what Minuchin and Fishman (1981) call the *stroke and kick* method. In the *stroke and kick* method, the supervisor encourages the supervisee as to strengths and successes, but also confronts the supervisee, such as encouraging the supervisee to implement an intervention in a session that he or she is hesitant to utilize due to fear of a negative reaction from the family. The supervisor guides and assists the supervisee but still allows the supervisee's strengths and talents as a therapist to develop and be refined as the supervisee gains experience in the field.

Learning Goals and Objectives

The supervisor encourages and guides the supervisee to begin developing a new system with the family: a system that involves the family and the therapist during their work in therapy. The supervisee will utilize joining and accommodation to begin this process. The supervisor will work with the supervisee to (a) support the family structure; (b) carefully track content and process in family interactions; (c) accommodate family members' styles and range of affect and interaction (through mirroring); and (d) be aware of the importance of joining without taking sides, without forgetting to be supportive when confronting (Vetere, 2001). The supervisor also encourages and guides the supervisee in assessing and exploring the family's structure specific to subsystems, boundaries, functions, relationships, and social supports.

Structural supervision focuses on assisting supervisees in becoming more aware of family system dynamics and patterns so they can become effective in knowing when and how to intervene to bring about change. Teaching the supervisee necessary tools and interventions is paramount to the supervision process. Structural supervision highlights the importance of structure, boundaries, and relational dynamics in both therapy and supervision. The overall goal of supervision is for the supervisee to develop skills and an awareness of the client's system and structure as well as of the supervisee's impact and interaction within that system.

Self of Supervisor

Self-disclosure should be done only after careful assessment of the reason for sharing these experiences, such as to encourage or normalize the supervisee's experiences. This reflection can provide important insight for the supervisor as well as facilitate discussion of the supervision relationship and can enhance one's growth as a supervisor. The importance of structure, boundaries, and hierarchy also applies in the supervision process. For example, it is imperative for the supervisor to develop a structure and establish boundaries in the supervision relationship. Not only is it necessary for the supervision relationship, it also models the use of the appropriate boundaries, hierarchy, and structure that are essential for the therapist to model with families.

The supervisor-supervisee relationship, just like the therapist-client relationship, affects the supervision and growth process. The supervision relationship has a profound impact on the supervisee's growth and development in the areas of empathic understanding, openness to change, commitment, communication, genuineness, and respect (Bernard & Goodyear, 2009; Falender & Shafranske, 2004; Nerdrum & Rønnestad, 2002; Young, Lambie, Hutchinson, & Thurston-Dyer, 2011). This can stimulate an isomorphic process wherein the supervisee utilizes the same stance of encouragement, structure, boundaries, and hierarchy when working with clients. The modeling that is utilized in supervision is typically also used in the supervisee's sessions with his or her own clients. This highlights the importance of understanding self-as-a-supervisor and self-as-therapist issues because they affect professional efforts and development in many ways. Because personal issues, values, and beliefs also affect supervision and therapy, it is important for both supervisor and supervisee to be readily aware of these during supervision. This awareness assists one in not pushing one's personal ideals, values, or beliefs on the supervisee or client. The supervisor should also be readily open to having discussions regarding outside influences and their effect on supervision and therapy in the context of developing ways to maintain appropriate boundaries and structure.

Strengths and Limitations of Structural Supervision

Working through the Structural supervision model has many advantages. First, it provides a framework wherein providing supervisees with skills, interventions, and direction is the focus. Second, this model works from a hierarchical standpoint, which reduces issues of power in the supervising relationship. Nevertheless, despite the many strengths of working with this model, there are limitations that should be noted. A majority of the supervision literature discusses working from a postmodern perspective (e.g., Anderson & Goolishian, 1990; Unger, 2006). The postmodern model of supervision emphasizes collaboration *without* a hierarchical structure, with the supervisor and the supervisee learning together and from one another (Unger, 2006). This model is appealing to many supervisors, due to a new collaborative wave in MFT and the desire for both supervisee and supervisor to work laterally. However, the Structural model of supervision does not function from a collaborative perspective, which can often make building the relationship between the supervisor and supervisee difficult.

Some would state that a shortcoming of the Structural supervision model is that it neglects the professional and personal growth of the supervisee (Todd & Storm, 2002). These authors view the absence of addressing personal growth of the supervisee as one of the largest limitations of the Structural supervision model. Structural supervision not only does not address the development of the supervisee, but also does not discuss development and growth of the supervisor. It is important to be aware of this limitation and make appropriate adjustments to provide an effective supervision process.

CASE EXAMPLE

Supervision of Brittney from a Structural standpoint will take into account Brittney's level of experience and stage of development as a therapist. Establishing a good supervision relationship would first begin with joining. Learning about Brittney and finding a way to connect with her would be of utmost importance. The next step would be a discussion of the goals and objectives of supervision. Establishing a structure and hierarchy in the supervision process is essential when working within this model. Dr. Bob would discuss protocols and the necessary steps that Brittney must follow regarding safety issues and other issues that may arise that she may need assistance with.

(continued)

Working from a hierarchical position and taking a teaching stance, Dr. Bob as the supervisor would take a leading role in the supervision process. This would include teaching Brittney the skills and tools she will need to provide effective therapy. Dr. Bob would focus on teaching the use of Structural Family Therapy techniques, such as joining with a family member, the use and purpose of the family map, how to facilitate enactments, dancing in and out of the system in a productive manner, and the use of skills such as reframing and challenging. Supervision would take the form of live supervision, with Dr. Bob observing Brittney working with the family, either from behind a one-way mirror or via video. If live supervision cannot be used, Dr. Bob would use a combination of video playback and case consultation. These forms of supervision would still allow for observation of Brittney's in-session work with the family and provide the opportunity to practice specific interventions with Brittney during her current work with the family.

Relating to the specific case regarding Brittney's work with the family of four (father, mother, and two boys), an initial case conceptualization would be processed during the supervision sessions.

> **Case conceptualization**: Analyzing the relationship and family dynamics, such as the marital dyad, overall parenting styles (i.e., who is the disciplinarian, who maintains rules and structure, permissive style, rigid or fluid, etc.), the parent-child relationships, any possible triangulations, what familial boundaries are present, if any alliances and coalitions exist among family members, completing a family map in order to illustrate these concepts, and outlining the family goals for therapy.

Dr. Bob would encourage a discussion regarding what theoretical approach Brittany will use with the family (e.g., Structural, strategic, postmodern, etc.) to assist in Brittney's theoretical orientation development. A discussion of ways in which Brittney could attempt to join with the family in order to understand patterns would be especially important to provide insight into the system's patterns and dynamics. Role-plays would be used in the supervision sessions to illustrate how to gently "dance" in and out of the family system while maintaining a firm hierarchical structure as well as practicing the use of enactments.

It will be necessary for Dr. Bob and Brittney to have a discussion regarding multiculturalism. This is essential to address in supervision

(*continued*)

for several reasons. First, the discussion of multiculturalism is important to have with each and every supervisee in order to understand their views and openness to the issues surrounding multiculturalism and diversity. Second, this conversation is important in order to ensure that the supervisee is comfortable with addressing any possible multicultural or diversity issues that are present in therapy sessions with clients. In regard to this case, Brittney is Caucasian and the family is African American, so it would be important for Brittney to address the cultural differences between her and the family and explore how the family feels and may react with regard to these differences. A role-play of this conversation would be extremely helpful between Dr. Bob and Brittney, so Brittney could feel confident and comfortable discussing these cultural differences with her clients.

Dr. Bob would provide Brittney with structure, support, and teaching of therapeutic skills so she feels competent in working with this family. He would address the importance of Brittney continuing to assess the mental and emotional safety of Travis and the other family members. This would include beginning a discussion about the impact the family's current life stressors have and any symptoms the family members may be experiencing.

Dr. Bob would encourage Brittney to discuss her thoughts and insights about the family in supervision. He would be interested in knowing what Brittney's thoughts are regarding the impact Travis's return from military duty may be having on the family (e.g., relationally, financially, occupationally, employment, emotional and mental health) and her perception of what the family's issues are. Dr. Bob would do this to encourage and coach Brittney to think outside the box regarding the possible impact this change has had on the family and to help Brittney become able to view familial problems from a systemic perspective. The discussion of Brittney's observations and hypotheses can assist in developing a discussion regarding what therapeutic interventions would be beneficial and useful in working with the family.

This case example explains the process of how Dr. Bob would structure supervision, teach, coach, and encourage Brittney as she works with her clients within a Structural framework. As illustrated, the Structural supervision process would focus on joining with the supervisee from a hierarchical, teaching stance. The Structural supervision model would provide Brittney with the guidance and tools for her to work effectively with the family, while also considering and addressing the necessary and apparent issues of diversity.

REFERENCES

American Association for Marriage and Family Therapy (AAMFT). (1999). *Approved supervisor designation: Standards and responsibilities.* Washington, DC: Author.

Anderson, H., & Goolishian, H. (1990). Beyond cybernetics: Comments on Atkinson and Heath's "Further thoughts on second-order family therapy." *Family Process, 29,* 157–163.

Bernard, J. M., & Goodyear, R. K. (2009). *Fundamentals of clinical supervision* (4th ed.). Needham Heights, MA: Allyn & Bacon.

Bowen, D. E., & Lawler, E. E. (1995). Empowering service employees. *MIT Sloan Management Review, 36,* 73–84.

Brodsky, A. M. (1980). Sex role issues in the supervision of therapy. In A. K. Hess (Ed.), *Psychotherapy supervision: Theory, research, and practice* (pp. 509–522). New York, NY: Wiley.

Brown, M., & Brown, J. (1995). Counselor supervision: Cross-cultural perspective. In J. G. Ponterotto, J. M. Casas, L. A. Suzuki, & C. M. Alexander (Eds.), *Handbook of multicultural counseling* (pp. 263–286). Thousand Oaks, CA: Sage.

Colapinto, J. (1988). The structural way. In H. Liddle, D. Breunlin, & R. Schwartz (Eds.), *Handbook of family therapy training and supervision* (pp. 17–37). New York, NY: Guilford Press.

Connell, C. (2010). Multicultural perspectives and considerations within structural family therapy: The premises of structure, subsystems, and boundaries. *Insight: Rivier Academic Journal, 6*(2), 1–6.

Cooklin, A., Asen, E., Ayers, M., & Glienke, O. (Producers/Directors). (2006). Inviting the family dance: Salvador Minuchin [DVD]. (Available from The Minuchin Center for the Family, 303 Fifth Avenue, Suite 603, New York, NY 10016)

Daft, R. L. (2008). *The leadership experience* (4th ed.). Mason, OH: South-Western Cengage Learning.

Falender, C. A., & Shafranske, E. P. (2004). Clinical supervision: A competency-based approach. Washington, DC: American Psychological Association.

Goldner, V. (1998). The treatment of violence and victimization in intimate relationships. *Family Process, 37,* 263–286.

Haley, J. (1974). Fourteen ways to fail as a teacher of family therapy. *Family Therapy, 1,* 1–8.

Hodgkinson, C. (1991). *Educational leadership: The moral art.* Albany, NY: SUNY Press.

Jones, C. W., & Lindblad-Goldberg, M. (2008). Ecosystemic structural family therapy: A primer. In K. Jordan (Ed.), *The quick theory reference guide* (pp. 331–347). New York, NY: Nova Science.

Katzenbach, J. R., & Smith, D. K. (1994). *The wisdom of teams: Creating the high-performance organization.* New York, NY: HarperBusiness.

Kouzes, J. M., & Posner, B. Z. (2007). *The leadership challenge* (4th ed.). San Francisco, CA: Jossey-Bass.

Lawless, J. L., Gale, J. E., & Bacigalupe, G. (2001). The discourse of race and culture in family therapy supervision: A conversation analysis. *Contemporary Family Therapy, 23*(2), 181–197.

Lee, R. & Everett, C. A. (2004). The integrative family therapy supervisor: A primer. New York, NY: Brunner-Routledge.

Liddle, H. A., & Halpin, R. J. (1978). Family therapy training and supervision literature: A comparative review. *Journal of Marriage and Family Counseling, 4*(4), 77–98.

Madanes, C., & Haley, J. (1977). Dimensions of family therapy. *Journal of Nervous and Mental Disease, 185,* 88–98.

McCollum, E. E. (1990). Integrating structural-strategic and Bowen approaches in training beginning family therapists. *Contemporary Family Therapy, 12*(1), 23–34.

McDowell, T., Goessling, K., & Melendez, T. (2012). Transformative learning through international immersion: Building multicultural competence in family therapy and counseling. *Journal of Marital and Family Therapy, 38*(2), 365–379. doi:10.1111/j.1752-0606.2010.00209.x

McInnes Miller, M., & Ivey, D. C. (2006). Spirituality, gender, and supervisory style in supervision. *Contemporary Family Therapy, 28*, 323–337. doi:10.1007/s10591-006-9012-0

Minuchin, S. (1974). Families and family therapy. Cambridge, MA: Harvard University Press.

Minuchin, S., & Fishman, H. C. (1981). *Family therapy techniques*. Cambridge, MA: Harvard University Press.

Minuchin, S., Montalvo, B., Guerney, B. G., Jr., Rosman, B. L., & Schumer, F. (1967). *Families of the slums: An exploration of their structure and treatment*. New York, NY: Basic Books.

Minuchin, S., & Nichols, M. P. (1993). *Family healing: Tales of hope and renewal from family therapy*. New York, NY: Free Press.

Montalvo, B. (1973). Aspects of live supervision. *Family Process, 12*, 343–359.

Nelson, T. (1991). Gender in family therapy supervision. *Contemporary Family Therapy, 13*(4), 357–369.

Nerdrum, P., & Rønnestad, M. H. (2002). The trainees' perspective: A qualitative study of learning empathic communication in Norway. *The Counseling Psychologist, 30*, 609–629. doi: 10.1177/00100002030004007

Rockinson-Szapkiw, A. J., Payne, L. Z., & West, L. C. (2011). Leadership lessons from Salvador Minuchin. *Family Journal: Counseling and Therapy for Couples and Families, 19(2)*, 191–197. doi:10.1177/1066480711400159

Simon, G. M. (1995). A revisionist rendering of structural family therapy. *Journal of Marital and Family Therapy, 21*, 17–26.

Stone, G. L. (1997). Multiculturalism as a context for supervision: Perspectives, limitations, and implications. In D. B. Pope-Davis & H. L. K. Coleman (Eds.), *Multicultural counseling competencies* (pp. 263–289). Thousand Oaks, CA: Sage.

Todd, T. C., & Storm, C. L. (2014). The complete systemic supervisor: Context, philosophy, and pragmatics, 2nd edition. New York, NY: Wiley.

Todd, T. C., & Storm, C. L. (2002). *The complete systemic supervisor: Context, philosophy and pragmatics*. New York, NY: Authors Choice.

Tolhurst-Christiansen, A., Thomas, V., Kafescioglu, N., Karakurt, G., Lowe, W., Smith, W., & Wittenborn, A. (2011). Multicultural supervision: Lessons learned about an ongoing struggle. *Journal of Marital and Family Therapy, 37*(1), 109–119.

Unger, M. (2006). Practicing as a postmodern supervisor. *Journal of Marital and Family Therapy, 32*, 59–71.

Vetere, A. (2001). Structural family therapy. *Child & Adolescent Mental Health, 6*(3), 133–139. doi:10.1111/1475-3588.00336

Vetere, A., & Cooper, J. (2000). Working systemically with family violence. In N. Singh, J. P. Leung, & A. N. Singh (Eds.), *International perspectives on child and adolescent mental health* (pp. 113–135). New York, NY: Elsevier Science.

Waldegrave, C. (2009). Cultural, gender and socioeconomic contexts in therapeutic social policy work. *Family Process, 48*(1), 85–101. doi:10.111/j.155-5300.2009.01269.x

Warren, R. L. (1977). *Social change and human purpose: Toward understanding and action*. Chicago, IL: Rand McNally.

Weaver, A., Greeno, C., Marcus, S., Fusco, R. A., Zimmerman, T., & Anderson, C. (2013). Effects of structural family therapy on child and maternal mental health symptomatology. *Research on Social Work Practice, 23,* 294–303.

Weiling, E., & Marshall, J. P. (1999). Cross-cultural supervision in marriage and family therapy. *Contemporary Family Therapy, 21*(3), 317–329.

Young, T. L., Lambie, G. W., Hutchinson, T., & Thurston-Dyer, J. (2011). The integration of reflectivity in developmental supervision: Implications for clinical supervisors. *The Clinical Supervisor, 30*(1), 1–18. doi:10.1080/07325223.2011.532019

NINE

Training the Strategic Therapist

Laura Tejada

TEACHING THEORETICAL CONTEXT

Strategic Therapy is a "deceptively simple" (Haley & Richeport-Haley, 2003, p. 1) approach to relational therapy that centers on a brief, highly focused course of therapy sessions aimed at reducing and eliminating problematic relational interaction patterns that are distressing to clients. Clients are viewed as being in distress due to these troublesome patterns in their relationships, not from mental or emotional shortcomings. Clients are also assumed to be actively working to solve the problems that have confronted them but to have become stuck along the way (Haley & Richeport-Haley, 2003; Nardone & Watzlawick, 2005). A Strategic Therapist conducts structured and collaborative therapy sessions during which presenting problems and specific treatment goals are clearly stated. Treatment goals and therapy interventions focus on disrupting troublesome relational and behavioral patterns and implementing more effective options.

Supervision of trainees within this model embodies these principles by making extensive use of *isomorphism:* that is, the framework for supervision of students is a direct reflection of what the Strategic Therapist would do in a client session. Thus, supervisees from the Strategic Therapy model are seen as capable of learning effective therapeutic skills, and the Strategic supervisor is to be an active and directive trainer (Haley, 1987).

This chapter outlines the therapy process from the perspective of Strategic Therapy and discusses its isomorphic relationship to the supervision process. It is hoped that this chapter will be the most help to supervisors who are not themselves experienced in Strategic Therapy but are supervising trainees who are in a clinical environment where they are receiving supervision from an on-site clinician who practices from this model. Providing effective supervision to trainees who may be working under two different supervisors is an issue at the heart of the Strategic Therapy supervision model. Strategic Therapy places emphasis on supervision of trainees who might be caught between supervisors with distinctly different theoretical orientations (Haley,

1987), such as might occur in an academic setting, or when trainees contract with a supervisor outside their agency in order to complete clinical supervision hours for credentialing purposes. (See Ungar and Costanzo [2007] for a related discussion of supervision that occurs outside the trainee's agency.)

Founders

Strategic Therapy evolved from a growing awareness in the 1950s and 1960s in the family therapy field of the importance of communication in maintaining behavioral patterns in relationships. It is not centered in one approach or philosophy of conducting Strategic Therapy but is most commonly associated with the models put forth by Jay Haley and the Mental Research Institute (MRI), both of which were influenced by the works of Gregory Bateson and Milton Erickson (Nardone & Watzlawick, 2005). Haley was a colleague of Gregory Bateson and John Weakland as part of The Palo Alto Group studying the double-bind and other communication patterns in the 1950s and 1960s (Haley, 1987; Nardone & Watzlawick, 2005). When this project ended, Don Jackson formed the Mental Research Institute, with a primary focus at that time of studying and conceptualizing the role of paradox in communication (Haley, 1987).

Strategic Therapy came into its own in the 1980s (Haley & Richeport-Haley, 2003), as Haley authored numerous books and articles from the late 1970s into the early 2000s and created a series of training videos. In the late 1980s, Haley wrote about training and supervision of clinicians in the Strategic Therapy model, stating the importance of live supervision at best and videotaped supervision at the least (Haley, 1987, 1996). Although Strategic Therapy models included children and adolescents as part of the relational system, Madanes (1984) highlighted the importance of viewing younger children as change agents in relational difficulties. In the 1990s, Strategic Therapy models and interventions were proposed in the areas of art therapy (Riley, 1990), in the treatment of children's concerns individually in school counseling (Schiff, 1990) and play therapy (Kissel, 1990), and relationally through Strategic Family Play Therapy (Ariel, 1992). As the field of relational counseling has transitioned to conform to the accountability standards for training and conducting therapy, core competencies have been identified for training in Strategic Therapy (Quick, 2012).

Philosophical Foundations

Haley directly credits the research into communication theory conducted by the MRI and the Palo Alto Group, the works of Milton H. Erickson and his perspectives on how change occurs, and the influence of Minuchin and other colleagues at the Philadelphia Child Guidance Clinic, along with Braulio Montalvo (Haley, 1980, 1987) as providing important foundational concepts for his approach to therapy and training (Haley, 1987). Each of these different perspectives supplies core foundational tenets that meld into the philosophy of Strategic Therapy.

The Communications School and the MRI Model

The elements of both groups of communication theorists (the Palo Alto Group and MRI) are incorporated into Strategic Therapy (Haley, 1990). Central contributions to Strategic Therapy from the works by John Weakland (1976) and Weakland, Fisch, Watzlawick, and Bodine (1974) emphasize that all communication must be considered within its cybernetic, relational context to be fully understood. Weakland (1976) states that, "There is no 'starting point' in an ongoing stream of interaction; the simply linear model of cause and effect is not appropriate. When we began to examine the more complex, yet highlight[ed] patterned repetitive interaction occurring in families, it became even clearer that the relevant epistemological model is one derived not from mechanics, but from cybernetics, where the focus is on the structure of interaction within some ongoing system" (p. 120). From this perspective, it is necessary to take a relational perspective (Nardone & Watzlawick, 2005) on how individuals and family units are motivated, how distressing behavior patterns arise and are maintained, and how change occurs. Central to this process is viewing both symptoms and clients positively and reframing them as active problem solvers seeking to bring stability to their family, "even if it is an unpleasant stability" (Haley, 1990, p. vii).

In relational systems, all verbal and nonverbal messages carry both a command and a report function, in which information may be imparted (the report function) along with a command for the receiver to change his or her behavior (Weakland, 1976; Weakland et al., 1974). The influence of these overt and covert messages in a relational system builds up over time; thus, "small signals may easily have large effects, and still further multiplication of effect can occur when one signal frames the interpretation of many others" (Weakland, 1976, p. 117). As the overt and covert verbal and nonverbal messages continue, distressing behavioral patterns evolve as family members attempt to cope with the often conflicting commands and reports being offered and received. Thus, the Communications School asserts "that specific behavior of all kinds is primarily an outcome of . . . communicative interaction within a social system; and . . . that 'problems' consist of persisting undesired behavior" (p. 121). From this perspective, distressing communication and behavioral patterns in family relationships should be the issues addressed in therapy, not the history of the problem or exploration of internal dynamics with "insight" as a goal (Haley, 1987, 1990).

Contributions of Milton Erickson

Once a relational therapist is charged with intervening in relational behavioral cycles in order to reduce client distress, rather than working to develop insight or explore the history of the presenting concerns, an active, goal-directed approach is necessary. The term *strategic* is used to describe the work of the therapist using this model, as defined by Milton Erickson, with whom Haley studied (Haley, 1973, 1990). Haley (1973) adopted Erickson's philosophy that "[t]herapy can be called strategic if the clinician initiates what happens during the therapy and designs a particular approach for each problem. . . . He must identify solvable problems, set goals, design interventions to achieve these goals, examine the responses he receives to correct his approach, and ultimately examine the outcome of his therapy to see if it has been effective" (p. 17). Because the family is assumed to be active problem solvers who are stuck in a first-order change process, the therapist has the responsibility to help the family gain a new way to behave and initiate second-order change to their family system (Nardone & Watzlawick, 2005). However, the therapist is also expected to empower clients, directly and indirectly, to bring about these changes. "The family therapist takes charge by laying down the general rules of the game for the family, and within this framework he places the family in charge of what is to happen in the room" (Haley, 1990, p. 172).

A significant philosophical foundation of Strategic Therapy influenced by Erickson is a positive view of clients as "changeable and curable" (Haley, 1996, p. 32). Also viewed positively is the hesitation clients have about attempting changes in their lives and relationships (Haley, 1973, 1996). Understanding the cybernetic principles of relational interactions gives the therapist an awareness of how hard it is for individuals and family systems to change and maintain change (Haley, 1990). This positive view of clients contributes to the philosophy of meeting clients where they are, working to understand their perspective of the presenting problem and the meaning they ascribe to it, and then using their language to craft interventions (Weakland et al., 1974).

Contributions From Minuchin

The Structural Family Therapy concepts of boundaries, hierarchy, and power are used to help Strategic Therapists frame interventions for the family (Haley, 1980, 1996), with the goal of behavior change and disruption of distressing cycles. Reorganization of the family system might be a potential outcome, but it is not the primary goal of the Strategic Therapy process. If the behavior cycle is disrupted and the family is able to stabilize in a less distressing pattern, they would be considered to have met their treatment goals, regardless of what other structural or relational concerns might be present. From this model, "the therapist . . . may also terminate treatment while still aware of other problems the patient may have" (Haley, 1990, p. 67).

Taken together, the philosophical foundations of Strategic Therapy model indicate a treatment goal of understanding the distressing behavioral patterns that are affecting the lives and relationships of clients, and then effectively assisting them to disrupt these cycles and develop more helpful interactions. A therapist is expected to be active in session and to craft interventions in collaboration with clients in order to interrupt problematic behaviors and the responses to them. Therapeutic success is determined to be the reduction and/or elimination of the distressing problems or symptoms, clients empowered to continue changes independently of the therapist, and maintenance of these gains after the termination of therapy (Haley, 1996; Haley & Richeport-Haley, 2001).

Process of Change

Four philosophical principles on how change occurs in Strategic Therapy are termed "heresies" by Nardone and Watzlawick (2005). These principles are:

- Passing from closed to open theoretical systems
- Focus on how rather than why
- The therapist is responsible
- Change comes before insight

Passing From Closed to Open Theoretical Systems

The process of change within the Strategic Therapy model is based upon a keen awareness of the cybernetic influences that shape our individual and relational behavior. From this perspective, change must occur within these relational interactions. To adopt this stance means that a Strategic Therapist must resolutely step away from the closed conceptualization of linear causality and understand presenting concerns and therapeutic interventions in systemic terms, moving from first-order interventions to the second-order interventions that engender lasting change (Nardone & Watzlawick, 2005; Watzlawick, 1978). A successful Strategic Therapist abandons the comfort of matching clients and presenting concerns to preconceived interventions, treatment models, and theories. Instead, a successful Strategic Therapist must be inventive and flexible in order to craft a unique intervention process for each client. This process is focused on the clients' manifestations of distressing relational patterns, not on the ideal of health presented by a theory or outcome measure (Haley, 1996; Nardone & Watzlawick, 2005).

How, Not Why

Taking a cybernetic, systemic view of relational patterns means that the focus in session is on process rather than content; that is, *how* the problem is described, experienced, and maintained, not *why* it developed or what factors are determined to be the cause. According to the Strategic Therapy school of thought, focusing on *why* eliminates the cybernetic, relational conceptualization of the presenting problem and related interaction cycles by reducing it to an individual, linear, cause-and-effect perspective. Weakland (1976) asserts that causal factors for relational difficulties should be considered only if there "exists clear and clearly relevant evidence—not just a possibility, or ambiguous signs—of some other significant causal factor such as organic pathology" (p. 121). But *how* these factors are considered is crucial to maintaining the nonpathologizing and problem-directed approach. The therapist must continue to focus on the observed behavior patterns by exploring *how* the family members and others in their realm label and judge these behaviors. "This distinction is important because in some cases—for instance, parents' overanxious concerns about ordinarily childish mischief—the judgment, more than the behavior it labels, makes the problem" (p. 122).

This perspective respects the power of the increasingly more intense cybernetic positive feedback cycles that build up in a relational unit over time. As family members attempt to reduce their distress by intervening behaviorally, the problem paradoxically escalates with each effort to resolve it. This is expressed cybernetically as "A makes more of B which makes more of A" (Weakland, 1976, p. 123). Thus, "the solution becomes the problem" as family members work harder to resolve the issue, which paradoxically escalates the more they attempt to eliminate it. The Strategic Therapist's central job is to interrupt these cycles and help family members explore alternative responses. This leads to the third "heresy."

The Therapist Is Responsible

The Strategic Therapist is responsible for working with the client to create unique intervention strategies to reduce the distress experienced by the family or couple. It is crucial that these interventions begin early in the therapy process because clients are under duress, which the therapist has the ethical responsibility to help alleviate (Haley, 1996; Haley & Richeport-Haley, 2001). The best way to meet this responsibility is to join with the client and focus on understanding their perspective of what the problem is, what they have already attempted as solutions, and how they will define success (Nardone & Watzlawick, 2005). This clear understanding is paramount in Strategic Therapy because interventions have to begin promptly yet be crafted for each individual client situation (Haley & Richeport-Haley, 2001).

The therapist has the responsibility to intervene directly to help the family resolve their distress (Haley & Richeport-Haley, 2001). Direct intervention does not mean "doing more of the same" by directing family members to stop what they are doing and do something else instead. From the perspective of Strategic Therapy, this is the same dynamic that brought the clients to the therapist in the first place. Instead, the therapist must use a strategically planned approach to helping families change their distressing behavioral patterns. This plan must consider the clients' perspective on their problems and possible solutions.

Change Comes Before Insight

Changing behavior creates opportunities for client and therapist alike to observe alternate perspectives on presenting concerns and client responses to troublesome behavior patterns (Nardone & Watzlawick, 2005). Crafting different behavioral experiences gives the family members an opportunity to experience change at a visceral and interactive level, rather than just talking about it (Haley & Richeport-Haley, 2001).

Role of the Therapist

The therapist using a Strategic model is an active, directive therapist, who is able to build rapport with a variety of clients and see their strengths. These therapists understand that people struggling with problems are caught in ineffective and often painful behavioral cycles in their relationships (Haley, 1996). The therapist must have and convey to the client "a genuine acceptance of the client's meaning system" (Duncan, 1992, p. 23) and accept the responsibility for intervening actively and effectively to reduce the distress clients are experiencing (Haley, 1990; Haley & Richeport-Haley, 2001; Weakland et al., 1974).

Key to being an effective Strategic Therapist is building a relationship with individuals and families (Duncan, 1992; Nardone & Watzlawick, 2005). This is not done to manipulate or gain power over the client. Instead, it is done in order to develop a clear understanding of the clients' distress, previous attempts at solutions, and their view of the meaning of their struggles. This clear understanding is crucial to developing effective intervention strategies that reflect the clients' situation and the meaning they give to it. "The effectiveness of the intervention depends on the meaning the client ascribes to it, and that meaning is acquired in the interactional context of the therapist and client" (Duncan, 1992, p. 22).

Part of building this rapport and understanding of the client's context is recognizing and identifying emotions. Emotion in Strategic Therapy takes on a different role in the therapy process than in more insight-oriented therapy approaches. Accordingly, a Strategic Therapist must be able to take on a new perspective about the role of emotions and insight in relation to change. Strategic Therapists "do not spend significant amounts of time merely talking about feelings or getting clients to recognize and own feelings; they concentrate instead on getting clients to express their feelings in ways that are more likely to lead to client satisfaction in daily life" (Kleckner, Frank, Bland, Amendt, & Bryant, 1992, p. 49). Worded more strongly, "Behavior causes how people feel" (Haley & Richeport-Haley, 2001). Emotions can thus be accessed and worked with behaviorally in concrete and pragmatic ways that contribute to client success in resolving the problematic behavioral cycles that result in them feeling distress.

Because individuals and families have been working directly to resolve their issues, they already know what they should or should not be doing (Weakland et al., 1974). They are assumed to have gotten good advice that they have not been able to implement successfully (Haley, 1987). Thus, when clients begin therapy, they are locked in a cycle of distressing behavior exacerbated by unsuccessful attempts at resolution. The cycle may be distressing, but it is a predictable way of life and brings a form of stability to the family. Therapists must understand and accept that changing these symptoms and behaviors means giving up a way of living for individuals and relational systems (Haley, 1990). Weakland (1976) wryly notes, "The therapist's job, then, is apt to be the unenviable task of getting people to change that which they are apt to be clinging to most strongly" (p. 125).

Process of Therapy

Once an initial rapport has been built, Strategic Therapists concentrate on establishing a clear description of the problem behavioral cycles from each member of the system present. The therapist's focus is on understanding the experience of family members; identifying their language, metaphors, and motivations; and crafting interventions designed to reflect their perspectives while pushing toward change. The following stages in the Strategic Therapy process are summarized from Weakland et al. (1974) and Nardone and Watzlawick (2005).

Problem Definition

The presenting concern is clearly stated in measurable terms. If there are multiple family members present, the input of each person is considered. It is possible that there may be more than one problem definition. This part of the therapy process may be the most time-consuming, due to the distress that clients may be experiencing, but it is important to invest this time. As Nardone and Watzlawick (2005) state, "a clear and concrete definition of the problem is of great help in finding the fastest and most effective solution, and thus the time spent in this so-called diagnostic phase will be made up for later" (p. 62).

How the Problem Is Continued

The cybernetic behavioral interactions that help maintain the problem behavior cycles are identified in concrete terms. Ways in which the different family members experience these interaction cycles are clarified, and the various perspectives of all family members are sought in order to have the most complete view of the interaction cycle possible. Ways that the clients have attempted to solve the troublesome behavior cycles are identified to reveal and clarify the positive feedback loop that family members may be trapped in.

Behavioral cycles may fall into three areas of "mishandling" (Watzlawick, Weakland, & Fisch, 2011, p. 38). First, the solution has been to deny that there is a problem. Action has not been taken when it is necessary. Second, attempts for change are directed at an unchangeable situation. Action was taken when it was not useful. Third, change attempts are made but are directed toward first-order change when second-order change is needed, or vice versa. For example, parents may be seeking the compliance of their teenagers with household chores but may be focusing on "bad attitude" (a second-order change) when completion of the chore by the children, no matter how sulkily, is the desired goal (first-order change).

Agreement on the Goals of Therapy

With a clear problem definition and identification of past attempts at resolution, the client and therapist may then develop a clear goal for therapy. Goals should be small and definitively stated so that progress toward these goals may be readily observed.

Development and Selection of Intervention Strategies

Strategies are considered and implemented with the goal of disrupting distressing behavioral cycles. Targets for interventions may be for significant behaviors but may also be focused on minor and perhaps unimportant details of the interaction cycle. This helps by not overwhelming the client and increases the odds that the client will work toward change and have success. Harnessing the power of the positive feedback loops in which clients are trapped means that small changes can reduce significant issues (Weakland, 1976; Weakland et al., 1974).

ETHICAL ISSUES IN STRATEGIC THERAPY

Ethical issues are important in Strategic Therapy. Because the therapist is active and directive, great care must be taken to fully understand the client's perspective on the problem, the attempted solutions, and the definition of successful treatment. It is then the therapist's job to work collaboratively within the client's framework to bring about change (Haley & Richeport-Haley, 2003). Therapists are also expected to keep in mind that they do have a position as "expert" by default because of their role as therapist (Weakland et al., 1974). Techniques to be used must be implemented within a framework that views the couple and family as capable of enacting change (Haley & Richeport-Haley, 2003). The implementation of techniques is done with the clients' best interest in mind and with the goal of reducing their distress quickly, effectively (Nardone & Watzlawick, 2005), and in a way that empowers the clients to be free of the therapist and the therapy process

(Haley & Richeport-Haley, 2001). Therapists are charged with monitoring their use of power when implementing techniques by keeping in mind that "[o]nly those therapy procedures will be used that the therapist is willing to experience or have his wife and children experience" (Haley & Richeport-Haley, 2003, p. 13).

Techniques to Be Used

Technique is a crucial aspect of Strategic Therapy, and isomorphically, to the supervision of Strategic Therapists. However, no technique is to be used simply for its own sake or without sufficient time spent developing a relationship with the client. Strategic Therapists must continually monitor the therapeutic rapport: "Technique and relationship are completely interdependent and cannot be separated" (Duncan, 1992, p. 22). A Strategic Therapist conducts therapy with active and typically directive use of specific interventions designed to help the family do or think something different (Haley, 1987; Nardone & Watzlawick, 2005; O'Hanlon, 1982). As the therapist responds differently to clients, clients then respond in different ways to the therapist and each other (Weakland, 1976). With different behaviors enacted, changes can occur. Small changes are considered to make large differences when introduced into the cybernetic feedback loops present, or when interrupting these cycles (Gardener, Burr, & Wiedower, 2006; Watzlawick et al., 2011; Weakland et al., 1974). In other words, the intervention and technique used need not be as "big" as the presenting concern (Weakland).

Directives

Strategic Therapists make consistent and skilful use of directives. *Directives* tell clients what to do. This is not "advice" or "commonsensible" interventions, but directives, both straightforward and indirect, which are designed to reduce client distress (Haley, 1996; Haley & Richeport-Haley, 2001). Straightforward directives are used when the therapeutic relationship is such that the individuals or families who are clients will follow the directive. Indirect directives are used when straightforward directives have not been successful, or the therapist does not have sufficient authority with the client (Haley, 1996).

Generic Interventions

Generic interventions (O'Hanlon, 1982), based upon the work of Milton Erickson, may be used to intervene both directly and indirectly in problematic behavioral patterns. These interventions are used without concern for the "cause" of problem behaviors and provide avenues of intervening in behavioral cycles via "personal and interpersonal patterns of perception, behavior, and experience" (p. 29). It is assumed that the client is able to make these changes and that disrupting these patterns will initiate different perspectives, behaviors, and experiences, which will in turn promote change. The generic interventions are:

1. Change the frequency/rate of the symptom/pattern.
2. Change the intensity of the symptom/pattern.
3. Change the duration of the symptom/pattern.
4. Change the time (hour/time of the day/week/month/year) of the symptom/pattern.
5. Change the location (in the world or the body) of the symptom/pattern.
6. Change the quality of the symptom/pattern.
7. Perform the symptom without the pattern.
8. Perform the pattern without the symptom.
9. Change the sequence of the elements in the pattern.
10. Interrupt or otherwise prevent the pattern from occurring.
11. Add (at least) one new element to the pattern.
12. Break up any previously whole element into smaller elements.
13. Link the symptom/pattern to another pattern/goal.

Communication-Based Interventions

Communication-focused interventions are important strategies in Strategic Therapy. They may be direct directives, such as the generic interventions, or indirect, such as reframing, metaphors, and paradox. Haley (1990) describes three general communication-based tactics that bring about change. First is ambiguous directives, such as "Discuss your feelings" (p. 175). Anything that is said or not said is complying. Second, emphasizing the positive helps motivate clients toward change, because it's not easy to resist someone who sees the positive in you. Third, encouragement of the usual behavior enables change because it is a different thing for a therapist (or supervisor) to say. This unexpected response may disrupt the family's anticipated response pattern because it disrupts the positive feedback loop in which they are stuck by not adding more "A" to "B."

The use of what Nardone and Watzlawick (2005) refer to as "nonordinary logic" (p. 22) is common in developing communication-based strategies to interrupt problematic behavioral cycles. This different logic is surmised to bypass the left-sided logic centers of the brain and engage the more creative and flexible right side of the brain. In avoiding a "logical" or "commonsensical" intervention, the expected pattern and response are short-circuited, allowing change to be introduced into interactional sequences (Watzlawick, 1978). For example, individuals and families may be given permission to continue doing what they are doing, since it will happen anyway (Haley, 1990). A family might be told that it is important for them to continue with their problematic behaviors, such as arguing, so that they have sufficient opportunity to observe this distressful behavioral interaction and give the most helpful information to the therapist at the next session.

Paradox

The use of nonordinary logic also invites the use of paradox in interventions. Paradoxes are helpful in circumventing the "commonsense" behavioral maintenance patterns that exacerbate distress. In the example of the family told to continue arguing because the most helpful thing they can do for themselves in the next week is to continue having their problem so they have good observational data for the next session, this gentle paradox creates a "benevolent ordeal" (Haley, 1990, p. 187). Because it is benevolent—that is, oriented toward a positive outlook toward the clients and accepting of the issues they bring—the clients can go along with it. If it is an ordeal, which continuing the distressing behavior could be, then the client can resist. But the benevolent ordeal puts the client in a double-bind of both resisting and complying. The nonordinary logic of the directive provides success either way for the client. Compliance gives more information for the problem definition and attempted solutions. Resistance means a reduction in the distressful behavior. Either way, change has resulted, either in behavior or in the perception of the behavior.

Reframing

Reframing is a central skill to be mastered in Strategic Therapy, as "effective intervention usually requires reframing" (Weakland, 1976, p. 127). *Reframing* is an indirect intervention that is not intended to change or explain the experience or behaviors of the relational system, only the *perception* of these experiences or behaviors (Nardone & Watzlawick, 2005; Weakland et al., 1974). Sometimes the problem is not the situation but the meaning we have attached to it (Weakland et al., 1974). Reframing problems and behaviors does not change the first-order reality for the family or couple, but it does change their perspective on the problem, which leads to second-order change (Watzlawick, 1978). Successful reframing only happens when the therapist has a good understanding of the clients' perceived realities and experiences because the most meaningful reframes must fit within the clients' worldview (Watzlawick).

Reframing may take the form of seeking a positive or helpful reason for a behavior, or "simply the useful relabeling of behavior" (Weakland et al., 1974, p. 155). For example, a man who argues with his wife when she begins to exhibit behavior typical of her depressive behavior patterns is reframed as attempting to energize or distract his wife from her unhappiness. The goal of therapy then becomes to interrupt this pattern so that the husband finds a less distressing way of helping his wife cope with depressive behaviors (Haley & Richeport-Haley, 2001).

Culture/Race and Gender Issues

Focusing on the initial presenting concerns and observed behaviors does not exclude discussion of issues such as gender, culture, ethnicity/race, and socioeconomic status with clients. These issues may be raised as part of the exploration of the problem presentation or the mechanisms by which the problematic interactions are maintained (Haley & Richeport-Haley, 2001). Issues related to culture, ethnicity, and gender are not seen as "the" problem, nor is insight into these issues considered a mechanism of change (Haley & Richeport-Haley, 2003). It is the Strategic Therapist's job to understand the client's perspective on these issues, not to apply broad generalizations about how families or couples from a specific demographic group may interact (Haley & Richeport-Haley, 2003; Quick, 2012). The supposed traits or cultural behaviors that are ascribed to a group of individuals or a cultural group as a whole are much less important to the effective Strategic Therapist than how those specific traits or cultural behaviors interact in unique ways within each relational client unit. As with any other "causal" condition, cultural variables are considered part of the clients' worldview. The general approach is to examine how cultural, ethnic, and gender variables have impacted the relational system's attempts to adapt to mainstream society and/or variables within individuals in the family, not to "resolve" the unresolvable issues of sociocultural beliefs, experiences, and perceptions (Haley & Richeport-Haley, 2003).

Supervisees from the Strategic Therapy model are expected to examine their own assumptions, culturally influenced beliefs and behaviors, and reactions (Quick, 2012). Therapists practicing from the Strategic approach need a keen eye for culturally influenced behavior patterns in the mainstream culture as well as minority cultures. Typical behaviors in the mainstream culture may also be part of the positive feedback loops that maintain distressing behavioral patterns, such as the culturally supported behavior of "cheering up" someone who is coping with sadness or depression by telling them to keep a positive outlook (Weakland et al., 1974). Because this sort of "look on the bright side" type of behavior is so deeply rooted in the problem-solving approaches that are approved of and expected in the mainstream culture, therapists need to be able to discern how these behaviors intertwine in positive feedback loops to fully understand the repetitive behavior cycles that are causing discomfort to their clients.

Strategic Therapists are also expected to listen for the details of problem-maintaining behavioral cycles from the client's cultural perspective and to incorporate these details into interventions and interactions in session. These interventions may fall into the categories of using the client's beliefs that differ from the mainstream culture to frame problem statements or treatment goals, using alternative healing beliefs in interventions, or referring to and/or collaborating with traditional cultural healers (Haley & Richeport-Haley, 2003). Trainees using the Strategic approach are expected to consider and respect the culturally influenced hierarchies and interaction patterns in place within families as part of meeting clients where they are and accepting what they bring to the therapy session (Quick, 2012). The culturally influenced language and problem conceptualization of the client will be accepted and used to develop problem statements and treatment goals. Previous culturally based strategies to eliminate a troublesome interaction pattern will be accepted and explored in the same way a mainstream-culture client family might discuss previous therapy visits and medication. A Strategic Therapist might give a directive that a family consult with informal healers in the client's ethnic community, if this is part of their belief system (Haley & Richeport-Haley, 2003). Conversely, a client may not view cultural factors as central to their distress and presenting concerns. If this is the case, the therapist practicing from a Strategic perspective is expected to respect the client's view and approach interventions from a different angle (Quick, 2012).

Finally, focus on the presenting problem keeps therapy oriented toward what the client wishes to address. This can protect the family's privacy, because the Strategic Therapist intervenes and inquires about only the issues that the clients present for therapy (Haley, 1987). Even if the therapist perceives that there are other issues, these are off limits unless the clients raise them as a concern. In short, if it ain't broke, don't fix it for the client (Haley & Richeport-Haley, 2001; O'Hanlon, 1982). An example might be a family in which aggression between the father and a teenaged son has become common when adults attempt to impose consequences on the teen. The therapist may notice that the parents have culturally influenced closely prescribed roles for male and female parents. Unless these roles are identified as part of the presenting concern that is bringing distress to the family, these culturally influenced gender roles would not be a target of interventions. Instead, interventions would target problematic behavioral interactions that create challenges for the family in adapting to the mainstream culture. Clients will have the opportunity to explore new roles in session and while implementing therapist directives. This is not done to force adaptation to the mainstream culture, but instead to "respect the client's culture and still resolve their problems" (Haley & Richeport-Haley, 2003, p. 19). An example is an immigrant father whose cultural values expect the mother to handle all school concerns; this father would be directed to attend a school meeting for his son as part of the role a father plays in teaching his son to adapt to the adopted country. This is an example of adding an element to the behavior pattern (O'Hanlon, 1982).

Research to Consider

Strategic Therapy has been criticized as not having a robust body of systematic, evidence-based research to support its effectiveness (Nardone & Watzlawick, 2005). Much of the evidence-based research available for Strategic Therapy focuses on models that combine Strategic Therapy with

Brief Therapy approaches, most notably Brief Strategic Family Therapy (Szapocznik, Hervis, & Schwartz, 2003), or cognitive behavioral approaches (Nardone & Watzlawick). However, this robust body of literature is beyond the scope of this chapter.

The fact that little research exists centering on Strategic Therapy belies the tradition of Strategic Therapy practitioners being very open to examining the effectiveness and efficacy of their approach. In keeping with the pragmatic focus on behavioral change, early practitioners of Strategic Therapy used action research to evaluate clients' perceptions of maintenance of gains following the termination of therapy, with no attention given to separating clients by presenting concerns or demographic variables (Haley 1987; Weakland et al., 1974). Former clients were called by an interviewer who was not the therapist to inquire if the specific treatment goal was met, if any other areas of improvement were noted by the client, and if other problem behaviors or interactional patterns had arisen since terminating therapy. This action research focused on rating outcomes in regard to the presenting concern as being successful, as producing significant improvement in reduction of symptoms, or as failures. Weakland et al. conclude that 72% of the 97 clients who had an average of seven sessions reported success or significant improvement in these follow-up studies. This pragmatic and hands-on method of evaluating treatment outcomes in Strategic Therapy continued to be used and recommended by Haley in the training of Strategic Therapists (Haley & Richeport-Haley, 2001).

This practice of action research with Strategic Therapy models has been continued by Nardone (1996). Client outcomes are assessed by categories based on presenting concerns and duration of therapy. Focusing on the efficacy of Strategic Therapy with phobic and obsessive disorders, Nardone reports that 79% of the former clients ($N = 152$) stated their issues were resolved at the time of follow-up, with another 7% reporting significant improvement. Of the clients surveyed, 87% had between four and 20 sessions, with the median number of sessions being 14. In 2005, Nardone and Watzlawick reported positive outcomes in 86% of 3,484 outpatient cases assessed over 10 years, with 52% of the sample attending 6 to 10 therapy sessions. Positive outcomes were identified in the presenting concerns of depressive disorders, eating disorders, obsessions and obsessive-compulsive disorders, sexual disorders, marital problems, and anxiety/phobic/panic disorders.

The pragmatic focus of Strategic Therapy lends itself well to case study research and conceptualizations exploring specific applications of the model. Case studies have focused on using Strategic Therapy in school settings to successfully address behavioral issues in children and teens through the use of strategic directives to parents/caregivers (Botman, 1997; Nelson, 2006; Schiff, 1990; Stone & Peeks, 1986) and in couples therapy when anorexia nervosa is a presenting concern (Murray, 2014).

Research examining the application of or effectiveness of Strategic Therapy supervision as a stand-alone model in the training of relational therapists is scarce. Much of the literature concerning therapist skills and other factors of the therapy process that would be of interest to a supervisor of relational therapy is centered on the Brief Strategic Family Therapy model and is beyond the scope of this chapter.

ROLE OF THE SUPERVISOR AND SUPERVISEE

The role of the supervisor in Strategic Therapy is isomorphic to that of the therapist. Haley (1996) notes that "it is possible to consider the techniques of therapy and the techniques of supervision as synonymous" (p. 163). Because the goal of therapy is to change clients by actively helping them find less distressing ways to interact in relationships, the goal of supervision in Strategic Therapy is to actively change therapists (Haley & Richeport-Haley, 2001) by helping them find more effective ways to work with clients. Thus, a Strategic supervisor approaches supervision with the goal of teaching therapy as a skill set to be learned (Haley, 1987, 1996). Trainees are given as much in-session practice and live supervision as possible, so that a supervisee will be able "to have maximum success" (Mazza, 1988, p. 93).

Supervisor–Supervisee Relationship

The supervisor–supervisee relationship in Strategic Therapy is isomorphic to that of the therapist–client relationship. The relationship between trainer and trainee is hierarchical, with the supervisor actively directing the trainee (Haley, 1996; Mazza, 1988). The same techniques are used with trainees as with clients (Haley, 1996): that is, direct and indirect directives, reframing, generic interventions to disrupt behavioral patterns, use of metaphors, and paradoxes. The goal of the Strategic supervisor is to help the supervisee learn the technical skills of relational therapy and case conceptualization by developing supervisory interventions for trainees so that they may provide productive therapy sessions in the same way trainees are expected to focus on the interventions needed to help a family have more productive interactions (Haley, 1996).

The supervisor is held responsible for the outcome of therapy in the same way the therapist is held responsible for outcomes. The unit of therapy for the supervisor is both the trainee and the client family (Haley, 1987, 1996). If there are difficulties for the trainee in working with a family or couple, Strategic supervisors are charged with intervening with the trainee in the same direct active way they would intervene were they the therapist (Haley 1987, 1996), with the goal of ensuring that clients receive competent care (Haley, 1987). This must be balanced with awareness of the hierarchy between therapist and client, and the supervisor is held accountable for intervening effectively with therapists so that they are able to maintain their authority and position in the hierarchy in therapy sessions, rather than be usurped by the supervisor (Mazza, 1988).

Ideal Structure of Supervision

The optimal structure for supervision under the Strategic model is live supervision led by an active supervisor with competent therapeutic skills and training in supervisory interventions (Haley, 1987). "Ideally the therapist learns to do therapy by doing it while guided by a supervisor at the moment therapy is happening" (Haley, 1987, p. 205). The supervisor must be both a skilled clinician and a teacher competent to impart both clinical skills and how to think strategically about human behavior and therapeutic change-focused interventions.

Isomorphic with the assumption of Strategic Therapy that talking with clients about their problems in therapy is less effective than working in the moment with observed behaviors in session, it is assumed that a supervisee cannot convey the full sense of the therapy session to a supervisor after the fact. Therefore, live supervision is the best way to evaluate the dynamics of the therapist-client interactions in session (Haley, 1987). Live supervision also allows the supervisor to assess whether what the trainee states about case conceptualizations and hypotheses correlates with the actions the supervisee is taking or not taking in session (Haley, 1987). If sessions are not all supervised live, videotaping may be used, albeit as a much-less-preferred second option (Haley, 1996).

Strategic Therapy supervision is considered to be best conducted in groups, because this increases the number of cases to which trainees are exposed and augments the number of suggested interventions (Haley, 1987). The group supervision experience is structured and active. Only the supervisor may make comments to or about the therapist, but the supervision group may ask questions of the supervisor and suggest interventions (Mazza, 1988). Trainees are expected to earn the privilege of observing and suggesting interventions by presenting their own clinical sessions for observation by their training cohort (Haley, 1987; Mazza, 1988).

Didactic content is balanced with live supervised clinical sessions. Content presented in supervision sessions should reflect the presenting needs of the clients the trainees are seeing (Haley, 1987). From this perspective, supervisees are highly motivated to learn both content and clinical skills when confronted with the immediate need to be effective with their current clients. Strategic Therapists are considered to be generalists; thus, the caseload of trainees and the training site should include a wide range of family configurations, populations, and presenting concerns, so that trainees may gain as much realistic experience as possible. A variety of clients, demographic groups, and presenting concerns increases the flexibility of the trainees, with the goal that they will be able to intervene effectively with many different issues and populations, and adapt to a variety of clinical settings (Haley, 1987).

Group supervision does not mean that supervision interventions are implemented en masse. The effective Strategic supervisor tailors supervisory interventions to trainees in the same way a therapist is expected to fit interventions to the clients (Haley, 1987). Students are expected to be flexible and meet the clients where they are. So must the supervisor meet the students where they are. Caseloads should be varied to provide opportunities to learn different skills as well as allowing each trainee to experience success and challenge in the best combination for each student (Haley, 1996).

Strategic Therapy devotes attention to ensuring that supervisees are not caught between competing supervisory philosophies when of necessity they must have multiple supervisors. Ideally, students should have only one supervisor during each segment of their training. Having more than one supervisor increases the likelihood that students will be caught between different theoretical and interventional perspectives, leaving the trainee trapped in a triangle in the same way a child might be snared between adult caregivers (Haley, 1987; Haley & Richeport-Haley, 2001). This situation may result in the trainee becoming confused or caught between conflicting instructions and expectations (Haley, 1987), ultimately rendering confused or conflicting therapy, with the student therapist functioning as the inept child in a parent-child triangle (Haley, 1996).

Ethics in the Supervisory Relationship

The Strategic Therapy training model focuses the supervisory interventions on therapy skills and case conceptualization. The students' personal socioemotional adjustment and "issues" are not considered to be part of the content of supervisory sessions or interventions (Haley, 1987, 1996). The importance of students having a supportive personal support network in their nonprofessional lives is recognized, but the specific aspects of the students' personal lives are not considered to be part of the supervisory content. As Haley points out, "A person's personal life is too important to be tampered with by teachers" (1987, p. 210). This is not to say that a Strategic supervisor is not supportive. Developing a good rapport and understanding with trainees is as important as it is for clients, but the support is in the supervisory training relationship, not in personal issues students may need to address (Haley, 1987).

The Structure of Training

Training of Strategic Therapists begins with developing a positive relationship with trainees in the same way that trainees will need to build a relationship with each client family. Haley (1996) proposes a staged model for training supervisees in Strategic Therapy, which is summarized in this section. More specific information for the content of these stages is given in the next section.

A Safe Learning Environment

First, students are helped to feel welcome in training by the supervisor. They are shown the training and supervisory rooms with the goal of having them feel comfortable in this space. Group members are welcomed by the supervisor and introduce themselves and their current work or training environment. This assists the supervisor in learning what other clinical and/or administrative supervisors students may be working under and what type of experience trainees might have. The supervisor then discusses the supervisory contract with the students, in which the goals of supervision and skills to be taught are clearly stated, as are expectations and responsibilities of both the trainee and the supervisor.

Specific Skills Are Taught and Practiced

Initial training sessions are structured with directives by the supervisor and are live supervised and videotaped. Trainees are directed to avoid making interpretations about past events and perceived negative motives, asking about feelings, or commenting on body language in initial sessions. They are taught how to greet clients, how to help them feel comfortable during the first interview, and how to concisely explain the therapy process and informed consent to family members. Next, trainees are taught different ways to structure the therapy sessions for success by attuning to the hierarchies present in the family. Finally, the initial intake information for the family is reviewed with the trainees to help them develop flexible session plans and initial interventions.

Planning Interventions for Trainees

Through live supervision and videotape review, the supervisor assesses the trainee's skills and develops interventions to improve skills and utilize the trainee's strengths. The trainee is considered to be able to make these changes in the same way a client family is expected to be able to enact change in their relationships (Haley, 1996). The supervisor makes clear, concise suggestions during live supervision via telephone or during an arranged break in the session for consultation with the team. In general, the supervisor is to give only one directive at a time to the student, to minimize the amount of call-ins or check-ins, and to focus the student on the immediate issue in the session.

Learning Goals/Objectives

In Strategic Therapy, the goal of the therapist is to not be needed (Haley & Richeport-Haley, 2001). Isomorphically, the Strategic supervisor is to aim for not being needed. Trainees need to be empowered to work independently of the supervisor while at the same time staying engaged in the learning process and open to the supervisor's interventions. This is the same delicate balance between supporting yet challenging clients to change that supervisees must navigate with clients. It is crucial that the Strategic supervisor maintain this balance, due to the "reciprocal relation between the supervisory structure and the family structure" (Haley, 1980, p. 62). The more disorganized the family structure, particularly the hierarchy, the more clear and direct the hierarchy must be between the Strategic supervisors and their trainees. The supervisor-trainee relationship and interventions focus specifically on the needs of the supervisee's client families, not general ideas about technique or theory (Haley, 1980), because the goal is to produce clinicians, not theoreticians (Haley, 1987). This mirrors the focus a supervisee is expected to maintain in therapy sessions on presenting concerns and maintenance patterns, not on past events, interpretation, or insight.

How to Build Relationships With Clients

Because the rapport is crucial to developing effective interventions, trainees must be taught how to engage with each member of the family presenting for therapy. This begins with teaching students how to welcome clients to the therapy setting, greet each member of the family, and help the family feel comfortable in the interview room. Trainees are taught to accept client communications with body language and the positioning of their bodies and chairs in space (Haley, 1996) and to refrain from interpreting these or seeking "insight" from these behaviors.

Next, the trainees are taught how to structure the first session so that they can effectively teach the family about the therapy process. They are taught how to introduce the recording equipment and the structure of therapy to clients in a concise manner that will provide full information for informed consent (Haley, 1996). Supervisees are then guided in speaking to each member of the family in order to set the expectation that each family member will be involved in the therapy process. Trainees are also taught to recognize their personal traits that help them engage effectively with clients so that they can make the best use of their individual engagement skills (Haley, 1996; Haley & Richeport-Haley, 2001). Finally, trainees are directed to answer personal questions that clients may pose to them directly and succinctly, and then return to the therapy process (Haley, 1996).

Learn What Information to Gather

Prior to the first session, the supervisor reviews the intake information with the trainee for the purpose of developing a plan for the information-gathering process. Background information that is helpful to know prior to delving into the specifications of the presenting problems are custody information if children are involved; if there has been previous therapy and/or if there are other therapists involved with individual family members at this time; if clients are court-ordered or otherwise compelled to attend therapy; the family composition; and who comes to therapy. In general, it is the supervisor and then later the advanced student who determines who comes to therapy (Haley, 1996).

Develop Treatment Plans

The supervisor actively assists trainees at all experience levels to develop treatment plans that focus on presenting problems and then to develop clear goals and effective interventions. Prior to each session, the supervisor meets with the trainee to discuss the therapeutic position on the following factors:

- Presenting problem
- Relational unit of intervention (parent subsystem, triangles, etc.)
- Sequence of problematic behavior and subsequent maintenance behaviors
- The hierarchy of the family
- Motivation for the problematic behaviors and the maintenance behaviors

During this planning session, the supervisor makes notes of trainee biases and identifies these for subsequent intervention in conjunction with what is observed in session. It is believed, in the training of Strategic Therapists, that dealing with these biases will be more effective if done in the moment by the supervisor or in tape review in which therapist interventions and strategies can be discussed, rather than dealt with by discussion of hypotheses at an abstract level (Haley, 1996).

Learn How to Motivate Clients for Change

Trainees need to be taught how to tell clients what to do. However, they need to be taught when and where to use these directives, and when to use indirect suggestions. The same holds for supervisors. When the trainee will follow straightforward directives, these are what are used. When these have not been successful, indirect techniques are used (Haley, 1996).

Trainees are taught how to engage clients into the therapy process by focusing on the specific needs that they request assistance for. They are then taught how to phrase clear, straightforward directives, or effective indirect directives, such as reframing or using the client's metaphors. Students are taught to work with reluctance to follow directives (Haley, 1987), viewing this as the clients' understanding that change is not easy, not that the client is "resistant." Above all, supervisees are taught to have a positive outlook toward their clients and to empower individuals and families to no longer need the therapy process to assist with reducing their problematic relational behavior cycles.

Supervisory Interventions

Supervisory interventions are isomorphic to therapeutic interventions (Haley, 1996; Mazza, 1988). Straightforward directives are crucial in the initial stages of training, when anxiety is high and students need a great deal of structure so that they can successfully manage both their own anxiety and that of the client. However, just like with clients, commonsensible directives are not helpful, especially with anxious early-stage trainees. How many of us have worked with students who became more anxious the more their fellow trainees told them to relax? For these students, it is helpful to use more indirect strategies to manage anxiety, such as reframing it as helping them to be sure they will indeed "do no harm" to their clients. Beginning trainees may be told that such anxiety is welcome in this training setting because it means they truly have the clients' best interest at heart. Students are then told to continue feeling anxious for as long as necessary to help them develop their skills.

Telling Them to Continue Doing What They Are Doing

Just as clients may be told to continue with problematic behaviors, so too may trainees be told to "do what they are going to do anyway," particularly with common rookie mistakes. A Strategic supervisor might inform trainees that the goal is to make one mistake on purpose in each session, and that this will be the only mistake that will be discussed in supervision. For the remainder of the supervision session, the focus would then be on identifying the strengths of the trainee and tying therapist skills to hypotheses and intervention strategies. Often the students will forget to make the requested mistake, which "prohibits" the supervisor from discussing mistakes during tape or session review. The Strategic supervisor is then free to discuss strategy and skills because these would not be considered mistakes. It is sometimes helpful to express disappointment that there are no mistakes to discuss because mistakes are important to the learning process; a dedicated Strategic supervisor is always hopeful that there will be many so that good learning will take place. In the author's experience, this supervisory stance opens up the opportunity for trainees to view their missteps as learning tools and to discuss them helpfully as learning experiences, not as shameful incompetence.

Directing Them to Do the Opposite of What They Think Good Therapy Is

Students who believe that therapy is "helping" people to "fix" themselves or to "correct faulty thinking" often fall into advice-giving or excessive use of closed questions designed to harry a client toward a specific, therapist-determined goal. Such students are directed to be "less helpful" or even to "stop helping." If this is not successful, the next supervisory intervention is indirect, in which the student's prior experience as a "helper" is commended for bringing him or her to this level of specialized training; this prior experience is then invited to "help the helper" by allowing new skills to be learned. For example, students who are successful and competent school teachers may do well throughout family therapy sessions, and then in the final moments, relapse into being teachers by "summarizing today's lesson." A Strategic supervisor might use an indirect intervention by commending the teacher side of the supervisee for waiting until the final moments of the session before interrupting the productive time on task that the trainee as counselor was completing. The teacher side of the trainee might be asked to help further by allowing the entire session time to be time on task with counseling skills, and then discussing the session with the supervisor afterward, not the client, in the same way they would discuss a successful classroom lesson with a colleague rather than the student. This example also illustrates the use of individual student language, metaphors, and professional competencies to build further success.

Gentle Paradox

With trainees who question or do not comply with the supervisor's methods or case conceptualization, it is helpful to compliment them on their refusal to accept anything at face value. Next, direct them to continue being skeptical of anything you might say. This places them in the position of doing what you request by continuing to be skeptical of you, which means they are complying with you.

A Benevolent Ordeal

Supervision may also be conceptualized as a benevolent ordeal (Haley, 1990). As supervisors, we bring the trainee a stressful experience with each supervision session, but one in which we are compassionate to the supervisees and believe that they are very capable of improving their therapeutic and case conceptualization skills.

CONCLUSION

Strategic Therapy and the accompanying supervision process are straightforward yet complex models for intervening in the distressing relational interaction cycles that bring clients to couple and family counseling. The Strategic supervisor is expected to view trainees as competent learners who are capable of building skills to intervene effectively and efficiently with client concerns, while respecting the recursive dynamics of relational patterns and the sociocultural context in which these concerns are embedded. The supervisory process is developmental, pragmatic, and isomorphic to the therapy process, creating a consistent framework in which supervisees may gain skills in intervention and case conceptualization for a wide variety of distressing issues that bring pain and discomfort to clients from all walks of life. Supervision from the perspective of Strategic Therapy values ethical relationships with supervisees by protecting their personal lives and individual temperaments from intervention by the supervisor while challenging trainees to examine how these personal beliefs and traits interact in the therapy process. The supervisor is held to the same standards of active, compassionate interventions in supervision that are expected of trainees with their clients. Above all, the Strategic Therapy model of supervision places a clear emphasis on respecting the reality that trainees may be working under two different supervision models at some point during their training process and seeks to minimize the impact of this on trainees.

The application of the skill and case conceptualization of a supervisor working from the Strategic Therapy model of supervision may be seen in the following case study.

CASE EXAMPLE

Supervisor Stance

To make use of the isomorphism between therapy and supervision in the Strategic Therapy model, Dr. Bob will approach supervision with Brittney in the same direct, compassionate, strengths-based manner he expects her to take with Travis, Tekia, and the two children. Because Brittney is a beginner, Dr. Bob is anticipating that Brittney will make missteps as a trainee and will view it as his job to ensure that she provides competent therapy to the family while at the same time empowering her to be as independent as possible. He is also assuming that she will be able to learn the skills of therapy and holds himself responsible for creating good learning conditions in the supervisory relationship, in the same way he will hold Brittney accountable for facilitating change in her clients. He will maintain an awareness of where she is developmentally as a therapist throughout her training and adjust his stance to reflect both her needs as a supervisee and the needs presented by Brittney's clients. As a Strategic supervisor, Dr. Bob's ultimate goal is to train Brittney sufficiently that she no longer needs his direction and support to be a competent relational therapist.

A Safe and Structured Training Environment

Training an intern to be independent of the supervisor begins immediately. Dr. Bob begins supervision with Brittney by establishing a safe training environment. He has given her a thorough orientation to his agency, ensuring that she is comfortable in the setting and can navigate the required paperwork and office procedures of the agency

(*continued*)

as independently as possible. He has inquired about her previous experience and training to gain an idea of what strengths she may have that she can build on as a therapist.

Clarifying roles. Next, they discuss the framework for Brittney's training. As a Strategic supervisor, Dr. Bob is keenly aware that both Brittney's learning and client care would be harmed if she is triangled between supervisors, leaving her unclear about how to proceed with clients. He discusses her previous preparation in her counselor education program and inquires about the supervision she is receiving through the university. This will help him reduce the possibility that Brittney may be caught between himself and the other supervisor should there be a difference in styles. He also clearly specifies the relationship and expectations between himself and other clinicians at the agency, so that all parties understand their roles in the training process. If there are other trainees at the site, supervision will be conducted in a group setting to maximize their exposure to different presenting concerns and relational configurations. Brittney's interest in gaining a broad range of experience is highlighted as a strength for her and will be presented as an indication that she has successfully landed at a training setting that will be of benefit to her.

In the first few weeks of the internship, Brittney gets explicit directions from Dr. Bob about what appear to be unimportant items. This attention to detail builds on the Strategic Therapy principle that therapy is a skill set that can be learned. Brittney is taught such concrete skills as how to prepare the counseling room prior to session, how to greet clients in the waiting room, and how to concisely yet thoroughly complete the initial informed consent process. Next, Dr. Bob discusses his expectations for how she will structure the session, giving her a rough guide to the first session, emphasizing the Strategic Therapy goal of establishing a clear behavioral treatment goal by the end of the initial session. This preliminary training is done efficiently so that students may begin seeing clients as soon as possible to maximize their in-session practice of therapy skills.

Pre-Session Preparation

Brittney's anxiety about meeting Travis, Tekia, and their children has been accepted by Dr. Bob. Rather than seeking to reduce her anxiety, he might commend her with a gentle paradox, telling her that these worries indicate she is appropriately concerned about doing a competent job with her clients, and to keep feeling anxious as long as

(*continued*)

necessary so she will remain sharp in session. Her feelings are not minimized, and the very real nature of her discomfort is validated.

Next, they review the case file together. Dr. Bob directs Brittney to speak to each family member to convey the expectation that all will be involved in the therapy process. Dr. Bob may give Brittney a short checklist with background information that is important to gather, particularly concerning what services Travis and Tekia might have received during his recovery process, what services are currently in place, their previous attempts to solve the "short fuse" problem, and safety assessment. He directs Brittney to mentally break the session into segments by giving a flexible structure to the session based upon the stages of Strategic Therapy (Nardone & Watzlawick, 2005; Weakland et al., 1974):

1. Problem definition
2. How the problem is maintained
3. Agreement on the goals of therapy
4. Development and selection of intervention strategies

During this initial consultation with Brittney, Dr. Bob refrains from discussing the history of the presenting concern, diagnoses that may be in place or hypothesized, or interpreting the case from a theoretical viewpoint. Instead, attention is focused on how to effectively use questioning and reframing to build a cybernetic, relational view of the presenting concern to gain a clear picture of what the family views as the core concern and what each family member views as a successful outcome of therapy. Dr. Bob also directs Brittney to listen for the family's metaphors and to build rapport by recognizing their current distress and commending past efforts to resolve their issues on their own.

Diversity. Dr. Bob does not use this initial consultation time to focus on Brittney's comfort with clients who are not of her culture and/or ethnicity, or the presenting concern, or what biases she might have. However, this is not to say that these issues are overlooked. Brittney is expected to learn and use culturally appropriate therapy interventions. But instead of applying broad generalizations about working with African American families, military families, families with small children, or families coping with injuries or financial stress, Dr. Bob instructs Brittney on how to use questions effectively to clarify the family's view of how and to what extent these issues affect their "short-fuse" problem. She is expected to follow the family's lead in this area and to respect their perspective. If Brittney strays from this

(*continued*)

task in session, Dr. Bob may direct her (through a phone-in or short consultation) to return to the discussion of the problematic behavioral cycle.

Dr. Bob focuses on diversity issues during group supervision with tape review, when Brittney and her fellow trainees are considered to be the most motivated to learn and apply this information. He may direct them to research specific information about relational interventions with African American families or military families, and/or about implications that Travis's specific injuries might have for the therapy process. This information will be applied to specific cases to give concrete examples of case conceptualization and interventions, and thus maximize the learning for all.

Initial interventions. The pre-session consultation ends with Dr. Bob recommending that Brittney give Travis and Tekia a directive in the final segment of the session. Brittney's eagerness to learn and her developmental stage as a beginner indicate that she is probably open to directives about therapeutic interventions from him. Isomorphically, he will prepare Brittney to give a direct intervention related to the presenting concern of Travis and Tekia, perhaps using the generic interventions of O'Hanlon (1982). Dr. Bob also prepares Brittney to give an indirect intervention if it appears that a direct intervention may not be warranted. For the purpose of this case study, no safety issues are determined to be present, so an indirect intervention might be for the family to continue to have "a short fuse" so that they have many observations to bring to the next session.

Treatment Planning

Following the initial sessions, and as needed in future sessions, Dr. Bob will assist Brittney to develop and fine-tune a treatment plan that focuses on the specific presenting problems that Travis and Tekia identify. If they determine that the issue is a parenting concern and not a marital issue, Brittney will be expected to respect this view and center the treatment plan on family interventions. Accordingly, if they report they want couples counseling, Dr. Bob will assist Brittney in constructing a corresponding treatment plan. It is at this point that Dr. Bob would open discussion of the information Brittany has gathered related to the specific concerns and diversity issues that Travis and Tekia identify as important to their relational interactions. Dr. Bob will also discuss with Brittney which relational units come to therapy, over time allowing her to make this decision more independently as she advances in her case conceptualization skills.

REFERENCES

Ariel, S. (1992). *Strategic family play therapy*. Chichester, UK: John Wiley & Sons. doi:10.1177/1066480710387370

Botman, J. A. (1997). A Strategic intervention for treating children's behavioral problems. *Behavioral Health Treatment, 2*(9), 8.

Duncan, B. L. (1992). Strategic Therapy, eclecticism, and the therapeutic relationship. *Journal of Marital and Family Therapy, 18*(1), 17–24. doi:10.1111/j.1752-0606.1992.tb00914

Gardener, B. C., Burr, B. K., & Wiedower, S. E. (2006). Reconceptualizing Strategic Family Therapy: Insights from a dynamic systems perspective. *Contemporary Family Therapy, 28*, 339–352. doi:10.1007/s10591-006-9007

Haley, J. (1973). *Uncommon therapy*. New York, NY: W.W. Norton.

Haley, J. (1980). *Leaving home*. New York, NY: McGraw-Hill.

Haley, J. (1987). *Problem-solving therapy*. San Francisco, CA: Jossey-Bass.

Haley, J. (1990). *Strategies of psychotherapy*. Rockville, MD: Triangle Press.

Haley, J. (1996). *Learning and teaching therapy*. New York, NY: Guilford Press.

Haley, J., & Richeport-Haley, M. (2001). *Learning & teaching therapy with Jay Haley*. [DVD]. New York, NY: Insight Media.

Haley, J., & Richeport-Haley, M. (2003). *The art of Strategic Therapy*. New York, NY: Brunner/Routledge.

Kissel, S. (1990). *Play therapy: A Strategic approach*. Springfield, IL: Charles C. Thomas.

Kleckner, T., Frank, L., Bland, C., Amendt, J. H., & Bryant, R. (1992). The myth of the unfeeling Strategic therapist. *Journal of Marital and Family Therapy, 18*(1), 41–51.

Madanes, C. (1984). *Behind the one-way mirror: Advances in the practice of Strategic Therapy*. San Francisco, CA: Jossey-Bass.

Mazza, J. (1988). Training Strategic therapists: The use of indirect techniques. In H. A. Liddle, D. C. Breunlin, & R. C. Schwartz (Eds.), *Handbook of family therapy training and supervision* (pp. 93–109). New York, NY: Guilford Press.

Murray, S. B. (2014). A case of Strategic couples therapy in adult anorexia nervosa: The importance of symptoms in context. *Contemporary Family Therapy, 36*, 392–397. doi:10.1007/s10591-014-9301-y

Nardone, G. (1996). *Brief Strategic solution-oriented therapy of phobic and obsessive disorders*. Northvale, NJ: Aronson.

Nardone, G., & Watzlawick, P. (2005). *Brief Strategic Therapy: Philosophy, techniques, and research*. Lanham, MD: Jason Aronson.

Nelson, J. A. (2006). For parents only: A Strategic Family Therapy approach in school counseling. *The Family Journal: Counseling and Therapy for Couples and Families, 14*(2), 180–183. doi:10.1177/1066480705285557

O'Hanlon, W. (1982). Strategic pattern interventions. *Journal of Strategic and Systemic Therapies, 4*, 26–33. doi:10.1037/e535522013-001

Quick, E. K. (2012). *Core competencies in the solution-focused and Strategic therapies: Becoming a highly competent solution-focused and Strategic therapist*. New York, NY: Routledge.

Riley, S. (1990). A Strategic Family Therapy approach to art therapy with individuals. *American Journal of Art Therapy, 28*(3), 71–79.

Schiff, N. P. (1990). Strategic Therapy for children's school problems. *Guidance & Counseling, 5*(3), 32.

Stone, G. P., & Peeks, B. (1986). The use of Strategic Family Therapy in the school setting: A case study. *Journal of Counseling and Development, 67*, 200–203. doi:10.1002/j.1556-6676.1988.tb02052

Szapocznik, J., Hervis, O., & Schwartz, S. (2003). *Brief Strategic Family Therapy for adolescent drug abuse* (National Institutes of Health Publication No. 03-4751). Bethesda, MD: U.S. Department of Health and Human Services.

Ungar, M., & Costanzo, L. (2007). Supervision challenges when supervisors are outside supervisees' agencies. *Journal of Systemic Therapies, 26*(2), 68–83. doi:10.1521/jsyt.2007.26.2.68

Watzlawick, P. (1978). *The language of change: Elements of therapeutic communication*. New York, NY: Basic Books.

Watzlawick, P., Weakland, J. H., & Fisch, R. (2011). *Change: Principles of problem formulation and problem resolution*. New York, NY: W.W. Norton.

Weakland, J. (1976). Communication theory and clinical change. In P. J. Guerin (Ed.), *Family therapy: Theory and practice* (pp. 111–128). New York, NY: Gardner Press.

Weakland, J., Fisch, R., Watzlawick, P., & Bodine, A. M. (1974). Brief therapy: Focused problem resolution. *Family Process, 13*(2), 141–168.

TEN

Training the Multigenerational Family Therapist

Heather Katafiasz

"There is a little schizophrenia in all of us."
—Murray Bowen

TEACHING THEORETICAL CONTEXT

Bowen Family Systems Theory attempts to explain life and the human phenomenon, rather than specific families (Friedman, 1991). Murray Bowen's theory has been termed a middle range theory, applying to the wide range between micro-level processes (families) and macro-level processes (life; Innes, 1996). According to Friedman (1991), Bowen found inspiration in the physical sciences (physics, astronomy, and biology) as well as deviations from reductionist thinking (study of the components rather than the whole system). He observed the parallel processes occurring throughout the physical sciences, in which theorists across these fields attempted to develop unified theories. These theories, like field theory and string theory, aimed to explain the interactions among the components of systems, rather than the components in isolation. Similarly, rather than focusing on how individuals develop dysfunction, Bowen attempted to create a unified theory that explained the interactions among people in a system, with health and dysfunction being possible consequences of those interactions. He believed that the constructs he identified were not unique to human emotional systems, but that those same processes (albeit in a simplified presentation) were also displayed in nonhuman systems. The ramification of Bowen's presentation of his theory as a unified clinical theory is that it should be applicable to all clients, regardless of background or symptom presentation (Friedman, 1991).

FOUNDER

Murray Bowen received his medical degree in 1937 from the University of Tennessee Medical School (Bowen, 1978). Although he was initially interested in surgery, after serving 5 years in the U.S. Army during World War II, he switched to a focus on psychiatry. He trained as a psychiatrist at the Menninger Foundation and later became a staff member there (Bowen, 1978).

As with many early family therapists, the development of Bowen's theory grew out of the decision to include family members in the treatment of patients diagnosed with schizophrenia (Kerr & Bowen, 1988). Bowen treated these patients and their families while at the Menninger Institute in the 1940s, and later at the National Institute of Mental Health in the 1950s, during which time he conducted a 5-year research project involving the hospitalization of the entire nuclear family along with the client (Kerr & Bowen, 1988). It was from these clinical experiences that Bowen identified the symbiotic mother-child relationship and its impact on symptom development in the client, with the subsequent recognition that the relational dynamics encompassed all family relationships (Bowen, 1978).

When he moved to Georgetown University in 1959, Bowen established the Georgetown Family Centre and began to apply the concepts of his theory to the treatment of families experiencing less severe psychological disturbances (Kerr & Bowen, 1988). During that time, Bowen recognized that the processes of families were similar, regardless of the severity of the symptoms, observing that "[t]here is a little schizophrenia in all of us" (Kerr & Bowen, 1988). Bowen's theory was derived not only from his clinical experiences, but also from his own experiences with his family of origin (Bowen, 1978). In 1967, at the Family Research Conference, he presented his own attempts to differentiate from his family of origin, demonstrating that a therapist should not only utilize a theory in therapy, but also live the theory of therapy (Bowen, 1978).

PHILOSOPHICAL FOUNDATION

Bowen Family Systems Theory is based on the need for intrapersonal and interpersonal balance, with chronic anxiety and differentiation serving as the foundational constructs for the theory. These constructs are opposing: the former causing symptoms and the latter acting as the antidote (Kerr & Bowen, 1988). What makes the two concepts similar is that humans continually struggle with both, struggling against succumbing to the former and attempting to achieve the latter. Influencing the levels of chronic anxiety and differentiation are two counterbalancing interpersonal forces: individuality and togetherness (Kerr & Bowen, 1988). Like gravity, the togetherness force pulls people toward one another; however, getting too close triggers anxieties about losing one's sense of self. In contrast, attempting to individuate from others can trigger anxieties about being alone. In other words, people are driven to be as close to others as possible, yet, at the same time, they need to find a way to maintain emotional or possibly physical distance without becoming too separate—a process often referred to as *distance regulation*. Bowen referred to the interdependent relationships in which this distance regulation occurs as *emotional systems* (sometimes also referred to as an emotional *field*, in recognition of the similarity between human and natural systems; Friedman, 1991). Furthermore, the automatic emotional and physical responses that occur in human interactions while attempting to balance these forces are known as *chronic anxiety* (Friedman, 1991). The chronic anxiety that permeates all situations may overwhelm us, blocking our ability to think and reflect rationally and wisely when under that emotional pressure (Kerr & Bowen, 1988).

All symptoms and problems that develop in individuals and families result from the inability to control this chronic anxiety. The antidote for these problems is the ability to act in a differentiated manner (Friedman, 1991). Interpersonally, *differentiation* has been defined as the capacity to not automatically respond to emotional pressures incurred by interacting with others (Kerr & Bowen, 1988). Intrapersonally, it has been described as the ability to be able to distinguish cognitions and emotions (Bowen, 1978), so that when people experience interpersonal imbalances in individuality and togetherness (relationships with high chronic anxiety), they are still able to make choices about how to react to the emotions and cognitions experienced. Bowen believed that differentiation exists on a continuum, with the majority of people lying somewhere in the middle (Kerr & Bowen, 1988), possibly similar to a bell curve, in a constant battle to react in a differentiated manner to the chronic anxiety.

Bowen's perspective on the continuum of differentiation has two implications. First, becoming differentiated is a continual, lifelong goal that all should strive for, but most are unlikely to achieve (Friedman, 1991). Those who claim to be "differentiated" do not fully understand the concept, as one cannot ever attain complete differentiation. Second, all emotional systems are symptomatic to some extent, although the development of symptoms, the severity of the symptoms that develop, and the interference with daily life may vary depending on the interaction between differentiation level and the intensity of stressful situations (Friedman, 1991). Lower differentiation increases the risk for the development of more severe symptoms, but it alone cannot cause symptom development. Bowen hypothesized that those with higher differentiation will be less likely to develop symptoms because their level of differentiation provides them with more resources to handle more stressful situations. However, those with lower differentiation, and fewer resources, may develop symptoms under only somewhat stressful situations (Friedman, 1991).

When confronted with a stressful situation, those with lower differentiation easily become overwhelmed by chronic anxiety, thereby experiencing emotional reactivity (Bowen, 1978). They can become so emotionally reactive in relationships that their individual identities become defined by who they are in that relationship, so when they experience chronic anxiety, they perceive any threat to the relationship as a threat to their individual identity (Bowen, 1978).

Getting too close (fusing) or too far away (cutting off) from loved ones is an emotionally reactive response that will temporarily reduce the impact of chronic anxiety. Interpersonally, *fusion* can be defined as inability to maintain an "I" stance in the face of an overwhelming "we"; *cutting off* can be defined as inability to maintain "we" while asserting an "I" stance. Although fusion and cutoff appear to be different, they are manifestations of the same inability to act in a differentiated manner; that is, not to become emotionally reactive when faced with high levels of chronic anxiety (Kerr & Bowen, 1988). Research has supported this relationship between differentiation and the use of fusion or cutoff as a distance regulator, such that higher differentiation is associated with less use of fusion and cutoff (Skowron & Schmitt, 2003).

In the formation of new emotional systems, Bowen believed that people will tend to be drawn to partners with levels of differentiation similar to their own, suggesting that their ability to deal with the chronic anxiety is similar, although he theorized that their emotionally reactive responses to the chronic anxiety tend to contrast (Bowen, 1978; Kerr & Bowen, 1988). Moreover, when two people form a new emotional system, their individual levels of differentiation coalesce to form a basic relational level of differentiation (Friedman, 1991). Therefore, it is not that one person has a certain level of differentiation or emotional reactivity, but that both simultaneously experience the same level of differentiation and are emotionally reactive (or unreactive) to each other. Pulling them apart to examine them individually does not allow identification of the relational processes at play (Kerr & Bowen, 1988).

Some research has supported Bowen's proposition that partners will have similar levels of differentiation (Miller, Anderson, & Kaulana-Keala, 2004), while other research has not (Bartle, 1993; Miller et al., 2004). A possible explanation for this contradictory evidence is that although their basic level of differentiation should be the same, Bowen hypothesized that in emotionally reactive couples, one member of the couple may develop a higher functioning level of differentiation than the other (Kerr & Bowen, 1988). This emotional reactivity may present in one of three forms of dysfunction: relational conflict, symptom development in one member of the couple, or symptom development in a child (Friedman, 1991). In the case in which one member of couple presents as the symptom bearer (alcoholic, adulterer, etc.) and appears to have lower differentiation, that person has become the "underfunctioner" for that emotional system. This presentation co-emerges in the context of the other member of the couple, who is not the symptom bearer, becoming the "overfunctioner" for the emotional system, and thus presenting with higher differentiation. In other words, their level of basic differentiation does not change, but their level of functional differentiation may look different in response to their relational processes in their emotional system (Kerr & Bowen, 1988).

The main mechanism through which people reduce chronic anxiety is to create an emotional triangle (Bowen, 1978). Bowen believed that the most stable form of social interaction is a triangle, like a three-legged stool; the dyad needs the extra leg to become stable (Friedman, 1991). A triangle usually involves bringing another person into the interaction, which disperses the chronic anxiety in the primary dyad into three separate dyads, allowing for all members to experience moderate amounts of chronic anxiety, rather than the primary couple experiencing an overwhelming amount of chronic anxiety. However, when these triangles become rigid and inflexible, it becomes *triangulation*, which can be problematic for families and lead to symptom development. Triangulation typically involves two people becoming closer and a third becoming the "enemy," but the third leg can also experience symptoms (Kerr & Bowen, 1988). In the cases where symptoms are triangulated, the anxiety in the primary dyad is reduced because the focus of the anxiety becomes solving the symptom and avoiding the underlying problem.

Bowen believed that relational processes, including levels of differentiation and responses to chronic anxiety, will be carried forward from previous generations and relived in the processes of current emotional systems; he called this the *multigenerational transmission of emotional processes* (Friedman, 1991). This implies that in any given family tree, there will be branches that have higher differentiation and branches with lower differentiation, as patterns from both higher- and lower-differentiated emotional systems will be transmitted. For the development of a well-functioning emotional system, the family members should be highly differentiated; the chronic anxiety and emotional reactivity should be low, as family members are able to adapt to changes in emotional and physical distance that accompany life transitions; and parents should be in good contact with their families of origin (Kerr & Bowen, 1988). However, remembering that Bowen believed that differentiation exists on a continuum and it is unlikely that any emotional system will always be able to react in a differentiated manner, it is possible that one of the three forms of dysfunctions may still develop.

Conversely, consistently high chronic anxiety in a family will form an undifferentiated ego mass, or an excessive amount of emotional reactivity in the family, such that the family members continually become fused or cut off from each other (Kerr & Bowen, 1988). When couples with lower differentiation have children, they will frequently triangulate their children to regulate the distance, usually doing so with one child more frequently than others. Bowen believed that sibling order affects a child's position in the emotional system and the likelihood that she or he will be triangulated (Friedman, 1991), although this has not been consistently supported by research (Cook, 2007; Miller et al., 2004). The extent to which certain children are triangulated more often than others will affect the levels of differentiation in all of the children. Those children not often triangulated will experience less chronic anxiety and may develop higher differentiation than their parents, leading to a higher-functioning branch of the family tree. Those children who are triangulated more often will have more of their parents' unwanted chronic anxiety projected onto them and may develop lower differentiation than their parents, leading to a more symptomatic branch of the family tree (Kerr & Bowen, 1988).

Bowen believed that continued declines in differentiation in subsequent generations would culminate in the presence of schizophrenia. In the same way that parents will develop "symptoms" as a way to reduce chronic anxiety and regulate their distance, triangulated children may also develop symptoms as a way to regulate the distance for the family. Eventually, when children raised by parents with lower differentiation attempt to leave their families, they may feel that their only course of action is to cut off, as they are unable to separate from their family without the family attempting to fuse again. As adults, they will have emotionally reactive responses (e.g., substance use, infidelity, triangulation, etc.) similar to those of their parents (or other significant relatives) to the chronic anxiety. This process of parents projecting their chronic anxiety, emotional reactivity, and differentiation onto their children is known as the *family projection process* (Kerr & Bowen, 1988).

Finally, Bowen believed that the multigenerational transmission of emotional processes exists on a much larger scale as societal emotional processes (Bowen, 1978). He hypothesized that humankind, as a species, is on a spiral of downward differentiation. Specifically, he stated that the population explosion and resulting limited physical space have created an imbalance in humankind's sense of individuality and togetherness, which have increased societal-level chronic anxiety; increased social and environmental problems are humankind's emotionally reactive responses to this anxiety. He suggested that the societal regression would come to a halt either when the consequences of the emotional reactivity are greater than the societal-level chronic anxiety (a global crisis) or the societal level of chronic anxiety dissipates (Bowen, 1978). Several contemporary theorists have discussed the issue of societal regression, specifically related to the mental health field and the perhaps false sense of security that licensing psychologists creates for the public in the United States (Smith, 2011), New Zealand (Tudor, 2011), and Italy (Mazzetti, 2011). While not specifically discussing Bowen's concept of societal emotional processes, other theorists have identified terrorism as exemplifying possible societal regression (Brenner, 2006; Volkan, 2002).

THE PROCESS OF CHANGE

Bowen Family Systems Theory is very insight-oriented. Therapy focuses on allowing clients the opportunity to learn more about themselves and their relationships, so they can assume responsibility for their own problems (Kerr & Bowen, 1988). The therapist's role is to be a coach and facilitator of the clients' understanding, rather than directing or pushing them. Regardless of the presenting symptom, the focus of therapy is on helping clients learn to manage their own chronic anxiety, separating their cognitions and emotions while in situations that provoke chronic anxiety (dysregulation in distance); this will allow them to react in a more differentiated manner and reduce the need for the symptom in the emotional system. Bowen believed that in order for clients to remain emotionally unreactive, the chronic anxiety in the room should be reduced as much as possible, so that clients can think as clearly as possible during the session (Kerr & Bowen, 1988).

Ideally, all members of the emotional system would be present for therapy. However, if the therapist is able to detriangulate the main triangle of the family, which generally involves the parents, then the entire emotional system can change (Kerr & Bowen, 1988). Friedman (1991) also suggested that if the differentiation level in the leader of the family, which may be the person with the highest level of differentiation or the greatest capacity for change in differentiation level, can be increased, the resulting interactional change may serve as a catalyst to increase the levels of differentiation among all members of the emotional system. Therefore, while family therapy is the preferred method of treatment for Bowenian therapists, therapy with an individual can also create the desired family change.

ROLE OF THE THERAPIST (THERAPEUTIC STANCE)

According to Friedman (1991), "the capacity of the therapist to apply Bowen theory is a function of the therapist's own differentiation" (p. 138). During therapy, the therapist forms a new emotional triangle with the client system but must remain emotionally unreactive and not become triangulated during the session (Kerr & Bowen, 1988). To prevent triangulation into the family's system, the therapist should take an "I" position, which is an autonomous position. Furthermore, the therapist must remain aware of his or her own emotional reactions in the room, while still being able to think clearly, so that he or she can use those reactions to help guide therapy (Kerr & Bowen, 1988). Friedman (1991) suggested that the therapist needs to balance the emotional and physical distance with clients and still be connected to them as humans. To accomplish this task, the therapist must have at the least a higher level of differentiation than the clients. Bowen believed that the client can only mature to the maturity level of the therapist; therefore, self-development of the therapist should necessarily be continual process. Furthermore, many Bowen therapists believe that therapists cannot simply learn the theory, but rather must live it by applying it to their lives, as Bowen did during his development of the theory. By modeling a more differentiated response to the client and with the client, clients are helped to learn to regulate their chronic anxiety and become more differentiated themselves, thereby enabling them to choose how to respond to their emotional and cognitive experience. Because the therapists' use of self in the therapy room is integral to helping clients become more differentiated, any technique has the possibility of being an effective intervention during the course of therapy (Friedman, 1991).

TECHNIQUES TO BE USED

The most frequently used technique of Bowen Family Systems Theory, and a multitude of other theories, is the *genogram*, a family tree that includes relationships, emotional triangles, and any other processes that can be passed from one generation to the next. A genogram usually goes back at least three generations (Kerr & Bowen, 1988). The therapist uses the genogram to show the client the family projection process, the emotional triangles, and the multigenerational transmission of emotional processes. By seeing these intergenerational patterns, clients can begin to detriangulate themselves from their family of origin and become more differentiated using relational experiments (Kerr & Bowen, 1988). This technique has become so widely utilized by therapists that several software programs have been created to assist with the drawing of a genogram (e.g., GenoPro), and books have been published detailing how to create and utilize a genogram in therapy (see, e.g., McGoldrick, 2011).

Another technique commonly utilized by Bowenian therapists is the *relationship experiment*, in which clients attempt to act in a differentiated manner in the context of the chronic anxiety that occurs in their emotional systems outside of therapy (Nichols, 2010). Although Bowen Therapy is a highly insight-oriented therapy, relationship experiments allow clients to begin to practice a more differentiated stance with those in their emotional systems and begin to detriangulate themselves from the emotionally reactive processes they are involved in. These experimental interactions are viewed as endeavors that may be successful or ineffective; regardless, the experience can then be processed in therapy for further understanding and adjustments (Nichols, 2010).

CULTURE/RACE AND GENDER ISSUES/RESEARCH TO CONSIDER

As stated previously, Bowen believed that his theory was describing life, not merely families (Bowen, 1978), arguing that the constructs of his theory were universal to all genders and cultures. The implication of this belief is that culture—including race, ethnicity, gender, and the specific beliefs, values, and traditions that vary between people—does not matter and does not affect the ability for change to occur in therapy. Bowen emphasized the emotional processes of the family, especially those transmitted multigenerationally, those that he believed were universal to all groups of people, and de-emphasized the factors that differentiate groups of people. These differentiating factors can be viewed as part of the client's content, in that they do not cause dysfunction, but they do illuminate the dysfunction in a manner in which the therapist is able to view it (Friedman, 1991). From this viewpoint, culture is discussed only to provide contextual information about the emotional processes (Friedman, 1991; Kerr & Bowen, 1988).

This de-emphasis on cultural considerations has spurred criticisms of Bowen's theory and inquiry into the cross-cultural applicability of the "universal" constructs (Skowron, 2004). Contemporary family theorists and therapists recognize the importance of understanding the influence of gender and multicultural factors in the application of therapy, yet few researchers have offered empirical examinations of the cross-cultural applicability of the Bowen theory or its constructs (Miller et al., 2004). One of the most criticized constructs, differentiation, has been interpreted as overly emphasizing individuation, both a stereotypically masculine characteristic and a Westernized value (Knudson-Martin, 1994; Tamura & Lau, 1992). Some theorists have offered modifications to Bowen's theory to answer the critiques, such as Knudson-Martin's (1994) feminist perspective on the applicability of the therapy with women and Tamura and Lau's (1992) multicultural perspective on the applicability of the therapy with the Japanese culture. Conversely, both of these researchers suggested modifications that have in turn been challenged by others suggesting that Bowen Theory has merely been misinterpreted as culturally or gender biased (Bregman, 1993; Horne & Hicks, 2002). McGoldrick and Carter (2001) address the cultural applicability of Bowen Theory, but through a slightly different lens, as they do not critique the concepts in Bowen's theory. Instead, they argue that the theoretical conceptualization of clients must be contextualized to account for differential experiences based on societal-level processes, specifically identifying that differentiation may be exceptionally difficult for those of minority background (racial, ethnic, sexual, or gender) to achieve because any attempts may be met with ostracization or even violence (McGoldrick & Carter, 2001). It is suggested that culture be assessed in the beginning stages of therapy; the information obtained can help the therapist determine how an individual family interprets and represents their cultural group (Thomas, 1998). The cultural genogram can be utilized to gather additional information about the cultural background of a particular symptom and aid the therapist in identifying interventions that will help that particular client system make progress toward the goal of increased differentiation (Thomas, 1998).

ROLE OF THE SUPERVISOR AND SUPERVISEE: LEARNING GOALS/OBJECTIVES

The goal of Bowenian supervision, similar to the goal of Bowenian therapy, is to assist supervisees in becoming as differentiated as possible, so that they will be able to maintain a differentiated stance and prevent emotional reactivity in therapy sessions with clients (Papero, 1988). Within this goal, supervisees must have a thorough understanding of the theory—but an academic understanding of the theory alone is not enough. Supervisees must also apply the theory to their daily lives; they must live the theory as Bowen attempted to (Papero, 1988). Therefore, supervision largely focuses on the self-of-the-therapist of the supervisee (Lee, 1997; McDaniel, Weber, & McKeever, 1983), in the belief that techniques can be developed later in the training process (McDaniel et al., 1983). This type of supervision is also referred to as *family-of-origin supervision*, with supervisors in the role of coach to the supervisees' self-exploration of their family-of-origin issues (McDaniel et al., 1983). However, the blending of supervision and therapy creates an ethical dilemma due to the dual relationships created, the possibility of exploitation of the supervisees, and the difficulties in maintaining objectivity in grading when supervisors are aware of personal information about supervisees (Donnelly & Gosbee, 2009; Liddle & Halpin, 1978; Winek & Shaw, 2000); nevertheless, Bowen viewed addressing family-of-origin issues as an essential part of clinical training (Friedman, 1991; Liddle & Halpin, 1978; Papero, 1988; Storm, 2000). Proponents of family-of-origin supervision point out that there are disadvantages experienced by supervisees who have not worked on family-of-origin issues (Donnelly & Gosbee, 2009). Specifically, these supervisees may experience decreased theoretical understanding of families as emotional systems, and the training progress of the supervisees as well as the therapeutic progress of clients may be impeded due to the parallel and isomorphic processes derived from unresolved family-of-origin issues (Donnelly & Gosbee, 2009). One way to protect the privacy of the supervisee (Brooks & Roberto-Forman, 2014) and obviate the potential ethical challenges of this form of supervision is to ensure that the family-of-origin material discussed is always case-related and to discuss the family-of-origin material only after a case has been presented and problem areas of the case have been identified (Munson, 1984).

The focus of supervision from a Bowenian perspective is on the supervisee's emotional reactivity to processes in client systems that resemble emotional processes in the supervisee's family of origin. When supervisees who have lower differentiation are confronted with the stressors of actually beginning to provide therapeutic services to clients, their own resources may be taxed, leading to emotionally reactive responses to clients and supervisors. Several family-of-origin interactional patterns have been identified that frequently reoccur as parallel processes in either the therapeutic or supervisory relationships and therefore should be addressed with supervisees: triangulation (Getz, 1994; Lawson, 1993; Protinsky & Keller, 1984), overfunctioning and underfunctioning (Getz, 1994; Protinsky & Keller, 1984), and pursuit and distancing (Getz, 1994). These processes and additional emotional processes based on Bowen Theory can be identified in family-of-origin supervision via the supervisees' completion of their genogram (Lawson, 1993). This can be especially helpful for students struggling to see the "schizophrenia" in all families as well as to accept that differentiation is not a destination, but rather a never-ending process of increasing their level of differentiation and attempting to respond in a more differentiated manner.

The supervisees' cultural history can also be integrated into the completion of the genogram (Kelly, 1990). Friedman (2000) states that the inclusion of cultural factors is not a necessary component of supervision from a Bowenian perspective, as the emotional processes described by the theory "transcend culture and natural systems" (p. 212). However, this stance has been contradicted by fellow Bowenian supervisees and supervisors, who concur with the inclusion of contextual factors in both therapy and supervision (McGoldrick, 2000). Additionally, when exploring the contextual makeup of the supervisee, the supervisee's experiences of and beliefs about gender, sexual orientation, religion, power dynamics should also be discussed. Hardy and Laszloffy (1995), as well as Shellenberger and colleagues (2007), have provided detailed recommendations on how to construct a cultural genogram for use in supervision.

Family-of-origin supervision generally consists of clinical case presentation or the presentation of the supervisee's family of origin (Lawson, 1993; McDaniel et al., 1983; Papero, 1988). When clinical cases are presented, they are frequently accompanied by raw data supervision, although this is not a strict requirement of Bowenian supervision (live or video; McDaniel et al., 1983). The goal of the case presentation is to apply the concepts from Bowen Theory and help supervisees identify how, in the next interaction they have with the members of the case, they can maintain a differentiated stance, remaining emotionally unreactive and therefore detriangulated.

Additional educational activities that can be utilized in family-of-origin supervision to enhance supervisees' understanding of their family-of-origin interaction patterns are reading additional theoretical materials, writing a detailed family autobiography, and completing family-of-origin standardized inventories (Lawson, 1993). Experiential activities that can elucidate family-of-origin processes are surrogate family sculpting and family-of-origin group work (Lawson, 1993). Winek and Shaw (2000) suggest conducting a family supervision session involving the supervisee's family members to further the supervisee's understanding of the reciprocal impact of clinical training on the family and of the family on clinical training. Although therapy is not specifically prescribed as a part of Bowenian supervision, it may be recommended to supervisees to help work toward their goal of becoming more differentiated.

SUPERVISOR-SUPERVISEE RELATIONSHIP

In family-of-origin supervision, the roles of the supervisor and supervisee are isomorphic to the roles of the therapist and client, respectively (McDaniel et al., 1983). Put another way, the supervisor-supervisee relationship often mirrors the supervisee-client relationship (Lee, 1997). Therefore, if an impasse is experienced in the supervisee-client relationship, it is likely that a similar impasse is being experienced in the supervisor-supervisee relationship (Lee, 1997). Impasses in this multigenerational clinical system will likely occur when the supervisor, supervisee, and client become triangulated (Lawson, 1993). A change in the supervisor-supervisee relationship can result in change in the supervisee-client relationship; however, at times the impasse may be difficult to identify (Lee, 1997). The supervisor must then become simultaneously an observer: of himself or herself, of the supervisor-supervisee relationship processes, and of the entire multigenerational supervisory system (Lee, 1997). To avoid impasses in therapy, the supervisor needs to be more differentiated than the supervisee, just as the therapist needs to be more differentiated than the client (Friedman, 1991; Liddle & Halpin, 1978). Additionally, the supervisor should be open to identifying self-of-the-supervisor issues that may affect the supervisor-supervisee relationship (Lee, 1997), including additional supervisor family-of-origin work (Liddle & Halpin, 1978; Papero, 1988) or gaining outside consultation and supervision (Lee, 1997).

Underlying the philosophy of self-of-the-therapist supervision is an assumption that, just as the therapy room must be perceived as a safe place for clients to openly discuss difficult topics, the supervision room must be perceived as a safe place for the supervisee to openly discuss both clinical struggles and successes (Lee & Everett, 2004). Trust in the supervisor-supervisee relationship (Braverman, 1984; Kaiser, 1992; Winek & Shaw, 2000) and amongst supervisees (Braverman, 1984) has been identified as key to family-of-origin supervision. Additionally, supervisees in family-of-origin supervision must be willing to participate in this form of supervision for the process to be productive and ethically sound (Braverman, 1984; Brooks & Roberto-Forman, 2014; Munson, 1984). To this end, supervisees should be granted the ability to set boundaries on what is discussed, based on their readiness to address a specific family-of-origin issue, and the supervisor should adhere to the boundaries set (Brooks & Roberto-Forman, 2014).

CASE EXAMPLE

A necessity of supervision from the perspective of Bowen Theory is that Brittany has a thorough knowledge about Bowen Family Systems Theory and an ability to apply its theoretical concepts to the case. Knowledge of the basic concepts of Bowen Theory is required to attend to the primary objective of supervision, which is for Brittany to improve her differentiation to a level that is higher than that of her clients.

To first begin to examine Brittany's basic knowledge of Bowenian concepts, Dr. Bob asked her to present a case in which she was currently providing therapeutic services. Often supervisees will choose cases that have self-of-the-therapist relevance to them, whether this is through a shared piece of content or perhaps due to something more process-related. For instance, Brittany chose to present this particular case because of the shared content of military experiences. What became apparent to Dr. Bob was that Brittany was responding to the family in emotionally reactive ways on issues surrounding this shared content. Specifically, Dr. Bob noticed that Brittany seemed to

(*continued*)

be telling the clients how they should feel about Travis's discharge from the military. Dr. Bob utilized this observation of the process to begin to address self-of-the-therapist issues with Brittany, specifically exploring in her presentation of her own genogram what this content meant to her and how that could be discerned and separated from what it meant to the client.

During Brittany's presentation of her genogram to Dr. Bob, Brittany revealed the source of her emotional reactivity to the military content: She had negative experiences both in her family's reaction to her enlisting and during her time of service. Brittany enlisted merely to gain access to the financial resources that would enable her to attend college and was relieved when she was discharged due to the extreme gender discrimination she experienced. Further, Brittany's parents also disagreed with her decision to enlist, causing a rift between her and her parents that she has not yet repaired. Dr. Bob identified that Brittany was having difficulty in the therapy room reacting in a differentiated manner with her clients and that her emotional reactivity was being displayed in both her assumptions about how the client should feel about the discharge from the military and her inability to focus on the clients' emotional processes. This content had become triangulated between Brittany and the clients, with her seeing it as something they should be happy to be rid of, whereas the clients were still grieving the loss of an important part of their lives. When Dr. Bob attempted to discuss his observations with Brittany, she began to become reactive to him as well. Brittany later identified that the emotional reactivity she had to Dr. Bob mirrored her emotional reactivity to her father, who was unsupportive of her military enlistment. Dr. Bob suggested that Brittany seek her own family-of-origin counseling so that she could work on her level of differentiation; if she is able to react in a differentiated manner with her parents on this evocative topic, she should be able to maintain a differentiated manner with her clients in the therapy room.

After processing these self-of-the-therapist issues with Brittany, Dr. Bob asked her to repeat her case presentation on this family highlighting the emotional processes she has observed. Brittany was then able to identify that the family is likely experiencing at least one of the forms of dysfunction hypothesized by Bowen: relationship conflict. She also identified that triangulation may also be occurring in the family, with Travis's "short fuse" and possible child behavior problems, described as parenting issues by the couple, becoming the triangulated symptoms. Brittany then suggested that the family's emotional system is struggling to balance the separateness and connectedness through the many transitions the family has

(*continued*)

experienced: Travis's military tours, his life-threatening injury and recovery, and his new and unstable work experiences. Brittany and Dr. Bob also explored other contexts of the family that may be relevant for the supervisor-supervisee as well as for the supervisee-client relationship: racial similarities and differences, gender roles and stereotypes, financial instability, emotional expression, relationship expectations, and parenting expectations.

REFERENCES

Bartle, S. (1993). The degree of similarity of differentiation of self between partners in married and dating couples: Preliminary evidence. *Contemporary Family Therapy, 15*(6), 467–484.

Bowen, M. (1978). *Family therapy in clinical practice*. New York, NY: Aronson.

Braverman, S. (1984). Family of origin as a training resource for family therapists. *The Clinical Supervisor, 2*(2), 37–47.

Bregman, O. (1993). Bowen theory. *Family Process, 32*, 377–378.

Brenner, I. (2006). Terror and societal regression: Does psychoanalysis offer insights for international relations? Panel report. *Journal of the American Psychoanalytic Association, 54*(3), 977–988.

Brooks, S., & Roberto-Forman, L. (2014). The transgenerational supervision models. In T. Todd & C. Storm (Eds.), *The complete systemic supervisor: Context, philosophy, and pragmatics* (2nd ed., pp. 186–207). West Sussex, UK: Wiley-Blackwell.

Cook, L. (2007). Perceived conflict, sibling position, cut-off, and multigenerational transmission in the family of origin of chemically dependent persons: An application of Bowen Family Systems Theory. *Journal of Addictions Nursing, 18*(3), 131–140.

Donnelly, H., & Gosbee, M. (2009). Family of origin supervision in the workplace: Impacts on therapist and team functioning. *Australian & New Zealand Journal of Family Therapy, 30*(4), 300–314.

Friedman, E. (1991). Bowen theory and therapy. In T. Sexton, G. Weeks, & M. Robbins (Eds.), *Handbook of family therapy: The science and practice of working with families and couples* (Vol. 2, pp. 134–170). New York, NY: Brunner/Routledge.

Friedman, E. (2000). Dear supervision committee: Contextual issues protest. *Readings in family therapy supervision: Selected articles from the* AAMFT Supervision Bulletin (pp. 212–213). Washington, DC: AAMFT.

Getz, H. (1994). Training marriage and family counselors: A family-of-origin approach. *Counselor Education & Supervision, 33*(3), 183.

Hardy, K., & Laszloffy, T. (1995). The cultural genogram: Key to training culturally competent family therapists. *Journal of Marital and Family Therapy, 21*, 227–237.

Horne, K., & Hicks, M. (2002). All in the family: A belated response to Knudson-Martin's feminist revision of Bowen theory. *Journal of Marital and Family Therapy, 28*(1), 103–113.

Innes, M. (1996). Connecting Bowen theory with its human origins. *Family Process, 35*, 487–500.

Kaiser, T. (1992). The supervisory relationship: An identification of the primary elements in the relationship and an application of two theories of ethical relationships. *Journal of Marital and Family Therapy, 18*, 283–296.

Kelly, G. (1990). The cultural family of origin: A description of a training. *Counselor Education & Supervision, 30*(1), 77.

Kerr, M., & Bowen, M. (1988). *Family evaluation: An approach based on Bowen theory.* New York, NY: W.W. Norton.

Knudson-Martin, C. (1994). The female voice: Applications to Bowen's Family Systems Theory. *Journal of Marital and Family Therapy, 20*, 35–46.

Lawson, D. (1993). Supervision methods for addressing triangulation issues with counselors-in-training. *The Family Journal, 1*(3), 260–268.

Lee, R. (1997). Seeing and hearing in therapy and supervision. *Journal of Family Psychotherapy, 8*(3), 51–57.

Lee, R., & Everett, C. (2004). *The integrative family therapy supervisor.* New York, NY: Routledge.

Liddle, H., & Halpin, R. (1978). Family therapy training and supervision literature: A comparative review. *Journal of Marital and Family Therapy, 4*, 77–98.

Mazzetti, M. (2011). Some reflections on "Licensing of psychotherapists in the United States: Evidence of societal regression?" *Transactional Analysis Journal, 41*(2), 147–150.

McDaniel, S. H., Weber, T., & McKeever, J. (1983). Multiple theoretical approaches to supervision: Choices in family therapy training. *Family Process, 22*, 491–500.

McGoldrick, M. (2000). More on the contextual learning objective debate: Objective supports our democratic ideals. In *Readings in family therapy supervision: Selected articles from the* AAMFT Supervision Bulletin (pp. 219–220). Washington, DC: AAMFT.

McGoldrick, M. (2011). *The genogram journey: Reconnecting with your family.* New York, NY: W.W. Norton.

McGoldrick, M., & Carter, B. (2001). Advances in coaching: Family therapy with one person. *Journal of Marital and Family Therapy, 27*(3), 281.

Miller, R., Anderson, S., & Kaulana-Keala, D. (2004). Is Bowen theory valid? A review of basic research. *Journal of Marital and Family Therapy, 30*(4), 453–466.

Munson, C. (1984). Uses and abuses of family of origin material in family therapy supervision. *The Clinical Supervisor, 2*(2), 61–74.

Nichols, M. (2010). *Family therapy: Concepts and methods* (10th ed.). Boston, MA: Pearson.

Papero, D. V. (1988). Training in Bowen theory. In H. A. Liddle, D. C. Breunlin, & R. C. Sehwartz (Eds.), *Handbook of family therapy training and supervision* (pp. 62–77). New York, NY: Guilford Press.

Protinsky, H., & Keller, J. (1984). Supervision of marriage and family therapy. *The Clinical Supervisor, 2*(2), 75–80.

Shellenberger, S., Dent, M., Davis-Smith, M., Seale, J., Weintraut, R., & Wright, T. (2007). Cultural genogram: A tool for teaching and practice (Report). *Families, Systems & Health, 25*(4), 367–381.

Skowron, E. (2004). Differentiation of self, personal adjustment, problem solving, and ethnic group belonging among persons of color. *Journal of Counseling & Development, 82*, 447–456.

Skowron, E., & Schmitt, T. (2003). Assessing interpersonal fusion: Reliability and validity of a new DSI fusion with other subscales. *Journal of Marital and Family Therapy, 29*(2), 209–222.

Smith, J. (2011). Licensing of psychotherapists in the United States: Evidence of societal regression? *Transactional Analysis Journal, 41*(2), 139–146.

Storm, C. (2000). Changing the line: An interview with Edwin Friedman. *Readings in family therapy supervision: Selected articles from the* AAMFT Supervision Bulletin (pp. 13–15). Washington, DC: AAMFT.

Tamura, T., & Lau, A. (1992). Connectedness versus separateness: Applicability of family therapy to Japanese families. *Family Process, 31,* 319–340.

Thomas, A. (1998). Understanding culture and worldview in family systems: Use of the multicultural genogram. *The Family Journal, 6*(1), 24–32.

Tudor, K. (2011). There ain't no license that protects: Bowen theory and the regulation of psychotherapy. *Transactional Analysis Journal, 41*(2), 154–161.

Volkan, V. (2002). 26th S.H. Foulkes annual lecture: Part one—September 11 and societal regression. *Group Analysis, 35*(4), 456–483.

Winek, J., & Shaw, L. (2000). Family supervision: Co-constructing basic therapy skills. In *Readings in family therapy supervision: Selected articles from the* AAMFT Supervision Bulletin (pp. 90–93). Washington, DC: AAMFT.

ELEVEN

Training the Cognitive Behavioral Therapist

Joanne Holbert
Katie Wootton

This chapter examines the theory of Cognitive Behavioral Therapy (CBT) and use of the theory as a model of supervision. The theory of CBT is reviewed by examining a philosophical foundation, techniques and interventions, the role of the therapist, the process of change, and cultural issues. Lastly, the significance of utilizing a CBT approach to therapy within the supervisor-supervisee relationship is examined. The authors review the supervisor-supervisee relationship, looking specifically at goals and challenges, and follow with a case example.

PHILOSOPHICAL FOUNDATION

Cognitive Behavioral Therapy is a brief therapy that can be utilized with a selection of mental and behavioral health populations. Cognitive Behavioral Therapy (CBT) can be used with adults, children, and older populations throughout an extensive continuum of mental and behavioral health diagnoses with couples, families, or individual concerns. CBT is an evidence-based treatment approach that has been researched within the context of a vast array of mental and behavioral disorders with various populations. This therapy is short-term, "present-orientated psychotherapy . . . directed toward solving current problems and modifying dysfunctional thinking and behavior" (Beck, 2011, p. 2).

CBT was founded by Aaron Beck in the 1960s and was originally called cognitive therapy (Beck, 2011). Beck's daughter, Judy Beck, followed him in incorporating CBT theory into practice. Beck incorporates a number of different theories and ideas from different foundational theorists, including Karen Horney, Alfred Adler, George Kelly, Albert Ellis, Richard Lazarus, and Albert Bandura, and utilizes different techniques and interventions to form a cognitive framework (Beck, 2011).

Cognitive conceptualizations of the client are used to formulate goals in CBT (Beck, Sarnat, & Barenstein, 2008). "Clients' difficulties, including both emotional and behavioral reactions, are understood in terms of their perceptions of situations, which are influenced by their basic understandings of themselves, others, and their worlds" (Beck et al., 2008, p. 58). Thoughts, feelings, and behaviors are often seen as a triangle where all three are connected to and affected by each other (Beck, 2011; Greenberger & Padesky, 1995). CBT focuses on changing the dysfunctional thoughts to allow and stimulate behavior changes and changes in feelings, which then affect how clients think about new situations. Therapy sessions focus on examining automatic thoughts that lead to underlying assumptions connected to core beliefs.

The Process of Change

CBT is collaborative, as the therapist and client are considered a team working on problems together (Beck, 2011). CBT is active and engages the client through a treatment relationship that encourages but also challenges. The process of change occurs through structured sessions followed by "booster sessions" (Beck, 2011) after termination of the structured meetings. When clients are able to examine the dysfunctional thoughts and trace these back to underlying assumptions and core beliefs, the clients will understand how the core beliefs affect their current thoughts and feelings. Underlying assumptions and core beliefs are often connected to an event or events that can be explored utilizing CBT techniques. The therapist then works as a "teammate" or a "business partner" with the client to utilize CBT techniques and interventions to promote change (Beck, 2011). Dysfunctional automatic thoughts are replaced with more adaptive thoughts that lead to more functional behavior and feelings.

Role of the Therapist

The therapist using a CBT approach serves as a "tour guide," working to navigate any automatic thoughts and feelings that the client may have and organizing them. The therapist works with the client to reorganize and make new meaning through restructuring distorted beliefs. The therapist examines the origins of these beliefs through Socratic questioning. The therapist and the client then collaborate, using structured sessions, to employ specific CBT techniques and interventions to look for evidence with which to test these beliefs. The therapist writes notes as the agenda is prepared and interventions are utilized. These notes are often copied and given to clients. The notes are also used as part of the clinical record to develop progress notes and treatment goals.

Case Conceptualization

In all therapeutic approaches, it is important to be able to formulate an idea or hypothesis about what issues the client is presenting in therapy. In Cognitive Behavioral Therapy, the case conceptualization is an important component of the structure and process of therapy. Kuyken, Padesky, and Dudley (2011) write extensively about the importance of case conceptualization in CBT. They emphasize the progressive development of the conceptualization through subsequent therapy sessions and describe three levels of case conceptualization. The first is *descriptive conceptualization,* which connects the client's presenting concerns with the language of CBT in behavioral and cognitive expressions. The second level of conceptualization seeks to understand what triggers and maintains these cognitions and behaviors; this is referred to as *cross-sectional conceptualization*. *Longitudinal conceptualization* is the third level of case conceptualization, utilized when core beliefs have a substantial context from previous experiences.

Techniques

CBT utilizes a structured format for sessions. Each session begins with setting an agenda, bridging back to the previous session, and setting a target for the current session. The therapist reviews the previous session and the homework assignments and then, with the client, establishes the goals for the current session by setting the agenda. CBT techniques are then applied to the target goal. Summarizing the information frequently, and then the session as a whole, helps to stay on task and to plan for homework assignments. The homework is checked for understanding, and feedback from the session is gathered from the client. Different techniques are part of the therapy process used within the CBT approach.

CBT utilizes a variety of successful, evidence-based techniques and interventions (Beck, 2011). Therapists working from this approach begin by working with the client to identify automatic thoughts. Once these thoughts are brought to the client's awareness, in a process of discovery through collaboration with the therapist, the client and therapist then use a variety of techniques as homework to promote change.

Automatic Thoughts

Automatic thoughts are a core concept within the CBT theory; these are thoughts that clients might notice affecting their everyday behaviors (Beck, 2011). Cully and Teten (2008) define *automatic thoughts* as follows.

> An automatic thought is a brief stream of thought about ourselves and others. Automatic thoughts largely apply to specific

> situations and/or events and occur quickly throughout the day as we appraise ourselves, our environment, and our future. We are often unaware of these thoughts, but are very familiar with the emotions that they create within us. (p. 44)

The therapist and the client work to identify and then challenge automatic thoughts. An automatic thought record helps clients begin to identify these destructive thoughts. An *automatic thought record* is a multicolumn table that connects situations, automatic thoughts, feelings, and alternate thoughts. A variety of automatic thought record formats are available online and in CBT texts (Beck, 2011; Padesky & Greenberger, 1995). Automatic thoughts stem from core beliefs, which are often formed in childhood or as the result of a stressful event. These learned experiences lead to underlying "rules," beliefs, or assumptions. The therapist and client work to identify cognitive distortions causing some of these negative thoughts. *Cognitive distortions* are errors in thinking, such as all-or-nothing thinking, overgeneralization, mind reading, labeling, catastrophizing, or emotional reasoning (see Beck, 2011, p. 181, for definitions and examples). A therapist might assign homework to a client to guide him or her in becoming more aware of these automatic thoughts and core beliefs through automatic thought records or logs, awareness exercises, and cognitive case diagrams (Beck, 2011).

Homework

Homework is essential within CBT to promote positive change. "[H]omework assignments provide opportunities . . . to educate . . . further (e.g., through bibliotherapy), to collect data (e.g., through monitoring . . . thoughts, feelings, and behavior), to test [clients'] thoughts and beliefs, to modify their thinking, to practice cognitive and behavioral tools, and to experiment with new behaviors" (Beck, 2011, p. 295). Homework allows change to occur outside the therapy room (Walker & Lampropoulos, 2014). The therapist may practice the homework with the client in session and will check for understanding and assess any barriers that the client may perceive as possibly impeding homework completion.

Haarhoff and Kazantzis (2007) examined obstacles in homework completion by clients, as there is evidence to support the importance of homework in CBT. Although symptom reduction is enhanced when clients adhere to homework assignments, the researchers found that CBT trainees often feel rushed or inadequate in the assignment of homework and may fail to follow through with homework exercises or give poorly planned assignments. Haarhoff and Kazantzis (2007) offer some recommendations for supervisors to enhance homework completion; this is discussed in the "Cognitive Behavioral Therapy Supervision" section of this chapter.

Problem Solving and Skills Training

CBT theory works to promote change in daily living. Clients spend the majority of time practicing CBT homework within the realm of daily living, using problem-solving skills to generate plans and goals and determine solutions (Vaske, Galyean, & Cullen, 2011). Promoting problem solving and skills training evokes changes for those clients with both strong and weak skills. Clients who elicit strong problem-solving skills might be negatively affected when using these skills, due to negative automatic thoughts intruding during use of their skills. The therapist and client can work through structured sessions to identify these negative thoughts, using problem-solving skills to develop alternate strategies.

Problem-solving skills include clearly defining the problem, brainstorming possible ways to solve the problem by looking at what has worked or not worked before, or possibly creating a list of new techniques to consider. Problem-solving skills also involve setting clear goals and accepting when situations cannot be changed. Examples of problem-solving techniques include (but are not limited to) making pros/cons lists, considering worst-case and best-case scenarios, and rating possible solutions based on a criterion important to the client.

Relaxation and Mindfulness

Relaxation and mindfulness techniques are used within the CBT approach to increase internal experiences and awareness and to decrease stress and tension that impact the client mentally, emotionally, and physically. "There are several kinds of relaxation exercises, including progressive muscle relaxation, imagery, and controlled breathing" (Beck, 2011, p. 264). Relaxation techniques are often paired with imagery of dysfunctional thoughts to help the client reduce anxious feelings connected to dysfunctional thoughts. Mindfulness is used to help clients better understand and become aware of different thoughts and emotions, in the hope that the client will be able to identify and accept these thoughts with nonjudgmental purpose. The goal of mindfulness is to interrupt those automatic thought processes and replace the focus. Instead of reacting to incoming stimuli, the client focuses on accepting and observing the stimuli without judgment (Woolhouse, Knowles, & Crafti, 2012).

Role-Playing

The role-playing technique is often used within the therapy process between client and therapist to better identify automatic thoughts and core beliefs. Role-playing allows the therapist and client to gauge and predict what social situations or responses to automatic thoughts and core beliefs might look like. Role-playing also allows the client to practice newly acquired skills. Learning new social skills, practicing new adaptive responses, and adjusting communication skills are also done through role-play (Beck, 2011). In role-play, the therapist can play the role of the client or the role of someone with whom the client would like to use a new skill outside of the session. The concept behind the role-play is to practice situations that may present outside of therapy.

Summary

Overall, a CBT therapist might utilize several or many of the various techniques and interventions available within the therapy process with clients. These techniques work together to promote positive change and instill new awareness of automatic thoughts and core beliefs. Training in CBT often has an individual and diagnostic focus. A plethora of books has been written on using CBT in working with anxiety, depression, substance abuse, and schizophrenia. Significant research has also been done in utilizing CBT techniques with social phobia in adolescents, bipolar disorder, eating disorders, LGBT clients, children, and the elderly. CBT techniques are also appropriate in work with families and couples.

COGNITIVE BEHAVIORAL THERAPY FROM A SYSTEMS PERSPECTIVE

CBT techniques can be utilized individually or in a systems setting when working with couples, families, or groups. Often a support system within the therapy setting is beneficial. CBT is a collaborative approach; thus, a systems perspective to therapy is usually beneficial to the client. For example, a family presents to therapy with two competing core beliefs: one from their family of origin and a generalized family core belief. Therapists examine the core beliefs from families and couples and look at the family of origin and what these clients learn from families (Dattilio, 2005). The systems technique of circular questioning may be used concurrently with CBT Socratic questioning (Nelson, Fleuridas, & Rosenthal, 1986). An example of this is: "What did it look like when mom and dad did their homework?"

Cognitive Behavioral Therapy is an effective technique to use with couples and families. In relationships, couples and families often develop schemata or core beliefs about family and couple relationships based on their own family of origin and the relationship of their parents or other significant couples in their lives. Sometimes the core beliefs come from what is perceived to be beliefs about families or couples in general. These core beliefs come from the clients' own families or families in general and affect how they think, feel, and behave within their current couple/family setting.

CBT for families is brief and solution-focused. A wide range of presenting problems, such as depression, anxiety, anger outbursts, acting-out behavior, obesity, addictions, and coping with physical illness, can be treated with couples and families utilizing CBT techniques. In systemic CBT, patterns are explored to identify the roots of the problem, especially regarding issues of relationship conflict. Agenda setting and case conceptualization help to establish the goals for the therapy. CBT couple/family work focuses on the family members' thinking and behaving more adaptively and learning to make better choices in efforts to get their needs met, which leads to a family environment that is more stable and peaceful. Insight, empathy, respect, and caring are enhanced with family members who are invested in the treatment.

When working with families, therapists may see members individually and/or together to get a full picture of the issues involved and to work more in depth with each individual regarding his or her role in or contribution to the family's problematic patterns. The family meetings are forums in which perspectives and experiences are shared, and increased understanding and problem solving occur. Treatment focuses on improving relationships and increasing adaptive functioning skills.

The relationship between the supervisor and the supervisee is a mutual relationship. Holloway (1995) emphasizes that supervision should be focused on giving power to both supervisor and supervisee, and that the relationship between the supervisor and the supervisee is at the center of the supervision process from a systemic perspective. The relationship is seen as mutual and involves both the supervisor and supervisee. Holloway (1995) describes several dimensions of supervision that are connected by this important relationship. The functions of and the tasks of supervision are two of the important components of the supervisory relationship and are at the foreground of the interaction. The other dimensions of the relationship include the client, the trainee, the supervisor, and the institution; these factors are unique contextual influences in the supervisory process (Holloway, 1995).

COGNITIVE BEHAVIORAL THERAPY SUPERVISION

Although Bernard and Goodyear (2009) described the supervision relationship as "evaluative and hierarchical," the collaborative and curious stance of CBT theory would suggest otherwise. The CBT supervisor may set the structure for supervision, but the supervisees should be encouraged to share supervision needs and their own goals for the supervision process. In establishing clear expectations for supervision, the supervisor and supervisee may brainstorm what the process will look like. An example of this CBT problem-solving technique would be creating a two-column template with headings for expectations for self (as supervisor or supervisee) and one with expectations for the supervisor/supervisee (see Figure 11.1). The supervisor completes a list of expectations of self as supervisor and expectations that he or she has for the supervisee. The supervisee also generates a list of expectations for self as therapist in training and what he or she expects from the supervisor in the supervision experience. The supervisor and supervisee may have expectations that are similar or different in theoretical orientation based on training or previous experiences in the supervision process. They may have similar or very different expectations of what the supervision experience should be. By presenting these expectations on a blank template, the supervisor and supervisee have an opportunity to write down their expectations of each other without influence from the other's thoughts. The supervisor and supervisee then compare their lists of expectations and develop a supervision contract based on this comparison.

Techniques used in CBT supervision are similar to those used in CBT and include setting an agenda for the supervision sessions, bridging content from previous supervision sessions, assigning homework to the supervisee, and providing frequent summaries by both the supervisor and the supervisee (Liese & Beck, 1997). CBT supervision is "structured, focused, and educational in nature, and the responsibility for the content and structure is shared by the supervisor and the supervisee" (Hays & Iwamasa, 2006, p. 269).

Role of the Supervisor

The goal in CBT supervision is to change dysfunctional thought patterns that are affecting the therapist's behaviors and emotions. The supervisor is

Supervisor		Supervisee	
WHAT DO I EXPECT FROM MYSELF?	WHAT DO I EXPECT FROM MY SUPERVISEE?	WHAT DO I EXPECT FROM MYSELF?	WHAT DO I EXPECT FROM MY SUPERVISOR?
1. Evaluate own automatic thoughts toward supervisee and presented cases	1. Present automatic thoughts regarding supervision process and presented cases	1. Present cases with challenging issues	1. Offer suggestions to solve problem issues
2. Utilize Socratic questioning	2. Effectively utilize and be open to using Cognitive Behavioral Therapy techniques	2. Demonstrate competency in therapy skills	2. Check in with me and have a plan for supervision
3. Set an agenda for each supervision session	3. Be able to verbalize what is needed from supervisor	3. Be able to articulate strengths and weaknesses as therapist	3. Role-play client situations for practice

FIGURE 11.1 Example of brainstorming expectations for supervision process.

a collaborator with the supervisee and relates as a colleague in CBT supervision. The supervisor may guide the supervisee to examine thought patterns that are negatively affecting the therapy system. Through the use of automatic thought records, the supervisor may need to examine his or her own thoughts about the supervisee or cases on which he or she is providing supervision. The supervisor may encourage the supervisee to visualize the case going in a direction the supervisee sees as appropriate. The supervisor may have the therapist-in-training envision what the client might look like or what the client might say when the supervisee utilizes a new or particular technique. The supervisor may have the supervisee visualize how the supervisee sees himself or herself looking when utilizing a specific technique. Socratic questioning may be modeled by the supervisor by asking for evidence to support/refute an idea or thought that the supervisee has, by asking about alternative explanations or viewpoints that the supervisee might see, or by assessing what the worst and best things that the supervisee could see happening are and how the supervisee might cope with these eventualities. The supervisor may question the effect of the supervisee believing an automatic thought, may assess what might be told to a friend/relative in a similar situation, or may question what he or she might do (Beck, 2011; Padesky & Greenberger, 1995).

The CBT supervisor should ask specific questions about the case and focus only on relevant historical and session information that is important to the specific question. Giving a time limit for this information (2 to 4 minutes) helps the trainee determine what is important to the specific concern. Planning an agenda or flow chart of important questions for the supervision session may help the supervisor stay focused. As an example, an agenda might include a case conceptualization for the client and the treatment plan, adherence to the treatment plan, CBT skills needed to implement the treatment plan, the client's response to the treatment, and what might be interfering with success (including the conceptualization of the case, the treatment plan, the therapeutic relationship, and self-of-therapist issues).

The CBT supervisor needs to be able to provide education regarding the concepts and teaching of the techniques of the CBT theoretical orientation, just as in other psychotherapy approaches. It is not necessarily expected that either the supervisor or the supervisee be "expert" in the provision of CBT; rather, through the collaborative relationship, both supervisor and supervisee gain knowledge through this psychoeducation process. The supervisor may also use psychoeducation on the process of CBT.

It is important for the supervisor to be able to bring out the supervisee's strengths, as focusing on negatives or being critical is counterproductive in CBT supervision. The strengths are used to work toward the goals of the supervision process. The CBT supervisor utilizes those observed cognitions and behaviors that affect the therapy process, paying special attention to how the supervisee identifies as a professional and reacts to the client (Haynes,

Corey, & Moulton, 2003). The supervisor helps the supervisee to develop a more curious stance about his or her work as a therapist and about the client.

Haarhoff and Kazantzis (2007) offer strategies for homework completion difficulties. They discuss how homework noncompliance is common, even though it is an important technique in CBT. Supervisors can be helpful to trainees in recognizing the impact of the trainees' beliefs and assist the trainees to gain self-awareness of this impact. Haarhoff and Kazantzis also discuss supervisors conceptualizing the experience of the client first and then conceptualizing the supervisee's experience with the client. Self-practice and self-reflection by the trainee in supervision, through the use of automatic thought records, role-plays, and conceptualizations of the trainee's schema, can lead to improvements in the thoughtfulness of homework and its effectiveness in therapy.

Reiser and Milne (2012) discuss a recursive cycle in CBT supervision that fosters a systemic orientation. Agenda setting at the beginning of each supervision session provides a structure for examining the developmental needs and objectives of the supervisee. The supervisor may need to demonstrate or model CBT techniques either directly, with client interventions, or through the supervision process. Discussion of the client, review of sessions, and problem-solving techniques are also a part of the CBT supervision cycle process. CBT supervision also is evaluative, in that clinical data, techniques, and therapy processes are closely monitored.

Although collaboration is ideal in CBT supervision, the supervisor may have to supersede a plan that could be dangerous to client welfare. When evaluation is part of the process of supervision for competency and gatekeeping, the collaborative nature of CBT supervision may be of lesser importance. Supervisors also need to monitor their own clinical role by maintaining and improving their own skills through consultation, readings, supervision of supervision, and continuing education.

Role of the Supervisee

To embrace making changes, supervisees have to feel a part of the problem solving (AAMFT, 2000). Supervisees may examine their own thought records about the conceptualization of the case, their perceptions of the way they are viewed as therapists, and/or their abilities as therapists. Trainees need to be able to develop some problem-solving skills through their own self-reflection.

Reiser and Milne (2012) also discuss the supervisee learning cycle that occurs in CBT supervision, which includes experiencing, reflecting, conceptualizing, and experimenting. The first phase of the cycle is *experiencing,* which includes awareness of problems, the changes occurring, and the processes taking place; the identification of affect, and the processing of assimilation of new schema. *Reflecting* is the phase in which summarizing and integrating the subjective material occurs. In the *conceptualizing* stage, the objective materials, or the theories and findings, are integrated and planning or decision making about actions occurs. The final stage, *experimenting*, involves trial-and-error learning through role-playing, reality checking, and the enactment of plans that are created both in and out of supervision.

Supervisor-Supervisee Relationship

Cognitive Behavioral supervision can be especially helpful when the trainee feels stuck. Self-reflection is essential and the supervisor-supervisee relationship should encourage and model this self-reflection (Prasko, Mozny, Novotny, Slepecky, & Vyskocilova, 2012; Prasko & Vyskocilova, 2010). Self-reflection (self of therapist) is continuous and critical for therapeutic growth and establishment of relationships that are therapeutic and appropriate. The supervisor can "be in the issue" with the supervisee and collaborate in assessment of the cognitions and behaviors affecting the therapeutic process.

Therapists are constantly experiencing new situations in Cognitive Behavioral Therapy through the expression and processes of their clients, and the supervisor-supervisee relationship employs the same learning process. Therefore, giving and receiving feedback is important in CBT supervision. The supervisees need to be aware of both their strengths and their weaknesses, and the supervisor needs to elicit feedback from the supervisees about strategies that are not beneficial and those that are beneficial to the supervision process. Both the supervisor and supervisee should be able to point out helpful strategies and events, and then be able to transfer that information to other situations. Supervision in CBT includes live observation and observation of taped material, in which the sessions are observed in their entirety, contributing to an intense relationship between the supervisor and trainee.

Reiser and Milne (2012) discussed common factors that supervisors and supervisees share in the supervision process: relating, collaborating, managing, and facilitating. In *relating*, the core conditions of the therapeutic process are discussed with the shared goal of having a restorative process occur. *Collaborating* refers to the alliance that the supervisor and supervisee form in working toward the goals of supervision and toward the client's goals. The building of technique and reduction of dysfunction and distress are the *managing* factor in the process. *Facilitating* involves eliciting the desirable emotions and promoting learning and insight.

Storm and Todd (1997) emphasize the importance of clear boundaries in the supervisory relationship. Clear roles should be established, and the supervisor needs to share responsibility with the trainee of the clients. Supervisors need to be sensitive to therapist trauma issues and how these issues fit with the trainee's schema. Storm and Todd also emphasize that the balance of the hierarchal supervision within the more collaborative approach be established with awareness of the supervisee's developmental level.

Learning Goals/Objectives

Newman (2013) suggests forming a contract with the supervisee that outlines expectations for the CBT supervision process. A contract might include how many clients are expected to be seen utilizing CBT techniques over a semester or given period of time. What should be included in CBT notes, and what should be presented for review each week at supervision, may also be addressed. Readings in the use of CBT with specific diagnoses or populations may also be included in a contract. A Cognitive Behavioral Therapy supervisee's contract could also include a plan to address one client thoroughly each week, examining the agenda setting, the mood check, the automatic thought records, and the homework results. The goals and objectives should adhere to ethical standards, attend to how safety is addressed, and outline the number of sessions that will be viewed.

Wright, Basco, and Thase (2006) developed a useful CBT Supervision Checklist that is available from their *Learning Cognitive-Behavior Therapy: An Illustrated Guide*. The checklist monitors and evaluates CBT competencies for each session and over the course of therapy. Important competencies that the supervisor evaluates are the trainee's relationship with the client in terms of collaboration, empathy, and understanding. Professional boundaries and flexibility in meeting the needs of the client are also evaluated. The use of CBT core techniques, such as the use of feedback, guided discovery, agenda setting, effective homework, use of automatic thought records, and modifications of schemas, are also assessed. Other competencies assessed include treatment plans, psychoeducation, and relaxation and stress management tools (Wright et al., 2006).

The Cognitive Therapy Rating Scale (CTRS), developed by Beck (Beck, 2011, pp. 368–374) can also be utilized as a supervision tool. The CTRS consists of 11 items that rate general therapeutic skills, conceptualization, strategy, and technique. There are also seven general considerations that look at special circumstances that arose and the overall skill of the clinician. The general therapeutic skills section evaluates how the agenda was set, how feedback was elicited from the client, the therapist's understanding of the client, the effectiveness of the therapist's interpersonal skills, the therapist's ability to collaborate with the client, and the use of time in the session (efficient or inefficient). The second part of the scale rates the therapist's use of guided discovery, focus on key cognitions and behaviors, strategy for change, application of cognitive behavioral techniques, and the incorporation of homework. The general considerations section examines whether special circumstances ought to be considered and the skill of the clinician in handling the situation, the overall skill of the clinician, the difficulty of the client, and additional comments or suggestions for the therapist.

Liese and Beck (1997) proposed weekly supervision sessions of 60 minutes in length where one client is discussed in depth. In addition, diverse learning groups may be beneficial to foster other perspectives and experiences.

Hays and Iwamasa emphasize that "culture serves as a filter through which individuals identify their thoughts, feelings, and behaviors" (2006, p. 277). Combining group supervision for 1 to 2 hours weekly with 1-hour in-depth weekly individual supervision may help to reduce multicultural problems.

CONCLUSION

This chapter examined the theory of Cognitive Behavioral Therapy (CBT) and use of that theory as a supervision model. The significant philosophical foundations of CBT, CBT techniques and interventions, and the process of change in CBT were reviewed. The role of the CBT therapist, the role of the CBT supervisor, and cultural issues to consider in CBT were discussed. CBT is effective with working with a variety of clients and supervisees at different developmental levels due to the specific, articulate interventions and techniques. The short-term, present-oriented qualities of CBT lead to effective change in both clients and supervisees. The discussion of theory and the application of CBT techniques through a case example promotes understanding in utilization of this approach.

CASE EXAMPLE

A. AAMFT Approved Supervisor—Dr. Bob
The supervisor is a 53-year-old Caucasian male. He is a seasoned supervisor who has been credentialed as an AAMFT Approved Supervisor for more than 20 years. He provides supervision for both students (in practicum and internship as an adjunct professor), and postgraduates, at his agency.

B. Supervisee—Brittney
The supervisee is a 33-year-old Caucasian female in her first semester of internship at a local nonprofit agency. She is particularly interested in doing family therapy and excited to have the opportunity to work with a broad range of families.

C. Family
Travis (29) and Tekia (27) are an African American couple with two children, Tyrone (5) and Tommy (4). Tekia has been a stay-at-home mom. Travis was in the military (as has been every male in his family) and "served two tours of duty in Afghanistan." He was injured when an improvised explosive device (IED) exploded and left him severely injured. He has spent many months in the hospital and rehabilitation. He was discharged from the military (which he had wanted to make his career) as a result of his

(continued)

injuries, and now has difficulty finding a job. He currently works two part-time jobs and is hoping to find a full-time job.

Presenting Problem:

- Couple relationship and parenting issues, as Travis reportedly has a "short fuse."
- There are no reported safety concerns.
- Continuing recovery from his injuries.
- Job issues—inability to find full-time employment/financial strain.

Travis's older brother is still in the military and was recently re-deployed to Afghanistan.

Case Conceptualization

Client Background:

- Client's diagnosis
- Current problems
- Dysfunctional thoughts/beliefs with problem
- View of self/world
- Attitudes, expectations, rules underlying beliefs
- Contributing stressors
- Early experiences that may have contributed
- Coping mechanisms

Supervisor Background:

- Trained in CBT theory
- Utilizes CBT theory with a variety of client populations, issues, and diagnoses
- Utilizes CBT theory within supervision orientation

Supervisee Background:

- PhD student
- Trained in CBT theory
- Utilizes CBT theory with a variety of client populations, issues, and diagnoses

Brittney was a 33-year-old graduate student presenting to individual supervision with Dr. Bob, an experienced clinical supervisor within systems theories and Cognitive Behavioral Therapy. During the initial supervision meetings, Brittney and her supervisor

(*continued*)

discussed supervision expectations and reviewed and completed an agreement for supervision.

The client that Brittney introduced during this initial supervision meeting was a family of four, with the dad, age 29; mom, age 27; son, age 5; and son, age 4. The dad was the identified client with trauma issues; however, mom and the children had exhibited some anxiety issues.

Brittney has been working with the family for three sessions, utilizing a CBT approach to treatment. Brittney presented a tape of completing an automatic thoughts log with mom and dad. Both mom and dad were asked to complete an automatic thoughts log for homework.

Brittney had been thoroughly exposed to and practiced CBT techniques and interventions throughout her program. Brittney was asked to complete a case conceptualization for each member of the system prior to the third supervision meeting. She was also asked to review Judy Beck's text on CBT (Beck, 2011).

Brittney's supervisor, Dr. Bob, reviewed her first three sessions thoroughly through video tape and case consultation. Dr. Bob observed that Brittney focused on the children's thoughts and feelings but focusing little on the parents. It was evident that mom was experiencing some anxiety, due to her reported "worries," and she met criteria for anxiety. Brittney's supervisor also observed some of Brittney's own anxiety within supervision sessions; Brittney had reported that she felt nervous confronting the parents in the system. The supervisor then set an agenda.

After thinking about the changes that Brittney might need to make and the guidance that she can provide during supervision, Dr. Bob planned the following supervision session:

Supervisor: [Checking in] *How are you doing today, Brittney?*

Supervisee: *I am doing well today. I am busy with seeing clients and trying to complete all of my required schoolwork.*

Supervisor: *That sounds very difficult to juggle all of that at once.*

Supervisee: *Yes, it is* [sigh]. *I will tell you that I have been quite anxious leading up to supervision today. I feel that I am "stuck."*

Supervisor: [Setting the agenda] *Should we add your own "anxieties" about this case to discuss on the agenda today for supervision?*

Supervisee: [Sigh] *Yes, that would be beneficial.*

Supervisor: *Great. Are there any other issues that have come up with seeing clients?*

(*continued*)

Supervisee: *I am currently seeing a family and feel that I am "stuck" with the client's progress.*

Supervisor: [Emphasizing an awareness of automatic thoughts and cognitions] *Let's look at where those feelings are stemming from.*

Supervisee: *I think that I am not connecting with the family.*

Supervisor: *From where are those thoughts stemming?*

Supervisee: *I think that the mom does not like me. She is not helping her children with her inconsistency at home and she does not complete homework.*

Supervisor: *Have you tried working with mom on her anxieties?*

Supervisee: *I am nervous about addressing any of her issues, as I don't feel she trusts me!*

Supervisor: *How about we practice reducing your anxiety with this case looking at those automatic thoughts? What are you thinking in terms of mom's reactions to you working with the family? Let's do a thought record to look at this* (see Figure 11.2).

Supervisee: *I think that she thinks I am too young. I think that she thinks that I do not know what I am doing or how to work with the family.*

Supervisor: [Challenging automatic thoughts] *You have worked well with other parents who are older than you in the past. What is the evidence to support that this is what mom is thinking?*

Supervisee: *That is true. I have worked well with clients older than me in the past.*

Supervisor: [Modifying negative thoughts; using self-disclosure] *I remember when I first started seeing clients at a young age and making the shift to work collaboratively with parents. I found that I had to really challenge those assumptions I had about working with parents, since I was not a parent at the time. What was going through your mind when you asked mom about her own anxieties* [referring to video clip]?

Supervisee: *She thinks that I have no idea where she is coming from. So I froze and went to work with the rest of the family, specifically the children.*

Supervisor: *Why don't we role-play what it would look like to work with mom on her anxieties? I will be the therapist and you will be the client.*

Supervisee: [Hesitantly] *Okay.*

Supervisor (as trainee): *Mom, I am really encouraged by the progress that I have seen in your children with making eye contact in the therapy session. I can't imagine the worry you must have when they*

(*continued*)

Situation	*I am Examining Mom's Reaction to Children's Anxiety*
Emotions	*Anxious (35%)* *Frustrated (50%)*
Negative Thoughts	*Mom is so inconsistent; it's no wonder the anxiety isn't improving!* *What do I know about parenting these children?* *She is probably rolling her eyes and talking about me after session that I don't know what I am doing.* *I don't think mom will ever trust me.*
Behavior	*Avoid addressing mom's anxieties.* *Focus on the children.* *Avoid addressing the need for consistency in the home.*
Evidence that supports the prominent negative thought	*Mom continues to let children sleep in her room.* *Children don't trust other adults.* *Mom doesn't follow through with homework.*
Evidence that does not support the prominent negative thought	*The family continues to attend sessions regularly.* *Mom has called for guidance during a crisis.* *Other mothers are respectful of my age and knowledge.*
Alternative thought, strength of belief and emotion	*I can point out positive changes seen in children (50%)* *I can empathize with mom's anxiety and worry for her children (50%)* *I could explore barriers to completing homework (60%)* *Anxious (30%)* *Frustrated (10%)*

FIGURE 11.2 Example of a thought record for supervision.

are engaging in one of their tearful episodes, like when you called me two weeks ago. I am wondering about times that you may have felt anxious like your sons, and what you did to help yourself through your own anxiety.

Supervisee (as client): *I guess I don't feel I handle my anxiety well at all. I am probably not a good example for my sons. I guess that is one reason we came to therapy, so they would have better tools. I am learning some ways to better handle my anxiety as a result of the things you have taught my sons.*

(*continued*)

Supervisor (as trainee): *That's great! Tell me about some of the tools that you are using.*

Supervisee (as client): *Well, I noticed that when we are in session, taking a deeep breath makes me feel less anxious. I am afraid that I won't be able to use it at home as well as you help us in session.*

Supervisor (as trainee): *I am wondering if there are some thoughts about not being able to continue what we learn in therapy that get in the way of completing homework?*

Supervisee (as client): *I guess I never thought of it that way, but now that you say that I am afraid that you will be disappointed in what we do at home, and therefore it is better to just not do it.*

Supervisor (as trainee): *Well, the good news is there are no right answers to the homework! The homework is a way to measure what techniques are helpful and what is working outside of session. Would it be helpful to call it something else, like a home adventure?*

Supervisee (as client): *That might be very helpful, as every day has a new adventure at our house!*

Supervisor: [Using guided imagery] *Now let's visualize the next therapy session going the direction you would envision following this role-play. Let's close our eyes and picture what the room would look like with the family. Where would you see mom sitting in session? How would you start the session? How would the family respond?*

Supervisee: [Visualizing] *I would start the session reviewing the homework, or "home adventure." I would do a check-in to see how the week went and then set the agenda based on this information. Mom would be sitting next to me so that we are collaborating on effective anxiety-reducing techniques. As mom begins to feel her anxiety increase, she would be comfortable talking about it and utilizing effective relaxation and mindfulness techniques herself in session as well as at home. I would feel confident in my skills to work through whatever issues the family presents.*

Supervisor: *On a scale of 0 to 100, how confident are you that the next session will go similar to this visualization?*

Supervisee: *About 80, I guess.*

REFERENCES

American Association of Marriage and Family Therapy (AAMFT). (2000). *Readings in family therapy supervision: Selected articles from the* AAMFT Supervision Bulletin. Washington, DC: Author.

Beck, J. S. (2011). *Cognitive behavior therapy: Basics and beyond* (2nd ed.). New York, NY: Guilford Press.

Beck, J. S., Sarnat, J. E., & Barenstein, V. (2008). Psychotherapy-based approaches to supervision. In C. A. Falender (Ed.), *Casebook for clinical supervision: A competency-based approach.* Washington, DC: American Psychological Association.

Bernard, J., & Goodyear, R. (2009). *Fundamentals of clinical supervision* (4th ed.). Boston, MA: Pearson Education.

Cully, J. A., & Teten, A. L. (2008). *A therapist's guide to brief cognitive behavioral therapy.* Houston, TX: Department of Veterans Affairs South Central MIRECC. Retrieved from http://www.mirecc.va.gov/visn16/docs/therapists_guide_to_brief_cbt manual.pdf

Dattilio, F. M. (2005). Restructuring schemata from family of origin in couple therapy. *Journal of Cognitive Psychotherapy: An International Quarterly, 20*(4), 359–373.

Greenberger, D., & Padesky, C. A. (1995). *Mind over mood.* New York, NY: Guilford Press.

Haarhoff, B., & Kazantzis, N. (2007). How to supervise the use of homework in cognitive behavior therapy: The role of trainee therapist beliefs. *Cognitive and Behavioral Practice, (14),* 325–332.

Haynes, R., Corey, G., & Moulton, P. (2003). *Clinical supervision in the helping professions: A practical guide.* Pacific Grove, CA: Brookes/Cole-Thomson.

Hays, P. A., & Iwamasa, G. (2006). *Culturally responsive cognitive-behavioral therapy: Assessment, practice, and supervision.* Washington, DC: American Psychological Association.

Holloway, E. (1995). *Clinical supervision: A systems approach.* Thousand Oaks, CA: Sage.

Kuyken, W., Padesky, C. A., & Dudley, R. (2011). *Collaborative case conceptualization: Working effectively with clients in cognitive-behavioral therapy.* New York, NY: Guilford Press.

Liese, B. S., & Beck, J. S. (1997). Cognitive therapy supervision. In C. E. Watkins, Jr. (Ed.), *Handbook of psychotherapy supervision* (pp. 114–133). New York, NY: John Wiley & Sons.

Nelson, T. S., Fleuridas, C., & Rosenthal, D. M. (1986). The evolution of circular questions: Training family therapists. *Journal of Marriage and Family Therapy, 12*(2), 113–127.

Newman, C. F. (2013). *Core competencies in cognitive-behavioral therapy.* New York, NY: Routledge.

Padesky, C. A., & Greenberger, D. (1995). *Clinician's guide to mind over mood.* New York, NY: Guilford Press.

Prasko, J., Mozny, P., Novotny, M., Slepecky, M., & Vyskocilova, J. (2012). Self-reflection in cognitive behavioural therapy and supervision. *Biomedical Papers of Medical Faculty of Palacky University, 156*(4), 377–384.

Prasko, J., & Vyskocilova, J. (2010). Countertransference during supervision in cognitive behavioral therapy. *Activitas Nervosa Superior Rediviva, 52*(4), 253–262.

Reiser, R. P., & Milne, D. (2012). Supervising cognitive-behavioral psychotherapy: Pressing needs, impressing possibilities. *Journal of Contemporary Psychotherapy, 42,* 161–171. doi:10.1007/s10879-011-9200-6

Storm, C. L., & Todd, T. C. (Eds.). (1997). *The reasonably complete systemic supervisor resource guide.* Boston, MA: Allyn and Bacon.

Vaske, J., Galyean, K., & Cullen, F. T. (2011). Toward a biosocial theory of offender rehabilitation: Why does cognitive-behavioral therapy work? *Journal of Criminal Justice, 39*(1), 90–102. doi:10.1016/j.jcrimjus.2010.12.006

Walker, J., & Lampropoulos, G. K. (2014). A comparison of self-help (homework) activities for mood enhancement: Results from a brief randomized controlled trial. *Journal of Psychotherapy Integration*, 24(1), 46–64. doi:10.1037/a0036145

Woolhouse, H., Knowles, A., & Crafti, N. (2012). Adding mindfulness to CBT programs for binge eating: A mixed-methods evaluation. *Eating Disorders: The Journal of Treatment & Prevention*, *20*(4), 321–339. doi:10.1080/10640266.2012.691791

Wright, J. H., Basco, M. R., & Thase, M. E. (2006). *Learning cognitive-behavior therapy: An illustrated guide.* Washington, DC: American Psychiatric.

TWELVE

Training a Postmodern Supervisor

Teresa Rogers
Megan Miranda

In this chapter, postmodern theory is explored in relation to supervision. Common assumptions, lead contributors, interventions, strengths and limitations are addressed. Effective and successful postmodern supervision techniques are examined, including supervisor self and multicultural issues. Lastly, a case example is explored through the eyes of a postmodern supervisor.

PHILOSOPHICAL FOUNDATION OF POSTMODERN SUPERVISION THEORY

Postmodern supervision theory is grounded in the postmodern movement and perspective. The postmodern perspective, as explained by Todd and Storm (1997), relies on the expanding appreciation of multiple viewpoints. This perspective does not acknowledge inflexible norms or limiting exploration of thoughts. Postmodern theorists believe in individual perspective and the usefulness of each separate experience (Todd & Storm, 1997). This theory also incorporates social constructionism; that is, meanings are developed through social interaction and social consensus. Whiting (2007) suggests that realities are constructed, subjective, and based on our language system. Further, he states, "knowledge is not only shared in interaction, it is created in interaction" (2007, p. 141). According to Todd and Storm, social constructionism requires that no meaning be static. Additionally, meaning is always new and a product of social interaction over time. Social constructionists assert that no communication is ever complete; there will always be new meaning or interpretation based on social interaction. Similarly, Whiting (2007) refers to *structural coupling*, meaning that as individuals interact, they influence one another to create more congruent views. In this way, individual perspectives tend to unite and create common values.

Current postmodern theories have common values that may have resulted from the process of structural coupling. Anderson (2001) refers to common values of most postmodern theories, which include taking a non-judgmental stance, utilizing and respecting the client's reality, emphasizing a collaborative framework, avoiding labeling, employing narratives to elicit personal perspective, and being more public with information and thoughts. Utilizing a postmodern perspective and respecting the postmodern values, Whiting (2007) notes that an isomorphic process occurs from supervision session to the therapy session. These common values can easily be applied to counseling or supervision sessions.

LEAD CO-CREATORS

Contributors to the postmodern movement emphasize that truth, reality, and knowledge are socially embedded. Social psychologist Kenneth Gergen states that postmodern theory is "principally concerned with explicating the processes by which people come to describe, explain or otherwise account for the world (including themselves) in which they live" (1985, p. 266). Lead co-creators of the postmodern movement include Tom Andersen, Harlene Anderson, Peggy Penn, Lynn Hoffman, Jaakko Seikkula, and Michael White (Anderson, 2001). These contributors emphasize that the client's presenting problem is not the individual, couple, or family; the problem stems from linguistic constructions and universal narratives (Todd & Storm, 1997). Anderson (2001) asserts that the aim of the postmodern therapist has shifted from changing the person to a philosophy focusing on the process for change.

Postmodern theory tends to be referred to as a philosophy of therapy. Anderson and Goolishian (1988) state that knowledge and language inform a worldview or "way of being" that does not separate personal from professional. Anderson and Goolishian emphasize the importance of conversational partners engaging in dialogical conversations and collaborative relationships. This stance requires the supervisor to constantly learn and engage with the supervisee as a co-learner. An isomorphic process is created when the therapist utilizes this stance with clients.

Anderson (2001) identifies other contributors, including Michael White and David Epston. White and Epston began working together in the early 1980s. Epston was interested in anthropology and storytelling, whereas White focused on interpretive methods inspired by Gregory Bateson. White and Epston developed what is now known as *narrative therapy* (Anderson, 2001). White and Epston (1990) assert that narrative therapy aims to give more possibility to each person's life by deconstructing culturally dominated stories and constructing new stories. White and Epston's theory can easily be applied to supervision, as it is the postmodern supervisor's role to focus on political and social agendas to encourage the supervisee to develop new meaning.

In sum, postmodern theory is more of a lifestyle than a concrete theory. Leaders in the creation of postmodern supervision theory focused on how truth and knowledge are embedded in society. Thus, the presenting problem is not the individual; it lies within social constructs and universal narratives. Addressing these contrasts and breaking down the social narratives is the job of the postmodern supervisor and therapist. Exploration of new meaning must take place to develop insight and elicit behavior change.

SUPERVISOR ROLE AND RESPONSIBILITIES

Key to postmodern supervision theory, as articulated by Unger (2006), is that the postmodern supervisor does not assume the expert role. The postmodern supervisor does not monopolize truth or the "correct way." He or she believes heavily in the power of collaboration. Per Unger, to avoid assuming the expert role, the postmodern supervisor will incorporate the supervisee's innate and learned tools into the supervision sessions. However, the postmodern supervisor must be cautious that the supervisee does not rely so heavily on past skill that the therapy session is negatively affected. The postmodern supervisor wants to aid the supervisee in appreciating multiple perspectives and developing new meanings to support clients' growth.

Anderson (2001) emphasizes the importance of learner-directed learning or collaborative learning. This means that the learner is in charge of the learning. She suggests that the role of the supervisor is to lead by following. Anderson states, "I still believe, that we could not teach a person how to be a therapist but we could provide an experience in which they could learn to be one" (2001, p. 357). Based on this statement, one role of the postmodern supervisor is to provide an open and collaborative space for the supervisee to grow.

Six Roles of Postmodern Supervisor

Unger (2006) describes six roles of the postmodern supervisor that encompass the supervision process that occurs throughout the supervision process. These roles include supporter, supervisor, case consultant, advocate, trainer/teacher, and colleague (Unger, 2006). In the *supporter* role, the supervisor supports the supervisee's emotional awareness by bringing his or her experiences with clients to a personal level. Through questioning and maintaining a non-expert stance, the supervisor is able to aid the supervisee by increasing the supervisee's self-efficacy as a therapist. This supporter role allows the supervisee to acknowledge possible self-as-therapist issues that may be affecting the therapeutic relationship. Next, the *supervisor* role fosters growth related to the clinical skills and abilities of the supervisee. The postmodern therapist will discuss case conceptualization with the supervisee, carefully noting the difference in individual perspectives. Additionally, the case consultant role offers advice on specific techniques based on the clinical experience of the supervisor. Assuming the *case consultant* role, the postmodern supervisor may ask the supervisee to explore techniques that could be effective for both supervisee and client. As the *advocate*, the supervisor actively supports the supervisee and/or the client system by encouraging the supervisee to take necessary actions for himself or herself or the client. This role includes supporting the supervisee to report child abuse, find appropriate services to meet client needs, or assisting supervisees in their professional growth. Next, the *trainer/teacher* role is fulfilled by educating the supervisee about specific techniques through instruction or practice. For example, the postmodern supervisor and supervisee may explore the fit of an intervention through a role-play together. Lastly, when taking on the *colleague* role, the postmodern supervisor acts as a peer to the supervisee. The postmodern supervisor works collaboratively alongside the supervisee while fulfilling needs of the client. These six roles have diffuse boundaries and postmodern supervisors will often find themselves taking on more than one role at a time (Unger, 2006).

Process of Change

In the postmodern perspective, *change* is a relative term and largely based on individual perception. In attempts to measure change, the postmodern supervision has achieved success if changes in the internal and cognitive processes of the supervisee are applied to therapy (Todd & Storm, 1997). Further, a behavioral change will be apparent in how the supervisee conducts each session from a not-knowing position. Anderson (2001) does not refer to "change" as Todd and Storm do; rather, she refers to the goal as "transformation." From a collaborative perspective, the supervisor or therapist is not a change agent; rather, all participants (the supervisor, supervisee, and client triad) embark on a process of mutual transformation. The supervisor uses his or her "expertise" to create space for such a transformation to take place. Todd and Storm assert that success can be measured if the supervisee is able to direct the therapeutic interaction and encourage clients to generate new ideas about themselves. If the supervisee can develop new understanding and empathy for the least-liked person in the system, the postmodern supervisor has achieved success (Todd & Storm, 1997; Unger, 2006).

Development of Supervisor Self

As noted, postmodern supervision theory is more of a "way of being" or life philosophy and less of a manualized, step-by-step supervision theory (Anderson & Goolishian, 1988). Employing a philosophy instead of a concrete theory, the postmodern supervisor will face several challenges with relation to the development of self-of-supervisor and, isomorphically, self-of-trainee. Liddle and Halpin (1978) compare family therapy supervision theories. Their writings indicate a split between supervision based on skill development, as utilized by Structural (Minuchin), Behavioral (Cleghorn and Levin), and Strategic (Haley) theories; and supervision based on self-development, as utilized by experientially oriented (Ferber and Mendelsohn) and psychodynamically based (Ackerman) theories. Skill-development supervision emphasizes the development of therapeutic skill and a way of intervening in dysfunctional systems. Conversely, self-development-based supervision relies on personal growth and the affective life of the trainee. Challenges arise, as postmodern supervision theory is neither purely skill nor purely self-development based.

Postmodern supervision theory encounters challenges that skill-based and self-development-based supervision theories address. Specifically, postmodern supervision theory is embedded in the relationship between supervisor and supervisee (Todd & Storm, 1997). This can be observed in a horizontal hierarchy, minimization of power differentials, and a "lead by following" stance (Todd & Storm, 1997). Liddle and Halpin (1978) report that a skill-development-based supervisor would challenge this approach, stating that the supervisor should aid the therapist by providing directives. Anderson (2001) reports that, in some cases, it is necessary for the postmodern supervisor to provide directives. However, providing directives does not maintain fidelity to the postmodern paradigm. Similarly, the postmodern supervisor may challenge the supervisee but must approach challenging in a tentative and curious manner (Anderson, 2001). From a skill-based perspective, Liddle and Halpin report that direct challenges are vital and help to organize and develop the relationship around the specific task. Providing directives and challenges is more time efficient and focused. Efficient and focused supervision is important; however, exploration and self-of-therapist development is vital for the supervisee.

Another gray area that may challenge the postmodern supervisor is the exploration and development of self through the supervisor's own family of origin. Liddle and Halpin (1978) indicate that a supervisor should work with his or her own family to increase development of self and to model for the supervisee. Self-based development supervisors and supervisees are pushed to define *self* in the context of their own families of origin. In contrast, a postmodern supervisor may have conversations related to family of origin as it affects the supervisor-supervisee-client relationship. However, a postmodern supervisor would not require the supervisee to invite family/spouses to the supervision session. Although that type of exposure may enhance the development of both the supervisor and the supervisee, a postmodern supervision perspective does not emphasize the importance of intense family-of-origin exploration (Liddle & Halpin, 1978).

Along with family-of-origin exploration, the self of a postmodern supervisor develops through mutual transformation as he or she engages in a relationship with a supervisee (Anderson, 2001). The journey of this relationship is what will impact and expand a sense of self. This transformation is intended not only for the supervisee/client, but for the postmodern supervisor as well.

Co-learning and open collaboration are vital to the development of a postmodern self-of-supervisor.

Summary

The postmodern supervisor is collaborative, open to the supervisee's expertise, and co-creates knowledge with the supervisee. There is no "correct way" of practicing therapy; thus, the postmodern supervisor encourages exploration of self and meaning. The postmodern supervisor may take on several roles while engaged in the supervision process. Each role assists the supervisee in developing himself or herself as a competent therapist. Along with encouraging the supervisee's growth, the postmodern supervisor is focused on his or her own growth. In this collaborative environment, both supervisee and supervisor are learning and creating together. This mutual transformation is the very essence of postmodern supervision theory.

TECHNIQUES

Techniques employed by the postmodern supervisor may not be as concrete or structured as those used under other theories. Postmodern supervision interventions aim to expand knowledge and explore differing perspectives (Todd & Storm, 2002). Similar to therapists asking questions of clients and using directives to bring about insight and change, the postmodern supervisor poses questions and uses exercises (such as the Cinderella Technique) to increase knowledge and growth and understanding of different experiences. Many techniques utilized in the counseling setting are applicable in the supervision setting (Wieling et al., 2001).

Todd and Storm (1997) indicate that similar to the counseling setting, the postmodern supervisor will establish goals with the supervisee based on the perceived need. Wieling et al. (2001) assert that the supervisor relies on live supervision, group supervision, and case report. *Reflecting teams* allow multiple perspectives and narratives to be developed and explored. Another intervention utilized by the supervisor is asking the supervisee to act "as if" he or she were a specific client in the system. Again, this technique helps to identify new meaning and explore new perspectives (Todd & Storm, 1997). Similar to acting "as if," Whiting (2007) suggests that the postmodern supervisor ask thought-provoking questions of the supervisee about each member in the system, so as to expand the supervisee's stories about the client system.

Isomorphic Process

According to Whiting (2007), a key technique for a postmodern supervisor is to rely on the isomorphic process that occurs between supervision and therapy. The aim of the postmodern supervisor is to create space to facilitate change in the behaviors and/or cognitive operations of the supervisee. Concurrently, the supervisee has the same goal in mind for his or her client system (Todd & Storm, 1997). The postmodern supervisor should identify the isomorphic processes when the roles of supervisor and supervisee replicate the roles of the client and therapist (Koltz, Odegard, Feit, Provost, & Smith, 2012).

Weir (2009) explains *isomorphism* as "when things take on the same or similar form" (p. 60). Weir advises the postmodern supervisor to always be aware of the isomorphic processes that are occurring in the supervision session and utilize them as a training tool. From a postmodern perspective, this can be done if the supervisor approaches the conversation with curiosity. If the supervisee can identify the isomorphic process that is occurring, he or she can take ownership, develop alternate meaning, and address that meaning effectively. However, Koltz et al. (2012) caution that the supervisor must consider whether discussing the isomorphic interaction will help the supervisee in the immediate moment. With such a heavy belief in the isomorphic process in postmodern theory, techniques will be utilized both in supervision sessions and in therapy sessions.

Collaborative Relationship

In addition to the isomorphic process, Anderson (2001) emphasizes the importance of conversational partners. The postmodern supervisor and supervisee engage in dialogical conversations and collaborative relationships. This conversation is a two-way, give-and-take, back-and-forth process. In this technique, the postmodern supervisor and supervisee are co-learners. Together they co-develop new meaning and clarify and expand constraints (Anderson, 2001). Key in this technique is respect; both must respect the richness and diversity in other's thoughts and narratives. Whiting (2007) suggests asking open-ended and thought-provoking questions about each member in the system. This may expand the supervisee's stories about his or her client system.

Not-Knowing Position

When asking open-ended and thought-provoking questions, the postmodern supervisor will always assume a not-knowing position so as to encourage the free flow of ideas (Todd & Storm, 1997). The goal of postmodern supervision is to open new possibilities within the current worldview. Simon (2006) emphasizes that the supervisee should be directed in an exploration of his or her personal worldview. This allows the supervisee to identify a theoretical model(s) that best accommodates that worldview. Anderson (2001) adds that a collaborative supervisor does not have superior knowledge; rather, he or she provides expertise. When providing expertise, the postmodern supervisor contributes in a tentative and curious manner (Anderson, 2001). The purpose of utilizing a tentative voice is to uphold the collaborative relationship and create room for all perceptions and thoughts. There will always be a level of uncertainty in postmodern supervision (Anderson, 2001). The postmodern supervisor who employs the not-knowing technique must do so with the intent of facilitating a collaborative relationship.

Transparency of Thought

As mentioned earlier, Anderson (2001) discusses the postmodern value of being public and transparent. She emphasizes that a postmodern supervisor's inner dialogue can become a monologue. Maintaining one's own private conversation collapses the very core of collaboration. Anderson says that she finds that making silent thoughts public increases rapport and enhances the conversational partnership. However, she does caution about oversharing and being mindful of the well-being of the supervisee/client. The postmodern supervisor should verbalize his or her inner voice as a model for the supervisee and to bolster the supervision relationship.

Power

The construct of power must be addressed. However, along with verbalizing his or her inner voice, the postmodern supervisor should also discuss power dynamics with the supervisee. Murphy and Wright (2005) provide techniques to address power in the postmodern supervision relationship, asserting that open discussions of power must take place between supervisor and supervisee. This includes using the term *power*, addressing power in the first supervision session, and revisiting power throughout the relationship. Further, Murphy and Wright offer the idea that empowerment and collaboration with the supervisee allows the supervisor to assert positive power. Supervisors can empower by encouraging supervisees to make decisions regarding their cases, affirming knowledge and providing the supervisees with choices. Conversely, the supervisee can exert power within the relationship by acknowledging that he or she is a consumer of supervision services and can join with peers to have a united voice. These topics related to power should be made overt in the postmodern supervisor-supervisee relationship.

Summary

Techniques employed by the postmodern supervisor are not as structured as in other theories. Most techniques are embedded within the supervisor-supervisee relationship. The postmodern supervisor must be well versed in and able to identify these processes. Attending to the isomorphic processes is a vital technique utilized by the postmodern supervisor. Additionally, maintaining a collaborative relationship, providing transparency of thought, and addressing power dynamics are all important techniques that yield a successful supervisory relationship.

RELEVANT RESEARCH

Unfortunately, as reported by Unger (2006), there are limited outcome data supporting the effectiveness of utilizing a postmodern supervision approach; there tends to be more research supporting the effectiveness of a postmodern therapeutic approach. However, as noted by Whiting (2007) both supervision and therapeutic theory rely heavily on the isomorphic process. As mentioned earlier, isomorphism is a vital piece of postmodern theory and one that can link supervision theory to therapeutic theory effectiveness. Storm, Todd, Sprenkle, and Morgan (2001) also assert that supervision is isomorphic to therapy and that all postmodern supervisors must examine these processes. Further, more research is needed to support the effectiveness of a postmodern supervision approach.

As research related to postmodern supervision theory is lacking, Storm et al., (2001) focused their research on a common-factors approach to supervision theory. The authors emphasize that there is no evidence that one approach is more effective than any other. However, they note that the field could benefit from the development of postmodern methods to fit in the supervision paradigm. Additionally, the authors' suggestions for best practice support the postmodern approach of making power issues transparent. Further, Storm et al. state that the context and expectations of the supervisee should be considered by the supervisor. Reflecting back on the philosophy of a postmodern epistemology, a continuous re-evaluation of information and avoiding attachment to any one source would support a postmodern supervision theory (Storm et al., 2001).

Selicoff (2006) further supports postmodern supervision theory, utilizing the motto of "Lead from one step behind" (p. 48). She states that the postmodern techniques of multiple viewpoints, reflecting teams, and allowing the supervisee to lead and be responsible for his or her own learning all result in continuous learning for both supervisor and supervisee. She emphasizes that good supervision is based on a trusting and respectful relationship. Whiting (2007) supports Selicoff's findings and reports that when a postmodern supervisor shows respect by seeking feedback from the supervisee, he or she is modeling an effective therapeutic technique of processing feedback with clients regarding the therapeutic process.

Future Research

The postmodern supervision theory could benefit from future research. Most postmodern authors note that further research is warranted (Gray & Smith, 2009; Selicoff, 2006; Storm et al., 2001; Todd & Storm, 1997; Unger, 2006;

Whiting, 2007). Storm et al., (2001) add a call to action regarding the supervisory regulations and designation of qualified supervisors. These authors note an erosion of the quality of supervision due to state laws. Further research supporting postmodern supervision theory may encourage professionals, professional organizations, and lawmakers to practice and enhance current standards, thus propelling the field into the future.

Summary

Current research is lacking regarding postmodern supervision theory. However, research supporting postmodern therapeutic theory is fairly prevalent. Current postmodern supervision literature emphasizes the importance of the isomorphic process as well as addressing power within the relationship. Additionally, current literature calls for more research on the effectiveness of postmodern supervision theory. Further support may enhance the profession as well as bolster the theory.

EFFECTIVENESS WITH DIVERSE SUPERVISEES

Postmodern supervisors address diversity in a similar manner as they address all supervisee matters: from a collaborative and open stance. Each individual supervisee maintains his or her own culture. The postmodern supervisor provides an open and safe environment to explore and discuss culture. (Tapp & Wright, 1996). This section addresses diversity as it relates to gender, ethnicity, and power.

Gender

A comprehensive review revealed that there is a gap in the literature specifically addressing gender issues from a postmodern perspective. There is, however, literature pertaining to gender issues in supervision that originate from a postmodern perspective, which could fill that gap and increase awareness of gender issues in supervision (Ancis & Marshall, 2010; Doughty & Leddick, 2007; Estrada, 2005). Postmodern supervisors accept the position that knowledge is socially constructed and that each individual experience creates worldview. This acceptance allows for and promotes the sensitivity and awareness needed to address the impact of gender on supervision and therapy. Although the literature did not define a postmodern perspective, much of the literature is developed from a paradigm similar to that of the postmodern supervisor (Ancis & Marshall, 2010). Being gender-sensitive to differing experiences is a natural position for the postmodern supervisor (Unger, 2006).

From a postmodern perspective, gender is negotiated as any issue would be addressed: from a not-knowing position. Paisley (1994) expresses that gender is one of the most descriptive characteristics; thus, it tends to be a sensitive area of personal exploration. Further, it is crucial to use the supervisory relationship to educate, confront, and model this conversation and development. Unger (2006) explains that it is the supervisor's responsibility to challenge the supervisee to develop new meaning or perspectives of a person different from himself or herself. Because postmodern theory relies on each person having a different perception of reality, gender and diversity are taken into consideration, as are many contextual factors. Tapp and Wright (1996) write that pluralism is a key focus in postmodernism. They encourage creating space to hear new perspectives and voices as related to gender and power. Todd and Storm (1997) describe how power, gender, and dominant culture can affect the supervisory relationship.

Shakeshaft, Nowell, and Perry (2000) state that the postmodern supervisor needs to acknowledge his or her own background and training as well as the impact of culture. Further, the postmodern supervisor must accept that he or she has no power over what is taught by society, school, and family. Similarly, Paisley (1994) reinforces the importance of awareness of self and society. From a postmodern perspective, the supervisor takes into consideration these past ideas and develops new and alternate meanings. Wheeler, Avis, Miller, and Chaney (1986) indicate that the postmodern supervisor has a responsibility to address gender in the relationship. One way to address gender is to challenge any stereotypical behavior and to encourage more balanced behavior and perspective. Postmodern theory provides space within the supervisory relationship to develop these new perspectives.

Diversity

From the postmodern perspective, diversity is approached with curiosity (Unger, 2006). Banks (2001) estimates that by 2040, members of nonmajority groups will become the predominant population. Further, she indicates that the postmodern supervisor should infuse cultural sensitivity and awareness into training and supervision of family therapists. In the postmodern spirit of minimizing hierarchy, Banks suggests that the supervisor and supervisee explore their cultural influences and their philosophy of multiculturalism within the supervision setting.

Further, Gray and Smith (2009) assert that the social constructionist model is congruent with the philosophy of multiculturalism. The emphasis on multiple realities and the assumption of "perception is truth" enhance a diverse worldview for the supervisee. However, Gray and Smith also mention a limitation to a postmodern stance with regard to multiculturalism. Assuming a "not knowing" and "supervisee as expert" stance may be in direct opposition to the views of some cultures. Different cultural groups may elevate professionals as experts, expecting guidance and solutions from them. One way to mitigate this issue is for the postmodern supervisor to convey that he or she has expertise in the process of supervision but will not direct the supervisee to engage in behaviors contrary to supervisee goals (Gray & Smith, 2009). Again, modeling this process in supervision will allow the supervisee to re-create this stance in work with clients.

A study conducted by Wieling and Marshall (1999) concludes that marriage and family therapists (MFTs) believe that a cross-cultural supervisory relationship provides the most potential for growth and self-awareness. Unfortunately, these relationships are rare in the MFT field. In addition, Taylor, Hernández, Deri, Rankin, and Siegel (2006) report that though a cross-cultural supervisory relationship may be ideal, the burden of cultural competency cannot be placed only on minority supervisors: All supervisors, across all supervision theories, must be culturally competent.

Morgan and Sprenkle (2007) write about a common-factors approach to supervision. They conclude that one role a supervisor fills is the "mentor role," which focuses on the development of the supervisee. The supervisor will focus on self-of-therapist issues as they relate to clinical work. Addressing cultural biases, norms, and learned behaviors are not only the responsibility of the postmodern supervisor but, as Morgan and Sprenkle indicate, a common factor throughout supervision theory. Additionally, Paris, Linville, and Rosen (2006) report that the personal life of the supervisee has an impact on clinical work. Difficult life experiences have been found to positively influence clinical abilities, as noted by therapist participants. Filling the mentor role provides the supervisee with a space in which to process these experiences (Morgan & Sprenkle, 2007).

Power

The supervisor-supervisee relationship is laden with power over both the supervisee and the clients (Todd & Storm, 2002). The addition of different cultural backgrounds can further compound this issue of power. This power differential works against the postmodern stance of horizontal hierarchy. Taylor et al. (2006) report that effective multicultural supervision requires that supervisors:

1. Take responsibility for holistically addressing multicultural issues within the supervisor-supervisee-client triad
2. Promote multiculturalism within their educational or clinic-based institutions
3. Recognize how their own worldview, as well as those of their supervisees and clients, impacts therapy
4. Include cultural issues (negative and positive values)
5. Develop specific, stage-by-stage multicultural competency goals for trainees (p. 6).

Utilizing these techniques of postmodern theory, a postmodern supervisor can meet the criteria presented by Taylor et al. In sum, Bernard and Goodyear (1998) state, "[T]he final stage of multicultural supervision must be social action for client, supervisees, supervisors and institutions" (p. 136).

Along with power related to diversity, power related to hierarchy consistently affects the supervision relationship. The postmodern supervisor attempts to minimize the hierarchical stance (Todd & Storm, 2002); however, in some settings, elimination of the hierarchy is impossible. Murphy and Wright (2005) report that supervisees naturally expect their supervisor to have—and exercise—a level of power. Further, supervisees preferred that the supervisor utilize that power in a positive manner. Murphy and Wright discuss the importance of empowerment and collaboration as positive uses of the supervisor's power. Supervisees reported that supervisors who utilized their power to uplift and encourage were most effective.

Summary

Postmodern supervision with relation to diversity is treated very similarly to postmodern supervision in general. The postmodern supervisor approaches every interaction with a not-knowing and curious stance. Gender is one of the most defining features of a human, and, often, dialogue about the impact of gender goes unsaid. The postmodern supervisor should address gender as it affects the supervisory relationship. Cultural differences also may occur within the supervisor-supervisee relationship. The postmodern supervisor is careful to address these differences and explore perceptions and meaning. Lastly, power dynamics with relation to diversity may be visible. The responsibility of the postmodern supervisor is to empower supervisees. All of these interactions enhance the supervision relationship and also model appropriate ways to discuss these sensitive issues with clients.

THEORETICAL LIMITATIONS

This theory does encounter dilemmas that are immanent in the supervision process. Wheeler et al. (1986) emphasize that the postmodern supervisor would attempt to reduce any hierarchical relationship, as it may limit the supervisee's own process. Unger (2006) discusses the idea that assigning grades or consequences is detrimental to the nonexpert position. The postmodern supervisor always wants to encourage cooperation, collaboration, and mutual generation of ideas with the supervisee. In addition, Wieling et al. (2001) assert that the postmodern perspective tends to be more of a lifestyle philosophy than a specific theory; this issue has created controversy among professionals. Another dilemma relates to the postmodern concept of exploration of multiple truths. This may lead to accepting that a different approach may be more useful for a particular client or supervisee (Todd & Storm, 1997).

As noted earlier, the postmodern supervisor does not allow or take a hierarchical stance during supervision (Todd & Storm, 1997; Unger, 2006; Wheeler et al., 1986). Ideally, the supervisor should not assign grades, give directives, or assume an expert role. This is somewhat problematic when working with beginning trainees in a training facility. Grading is a part of the trainee experience; thus, the supervisor must compromise theory to accommodate the practicality of the situation. A balance can be developed if the supervisor is proactive and aware. One way to mitigate this potential dilemma regarding evaluation is to complete an evaluation of the student's progress with the student in supervision, maintaining transparency (Unger, 2006). Wheeler et al. (1986) suggest that the postmodern supervisor allow for the student's input and development of new meaning during the evaluation. This will help the student to take more ownership of his or her strengths and growth edges and alleviate the need for the postmodern supervisor to assume an authoritarian/expert role.

Along with assuming the expert role, giving directives to the supervisee presents a contradiction to postmodern theory. Again, in a training facility, giving directives is part of the environment. If issues of safety or ethics are present, per the AAMFT Code of Ethics the supervisor would be acting unethically if he or she did not provide clear instruction to the supervisee (AAMFT, 2012). Any directive will be processed with the supervisee and new meaning or perspective will be developed after the situation has been handled and there is no imminent risk to client or therapist (Wheeler et al., 1986). This permits the supervisee to take ownership and perhaps act differently in future settings.

The last area in which postmodern supervision theory can be problematic in a training facility setting is related to developmental stages. As stated by Todd and Storm (1997), postmodern theory does not rely on a supervisee's developmental stages, as this can box the supervisee in. Though developmental stages may not be linear or exactly the same for each supervisee, there is a difference between a supervisee who has never seen a client before and a supervisee who has 1,000 direct client contact hours. The supervisee with no experience is looking for an expert and for directives from his or her supervisor (Stoltenberg, 1981). Similar to grading and directives, a postmodern supervisor can remain true to the theory by encouraging the supervisee to reflect back to past experiences and use past knowledge to develop his or her own ideas. As reported by Todd and Storm (1997), the postmodern supervisor does not have to subscribe to the notion of developmental stages per se. However, Melchent, Hays, Wiljanen, and Kolocek (1996) imply that the postmodern supervisor must be aware of the struggles a new clinician and a seasoned clinician may encounter. This awareness is vital to the growth of the supervisee.

In sum, there are several conflicting forces that affect the postmodern supervisor. Specific dilemmas result from the inability to maintain a nonhierarchical stance and the notion that postmodernism is a lifestyle rather than a theory. Most limitations occur in the training facility setting, as postmodern supervisors must assign grades, give directives, and acknowledge developmental differences. These act in direct contrast to pure postmodern theory. However, if the postmodern supervisor is transparent and invites the supervisee to explore and develop new meaning throughout, the process fidelity to the model is maintained.

CONCLUSION

This comprehensive chapter described the perspectives, attributes, experiences, beliefs, roles, and values of the postmodern supervisor, which revolve around the central epistemology that knowledge is socially constructed and that there is no singular absolute truth (Todd & Storm, 2002). The six key roles of the postmodern supervisor, as described by Unger (2006) include those of supporter, supervisor, case consultant, trainer or teacher, colleague, and advocate. The postmodern supervisor constantly oscillates throughout the different roles while keeping his or her primary objective of being a collaborative learner. The postmodern supervisor encourages the exploration of the supervisee's accomplishments, ideas, and experiences.

Postmodern supervision theory attempts to address the needs of the supervisor, supervisee, and client; however, there are limitations. Further research is necessary to address the use of the postmodern theory of supervision when working with diverse populations and gender-related issues. In addition, research should be conducted regarding the effectiveness of the postmodern theory in general. Postmodern supervision theory supports the perceptions, experiences, and collaborative growth of both supervisor and supervisee.

CASE EXAMPLE

Case Information

A. AAMFT Approved Supervisor—Dr. Bob
The supervisor is a 53-year-old Caucasian male. He is a seasoned supervisor who has been credentialed as an AAMFT Approved Supervisor for more than 20 years. He provides supervision for both students (in practicum and internship as an adjunct professor), and postgraudates, at his agency.

(*continued*)

B. Supervisee—Brittney
The supervisee is a 33-year-old Caucasian female, in her first semester of internship at a local nonprofit agency. She is particularly interested in doing family therapy and is excited to have the opportunity to work with a broad range of families.

C. Family
Travis (29) and Tekia (27) are an African American couple with two children, Tyrone (5) and Tommy (4). Tekia has been a stay-at-home mom. Travis was in the military (as has been every male in his family) and "served 2 tours of duty in Afghanistan." He was injured when an improvised explosive device (IED) exploded and left him severely injured. He has spent many months in the hospital and rehabilitation. He was discharged from the military (which he had wanted to make his career) as a result of his injuries and now has difficulty finding a job. He currently works two part-time jobs, hoping to find a full-time job.

Presenting Problem:

- Couple relationship and parenting issues, as Travis reportedly has a "short fuse."
- There are no reported safety concerns.
- Continuing recovery from his injury.
- Job issues—inability to find full-time employment/financial strain.

Travis's older brother is still in the military and was recently re-deployed to Afghanistan.

Application of Case

As a postmodern supervisor, Dr. Bob would hold a non-expert and curious stance. Dr. Bob would encompass Unger's (2006) six roles as supporter, supervisor, trainer/teacher, case consultant, colleague, and advocate, oscillating among the different roles throughout his supervision sessions with Brittney. Initially, Dr. Bob and Brittney would establish goals of supervision based on Brittney's perceived need. Dr. Bob may live-observe sessions to be a support and advocate for therapist and clients. Dialogue between Dr. Bob and Brittney would be open and exploratory. Dr. Bob may ask open-ended questions to better understand her perspective regarding not only the case, but also life in general. Questions, thoughts, and feelings would be encouraged and explored by Dr. Bob and Brittney. In this

(*continued*)

way, both would be learning and creating new meaning based on dialogue.

As the case consultant and trainer/teacher, Dr. Bob would have his ideas about this case and would share his thoughts regarding ethical decision making. As the supporter, Dr. Bob may ask if the details of the case impact Brittney on a personal level. Any self-as-therapist issues that may arise will be addressed and investigated. An area that Dr. Bob would explore as the supporter is the cultural and gender dynamics that may be affecting the therapy process. Through a not-knowing and curious stance, Dr. Bob, as a postmodern supervisor, would ask Brittney to share her perspective based on her cultural identity. Additionally, Dr. Bob and Brittney would identify Tekia's and Travis's perspectives from their backgrounds in order to build further cultural sensitivity and awareness for both Dr. Bob and Brittney as collaborative learners.

In the role of the advocate, Dr. Bob would advocate for the client system and Brittney by pointing out red flags that could lead to ethical dilemmas with this case, such as Travis's "short fuse." Because the postmodern supervisor knows that supervisee growth occurs through the use of collaborative dialogue in a safe environment, Dr. Bob would ask Brittney to share her perspectives of possible ethical dilemmas as well as her case conceptualization based on theoretical orientation. Dr. Bob would pose questions to elicit mutual understanding of the differing possible perspectives of Travis, Tekia, and Brittney by using the acting "as-if" technique. For example, Dr. Bob might ask Brittney, "Act as if you are Tekia and you are witnessing Travis becoming angry. How do you feel? What actions do you take?" This will provide Brittney with insight into individual perspectives.

The postmodern supervisor instills a sense of safety and security through the process of mutual learning and support (Unger, 2006). Through collaborative dialogue, Dr. Bob and Brittney would discuss the possible safety issues of Travis's "short fuse" and possible post-traumatic stress disorder symptoms. If the supervisee does not bring this up as a concern, as the advocate and teacher, Dr. Bob would address this issue. Dr. Bob may begin a dialogue about ethical dilemmas and ethical decision-making models. Through conversational discourse, he would better understand Brittney's perspective. Similarly, Brittany may start to identify this as a potential ethical issue. Dr. Bob may guide Brittney through the process of ethical decision making. The collaborative nature of this approach allows Brittney to develop her own insight and apply it effectively.

(*continued*)

While observing, and through case conceptualization with Brittney, Dr. Bob will be noting any isomorphic processes. For example, Travis may be angry or "short-fused" in session without Brittney acknowledging the behavior. In the supervision session, Brittney may be angry or short with Dr. Bob. If Dr. Bob does not acknowledge the isomorphic process that is occurring, he will be reinforcing Travis's angry outbursts and not model for Brittney how to address these behaviors in session. To address this isomorphic process, Dr. Bob will come from a tentative stance and allow Brittany to develop her own understanding of the underlying processes at work in the supervision and therapy sessions.

Power is a common underlying process in the supervision relationship. Dr. Bob would discuss the idea of power and model appropriate uses of power for Brittney. Dr. Bob would empower Brittney through affirming her knowledge and perspectives as being accurate. He would foster her efficacy as a therapist by providing her the freedom to make decisions regarding case conceptualization and client care. Dr. Bob would treat Brittney with the respect of a colleague and expect that both he and Brittney would grow through their relationship as horizontal learners. Dr. Bob would not view Brittney as being "less than" due to their age differential or her developmental stage as a therapist. He would co-create knowledge along with her and eliminate the power differential by minimizing hierarchies in their relationship. This postmodern relationship allows each participant to learn, experience, and grow together.

REFERENCES

American Association for Marriage and Family Therapy (AAMFT). (2012). *AAMFT code of ethics.* Retrieved from http://www.aamft.org/resources/lrm_plan/Ethics/ethicscode2012.asp

Ancis, J. R., & Marshall, D. S. (2010). Using a multicultural framework to assess supervisees' perceptions of culturally competent supervision. *Journal of Counseling & Development, 88,* 277–284.

Anderson, H. (2001). Postmodern collaborative and person-centered therapies: What would Carl Rogers say? *Association for Family Therapy and Systemic Practice, 23,* 339–360.

Anderson, H., & Goolishian, H. A. (1988). Human systems as linguistic systems: Evolving ideas about the implications for theory and practice. *Family Process 27,* 371–393.

Bernard, J., & Goodyear, R. (1998). *Fundamentals of clinical of supervision* (2nd ed.). Needleham Heights, MA: Allyn & Bacon.

Doughty, E. A., & Leddick, G. R. (2007). Gender differences in the supervisory relationship. *Journal of Professional Counseling: Practice, Theory & Research, 35,* 17–30.

Estrada, D. (2005). Supervision of cross-cultural couple's therapy: Giving voice to the code of silence in the supervision and therapy room. *Journal of Family Psychotherapy, 16,* 17–30. doi:10.1300/J085v16n04_02

Gergen, K. (1985). The social constructionist movement in modern psychology. *American Psychologist, 40,* 255–275.

Gray, S. W., & Smith, M. S. (2009). The influence of diversity in clinical supervision: A framework for reflective conversations and questioning. *Clinical Supervisor, 28,* 155–179. doi:10.1080/07325220903324371

Melchent, T., Hays, V., Wiljanen, L., & Kolocek, A. (1996). Testing models of counselor development with a measure of counseling self-efficacy. *Journal of Counseling and Development, 74,* 640–645.

Morgan, M., & Sprenkle, D. (2007). Toward a common-factors approach to supervision. *Journal of Marital and Family Therapy, 33,* 1–17.

Murphy, M., & Wright, D. (2005). Supervisees' perspectives of power use in supervision. *Journal of Marital and Family Therapy, 31,* 283–295.

Paisley, P. (1994). Gender issues in supervision. *ERIC Digest, 13.* Retrived from http://www.ericdigests.org/1995–1/issues.htm

Paris, E., Linville, D., & Rosen, K. (2006). Marriage and family therapist interns' experiences of growth. *Journal of Marital and Family Therapy, 32,* 45–57.

Selicoff, H. (2006). Looking for good supervision: A fit between collaborative and hierarchical methods. *Journal of Systemic Therapy, 25,* 37–51.

Shakeshaft, C., Nowell, I., & Perry, P. (2000*). Educational leadership: Gender and supervision.* San Francisco, CA: Jossey-Bass.

Simon, G. (2006). The heart of the matter: A proposal for placing the self of the therapist at the center of family therapy research and training. *Family Process, 45,* 331–344.

Storm, C. L., Todd, T. C., Sprenkle, D. H., & Morgan, M. M. (2001). Gaps between MFT supervision assumptions and common practice: Suggested best practices. *Journal of Marital and Family Therapy, 27,* 227–240

Tapp, D., & Wright, L. (1996). Live supervision and family systems nursing: Postmodern influences and dilemmas. *Journal of Psychiatric and Mental Health Nursing, 3,* 225–233.

Taylor, B. A., Hernández, P., Deri, A., Rankin, I. R., & Siegel, A. (2006). Integrating diversity dimensions in supervision: Perspectives of ethnic minority aamft approved supervisors. *Clinical Supervisor, 25,* 3–21. doi:10.1300/J001v25n01•02

Todd & Storm (1997). The complete systemic supervisor. Boston: Allyn & Bacon.

Todd, T. C., & Storm, C. L. (2002). *The complete systemic supervisor: Context, philosophy and pragmatics.* New York, NY: Authors Choice.

Unger, M. (2006). Practicing as a postmodern supervisor. *Journal of Marital & Family Therapy, 32,* 59–71.

Weir, K. N. (2009). Countering the isomorphic study of isomorphism: Coercive, mimetic, and normative isomorphic trends in the training, supervision, and industry of marriage and family therapy. *Journal of Family Psychotherapy, 20,* 60–71. doi:10.1080/08975350802716517

Wheeler, D., Avis, J., Miller, L., & Chaney, S. (1986). Rethinking family therapy education and supervision: A feminist model. *Journal of Psychotherapy & the Family, 4,* 53–71.

White, M., & Epston, E. (1990). *Narrative means to therapeutic ends.* New York, NY: Norton.

Whiting, J. B. (2007). Authors, artists, and social constructionism: A case study of narrative supervision. *American Journal of Family Therapy, 35*(2), 139–150. doi:10.1080/01926180600698434

Wieling, E., & Marshall, J. (1999). Cross-cultural supervision in marriage and family therapy. *Contemporary Family Therapy, 21*, 317–329.

Wieling, E., Negretti, M., Stokes, S., Kimball, T., Bateman, C., & Bryan, L. (2001). Postmodernism in marriage and family therapy training: Doctoral students' understanding and experiences. *Journal of Marital and Family Therapy, 27*, 527–533.

THIRTEEN

Training the Systemic Cognitive-Developmental Therapist

Michael Redd

The field of family therapy has been moving ever onward toward attempting to address the complex, multidimensional, diverse, and multicultural needs of the profession (Rigazio-DiGilio, Gonçalves, & Ivey, 1996; Walsh, 2012). Woolley (2010) and Norcross (2005), both summarizing large recent studies that include marriage and family therapists (MFTs), indicate that not only are a sizeable minority of therapists now practicing from an integrative or eclectic orientation, there is significant variability in approaches used. Supervisors need to be able to address the additional training needs of these therapists and help them further their grounding in case conceptualization and intervening intentionally (Woolley, 2010). Further, supervisors in the context of clinical training programs are often required to supervise trainees who are learning and using a broad range of theoretical orientations (Storm & Sprenkle, 2002). More and more is required of trainers and supervisors in order to meet the needs of therapists practicing from diverse theoretical orientations.

Facilitating intervention and training from a theoretical orientation is only one of many purposes of supervision. Other purposes include protecting client welfare, facilitating client progress, serving a gatekeeping function for the profession, and facilitating the professional growth and development of the supervisee (Rigazio-DiGilio & Anderson, 1995a; Storm, Todd, Sprenkle, & Morgan, 2001). These many purposes present a significant amount of complexity. Supervision models, though still largely lacking research on outcomes (see Morgan & Sprenkle, 2007), can present one important way of organizing the complexity inherent in the supervision of family therapists.

The use of integrative supervision models is indicated to manage this complexity and adequately allow for both flexibility and increased access to a theoretical landscape that will address varied clinical and training needs (Rigazio-DiGilio & Anderson, 1995a). Systemic Cognitive-Developmental Supervision (SCDS) is one such integrative supervision model, which presents a metatheoretical approach that can be applied by supervisors who use different supervisory training approaches with supervisees who have training in and are working under different theories (Rigazio-DiGilio & Anderson, 1995a). This chapter presents a summary of SCDS and integrates a case study to illustrate basic concepts and use of the SCDS supervision model.

TEACHING THEORETICAL CONTEXT

Founders

Systemic Cognitive-Developmental Supervision is a supervision model that was developed upon the same theoretical foundations as Systemic Cognitive-Developmental Therapy (SCDT). Both SCDT (Rigazio-DiGilio,1997, 2000, 2012) and SCDS (Rigazio-DiGilio, 2014a; Rigazio-DiGilio & Anderson, 1995a) were developed primarily by Dr. Sandra Rigazio-DiGilio. Rigazio-DiGilio and McDowell (2007) locate the development of SCDT within the context of other systemic therapies as being in the fourth force, or integrative/metaframework models that followed and built upon concepts from other approaches to therapy and supervision, including psychodynamic/historical (e.g., Bowenian theory), cognitive-behavioral/interactional (e.g., Minuchin and Haley's structural/strategic approaches), existential/humanistic (e.g., Whitaker's symbolic/experiential or Satir's communication approach), and ecological/postmodern approaches (e.g., social constructivism or narrative approaches). These later approaches, such as the Integrative Problem-Centered Metaframeworks approach (Breunlin, Pinsof, Russell, & Lebow, 2011) and SCDT (Rigazio-DiGilio, 1997), all share an attempted integration and organization of previous theoretical approaches and interventions to broaden clinical options (Rigazio-DiGilio, 2014a). SCDT departs from many traditional models by moving beyond locating dysfunction solely within individuals or families (Rigazio-DiGilio, 2000). Indeed, "the territory of investigation and intervention, then, includes all active participants (e.g., individuals, family units, personnel from community institutions and agencies), the actual symptoms, and the multiple ways participating members construct their interpretations and reactions to these symptoms" (p. 1018).

SCDT has been described as a metatheoretical approach (Rigazio-DiGilio, 2000) that has its theoretical roots in Developmental Counseling and Therapy (DCT), developed by Dr. Allen Ivey (Ivey, 1986; Ivey & Ivey, 1999; Ivey, Ivey, Myers, & Sweeney, 2005; Ivey, Ivey, & Rigazio-DiGilio, 2011). The metatheory does not dictate specific interventions, but rather provides a framework that guides therapists in choosing from strategies and interventions they are trained in to co-create treatment plans (Rigazio-DiGilio, 1997). DCT is also a metatheoretical approach that seeks to integrate multicultural, developmental, and traditional counseling theories (Ivey, 1986). SCDT extended DCT and its concepts for working with families and wider systems. SCDT seeks to provide a comprehensive system that combines individual, family, and network assessment and intervention (Rigazio-DiGilio, 2000).

Philosophical Foundation

SCDS is a supervision model that is built upon the theory of SCDT and is grounded in integrative, developmental, co-constructive, holistic, and systemic assumptions. The integrative assumption has already been discussed and the co-constructive assumption will be discussed later in the section on culture/race and gender issues. The others are briefly discussed here.

Developmental

As noted by Sansbury (1982), stage-like developmental theories of supervision and research contribute to understanding salient issues that clinicians with varying degrees of experience may have. Stoltenberg's Integrative Developmental Model is an example of this kind of stage-like developmental model (Stoltenberg, Bailey, Cruzan, Hart, & Ukuku, 2014; Stoltenberg, McNeill, & Delworth, 1998). SCDS expands on developmental assumptions that allow one to understand supervisee growth by conceptualizing development as nonstage-like and to see supervisees as presenting with unique worldviews, developmental and contextual histories, and idiosyncratic needs, capabilities, and limitations (Rigazio-DiGilio, 2014a; Rigazio-DiGilio & Anderson, 1995a). Thorough assessment of supervisee development is integral to effectively tailoring supervision to this diversity. This developmental assessment is then used as part of a co-constructive process to develop and address supervisory goals and objectives. Differences between even beginning supervisees—in terms of lived experience, experience with helping relationships, skill and perceptual abilities, espoused theoretical orientation, inherent theory of human development, and access to various information processing styles (discussed hereafter)—reinforce the need to tailor supervision with a more complex and nuanced view, rather than using a nondevelopmental model or making assumptions about development based only on years of experience. SCDS provides one version of a nuanced view of development that is intricately connected to supervisory modalities, approaches, interventions, and roles (Rigazio-DiGilio, 2014a, 2014b). The underlying theory is also connected to an understanding of culture and worldviews (Rigazio-DiGilio, 2014a ; Rigazio-DiGilio & Ivey, 1995), which provides entry points into cultural identity development as well (see Ivey et al., 2005, for DCT integration with cultural identity development). A strong focus on supervisees and their development as competent therapists can also be seen as critical for preserving client welfare in the future following independent licensure.

Systemic

Isomorphism, or repetition of patterns at different systemic levels (Todd & Storm, 2014), is an important therapeutic and supervisory principle. Isomorphism is built into the SCDS conception of development, or the goal of being able to address flexible information-processing styles at the client level, self-of-therapist level, therapeutic process level, and supervisory process level (Rigazio-DiGilio & Anderson, 1995a). SCDS is built upon SCDT, which is an ecosystemic model addressing not just family systems but also individual and community systems (Rigazio-DiGilio, 2000). SCDT and DCT go beyond many therapy models and uniquely focus on wider systems and networks to address environmental resources, constraints, and contributors to the distress of families (Ivey et al., 2005; Rigazio-DiGilio, 2000). Thus, within the supervision context, larger systems such as the environment (of supervision, of therapy, and of the clients), training context, rules/regulations, and so forth, are important to assess and be aware of, as they also contain both constraints and competencies/resources that affect supervision and therapy. Supervisors can facilitate perceptual skills on these different levels by bringing attention to them and by modeling switching to different systemic levels while staying within one information-processing style. Supervisors can bring attention to isomorphic patterns when they occur at different levels. This would mean that a male supervisor discussing with a female supervisee how she could be more direct in bringing up gender with a couple she was working with would also need to be more direct about addressing the possible effects of gender in the supervisory realm. This not only models the use of these perceptual skills, but also encourages consistency and the need to be aware of different systemic levels that affect the work of therapy. Supervisors who show a willingness to demonstrate and do what they are asking supervisees to do (an isomorphic process) thus model respect and co-construct supervision in a way that better balances the power differential in supervision and therapy.

Cognitive-Emotional Information-Processing Styles

In SCDS, information-processing styles are one way that development is operationalized. The construct serves to anchor supervisory assessment and intervention to the here and now of the supervisory environment. Cognitive-emotional information-processing styles in SCDS are predicated upon a reinterpretation of Piagetian cognitive developmental stages. There are four SCDT information-processing styles (sensorimotor/elemental, concrete/situational, formal/reflective, and dialectic/systemic), each of which represents "a partial, distinct, and viable frame of reference and a range of cognitive, emotional, and behavioral resources used in the construction of worldviews" (Rigazio-DiGilio, 2000, p. 1023). These styles are essentially modes of processing information that affect our experience, understanding, and ways of functioning in the world (Rigazio-DiGilio, 2012). Supervisees who are using primarily a sensorimotor/elemental style process events in therapy by using sensory experience and may be able to more easily access the emotional climate in sessions and more clearly read nonverbal cues and clues (Rigazio-DiGilio, 2014b). However, supervisees constrained in this style may be overwhelmed by the emotions of the moment (such as anxiety, frustration, or anger) or have difficulty with transference/countertransference and boundaries. Supervisees constrained in this orientation may also rush decision making, deciding based on gut feelings (i.e., what feels "right"). Supervisees who are using primarily a concrete/situational style to process information typically have a linear perspective (cause and effect) and may be able to act with more predictability. Supervisees constrained in this style may have difficulty moving beyond linear hypotheses (i.e., get caught in black/white thinking) to look at and act on abstract principles, patterns, or systems dynamics. Supervisees who are using primarily a formal/reflective style make sense of the world mostly through pattern identification, focusing on abstractions and multiple perspectives. Supervisees constrained in this style may be overly abstract in their reasoning, which may lead to them over-intellectualizing, neglecting important concrete data, and having difficulty actually applying abstract principles to direct interventions in therapy. Supervisees who are using primarily a dialectic/systemic style to process information are able to become aware of and deconstruct underlying assumptions, rules, and broader contextual and historical influences on the presenting issues. Supervisees who are constrained in this style may become bogged down with overanalysis, get stuck in over-intellectualizing, be unable to choose from many identified alternative courses of action, and take a concrete perspective (Rigazio-DiGilio, 2014b).

Information-processing styles, unlike Piaget's constructs, are not hierarchical (i.e., moving upward in complexity) or linear (i.e., advancing sequentially) (Rigazio-DiGilio, 2012). According to Rigazio-DiGilio and Ivey (1995), development is toward broader application and ability to access the resources inherent in these different styles. Rigazio-DiGilio and Anderson (1995a) postulate that supervisees who are able to more flexibly and broadly access different information-processing styles have access to greater resources and are better able to meet client and clinical needs. The SCDS supervisor assesses supervisees' primary and ancillary information-processing styles through their language use in the supervisory session. Supervisors pay attention to the "linguistic cues" that signal the use of different information-processing styles (Rigazio-DiGilio, 2014b). For example, a supervisee who tends to use detailed and concrete descriptions of therapy sessions ("stories") from a linear perspective would be primarily operating from a concrete/situational information-processing style. When supervisors are able to assess, without prematurely influencing, the supervisee's frame of reference, they are able to open a window on how the supervisee makes meaning and understands. The concepts of horizontal development (matching the supervisee's primary style and working to build and expand the use of resources and address constraints within that style) and vertical development (style-shifting to ancillary styles to facilitate processing, skill building, and use of resources in less-used styles) are implemented to facilitate supervisee development. Within the supervisory model, horizontal development is also conceptualized as being better able to focus on and use one style effectively to address the different client, self-of-therapist, therapeutic process, and supervisory process issues. Further developmental concepts derived from Piaget and adopted by SCDS, such as accommodation and assimilation, give further guidance to the supervisor in assessing the degree to which supervisees are balancing the integration of new information into their existing worldviews (Rigazio-DiGilio, 2012).

The Process of Change

The process of change in both SCDT and SCDS focuses on developmental conceptualizations: that is, increasing the ability of clients and supervisees to adapt to internal and external demands for change and meet developmental needs by more broadly and flexibly using the resources inherent within each information-processing style (Rigazio-DiGilio, 2000, 2012). This also means addressing the constraints clients and supervisees may have in a particular style as well as accessing competencies to enable them to flexibly use the range of styles in a more intentional manner.

Culture/Race and Gender Issues

Although no current research has examined the extent to which SCDS is effective when used with diverse supervisees and/or clients, there are some important elements that point to SCDS being a culturally centered approach that has the foundational elements needed to be culturally sensitive. A culturally centered approach to supervision is needed to account for supervisor-supervisee interactions in the here-and-now of the supervisory session (including the impact of assumptions, beliefs, attitudes, behaviors, and feelings), address how to co-construct a dialogue about supervisory issues rather than dictate them, address power, seek to understand supervisee and client development and context, and seek to respect and seek out supervisee worldviews (Rigazio-DiGilio & Anderson, 1995a; Rigazio-DiGilio & Ivey, 1995). SCDS sees culture as primary and suggests that there is a dialectic and co-constructive process between individuals and their social and cultural environments (Rigazio-DiGilio & Ivey, 1995). Inherent in a culturally sensitive view of supervision is the anticipation of differences or conflicts in worldview between the supervisor and supervisee (or supervisee and client or supervisor and client), the need to assess the saliency of different worldview positions, and to work through relational and supervisory issues in a co-constructive fashion (Brown & Landrum-Brown, 1995). SCDS provides a unique way to address differences in worldviews by focusing on co-construction and provides a helpful heuristic to describe worldview differences.

Supervisors need to continue to remain open to learning more and hearing disparate voices in the context of supervision (Aducci & Baptist, 2011; Rambo & Shilts, 2002). They must recognize that supervision models are as vulnerable to gender and ethnocentric biases as are the theoretical orientations upon which they are based (Banks, 2001; see also Rigazio-DiGilio & Anderson, 1995b). SCDS does focus on a nonpathological, developmental context that takes into account unique perspectives and contextual history, allowing for the examination of different perspectives (Rigazio-DiGilio, 2000). It has been noted that being able to strike a balance between collaborative and hierarchical methods in supervision is important to take into account as well as developmental level of supervisee-therapists (Prouty, Thomas, Johnson, & Long, 2001). Using more collaborative methods increases the chance that supervision will be power-sensitive and will not be oppressive or exploitative. One strength of the SCDS model is the range of supervisory styles and interventions that can address both more directive and more collaborative methods. Another is its conceptualization about the co-constructive nature of supervision environments.

The intersecting domains of class, gender, race, ethnicity, ability, sexual orientation, spirituality, and so forth also provide important context to supervision and therapy. Multicultural understanding, awareness, and competence are seen as a developmental process that is ongoing (with no end point). It is a continuous process largely due to the inherent difficulty involved in attempting to see the world from another's perspective. Because each new client family presents a unique intersection of levels of diversity, there is also significant complexity involved in these intersections between race, ethnicity, age, gender, able-bodiedness, spirituality, language, and so forth. Although the general SCDS model provides an important framework as a beginning, value is added by a knowledge of the developing literature on culturally sensitive therapy and supervision (e.g., Christiansen et al., 2011). To become more multiculturally competent, therapists and supervisors must continue to have conversations, discussions, and ways of challenging themselves both in and out of supervision and therapy. One important way this is done is through paying attention to social justice and equity issues (see Constantine, Hage, Kindaichi, & Bryant, 2007; Haddock, Schindler Zimmerman, & MacPhee, 2000) and the process of development of critical consciousness, or cultural self-awareness (see Garcia, Kosutic, McDowell, & Anderson, 2009; Pitner & Sakamoto, 2005). Consistent with SCDS's integrative view, it is asserted that there are both cognitive and affective barriers that must be addressed on the path of developing critical consciousness (including identity development; how one's worldview may obscure difference; automaticity of stereotypes; cognitive load; and fear, anger, anxiety, and hostility that may come when self-image is challenged) (Pitner & Sakamoto, 2005).

Research to Consider

Morgan and Sprenkle (2007) have critiqued supervision models that are based on therapy models (saying that they do not take into account the significant differences between therapy and supervision) as well as supervision models based on developmental theories (saying that there is limited empirical support for the developmental assumptions underlying these stage-models). However, their discussion does not specifically speak to the unique aspects of SCDS, which is in essence a model based on a broad integrative therapy theory *and* a model that has atypical developmental assumptions, due to the nonlinear and nonhierarchical formulations upon which it is based. This being said, the overall message and critique of their article is that supervision models in general lack empirical support and outcomes have mostly not been empirically tested (see also Rigazio-DiGilio, 2014c). Rigazio-DiGilio and Anderson (1995a) reference unpublished preliminary research, which they describe as showing that trainees' information-processing styles can be rated reliably using the SCDS assessment framework. Preliminary support does have to be expanded upon, and similar to supervision models in general, SCDS should be further researched, including testing its general assumptions and a contextualized view of outcomes that is linked to training, supervisee practice, and client outcomes (Rigazio-DiGilio, 1998).

ROLE OF THE SUPERVISOR AND SUPERVISEE

Supervisor-Supervisee Relationship

SCDS values co-construction, involving explicit goal setting, meeting supervisees where they are, and working with them rather than prescribing to them (Rigazio-DiGilio & Anderson, 1995a). Supervisors using this model have to not only address limitations and skills deficits, but also build upon strengths and enable collaboration. Again, as in many components of this model, the level of collaboration versus directedness is not one-size-fits-all; it requires significant flexibility from the supervisor. This flexibility is important, although it makes it difficult to characterize a typical supervisor-supervisee relationship. Co-construction of the supervisory environment occurs in all cases, though, and is also facilitated by seeking collaboration from the supervisee throughout the supervisory experience, from contracting and deciding on supervisory goals, to evaluating progress toward those goals (Rigazio-DiGilio, 2014a). Using clear supervision contracts also helps to establish the boundaries and expectations of the supervisory relationship and is essential (Rigazio-DiGilio, 2014a).

Learning Goals/Objectives

Learning goals and objectives are specific to each case and context and must take into account the supervisory needs and setting, supervisee developmental needs, client needs, and how those needs intersect in supervision and therapy as well as outside contexts (culture, practice environment, etc.). Specific supervisee developmental needs are informed by developmental assessment. Supervisors must work broadly and bring in more resources to bear on solutions. Supervisors need to be able to work within all of the cognitive-emotional information-processing styles; to be more experiential at times and work with supervisee emotions; to work concretely as well as reflectively; and to conceptualize ecosystemically, focusing on culture, community, challenging assumptions, and so forth. Supervisors who use an integrative therapy orientation for their clinical work may more easily be able to bring their understanding of multiple ways of intervening and differing worldviews to help expand perspectives and facilitate flexibility.

Supervisor Roles and Interventions

Being able to tailor supervision to the unique development and worldview of supervisees includes being able to assume different roles as a supervisor. Flexibly adapting to different supervisory roles facilitates certain environments that can match, expand, and use supervisee competencies, and then facilitate supervisee growth and address supervisee constraints (Rigazio-DiGilio & Anderson, 1995a). Supervisor roles can include a more directive role, to provide structure to explore sensorimotor experience; a coaching role, to work in a concrete fashion to provide skill building within a linear perspective; a consultation role, to facilitate reflection on patterns; and a collaboration role, to facilitate metacognitive processes and ability to examine the interface between the client and the environment (culture, community, school/work, family-of-origin, etc.). Further descriptions of these roles, including supervision objectives, how to address competencies and constraints in these information-processing styles, and sample questions that supervisors working from these roles might ask are included in the literature (see Rigazio-DiGilio & Anderson, 1995a; Rigazio-DiGilio, 2014b). Put simply, knowing how to categorize different types of questions can allow supervisors, for example, to ask more concrete questions when trying to facilitate a concrete processing style (e.g., "What happened next in the session after the father said that?" or "What did the mother actually say?"), or formal/reflective questions when trying to facilitate a more reflective style (e.g., "What do you see as the basic dysfunctional pattern that has played out across your sessions with the family?" or "What typically does the rest of the family do in sessions when the mother is talking to you?"). Questions centered on here-and-now experiencing would be used to facilitate a sensorimotor/elemental style (e.g., "How are you feeling right now as we discuss this case?"), and questions regarding underlying assumptions/rules would be used to facilitate a dialectic/systemic style (e.g., "What rules does this family operate from in this interaction?" or "What cultural influences appear to be operating in the husband's decision?"). A basic assumption here is that the kinds of questions, conversations, and interventions that a supervisor facilitates/co-facilitates with a supervisee can create specific learning environments, which themselves "foster conceptual, perceptual, and technical competencies within each of the four [information-processing styles]" (Rigazio-DiGilio & Anderson, 1995a, p. 97).

As noted earlier, SCDS builds upon an integrative, metatheoretical approach that provides a framework that supervisors can use to guide selection of different modalities, strategies, and interventions, many of which they may already be familiar with and may be included in different supervision models. Supervisory modalities and interventions are thus chosen from among a range of possibilities depending on supervisee information-processing styles and intended clinical and supervision objectives. For example, in facilitating a structuring environment to style-match or style-shift to a sensorimotor/elemental style, a supervisor can use modalities such as "cotherapy teams, live supervision with bug-in-the-ear guidance, and role-plays" (Rigazio-DiGilio, 2014a, p. 270). A coaching environment for style-matching or style-shifting to a concrete/situational style could be facilitated, among other ways, by direct coaching using live supervision with coaching prior to the session, coaching during a midsession break, and coaching postsession to enhance intentional use of interventions and skill building. A consulting environment for style-matching or style-shifting to a formal/reflective style could be facilitated by tracking patterns across cases using video clips, the identification of isomorphic patterns, or the use of homework exercises that require self-analysis and self-reflection. A collaborating environment for style-matching or style-shifting to a dialectic/systemic style could be facilitated, for example, by dialogues in supervision that uncover and deconstruct hidden assumptions, rules, biases, and impacts of broader systems on the supervisee, supervision process, and therapy (Rigazio-DiGilio, 2014a). A more comprehensive list of sample supervision modalities and interventions is found in Rigazio-DiGilio (2014b).

CONCLUSION

SCDS is a promising integrative supervision model that can provide significant direction to the supervisor in the supervisory context; it also provides a way for supervisees, no matter where they are in their development, to use clinical issues and the supervisory environment to continue to expand their development and worldview. Future research into SCDS outcomes and diversity applications will continue to build upon its current research foundations. It is hoped that this field will continue to move forward, making use of and refining these kinds of integrative supervision models to meet the ever-changing demands of an evolving profession.

CASE EXAMPLE

A case example was provided to the authors of the chapters in this book in order to give readers a reference point to examine similarities as well as differences in the different supervision models. In this example, the AAMFT Approved Supervisor, Dr. Bob, as an experienced supervisor for more than 20 years, provides supervision for a range of experience levels (from students in practicum and internship to therapists in an agency setting). One of the strengths of the SCDS approach to supervision is its way of organizing many of the familiar supervision interventions and strategies within a comprehensive framework (Rigazio-DiGilio, 2014b). Experienced supervisors such as Dr. Bob who learn this approach may find this comprehensive framework an effective way of putting together interventions learned from a variety of different places and contexts over their career as supervisors. In addition, Dr. Bob's situation, as a supervisor who works with supervisees from a range of settings, experience, and needs, fits well with SCDS and its integrative framework.

The supervisee in this example is Brittney, who is an older-than-traditional student, being in the Master's program at age 33. Many individuals who come to therapy training later in life or as a second career often have a wealth of experience and strengths that cannot be neatly categorized based on stage-like developmental theories. The more nuanced developmental view of SCDS, as described earlier, allows for targeted assessment and intervention. This tailored approach can take into account the wide range of competencies and constraints that each supervisee brings, even though supervisees are all "beginning therapists."

When starting to supervise Brittney, Dr. Bob, working from an SCDS orientation, would spend time to develop some rapport with her and to contract for supervision. This contracting is important to clarify the boundaries and expectations of this particular context (Rigazio-DiGilio, 2014a). This would also involve a discussion of supervisory goals. This should be a collaborative endeavor with input from Brittney to develop goals that will be relevant to Brittney's development as a therapist. The input regarding goals from Brittney's perspective may provide some linguistic cues regarding the information-processing style that Brittney is using to make sense of the supervision environment. For example, as Brittney talks about goals, she discusses how she has had difficulty in her practicum experience moving from conceptualization of the family, using her theory of therapy, to knowing which interventions to apply. This

(*continued*)

conversation gives some initial clues that Brittney may primarily use a formal/reflective style, but that she is constrained by her underdeveloped access to a concrete/situational style. Dr. Bob would continue to assess Brittney's ability to flexibly access different cognitive-emotional information-processing styles, both through informal assessment of the natural dialogue as well as through using a question matrix to guide the assessment (see Rigazio-DiGilio, 2014b). By using different kinds of questions, Dr. Bob can see how able Brittney is in discussing supervision goals, the supervision context, the therapy environment and context, and the client family she is working with from different information-processing styles. The following is an example of a part of this process.

Dr. Bob (to Brittney): *So tell me what comes up for you when you think of Travis and Tekia?* [A more neutral open-ended question to see which information-processing style Brittney will go to first. He could have also used something such as "What do you need?"]

Brittney: *They really seem to be motivated to get help. They are coming despite their difficulty in getting care for their children for each session and are engaged every session. I do find, though, that I am getting stuck with them in the same place I typically get stuck with my clients.*

Dr. Bob: *What are you seeing?*

Brittney: *My concern has been that Tekia always seems to speak for Travis in sessions. Travis is withdrawn typically, only getting engaged when he seems to agree strongly with Tekia and is getting angry. I see the dynamics happening in session, but I'm not sure what to do to change them.*

Dr. Bob noted the initial linguistic cues that signal Brittney was processing from a formal/reflective style and started to ask some more reflective questions to assess her ability to broadly access this style. He could continue on this line of questioning, asking more about patterns of the family in session, switching to asking about other perspectives or ways of looking at what was going on in session, or perhaps asking about the particular patterns she has found herself doing in session with this family. After spending some time assessing Brittney's ability to access a formal/reflective style, he would ask questions in other styles to assess her ability to understand the family from different styles. He might go to a more concrete style first, as that was indicated as possibly problematic in the supervisory goal setting.

(*continued*)

Dr. Bob: *Tell me about what specifically happened in the last session when you went over the homework with the couple.* [Asking a concrete question]

Brittney: *They didn't understand the assignment from the last session and so we moved on to the rest of the session.*

Dr. Bob: *Okay, so take me through specifically what happened. You asked them to take out the homework?*

Brittney: *Yes, I said "How did the homework go?" and they were frustrated like they have been in other sessions.*

Dr. Bob: *What exactly did they say?*

Brittney: *They said they didn't really understand, which has been typical.*

Dr. Bob: *Take me through this situation step by step, just describing what happened. Tell me who spoke first, and what they said, then who spoke next, and so forth.*

Here Dr. Bob continues to ask concrete questions that encourage Brittney to access the concrete/situational information-processing style. This dialogue gives some clues that Brittney may be constrained in the formal/reflective information-processing style and has less access to a concrete/situational style. This is shown by her continuing to revert to a formal/reflective style (referencing patterns) and having some difficulty telling the story of what happened using concrete details and a linear perspective. When focused on for long enough, she is able to become more concrete, though the concrete style is not one that comes easily to her.

Further questioning reveals that Brittney also has a relatively strong ability to access a dialectic/systemic processing style. She is able to use the resources inherent in the style to make sense of the family cultural context of Travis's injuries and discharge from the military, the sociocultural issues related to the military and seeking emotional/relationship help, and the family rules around "failure" to provide, among other issues. However, Brittney tends to get overwhelmed with the possibilities and analysis when she accesses this style and does not seem to know what to do with this information and reflective process. Also, similar to her use of the formal/reflective style, Brittney becomes constrained when she gets stuck in these information-processing styles and cannot access or switch effectively to the concrete/situational style. This is most evident in her treatment plans, which show little connection to her understanding of the systemic context and her difficulty in translating her abstract reasoning/processing into skillful execution of specific interventions in session that connect to this understanding.

(*continued*)

After a thorough assessment process, Dr. Bob assesses that Brittney's primary information-processing style is formal/reflective, her secondary style is dialectic/systemic, and her ancillary styles are concrete/situational and sensorimotor/elemental. An initial supervisory plan might initially style-match to Brittney's primary style of formal/reflective in order to build a stronger foundation in that style. This plan could involve facilitating a consulting environment through relevant questions in order to analyze and more fully describe patterns and the context involved (including the therapy environment). The plan might also involve listening to audiotapes of sessions to identify themes and patterns. Facilitating Brittney's ability to broadly and flexibly access the formal/reflective style through style matching would allow the supervisor first to join with the supervisee in her primary meaning-making and then to help her better access the resources inherent in the style in a flexible way that contrasts with more rigid use of the style.

After a stronger foundation in her primary style had been established, a coaching environment would be facilitated to strengthen Brittney's ability to access the concrete/situational style. The supervisor would use a semi-directive questioning style that focused on directing attention to the cause-effect linear perspective, and especially to describing Brittney's own actions in therapy and their effects. This plan would involve skill building through practice and possibly role-plays, strengthening the connection between intention and action (through step-by-step video replays), and live supervision with coaching before and after the session. Facilitating Brittney's ability to access the concrete/situational style through style shifting will allow her to better access the resources inherent in the concrete/situational style, such as better tracking skills, a more accurate understanding of patterns (as they will be more embedded in actual behavioral data from the sessions), and more intentional interventions (see Rigazio-DiGilio, 2014b). This example showcases a part of the overall supervision plan, which will continue to evolve as supervisee, client, and supervisory needs are met or change. As noted earlier, the overall goal is to help Brittney develop competencies by being able to more broadly and flexibly access the different information-processing styles. This involves strengthening current competencies, addressing constraints, and building underdeveloped abilities. The assumption is that supervisees who are able to broadly and flexibly access these different information-processing styles have many more options available to effect change in meeting the varied clinical and supervision needs (Rigazio-DiGilio, 2014b).

REFERENCES

Aducci, C. J., & Baptist, J. A. (2011). A collaborative-affirmative approach to supervisory practice. *Journal of Feminist Family Therapy, 23*(2), 88–102.

Banks, A. (2001). Tweaking the Euro-American perspective: Infusing cultural awareness and sensitivity into the supervision of family therapy. *The Family Journal, 9*(4), 420–423.

Breunlin, D., Pinsof, W. M., Russell, W., & Lebow, J. (2011). Integrative problem-centered metaframeworks (IPCM) therapy I: Core concepts and hypothesizing. *Family Process, 50,* 293–313.

Brown, M. T., & Landrum-Brown, J. (1995). Counselor supervision: Cross-cultural perspectives. In J. Ponterotto, J. Casas, L. Suzuki, & C. Alexander (Eds.), *Handbook of multicultural counseling* (pp. 263–286). Thousand Oaks, CA: Sage.

Christiansen, A. T., Thomas, V., Kafescioglu, N., Karakurt, G., Lowe, W., Smith, W., & Wittenborn, A. (2011). Multicultural supervision: Lessons learned about an ongoing struggle. *Journal of Marital and Family Therapy, 37*(1), 109–119.

Constantine, M. G., Hage, S. M., Kindaichi, M. M., & Bryant, R. M. (2007). Social justice and multicultural issues: Implications for the practice and training of counselors and counseling psychologists. *Journal of Counseling & Development, 85*(1), 24–29.

Garcia, M., Kosutic, I., McDowell, T., & Anderson, S. A. (2009). Raising critical consciousness in family therapy supervision. *Journal of Feminist Family Therapy, 21*(1), 18–38.

Haddock, S. A., Schindler Zimmerman, T., & MacPhee, D. (2000). The power equity guide: Attending to gender in family therapy. *Journal of Marital and Family Therapy, 26*(2), 153–170.

Ivey, A. (1986). *Developmental therapy: Theory into practice.* San Francisco, CA: Jossey-Bass.

Ivey, A. E., & Ivey, M. B. (1999). Toward a developmental diagnostic and statistical manual: The vitality of a contextual framework. *Journal of Counseling & Development, 77*(4), 484-490.

Ivey, A. E., Ivey, M., Myers, J., & Sweeney, T. (2005). *Developmental counseling and therapy: Promoting wellness over the lifespan.* Boston, MA: Lahaska Press.

Ivey, A. E., Ivey, M. B., & Rigazio-DiGilio, S. A. (2011). Developmental counselling and therapy: A brain-based approach to counselling and therapy. *Counselling Australia, 11,* 16–24.

Morgan, M. M., & Sprenkle, D. H. (2007). Toward a common factors approach to supervision. *Journal of Marital and Family Therapy, 33*(1), 1–17.

Norcross, J. C. (2005). A primer on psychotherapy integration. In J. C. Norcross & M. R. Goldfried (Eds.), *Handbook of psychotherapy integration* (2nd ed., pp. 3–23). New York, NY: Oxford University Press.

Pitner, R. O., & Sakamoto, I. (2005). The role of critical consciousness in multicultural practice: Examining how its strength becomes its limitation. *American Journal of Orthopsychiatry, 75*(4), 684–694.

Prouty, A. M., Thomas, V., Johnson, S., & Long, J. K. (2001). Methods of feminist family therapy supervision. *Journal of Marital and Family Therapy, 27*(1), 85–97.

Rambo, A. H., & Shilts, L. (2002). Four supervisory practices that foster respect for differences. In T. Todd & C. Storm (Eds.), *The complete systemic supervisor: Context, philosophy, and pragmatics* (pp. 83–91). New York, NY: Authors Choice Press.

Rigazio-DiGilio, S. A. (1997). Systemic cognitive-developmental therapy: A counseling model and an integrative classification schema for working with partners and families. *International Journal for the Advancement of Counselling, 19,* 143–165.

Rigazio-DiGilio, S. A. (1998). Toward a reconstructed view of counselor supervision. *Counselor Education and Supervision, 38*(1), 43–51.

Rigazio-DiGilio, S. A. (2000). Relational diagnosis: A coconstructive–developmental perspective on assessment and treatment. *Journal of Clinical Psychology, 56*, 1017–1036.

Rigazio-DiGilio, S. A. (2012). Developmental therapy and systemic cognitive-development therapy: Postmodern implications and beyond. *Japanese Journal of Microcounseling, 7*(1), 2–25.

Rigazio-DiGilio, S. A. (2014a). Common themes across systemic integrative supervision models. In T. Todd & C. Storm (Eds.), *The complete systemic supervisor: Context, philosophy, and pragmatics* (2nd ed., pp. 231–282). London, UK: Wiley-Blackwell.

Rigazio-DiGilio, S. A. (2014b). Assessing styles and tailoring systemic supervisory interventions. In T. Todd & C. Storm (Eds.), *The complete systemic supervisor: Context, philosophy, and pragmatics* (2nd ed.). London, UK: Wiley-Blackwell. www.wiley.com/go/todd

Rigazio-DiGilio, S. A. (2014c). Supervising couple and family therapy practitioners. In C. E. Watkins & D. Milne (Eds.), *The Wiley international handbook of clinical supervision* (pp. 622–647). London, UK: Wiley-Blackwell.

Rigazio-DiGilio, S. A., & Anderson, S. A. (1995a). A cognitive-developmental model for marital and family therapy supervision. *The Clinical Supervisor, 12*, 93–118.

Rigazio-DiGilio, S. A., & Anderson, S. A. (1995b). Gender-aware supervision in marriage and family counseling and therapy: How far have we actually come? *Counselor Education & Supervision, 34*, 344–355.

Rigazio-DiGilio, S. A., Gonçalves, O. F., & Ivey, A. E. (1996). From cultural to existential diversity: The impossibility of psychotherapy integration within a traditional framework. *Applied and Preventive Psychology, 5*(4), 235–247.

Rigazio-DiGilio, S. A., & Ivey, A. E. (1995). Individual and family issues in intercultural therapy: A culturally centered perspective. *Canadian Journal of Counselling, 29*, 244–261.

Rigazio-DiGilio, S., & McDowell, T. (2007). Family therapy. Chapter in M. Spiegler and J. Frew (Eds.) *Counseling and psychotherapy for a world of diversity: Theory and practice.* Boston: Houghton Mifflin.

Sansbury, D. L. (1982). Developmental supervision from a skills perspective. *The Counseling Psychologist, 10*(1), 53.

Stoltenberg, C. D., Bailey, K. C., Cruzan, C. B., Hart, J. T., & Ukuku, U. (2014). The integrative developmental model of supervision. In C. E. Watkins & D. L. Milne (Eds.), *The Wiley international handbook of clinical supervision*. Oxford, UK: John Wiley & Sons.

Stoltenberg, C. D., McNeill, B., & Delworth, U. (1998). *IDM supervision: An integrated developmental model of supervising counselors and therapists.* San Francisco, CA: Jossey-Bass.

Storm, C. L., & Sprenkle, D. H. (2002). The ivory tower and the institute. In T. Todd & C. Storm (Eds.), *The complete systemic supervisor: Context, philosophy, and pragmatics* (pp. 93–111). New York, NY: Authors Choice Press.

Storm, C. L., Todd, T. C., Sprenkle, D. H., & Morgan, M. M. (2001). Gaps between MFT supervision assumptions and common practice: Suggested best practices. *Journal of Marital and Family Therapy, 27*, 227–239.

Todd, T. C., & Storm, C. L. (2014). *The complete systemic supervisor: Context, philosophy, and pragmatics* (2nd ed.). London, UK: Wiley-Blackwell.

Walsh, F. (2012). Clinical views of family normality, health, and dysfunction: From a deficits to a strengths perspective. In F. Walsh (Ed.), *Normal family processes: Growing diversity and complexity* (4th ed., pp. 27–54). New York, NY: Guilford Press.

Woolley, S. R. (2010). Purposes, diversities, and futures in MFT doctoral education. *Journal of Marital and Family Therapy, 36*(3), 282–290.

FOURTEEN

Training the Contextual Therapist

Rikki Patton

Few guides exist that outline the use of contextual therapy theory as a supervision model for training systemic therapists. By sharing how I understand the integration of contextual therapy theory as a supervision model, I hope to contribute a practical interpretation of using contextual therapy theory as a guide for supervision. This model of supervision is based not only on prior research and conceptual literature regarding contextual theory as a therapeutic model, and on clinical supervision practice in general, but also on my own experiences of training as a contextual family therapist, being supervised by a contextual family therapist and supervisor, and supervising students using Contextual Family Therapy as a lens. To help guide the reader, this chapter includes an overview of contextual therapy theory and its application to supervision, the role of the supervisor and supervisee, and the application of this supervision model to the given case example.

TEACHING THEORETICAL CONTEXT

Founder(s)

Ivan Boszormenyi-Nagy, a trained psychiatrist, was the founder of Contextual Therapy (CT; Goldenthal, 1993). Coming from a family with three generations of judges (Ruzgyte, 2011), Nagy was forced to leave his home country of Hungary in 1948 due to his unwillingness to "submit" to political pressures and lived in Austria as a refugee for 2 years before moving to the United States in 1950 (Goldenthal, 1993; Ruzgyte, 2011). It was during the 1950s that Nagy began developing the contextual therapy model (Soyez, Tatrai, Broekaert, & Bracke, 2004). Relatively scant information was published on the topic until the 1980s, when many major books and articles on Contextual Therapy were published (Soyez et al., 2004; Wilburn-McCoy, 1993). Despite the increase of literature in the past few decades, a systematic conceptualization of using contextual therapy theory as a lens for clinical supervision is lacking. By understanding the philosophical foundations of the contextual model, a supervision model based on contextual therapy theory principles can be discerned.

Philosophical Foundation and Application to Supervision

The contextual therapy approach assumes an integrative, intergenerational stance, positing that both individual and relational realities constitute human existence (Boszormenyi-Nagy & Krasner, 1986; Goldenthal, 1993). The contextual approach assumes that individuals do not exist in isolation, but rather that people impact, and are impacted by, past and present relational factors (Ruzgyte, 2011). It is suggested that this focus on both the individual and the system sets the contextual model apart from other systemic models (Goldenthal, 1993). Hence, within this paradigm, the contextual supervisor attends to both individual and systemic processes within the supervision relationship.

In order to examine both individual and relational realities in the supervision relationship, the contextual supervisor should incorporate the four basic tenets of the contextual therapy model into the supervision approach: existential facts/biology, individual psychology, transactional relational patterns, and relational ethics. It is argued that understanding these four dimensions is crucial to understanding the meaning of one's experience (Boszormenyi-Nagy & Krasner, 1986); a contextual supervisor will not only help supervisees identify these four dimensions when working directly with clients, but will also aid supervisees in discerning how these four dimensions affect their clinical work and will explore how the basic tenets affect the supervisors themselves. Just as it is important to gather information about clients regarding the four dimensions inherent in CT, it is crucial that these four dimensions be considered in regard to the clients, the therapist, the supervisor, the client-therapist system, and the therapist-supervisor system, along with other contexts in which these systems exist. This conceptualization is similar to findings from White and Russell (1995), in which supervisor variables, supervisee variables, supervisor-supervisee relationship variables, supervisory interaction variables, and contextual variables must be considered as part of MFT supervision. Although it is easiest to explain each of the dimensions individually, it should be noted that the dimensions are highly interrelated (Goldenthal, 1996).

Dimension I: Existential Facts

Facts include the details—*facts*—about someone's life, such as biological, physical, and economic issues (Bernal, Rodriguez, & Diamond, 1990); preexisting factors (such as genetic makeup and family history); and unavoidable existential familial conflicts (including childrearing; Boszormenyi-Nagy & Krasner, 1986). These are considered objective measures that are difficult to change (Ruzgyte, 2011). Further, facts are considered important in guiding the therapeutic pathway (Boszormenyi-Nagy & Krasner, 1986).

Applying Dimension I to supervision: Existential facts affect supervision and, relatedly, clinician development. As supervisor, it is important to recognize the facts about the supervisee and how those facts may impact his or her role as therapist and supervisee. Contextual supervisors help trainees identify client facts as part of the assessment procedure. Additionally, trainees working with a contextual supervisor are encouraged to define their own existential facts and explore how those self-identifiers affect their relationships with clients, colleagues, and their supervisor as well as their own clinical development. Further, the supervisor needs to consider how the facts about himself or herself may affect the supervisory relationship and how the facts about the client may affect the therapist and supervisor; contextual supervisors are cognizant of their own self-identifiers and the role of those existential facts in the supervision relationship.

Existential facts include many diversity issues, such as gender, race/ethnicity, age, and multiculturalism. Therefore, it is necessary to examine the existential facts for each part of the supervision system (including the supervisor, trainee, and client) through a lens of multicultural competence. Prior research suggests that multicultural competence among supervisors affects the supervisor-supervisee alliance, supervision satisfaction, and supervisee multicultural competence (Inman, 2006), which underscores the need to account for how issues of diversity impact therapy and supervision within the contextual supervision model.

Dimension II: Individual Psychology

The contextual approach accounts for the self in the system by addressing individual psychology. This dimension incorporates an understanding of individual cognitive and emotional functioning for all members of the system (Goldenthal, 1993), including cognition, personal motivations, personality, coping styles, presence of mental illness, developmental stage, life goals, and motivations that affect people's lives (Boszormenyi-Nagy & Krasner, 1986; Goldenthal, 1993; Lyness, 2003).

Applying Dimension II to supervision: The cognitions, personal motivations, personality, coping styles, presence of mental illness, and developmental stage of each member of the supervision system (supervisor, supervisee, and clients) all affect the supervision relationship. Within a contextual supervision model, the contextual supervisor not only guides the trainee in identifying the individual psychology of each member of the client system being worked with, but also in taking note of his or her own psychological characteristics and how those characteristics affect how she or he approaches clinical work. Contextual supervisors also examine their own individual psychology and the ways in which those attributes may influence their supervisory work.

Dimension III: Transactional Relational Patterns

Transactional patterns incorporate concepts from general systems theory and classical family therapy to explain patterns that develop through the transactions within a system (Boszormenyi-Nagy & Krasner, 1986; Goldenthal, 1993).

Applying Dimension III to supervision: Contextual supervisors guide clinicians to identify the relational patterns present within the systems they are working with. Additionally, a supervisor using a contextual framework helps trainees identify relational patterns in their own lives and examine how those patterns may influence how they interact with clients. Further, contextual supervisors also explore how their own relational patterns and processes affect their role as supervisor.

Because family history is considered integral to development and change within the contextual approach, it will be important for the supervisor to be aware that potential family issues of the therapist may affect the supervisee both within the therapy room and in supervision. For instance, a supervisee whose mother was withdrawn and depressed may react differently to a client mother who expresses depressive symptoms compared to a client mother who is overly involved with her child. Similarly, the supervisor's own family history may influence his or her interactions with the supervisee or the interpretation of the client. Thus, accounting for the impact of family history on therapy, as appropriate for supervisory purposes, will be an integral part of supervision (Todd & Storm, 1997). It is also important for the contextual supervisor to observe transactions as they occur within all of the subsystems of the supervisory relationship. Effective supervision involves the creation of a safe place in which the supervisee can learn and grow as a trainee (Todd & Storm, 1997).

Dimension IV: Relational Ethics

The fourth dimension of contextual theory, relational ethics, is the foundational dimension of the contextual theory (Boszormenyi-Nagy & Krasner, 1986; Goldenthal, 1996). *Relational ethics* refers to "a basic existential human concern for fairness, reciprocity, and equitability" (Adams & Maynard, 2004). Thus, people develop an understanding of themselves in relation to their world based on the process of give and take within significant relationships. Although there are numerous concepts associated with relational ethics, which have been covered in detail elsewhere (Boszormenyi-Nagy, 1987; Boszormenyi-Nagy & Krasner, 1986; Goldenthal, 1993, 1996), the concepts of fairness, trust, loyalty, and entitlement play a particularly important role in the supervisory relationship and, thus, are discussed in detail here. Although there is significant overlap within the key variables, they are presented here separately for clarity in definition.

Relational fairness. Fairness is a core component that differentiates the contextual model from other integrative models (Gangamma, Bartle-Haring, &

Glebova, 2012); it is defined as the right to both give and receive in relationships (Goldenthal, 1993). Perceived unfairness is believed to be at the core of relationship distress (Boszormenyi-Nagy, Grunebaum, & Ulrich, 1991; Gangamma et al., 2012) and can lead to trust violations and the development of destructive entitlements (Ruzgyte, 2011).

Contextual supervisors aim to guide trainees to identify the perceived imbalance of give and take in clients' relationships in order to determine where to effectively intervene. Additionally, the contextual supervisor suggests that supervisees explore their own perceptions of fairness, both abstractly and within their own relationships, to determine how their own perceptions play a role in the therapeutic process. Further, the contextual supervisor pays attention to the balance of give and take within the supervisory relationship, noting that the amount of give and take will shift over time as the supervisee advances from a trainee to a peer. The contextual supervisor also considers the impact of his or her own perceptions and experiences of fairness in the supervisory relationship.

Trustworthiness. The presence of relational trustworthiness is considered the underlying premise of healthy relationships in contextual theory (Adams & Maynard, 2004; Boszormeny-Nagy & Krasner, 1986; Mauldin, 2003; Ruzgyte, 2011). Trustworthiness is a foundational tool that individuals use to learn what it means to relate to other people (Hargrave & Pfitzer, 2004) and is considered a prerequisite for the occurrence of authentic dialogue within relationships (Lyness, 2003).

Contextual supervisors help trainees identify intact and/or breached trustworthiness within their client systems. Supervisors working from a contextual lens also guide supervisees in exploring their own understanding of trustworthiness, and its role in relational fairness, so as to avoid potential projection into the therapeutic relationship. Further, the supervisor attends to her or his own experiences of trustworthiness in relationships and pays particular attention to maintenance and mutuality of trustworthiness in the supervisory relationship.

Fairness and trust in the supervision relationship should include elements of respect, safety, empathy, clear communication, and mutual understanding (Kaiser, 1992). A fair supervision relationship should account for the greater power and authority of the supervisor. It is important for the supervisor to create a safe space for the supervisee to be vulnerable within the supervision setting; this appropriate use of power transfers into a trusting supervisor-supervisee relationship.

Loyalty. Loyalty, as a relational ethic, involves the preference of one relationship over another (Boszormenyi-Nagy & Krasner, 1986; Goldenthal, 1993; Ruzgyte, 2011) and allows an individual to blame others for his or her misgivings (Goldenthal, 1993, 1996). When supervising through a contextual lens, the supervisor aims to guide the trainee in identifying the various loyalties present within the client system and within the trainee's own relationships. The contextual supervisor also attends to his or her own experiences of loyalty and pays attention to how loyalty may affect the supervisory relationship.

Entitlement. There are two dimensions of entitlement: constructive and destructive. *Constructive entitlement* is earned through (1) giving and (2) acting in a way that demonstrates taking accountability for how one's actions impact others (Goldenthal, 1993; Lyness, 2003; Ruzgyte, 2011). *Destructive entitlement* develops from perceived unfairness/injustice in one's own life and is defined as a lack of caring for other's needs and wants, which can include over-taking or over-giving behaviors (Goldenthal, 1993). As a result, individuals who display destructive entitlements do not take accountability for their own actions/inactions.

The contextual supervisor guides the trainee in identifying the entitlements, and related accountability issues, present in the client system. It is also the contextual supervisor's responsibility to role-model and guide supervisees toward applying accountability as an integral part of their clinical practice (Kaiser, 1992). Specifically, the supervisees must understand themselves with regard to entitlement and accountability and feel safe expressing that understanding; the supervisor aims to guide the trainee in this way (Kaiser, 1992). Finally, the contextual supervisor must explore his or her own entitlements and accountability and be cognizant of how the supervisory relationship is potentially affected by the supervisor's entitlements.

Applying Dimension IV to supervision. Of greatest importance within this philosophy of contextual supervision is the supervisors' accounting for and acknowledging the relational ethics of themselves, their supervisees, and the clients in the cases they are supervising. It is crucial that the supervisor process how his or her own relational ethics plays a role in the supervisor's ability to effectively supervise (Wosket, 2009). It is also important for supervisors to be able to discern the relational ethics of their supervisees and how the supervisees' ethics may be affecting their role as a therapist and as a supervisee. It is suggested that the supervision relationship should be based on an authentic, quality emotional connection and a strong working alliance between the supervisor and supervisee, thereby creating a milieu of trust and fairness (Wosket, 2009).

The supervision relationship is a place where issues related to relational ethics can emerge and be explored (Wosket, 2009), if clinically relevant. As the supervisor, one needs to be sensitive to behaviors and other communications by the supervisee that might suggest the need to attend to an ethical issue (Wosket, 2009). The supervision relationship should also incorporate specific components related to ethics, such as fairness, trust, loyalty, and entitlement.

Summary

While the focus of contextual theory centers around the relational ethics dimension, the four dimensions of relational reality are assumed to exist simultaneously, and all four dimensions are considered key in conceptualizing and intervening with client systems (Goldenthal, 1993) as well as with trainees within a supervision relationship. There are key components of relational ethics that play a particularly important role in the supervision relationship: fairness, trust, loyalty, and entitlement. The supervisor acts as a guide for the trainee in identifying the relational ethics within the client system, and within the trainee's own experience, in order to avoid projection of the trainee's own relational ethics onto the client system. The contextual supervisor also plays particular attention to the relational ethics present in the supervision relationship and attends to self-of-the-supervisor issues by being cognizant of his or her own relational ethics.

CONTEXTUAL SUPERVISION PROCESS

According to contextual theory, dysfunction occurs when there is an imbalance in relational fairness in current relationships, possibly due to unresolved imbalances from past relationships, including the family of origin (Wilburn-McCoy, 1993). When this fairness is imbalanced, trustworthiness diminishes and authentic dialogue within the system is often discarded. Change occurs by rebalancing fairness and re-establishing dialogue as a tool for rebuilding trust in the form of "talking, listening, and validating" (Wilburn-McCoy, 1993, p. 395). Based on these mechanisms of change, goals are focused on rebalancing fairness and trustworthiness through promoting authentic dialogue among family members (Goldenthal, 1993; Ruzgyte, 2011). Goals are set via the dialogue of the family members; the therapist does not set the goals but simply guides the dialogue so that the family can determine its own goals (Ruzgyte, 2011).

The contextual supervisor guides the trainee in identifying the perceived imbalance in fairness and deciding how to intervene to rebalance fairness and rebuild trust by guiding clients toward a fair and just relationship. Within the supervision relationship, the contextual supervisor attends to perceptions of unfairness and lack of trustworthiness displayed by the supervisee or by the supervisor and addresses those perceptions with regard to the supervision relationship.

Role of the Supervisor and Supervisee

"Contextual therapists have only recently begun to articulate guidelines and clinical techniques that would help clinicians translate theory into practice."
—Adams & Maynard, 2004

Specific frameworks for contextual supervision are sparse; only two publications were found that mentioned supervision and Contextual Therapy/relational ethics. Kaiser (1992) applied general supervisor-supervisee relationship elements specifically to Contextual Therapy. These elements included self-of-the-therapist issues, such as accountability and personal awareness, trust, and power and authority. Although Kaiser's work can be used as an initial guide to examine supervision within the contextual framework, Kaiser focused on fitting contextual components within a general supervision paradigm, thereby missing the incorporation of the four dimensions of relational ethics into the supervision model.

Wosket (2009) contributed a chapter titled "Relational Ethics in Supervision" to a book examining relational ethics within the clinical context. Using numerous case examples of supervisee relationships, Wosket was able to exemplify how relational ethics, broadly defined, might look in a supervisor-supervisee relationship. However, Wosket's work did not use relational ethics specifically as defined within the Contextual Therapy paradigm and therefore does not offer much guidance on how to apply specific ethical constructs viewed through the Contextual Therapy lens, or Contextual Therapy as a larger framework to the supervision relationship. Though Kaiser's and Wosket's work both have limitations, their presuppositions will be incorporated as part of the current conceptualization of supervision through a contextual lens.

A misconception about contextual theory, and the relational ethics dimension in particular, is the assumption that the therapist will drive the client toward a moralistic notion of right and wrong. In actuality, the role of the contextual therapist is simply to guide the clients toward considering their own ethics and, consequently, consideration of the role of fairness in their relationships (Goldenthal, 1993; Ruzgyte, 2011). The contextual supervisor's job is to affirm the notion that the therapist's role is that of a guide, not a moral compass. The contextual supervisor aims to help trainees understand their role by helping the trainees understand the following components of the contextual therapist's role, as outlined by Ruzgyte (2011): (1) limiting therapist goals to the perceptions of fairness and needs/wants of the client; (2) using multidirected partiality to show empathy to each member of the system who is affected by the therapy process, including those who are not in the room; (3) involving all family members in the session, with the goal of having each person recognize and share his or her own relational stance in the system and listen to the response of other members; (4) parsing out the four dimensions of relational reality for the client system, focusing on relational ethics; (5) refraining from judgment; (5) being active and precise, rather than prescriptive or directive, in guiding clients; and (6) role-modeling listening to the clients, thereby inviting them to think about issues of fairness and trustworthiness (Ruzgyte, 2011). The contextual supervisor should also incorporate these components into how he or she approaches the role of supervisor by limiting supervision goals to the trainee's perceptions of fairness, using multidirected partiality, parsing out the supervisee's relational reality, being active and precise instead of prescriptive, and role-modeling listening.

Techniques to Be Used

Techniques specific to contextual theory include multidirected partiality, developing a trustworthy relationship, assessing fairness issues, and acknowledgment (Goldenthal, 1993). Each of these techniques should be taught to trainees and used by contextual supervisors within the supervision relationship.

Multidirected Partiality

Multidirected partiality is the main tool that a contextual therapist (and a contextual supervisor) has in the proverbial toolbox. It differs from therapeutic neutrality and unidirected partiality in that the therapist aims to understand each system member's perspective, including all those who may be affected through the course of therapy (Goldenthal, 1993). Multidirected partiality includes a multitude of interventions within this technique, including empathy, crediting rather than blaming, lending weight to less empowered members, giving room for members to spontaneously acknowledge, taking accountability, and acknowledgment (Goldenthal, 1993; Ruzgyte, 2011). As a key component of contextual supervision, just as within Contextual Therapy, the contextual supervisor aims to model multidirected partiality in the supervisory relationship.

Further, the supervisor uses multidirected partiality in supervision to promote a trustworthy relational environment by (1) modeling a nonjudgmental view of each member of the system, (2) being accepting and transparent about observations of the supervisee's own countertransference issues, and (3) modeling how to hold every member of the system accountable (Kaiser, 1992).

Developing a Trustworthy Relationship

Developing a trustworthy relationship is akin to building a therapeutic alliance. The contextual supervisor guides the trainee to understand that in order to build a trustworthy relationship, the therapist does not ask clients to just "trust" in the process; rather, in the beginning of the therapeutic relationship, the therapist gathers information and assesses for the appropriateness of treatment, while encouraging the clients to use the time to conduct their own assessment regarding the appropriateness of treatment for them (Goldenthal, 1993). A similar process should be modeled within the supervision relationship.

Assessing Fairness Issues

Assessing for fairness issues includes exploring perceived injustice and destructive entitlements (Goldenthal, 1993). Using a genogram to map out issues of fairness in past and current relationships can allow the therapist a better conceptualization of the imbalances in fairness and potentially guide the clients toward a better understanding of their own relational ethics in context (Ruzgyte, 2011; Wilburn-McCoy, 1993). Similarly, through supervision, the contextual supervisor pays particular attention to perceived fairness within the supervisory relationship, the therapist-client relationship, and the client's own system. The contextual supervisor aims to role-model assessment of fairness issues within these multiple relationships.

Acknowledgment

Acknowledgment is defined as recognizing and acknowledging the giving within the relationship and giving credit to the other person. Acknowledgment takes two forms: acknowledging a considerate act and acknowledging an injustice. By acknowledging one another, members within a client system can begin working toward a trustworthy relationship (Goldenthal, 1993). The contextual supervisor role-models the use of acknowledgment through interacting with the supervisee and guides the supervisee toward role-modeling acknowledgment for the clients.

Culture/Race and Gender Issues/Research to Consider

Evidence drawn from the research available suggests that contextual theory as a therapeutic model has been applied to a variety of populations and presenting problems, including substance-using populations (Bernal et al., 1990; Soyez et al., 2004); couples therapy processes (Adams & Maynard, 2004; Gangamma et al., 2012; Mauldin, 2003); sexual difficulties (Goldenthal, 1996) and couple infidelity (Mauldin, 2003); relationship conflict and resolution (Beckenbach, Patrick, & Sells, 2010); postdivorce families (Cole & Cole, 1999); marital satisfaction (Gangamma et al., 2012; Grames, Miller, Robinson, Higgins, & Hinton, 2008); grandparents raising grandchildren (Brown-Standridge & Floyd, 2000); families making end-of-life decisions (Foster & McLellan, 2002); trauma survivors (Gold, 2000); families who experienced homicide (Temple, 1997); mental health issues (Grames et al., 2008), physical health problems (Grames et al., 2008); and examination of earned constructive entitlement (Horowitz, 2009). While there are numerous articles covering varied topics and populations using contextual theory in therapy, guidance for directly applying contextual theory within the supervision relationship, including implications of diversity in supervision, is lacking.

Further, little work is available that has explicitly examined culture/race and gender issues through the Contextual Therapy lens. In a study examining the impact of relational ethics on relationship satisfaction among couples, findings showed that the relationship between fairness and relationship satisfaction is affected by gender (see Gangamma et al., 2012). Although Gangamma et al.'s findings suggest that there are gender issues to consider within Contextual Therapy, more work is needed to further examine other diversity issues through this lens in both therapy and supervision.

The overall lack of guidance regarding contextual theory and diversity issues has several implications for clinical supervision through the contextual lens. First, supervisors need to help supervisees understand the limitations of the contextual model, due to the lack of information about its utility with diverse populations. Additionally, supervisors need to attend to the multiple levels of diversity issues present within the supervisory relationship by exploring the impact of diversity issues that are presented within client systems, from the supervisee, from the supervisor himself or herself, and the interactions among these different systems. Wosket (2009) suggests that supervisors can use the conceptualization of relational ethics in supervision as a tool to guide supervisees in working with issues of difference, which is a global way in which Contextual Therapy supervision can incorporate diversity issues into the training process.

Learning Goals/Objectives

The goals for training a contextual therapist mirror the goals put forth for the therapeutic relationship. As therapist, the goal is to help the client system rebalance fairness and trust in the relationship (Wilburn-McCoy, 1993). Fairness and trustworthiness are gained through promoting authentic dialogue among family members (Goldenthal, 1993; Ruzgyte, 2011), which is achieved by using guiding questions rather than offering directives (Ruzgyte, 2011). Supervision using a Contextual Therapy theory lens should also be strengths-based, with a focus on guiding supervisees toward understanding of their role as therapists and competence as therapists. The supervisor asks supervisees guiding questions about the therapy process, rather than offering straightforward directives, in order to help supervisees better understand the client and themselves (Wosket, 2009).

CASE EXAMPLE

In the case example shared in this book, Dr. Bob is an AAMFT Approved Supervisor with more than 20 years' experience supervising students and clinicians. If Dr. Bob were training using a Contextual Therapy framework for supervision, he would likely be attuned to his own relational ethics and how those ethics affect how he understands supervisees as well as client systems.

(*continued*)

Dr. Bob would also be able to conceptualize the relational reality of the supervisee, Brittney, through the ongoing supervisory relationship. One of Dr. Bob's initial goals while working with Brittney would include engaging in authentic dialogue with Brittney to gather information about her relational reality as a tool for understanding how her own facts, individual psychology, transactional patterns, and relational ethics affect both the supervision relationship and her work with clients. A substantial part of gathering such information would include questions guiding Brittney toward insight about her relational reality and how such reality affects her role as therapist and supervisee.

Dr. Bob, as supervisor, would also want to clearly conceptualize each client system through the contextual lens. By understanding the client system through the contextual lens, the supervisor would be better able to provide guidance and feedback to the supervisee for working with and conceptualizing each client system. More importantly, though, the supervisor wants to guide the supervisee toward independent conceptualization of each client system through the contextual lens by asking guiding questions that help the supervisee connect the proverbial dots between each dimension of relational reality and how that information can be used to intervene in the presenting problem. Following include some considerations related to information from the case example.

Facts. The case example presents biological and racial/cultural facts about the supervisor, supervisee, and client system. Dr. Bob is a 53-year-old Caucasian male who is supervising Brittney, a 33-year-old Caucasian female. Brittany is seeking supervision for work with a family system that includes Travis and Tekia, ages 29 and 27, respectively, an African American couple with two young children. The intersections of gender, age, and race in the supervisor-supervisee-client system should be considered. As Goldenthal (1993) states, clearly conceptualizing the facts for each member in the system can help clarify how relational ethics affect current relationships. In this case, the impact of power and related unfairness, injustice, or perceived untrustworthiness, as it relates to diversity issues such as gender, age, and race, should be considered as potentially influential on the overall supervisory and therapeutic relationship.

In addition, the biological fact of Travis experiencing severe injury while deployed is an important fact to consider, as it directly impacted other facts in his life, including being discharged from the military and lacking full-time work. These facts likely relate to the

(*continued*)

other three dimensions of relational reality for Travis, his partner, and his children—the family system. Further, the fact that his older brother is an active military member who is currently deployed may relate to the other dimensions as well. The supervisor's and supervisee's connections to the military, via their own involvement or through relationships with active or veteran military members, may have to be explored, as the facts from Dr. Bob and Brittney may intersect with the family's fact of military involvement. Furthermore, it is a fact for this system that Travis and Tekia have two young children. Although the parenting status of either Dr. Bob or Brittney is currently unknown, the facts from the client and therapist may intersect and lead to possible misperceptions by either part of the therapeutic system.

While many facts can be identified from the case example, Dr. Bob would encourage Brittney to incorporate a more comprehensive assessment structure to gather more information about other facts related to this system (e.g., relationship status, personal history, etc.). In addition to a more thorough assessment, Dr. Bob would guide Brittney to (1) identify the multitude of facts that the family presents with; (2) recognize her own facts and how the client system may construe them; (3) consider how her own facts may be related to her own relational ethics and, consequently, how she views the family system; and (4) conceptualize how the client's facts may interact with the other dimensions of their lives, especially relational ethics.

Individual psychology. Little information about individual psychology is provided as part of the case example; the only direct evidence regarding individual cognitive or emotional functioning for the members of the system was related to Travis's "short fuse." Hence, Dr. Bob would guide Brittney toward incorporating questions about individual psychology into a more thorough assessment process.

Despite the lack of direct reports related to the individual psychology dimension, information was provided within the case example that Dr. Bob could help Brittney recognize as possible avenues worth more exploration within this domain. For instance, Travis's involvement in active duty, and resulting injury, could be related to possible mental health issues, such as post-traumatic stress disorder, as evidenced in prior literature (e.g., Sabella, 2012). His premature discharge from the military as a result of his injury could be related to feelings of loss and grief, which have been connected to depression and post-traumatic stress disorder in past research (e.g., Copeland et al., 2011; Mojtabai, Rosenheck, Wyatt, & Susser, 2003). Tekia's role as a military wife, and the change from her husband being deployed

(*continued*)

to being home, could impact her individual psychology (e.g., Whitt, Cabrera, McGurk, & At, 2008). Further, the presence of a chronic medical issue has been shown to affect individual and couple functioning (e.g., Martire, Schulz, Helgeson, Small, & Saghafi, 2010).

Information about Dr. Bob or Brittney regarding this dimension was also missing from the case example. It would be important for Dr. Bob to continually reflect on his own individual psychology and how it may affect the supervisory relationship. Further, it would be important for Dr. Bob to help guide Brittney toward (1) identifying the individual psychology of all family members; (2) identifying how her own individual psychology may affect her work with this client system; (3) recognizing how her own individual psychology may be construed by the client; and (4) conceptualizing how the clients' individual psychologies are related to the other dimensions of relational reality.

Family transactions and power. Based on the client report that the presenting problem includes couple and parenting issues, it is suggested that this client system may be experiencing an imbalance in family transactional patterns, such as communication patterns, boundaries, and family roles and rules. As with the facts and individual psychology dimensions, Dr. Bob would guide Brittney in examining the systemic process within this family to get a better picture of the transactional patterns and issues of power within the client system. Dr. Bob would also ask Brittney guiding questions to help her identify her own experiences and perceptions of transactional patterns within her own relationships and how those experiences may affect how she conceptualizes her client system.

Relational ethics. Dr. Bob will approach supervision having already explored (and continuing to assess) his own relational ethics and the potential impact of those ethics on the supervisory relationship and his conceptualization of both supervision and the client system. By having some awareness about his own relational ethics, Dr. Bob will be better able to bracket his own biases and be better able to engage in authentic dialogue with Brittney; thus, he can offer her more neutral guidance toward self-of-the-therapist insight and client conceptualization through a contextual lens.

Connecting the dots. In addition to helping Brittney identify the four dimensions of relational reality in the client system, and within herself, Dr. Bob also needs to guide Brittney toward using the information gathered about the four dimensions to develop a treatment

(*continued*)

plan to assist the client system to move away from perceived unfairness and lack of trustworthiness toward viewing their relationships as fair and trustworthy. Dr. Bob will role-model the use of guiding questions, multidirected partiality, and acknowledgment within the supervision relationship and will guide Brittney to use these techniques directly with the client system. The overall goals for Dr. Bob include (1) guiding Brittney toward insight in how components of contextual theory play out in her own life in order to bracket potential projection into the therapeutic relationship; and (2) guiding Brittney toward confident use of the contextual model to assess and intervene with the client system. In doing so, Dr. Bob is helping Brittney develop into a skilled contextual therapist.

REFERENCES

Adams, J. F., & Maynard, P. E. (2004). Contextual therapy: Applying the family ledger to couple therapy. *Journal of Couple & Relationship Therapy, 3*, 1–11.

Beckenbach, J., Patrick, S., & Sells, J. (2010). Relationship conflict and restoration model: A preliminary exploration of concepts and therapeutic utility. *Contemporary Family Therapy, 32*(3), 290–301.

Bernal, G., Rodriguez, C., & Diamond, G. (1990). Contextual therapy: Brief treatment of an addict and spouse. *Family Process, 29*, 59–71.

Boszormenyi-Nagy, I. (1987). *Foundations of contextual therapy. Collected papers of Ivan Boszormenyi-Nagy.* New York, NY: Brunner.

Boszormenyi-Nagy, I., Grunebaum, J., & Ulrich, D. (1991). Contextual therapy. *Handbook of Family Therapy, 2*, 200–238.

Boszormenyi-Nagy, I., & Krasner, B. (1986). *Between give and take.* New York, NY: Brunner.

Brown-Standridge, M. D., & Floyd, C. W. (2000). Healing bittersweet legacies: Revisiting contextual family therapy for grandparents raising grandchildren in crisis. *Journal of Marital and Family Therapy, 26*(2), 185–197.

Cole, C. L., & Cole, A. L. (1999). Boundary ambiguities that bind former spouses together after the children leave home in post-divorce families. *Family Relations, 48*(3), 271.

Copeland, L. A., Zeber, J. E., Bingham, M. O., Pugh, M. J., Noël, P. H., Schmacker, E. R., & Lawrence, V. A. (2011). Transition from military to VHA care: Psychiatric health services for Iraq/Afghanistan combat-wounded. *Journal of Affective Disorders, 130*(1), 226–230.

Foster, L. W., & McLellan, L. J. (2002). Translating psychosocial insight into ethical discussions supportive of families in end-of-life decision-making. *Social Work in Health Care, 35*(3), 37–51.

Gangamma, R., Bartle-Haring, S., & Glebova, T. (2012). A study of contextual therapy theory's relational ethics in couples in therapy. *Family Relations, 61*, 825–835.

Gold, S. N. (2000). *Not trauma alone: Therapy for child abuse survivors in family and social context.* Philadelphia, PA: Brunner/Routledge.

Goldenthal, P. (1993). *Contextual family therapy: Assessment and intervention procedures.* Sarasota, FL: Professional Resource Press.

Goldenthal, P. (1996). *Doing contextual therapy: An integrated model for working with individuals, couples and families.* New York, NY: Norton.

Grames, H. A., Miller, R. B., Robinson, W. D., Higgins, D. J., & Hinton, W. J. (2008). A test of contextual theory: The relationship among relational ethics, marital satisfaction, health problems, and depression. *Contemporary Family Therapy, 30*(4), 183–198.

Hargrave, T. D., & Pfitzer, F. (2004). *The new contextual therapy: Guiding the power of give and take.* New York, NY: Brunner/Routledge.

Horowitz, H. (2009). The healing power of giving: A contextual therapy case study. *Journal of Spirituality in Mental Health, 11*(3), 213–217.

Inman, A. G. (2006). Supervisor multicultural competence and its relation to supervisory process and outcome. *Journal of Marital and Family Therapy, 32,* 73–85.

Kaiser, T. L. (1992). The supervisory relationship: An identification of the primary elements in the relationship and an application of two theories of ethical relationships. *Journal of Marital and Family Therapy, 18*(3), 283–296.

Lyness, K. P. (2003). Extending emotionally focused therapy for couples to the contextual realm: Emotionally focused contextual therapy. *Journal of Couple & Relationship Therapy, 2*(4), 19–32.

Martire, L. M., Schulz, R., Helgeson, V. S., Small, B. J., & Saghafi, E. M. (2010). Review and meta-analysis of couple-oriented interventions for chronic illness. *Annals of Behavioral Medicine, 40*(3), 325–342.

Mauldin, G. R. (2003). "Forgive and forget": A case example of contextual marital therapy. *The Family Journal, 11*(2), 180–184.

Mojtabai, R., Rosenheck, R. A., Wyatt, R. J., & Susser, E. S. (2003). Use of VA aftercare following military discharge among patients with serious mental disorders. *Psychiatric Services, 54*(3), 383–388.

Ruzgyte, E. (2011). Contextual Family Therapy. In Metcalf, L. *Marriage and family therapy: A practice-oriented approach.* Springer Publishing Company.

Sabella, D. (2012). PTSD among our returning veterans. *American Journal of Nursing, 112,* 48–52.

Soyez, V., Tatrai, H., Broekaert, E., & Bracke, R. (2004). The implementation of contextual therapy in the therapeutic community for substance abusers: A case study. *Journal of Family Therapy, 26,* 286–305.

Temple, S. (1997). Treating inner-city families of homicide victims: A contextually oriented approach. *Family Process, 36*(2), 133–149.

Todd, T. C., & Storm, C. L. (1997). *The complete systemic supervisor: Context, philosophy, and pragmatics.* Lincoln, __: Allyn and Bacon.

White, M. B., & Russell, C. S. (1995). The essential elements of supervisory systems: A modified Delphi study. *Journal of Marital and Family Therapy, 21*(1), 33–53.

Whitt, A. A., Cabrera, C. O. A., McGurk, M. D., & At, M. S. U. S. (2008). Prevalence of mental health problems, treatment need, and barriers to care among primary care-seeking spouses of military service members involved in Iraq and Afghanistan deployments. *Military Medicine, 173,* 1051.

Wilburn-McCoy, C. (1993). Rediscovering Nagy: What happened to contextual therapy? *Contemporary Family Therapy, 15,* 395–404.

Wosket, V. (2009). Relational ethics in supervision. In L. Gabriel & R. Casemore (Eds.), *Relational ethics in practice: Narratives from counselling and psychotherapy* (pp. 44–57). New York, NY: Routledge.

FIFTEEN

Training the Narrative Therapist

Jerry Gale
D. Bruce Ross III
Megan A. McCoy

POSITIONING THE AUTHORS

In composing a chapter about Narrative Therapy, it is appropriate to first begin by situating ourselves to the reader. This information is included so as to be transparent about effects in our lives that have helped us write this chapter as well as to note our own limitations and constraints.

Jerry Gale has self-identified as a family therapist for more than 35 years. Influenced by many traditions and leaders in the field, Jerry did a week-long training with Michael White in 1999. This was very significant to him and led to an ongoing appreciation of and sensitivity to issues of social justice. Jerry himself has benefited from many contexts of privilege, including being Caucasian, heterosexual, raised in a middle-class family, and employed as an academic. While personally beneficial, these privileges also blinded him to the challenges and inequities many people face.

D. Bruce Ross has been a marriage and family therapist for the past 5 years. He completed a master's degree at the University of Maryland in its Couples and Family Therapy program, where there was an emphasis on understanding and addressing how diversity and social discourses play a role in our lives and therapeutic experience. He continued this experience at the University of Georgia, earning a doctorate in Human Development and Family Science with an emphasis in Marriage and Family Therapy. Bruce works to understand the systemic discourses affecting clients' lives through his work as a traditional family therapist and a supervisor in training as well as a financial practitioner for low-income populations.

Megan McCoy has been doing therapy for almost a decade. She was trained in the Couple and Family Therapy program at Drexel University, a program known for its emphasis on social justice. This focus on exploring dominant discourses and their impact on all lives continued into her doctoral education in the University of Georgia. Megan is a strong practitioner of narrative theory in the therapy room. However, she recently had her first foray into the world of supervision through the doctoral program and was fascinated with how her narrative therapy skills translated into the supervision room. She was also surprised how becoming a supervisor-in-training made her aware of her position of privilege as a supervisor and doctoral candidate.

AN INTRODUCTION TO NARRATIVE THERAPY

Narrative Therapy was first introduced in the early writings of Michael White and David Epston in the early 1990s (White, 2007; White & Epston, 1990, 1992). Narrative Therapy, in its beginning, initially drew from a cybernetic framework, as White was strongly influenced by Gregory Bateson (1972) and his writings about constraints. Epston came to the narrative approach through his study of the narrative metaphor in anthropology (Epston, 1989; White & Epston, 1990). The expression of the narrative metaphor, which became an integral aspect of Narrative Therapy, was a radical departure from other clinical approaches. Rather than focusing on the problem as psychologically within the client and treating the person as something to solve or fix, Narrative Therapy works to help clients achieve new narratives of their lives and relationships. Narrative theory (Brunner, 1994) presents stories as the performance of identity, not as representative of identity. This is a key point: It means that talk and interactions are not viewed as representative or as corresponding to internal psychological constructs or schemas. Rather, our stories and our telling of these stories are generative of the meaning of our experiential states of being (Gale, 2010).

This is both a collaborative and a directive process. While the therapist and client(s) co-construct the preferred narrative of the clients' lives, the clinician also carefully attends to what is said and not said by the clients in order to guide the conversation with questions. These questions explore such things as: (a) how the person came to particular understandings of his or her situation; (b) finding out who in the person's life are influential in supporting new narratives; (c) finding sparkling moments when the person did experience different narratives of being; and (d) finding times when the person resisted dominant narratives presenting particular norms of identity. Narrative Therapy is premised on a poststructural view that identity is constantly being created, challenged, or changed within the meanings imposed through cultural and societal discourses. When we are born, we enter into the world of these cultural discourses, which present norms, values, and preferred and dis-preferred ways of living our lives. The narratives that we say to others and to ourselves are not simple accounts of events, but rather organize and define our identity. A saying of Narrative Therapy is "the client is not the problem, the problem is the problem." This shift of narrative is not simply a rhetorical device, but rather serves to change the client's phenomenological experience. It is the narrative therapist's work to unpack the cultural and societal influences to separate the problem from the person, creating space for the story (and thereby the identity of the client) to become thicker and more multi-stranded (Freedman & Combs, 1996). Thicker and multi-stranded stories mean more complexity of details and description and characterization of participants and provide richer reflections on the past and on future possibilities. Although Narrative Therapy has evolved over the past 25 years, this remains a foundational core of the approach.

Narrative Therapy has been a very popular technique for years. As stated in Etchison and Kleist (2000), "Hevern (1999) reported more than 2,000 bibliographic narrative therapy resources of scholarly articles, book chapters and full texts, and doctoral dissertations, and this number has only grown over the past 15 years" (p. 61). Despite its popularity, until recently it has been difficult for Narrative Therapy to provide empirical support through randomized controlled research. There are several reasons for this reluctance. First, the postmodern wave of the early 1990s, of which Narrative Therapy was a part, was mistrusting of research claims presenting grand narratives (with a level of certainty) explaining human behavior and reducing human experience to numeric generalizations (Strong & Gale, 2013). A second reason was the challenge of achieving valid randomized controlled trials, which require manualization and controlled replication by independent investigators. This is a difficult process for Narrative Therapy, as it is language-based, client-directed, and focused on relational process rather than step-by-step operational techniques (Larner, 2004). However, this skepticism of research has begun to shift with the development of research that is contextually sensitive, a continued refinement of narrative models, and clinical manuals that are more flexible and context sensitive (Strong & Gale, 2013).

In 2000, Etchison and Kleist highlighted the dearth of literature by reviewing the current state of Narrative Therapy. They found empirical support for Narrative Therapy in four studies. For example, Besa's (1994) use of a single-case research design demonstrates effectiveness in reducing parent–child conflict for five out of six families. The families that showed improvement ranged from an 88% to a 98% decrease in parent-child conflicts with Narrative Therapy. No improvements were observed in the absence of Narrative Therapy. In addition, St. James-O'Connor, Meakes, Pickering, and Schuman (1997) examined families' perceptions of their Narrative Therapy experience and the meaning that these families attributed to this experience. The eight families who participated in the study reported some reduction in the presenting problem. The reduction of problems was greater in families involved in Narrative Therapy for longer periods than in families involved for shorter periods. The authors conclude that the results support the view of Narrative Therapy as empowering personal agency in family members.

Since this review was published, more empirical support has been published. Weber, Davis, and McPhie (2006) conducted a study on narrative interventions with seven women who identified themselves as experiencing depression as well as an eating disorder. A comparison of pre- and post-group tests demonstrates a reduction in depression scores and eating disorder risk. Furthermore, all the participants reported a change in daily practices and less self-criticism. Vromans and Schweitzer (2011) investigated depressive symptom and interpersonal relatedness outcomes from eight sessions of manualized Narrative Therapy for 47 adults with major depressive disorder. After therapy, depressive symptom improvement was seen ($d = 1.36$) and 74% of the clients achieved reliable improvement, 61% moved to the functional population, and more than half showed clinically significant improvement (53%). These findings are comparable to benchmark research outcomes. Lopes et al. (2014) completed a study with 63 depressed clients who were assigned to either Narrative Therapy or Cognitive Behavioral Therapy. The authors find a significant symptomatic reduction and pre- to post-treatment effect sizes for both groups superior to the wait-listed control groups.

Despite the effectiveness of the narrative approach in therapy, the literature show a deficit on how to supervise a narrative therapist when working with a narrative lens. There are numerous methods of conducting supervision but few concrete guidelines on what constitutes "good" supervision (Anderson, Schlossberg, & Rigazio-DiGilio, 2000). Most supervisors have tended to apply their own model of therapy to how they supervise (Clarkin & Glick, 1982; Nichols, Nichols, & Hardy, 1990), with the underlying assumption that the supervision process is equivalent or parallel to the therapeutic process. However, the assumption that supervision is isomorphic to therapy and can contribute to supervision efficacy has not received any empirical support (Storm, Todd, Sprenkle, & Morgan, 2001). The purpose of this chapter is to provide perspective on the supervisory process from a narrative framework through a discussion of Narrative Therapy in the current contextual environment; it also discusses how to incorporate a poststructural approach in supervision sessions; how to engage therapists-in-training in conversations about the impact of societal discourses, context, and issues of privilege and diversity; and provision of supervisory techniques to engage narrative therapists-in-training to be effective practitioners. It is not a step-by-step recipe of how to do "narrative supervision," but rather provides a framework for approaching supervision from this theoretical standpoint.

NARRATIVE THERAPY IN TODAY'S WORLD

The uniqueness of Narrative Therapy in today's society is that it emphasizes and brings in a different perspective about looking at and situating human beings in their environment. Building on Foucault's ideas about relational power, there is always a tension in the construction of meaning, identities, and agency. We are always situated within cultural contexts in which discourses are shaping and constraining personal and interpersonal narratives.

With a strong appreciation for social justice, Narrative Therapy does attend to inequities of gender, race, economics, sexual orientation, and more. The ways most people in a society hold, talk about, and act on a common, shared viewpoint are part of and sustain the prevailing discourses (Hare-Mustin, 1994). These discourses reflect the ideologies of those who have the most power in society (Combs & Freedman, 2012). Hare-Mustin (1994) highlights some of the different discourses that may be negative (e.g., "men are not in control of their sexual impulses" [p. 24]) or sometimes even contradictory (e.g., "marriage-between-equals discourse that conceals male domination" [p. 29]). However, the narrative therapeutic approach seeks to disentangle the societal and cultural discourses that start to shape our identity and separate them from constraining our identities.

The development of the preferred narrative allows individuals to not accept the societal definition of their identity and start to reclaim their own narratives, voices, and meaningful identities. This provides a different approach for understanding individuals, families, and communities through the lens of social justice. Understanding the meaning and the context behind the language that clients, therapists, and supervisors use is an important aspect of the narrative process. Consequently, helping therapists understand and find the preferred language to use with clients is crucial to the narrative therapeutic process. For instance, language such as homosexuality and bisexuality has contextual roots in the discourse of psychopathology, and thus may have negative connotations for the client (Neal, 2000). In today's culture, language such as "trans," "queer," or "gay" may be more readily used or even preferred; however, it is important to understand the meaning-making behind the language used for the client(s), therapist, and supervisor. Narrative Therapy provides space for clients to redefine the discourses and challenge the ideologies of those who hold the power in our society.

A parallel process occurs between the narrative therapist-client relationship and the narrative supervisor-therapist relationship. As the therapist helps the clients to disentangle the societal discourses afflicting their problem story, so too does the supervisor help the therapists to understand and realize the impact of their own identity stories and the prevailing societal and cultural discourses shaping their thoughts, behaviors, and values. This parallel process requires both the therapist and supervisor to be mindful and responsive to the various contexts that influence the course of the narrative therapeutic process. As Rachel Hare-Mustin (1994) discusses in her seminal paper, the narratives in the therapy room are not just the words spoken, but also the cultural discourses shaping the clients' understanding and actions. The challenge for the clinician is not only to hear what is said in session, but also to attend to unspoken cultural discourses shaping these stories. These discourses can play out at different levels of context, including: self-of-the-individual (both client and clinician experiences of such things as race, sexual orientation, gender bias, etc.); the relationship between therapist and supervisor, and client(s)/therapist/supervisor (relational power, gatekeeping, limiting preferred stories, etc.); the institutional setting (e.g., the norms, rules, and policies of private practice, agency, university, nonprofit, etc.); issues of the profession and state (e.g., ethical and legal codes, policy, etc.); the local community level (racism, violence, incarceration, unemployment, etc.); cultural level (e.g., norms and values around gender, race, sexual orientation); and even cross-cultural and international influences that may be at play. It is helpful to reflect upon the level(s) at which all participants may be impacted or constrained in the generation of alternative narratives and address the congruencies, contradictions, and isomorphism between levels in order to uncover discourses and relational power dynamics that may be at play.

CONSIDERATIONS BETWEEN A NARRATIVE THERAPIST AND TRAINING A NARRATIVE THERAPIST

An important consideration in conducting narrative supervision is the therapist's relationship to the supervisor. Just as the relational stance and the meaning derived from it are integral to the effectiveness of therapy for a narrative therapist, the way in which our cultural, societal, and family narratives influence our perspectives is just as important in the supervisor-trainee relationship. The meaning attached to these perspectives position supervisor and training therapist alike in the relationship and the working alliance that is created. Consequently, the narrative supervisor must be just as aware of his or her own narratives and how they shape the supervisory experience and direction as well as the relationship between supervisor and trainee. As a supervisor, it is critical to maintain multiple perspectives of how each person's understanding can open or constrain the narratives of the clients and narratives of how to proceed.

Understanding the Role of the Narrative Therapist in Training

The process of supervising and training a narrative therapist parallels the process in the therapist-client relationship. Supervisors should adopt a co-constructionist perspective, viewing the therapist as an expert in his or her own life and abilities, while also offering guidance for the clinician to create his or her own narrative as an effective therapist. Consequently, the narrative supervisor will ask many guiding questions, rather than predominately providing answers. The supervisor may be directive but more so for helping the therapist expand his or her narrative rather than for imposing an expert knowledge of prescribed meaning or behaviors. Narrative questioning has the intent of uncovering meaning and generating experience rather than imposing prescriptive information. Narrative therapists use different kinds of questions to achieve these effects (Freedman & Combs, 1996). Several books and articles have been written exploring the use of narrative questions. The list of types of narrative questions has grown in response (for more information, see McCoy, Stinson, Bermudez, & Gladney, 2013; Shapiro & Ross, 2002). Yet, the crux of narrative questions is to deconstruct the dominant discourses that are internalized within the clients so they can externalize the problem and fight against the problem (McCoy et al., 2013). Within the supervisory context, the goal of narrative questions is to deconstruct the discourses that are prohibiting the supervisee from seeing the larger clinical picture beyond the presenting internalized issues.

Special consideration should be applied to issues of privilege, self-of-the-therapist, self-of-the-supervisor, and collaborative factors. For instance, questions that are asked around privilege may include: Whose voice is being privileged in the therapist-client relationship? And what effect does that privilege have on the relationship and therapeutic process? Questions addressing self-of-the-therapist and self-of-the-supervisor include: How does your own narrative, as a therapist or individual, affect the process of therapy? And how does your own narrative affect the client's narrative? Questions around collaboration may include: How are you, as a therapist, fostering collaboration between you and the client? Who are you collaborating with in session? Whose voice is not being heard in session? What effect does this collaboration have on the therapeutic process? Suggesting this line of questioning allows the narrative therapists-in-training to understand their own impact on the therapeutic process and the implications of their respective narratives, not only on how they conduct therapy, but also on how it guides their own lines of questioning, which clients they align with, how they negotiate power imbalances, and how they begin to understand the therapeutic process overall.

This line of questioning highlights the intimacy of narrative-based supervision. White (1992) encouraged supervisors to explore how their understanding of therapy and the construction of their personal identity is developing through the supervision process. This allows for development of both the supervisees' skills and their personal identities. The narrative supervisor will be curious about new developments noticed by the therapist. The knowledge, experience, and practices we use in supervising a therapist will affect the therapists' construction of the stories of their experiences (Foucault, 1980; White, 1997). Carlson and Erikson (2001) explored the accountability this places on supervisors and highlighted the very personal implications of this accountability. This level of intimacy can be an invaluable experience for both supervisor and supervisee. Yet, it is important for the supervisor to practice in a way that holds himself or herself accountable for protecting this connection. Carlson and Erikson (2001) believe that "when our knowledges, skills, beliefs, values, commitments, and desires as therapists are intimately connected to our personal lived experience[,] this allows our work and our personal lives to be mutually beneficial and sustaining of one another" (p. 218). This chapter expands on their premise and explores other practices of narrative supervision that can ensure the supervisor remains accountable to both his or her supervisee and the supervisee's client systems.

Attending to Issues of Power, Privilege, and Diversity

A primary goal of any supervisor training a narrative therapist is to develop the supervisee's ability to appreciate and attend to a multitude of diverse viewpoints and perceptions. The supervisee must learn to understand not only the meaning the client has created behind these viewpoints and perceptions, but also the meaning created by the supervisee himself or herself about these viewpoints and perceptions. As therapeutic sessions are often fraught with issues of power, privilege, and diversity, narrative supervisors must focus directly on how these therapeutic issues shape the course of treatment, influence the roles within sessions, direct the questions asked, and change the meaning of the problem for both the client and the supervisee. Consequently, because of the multiple viewpoints that may exist within session, questions are generally open-ended but also pointed. Questions beginning with "did you," "have you," or "are you" often lead to yes or no answers, and thus do not allow for the possibility of multiple viewpoints. Instead, questions should center on beginning with "what," "where," "when," or "how," as well as tentative language of possibility. Examples of questions asked in supervision are: "What are some different possibilities for considering how the client is understanding his or her situation?" "What informs your understanding of the client's situation?" "What are the possible cultural and/or gender influences that impact the family's interactions?" "How might your assumptions impact your interaction with the client and how do your assumptions impact the client's interaction with you?" This frame of questioning invites both the client in session and the supervisee in supervision to engage in multiple viewpoints during the course of the therapeutic discussion.

Another therapeutic issue often addressed in narrative supervision is the role of hierarchy. Hierarchies can exist across many different levels within systems. Within the clinical system, there are multiple system and personal levels that extend through society, culture, gender, and family roles, as well as the client-therapist-supervisor levels, and even the supervisor-work context level. These different levels are often intersecting, contradictory, and overlapping. Thus, it becomes the supervisor's role to help the therapist-in-training understand the different viewpoints occurring and how they shape the experience of the client and therapeutic process. Because Narrative Therapy assumes the position of *unknowing* the experience of another person, and instead getting rich descriptions of their narratives, there are no experts imposing meaning upon another in the supervisory process. The skills and *expertise* of supervisors are articulated through their knowledge, skills, and experiences to develop effective therapists through their questions and suggestions. The dualistic assumption or account of identity, where the knowledge and power that exist are revealed or not, is uncovered in the process of the *unknowing* approach (Brown, 2007), because the therapist has the power and expertise; consequently, the supervisor must then deny his or her own power. Through positioning themselves to maximize collaborative dialogue, narrative supervisors teach through encouraging and generating new ideas and meaning-making with their supervisees. In turn, a parallel process occurs in therapy, as the therapists-in-training use the same process with their clients. This allows parties to work on a more collaborative level of interaction, instead of one position (i.e., the supervisor) having a more expert or superior viewpoint, which would result in a defined hierarchy.

Supervision involves a newer member in the field consulting with a more experienced colleague to be evaluated and managed (Behan, 2003). There *is* a power differential in the supervisory relationship, and that differential is coupled with the supervisee's race, culture, gender, and multiple other identity differences (Shachar et al., 2012). It is the role of narrative supervisors to flatten this hierarchy, as this is a tenet of social constructionism (Fife & Whiting, 2007). However, the structure of supervision makes it difficult to eradicate the hierarchy completely, so in addition to flattening the hierarchy as much as possible, the supervisor's role is to make this hierarchy overt in the relationship. It is essential that the hierarchy of power within the supervisory system be acknowledged to avoid the professional expertise of both the supervisor and the therapist from diminishing the clients/supervisee's voices and knowledge (Whiting, 2007). There may be risks associated with oversimplifying the client's knowledge or expert role in the therapeutic process. The therapeutic approach of taking a *not-knowing* stance is not to negate the therapists' own experience or knowledge base, but rather to seek to listen in a manner in which their pre-experience and knowledge do not limit them in understanding the full meaning of the client's experience (Anderson & Goolishian, 1992). Similarly, an isomorphic process occurs in supervision, where the supervisor does not negate the expertise, experience, and knowledge of the therapist-in-training. The power differential places supervisors directly in the path of temptation of being the expert. When playing "expert," supervisors may inadvertently recruit the therapist to one "right way" of seeing: inviting traditional, diagnosis-like statements, and imposing their knowledge, life views, and internalized dominant discourses on the supervisee (and in turn the supervisee's clients) (Shachar et al., 2012; Whiting, 2007; Whiting, Nebeker, & Fife, 2005).

Although the relationship is "never flat" (Fine & Turner, 1997, p. 230), supervisors can reduce hierarchical differences by presenting ideas using tentative language (Gardner, Bobele, & Biever, 1997). A narrative supervisor should be cautious about making pronouncements, instead retaining a tentative and collaborative approach that encourages the therapist to take the lead. This tentative and curious approach facilitates the supervision process as a dialogic process of exploring possibilities. When supervisors are curious about the knowledge of the therapist, they can begin to inquire about differences between themselves and therapists, and between therapists and clients, according to social constructions such as gender, class, race, culture, and religion (Kaiser, 1997; Lappin & Hardy, 1997). Hair and Fine (2012) explain how supervision questions beginning with "I wonder . . ." or "I am curious about . . ." can help stimulate collaborative knowledge production. Questions that begin with this tentative stance can explore what aspects of the supervisee's identity are dominant or not in order to uncover discourses that have been silently endorsed in the therapy room and the therapist's life (Anderson, 2000; Murphy & Wright, 2005; Storm et al., 2001). Through this tentative and curious approach, the supervisee's voice and knowledge will become privileged in the supervision room. This creates opportunity for growth and self-efficacy to develop in the therapist as well as the supervisor—a beautiful outcome. Yet, in addition, in an isomorphic process, the supervisee will then begin to privilege the clients' voice and knowledge in the therapy room, thus creating change in all levels of intervention and interaction (Hair & Fine, 2012; Whiting, 2007).

As a caveat, we do not dismiss the fact that there are times when the collaborative and tentative stance espoused by Narrative Therapy has to be put to the side temporarily. For instance, if the clients are at risk of harm to self or others, or the therapist is at risk of harming the clients, supervisors need to be less collaborative and more directive, informing therapists of a particular plan of action, even if it is contrary to the preferences of the therapist (Morgan & Sprenkle, 2007). However, this can be done in a transparent and dialogical way to support the therapists rather than depriving them of autonomy in their role (Hair & Fine, 2012; Selicoff, 2006; Storm et al., 2001).

In presenting Narrative Therapy in today's world as it relates to meaning around culture and diversity, we would be remiss if we did not discuss and attend to the influence of social media and other electronic media. Knowledge and information are now at the tips of your fingers as well as instantaneous and consequently can have an impact on meaning. The rise of social media applications brings instant feedback, from both personal and anonymous sources, which can shape the value, meaning, and even perceived failures of our experiences and thoughts that are shared. As a narrative supervisor, it is necessary to discuss, understand, and at times even challenge the pervasive impact of media influences on our meaning.

Self-of-Therapist Considerations

Given the emphasis on meaning-making in the narrative approach to therapy, it is important that the narrative supervisor elicits discussion and ideas centered on the therapist's own meaning-making and the implications derived from it (which are, in turn, centered on the clients' meaning-making). As we are impacted by the different contexts within our lives on a continuous basis, we must be aware of how those contexts influence our therapeutic processes. This is consistent with the ideas of Bakhtin (Morson & Emerson, 1990) of polyphony and engaging with multiple voices in dialogue while maintaining agency of each voice. Thus, for a supervisor training a narrative therapist, attention must be placed on fostering the therapist's own understanding of his or her narrative and the contextual factors that influence it. Narrative therapists must be able to understand their own narratives, and their processes of constructing those narratives, in order to effectively elicit the narratives of their clients. Thus, discussions of the therapist's upbringing, family values, and belief systems that construct his or her own narrative must occur in order to discover implicit biases, parallels held with client systems, and how the societal discourses affect how the therapist draws meaning in the world.

Many of the self-of-the-therapist supervision approaches tend to be deficit or pathology based. The approach centers on encouraging new therapists to *fix* their personal perspectives so they can be nonreactive and avoid bringing their personal attitudes, perceptions, and possible biases into the therapeutic relationship and process (Carlson & Erickson, 2001). This should not be the experience in narrative supervision. Instead, Carlson and Erickson (2001) explain that the exploration of self in Narrative Therapy is more aligned with the metaphor of the "rite of passage." Novice therapists need not be fixed in order to do therapy. Instead, trainees have valuable lived experiences, knowledge, skills, and desires that have brought them into this helping field. These aspects within the self-of-the-therapist process can be explored to help therapists develop confidence in their abilities and help them experience personal agency in regard to their work (Carlson & Erickson, 2001).

Teaching Techniques Through a Narrative Lens

At its core, the process of training a narrative therapist through a narrative supervisory lens entails the utilization of a collaborative approach. Though it is often tempting for supervisors to offer suggestions or direct guidance in the supervisory process, supervisory efforts are shifted to asking guiding questions that help the therapist to think through the complexity of the client's treatment plan rather than providing answers. Collaboration is a two-way process in which a dialogue develops when the supervisor neither offers expert answers nor simply elicits ideas from the therapist-in-training. Consequently, the supervision process concentrates on sharing the knowledge and experiences of both supervisor and trainee in a collaborative conversation (which also is isomorphic to the dialogue between therapist and clients).

A technique to enhance the collaborative nature of narrative supervision is the utilization of reflecting teams. Tom Andersen (1997) presented reflecting teams as a supervisory team (composed of supervisor(s) and/or other therapists) who watch a session, noticing how the clients' contexts shape their narratives, and then provide feedback for both the primary therapist and the clients. Reflecting teams allow more people to join the conversation and offer insight into the processes of the client system. Reflection can occur facing the clients, talking with the therapist while the clients watch, or talking among only other members of the reflecting team while the therapist and clients watch. Narrative therapists have adapted this technique to be more aligned with narrative theory. In a narrative approach to reflecting teams, reflecting the team members' perspectives permits each voice to be heard, allowing for "the postmodern notions of multiple perspectives, horizontal, collaborative relationships, and transparency into action" (Freedman & Combs, 1996, p. 171). Although narrative theorists shy away from being prescriptive, Freedman and Combs (1996) provided three primary tasks that are helpful in organizing supervisees in the reflection process. They recommend that the focus be on joining with the family, supporting the new preferred narrative, and deconstructing the problem-saturated descriptions. The reflection team provides an opportunity to notice what beliefs, ideas, and contexts continue to support the problem-saturated descriptions. Reflecting teams are most effective when there is a meaningful conversation that engages the narrative of the client with the acknowledgment of the context factors and discourses effecting the problem story.

The reflection team can also serve a secondary function. Often, during the reflection process, the reflecting team members will share their own personal experiences as they reflect on clients' problems. Through team members' use of self-disclosure during reflection teams, reflecting teams can flatten the hierarchy between the therapeutic team and the clients (Cole, Demerrit, Shatz, & Sapoznik, 2001). The narrative perspective on reflecting teams allows new and meaningful narrative to emerge through the process.

The utilization of definitional ceremonies in the therapeutic process also enhances the reflecting team's effectiveness. Definitional ceremonies are typically provided to underserved populations whose personal narratives are affected by issues of diversity, privilege, power, and marginalization (White, 1995). These underserved populations often feel invisible within the larger society. Definitional ceremonies provide an opportunity for every person to be seen on his or her own terms and have witnesses to the fact of their worth and identity. The purpose of the reflecting team is to bear outside witness to the claims of their new significant identity and ability to shape their lives on their own terms. The supervisor can also offer definitional ceremony to the clinician to celebrate and honor the therapist's own work.

There are several other practices derived from Narrative Therapy that can be used in narrative supervision. One is the "re-membering" conversation. Re-membering conversations are based upon the premise that an individual's identity is influenced by significant figures in that person's life, and that these figures construct particular associations that constitute the person's narrative of his or her life and his or her identity (Little, Hartman, & Ungar, 2008; Redstone, 2009; White, 2007). Re-membering is more than thinking about or remembering significant people. It is an active process of bringing these relationships back into one's life and especially of re-experiencing how these relationships invite one to feel and experience oneself as a person at an emotional level (Carlson & Erickson, 2001). Behan (2003) explores how assisting his supervisees in re-membering people in their lives helped the supervisees have a stronger sense of identity as therapists and counteracted any potential for caregiver fatigue.

Another practice is the use of therapeutic documentation. Behan (2003) believes that supervisors can become the keepers of documents that they or their supervisees have contributed over the years—certificates, letters, identity documents, collections of knowledge, and so on—which can be shared with supervisees and passed on to their clients. Through therapeutic documentation, the supervisor and supervisee can avoid being drawn into believing social discourses. Therapeutic documentation is another way to create dialogues that deconstruct these discourses, making them visible, evaluable, chosen, or rejected (Behan, 2003; Redstone, 2009). In addition, the supervisor can provide therapeutic documentation in his or her practice. The use of end-of-semester ceremonies or rituals to explore the growth that occurred over the semester can re-engage and reignite supervisees' passions.

Narrative therapists have always applauded the use of outsider witnesses in their work (Freedman & Combs, 1996; Madigan & Epston, 1995; White, 1995, 1997). New and different therapeutic stories cannot just be told in the therapy room; they have to be lived and performed to be internalized (Freedman & Combs, 1996; White & Epston, 1990). Communities of support should be available for the supervisees in narrative supervisors. Carlson and Erickson (2001) encourage supervisors to create communities of concern for new therapists.

HOW TO GUIDE NARRATIVE THERAPISTS OUT OF AN IMPASSE DURING THE TREATMENT PROCESS

Over the course of the supervisory period, many therapists-in-training will find themselves at an impasse or feel "stuck" with a particular client or problem. Although this is not an unusual experience, especially during the learning process, it may bring up feelings of incompetence or anxiety about treating the particular client system. While this may feel like a challenge for the therapist-in-training, it also presents a unique area of growth for the therapist in understanding his or her own identity story. This situation can emphasize how the client understands, overcomes, and accepts the challenges that may not be within his or her control related to the cultural and environmental discourses that he or she was raised with and internalized into his or her own life.

A supervisor working through a narrative lens will not assume a "right" answer and will resist temptations of certainty. Through taking a not-knowing stance, the supervisor will be able to elicit rich and thoughtful answers from the therapist through a collaborative discussion. Supervisor questions should ask the therapist to evaluate the current situation and name the problem, then consider the relation it has to the client, the problem, therapy, and to the therapist. The supervisor would help the therapist choose a particular direction and relate more satisfying stories incorporating that direction. Evaluation of the effectiveness of that new, alternative story would help the client system understand a new preferred identity. Narrative supervisors aspire to minimize the imposition of their own knowledge, life views, and cultural beliefs on supervisees (Whiting, 2007). During times of impasse, it is a natural desire in many supervisors to become the expert and help the supervisee. However, it is important to avoid doing this. The expert stance will create a sole "truth" or a sole direction for treatment. It does not honor the supervisees' voice and knowledge, but rather produces new dominant discourse in the supervision room that marginalizes other truths (Shachar et al., 2012).

FUTURE DIRECTIONS

There is a dearth in the literature of material providing insight on narrative approaches to clinical supervision. Moreover, little is known about how the social justice initiatives embedded within narrative therapeutic practices are translated into a narrative supervision process (Crocket, 2002). The primary focus has been on aiding supervisees in making the epistemological shift from a modernist epistemological orientation to a postmodern narrative stance. As a result, the narrative supervision process can be considered a context for the "re-authoring of stories about therapy and the participant's life" (White, 1992, p. 76). Unfortunately, only a limited number of articles speak to the pedagogical initiatives of training supervisees in the art of narrative skill acquisition. In addition, the majority of Narrative Therapy books and articles are addressed to experienced clinicians, giving the impression that Narrative Therapy should not be taught to beginning therapists or students (Carlson & Erickson, 2001). Future directions for research and writing include exploring how to aid supervisors in the isomorphic translation of Narrative Therapy into supervision, support for flattening the power hierarchies in the supervision room and process, and providing resources for supervisors of novice therapists.

CASE EXAMPLE

In presenting a case study, it is important to note that the following exemplar is just one possible way of implementing a narrative approach to supervision. Furthermore, Narrative Therapy recognizes that the therapist should not make assumptions based on the limited information that is often provided in the initial description of a case. How each story is presented has certain consequences to how it is viewed by others. Thus, if it is presented by the client system to the therapist, it may be viewed one way—possibly differently by the client(s) and the therapist. If, in turn, the therapist presents the case

(*continued*)

to the supervisor, it may be viewed in yet a different manner. Furthermore, if the supervisor views the session video or watches the case live, he or she may develop still another view of the case. Consequently, when viewing a story through a Narrative Therapy lens, it is important not only how a case is presented, but also how it is considered by those within and a part of it. How we ask questions, and even the questions chosen to be asked, by the client system, therapist, and supervisor, about the assumptions made in the case will shape and change the course of the client's story and process.

In the current case example, there are six different characters, resulting in six different voices, six different perspectives, six different sets of assumptions about the case, six different stories contributing to the current situation, and multiple permutations of shared narratives. There is the supervisor, Dr. Bob; the supervisee/therapist, Brittney; and four members of the client system: the father, Travis; the mother, Tekia; the 5-year-old son, Tyrone; and the 4-year-old son, Tommy. The role of the supervisor is to collaboratively aid the supervisee in understanding the client system's own problematic story and helping to find and adopt a new preferred narrative in the clients' lives.

It is a difficult process for a beginner therapist to remain open to multiple narratives. The supervisor must remain empathetic around the difficulties of maintaining curiosity and ambiguity with the multiple perspectives and storylines of all the family members (Sutherland, Fine, & Ashbourne, 2013). This is one of the more challenging components of being a Narrative Therapy supervisor, yet also the most essential. By generating a polyphonic stance in the supervision room, you are modeling for the supervisee how to remain a *family* therapist while doing Narrative Therapy (Neal, 1996).

For a supervisor training a narrative therapist, the first task is to understand the supervisor-supervisee relationship. A narrative supervisor would seek to understand the similarities and the differences between his or her own story and the story of the supervisee. It is crucial to understand how these similarities and differences shape the supervisory experience. This requires each member of this system to attend to aspects of gender, race/ethnicity, sexual orientation, culture, backgrounds, skill sets, and life experience, and how each of these aspects shapes the direction of and relationship in supervision. Similarly, the second task for the supervisor is to understand how his or her own story relates to the case and how his or her own assumptions about the clients and their story influence the direction of supervision. For instance, how does Dr. Bob perceive the military? How does Dr. Bob perceive those with disability? The supervisor also collaboratively aids the narrative therapist (Brittney) in uncovering and understanding her own story, and how it relates to the clients' story, as well as how it shapes the treatment process. In the current case, it is important to pay attention to the differences in gender, culture (e.g., military vs. civilian), disability, marital status, family structure, and possible financial strain.

For the therapist, the beginning of treatment consists of a focus on creating the therapeutic alliance with the family. The therapist does not focus directly on the problem, but rather asks the couple how they met, what drew them to one another, what kept the family together while Travis served his tour of duty in Afghanistan, how they each experienced Travis's reentry into civilian life, and what they do for fun together. The therapist also lends voice to the children, Tyrone and Tommy, and asks about their perspectives of each other and their parents, what they do as a family together, and how family members interact in the household. The therapist also highlights the family's strengths and the resources they have. As the supervisor, Dr. Bob facilitates this process by asking Brittney how she is conceptualizing the family's story. In traditional approaches to therapy, the therapist is lauded as the expert; within the narrative context, however, the client is the expert. Consequently, the family system already has the strengths and resources to overcome the externalized problem. It is Brittney's job to ask questions that highlight the family's pre-existing resourcefulness, but it is Dr. Bob's role to ensure that Brittney is internalizing the family's potentially new and fragile new story of resilience (Sutherland et al., 2013).

Questions asked may be about Travis's perceptions of the military in general and perceptions of his experience in the military, and then reentering civilian life. Given his disability from active service, questions may be asked about his connection with Veterans Affairs (VA) and how he is supported for his service. Similar questions would be asked of each of the family members as well, so that each voice is heard and given meaning. Clients are also each asked about connection to community and family support, in addition to understanding the family histories and backgrounds the clients originate from. As the narrative unfolds, the therapist will begin to understand the specific issues affecting the problem-saturated story and deconstruct the powerful discourses that are causing or maintaining the problem-saturated story. This is a time where Dr. Bob's role may become more active. As Brittney begins helping the clients deconstruct the cultural discourses, she may be unaware of the culturally dominant ideas that she may be holding on to as true. Dr. Bob's role is to help observe the role discourses play in dictating the power dynamics in both Brittney's therapy room and Brittney's own life.

In summary, despite the fact that it is difficult to provide prescriptions for Dr. Bob within the narrative approach, there are some general orientations that can be highlighted (Sutherland et al., 2013). Dr. Bob's job is to ensure that the "lived experiences of therapists are privileged through the exploration of their stories and ideas about competent and helpful (to their clients) practice" (Sutherland et al., 2013, p. 374). The questions that Brittney chooses to ask will shape the course of the clients' stories. The questions that Dr. Bob chooses to ask will shape the course of the clients' stories, Brittney's story for her clients, and Brittney's own personal story (as well as Dr. Bob's own story).

REFERENCES

Andersen, T. (1997). Researching client-therapist relationships: A collaborative study for informing therapy. *Journal of Systemic Therapies, 16*(2), 125–133.

Anderson, H. (2000). "Supervision" as a collaborative learning community. In The American Association for Marriage and Family Therapy (Ed.), *Readings in family therapy supervision: Selected articles from the* AAMFT Supervision Bulletin (pp. 8–11). Washington, DC: AAMFT.

Anderson, H., & Goolishian, H. (1992). The client is the expert: A not-knowing approach to therapy. *Therapy as Social Construction,* 25–39.

Anderson, S. A., Schlossberg, M., & Rigazio-DiGilio, S. (2000). Family therapy trainees' evaluations of their best and worst supervision experiences. *Journal of Marital and Family Therapy, 26*(1), 79–91.

Bateson, G. (1972). Steps to an ecology of mind: Collected essays in anthropology, psychiatry, evolution, and epistemology. University of Chicago Press.

Behan, C. P. (2003). Some ground to stand on: Narrative therapy. *Journal of Systemic Therapies, 22*(4), 29–33.

Besa, D. (1994). Evaluating narrative therapy using single-system research designs. *Research on Social Work Practice, 4,* 309–325.

Brown, C. (2007). Situating knowledge and power in the therapeutic alliance. In C. Brown, T. Augusta-Scott (Eds.), *Narrative therapy: Making meaning, making lives* (pp. 3–22). Thousand Oaks, CA: Sage Publications.

Brunner, D. D. (1994). *Inquiry and reflection: Framing narrative practice in education.* Albany, NY: SUNY Press.

Carlson, T., & Erickson, M. (2001). Honoring and privileging personal experience and knowledge: Ideas for a narrative therapy approach to the training and supervision of new therapists. *Contemporary Family Therapy, 23*(2), 199–220.

Clarkin, J. F., & Glick, I. D. (1982). Supervision of family therapy. In M. Blumenfield (Ed.), *Applied supervision in psychotherapy* (pp. 87–106). New York, NY: Grune & Stratton.

Cole, P. M., Demerrit, L. A., Shatz, K., & Sapoznik, M. (2001). Getting personal on reflecting teams. *Journal of Systemic Therapies, 10*(2), 74–87.

Combs, G., & Freedman, J. (2012). Narrative, poststructuralism, and social justice current practices in narrative therapy. *The Counseling Psychologist, 40*(7), 1033–1060.

Crocket, K. (2002). Introducing counsellors to collaborative supervision. *The International Journal of Narrative Therapy and Community Work, 4,* 19–24.

Epston, D. (1989). *Collected works.* Adelaide, Australia: Dulwich Centre Publications.

Etchison, M., & Kleist, D. M. (2000). Review of narrative therapy: Research and utility. *Family Journal, 8,* 61–66.

Fife, S. T., & Whiting, J. B. (2007). Values in family therapy practice and research: An invitation for reflection. *Contemporary Family Therapy, 29*(1–2), 71–86.

Fine, M., & Turner, J. (1997). Collaborative supervision: Minding the power. In T. C. Todd & C. L. Storm (Eds.), *The complete systemic supervisor: Context, philosophy, and pragmatics* (pp. 229–240). Needham Heights, MA: Allyn & Bacon.

Foucault, M. (1980). *Power/knowledge: Selected interviews and other writings.* New York, NY: Pantheon Books.

Freedman, J., & Combs, G. (1996). *Narrative therapy: The social construction of preferred realities*. New York, NY: W.W. Norton.

Gale, J. (2010). Discursive analysis: A research approach for studying the moment-to-moment construction of meaning in systemic practice. *Human Systems: The Journal of Systemic Consultation and Management, 21*, 27–37.

Gardner, G. T., Bobele, M., & Biever, J. L. (1997). Postmodern models of family therapy supervision. In T. C. Todd & C. L. Storm (Eds.), *The complete systemic supervisor: Context, philosophy, and pragmatics* (pp. 217–228). Needham Heights, MA: Allyn & Bacon.

Hair, H. J., & Fine, M. (2012). Social constructionism and supervision: Experiences of AAMFT supervisors and supervised therapists. *Journal of Marital and Family Therapy, 38*(4), 604–620.

Hare-Mustin, R. T. (1994). Discourses in the mirrored room: A postmodern analysis of therapy. *Family Process, 33*(1), 19–35.

Hevern, V. W. (1999). *Narrative psychology: Internet and resource guide*. Syracuse, NY: Author. Retrieved from http://maple.lemoyne/narpsych.html

Kaiser, T. L. (1997). *Supervisory relationships: Exploring the human element*. Pacific Grove, CA: Brooks/Cole.

Lappin, J., & Hardy, K. V. (1997). Keeping context in view: The heart of supervision. In T. C. Todd & C. L. Storm (Eds.), *The complete systemic supervisor: Context, philosophy and pragmatics* (pp. 41–58). Needham Heights, MA: Allyn & Bacon.

Larner, G. (2004). Family therapy and the politics of evidence. *Journal of Family Therapy, 26*, 17–39.

Little, A., Hartman, L., & Ungar, M. (2008). Introducing the narrative construal of reality and the "club of life." *International Journal of Narrative Therapy and Community Work, 1*, 46–48.

Lopes, R. T., Gonçalves, M. M., Machado, P. P. P., Sinai, D., Bento, T., & Salgado, J. (2014). Narrative therapy vs. cognitive-behavioral therapy for moderate depression: Empirical evidence from a controlled clinical trial. *Psychotherapy Research, 24*(6), 662–674. doi:10.1080/10503307.2013.874052

Madigan, S. P., & Epston, D. (1995). From "spy-chiatric gaze" to communities of concern: From professional monologue to dialogue. In S. Friedman (Ed.), *The reflecting team in action: Collaborative practice in family therapy* (pp. 257–276). New York, NY: Guilford.

McCoy, M., Stinson, M., Bermudez, J. M., & Gladney, L. (2013). Utilizing a narrative approach to increase intimacy after prostate cancer. *Journal of Creativity in Mental Health, 8*(1), 48–69.

Morgan, M., & Sprenkle, D. (2007). Toward a common-factors approach to supervision. *Journal of Marital and Family Therapy, 33*(1), 1–17.

Morson, G. S., & Emerson, C. (1990). Mikhail Bakhtin: Creation of a prosaics. Stanford, CA: Stanford University Press.

Murphy, M. J., & Wright, D. W. (2005). Supervisees' perspectives of power use in supervision. *Journal of Marital and Family Therapy, 31*, 283–295.

Neal, C. (2000). *Issues in therapy with lesbian, gay, bisexual and transgender clients* (Vol. 3). Philadelphia, PA: McGraw-Hill International.

Neal, J. H. (1996). Narrative therapy training and supervision. *Journal of Systemic Therapies, 15*(1), 63–77.

Nichols, W. C., Nichols, D. P., & Hardy, K. V. (1990). Supervision in family therapy: A decade restudy. *Journal of Marital and Family Therapy, 16*(3), 275–285. doi:10.1111/j.1752-0606.1990.tb00848.x

Redstone, A. (2009, October). Narrative practice and supervision: The re-population of identity. *Context*, 21–24.

Selicoff, H. (2006). Looking for good supervision: A fit between collaborative and hierarchical methods. *Journal of Systemic Therapies, 25,* 37–51.

Shachar, R., Nasim, R., Leshem, T., Rosenberg, J., Schmidt, A., & Schmuely, V. (2012). Power hierarchy, multiple truth, and innovations in narrative supervision. *Journal of Systemic Therapies, 31*(4), 34–48.

Shapiro, J., & Ross, V. (2002). Applications of narrative theory and therapy to the practice of family medicine. *Family Medicine, 34,* 96–100.

St. James-O'Connor, T., Meakes, E., Pickering, M., & Schuman, M. (1997). On the right track: Client experience of narrative therapy. *Contemporary Family Therapy: An International Journal, 19,* 479–495.

Storm, C. L., Todd, T. C., Sprenkle, D. H., & Morgan, M. M. (2001). Gaps between MFT supervision assumptions and common practice: Suggested best practices. *Journal of Marital and Family Therapy, 27*(1), 227–239. doi:10.1111/j.1752-0606.2001.tb01159.x

Strong, T., & Gale, J. (2013). Postmodern clinical research: In and out of the margins. *Journal of Systemic Therapies, 32,* 46–57.

Sutherland, O., Fine, M., & Ashbourne, L. (2013). Core competencies in social constructionist supervision. *Journal of Marital and Family Therapy, 39*(3), 373–387. doi:10.1111/j.1752-0606.2012.00318.x

Vromans, L. P., & Schweitzer, R. D. (2011). Narrative therapy for adults with major depressive disorder: Improved symptom and interpersonal outcomes. *Psychotherapy Research, 21*(1), 4–15. doi:10.1080/10503301003591792

Weber, M., Davis, K., & McPhie, L. (2006). Narrative therapy, eating disorders and groups: Enhancing outcomes in rural NSW. *Australian Social Work, 59*(4), 391–405. doi:10.1080/03124070600985970

White, M. (1992). Family therapy training and supervision in a world of experience and narrative. In D. Epston & M. White (Eds.), *Experience, contradiction, narrative, and imagination* (pp. 75–95). Adelaide, Australia: Dulwich Centre Publications.

White, M. (1995). *Re-authoring lives: Interviews and essays.* Adelaide, Australia: Dulwich Centre Publications.

White, M. (1997). *Narratives of therapists' lives.* Adelaide, Australia: Dulwich Centre Publications.

White, M. (2007). *Maps of narrative practice.* New York, NY: W.W. Norton.

White, M., & Epston, D. (1990). *Narrative means to therapeutic ends.* New York, NY: W.W. Norton.

White, M., & Epston, D. (1992). *Experience, contradiction, narrative, and imagination.* Adelaide, Australia: Dulwich Centre Publications.

Whiting, J. B. (2007). Authors, artists, and social constructionism: A case study of narrative supervision. *American Journal of Family Therapy, 35,* 139–150.

Whiting, J. B., Nebeker, R. S., & Fife, S. T. (2005). Moral responsiveness and discontinuity in therapy: A qualitative study. *Counseling and Values, 50,* 20–37.

SUGGESTED ADDITIONAL RESOURCES

www.dulwichcentre.com.au/catalogue.pdf
www.dulwichcentre.com.au/publications.html
therapeuticconversations.com/
www.narrativetherapychicago.com/

SIXTEEN

Training the Systems/Dialectical Therapist

Thomas W. Roberts

The systems/dialectical model of supervision is based in emotional restructuring and isomorphic and developmental components. It consists of three simultaneously occurring stages or processes in which emotional, cognitive, and behavioral components interact to produce change: (1) building relationships, (2) breaking impasses, and (3) orchestrating change. During the process of working with a family, the supervisee becomes aware that his or her own experiences are isomorphic with those of the family and that he or she undergoes a process of change similar to the family's process.

DEVELOPMENT OF THE SYSTEMS/DIALECTICAL MODEL

The systems/dialectical model of Marriage and Family Therapy (MFT) supervision was developed in a master's-level program in order to apply a generic systems approach to supervision that would be compatible with various systemic approaches (Roberts, Winek, & Mulgrew, 1999). It was intended to provide a generic systemic model of MFT supervision to counter the proliferation of model-based supervision methods. Theory-based models tended to focus on learning principles of the model and how to use concepts of the model. This overdeveloped emphasis on specific models reflected a kind of competition between theories and models as to which was the most effective. In such a competitive atmosphere, it is no wonder that the use of integrative factors across models was slow to develop. In contrast, a systems/dialectical approach is a process-based model that emphasizes how the supervisee and the couple or family in therapy change. It incorporates both systemic and developmental concepts and allows a comprehensive integration of systemic principles across various models. It posits a process approach that can be used with different models of therapy.

The literature has expressed a growing concern about the lack of integrative models in MFT and supervision of trainees (Sprenkle, 1999). Common factors across different schools of therapy have been considered outliers to the essential elements of therapy. Recently, with less charismatic leaders in the MFT field and more focus on the process of change, there is more effort both to identify the common factors and to delineate a new direction in understanding how clients change in therapy.

White and Russell (1995) provided a number of coordinated principles that were useful in developing this integrative MFT model of supervision, including factors related to the supervisor, supervisee, the relationship between the supervisor and the supervisee, the supervisory interaction, and the context in which the supervision takes place. The level of training, personal traits, and family relationships affect the supervisor dimension. The supervisee dimension consists of the cognitive and emotional functioning of the supervisee and his or her capacity for secure attachment relationships. The supervisor-supervisee dimension includes the structure and the quality of the supervisory relationship. The ability to receive and use feedback in supervision makes up the supervisory interaction. *Context* refers to both the present setting of supervision and past significant relationships with their unresolved issues.

Philosophical Foundations

Two guiding principles undergird the systems/dialectical model: the process of supervision is isomorphic with the therapy process, and developmental change occurs in both the supervisee and the couple or family (Roberts et al., 1999). According to Liddle (1988), *isomorphism* is an essential component of supervision from a systems perspective. It is defined as a parallel process occurring between the supervisee in training and the couple or family in therapy (Everett & Koerpel, 1986; Kerlinger, 1986). Liddle (1991) proposes that the isomorphic dimension of supervision goes beyond mere parallelism and can represent a technique whereby the supervisor can use isomorphism to guide the supervisee's interventions to cause change in the family. Some suggest that a focus on isomorphism can be too narrow and reduce supervision to a parallel process (Mead, 1990). The differences between supervision and therapy also create ample dimensions of supervision that would be lost if the only focus was isomorphism.

This author contends that the systems theory concept of interconnections between elements in a system is the basis of understanding isomorphism in supervision. The supervisory experience between supervisor and supervisee, like the experience of living and interacting in a family, takes place through a process that will both aid and impede change. This isomorphic process affects the supervisor and the supervisee equally. Resolved and unresolved issues for both intermingle in the course of their relationship and affect the outcome of supervision. Concurrent with the supervisee's development as a therapist is the supervisor's own growth and development. Unresolved issues in the supervisor's life, including family or marital problems, financial or life course issues, and dissatisfaction with the role of supervisor, will all affect the supervisory process.

Although the isomorphic or parallel process is important, the systems/dialectical model does not reify it as the *only* important process. It is an important aspect particularly when impasses are experienced. It is at these times that the clinical relationship most nearly reflects that of the supervisory relationship. It must be kept in mind that supervision also includes didactic learning, self-reflection, application of theory to clinical interventions, and evaluation and assessment by the supervisor of the clinical strengths of the supervisee. The specific role of the isomorphic process is explained in more detail when the stages of the process are discussed.

The second underlying principle of the systems/dialectical perspective is developmental theory. The idea that supervisees grow in their clinical skills, as evidenced by more in-depth cognitive and emotional responses to their clients and supervisor, has been proposed for some time. Psychotherapy models and Marriage and Family Therapy models have espoused the growth of the supervisee in developmental terms. Some have suggested that supervision should be geared specially to the supervisee's developmental level, since beginning therapists have different needs than more advanced therapists (Flemons, Green, & Rambo, 1996). From this perspective, supervisors must be flexible and able to shift in interventions to accommodate the supervisee's developmental level (Carlson & Lambie, 2012). The individual needs of each of the supervisees when they begin supervision and as they continue in supervision must be addressed. Some researchers have found that to attain maximum benefit of supervision, supervisors must tailor their interventions to meet specific learning and emotional needs of each supervisee (Hart & Nance, 2003). As the supervisee changes, the nature of the relationship with the supervisor also changes (Russell-Chapin, 2007).

Developmental models differ in structure but tend to have a sequence of stages. For example, Stoltenberg, McNeill, and Delworth (1998) propose a model of developing competence from beginner to master therapist. Another developmental model, put forward by Hersey, Blanchard, and Johnson (2000), calls for different supervisor responses for each developmental level of the supervisee. The problem with these models is that the relationship between the supervisee and the supervisor is minimized. It does not depict how this relationship affects the growth in the supervisee, supervisor, and clients. A developmental model should be consistent for all parties because it is should be considered co-development. Like therapy, supervision is a fluid process that simply cannot be reduced to specific maneuvers. Like therapy, there are unexpected occurrences, spontaneity, and serendipity. The more the steps are determined, the more rigid and less like real life supervision (and therapy) becomes.

In contrast to stage models, process models of supervision allow for reflection and both cognitive and emotional learning. Loganbill, Hardy, and

Delworth (1982) postulate a process model based on three stages: stagnation stage, confusion stage, and integration stage. This process model gives a deeper understanding of the relationship between the supervisee and the supervisor, but also tends to be linear. This model does not seem to consider that supervision like therapy, happens unevenly and sometimes occurs nonconsciously. Supervisees may be aware at one time that they have a greater feeling of competency, only to experience a sense of confusion and ignorance with the next case. Like clients in therapy, supervisees will experience ongoing visitations of those emotions and lack of cognitive acuity that predated their supervisory experience. Generally, neither stage nor process models of supervision have accounted for this lingering predisposition to return to presupervisory levels of development.

In contrast to linear stage and process models of development, the systems/dialectical model is based on a nonlinear process outlook that is steeped in ideas from systems theory and child development. Development is viewed as a process of continuing change in which each change is influenced by present environmental circumstances. However, the past influences are never eliminated and continue to influence the supervisee in emotions, cognition, and behaviors. The earliest attachment experiences, though amenable to change, are powerful influences on emotional development throughout life. Applied to supervision in MFT, predetermined stages characterized by concomitant interventions by the supervisor do not paint an accurate description of how change occurs. Bowlby (1973) reflects this view in his writings on attachment, which he views as resulting from the dynamic interaction of past experiences and present experiences. Present or new experiences can change past experiences but are also partly formed by past experiences. Developmental change is an active process that does not follow a strictly linear pathway.

This nonlinear view of development suggests that the attachment style in childhood sets an initial pattern that can be either supported or altered by subsequent experiences. This means that the outcome of development is a cumulative process and not merely an expression of the early attachment style. Experiences over the life course either support the initial pattern or aid in changing it. Because developmental change is cumulative, the past is always influential; it is never entirely wiped out and tends to be maintained after change has occurred. Bowlby (1973) labels the tendency to return to prior levels of development as "homeorhetic." Consequently, even after discernible change, older patterns of development will continue (Sroufe, 2005). Parental interactions aside from attachment behaviors, along with peer interactions, can have an effect on developmental outcomes. In terms of developmental change, the supervisory process parallels the therapy experience because the change that occurs fundamentally alters the supervisee and the couple or family. It is also assumed that both cognitive and emotional functions are altered, resulting in greater integration of brain functions.

Change

The systemic and developmental perspectives of change in this model were influenced by many writers, but a few were especially significant. First, the writings of Carl Whitaker (Whitaker & Abroms, 1974) emphasize the primacy of emotion over cognition, the concept of the selfhood of the therapist, and an understanding of boundaries between the therapist and clients. Whitaker was one of the first systemic therapists and writers to understand the underlying significance of the therapist's use of self in relating to the client. He understood that therapeutic change is related to how the therapist responds internally to the couple or family and then uses this reaction to guide therapeutic interventions. Brain researchers over the past two decades no doubt would label Whitaker the first therapist to utilize mirror neurons in therapy. He clearly demonstrated that what he saw and heard from his clients he felt as completely as they did and could respond by monitoring his internal reactions. He also understood that techniques alone were of little value in changing clients. Though he did not directly state that the quality of the therapist-client relationship is the main curative factor in therapy, his interactions with clients support this conclusion. Brain researchers have found evidence that clients are much more strongly affected by the quality of the therapeutic relationship than by any techniques used by the therapist. When asked to recall specific things that helped them in therapy, they were unable to remember any specific techniques but could recall very vividly how they felt about the relationship with the therapist (Lyons-Ruth, 1998).

Although Whitaker focused on therapy and not supervision, the lesson can easily be applied to supervision. If the supervisory relationship is parallel to or isomorphic with the therapeutic relationship, it can be assumed that the relationship between the supervisor and the supervisee is an influential factor in supervision. In fact, the relationship between the supervisee and the supervisor is primary and especially significant in helping the supervisee work through impasses.

Likewise, therapy can be viewed as a developmental process in which the client's brain structure and functioning may be altered in a way that results in better adjustment and more satisfying interactions with others. There is recent evidence from brain researchers that therapy can change the cellular makeup of the brain (Koch, 2009, as cited in Kift, 2009). Such a change as this demonstrates that therapy is a *developmental* change, because it rewires the brain in a new and more functional way. As a parallel process to therapy, supervision is viewed as a rewiring of the supervisee's brain. This rewiring is an emotional restructuring that changes how the supervisee relates to others. As noted earlier, though there may be change or rewiring of the brain circuits, the past entrenched patterns still can emerge at times of stress or particular circumstances. This model does not posit a linear or teleological process whereby the supervisee reaches a level of expertise that is fundamentally different from the beginning. How this process occurs is discussed in another section of this chapter.

Second, the concept of co-constructed reality, developed by a number of influential writers in constructivism, such as Anderson (1987, 1991) and Watzlawick (1984, 1990), was instrumental in forming this model's understanding of change. According to constructivism, reality is the product of interaction between or among persons. While this view of reality differs significantly from the view of reality as comprising objectivity residing outside the individual, constructivist authors, particularly Simon (1992), posit that a "both/and" orientation is needed rather than an "either/or." A both/and position is essential in moving on from dialectical opposites experienced by the supervisee in supervision and the client in therapy. A drawback in the constructivist position is the reliance on cognitive processes in ascertaining reality. Systems concepts rely on a view of reality as based in subjective construction, which means that the context defines the reality.

Third, the little-known child development theory of Klaus Riegel (1976, 1979) was helpful in understanding how dialectics create the process for developmental change. Riegel believes that development occurs through the interaction of four dimensions: the (1) inner/biological, (2) individual/psychological, (3) outer/physical, and (4) social/cultural. Change occurs as the individual experiences discordant events that force a synthesis, a process that repeats itself continually.

According to Riegel (1976), development occurs as the individual experiences contradictions, or crises, among the four dimensions. This developmental process is not linear with stages that build on each other. From this perspective, change is fluid and follows from the type of confrontation experienced and the synthesis of the dialectic. Applied to supervision, it means that supervisees do not progress along a neat line from novice to advanced therapist. Rather, the interaction of various dimensions represents uneven change and, because of past entrenched styles of interacting, regression may take place.

A strong influence in the development of the systems/dialectical model is recent findings in neuroscience in the areas of memory, learning, and emotions. Researchers have demonstrated that psychotherapy changes both brain structure and function by altering the neurochemistry of the brain (Kandel, 1999). For example, researchers have established a link between social isolation and structural and functional changes in the brains of primates (Amini et al., 1996). The effects of attachment relationships on the brain can be far-reaching. The affective system of humans develops prior to the formation of explicit memory and forms a filter through which the person views the world.

Learning is also a key factor in therapy or supervision. Recent brain research in learning reveals that synaptic connections are strengthened and, therefore, more likely to be retained when something is learned. New experiences must be incorporated into preexisting beliefs systems, which take significant effort and repeated trials to change. Learning is accomplished through short-term memory being transformed into long-term memory. The more specifically patterns are reinforced, the more likely it is that they will be retained. Most emotional learning takes place outside of conscious memory.

THE ROLE OF THE THERAPIST

A systems/dialectical model proposes that the role of the therapist is to bring about change in the client using both implicit and explicit processes. Because many issues that form blocks and contradictions seem to arise from implicit or nonconscious formations, the therapist must understand how to facilitate change by addressing implicit memory systems. Past experiences and attachment issues have set a nonconscious pattern that repeats. By focusing on how these implicit processes repeat, the therapist more effectively intervenes in communication patterns and other automatic responses that cause blocks and impasses. Past attachment issues should be addressed through developing new experiences.

The therapist's role is to help revise the attachment styles of the couple or family members that will emerge as therapy unfolds. Experiencing these issues in the present is viewed as extremely effective because this is where change is most likely to occur. Borrowing from Minuchin (1974) the idea of increasing the stress level therapeutically, the therapist wants to set the best stage for change. Recent brain research supports the idea of increasing or overloading the brain with emotional content, so that the left brain can reorganize the input. This working-together of the two hemispheres reintegrates the brain and improves functioning. It is suggested that the therapist use multiple means of interaction so that new learning can take place. Nonverbal responses can be a powerful form of change in the therapeutic process. The therapist must be prepared for the spontaneous emergence of impasses as the context of the problem unfolds.

Techniques

This model tends to reduce the importance of techniques per se. As mentioned earlier, the therapist-client relationship and the supervisor-supervisee relationship is the main basis for change. Of primary importance is the understanding of the relationship of emotions and cognitions and maintaining a both/and understanding of their interaction in problem formation and resolution. The rational centers of the brain and the emotional center work together to produce impasses and blocks and thus must work together to resolve them. Recent brain research does not place either cognitions or emotions in a position superior to the other but acknowledges that emotions and rational thinking overlap (Damasio, 1999). The purpose of the rational brain is not to control the emotional brain. Rather, emotions provide assistance in making rational choices. In fact, according to Damasio, one cannot make a rational decision without input from the emotional centers of the brain. Emotions provide a useful tool in changing blocks and impasses. The therapist should understand that emotions always play a role in informing the rational brain how to make decisions.

Brain researchers have found that right-brain-dominant persons are more likely to be depressed and thus experience more impasses. Therefore, activation of the left hemisphere permits more positive functioning. Humor has been infrequently suggested as a technique to reduce impasses, except for Carl Whitaker (Whitaker & Abroms, 1974), but researchers have found that humor activates and integrates different areas of the brain. For example, humor in the form of a joke is first processed in the left hemisphere and the left prefrontal lobe. As the joke is being recognized, the right hemisphere and the occipital lobe are activated. Increasing the experience of humor can reduce the preoccupation with negative experiences. The therapist should take cues from the couple or family as to opportunities for the appropriate interjection of humor.

From a systems/dialectical model, techniques all tend to relate to reducing memory work. For example, without explicit or implicit memory there would be little emotional distress. As has been noted, much distress comes from implicit memory through the formation of attachment styles. Mindfulness techniques can both increase positive experiences and decrease the strength of negative memories. Areas of the brain affected by mindfulness techniques act to prevent stress.

Finally, though generally not raised in couples therapy, love and romantic feelings are powerful emotions that demonstrate the interplay of emotions and cognitions. Recent research demonstrates that romantic love stimulates positive centers of the brain, whereas loss of romantic love stimulates the pain center of the brain (Acevada, Aron, Fisher, & Brown, 2011). This may be why many couples are unable to continue their relationship even though they have few arguments. When love feelings are reduced, the presence of the other person is enough to set off the pain center of the brain, leading to unresolved impasses.

Cultural and Gender Issues

According to most research on the topic, cross-cultural/racial therapy connections tend to create concerns (Sue & Sue, 2008). One of the major concerns is that differences in values may result in misunderstandings. Nonverbal gestures, use of certain words, and tone of voice may carry different meanings for persons from different cultural or racial groups (Kim, Ng, & Ahn, 2005). The level of discomfort is felt by both clients and therapists. Therapists tend to steer away from racial discussion in therapy, which reduces open and honest communication and impedes spontaneity (Norton, Sommers, Apfelbaum, Pura, & Ariely, 2006). Although cross-racial and cultural therapy raises concerns expressed by clients and therapists alike, no distinct way of addressing these concerns has been initiated (Tang & Gardner, 2006). Central to this discomfort is transference and countertransference of racial stereotypes (Tang & Gardner, 2006). According to Riegel (1979), race is an integral, or constitutive, aspect of the personality and therefore must be addressed along with the presenting problems (Hamer, 2002). Because race and culture are constitutive to the client, it is necessary to provide an avenue for expression (Chang & Berk, 2009). Although research suggests that clients view cross-racial therapy as potentially discordant, they nevertheless believe that it can be successful (Zang & Burkhard, 2008). Some related factors, such as the degree of cultural identity, provide some clarity as to whether these issues should be raised in therapy.

In addition to cultural and racial components of therapy, gender is also an indisputable factor—gender differences are rooted in the brain. A male therapist and a female client will experience different concerns in therapy. Gender role expectations and stereotypes can become problematic. Sometimes the choice of the therapist's gender is made because of the type of presenting problem; for example, a woman whose husband is cheating on her may prefer a female therapist. Generally, researchers have found that a key aspect is whether there is need for support or identification with the therapist. Research has not found compelling evidence that a distinct difference exists in the effectiveness of same-gender therapy, except in cases in which biases and specific expectations already exist (Chang & Berk, 2009).

A systems/dialectical perspective of cultural, racial, and gender factors in cross-cultural therapy is dialectical in nature. It suggests that cross-cultural and gender therapy allows the exploration of opposites. In some therapy situations, there is an increase in effectiveness, while in others—the dialectic of White therapist and Black client, for example—may raise past unresolved cognitive and emotional responses that will have to be resolved. Like other blocks that must be addressed in therapy, it becomes imperative that issues of culture or race be raised. These issues should be raised in such a way as to give the client, rather than the therapist, the initiative to move it forward. Most literature confirms that the way the issue is usually raised is from the perspective of the therapist. Allowing the client to have a voice and express concerns and feelings about this issue would greatly enhance the therapeutic alliance.

ROLE OF SUPERVISOR AND SUPERVISEE

Supervisor-Supervisee Relationship

According to Lambert (1992, as cited in Sprenkle, 1999), approximately 30% of clients' improvement in therapy is due to the relationship with the therapist. Likewise, the improvement in supervision has much to do with the relationship between the supervisee and the supervisor. Literature on supervision in related fields, such as in psychiatric training, supports the dynamic relationship between the supervisor, supervisee, and clients as a key component of change (Binder & Strupp, 1997). Although the quality of the therapist-client relationship is somewhat taken for granted, it deserves more attention than it is generally given. Too often the focus in therapy is on the techniques used by the therapist rather than the quality of the relationship between therapist and client. There is some evidence that clients fail to be impressed by the dazzling array of therapy techniques when they recall their therapeutic experience. They are much more likely to remember the warmth of the therapist and their general feeling about the therapist than any specific techniques (Lyons-Ruth, 1998).

Because the supervisory experience closely parallels the therapy experience, the quality of the relationship with the supervisee will account for much of the growth in supervision. A study by Frankel and Piercy (1990, as cited in Sprenkle, 1999) finds that both supervisee and client have better outcomes when the supervisor exhibits support behavior in the supervisory relationship. Although the relationship with the supervisee seems to beg the question, much investigation is still needed to determine the specific behaviors that are helpful.

Learning Goals and Objectives

The systems/dialectical model of supervision is a process model based on the dynamic interaction of the supervisor, supervisee, and client(s). The primary focus is on the emotional interplay of these principal participants, rather than on cognitive or didactic processes. The emphasis is on emotional learning brought about by conflicts and contradictions that are encountered in the interaction process. This model views both the supervisor and the supervisee as developing organisms in which the four dimensions of development—the biological, the social, the physical, and the psychological—interact, resulting in conflicts and contradictions (Roberts et al., 1999). It is through resolving these contradictions that growth and change occur. The outcome of supervision, like therapy, can be varied. The supervisory experience can result in better emotional, cognitive, and behavioral functioning; or, in contrast, can lead to impasses, blocks in understanding, and poorer emotional well-being.

Because of the influence of past emotional experiences, including the family of origin and the development of attachment styles, the supervisor must uncover these relationships. This model assumes that some of the impasses with clients will emanate from these past attachment experiences. The supervisor must be acutely aware of how the current blocks in working with a couple or family reignites the supervisee's past issues. It is assumed that the issues emerge from the process of doing therapy and are discussed only in that context. The purpose of addressing past emotional experiences and attachment is not to do "therapy" with the supervisee, but to acknowledge that much of the supervisee's confusion, blind spots, and feelings of incompetence with specific cases have roots in the supervisee's implicit memory system, which is beneath conscious awareness.

The systems/dialectical model views time as a dialectical phenomenon (Roberts et al., 1999). Generally, clients are too concerned about either the past or the future, with the present overwhelmed or ignored. The present time is the midpoint between the past and the future and is the only time dimension in which change can take place. Understanding the present dimension of interaction between the supervisee and supervisor, through verbal, nonverbal, and behavioral components, is necessary to resolve past and future issues. The present time also includes the daily lives of the supervisee, the supervisor, and the client couple or family. Breaking up with a significant other, a midlife crisis, financial difficulty, a poor grade on an exam, a change of residence, and the like can all have negative effects on both the clinical work of the supervisee and the supervisory experience.

Though this model shies away from specific stepwise goals, it does outline a specific process whereby emotional, cognitive, and behavioral growth can take place. It incorporates a reflecting team approach (Priest, Darden, & Keller, 1990) that makes possible full exploration of isomorphic and developmental components. Videotapes are made of all sessions with couples and families, with the supervisor and another student acting as a reflecting team behind a one-way mirror (Roberts et al., 1999). The reflecting team may phone the student/supervisee during the session and ask for a consultation. After the session, the student is asked to present a portion of the videotape to the other supervisors and students. The supervisor and student who observed the session give feedback to the group, followed by other students and supervisors commenting on this feedback. Next, the supervisee and the reflecting team discuss the feedback from the other students and supervisors. The feedback addresses what is taking place between the supervisee and the couple or family and maintains a present context. Generally, questions such as "What are you feeling here?" are asked rather than questions about the content of the session. This process of reflecting allows the supervisee to respond to the comments that are most helpful. The model posits that the isomorphic and developmental issues are addressed though the simultaneous interaction of three processes: developing relationships, breaking impasses, and orchestrating change.

As the supervisee experiences the three interlocking processes, he or she is isomorphic with the changes that the couple or family is experiencing in therapy (Roberts et al., 1999). Like their clients, the supervisees will experience emotional, cognitive, and behavioral restructuring. The outcome of experiencing the three processes will be a change in the supervisee's identity from a student to a therapist. The supervisory experience can become a model for how the supervisee addresses impasses in the therapy session. Consequently, the supervisee can identify his or her own growth and witness the parallel growth taking place in the clients. As impasses arise and resolutions are accomplished in supervision, developmental and attachment issues surface and can be addressed as they occur.

The structure of the feedback sessions permits the supervisees to listen and reflect on a conversation about their therapy session. The supervisee is not allowed to respond directly to the feedback of the group, which inhibits the defensive mechanism of defending one's actions. The supervisee must take in the comments and process them internally. This procedure initiates an implicit, or nonconscious, processing of the feedback. At a later time, the supervisee meets with the supervisor to reflect on the feedback received from the group. The supervisee is guided in reflecting emotionally on the group feedback. The supervisor takes on an empathic, nonjudgmental, and caretaking role that aids nonconscious processes, such as in reattachment. The supervisee is able to see how his or his own developmental change parallels the change occurring in clients.

SYSTEMS/DIALECTICAL MODEL APPLIED TO THE CASE EXAMPLE

Developing Relationships

Although developing relationships is discussed first, and no doubt extremely important for the overall outcome in therapy, this model does not set in motion or follow a linear process. Aspects of relationship development and maintenance are ongoing and never considered complete or a stage to be completed before moving on to other stages. In this model, the process of maintenance of the relationship is just as important as establishing rapport. The focus on building relationships is on honest, open, and ethical communication that includes taking charge of the direction of therapy, providing empathy and encouragement, and appropriately challenging when needed. The major dialectical dilemma for building relationships is objectification versus overidentification (Roberts et al., 1999). When the supervisee objectifies, he or she fails to join with the couple or family. The outcome is taking on too much of the therapeutic work and feeling frustrated and angry. The other side of the dialectic is overidentification with the couple or family. A typical scenario for beginning supervisees is to identify with the partner or family member who is seen as the victim or scapegoat. Overidentification may reflect blocks in development that the supervisee must resolve. Investigation of unresolved family-of-origin issues would be helpful to resolve these issues.

Applying this process of relationship building to the case example requires some speculative thinking, as the case information gives little detail. First, we can assume that the supervisee was raised in a traditional family, is in her mid-thirties, of the White race, and never married. The family is Black and displays traditional work roles of the husband working outside the home and the wife not working. The most obvious dialectic is the cross-racial connection. If the supervisee is unable to form a collaborative bond, she will feel frustrated and even consider transferring the case to someone else.

We can assume that both the supervisee and the family have some initial concerns, and the family may prefer a Black therapist. The first videotape with this family may reflect this discomfort. If the supervisee appears tense and stressed, it might reflect objectification of the family and a failure to initially establish a collaborative alliance. The reflecting team might speculate about this and wonder if the therapist's past relationships included cross-racial friendships. It could be that the supervisee has little experience with Black people and responds from stereotypes. However, if the supervisee overidentifies with the family on the basis of her past associations with Black people, she may see them as victims, particularly Travis, who was injured in the war. She could believe that the government is not genuine in recruiting poor or ethnic minority individuals into the army. Her overidentification with the family would be an issue raised by the reflecting team. Follow-up sessions with her supervisor could explore in greater detail her internalized view of Black families. Her resolution of this issue is necessary to be able to therapeutically intervene in this family.

Breaking Impasses

Breaking impasses occur throughout the supervisory experience, just as it does in therapy. The major dialectic in breaking impasses is the tension between *homeostasis*, or stability, and *morphogenesis*, or change (Roberts et al., 1999). The model assumes that the presenting problems are symptoms of deeper blocks or impasses related to the social, cognitive, psychological, or physical dimensions of development. Further, the model posits that addressing these impasses requires an emotional restructuring.

In starting with the presenting problem of Travis's "short fuse," care must be taken not to get bogged down in content. During the first session, the supervisee would need to gauge her own reactions to the description of this problem and be aware of the feelings behind the description. Travis's goal of making the military a career was destroyed by his injury, a loss that is now further compounded because of the inability to find a well-paying full-time job. Perhaps the sense of failure, loss of a goal, and financial stress reduce Travis's behavioral coping mechanism. The dialectic of stability versus change is the underlying impasse with this problem. For example, holding on to the sense of loss and failure may be an easier choice than letting go, embracing the changes, and setting new goals.

As therapy progresses, it could become clear that Travis will not be able to find a job that meets the family's needs. The supervisee may discover that Tekia does not want to work because the children are preschool age, and even when they reach school age she wants to be home when they come home from school. If Brittney begins to push toward a particular solution, it could be due to her own impasse from an unresolved past experience in the family of origin. The supervisor and the reflecting team would need to be observant of Brittney's processing of this problem. Parallel processes would have to be identified. As Brittney addresses how financial stress derailed her own family of origin and connects the parallel dimensions with this family, she will be less inclined to "fix" the problem. Instead, she will be more empathic and understanding.

Orchestrating Change

As stated earlier, the stage of orchestrating change is integral to the others and not sequential. Even in the building relationships stage, change is beginning to take place because the relationship with the couple or family is forming the basis for more functional relationships. A systems/dialectical model hypothesizes that change, like learning, can be nonconscious and beneath awareness. For this reason, the emotional dimensions of the intervention are highlighted. Persons change when they are in a relationship that modifies their attachment styles, which results largely from nonconscious processes. To be truly present with this family, Brittney must address the parallel issues that block her from freely experiencing this family. These blocks may include addressing racial stereotypes and past familial experiences that hinder her emotional openness to this family, then looking for times in which the family uses humor and reinforce the relationship by creating additional opportunities for humor.

There seems to be some inflexibility in this couple. It could be true that Travis's long months of recovery from his injuries and the family's financial situation have reduced their feeling of love and support for each other. Tekia may have shifted her view of Travis as a strong, virile man, to one whose physical performance in their intimate relationship is lacking and who is emotionally weak. These are unspoken feelings that may be difficult to explicitly acknowledge. Making these concerns explicit would enhance the overall functioning of this couple.

Emotional restructuring may be more effective with this couple through implicit means, that is, indirect interventions. The reflecting team may notice this inflexibility and suggest ways to introduce new experiences. Some new experiences might include having date nights, taking walks together, or beginning a home project, along with other things. New positive experiences can change the emotional dynamic of the relationship and produce a change. These new experiences must be reinforced and practiced to become part of the clients' ongoing repertoire.

CONCLUSION

In summation, the systems/dialectical approach views supervision and therapy as similar in how change is manifested. Impasses occur in both the supervisee and the client couple or family. Through the ongoing process of reflection, the supervisee is able to resolve the impasses that occur. The change experienced by the supervisee and the family is a developmental change. As discussed previously, some issues reappear because past experiences are never completely eliminated. Reinforcing the change is necessary so that the regression to past emotional, cognitive, and behavioral manifestations does not dominate.

REFERENCES

Acevada, B., Aron, A., Fisher, H., & Brown, L. L. (2011). Neural correlates of long-term intense romantic love. *Social, Cognitive, and Affective Neuroscience* [online publication]. doi:10.1093/scan/nsq092

Amini, F., Lewis, T., Lannon, R., Louie, A., Baumbacher, G., McGuinness, T., & Schiff, E. Z. (1996). Affect, attachment, memory: Contributions toward psychobiologic integration. *Psychiatry, 59,* 213–239.

Anderson, T. (1987). The reflecting team: Dialogue and meta-dialogue in clinical work. *Family Process, 27,* 415–428.

Anderson, T. (1991). *The reflecting team: Dialogues and dialogues about dialogues.* New York, NY: W.W. Norton.

Binder, J. L., & Strupp, H. H. (1997). Supervision of psychodynamic psychotherapies. In C. E. Watkins, Jr. (Ed.), *Handbook of psychotherapy supervision* (pp. 44–62). New York, NY: Wiley.

Bowlby, J. (1973). *Attachment and loss—Vol 2. Separation: Anxiety and anger.* New York, NY: Basic Books.

Carlson, R. G., & Lambie, G. W. (2012). Systemic-developmental supervision: Clinical supervisory approach for family counseling student interns. *The Family Journal, 20*(1), 29–36. doi:10.1177/1066364807114198O9

Chang, D. F., & Berk, A. (2009). Making cross-racial therapy work: A phenomenological study of clients' experiences of cross-racial therapy. *Journal of Counseling Psychology, 56*(4), 521–536. doi:10 1037/a0016905

Damasio, A. (1994). *Descartes' error: Emotion, reason and the human brain.* New York, NY: Avon Books.

Damasio, A. (1999). *The feeling of what happens: Body and emotion in the making of consciousness.* Orlando, FL: Harcourt Brace.

Everett, C. A., & Koerpel, B. J. (1986). Family therapy supervision: A review and critique of the literature. *Contemporary Family Therapy, 8,* 62–74.

Flemons, D. G., Green, S. K., & Rambo, A. H. (1996). Evaluating therapists' practices in a postmodern world: A discussion and a scheme. *Family Process, 35,* 43–56.

Hamer, F. M. (2002). Guards at the gate: Race, resistance, and psychic reality. *Journal of the American Psychoanalytic Association, 50*(4), 1219–1236. doi:10.1177/0003065 1020500041301

Hart, G. M., & Nance, D. (2003). Styles of supervision as perceived by supervisors and supervisees. *Counselor Education and Supervision, 43*(2), 146–158.

Hersey, P., Blanchard, K. H., & Johnson, D. (2000). *Management of organizational behavior: Leading human resources* (8th ed.). Englewood Cliffs, NJ: Prentice-Hall.

Kandel, E. C. (1999). Biology and the future of psychoanalysis: A new intellectual framework for psychiatry revisited. *American Journal of Psychiatry, 156,* 505–524.

Kerlinger, F. N. (1986). *Foundations of behavioral research* (3rd ed.). New York, NY: Holt, Rinehart & Winston.

Kift, L. (2009). Can therapy really change your brain? *Psych Central*. Retrieved from http://psychcentral.com/blog/archives/2009/11/25/can-therapy-really-change-your-brain/

Kim, B. S. K., Ng, G. F., & Ahn, A. J. (2005). Effects of client expectation for counseling success, client-counselor worldwide view match and client adherence to Asian and European American cultural values on counseling process with Asian Americans. *Journal of Counseling Psychology, 52*, 67–76. doi:10.1037/0022-0167.52.1.67

Liddle, H. A. (1988). Systemic supervision: Conceptual overlays and pragmatic guidelines. In H. A. Liddle, D. C. Breunlin, & R. C. Schwartz (Eds.), *Handbook of family therapy training and supervision* (pp. 153–171). New York, NY: Guilford.

Liddle, H. A. (1991). Training and supervision in family therapy: A comprehensive and critical analysis. In A.S. Gurman & D.P. Kniskern. *Handbook of family therapy Vol II* (pp. 638–697). New York, NY: Brunner/Mazel.

Loganbill, C., Hardy, E., & Delworth, U. (1982). Supervision: A conceptual model. *Counseling Psychologist, 10*, 3–42.

Lyons-Ruth, K. (1998). Implicit relational knowing: Its role in development and psychoanalytic treatment. *Infant Mental Health Journal, 192*, 282–289.

Mead, D. E. (1990). *Effective supervision: A task-oriented model for the mental health professions.* New York, NY: Brunner/Mazel.

Minuchin, S. (1974). *Families and family therapy.* Cambridge, MA: Harvard University Press.

Norton, M. I., Sommers, S. R., Apfelbaum, E. P., Pura, N., & Ariely, D. (2006). Colorblindness and interracial interaction: Playing the political correctness game. *Psychological Science, 17*(11), 949–953. doi:10.1111/j1467-9280.2006.01810x

Priest, L., Darden, E., & Keller, J. (1990). The fly on the wall reflecting team supervision. *Journal of Marital and Family Therapy, 16*, 265–273.

Riegel, K. F. (1976). The dialectics of human development. *American Psychologist, 30*, 689–700.

Riegel, K. F. (1979). *Foundations of dialectical psychology*. New York, NY: Academic Press.

Roberts, T. W., Winek, J., & Mulgrew, J. (1999). A systems/dialectical model of supervision: A symbolic process. *Contemporary Family Therapy, 21*(3), 291–302.

Russell-Chapin, L. A. (2007). Supervision: An essential for professional counselor development. In J. Gregoire & C. M. Jurgers (Eds.), *The counselor's companion: What every beginning counselor needs to know* (pp. 79–80). Mahwah, NJ: Lawrence Erlbaum.

Simon, G. (1992). Having a second order mind while doing first order therapy. *Journal of Marital and Family Therapy, 18*, 377–387.

Sprenkle, D. H. (1999). Toward a general model of family therapy supervision: Comment on Roberts, Winek, and Mulgrew. *Contemporary Family Therapy, 21*, 309–315.

Sroufe, L. A. (2005). Attachment and development: A prospective, longitudinal study from birth to adulthood. *Attachment and Human Development, 7*(4), 349–357.

Stoltenberg, C. D., McNeill, B., & Delworth, U. (1998*). IDM supervision: An integrated developmental model for supervising counselors and therapists.* San Francisco, CA: Jossey-Bass.

Sue, D. W., & Sue, D. (2008). *Counseling the culturally diverse: Theory and practice* (5th ed.). Hoboken, NJ: John Wiley & Sons.

Tang, N. M., & Gardner, J. (2006). Interpretation race in the transference: Perspectives of similarity and differences in the patient/therapist dyad. In R. Moodley & S. Palmer (Eds.), Race, culture and psychotherapy: Critical perspectives in multicultural practices (pp. 89–99). New York, NY: Routledge/Taylor and Francis.

Watzlawick, P. (1984). *The invented reality: How do we know what we believe we know (Contributions to constructionism)*. New York, NY: Norton.

Watzlawick, P. (1990). *Munchausen's pigtails: Or psychotherapy and reality*. New York, NY: Norton.

Whitaker, C. A., & Abroms, G. M. (1974). New approaches to residency training in psychiatry. In G. Farwell, N. Gamsky, & P. Mathier-Couglan (Eds.), *The counselor's handbook* (pp. 33–52). New York, NY: Intext Educational.

White, M. B., & Russell, C. S. (1995). The essential elements of supervisory systems: A modified Delphi study. *Journal of Marital and Family Therapy, 21*, 33–54.

Zang, N., & Burkhard, A. W. (2008). Discussions of racial differences and the effects on client ratings of counselor creditability. *Journal of Multicultural Counseling and Development, 36*, 77–88.

SEVENTEEN

Training the Emotionally Focused Therapist

Scott R. Woolley
George Faller
Lisa Palmer-Olsen
Angela DeCandia Vitoria

TEACHING THEORETICAL CONTEXT

Founders

Emotionally Focused Therapy (EFT) was developed by Sue Johnson and Les Greenberg and was first published as a couple therapy outcome study based on Johnson's dissertation (Johnson & Greenberg, 1985), which Greenberg chaired. The couple and family version has been developed primarily by Johnson and colleagues since that time. Greenberg and colleagues have developed a model they call Emotion-focused Therapy (Elliott, Watson, Goldman, & Greenberg, 2004). Because the couple and family EFT outcome and process research to date has primarily been done on Emotionally Focused Therapy (Greenman & Johnson, 2013; Halchuk, Makinen, & Johnson, 2010), this chapter is focused on supervising from an Emotionally Focused Therapy perspective, and EFT will refer to Emotionally Focused Therapy.

The EFT supervision model was first developed and articulated by Palmer-Olsen, Gold, and Woolley (2011) and was based on a study of how certified EFT therapists developed competency in EFT. The EFT supervision model is the first known empirically derived model of supervision in the field of couple and family therapy. This chapter presents the basics of the EFT supervision model, along with additional insights that have been gained by the authors since 2011.

Philosophical Foundation

Emotionally Focused Therapy is a humanistic, systemic, experiential model which posits that people and relationships can grow and change (Johnson, 2004). The model was developed and has evolved through careful study of what is effective in therapy and integration of this knowledge in well-grounded theory and research. EFT integrates Rogerian, Systemic, and Experiential models and is based in Attachment Theory, which serves as an overarching theory of love and human development.

From an EFT perspective, couples typically get together for emotional reasons and separate for emotional reasons. Couple and family relationships contain powerful emotional bonds that are logical and understandable when viewed in their proper context. EFT helps supervisors, clinicians, and the general public understand the logic of these emotional bonds and provides a road map for change.

THE PROCESS OF CHANGE: BETWEEN AND WITHIN

"All knowledge is experience, everything else is just information."
—Albert Einstein

The primary change mechanism in Emotionally Focused Therapy is experiential (Johnson, 2004; Johnson et al., 2005). Happy, strong couple and family relationships are characterized by positive, intimate, safe connections (Johnson, 2013). Secure lovers turn to each other in times of vulnerability and distress for comfort, safety, and security, which helps form powerful bonds of love and commitment. Parents, when seeing their newborns, often talk of falling in love with their children. As couples attune with each other and with their children, strong attachment bonds develop that can last a lifetime. However, as couples and families move through life cycle transitions, it is easy to misattune, misunderstand, miscommunicate, and even knowingly or unknowingly betray each other, which can result in emotional pain, insecurity, and fears of disconnection. If these misses are not repaired, couples and families get caught in negative patterns or cycles of interaction. Rather than turning to each other for comfort, intimacy, and connection, they tend to get reactive and anxiously pursue or withdraw. These responses tend to form patterns or cycles of interaction that become rigid, painful, and very destructive. The relationship becomes dangerous rather than secure and feelings of fear and helplessness replace security and safety in the relationship. It is the goal of the EFT therapist to identify the patterns, access and reprocess the emotions and attachment longings that drive the cycle, and create bonding events to fundamentally change the cycle from one of negativity to one that is positive, nurturing, and secure.

The EFT change processes has been divided into nine interactive steps (Johnson, 2004), which are in turn divided into three fundamental stages. Each of these steps and stages interactively builds upon the others. Consequently, more advanced steps and stages require the successful, and to some degree ongoing, implementation of previous steps.

Stage I: Assessment and Cycle De-Escalation (Steps 1–4)

In stage I, the therapist joins with members of the couple or family to identify the negative relationship patterns, access and process the underlying emotions, and reframe the problem as being about the cycle and the underlying attachment-related emotions. Through this process, the couple learn not only to see the problem as the cycle, but also to stay out of the cycle or revisit and repair rocky moments (Johnson, 2008).

Step 1: Create a safe therapeutic alliance and identify core struggles. Here the therapist works to establish a strong, safe working alliance with each member of the couple or family and do a basic assessment.

Step 2: Identify the negative interaction cycle and each partner's position in that cycle. The assessment in step 1 leads to identification of the negative interactive cycle, which characterizes almost all distressed relationships. The negative cycle is often referred to as a *dance* in popular literature (Johnson, 2008, 2013) and, over time, undermines the safety of each person in the relationship and eventually leads to relationship dissolution.

There are five levels of the cycle (Figure 17.1). The top layer consists of behaviors, which are patterned and typically involve some form of pursue-withdraw. The second layer consists of each partner's perceptions or attributions of self, the other, and the relationship itself. The third level consists of the secondary/reactive/harder emotions that occur in response to the more vulnerable primary emotions. The secondary reactive emotions, such as anger, resentment, jealousy, and anxiety, block intimacy and safe connection and tend to evoke fear, negative appraisals, and more reactive behaviors in the partner. These aspects of the cycle (behaviors, perceptions, secondary emotions) are all above the line because they are readily apparent. Clients usually talk about and show these aspects of the cycle in therapy.

The below-the-line parts of the cycle consist of primary emotions and attachment needs. *Primary* emotions are the more vulnerable emotions, such as fear, sadness, and loneliness. Attachment needs or longings are the basic human needs to be safely connected with others and feel loved, valued, and respected. These vulnerable feelings and needs, when expressed, tend to evoke empathy and draw people close.

Step 3: Access, crystallize, and reprocess underlying attachment-related emotions. In this step the therapist works to identify, access, clarify, and reprocess both secondary and primary emotions associated with moments of connection and disconnection (attachment-related aspects of the cycle).

Step 4: Reframe the problem in terms of the negative cycle, underlying emotions, and attachment longings. In this step the therapist reframes the problem in terms of the cycle (identified in step 2) and underlying

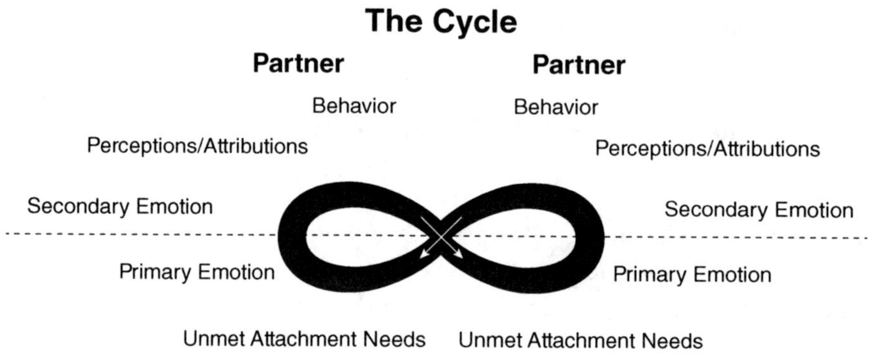

FIGURE 17.1 The Emotionally Focused Therapy (EFT) cycle.

Note: This figure may be used in supervision and therapy as long as the copyright notice and author are included. The figure can be downloaded from www.trieft.org under Forms. © Scott R. Woolley.

attachment-related emotions (identified and accessed in step 3). Negative behaviors, such as angry pursuit or cold withdrawal, are framed as either ineffective protests against the lack of safe connection or ineffective attempts to create connection through pursuing for connection, trying to correct problems through criticism, or trying to protect the relationship from unproductive conflict through withdrawal. These reframes typically start as the cycle is identified and primary emotions are accessed and continue to be reinforced through the rest of therapy.

Stage II: Changing Interactional Positions and Creating New Bonding Events (Steps 5–7)

Stage II involves fundamentally changing the cycle or dance and replacing it with one in which each member of the couple feels safe enough to share vulnerability and intimate connection and ask directly for attachment needs and desires to be met. Successful completion of stage II of EFT is associated with positive outcomes (Johnson, 2004).

Step 5: Promote identification with disowned attachment longings and aspects of self and integrate these into relationship interactions. In this step the therapist helps all clients to own and identify with their deeper emotional attachment needs for connection, belonging, and safety. This step often involves "parts work," particularly with traumatized clients who internally compartmentalize and disown aspects of self that were abandoned, abused, and traumatized. As clients come to own their deeper attachment needs, emotions, and aspects of self, they are encouraged to share them directly with the partner, which helps set the stage for step 6.

Step 6: Promote acceptance of the partner's attachment longings and aspects of self. In this step, the therapist works to help each partner accept, respect, and attune to the other partner's attachment-related needs, longing, and aspects of self that were shared in step 5. Steps 5 and 6 typically go together as the therapist works to help partners share and accept the deeper, more vulnerable attachment fears, longing, and needs.

Step 7: Facilitate the direct expression of needs and wants to create emotional bonding and connection. In this step, the therapist works to get each member of the couple to ask directly for attachment needs of safe closeness and connection to be met. This step builds on but is different from the work in steps 5 and 6 in that asking directly for needs to be met, by an engaged and open partner, is more vulnerable and also more powerful than just letting the partner know about the needs. In step 5 the partner may say, "I have a deep longing to know I am loved and not alone." In step 7, the same partner may say, "Come and be with me and let me know you love me." These kinds of direct expressions of needs and wants help create powerful bonding events that are the basis of lasting relational change.

Stage III: Consolidation (Steps 8–9)

The final stage of EFT involves helping couples past particular difficult issues and guiding them to consolidate their gains to instill lasting change.

Step 8: Facilitate the emergence of new solutions to old relationship problems. In this step, the therapist supports the couple as they work to resolve long-standing issues. Once couples and families have the kinds of bonding events and connects developed in stage II, they solve many of their differences on their own. However, there may be particularly painful, long-standing issues that still have to be addressed. Sometimes step 8 sessions involve breakthrough solutions. In other cases, there are emotional shifts that help the couple live with and even appreciate differences.

Step 9. Consolidate new positions/cycles of attachment behaviors. In this step, the therapist helps the couple/family recognize the changes they have made and discuss ways of staying connected. The therapist reviews the changes to help make the new story of the relationship explicit and concrete.

The nine steps build on each other and are additive. Once a therapist successfully implements the processes of a step, the therapist continues to integrate that work into the present processes of therapy. So, for example, the joining and assessment in step 1 continues to some degree throughout therapy. Once the cycle is identified in step 2, it is worked with and used from then on in therapy. When underlying emotions are accessed (step 3), the therapist continues to work with and use those emotions throughout therapy. The reframes of step 4 are continually emphasized and worked with for the rest of therapy. This same process of using and building on each step continues through all of the steps and stages of the therapy.

ROLE OF THE THERAPIST

The EFT therapist takes a collaborative stance and assumes that client emotions, experiences, and behaviors make sense in their context. Clients are seen as having expertise in their own life experiences, and therapists are seen as having expertise in the processes of change. The EFT therapist honors both of these areas of expertise and collaboratively combines them to facilitate change. EFT therapists are typically very active in the therapy room while staying closely attuned to all therapy participants. The therapist works to be a temporary secure base in the therapy process and actively facilitates change through the interventions and stages/steps of change. As a safe attachment figure, the therapist engages in the reciprocal process of creating new experiences through the co-regulation of affect and the co-creation of meaning. To engage in this "intersubjective" exchange, therapists need awareness of their own emotional responses, triggers, and vulnerabilities (Stern, 1985).

Techniques

Emotionally focused therapists use a variety of interventions drawn primarily from EFT's theoretical roots. To identify cycles of interaction, EFT therapists use circular questions, assessment enactments, and observation of present processes. Interventions focused on accessing emotions emphasize the reciprocal nature of attunement. These two-way interventions are active attempts by the therapist to "feel" the clients' world and allow the clients to feel their own experience.

Access, Expand, and Reprocess Emotional Experience

An important part of EFT involves accessing, expanding, and reprocessing emotional experience. The following interventions are key and are used throughout EFT.

Empathic reflection
EFT uses empathic reflection to help build the alliance, to help clients feel understood, to attune with clients, to bring focus to what a client has said, to help clients order their experience and hear themselves, and to slow the process of therapy down (particularly helpful with high-conflict couples).

Validation
Validating client realities helps normalize and put client experiences into a context. In couple and family therapy, the EFT therapist often validates contradictory positions and experiences of different clients through validating the experience while not taking a position on whose experience is accurate.

Empathic conjecture
Empathic conjecture involves tentatively conjecturing about a person's emotions or experience. For example, "Help me if I am wrong, but I am getting that when he turns away and shuts down, there is a part of you that fears you two will never really connect. Is that right?" Empathic conjecture is a way of helping clients name their emotional experiences, feel understood and safe

with the therapist, and focus on and identify with their deeper emotional experiences. If the clients accept the interpretation, they are encouraged to put it into their own words. If the client rejects the conjecture, the therapist sees this as helping the therapist attune with greater accuracy.

Evocative questions
An evocative question is an open-ended question about emotions related to some stimulus (such as what a family member just said) or a bodily response. Evocative questions are designed to clarify and deepen emotional experience.

Heightening
Heightening involves intensifying and focusing on emotional experience through repeating and enacting, The letters *RISSSC* stand for a set of procedures that are used especially in heightening.

- Repeat—Repeat clients' words, particularly emotionally charged words or metaphors
- Images—Use images around emotional content and action tendencies
- Simple—Use simple words and phrases
- Slow—Slow down to allow clients to feel
- Soft—Use a soft voice, particularly around emotional content
- Client words—Use client words as much as possible

Restructuring Interactions

In order to change interaction patterns and create bonding events, the EFT therapist tracks and reflects the process of interaction and uses enactment to create new experiences and interactive patterns.

Track and reflect process of interaction
The therapist works to illuminate and restructure interactions through tracking and reflecting the process of interaction, both as reported by the couple and as seen in the therapy room. The therapist uses the knowledge of these interactions to help create new interactions.

Enactments
In the process of enactments, the couple or family talks directly to each other. Enactments are typically used to create and intensify new interactions and create bonding events. They are used throughout EFT but particularly in stage II as the therapist works to create new experiences. It is important to note that enactments are not used only to create insight, but rather are used to create security through corrective emotional experiences.

CULTURE/RACE AND GENDER ISSUES/RESEARCH

Basic attachment needs are wired into the brains of humans and are therefore universal. All humans need to feel loved, valued, respected, and safely connected with others. However, the way safe connections are formed and maintained and the way attachment needs are expressed and understood and responded to are often different in different cultures. The EFT therapist is sensitive to these differences and seeks to curiously tune into and track culturally specific attachment strategies and processes that result in safe connections (van Ijzendoorn & Sagi-Schwartz, 2008).

All forms of discrimination and oppression and the associated traumas can create insecurity and a heightened need for secure attachment. However, discrimination and oppression can also make it more difficult to develop and maintain secure attachments. For example, discrimination can lead to economic oppression, which may necessitate frequent moves and long periods of separation and disconnection. Additionally, the experiences of discrimination and oppression inform internal working models of self and others and can make it more difficult to form and maintain safe attachment bonds. The EFT therapist is sensitive to various forms of discrimination and oppression and their impact on attachment needs, models, and strategies.

EFT has been used with, written about, and found to be effective with a variety of populations and problems. EFT has been shown to be effective with couples in North America and in the Middle East, in treating trauma (including child abuse), in changing brain processes associated with threat, in changing attachment processes, and in treating depression (Dalton, Greeman, Classen, & Johnson, 2013; Dessaulles, Johnson, & Denton, 2003; Johnson, Hunsley, Greenberg, & Schindler, 1999; Mehr et al., 2014). For a current list of outcome studies, process studies, and key articles, see http://www.iceeft.com/images/PDFs/EFTResearch.pdf. An EFT casebook illustrates the use of EFT with same-gender couples; couples facing cancer, sexual issues, addictions, and trauma; culturally diverse couples; and military couples (Furrow, Johnson, & Bradley, 2011).

ROLES OF THE SUPERVISOR AND SUPERVISEE

Supervisor-Supervisee Relationship

The EFT supervisor-supervisee relationship is one of collaboration and respect. Palmer-Olsen et al. (2011) found that a secure supervisory alliance was identified as a key factor for therapists learning EFT. The supervisor respects and honors the talents, life experiences, and agency of the therapist while guiding the therapist to greater expertise in the understanding and implementation of EFT. Learning to do therapy, particularly an experiential form of therapy, is a vulnerable process. The supervisor works to be transparent about his or her own experiences and works to provide a secure base for the therapist. The EFT supervisor's stance is characterized by acceptance, empathy, authenticity, curiosity, and responsiveness.

Learning Goals/Objectives

For the purpose of understanding and teaching EFT supervision, the process has been broken into four primary goals: The acronym ACES (Alliance, Conceptual, Experiential, Self-of-the-Therapist) is a simple way to remember these primary goals.

Goal One: Alliance

Developing a secure supervisory alliance is the basic building block of EFT supervision; it provides the solid foundation for the remaining three goals. Starting with the paramount importance of a secure alliance helps put the other goals in context. The person of the supervisee is the most important component, not the issues the supervisee presents with. Palmer and Johnson (2002) summarize well the elements that are necessary for establishing the quality of alliance that best facilitates supervisees' learning of EFT: "The supervisory relationship needs to be characterized by mutuality, collaboration and respect in order for the learning to take place in an atmosphere that allows for creativity and risk taking. . . . The novice therapist needs then to feel not only heard but also held emotionally and resonated with by the supervisor" (p. 18).

An authentic and transparent EFT supervisor creates a collaborative atmosphere that emphasizes journeying together as the best path toward discovering truth. The stance of an EFT supervisor is playful, accepting, curious, and engaging (Hughes, 2007). In order to establish the safety necessary for safe exploration, the supervisor must provide structural and emotional support to the supervisee. This includes ongoing assessment and a supervision contract. Identifying and being explicit about the primary goals of supervision, the developmental stage of the therapist, evaluation of progress, and strategies for handling conflict are essential for creating safety and reducing supervisees' anxiety.

Goal Two: Conceptual Development and Integration

To achieve this goal, the EFT supervisor adapts a more traditional, didactic stance that focuses on understanding the model. In pursuit of this goal, the supervisor is mostly imparting information so the supervisee can learn to integrate attachment theory and EFT into practice. Reading materials are used to help the supervisee cognitively grasp the EFT process and concepts. The supervisor provides answers to the supervisor's theoretical questions in a safe way to build trust and develop the same cognitive map of therapy.

Goal Three: Experiential Focus

Here the emphasis is on the supervisee's ability to work with emotion and facilitate couples' bonding events. Moving beyond merely understanding concepts, an experience focus involves putting the theory into practice. Helping supervisees develop the skill set necessary to empathetically attune and stretch their experimental edge is critical to effective EFT. Palmer-Olsen et al. (2011) find that inability to manage or utilize emotions effectively is a major contributor to blocks in both therapy and supervision. Just accumulating information on how to do EFT is akin to learning how to swim by just reading a book. Consequently, movement toward the third goal focuses on experiential, bottom-up learning through role-plays, exercises, video review, and real-time supervision. The supervisor is teaching how to attune and regulate emotion by doing it with the supervisee and having the supervisee practice. Providing moment-by-moment feedback with support and encouragement enables the supervisee to gain experiential knowledge and skill.

Goal Four: Self-of-the-Therapist

The EFT supervisory experience is isomorphic to the EFT therapeutic process. For supervisees to effectively assist clients in identifying and processing their own emotions and attachment needs related to therapy and learning EFT, supervisees must successfully identify and process their own emotions and attachment needs and how those needs get triggered and affect therapy (Palmer-Olsen et al., 2011). When the supervisee's own emotional blocks get triggered, they create barriers to attunement and negatively affect the therapeutic process. The two most common therapist action tendencies associated with their own issues involve exiting the emotional process in session (withdrawal) or pushing too hard to fix issues (pursuit). When supervisors see these behaviors, it indicates that self-of-the-therapist issues are likely to be involved.

Creating the safety for supervisees to risk exploring their own emotions is necessary before shifting the focus to self-of-the-therapist issues. If a supervisee is not in a place to examine self-of-the-therapist issues, the supervisor respects that position. Ultimately, it is counterproductive to press for self-disclosure without the supervisee being open. Good supervision models the "how to" in connection as the supervisor teaches the supervisee through the supervisory dyad. When the supervisee is ready to delve into self-of-the-therapist issues, the supervisor deepens Goal Four through the HEARTS process, which stands for Here-and-now focus, Explore/energize blocks, Accept the function of the block, Resolve the block.

HEARTS Process of Self-of-the-Therapist Supervision

Here-and-Now Focus

The supervisor creates a safe environment to explore, without judgment, whatever is emerging for the supervisee in real time, both intrapsychically and interpersonally. The supervisor must be clear on the intended purpose of exploring the supervisee's self-of-the-therapist issues and gain explicit permission to proceed to actually exploring blocks.

Explore/Energize Blocks

After creating a safe, collaborative environment, the supervisor helps the supervisee discover and identify the block or trigger that is impeding attunement to a client in the therapeutic process. For example, if a silent client triggers anxiety in the therapist, the therapist may soothe his anxiety by pushing the client to open, asking the client more questions and invariably causing more silence. This leaves the therapist more anxious, the client more silent and withdrawn, and both client and therapist completely misattuned. However, recognizing this feedback loop of misattunement, and gaining insight into how the block interferes with the therapist's ability to stay with the client's experience, is not enough for actually working through the self-of-the-therapist block. Real change can only come from creating neural pathways through new experiences for the supervisee, not from simply recognizing the miss. To evoke a new experience for the therapist, the supervisee must activate the block and engage it in the here and now. The supervisor helps the block come alive by having the supervisee visualize the precise moment of the trigger—client's words, tone, facial expression, body posture—and deepens the activation by exploring the therapist's feelings around the trigger.

Bruce Ecker and colleagues' work on memory reconsolidation describes well how neuroplasticity of the brain actually works. To change the brain, the synapses that store a memory of an event must be unlocked and a new, corrective experience must be attached to the unlocked synapse (Ecker, Ticic, & Hulley, 2012). Regardless of the theoretical orientation of the therapist, the brain changes through this simple process. As the supervisor activates the block, the supervisee's old "protected" response emerges. Engaging the block brings the therapist's action tendency to protect herself from the threat of the trigger into the room.

Acceptance of the Function of the Block and Accessing Unmet Needs

Attuning to the block begins with honoring the function of the supervisee's protection. Highlighting the block often triggers shame and feelings of inadequacy in the therapist. Letting the therapist know with empathy and acceptance that his behavior and defenses make sense creates a powerful antidote to shame. After validating the therapist's good reasons for her responses, it is necessary to expand the frame and help the therapist access the unmet needs beneath the block. If a therapist exits the process or pushes a client too hard for a response due to the therapist's own survival strategies, no one, including the therapist, is seeing or comforting the therapist's underlying fears of rejection and failure. The action tendency of the block (pursue/withdrawal) to avoid being hurt (unmet needs) often ensures that therapists preemptively reject or hide themselves to avoid the possibility of hurt. Recognizing the trap reveals the costs of the blocks and allows the therapist to access the vulnerable parts of self previously hidden. A helpful image is therapists asking their loyal soldier (the block) to move aside so they can see the little child hiding behind (unmet needs).

Resolving the Block With a Resonating, Healing Response to Unmet Needs

Once the unmet needs are revealed, the timing is right for the supervisee to receive a healing, corrective emotional response that removes, reduces, or redeems the block. This reparative response from the supervisor, a member of the supervision group, or the supervisee's self arises from a place of deep connection called *resonance*. If attunement is the process of joint exploration and coordination of intentions, then resonance is the destination of successful attunement.

Joseph Pearce used an electroencephalograph (EEG) to study the brain waves of two people working together on activities with a singular purpose (McTaggart, 2011). He discovered that the two brains synchronize into brain-wave "entrainment" or resonance. In brain-wave entrainment, the electrical signaling in the brains of two people quickly coordinate. The frequencies, amplitudes, peaking, and troughing of the brain waves all begin to resonate in tandem: Literally, two separate brains reach beyond the borders of each mind and join together into one. This kind of resonance is not limited to just the brain. Many researchers are finding that a large number of physiological processes—heart waves, blood flow, respiration, body language, vocal prosody—all can come into synchrony in separate individuals (McTaggart, 2011, p. 63). It seems that our biology is the product of synchronizing with the world and the people around us (p. 69).

The supervisor guides the supervisee toward this place of clarity, new awareness, communion, synchrony, truth, matched affect, core states, and resilience. To satisfy our deepest needs, we merge with others, constantly seeking synchrony (McTaggart, 2011, p. 70). The felt sense of body and mind coming into synchrony is what we call *resonance*, which is marked by feelings of calmness and harmony. It is a sacred space of direct connection between two people from and within their most vulnerable states. No one is in control because the space between is shared.

The corrective emotional experience occurs here when supervisee-therapists get a new response to their previously hidden attachment needs. The therapist's receiving a loving response attaches a new experience to the unlocked synapse of the block, rewiring it as a positive experience and erasing the negative. Imagine what a radically different outcome a supervisee can achieve if, instead of beating himself up for failing to attune to a client, he instead acknowledges and releases his "soldier" (blocks) and comforts his scared little boy with empathy. Transformation and positive affect is the natural aftermath of resonance and corrective emotional experience. The supervisee experiencing stated feelings of joy, relief, lightness, peacefulness, and freedom are predictable physical markers of successful block resolution. There is no reason to guess; positive affect is evidence of a shift. Taking a few moments to celebrate these new positive shifts explicitly with the supervisee helps consolidate the progress made.

Tie-Back to the Therapeutic Process

Having their needs met, experientially, enables the supervisees to return to the therapeutic process of their clients and lean into the direction that was previously blocked. Liberated from the need to protect and survive, the therapist can unleash curiosity and explore new territories. If a supervisee is comforted in her fear, then attempting to attune to a client's fear become less scary. A therapist demonstrating an ability to attune and co-regulate with a client to collaboratively access emotions previously impeded is proof of effective block resolution.

Synthesize Experience Into Coherent Narrative

After celebrating the positive affect accompanying successful tracking-back (block resolution), it is important to install the whole process into a coherent narrative. Organizing the experience into a story fosters integration and replication. Neurons that fire together wire together. Putting the corrective emotional experience into a more global perspective allows the gains to be more far-reaching. Reinforcing what works increases the likelihood of similar results in the future.

CASE EXAMPLE

Brittney entered her supervision session with her supervisor, Dr. Bob, feeling disappointed about her first couple session with Travis and Tekia. Brittney described feeling lost and ineffective as Travis and Tekia argued about multiple topics. The more Brittney tried to gently interrupt the fighting and help them listen to each other, the more the fighting escalated. At the end of the session, Travis announced that he was not going to listen to any more of Tekia's criticism about "him being just like his father and not finding work" and angrily left the therapy room. Afterward, Brittney talked to Travis on the phone and the couple agreed to come back for another session. Brittney was anxious not to have another "out of control" session where she felt like a terrible therapist.

Dr. Bob noticed Brittney's discouragement and normalized how difficult couples can trigger feelings of frustration, helplessness, and inadequacy. He empathized with her struggle and validated her willingness to jump in to help. Focusing on the first goal of EFT supervision, building an alliance, Dr. Bob stressed connecting with Brittney before trying to provide advice. Dr. Bob shared his "nightmare"

(*continued*)

therapy session as a novice therapist in an attempt to let Brittney know he understands how badly she feels. In doing this, Dr. Bob role-modeled vulnerability and attempted to join Brittney in her struggle. Brittney laughed, expressing gratitude by saying, "I guess it's true that misery loves company."

Feeling understood and not judged, Brittney then asked Dr. Bob for his help so she could be more effective in the next session. Dr. Bob asked her to describe a piece of the session where she felt most stuck. Brittney depicted the end of the session where Tekia told Travis he needed to work harder, unlike his father, in his job search and send out more applications. Travis replied that she needed to work harder disciplining their kids. Brittney didn't know where to go as each partner was firing off complaints. Dr. Bob used this example to help Brittney shift focus from the surface content to the underlying attachment process. Providing a map to conceptualize the unfolding attachment drama between partners is the second goal of EFT supervision.

Dr. Bob encouraged Brittney to try and track the recurring sequences of interactions between the partners that maintain the couple's distress. Escalating emotional tensions lead both partners to utilize adaptive protective strategies, the two most common being pursual (pursuit), which is an intense hyperactivating emotional response (moving toward the partner); or withdrawal, which is a deactivating emotional response (moving away from the partner). The therapist's primary intention is first to assist each partner in identifying their protective strategies and then to expand the couple's frame so each partner can recognize how their protective responses pull defensive strategies from their partner, mutually constructing an interdependent negative cycle. Couples become stuck in predictable patterns of reactivity marked by increased emotional distance and decreased responsiveness.

Mapping out the moves between Travis and Tekia reveals a common negative cycle of pursuit and withdrawal. Tekia is anxious about the family's financial situation and reaches out to Travis for information in the hopes that he can calm her fears. She moves toward Travis to face the challenge together. Unfortunately, Travis, who is feeling bad about his lack of progress in finding a job, hears Tekia's questions as criticism. This criticism is compounded by cultural messages of the challenges African American males face in finding employment. He doesn't notice Tekia's bid for connection; rather, he experiences it as an attack. Travis's underlying fear of failure is reinforced with each of Tekia's suggestions and attempts to help with advice. Every word points out either what he is not doing or what he is doing wrong. His specific failures in the family further reinforce his cultural perspective that the system is setting him up to fail. Real opportunities are

(*continued*)

unavailable and following the rules gets him nowhere. Despite his efforts, he keeps finding himself stuck on the bottom. Travis's mounting pressure is exacerbated by Tekia's attempts at conversation and he withdraws to reduce the tension and avoid escalation. When Travis withdraws, Tekia feels rejected and dismissed. Her anxiety increases as Travis refuses to engage. She follows Travis into another room and continues to force the conversation, letting Travis know he needs to do more. As Tekia's unrelenting messages of disappointment, doubt, and failure bombard Travis, he feels trapped. The criticism he hoped to escape is following him. With nowhere to turn, he lashes out with anger, screaming at her about her parenting.

Dr. Bob helped Brittney see the negative cycle running rampant. Trying to connect and empathize with each partner's struggle, including awareness of how cultural messages reinforce interpersonal reactivity, is crucial for building a healthy alliance. Both partners are protecting themselves and are unaware of how their protection is truly traumatic for their partner. If Travis is discouraged in his role as a provider, he desperately needs encouragement and Tekia's assistance in reducing the mounting pressure. Instead, he gets advice, criticism, heightened pressure, and reinforcement of his failures, all of which increase the likelihood of his withdrawal. If Tekia needs Travis by her side to share her fears and empower her to have some influence in the family's financial future, then his disengagement is the worst-case scenario. His withdrawal fuels her anxiety, helplessness, and loneliness, which drive Tekia to protest further. For Tekia, a negative response is better than no response at all. Making explicit these negative patterns that fuel reactivity and replacing them with positive cycles that foster responsiveness and secure attachment is the main objective of EFT.

Possessing a map doesn't just show someone where they are (identifying the negative cycle); more importantly, it highlights where they need to go (positive cycle). It is crucial that the therapist possess both a starting point and a destination. Dr. Bob gave Brittney a glimpse of how Travis and Tekia's conversation could have ended differently. What if each partner explicitly shares their attachment needs instead of their protective strategies? Imagine Tekia telling Travis, "I know you are trying your best, this is about my fears, not that you are doing anything wrong. I'm just really scared our financial situation is going to get worse and maybe if we keep fighting you will not want to be with me anymore." This bid for connection is direct and Travis replies, "Oh my goodness, I never knew you were scared I would leave you. You are the best thing in my life and we'll figure it out [reaching over to hug her]. Let's talk and come up with a plan." This vulnerable conversation pulls Travis closer as opposed to Tekia's reactivity that pushes him away.

(*continued*)

It is important to realize that Travis is equally empowered to initiate a vulnerable conversation. Envision Travis saying, "I know you are nervous and justifiably want information to calm your fears. Yet I'm trying my best and I hear your questions as evidence that I'm a loser. I get so discouraged and weighted down that I just want to escape. I need help to lift my spirits but I don't know how to ask for help." Another direct expression of a vulnerable need, to which Tekia responds, "Wow, I didn't realize my questions make you feel like you are a loser. I'm so sorry. I'll be your biggest cheerleader [reaching over to hug him]. We can do it together."

The attachment map was really helpful for Brittney in understanding the couple's dynamics. Next she wanted help in implementing the map. Dr. Bob set up a role-play where he played the role of Travis and directed Brittney to practice finding her words to get alongside Travis. The third goal of EFT supervision is the therapist working with real-time emotion. Brittney needs practice finding her words to go deeper emotionally with her clients. Like the theory of change in EFT, isomorphic corrective change arises from new experience, not just insight. Insight is essential to create the safety needed to take risk, but it is the reaching for and responding back that fosters secure connections.

In the role-play, Brittney did a great job validating Travis's vulnerabilities and fears of failure. She clearly empathized with his pain of not being able to amply provide for his family. A few times Brittney needed help finding the right word, but Dr. Bob provided assistance and she quickly got back into a groove. In the role of Travis, Dr. Bob used his own sense of attunement to feel if Brittney was alongside, and he provided moment-by-moment feedback to help her stay close. When she got out too far ahead or fell too far behind where Travis was in the moment, Dr. Bob facilitated minor adjustments. EFT supervisors emphasize that the best way to learn emotional engagement with others is by hands-on learning, stressing less supervisor talking and more supervisee doing. To become good at emotional work, one needs corporeal practice, not abstract ideas.

Brittney did an excellent job of co-regulating Travis's implicit vulnerable attachment needs. Travis (Dr. Bob) felt new awareness and understanding. Brittney then asked Travis what it would be like to tell Tekia about his vulnerable needs for support. Immediately, Travis balked and started to get frustrated. This is a common reaction for a partner who is touching a vulnerable part of himself or herself, because it is unfamiliar and they mistrust their partner's response. There is too much risk of failure. Brittney lost attunement with Travis's frustration and pushed Travis to talk about his fears. When Travis grew more frustrated, Brittney went silent. Dr. Bob paused the

(*continued*)

role-play to explore what was happening in Brittney, and whether she was needing a little help finding a word to attune to the frustration (third goal) or if she was being triggered by her own experiences.

The fourth goal of EFT supervision is temporarily putting aside the therapeutic process and opening a space to explore self-of-the-therapist blocks. In this situation, Brittney was freezing, unsure how to lean into Travis's anger. She was fluent in his vulnerability but uneasy with his frustration. Dr. Bob and Brittney agreed to discuss the possibility of using the HEARTS process to help Brittney work through her block of attuning to Travis's anger.

H = Here-and-now focus; expresses clear intentions about the process and an emphasis on taking a curious, nonjudgmental stance. It is critical for the supervisor to empower the supervisee with decision-making power. Dr. Bob explained to Brittney that blocks to attunement with clients are normal to the therapeutic process and if Brittney wants to explore the very good reasons why Travis's anger is so challenging, they could investigate it together. If she was uncomfortable shifting focus toward her own personal blocks, then they could continue to focus on the case. Dr. Bob also explained that if Brittney decided to explore the block, she could stop at any point. Brittney cautiously agreed.

E = Engage/energize the block. Dr. Bob had already identified the block as Brittney's withdrawal response to Travis's anger; now it was time to activate the block and make it explicit. Dr. Bob asked Brittney to describe Travis's face, body language, tone, and words. As Brittney re-experienced the scene, she responded to Travis's "mean eyes" with fear and a feeling of wanting to run away. With her block in regard to Travis's anger energized and the synapse where it was stored unlocked, Dr. Bob moved to understanding her protection.

A = Attune to the function of the block and the supervisee's unmet needs. With compassion and curiosity, Dr. Bob explored Brittney's strong response to anger and wondered what was familiar about wanting to run away from anger. The basic tenet of EFT and EFT supervision is that the behavior always makes sense when you understand the attachment context. Brittney described growing up with an alcoholic, abusive dad who would explode; she constantly had to hide to avoid violence. Everyone in her family walked on eggshells, although no one ever talked about their fears. Brittney's only option to deal with her father's anger was withdrawal. There was no room for her to fight back or to question what was bothering her dad. Her hiding kept her safe. Dr. Bob validated her great reasons for withdrawing in the face of

(*continued*)

anger and normalized how her body was responding in the same way with Travis. Brittney's nervous system registers threat and retreats. Dr. Bob also opened up space for Brittney to see how her withdrawal hides her own feelings of terror and massive helplessness. As Dr. Bob validated her protection, he expanded the frame so that Brittney could see how her short-term avoidance ensures that no one sees the vulnerable part of her that is so afraid. Her protection guarantees that her need for comfort goes unmet. She is stuck in a double bind where withdrawal is necessary, but as long as she avoids no one can ever see and respond to her sadness. Dad's failure to attune to Brittney created a survival coping strategy of avoidance that caused Brittney to be unable to attune to her own vulnerability.

R = Resolve the block. As Brittney was able to identify both her block and her unmet attachment needs, it became possible to provide Brittney with a corrective emotional experience attached to the old block. From a place of resonance and understanding, Dr. Bob was able to respond to Brittney's unmet needs. He expressed sadness that no one ever acted to protect Brittney, and as a result Brittney had no option but to hide. He noted how hard it was for her, explaining that when Brittney was triggered with threat, her withdrawal guarantees that she remains stuck in herself with the same lack of responsiveness she experienced with her parents. Dr. Bob conveys his hope for Brittney to learn how to be present and fight for a new response. He encouraged Brittney to imagine herself as a little girl hiding in the closet and feeling the unfairness of the situation. Not only was this little girl all alone, but no one came to her rescue. Then Dr. Bob asked Brittney to imagine herself as an adult opening up the closet door and rescuing the little girl. This powerful image created permission for Brittney to experience self-compassion and comfort. Brittney began to cry tears of relief.

Slowly and gently, Dr. Bob reinforced the new, positive response by telling Brittney how proud he is of her efforts and how he too wants to fight for her. He lets her know that she deserves comfort: we all do. Positive affect and transformation are predictable outcomes of successfully resolving a block. Brittney described feeling lighter, calmer, hopeful, and freer—all physical markers of expansion into new areas previously blocked. Dr. Bob helped Brittney put words to her new experience of responsiveness and stressed how the positive affect is proof of her getting to the other side of her block.

T = Track back. With the block being identified and worked through, it was possible to track back to the therapeutic setting and see

(*continued*)

if Brittney could lean into the direction that was previously obstructed. Before, Brittney withdrew from Travis's anger. Dr. Bob wanted to see if she can move toward Travis's anger. Learning to match affect is essential for attunement. If clients are angry, they signal their anger by demanding that the therapist pay attention and listen. Brittney's avoidance of the anger, signaled through her own withdrawal, intensified Travis's frustration. A calmer, less afraid Brittney, armed with the awareness of how her avoidance triggers Travis's anger, wanted another chance to role-play attuning to the anger.

Brittney recognized Travis (Dr. Bob) from a more empathic place: "It must really frustrate you when Tekia offers advice and sends the message you aren't doing enough. You know how hard you try and her words don't capture your real effort. Correct me if I am wrong, but that must feel pretty unfair." Dr. Bob smiled in the role of Travis. He felt his body relax and his anger diminish as Brittney conveyed attunement and understanding. Brittney's validation of Travis's anger and the good reasons for his defensiveness, and effectively giving Travis permission for his protection, he no longer needed to be angry.

S = Synthesize the experience. With Dr. Bob's help, Brittney organized her story into segments; identifying her block with regard to anger, honoring the valid motives for the block, recognizing the unmet needs of fear and loneliness underneath the block, working through the block by having a new experience of self-compassion and receiving comfort for her needs, tracking back to the therapeutic relationship, and learning to attune to Travis's anger. Dr. Bob and Brittney celebrated a job well done.

CONCLUSION

Like EFT, EFT supervision embodies a humanistic, experiential foundation. Through a secure supervisory alliance grounded in attachment theory and experiential process, including the self-of-the-therapist exploration, EFT supervisees are given the unique opportunity to gain invaluable perspective on their own emotional experience, which is always part of the therapeutic process. Attuning to supervisees, like attuning to clients, demands flexibility and self-awareness. Recognizing the many levels of intervention—alliance, conceptual, experiential, and self-of-the-therapist (ACES)—provides a broad framework for the supervisor to meet the supervisee's varying needs. EFT supervision allows the supervisor, supervisee, and client to experience growth and change in the process of healing and the practice of Emotionally Focused Therapy.

REFERENCES

Dalton, J., Greeman, P., Classen, C., & Johnson, S. M. (2013). Nurturing connections in the aftermath of childhood trauma: A randomized controlled trial of emotionally focused couple therapy (EFT) for female survivors of childhood abuse. *Couple and Family Psychology: Research and Practice, 2*(3), 209–221.

Dessaulles, A., Johnson, S. M., & Denton, W. (2003). Emotion focused therapy for couples in the treatment of depression: A pilot study. *American Journal of Family Therapy, 31*, 345–353.

Ecker, B., Ticic, R., & Hulley, L. (2012). *Unlocking the emotional brain: Eliminating symptoms at their roots using memory reconsolidation*. New York, NY: Routledge.

Elliott, R., Watson, J., Goldman, R., & Greenberg, L. S. (2004). *Learning emotional-focused therapy: The process-experiential approach to change*. Washington, DC: American Psychological Association.

Furrow, J., Johnson, S., & Bradley, B. (2011). *The emotionally focused casebook*. New York, NY: Brunner/Routledge.

Greenman, P., & Johnson, S. (2013). Process research on EFT for couples: Linking theory to practice. *Family Process, 52*(1), 46–61.

Halchuk, R., Makinen, J., & Johnson, S. M. (2010). Resolving attachment injuries in couples using emotionally focused therapy: A 3-year follow-up. *Journal of Couple and Relationship Therapy, 9*, 31–47.

Hughes, D. (2007). *Attachment focused family therapy*. New York, NY: Norton.

Johnson, S. M. (2004). *The practice of emotionally focused couple therapy: Creating connections*. New York, NY: Brunner-Routledge.

Johnson, S. M. (2008). *Hold me tight: Seven conversations for a lifetime of love*. New York, NY: Little, Brown.

Johnson, S. M. (2013). *Love sense: The revolutionary new science of romantic relationships*. New York, NY: Little, Brown.

Johnson, S. M., Bradley, B., Furrow, J., Lee, A., Palmer, G., Tilley, D., & Woolley, S. R. (2005). *Becoming an EFT therapist: The workbook*. New York, NY: Brunner-Routledge.

Johnson, S. M., & Greenberg, L. (1985). The differential effectiveness of experiential and problem solving interventions in resolving marital conflict. *Journal of Consulting & Clinical Psychology, 53*, 175–184.

Johnson, S. M., Hunsley, J., Greenberg, L., & Schindler, D. (1999). Emotionally focused couples therapy: Status and challenges (a meta-analysis). *Journal of Clinical Psychology: Science and Practice, 6*, 67–79.

McTaggart, L. (2011). *The bond: Connecting through the space between us*. New York, NY: Free Press.

Mehr, S. E., Bahrami, F., Karami, B., Eslami Mehr, Y., Hedayati, A. M., Ahmadi, S., & Rozeyan, A. (2014). Studying the effect of emotion focused therapy on couples' attachment styles. *MAGNT Research, 2*(5), 595–602.

Palmer, G., & Johnson, S. M. (2002). Becoming an emotionally focused couple therapist. *Journal of Couple and Relationship Therapy, 1*(3), 1–20.

Palmer-Olsen, L., Gold, L., & Woolley, S. (2011). Supervising emotionally focused therapists: A systematic research-based model. *Journal of Marital and Family Therapy, 37*(4), 411–426.

Stern, D. N. (1985). *The interpersonal world of the infant*. New York, NY: Basic Books.

van Ijzendoorn, M., & Sagi-Schwartz, A. (2008). Cross-cultural patterns of attachment: Universal and contextual dimensions. In J. Cassidy & P. R. Shaver (Eds.), *Handbook of attachment: Theory, research, and clinical applications* (2nd ed., pp. 880–905). New York, NY: Guilford Press.

EIGHTEEN

Training Feminist Family Therapists

Anne M. Prouty
Markie L. C. Twist

Feminist family therapy is a meta-philosophy (often referred to in the literature as a *feminist-informed lens*): a group of feminist perspectives that can focus and expand psychotherapy models. Initial feminist critiques focused on noticing sexist assumptions within systems theory and therapy models and quickly moved to tracking power dynamics in relationships, families, communities, and societies (e.g., Avis, 1988; Comas-Díaz & Greene, 1994; Goldner, 1989; Goodrich, Rampage, Ellman, & Halstead, 1988; Hare-Mustin, 1978, 1989). It is useful for feminist trainers to be familiar with several types of feminist perspectives because each is helpful in its own way and because therapists will vary as to which perspectives interest them. Therefore, in this chapter we review several feminist perspectives and how each can inform a feminist family therapy lens. We then discuss the role of feminist clinicians, the process of change, specific feminist-informed conceptual and executive techniques, and the role of feminists as supervisors and trainers, and conclude with applying feminist family therapy to the book's common case example.

TEACHING THEORETICAL CONTEXT: FEMINIST PERSPECTIVES

Several kinds of feminism have informed family therapist scholars and trainers. Quite often neither a scholar nor her writing falls neatly into only one form of feminism. We have found it helpful, however, in our work training therapists to help them understand that there is not only one kind of feminist, and that feminists do not all agree. To use a systems theory concept, different feminist foundations *punctuate* different aspects of social process and power. Understanding the different feminist perspectives can explain how perspectives on topics like pornography and marriage can differ and yet both be feminist.

Radical Feminist Foundations in Family Therapy

Radical feminism is probably the most widely known, as this perspective focuses most keenly on the disproportionate allocation of social power between women and men and the vehement proposal that society's only option is for transformation to equal power. The examination of gendered power in family therapy first appeared in Hare-Mustin's (1978) critique of systems theory. Interventions based on neutrality were seen as passive agreement with the status quo that disproportionately favored and empowered white, heterosexual, stereotypically gender identifying men. Several feminist therapists and scholars (e.g., Goodrich et al., 1988; Walters, Carter, Papp, & Silverstein, 1988) formed groups to support each other in the re-envisioning of family therapy models (e.g., Ault-Riché, 1986; Dankoski & Deacon, 2000; Knudson-Martin, 1994; Leupnitz; 1988; Walsh & Schneikman, 1988) and re-envision, design, and implement interventions that would dismantle gendered power in families and within the therapy room.

It was the initial group of radical feminists who first critiqued systems theory as being too often applied to people without considering social power inequities. Radical feminists paid most of their attention to gender but also acknowledged that racism (McGoldrick, Garcia-Preto, Hines, & Lee, 1989; Pinderhughes, 1986), classism (Kliman, 1998; Laszloffy, 2008; Walsh, 1989), and heterosexism (Laird, 1988) also affected people in couples and families differently, and argued that overt attention to the origins and outcomes of power dynamics should be incorporated into the therapist's work. Radical feminists in family therapy focused on couple dynamics and urged family therapists to promote egalitarian couple relationships that would include shared parenting, negotiated workloads, and avoidance of mother-blaming and father-praising in therapy (Ault-Riché, 1986; Avis, 1985; Haddock, Zimmerman, Current, & Harvey, 2003; Prouty & Lyness, 2011; Tamasese, 2003).

Finally, the discussion of gendered power expanded how we thought about emotion as an adult skill (Goldner, 1989; Knudson-Martin, 1994) and an important aspect of training the person-of-the-therapist (Aponte & Kissil, 2012). Feminist supervisors focus on training therapists to elicit, validate, and effectively use emotion as a "way of knowing" (Belenky, Clinchy, Goldberger, & Tarule, 1997) and understanding the social world and experience (Prouty, 2001). Feminist trainers often see emotional depth, understanding, and congruency as adult skills that are equally important to cognitive ways of knowing as well as an indication of a flexible mind more capable of managing relationships and life challenges. Feminists stress the importance of validating a wide variety of clients' and therapists' emotions, including those sometimes restricted by culture, like women's anger (Flemke & Allen, 2008). Feminists have also examined the development of complex and nuanced *emotional processes* as a key aspect of *moral reasoning* and have stressed that therapists should move away from the superficial definition of change as a way to feel better and instead to think of change as a way to *be* better

(Goldner, 1989) and to "struggle well" (Walsh, 1998). Feminists have urged the social contextualization of people's emotional experiences and responses, especially in regard to gendered roles in couple therapy (Goodrich, 1991; Vatcher & Bogo, 2001).

Queer-Informed Radical Feminism

Scholarly understandings of "gender" around the globe are now bringing about a newly emergent critique of feminism, and thus influencing each one of these feminist meta-perspectives. Indeed, just as radical feminists critiqued systems theory and theorists as being applied to people without consideration of social power inequities, some scholars are critiquing many forms of feminism as being applied to cisgender people without consideration of nonbinary (male or female) gender-identifying or sex-identifying persons (Ansara & Hegarty, 2014; Blumer, Ansara, & Watson, 2013); they urge deconstruction of the concept of gender and cisgender (Ansara & Hegarty, 2014; Blumer et al., 2013). Ansara's global perspective of gender has enabled him to deconstruct gender in ways that have added to critiques of gender and relatedly of gender-focused feminism. Ansara and colleagues have noted that there are places on the globe where there are three to five gender classifications (e.g., Bartlett & Vasey, 2006; Peletz, 2009) and cultures where gender shifts over one's life span. In these global contexts of understanding gender, gender is viewed as being more than binary or more than cisgender. Hence, feminist perspectives informed by the critiquing of cisgenderism, and the privileging of patriarchies rooted in binary gender constructions, have revealed that the very conceptualization of gender by Westerners has been rooted in cultures with ethnocentric and reductionistic perspectives (Ansara & Hegarty, 2014). Thus, post-gender feminist trainers enable therapists to recognize the existence of people whose assigned gender differs from their gender self-designation as well as to recognize people who have nonbinary identities (e.g., gender queer, bigender, transgender, two-spirit, polygender, or agender).

Cultural Feminist Perspectives

Cultural feminism arose concurrently with the radical feminist critique and quickly brought into discussion the construction of social power as it intersected with class, ethnic identity (including religion; e.g., Almeida, 1993; Ault-Riché, 1986; Boyd-Franklin & Garcia-Preto, 1994; McIntosh, 1988), gender, age, physical ability, and soon thereafter sexual orientation (Carlson, McGeorge, & Toomey, 2013; Greene, 1994; Long & Serovich, 2003). Cultural feminist family therapists have helped the family therapy field to place therapists' own cultural identities, as well as those of their clients, in the forefront of clinical and research training. For example, McGoldrick and colleagues

(McGoldrick, Giordano, & Garcia-Preto, 2005; McGoldrick, Giordano, & Pearce, 1996; McGoldrick, Pearce, & Giordano, 1982) have provided key textbooks to aid therapists in their self-examination of their own ethnic values and assumptions as well as the essential need to be respectful, curious, and admiring of the complexities of ethnic identity and cultural systems. Cultural feminists have encouraged systemic trainers to help novice therapists notice unexamined and unearned privileges: for instance, white privilege, which might be expressed in a phrase such as "race doesn't matter because people are all the same deep down"; and social class privilege, which might be expressed via unexamined ideas like "we are all mostly middle class." Recently, feminists like Walsh (1999), and Helmeke and Sori (2006) have emphasized that religion and spirituality are at the heart of many people's cultural identities. Religion often structures power dynamics within families as well as the daily and yearly rhythm of families' lives (Imber-Black & Roberts, 2003). In addition to being the heart of cultural identity, religion and religious traditions also provide paths of resistance to many forms of racism and xenophobia.

Within family therapy, the socialist feminist perspective has largely been woven into the larger cultural feminist perspective. The socialist feminist lens focuses on the effects of class and money on people's access to political power, resource power, and the power of different types of knowledge (from how things are made to formal education). In her essay entitled "For Love or Money," McGoldrick (1991) rocked the field by suggesting that couple therapy not take place until the couple's assets had been equally divided. Feminist therapists informed by the socialist perspective are trained to pay attention to how work history, skills training, education, and other monetary assets like inheritances influence the power dynamics in couples, families, and communities. This perspective also examines the complexities of class as going beyond money to look at how class informs various aspects of the human condition including life trajectories, the definition of *family*, and mental health. Aponte's classic eco-structural book, *Bread and Spirit* (1994) expanded structural family therapy with a simultaneous inclusion of identity, hope, and communion while sensitively examining the positive and negative effects of intergenerational poverty on families. Relatedly, one of the most widely talked about topics among feminist family therapists is the positive and negative effects of parents who work outside of the home and the distribution of work inside the home (Zimmerman, 2001). Balancing work and family depends a great deal on resources that are familial, social, geographical, and class dependent. In addition, cultural and gender dynamics inform families' priorities and levels of guilt (Haddock et al., 2003).

Liberal Feminist Perspectives

Liberal feminism has been more lively within the sister discipline of family studies, contributing to the specialty area of family policy. Within family therapy, feminists who endorse a liberal feminist perspective learn to examine how laws, rules, and public policies affect the lives of people and families differently. Some of the most commonly talked about topics among liberal feminists are equal pay for equal work, marriage, divorce, and custody. Liberal feminists (e.g., Blumer, Haym, Zimmerman, & Prouty, 2014; Prouty-Lyness, 2005; Twist, 2006) examine intended and unintended consequences and bring to light those laws and policies that oppress people, inhibit flexibility of gender roles, and limit family options for healthy development over the life span (e.g., parental leave, family leave, tax codes, marriage and divorce laws, and health care laws). Mostly informed by cultural and liberal feminist foundations, perceptual and intervention skills enable the feminist therapist to help clients deal with the effects of public policies and laws in their lives. This is especially important in the areas of family law, medical law, financial laws, and immigration law and policies.

Postcolonial Feminist Contributions

Postcolonial feminists have expanded feminist family therapy to examine cultural and community expertise as extremely important ways of creating knowledge and contributing to mental health (Almeida, Dolan-Del Vecchio, & Parker, 2007; Tamasese & Waldegrave, 1994; Waldegrave, 2005). People's identities are extremely complex, and how people experience and prioritize different aspects of their identities often depends upon the social context (Hernández, Taylor, & McDowell, 2009; Kirk & Okazawa-Rey, 2003; Hernández & McDowell, 2010). In addition to prioritizing and experiencing different aspects of one's personal identity in different ways in different situations and in different locations, we also experience ourselves differently in relation to others around us. Kirk and Okazawa-Rey have talked about people's various identities as "social locations" (e.g., being heterosexual, Christian, a parent, a daughter, married, etc.) that socially construct identities in relation to each other and in varying contexts. Exploring the various intersections of one's social locations can help one better understand how those locations interact with and are perceived by those around them (for example, use the Intersectionality Grid developed by Prouty, Bermúdez, Helmeke, & Ko, 2011; see also Prouty, 2014). For instance, some social locations may hold privilege while others are minoritized or oppressed within the same context, and locations that are privileged in one context may be oppressed in another.

Postcolonial feminists have contributed much to the conversation and design of therapists' training in the 21st century. Almeida, Hernandez-Wolfe, and Tubbs (2011) encouraged trainers and trainees to build a critical consciousness and to guard against using cross-cultural/national experiences as even unintended and gentle forms of colonization that use knowledge as the commodity (Bacigalupe, 1998). Hence, postcolonial feminists are also leading the way in developing cross-cultural and international training, with a focus on not colonizing, but rather expanding family therapy training ideas and methods (e.g., Bacigalupe, 1998; Fang, McDowell, & Holland, 2006). They have called for challenging cultural encapsulation and sociocentrism, examining how cross-border social policies and economics influence the development of family therapy and mental health.

Technology can help systemic therapists to share and export more knowledge embedded in the experiences of local people and therapists, rather than using only imported ideas. McDowell and colleagues (e.g., Fang et al., 2006; McDowell, Goessling, & Melendez, 2012) have shared their methods of internationalizing courses and training workshops that involved co-creation of what defined both problems and solutions, ideas for working with families, and methods of sharing ideas across cultures. In our global classrooms and therapy rooms, postcolonial feminists are thereby emphasizing the importance of discovering people's perspectives, priorities, and ways of living so as to enhance our training, practice, and research.

THE PROCESS OF CHANGE

Feminist perspectives inform models of therapy, and each therapist connects with one or more of the branches of feminism in relation to who they are as a person and with what their clients present. Supervisors can help therapists to learn to focus on different goals of change, which can vary depending on the type of feminism espoused. In addition, feminist therapists believe that the only tool we ultimately have in therapy is ourselves, and that change occurs through use and recognition of the self-of-the-therapist (Aponte, 2009; Aponte & Kissil, 2012; Aponte & Winter, 1987). Thus, a strong trainer-therapist-client alliance is important (Brown, 2006; Prouty, 2001).

Supervisors can help therapists to develop their feminist lens through increasing their ability to notice power dynamics and how power intersects with identity variables such as gender and ethnic identity. Live supervision, review of digital video footage of clinical work, and movie reviews (Blumer, 2010) are all methods that trainers use to help new therapists learn to recognize dynamics related to gender and power (see, for example, Keeling, Butler, Green, Kraus, & Palit, 2010). Reviewing movies and digital footage of clinical and supervisory work is particularly useful when learning to recognize microaggressions, because the digital footage can be paused for discussion. Because emotional competency is important to many feminists, self-reflection is commonly integrated into self-of-the-therapist training, because learning to respect one's own emotional intelligence is essential to helping clients to do the same.

ROLE OF THE THERAPIST

The specific role of the feminist therapist will vary slightly based upon the feminist perspective from which the clinician is operating. Nevertheless, there are some commonalities in the role of the feminist family therapist. For example, feminist family therapists develop noticing skills so that covert power dynamics in therapy sessions do not continue to be normalized and perpetuated. The Power Equity Scale provides the trainer with a way to scaffold therapists' learning to notice and develop these interpersonal skills (Haddock, Zimmerman, & McPhee, 2000).

Therapists are different, yet trainers can help a feminist therapist to develop a style and several interventions by which to interrupt clients' covert attempts to reinforce various stereotypes and destructive power dynamics, so as to encourage human dignity and empowerment. For example, a therapist might state that she is "confused and needs just a bit of clarification" when a man states that he babysat his children or washed his partner's dishes. Therapists aware of cisgenderism might start all sessions with first-time clients by introducing themselves not only by their names, but also with their preferred gender pronouns (e.g., he, her, she, him, them, they, their, zie, hur), and invite their clients to do the same, so as not to assume the gender of any of the clinical co-participants (Blumer et al., 2013). This latter point is important because if one does not know the self-designated gender of the clinical co-participants with whom one is working, then a full picture of the role that gender and power are playing in their relational dynamics and presenting problems may not be clear (Blumer et al., 2013).

There are several ways in which trainers can assist therapists with improving their noticing skills regarding power and gender dynamics. They can encourage therapists to become more aware and take note of the way that power and gendered dynamics are occurring in their own lives, in the media, and in the world at large. Seeing these dynamics in themselves and their own lives can help them to better recognize similar dynamics in their clients. Using self-disclosure (Cheon & Murphy, 2007) about this awareness of said dynamics with one's clients, when appropriate, can also help with another common role of the feminist clinician: serving as a role model (Blumer & Barbachano, 2008; Green & Dekkers, 2010; Mowbray, Lanir, & Hulce, 1984).

There are also several ways in which trainers can encourage clinicians to go beyond noticing these dynamics to uncovering them by bringing voice to them (Turner & Avis, 2003). This can occur through creating collaborative working relationships in supervision (Blumer, Green, Compton, & Barrera, 2010) as well as in therapy (Brown, 2006). Collaborative relationships can be established for creating the space for therapists and their clients to recognize themselves as the experts on their own lives and encouraging their use of this expertise (Blumer et al., 2010). This can be particularly important when working with female-identifying therapists, as women who have been discouraged from owning expertise and opinions may need more support and encouragement both in the supervisory and in the therapy room. This dynamic can play out isomorphically in therapy, as both therapists and clients may need to relearn encouragement of women's expertise and opinions. Indeed, a study by Werner-Wilson, Price, Zimmerman, and Murphy (1997) revealed that female clients were interrupted with greater frequency than their male counterparts.

TECHNIQUES

Facilitating a Curious and Respectful Examination of Human Diversity and Power Dynamics

One of the most important first skills that a feminist therapist must learn is how to facilitate authentic and genuine dialogues. Several feminists have written about the importance of facilitating genuineness through having conversations about the similarities and differences between supervisees in training groups, as well as the supervisor and supervisee, early on in the professional relationship (Blumer & Barbachano, 2008; Turner & Avis, 2003). Areas to attend to include paying attention to gender (Helmeke, 1994; Keeling et al., 2010; Roberts, 1991), power (Haddock et al., 2000), ethnicity (Almeida, 1993; Boyd-Franklin & Garcia-Preto, 1994), sexual orientation (Rock, Carlson, & McGeorge, 2010), self-of-the-supervisor and -supervisee concerns (Prouty, 2001), and personal aspects of the supervisor and supervisee that can be used in an active and purposeful way to further the goals of supervision (Korinek & Kimball, 2003; Prouty, 2014). Such sharing can be accomplished through the use of a background questionnaire (Blumer & Barbachano, 2008), the "intersectionality grid" (see Prouty, 2014), or following Aponte and Kissil's (2012) person-of-the-therapist model of supervision and training.

Areas of importance from several feminist perspectives can be covered in a way that normalizes the examination of similarities and differences and provides the supervisee a structured means of communicating about his/her/their background and ways in which different aspects of her/his/their experience facilitates and provides barriers to power. Topics to be covered can include regions of residence, migration history, culture and ethnicity, socioeconomic experiences, military history, parenting experience, relevant professional experience (clinical, research, teaching), previous supervision experience (format, setting length, likes, dislikes), strengths and areas of growth of the supervisee, supervisee goals of supervision, and perception of power in the supervisory relationship (expectations of supervisory process with other supervisees, evaluation procedures). As the supervisee shares his/her/their individual background, the supervisor has a chance to self-disclose about her/his/their background as well—highlighting areas of similarity and difference for discussion (Blumer & Barbachano, 2008). This can also be useful in small-group supervision, as it allows for practicing respectful and curious questions. Additional benefits of such discussion, besides assistance in the development of a strong working alliance, are that it can be a means of reducing power conflicts (Murphy & Wright, 2005) and can lead to higher levels of satisfaction with supervision, particularly when supervisors facilitate conversation regarding culture and express interest in the culture of the supervisee (Cheon, Blumer, Shih, Murphy, & Sato, 2009; Duan & Roehlke, 2001; Gatmon et al., 2001; Green & Dekkers, 2010).

Recognizing Gendered Power in Clients' Lives

Supervisors and trainers who need a basic evaluative tool to help therapists learn to implement feminist ideas in therapy can use the Feminist Family Therapy Scale by Black and Piercy (1991); the Feminist Family Therapy Checklist by Chaney and Piercy (1988); the Feminist Couple Therapy Scale by McGeorge, Carlson, and Toomey (2015b); or the Power and Equity Guide developed by Haddock and colleagues (2000). Looking more closely at the latter illuminates the themes common across all three, as the Power and Equity Guide facilitates three therapeutic goals. The first goal is to consciously manage the power differentials between the therapist and the clients. This is not to say that the therapist is always in a position of more power in all aspects, as this depends on the identity characteristics of all involved. But feminists do wish therapists to acknowledge their skills and abilities and to acknowledge the business dynamics as well as the vulnerability of clients in pain and under duress. The second goal is to reduce or eliminate the power inequities between client partners. This goal encompasses all aspects of the intimate relationship(s), including communication, money, sex, work within and outside of the home, emotional care, parenting, and in taking responsibility for the relationship. In the area of interpersonal violence, the Power and Control wheel developed by Almeida and colleagues (Almeida & Lockard, 2005) is a tool that can help therapists to understand how interpersonal violence and destructive power dynamics intersect with gender, sexual orientation, racism, and economic resources.

ROLES OF THE SUPERVISOR AND SUPERVISEE

Supervisor-Supervisee Relationship

Although there are myriad ways for a supervisor to create a strong working alliance with a supervisee, feminist supervisors use contracting to promote safety and clarity in the supervisor-supervisee relationship (Prouty, Thomas, Johnson, & Long, 2001). Through discussion of mutual and individual goals, contracting assists in the creation of an opportunity to focus on strengths and build the therapist's skills in an empowering manner. A focus on strengths and empowerment may be particularly important in working with supervisees who are ethnic/citizen/sexual orientation/gender identity minority persons, especially when they have already had experiences with marginalization and discrimination. Feminist supervisors take time to establish a personal relationship that focuses on both current and long-term learning goals of the therapist while modeling attention to the supervisor's learning goals, skills, knowledge, and limitations. Care is taken to establish how the therapist can have access to the supervisor as both a trainer and a professional mentor.

Researchers advocate that clinical program personnel create space for dialogue, including fostering an environment where clinical interns and supervisees who may have minoritized identities feel comfortable choosing to be "out" (Long & Serovich, 2003). One way to help create a space for such dialogue is through first establishing a strong working relationship (Blumer & Barbachano, 2008). Indeed, recent research in the family therapy field has shown that it is the supervisory relationship itself—specifically the working alliance—that most contributes to supervisee satisfaction in supervision (Cheon et al., 2009). In looking at this safety dynamic from both the supervisee and trainer perspectives, the research team of McGeorge, Carlson, and Toomey (2015a) developed affirmative training guides for both therapists and trainers.

Balancing Directive Teaching With Collaborative Teaching

Proportions of directive and collaborative training interventions are based on the developmental level of the therapist, the context in which the supervision contract exists, the supervisee's therapy model, and the supervisor's philosophy of supervision. For example, feminist-informed structural supervisors may use more directive call-ins than feminist-informed narrative supervisors. Regardless, one of the hallmarks of feminist supervision is the use of transparency: Feminists may use power-over (directives) in a teachable moment but will debrief with the therapist later in an effort to translate their reasoning and create an opportunity for empowering (power-to) the therapist for future events. The goal is to focus on process scenarios for greatest generalization of learning, and thereby empowerment, via the debriefing of teachable moments.

Mentorship as an Overt Aspect of Training Feminist Therapists

Through research with recent graduates of systemic training programs, four types of mentorship have been found to be common (psychosocial, career, clinical, and research mentorship), and around 70% of trainees report receiving some form of mentorship (Prouty, Helmeke, & Fischer, in press). Feminist systemic trainers have been discussing the importance of mentorship to their training for the past decade (e.g., Blumer et al., 2010; Prouty-Lyness & Helmeke, 2008). Prouty-Lyness and Helmeke found that feminist supervisors included mentorship within their clinical supervision relationships. Blumer et al. (2010) have provided a detailed description of their profound appreciation for the deep and long-reaching influence of their feminist mentors on their careers and professional identities, emphasizing the intergenerational nature of feminist mentorship within the profession of family therapy. Hence, it is important to discuss with trainees their needs and expectations for mentorship during and after clinical training.

Learning Goals/Objectives

Learning goals and objectives can vary with the type of feminist lenses familiar to the supervisor and requested by the feminist therapist. Additionally, like goals and objectives of feminist family therapy supervision can be extrapolated from those outlined as foundational in family therapy practices. Thus, the components of feminist family therapy and supervision include, but are not limited to, the following: tailoring treatment to the client/supervisee; collaboration with the client/therapist on goals of therapy/supervision; creation of a strong working alliance (Brown, 2006); use of self and self-disclosure (Allen, 1995; Baber & Murray, 2001); reflexivity (Baber & Murray, 2001; Hare-Mustin & Marecek, 1990); practice of gender sensitivity; facilitation of social change; emphasis on the personal as political; perception of therapy/supervision as value based; and viewing the clinician/supervisor as a role model in the lives of clients/supervisees (DeVoe, 1990; Kaschak, 1981; Mowbray et al., 1984; Sherman, Koufacos, & Kenworthy, 1978).

CONCLUSION

In summary, feminist family therapists have critiqued and stretched family therapy by analyzing power dynamics as they occur interpersonally, socially, culturally, and cross-nationally. Feminists examine, question, and encourage people to recognize barriers and options to access personal and social power within their own lives; balance and share power interpersonally; and both respect and celebrate the diversity we experience in our human identities and cultures. Feminist perspectives can be used to enhance any therapeutic model and apply to all of our lives.

CASE EXAMPLE

The Supervisor-Supervisee Relationship

A. AAMFT Approved Supervisor—Dr. Bob
The supervisor is a 53-year-old Caucasian, gay, and male-identifying. He is a seasoned supervisor who has been credentialed as an AAMFT Approved Supervisor for more than 20 years. He provides supervision for both students (in practicum and internship as an adjunct professor) and postgraduates at his agency.

(*continued*)

B. Supervisee—Brittney
The supervisee is a 33-year-old Caucasian, heterosexual, female-identifying from British Columbia, Canada, and is in her first semester of her master's programs internship in a local nonprofit agency. She is particularly interested in doing feminist family therapy and is excited about the opportunity to work with a broad range of families.

Dr. Bob would begin working with Brittney by introducing her to the different feminist perspectives emphasized within family therapy. Brittney could begin with one of them and then learn how to focus on the key social constructs and interpersonal dynamics emphasized by that branch of feminist family therapy. Because all branches of feminist family therapy include analysis of human identity (like gender) and access/barriers to social power, the Intersectionality Grid could be used to help Brittney learn how to notice and discuss similarities and differences between her and her supervisor, Dr. Bob. Prior to receiving the case assignments, Dr. Bob begins the feminist training by discussing how diverse people's identities are, and how each aspect of a person's identity provides access to power in some circumstances and might reduce or prohibit it in others. Dr. Bob works with Brittney to identify as many types of identity aspects as she can think of and adds a few that prior therapists have thought of, if he can. Dr. Bob then models self-of-the-therapist self-evaluation by discussing how he has experienced himself with regard to each of the identified aspects. He encourages Brittney to journal about her own sense of identity and then discuss it in supervision. Brittney begins to gain a larger understanding of the complexities of how to respectfully and curiously understand how people's access to power changes based upon not only who they are, but the intersections of who they are and the social context in which they reside.

Brittney could then move on to interviewing some peers and practicing being curious about her similarities and differences with them, and how experiences of being in the world and in different contexts influenced them. Bob would then talk with Brittney about how she could integrate these ideas into her model of therapy.

The Case

Family

Travis (29 years old) and Tekia (27 years old) are an African American couple with two children, Tyrone (5 years old) and Tommy (4 years

(*continued*)

old). Tekia has been a stay-at-home mom. Travis was in the military (as has been every male in his family) and "served 2 tours of duty in Afghanistan." He was injured when an improvised explosive device (IED) exploded and left him severely injured. He has spent many months in the hospital and rehabilitation. He was discharged from the military (which he had hoped to make his career) as a result of his injuries and now has difficulties finding a job. He currently works two part-time jobs but is hoping to find a full-time job.

Presenting Problem:

- Couple relationship and parenting issues, as Travis reportedly has a "short fuse."
- There are no reported safety concerns.
- Continuing recovery from his injury.
- Job issues—inability to find full-time employment/financial strain.

In supervising this case, the supervisor training a feminist therapist would first start with assessment of the therapist's developmental level in relation to these clients' safety status, presenting problems, and familiarity with clients' cultural identities and family configuration. Brittney identifies the couple as low suicidal and homicidal risk but is unclear as to the risk for interpersonal violence and type of parental discipline that underlies the couple's description of Travis's "short fuse."

When Travis and Tekia's case is assigned to Brittney, Dr. Bob determines that because of Brittney's moderately advanced developmental level and the low risk for client and therapy room safety concerns, he will take a more collaborative approach to supervision. He asks Brittney to describe the clients' strengths and challenges, their gendered power interactions, their approach to resisting racism within and outside of the military, and their coping strategies related to Travis's physical changes and challenges since his military service. Brittney uses the Intersectionality Grid to do this, and then brainstorms respectful and curious questions that will enable her to better understand her clients' unique identities and experiences with interpersonal power, barriers to power, and resiliency strategies utilized by each member of the couple (their families of origin if she likes intergenerational therapy models) and the couple as a team.

Examples of Questions From Radical Feminist Lens

1. In our generation, many of us are able to experience and express gender more freely than did our parents—how do each of you identify?

(continued)

2. What does being female or male (or however the clients answered Q1) play out in your couplehood? As parents? In your families of origin? In practicing your religion? With friends? At work?
3. Do you feel like each of you are encouraged to show a wide range of emotions, or do you sometimes feel like some are less okay to feel and express (making sure both vulnerability and anger are included in discussion)?

Examples of Questions From a Cultural Feminist Lens

1. So, I'm unmarried, White, and Canadian. I've done some reading and interviewed some local colleagues about their experiences being African American and parents here in [state]. But each of us is very different as we draw upon our ethnic cultures, religions, and families of origin. Could you both give me a picture of how you have experienced your ethnic identities, including religions? How have these informed both of your identities?
2. Did these (ethnic cultures, religions, and families of origin) influence whether/how you negotiated being a couple?
3. Did these (ethnic cultures, religions, your access to money, and examples from your families of origin) influence how you disagree and resolve arguments? (Using ideas from the Power and Control wheel)
4. Did these (ethnic cultures, religions, and families of origin) influence how each of you would be parents?
5. What type of resources, money and other types, do you have as a couple? How do you interact with each other about these resources? How are financial decisions made (including insurance)?
6. What has it been like to be an active military family and now a wounded warrior veteran?
7. How do you think your various identities have provided options in your lives?
8. If you could design a family crest that displayed your family's resiliency, what would it be?

Examples of Questions From a Liberal Feminist Lens

1. What went into your decision to legally marry? Do you have powers of attorney and wills?
2. Were there military regulations that affected your couple and family decisions?
3. What forms of disability and health care does your family have access to now?
4. Have you been provided with other support services from the military, the state, and/or your religious community?

(*continued*)

5. What supports do you wish you had for the challenges you are now facing?
6. Do you have to play down or hide your marital/family status in any context?
7. Do you have to play down or hide your religious affiliation/ethnic identity/ physical status in any context?
8. Are your home and work environments safe and accessible to you and your family members?

Examples of Questions From a Postcolonial Feminist Lens

1. Sometimes our context influences which aspects of ourselves are emphasized. Sometimes we adjust our emphasis and sometimes the context does this. Have either of you experienced this?
2. Have there been times in either of your lives when you felt you were able to have control/not have as much control over which aspects of yourselves were emphasized?
3. When others want to label you with something that doesn't fit, how have you resisted?
4. How have you dealt with these experiences as a couple?
5. How do you help your children to experiment with identities, their sense of belonging to family and communities, and being able to be who they each wish to be?
6. Have your military experiences and/or your experiences within medical settings influenced this process?

As one can see from this long list of potential questions, Brittney would have to choose questions to fit her theory of therapy and the goals she and the couple wish to focus upon first. Brittney might even choose to integrate the questions into a genogram or an ecogram, or have the couple experientially sculpt their experiences around a topic (Duhl, Kantor, & Duhl, 1973). Dr. Bob could help Brittney to gain ease and skill with using feminist perspectives in her work. Any feminist perspectives can be molded to enhance both modern and postmodern therapy models.

REFERENCES

Allen, K. R. (1995). Opening the classroom closet: Sexual orientation and self-disclosure. *Family Relations, 44*(2), 136–141. Retrieved from http://www.jstor.org/discover/10.2307/584799?sid=21105805783723&uid=2&uid=4

Almeida, R. V. (1993). Creating collectives of liberation. In L. B. Silverstein & T. J. Goodrich (Eds.), *Feminist family therapy: Empowerment and social context* (pp. 293–306). Washington, DC: American Psychological Association.

Almeida, R. V., Dolan-Del Vecchio, K., & Parker, L. (2007). *Transformative family therapy: Just families in a just society.* New York, NY: Pearson.

Almeida, R. V., Hernandez-Wolfe, P., & Tubbs, C. (2011). Cultural equity: Bridging the complexity of social identities with therapeutic practices. *International Journal of Narrative Therapy & Community Work, 3*, 43–56.

Almeida, R. V., & Lockard, J. (2005). The cultural context model: A new paradigm for accountability, empowerment, and the development of critical consciousness against domestic violence. In N. Sokoloff (Ed.), *Domestic violence at the margins: Readings on race, class, gender, and culture* (pp. 301–320). New Brunswick, NJ: Rutgers University Press.

Ansara, Y. G., & Hegarty, P. (2014). Methodologies of misgendering: Recommendations for reducing cisgenderism in psychological research. *Feminism & Psychology, 24*(2), 259–270. doi:10.1177/0959353514526217

Aponte, H. J. (1994). *Bread and spirit: Therapy with the new poor.* New York, NY: W.W. Norton.

Aponte, H. J. (2009). Introduction for special section on training and supervision. *Journal of Marital and Family Therapy, 35*(4), 369. doi:10.111/j.1752-0606.2009.00128.x

Aponte, H. J., & Kissil, K. (2012). "If I can grapple with this I can truly be helpful in the therapy room": Using the therapist's own emotional struggles to facilitate effective therapy. *Journal of Marital and Family Therapy, 40*, 1–13. doi:10.1111/jmft.12011.

Aponte, H. J., & Winter, J. E. (1987). The person and practice of the therapist: Treatment and training. *Journal of Psychotherapy and the Family, 3*(1), 85–111. doi:10.1300/J287v03n01_10

Ault-Riché, M. (1986). *Women and therapy*. Rockville, MD: Aspen.

Avis, J. M. (1985). The politics of functional family therapy: A feminist critique. *Journal of Marital and Family Therapy, 11*(2), 127–138. doi:10.1111/j.1752-0606.1985.tb00601.x

Avis, J. M. (1988). Deepening awareness: A private study guide to feminism and family therapy. *Journal of Psychotherapy and the Family, 3*(4), 15–46. doi:10.1300/J287v03n04_03

Baber, K. M., & Murray, C. I. (2001). A postmodern feminist approach to teaching human sexuality. *Family Relations: An Interdisciplinary Journal of Applied Family Studies, 50*(1), 23–33. doi:10.1111/j.1741-3729.2001.00023.x

Bacigalupe, G. (1998). Cross-cultural systemic therapy training and consultation: A postcolonial view. *Journal of Systemic Therapies, 17*, 31–44. Retrieved from http://psycnet.apa.org/psycinfo/1998-02863-003

Bartlett, N. H., & Vasey, P. L. (2006). A retrospective study of childhood gender-atypical behavior in Samoan fa'afafine. *Archives of Sexual Behavior, 35*(6), 659–666. doi:10.1007/s10508-006-9055-1

Belenky, M. F., Clinchy, B. M., Goldberger, N. R., & Tarule, J. M. (1997). *Women's ways of knowing: The development of self, voice, and mind.* New York, NY: Basic Books.

Black, L., & Piercy, F. P. (1991). A feminist family therapy scale. *Journal of Marital & Family Therapy, 17*, 111–120. doi: 10.1111/j.1752-0606.1991.tb00874.x

Blumer, M. L. C. (2010). "And action!" Teaching and learning through film. *Journal of Feminist Family Therapy: An International Forum, 22*(3), 225–235. doi:10.1080/08952833.2010.499703

Blumer, M. L. C., Ansara, Y. G., & Watson, C. M. (2013). Cisgenderism in family therapy: How everyday practices can delegitimize people's gender self-designations. *Journal of Family Psychotherapy, 24*(4), 267–285. doi:10.1080/08975353.2013.849551

Blumer, M. L. C., & Barbachano, J. M. (2008). Valuing the gender-variant therapist: Therapeutic experiences, tools, and implications of a female-to-male trans-variant clinician. *Journal of Feminist Family Therapy: An International Forum, 20*(1), 46–65. doi:10.1080/0895280801907135

Blumer, M. L. C., Green, M. S., Compton, D., & Barrera, A. M. (2010). Reflections on becoming feminist therapists: Honoring our feminist mentors. *Journal of Feminist Family Therapy: An International Forum, 22*(1), 57–87. doi:10.1080/08952831003652319

Blumer, M. L. C., Haym, C., Zimmerman, K., & Prouty, A. (2014). What's one got to do with it?: Considering monogamous privilege. *Family Therapy Magazine*, 28–33.

Boyd-Franklin, N., & Garcia-Preto, N. (1994). Family therapy: The cases of African American and Hispanic women. In L. Comas-Díaz & B. Greene (Eds.), *Women of color: Integrating ethnic and gender identities in psychotherapy* (pp. 239–264). New York, NY: Guilford Press.

Brown, L. S. (2006). Still subversive after all these years: The relevance of feminist therapy in the age of evidence-based practice. *Psychology of Women Quarterly, 30*, 15–24. doi:10.1111/j.1471-6402.2006.00258.x

Carlson, T. S., McGeorge, C. R., & Toomey, R. B. (2013). Establishing the validity of the Affirmative Training Inventory: Assessing the relationships between lesbian, gay, and bisexual affirmative training and students' clinical competence. *Journal of Marital and Family Therapy, 39*, 209–222. doi:10.1111/j.1752-0606.2012.00286.x

Chaney, S. E., & Piercy, F. P. (1988). A feminist family therapy behavior checklist. *American Journal of Family Therapy, 16*, 305–318. doi:10.1080/01926188808250736

Cheon, H. S., Blumer, M. L. C., Shih, A., Murphy, M. J., & Sato, M. (2009). The influence of supervisor and supervisee matching, role conflict, and supervisory relationship on supervisee satisfaction. *Contemporary Family Therapy, 31*(1), 52–67. doi:10.1007/s10591-008-9078-y

Cheon, H. S., & Murphy, M. J. (2007). The self-of-the-therapist awakened: Postmodern approaches to the use of self in marriage and family therapy. *Journal of Feminist Family Therapy, 19*(1), 1–16. doi:10.1300/J086v19n01_01

Comas-Díaz, L., & Greene, B. (Eds.). (1994). *Women of color: Integrating ethnic and gender identities in psychotherapy*. New York, NY: Guilford Press.

Dankoski, M. E., & Deacon, S. A. (2000). Using a feminist lens in contextual therapy. *Family Process, 39*, 51–66. doi:10.1111/j.1545-5300.2000.39107.x

DeVoe, D. (1990). Feminist and nonsexist counseling: Implications for the male counselor. *Journal of Counseling and Development, 69*(1), 33–36. doi:10.1002/j.1556-6676.1990.tb01451.x

Duan, C., & Roehlke, H. (2001). A descriptive "snapshot" of cross-racial supervision in university counseling center internships. *Journal of Multicultural Counseling and Development, 29*(2), 131–146. doi:10.1002/j.2161-1912.2001.tb00510.x

Duhl, F. S., Kantor, D., & Duhl, B. S. (1973). Learning space and action in family therapy: A primer of sculpting. In D. Bloch (Ed.), *Techniques of family psychotherapy: A primer.* New York, NY: Grune & Stratton.

Fang, S., McDowell, T., & Holland, C. (2006). Internationalizing family science programs: A spherical expansion of inclusive perspectives. In R. R. Hamon (Ed.), *International family studies: Developing curricula and teaching tools* (pp. 1–20). Binghamton, NY: Haworth Press.

Flemke, K., & Allen, K. R. (2008). Women's experience of rage: A critical feminist analysis. *Journal of Marital and Family Therapy, 34*(1), 58–74. doi:10.1111/j.1752-0606.2008.00053.x

Gatmon, D., Jackson, D., Koshkarian, L., Martos-Perry, N., Molina, A., Patel, N., & Rodolfa, E. (2001). Exploring ethnic, gender, and sexual orientation variables in supervision: Do they really matter? *Journal of Multicultural Counseling and Development, 29*(2), 102–113. doi:10.1002/j.2161-1912.2001.tb00508.x

Goldner, V. (1989). Generation and gender: Normative and covert hierarchies. In M. McGoldrick, C. M. Anderson, & F. Walsh (Eds.), *Women in families: A framework for family therapy* (pp. 42–60). New York, NY: W.W. Norton.

Goodrich, T. J. *Women and power.* (1991). New York, NY: W.W. Norton.

Goodrich, T. J., Rampage, C., Ellman, B., & Halstead, K. (1988). *Feminist family therapy: A casebook.* New York, NY: W.W. Norton.

Green, M. S., & Dekkers, T. D. (2010). Attending to power and diversity in supervision: An exploration of supervisee learning outcomes and satisfaction with supervision. *Journal of Feminist Family Therapy, 22*(4), 293–312. doi:10.1080/08952833.2010.528703

Greene, B. (1994). Lesbian women of color: Triple jeopardy. In L. Comas-Díaz & B. Greene (Eds.), *Women of color: Integrating ethnic and gender identities in psychotherapy* (pp. 389–427). New York, NY: Guilford Press.

Haddock, S. A., Zimmerman, T. S., Current, L. R., & Harvey, A. (2003). Parenting practices of dual-earner couples who successfully balance family and work. *Journal of Feminist Family Therapy, 14*(3/4), 37–56. doi:10.1300/J086v14n03_03

Haddock, S. A., Zimmerman, T. S., & McPhee, D. (2000). The power equity guide: Attending to gender in family therapy. *Journal of Marital and Family Therapy, 26,* 153–170. doi:10.1111/j.1752-0606.2000.tb00286.x

Hare-Mustin, R. (1978). A feminist approach to family therapy. *Family Process, 17*(2), 181–194. doi:10.1111/j.1545-5300.1978.00181.x

Hare-Mustin, R. (1989). The problem with gender in family therapy theory. In M. McGoldrick, C. M. Anderson, & F. Walsh (Eds.), *Women in families: A framework for family therapy.* New York, NY: W.W. Norton.

Hare-Mustin, R. T., & Marecek, J. (Eds.). (1990). *Making a difference: Psychology and the construction of gender.* New Haven, CT: Yale University Press.

Helmeke, K. (1994). Fostering a safe atmosphere: A first step in discussing gender in family therapy training programs. *Contemporary Family Therapy, 16(6),* 553–519. doi:10.1007/BF02196844

Helmeke, K., & Sori, K. (Eds.). (2006). *The therapist's notebook for integrating spirituality in counseling I & II.* New York, NY: Haworth Press.

Hernández, P., & McDowell, T. (2010). Intersectionality, power, and relational safety in context: Key concepts in clinical supervision. *Training and Education in Professional Psychology, 4,* 29–35. doi:10.1037/a0017064

Hernández, P., Taylor, B. A., & McDowell, T. (2009). Listening to ethnic minority AAMFT approved supervisors: Reflections on their experiences as supervisees. *Journal of Systemic Therapies, 28,* 88–100. doi:10.1521/jsyt.2009.28.1.88

Imber-Black, E., & Roberts, J. (2003). Rituals in families and family therapy. New York, NY: W.W. Norton.

Kaschak, E. (1981). Another look at sex bias in students' evaluations of professors. *Psychology of Women Quarterly, 5*(s5), 767–772. doi:10.1111/j.1471-6402.1981.tb01100.x

Keeling, M. L., Butler, J., Green, N., Kraus, V., & Palit, M. (2010). The Gender Discourse in Therapy questionnaire: A tool for training in feminist-informed therapy. *Journal of Feminist Family Therapy, 22*(2), 153–169. doi:10.1080/08952831003787883

Kirk, G., & Okazawa-Rey, M. (2003). Identities and social locations: Who am I? Who are my people? In G. Kirk & M. Okazawa-Rey (Eds.), *Women's lives: Multicultural perspectives* (3rd ed., pp. 59–110). New York, NY: McGraw-Hill.

Kliman, J. (1998). Social class and relationship: Implications for family therapy. In M. McGoldrick (Ed.), *Re-visioning family therapy: Race, culture, and gender in clinical practice* (pp. 50–61). New York, NY: Guilford Press.

Knudson-Martin, C. (1994). The female voice: Applications to Bowen family system's theory. *Journal of Marital and Family Therapy, 20*, 35–46. doi:10.1111/j.1752-0606.1994.tb01009.x

Korinek, A. W., & Kimball, T. G. (2003). Managing and resolving conflict in the supervisory system. *Contemporary Family Therapy, 25*(3), 295–310. doi:10.1023/A:1024559321634

Laird, J. (1988). Enactments of power through ritual. In T. J. Goodrich (Ed.), *Women and power* (pp. 123–147). New York, NY: W.W. Norton.

Laszloffy, T. A. (2008). Social class: Implications for family therapy. In M. McGoldrick & K. Hardy (Eds.), *Re-visioning family therapy: Race, culture, and gender in clinical practice* (2nd ed., pp. 48–60). New York, NY: Guilford Press.

Leupnitz, D. A. (1988). *The family interpreted: Feminist theory in clinical practice.* New York, NY: Basic Books.

Long, J. K., & Serovich, J. M. (2003). Incorporating sexual orientation into MFT training programs: Infusion and inclusion. *Journal of Marital and Family Therapy, 29*, 59–67. doi:10.1111/j.1752-0606.2003.tb00383.x

McDowell, T., Goessling, K., & Melendez, T. (2012). Transformative learning through international immersion: Building multicultural competence in family therapy and counseling. *Journal of Marital and Family Therapy, 38*, 365–379. doi:10.1111/j.1752-0606.2010.00209.x

McDowell, T., & Hernández, P. (2010). Decolonizing academia: Intersectionality, participation, and accountability in family therapy and counseling. *Journal of Feminist Family Therapy, 22*, 93–111. doi:10.1080/08952831003787834

McGeorge, C., Carlson, T. S., & Toomey, R. B. (2015a). Assessing lesbian, gay, and bisexual affirmative training in couple and family therapy training: Establishing the validity of the faculty version of the Affirmative Training Inventory. *Journal of Marital and Family Therapy, 41*, 57–71. doi:10.1111/j.1752-0606.2012.00286.x

McGeorge, C., Carlson, T. S., & Toomey, R. B. (2015b). Establishing the validity of the feminist couple therapy scale: Measuring therapists' use of feminist practices with heterosexual couples. *Journal of Couple and Relationship Therapy, 12*, 3–21. doi:/10.1080/15332691.2013.749760

McGoldrick, M. (1991). For love or money. In T. J. Goodrich (Ed.), *Women and power: Perspectives for family therapy* (pp. 239–244). New York, NY: W.W. Norton.

McGoldrick, M., Garcia-Preto, N., Hines, P., & Lee, E. (1989). Ethnicity and women. In M. McGoldrick, C. M. Anderson, & F. Walsh (Eds.), *Women in families: A framework for family therapy* (pp. 169–199). New York, NY: W.W. Norton.

McGoldrick, M., Giordano, J., & Garcia-Preto, N. (Eds.). (2005). *Ethnicity and family therapy* (3rd ed.). New York, NY: Guilford Press.

McGoldrick, M., Giordano, J., & Pearce, J. (Eds.). (1996). *Ethnicity and family therapy* (2nd ed.). New York, NY: Guilford Press.

McGoldrick, M., Pearce, J. K., & Giordano, J. (Eds.). (1982). *Ethnicity and family therapy.* New York, NY: Guilford Press.

McIntosh, P. (1998). White privilege: Unpacking the invisible knapsack. In M. McGoldrick (Ed.), *Re-visioning family therapy: Race, culture, and gender in clinical practice* (pp. 147–152). New York, NY: Guilford Press.

Mowbray, C. T., Lanir, S., & Hulce, M. (Eds.). (1984). *Women and mental health: New directions for change.* New York, NY: Psychology Press.

Murphy, M. J., & Wright, D. W. (2005). Supervisees' perspectives of power use in supervision. *Journal of Marital and Family Therapy, 31*(3), 283–295. doi:10.1111/j.1752-0606.2005.tb01569.x

Peletz, M. G. (2009). *Gender pluralism: Southeast Asia since early modern times.* New York, NY: Routledge.

Pinderhughes, E. (1986). Minority women: A nodal position in the functioning of the social system. In M. Ault-Riché (Ed.), *Women and family therapy* (pp. 51–63). Rockville, MD: Aspen.

Prouty, A. M. (2001). Experiencing feminist family therapy supervision. *Journal of Feminist Family Therapy: An International Forum, 12*(4), 171–203. doi:10.1300/J086v12n04_01

Prouty, A. M. (2014). Incorporating feminist practice in supervision. In C. L. Storm & T. C. Todd (Eds.), *The systemic supervisor electronic resources.* London, UK: Wiley-Blackwell.

Prouty, A. M., Bermúdez, J. M., Helmeke, K., & Ko, M.-J. (2011, September). *Intersections of human diversity within MFT training and supervision.* Three-hour presentation at the American Association for Marriage and Family Therapy annual conference, Charlotte, NC.

Prouty, A. M., Helmeke, K.B., & Fischer, J. (in press). Mentorship in family therapy training programs: Students' and new graduates' perspectives. *Journal of Family Psychotherapy, 27.*

Prouty, A. M., & Lyness, K. P. (2011). Feminist couple therapy. In J. L. Wetchler (Ed.), *Handbook of clinical issues in couple therapy* (pp. 271–289). New York, NY: Routledge.

Prouty, A. M., Thomas, V., Johnson, S., & Long, J. K. (2001). Methods of feminist family therapy supervision. *Journal of Marital and Family Therapy, 27,* 85–97. doi:10.1111/j.1752-0606.2001.tb01141.x

Prouty-Lyness, A. M. (Ed.). (2005). *The politics of the personal in feminist family therapy: International examinations of family policy.* New York, NY: Haworth Press.

Prouty-Lyness, A. M., & Helmeke, K. B. (2008). Clinical mentorship: One more aspect of feminist supervision. *Journal of Feminist Family Therapy: An International Forum, 20*(2), 166–199. doi:10.1080/08952830802023318

Roberts, J. M. (1991). Sugar, spice, toads, and mice: Gender issues in family therapy training. *Journal of Marital and Family Therapy, 17,* 121–133. doi:10.1111/j.1752-0606.1991.tb00875.x

Rock, M., Carlson, T. S., & McGeorge, C. R. (2010). Does affirmative training matter? Assessing CFT students' beliefs about sexual orientation and their level of affirmative training. *Journal of Marital and Family Therapy, 36*(2), 171–184. doi:10.1111/j.1752-0606.2009.00172.x

Sherman, J., Koufacos, C., & Kenworthy, J. A. (1978). Therapists: Their attitudes and information about women. *Psychology of Women Quarterly, 2*(4), 299–313. doi:10.1111/j.1471-6402.1978.tb00510.x

Tamasese, K. (2003). Gender and culture: Together. In C. Waldegrave, K. Tamasese, F. Tuhaka, & W. Campbell (Eds.), *Just therapy—A journey: A collection of papers from the Just Therapy Team* (pp. 203–206). Adelaide, Australia: Dulwich Centre Publications.

Tamasese, K., & Waldegrave, C. (1994). Cultural and gender accountability in the "Just Therapy" approach. *Journal of Feminist Family Therapy, 5,* 29–45. doi:10.1300/J086v05n02_03

Turner, J., & Avis, J. M. (2003). Naming injustice, engendering hope: Tensions in feminist family therapy training. In L. B. Silverstein & T. J. Goodrich (Eds.), *Feminist family therapy: Empowerment in social context* (pp. 365–378). Washington, DC: American Psychological Association.

Twist, M. (2006). A response to the Babies and Bosses Report: The effects of policy on therapy and the influence of therapists on politics. *Journal of Feminist Family Therapy: An International Forum, 17,* 67–77. doi:10.1300/J086v17n03_04

Vatcher, C. A., & Bogo, M. (2001). The feminist/emotionally focused therapy practice model: An integrated approach for couple therapy. *Journal of Marital and Family Therapy, 27*(1), 69–83. doi:10.1111/j.1752-0606.2001.tb01140.x

Waldegrave, C. (2005). "Just Therapy" with families on low incomes. *Child Welfare, 84,* 265–276.

Walsh, F. (1989). Reconsidering gender in the marital quid pro quo. In M. McGoldrick, C. M. Anderson, & F. Walsh (Eds.), *Women in families: A framework for family therapy* (pp. 267–285). New York, NY: W.W. Norton.

Walsh, F. (1998). *Strengthening family resilience.* New York, NY: Guilford Press.

Walsh, F. (Ed.). (1999). *Spiritual resources in family therapy.* New York, NY: Guilford Press.

Walsh, F., & Schneikman, M. (1988). (Fe)male: The hidden gender dimension in models of family therapy. In M. McGoldrick, C. M. Anderson, & F. Walsh (Eds.), *Women in families: A framework for family therapy* (pp. 16–41). New York, NY: W.W. Norton.

Walters, M., Carter, B., Papp, P., & Silverstein, O. (1988). *The invisible web: Gender patterns in family relationships.* New York, NY: Guilford Press.

Werner-Wilson, R. J., Price, S. J., Zimmerman, T. S., & Murphy, M. J. (1997). Client gender as a process variable in marriage and family therapy: Are women clients interrupted more than men clients? *Journal of Family Psychology, 11*(3), 373. doi:10.1037/0893-3200.11.3.373

Zimmerman, T. S. (Ed.). (2001). *Balancing work and family: Special considerations in feminist therapy.* New York, NY: Haworth Press.

SECTION IV

Population-Specific Supervision

NINETEEN

Training the Medical Family Therapist in an Integrated Care Setting

Jenny Speice
Susan McDaniel

In recent years, a dramatic transformation of our nation's health care system has begun, with implications for training medical family therapists. Implementation of the Affordable Care Act (ACA, enacted in 2010) is expanding access to care, shifting reimbursement for services, and realigning health care delivery in new and creative models. Patient[1]- and family-centered care is growing in a meaningful way. Patient and family preferences are more actively incorporated to enhance satisfaction with the care received and to improve the treatment and health outcomes. As part of improving outcomes, cost, and satisfaction, interdisciplinary approaches to care—integrated in shared locations—are on the rise. Medical Family Therapists (MedFTs) should have a central role in this integrated health care team, as they have unique skills to contribute.

In this chapter, the authors provide a theoretical background for Medical Family Therapy (MedFT) as the systemic approach to integrated care as well as unique considerations for supervision in the integrated care settings where MedFTs typically practice. A case example offers a frame and illustration for the application of this important approach.

TEACHING THEORETICAL BACKGROUND

With the publication of *Medical Family Therapy* (McDaniel, Hepworth, & Doherty, 1992), and the second edition, *Medical Family Therapy and Integrated*

[1]We use the word *patient* in Medical Family Therapy, rather than *client*, given the convention in the health care setting.

Care (McDaniel, Doherty, & Hepworth) in 2014, a new way of conceptualizing clinical practice for family therapists was launched. In addition to solid skills in providing traditional family therapy, MedFTs bring expertise in serving patients and their families where they access care most often—in their primary care providers' offices and medical specialty centers. Patients and families might be referred for traditional mental health, individual and family life transition concerns, needed health behavior change, or assistance in navigating acute or terminal illness issues. Not infrequently, health crises surface unresolved individual, couple, or family issues, offering the opportunity for better resolution through Medical Family Therapy. Often families present while managing chronic illnesses that require rallying a family's resources for the long haul over many life cycle transitions and developmental milestones.

Philosophical Foundation

The two central underpinnings of Medical Family Therapy are the Biopsychosocial Model (Engel, 1977, 1980) and Family Systems Theory, both of which have origins in General Systems Theory (von Bertalanffy, 1968). In the early 1970s, these approaches were developed separately in pioneering work by two University of Rochester medical faculty: internist George Engel, MD, and family psychiatrist Lyman Wynne, MD, PhD. Medical Family Therapy brought these two theories together in what was termed a *biopsychosocial systems approach* (McDaniel et al., 1992). The integration of these theories brings together the individual's internal experience (from the precellular level to organ systems and physiology) and the patient's external experience in a family and community. Practically, this means that a MedFT considers the many aspects of a patient and family system, which include one's own body, family, neighborhood, community, religious and cultural contexts, and health care team. By incorporating the patient's biology with relationships, the MedFT avoids the classic mind-body split that artificially relegated these different parts of a patient's life to separate spheres of influence and treatment.

In addition to biopsychosocial systems theory, three other metaframeworks are relevant for Medical Family Therapy: transgenerational, developmental, and collaborative approaches. Given the demonstrated family influences on health and well-being, a transgenerational, family-of-origin approach, pioneered by Murray Bowen (Kerr & Bowen, 1988), acknowledges that in addition to biological and genetic connections within many families, we all grow up with legacies of health and illness whether or not biological ties exist. The transgenerational transmission of health behaviors, beliefs, interactions with the health care system, and encounters with death and dying all influence a patient's experience of a current symptom or illness episode. Meanings that a patient ascribes to a bodily sensation, anticipated test results, routine doctor's visits, new diagnoses, or relapsing illness episodes powerfully influence the patient's psychological health as well as interactions with family members and significant others. Furthermore, those relationships and psychological vulnerabilities recursively influence the person's physical health and abilities to initiate or sustain important health behavior change, manage medications and other treatment recommendations, and even to attend routine health care visits.

All of this takes place within a developmental trajectory for the individual, couple, family, and for any illness experiences. MedFTs play a central role on a treatment team and can encourage recognition of the unique patient and family experience, reducing the potential of feeling treated "like just another cancer patient." Family psychiatrist John Rolland (1984) proposed a typology-of-illness model that accounts for four distinct and overlapping categories: onset, course, outcome, and degree of incapacitation. He further overlaid the illness typology onto the individual and family life cycle stages to demonstrate how an illness experience can uniquely affect a patient or family at different life cycle points and transitions and is likely to change the family's previously planned course (Rolland, 1988). During the normative family life cycle, there are periods of relative closeness (births, marriages, end of life) and separation (adolescence and young adulthood) in all families (Combrinck-Graham, 1985). Illness and caregiving needs might disrupt the typical family undertakings for connection or independence depending on the timing of an illness presentation or exacerbation in a particular family. Additionally, Reiss and Kaplan-DeNour (1989) acknowledge that phases of an illness life cycle (acute, chronic, terminal) require that patients, families, and the health care team relate to each other differently, with unique challenges at each phase. A MedFT helps a patient, family, and care team navigate these fluid and intersecting phases.

A commitment to this collaborative approach to care is another central meta-framework underpinning Medical Family Therapy (McDaniel et al., 2014). Collaboration includes partnering among interdisciplinary health care team members (physicians, nurses, medical assistants, and other health and behavioral health practitioners) as well as the patient and family (Seaburn, Lorenz, Gunn, Gawinski, & Mauksch, 1996). MedFTs bring systemic skills to this team, incorporating the diverse perspectives from team members and bridging gaps in worldview, language, and culture to achieve more effective outcomes for a patient and her family. At times this collaboration occurs in joint visits that include the MedFT, the physician and/or other health professional, and the patient and family. At other times, it looks more like shuttle diplomacy, with the MedFT gaining a critical understanding of the illness experience and treatment needs from each perspective, helping to translate or transmit that understanding to the other parties in a way that helps everyone to address concerns directly with each other, and developing a mutually agreed-upon treatment plan.

THE PROCESS OF CHANGE

A change in any part of the system (including the physical health of a family member) will have important, and unpredictable, effects on the other parts of the system. In the context of an interdisciplinary, collaborative team, a

MedFT helps to identify the strengths and resilience of a patient and family.[2] MedFTs help to uncover and encourage a patient and family's sense of agency (the ability to control what is within one's control) and communion (the sense of connection with important personal and professional support people) (Bakan, 1966; Totman, 1979). Sometimes this involves recognizing the transgenerational legacies of strength, resilience, and advocacy in the face of illness or loss; providing psychoeducation; and/or connecting patients and families to community groups or online supports with people facing similar diagnoses and struggles. For traditionally trained psychotherapists, navigating the foreign land of health care can be overwhelming and confusing, given all of the complexities inherent in the system. It is particularly easy for a patient or family to feel lost, abandoned, or hopeless, left without a sense of agency and communion. MedFTs can play an essential role in ensuring that patients and families are active members of their own health care team.

In Medical Family Therapy, change can take many different forms. For some, a brief focused behavioral intervention, such as to stop smoking or develop a plan to maintain blood glucose levels, is the critical goal. Introducing a relational lens to this brief episode of biobehavioral health care provides greater potential for enduring change. For others, releasing the unresolved grief of losing a mother to early breast cancer addresses a patient's avoidance of her own routine mammography screening as well as resolving a lengthy depressive episode. For those living with chronic illness, a MedFT helps patients and families to "put their illness in its place" (Gonzalez, Steinglass, & Reiss, 1989) and return to important routine life cycle challenges such as launching young adults rather than living bound by relentless illness demands.

It is also important to take an illness history for any presenting problem. Patients often do not recognize the impact of health and illness events on individual and relational dynamics. Timelines and genograms can reveal significant connections if the history includes physical, emotional, and interpersonal experience of the patient and his or her significant others.

ROLE OF THE MEDICAL FAMILY THERAPIST/BEHAVIORAL HEALTH CONSULTANT

Within this integrative, collaborative frame, a MedFT normalizes the patient and family experience of health and illness, respects natural defenses, highlights strengths and resilience, and increases their sense of agency and communion. For many, MedFTs facilitate a healing process of meaning-making, while helping to remove blame that patients or families may experience.

MedFTs master the elements of a traditional mental health intake to assess any psychopathology (mental illness and substance use), history of past

---

[2]These very same skills are also applied to the health care team to facilitate team functioning.

treatments, use of medications, family and social history, and relational dynamics that influence the exacerbation or maintenance of individual, relational, and/or health problems. In addition, MedFTs are trained to explore the "why now" in a broader frame:

- What is the biopsychosocial view of the presenting problem or problems?
- What are the strengths and demonstrated resilience that individuals and this family bring? What are the individual and family vulnerabilities?
- What are the life cycle influences at this moment in time for the family (centripetal versus centrifugal)?
- For anyone with an identified illness, understanding the type, course, level of incapacitation, treatment, and prognosis of the illness is essential. Then, how does that intersect with the individual and family life cycle needs?
- What is this family's sense of agency and communion for navigating similar experiences?
- What meaning do they make of the illness?
- What are the transgenerational legacies for health and illness?
- What past experiences in treatment shape this current episode of care?

Additionally, the MedFT integrates another system layer outward and explores this patient's and family's interactions with the health care team (physicians, physician assistants, nurse practitioners, nursing staff, and medical assistants). This is made easier by the fact that many health care organizations are moving in the direction of patient- and family-centered care, with an emphasis on enhancing the patient's and family's satisfaction with care. Ultimately, by partnering with patients and their families, the treatment team better understands patient and family preferences, abilities, resources, and barriers to effective care. The result is an evolving treatment plan that improves the care and health of the patient. Some essential questions for the therapist to explore about the treatment and the team from the patient/family perspective include:

- What is your experience of relationships with doctors, nurses, and other health professionals? How well do you think they know you, your family, and what you are going through?
- How does this care experience improve your ability to manage your illness and treatment needs? How does it make for more challenges?
- What are the barriers that get in the way of being able to manage your illness and treatment needs? Are you able to attend your appointments, take medications as prescribed, follow through on recommendations for your diet, exercise, or other lifestyle changes? Are there concerns you have about your health insurance or the financial burden of your health care needs?
- How do your health care clinicians incorporate important parts of your culture into your care? This could be as simple as making sure that visits and materials are available in a patient and family's primary language,

incorporating a same-sex partner, or asking about the use of prayer and religious leaders/communities as a healing support.
- What other experiences has your family had with health care that make it easier or more challenging for you to trust your health care team now?

After completing a comprehensive biopsychosocial systems assessment, the MedFT considers the most appropriate collaborative care treatment plan for this patient and family.

MedFT TECHNIQUES

In integrated care settings, a course of treatment typically unfolds episodically (as with the patient's relationship with the primary care physician or nurse practitioner) because the patients and their family members will have onsite access to return to psychotherapy when the need arises again. MedFTs make use of brief, evidence-informed therapies in an effort to provide effective care to the largest number of patients in the practice. Problem-solving therapy, developed for use in primary care and manualized for the IMPACT study by Mark Hegel and Patricia Arean (2003), shows promise as a brief (6- to 10-session) treatment for depression. Likewise, Cognitive Behavioral Therapy approaches provide a useful structure for brief intervention. Another common starting point is the use of motivational interviewing (Miller & Rollnick, 2002) to explore and resolve ambivalence regarding behavior change. Using the techniques of expressing empathy, developing the discrepancy between present behavior and broader goals and values, rolling with resistance, and supporting self-efficacy, MedFTs work to engage patients in positive changes that align with their own values. Over the course of several years, and caring for multiple family members in the same setting, a MedFT has the advantage of weaving together larger stories and themes for a patient and family over time. Therefore, the discrete episodes of psychotherapy that occur within this larger frame can be efficient and focused on addressing the presenting symptoms and functional needs, always informed by the larger biopsychosocial transgenerational systems lens.

For those who present with a diagnosed illness, the principal techniques of Medical Family Therapy include normalizing the illness experience as an adaptation that anyone would struggle to come to terms with as well as building the patient's and family's sense of agency and communion. Patients and their family members are encouraged to "put the illness in its place" (Gonzalez et al., 1989) as a way to accomplish their important developmental individual and family life cycle tasks while appropriately attending to the medical regimen. MedFTs help to reinforce the family's nonillness identity, and tend to their emotional life and other routines apart from the necessary demands of the illness, in a way very similar to "externalizing the illness" (White & Epston, 1990). When families are struggling with making sense of the "why us?" question (McDaniel, Hepworth, & Doherty, 1997), the MedFT creates space for these important conversations, along with extended family, religious leaders and spiritual guides, and other significant people in the family's life (Hodgson, Lamson, Mendenhall, & Crane, 2014). Often there are peer supports available in the community or online who genuinely relate to the family's struggles and can be valuable sources of information and strength during the long haul of illness.

CULTURAL CONSIDERATIONS

If unfamiliar with a medical setting, one of the primary goals for a MedFT is to adjust to the sights, sounds, and smells of health care. Acclimating to what can seem like a foreign medical culture is a common theme in supervision. Supervisees explore their reactions to blood, needles, and the frail or wounded bodies of patients as well as their own vulnerabilities. Other challenges arise when patients express frustration with their health care team or have additional burdens of managing illnesses accompanied by social stigma or the disconfirmation of not appearing sick. Additionally, MedFT supervisees often need time in supervision to grapple with the differences in hierarchy, jargon, and pace of medical culture. Training in an integrated care setting might be the first exposure to larger public health considerations, barriers to accessing care, and the dramatic disparities in care experienced by racial and ethnic minorities (Smedley, Stith, & Nelson, 2003). Most patients are cared for in health care settings where the clinical staff does not adequately match the racial or ethnic diversity represented in the patient population. If unaware of their own privilege and social location, MedFTs might unknowingly perpetuate the long history of racial and ethnic discrimination in health care (Shavers, Klein, & Fagan, 2012).

Similarly, there is a long history of medicine as a patriarchal discipline. That is changing dramatically, with medical schools (like the mental health disciplines) typically admitting more women than men. Still, more women seek health care than men, and women are often the family's informal health provider, so gender issues in health care and in Medical Family Therapy are important to monitor (McDaniel & Cole-Kelly, 2003; McDaniel & Hepworth, 2003).

RESEARCH TO CONSIDER

Much of the research on integrated care models focuses on engaging patients in their own health and behavior change or managing chronic health conditions (Bodenheimer, MacGregor, & Shafiri, 2005), often using brief interventions with a problem-solving focus (Hegel & Arean, 2003). Recent work has focused on team-based outcomes involving a range of clinicians from a range of disciplines and nonclinical staff all coordinating their efforts (Ghorob & Bodenheimer, 2012). In addition, all MedFTs should be familiar with the ever-increasing number of studies demonstrating lifelong health consequences of adverse childhood experiences, built on the seminal work of Felitti and colleagues' study of more than 17,000 patients (Felitti et al., 1998). MedFT supervisors will orient supervisees to the latest developments in health outcomes literature, patient- and family-centered care, or team-based initiatives in their setting. MedFTs bring unique skills to support team functioning and advocate for patient and family voices as part of the team.

ROLE OF THE SUPERVISOR AND SUPERVISEE

In the beginning of supervision, the supervisor and supervisee co-create a specific supervision contract that establishes a range of important aspects of their supervision. Most notably, the contract covers the routine meeting times, agenda setting for meetings, arranging live/video supervision, how to handle crises while onsite or offsite, whether backup supports exist if unable to reach the supervisor immediately, documentation expectations, competencies, and professionalism in a collaborative integrated care environment.

For most MedFTs, clinical care and supervision occur in an integrated health care setting. Typically the supervisor is a behavioral health clinician who has served in that setting for many years and is well integrated into the clinical milieu. Supervisors focus on the essentials of family therapy supervision, along with helping supervisees conduct broad biopsychosocial assessments, and facilitate treatment that incorporates relevant clinical aspects, often episodically. In integrated primary care settings, for example, patients and their families often are served over the course of many years, throughout a number of different episodes of care, often with different family members presenting as the "identified patient." MedFTs who practice in these settings appreciate the privilege of knowing families over time, through various life cycle transitions, periods of illness, relative health, crisis, and comparative normalcy.

Another essential element in Medical Family Therapy supervision is exploring self-of-the-therapist issues. As in all supervision, it is best when these conversations happen in the context of a safe and validating supervision relationship. In our setting, we incorporate a basic genogram in early supervision conversations and ask supervisees to consider the relevant health and illness narratives from their families of origin. We acknowledge that most therapists who choose Medical Family Therapy as a specialty come to this work with their own individual or family experiences with illness, caregiving, or family legacies in the health professions. For those for whom it is less obvious in their family's life to date, we know that family health events are inescapable in the years ahead, whether a therapist confronts a new diagnosis of infertility or pregnancy loss, a parent with a new-onset debilitating illness, or a partner with more salient family illness caregiving stories. As supervisors, we ask our supervisees to examine the meta-themes that they bring to Medical Family Therapy in general, and then address any specifics related to the current case. We also have the supervisee read several chapters from the casebook accompanying *Medical Family Therapy and Integrated Care*, called *The Shared Experience of Illness* (McDaniel et al., 1997), in which MedFTs describe their own approach to the work and how it has been shaped by personal or family experience with illness or disability. Each trainee may review the following:

- While drawing or reviewing a prepared genogram, consider the health and illness experiences of each family member, including the causes of death for those who have died.
- What themes of health and illness were present in your family growing up? What were your earliest memories of you or someone else being sick, and how did the important people in your family relate about being sick? What are the positive and negative family experiences with chronic diseases and acute episodes or trauma requiring medical care?
- Who else in your family has roles in health care or caregiving, either formally (as physicians, nurses, or allied health professionals) or informally (the informal health expert in the family)? How did you find your way to becoming a MedFT?
- What are your own personal experiences with illness and health behavior changes? What challenges have you experienced, and what insights or resources have you garnered?
- Are there specific presenting problems or life cycle transitions that might be particularly relevant given your personal or family-of-origin history (childhood-onset diseases or injuries, adolescent or early adulthood management of chronic conditions and transition in roles toward more independence, fertility/infertility/pregnancy loss, new adult-onset diagnoses, aging or later-life medical comorbidities, and end-of-life issues)?

Another aspect of learning about MedFT as a new supervisee is documenting in an integrated electronic health record (EHR). This is the norm for most integrated health care settings, given federal legislation. For those who were not trained on EHRs in mental health settings, training for documenting electronically, and in an integrated system, is an important focus in early supervision. Electronic health records have distinct advantages for coordination of care, reducing medication errors and decreasing redundancies in data collection across settings. For MedFTs, there are also professional and ethical responsibilities for documenting in a patient's chart, including how to reference important family relationships, dynamics, or events and any other family members who might be present in the sessions. Although patient privacy and security protections are in place in EHRs, one unique documentation challenge in EHRs is that these records exist in perpetuity: patients (or their health care designees) may have routine access to these notes in the not-so-distant future. For example, a 35-year-old woman might access her record during routine prenatal care and discover family therapy notes from her teenage years. If not considered thoughtfully by the therapist at the time, these notes might include information formerly unknown to the patient (e.g., about a messy divorce or parental infidelity).

LEARNING GOALS/OBJECTIVES

The supervisor and supervisee establish developmentally appropriate learning goals for the duration of their contract. These are reviewed and renegotiated at regular intervals that are appropriate for the level of experience and the context of the relationship (training program versus employment, for example). If a novice MFT, the supervision contract initially focuses on establishing basic competency in assessment, treatment planning, intervention, safety and crisis management, all within this biopsychosocial, family systems, and collaborative care framework. If the trainee has already been socialized and trained as another type of health or mental health professional, the contract may focus more on the biopsychosocial, family systems, and collaborative care framework and the special theoretical, technical, ethical, and professional aspects of Medical Family Therapy.

The supervisor supports the new supervisee's process of becoming a part of the context and attends to any relevant self-of-the-therapist aspects that might arise. Typically this begins with some shadowing experiences of medical and nursing colleagues, often observing routine medical exams or office surgical procedures with patient consent. Observing other health care colleagues on the team (including nutritionists, pharmacists, care managers, medical assistants, and medical secretaries) helps the MedFT begin to acclimate to the medical culture, if it is unfamiliar to the supervisee. In addition to the outpatient setting, it is useful to have the trainee shadow a medical professional or trainee who is taking call overnight. These experiences serve to educate the trainee about both the needs of the medical professional who makes a referral and the experience of the patient. It also serves as an opportunity to build initial rapport with the interdisciplinary team members and better appreciate the pace, language, and "culture" of the environment. All of these experiences will help to inform their future collaboration when therapists are writing notes in the chart that will be relevant for other treating clinicians, conducting a joint Medical Family Therapy appointment with a medical or nursing colleague, or engaging in a brief hallway consult in between patient visits. MedFTs need to become skilled in and comfortable with brief, focused, clear statements or requests that are free from the typical mental health or family therapy jargon when they are interacting with their physician and nursing colleagues. In addition, these shadowing experiences allow novice MedFTs to become more comfortable managing their anxiety in medical settings, particularly if they purposefully avoided medical school because of an aversion to "blood and guts." Acclimating to the sights and smells of a medical environment is an early priority for MedFTs, who may also need the safety of their supervisory relationship to address feelings they have about the power and hierarchy that are clearly evident in medical settings. Understandably, supervisees also need time for reflection on their role as witnesses to human suffering and their feelings of inadequacy about "fixing it," or providing some relief.

Other supervisee goals can include learning about common primary care screeners for depression, anxiety, or substance abuse; enhancing motivational interviewing skills; discovering brief behavioral health consultation or therapy interventions useful in integrated care settings; or expanding knowledge about particular diagnoses, medications, or patient populations. In addition, the new supervisee will shadow the supervisor in a variety of roles: as a therapist, as a collaborator with the other clinicians, and as a consultant to the clinical team. In some settings, the MedFT also serves as an educator for physicians-in-training, as is the case in our context. This is yet another developmental milestone, for supervisees who understand the roles and needs of others on the teams. For example, the MedFT might provide a "lightning talk" to his or her team before an afternoon "huddle" on depression, its effect on a spouse, and the importance of including spouses in the treatment of depression when the patient is married or committed. The multiple roles can be both stimulating and daunting for a new clinician to undertake. Accordingly, this often is a developmental process that occurs over time, with pacing guided by the supervisor.

In the following example, the case vignette is adapted for an integrated care setting for the purposes of demonstrating a typical course of Medical Family Therapy. All members of the family presented receive their health care in this setting, as is common for a family medicine practice. The clinic uses an electronic health record (EHR), and patients routinely provide consent for sharing of information between treating clinicians.

CONCLUSION

Supervising a novice MedFT in an integrated care setting highlights the significance of context in culturally informed patient- and family-centered care. Supervisors attend to MedFT supervisees' development of clinical competencies for patient, family, and team collaboration, while being sensitive to important self-of-the-therapist needs. Exposure to Medical Family Therapy skills is important for all family therapists, as illness occurs in all families and affects and is affected by relationships. In addition, exposure—if not experience and expertise—is important for family therapy to participate in the transformation of health care and team-based care in the United States and around the world.

CASE EXAMPLE

Supervisor/Supervisee/Setting

Dr. Bob, a 53-year-old Caucasian MedFT and AAMFT Approved Supervisor, has practiced at the Family Medicine Center (FMC) for the past 20 years, having completed his own postdoctoral fellowship training at FMC. Brittney is a 33-year-old Caucasian MedFT student who is new to the site. It is her first clinical placement.

Patient/Family/Presenting Concerns

Travis (29) and Tekia (27) are an African American couple with two children, Tyrone (5) and Tommy (4). Tekia has been a stay-at-home mom. Like most of the men in his family, Travis was in the military. He served 2 tours of duty in Afghanistan and was injured when an improvised explosive device (IED) exploded and left him severely disabled. He spent many months in the hospital and rehabilitation. To his disappointment, he was discharged from the military as a result of his injuries and now has difficulty finding a job that is rewarding and financially satisfying. He works two part-time jobs but is hoping to find a full-time job. Travis's older brother is still in the military and was recently re-deployed to Afghanistan. Travis and Tekia present to therapy about their relationship and parenting issues, as Travis reportedly has a "short fuse." They also report needing help as he continues recovery from his injury and navigates challenges with employment and finances.

Preparing for the Intake

As is the common practice at FMC, Brittney reviewed the referral from the primary care practitioner, which came to the in-basket of her EHR; she scanned the medical chart for a brief history. Brittney ran into the physician in the hallway and they ducked into an exam room to have a brief chat to learn more about the reason for referral. Occasionally, referrals happen within the context of a medical visit, and the physician or nurse practitioner can introduce the patient to the MedFT in a "warm hand-off" that makes follow-up likely. In this case, the referral was made electronically. In supervision, Dr. Bob and Brittney reviewed the presenting problems and reason for referral, went over the background information collected from the chart and collaborative contact, and made a plan for the first session. In addition to the traditional first session joining, and obtaining basic intake information, they prioritized more extensive risk assessments for self-injury, partner violence, and harm to children in the home given

(*continued*)

the concerns about Travis's reported "short fuse." Brittney planned to explore any current crises as well as the more chronic, long-term adjustments to the injury and resultant challenges.

From her chart review, Brittney learned that Travis suffered a severe injury to his right leg, leading to amputation above the knee. There was some indication about preliminary evaluations for traumatic brain injury (TBI) from the explosion as well. Brittney was also eager to identify the strengths and resources of the patient and his family and to highlight the resilience they have shown in recent months and over time. After addressing specific risks and acute needs, she decided that she will explore the range of biopsychosocial (and spiritual) aspects to their recovery from this devastating loss. She would obtain an extensive history of the physical injuries, resulting disabilities, and rehabilitation efforts. She would have Travis and Tekia recount the timeline of their lives together, including where they were relationally as a couple and young family when the IED "exploded" in their lives. Brittney would also explore the similar and different family-of-origin experiences for Travis and Tekia related to military life, employment, illness, disability, marriage, parenting, caregiving, and loss. Dr. Bob cautioned that several sessions are usually required to gain a comprehensive picture, and MedFTs prepare with this scaffolding to help guide the patient and family storytelling.

Dr. Bob integrated relevant learning goals, developmental needs, and self-of-the-therapist issues for Brittney as they prepare to care for this family. He explored her background experiences with military family life and trauma and directly addressed her cultural sensitivity about caring for an African American disabled veteran and his young family. He also attended to important collaborative needs for the treating health professionals (e.g., some mental health referrals are motivated partially by primary care clinicians' frustrations at a patient's noncompliance with medical recommendations or difficulty managing family dynamics that interfere with treatment adherence). Brittney acknowledged her lack of personal and family-of-origin experience with military service, living with a physical disability or racial discrimination, and demonstrated openness to exploring these themes in ongoing supervision.

Ongoing Supervision for Psychotherapy and Collaborative Care

As a supervisor for a novice MedFT, Dr. Bob kept close watch on the evolving assessment for this family and Brittney's collaboration with

(*continued*)

the team and provided specific guidance on her documentation. He also checked in regularly about how she was managing the emotional intensity of the care and offered support through live supervision as well as joining a session to conduct a more thorough risk assessment. Brittney and Dr. Bob decided together about how to sequence individual, couple, and family sessions to address the relevant treatment plan goals: monitor Travis's PTSD symptoms and maintain safety; improve marital functioning and quality; enhance parenting skills; develop more active coping to reduce angry outbursts; increase a sense of agency and communion for living with the injury/disability; and grieve the many losses for this man, couple, and family.

Over the course of the first 10 sessions, Brittney created an effective working alliance with Travis and Tekia and helped them begin to establish an increased sense of agency and communion relating to their "new normal" of living with disability. The couple agreed upon a number of new routines in their home to help structure their days to meet the various needs for Travis's care as well as family life with their energetic young sons. The couple navigated a few role shifts in the process, including Travis being responsible for the family breakfast routine because he is an early riser, which allows Tekia time for a morning run, something the couple agreed reduces the tension in the family enormously. Travis was proud of his new culinary talents and efforts to creatively incorporate healthy choices into his sons' diets within a limited budget. This led to the exploration of a more central role for meal planning and preparation in the family, as well as potential career options, for Travis. The couple also organized a new calendar system to keep track of the many appointments for physical therapy and follow-up medical care as well as Travis's medications. They were pleased to have rallied the support of Travis's mother and Tekia's sister, who had offered to "help" in the past, but without a specific plan their offers went unaccepted. Now the extended family has a schedule, with planned needs for child care or rides, and could readily follow through on their desires to be useful. After careful coordination and support of the care team, the once-chaotic process of managing medications and appointments became an established routine with rare problems.

With newfound energy, the couple decided to tackle their ongoing challenges with the complex Veterans Administration systems for disability benefits. In frustration at past attempts to navigate the complexities, they had missed several deadlines for paperwork processing. Brittney advocated for Travis to reconnect with resources through the Veterans Outreach Center, where he also attended a

(*continued*)

group. In recent weeks, they met another disabled veteran and his wife and young son, who share similar challenges. The two couples planned to bring their families together a few times and reduce their sense of isolation from their old friends and former lifestyles. By increasing a sense of agency and communion, Travis and Tekia feel more in control, are less tense, have fewer angry outbursts, and are ready to turn to nurturing more intimacy in their marriage, including a focus on adapting their sexual expression with his amputation. The couple also initiated tentative conversations about how to balance their roles for paid work and household responsibilities in ways that demonstrate more respectful consideration for how their lives have changed from their early ideas about their partnership.

Dr. Bob acknowledged the clear progress this couple has made in treatment and offered Brittney caution regarding the next phase of treatment, just at a time when she was considering tapering off the sessions. He pointed out that they are nearing the anniversary of the IED explosion, and it was possible there might be some exacerbation of stress and symptoms and a need to keep more regular appointments. Brittney reached out to her collaborative team, including the facilitator for the group that Travis attends with other veterans who are disabled. Indeed, the group facilitator had expressed concern that Travis has seemed more agitated in the group in recent weeks and had been yelling angry complaints to the staff about the lack of progress in obtaining disability payments. Tekia arrived at the next session in tears, reporting that Travis has been experiencing flashbacks and night terrors again. She expressed worries about his increased drinking and finding him one night holding a gun. Travis was quiet and withdrawn but visibly shaky. Brittney was grateful for Dr. Bob's anticipatory guidance and acknowledged the anniversary timing to Tekia and Travis. In the context of a comprehensive assessment of the crisis, Brittney gently explored concerns for harm to Travis or anyone in the family. After ruling out imminent risk, she made a safety plan to mitigate the elevated risk, including removing access to his guns, which is often a critical and sensitive issue for veterans. Travis had one gun that he used for hunting. He agreed for Tekia to take it to her brother's. Brittney scheduled more frequent individual sessions with Travis and coordinated ongoing contact with Tekia. Together they considered if and when a higher level of care might be briefly needed to help Travis, and the family, through this crisis.

After documenting the session progress note and Travis's safety plan, Brittney sent it to his physician's EHR in-basket, so that he could follow up at Travis's medical appointment later in the week. In

(*continued*)

the next supervision session, Dr. Bob and Brittney debriefed the session and Brittney revealed that she was very worried about Travis's risk and felt unskilled to manage the crisis. Dr. Bob watched a portion of the risk assessment from the session DVD with Brittney. They discussed the option for Brittney to join Travis's medical visit later that week to re-assess any safety concerns; enact his safety plan with Tekia and Travis's mother; and to reach out to her collaborative team, including the group facilitator, to help Travis safely navigate this crisis. In the course of their supervision session, Brittney unpacked her reactions with Dr. Bob. Clearly she was relying on their combined clinical judgment, but she also became aware of her own internal fear in response to the elevated suicide risk in a veteran with access to firearms and struggling with comorbidities of PTSD, TBI, leg amputation, disability, family stressors, racism, and increased substance use. Together, Dr. Bob and Brittney made a plan for live supervision of the next session, and a plan to continue to monitor how Brittney was managing her own anxiety and self-care, as well as to increase the supports from her collaborative care team.

Brittney went to work enacting the plan for care that she had developed with Dr. Bob. In a series of individual sessions, combined with phone updates from Tekia, Brittney and Travis addressed the current crisis. Travis was agreeable to attending a substance use evaluation and additional group work targeting the PTSD symptoms, as he was motivated to return to the connection he was beginning to feel to Tekia, the boys, and their new family routine. Brittney also encouraged Travis to invite another disabled veteran from his unit in Afghanistan into a session; this was tremendously valuable given her lack of personal or family experience in the military. Expanding the collaborative team is a critical step in managing crises, and Brittney felt the support and relief as well as more hope for Travis's recovery and this family's return to the new path they were making together.

As the end of the summer approached, Brittney was pleased to see that Tyrone and Tommy, the couple's young sons, were also scheduled for well-child visits anticipating their re-entry into school. Given the crisis with Travis, most of the recent therapy sessions had focused on stabilizing his symptoms and functioning and monitoring for safety. Brittney appreciated that the same family physician cared for the whole family and would see the boys in their upcoming well-child check. They could assess together whether the boys would benefit from some focused play therapy related to the school transition or any of their household stress and worries about their dad. Brittney had planned to do her own assessment in a family session in

(*continued*)

the coming weeks, but the annual school physical provided a timely opportunity to check in on their well-being.

After the well-child visits revealed some sleep disturbances and more than typical behavioral issues for both boys, their physician, Brittney, and Dr. Bob consulted. Given all of the recent stressors, and upcoming school transition, they decided to intersperse a couple of family sessions now that Travis's current crisis was resolving. Clearly this crisis had rocked the family, and they all remained on high alert and edgy even though Travis and Tekia were trying to re-establish the morning breakfast and other routines that had become important to a sense of coherence as a family. Anticipating school transitions is challenging for all families with young children, and this family was vulnerable to having a more difficult time.

The family physician also reported to Brittney that Tekia had made a passing remark about how Travis's TBI work-up and disability qualification process would have gone faster if he were not "just another black man." Although Brittney and Dr. Bob had discussed whether to address race directly with Travis and Tekia, Brittney had never brought it up directly in session. She explored her hesitation in the next supervision session. Together they considered the pros and cons of inviting or initiating the conversation with Tekia and Travis and the appropriate timing. Dr. Bob also revisited their earlier supervision conversations about race relative to other patients or families where the topic was more openly discussed.

As treatment continued, with the safety issues resolved for the moment, Travis improving, and the boys launched to school, Brittney could revisit the progress the couple had made prior to being destabilized by the anniversary crisis. She reflected with the couple about their resilience through all the fallout since the IED explosion at a time in their lives when they were filled with hope and expectations about their future. Brittney sensitively acknowledged that this trauma and the resultant long recovery had brought them into care with her, a young White therapist, and at times they must experience a gap in her understanding of their lives. Travis was silent. Tekia tearfully agreed that she and Travis had been "doing just fine" before all of this and that re-entering the medical system reminded her of how she was treated while in childbirth with their youngest son, Tommy. Travis was deployed, and although she had the support of both families during the pregnancy and birth, she was subjected to insensitive hospital staff comments about her son's "absent father." She acknowledged that she was initially wary of Brittney's abilities to connect given her age, race, and, more painfully, Tekia's attributions

(continued)

about Brittney's "perfect family life" with an able-bodied husband and happy children living an affluent lifestyle. Brittney was initially taken aback, silent about her lack of a partner or children and the "perfect family life." She acknowledged Tekia's courage in sharing that story and her beliefs about Brittney and recommitted that, although their lives were quite different, and separated by important health and cultural differences, she felt able to be of use to their family, along with the rest of the health care team. Brittney debriefed with Dr. Bob after the initial challenging conversation and was eager to update him several sessions later about how the candid dialogue seemed to deepen the therapeutic alliance. This increased safety allowed Tekia and Travis to move into sensitive and direct conversations about their sexual functioning, a taboo subject for both of them since the amputation.

As is common in integrated care settings, Travis and Tekia completed one episode of psychotherapy after about 6 months, with the understanding that they could access Brittney or a colleague of hers as needed in the future. Both Travis and Tekia had made solid steps toward accomplishing their goals and were well aware of clues that might warn of any potential crises brewing. Prior to an official discharge of Travis's mental health care, Brittney coordinated with each of the members of the collaborative team to review the treatment goals that had been accomplished and progress toward the remaining goals and to determine if there were any considerations for not ending psychotherapy at this time. Dr. Bob helped Brittney to construct the discharge summary in a way that would be brief and concise, yet include essential elements for other members of her collaborative team as well as future clinicians.

REFERENCES

Bakan, D. (1966). *The duality of human existence: An essay on psychology and religion*. New York, NY: Rand McNally.

Bodenheimer, T., MacGregor, C., & Shafiri, C. (2005). *Helping patients manage their chronic conditions*. Oakland, CA: California HealthCare Foundation.

Combrinck-Graham, L. (1985). A developmental model for family systems. *Family Process, 24*, 139–150.

Engel, G. L. (1977). The need for a new medical model: A challenge for biomedicine. *Science, 196*, 129–136.

Engel, G. L. (1980). The clinical application of the biopsychosocial model. *American Journal of Psychiatry, 137*, 535–544.

Felitti, V. J., Anda, R. F., Nordenberg, D., Williamson, D. F., Spitz, A. M., Edwards, V. . . . Marks, J. S. (1998). Relationship of childhood abuse and household dysfunction to many of the leading causes of death in adults: The Adverse Childhood Experiences (ACE) Study. *American Journal of Preventive Medicine, 14,* 245–258.

Ghorob, A., & Bodenheimer, T. (2012). Sharing the care to improve access to primary care. *New England Journal of Medicine, 366*(21), 1955–1957.

Gonzalez, S., Steinglass, P., & Reiss, D. (1989). Putting the illness in its place: Discussion groups for families with chronic medical illnesses. *Family Process, 28,* 69–87.

Hegel, M., & Arean, P. A. (2003). Problem-solving treatment for primary care (PST-PC): A treatment manual for depression. Lebanon, NH: Project IMPACT, Dartmouth College.

Hodgson, J., Lamson, A., Mendenhall, T., & Crane, D. R. (Eds.). (2014). *Medical family therapy: Advanced applications.* New York, NY: Springer.

Kerr, M. E., & Bowen, M. (1988). *Family evaluation.* New York, NY: W.W. Norton.

McDaniel, S. H., & Cole-Kelly, K. (2003). Gender, couples, and illness: A feminist analysis of medical family therapy. In T. J. Goodrich & L. Silverstein (Eds.), *Feminist family therapies* (pp. 267–280). Washington, DC: American Psychological Association.

McDaniel, S. H., & Hepworth, J. (2003). Family psychology in primary care: Managing issues of power and dependency through collaboration. In R. Frank, S. H. McDaniel, J. Bray, & M. Heldring (Eds.), *Primary care psychology* (pp. 113–132). Washington, DC: American Psychological Association.

McDaniel, S. H., Doherty, W. J., & Hepworth, J. (2014). *Medical family therapy and integrated care* (2nd ed.). Washington, DC: American Psychological Association.

McDaniel, S. H., Hepworth, J., & Doherty, W. J. (1992). *Medical family therapy: A biopsychosocial approach to families with health problems.* New York, NY: Basic Books.

McDaniel, S. H., Hepworth, J., & Doherty, W. J. (1997). *The shared experience of illness: Stories of patients, families and their therapists.* New York, NY: Basic Books.

Miller, W., & Rollnick, S. (2002). *Motivational interviewing: Helping people change* (2nd ed.). New York, NY: Guilford Press.

Reiss, D., & Kaplan De-Nour, A. (1989). The family and medical team in chronic illness: A transactional and developmental perspective. In C. J. Ramsey (Ed.), *Family systems in medicine* (pp. 435–444). New York, NY: Commonwealth Fund.

Rolland, J. S. (1984). Toward a psychosocial typology of chronic and life-threatening illness. *Family Systems Medicine, 2,* 245–262.

Rolland, J. S. (1988). Family systems and chronic illness: A typological model. In F. Walsh & C. Anderson (Eds.), *Chronic disorders and the family* (pp. 148–163). New York, NY: Hawthorn Press.

Seaburn, D. B., Lorenz, A. D., Gunn, W. B., Jr., Gawinski, B. A., & Mauksch, L. B. (1996). *Models of collaboration: A guide for mental health professionals working with healthcare practitioners.* New York, NY: Basic Books.

Shavers, V. L., Klein, W. M., & Fagan, P. (2012). Research on race/ethnicity and health care discrimination: Where we are and where we need to go. *American Journal of Public Health, 102*(5), 930–932.

Smedley, B. D., Stith, A. Y., & Nelson, A. R. (Eds.). (2003). *Unequal treatment: Confronting racial and ethnic disparities in health care.* Washington, DC: National Academies Press.

Totman, R. (1979). *Social causes of illness.* New York, NY: Pantheon Books.

von Bertalanffy, L. (1968). *General systems theory: Foundations, development, applications.* New York, NY: George Braziller.

White, M., & Epston, D. (1990). *Narrative means to therapeutic ends.* New York, NY: W.W. Norton.

trauma in
the comm: & culture
's trauma

TWENTY

Training the Trauma-Informed Marriage and Family Therapist

Karin Jordan

Trauma occurs on various levels, such as child abuse, spouse/partner abuse, street violence, school and job violence, large-scale transportation accidents, human-generated (e.g., war, terrorism, cyberterrorism, etc.) disasters, and natural disasters (e.g., tornado, hurricane, tsunami, wildfire, etc.). Each level and occurrence has implications for training trauma-informed family therapists. The field of traumatology is growing rapidly, and research such as the Adverse Childhood Experiences (ACE) study (Anda et al., 2006) is showing the long-term effects of childhood trauma, both psychological and physiological. Other research studies have shown how trauma causes DNA changes in the survivor, and that these DNA changes are transmitted to future offspring by the mother (Nauert, 2010). In addition, military personnel returning to the United States are often dealing with post-traumatic stress disorder (PTSD), depression, suicidal ideation, anxiety, reintegration problems, traumatic brain injury (TBI), limb loss, and other disabilities. Trauma-informed marriage and family therapy has proven to provide a buffering effect and help military personnel to reintegrate into civilian life. Marriage and Family Therapy (MFT) should have a central role in providing trauma-informed therapy, as MFTs have an understanding of how to work from a systemic perspective.

It also is important that MFTs understand self-of-therapist, given the complexity of working with trauma-affected clients. Self-of-therapist includes three levels, all of which must be considered: (1) the immediate, visceral clinical encounter, in which the therapist hears clients' trauma experience; (2) the MFT's life experience of trauma; and (3) the trauma embedded in the culture and the community. There are different types of self-of-therapist: (1) matched self-of-therapist, when the MFT feels the same feelings as the trauma survivor; and (2) complementary self-of-therapist, when the MFT takes on the role(s) that the trauma survivor places on him or her. Therefore, it is important for MFTs to maintain appropriate boundaries with the trauma survivor when they become witness to the trauma experience(s) of the survivor, without getting immersed in the story. Also, self-of-therapist issues experienced with the couple and/or family of the trauma survivor may be more complex than those encountered in individual work with a trauma survivor.

There is a growing recognition that vicarious trauma is a job hazard for MFTs working with trauma survivors. *Vicarious traumatization* can best be described as the MFT's "cost of caring" for trauma survivors (Basham, 2008, p. 87); it differs from burnout in that burnout does not impact the MFT's trust, issues of intimacy, or safety concerns, or include intrusive imagery (Rosenbloom, Pratt, & Pearlman, 1995). Today, there is increased awareness that family therapists who provide trauma-informed therapy need to be aware of the impact that working with this population can have on them, and the importance for supervisors to be knowledgeable about vicarious trauma and what aspects of the work contribute to family therapists developing vicarious traumatization.

Organizational factors refers to the MFT's workplace experience. More specifically, it refers to the number of trauma survivors the MFT sees and the balance between trauma and non-trauma survivors (Harrison & Westwood, 2009). MFTs seeing only trauma survivors are believed to be at higher risk of developing vicarious traumatization, as survivors often deal with safety issues and self-destructive behavior, which can create additional stress for the MFT (Fontana & Rosenheck, 2008). In addition, it is helpful if the agency/center has emergency services accessible, rather than having the MFT also being on call (on weekends and evenings) for emergencies, as this can compound to the stress. Furthermore, it is believed that MFTs who isolate themselves—who do not reach out, and consequently receive only limited peer and administrative support—are more likely to develop vicarious traumatization (Courtois & Ford, 2009; Harrison & Westwood, 2009). When working with trauma survivors, it is helpful to know other MFTs who are also working with trauma survivors to serve as a support system (Abu-Bader, 2000). Furthermore, training and education in traumatology are important to support the MFT in his or her work and is also believed to assist in the MFTs work satisfaction.

In this chapter, the author provides a theoretical background for Trauma-Informed Marriage and Family Therapy, as well as unique considerations regarding training and experience, self-of-therapist, vicarious traumatization, and organizational factors, and the Ecosystemic Developmental Trauma Model for supervision where Trauma-Informed Marriage and Family Therapy is practiced. A case example offers a framework for and illustration of the application of this approach.

TEACHING THEORETICAL BACKGROUND

The field of traumatology has been growing rapidly over the past decade, yet little has been written in the area of traumatology and MFT. Seminal writings in the field of traumatology include *Secondary Traumatic Stress: Self-Care for Clinicians, Researchers, and Educators* (Stamm, 1995); *Treating Compassion Fatigue* (Figley, 2002); *Trauma and the Therapist: Countertransference and Vicarious Traumatization in Psychotherapy with Incest Survivors* (Pearlman & Saakvitne, 1995); *The Body Remembers: The Psychophysiology of Trauma and Trauma Treatment* (Rothchild, 2000; van der Kolk & Fisler, 1994). In the field of MFT, the research on traumatology has been more limited but includes important contributions such as *Helping Traumatized Families* by Charles Figley and Laurel Kiser (2012) as well as *Emotion Focused Couple Therapy with Trauma Survivors: Strengthening Attachment Bonds* by S. M. Johnson (2002).

MFTs serve as experts helping survivors of various trauma events and their families. Survivors and families might be referred for traditional mental health, individual matters, and family life transition concerns. MFTs are needed to address the trauma event and the emotional, psychological, and physiological changes the survivor is experiencing. It is not unusual to see survivors dealing with a sense of a foreshortened future; this is a symptom of PTSD, which causes trauma survivors to feel that they will die prematurely without ever reaching certain developmental milestones, such as going to college, achieving a certain career, having a marriage/committed relationship, having children, and so forth. As the severity of this symptom can vary from mild to severe, so can the person's behavior. For example, a survivor might decide that she wants to have the child now that she had wanted to have in five years, or a survivor may be so depressed and hopeless that he stops participating in life. Other frequent challenges for survivors, especially in Western society, are changes in their core belief systems; beliefs about themselves, others, and the world; and their spiritual beliefs/values. Other survivors experience sadness and depression and are so overwhelmed that they struggle with suicidal ideations. Family support for survivors as they work through their trauma experience requires family resources, resiliency, and stress buffers, often over multiple life cycle transitions.

Philosophical Foundation

The central underpinning of MFT Trauma Therapy is the Ecosystemic Developmental Trauma Model, which is rooted in Bronfenbrenner's (1987) Ecological Systems Theory, in which an ecological paradigm is used to better understand the development of a person in the context of a trauma event.

$$TRt = (t[t\text{-}p][PE][t\text{-}p]) + TE$$

This formula defines trauma reactions as a set of processes through which properties of the person and the environment interact to produce constancy and change in the characteristics of the person over the life course, with the addition of the trauma event. This formula defines development as a set of processes through which properties of the person and the environment interact to produce constancy and change in the characteristics of the person over the life course.

The formula's symbols are defined as follows:

D = Development/outcome at a specific time
t = Time at which a developmental outcome is observed
P = Person
E = Environment
t-p = a joint force of person and environment over time, to produce the outcome that is being observed
Dt = t(t-p)(PE)(t-p)
TE = Trauma Event
TR = Trauma Reaction

Developmental outcomes of today shape the developmental outcomes of tomorrow. (This is most easily seen in early childhood development; for example, attachment is impacted by consistent love and care versus neglect and will affect later abilities to cope.) Likewise, trauma events of today also shape the developmental outcome of tomorrow. The ecological paradigm involves:

- The interrelatedness of a person having experienced a traumatic event
- The environment where the trauma event occurred
- The environment where the recovery from the traumatic event occurs
- The developmental tasks that must occur

When working with a trauma survivor in the ecological system (Bronfenbrenner, 1987), the taxonomy consists of a hierarchy of four systems (see Figure 20.1): the *microsystem* (interactions among the person's individual factors, including age, gender, ethnicity, etc.); the *mesosystem* (relations among the settings in which the developing person is involved at any particular life stage, such as home and school); the *exosystem* (encompasses the linkage and processes taking place between two or more settings, at least one of which does not ordinarily contain the developing person, but in which events occur that influence processes within the immediate setting that does contain that person); and the *macrosystem* (the societal/cultural values and structure as well as historical underpinnings that involve the more proximal values and behaviors of families, including factors such as poverty, socioeconomic class, etc.).

The integration of these four systems brings together the individual's internal experience (trauma event[s]) and the survivor's external experience in a family and community. Practically, this means that a Trauma-Informed Marriage and Family Therapist (T-IMFT) considers the many aspects of a person and family system, which include the individual trauma survivor's own aspects, risk factors, trauma experience, services received, protective factors, family, neighborhood, community, religious and cultural contexts, and larger

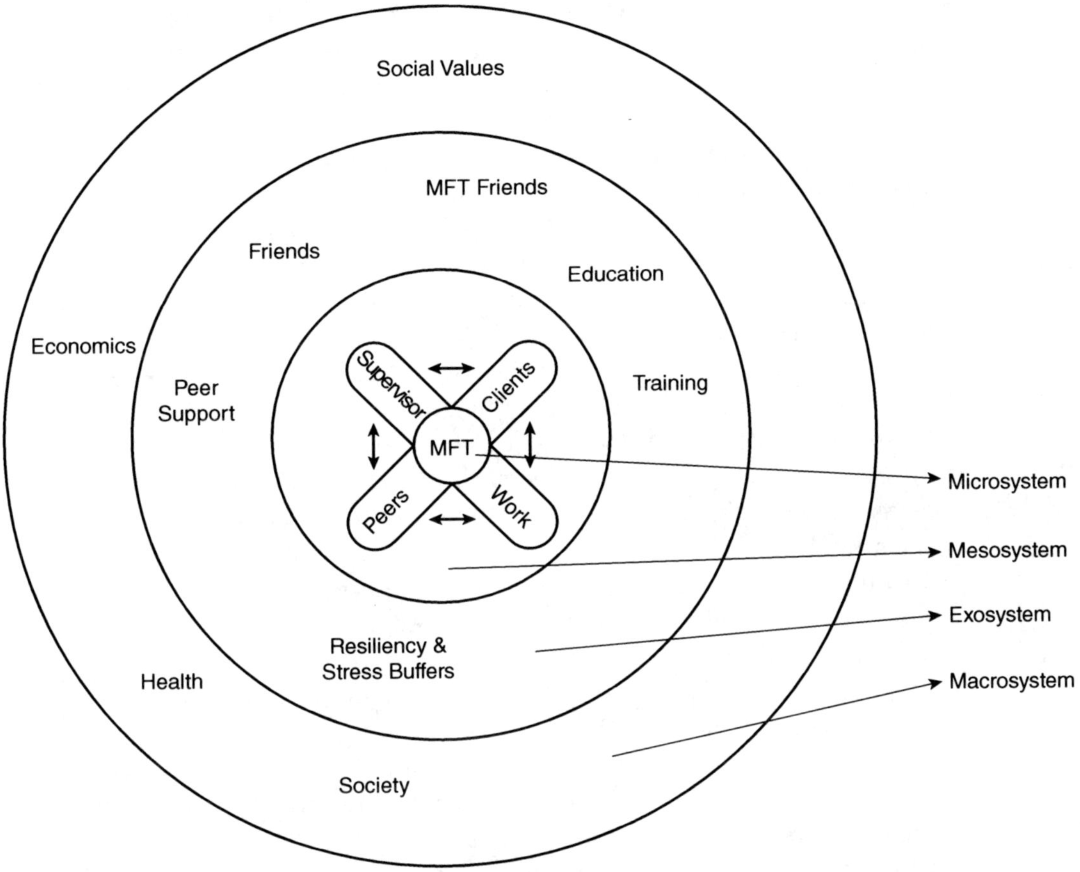

FIGURE 20.1 Ecosystemic Developmental Trauma Model.

social cultural aspects. By incorporating the survivor's individual factors with relationships and larger sociocultural aspects, the T-IMFT avoids the classic focus on the survivor alone, which often ignores these different parts of a survivor's life that greatly influence their function and treatment.

There is a broad range of trauma events that a person might have experienced. The different types include:

- Natural disaster—Can best be defined as a large-scale, not directly human-caused injury- or death-producing environment that adversely affects a significant number of people. In the United States, between 13% and 30% of the population will be exposed to one or more natural disasters in their lifetime (Briere & Elliotte, 2000).
- Mass interpersonal violence—Intentional violence that involves high numbers of injuries or casualties but does not occur in the context of war. This is a newer category in the trauma field and includes such things as terrorist attacks and mass human rights abuses by totalitarian regimes.
- Large-scale transportation accidents—Transportation accidents such as airline crashes, train derailments, and shipwrecks.

- House or other domestic fires—House/structure fires, often caused by short circuits, gas explosions, leaking propane tanks, stoves, or heaters, and often resulting in deaths, serious burns and smoke inhalation injuries, and loss of property.
- Motor vehicle accidents—About 20% of the population in the United States have experienced a serious motor vehicle accident.
- Rape and sexual assault—*Rape* means nonconsensual oral, anal, or vaginal penetration. *Assault* is forced sexual contact. In the United States, the rate of rape and sexual assault against women ranges from 14% to 20% and against men from 2% to 4%.
- Stranger physical assault—Includes mugging, beating, stabbing, shooting, attempted strangulation, and other violent acts against a person committed by someone not well known.
- Partner battery—Also known as *wife battery, spousal abuse*, or *domestic violence*; includes physical and sexual abuse. In the United States, occurrence rates are 9% to 15%.
- Torture—Includes "any act by which severe pain or suffering, whether physical or mental, is intentionally inflicted on a person for such purposes as obtaining information or confession, or punishing for an act committed or is suspected of having committed such act, or intimidating him or a third person" (Vesti & Kastrup, 1995, p. 214).
- War—Involves a wide range of violence and traumatic experiences, threat of death and injury to self and others, and being involved in injury or killing of others.
- Child abuse—Emotional, physical, and/or sexual abuse. In the United States, approximately 25% to 35% of women and 10% to 20% of men were abused as children.

Trauma events are a particular kind of crisis event that are extreme, powerful, and threatening, so that they overwhelm the person's sense of safety and security and can affect the person's ability to grow and change in expected ways (cognitive, emotional, behavioral). Trauma events are seen as a time when normal coping and problem-solving strategies do not work; they are generally time limited (minutes, hours, days, weeks, months, to years). A specific trauma situation or trauma event is the catalyst for feeling overwhelmed, as normal coping skills are exceeded. More specifically, Slaikeu (1990, p. 15) describes this experience as a "temporary state of upset and disorganization, characterized chiefly by an individual's inability to cope with a particular situation using customary methods of problem-solving, and by the potential for a radically positive or negative outcome." Trauma events can impact a person's core beliefs about himself or herself, the world, others, and religious/spiritual beliefs. This impact is also called "Shattered Assumption Theory" (Fernando, 2004), a Western-culture phenomenon. Trauma survivors can deal with the trauma event through accommodation (in which they incorporate new information into their core values about themselves, others, the world, and their spiritual/religious beliefs and values) or assimilation (in which they change their core values about themselves, others, the world, and their spiritual/religious beliefs and values).

Trauma can occur as a single or ongoing event, and some people go through multiple trauma events. Studies such as the ACE study (Anda et al., 2006) indicate that there are long-term effects of trauma, and therefore trauma should be placed within a developmental trajectory for the individual, couple, and/or family. The T-IMFT plays a central role in treatment of a broad range of trauma survivors and their families, especially as part of a treatment team, and can encourage recognition of the unique trauma survivor and family experience, reducing the potential of feeling treated "like just another survivor." The trauma model accounts for four distinct and overlapping categories: onset (trauma event), course of recovery, outcome (full or partial recovery), and degree of trauma impact. The therapist needs to do an overlay of the trauma event onto the individual and family life cycle stages to demonstrate how a trauma experience can uniquely affect an individual or family at different life cycle transitions and is likely to change the family's previously planned course. During the normative family life cycle, there are periods of relative closeness (births, marriages, end of life) and separation (adolescence and young adulthood) in all families (Combrinck-Graham, 1985). Trauma events can impact an individual, an entire family, or even a whole community and might disrupt the typical family tasks for connection or independence, depending on the onset and duration of the trauma event. Additionally, the individual and family needs during recovery phases of a trauma event might be quite different, based on the potential for ongoing threat to safety (unsafe home environment, redeployment, etc.) and for displacement and losses (e.g., loved ones, neighbors, property, community, etc.). Survivors have unique needs in each of these situations, so a T-IMFT has to attend uniquely to these unique challenges.

A commitment to meeting the trauma survivor and his or her family "where they are" is another central meta-framework underpinning Trauma-Informed Marriage and Family Therapy. Meeting the trauma survivor where she or he is includes a thorough assessment about the survivor's five factors:

- *Individual and personal status factors*—age; gender; ethnicity; race; military branch; active duty, enlisted or reserved; single- or dual-service family; married or divorced; daughter/son, father/mother, husband/wife
- *Predisposing factors*—past experience of trauma, such as exposure to aggression and violence (actual vs. vicarious, acute vs. chronic), personal and/or professional, personality disorders, poor coping skills and strategies, difficulty learning from previous experiences, impulsivity, low self-esteem, unstable work history, lack of finances, legal problems, chronic mental health issues (obsessive compulsive disorder, anxiety disorder, paranoia, PTSD, depression)
- *Peridisposing factors*—duration of trauma events, from minutes to hours to days or weeks, with risk of injury or death of self and others; proximity/duration/severity of the potential hazard (exceeds normal coping skills)

- *Postdisposing factors*—resources (water, food, shelter, safety, family, other support, and information)
- *Protective factors*—resiliency/postdisaster growth; having the ability to cope, bounce back, and grow emotionally and psychologically despite multiple losses (lives and property) and destruction (community and home) (Lambert & Lawson, 2013); assistive factors and stress buffers such as positive automatic support (PAT; Lightsey, 1994), physical activity (Klaperski, Seeling, & Fuchs, 2013), physical health, mental health, self-determination (Parker, Jimmieson, & Amiot, 2010), and religiosity (Lechner, Tomasik, Silbereisen, & Wasilewski, 2013)

Assessment includes gathering information regarding all of the five factors, with the T-IMFT gaining a critical understanding of the trauma experience and treatment needs from each survivor and family member and developing a mutually agreed-upon treatment plan.

In addition to the Ecosystemic Developmental Trauma Model, two other meta-frameworks are relevant for Trauma-Informed Marriage and Family Therapy: transgenerational (Schwartz, 2010) and collaborative (CMFT; Madsen, 2007) approaches. The transgenerational, family-of-origin approach, pioneered by Murray Bowen (Kerr & Bowen, 1988), acknowledges that in addition to families being biologically and genetically connected, some families have a multigenerational history of trauma. More specific focus is given to: (1) the transgenerational transmission of trauma, (2) how to achieve trauma recovery, (3) to what level recovery is achieved, and (4) how to seek out mental health and other support services. Meanings that a client ascribes to signs and symptoms often experienced by trauma survivors (such as flashbacks, hypervigilance, nightmares and night terrors, PTSD, foreshortened sense of the future, etc.) all influence the survivor's emotional and psychological well-being, which can lead to physiological problems and often influence interpersonal relationships with family members and significant others. These emotional, psychological, physical, and relationship vulnerabilities are recursive and can influence the overall well-being of the survivor.

All of this takes place within a developmental trajectory for the individual, couple, and family and for any trauma experiences. T-IMFTs play a central role in treating survivors and their families, looking at present individual and family structure and function as well as intergenerational trauma transmission and coping skills. A trauma event and a past trauma history affect the individual and family life cycle stages and development. Trauma can impact the survivor and his or her family at different life cycle transitions and is likely to change the survivor's and often his or her family's previously planned course. As noted earlier, during the normative family life cycle, there are periods of relative closeness (births, marriages, end of life) and separation (adolescence and young adulthood) in all families (Combrinck-Graham; 1985). Trauma exposure and recovery might disrupt the typical family tasks for connection or independence depending on the timing of a single or multiple trauma event impacting one or more family members in a particular family. Additionally, trauma recovery requires that the trauma survivor work through the trauma experience, which can be challenging if the trauma event is ongoing and no resources and support are available.

THE PROCESS OF CHANGE

Change in any part of the system (including the trauma survivor) will have important and unpredictable effects on the other parts of the system. In the context of a trauma survivor, a T-IMFT helps to identify the post-traumatic growth of the survivor and the strengths and resilience of the family. The T-IMFT helps the survivor tell his or her story and deal with the grief and loss issues. T-IMFTs also focus, with the survivor, on the foreshortened sense of the future. In addition, the T-IMFT helps the survivor accommodate or assimilate the trauma event as well as find ways to gain a sense of connection with family and friends. Often this involves recognizing the transgenerational legacies of strength, resilience, and post-traumatic growth as well as stress buffers in the face of trauma recovery and loss; providing psychoeducation; and/or connecting survivors and/or their families to community groups or online support with other trauma survivors and families of trauma survivors. For traditionally trained MFTs, navigation of the unfamiliar field of traumatology can be foreign and confusing. Survivors and their families are also often overwhelmed when trying to access and navigate trauma resources (e.g., dealing with FEMA and how to apply for resources after a disaster; military personnel trying to reintegrate into civilian life; wounded veterans trying to use the VA system). T-IMFTs can play an essential role in ensuring that survivors and their families know how to navigate the system to get their needs met.

In Trauma-Informed Marriage and Family Therapy, change can be dealt with through many different forms. It can include various theoretical approaches, including Trauma-Informed Cognitive Behavioral Therapy (TI-CBT; Cohen, Mannarino, & Deblinger, 2006). Narrative Family Therapy has been used with veterans (Freedman & Combs, 1996); Structural Family Therapy with childhood sexual abuse (Gelles & Maynard, 1987); and Integrative Family Therapy (Mikesell, Lusterman, & Daniels, 1995) or Intergenerational Family Therapy with spouse/partner abuse (Schwartz, 2010). The goal is to assist survivors to put the trauma event in its place through assimilation or accommodation and return to important daily routines as well as surmounting routine life cycle challenges, such as leaving home and going to college, rather than dealing with a foreshortened sense of the future and bypassing life cycle transitions out of fear that tomorrow might happen.

It is also important to take a history of previous trauma. Survivors often do not recognize the impact of previous trauma and how trauma histories have affected individual and relational dynamics. The trauma genogram timelines can reveal significant connections if the history includes physical, emotional, and interpersonal experience of the trauma survivor and his or her significant others.

ROLE OF THE TRAUMA-INFORMED MARRIAGE AND FAMILY THERAPIST

Within this trauma-informed frame, a T-IMFT normalizes the survivor and family experience of trauma, acknowledges loss and grief issues, and assists in getting basic needs met (food, water, safety, and shelter). In addition, the T-IMFT highlights stress buffers and resilience/post-traumatic growth. T-IMFTs facilitate a healing process of meaning-making while helping to integrate or accommodate the trauma event, allowing the trauma survivor and his or her family to heal and grow together.

T-IMFTs master the elements of a traditional MFT intake to assess for any psychopathology (mental illness and substance use), history of past therapy, trauma history, use of medications, family and social history, and relational dynamics that influence the survivor's relational, grief and loss, displacement, and/or health problems. In addition, T-IMFTs are trained to explore the trauma impact in a broader context:

- What were the magnitude and duration of the trauma event, including the loss of life, property, and community?
- What psychoeducational materials were provided for the survivor, family, and community after the trauma event?
- What help was provided for the survivor and/or family right after the disaster event?
- What impact did the trauma event have on the survivors' community? What community resources were lost?
- What are the strengths and demonstrated resilience that individuals and this family bring? What are the individual and family vulnerabilities?
- What are the life cycle influences at this moment in time for the family (centripetal versus centrifugal)?
- Those with a history of trauma understand the seriousness, level of incapacitation, loss and grief issues, foreshortened sense of the future, and physiological effects (short- and long-term). How does that intersect with the individual and family life cycle needs?
- What is this family's ability to serve as a support system for the survivor?
- What meaning do they make of their trauma experience(s)?
- What are the transgenerational legacies for trauma?
- What past experiences in treatment affect present service requests?

Additionally, the T-IMFT integrates another system and explores this survivor's and family's interactions with first responders and trauma resources from other agencies. Many of these initial first responders and organizations are moving in the direction of "survivor- and family-centered care," with an emphasis on stabilizing the survivor and family system. Ultimately, this is done by providing crisis therapy, debriefings, and defusing with survivors, providing support and information as well as education for their families. Initial services help stabilize the survivor, with the assumption that survivors and their families will seek out long-term Trauma-Informed Marriage and Family Therapy, focusing on understanding the trauma survivor and his or her family's history, needs, abilities, resources, and barriers to effective therapy. The result is an evolving treatment plan that improves the therapy as well as the assimilation or accommodation of the trauma event. Here are some essential questions for the T-IMFT to explore regarding the initial treatment from the patient/family perspective:

- What resources were made available to the survivor right after the trauma event? What services were made available for the family? How well did these early resources meet your needs?
- Did you access crisis therapy, debriefings, or defusing? How did those services improve your ability to manage the trauma event? How does it make for more challenges?
- What barriers get in the way of dealing with the trauma event and eventually achieving accommodation or assimilation of the trauma experience? Are you dealing with a foreshortened sense of the future? Are there concerns about your ability to deal with the grief and loss you have experienced?
- How did first responders and initial agency representatives incorporate important parts of your culture into your care? (Were resources made available in the survivor's and his or her family's primary language, incorporating a same-sex partner, or asking about the use of prayer and religious leaders/communities as a healing support?)
- What other experiences has your family had with first responders and other agencies that make it easier or more challenging for you as survivor to trust and access Trauma-Informed Marriage and Family Therapy now?

After completing a comprehensive trauma history assessment, the T-IMFT considers the most appropriate trauma-informed treatment plan for the survivor and his or her family.

TRAUMA-INFORMED MARRIAGE AND FAMILY THERAPY TECHNIQUES

In Trauma-Informed Marriage and Family Therapy settings, the course of therapy typically unfolds episodically (as does the survivor's relationship with his or her support system [family and friends]), because the survivor and his or her family members are engaged in Trauma-Informed Marriage and Family Therapy as needed. T-IMFTs make use of a broad range of family therapy approaches in an effort to provide effective care to the various trauma-affected survivors. Focus is upon the trauma-affected person's (1) individual factors (e.g., age, gender, ethnicity, race, etc.), (2) predisposing factors (e.g., previous trauma history, mental health problems, life cycle transition, history of legal problems, etc.), (3) peridisposing factors (e.g., trauma event, potential for injury or death of self or others, duration of the trauma event exposure, etc.), (4) postdisposing factors (e.g., educational services, resources, pamphlets, etc.), and (5) protective factors (e.g., resiliency, stress buffers, etc.). After an assessment has been conducted, focusing on the five factors and assessing and attending to the survivor's physical (e.g., sleeping problems, headache, stomachache, weight loss or gain, etc.), emotional (e.g., depression, sadness, anxiety, flashbacks, hypervigilance, PTSD, etc.), and psychosocial (e.g., lack of family support, relationship problems, divorce, etc.) issues, trauma survivors are also asked to tell their trauma story, having the family become witnesses (and a support system to the survivor) by listening and supporting the survivor. Trauma survivors and their family members are encouraged to put the trauma experience in its place, through assimilating or accommodating the experience as a way to accomplish their important developmental individual and family life cycle tasks while appropriately attending to the trauma experience.

In addition, symptom-specific issues are addressed as they arise. Communication between the survivor and his or her family are addressed through communication skills development. Safety issues (e.g., suicidal ideations, dissociation, arousal level, and bodyworks, which includes self-destructive behaviors such as substance abuse and/or drug use) are assessed on an ongoing basis, whereas mental health (e.g., depression, anxiety, etc.) and physical health issues (e.g., traumatic brain injury, loss of limbs, etc.) are addressed by collaborating with health professionals. A T-IMFT has the advantage of being able to weave together complex stories and themes through a larger frame using the Ecosystemic Developmental Trauma Model for trauma survivors and their families. Therefore, T-IMFTs who use this larger frame focus on addressing the trauma event and functional needs of the trauma survivor and his or her family. More specifically, the T-IMFT is aware of the importance of seeing the trauma survivor within the context of the whole family and recognizing the importance of looking at the functioning of larger transgenerational systems.

T-IMFTs help to reinforce the family's post-trauma function and non-trauma identity, tending to their emotional life and other routines apart from the necessary demands of any physical needs after a trauma event, in a way very similar to externalizing the trauma experience. When trauma survivors and their families are struggling with making sense of the trauma event and the "why me/us?" question, the T-IMFT creates space for these important conversations with the trauma survivor along with his or her family. Depending on the trauma event (family violence, community violence, large-scale transportation accidents, war, terrorism, and natural disasters), peer support is often available in the community or online from persons who genuinely relate to the trauma survivor's struggles and can be valuable sources of information and strength during the long haul of healing.

CULTURAL CONSIDERATIONS

If unfamiliar with the various forms of trauma events that clients may encounter, T-IMFTs need to become well versed in the various trauma events, as there is a big difference in working with a survivor depending on the trauma event (child abuse, spouse/partner abuse, bullying, gang activities, work violence, terrorism, combat, sexual abuse during deployment, natural disaster, etc.). Because no formal training is provided in Commission on Accreditation of Marriage and Family Therapy Education (COAMFTE) accredited programs for this range of trauma events, T-IMFTs need to adjust to working with these often complex problems. T-IMFTs need to know that working with trauma survivors will impact the therapist. *Vicarious trauma,* also called the *cost of caring* (Figley, 2002) is a job hazard of working with trauma survivors. Acclimating to what can seem like a foreign trauma culture is a common theme in supervision. Supervisees explore their reactions to the trauma stories of survivors as well as their own vulnerabilities. Other challenges arise when trauma survivors are in ongoing trauma situations; when there are multiple trauma events; when the system (e.g., family friends, community/community leaders, the Veterans Administration, etc.) the trauma survivor reached out to is passive and no assistance is offered; and when trauma survivors are displaced (e.g., shelters, foster care, etc.) and/or separated from their families. Most trauma survivors find that the available service providers do not adequately match their racial or ethnic diversity. If unaware of their own privilege and social location, T-IMFTs might unknowingly perpetuate a long history of racial and ethnic discrimination when working with trauma survivors in the field of traumatology.

Similarly, there is a long history of traumatology as a patriarchal discipline. That is changing gradually, with mental health professionals in counseling psychology, counseling, and MFT contributing more and helping to shape the field, as these professions typically have more women than men. Still, more women seek trauma-informed therapy than men. Women are frequently the family's psychosocial monitors and are often the ones who will seek out mental health services, so it is important to monitor gender issues in mental health services generally and in Trauma-Informed Marriage and Family Therapy in particular.

RESEARCH TO CONSIDER

Much of the research on trauma-informed mental health care has focused on the trauma survivor (Ford, Chapman, Mack, & Pearson, 2006; Lonigan, Shannon, Finch, Daugherty, & Taylor, 1991; Steele & Malchiodi, 2011) and the importance of taking an ecosystemic developmental trauma perspective. The use of integrated trauma models is only gradually growing, whereas various marriage and family therapy models such as Narrative Therapy (telling their trauma narrative) and the intergenerational approach (intergenerational trauma transmission) have been used in the past. Recent work has focused on evidence-based models such as Cognitive Behavioral Therapy (CBT). In addition, all T-IMFTs should be familiar with the Adverse Childhood Experiences study (ACE Study; Anda & Felitti, 2006) and the increasing amount of literature on the long-term effects of trauma (physiological, psychological, and psychosocial). Trauma-Informed Marriage and Family Therapy supervisors should orient supervisees to the latest developments in the traumatology literature, trauma survivor and family-centered care, and intergenerational trauma transmission. T-IMFTs bring unique skills for assisting trauma survivors and their families and understand the importance of doing so through an ecosystemic developmental lens.

ROLE OF THE SUPERVISOR AND SUPERVISEE

Supervisors and supervisees begin their professional relationship by co-creating a specific supervision contract that cover a range of important aspects of their supervision. These items should include (among many possible others) frequency and duration of meeting times, agenda setting for meetings, arranging live/video supervision, fees, theoretical orientations, crisis management, backup system, duration of the contract, documentation, competencies, professionalism, trauma-informed MFT, and vicarious trauma.

For most T-IMFTs, clinical care and supervision may occur in a variety of settings. Typically, the supervisor is a licensed T-IMFT who has served as such for many years, is well versed in both theory and practice, and has integrated traumatology into the MFT clinical competencies. Supervisors focus on the essentials of MFT supervision, along with helping supervisees conduct specifically Trauma-Informed Marriage and Family Therapy. Trauma-informed services are provided in a variety of settings: private practice, for-profit and nonprofit agencies, Veterans Administration facilities, hospitals, trauma centers, and so on. Services can range from individual to couple to family to group therapy. T-IMFTs who practice in these settings understand that trauma experiences can be triggered through various life cycle transitions, which might require additional therapy.

Another essential element in Trauma-Informed Marriage and Family Therapy supervision is discussion of vicarious trauma, which can best be explored through self-of-the-therapist approaches. These issues are most appropriately explored when the supervisor and supervisee have developed a safe and validating supervision relationship. It is helpful to incorporate a trauma genogram early in the supervision process to explore the supervisee's own trauma history and trauma narrative from the family of origin. It is not unusual for T-IMFTs who chose this specialization to come to this work with a personal or family-of-origin trauma history, which should be examined to ensure that it will not interfere with the supervisee's work with trauma survivors. Trauma-Informed Marriage and Family Therapy supervisors need to ask their supervisees to examine the meta-themes that they bring to Trauma-Informed Marriage and Family Therapy in general, and then address any specifics related to the current case. Supervisees are also encouraged to read more about trauma therapy as well as seek their own personal therapy should their own trauma history be triggered or they find themselves dealing with countertransference Each trainee may review the following:

- While drawing or reviewing a prepared genogram, consider the trauma experiences of each family member, including the causes of trauma, duration and frequency (single versus multiple trauma events) and the impact of the trauma event.
- What trauma themes were present in your family growing up? What were your earliest trauma memories? What memories do you have of your family's trauma history and the impact it had on them? What coping skills were utilized by you and other family members? What about resiliency and stress buffers? What services were accessed/provided?
- Who else in your family has roles in trauma caregiving, either formally (as mental health provider, crisis counselor, first responder, medical person) or informally (the informal counselor in the family)? How did you find your way to becoming a T-IMFT?
- Are there trauma issues or life cycle transitions that might be particularly relevant given your personal or family-of-origin trauma history? Are there unresolved trauma issues that might interfere with your effectiveness as a T-IMFT?

Another aspect of learning about Trauma-Informed Marriage and Family Therapy as a new supervisee is learning about vicarious trauma. As described earlier, *vicarious traumatization* is the process through which the T-IMFT's inner experience is negatively transformed through empathic engagement with the client's trauma material (McCann & Pearlman, 1990). Supervisees need to be aware of the signs and symptoms of vicarious traumatization, which include: (1) hypervigilance; (2) suspicion about people's motives and behaviors; (3) difficulty sleeping, nightmares; (4) intrusive images; (5) anxiety; (6) numbness; (7) inability to experience pleasure; (8) excessive caregiving (on and off the job); (9) cynicism; (10) anger and irritability; (11) depression; (12) exhaustion; (13) increased self-criticism; (14) feeling estranged from others; (15) shattered ability to believe in oneself (I can keep myself safe) and others (humans are basically good, and the world is a safe, just, controllable, and meaningful place); (16) carry the burden of the trauma stories the T-IMFT has become witness to; and (17) re-examination of the T-IMFT's life priorities (can be good but disruptive). The supervisor also needs to check in with the supervisee to see if the T-IMFT is experiencing vicarious traumatization. In addition, the supervisee needs to understand that vicarious traumatization can best be described as the trauma survivor's powerlessness during the trauma event, which parallels the T-IMFTs' powerlessness if they feel they need to "fix" their clients. Supervisors need to assess whether the supervisee believes that self-care is less important than the care she or he provides for others. This kind of thinking should be addressed as part of the Trauma-Informed Marriage and Family Therapy supervision, as does self-care, to reduce stress and transform the T-IMFT's potential despair that comes with having a shattered perception of self, others, the world, and religious/spiritual values/beliefs. Trauma-Informed Marriage and Family Therapy supervisors do so by normalizing and not pathologizing the supervisee's behavior. Trauma-Informed Marriage and Family Therapy supervision should focus on exploring ways that in which supervisees can engage in self-care, which means becoming aware of their own needs, limits, emotions, and resources, as well as finding a balance in their work and play (taking care of others *and* taking care of oneself). It also means that supervisees need to stay connected with themselves, others, and to something larger. Because of the relationship that the supervisor and supervisee have established, it creates an opportunity for the T-IMFT to ask for help if she or he needs it.

LEARNING GOALS/OBJECTIVES

Developmentally appropriate learning goals are established early on in the supervisor-supervisee relationship as part of the supervision contract. These learning goals are assessed periodically and updated as needed for the level of experience and the context of the relationship (student versus employee, for example). Focus is on the developmental stage of the supervisee and his or her readiness to move beyond basic competency. Assessment, treatment planning, intervention, safety, and crisis management are all addressed within this ecosystemic developmental trauma-focused framework. If the supervisee has already been trained in another type of trauma therapy, the contract may focus more on the ecosystemic developmental framework and the special assessment focusing on the five aspects (physical, emotional, and psychosocial symptoms, ethics, and professionalism) of Trauma-Informed Marriage and Family Therapy.

Generally, it is appropriate to have the supervisor support the new supervisee's process of becoming comfortable with serving trauma survivors, hearing their trauma stories, and becoming witness to their trauma events. Any relevant self-of-the-therapist issues that might arise with the supervisee must be addressed. Typically this begins with the supervisees getting consultation, video or live supervision, and encouragement to look at their own trauma genograms. T-IMFTs see trauma survivors in various settings and at times have the opportunity to engage in co-therapy with a more seasoned T-IMFT. In addition, it is useful to have the supervisee observe another T-IMFT. Additional training, such as workshops and seminars, serve as a way to educate the supervisee about the need for trauma-informed therapy. It also serves as an opportunity to help supervisees see how the Ecosystemic Developmental Trauma Model is used with a range of trauma events and how to include the survivor's partners and/or family. Focus is on how to build initial rapport, especially when trust has been broken, and learning to understand and speak the different languages and "cultures" of the trauma events (e.g., bullying in high school versus three-time combat deployment, or spouse/partner abuse). All of these experiences help inform the T-IMFT's future trauma work with a broad range of trauma survivors and their spouses/partners and/or families. T-IMFTs need to become skilled and comfortable with PTSD and other trauma symptomology as well as post-traumatic growth (PTG). In addition, co-therapy allows novice T-IMFTs to become more comfortable managing their anxiety while working with trauma survivors; these T-IMFTs may also need the safety of their supervisory relationship to address vicarious traumatization. It is therefore important to encourage supervisees to take time for reflection on their role as witnesses to human suffering and their feelings of helplessness about "fixing it." In addition, T-IMFTs need peer support, talking with other T-IMFTs one-on-one or in a peer consultation group. In addition, they need to find balance in seeing trauma survivors and other clients and get trauma-informed supervision to address the unique supervision needs of the T-IMFT. This should include teaching supervisees about the various trauma assessment tools available; assessing for PTSD, depression, anxiety, and substance abuse; enhancing interviewing skills; and looking at the various trauma therapy interventions useful in working with a broad range of trauma survivors. It also means expanding the T-IMFT's knowledge about particular trauma events, such as spouse/partner abuse, natural disasters, military combat deployment, and so on. In addition, attention should be paid to vicarious traumatization and its signs and symptoms. T-IMFTs need to learn that self-care, education, and a balanced caseload (trauma survivors and other clients), as well as trauma supervision and peer support, all serve as buffers against vicarious traumatization. More specifically, it means recognizing that these things may not prevent the T-IMFT from developing vicarious traumatization, but rather are intended to help the T-IMFT to deal more effectively with the vicarious traumatization that does occur.

Furthermore, the T-IMFT serves as therapist and educator to the survivor and his or her family. If other professionals, such as psychiatrists and/or other medical doctors, are involved in the survivor's care, the T-IMFT serves as a collaborator, coordinating the survivor's and family's treatment, and as consultant to other systems (such as the courts, insurers, etc.). In some situations, the T-IMFT also serves to educate physicians and other service providers. T-IMFT supervisors must understand the various roles T-IMFTs play in coordinating services and educating others to provide quality care for trauma-affected clients. The coordination of services can be challenging to the T-IMFT and supervisory guidance in this area can be especially helpful.

The following case vignette demonstrates supervision of a typical course of Trauma-Informed Marriage and Family Therapy using the Ecosystemic Developmental Trauma Model.

CONCLUSION

Supervising a novice T-IMFT working with a survivor and his or her family highlights the significance of context in culturally informed survivor- and family-centered trauma care. Supervisors attend to T-IMFT supervisees' development of clinical competencies for trauma survivors and their families from an ecosystemic developmental trauma perspective, while being sensitive to important self-of-the-therapist needs and the development of vicarious traumatization. Education (instruction and clinical) about Trauma-Informed Marriage and Family Therapy skills is essential, as all MFTs will eventually work with some kind of trauma survivors and/or their families. Necessary training should become part of all accredited MFT programs. Currently training is a shortcoming in preparing MFTs for Marriage and Family Therapy to participate in helping survivors and their families to heal, in an era when trauma from war (service men and women), terrorism, and large-scale natural disasters, as well as from family, street, and school/workplace violence, is so prevalent in the United States.

CASE EXAMPLE

Supervisor/Supervisee/Setting

Dr. Bob, a 53-year-old Caucasian T-IMFT and AAMFT Approved Supervisor, has specialized in traumatology for the past 20 years, having completed additional workshops and coursework in traumatology and received trauma-informed supervision. Brittney is a 33-year-old Caucasian Trauma-Informed Marriage and Family Therapy student who is new to the site. It is her first clinical placement.

Trauma Survivor/Family/Presenting Concerns

Travis (29) and Tekia (27) are an African American couple with two children, Tyrone (5) and Tommy (4). Tekia has been a stay-at-home mom. Like most of the men in his family, Travis was in the military. He served 2 tours of duty in Afghanistan and was injured when an improvised explosive device (IED) exploded and left him severely disabled. He spent many months in the hospital and rehabilitation.

(*continued*)

To his disappointment, he was discharged from the military as a result of his injuries and now has difficulty finding a rewarding job. He currently works two part-time jobs but hopes to find full-time work. Travis's older brother is still in the military and was recently re-deployed to Afghanistan. Travis and Tekia present to therapy with relationship and parenting issues, as Travis reportedly has a "short fuse." They also report needing help as he continues recovery from his injury and assistance in navigating challenges with employment and finances.

Preparing for the Intake

Brittney reviews a self-referral initiated by Tekia, requesting couple therapy. A phone intake had already been done, which Brittney reviewed to see what the presenting issues were and if there were any safety concerns. None were reported. Brittney saw that there were some reported medical issues for Travis as a result of his injuries during his second tour of duty and that the couple was struggling with marital issues, which was the reason for seeking the MFT services. It is not unusual for service men and women to have relationship issues after deployment, especially if they had multiple tours of duty and were deployed to combat zones. It is believed that couples grow apart, and both partners change, as they are separated for months at a time (Jordan, 2011).

In supervision, Dr. Bob and Brittney review the presenting problems and reason for self-referral, as well as the phone interview, and make a plan for the first session. In addition to the traditional first-session joining, information focusing on the five factors: (individual, predisposing, peridisposing, postdisposing, and protective) will be assessed. The assessment is service man/woman and spouse/partner specific. Travis's military training, branch of the military, age first deployed, and so on (personal factors), and his combat experience (peridisposing factors) are focused upon. For Tekia, the focus is on history of previous trauma, life cycle transitions, and other stressors (predisposing factors). Present functioning will be addressed, focusing on risk factors such as self-injury, self-destructive behavior, partner violence/abuse, and harm to children in the home, as Travis reportedly has a "short fuse." Brittney plans to find out more information regarding Travis's disability to get a better understanding of what kind it is, if it is temporary or chronic/permanent, what the short- as well as long-term adjustments will be needed, and the resultant challenges. She will get an extensive history and prognosis.

(*continued*)

After that assessment has been completed, focus will be upon both Travis's and Tekia's resiliency and stress buffers. They will do a basic genogram, looking at family-of-origin (FOO) experiences, relationship history, employment including military experience, and history of injury resulting in disability, and add a trauma timeline for both Travis and Tekia. Focus will be on FOO experiences for Travis and Tekia related to military life, employment, trauma experiences and survival, grief and loss, disability, marriage, communication, and conflict management. Dr. Bob is supportive of the plan, cautioning Brittney to assess for safety issues, and reminds her that the assessment uses several questions to gain a more comprehensive picture of the situation and serve as a springboard to determine the treatment. This assessment process is done through intake questionnaires, the trauma genogram, and story telling.

Supervision with Dr. Bob focuses on relevant learning goals, developmental needs, and self-of-the-therapist issues for Brittney as she prepares to provide trauma-informed couple therapy. He explores her trauma history and self-of-therapist by having Brittney develop her own genogram. When exploring the genogram, Dr. Bob asks questions about her background experiences with military family life and trauma. He also asks her about cultural awareness and sensitivity about caring for an African American disabled veteran and his young family. He attends to important potential vicarious traumatization issues and other signs of trauma affect. In addition, he explores what kind of caseload Brittney has as well as her commitment to self-care and continuing education on traumatology. Brittney acknowledged her own childhood trauma experiences as well as grief and loss issues. She identified a personal and family-of-origin experience with service men and women. Her father was a Vietnam veteran and, after many struggles upon returning from combat, developed a drinking problem and eventually committed suicide. Brittney felt somewhat prepared to address issues of racial discrimination and demonstrated openness to exploring these themes in ongoing supervision.

Ongoing Supervision for Psychotherapy and Trauma-Informed Care

As a supervisor for a novice T-IMFT, Dr. Bob keeps close watch on the evolving assessment of this couple. Brittney consults regularly with Dr. Bob and he provides specific guidance on her documentation. He also checks in regularly to see how Brittney is handling the trauma stories and managing the emotional intensity of the care and offers support through live supervision as well as joining a session

(*continued*)

to conduct a more thorough risk assessment when Travis presents with suicidal ideations. Brittney and Dr. Bob decide together about how to sequence individual, couple, and family sessions to address the relevant treatment plan goals: monitor Travis's PTSD symptoms and maintain safety; improve marital functioning and quality; enhance coping to reduce angry outbursts; assist Travis with the injury/disability and grieve the many losses he has experienced; help Tekia to grieve the old couple relationship that no longer exists, and embrace the new changed Travis and the relationship.

Over the course of the first three sessions, focus is on Travis and support services, such as the emergency room crisis number and the suicide prevention hotline. A safety plan is put in motion each session. Travis is assessed by a psychiatrist from the Veterans Administration, who determines that an antidepressive medication will be most helpful for Travis. During these sessions, Brittney created an effective working alliance with Travis and Tekia and helped them begin to establish a changed "new normal" couple relationship with new routines to manage their daily life and attend to Travis's care as well as their couple relationship. Both Travis and Tekia engaged in a ritual in which they grieved the couple relationship they once had (which had changed) through the deployments that both experienced: (1) Travis experienced combat twice and saw fellow service men and women being injured and killed. He was injured and struggled with adjusting to a disability and to what he had seen in combat. (2) Tekia learned to be more self-reliant, as she had to manage their family finances, home, children, and other responsibilities by herself, but also had been dealing with anxiety, wondering if Travis would make it back home safe. Travis and Tekia are dealing with their finances together, which seems to work well. Travis, as he gradually gets better, has started several small home improvement projects, which have helped make him feel like he is contributing to the couple's well-being.

Travis is often tense when he is in public by himself, as he feels overwhelmed with all the sounds and activity, and was able to get a service dog to help him. He had been relying on other service men and seldom felt safe but has made great improvement through being with his service dog, Jacky. Tekia, who was at first hesitant to have a service dog in the house, has seen great improvements over the last couple of months and sees a strong bond forming between Travis and Jacky. The Veterans Administration (VA), at times difficult to work with, has been very helpful in the rehabilitation process and supported Travis when he got his service dog. Travis and Tekia reported a gradual improvement in their relationship. Travis has been less tense,

(*continued*)

with fewer angry outbursts, and is more relaxed. Their relationship, although different from when they first met pre-deployment, is once more nurturing, with more intimacy. His injuries were painful, and the healing as well as the physical therapy was a slow, drawn-out process, which had impacted their sexual relationship. As the pain gradually improved, so did their sexual relationship. The couple was successful in finding their "new normal."

Dr. Bob, who guided Brittney in her work with Tekia and Travis, acknowledged the progress that the couple made. Jacky attended the session with the couple and also became an integral part of the therapy process. Even though the couple made good progress, at one point Travis learned that one of his friends, Darrell, a service man he was deployed with, had committed suicide. It was a difficult time for him and Brittney, and after carefully assessing him to rule out safety issues, Brittney was still worried and requested that no weapons be kept in the home. Travis was at first hesitant but agreed to it. Tekia was visibly relieved, as she was very worried as well and feared that Travis might, in a moment of grief, hurt himself. Couple therapy over time became a combination of individual and couple sessions. Travis, at first hesitant, talked about his experiences in Afghanistan. Tekia not only learned about Travis's experiences, but also, with Brittney's help, assisted him in the healing process by supporting and comforting him. Travis also got support from his extended family and community, who both helped with some of the more complex home improvement projects and also raised funds to help the couple cover the many medical bills and expenses they had accumulated after Travis returned from Afghanistan. Travis tried for a couple of weeks to escape all of the pressures he was experiencing, trying to reintegrate into civilian life, but started drinking heavily. He stopped when Tekia threatened to leave him, and started attending an AA group, which he found helpful. Travis also found support in talking with Tekia's cousin, who had also been deployed to Afghanistan and had experienced combat.

Travis, however, was clear that Jacky had "given him his freedom back." The bond between the two of them grew stronger and stronger. At first, it seemed that Tekia was somewhat jealous, but she soon saw the influence Jacky had in Travis's healing and recovery process. After several months of weekly sessions, the couple reported and demonstrated a better level of functioning and relationship satisfaction, so therapy was spread out to every other week, then to every three weeks, and then to once a month, and eventually the couple indicated their readiness to terminate. A six-month follow-up interview indicated that the couple was continuously making progress

(*continued*)

and, although they had encountered some challenges, had successfully navigated though them and are continuing to grow and change.

Brittney reached out to her supervisor and some of her peers for consultation and support. She was unclear about how to attend to Travis when he struggled with his friend's death and was glad that Dr. Bob provided clear instructions on how to support Travis. She scheduled more frequent individual sessions, as a higher level of care was needed on a short-term basis. With Dr. Bob, Brittney also explored her own reactions and feelings when working with the couple. They scheduled a combination of live and consultation sessions, based on what was going on with the client(s). Dr. Bob regularly checked in with Brittney regarding her ability to deal with the trauma stories to which she became witness. Dr. Bob and Brittney talked about vicarious trauma and how it is a job hazard when working with survivors. Focus was on Brittney's self-care and how she was dealing with her own anxiety. Peer support and consultation became a valuable addition to her supervision with Dr. Bob; also, she was able to engage in regular tai chi sessions, which she found particularly helpful for her self-care. Dr. Bob also encouraged Brittney to get additional training in trauma-informed MFT, education about working with service men and women after combat deployment, and assistance in creating a balanced work load of trauma- and non-trauma-affected clients.

As is common in most MFT settings in preparation of termination, sessions were spaced out with Travis and Tekia and termination was conducted by focusing on Travis's and Tekia's achievements in T-IMFT. Both clients were informed that booster sessions are available as needed by contacting Brittney or a colleague of hers as needed. Prior to the official termination, Brittney reviewed with Travis and Tekia the treatment goals that had been accomplished and progress toward the remaining goals, so as to determine if it was the appropriate time to terminate. Dr. Bob helped Brittney to prepare for termination and guided her on how to do the reflection with her clients. He also reviewed her termination paperwork in a way that would be brief and concise, yet include essential elements for other T-IMFTs if Travis and Tekia decided to resume therapy.

REFERENCES

Abu-Bader, S. H. (2000). Work satisfaction, burnout, and turnover among social workers in Israel: A clinical diagram. *International Journal of Social Welfare, 9,* 191–200.

Anda, R. F., & Felitti, V. J. (2006, April 18). *Adverse childhood experiences as a national public health problem* (congressional briefing). Sponsored by the American Academy of Pediatrics and The Family Violence Prevention Fund. Washington, DC.

Anda, R. F., Felitti, V. J., Brown, D. W., Chapman, D., Dong, M., Dube, S. R., . . . Giles, W. H. (2006). Insights into intimate partner violence from the Adverse Childhood Experiences (ACE) study. In P. R. Salber & E. Taliaferro (Eds.), *The physician's guide to intimate partner violence and abuse.* Volcano, CA: Volcano Press.

Basham, K. (2008). Homecoming as safe haven or the new front: Attachment and detachment in military couples. *Journal of Clinical Social Work, 36,* 83–96.

Briere, J., & Elliotte, J. (2000). Trauma orientation and detection of violence histories in the psychiatric emergency services. *Journal of Nervous and Mental Disease, 188,* 622–624.

Bronfenbrenner, U. (1987). Interacting systems in human development: Research paradigms—Present and future. In N. Bolger, A. Caspi, G. Downey, & M. Morehouse (Eds.), *Person in context: Developmental processes* (pp. 25–49). Cambridge, UK: Cambridge University Press.

Cohen, J. A., Mannarino, A. P., & Deblinger, E. (2006). *Treating trauma and traumatic grief in children and adolescents.* New York, NY: Guilford Press.

Combrinck-Graham, L. (1985). A developmental model for family systems. *Family Process, 24*(2), 139–150.

Courtois, C. A., & Ford, J. D. (2009). *Treating complex traumatic stress disorders: An evidence-based guide.* New York, NY: Guilford Press.

Fernando, G. A. (2004). Working with survivors of war in non-western cultures: The role of the clinical psychologist. *Intervention: International Journal of Mental Health, Psychosocial Work and Counseling in Areas of Armed Conflict, 2,* 108–117.

Figley, C. R. (2002). *Treating compassion fatigue.* New York, NY: Brunner Routledge.

Figley, C. R., & Kiser, L. J. (2012). *Helping traumatized families.* New York, NY: Routledge/Taylor & Francis Group.

Fontana, A., & Rosenheck, R. (2008). Treatment-seeking veterans of Iraq and Afghanistan: Comparison with veterans of previous wars. *New England Mental Illness Research, Education and Clinical Center, 196*(7), 513–521.

Ford, J. D., Chapman, J., Mack, M., & Pearson, G. (2006). Pathways from traumatic child victimization to delinquency: Implications for juvenile and permanency court proceedings and decisions. *Juvenile and Family Court Journal, 2,* 13–23.

Freedman, J., & Combs, G. (1996). *Narrative therapy: The social construction of preferred realities.* New York, NY: W.W. Norton.

Gelles, R. J., & Maynard, P. E. (1987). A structural family systems approach to intervention in cases of family violence. *Family Relations, 36*(3), 270–275.

Harrison, R. I., & Westwood, M. J. (2009). Preventing vicarious traumatization of mental health clinicians: Identifying protective practices. *Psychotherapy: Theory, Research, Practice, Training, 46,* 203–219.

Johnson, S. M. (2002). *Emotion focused couple therapy with trauma survivors: Strengthening attachment bonds.* New York, NY: Guilford Press.

Jordan, K. (2011). Counselors helping service veterans re-enter their couple relationship after combat and military service: A comprehensive overview. *The Family Journal, 19*(3), 1–11. doi:10.1177/1066480711406689

Kerr, M. E., & Bowen, M. (1988). *Family evaluation: An approach based on Bowen theory.* New York, NY: W.W. Norton.

Klaperski, S., Seeling, H., & Fuchs, R. (2013). Physical activity as a stress buffer. *Zeitschrift fur Sportpsychologie, 19*(2), 80–90.

Lambert, S. F., & Lawson, G. (2013). Resilience of professional counselors following Hurricanes Katrina and Rita. *Journal of Counseling and Development, 91,* 261–268.

Lechner, C. M., Tomasik, M. J., Silbereisen, R. K., & Wasilewski, J. (2013). Exploring the stress-buffering effects of religiousness in relation to social and economic change: Evidence from Poland. *Psychology of Religion and Spirituality, 5*(3), 145–158.

Lightsey, O. R. (1994). Thinking positive as a stress buffer: The role of positive automatic cognition in depression and happiness. *Journal of Counseling Psychology, 4*(3), 325–334.

Lonigan, C., Shannon, M., Finch, A., Daugherty, T., & Taylor, C. (1991). Children's reactions to a natural disaster: Symptom severity and degree of exposure. *Advances in Behavioral Research and Therapy, 13,* 135–154.

Madsen, W. C. (2007). *Collaborative therapy with multi-stressed families.* New York, NY: Guilford Press.

McCann, I. L., & Pearlman, L. A. (1990). Vicarious traumatization: A contextual model for understanding the effects of trauma on helpers. *Journal of Traumatic Stress, 3*(1), 131–149.

Mikesell, R. H., Lusterman, D. D., & Daniels, S. (1995). *Integrating family therapy: Handbook of family psychology and systems theory.* Washington, DC: American Psychological Association.

Nauert, R. (2010). Can psychological trauma be inherited? *PsyCentral.* Retrieved from http://psychcentral.com/news/2010/09/09/can.psychological-trauma-be-inherited/1790/html

Parker, S. L., Jimmieson, N. L., & Amiot, C. E. (2010). Self-determination as a moderator of demands and control: Implication for employee strain and engagement. *Journal of Vocational Behavior, 76*(1), 52–67.

Pearlman, L. A., & Saakvitne, K. W. (1995). Trauma and the therapist countertransference and vicarious traumatization in psychotherapy with incest survivors (pp. 451). New York, NY: W.W. Norton.

Pearlman, L. A., & Saakvitne, K. W. (1995). Treating therapists with vicarious traumatization and secondary traumatic stress disorder. In C. R. Figley (Ed.), *Compassion fatigue: Coping with secondary traumatic stress disorder in those who treat the traumatized* (pp. 150–177). New York, NY: Brunner-Mazel.

Rosenbloom, D. J., Pratt, A. C., & Pearlman, L. A. (1995). Helpers' responses to trauma work: Understanding and intervening in an organization. In B. H. Stamm (Ed.), *Secondary traumatic stress: Self-care issues for clinicians, researchers and educators* (pp. 65–79). Lutherville, MD: Sidran Press.

Rothchild, B. (2000). *The body remembers: The psychophysiology of trauma treatment.* New York, NY: W.W. Norton.

Schwartz, G. (2010). *Haunting legacies: Violent histories and transgenerational trauma.* New York, NY: Columbia University Press.

Slaikeu, K. A. (1990). *Crisis intervention: A handbook for practice and research.* 2nd ed. (p. 519). Boston, MA: Allyn & Bacon.

Stamm, B. H. (ed.) (1995a). Secondary traumatic stress: Self-care issues for clinicians, researchers and educators (p. 279). Lutherville, MD: Sidran Press.

Stamm, B. H. (1995b). *A process approach to the scientific method.* Paper presented at the Trauma, Loss and Dissociation Conference, Washington, DC.

Steele, W., & Malchiodi, C. (2011). *Trauma informed practices for children and adolescents.* New York, NY: Routledge/Taylor & Francis Group.

van der Kolk, B., & Fisler, R. (1995). Dissociation and the fragmentary nature of traumatic memories: Overview and exploratory study. *Journal of Traumatic Stress, 8*(4), 505–525.

Vesti, P., & Kastrup, M. (1995). Refugee status, torture, and adjustment. In J. R. Freedy & S. E. Hobfoll (Eds.), *Traumatic stress: From theory to practice* (pp. 213–232). New York, NY: Plenum Press.

Index